# *AIRCRAFT DYNAMICS AND AUTOMATIC CONTROL*

DUANE McRUER
IRVING ASHKENAS
DUNSTAN GRAHAM



PRINCETON UNIVERSITY PRESS
PRINCETON, NEW JERSEY

L.C. Card: 73-134250
ISBN: 0-691-08083-6
ISBN: 0-691-02440-5, pbk.
First Princeton Paperback printing, 1990
10
10 9 8 7 6 5 4 3 2 1

This book has been composed in Monotype Modern 8A
Printed in the United States of America by
Princeton University Press

# *PREFACE*

Flight control is a systems discipline that brings together the component dynamic characteristics of aircraft and flight controllers to form the system dynamic characteristics of the vehicle under the action of feedback control. Unfortunately, it has seemed to us that, by and large, the texts, monographs, and courses of instruction that treat these topics have tended to emphasize their disparities. There is certainly no lack of books on aircraft stability and control nor on feedback control systems. Our conviction, however, is that there is a field that comprises both of these subjects and that an understanding of either one can help to illuminate the other.

The purpose of this book is to present an integrated, analytical treatment of the dynamics of the vehicle (the controlled element) and its flight control systems. The book has been written by and for engineers concerned with the analysis of aircraft dynamics and the synthesis of aircraft flight controls. Such studies are at least as old as powered flight itself and they seem likely to remain pertinent as long as there are new and more advanced aeronautical vehicles.

Not long ago, the intellectual mathematical equipment of skilled stability and flight control analysts generally exceeded their physical ability to perform all the design and tradeoff calculations that might be needed or desired. Nowadays quite the opposite situation exists because advances in both analog and digital computation allow the consideration of problems that at one time would have been rejected as being too time-consuming. As a consequence, the analyst's physical means now often exceed his mental grasp; what he can compute may, possibly, far exceed his understanding or appreciation. This can lead to an excessively empirical approach to design which is similar to the one used by "practical" designers 30 or more years ago. Then airplane stability and control properties were evaluated only in flight test, and flight control equipment was also "designed" with the aid of extensive full scale testing. A difference, of course, lies in the abstractions involved, for, regardless of the detail and complexity of our mathematical models, they remain just that; whereas the physical equipment and the aircraft are the objects of our abstractions. Viewed in these terms, too great a reliance on a numerical-empirical approach to design is no better and may even be worse than the physical empiricism of earlier days. Inundated by computer printouts and strip chart recordings, the analyst is confronted with a crucial problem—what is the essence, what does it all mean?

For this reason we have strongly emphasized an analytical approach to flight control system design and have summarized an eclectic collection of efficient, neatly interconnected techniques that *inherently* and readily display the essential aspect of complex system problems. When skillfully applied, either with pencil and paper or using computer aids, these techniques enable one to attain a high level of insight and physical understanding with a minimum of effort. They are suitable for the establishment of nominal system designs, for the forecast of off-nominal problems, and for the diagnosis of the root causes of the problems that almost inevitably occur in the course of the design process.

While we have tried to be as definitive as possible on the subject of aircraft and flight control system dynamics and the procedures that are employed to accomplish automatic flight control system designs, the scope of our work has had to be limited to keep within the confines of one volume. The necessary limitation has been accomplished primarily by considering the aircraft only as a rigid body, and by almost exclusively emphasizing the theory of linear constant coefficient systems. The decisions on both these limitations were made somewhat reluctantly, since the flexible airframe and nonlinear features of flight control are always fascinating academically and are often important practically. We should hasten to remark, however, that regardless of the number of modes or nonlinearity of a problem, linearized solutions to comparatively low order problems almost always give reasonable approximate answers. They provide, as it were, a most useful species of limiting case solution, and limiting cases are, in general, the basis for much of our physical understanding of complex phenomena. With a solid grounding in linear theory, the extension of the results to nonlinear problems, especially of stability, is ordinarily rewarding and effective. Thus linear theory is, very generally, a theory of a first approximation which has the great virtue that it can be conceptually assimilated in its entirety. Further, as a practical matter, it is our observation that the great majority of the physical problems of aircraft flight control which are susceptible to mathematical treatment are, in fact, handled to a very good first approximation by linear treatments.

This book has a genealogy. Its immediate predecessors are the series of Bureau of Aeronautics volumes prepared in the early 1950s at Northrop Aircraft. The considerable success and the reputation of these volumes in industrial design departments, government laboratories, and in engineering schools prompted the original intent of the Naval Air Systems Command in sponsoring a large portion of the present work (Contract NOw 62-0781-c), so as to provide revisions and an updating for two of those volumes:

"Dynamics of the Airframe," Bureau of Aeronautics Report AE-61-4II, September 1952; "Automatic Flight Control Systems for Piloted Aircraft," Bureau of Aeronautics Report AE-61-4VI, April 1956. Although some of the numerical data and examples from these earlier volumes have been used here, we present an essentially new effort rather than a revision. Furthermore, in order to provide an integrated treatment, we have included material that partially revises the first of the Bureau of Aeronautics volumes: "Methods of Analysis and Synthesis of Piloted Aircraft Flight Control Systems," BuAer Report AE-61-4I, March 1952. Consequently, this book will, in the main, supersede the above three volumes of the Bureau of Aeronautics series.

We are indebted to many people and organizations for their assistance in the preparation of this book. First and foremost is the Naval Air Systems Command (NASC), which sponsored the preparation of much of the manuscript. The NASC project monitor, Jack Crowder, was an ideal supporter, continually interested and anxious to get the job done, yet patient and understanding, in spite of the inevitable delays that projects of this sort seem to incur. We also owe major debts to our colleagues, at Systems Technology and elsewhere, who have critically reviewed various versions and portions of the manuscript and have offered constructive criticisms and suggestions for its improvement. First in this group is Robert L. Stapleford of Systems Technology, who has gone through the book several times, exercising his penchant for clarity and his keen eye for error. Robert J. Woodcock of the Air Force Flight Dynamics Laboratory, who thoroughly reviewed several chapters, was also a great help. H. R. Hopkins of the Royal Aircraft Establishment, Farnborough, reviewed Chapter 1, making many helpful suggestions and very graciously offered us the use of his own extensive material on the history of flight control. Dr. Malcolm J. Abzug of Thompson-Ramo-Wooldridge also made a number of correcting and clarifying remarks related to the history presented in Chapter 1, for which we are very grateful; Ronald O. Anderson of the Air Force Flight Dynamics Laboratory made available to us his bibliography on the history of feedback controls. Gary Teper of Systems Technology was responsible for the collection and presentation of the data contained in Appendix A. Particular acknowledgment is further due to the publication staff of Systems Technology, who labored long and hard to prepare the manuscript for publication, and especially to their Publications Manager, Junichi Taira, whose meticulous attention to every detail is revealed on each page of the book.

Besides those who helped directly, there are others in the background. Most important, of course, are our many colleagues in the flight control and automatic control community whose original work is reflected here. We have tried to acknowledge them throughout the book with pertinent references to the published literature. As is evident from these footnotes, a great deal of the work summarized here was originally accomplished for the Control Criteria Branch of the Air Force Flight Dynamics Laboratory. In fact, some of the material appearing here for the first time is based on unpublished notes prepared in the course of work sponsored by the United States Air Force. We must also acknowledge our former colleagues at Northrop Aircraft, Warren Koerner and Robert E. Trudel, who were among the authors of the old BuAer "Dynamics of the Airframe" volume on which parts of Chapters 4, 5, and 6 are based. Finally, we wish to acknowledge our present or past Systems Technology co-workers, J. J. Best, T. S. Durand, D. E. Johnston, H. R. Jex, W. A. Johnson, L. G. Hofmann, J. D. McDonnell, R. A. Peters, R. J. Wasicko, D. H. Weir, and J. Wolkovitch, for their several original contributions to the material presented in the following pages.

The merits that this book may possess can, in large part, be attributed to all of these people. Any faults are the responsibility of the authors. We hope that this work will prove both instructive and useful to others who, like ourselves, wish to help solve the flight control system design problems of future generations of aircraft.

Duane McRuer
*Hawthorne, California*

Irving Ashkenas
*Hawthorne, California*

Dunstan Graham
*Princeton, New Jersey*

February 1971

# CONTENTS

# *LIST OF FIGURES*

# LIST OF TABLES

*AIRCRAFT DYNAMICS AND AUTOMATIC CONTROL*

CHAPTER 1

# *INTRODUCTION AND ANTECEDENTS*

> We now know a method of mounting into the air, and, I think, are not likely to know more. The vehicles can serve no use till we can guide them; and they can gratify no curiosity till we mount with them to greater heights than we can reach without; till we rise above the tops of the highest mountains.[1]

The economic or military value of any vehicle depends fundamentally on its ability to traverse a controllable path between its point of departure and its destination or "target." Abstractly, the vehicle is a velocity vector in space. It has a direction in which it is going and a speed with which it is going there. The time integral of the velocity vector is the path. Each type of vehicle, however, is made to move and carry in a certain medium and its motions may be subject to constraints. Means for the control of the path vary widely and depend on the constraints. Thus a train, for example, is constrained to move along a track; the control that is provided is merely a speed control. The train is not steered. An automobile or a ship, on the other hand, while constrained to move on the surface of the land or the sea, must be steered as well. Aircraft share with submarines and torpedos an unusual freedom from constraints, and the problems of the control of aircraft are of unusual complexity. We do indeed "know a method of mounting into the air," but the solution of the problems of control still requires both sensibility and diligence.

An aeronautical vehicle or weapon system contains spatial sensors, and guidance and control devices (possibly all subsumed in the human pilot) whose purpose it is to develop three-dimensional flight path commands appropriate to steering so as to reach a destination or target, and then to execute those commands by maintaining or modifying the forces on the vehicle so as to maintain or modify the velocity vector. This allows an intended purpose or "mission" to be accomplished.

Qualities of an aircraft that tend to make it resist changes in the direction or magnitude of its velocity vector are referred to as *stability*, while the ease and expedition with which the vector may be altered are referred

[1] Samuel Johnson, "A Dissertation on the Art of Flying," Chapter 6 in *History of Rasselas*, originally published in 1759, republished by Clarendon Press, Oxford, 1931.

to as the qualities of *control*. Stability makes a steady unaccelerated flight path possible; maneuvers are made with control. The path of an aircraft, however, is never stable of itself; whether through the intervention of the human pilot or by means of automatic control, stability is actually secured with the mechanism of *feedback*, a principle by which cause and effect systems are modified to secure certain desirable properties. Information about the effect (or output) is fed back (or returned) to the input and is used to modify the cause. Typical of feedback control is its speed of response and its accuracy in following commands and in suppressing the effects of disturbances. Also typical, however, is its tendency to "hunt" or oscillate. The particular advantages of feedback are enhanced by high gain, but this is inimical to dynamic stability, and high gain also increases the susceptibility of the system to spurious signals or "noise." Therefore, a designer intending to exploit the potential advantages of feedback is compelled to strike a fine balance between the desirable properties that might be secured and the pressing danger of disastrous performance.

The earliest aeronautical experimenters had hoped to achieve "inherent" stability (i.e., without feedback), and while many, such as Cayley, Penaud, Lilienthal, Chanute, and Langley, pursued this goal and discovered how to set the incidence of the tailplane so as to achieve longitudinal stability with respect to the relative wind, and to use wing dihedral so as to achieve "lateral stability," it gradually became clear that configurations with a large amount of such inherent stability were particularly, and distressingly, susceptible to being upset by gusts.

Speaking before the Western Society of Engineers in 1901, Wilbur Wright said: "Men already know how to construct wings or aeroplanes, which when driven through the air at sufficient speed, will not only sustain the weight of the wings themselves, but also that of the engine, and of the engineer as well. Men also know how to build engines and screws of sufficient lightness and power to drive these planes at sustaining speed .... Inability to balance and steer still confronts students of the flying problem .... When this one feature has been worked out, the age of flying machines will have arrived, for all other difficulties are of minor importance."[2]

While this statement was somewhat optimistic with respect to the state of knowledge concerning airfoils and propellers, as the Wright brothers themselves soon discovered, it was correct in its essentials, and there is no doubt at all that suitable stability and control characteristics were the very last features of the first successful airplane to be developed. It is now generally agreed that the principal contribution of Wilbur and

[2] M. W. McFarland, ed., *The Papers of Wilbur and Orville Wright*, Vol. 1, McGraw-Hill, New York, 1953, pp. 99–100.

Orville Wright was their recognition that the frustrating search for inherent stability might well be abandoned if only the operator were provided with sufficiently powerful controls with which to balance and steer, i.e., that the human pilot, operating on feedback signals, could use the controls to stabilize a neutrally stable or an inherently *unstable* aircraft.[3] Of course, the Wright brothers did not use this language, and indeed the recognition of the essential character of the airplane as an element in a feedback control loop came comparatively recently.

While the first *automatic* feedback control system for an airplane actually antedated the first successful flight by more than a decade, and the demonstration of completely automatic control of an airplane in full flight took place more than 50 years ago—in 1914, the means employed to secure satisfactory flying qualities of the aircraft themselves and to develop artificial stabilizers and automatic pilots were, at first, largely empirical arts. They seem to have made progress with a minimum amount of mathematics until after the end of the 1939–1945 war.

The modern view of the dynamics of aircraft and their control systems, in terms of the stability and response of the entire closed-loop (feedback) system, can be traced from its sources by way of three separate branches of technical knowledge, their confluence, and the recent advance and augmentation of the subject (see Fig. 1-1). During roughly the first 50 years of aviation's history, the study of the dynamics of aircraft and their control systems was of negligible interest to designers, who learned to get by with rules of thumb for proportioning the stabilizing and control surfaces and to develop automatic feedback controls by cut-and-try methods. This was in spite of the fact that a mathematical theory for the stability of the unattended motion and of the aircraft's response to control was developed at an early date. On the other hand, design trends since World War II, which have greatly extended the flight envelope of fixed-wing airplanes and introduced new types of vehicles such as helicopters, VTOL airplanes, ground effect machines, hydrofoil boats, winged missiles, and space launchers, have so enormously multiplied the number and type of problems that the techniques formerly employed in practice would have been totally inadequate. Very fortunately, wartime pressures produced two developments that fundamentally altered techniques for the design of automatic flight control systems. The first of these was the theory of servomechanisms; the second was the electronic computer. Analysis and simulation are today the twin pillars on which the entablature of aircraft flight control system design stands.

There has been an explosive growth in the practice of "experimenting" with mathematical models. It has been urged by both the expanding complexity of the problems and the increasing availability of appropriate

[3] C. S. Draper, "Flight Control," *J. Roy. Aeron. Soc.*, 59, 451–477 (July 1955).

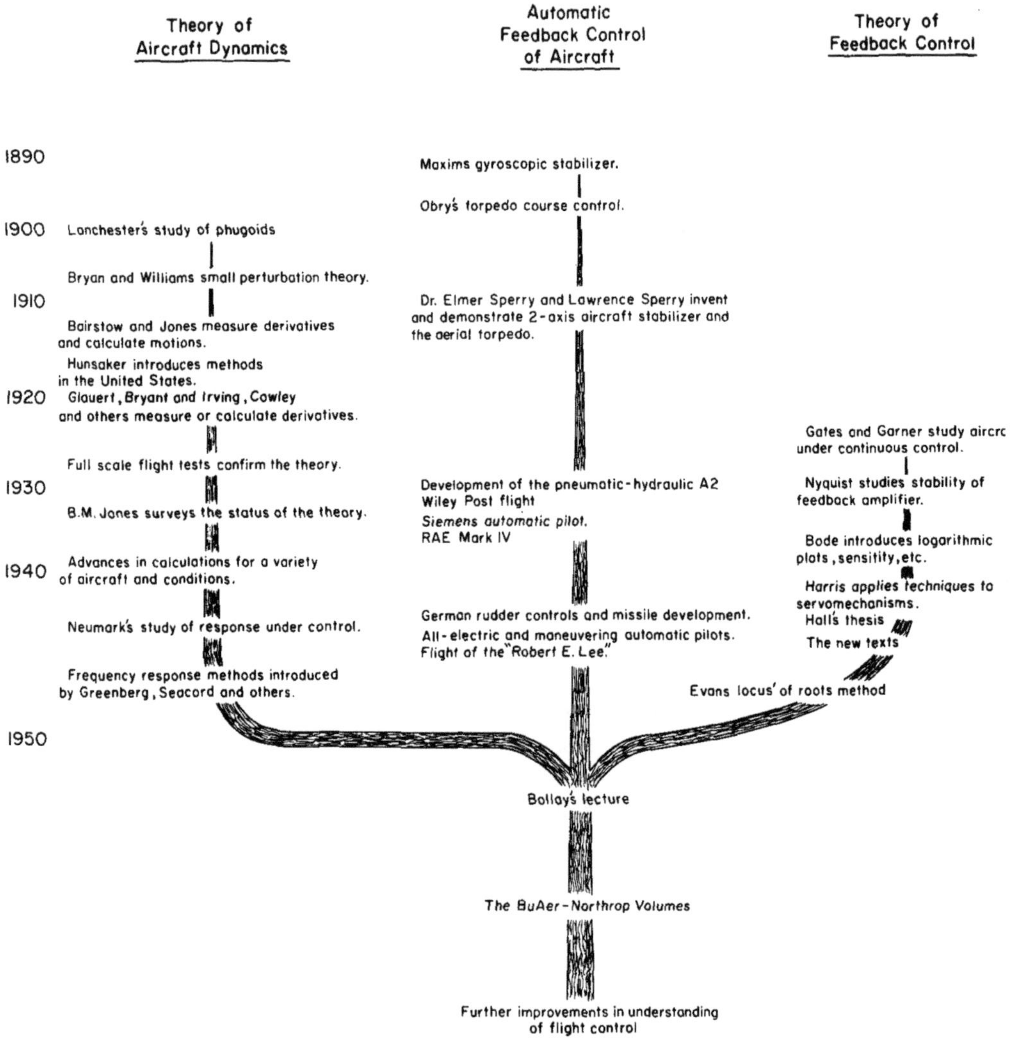


Fig. 1-1. Confluence and augmentation of the theory and practice of automatic feedback control of aircraft.

methods and techniques. Further, the mathematical theory has served for the classification, interpretation, and extrapolation of the growing number of results of physical experiments.

It is to the development, exposition, and demonstration of methods of analysis and synthesis for aircraft automatic flight control systems that this monograph is addressed. It is not a text on design but is rather a guide to the consideration of the effects of vehicle and equipment features on the dynamic performance of the system. Where possible, the emphasis in treating the elements of the system is on the largest entities. Thus attention is directed to the response of the airplane to elevator motion, rather than to the change in airflow over the tail, and to the input/output characteristics of a rate gyro, rather than to detailed consideration of the torques acting on the gimbal. The vehicles considered are the ones that are heavier than the fluid in which they operate but which are acted on by significant fluid dynamical forces. This class includes at least the following types of vehicles:

Airplanes
Helicopters
Vertical takeoff and landing aircraft
Ground effect machines
Hydrofoil boats

Control, as somewhat distinct from guidance, is taken to be the subject of interest. For this reason it will ordinarily be possible to consider the motions in moving coordinate systems fixed in the vehicle and to avoid the coordinate axis transformations required to obtain the vehicle motion in, for example, a coordinate system fixed in the earth. When the origin of the moving coordinate system is in an "equilibrium" state of motion along a nominal trajectory, the equations of motion of the vehicle can be linearized for small perturbations and the linearized equations will have constant coefficients. Then it is possible to use the convenient *transfer function* models for the dynamics of the vehicle, and all the analytical techniques for the study of linear feedback systems can be brought to bear on the problem.

Although there are a number of modern treatments of the stability and control of aircraft,[4] all of which emphasize the same approach to the

[4] Among the more recent texts are C. D. Perkins and R. E. Hage, *Airplane Performance, Stability, and Control*, Wiley, New York, 1949; W. J. Duncan, *Control and Stability of Aircraft*, Cambridge University Press, London and New York, 1952; B. Etkin, *Dynamics of Flight*, Wiley, New York, 1959; W. R. Kolk, *Modern Flight Dynamics*, Prentice-Hall, New York, 1961; A. W. Babister, *Aircraft Stability and Control*, Pergamon Press, New York, 1961; R. L. Halfman, *Dynamics*, Vol. 1, *Particles, Rigid Bodies, and Systems*, Addison-Wesley, Reading, Mass., 1962; E. Seckel, *Stability and Control of Airplanes and Helicopters*, Academic Press, New York, 1964.

linearized dynamics that is to be adopted here, and there is also a very wide selection of both introductory texts and more advanced treatises on automatic feedback control,[5] there has been a conspicuous lack of any significant treatment of these subjects in concert and therefore no proper introduction to the area between these fields. It is a fact that the methods of servomechanism analysis can be used as a powerful tool in the study of aircraft dynamics, and, additionally, that the characteristics of aircraft and their control systems provide a series of both subtle and complex problems that are likely to carry the student of feedback systems beyond what he may have learned in connection with the customary examples of remote position control, speed regulation, process control, and instrumentation. The discussion that follows will serve to bridge a gap between existing technical disciplines and to make more readily available some of the results contained in a scattered engineering report literature which is now familiar only to a small group of specialists.

The authors have adopted an eclectic view, taking from several fields what best appeals and suits but attempting, at the same time, to provide a unified treatment. Where a completely unified view is not feasible, the dominant theme is stated and the minor theme is contraposed.

It is the conviction of the authors that only the most thorough understanding of the dynamics of each element is a suitable basis for system synthesis. While digital and analog computers are now generally available to produce "solutions," even a sheaf of solutions may not clearly show the designer how to obtain the most satisfactory behavior and to avoid unpleasant surprises when the machinery is built. It is for this reason that the mathematical analysis of aircraft feedback control systems is emphasized throughout the treatment here. Of course, simulation and flight testing are valuable tools in the development of aircraft control systems, but, to an extent, a good theory is a summary of, and substitute for, experience, and the understanding which is conferred by analysis is a short-cut to the best results. It may seem, however, that a linearized theory is unrealistic because practical aircraft feedback control systems inevitably include nonlinear elements. The results that are achieved justify its use. Restrictions that are implicit in the use of linear theory are

[5] See, e.g., H. M. James, N. B. Nichols, and R. S. Phillips, *Theory of Servomechanisms*, McGraw-Hill, New York, 1947; H. S. Tsien, *Engineering Cybernetics*, McGraw-Hill, New York, 1954; J. G. Truxal, *Automatic Feedback Control System Synthesis*, McGraw-Hill, New York, 1955; H. Chestnut and R. W. Mayer, *Servomechanisms and Regulating System Design*, 2nd edn., Vol. 1, McGraw-Hill, New York, 1959; J. J. D'Azzo and C. H. Houpis, *Feedback Control System Analysis and Synthesis*, McGraw-Hill, New York, 1960; R. N. Clark, *Introduction to Automatic Control Systems*, Wiley, New York, 1962; E. C. Barbe, *Linear Control Systems*, International Textbook, Scranton, Pa., 1963; I. M. Horowitz, *Synthesis of Feedback Systems*, Academic Press, New York, 1963; C. J. Savant, Jr., *Control System Design*, McGraw-Hill, New York, 1964.

nowhere nearly as severe as might be imagined. In part, this is because linear approximations often have a substantial validity; in part it is so because feedback, in itself, tends to "linearize" the system.

Finally, it may or may not be true, as George Santayana said, that "those who cannot remember the past are condemned to repeat it," but there is enough truth there so that the history of the present subject can be studied with considerable profit. It is evident upon knowledgeable consideration that some costly mistakes might have been avoided with a better appreciation of the difficulties that confronted previous investigators of the problems of flight control.

## 1-1. Outline of the Volume: A Guide for the Reader

The subject of the feedback control of flight has a considerable scope and variety, and there is no canonical approach to its understanding. Its students will typically have acquired a considerable knowledge of the theory of linear feedback systems, and of the dynamic stability of aircraft and their response to control, as substantially independent subjects. The background of the typical reader will probably include some knowledge of operational or Laplace transform techniques for the solution of ordinary linear differential equations with constant coefficients, conventional servo analysis techniques such as the root locus and frequency response methods, response calculations with either deterministic or random inputs, and the describing function method for the treatment of common control system nonlinearities. While many of these matters are reviewed here before they are applied, the pace is brisk and the treatment is not intended as an introduction to the elements of the theory. The reader is further presumed to have some acquaintance with the dynamics of rigid bodies, although it is not, strictly speaking, necessary to have studied the dynamics of aircraft. Again, the latter subject is treated here *ab initio* but with a purpose not shared with the conventional texts cited in note 4.

Figure 1-2 is a graphical representation of the outline for this volume. The book begins, in this first chapter, with a definition of control appropriate to aeronautical vehicles and a distinction between control and *guidance*. This is followed by a brief summary of the advantages of feedback for control and an introduction to some of the earliest examples of feedback control. Historical sketches of the development of aircraft dynamic stability and control, practical automatic flight control systems, and feedback system analysis complete the introduction.

Chapter 2 comprises a review of those aspects of applied mathematics pertinent to the construction and use of linear mathematical models of aircraft and their control systems. The Laplace transform method and

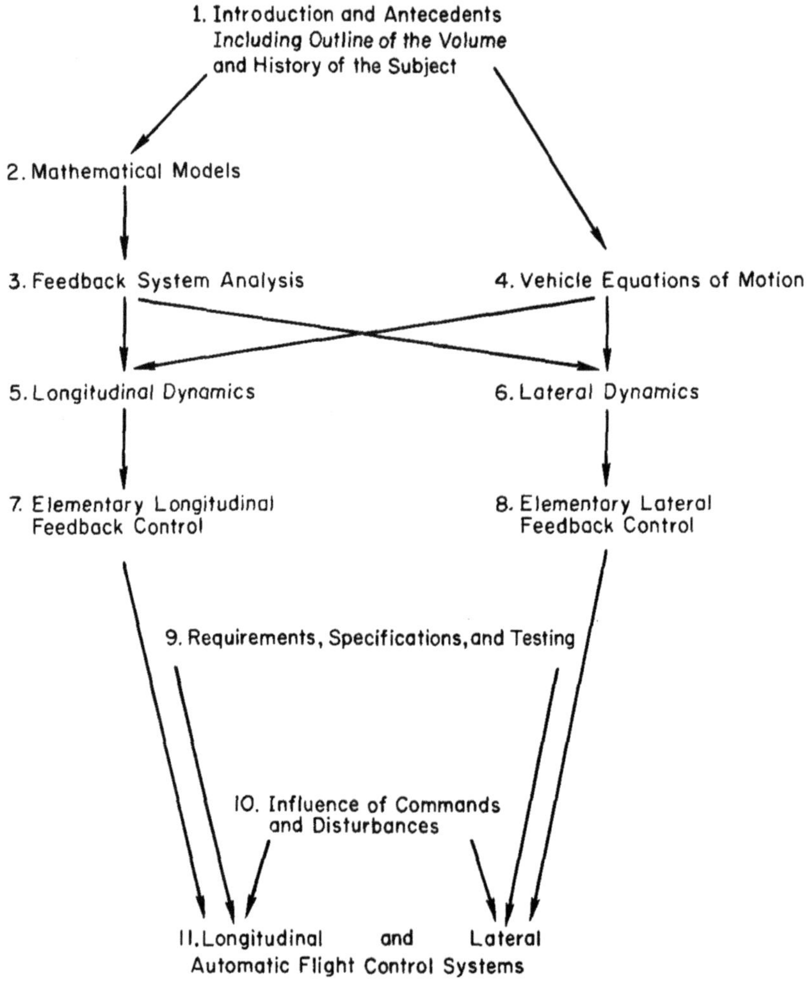


Fig. 1-2. Graphical outline of the volume showing interrelationship between topics.

the transfer function model, which play such a prominent part later, are discussed in detail, and considerable emphasis is placed on graphical representations and graphical constructions. While the typical reader is assumed already to have a considerable familiarity with this material so that he should be able to move ahead rapidly, he is likely to find that certain matters such as time vectors and the steady-state response to polynomial inputs are treated here in a unique way that provides a background for subsequent developments.

The material of Chapter 3 is a condensed account of the particular topics in feedback system analysis on which the remainder of the monograph strongly depends. Here the reader will find not only a review of the root locus method and the conventional open-loop/closed-loop logarithmic frequency response methods but also their presentation as elements of a unified servoanalysis method that is a complete generalization of the semigraphical analytical techniques. The reader will also find here an exposition of multiloop analysis procedures particularly appropriate to the study of vehicular control systems and, finally, a discussion of sensitivity, including the connection between gain sensitivity and the modal response coefficients (time vectors or eigenvectors) of the system response. This chapter is one of the most unusual features of the volume because many of the techniques, and especially their highly organized connections, are not explained in the conventional textbooks on linear feedback system analysis.

The main issue is discussed in Chapter 4, in which the equations of motion of aeronautical vehicles are developed from first principles. It is shown there that these equations can be linearized about a nominal flight path and that, when this flight path lies in the plane of symmetry of the unperturbed vehicle, the equations can normally be separated into two independent sets, the longitudinal equations and the lateral equations. This simplification is the basis for the division of the greater part of the balance of the discussion.

Still with the intention of studying the aircraft under active control, the longitudinal dynamics of the aircraft-alone are explored in Chapter 5. The transfer functions for the aircraft's response to control are evolved from the equations of motion, and *approximate factors* for the numerators and denominators are presented in terms of coefficients in the equations. Although approximate factors for parts of the characteristic functions (denominators) of airplanes have been known for about 40 years, it was only a few years ago that a similar understanding of the numerators was developed, and a similar approach to VTOL aircraft has only been successful even more recently. Here again, the presentation in Chapter 5 departs considerably from the conventional practice because little or no attention is paid to transfer function factors in the existing texts, and the developments summarized there represent a part of the novel approach of this volume.

Chapter 6 does for the lateral motions what Chapter 5 does for the longitudinal motions. The treatment is exactly parallel, although the results are different because of the distinction between the typical motions in the several degrees of freedom.

In Chapters 7 and 8 the discussion finally turns to the feedback control of airplanes and helicopters. The stability and response of vehicles under continuously active control are considered with the assumption of ideal proportional control, i.e., no account is yet taken of the practical imperfections such as lags, which inevitably are associated with real sensors, amplifiers, and actuators. The possible ideal feedback systems for control of the longitudinal motions are canvassed in Chapter 7, while a similar presentation on ideal feedback systems for control of the lateral motions is made in Chapter 8.

Chapter 9 is on the subject of general requirements, specifications, and testing. These subjects are presented in the context of a design process outline. Emphasis is placed on the sources of operational requirements and the logical evolution of the requirements from these origins. The requirements that derive from a consideration of flight control systems as feedback devices are also treated at length.

In Chapter 10 the effects of inputs and disturbances are treated as a performance consideration. Up to that point, the inputs to the system are considered to be relatively simple test signals such as an impulse or a sine wave. Now the influence on design of considering the structure of the inputs and disturbances is introduced for the first time. Actually, the inputs and disturbances are approximated by either deterministic signals more complicated than the ones previously considered, or, where their nature demands it, in probabilistic terms. The first probability density function and the second probability distribution function are reviewed; their use in system performance calculations is explained in some detail for the cases in which the signals have a Gaussian distribution and are stationary. In that case convenient calculations of the performance of linear systems can be carried out in the frequency domain by making use of the power and cross-spectral density functions. The "transient analog" and adjoint technique, which underlie the computer approach to more complex problems, are also introduced.

Finally, much of the material of all the previous chapters is used in discussions of longitudinal and lateral automatic flight control systems in Chapter 11. The influence of requirements and of imperfections in the components is particularly pointed out. Multiloop flight control systems of several types are treated as illustrative examples.

At the end of the book are two appendices and a bibliography. The bibliography supplements this book by providing references to those aspects of aircraft dynamics and automatic control that are not extensively treated here. It covers vehicle flexibility and other higher order dynamic effects, components, and descriptions of flight control systems. The first appendix presents tabulations of dynamic characteristics for some representative aircraft, and the second serves as a brief introduction to probability theory.

## 1-2. A Definition of Flight Control

It is not surprising that, when considered in detail, the abstract or physical attributes of an aeronautical vehicle or weapon system and its elements are so interrelated as almost to preclude discussion of any one aspect of the system without simultaneously treating most of the others. Still, it ultimately becomes necessary to stake out definite domains that can be treated more or less individually. This can be accomplished with some generality if other factors and entities in the system can be considered either precursory or *by definition* separated from the subject of special attention.

As a first step in separating the automatic flight control area from other aspects of the overall aeronautical vehicle or weapon system, it is necessary to distinguish control from guidance. Unfortunately, the boundary between these two areas is seldom inherently sharp because of basic functional, operational, and equipment interactions that they may share. As a practical matter, however, the following definitions can ordinarily be used:

Guidance: The action of determining the course and speed, relative to some reference system, to be followed by a vehicle.

Control: The development and application to a vehicle of appropriate forces and moments that

1. Establish some equilibrium state of vehicle motion (*operating point control*).
2. Restore a disturbed vehicle to its equilibrium (operating point) state and/or regulate, within desired limits, its departure from operating point conditions (*stabilization*).

To apply these definitions to a specific example, consider the air to surface missile system shown in Fig. 1-3. In this figure the blocks inscribed with capital letters in square brackets are not simple transfer functions relating outputs to inputs, but instead are matrix operations. It is readily apparent that the complete system, when viewed in the large, is complicated and analytically intractable. However, two major types of loops are seen to be present: one a series of inner loops involving the feedback of airframe motion quantities; the other an outer loop containing the kinematic transformations required to generate the relative orientation between target and vehicle, and closed through a geometry sensor and computer that generates flight path commands. By use of the definitions given above, it is now possible to separate the guidance and the control areas, at least in terms of the matrix operators shown in the block diagram.

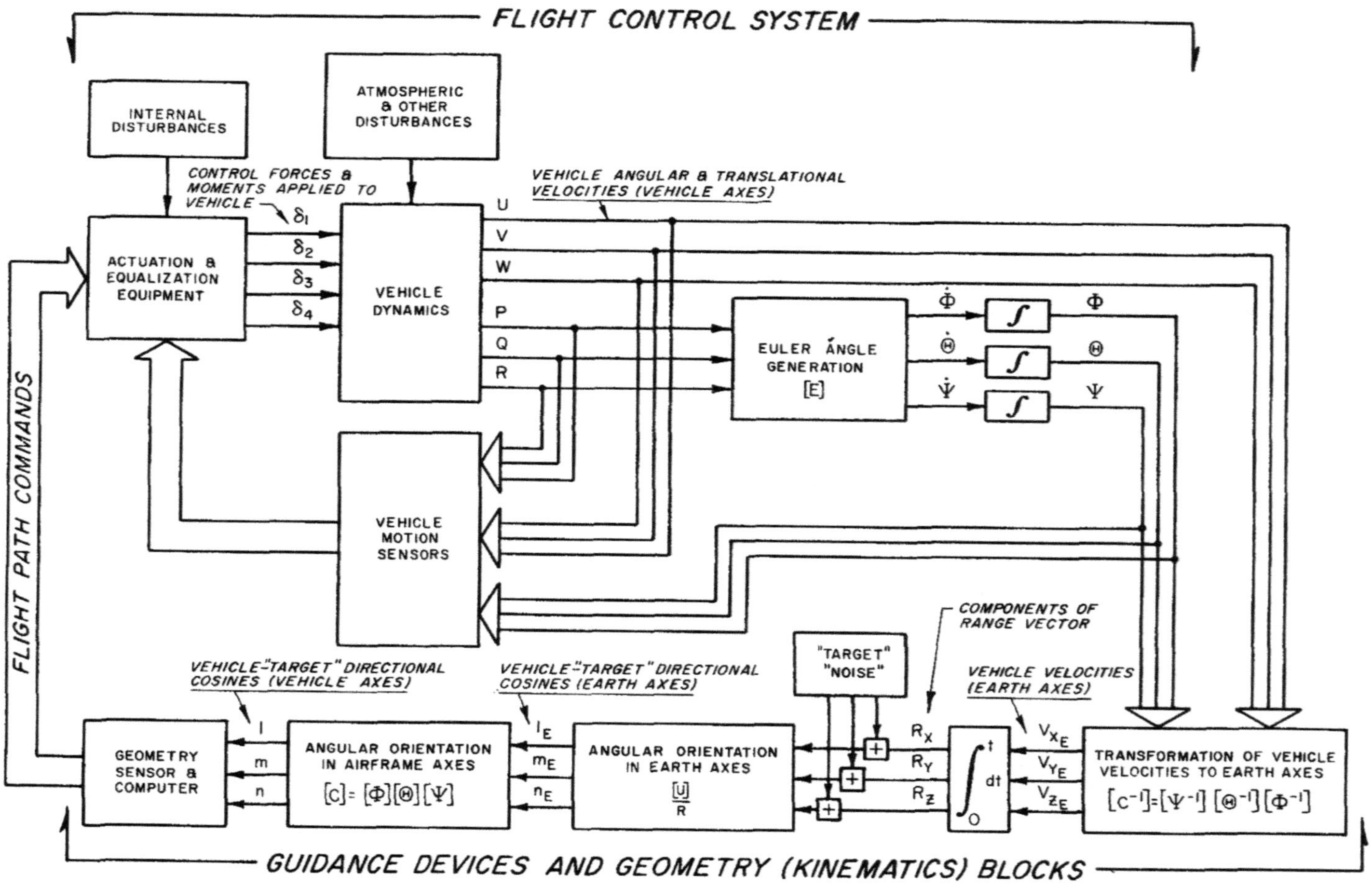


Fig. 1-3. Air to surface missile system block diagram.

Note, parenthetically, that an abstract or functional picture, rather than one drawn in terms of physical equipment, is preferred at this stage. If, e.g., in Fig. 1-3, the Euler angles, $\Phi$, $\Theta$, and $\Psi$, used as measures of vehicle motion, were obtained from a stable platform, this equipment would have to be considered a part of the flight control system; yet to many people the very words "stable platform" imply an item of guidance equipment.

On a physical basis Fig. 1-3 makes apparent an important distinction between the two types of loops. The flight control loop is concerned only with vehicle motion quantities measured in the aircraft (although two reference axis sytems are necessary), while the guidance loop involves axis system transformations that put the vehicle and target on comparable terms. For many systems this distinction is quite helpful in separating guidance from control. There is little doubt that the control of aircraft attitude angles is one of the functions of flight control, while the control of the *path* is, strictly speaking, a guidance function. Later it will become clear, however, that there are pseudo path variables such as pressure altitude and heading which are measured in the aircraft, and whose control, therefore, is logically considered to be a part of the domain of flight control. Further, it is often possible to formulate guidance problems such as terrain avoidance and approach to a runway on a localizer beam without involving more than linear approximations to the kinematic transformations in the guidance loop; then, with a single notable exception, guidance problems can be considered as minor extensions to the problem of flight control.

The exception is in those cases in which there are important dynamic interactions between the control and guidance loops. The complex diagram of Fig. 1-3 can be simplified by specifying the general type of guidance to be used and defining ideal steady-state "trajectories." The desired steady-state conditions can then be used as operating points, and all of the equations indicated by the block diagram of Fig. 1-3 can be *linearized* about these operating points. A simplified block diagram, emphasizing the system dynamics in a form suitable for dynamic analysis, can finally be drawn. Figure 1-4 shows linearized block diagrams (derived from Fig. 1-3) that relate perturbed quantities when the vehicle is on a straight line collision course with the target and is operating about straight and level flight condition.[6] Figure 1-5(a) results when the longitudinal control system block diagram is redrawn so as to use flight path angle, $\gamma$, instead of pitching velocity, $q$, and plunging velocity, $w$, as the motion

[6] While the implied assumption is surely a tremendously simplifying one, aeronautical vehicles do, in fact, spend most of their time in the air in straight and level flight, and the control system must be made to work for that flight condition first. The choice of operating point, however, is illustrative and is not necessary to the argument.

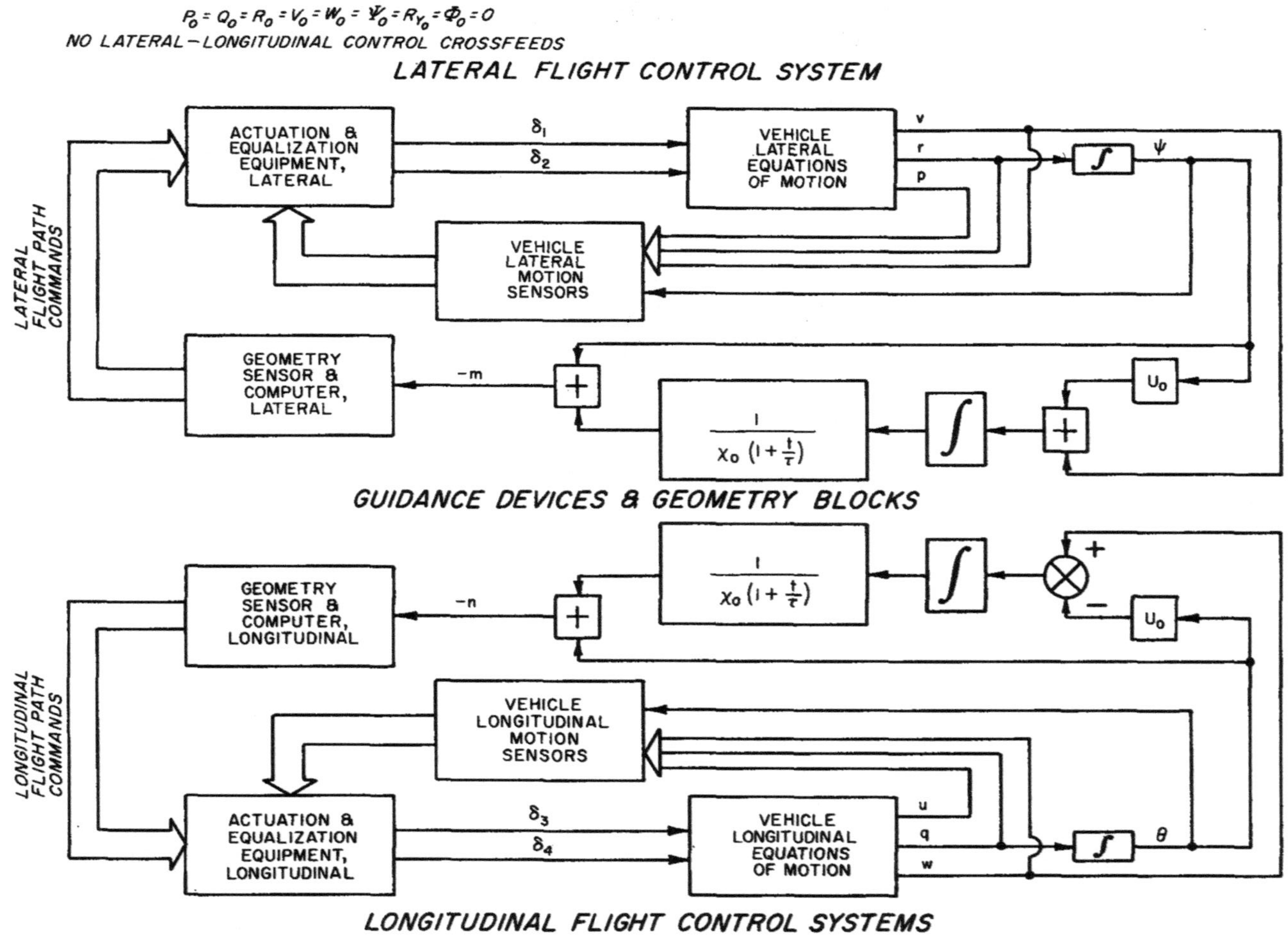


Fig. 1-4. Air to surface missile system linearized block diagram.

variables. Here the geometry relationships are shown in a single block, while the flight control system portion of the diagram is separated into functional divisions. Figure 1-5(b) goes one step further and shows a single closed-loop flight control system block with the geometry block broken into two parallel channels. Both diagrams in Fig. 1-5 assume unity dynamics for the geometry sensor and the computer.

Figure 1-5 emphasizes the fact that the geometry block contains a time-varying parameter $(1 - t/\tau)$, where the time variable, $t$, appears explicitly. The magnitude of the parameter defines the relative degree of dynamic interaction between the flight control and the guidance. When the ratio time:time-to-go, $t/\tau$, is very small, the sole dynamic effect of the guidance elements is to add a unity feedback path to the closed-loop flight control system. In most cases this effect, while certainly worthy of consideration, does not complicate the problem. *It can easily be taken into account as just another loop in the flight control system.* On the other hand, as $t/\tau$ approaches unity, the geometry block gain approaches infinity and the dynamic interactions between guidance and flight control may become the most prominent feature of system performance.

If the considerations exemplified in the discussions above are generalized for a variety of guidance system types, it is found that guidance and control can interact in only three ways. These are illustrated in Fig. 1-6 as connections between the two parts of a guidance and control dichotomy. The interaction with *operating point control*, as illustrated in the example above, is the starting point in the development of diagrams, such as the ones shown, which emphasize the dynamics. Guidance system effects on *limiting* involve the characteristics of particular physical mechanizations and, in any event, relate only to conditions outside the realm of linear theory. Finally, *dynamic interaction* between guidance and control occurs only in homing guidance, and even then the interaction is slight until the target is close. On these bases it should be clear that linear constant approximations to flight control and guidance systems can be treated completely separately, as far as their dynamics are concerned, for all guidance modes except final "homing" maneuvers. Therefore, if a particular operating point—straight and level flight—is assumed and a possible interest in homing maneuvers is neglected, the subject of control can be separated from guidance and the synthesis of automatic flight control systems can be studied in its own right.

## 1-3. Why Feedback?

The flight control systems in Figs. 1-3 to 1-5 are, quite apparently, shown as *feedback* systems, in which a portion of each output is fed back

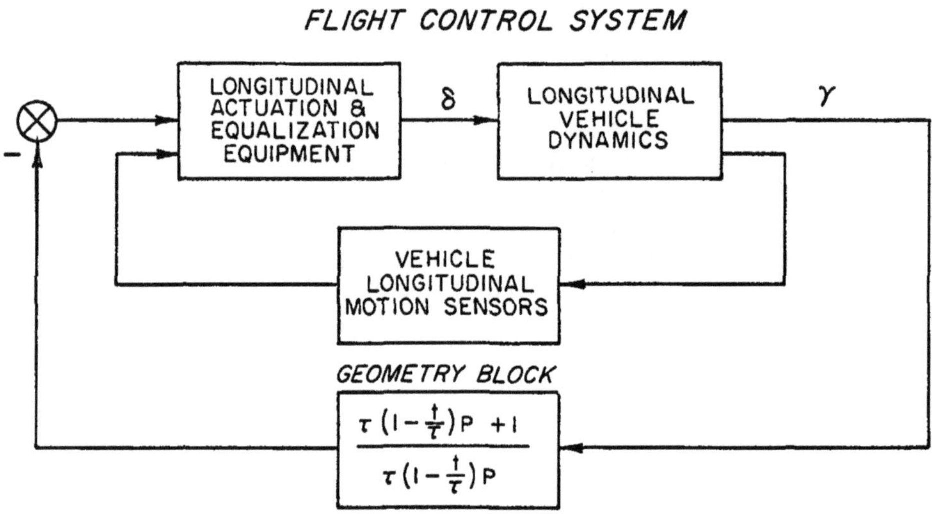


(a)

*Note: P is the operator,* $\frac{d}{dt}$

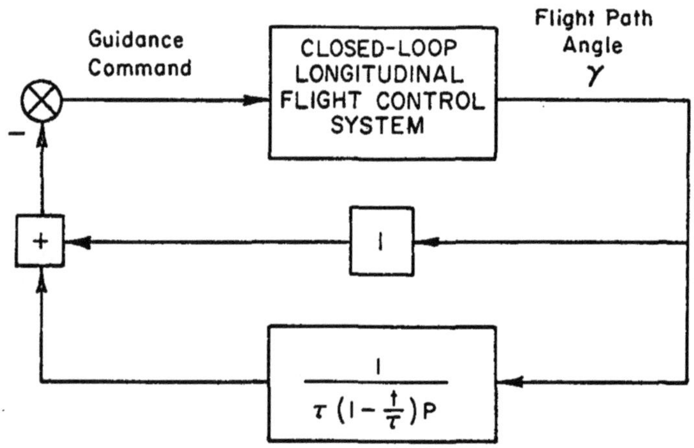


(b)

Fig. 1-5. Simplified longitudinal collision course guidance and control system.

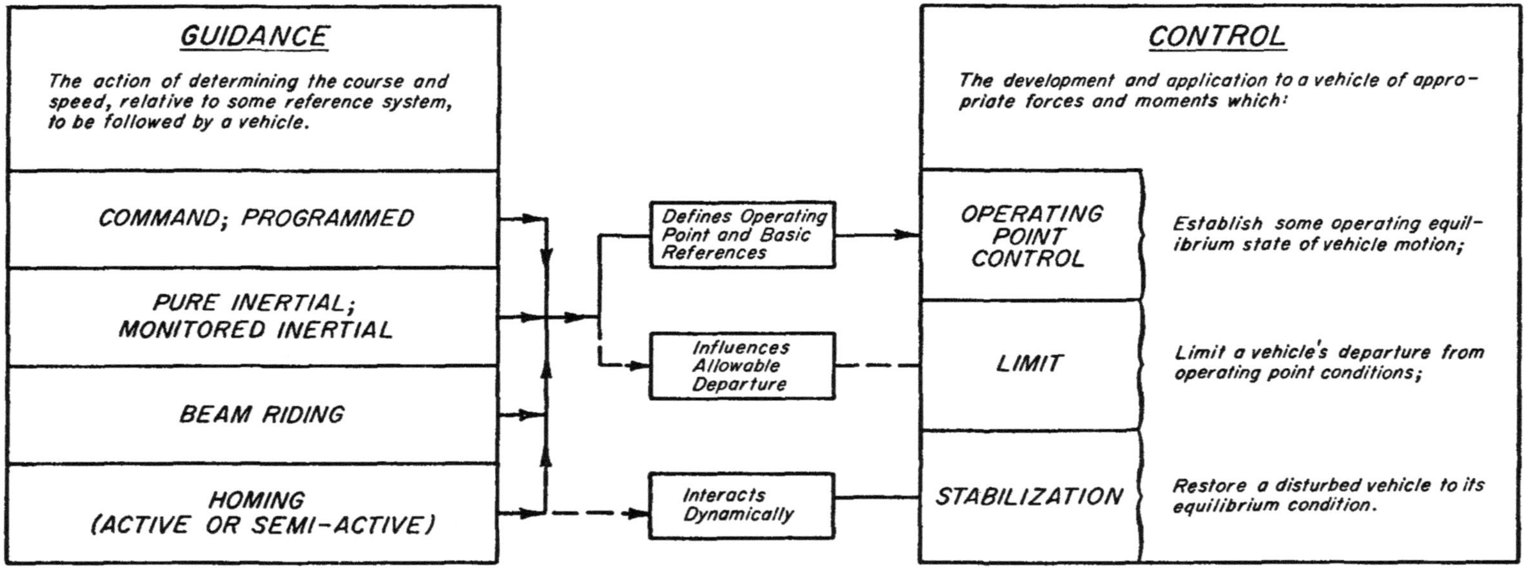


Fig. 1-6. Guidance/control dichotomy.

so as to modify the input. Effective flight control systems invariably are feedback systems, for a number of important reasons.

Even if invention had ultimately produced a satisfactory, inherently stable aircraft, the disadvantage of open-loop, i.e., nonfeedback, control would probably still preclude its use for the control of flight. Open-loop controls are programmed and calibrated. Their proper operation depends on the computation of an appropriate program and on maintaining the calibration of the controlled element or object of control. In flight control applications the appropriate program is often of considerable complexity, and a most notable feature of the aircraft's response to control is that it changes markedly with speed, altitude, and loading.

The advantages of feedback control are:

1. The provision of stability.
2. The adjustment of dynamic response, including
   a. Reduction of lags.
   b. Provision of desired or specified command/response relationships, especially as regards the improvement of linearity and the reduction of the effect of vehicle cross-coupling forces.
3. The suppression of unwanted inputs and disturbances.
4. The suppression of the effects of variations and uncertainties in the characteristics of the controlled element.

Feedback can make an unstable system stable. (It can, of course, also make a stable system unstable, a subject to which we shall have frequent occasion to return later.) Unfortunately, aircraft are never stable by themselves. At the very best they are neutrally stable in heading and altitude; continuous corrections must be made in order to fly a straight and level course. Otherwise, and this is especially true of modern configurations, a disturbance may start an aperiodically divergent motion such as the one pilots sometimes call the "graveyard spiral," or a similar disturbance may initiate weakly damped or perhaps divergent oscillations in the rolling, yawing, and pitching degrees of freedom. It is for the repair of any such deficiencies in stability that the classes of automatic flight control systems known as stability augmenters and automatic pilots are principally useful.

Feedback can improve the speed of response and may be used so as to enforce some desired correspondence between the input and output of the system. The series of figures that have been presented have served to emphasize the fact that one of the purposes of the flight control system is to follow the commands generated by the guidance system. Rapid and accurate response to commands, so that the commanded flight path is matched by the actual flight path, for example, is made possible or is largely enhanced by the feedback of aircraft motion quantities. These feedbacks have been illustrated and are defined in the figures to comprise the flight control system.

It is not only the case that the (feedback) flight control system will improve the speed of response and accuracy in following commands but it will also tend to suppress the effect of disturbances such as the atmospheric (gust) disturbances illustrated in Fig. 1-3, as well as the effects of changes in the characteristics of the vehicle's response to control. These are not the least reasons for employing feedback. The aircraft must typically fly in atmospheric turbulence which tends to upset it and to alter its flight path, and the response to control may very well be substantially changed by the consumption of fuel, the release of stores, and changes in the flight speed or altitude.

Some of the earliest inanimate feedback controls, aside from water level controls which were known in antiquity, were speed regulators for prime movers.[7] These were primarily designed to *regulate* against changes in speed because of disturbances such as changes in the load, or changes in the response of the machine to control such as a change in speed at the same throttle setting because of an increase in steam pressure. The early governors secured some of the very practical advantages of feedback but they also tended to display the largest disadvantage: a tendency to hunt or oscillate. The phenomenon of hunting of engine governors motivated a number of authors to study the stability of feedback control systems and to lay the foundations of a mathematical theory of the subject. Among the earliest of these investigators was the physicist J. C. Maxwell,[8] who in his own paper on the subject conceded his inability to discover the criteria for the stability of higher order systems. Later, he was one of the examiners who set the subject "The Criterion of Dynamical Stability" for the Adams Prize Essay Contest in 1877. The prize was won by E. J. Routh,[9] who considered not only governors but the stability of general motion of rigid bodies. His studies in that field became the basis for the investigation of the dynamic stability of aircraft and for many years provided the principal tool for the study of feedback control systems.

## 1-4. Early History of the Subject of Aircraft Dynamics

F. W. Lanchester was the first to investigate analytically the dynamic stability of aircraft. Before the turn of the century he experimented with glider models and studied the properties of the solutions to a simplified

[7] James Watt is commonly credited with the invention of the flyball governor, about 1784, but it seems likely that these were in use on windmills before his time. (See A. Wolf, *A History of Science, Technology, and Philosophy in the XVIIIth Century*, Macmillan, New York, 1939.)

[8] J. C. Maxwell, "On Governors," *Proc. Roy. Soc.* series A. No. 100, 270–283, 1868.

[9] E. J. Routh, *Stability of a Given State of Motion*, Macmillan, London, 1877.

set of equations for motion in the plane of symmetry.[10] He called the resulting flight paths "phugoids," a name that persists to this day.

In the year of the first powered flight, 1903, Bryan and Williams, using more conventional mathematical methods, introduced the linearized equations of motion that have been the foundation of studies of dynamic stability and response to control ever since.[11] Later, the theory of both the longitudinal and lateral motions was presented by Bryan.[12] The six Euler equations for the general motion of a rigid body were considered for "small" departures from steady straight flight of an airplane with a plane of symmetry. Under these assumptions the equations were shown to be separable into two groups of three each. One group related the motion variables in the plane of symmetry, while the other group related the motion variables out of the plane of symmetry. Neither group contained any variable occurring in the other, so that they could be treated entirely separately. The separate groups of equations were called the "symmetric" or *longitudinal* and the "asymmetric" or *lateral* equations. A further consequence of the assumption of small perturbations was that the air forces on the airplane could be shown to depend on certain *constants* or "stability derivatives," as they were called; Bryan suggested that these might be determined experimentally.

As early as 1912, Bairstow and Melvill Jones, at the National Physical Laboratory in Great Britain, had taken up Bryan's suggestion and had developed some of the wind tunnel techniques for measuring the stability derivatives of models. They reported the results of their initial effort the following year, showing how features of the motion could be recognized in the mathematical solutions for the free response of a hypothetical airplane for which they had measured or calculated all the derivatives.[13] In both language and notation this report is thoroughly modern; it might be used as a text on the dynamic stability of airplanes today.

The theory and the experimental practice were subsequently extended by the original investigators and others. Bairstow considered the stability of more complicated motions such as circling flight and treated the motion of dirigible airships. He provided a comprehensive account of the subject in 1920.[14] Hunsaker, who had visited the National Physical Laboratory

[10] F. W. Lanchester, *Aerodonetics*, Archibald Constable, London, 1908. See also B. Melvill Jones, "Dynamics of the Aeroplane," in W. F. Durand, ed., *Aerodynamic Theory*, Vol. 5, Durand Reprinting Committee, Pasadena, Calif., 1943; republished (Vols. 5 and 6 bound in one volume) by Dover, New York, 1963, pp. 2–3, 169.

[11] G. H. Bryan and W. E. Williams, "The Longitudinal Stability of Aerial Gliders," *Proc. Roy. Soc.* Series A, 73, No. 489, 100–116 (1904).

[12] G. H. Bryan, *Stability in Aviation*, Macmillan, London, 1911.

[13] L. Bairstow, B. Melvill Jones, and B. A. Thompson, "Investigation into the Stability of an Airplane," ARC R & M 77, 1913.

[14] L. Bairstow, *Applied Aerodynamics*, Longmans Green, London, 1st edn., 1920; 2nd edn., 1939.

in 1914, introduced Bairstow's wind tunnel techniques and the method of Bryan and Bairstow for the calculation of dynamic stability in the United States. He collaborated on the first report of the United States National Advisory Committee for Aeronautics which was concerned with the response of aircraft to gusts.[15] Glauert calculated the stability derivatives of a running propeller and the motions of an airplane with the elevator free.[16] The model measurement[17] or calculation[18] of particular stability derivatives continued to attract attention, and a considerable effort was made to measure derivatives, free motions, and the response to controls[19] in full scale flight tests. The references given in the footnotes are only typical, not comprehensive. More details are given in the historical sketch appended to the 1947 paper by Milliken[20]; anecdotal accounts of the work of the British pioneers are presented in several contributions to the "Centenary Journal" of the Royal Aeronautical Society,[21] particularly the ones by A. V. Stevens, Sir Harry Garner, J. L. Mayler, and R. W. McKinnon Wood.

By 1935, when the survey by B. Melvill Jones appeared in Durand's *Aerodynamic Theory*, the classical approach of Bryan and Bairstow was well established but was very little used. Results of the full scale experiments had led to the conviction that the theory of infinitesimal motions was practical for the prediction of the stability of motion, the time history of the motion following a disturbance, and the response to the application of control. The effect of variations in the configuration of a typical airplane had been traced via their influence on the derivatives to the result in terms of stability of motion. Furthermore, these results were

[15] J. C. Hunsaker, "Experimental Analysis of Inherent Longitudinal Stability for a Typical Biplane," NACA TR 1, Pt. I, 1915. See also "Dynamic Stability of Aeroplanes," *Smithsonian Misc. Collection*, 1916.

[16] H. Glauert, "The Stability Derivatives of an Airscrew," ARC R & M 642, Oct. 1919; see also "The Longitudinal Stability of an Airplane," ARC R & M 638, 1919.

[17] L. W. Bryant and H. B. Irving, "Apparatus for the Measurement of $M_q$ on a Complete Model Airplane," ARC R & M 616, 1919.

[18] W. L. Cowley, "The Effect of the Lag of Downwash on the Longitudinal Stability of an Aeroplane," ARC R & M 718, Feb. 1918.

[19] H. Glauert, "Analysis of Phugoids Obtained by a Recording Airspeed Indicator," ARC R & M 576, Jan. 1919; E. P. Warner and F. H. Norton, "Preliminary Report on Free Flight Tests," NACA TR 70, 1919; F. H. Norton, "Practical Stability and Controllability of Airplanes," NACA TR 120, 1921; see also NACA TR 112, TR 167, and TR 170; M. A. Garner and S. B. Gates, "The Full-Scale Determination of the Lateral Resistance Derivatives of a Bristol Fighter," ARC R & M 987, Aug. 1925; see also ARC R & M 1068 and 1070; H. A. Soule, and J. B. Wheatley, "A Comparison Between the Theoretical and Measured Longitudinal Stability Characteristics of an Airplane," NACA TR 442, 1933.

[20] W. F. Milliken, Jr., "Progress in Dynamic Stability and Control Research," *J. Aeron. Sci. 14*, No. 9, 493–519 (Sept. 1947).

[21] "Centenary Journal, Royal Aeronautical Society 1866–1966," *J. Roy. Aeron. Sci. 70*, No. 661, pp. 71–78, 82–84, 89–90, 95–97 (Jan. 1966).

appreciated not only in terms of the solutions to specific numerical examples but more generally, at least in part, as approximate solutions given in terms of the dominant literal stability derivatives.

Melvill Jones himself, speaking of complete solutions to the equations of motion and of approximate solutions to the stability quartic equations, evaluated the state of affairs in the following words:

> In spite . . . of the completeness of the experimental and theoretical structure . . . it is undoubtedly true that, at the time of writing, calculations of this kind are very little used by any but a few research workers. It is in fact rare for anyone actually engaged upon the design and construction of aeroplanes to make direct use of [such] computations . . . , or even to be familiar with the methods by which they are made . . . . In my own opinion it is the difficulty of computation . . . which has prevented designers of aeroplanes from making use of the methods.
>
> Though the process . . . will, if worked correctly, give the final answer required, it is so involved that it is not easy to trace the connection between the final answer and the separate characteristics of the airplane which are represented by the various derivatives included in the equation of motion.
>
> With regard to the response to specific disturbances no convenient means of tracing this connection has yet been devised; but when . . . the form of the solution of the quartic for $\lambda$ [i.e., the frequencies and damping factors of the free modes] in normal flight is all that is required, the omission of certain terms, which are then relatively unimportant, allows such drastic simplifications to be made that the relation between cause and effect can be displayed with comparative ease.[22]

The situation was hardly altered during the next ten years. In spite of the introduction of the method of operators,[23] which *did* reduce the labor of computation, and in spite of earnest efforts to make the techniques as simple and general as possible by introducing a nondimensional

[22] B. Melvill Jones, "Dynamics of the Aeroplane," in W. F. Durand, ed., *Aerodynamic Theory*, Vol. 5, Durand Reprinting Committee, Pasadena, Calif., 1943, republished (Vols. V and VI bound in one volume) by Dover Publications, New York, 1963, pp. 2–3, 169.

[23] L. W. Bryant and D. H. Williams, "The Application of the Method of Operators to the Calculation of the Disturbed Motion of an Airplane," ARC R & M 1346, July 1930; R. T. Jones, "A Simplified Application of the Method of Operators to the Calculation of the Disturbed Motion of an Airplane," NACA TR-560, 1936; see also "Calculation of the Motion of an Airplane under the Influence of Irregular Disturbances," *J. Aeron. Sci.*, 3, No. 12, 419–425 (Oct. 1936); A. Klemin and B. F. Ruffner, "Operator Solutions in Airplane Dynamics," *J. Aeron. Sci.* 3, No. 7, 252–255 (May 1936).

notation,[24] and by summarizing information on the stability factors in convenient charts,[25] and, further, in spite of hortatory expositions of the theory,[26] designers of airplanes continued to disdain dynamic stability analysis.

Nevertheless, research continued at a pace that was accelerated by the advent of the war, and some improvements were made in the understanding of, e.g., the importance of wing/fuselage interference and power effects on the stability derivatives, the effect of closely balanced free controls on the motion, the response to particular motions of particular controls such as spoilers, and the influence of changes in design on the character of the motions.

The point of view then current, however, did not permit one (with very rare and soon forgotten exceptions) to consider the response of the airplane under the continuous action of the controls, i.e., as a feedback system. The controls were, almost invariably, considered as:

1. Fixed, as in the earliest studies.
2. Free, i.e., restrained only by aerodynamic hinge moments (or later by friction as well).[27]
3. Programmed, i.e., moved as a simple function of time, such as a step or ramp function or a smooth pulse.[28]

It may have been not only the fact that the calculations were laborious but also that the assumptions of the analysis appeared unrealistic, which discouraged their use in design. The stability of unattended motion with the controls fixed or free and the response to programmed control motions were and are, indeed, of some interest in connection with the dynamics of an airplane, but it is evident that the human or an automatic pilot flies by operating the controls more or less continuously. The airplane plainly is an element in a system that includes a human pilot or an automatic

[24] H. Glauert, "A Non-Dimensional Form of the Stability Equations of an Aeroplane," ARC R & M 1093, 1927.

[25] S. B. Gates, "A Survey of Longitudinal Stability below the Stall, with an Abstract for Designers' Use," ARC R & M 1118, July 1927; C. H. Zimmerman, "An Analysis of Longitudinal Stability in Power-Off Flight with Charts for Use in Design," NACA TR-521, 1935; also "An Analysis of Lateral Stability in Power-Off Flight with Charts for Use in Design," NACA TR 589, 1937.

[26] O. C. Koppen, "Happier Landings," *Aviation*, Sept. 1934; "Control Sensitivity," *Aviation*, Oct. 1935; "Smart Airplanes for Dumb Pilots," paper presented to the SAE, Detroit, Mich., Jan. 1936; "Airplane Stability and Control from the Designer's Point of View," *J. Aeron. Sci.*, 7, No. 4, 135–140 (Feb. 1940).

[27] H. Glauert, "The Longitudinal Stability of an Aeroplane," ARC R & M 638, 1919; R. T. Jones and D. Cohen, "Analysis of the Stability of an Airplane with Free Controls," NACA TR 709, 1940.

[28] R. T. Jones, "A Simplified Application of the Method of Operators to the Calculation of the Disturbed Motion of an Airplane," NACA TR 560, 1936; K. Mitchell, "Lateral Response Theory," RAE Rept. Aero. 1952, Mar. 1944.

pilot. This view did not come to be generally accepted until after the war, and the understanding of a convenient means of tracing the connection between the response to specific disturbances, such as the operation of the controls, and the characteristics of the airplane that are represented by the various derivatives was of an even later date.

## 1-5. Early History of Automatic Flight Control

The development of automatic flight, like the development of airplanes themselves, proceeded for a long time with the benefit of very little theoretical knowledge.

Sir Hiram Maxim (1840–1916) was a prodigious inventor. In 1891, when he turned his attention to the design and construction of a heavier than air flying machine, he proposed to secure its longitudinal stability by means of a servo drive and automatic feedback. The devices that are described in Maxim's book and are illustrated there by a photograph of the installation in an airplane are surprisingly modern in concept and execution.[29]

A steam-driven, pendulous, vertical gyroscope was made to operate a valve that ported steam to a servo cylinder.[30] The motion of the piston drove the elevators, and the feedback link from the piston repositioned the gyro-operated valve body so as to close the valve. In principle, this "gyroscopic apparatus for automatically steering [the] machine in a vertical direction" is indistinguishable from the elevator control portion of automatic pilots of much more recent date. Easily recognizable are the elements of any automatic flight control system: the sensor (gyroscope), the amplifier (valve), and the control surface positioning servo. Unluckily, tests of the flying machine for which the gyroscopic control was designed ended in disaster when the aircraft lifted off the tracks designed to restrain it, turned over, and was destroyed. Maxim, who felt that his special contribution was to be the development of lifting surfaces and power plants, thought that the point about lift and power had been proved, and the inventor turned his energies in other directions.

Still in the nineteenth century, however, successful gyroscopic feedback control of the "flight" path was demonstrated by Ludwig Obry, an Austrian, who in 1894 introduced a course-keeping gyro as an improvement for the naval torpedo invented by Robert Whitehead in 1866. (Whitehead's torpedo had an automatic depth control.) A little later the

[29] H. S. Maxim, *Artificial and Natural Flight*, Whittaker, London, 1908, pp. 92–94.

[30] The principles of steam and hydraulic servomotors had already been known for some time. See A. B. Brown, British Patent 2253, 1871, and J. Farcot, *Le Servo Moteur ou Moteur Asservi. Gouvernails à Vapeur Farcot, Description Théoretique et Pratique*, J. Baudry, Paris, 1873. Among early applications to vehicle control were steering engines for steamships.

principle of gyroscopic stabilization of ships was introduced, and, although depending on a completely different principle, was the model for the next attempt at gyroscopic control of an aircraft. In 1909–1910 Dr. Elmer Sperry attempted to make a gyroscopic "stabilizer" for an airplane. This was a rigidly mounted engine-driven wheel with its spin axis vertical. It would have opposed rolling motions with a pitching torque, and vice versa. It was apparently never brought to a test because of the lack of success of the airplane in which it was installed but it served as the inspiration for further trials.[31]

Between 1910 and 1912 Dr. Sperry and his son Lawrence developed and installed, in an airplane belonging to Glenn H. Curtiss, an all-electric, two-axis automatic pilot. Roller contacts on a gyro platform, measuring the bank and pitch angles, actuated solenoid clutches which connected the ailerons and elevator to a propeller-driven "air turbine," and motion of the surfaces repositioned the contactor segments[32] (e.g., see Fig. 1-7). The machine was announced to the public in October 1912. In 1914 the aircraft and its automatic pilot were entered in a safety contest sponsored by the Aero Club of France. Lawrence Sperry made a dramatic demonstration of automatic flight as he flew at low altitude along the Seine in the vicinity of Paris, standing upright in the cockpit of the Curtiss flying boat,

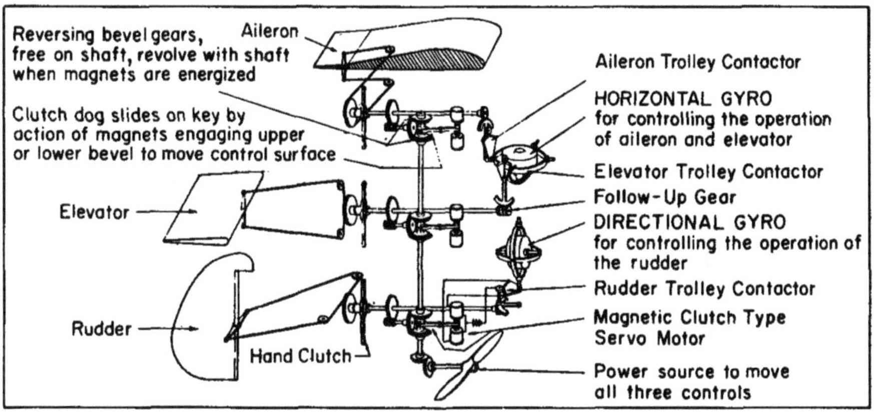


Fig. 1-7. Diagrammatic plan of Sperry automatic pilot.

[31] C. S. Draper, "Flight Control," *J. Roy. Aeron. Soc.*, 59, 451–477 (July 1955).

[32] One observer vividly recalls the loud groaning noise that this type of automatic pilot made. Presumably, the noise came from the grinding of the toothed clutch faces. See the discussion by Dr. A. L. Rawlings in the article by F. W. Meredith and P. A. Cooke, "Aeroplane Stability and Automatic Control," in *J. Roy. Aeron. Soc.*, 61, No. 318, 415–436 (June 1937).

Table 1-1. Selected early inventions in the feedback control of aircraft (adapted from F. Haus, "Automatic Stabilization," NACA TM 802, Aug. 1936).

| Feedback variable | Control | Inventor and date | Actuating means |
|---|---|---|---|
| Speed, U | Elevator deflection, $\delta_e$ | Budig 1912<br>Etévé 1914 | Mechanical connection to sensor |
| Incidence, $\alpha$ | Elevator deflection, $\delta_e$ | Etévé 1910 | Mechanical connection to sensor |
| Inclination, $\theta$ | Elevator deflection, $\delta_e$ | Regnard 1910<br>Sperry 1912<br>RAE 1927 | Electric type of servo.<br>Air-turbine-driven clutch servo.<br>Pneumatic servo |
| Angular velocity, $\dot{\theta}$ | Elevator deflection, $\delta_e$ | Lucas 1929<br>Girardville 1910 | Mechanical connection to sensor |
| Direction of apparent gravity, $g \sin \theta + dU/dt$ | Elevator deflection, $\delta_e$ | Moreau 1912 | Electric-motor-driven clutch servo |
| Speed, U, and incidence, $\alpha$ | Elevator deflection, $\delta_e$ | Etévé 1914 | Mechanical connection to sensor |
| Speed, U, and direction of apparent gravity, $g \sin \theta + dU/dt$ | Elevator deflection, $\delta_e$ | Doutre 1911 | Pneumatic servo |
| Speed, U, and magnitude of apparent gravity, $a_z$ | Elevator deflection, $\delta_e$ | Doutre 1913 | Pneumatic servo |
| Speed, U, and inclination, $\theta$ | Elevator deflection, $\delta_e$ | Marmonier 1909 | Unknown type of servo |
| Speed, U, and angular velocity, $\dot{\theta}$ | Elevator deflection, $\delta_e$ | Boykow 1928 | Hydraulic servo |
| Sideslip, $\beta$ | Aileron deflection, $\delta_a$ | Constantin 1920 | Mechanical connection to sensor |
| Bank angle, $\varphi$ | Aileron deflection, $\delta_a$ | Sperry 1912 | Air-turbine-driven clutch servo |

Table 1-1 (Continued).

| Feedback variable | Control | Inventor and date | Actuating means |
|---|---|---|---|
| Heading, $\psi$ | Rudder deflection, $\delta_r$ | RAE 1927 | Pneumatic servo |
| Yawing velocity, $r$, and side acceleration, $a_y$<br>Side acceleration, $a_y$, and yawing velocity, $r$ | Rudder deflection, $\delta_r$<br>Aileron deflection, $\delta_a$ | Mazade and Aveline 1922 | Pneumatic servos |
| Bank angle, $\varphi$<br>Heading, $\psi$ | Aileron deflection, $\delta_a$<br>Rudder deflection, $\delta_r$ | Sperry 1932 | Air-turbine-driven clutch servos |

holding his hands over his head, while his mechanic walked out along the wing. A photograph of this unusual event has been reproduced by Bollay and by Richardson, among others.[33] A similar demonstration was planned for early the next year in New York, but there the aircraft was overturned and extensively damaged by wind before the demonstration of its performance could be satisfactorily completed.[34]

This first automatic pilot was intended as an aircraft stabilizer. In other words, it was intended to supply stability, as we now say, "artificially," to aircraft that were often deficient in this regard. Other inventors were pursuing the same goals by the same and other means. The feedback of angle of attack and angle of sideslip, speed, longitudinal, side, and normal acceleration, lift, and body axis rates, as well as attitude angles, were all tried singly and sometimes in combination. Clarke[35] in an early paper described some of his own experiments in Great Britain, while Haus[36] has sketched some of the history of early developments on the continent of Europe (see Table 1-1). However, none of these original inventors was successful enough for his device to pass immediately into common use.

The design of aircraft made giant strides during the 1914–1918 war and it was found that sufficient stability for the human pilot's use could be supplied by suitable choice of the size and shape of the aerodynamic surfaces. Actually, many aircraft were still unstable but not dangerously so, and, with reference to the ground, the human pilot performed the stabilizing and control functions of the (feedback) control and guidance

[33] W. Bollay, "Aerodynamic Stability and Automatic Control," *J. Aeron. Sci.*, 18, No. 9, 569–624 (Sept. 1951); K. I. T. Richardson, *The Gyroscope Applied*, Philosophical Library, New York, 1954.

[34] "The Sperry Gyroscopic Stabilizer," *Flight*, 7, No. 5, 74–76 (Jan. 29, 1915).

[35] T. W. K. Clarke, "Auto-mechanical Stability," *Aeron. J.*, 101–115 (Apr. 1912).

[36] F. Haus, "Automatic Stability of Airplanes," NACA TM 695, Dec. 1932. *Automatic Stabilization*, NACA TM 802, Aug. 1936; 815, Dec. 1936.

systems. Neither artificial stabilizers nor automatic pilots were found to be particularly useful on the manned warplanes. They, in effect, disappeared from view. Under the cover of military secrecy, however, the development of the automatic pilot was continued for possible application to pilotless aircraft; in 1917–1918 Lawrence Sperry completed the construction and test of an aerial torpedo for the United States Navy.[37] In an advanced version the aerial torpedo was even remotely controlled by radio. The success of the project, however, came too late for the use of a "flying bomb" in World War I. Following the war, and turning to more prosaic applications, the Sperry Gyroscope Company had, by 1932, developed an automatic pilot for possible commercial transport use[38] (see Fig. 1-7). Except for the gyroscopic references that comprised the then new directional and vertical gyroscopes and the fact that it provided for control about all three aircraft axes, this automatic pilot retained many of the features of the original one of 1910–1915. Better results, however, were just around the corner.

In 1933 the prototype of the A2 automatic pilot was under construction. It featured panel-mounted gyroscopes with pneumatic pickoffs and three-axis control with proportional hydraulic servos. When Wiley Post, visiting the Sperry factory, saw it, he insisted that it be installed in his Lockheed Vega 5-C.[39] During the period July 15–22, 1933 Post, flying alone, set a round-the-world record of 7 days, 18 hours. The performance and reliability of the automatic pilot, which allowed the human pilot to perform the navigator's function and even to nap in flight, played a considerable role in this feat. The prototype automatic pilot used by Wiley Post, together with his airplane, the "Winnie Mae," are in the National Aeronautical Collection of the Smithsonian Institute in Washington.

The A2 automatic pilot came into widespread use among the airlines during the 1930s. This was due partly to its demonstrated reliability, partly because the panel-mounted instruments then coming into extensive use for routine operations under low visibility conditions supplied its gyroscopic references (an obvious economy in cost and weight), and partly because airplanes had achieved a range performance that made pilot relief attractive. The A2 was first introduced to airline service on the Boeing Model 247 in 1934. Its defects, if any, were its virtues; it flew straight and level. It was not designed for maneuvering.

In effect this automatic pilot provided for control surface deflections

[37] P. R. Bassett, "Instruments and Control of Flight," *Aeron. Eng. Rev.*, 12, No. 12, 118–123, 133 (Dec. 1953).

[38] E. A. Sperry, Jr., "Description of the Sperry Automatic Pilot," *Aviation Eng.*, 6, No. 1, 16–18 (Jan. 1932); see also E. S. Ferry, *Applied Gyrodynamics*, Wiley, New York, 1932, pp. 123–125.

[39] P. R. Bassett, "Instruments and Control of Flight," *Aeron. Eng. Rev.*, 12, No. 12, 118–123, 133 (Dec. 1953).

in three axes that were proportional to the departures from the reference attitude. A description of its operation is given by Richardson,[40] among others. It was as if the surfaces were "geared" to the instrument (see Fig. 1-8). This concept of gearing was frequently employed in some of the early mathematical studies of automatic control of aircraft but it lacks the generality offered by the concept of feedback.

During roughly the same period of time (1922–1937), somewhat similar automatic pilot developments were underway in Great Britain, although the aim there was the satisfaction of military requirements and the work was, at first, carried out in secrecy.[41] Interestingly, in both the earlier (Mark I) and later (Mark VII and Mark VIII) versions, the British pursued the design of two-axis controls with great diligence. The Mark I used a single free gyroscope measuring heading and pitch to control the rudder and elevators by means of low pressure pneumatic servos. An account of the results of the early efforts was presented by the pioneers Meredith and Cook of the Royal Aircraft Establishment (RAE) in England in 1937.[42] This account describes the use of the automatic pilot in aerial map-making and suggests its superiority over the Sperry three-axis design (Fig. 1-9) for applications requiring maneuvers. It also shows the considerable acquaintance of the authors with the theory, methods, and conclusions of their colleagues at the Royal Aircraft Establishment who had been engaged in the study of the dynamic stability of airplanes. The action of the automatic pilot was clearly explained in those terms. In fact the methods of dynamic stability analysis had been applied in the design of the RAE flight control equipment from 1924 on, but very few results were ever published in the open literature.

Following in the footsteps of Bairstow[43] and Glauert[44] and using a method developed by S. B. Gates,[45] who employed it in the obscure RAE

[40] K. I. T. Richardson, *The Gyroscope Applied*; see also P. R. Bassett, "Development and Principles of the Gyropilot," Instruments, 9, No. 9, 251–254 (Sept. 1936); *The Sperry Aircraft Gyropilot*, Sperry Gyroscope Co. Publication 15-665, July 1940.

[41] A comprehensive bibliography of British and foreign work on automatic flight control, both theoretical and experimental, from 1903–1957 has been prepared by the RAE. See R. C. Wright, A. T. E. Bray, and H. R. Hopkin, "List of Published and Unpublished References on the Remote and Automatic Control of Aircraft and Missiles, Pitotless Target Aircraft, Autopilots and Gyroscopic Flight Instruments, Inertial Guidance and Automatic Landing of Aircraft," RAE Library Bibliography No. 224, Ministry of Aviation, Sept. 1960.

[42] F. W. Meredith and P. A. Cooke, "Aeroplane Stability and the Automatic Pilot," *J. Roy. Aeron. Soc.*, 61, No. 318, 415–436 (June 1937).

[43] L. Bairstow, *Applied Aerodynamics*, 1st edn., Longmans Green, London, 1920.

[44] H. Glauert, "Summary of the Present State of Knowledge with Regard to Stability and Control of Aeroplanes, ARC R & M 710, Dec. 1920.

[45] S. B. Gates, "Notes on the Aerodynamics of Automatic Directional Control," RAE Rept. No. BA 487, Feb. 19, 1924; "Notes on the Aerodynamics of an Altitude Elevator Control," RAE No. BA 494, Mar. 19, 1924. (The latter report discusses the instability of altitude control with elevator at speeds below the speed for minimum power required.)

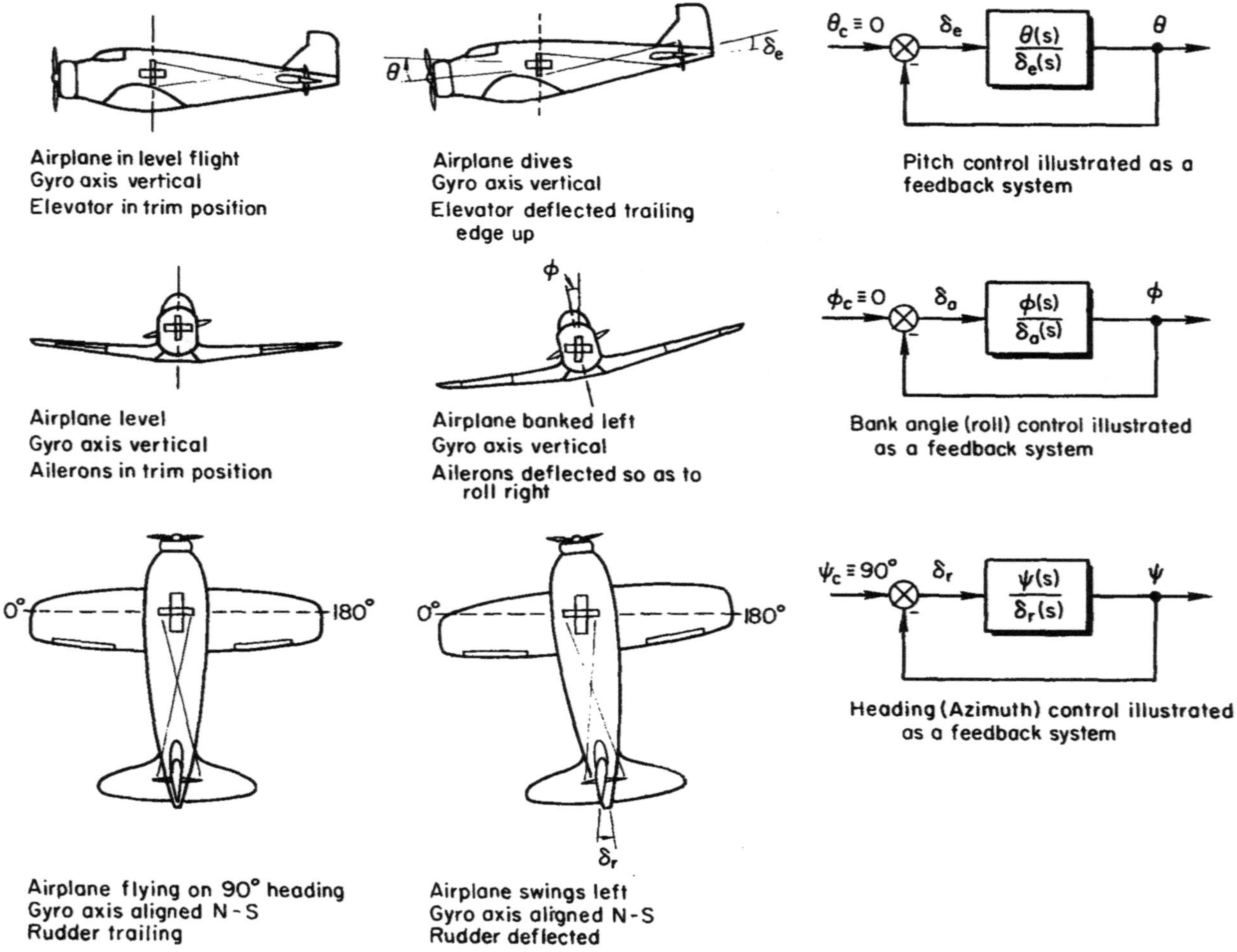


Fig. 1-8. Simple three-axis attitude control illustrating the concept of gearing.

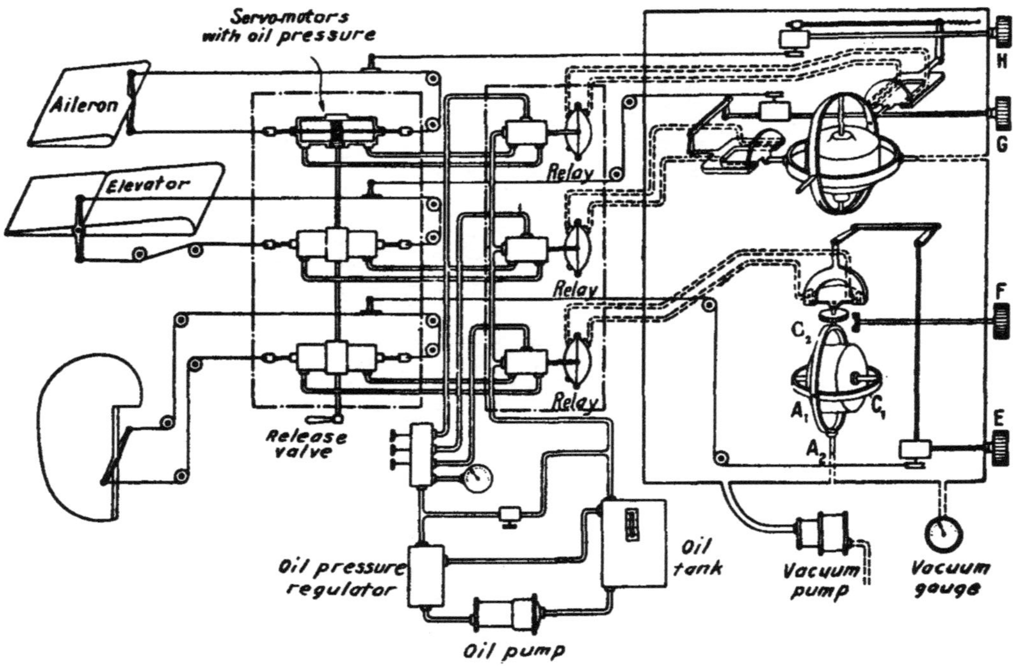


Fig. 1-9. Assembly sketch of Sperry stabilizer.

Reports BA 487 and 494, Garner,[46] in 1926, made an analysis of the lateral/directional motion of an airplane under the influence of feedback control. Gates assumed that the controls were moved according to certain laws, e.g., in proportion to certain output variables and their derivatives. He also stressed that good stability was not enough, since it was essential also to consider the amplitudes of the several modes of motion. With similar assumptions, Garner calculated the stability of the undisturbed motion and the transient motion, following an initial disturbance, under the influence of the feedback control system. It was specifically pointed out that the movements of the controls might be regarded as made either by the (human) pilot or by some mechanical means. Garner then further had the wit and vision to provide in the theoretical treatment for "lag" in the application of controls, and was able to point to a qualitative correspondence between his analytical results and flight tests of an RAE (automatic) rudder control that had appreciable reaction lag. Shortly after Garner's report, a further contribution by Cowley[47] appeared which proposed more elaborate methods of taking into account the time lag in the application of control. Both a pure time delay and a second-order lag were successfully treated.

It now seems surprising that these papers are not given more prominence in accounts of the development of the theory of automatic control systems. Perhaps they were simply too far ahead of their time. Perhaps, on the other hand, it was only in Great Britain, where automatic flight control system development was the responsibility of a government research establishment, that it was thought to be desirable to make response calculations in connection with the design of practical systems. In spite of an apparently adequate theory, however, stability difficulties attended the early flight trials of the RAE Mark IV automatic pilot about 1934.[48] A solution to the problem was apparently not found by analysis or simulation. The problem disappeared when the autopilots were installed in the larger aircraft for which they were intended, no doubt because of the larger inertia and slower response of the multi-engined bombers.

Comprehensive details of subsequent British automatic pilot development (1937–1947), as well as comments on American and German efforts, are set forth in the report by Hopkin and Dunn.[49] Included there is the

[46] H. M. Garner, "Lateral Stability with Special Reference to Controlled Motion," ARC R & M 1077, Oct. 1926.

[47] W. L. Cowley, "On the Stability of Controlled Motion," ARC R & M 1235, Dec. 1928.

[48] H. R. Hopkin and R. W. Dunn, *Theory and Development of Automatic Pilots, 1937–1947*, RAE Rept. IAP 1459, Monograph 2.5.03, Aug. 1947.

[49] Hopkin and Dunn, however, omit any mention of the Pollock/Brown all-hydraulic automatic pilot. This again was a two-axis unit, driving the elevators and rudder. It is described in the article "A New Automatic Pilot," *Flight*, 27, No. 1360, Mar. 14, 1935.

story of the uniquely conceived Mark VII autopilot. In this device the elevator was moved in response to airspeed error and error rate, while the ailerons were actuated by a combination of yaw and roll signals detected by a free gyroscope. The rudder was left free. Calculations showed that the stability properties of this arrangement should have been satisfactory, as indeed they were. Unfortunately, although the performance in average weather was good, in very rough air and in some aircraft at low speed, the elevator, responding to detected changes in the airspeed and airspeed rate, caused violent changes in pitch attitude. These were large enough in some cases so that the acceleration on the fuel system caused the engines to stop momentarily. Later, the matter was investigated theoretically by Neumark[50] and by Sudworth and Hopkin.[51] They were quite able to identify the source of the difficulty and to show that the calculation of damping factors alone was not sufficient to insure satisfactory performance. At the Royal Aircraft Establishment similar calculations were made both by hand and with a mechanical differential analyzer. Very few of the results, however, were published. The understanding of response to *specific* disturbances that Gates had stressed, and for which B. Melvill Jones had called, was thus only modestly enhanced.

Interestingly, by 1935 the German firm of Siemens had developed an elevator control that successfully used an airspeed reference and a rudder control with one of the earliest magnetic compass tie-ins.[52] Hydraulic positioning servos were employed to move the surfaces, but the key feature in both cases was the use of a rate gyro feedback. The rudder "course control" was an independent unit. In Germany it was argued that only the largest and heaviest airplanes would require a complete automatic pilot but that practically all small and medium machines could make good use of a course control.[53]

The firms of Siemens and Askania both developed three-axis automatic pilots that included an independent course control,[54] but during the war they concentrated on the production of the single-axis units. Eventually, more than 80 percent of the aircraft in the German Air Force were equipped with similar automatic stabilizers. A schematic diagram of the

[50] S. Neumark, "The Disturbed Longitudinal Motion of an Uncontrolled Aeroplane and of an Aeroplane with Automatic Control," ARC R & M 2078, Jan. 1943.

[51] J. Sudworth and H. R. Hopkin, "Influence of Automatic Pilots in Stabilization and Dynamic Stability in Pitch," RAE Tech. Note Instn. 775, July 1943.

[52] F. Haus, "Automatic Stabilization," NACA TM 802, Aug. 1936. See also "Siemens Autopilot," *Flight*, 27, No. 1359, 41–42 (Jan. 10, 1935).

[53] G. Klein, "Bedeutung automatischer Flugzeugsteuerungen für den Flugzeugbau," *Jahrbuch 1938 der Deutscher Luftfahrtforschung, Ergänzungsband*, R. Oldenbourg, Munich and Berlin, 1938, pp. 237–242.

[54] E. Fischel, "Vefahren und Bauglieder automatischer Flugzeugsteuerungen, *Jahrbuch 1938 der Deutscher Luftfahrtforschung, Ergänzungsband*, R. Oldenbourg, Munich and Berlin, 1938, pp. 231–236.

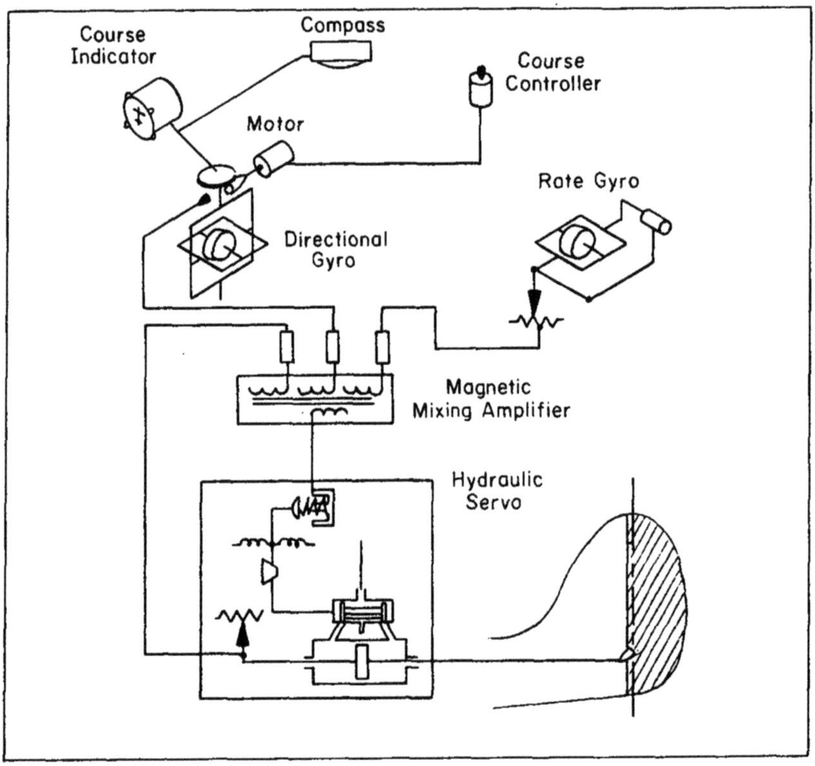


Fig. 1-10. Course control K-12.

Siemens K-12 unit is shown in Fig. 1-10. It was recognized at an early date that such a control could be used to supplement any deficiency in the aerodynamic damping about the yaw axis,[55] and that the course-holding feature, providing the airplane with a heading sense, would permit unattended operation, to a degree, so that even a pilot inexperienced in instrument flying could conduct operations in instrument weather in comparative safety. It may be noted that the methods of mathematical analysis that were introduced by Oppelt[56] for the study of automatic course-holding were rather sophisticated for their time. While he used much simplified linear equations to represent the aircraft in the 1937

[55] K. Wilde, "Über neuere Arbeiten auf dem Gebiet der automatischer Steuerungen," *Jahrbuch 1938 der Deutscher Luftfahrtforschung, Ergänzungsband*, R. Oldenbourg, Munich and Berlin, 1938, pp. 243–247.

[56] W. Oppelt, "Die Flugzeugkursteuerung im Geradeausflug," *Jahrbuch 1937 der Deutscher Luftfahrtforschung*, R. Oldenbourg, Munich and Berlin, 1937, pp. 3-22 to 3-34; also "Comparison of Automatic Control Systems," NACA TM 966, Feb. 1941.

paper, he also used phasor diagrams and approximate describing functions for friction and hysteresis to explore the deleterious effects on the action of the automatic control of these and other imperfections in the system such as quantized signals, and he pointed to the use of a rudimentary simulator as a means for exploring the effect of nonideal equipment characteristics in practice.

Later, during the war, the Germans introduced the rate-rate principle in the Siemens K-23 and Askania PKS-12 fighter course controls, and also in the experimental Patin three-axis automatic pilot. (Here the surfaces were made to move at a rate proportional to the rate of body axis rotation measured by a rate gyro. Damping was insured by electrical differentiation of the signal, in the case of the Siemens and Askania units, and by special design of the gyros to pick up a component of angular acceleration in the case of the Patin design. No feedback units measured the output of the servos.) All three of these control systems were all-electric.

It can be appreciated, even from this abbreviated account, that airplane automatic pilot development proceeded quite independently in Germany from its course in the United States and in Great Britain. Some idea of how it might have evolved can be obtained from Dudenhausen's[57] description of a three-axis, rate-rate automatic flight control system actually built in 1955 but based to a large extent on developments carried out just before the final collapse of Hitler's armies. (It is further amusing to note in the same issue of *Luftfahrttecknik* in which the Dudenhausen article appears that a speaker from an American company, tracing the history of automatic pilot development, says that course control was easily added after the more difficult problem of stabilizing the aircraft in pitch and roll was accomplished. His German translator felt constrained to correct him. From the German point of view the course control came first.[58])

The Germans were also, of course, very active in the development of pilotless aircraft and missiles.[59] The V-1 flying bomb had a conventional two-axis (elevator and rudder) automatic pilot with altitude and compass tie-ins (see Fig. 1-11). An air mileage counter determined when the final dive should begin. In spite of its, in many ways, very advanced technology the V-2 (A-4) simply used two free gyros: the master control

[57] H. J. Dudenhausen, "Dreiachsen-Flugregelung für Hochleistungsflugzeuge mit Integrations-Wendekreiseln als Hauptrichtgeber," *Luftfahrttecknik*, 4, No. 3, 49–58 (Mar. 18, 1958).

[58] J. F. Wiren, "Geschichtliches zur Entwicklung der Flugregelungen," *Luftfahrttecknik*, 4, No. 3, 46–68 (Mar. 18, 1958), and notes by the translator, Dpl. Ing. Manteufel.

[59] T. H. Benecke and A. W. Quick, eds., *History of German Guided Missiles*, Verlag E. Appelhans, Brunswick, Germany, 1959; J. N. Thiry, "Control Projects in the German Army, Air Force, and Navy," unpublished translation of a German report with the same title written in Aug. 1944, Sept. 10, 1958.

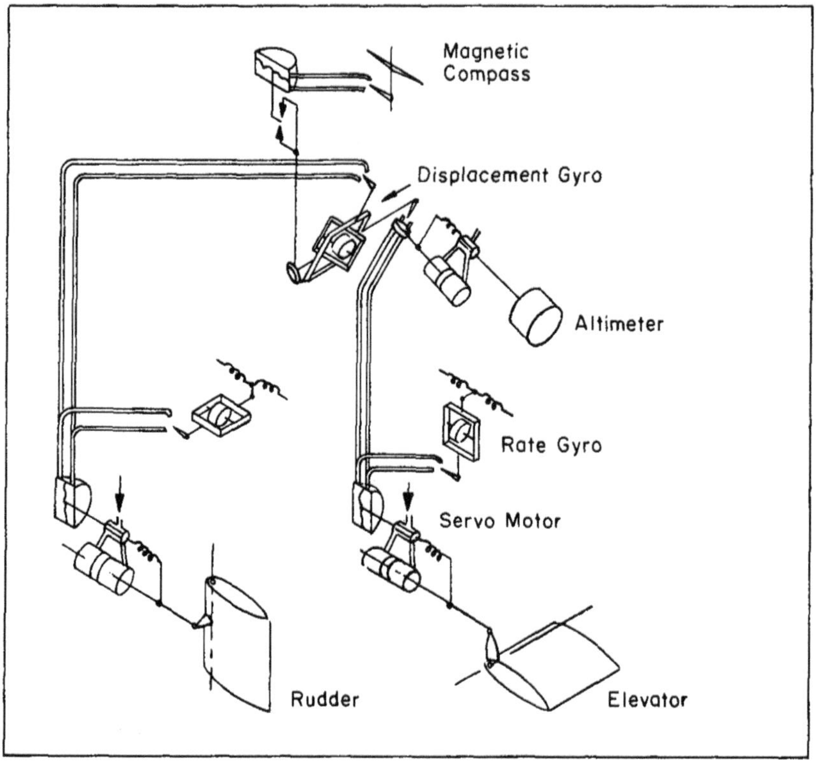


Fig. 1-11. The V-1.

gyro to control yaw and roll, and the "verticant" to control pitch by means of hydraulic servo-driven vanes in the exhaust blast of the rocket engine (see Fig. 1-12). A pitch maneuver was preprogrammed and thrust was cut off by an integrating accelerometer. Provision was made for monitoring and correcting the course with a radio beam. The gyro and vane arrangement was somewhat similar to the one evolved some time earlier (1932) by the pioneer R. H. Goddard for the control of his rockets.[60]

Beginning in 1941 there was a considerable amount of activity in the United States aimed at the development of electric automatic pilots ultimately capable of accepting maneuvering commands, either from the human pilot or from some other source of guidance information such as a bombsight. The first of these "all-electric" automatic pilots was the

[60] M. Lehman, *This High Man: The Life of Robert H. Goddard*, Farrar Straus, New York, 1963, pp. 202 ff.

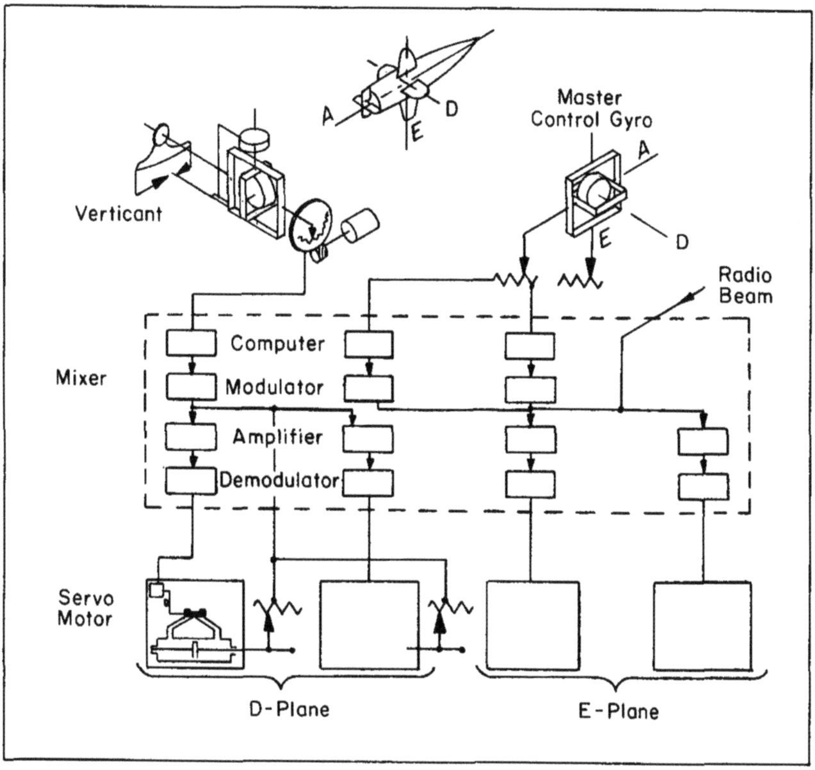


Fig. 1-12. The V-2.

C-1, built by the Minneapolis-Honeywell Regulator Company.[61] It was installed in all the American four-engined bombers such as the B-17, the B-24, and the B-29. In the C-1 deviations from the reference attitude in pitch, roll, and yaw were measured with vertical and directional gyroscopes, as in the Sperry A2 design, but the amplified signals were applied to electrical positioning servomotors driving the elevators, ailerons, and rudders. The automatic pilot unit was specifically designed to subject the aircraft to commands originating in the bombardier's operation of the bombsight, but a single-knob turn control, and later a "formation stick," was also provided for the pilot. Erection cutout in turns was one of the novel features of this automatic pilot. The electrically driven gyroscopes were not intended to be used as flight instruments. Later, in the C-1A

[61] W. H. Gille and R. J. Kutzler, "Application of Electronics to Aircraft Flight Control," *Trans. AIEE*, 63, 849–853 (Nov. 1944); see also W. H. Gille and H. T. Sparrow, "Electronic Autopilot Circuits," *Electronics*, 17, 110–117 (Oct. 1944).

(1945), a number of improvements were introduced, including the addition of a yaw rate gyro signal to the rudder.

This unit was closely followed in time by the General Electric design,[62] which was similar to the Sperry pneumatic/hydraulic A2 unit in its functions, and which differed from the C-1 in its design mainly in that the electrically driven gyroscopes were also the panel-mounted flight instruments, and in that electrohydraulic positioning servomotors provided the final stage of power amplification. At first the controls were merely trim knobs, one for each axis, but later a single-knob turn control was introduced.[63] This automatic pilot was notable for its light weight (74.5 lb), so that it was suitable for installation in fighters and light attack aircraft.

The Sperry Gyroscope Company also developed an electric automatic pilot during the war, the A5. It had a number of novel features such as altitude control, automatic elevator trim by means of an additional trim tab servo, and, notably, equalizing circuits that provided for phase advance of the servo actuating signals. The servos were electrohydraulic units with *force* feedback. This was thought to be a desirable feature in that, since control surface effectiveness and aerodynamic hinge moment vary in much the same way with speed and altitude, the closed-loop response with force feedback should tend to be invariant with flight condition.[64] Early flight tests in a Fairchild 22 were encouraging. Unfortunately, as it turned out in practice, the deleterious effect of control cable friction made it extremely difficult to secure satisfactory operation in the large aircraft,[65] such as the B-24E "Liberator," for which this automatic pilot was intended.

The Eclipse-Pioneer Division of the Bendix Corporation began work on the "flux-gate" compass in 1939 and, about 1943, introduced automatic heading control from the flux-gate compass in the all-electric P-1 (A-10) autopilot. This equipment obviated the very tiresome necessity of frequently resetting the directional gyro by reference to the magnetic compass. The P-1 also featured a yaw rate gyro signal fed to the rudder, automatic synchronization so that it could be engaged in any attitude, and a computed up-elevator signal in turns.

Somewhat later, considerable effort was expended in designing automatic pilots specifically for fighter aircraft; Lear introduced the F-5

[62] C. M. Young, E. E. Lynch, and E. R. Boynton, "Electrical Control in Automatic Pilots," *Trans. AIEE*, 63, 939–943 (Nov. 1944).

[63] H. R. Hopkin and R. W. Dunn, "Theory and Development of Automatic Pilots, 1937–1947," RAE Rept. IAP 1459, Monograph 2.5.03, Aug. 1947.

[64] P. Halpert and O. E. Esval, "Electric Automatic Pilots for Aircraft," *Trans. AIEE*, 63, 861–866 (Nov. 1944).

[65] B. Levine, Discussion of paper by P. Halpert and O. E. Esval, *Trans. AIEE*, 63, 1501 (Nov. 1944).

automatic pilot about 1950.[66] In the F-5 the problem of supplying power amplification for the control surface positioning servos was solved with magnetic powder clutches. The use of these units eliminated much relatively heavy equipment such as electronic or rotating power amplifiers and provided for high performance servomechanisms in a compact and lightweight package. The gyroscopic references, however, were the conventional vertical and directional gyroscopes, and the maneuvers that could be made under automatic control were limited by the phenomenon of "gimbal lock." This problem was attacked by Westinghouse[67] and the Instrumentation Laboratory at the Massachusetts Institute of Technology (MIT),[68] both of whom constructed laboratory models of fighter airplane all-attitude sensors, using different kinds of single degree of freedom gyroscopes.

Improvements were also made in providing for guidance tie-ins. Immediately after the war the Sperry Gyroscope Company brought out the A-12 automatic pilot,[69] and Bendix introduced the PB-10.[70] Both of these were equipped with approach couplers, and the Bendix system had automatic throttle controls for the control of airspeed on the approach to the runway. In England the Smith's firm brought out the all-electric, rate-rate SEP-2 automatic pilot and approach coupler.[71]

All the elements of a modern automatic pilot were now at hand, and in 1947 the United States Air Force All-Weather Flying Division's C-54 "Robert E. Lee," equipped with a Sperry automatic pilot and approach coupler and Bendix throttle controls, made a dramatic demonstration of completely automatic flight. Taking off from Stephenville, Newfoundland, on the evening of September 21, it flew through the night across the Atlantic and landed the next day at Brize-Norton in England. From the time the brakes were released for the takeoff roll until the landing roll was complete, no human hand touched the controls. The selection of course, radio station, speed, flap setting, landing gear position, and the final application of wheel brakes were all accomplished automatically from a

[66] J. Harper, "George Turns Tiger," *Flying Safety*, Jan. 1956.

[67] C. R. Hanna, K. A. Oplinger, and G. R. Douglas, "Automatic Flight Control System Using Rate Gyros for Unlimited Maneuvering," *Electrical Engineering*, 73, No. 5, 443–448 (May 1954).

[68] H. P. Whitaker, J. A. Gautraud, and S. A. Wingate," *Flight Test Evaluation of the MIT Automatic Control System for Aircraft*," MIT Instrumentation Lab. Rept. R-55, 1953.

[69] P. Halpert, *The A-12 Gyropilot*, paper presented to the SAE, New York, (Apr. 13–15, 1948).

[70] J. C. Owen, "Automatic Pilots," *Electrical Engineering*, 67, No. 6, 551–561 (June 1948); P. A. Noxon, "Flight Path Control," *Aeron. Eng. Rev.*, 7, No. 8, 36–45 (Aug. 1948).

[71] F. W. Meredith, "The Modern Autopilot," *J. Roy. Aeron. Soc.*, 409–428 (May 1949); see also W. H. Coulthard, *Aircraft Instrument Design*, Pitman, New York, 1952.

program stored on punched cards. The complete automation of aircraft flight seemed to be at hand. Anast described the performance and the prospects early in 1948.[72]

While the development of automatic pilots up to 1950 had, in general, been responsive to the needs of potential users, and while such features as single-knob turn control, erection cutout, automatic trim, altitude control, synchronizers, rate gyro feedbacks, compass tie-in, and approach coupling served useful purposes in some applications, in many cases they were introduced only as the technology became readily available and certainly not because they were *required* for any particular airplane. The automatic pilot was almost universally regarded as a useful but hardly an essential item of equipment; the day had not yet arrived when an automatic pilot would be designed for a specific airplane by taking into account, right from the beginning, particular and peculiar features of the mission of the airplane and of the design of its control system.

Perhaps because of the obvious necessity for a special design of the flight control system in a missile and the successes achieved with that approach, and certainly, in part, because the postwar generation of jet airplanes almost invariably needed some form of stability augmentation whose particular nature was often dictated by the unique configuration or mission of the vehicle, the most common design practice later came to be based on a careful enumeration of *requirements* to be met and *functions* to be performed by each system. This alteration in the methods of design was only one of several that, together, radically changed the nature of work in automatic flight control.

## 1-6. The Joining of Control Technology and Dynamic Analysis

While the mathematical tools for performing analyses of automatic flight control systems for aircraft had existed in at least a rudimentary form before the war of 1939–1945, there has been occasion to remark that they did not seem to be much used. The work of Gates, Garner, and Cowley[73] seems to have been nearly forgotten. Oppelt,[74] even though translated into English, does not seem to have attracted much attention, while Minorsky's paper[75] on the steering of ships was fairly widely known but did not seem to inspire other workers to follow similar lines. There had

[72] J. L. Anast, "Automatic Aircraft Control," *Aeron. Eng. Rev.*, 7, No. 7, 20–24 (July 1948).

[73] See notes 42–47.

[74] W. Oppelt, "Comparison of Automatic Control Systems," NACA TM 966 Feb. 1941.

[75] N. Minorsky, "Directional Stability of Automatically Steered Bodies," *J. Amer. Soc. of Naval Eng.*, 34, 280–309 (1922).

appeared, here and there, papers or monographs on the theory of servomechanisms,[76] the regulation of prime movers,[77] process control,[78] the dynamics of instruments,[79] and the Cauchy/Heaviside operational calculus applied to the dynamic response of aircraft,[80] but apparently the need to study the feedback control of aircraft for practical reasons was not yet felt.

The subject of the theory of automatic control of aircraft did receive some attention in the universities, technical institutes, and research laboratories, however. Longitudinal stability of an airplane under the action of a control system of the Sperry A2 or A3 automatic pilot type was investigated by Klemin, Pepper, and Wittner[81] at New York University, and H. K. Weiss[82] at Massachusetts Institute of Technology performed a comprehensive study of the stability of an automatically controlled airplane, including both the free longitudinal and lateral motions and the response to gusts, as his thesis research for the Master's degree. There was also a very original contribution from Imlay,[83] who explored the problem of selecting "optimum" gearings; but all this barely

[76] H. L. Hazen, "Theory of Servomechanisms," *J. Franklin Inst.*, 218, No. 3, 279–331 (Sept. 1934); see also "Design and Test of a High Performance Servomechanism," *J. Franklin Inst.*, 218, No. 5, 543–580 (Nov. 1934).

[77] A. Stodola, *Steam and Gas Turbines*, Vol. 1, translated from German 6th edn. by C. L. Loewenstein, McGraw-Hill, New York, 1927; W. Trinks, *Governors and the Governing of Prime Movers*, Van Nostrand, New York, 1919; H. K. Weiss, "Constant Speed Control Theory," *J. Aeron. Sci.*, 6, No. 4, 147–152 (Feb. 1939).

[78] G. Wünsch, *Regler für Druck und Menge*, R. Oldenbourg, Munich, 1930; A. Ivanoff, "Theoretical Foundations of the Automatic Regulation of Temperature," *J. Inst. of Fuel* (London), 7, 117–130 (Feb. 1934); S. D. Mitereff, "Principles Underlying the Rational Solution of Automatic Control Problems," *Trans. ASME*, 57, No. 4, 159–163 (May 1935); E. S. Smith, Jr., "Automatic Regulators, Their Theory and Application," *Trans. ASME*, 9, No. 4, 159–163 (May 1936); see also *Automatic Control Engineering*, McGraw-Hill, New York, 1944; A. Callender, D. R. Hartree, and A. Porter, Time Lag in a Control System," *Phil. Trans. Roy. Soc.* A235, 415–444 (1936); C. E. Mason and G. A. Philbrick, "Automatic Control in the Presence of Process Lags," *Trans. ASME*, 62, 295–308 (1940).

[79] C. S. Draper and G. V. Schliestett, "General Principles of Instrument Analysis," *Instruments*, 12, No. 5, 137–142, (May 1939); C. S. Draper and G. P. Bentley, "Design Factors Controlling the Dynamic Performance of Instruments," *Trans. ASME*, 62, No. 5, 421–432 (July 1940).

[80] L. W. Bryant and D. H. Williams, "The Application of the Method of Operators to the Calculation of the Disturbed Motion of an Airplane," ARC R & M 1346, July 1930. R. T. Jones, "A Simplified Application of the Method of Operators to the Calculation of the Disturbed Motion of an Airplane," NACA TR 560, 1936; see also "Calculation of the Motion of an Airplane Under the Influence of Irregular Disturbances," *J. Aeron. Sci.*, 3, No. 12, 419–425 (Oct. 1936). A. Klemin and B. F. Ruffner, "Operator Solutions in Airplane Dynamics," *J. Aeron. Sci.*, 3, No. 7, 252–255 (May 1936).

[81] A. Klemin, P. A. Pepper, and H. A. Wittner, "Longitudinal Stability in Relation to the Use of an Automatic Pilot," NACA TN 666, Sept. 1938.

[82] H. K. Weiss, "Theory of Automatic Control of Airplanes," NACA TN 700, Apr. 1939.

[83] F. H. Imlay, "A Theoretical Study of Lateral Stability with an Automatic Pilot," NACA TR 693, 1940.

carried the matter further than the state in which it had been left by the British authors nearly 15 years before. The difficulty, as Weiss pointed out, lay in the necessity for factoring characteristic functions of the fifth, sixth, and seventh degree. The same problem, of course, plagued students of other feedback control devices, and a considerable effort was made to find convenient methods for accomplishing the tedious algebra.[84]

In 1944, discussing the subject of automatic flight and airplane stability (which he treated separately), Zand[85] still found the situation very unsatisfactory. He wrote,

> a thorough knowledge of the stability of the airplane is a prerequisite toward the successful solution of the problem of automatic flight. We have sketched the difficult path which led airplane designers toward the understanding of fundamentals of dynamic stability. An equally thorny road full of obstacles has been conquered by the instrument engineer who succeeded against such odds as space and weight limitations, lack of power, etc. . . . In many instances the airplane engineering field and the instrument engineering fraternity have worked independently on the problem which concerns both branches directly. Pooling the knowledge of dynamic stability with the knowledge of instrument design for the general betterment of aeronautics is essential . . . . Experience has shown that if the matching is performed theoretically first the number of experimental flying hours will be reduced to a minimum and the results obtained superior.

Indeed, the theoretical matching of the two subjects was shortly to become not merely a desideratum but an absolute necessity.

The war had seen the advent, on both sides, of the turbojet engine, so that suddenly the limits of the flight envelope were enormously extended in both speed and altitude, with concomitant configuration changes involving increased wing loadings, mass distributions concentrated in long thin fuselages, and the aerodynamic benefits of short span, swept wings. These changes led to a marked defect in the damping of the longitudinal short period and dutch roll oscillations in the airplanes of the immediate postwar period. However, it was not only the "classical" modes that were deficient in stability, previously unknown coupled modes made their

[84] Y. J. Liu, "Servomechanisms, Charts for Finding Their Stability and for Finding the Roots of Their Third and Fourth Degree Characteristic Equations," Dept. of Electrical Eng., Massachusetts Institute of Technology, Oct. 1941; Shi-Nge Lin, "Method of Successive Approximations of Evaluating the Real and Complex Roots of Cubic and Higher Order Equations," *J. Math. and Phys.*, 20, 231–242 (Aug. 1941); A. Porter and C. Mack, "New Methods for the Numerical Solution of Algebraic Equations," *Phil. Mag.* 40, No. 304, 578–585 (May 1949); H. R. Hopkin, "Routine Computing Methods for Stability and Response Investigations on Linear Systems," ARC R & M 2392, Aug. 1946.

[85] S. J. Zand, "Automatic Flight and Airplane Stability," *Aviation*, 43, No. 6, 140–141, 290–296 (June 1944).

appearance. Among these were fuel slosh[86] and the rolling instability.[87] Furthermore, thinner wings and finer fuselages combined with advances in materials and manufacturing processes made for increased structural flexibility. Power-boosted controls[88] had also come into use to handle the large hinge moments of the control surfaces. Early hydraulic power units had stability difficulties of their own,[89] and the inevitable lag was inimical to the stability of automatic flight control systems. All these trends were bad news for the automatic flight control system designer, who now desperately wanted analytical help.

The joining of control technology and vehicle dynamic analysis which would, no doubt, have come about in any event, was forced by the marked deficiencies in stability of the new jet aircraft and also by the advent of the guided missile, in which it was obviously essential to match the dynamics of the airframe and the control system from the first flight on.[90] One of the first results was the "stability augmenter," a feedback control designed to modify the inherent aerodynamic stability of the airframe, augmenting one or more of the stability derivatives by imposing forces or moments through actuation of the controls in response to motion variables. Thus, in short order, the following devices were invented or reinvented:

Yaw damper
Sideslip stability augmenter
Pitch damper
Roll damper
Transonic trim shifter
Autothrottle

[86] A. Schy, "A Theoretical Analysis of the Effects of Fuel Motion on Airplane Dynamics," NACA Rept. 1080, 1952; H. Luskin and E. Lapin, "An Analytical Approach to the Fuel Sloshing and Buffeting Problems of Aircraft," *J. Aero. Sci.*, 19, No. 4, 217–228 (Apr. 1952).

[87] F. D. Graham and R. C. Uddenberg, "The Dynamic Stability and Control Problem of a Pivoted-Wing Supersonic Pilotless Aircraft," Boeing Airplane Co., Document D-8810, Feb. 1948; W. H. Phillips, "Effect of Steady Rolling on Longitudinal and Directional Stability," NACA TN 1627, June 1948.

[88] D. J. Lyons, "Present Thoughts on the Use of Powered Flying Controls in Aircraft," *J. Roy. Aeron. Soc.*, 53, 253–277 (Mar. 1949); T. A. Feeney, "Powered Control System Design Practice at Northrop Aircraft," Proc. Bureau of Aeronautics Symposium on Analysis and Design of Power Boosted and Power Operated Surface Control Systems, Oct. 6–7, 1949.

[89] D. T. McRuer, "An Analysis of Northrop Aircraft Powered Flight Controls," Proc. Bureau of Aeronautics Symposium on Analysis and Design of Power Boosted and Power Operated Surface Control Systems, Oct. 6–7, 1949.

[90] While it seems surprising that the developers of the V-2 were able to make do with stability diagrams and a rudimentary simulator, they did not entirely neglect analysis in the design of the control and guidance equipment. The frequency response method and Nyquist stability criterion were known but were not "popular." See O. Müller, "The Control System of the V-2," in T. Benecke and A. W. Quick, eds., *History of German Guided Missile Development*, Verlag E. Appelhans, Brunswick, Germany, 1959.

Since both the problem and the solution were inevitably connected with a particular aircraft and its control system, the old methods of designing general purpose equipment were totally inadequate. Extensive analysis and simulation for each application to a piloted aircraft or missile were found to be required.

In the interim, wartime pressures for very high performance servomechanisms and regulators for such uses as servo-controlled aircraft wing spar milling machines,[91] process controls in the manufacture of fissionable material,[92] and particularly for fire control and navigation computers,[93] as well as antenna drives for tracking radars,[94] had led to the widespread adoption of analytical design techniques originally developed for long distance telephone amplifiers.[95] While James, Nichols, and Phillips credit John F. Taplin, at the Massachusetts Institute of Technology, with frequency response technique work for servomechanisms as early as 1937, substantially nothing concerning the matter appeared in public print until after the war was over. It seems clear that the same or very similar ideas were shared by widely separated investigators. At both the Bell Telephone Laboratories and the Massachusetts Institute of Technology, classified memoranda were prepared on the eve of the United States involvement.[96] As far as the present authors know, these historic documents have still not been released. Later, at the Massachusetts Institute of Technology, Hall[97] prepared a dissertation that was initially classified but was released in 1947. The effort to keep all this information classified did not prevent the duplicate development and use of the ideas

[91] *Electronics*, Oct. 1944.

[92] H. Smyth, *Atomic Energy for Military Purposes*, Sections 7 and 27 and Appendix 4, Princeton University Press, Princeton, N.J., 1945.

[93] I. A. Greenwood, Jr., J. V. Holdam, Jr., and D. MacRae, Jr., *Electronic Instruments*, McGraw-Hill, New York, 1948.

[94] H. M. James, N. B. Nichols, and R. S. Phillips, *Theory of Servomechanisms*, McGraw-Hill, New York, 1947.

[95] H. Nyquist, "Regeneration Theory," *Bell Systems Tech. J.*, 11, No. 1, 126–147 (Jan. 1932); see also "The Regeneration Theory," *Trans. ASME*, 76, No. 8, 1151 (Nov. 1954); H. S. Black, "Stabilized Feedback Amplifiers," *Bell System Tech. J.*, 13, No. 1, 1–18; (Jan. 1934); see also United States Patent 2,102,671, Dec. 1937; E. Peterson, J. G. Kreer, and L. A. Ware, "Regeneration Theory and Experiment," *Bell System Tech. J.*, 13, No. 10, 680–700 (Oct. 1934); H. W. Bode, *Amplifiers*, United States Patent 2,123,17, July 12, 1938; see also "Relations Between Attenuation and Phase in Feedback Amplifier Design," *Bell System Tech. J.*, 19, No. 3, 421–454 (July 1940); *Network Analysis and Feedback Amplifier Design*, Van Nostrand, New York, 1945; "Feedback—The History of an Idea," in *Active Networks and Feedback Systems*, Polytechnic Press, Brooklyn, New York, 1961.

[96] D. C. Bomberger and B. T. Weber, "Stabilization of Servomechanisms," Bell Telephone Laboratories Restricted Publication M.M.-41-110152, Dec. 10, 1941; H. Harris, Jr., "The Analysis and Design of Servomechanisms," OSRD, NDRC (Section D-2), Rept. 454, Jan. 1942. (Brown and Campbell give the date of this report as Dec. 1941, but other authorities agree on the one given here.)

[97] A. C. Hall, *The Analysis and Synthesis of Linear Servomechanisms*, Technology Press, Cambridge, Mass., 1943.

on several sides. Almost simultaneously with Hall's dissertation, Profos[98] contributed his own at Zurich; a little later the book that we know in translation as *The Dynamics of Automatic Control*[99] was published in Munich. The system was not even airtight on our own side. The earliest published work in English that refers to the frequency response method in control seems to have been the 1944 paper by Prinz.[100]

Immediately upon the cessation of hostilities there appeared a rash of papers by the original contributors, and others.[101] Almost simultaneously with the papers a growing number of books began to appear, many of which are almost as valuable today as when they were first published.[102] Typically, these books not only expounded the new theory of the frequency response of automatic control systems but further connected it to the performance in the time domain via the correspondence between the transfer function and the transient response as revealed by the Laplace transform method. Operational techniques were not new but their rigorous and respectable foundation in the Laplace transformation was, at that time, a comparatively recent development.[103]

[98] P. Profos, *Die Behandlungen von Regel Problemen vermittels des Frequenzganges des Regelkreises*, Ph.D. dissertation, A. G. Gebr. Leeman, Zurich, 1943; also "A New Method of Regulating System Design," *Sulzer Tech. Rev.*, No. 2 (1945).

[99] R. C. Oldenbourg and H. Sartorius, *Dynamik Selbsttätiger Regelung*, R. Oldenbourg, Munich, 1944; see also *The Dynamics of Automatic Control*, ASME, New York, 1948.

[100] D. G. Prinz, "Contributions to the Theory of Automatic Controllers and Followers," *J. Sci. Instrum.* 21, No. 4, 53–64 (Apr. 1944).

[101] D. P. Campbell, "Theory of Automatic Control Systems," *Industrial Aviation*, 62–64, 94, 95 (Sept. 1945); E. B. Ferrell, "The Servo Problem as a Transmission Problem," *Proc. IRE*, 33, No. 11, 763–767 (Nov. 1945); G. S. Brown and A. C. Hall, "Dynamic Behavior and Design of Servomechanisms," *Trans. ASME*, 68, 503–524 (1946); A. C. Hall, "Application of Circuit Theory to Design of Servomechanisms," *J. Franklin Inst.*, 242, No. 4, 279–307 (Oct. 1946); see also "Early History of the Frequency Response Field," *Trans. ASME*, 76, No. 8, 1153–1154 (Nov. 1954); H. Harris, Jr., "Frequency Response of Automatic Control Systems," *Trans. AIEE*, 65, 539–545 (1946); R. E. Graham, "Linear Servo Theory," *Bell System Tech. J.*, 25, No. 4, 616–651 (Oct. 1946); H. T. Marcy, "Parallel Circuits in Servomechanisms," *Trans. AIEE*, 65, 521–529 (1946).

[102] L. A. McColl, *Fundamental Theory of Servomechanisms*, Van Nostrand, New York, 1945; H. Lauer, R. Lesnick, and L. E. Matson, *Servomechanism Fundamentals*, McGraw-Hill, New York, 1947; H. M. James, N. B. Nichols, and R. S. Phillips, *Theory of Servomechanisms*, McGraw-Hill, New York, 1947; W. R. Ahrendt and J. F. Taplin, *Automatic Regulation*, Vol. I, Ahrendt and Taplin, Washington, D.C., 1947; I. A. Greenwood, Jr., J. V. Holdam, Jr., and D. MacRae, Jr., *Electronic Instruments*, McGraw-Hill, New York, 1948; G. S. Brown and D. P. Campbell, *Principles of Servomechanisms*, Wiley, New York, 1948.

[103] G. Doetsch, *Theorie und Anwendung der Laplace Transformation*, Springer-Verlag, Berlin, 1937; N. W. MacLachlan, *Complex Variable and Operational Calculus*, Cambridge University Press, London, 1939; H. S. Carslaw and J. C. Jaeger, *Operational Methods in Applied Mathematics*, Clarendon Press, Oxford, 1941; M. F. Gardner and J. L. Barnes, *Transients in Linear Systems*, Vol. I, Wiley, New York, 1942; R. V. Churchill, *Modern Operational Mathematics in Engineering*, McGraw-Hill, New York, 1944.

There also appeared, immediately after the war, accounts of the improved mechanical analog computer developed at the Massachusetts Institute of Technology[104] and of the digital scientific calculators developed at Harvard University,[105] at the Bell Telephone Laboratories,[106] and at the University of Pennsylvania.[107] These machines had originally been employed primarily to compute ballistic tables but their potentialities for the solution of other problems, including the design of feedback control systems for aircraft, was quite plain. Furthermore, requirements for fire control computers had led to the development of a variety of new or improved components. Among these was the d-c or operational amplifier. In a prophetic paper published in 1947, Ragazzini, Randall, and Russell[108] pointed out that these might be used in a general-purpose machine for solving differential equations, an electronic analog computer. A number of firms almost immediately developed such machines for sale or for their own use, and by 1950 they were fairly common. Several universities also developed their own machines; among the first was the University of Michigan. In an early report on the feasibility of electronic analog computation,[109] autopilot control of the longitudinal motion of an airplane was given as one of the illustrative examples.

Knowledge of the development of the new methods of analysis and of the newly available computers spread very rapidly, and one could almost say that a new branch of the engineering profession came suddenly into being: men were proud to call themselves feedback systems engineers, or "systems engineers" for short, some of them became aircraft control system engineers. Application of the frequency response (or transfer function) technique to the design of aircraft and their control systems

[104] V. Bush and S. H. Caldwell, "A New Type of Differential Analyzer," *J. Franklin Inst.*, 240, No. 4, 255–326 (Oct. 1945).

[105] H. H. Aiken and G. M. Hopper, "The Automatic Sequence Controlled Calculator—I," *Elec. Eng.*, 65, No. 8–9, 384–391 (Aug.–Sept. 1946); "The Automatic Sequence Controlled Calculator—II," *Elec. Eng.*, 65, No. 10, 449–454 (Oct. 1946); "The Automatic Sequence Controlled Calculator—III," *Elec. Eng.*, 65, No. 11, 522–528 (Nov. 1946).

[106] F. L. Alt, "A Bell Telephone Laboratories Computing Machine—I," *Math. Tables and Other Aids to Computation*, 3, No. 21, 1–13 (Jan. 1948); "A Bell Telephone Laboratories Computing Machine—II," *Math. Tables and Other Aids to Computation*, 3, No. 22, 69–84 (Apr. 1948).

[107] H. H. Goldstine and A. Goldstine, "The Electronic Numerical Integrator and Computer (ENIAC)," *Math. Tables and Other Aids to Computation*, 2, No. 15, 97–110 (July 1946).

[108] J. R. Ragazzini, R. H. Randall, and F. A. Russell, "Analysis of Problems in Dynamics by Electronic Circuits," *Proc. IRE*, 35, No. 5, 442–452 (May 1947).

[109] D. W. Hagelbarger, C. E. Howe, and R. M. Howe, "Investigation of the Utility of an Electronic Analog Computer in Engineering Problems," External Memo. 28, Eng. Res. Inst., University of Michigan, Apr. 1, 1949.

was pointed out early,[110] and it quickly became a part of the design process for actual aircraft (or missile) control systems.[111]

Further improvements and extensions of the analytical techniques were also discovered. By no means the least of these, Evans' locus of roots method[112] was inspired by consideration of the problems of aircraft and missile flight control. In the classic Fourteenth Wright Brothers Lecture for the year 1950, Bollay summarized the then existing state of the art and pointed to the use of the Laplace transformation, frequency response techniques, the Nyquist stability criterion, the root locus method, analog computers, and other tools of the systems engineer in the design departments of the major aircraft manufacturers.[113]

At Northrop Aircraft, based in part on the experiences gained with the power controls and stability augmentation system developments needed for tailless and other advanced designs,[114] and in a comprehensive study of the F-5 automatic pilot for the F-89A aircraft,[115] an attempt was made to summarize the most useful aspects of the new knowledge of aircraft control system engineering in a series of seven volumes, prepared for the U.S. Navy Bureau of Aeronautics.[116] These volumes began to appear in

[110] W. F. Milliken, Jr., "Progress in Dynamic Stability and Control Research," *J. Aeron. Sci.*, 14, No. 9, 493–519 (Sept. 1947); J. B. Rea, *Automatic Tracking Control of Aircraft*, Sc.D thesis, Massachusetts Institute of Technology, 1947; H. Greenberg, "Frequency-Response Method for Determination of Dynamic Stability Characteristics of Airplanes with Automatic Controls," NACA TN 1229, Mar. 1947; R. C. Seamans, Jr., B. G. Bromberg, and L. E. Payne, "Application of the Performance Operator to Aircraft Automatic Control," *J. Aeron. Sci.*, 15, No. 9, 535–555 (Sept. 1948); J. R. Moore, "Application of Servo Systems to Aircraft," *Aeron. Eng. Rev.*, 8, No. 1, 32–43, 71 (Jan. 1949); C. L. Seacord, "Application of Frequency Response Analysis to Aircraft Autopilot Stability," *J. Aeron. Sci.*, 17, No. 8, 481–498 (Aug. 1950).

[111] P. A. Noxon, "Flight Path Control," *Aeron. Eng. Rev.*, 17, No. 8, 38–45 (Aug. 1948); R. J. White, "Investigation of Lateral Dynamic Stability in the XB-47, Airplane," *J. Aeron. Sci.*, 17, No. 3, 133–148 (Mar. 1950).

[112] W. R. Evans, *Servo Analysis by Locus of Roots Method*, North American Aviation Rept. AL-787, Nov. 1, 1948; "Graphical Analysis of Control Systems," *Trans. AIEE*, 67, 547–551 (1948); "Control System Synthesis by the Root Locus Method," Trans. *AIEE*, 69, 66–69 (1950); "The Use of Zeros and Poles for Frequency Response or Transient Response," *Trans. ASME*, 76, No. 8, 1335–1342 (Nov. 1954); *Control System Dynamics*, McGraw-Hill, New York, 1954.

[113] W. Bollay, "Aerodynamic Stability and Automatic Control," *J. Aeron. Sci.* 18, No. 9, 569–624 (Sept. 1951).

[114] D. T. McRuer, "An Electronic Tail for the Flying Wing," *Flight Lines*, (Northrop Aircraft, Inc) Nov. 1950.

[115] "Analysis Final Report: Analysis of Type F-5 Automatic Pilot Applied to the Type F-89 Aircraft and Control System," Northrop Aircraft, Servomechanisms and Dynamics Section, Rept. SMD-3, Sept. 13, 1950.

[116] "Fundamentals of Design of Pilot Aircraft Flight Control Systems," Bureau of Aeronautics Rept. AE-61-4; Vol. 1, "Methods of Analysis and Synthesis of Piloted Aircraft Flight Control Systems," Mar. 1952; Vol. 2, "Dynamics of the Airframe," Sept. 1952; Vol. 3, "The Human Pilot," Aug. 1954; Vol. 4, "The Hydraulic System," Mar. 1953; Vol. 5, "The Artificial Feel System," May 1953; Vol. 6, "Automatic Flight Control Systems for Piloted Aircraft," Apr. 1956; Vol. 7, "Methods of Design and Evaluation of Interceptor Fire Control Systems," Oct. 1959.

March 1952. The genealogy of the present work can be traced directly to several of the "Northrop Volumes." In fact, it began, about 10 years after the initial summary, as an effort to revise and update Volumes 2 and 6 and to provide in one volume a comprehensive account of the theory and application of analytical techniques in the design of automatic flight control systems.

CHAPTER 2

# *MATHEMATICAL MODELS OF LINEAR SYSTEM ELEMENTS*

## 2-1. Introduction

A major task in systems analysis is the estimation of system response to commands or disturbances. The most concrete way to determine behavior is to test the actual system. This direct experimental approach is precluded in the early phases of design, when the "system" may be but one of a number of competing possibilities, or when the physical system may be unavailable. Fortunately, many of the potential results of actual physical measurements can be foreseen by performing experiments utilizing various models of the system.

As the underlying basis for system models consider the block diagram representation of Fig. 2-1. The input, stimulus, command, disturbance, or forcing function elicits an output or response from the system. The system might be one of a very large number of elements including a human being, an airplane, or a society, and the words appropriate to the several portions of the diagram of Fig. 2-1 are quite different in these different contexts. Nevertheless, it is assumed here that we are dealing with cause and effect elements, perhaps combined into larger systems; and that an input and output of a particular element can always be identified. It is further assumed that the relationship between the input and output can be represented by one or more ordinary differential equations. These equations become the mathematical attorneys for the physical elements or systems, and it is through them as intermediaries that the transactions in which we are to engage will ordinarily be conducted.

Many physical elements and systems are practically linear and time-invariant; that is, they can be described adequately over a limited range by linear differential equations with constant coefficients. Examples of linearization abound in the literature,[1] and Chapter 4 presents an example of the assumptions and techniques that are employed, in this case, to

[1] D. Graham and D. McRuer, *Analysis of Nonlinear Control Systems*, Wiley, New York, 1961, pp. 9–12, 445–454; R. W. Jones, "Stability Criteria for Certain Non-linear Systems," in A. Tustin, ed., *Automatic and Manual Control*, Butterworths London, 1952, pp. 319–324.

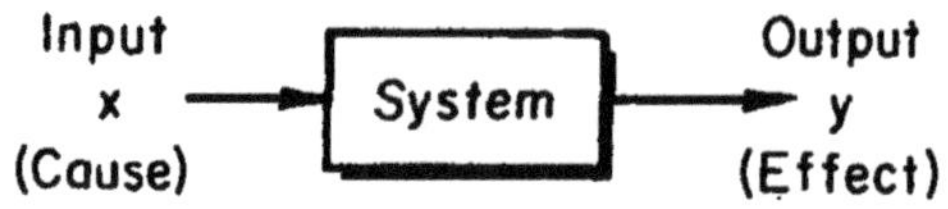


Fig. 2-1. A pattern.

linearize the equations of motion of an aircraft. It can be demonstrated that feedback control itself has the property of linearizing the performance of the systems or elements to which it is applied,[2] so that while all systems and elements are in fact nonlinear, the assumptions of our analyses are often not so restrictive as they may at first appear.

Proceeding on the assumption, for the moment, that we need only be concerned with systems that are linear with constant coefficients, or may legitimately be linearized, we shall introduce in this chapter the powerful and convenient concepts of the Laplace transformation. The rudiments of response calculations are first reviewed, including approximate calculations and modal response ratios. Following this is a discussion of the system descriptors: the weighting function or impluse response and its transform, and the transfer function. Graphical representations for both of these functions are emphasized: time vectors for the weighting function; pole-zero plots, $j\omega$-Bode diagrams, and $\sigma$-Bode diagrams for the transfer function.

## 2-2. Laplace Transformation

The system of Fig. 2-1 is, according to our assumptions, described by the equation

$$\left(\frac{d^{m+n}}{dt^{m+n}} + b_1 \frac{d^{m+n-1}}{dt^{m+n-1}} + \cdots + b_{m+n-1}\frac{d}{dt} + b_{m+n}\right)y(t)$$

$$= \kappa\left(\frac{d^n}{dt^n} + a_1\frac{d^{n-1}}{dt^{n-1}} + \cdots + a_{n-1}\frac{d}{dt} + a_n\right)x(t) \quad (2\text{-}1)$$

For any physical system, $m \geqq 1$. We are interested in discovering certain aspects of the performance of the system such as the stability, accuracy, and speed of response of the output for certain inputs. The *analysis problem* is defined as follows: Given the input and the differential equation, find the ouput. If the analysis problem were solved for all the inputs to which the system might be subjected and these solutions were tabulated

[2] J. C. West, *Analytical Techniques for Non-linear Control Systems*, English Universities Press, London, 1960, pp. 16–23.

as input-response pairs, the analyst would have a complete description of the performance of the system. Luckily, this turns out to be unnecessary for linear systems.

In modern engineering analysis an equation such as Eq. 2-1 is most often solved, if it has to be, by a digital or an analog computer. On the other hand, a great deal of information concerning the nature of the solutions for a variety of inputs can be found without solving the equation itself. The techniques that are commonly employed, however, are intimately related to the method of solving Eq. 2-1 by means of the Laplace transformation.

The Laplace transformation of a function of time is defined as

$$\mathcal{L}[f(t)] = F(s) = \lim_{\substack{T_2 \to \infty \\ T_1 \to 0}} \int_{T_1}^{T_2} f(t)e^{-st}\,dt \tag{2-2}$$

where $s$ is the complex variable $s = \sigma + j\omega$. Application of the definition allows the construction of tables of the transforms of operations such as differentiation and integration, and tables of the transforms of functions such as an impulse or a sine wave. The inverse transformation [i.e., $f(t)$, given $F(s)$] is usually carried out by finding the appropriate pair in a table so arranged that $f(t)$ can be associated with a particular $F(s)$.

The most interesting and useful properties of the Laplace transformation are summarized (without proof) in Table 2-1.[3]

Common transform pairs for the time functions that occur in the analysis and testing of feedback control systems are catalogued in the short illustrated table of transform pairs presented in Table 2-2.[4] In many cases two alternative forms are given for the transform. In some problems one form is more suitable than the other, so they are used interchangeably as convenience may dictate. To emphasize the physical interpretations of the

[3] For more details on the properties of the transformation and its uses, see: M. F. Gardner and J. L. Barnes, *Transients in Linear Systems*, Wiley, New York, 1942; J. A. Aseltine, *Transform Method in Linear System Analysis*, McGraw-Hill, New York, 1958; R. V. Churchill, *Operational Mathematics*, 2nd edn., McGraw-Hill; New York, 1958; W. Kaplan, *Operational Methods for Linear Systems*, Addison-Wesley, Reading, Mass., 1962; G. Doetsch, *Guide to the Applications of Laplace Transforms*, Van Nostrand, London, 1961.

[4] A table of transform pairs of particular value for vehicular control problems is S. Neumark's *Operational Formulae for Response Calculations*, ARC Tech. Rept. R & M 3075, Her Majesty's Stationery Office, London, 1958. This report comprehensively treats operational fractions of the first, second, third, and fourth order and has additional tables on the reduction of every fraction of fifth or sixth order to a combination of fractions of lower order. Other useful tables of transform pairs are the references on the Laplace transformation and its uses cited previously; see also F. E. Nixon, *Principles of Automatic Controls*, Prentice-Hall, New York, 1953 (the same table has also been published separately as *Handbook of Laplace Transformations*, Prentice-Hall, New York, 1960); A. Erdélyi, F. Oberhettinger, and F. G. Tricomi, *Tables of Integral Transforms*, Vol. 1, McGraw-Hill, New York, 1954.

Table 2-1. Fundamental properties of the Laplace transformation.

| Item | $F(s)$ | $f(t)$ | Special restrictions and remarks |
|---|---|---|---|
| Laplace transformation | $F(s)$ | $f(t)$ | $t$ is a real variable; $f(t)$ is known and single-valued almost everywhere for $0 \leqq t$ $\lim_{\substack{t_2 \to \infty \\ t_1 \to 0}} \int_{t_1}^{t_2} \lvert f(t) \rvert e^{-\sigma t}\, dt < \infty$ for some real number $\sigma$. |
| Differentiation | $sF(s) - f(0+)$ | $\frac{df(t)}{dt}$ | $f(t)$ and $df(t)/dt$ are Laplace transformable; |
| | $s^2F(s) - sf(0+) - \frac{df}{dt}(0+)$ | $\frac{d^2f(t)}{dt^2}$ | $f(t)$, $df(t)/dt$ and $d^2f(t)/dt^2$ are Laplace transformable; |
| | $s^nF(s) - \sum_{k=1}^{n} s^{n-k} \frac{d^{k-1}f}{dt^{k-1}}(0+)$ | $\frac{d^nf(t)}{dt^n}$ | $f(t)$, $df(t)/dt, \cdots d^nf(t)/dt^n$ are Laplace transformable. |
| Integration | $\frac{F(s)}{s} + \left.\frac{\int f(t)\, dt}{s}\right]_{t=0^+}$ | $\int f(t)\, dt$ | $f(t)$ is Laplace transformable; |
| | $\frac{F(s)}{s^k} + \sum_{n=1}^{k} \frac{f^{-n}(0+)}{s^{k-n+1}}$ | $\int\int \cdots \int f(t)(dt)^k$ | $f(t)$ is Laplace transformable. |
| Linearity | $aF(s)$ | $af(t)$ | $a$ is a constant or a variable that is independent of $t$ and $s$; $f(t)$ is Laplace transformable; |
| | $F_1(s) + F_2(s)$ | $f_1(t) + f_2(t)$ | $f_1(t)$ and $f_2(t)$ are Laplace transformable. |

| | | | |
|---|---|---|---|
| Scale change | $aF(as)$ | $f\left(\frac{t}{a}\right)$ | $f(t)$ is Laplace transformable; $a$ is positive and is a constant or variable independent of $t$ or $s$. |
| Convolution or complex multiplication | $F_1(s)F_2(s)$ | $\int_0^t f_1(t-\tau)f_2(\tau)\,d\tau$ or $f_1(t) * f_2(t)$ | $f_1(t)$ and $f_2(t)$ are Laplace transformable. |
| Real multiplication | $\frac{1}{2\pi j}\int_{C_2-j\infty}^{C_2+j\infty} F_1(s-\lambda)F_2(\lambda)\,d\lambda$ or $F_1(s) * F_2(s)$ | $f_1(t)f_2(t)$ | $f_1(t)$ and $f_2(t)$ are Laplace transformable. $\sigma_{a1} + \sigma_{a2} < \sigma$; $\sigma_{a1} < C_2 < \sigma - \sigma_{a2}$ |
| | $\sum_{i=1}^{n} \frac{N_1(s_i)}{\left[\frac{dD_1(s)}{ds}\right]_{s=s_i}} F_2(s-s_i)$ | | In the special case where $F_1(s) = N_1(s)/D_1(s)$, a rational fraction having no higher order and $k$ first-order poles, $s_1, s_2, \ldots, s_n$ |
| Real translation | $e^{-\tau s}F(s)$<br>$e^{\tau s}F(s)$ | $f(t-\tau)$<br>$f(t+\tau)$ | $f(t)$ is Laplace transformable; $\tau$ is nonnegative real number; and $f(t-\tau) = 0; \quad 0 < t < \tau$; $f(t+\tau) = 0; \quad -\tau < t < 0$ |
| Complex translation | $F\left(s + \frac{1}{T}\right)$ | $e^{-t/T}f(t)$ | $f(t)$ is Laplace transformable; $T$ has a nonnegative real part. |
| | $F\left(s - \frac{1}{T}\right)$ | $e^{t/T}f(t)$ | |
| | $F(Ts - b)$ | $\left(\frac{1}{T}\right)e^{bt/T} f\left(\frac{t}{T}\right)$ | |

Table 2-1 (Continued)

| Item | $F(s)$ | $f(t)$ | Special restrictions and remarks |
|---|---|---|---|
| Final value | $\lim_{s\to 0} sF(s)$ | $\lim_{t\to\infty} f(t)$ | $f(t)$ and $df(t)/dt$ are Laplace transformable; $sF(s)$ is analytic in the right half-plane and on the axis of imaginaries. |
| Initial value | $\lim_{s\to\infty} sF(s)$ | $\lim_{t\to 0} f(t)$ | $f(t)$ and $df(t)/dt$ are Laplace transformable. $\lim_{s\to\infty} sF(s)$ exists. |
| Complex differentiation | $-\dfrac{d}{ds}F(s)$ | $tf(t)$ | $f(t)$ is Laplace transformable. |
| | $(-1)^n \dfrac{d^nF(s)}{ds^n}$ | $t^nf(t)$ | |
| Complex integration | $\int_s^\infty F(s)\,ds$ | $\left(\dfrac{1}{t}\right)f(t)$ | $F(t)$ and $f(t)/t$ are Laplace transformable $\int_s^\infty F(s)\,ds$ exists. |
| Periodic functions | $\dfrac{\int_0^T e^{-st}f(t)\,dt}{1-e^{-Ts}}$ | $f(t)$ | $f(t)$ is Laplace transformable. $f(t) = f(t+T)$ |
| Antiperiodic functions | $\dfrac{\int_0^T e^{-st}f(t)\,dt}{1+e^{-Ts}} = F(s)$ | $f(t)$ | $f(t)$ is Laplace transformable. $f(t) = -f(t+T)$ |
| | $\dfrac{F(s)}{1-e^{-Ts}}$ | Half-wave rectification of $f(t)$. | |
| | $F(s)\coth\left(\dfrac{Ts}{2}\right)$ | Full-wave rectification of $f(t)$. | |

Table 2-2. Common transformation pairs and properties of their time response.

| | F(s) | f(t) | TIME RESPONSE PROPERTIES |
|---|---|---|---|
| UNIT IMPULSE | 1 | $\delta(t)$ | 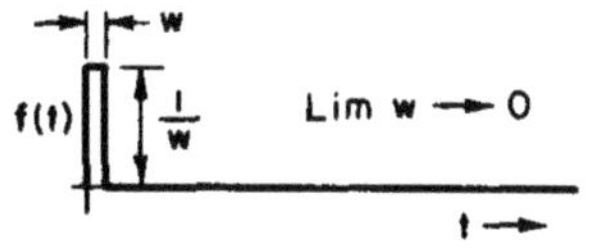 |
| UNIT DOUBLET | s | $\dot{\delta}(t)$ | 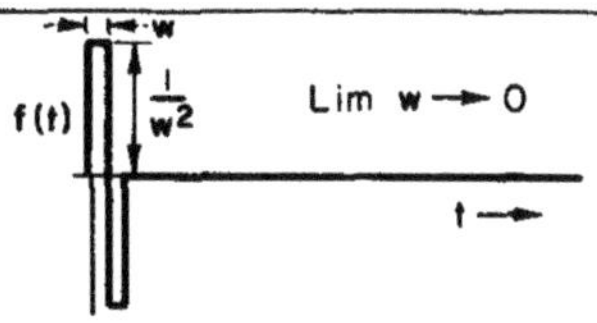 |
| STEP FUNCTION POSITION | $\frac{1}{s}$ | 1 or u(t) |  |
| STEP FUNCTION VELOCITY | $\frac{1}{s^2}$ | t | 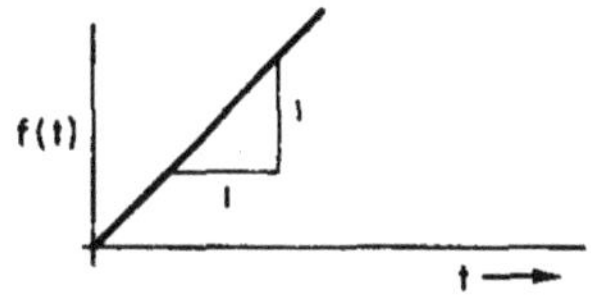 |
| STEP FUNCTION ACCELERATION | $\frac{1}{s^3}$ | $\frac{1}{2}t^2$ | 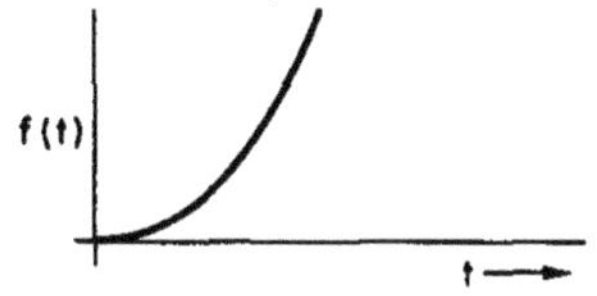 |
| $n^{th}$ ORDER STEP FUNCTION | $\frac{n!}{s^{n+1}}$ | $t^n$ | 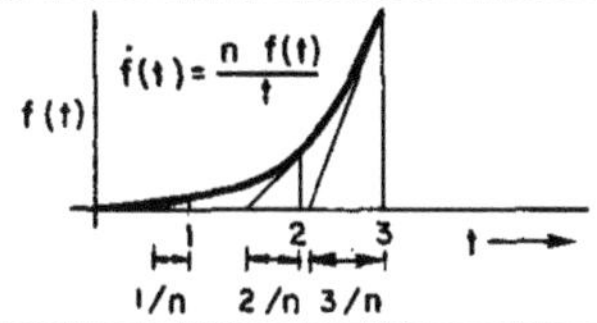 |
| PURE TIME DELAY | $e^{-\tau s}F(s)$ | $f(t-\tau)$ Where: $f(t-\tau)=0, t<\tau$ | 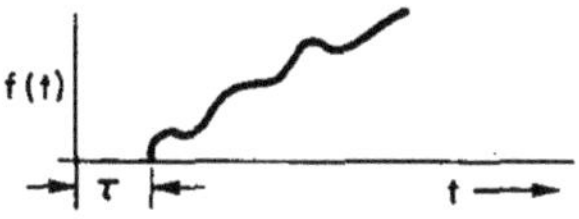 |

Table 2-2. (Continued)

| | F(s) | f(t) | TIME RESPONSE PROPERTIES |
|---|---|---|---|
| FIRST ORDER LAG CONVERGING | $\frac{T}{Ts+1}$ *or* $\frac{1}{s+\alpha}$ $0 < \alpha = \frac{1}{T}$ | $e^{-t/T}$ $e^{-\alpha t}$ | 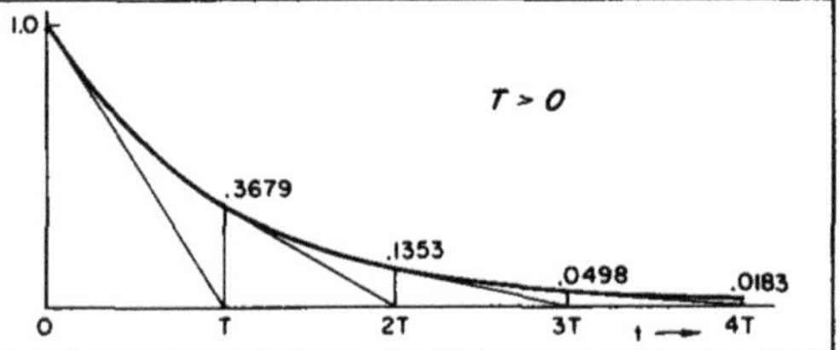  |
| FIRST ORDER LAG CONVERGING *and* FIRST ORDER POLE AT ORIGIN | $\frac{1}{s(Ts+1)}$ *or* $\frac{\alpha}{s(s+\alpha)}$ $0 < \alpha = \frac{1}{T}$ | $1-e^{-t/T}$ $1-e^{-\alpha t}$ | 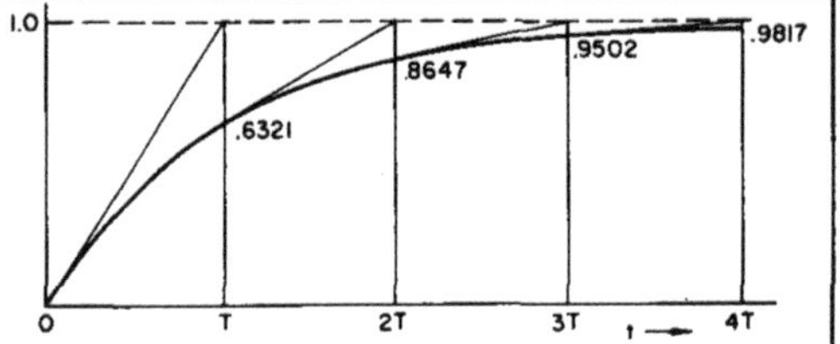  |
| FIRST ORDER LAG DIVERGING | $\frac{-T}{-Ts+1}$ *or* $\frac{1}{s-\alpha}$ $\alpha = \frac{1}{T}$ | $e^{t/T}$ $e^{\alpha t}$ | 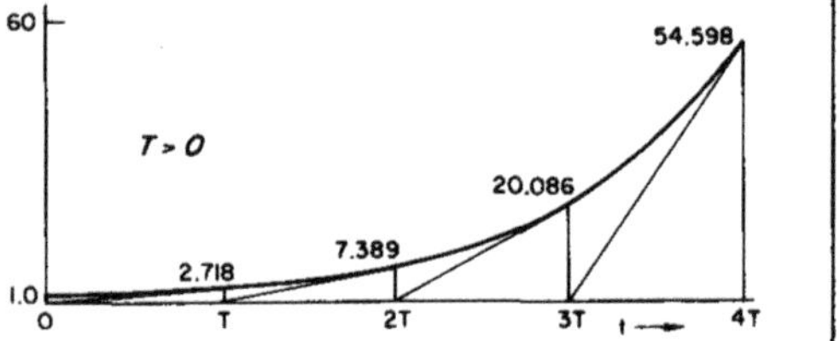  |
| FIRST ORDER LAG DIVERGING *and* FIRST ORDER POLE AT ORIGIN | $\frac{1}{s(-Ts+1)}$ *or* $\frac{-\alpha}{s(s-\alpha)}$ $\alpha = \frac{1}{T}$ | $-1+e^{t/T}$ $-1+e^{\alpha t}$ | 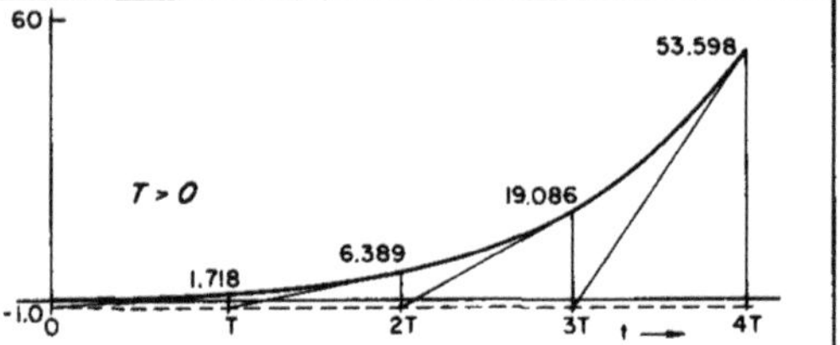  |
| MULTIPLE FIRST ORDER | $\frac{T^n}{(Ts+1)^n}$ *or* $\frac{1}{(s+\alpha)^n}$ $\alpha = \frac{1}{T}$ | $\frac{1}{(n-1)!}t^{n-1}e^{-t/T}$ $\frac{1}{(n-1)!}t^{n-1}e^{-\alpha t}$ | 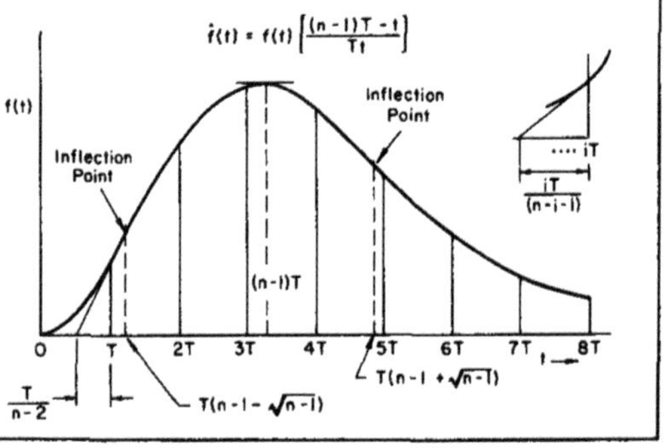  |

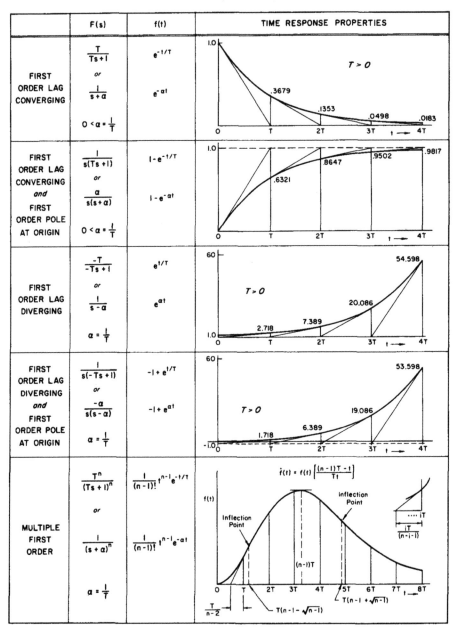

Table 2-2. (Continued)

| | F(s) | f(t) | TIME RESPONSE PROPERTIES |
|---|---|---|---|
| UNDAMPED SECOND ORDER | $\dfrac{\frac{1}{\omega_n}}{\frac{s^2}{\omega_n^2}+1}$ *or* $\dfrac{\omega_n}{s^2+\omega_n^2}$ | $\sin \omega_n t$ | 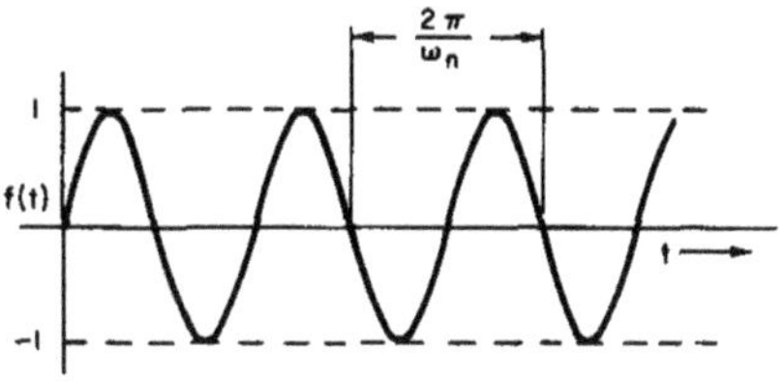  |
| | $\dfrac{\frac{s}{\omega_n^2}}{\frac{s^2}{\omega_n^2}+1}$ *or* $\dfrac{s}{s^2+\omega_n^2}$ | $\cos \omega_n t$ | 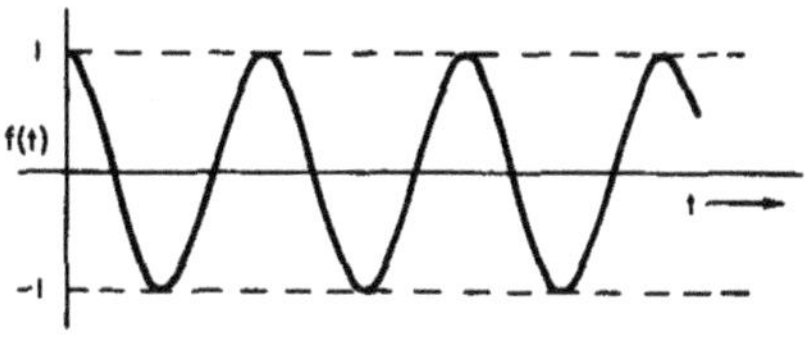  |
| CONVERGING SECOND ORDER | $\dfrac{1}{s^2+2\zeta\omega_n s+\omega_n^2}$ *or* $\dfrac{1}{(s+a)^2+\beta^2}$ | $\dfrac{1}{\omega_n\sqrt{1-\zeta^2}} e^{-\zeta\omega_n t} \cdot \sin \omega_n\sqrt{1-\zeta^2}\,t$ *or* $\dfrac{1}{\beta} e^{-at} \sin \beta t$ Where: $a = \zeta\omega_n$, $\beta = \omega_n\sqrt{1-\zeta^2}$ | 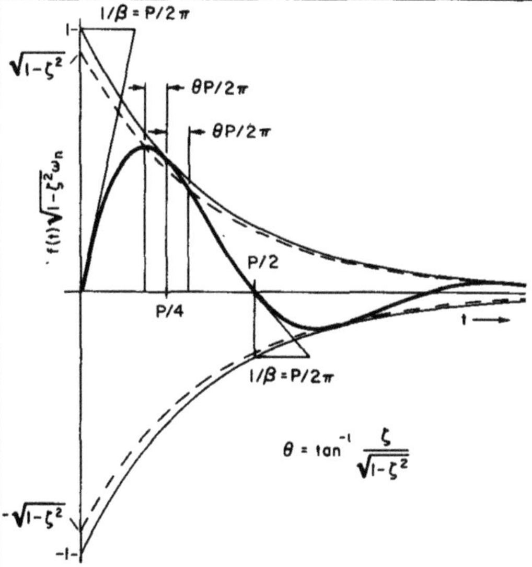  |

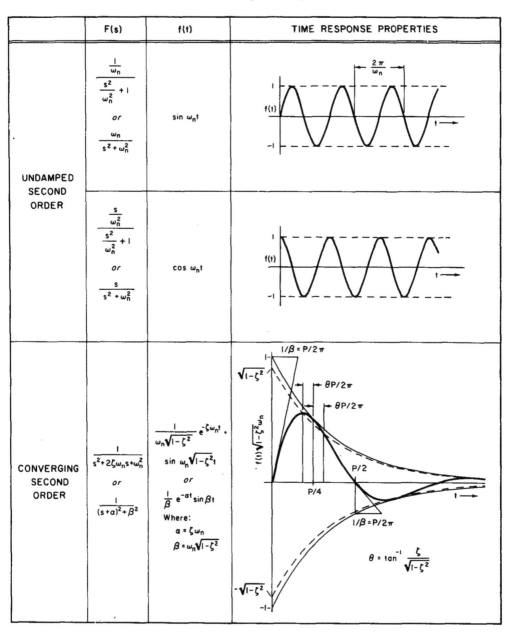

Table 2-2. (Continued)

<table>
<tr><th></th><th>F(s)</th><th>f(t)</th><th>TIME RESPONSE PROPERTIES</th></tr>
<tr>
<td>CONVERGING SECOND ORDER</td>
<td>$\dfrac{s}{s^2 + 2\zeta\omega_n s + \omega_n^2}$<br><br>*or*<br><br>$\dfrac{s}{(s+a)^2 + \beta^2}$<br>$a = \zeta\omega_n$<br>$\beta = \omega_n\sqrt{1-\zeta^2}$</td>
<td>$\dfrac{e^{-\zeta\omega_n t}}{\sqrt{1-\zeta^2}}\cos\left[\omega_n\sqrt{1-\zeta^2}\,t + \theta\right]$<br><br>$\dfrac{\sqrt{a^2+\beta^2}}{\beta}\,e^{-at}\cos(\beta t + \theta)$<br>*where:*<br>$\theta = \tan^{-1}\dfrac{a}{\beta} = \tan^{-1}\dfrac{\zeta}{\sqrt{1-\zeta^2}}$</td>
<td>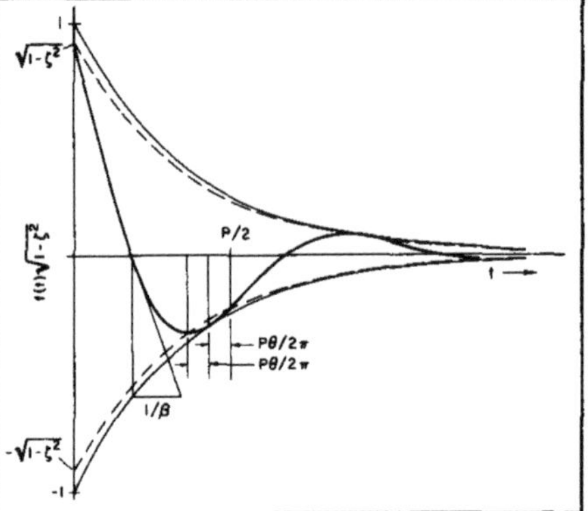
</td>
</tr>
<tr>
<td>CONVERGING SECOND ORDER</td>
<td>$\dfrac{\omega_n^2}{s(s^2 + 2\zeta\omega_n s + \omega_n^2)}$<br><br>*or*<br><br>$\dfrac{a^2+\beta^2}{s\left[(s+a)^2 + \beta^2\right]}$<br>$a = \zeta\omega_n$<br>$\beta = \omega_n\sqrt{1-\zeta^2}$</td>
<td>$1 - \dfrac{e^{-\zeta\omega_n t}}{\sqrt{1-\zeta^2}}\cos\left[\omega_n\sqrt{1-\zeta^2}\,t - \theta\right]$<br><br>$1 - \dfrac{\sqrt{a^2+\beta^2}}{\beta}\,e^{-at}\cos(\beta t - \theta)$<br>*where:*<br>$\theta = \tan^{-1}\dfrac{\zeta}{\sqrt{1-\zeta^2}}$</td>
<td>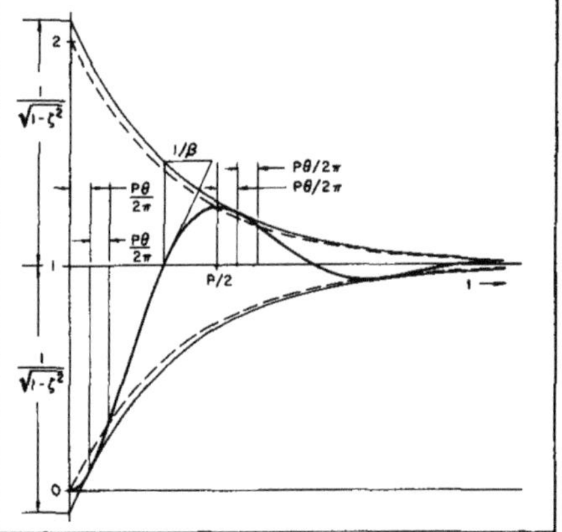
</td>
</tr>
<tr>
<td>DIVERGING SECOND ORDER</td>
<td>$\dfrac{1}{s^2 - 2\zeta'\omega_n s + \omega_n^2}$<br><br>*or*<br><br>$\dfrac{1}{(s-a')^2 + \beta^2}$<br>$a' = \zeta'\omega_n$<br>$\beta = \omega_n\sqrt{1-\zeta'^2}$</td>
<td>$\dfrac{e^{\zeta'\omega_n t}}{\omega_n\sqrt{1-\zeta'^2}}\sin\omega_n\sqrt{1-\zeta'^2}\,t$<br><br>$\dfrac{1}{\beta}\,e^{at}\sin\beta t$</td>
<td>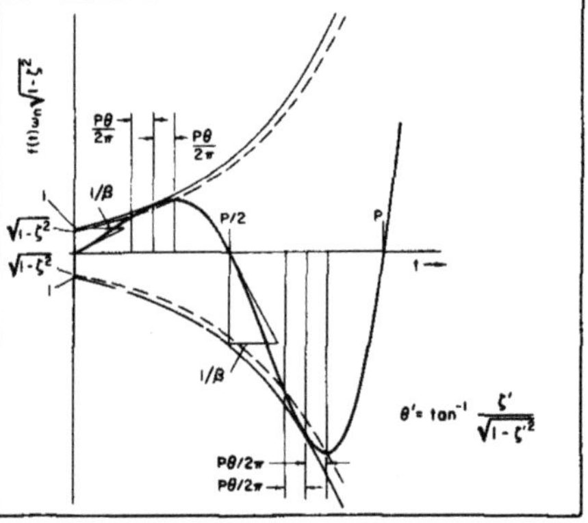
</td>
</tr>
</table>

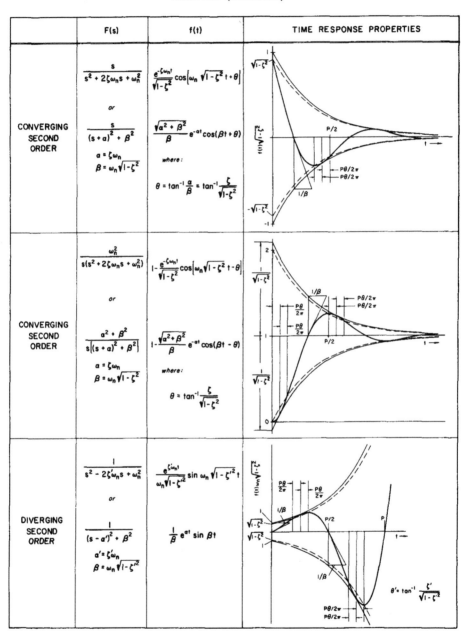

$f(t) - F(s)$ dichotomy, graphical representations of $f(t)$ are given as an integral part of the table. A knowledge of the details of these time histories enables the analyst to picture, either mentally or by a sketch, the behavior of a system element characterized by a given $F(s)$ or $f(t)$.

It is also necessary to point out that the integral that defines the Laplace transformation may fail to converge. For transformable functions this situation is avoided by defining an abscissa of absolute convergence, $\sigma_a$, which is set just large enough to assure the convergence of the transform integral. This is the minimum value that the real part of the complex variable $s = \sigma + j\omega$ may take. In the case of a function, such as $f(t) = t^t$, no value of $\sigma_a$ can be found that will assure convergence of the integral. The function is then said to be *not* Laplace *transformable.*

## 2-3. Response Determination

When the definition of the transform for the operation of differentiation is applied to Eq. 2-1, if it is assumed that *all initial conditions are zero*, the transformed equation is

$$(s^{m+n} + b_1 s^{m+n-1} + \cdots + b_{m+n-1}s + b_{m+n})Y(s)$$
$$= \kappa(s^n + a_1 s^{n-1} + \cdots + a_{n-1}s + a_n)X(s) \qquad (2\text{-}3)$$

The Laplace transformation has reduced the linear differential equation with constant coefficients to an algebraic equation in the transform variable, $s$. For any transformable input $x(t)$, which has the transform $X(s)$, Eq. 2-3 may be solved for $Y(s)$:

$$Y(s) = \kappa \frac{(s^n + a_1 s^{n-1} + \cdots + a_{n-1}s + a_n)}{(s^{m+n} + b_1 s^{m+n-1} + \cdots + b_{m+n-1}s + b_{m+n})} X(s) \qquad (2\text{-}4)$$

In principle, then, the inverse transformation yields $y(t) = \mathcal{L}^{-1}[Y(s)]$.

Example

The procedure can be illustrated with an elementary example. Consider the (rotary) spring-mass-damper system of Fig. 2-2. This device is the all-mechanical analog of a simple servomechanism. The equation of motion, obtained by summing torques on the wheel, is:

$$I\ddot{y} + B\dot{y} + Ky = Kx \qquad (2\text{-}5)$$

If this ordinary differential equation with constant coefficients is Laplace-transformed, it may be written as:

$$[Is^2 + Bs + K]Y(s) = KX(s) + [Isy(0+) + I\dot{y}(0+) + By(0+)] \qquad (2\text{-}6)$$

where $y(0+)$ and $\dot{y}(0+)$ are the initial conditions. Note that the action of

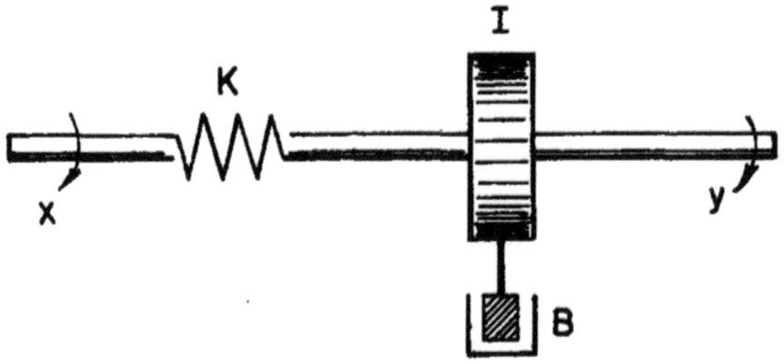


Fig. 2-2. Spring-mass-damper system.

the initial conditions is equivalent to that of an input made up of delta functions and higher order delta functions. That is, $Isy(0+)$ amounts to a doublet of weight $Iy(0+)$, while $I\dot{y}(0+)$ and $By(0+)$ are equivalent to a delta function input with weight $[I\dot{y}(0+) + By(0+)]$. Solving for $Y(s)$,

$$Y(s) = \frac{(K/I)X(s)}{s^2 + (B/I)s + K/I} + \frac{sy(0+) + [\dot{y}(0+) + (B/I)y(0+)]}{s^2 + (B/I)s + K/I} \tag{2-7}$$

Letting $x(t) = \delta(t)$, the unit impulse, so that $X(s) = 1$; setting the initial conditions to zero; and defining an "undamped natural frequency," $\omega_n = \sqrt{K/I}$, and a "damping ratio," $\zeta = B/(2\sqrt{KI})$, we find that

$$Y(s) = \frac{\omega_n^2}{s^2 + 2\zeta\omega_n s + \omega_n^2} \tag{2-8a}$$

or

$$Y(s) = \frac{1}{(s^2/\omega_n^2) + (2\zeta s/\omega_n) + 1} \tag{2-8b}$$

The inverse transformation of this function can be carried out by recognizing that the right-hand side of Eq. 2-8 is a pair in Table 2-2 and that:

$$y(t) = \frac{\omega_n}{\sqrt{1-\zeta^2}} e^{-\zeta\omega_n t} \sin \omega_n\sqrt{1-\zeta^2}\, t; \qquad \zeta < 1 \tag{2-9}$$

When the response transform has a more complex denominator, the transform can be broken down into a sum of partial fractions before inverse transforming. Suppose that the algebraic solution for the transform of the response is given in the form

$$Y(s) = \kappa \frac{(s^n + a_1 s^{n-1} + \cdots + a_{n-1}s + a_n)}{(s^{m+n} + b_1 s^{m+n-1} + \cdots + b_{m+n-1}s + b_{m+n})} \tag{2-10}$$

where $m \geqq 1$. The polynomials can be factored. Then,

$$Y(s) = \frac{N(s)}{D(s)} = \kappa \frac{\prod_{j=1}^{n} (s + z_j)}{\prod_{i=1}^{m+n} (s + q_i)} = \kappa \frac{\alpha(s)}{\beta(s)} \tag{2-11}$$

or

$$Y(s) = \frac{N_1(s)}{D_1(s)} = \frac{K \prod_{j=1}^{n} [(s/z_j) + 1]}{s^k \prod_{i=1}^{m+n-k} [(s/q_i) + 1]} = K \frac{A(s)}{B(s)} \tag{2-12}$$

and the numerator and denominator roots, $-z_j$ and $-q_i$, are called *zeros* and *poles*, respectively. For reasons that will appear later, the first style of the transform factors is called the *root locus form* and the second style is called the *Bode form*. Both are used extensively.

In principle, the right-hand side of Eq. 2-11 or 2-12 can be reformulated in a partial fraction expansion; elementary inverse Laplace transformations can then be carried out on each term. This yields the time response, $y(t)$.

$$Y(s) = \sum_{i=1}^{m+n} \frac{C_i}{s + q_i} \tag{2-13}$$

$$y(t) = \mathcal{L}^{-1}[Y(s)] = \sum_{i=1}^{m+n} C_i e^{-q_i t} \tag{2-14}$$

Thus, any response transform that is a ratio of rational polynomials with real coefficients results in a time domain response which is a sum of real or complex conjugate exponentials. The amount or magnitude of each mode that is present in the total response is indicated by the partial fraction coefficient, $C_i$.

When the response transforms have either of the equivalent rational proper fraction forms

$$Y(s) = \frac{N(s)}{D(s)} = \kappa \frac{(s + z_1)(s + z_2) \cdots}{(s + q_1)(s + q_2) \cdots} = \frac{C_1}{s + q_1} + \frac{C_2}{s + q_2} + \cdots \tag{2-15}$$

or

$$Y(s) = \frac{N_1(s)}{D_1(s)} = K \frac{(T_a s + 1)(T_b s + 1) \cdots}{(T_1 s + 1)(T_2 s + 1) \cdots} = \frac{C_1'}{T_1 s + 1} + \frac{C_2'}{T_2 s + 1} + \cdots \tag{2-16}$$

the coefficients $C_i$ or $C_i'$ may be evaluated as shown in Table 2-3.

If the explicit differential equation that describes the system is of low order, the polynomials in the system transfer function are of the same low degree. Then the task of factoring the polynomials and of finding the partial fraction coefficients can be carried out without difficulty.[5] On the

[5] Concise summaries of preferred methods appear in J. J. D'Azzo and C. H. Houpis, *Feedback Control System Analysis and Synthesis*, McGraw-Hill, New York, 1960 (Appendix B); Y. H. Ku, *Analysis and Control of Linear Systems*, International Textbook, Scranton, Pa., 1962, pp. 171–181.

Table 2-3. Partial fraction coefficients.

First-Order Poles

General: The coefficient of the component involving $1/(s + q_1)$ or $1/(T_1 s + 1)$ will be

$$C_1 = \left[\frac{N(s)(s + q_1)}{D(s)}\right]_{s=-q_1}; \qquad C_1' = \left[\frac{N_1(s)(T_1 s + 1)}{D_1(s)}\right]_{s=-1/T_1}$$

First-order pole at the origin: For this case,

$$X(s) = \frac{N(s)}{sD(s)} \quad \text{or} \quad \frac{N_1(s)}{sD_1(s)}; \qquad C_0 = \frac{N(0)}{D(0)} = \frac{N_1(0)}{D_1(0)}$$

First-order poles on the imaginary axis (at $s = \pm j\omega_1$): The inverse transform of this component of $X(s)$ will be

$$\text{Re}\left[\frac{(s^2 + \omega_1^2)}{s}\frac{N(s)}{D(s)}e^{st}\right]_{s=j\omega_1} = \text{Re}\left[\frac{-s[(s^2/\omega_1^2) + 1]N_1(s)}{D_1(s)}e^{st}\right]_{s=j\omega_1}$$

One conjugate-complex pair with $N$ real first-order poles:

$$\frac{N(s)}{D(s)} = \frac{a_{N+1}s^{N+1} + a_N s^N + \cdots + a_1 s + a_0}{[\Pi_{i=1}^N (s + q_i)](s^2 + bs + c)} = \sum_{i=1}^{N}\frac{C_i}{s + q_i} + \frac{As + B}{s^2 + bs + c}$$

$$A = a_{N+1} - \sum_{i=1}^{N} C_i; \qquad B = \frac{a_0}{\Pi_{i=1}^N q_i} - c\sum_{i=1}^{N}\frac{C_i}{q_i}$$

Higher Order Poles

$D(s)$ will contain terms such as $(s + q_1)^m$, and $D_1(s)$ will contain terms such as $(T_1 s + 1)^m$. The coefficients $C_{m-k}$ and $C'_{m-k}$, for the components $1/(s + q_1)^{m-k}$ and $1/(T_1 s + 1)^{m-k}$ of the $X(s)$ expansion, will be

$$C_{m-k} = \frac{1}{k!}\left[\frac{d^k}{ds^k}\frac{(s + q_1)^m N(s)}{D(s)}\right]_{s=-q_1}; \qquad C'_{m-k} = \frac{1}{k!\,T_1^k}\left[\frac{d^k}{ds^k}\frac{(T_1 s + 1)^m N_1(s)}{D_1(s)}\right]_{s=-1/T_1}$$

where $m - k \geqq 1$.

other hand, if the characteristic (denominator) polynomial is of the fourth degree or higher, the algebraic factoring of the polynomial and the determination of the partial fraction coefficients may be excessively tedious. It was partly to avoid most of the tedious labor involved in the solution of the equations of motion of linear systems that the semigraphical methods of linear feedback system analysis to be discussed in Chapter 3 were developed.

## 2-4. Simplified Methods to Obtain an Approximate $f(t)$ from its Unfactored Transform $F(s)$

A major source of practical difficulty in solving linear constant coefficient differential equations arises in finding the time response, $f(t)$, after its transform, $F(s)$, is known. This occurs because the poles of $F(s)$ must usually be found as a necessary preliminary to the partial fraction expansion or to entry into a transform table. Factoring the polynomials of $F(s)$ to find its poles and zeros may be more time-consuming than warranted. Therefore, schemes that avoid this operation, yet still allow the extraction of some information about the time response, $f(t)$, directly from its transform, $F(s)$, are valuable techniques for many applications.

In this section three ways are described to find information about a time function, $f(t)$, from its transform, $F(s)$, in unfactored form. The first two are fundamental properties of the Laplace transformation, the initial and final value theorems. Application of these properties gives the values of $f(t)$ at $t = 0$ and as $t \to \infty$. In essence, these theorems state an equality between two particular values of $f(t)$ and two particular values of its transform, $F(s)$. The initial value theorem can also be used to obtain the derivatives of the initial time response, thereby permitting its expansion in a Maclaurin series. The third method involves the expansion of the response transform in a Maclaurin series, which has a degree of validity in the steady state after transients have become insignificant, or in cases in which the input can be approximated as a power series. This series is the basis of the error coefficients so valuable in some phases of servo design.

### *The Initial Value Theorem*

The initial value theorem is a basic property of the Laplace transformation that allows the value of $f(t)$ and its derivatives at $t = 0$ to be found from its transform, $F(s)$. Specifically, if $f(t)$ and its first derivative have Laplace transforms and the limit as $s \to \infty$ of $sF(s)$ exists, where $F(s)$ is the transform of $f(t)$, then

$$\lim_{s \to \infty} sF(s) = \lim_{t \to 0} f(t) \tag{2-17}$$

For example, if $F(s) = K/(s^3 + as^2 + bs + c)$, then

$$\lim_{t \to 0} f(t) = \lim_{s \to \infty} \frac{Ks}{s^3 + as^2 + bs + c} \tag{2-18}$$

and

$$f(0) = 0$$

If the derivatives of $f(t)$ are Laplace transformable, they have Laplace transforms given by

$$\mathcal{L}[\dot{f}(t)] = sF(s) - f(0+) \tag{2-19}$$

$$\mathcal{L}[\ddot{f}(t)] = s^2F(s) - sf(0+) - \dot{f}(0+) \tag{2-20}$$

$$\mathcal{L}[\dddot{f}] = s^3F(s) - s^2f(0+) - s\dot{f}(0+) - \ddot{f}(0+) \tag{2-21}$$

$$\mathcal{L}[\overset{\text{IV}}{f}] = s^4F(s) - s^3f(0+) - s^2\dot{f}(0+) - s\ddot{f}(0+) - \dddot{f}(0+) \tag{2-22}$$

and so forth. Use of the initial value theorem to evaluate the derivatives for the example proceeds as follows.

Since $f(0+) = 0$,

$$\lim_{t\to 0} \dot{f}(t) = \lim_{s\to\infty} s[sF(s)] = \lim_{s\to\infty} s\,\frac{Ks}{s^3 + as^2 + bs + c} = 0 \tag{2-23}$$

and similarly, since $f(0+)$ and $\dot{f}(0+)$ are zero,

$$\lim_{t\to 0} \ddot{f}(t) = \lim_{s\to\infty} s[s^2F(s)] = \lim_{s\to\infty} s\,\frac{Ks^2}{s^3 + as^2 + bs + c} = K \tag{2-24}$$

Because $\ddot{f}(0) = K$, the initial value of the third derivative will be

$$\begin{aligned}\lim_{t\to 0} \dddot{f}(t) &= \lim_{s\to\infty} s[s^3F(s) - \ddot{f}(0+)]\\ &= \lim_{s\to\infty} s\left[\frac{Ks^3}{s^3 + as^2 + bs + c} - K\right]\\ &= \lim_{s\to\infty}\left[-\frac{Ks(as^2 + bs + c)}{s^3 + as^2 + bs + c}\right] = -Ka\end{aligned} \tag{2-25}$$

For the initial value of the fourth derivative,

$$\begin{aligned}\lim_{t\to 0} \overset{\text{IV}}{f}(t) &= \lim_{s\to\infty} s[s^4F(s) - s\ddot{f}(0+) - \dddot{f}(0+)]\\ &= \lim_{s\to\infty} s\left[\frac{Ks^4}{s^3 + as^2 + bs + c} - sK + Ka\right]\\ &= \lim_{s\to\infty} K\left[\frac{(a^2 - b)s^3 + (ab - c)s^2 + acs}{s^3 + as^2 + bs + c}\right] = K(a^2 - b)\end{aligned} \tag{2-26}$$

and so forth.

The derivatives evaluated in the fashion illustrated above are valuable as checkpoints and to enhance one's physical grasp of the initial character of a response. Also, several derivatives can be combined into a Maclaurin series to give an approximation to the initial response. The Maclaurin series, in general, is

$$f(t) = f(0) + t\dot{f}(0) + \frac{t^2}{2!}\ddot{f}(0) + \frac{t^3}{3!}\dddot{f}(0) + \cdots \tag{2-27}$$

For the example above this series would be

$$\begin{aligned}f(t) &= K\left[\frac{t^2}{2} - a\frac{t^3}{6} + (a^2 - b)\frac{t^4}{24} + \cdots\right]\\ &= \frac{Kt^2}{2}\left[1 - \frac{at}{3} + \frac{(a^2 - b)t^2}{12} + \cdots\right]\end{aligned} \tag{2-28}$$

This initial response version of the Maclaurin series supplements the Maclaurin series developed later for the steady-state response.

*The Final Value Theorem*

The final value theorem equates the value of a time function, $f(t)$, as $t$ approaches infinity to that of the function $sF(s)$ as $s$ approaches zero. Obviously, there must be more restrictions on the application of this theorem than there were on that for the initial value. For example, an $F(s)$ that has poles in the right half-plane, or on the axis of imaginaries (both allowed in the initial value theorem), gives rise to an $f(t)$ that has no final value. So, in addition to the requirement shared with the initial value theorem, i.e., that the function $f(t)$ being evaluated at $t = 0$ and its derivative are Laplace transformable, it is also necessary to specify that the function $sF(s)$ is analytic on the axis of imaginaries and in the right half-plane. Then, with these restrictions,

$$\lim_{s \to 0} sF(s) = \lim_{t \to \infty} f(t) \tag{2-29}$$

Using the previous example and assuming that $F(s)$ has no poles in the right half-plane or on the imaginary axis, we obtain

$$\lim_{t \to \infty} f(t) = \lim_{s \to 0} s \frac{K}{s^3 + as^2 + bs + c} = 0 \tag{2-30}$$

Similarly, all higher derivatives also have zero final values.

As another example, consider

$$F(s) = \frac{K}{s^2(s + \alpha)} \tag{2-31}$$

The initial value theorem gives $f(0) = 0$, but the application of the final value theorem to find $f(t)]_{t \to \infty}$ is not possible because $sF(s)$ has a pole at the origin (on the imaginary axis). However, if $\alpha$ is positive, the derivative $\dot{f}(t)$ does have a final value, i.e.,

$$\lim_{t \to \infty} \dot{f}(t) = \lim_{s \to 0} s \frac{Ks}{s^2(s + \alpha)} = \frac{K}{\alpha} \tag{2-32}$$

Further, since the theorem would show that $\ddot{f}(t)$, $\dddot{f}(t)$, and all the higher derivatives are zero as $t$ goes to infinity, it is apparent that $f(t)$ can be approximated for large values of $t$ by

$$f(t) \doteq \frac{K}{\alpha} t \tag{2-33}$$

In addition to giving the type of particular answers illustrated above, the initial and final value theorems may be regarded as a basis for establishing the intuitive feeling that the steady-state time response is determined largely by the behavior of $F(s)$ at small values of the complex frequency, $s$, and that the time response at small values of time depends largely on $F(s)$ at large values of $s$. This is often a helpful concept (although the restrictions of the two underlying theorems should be kept thoroughly in mind whenever it is used).

As a case in point, consider a distinction between the so-called Bode and root locus gains. A response transform, or one of its derivatives or integrals if more than one power of free $s$ is present, may be written either as

$$Y(s) = \kappa \frac{(s^n + a_1 s^{n-1} + \cdots + a_{n-1}s + a_n)}{s(s^{m+n} + b_1 s^{m+n-1} + \cdots + b_{m+n-1}s + b_{m+n})} \tag{2-34}$$

or as

$$Y(s) = K \frac{[(s^n/a_n) + (a_1/a_n)s^{n-1} + \cdots + (a_{n-1}/a_n)s + 1]}{s[(s^{m+n}/b_{m+n}) + (b_1/b_{m+n})s^{m+n-1} + \cdots + (b_{m+n-1}/b_{m+n}) + 1]} \tag{2-35}$$

As introduced by Eq. 2-10, et seq., the first style (Eq. 2-34) is in the root locus form, with a root locus gain, $\kappa$; the second (Eq. 2-35) is in Bode form, with a Bode gain, $K$. Assuming that the final value theorem is applicable, it is easy to see that the Bode gain is the final value, i.e.,

$$\lim_{t\to\infty} y(t) = \lim_{s\to 0} [sY(s)] = K \tag{2-36}$$

The root locus gain, on the other hand, is connected with the initial response. It is, in fact, the first nonzero derivative at $t = 0$, i.e.,

$$\left.\frac{d^m y(t)}{dt^m}\right|_{t\to 0} = \lim_{s\to\infty} s[s^m Y(s)] = \kappa \tag{2-37}$$

*Steady-State Response Calculations by a Maclaurin Series*

In the first article of this section a Maclaurin series having time as the variable was used to develop an expression for the initial response of a system. In this article a Maclaurin series will again be used, but in the transform domain and to evolve an approximation to the steady-state response. As a result of the two developments, approximate expressions for both the initial and final phases of response can be obtained directly from the polynomial ratio form of $F(s)$ without factoring.

As shown in previous parts of this chapter, the output of an element when excited by some input is given by

$$\begin{aligned} Y(s) &= \frac{N(s)}{D(s)} X(s) \\ &= W(s)X(s) \end{aligned} \tag{2-38}$$

where $Y(s)$ is the transform of the output response, $X(s)$ that of the input and $W(s) = N(s)/D(s)$ is a ratio of polynomials in $s$ containing all the system characteristics. If $W(s)$ is expanded in a Maclaurin series in $s$, the series will converge for small values of $s$.

$$W(s) = W(0) + s\left.\frac{dW(s)}{ds}\right|_{s=0} + \frac{s^2}{2!}\left.\frac{d^2W(s)}{ds^2}\right|_{s=0} + \cdots + \frac{s^n}{n!}\frac{d^nW(s)}{ds^n} + \cdots$$

$$= C_0 + C_1s + C_2s^2 + \cdots + C_ns^n + \cdots \tag{2-39}$$

where

$$C_n = \frac{1}{n!}\left.\frac{d^nW(s)}{ds^n}\right|_{s=0} = W^{(n)}(0)$$

Multiplying the series by $X(s)$ to give the output transform, $Y(s)$, we obtain

$$Y(s) = C_0X(s) + C_1sX(s) + C_2s^2X(s) + \cdots + C_ns^nX(s) + \cdots \tag{2-40}$$

This expression is valid in the region near $s = 0$, where the series is convergent. Recognizing that

$$\mathcal{L}\left[\frac{d^nx(t)}{dt^n}\right] = s^nX(s) - s^{n-1}x(0+) - s^{n-2}\dot{x}(0+) - \cdots - \frac{d^{n-1}x}{dt^{n-1}}(0+)$$

we find that the series for the output transform reduces to

$$\begin{aligned} Y(s) = C_0\mathcal{L}[x(t)] + C_1\mathcal{L}[\dot{x}(t)] + C_2\mathcal{L}[\ddot{x}(t)] + \cdots + C_n\mathcal{L}\left[\frac{d^nx}{dx^n}\right] \\ + \cdots + \{x(0+)[C_1 + C_2s + \cdots]\} \\ + \{\dot{x}(0+)[C_2 + C_3s + \cdots]\} + \cdots \end{aligned} \tag{2-41}$$

If an infinity of higher order impulse functions (which occur at $t = 0$ and hence have no effect on large $t$) are ignored, the inverse transform is given by

$$y(t) = W(0)x(t) + W'(0)\dot{x}(t) + \frac{1}{2!}W''(0)\ddot{x}(t) + \cdots + \frac{1}{n!}W^{(n)}(0)\frac{d^nx}{dt^n} + \cdots$$

$$= C_0x(t) + C_1\dot{x}(t) + C_2\dot{x}(t) + \cdots + C_n\frac{d^nx}{dt^n} + \cdots \tag{2-42}$$

where the primes denote differentiation with respect to $s$. This series is valid only at those times corresponding to $s \to 0$, i.e., in the neighborhood governed by the final value theorem, and all of the restrictions of the theorem apply. As a practical matter the series is suitable to define the steady-state components of a system response, presuming either that the transients have died away or that they have been otherwise removed.

The $C_n$'s shall be called *general output response coefficients.* In the special case in which $x(t)$ is the system command input and $y(t)$ is the system error, these coefficients are the well-known error coefficients of conventional servo analysis.

As might be expected, the result given by Eq. 2-42 is the forced solution for a system subjected to a power series input. The transient component of the solution is not obtained, although all the output terms having time variations identical with those of the power series input are given by the relation. Because power series are handy devices to describe such things as idealized command signals derived from empirical data, average effects of random functions which have stationary characteristics about a time-varying mean, etc., the response series has many uses.

To complete the discussion of the response series, the first few output response coefficients will be developed for a general system. For this purpose let $W(s)$ have the form

$$W(s) = \frac{b_0 + b_1 s + b_2 s^2 + \cdots + b_n s^n}{a_0 + a_1 s + a_2 s^2 + \cdots + a_n s^n}$$
$$= \frac{b_0 + b_1 s + b_2 s^2 + \cdots + b_n s^n}{a_0[1 + (s/a_0)(a_1 + a_2 s + a_3 s^2 + \cdots + a_n s^{n-1})]} \tag{2-43}$$

For most input-output combinations the order of the numerator will be less than that of the denominator, so $b_n$ and perhaps other $b_i$'s will be zero. In Eq. 2-43 the numerator is allowed to be the same order as the denominator to include such important cases as input-error, closed-loop response functions.

Perhaps the simplest way to generate the required Maclaurin series for functions like those of Eq. 2-43 is simply to divide the denominator into the numerator. Because the denominator is of the form $[1 + z(s)]$, this can be readily accomplished by expanding $[1 + z(s)]^{-1}$ as $1 - z(s) + z^2(s) - \cdots$, and then multiplying by the numerator polynomial. Proceeding in this way, we find that

$$W(s) = \frac{1}{a_0}(b_0 + b_1 s + b_2 s^2 + \cdots + b_n s^n)$$
$$\times \left\{1 - \frac{s}{a_0}(a_1 + a_2 s + \cdots + a_n s^{n-1}) + \left(\frac{s}{a_0}\right)^2 (a_1 + a_2 s + \cdots + a_n s^{n-1})^2 - \left(\frac{s}{a_0}\right)^3 (a_1 + a_2 s + \cdots + a_n s^{n-1})^3 + \cdots\right\}$$

$$= \frac{1}{a_0}(b_0 + b_1 s + b_2 s^2 + \cdots + b_n s^n)$$

$$\times \left\{1 - \frac{a_1}{a_0} s + \left[\left(\frac{a_1}{a_0}\right)^2 - \frac{a_2}{a_0}\right] s^2 - \left[\left(\frac{a_1}{a_0}\right)^3 - \frac{2a_1a_2}{a_0^2} + \frac{a_3}{a_0}\right] s^3 + \cdots\right\}$$

$$= \frac{1}{a_0}\left[b_0 - \left(\frac{a_1}{a_0} b_0 - b_1\right) s + \left\{\left[\left(\frac{a_1}{a_0}\right)^2 - \frac{a_2}{a_0}\right] b_0 - \frac{a_1}{a_0} b_1 + b_2\right\} s^2\right.$$

$$\left. - \left\{\left[\left(\frac{a_1}{a_0}\right)^3 - \frac{2a_1a_2}{a_0^2} + \frac{a_3}{a_0}\right] b_0 - \left[\left(\frac{a_1}{a_0}\right)^2 - \frac{a_2}{a_0}\right] b_1 + \frac{a_1}{a_0} b_2 - b_3\right\} s^3 + \cdots\right] \tag{2-44}$$

Putting this result in a somewhat different form, which is often easier to work with, we obtain

$$W(s) = \frac{b_0}{a_0} + \frac{a_0b_1 - a_1b_0}{a_0^2} s + \left[\frac{a_0b_2 - a_2b_0}{a_0^2} - \left(\frac{a_0b_1 - a_1b_0}{a_0^2}\right)\frac{a_1}{a_0}\right] s^2$$

$$+ \left\{\frac{a_0b_3 - a_3b_0}{a_0^2} + \frac{a_0b_1 - a_1b_0}{a_0^2}\left[\left(\frac{a_1}{a_0}\right)^2 - \frac{a_2}{a_0}\right] - \frac{a_0b_2 - a_2b_0}{a_0^2}\frac{a_1}{a_0}\right\} s^3 + \cdots$$

$$= C_0 + C_1 s + C_2 s^2 + C_3 s^3 + \cdots \tag{2-45}$$

The dependence of successive output response coefficients upon preceding coefficients can be deduced by properly associating various combinations of terms in Eq. 2-45, namely:

$$\begin{aligned} C_0 &= \frac{b_0}{a_0} \\ C_1 &= \frac{b_1}{a_0} - \frac{a_1}{a_0} C_0 \\ C_2 &= \frac{b_2}{a_0} - \frac{a_2}{a_0} C_0 - \frac{a_1}{a_0} C_1 \\ C_3 &= \frac{b_3}{a_0} - \frac{a_3}{a_0} C_0 - \frac{a_2}{a_0} C_1 - \frac{a_1}{a_0} C_2 \end{aligned} \tag{2-46}$$

This form, which is readily extended by inspection, provides a convenient algorithm for the computation of output response coefficients in a sequential fashion.

## 2-5. Partial Fraction Coefficient Ratios

Higher order equations come about through multidegree of freedom systems. In many cases the responses of more than one degree of freedom are desired. These can be found using the technique already described, i.e., decompose the response transforms for each degree of freedom into partial fraction expansions, and then inverse-transform term by term. With this procedure the labor is increased in proportion to the number of degrees of freedom for which information is desired.

A far more efficient way to determine responses for the several degrees of freedom starts with the original equations of motion for the system. In these the degrees of freedom are dependent variables in a set of simultaneous constant-coefficient differential equations. Transformation of the simultaneous differential equations simplifies them to a set of linear algebraic equations. At this point, determinants or any equivalent method such as the elimination of variables between equations can be applied to find the transform of each dependent variable. In the technique to be described the partial fraction coefficients are determined for one degree of freedom. For all the other degrees of freedom, ratios of partial fraction coefficients are obtained from the transformed equations of motion. The partial fraction coefficient ratios can then be used with the partial fraction coefficients for the first degree of freedom to determine the coefficients appropriate to the other degrees of freedom.

The procedure outlined above is best understood with the aid of an illustration. Consider the following set of three simultaneous equations:

$$a_{11}(s)X(s) + a_{12}(s)Y(s) + a_{13}(s)Z(s) = b_1(s)\delta_1(s) \tag{2-47}$$

$$a_{21}(s)X(s) + a_{22}(s)Y(s) + a_{23}(s)Z(s) = b_2(s)\delta_2(s) \tag{2-48}$$

$$a_{31}(s)X(s) + a_{32}(s)Y(s) + a_{33}(s)Z(s) = b_3(s)\delta_3(s) \tag{2-49}$$

Here, $x(t)$, $y(t)$, and $z(t)$ are the dependent variables, the $\delta$'s are the input forcing functions, and the equations are in the transformed state so that the coefficients $a_{ij}(s)$ are polynomials in $s$. Equations 2-47 to 2-49 are simple linear equations and can be solved by determinants or an equivalent procedure. Then $X(s)$ becomes

$$X(s) = \frac{\begin{vmatrix} b_1\delta_1 & a_{12} & a_{13} \\ b_2\delta_2 & a_{22} & a_{23} \\ b_3\delta_3 & a_{32} & a_{33} \end{vmatrix}}{\Delta(s)} = \frac{N_x(s)}{\Delta(s)} \tag{2-50}$$

and similarly, $Y(s)$ and $Z(s)$ are given by

$$Y(s) = \frac{\begin{vmatrix} a_{11} & b_1\delta_1 & a_{13} \\ a_{21} & b_2\delta_2 & a_{23} \\ a_{31} & b_3\delta_3 & a_{33} \end{vmatrix}}{\Delta(s)} = \frac{N_y(s)}{\Delta(s)} \tag{2-51}$$

and

$$Z(s) = \frac{\begin{vmatrix} a_{11} & a_{12} & b_1\delta_1 \\ a_{21} & a_{22} & b_2\delta_2 \\ a_{31} & a_{32} & b_3\delta_3 \end{vmatrix}}{\Delta(s)} = \frac{N_z(s)}{\Delta(s)} \tag{2-52}$$

where

$$\Delta(s) = \begin{vmatrix} a_{11} & a_{12} & a_{13} \\ a_{21} & a_{22} & a_{23} \\ a_{31} & a_{32} & a_{33} \end{vmatrix}$$

The denominator, $\Delta(s)$, is the *characteristic function* of the system. When it is equated to zero to make the *characteristic equation* $\Delta(s) = 0$, the separate factors yield the roots, $s = -q_i$, which determine the nature of the individual exponential motions or *modes* of motion. These equations can be solved for $x(t)$, $y(t)$, and $z(t)$ by performing the following conventional steps: Expand the determinant, $\Delta(s)$, and find its roots. These roots and the poles introduced by $\delta_i(s)$ are the poles shared by $X(s)$, $Y(s)$, and $Z(s)$. In the illustrative case the poles will be denoted by subscripts 1, 2, 3, . . . $n$.

$$\kappa_\Delta(s + q_1)(s + q_2)(s + q_3) \cdots (s + q_n) = \Delta(s) = 0$$

or

$$\Delta(-q_1) = \Delta(-q_2) = \Delta(-q_3) = \Delta(-q_n) = 0$$

Expand $X(s)$, $Y(s)$, and $Z(s)$ as partial fractions and find the partial fraction coefficients, as shown in the last section. Examples of this for $X(s)$ and $Y(s)$ are given below. The expressions for $X(s)$ and $Y(s)$ are continued to allow for poles of $\delta_1(s)$, $\delta_2(s)$, and $\delta_3(s)$, which are left as arbitrary functions.

$$X(s) = \frac{x_1}{s + q_1} + \frac{x_2}{s + q_2} + \cdots \frac{x_n}{s + q_n} + \cdots$$

$$Y(s) = \frac{y_1}{s + q_1} + \frac{y_2}{s + q_2} + \cdots \frac{y_n}{s + q_n} + \cdots$$

As the final step, perform the inverse transformation term by term by utilizing Table 2-2.

$$x(t) = x_1e^{-q_1t} + x_2e^{-q_2t} + \cdots + x_ne^{-q_nt} + \cdots$$

$$y(t) = y_1e^{-q_1t} + y_2e^{-q_2t} + \cdots + y_ne^{-q_nt} + \cdots$$

In the above procedure the partial fraction coefficients $x_k$, $y_k$, $z_k$ for each dependent variable are found separately. These coefficients will, of course, be partially determined by the particular type of input, requiring some recalculation if the forms of the inputs are changed.

The simplification promised at the outset of this section is made by finding the ratios of the partial fraction coefficients that describe a particular mode, i.e., the component of each dependent variable characterized by a particular root of $\Delta(s) = 0$. These ratios, say $y_1/x_1$ or $z_k/x_k$, are *independent* of the input $\delta$'s, and can therefore be found in general terms.

To illustrate the lack of dependence on input, consider the ratio of $x_k$ to $y_k$. For the mode characterized by the roots $s_k$,

$$x_k = \left[\frac{N_x(s)}{\Delta(s)}(s + q_k)\right]_{s=-q_k}; \qquad y_k = \left[\frac{N_y(s)}{\Delta(s)}(s + q_k)\right]_{s=-q_k} \tag{2-53}$$

Dividing $x_k$ by $y_k$, we obtain

$$\frac{x_k}{y_k} = \left[\frac{N_x(s)}{\Delta(s)}(s + q_k)\right]_{s=-q_k}\left[\frac{\Delta(s)}{N_y(s)(s + q_k)}\right]_{s=-q_k} = \left[\frac{N_x(s)}{N_y(s)}\right]_{s=-q_k} \tag{2-54}$$

For a set of arbitrary input $\delta$'s, say $\delta_1$, $\delta_2$, and $\delta_3$, the $N_x(s)/N_y(s)$ ratio becomes

$$\frac{N_x(s)}{N_y(s)} = \frac{b_1\delta_1\Delta_{11} - b_2\delta_2\Delta_{21} + b_3\delta_3\Delta_{31}}{-b_1\delta_1\Delta_{12} + b_2\delta_2\Delta_{22} - b_3\delta_3\Delta_{32}} \tag{2-55}$$

where the $\Delta_{ij}$ are the minors of the determinant $\Delta(s)$. For example, $\Delta_{11}$ is obtained from $\Delta(s)$ by crossing out the row and the column in which $a_{11}$ appears. For another set of arbitrary inputs, $\delta_a$, $\delta_b$, and $\delta_c$, the ratio becomes

$$\frac{N_x(s)}{N_y(s)} = \frac{b_1\delta_a\Delta_{11} - b_2\delta_b\Delta_{21} + b_3\delta_c\Delta_{31}}{-b_1\delta_a\Delta_{12} + b_2\delta_b\Delta_{22} - b_3\delta_c\Delta_{32}} \tag{2-56}$$

The various $\delta$'s are arbitrary, so $[N_x(s)/N_y(s)]_{1,2,3}$ is not equal to $[N_x(s)/N_y(s)]_{a,b,c}$ in general. However, the ratio may conceivably be equal for some values of $s$. To find these particular values, the two equations are set equal to one another. Then,

$$\frac{b_1\delta_a\Delta_{11} - b_2\delta_b\Delta_{21} + b_3\delta_c\Delta_{31}}{-b_1\delta_a\Delta_{12} + b_2\delta_b\Delta_{22} - b_3\delta_c\Delta_{32}} = \frac{b_1\delta_1\Delta_{11} - b_2\delta_2\Delta_{21} + b_3\delta_3\Delta_{31}}{-b_1\delta_1\Delta_{12} + b_2\delta_2\Delta_{22} - b_3\delta_3\Delta_{32}} \tag{2-57}$$

If we multiply the means by the extremes and combine some terms,

$$\begin{aligned} &b_1b_2(\Delta_{11}\Delta_{22} - \Delta_{12}\Delta_{21})(\delta_2\delta_a - \delta_1\delta_b) \\ &\quad + b_1b_3(\Delta_{11}\Delta_{32} - \Delta_{12}\Delta_{31})(\delta_1\delta_c - \delta_3\delta_a) \\ &\quad + b_2b_3(\Delta_{31}\Delta_{22} - \Delta_{21}\Delta_{32})(\delta_2\delta_c - \delta_3\delta_b) = 0 \end{aligned} \tag{2-58}$$

Now the difference in products of the minors can be identified as

$$\begin{aligned}\Delta_{11}\Delta_{22} - \Delta_{12}\Delta_{21} &= a_{33}\Delta(s)\\ \Delta_{11}\Delta_{32} - \Delta_{12}\Delta_{31} &= -a_{23}\Delta(s)\\ \Delta_{22}\Delta_{31} - \Delta_{21}\Delta_{32} &= a_{13}\Delta(s)\end{aligned}$$

so that Eq. 2-58 becomes

$$\Delta(s)[a_{33}b_1b_2(\delta_2\delta_a - \delta_1\delta_b) - a_{23}b_1b_3(\delta_1\delta_c - \delta_3\delta_a) + a_{13}b_2b_3(\delta_2\delta_c - \delta_3\delta_b)] = 0$$

or (2-59)

$$\Delta(s) = 0$$

Since $\Delta(s)$ is and can be zero only at the roots $s = -q_1, -q_2, -q_3, \ldots -q_n$, the ratio $x_k/y_k = [N_x(s)/N_y(s)]_{s=-q_k}$ is independent of the input $\delta$'s.

Because of this lack of dependence on inputs, the various ratios $y_k/x_k$ and $z_k/x_k$ can be computed for any input. Ordinarily, the simplest results are obtained when all the $\delta$'s but one are set equal to zero. Thus the modal response ratio $y_k/x_k$ can be found from any of the ratios of minors given below, evaluated at $s = -q_k$:

$$\frac{y_k}{x_k} = -\left.\frac{\Delta_{12}}{\Delta_{11}}\right]_{s=-q_k} = -\left.\frac{\Delta_{22}}{\Delta_{21}}\right]_{s=-q_k} = -\left.\frac{\Delta_{32}}{\Delta_{31}}\right]_{s=-q_k} \qquad (2\text{-}60)$$

The mode of motion corresponding to $q_k$ will be represented in the several degrees of freedom by the terms

$$\begin{aligned} & x_k e^{-q_k t} && \text{in } x(t)\\ & \left(\frac{y_k}{x_k}\right) x_k e^{-q_k t} && \text{in } y(t)\\ & \left(\frac{z_k}{x_k}\right) x_k e^{-q_k t} && \text{in } z(t)\end{aligned} \qquad (2\text{-}61)$$

where $x_k$ is presumed to be the partial fraction coefficient that is individually computed for the particular input of interest. The total responses are obtained by summing the responses for the several modes, e.g., for $y(t)$

$$y(t) = \left(\frac{y_1}{x_1}\right) x_1 e^{-q_1 t} + \left(\frac{y_2}{x_2}\right) x_2 e^{-q_2 t} + \cdots + \left(\frac{y_n}{x_n}\right) x_n e^{-q_n t} + \cdots \qquad (2\text{-}62)$$

Modal response ratios are, in general, complex quantities and can be considered as plane vectors. When the components for any one mode are inserted into the equations of motion, each term in the equations becomes, in general, complex and can also be treated as a plane vector. Thus, when the components from Eq. 2-61 are inserted into the homogeneous form of Eq. 2-47,

$$\left[1a_{11}(s) + \left(\frac{y_k}{x_k}\right)a_{12}(s) + \left(\frac{z_k}{x_k}\right)a_{13}(s)\right]_{s=-q_k} = 0 \qquad (2\text{-}63)$$

Interpreted graphically as a vector diagram, Eq. 2-63 amounts in general to a closed polygon. This provides a convenient check on the calculation of modal response ratios. As will appear later, both the modal response ratios and the vector polygons play a central role in the description and physical interpretation of vehicle motion characteristics.

## 2-6. Weighting Function and Modal Response Coefficients

The preceding sections have reviewed techniques for finding the transient response of an element when it is subjected to general types of analytical input functions. The definition of such transient responses for all system-dependent variables, together with the input functions that cause them, is one reasonable way to characterize the system. It has great virtue as a direct prediction of expected system behavior when the inputs used are representative of those to which the system will be subjected in practice.

On the other hand, the calculation of a catalog of input-response pairs can be a lot of trouble, and can sometimes tend to overcomplicate the physical picture. For instance, when the input is complex, the part of the response dominated by the system characteristics is usually difficult to separate from that part which depends primarily on the input. To alleviate the labor and to circumvent potential confusion, another way is needed to characterize the transient response. Two special transient responses to simple input forms are commonly used for this purpose. One is the *weighting function*, or system response to a unit impulse; the other is its integral, the *indicial response* or indicial admittance, which is the system response to a unit step function. Because the inputs in both cases are simple, these standard responses exhibit the characteristic modes of the system in the simplest ways possible without contamination due to complex input shapes. Also, with the weighting function or indicial response known, the time response, $y(t)$, of the system to any input, $x(t)$, is readily found by the use of superposition. For the weighting function the *convolution* integral is used, i.e.,

$$y(t) = \int_0^t w(t - \tau)x(\tau)\,d\tau, \qquad t \geqq 0$$

A modified form of this integral, called *Duhamel's integral*, is appropriate when the indicial response, $I(t)$, is available,

$$y(t) = x(0+)I(t) + \int_0^t I(t-\tau)\frac{dx}{d\tau}(\tau)\,d\tau \tag{2-64}$$

As a consequence of these equations, all response calculations can be carried out directly in the time domain if desired.[6]

More often than not, the weighting functions or indicial responses are used as ends in themselves to exhibit in a standard way the transient characteristics of the system, and not as intermediaries in response calculations using the convolution or Duhamel integral. The latter procedure is seldom followed because transient responses will usually be more easily obtained by working in the transform domain, i.e., by transforming the convolution to the algebraic equation

$$Y(s) = W(s)X(s) \tag{2-65}$$

expanding into partial fractions, and inverting to the time domain. There are cases, too, in which one finds a convolution integral itself of value. These applications include the following situations:

1. It is easier to integrate directly than to go through the other transformation processes.
2. The input function or weighting function does not possess a rational Laplace transform.
3. The input function or weighting function is so complex that taking its Laplace transform is impractical.
4. The input function or weighting function is known only graphically or experimentally.

Thus the use of the convolution approach is often a practical necessity.

Like other transient responses, the impulse response or weighting function will have a transform that can be resolved into partial fractions. In this case the partial fraction coefficients are called *modal response coefficients* and are accorded a special symbol, $Q_i$. Thus the transformed weighting function will be

$$W(s) = \sum_{i=1}^{m+n} \frac{Q_i}{s+q_i}$$

[6] There are convenient algorithms for numerical convolution, when this may be required. See, e.g., J. G. Truxal, *Automatic Feedback Control System Synthesis*, McGraw-Hill, New York, 1955, pp. 63–71; A. Tustin, "A Method of Analyzing the Behavior of Linear Systems in Terms of Time Series," *J. IEEE*, 94, Pt. IIa, 152–160 (1947).

where the roots $-q_i$ are those of the system's characteristic equation $\Delta(s) = 0$. Upon inverse transforming, the weighting function becomes

$$w(t) = \sum_{i=1}^{m+n} Q_i e^{-q_i t}$$

Splitting $W(s)$ into partial fractions is equivalent to replacing the $(m + n)^{th}$-order differential equation for $w(t)$ by $m + n$ first-order differential equations of the form

$$\dot{w}_i(t) + q_i w_i(t) = Q_i \delta(t), \qquad i = 1, 2, \ldots, m + n$$

where $w_i(0) = 0$ and the total weighting function is

$$w(t) = w_1(t) + w_2(t) + \cdots + w_i(t) + \cdots w_{m+n}(t)$$

Thus the total weighting function is equivalent to the summed responses of $m + n$ first-order systems excited by an impulse input, as shown in Fig. 2-3. The modal response coefficients can also be thought of as initial conditions on unity-gain elemental systems (the $Q_i$'s in Fig. 2-3 replaced by 1's) which, with no other excitation (no impulse input to the system), results in a system output equal to the weighting function. For this interpretation the elemental differential equations would be

$$\dot{w}_i(t) + q_i w_i(t) = 0, \qquad i = 1, 2, \ldots, m + n \tag{2-66}$$

where

$$w_i(0) = Q_i$$

The output response coefficients developed in Section 2-4 can be interpreted as time-weighted moments of the weighting function. The Laplace transform of the $n^{th}$ time moment, $t^n y(t)$, of the output $y(t)$ is

$$\mathcal{L}[t^n y(t)] = (-1)^n \frac{d^n Y(s)}{ds^n} \tag{2-67}$$

so the transform of the integral of $t^n y(t)$ will be

$$\mathcal{L}\left[\int_0^t \tau^n y(\tau)\, d\tau\right] = \frac{(-1)^n}{s} \frac{d^n Y(s)}{ds^n} \tag{2-68}$$

Under conditions in which the final value theorem will apply, i.e., where $\lim_{t\to\infty} \int_0^t \tau^n y(\tau)\, d\tau$ exists,

$$\lim_{t\to\infty} \int_0^t \tau^n y(\tau)\, d\tau = \lim_{s\to 0} s \frac{(-1)^n}{s} \frac{d^n Y(s)}{ds^n} = (-1)^n \left.\frac{d^n Y(s)}{ds^n}\right|_{s\to 0}$$

$$= (-1)^n \left[\frac{d^n}{ds^n} W(s) X(s)\right]_{s\to 0} \tag{2-69}$$

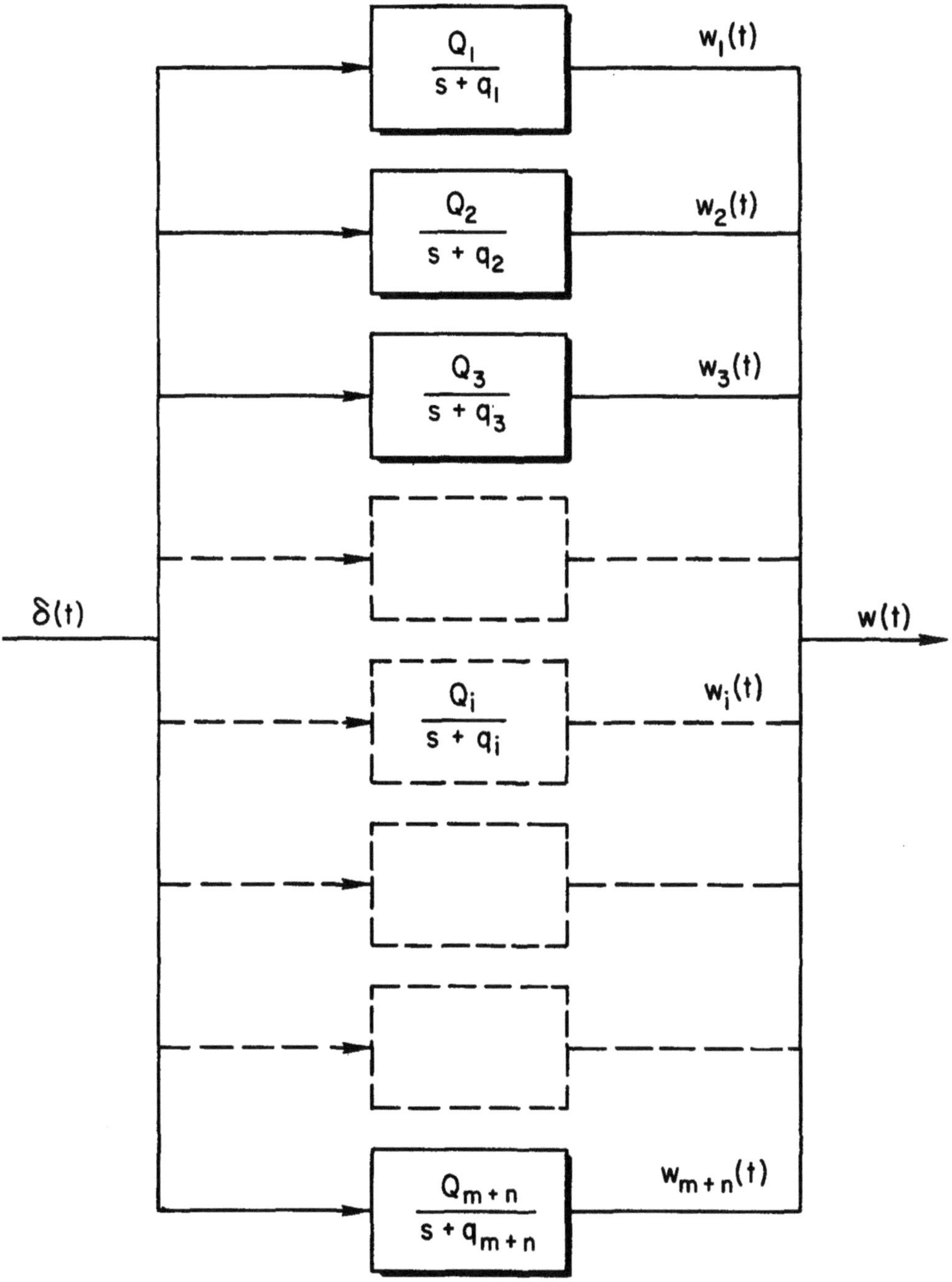


Fig. 2-3. Elemental first-order systems corresponding to components of the weighting function.

If the input, $x(t)$, is taken to be a unit impulse, then $X(s) = 1$ and the output, $y(t)$, will be the system weighting function. When these conditions are inserted into Eq. 2-69, the result becomes

$$\int_0^\infty \tau^n w(\tau)\, d\tau = (-1)^n \left[\frac{d^n W(s)}{ds^n}\right]_{s\to 0} \tag{2-70}$$

Using Eq. 2-39, we find that the output response coefficients $C_n$ can be identified as

$$C_n = \frac{1}{n!}\frac{d^n W(s)}{ds^n}\bigg|_{s\to 0} = \frac{(-1)^n}{n!}\int_0^\infty \tau^n w(\tau)\, d\tau \tag{2-71}$$

This relationship between output response coefficients and the time moments of the weighting function is helpful for physical interpretation and also provides the basis for simple measurement of the response coefficients using computer techniques.

Other useful connections between the output response coefficients and weighting function parameters are relationships involving the modal response coefficients, $Q_i$. It can be shown that[7]

$$\begin{aligned} C_0 &= \sum_{i=1}^{m+n} \frac{Q_i}{q_i} \\ C_1 &= -\sum_{i=1}^{m+n} \frac{Q_i}{q_i^2} \\ C_2 &= \sum_{i=1}^{m+n} \frac{Q_i}{q_i^3} \\ &\;\;\vdots \qquad\quad \vdots \\ C_k &= -\sum_{i=1}^{m+n} \frac{Q_i}{(-q_i)^{1+k}} \end{aligned} \tag{2-72}$$

## 2-7. Time Vector Representations for the Weighting Function

The weighting function is most conventionally shown as a time history. This gives a general view of system response characteristics and is often all that is required on low order systems. For higher order systems, however, many modes make up the composite motions, some more dominant than others. Also, a given mode is, in general, reflected with

[7] D. T. McRuer and R. L. Stapleford, "Sensitivity and Modal Response for Single-Loop and Multiloop Systems," ASD-TDR-62-812, Jan. 1963, pp. 19–21.

different scales into each of the several degrees of freedom. For these reasons attention must be focused on the modal components of the system weighting functions if a complete picture is to be obtained. This is accomplished by application of the principles already described in the discussion of modal response ratios as enhanced by the use of a graphical interpretation using the *time vectors*.

The method of time vectors is based on the concept of rotating vectors to represent component or total motion quantities. It is particularly useful in representing the amplitude and phasing relationships between such quantities in oscillatory motion. The concepts of time vectors stem from harmonic motion analysis and alternating current theory, with minor modifications to handle time-variable amplitudes.[8] The basic ideas are readily grasped with the aid of simple examples, so this procedure will be adopted here.

Consider the second-order system described by the differential equation

$$\frac{d^2x}{dt^2} + 2\zeta\omega_n \frac{dx}{dt} + \omega_n^2 x = \omega_n\sqrt{1-\zeta^2}\, f(t) \tag{2-73}$$

If the desired response, $x$, is to be the weighting function, $w(t)$, then the forcing function, $f(t)$, is replaced by the unit impulse, $\delta(t)$, and Eq. 2-73 becomes

$$\frac{d^2w}{dt^2} + 2\zeta\omega_n \frac{dw}{dt} + \omega_n^2 w = \omega_n\sqrt{1-\zeta^2}\, \delta(t) \tag{2-74}$$

The right-hand side of this equation is zero for $t > 0$, and most of our interest will be centered on the solutions in this region. When Laplace transformed, Eq. 2-74 becomes

$$\begin{aligned} W(s) &= \frac{\omega_n\sqrt{1-\zeta^2}}{s^2 + 2\zeta\omega_n s + \omega_n^2} \\ &= \frac{x_1}{s + (\zeta\omega_n - j\omega_n\sqrt{1-\zeta^2})} + \frac{x_2}{s + (\zeta\omega_n\sqrt{1-\zeta^2})} \end{aligned} \tag{2-75}$$

where $x_1 = 1/2j$ and $x_2 = -1/2j$. The inverse transform can be written either as a real function or as a sum of the two modes involved, represented as complex numbers. Both forms are useful for time vector considerations; each one is given below.

$$w(t) = e^{-\zeta\omega_n t} \sin \omega_n\sqrt{1-\zeta^2}\, t \qquad t \geqq 0 \tag{2-76}$$

[8] R. K. Mueller, "A Graphical Solution of Stability Problems," *J. Aeron. Sci.*, June 1937; M. F. Gardner and J. L. Barnes, *Transients in Linear Systems*, Wiley, New York, 1942, pp. 174ff; K. H. Doetsch, "The Time Vector Method for Stability Investigations," ARC R & M 2945, 1957; W. O. Breuhaus, "Résumé of the Time Vector Method as a Means for Analyzing Aircraft Stability," WADC-TR-52-299, Nov. 1952.

or

$$w(t) = w_1(t) + w_2(t)$$
$$= x_1 e^{-\zeta\omega_n t} e^{j\omega_n\sqrt{1-\zeta^2}t} + x_2 e^{-\zeta\omega_n t} e^{-j\omega_n\sqrt{1-\zeta^2}t} \qquad t \geqq 0 \quad (2\text{-}77)$$

The two complex modes in Eq. 2-77, when combined, become the damped real oscillation of Eq. 2-76. Either mode can be used to represent the real weighting function by considering only their real parts, i.e.,

$$w(t) = 2 \operatorname{Re} w_1(t) = 2 \operatorname{Re} w_2(t) \qquad (2\text{-}78)$$

To remove the $t \geqq 0$ restriction on the weighting function given as Eq. 2-76, the form there can be multiplied by the unit step function, $u(t)$. This makes $w(t)$ zero for $t < 0$ and equal to the damped oscillation thereafter. This will be of no consequence for the time vector representation, although it is required in order that $w(t)$ satisfy Eq. 2-74. Using this form, the weighting function and its first and second derivatives become

$$w(t) = [e^{-\zeta\omega_n t} \sin (\omega_n\sqrt{1-\zeta^2}\, t)]u(t) \qquad (2\text{-}79)$$

$$\dot{w}(t) = \left[\omega_n e^{-\zeta\omega_n t} \sin \left(\omega_n\sqrt{1-\zeta^2}\, t + \frac{\pi}{2} + \theta\right)\right]u(t) \qquad (2\text{-}80)$$

$$\ddot{w}(t) = [\omega_n^2 e^{-\zeta\omega_n t} \sin (\omega_n\sqrt{1-\zeta^2}\, t + \pi + 2\theta)]u(t) + \omega_n\sqrt{1-\zeta^2}\, \delta(t) \qquad (2\text{-}81)$$

where $\theta$, the so-called *damping angle*, is given by the equivalent expressions

$$\theta = \sin^{-1} \zeta$$
$$= \cos^{-1} \sqrt{1-\zeta^2} \qquad (2\text{-}82)$$
$$= \tan^{-1} \frac{\zeta}{\sqrt{1-\zeta^2}}$$

The insertion of Eqs. 2-79 through 2-81 into the differential equation for the weighting function satisfies the latter identically. Without the unit step function multiplier, the $\delta$ function term in $\ddot{w}(t)$ (Eq. 2-81) would not have arisen and Eq. 2-74 would then be satisfied only for $t > 0$. Now that this point has been made, we shall drop the $u(t)$ multipliers and the $\delta$ function in $\ddot{w}$ and consider only those times greater than zero.

The first mode of Eq. 2-77 and its derivatives are

$$w_1(t) = x_1 e^{-\zeta\omega_n t} e^{j\omega_n\sqrt{1-\zeta^2}t} \qquad t \geqq 0 \qquad (2\text{-}83)$$

$$\dot{w}_1(t) = \omega_n x_1 e^{-\zeta\omega_n t} e^{j[\omega_n\sqrt{1-\zeta^2}t + \pi/2 + \theta]} \qquad t \geqq 0 \qquad (2\text{-}84)$$

$$\ddot{w}_1(t) = \omega_n^2 x_1 e^{-\zeta\omega_n t} e^{j[\omega_n\sqrt{1-\zeta^2}t + \pi + 2\theta]} \qquad t \geqq 0 \qquad (2\text{-}85)$$

Comparing these three equations with their equivalents for the real oscillation provides the basis for a very useful and simple rule, to wit: When any derivative of a component of motion is differentiated, the amplitude is multiplied by $\omega_n$ and the phase angle is increased by $\pi/2 + \theta$ (when applying the rule to the second mode, the phase angle would be considered as all of the exponent multiplied by $-j$; thus, $\pi/2 + \theta$ is added to $\omega_n\sqrt{1 - \zeta^2 t}$).

The fundamental concept of time vectors is that either periodic or aperiodic motions can be considered as generated by a time vector rotating with constant angular velocity about a fixed point. When the motion is a constant amplitude oscillation, the time vector is of fixed length, whereas its length varies with time for subsiding or diverging oscillations and aperiodic motions. The generating motion of the time vector for the damped sinusoidal weighting function is illustrated in Fig. 2-4 for a damping ratio of 0.3. As the time vector rotates at a constant velocity, $\omega_n\sqrt{1 - \zeta^2}$, its amplitude decreases exponentially, so that the trace of the tip is a logarithmic spiral. At any time, the angle between the tangent and the normal to the radius vector of the spiral is just the damping angle, $\theta$.

The connection between the generating vector and the weighting function can be appreciated by considering the projection of the vector onto the vertical axis. At each instant the projection is equal to the value of the weighting function at that time. The derivatives of the weighting function can be considered in a similar fashion.

As illustrated in Fig. 2-4, $\dot{\bar{w}}$ is displaced by $\pi/2$ plus the damping angle from $\bar{w}$, and $\ddot{\bar{w}}$ is further displaced by this same increment from $\dot{\bar{w}}$. Also, the generating vectors are longer than that for the weighting function by the factors $\omega_n$ and $\omega_n^2$, respectively. If there were no damping, $\theta$ would be zero and the system would be a simple harmonic oscillator, wherein the velocity and acceleration are, respectively, 90 and 180° out of phase with the displacement.

Figure 2-5 reproduces the time vectors for the weighting function and its derivatives and also indicates the scaling of quantities in the differential equation proportional to these terms. When these individual components of the differential equation are added together, they form a time vector diagram that represents in graphical fashion the equation and its components. For the second-order system this is always an isosceles triangle. The vertex angle is equal to $2\theta$, which indicates the degree of damping in the system. The ratio of the altitude of the triangle to an isosceles side is equal to the cosine of $\theta$ and is therefore a measure of the frequency of the damped motion, $\omega_n\sqrt{1 - \zeta^2}$, as compared with that of the undamped motion, $\omega_n$. The length of each time vector in the triangle, and thus the entire triangle, shrinks at the same rate as the parameters of motion. The

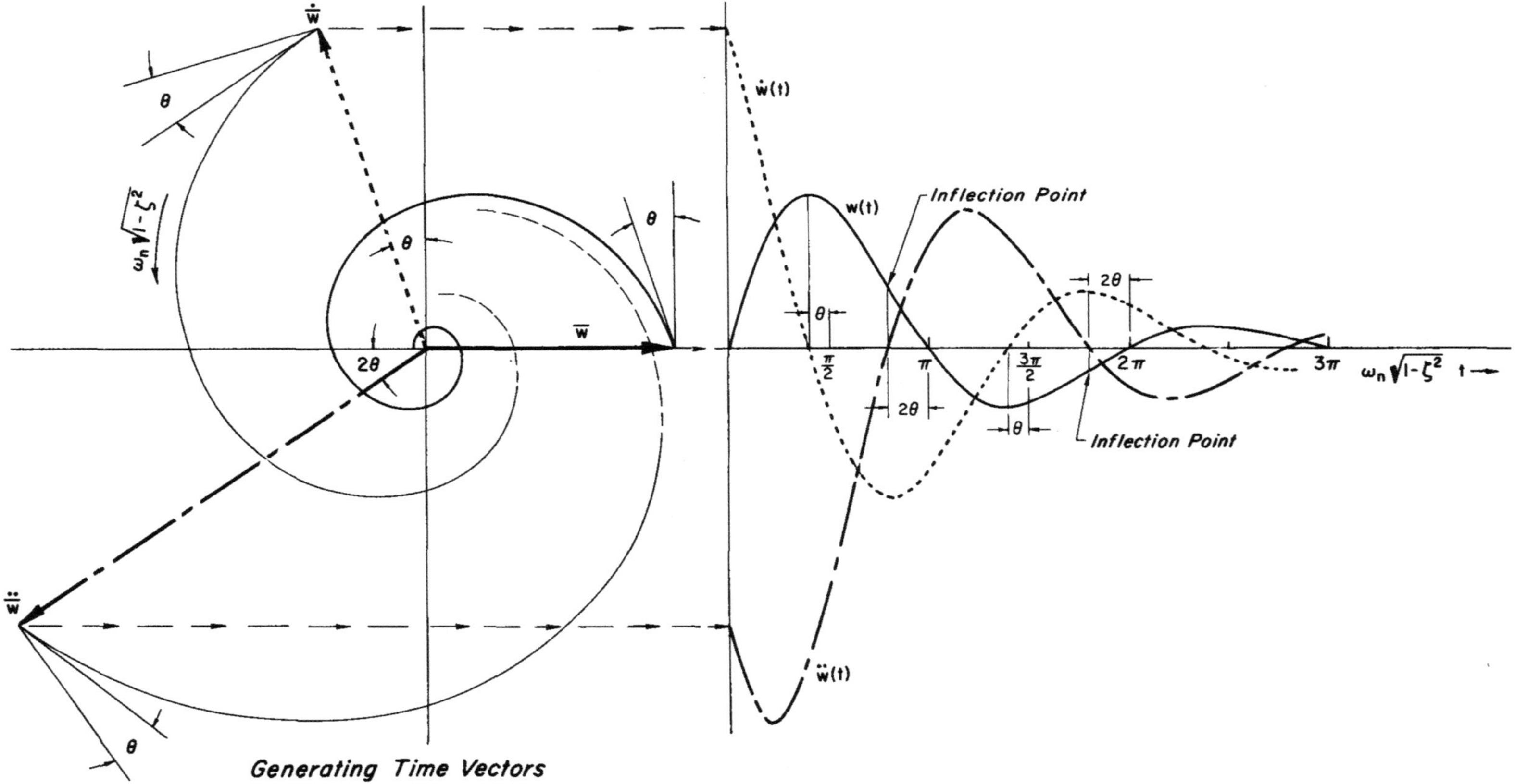


Fig. 2-4. Time histories and their generating time vectors for a second-order system.

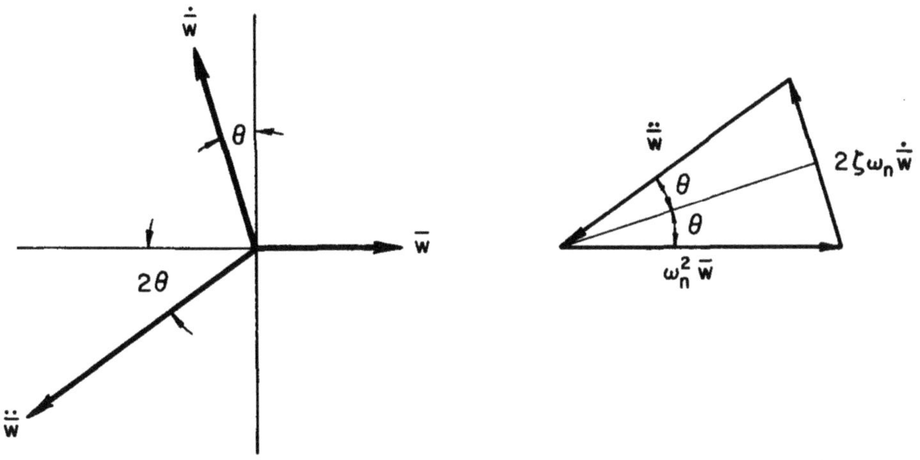


Fig. 2-5. Time vectors and a polygon of forces for second-order systems.

relative relationships, however, are unmodified, so it is usual to consider the time vector triangle to be frozen at a particular instant.

These explanations have been carried out using the real form of the weighting function, although the $w_1$ complex component could have been used just as well. For that matter, the $w_2$ component can also serve, although conventions would have to be changed because its direction of rotation is opposite that assumed in the figures.

The illustrative problem has thus far been treated as a one degree of freedom system. Since it is second-order, however, it can as well be considered a two degree of freedom system and thus serve as the simplest example of the use of modal response ratios in construction of time vector diagrams. A two degree of freedom system having the weighting function already described is given by

$$(s + \zeta\omega_n)X(s) - \omega_n\sqrt{1 - \zeta^2}\,Y(s) = 0 \qquad \text{(2-86)}$$
$$\omega_n\sqrt{1 - \zeta^2}\,X(s) + (s + \zeta\omega_n)Y(s) = F(s)$$

The characteristic function is, of course, $\Delta = s^2 + 2\zeta\omega_n s + \omega_n^2$ and the numerators of the $X$ and $Y$ response transforms are

$$N_x(s) = F(s)\omega_n\sqrt{1 - \zeta^2} \qquad \text{(2-87)}$$
$$N_y(s) = F(s)(s + \zeta\omega_n)$$

The modal response ratios are given by

$$\frac{y_1}{x_1} = \left.\frac{s + \zeta\omega_n}{\omega_n\sqrt{1 - \zeta^2}}\right]_{s=-\zeta\omega_n+j\omega_n\sqrt{1-\zeta^2}} = j$$

$$\frac{y_2}{x_2} = \left.\frac{s + \zeta\omega_n}{\omega_n\sqrt{1 - \zeta^2}}\right]_{s=-\zeta\omega_n-j\omega_n\sqrt{1-\zeta^2}} = -j \tag{2-88}$$

The time vectors for $x_1$ and $\dot{x}_1$ and the modal response ratio, $y_1/x_1$, considering the first mode only, are shown in Fig. 2-6(a). Also given there are scaled quantities involved in the two equations of motion. The two vector triangles shown in Figs. 2-6(b) and 2-6(c) illustrate the time vector diagram for the two equations of motion. The point illustrated here over and above those described previously is that the terms involving $y$ are derived from $x_1$ or $\dot{x}_1$ by using the modal response ratio $y_1/x_1$. Because all variables have the same phase angle $(\pi/2 + \theta)$ and the same multiplying factor $(\omega_n)$ between successive derivatives, the ratio of two derivatives of different components is not affected by increasing or lowering the order of differentiation simultaneously for both components, i.e.,

$$\frac{y_1}{x_1} = \frac{\dot{y}_1}{\dot{x}_1} = \frac{\ddot{y}_1}{\ddot{x}_1} \quad \text{or} \quad \frac{\dot{y}_1}{x_1} = \frac{\ddot{y}_1}{\dot{x}_1} \qquad \text{etc.}$$

Thus the modal response ratio $y_1/x_1$ can be used to obtain either $y_1$ or $\dot{y}_1$ by multiplying by $x_1$ and $\dot{x}_1$, respectively.

While the above developments have used the simplest possible example to provide clarity of explanation, the greatest benefits of the time vector method appear for higher order systems with several degrees of freedom. The vector diagrams in these cases become polygons of forces or moments, with, for each mode, one polygon per equation of motion. Because of cross-coupling forces or moments linking the different degrees of freedom, these polygons are generally more complicated than simple triangles. The graphical nature of the polygons, however, still allows the ready visualization of key physical relationships. Thus the diagrams show the relative amplitudes and phasings between the different variables of oscillatory motion modes and provide a direct physical appreciation for the effects of individual parameters on the motion.

## 2-8. Transfer Function Models

The discussion of mathematical models to this point has emphasized the transient response performance of an element when subjected to various inputs. This has been done because the usual end result of a system study is a prediction of the physical performance of a system, and transient responses are as physical a result as can be attained. However,

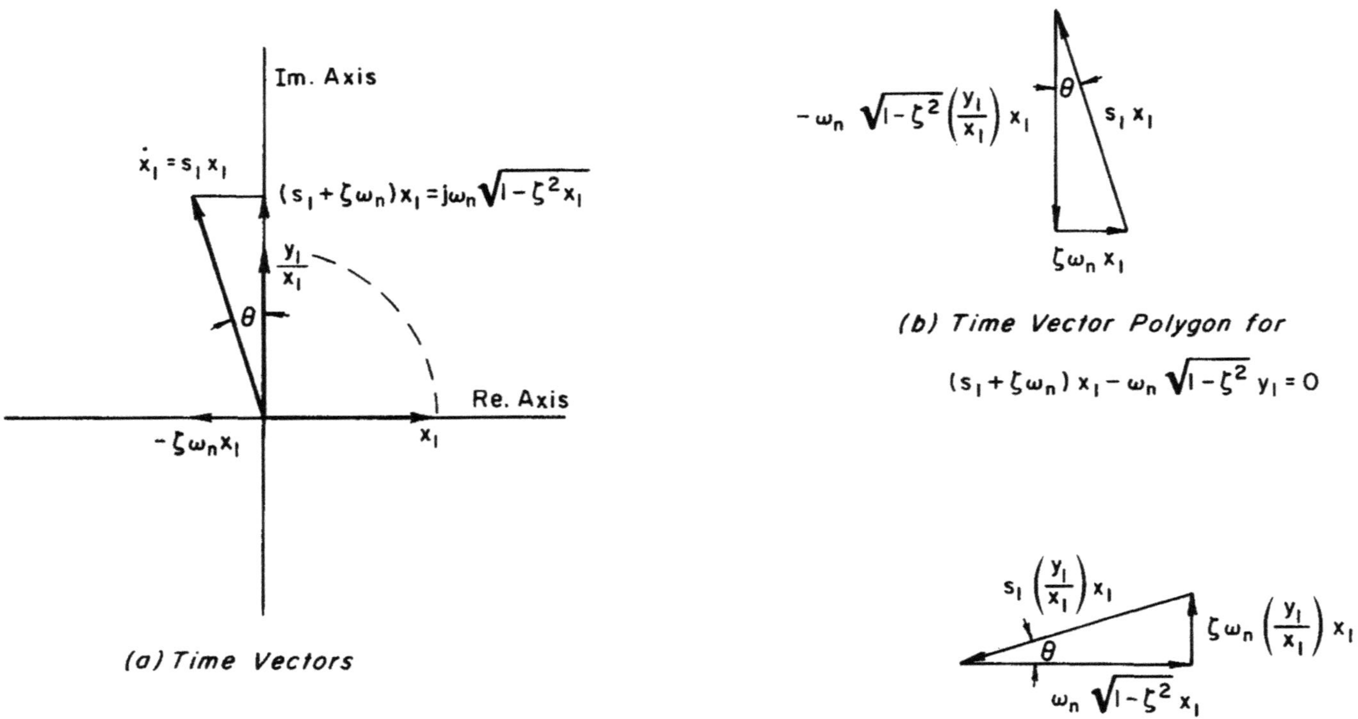


*(a) Time Vectors*

*(b) Time Vector Polygon for*

$$(s_1 + \zeta\omega_n)x_1 - \omega_n\sqrt{1-\zeta^2}\,y_1 = 0$$

*(c) Time Vector Polygon for*

$$\omega_n\sqrt{1-\zeta^2}\,x_1 + (s_1 + \zeta\omega_n)y_1 = 0$$

Fig. 2-6 Time vectors and vector polygons for second-order systems considered as a two-degree-of-freedom system.

analytical models should not be restricted to those of the transient response variety alone because transient response models have several defects. First, the transient response to a particular input is usually dominated by one or two modes, even though the system may be of higher order, because of differences in time and amplitude scale factors among the several modes. Modes that might be important in responses to different inputs, or that might have pronounced effects on performance if system parameters were slightly changed, may be suppressed to a large extent.

Second, it is seldom easy to combine directly the transient response models of several complex elements into a single one describing the combination of the elements in a system. Even to combine two or more series elements having known weighting functions into an overall weighting function using repeated convolution is an irksome computational task.

Third, and finally, it is sometimes troublesome, using the convolution integral, to modify the transient response model obtained for simple inputs to one consistent with other more complex inputs. To overcome these difficulties a different form of mathematical model is desired that

1. Defines the element as an entity by exhibiting with equal emphasis all of the element characteristic parameters and/or modes.
2. Allows the models of individual elements to be combined simply into those of the combination.
3. Can be directly applied in the transform domain for the intermediate stages of transient response calculations.

The transform of the weighting function, which is the transform of the output (with all initial conditions zero), divided by the transform of the input, is a model that fulfills these requirements. This function has a supreme theoretical and practical importance, and again a special name. It is called a *transfer function* of the system. For the general system of Eqs. 2-1 through 2-4 the transfer function will be

$$\frac{Y(s)}{X(s)} = W(s) = \frac{\kappa(s^n + a_1 s^{n-1} + \cdots + a_{n-1}s + a_n)}{(s^{m+n} + b_1 s^{m+n-1} + \cdots + b_{m+n-1}s + b_{m+n})} \tag{2-89a}$$

$$= \kappa \frac{\prod_{j=1}^{n} (s + z_j)}{\prod_{i=1}^{m+n} (s + q_i)} \tag{2-89b}$$

$$= \sum_{i=1}^{m+n} \frac{Q_i}{s + q_i} \tag{2-89c}$$

Although the transfer function is one step removed from a transient response model, it still has a direct interpretation in transient response terms. All the information necessary for the construction of weighting function time responses or time vector diagrams is implicitly contained in the transfer function in all its forms. This information is explicit in only the third form given above (Eq. 2-89c). The factored form (Eq. 2-89b) provides all the time history information, i.e., the system characteristic roots, $-q_i$, but the modal response coefficients are one step away. Finally, the first form (Eq. 2-89a) cannot, in general, be interpreted in time response terms without additional operations. Of particular interest in this connection is the application of tests that can describe the general position of the poles without factoring the polynomial. These tests will be referred to later when the subject of stability is discussed.

The transfer function contains only parameters that stem from the system; it does not depend on the inputs. In this sense the system as an independent entity is equivalent to the $(m + n)r$ transfer functions corresponding to its $m + n$ independent degrees of freedom when subjected to $r$ independent inputs. Alternatively, it can be stated that the gains ($\kappa$), zeros ($-z_j$), and poles ($-q_i$) of the transfer functions completely define the system.

Because input transforms are converted to output transforms by multiplication with the appropriate transfer functions, it is convenient to characterize the input and output of the "pattern" of Fig. 2-1 by their Laplace transforms and to represent the system by its transfer functions. The "pattern" of Fig. 2-1 in the case of the example system of Fig. 2-2 then becomes the *block diagram* of Fig. 2-7.

The ease of combining element transfer functions to form the transfer function of the combination can be illustrated by considering two elements in series. If the first has a transfer function

$$W_1(s) = \frac{Y(s)}{X(s)} \tag{2-90}$$

and the second a transfer function

$$W_2(s) = \frac{Z(s)}{Y(s)} \tag{2-91}$$

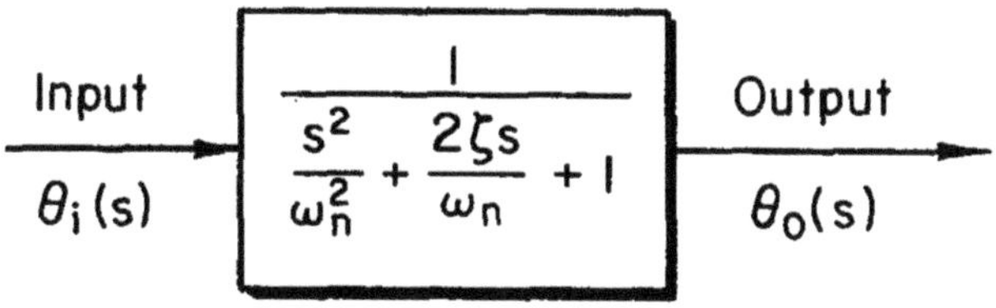


Fig. 2-7. Block diagram of the spring-mass-damper system.

then the combination has a transfer function that is the product of the individual transfer functions

$$W(s) = \frac{Z(s)}{X(s)} = \frac{Y(s)Z(s)}{X(s)Y(s)} = W_1(s)W_2(s) \tag{2-92}$$

Contrast this with the same operation, using the time domain models. Since the weighting function is the inverse transform of the transfer function

$$\mathcal{L}^{-1}[W(s)] = w(t)$$

and

$$\mathcal{L}^{-1}[W_1(s)] = w_1(t)$$

$$\mathcal{L}^{-1}[W_2(s)] = w_2(t)$$

then, by the convolution theorem of the Laplace transform,

$$\begin{aligned} w(t) = w_1(t) * w_2(t) &= \int_0^t w_1(\tau)w_2(t-\tau)\,d\tau \\ &= \int_0^t w_2(\tau)w_1(t-\tau)\,d\tau \end{aligned} \tag{2-93}$$

Because convolution is ordinarily much more difficult to perform than simple multiplication, the transfer function is generally the most convenient form of mathematical model to use for manipulations of this type.

Although the transfer function is fundamentally a *system* descriptor, it does have a physical interpretation as *response* components in the output response to the hypothetical signal $e^{s_1 t}$. When such an input is applied, the differential equation of the system takes the form

$$\begin{aligned} &\frac{d^{m+n}x}{dt^{m+n}} + b_1\frac{d^{m+n-1}x}{dt^{m+n-1}} + \cdots + b_{m+n-1}\frac{dx}{dt} + b_{m+n}x \\ &\qquad = \kappa\left(\frac{d^n}{dt} + a_1\frac{d^{n-1}}{dt^{n-1}} + \cdots + a_{n-1}\frac{d}{dt} + a_n\right)e^{s_1 t} \end{aligned} \tag{2-94}$$

By transforming with initial conditions zero, this becomes

$$\begin{aligned} X(s) &= \frac{\kappa(s^n + a_1 s^{n-1} + \cdots + a_{n-1}s + a_n)}{(s^{m+n} + b_1 s^{m+n-1} + \cdots + b_{m+n-1}s + b_{m+n})(s - s_1)} \\ &= \sum_{i=1}^{m+n}\frac{C_i}{s+q_i} + \left[\frac{\kappa(s^n + a_1 s^{n-1} + \cdots + a_{n-1}s + a_n)}{(s^{m+n} + b_1 s^{m+n-1} + \cdots + b_{m+n-1}s + b_{m+n})}\right]_{s=s_1}\left(\frac{1}{s-s_1}\right) \\ &= \sum_{i=1}^{m+n}\frac{C_i}{s+q_i} + \frac{W(s_1)}{s-s_1} \qquad \text{for } s_1 \neq -q_i \end{aligned} \tag{2-95}$$

or, in the time domain,

$$x(t) = \sum_{i=1}^{m+n} C_i e^{-q_i t} + W(s_1)e^{s_1 t} \tag{2-96}$$

The term in brackets in Eq. 2-95 is seen to be the transfer function evaluated at $s = s_1$. In essence, this is the partial fraction coefficient for the forced mode of the system, i.e., that mode having a time history identical to the input $e^{s_1 t}$. The summation represents the effect on the output of the system's natural modes as excited by the forcing function.

It is instructive to consider Eq. 2-96 for several special values of $s_1$. First, if $s_1$ is a real number, either $+\sigma$ or $-\sigma$, then the forced component in the time response is either $W(+\sigma)e^{+\sigma t}$ or $W(-\sigma)e^{-\sigma t}$. Here, the amplitude of the forced component is the transfer function for the particular values of $s$ represented by $+\sigma$ or $-\sigma$.

More generally, $s_1$ will be complex, as will $W(s_1)$. In this event, $e^{s_1 t}$ is one of the two complex conjugate components of a damped sinusoidal or damped cosinusoidal wave. Alternatively, $e^{s_1 t}$ may be considered as a generating vector, resulting in a damped sinusoidal or cosinusoidal time history. In either event, the real response to, say, a damped sinusoidal input, $e^{-\sigma_1 t} \sin \omega_1 t$, can be seen to be

$$x(t) = \sum_{i=1}^{m+n} C_i e^{-q_i t} + |W(s_1)|e^{-\sigma_1 t} \sin [\omega_1 t + \measuredangle W(s_1)]$$

where

$$s_1 = -\sigma_1 + j\omega_1$$

Here, the transfer function $W(s_1)$ appears as a magnitude and phase angle in the forced component of the response.

In all of the cases examined above, the transfer function could, in principle, be measured from the responses to various $e^{s_1 t}$ input forms to the extent that the forced response can be separated from the transients represented by the summation. Ordinarily, this is frustratingly difficult or practically impossible. However, for stable or just slightly unstable systems the separation is readily accomplished for the special case $s = j\omega$. Then the forced response is a sinusoid that is separated from the total response by the simple expedient of waiting for the transients to become insignificant for the stable case, or by subtracting out the slightly divergent mode(s) in the unstable case. The motion component remaining consists of a sinusoidal oscillation at frequency $\omega$, characterized by an amplitude equal to the magnitude of the transfer function evaluated at $j\omega$, and a phase angle relative to the input given by the angle of $W(j\omega)$. The special transfer function form, $W(j\omega)$, obtained in this manner, is called the *frequency response* because of its connection with the system response to oscillatory inputs.

## 2-9. Representations of Transfer Functions

The transfer functions of systems or elements can be represented in several useful ways. Among the more prominent of these are the algebraic representation, the pole-zero plot, and several varieties of the Bode diagram.[9] These all explicitly reflect part or all of the system's pole and zero characteristics. Other possible representations, such as the polar plot of the frequency response function, implicitly contain the same information, but pole and zero data are more difficult to dissect from the total representation. Because our emphasis is on transfer function representations which maximize the easily extractable information, such forms will not be considered here.

The transfer functions of concern are, as we have seen, ordinarily ratios of rational polynomials in which the denominator is of a higher degree than the numerator. When the polynomials are factored, they may be written in the alternative forms:

$$G(s) = \kappa \frac{\alpha(s)}{\beta(s)} = \kappa \frac{\prod_{j=1}^{n} (s + z_j)}{\prod_{i=1}^{m+n} (s + q_i)} = \kappa \frac{\prod_{j=1}^{n} (s + z_j)}{s^k \prod_{i}^{m+n-k} (s + q_i)} \tag{2-97}$$

$$G(s) = K \frac{A(s)}{s^k B(s)} = K \frac{\prod_{j=1}^{n} [(s/z_j) + 1]}{s^k \prod_{i=1}^{m+n-k} [(s/q_i) + 1]} \tag{2-98}$$

Equations 2-97 and 2-98 display the *algebraic* representation of the transfer functions. The $s^k$ style is appropriate when some ($k$) of the poles are zero. The $s^k$ style can also serve for zeros if the $k$ in the denominator product superscript is shifted to the numerator.

The function $G(s)$ is a function of a complex variable, which can be characterized by its poles and zeros. Then $G(s)$ itself can be represented by plotting its poles and zeros on the complex $s$ plane. This would constitute a *pole-zero plot*. The pole-zero plots of the most common transfer functions or transfer function factors are shown in Table 2-4, where an open circle represents a zero and a cross represents a pole.

Because $G(s)$ for a particular value of $s$ is a complex number, $A + jB$, it can be represented by a modulus or magnitude, $|G| = \sqrt{A^2 + B^2}$, and an argument or phase angle, $\measuredangle G = \tan^{-1} (B/A)$. It is particularly

[9] H. W. Bode, *Network Analysis and Feedback Amplifier Design*, Van Nostrand, New York, 1945; N. L. Kusters and W. J. M. Moore, "A Generalization of the Frequency Response Method for the Study of Feedback Control Systems," *Automatic and Manual Control*, Butterworths, London, 1952; D. T. McRuer, "Unified Analysis of Linear Feedback Systems," ASD-TR-61-118, Wright-Patterson Air Force Base, Ohio, July 1961; F. P. De Mello, "Evaluation of Transient System Response," *AIEE Trans.*, 78, Part 2, 177–186 (Sept. 1959).

convenient for the purpose of graphical constructions to plot $20 \log_{10} |G|$ and $\measuredangle G$ against $\log_{10} |s|$, while letting the complex variable, $s = \sigma + j\omega$, take on special values in the $s$ plane. The simplest special values of $s$ are obtained by letting $s$ vary along a straight line in the $s$ plane. When this is done, the transfer function plot comprising $20 \log_{10} |G|$ and $\measuredangle G$ versus $\log_{10} |s|$ is called a ***generalized Bode diagram.*** The term $20 \log_{10} |G|$ has the dimensions of decibels (dB) and is abbreviated as $|G|_{dB}$. A chart for converting magnitudes to dB, and vice versa, is given as Fig. 2-8.

There are four types of generalized Bode diagrams that are often used, corresponding to four different $s$-plane pathways along which $s$ is compelled to vary. First is the one in which $s = 0 + j\omega$, called the $j\omega$-Bode diagram, or simply Bode diagram. This is the classical plot introduced by Bode. Second is the diagram for which $s = +\sigma$ or $s = -\sigma$. These are called $\sigma$-Bode diagrams. For the $j\omega$-Bode and $\pm\sigma$-Bode diagrams, $s$ takes on its simplest values as either an imaginary or a real number, and the $s$ plane pathways are along the imaginary or real axes (Fig. 2-9). These are both special cases of general radial pathways along which $s = |s| \, (-\xi + j\sqrt{1 - \xi^2})$, i.e., for $s = j\omega$, $\xi = 0$, whereas for $s = \pm\sigma$, $\xi = \mp 1$. Third are the type of Bode diagrams based on these general radials, called $\xi$-Bodes. These are useful in concept and principle, but practical constructions require a very large number of $\xi$-templates as graphical aids. Because most of the information obtainable, using $\xi$-Bodes, can be found just as well with other types of conventional templates, we shall not consider $\xi$-Bode plots further here.[10] Fourth, and finally, is the diagram in which the complex variable has a constant real part, $\sigma_i$, and a variable imaginary part, $s = \sigma_1 + j\omega$. Such diagrams are called shifted Bode diagrams. The $j\omega$-Bode diagrams and the sigma diagrams for the common transfer function factors are also summarized in Table 2-4.

As is apparent in Table 2-4, the Bode magnitude diagrams are closely approximated for wide ranges of the independent variable by straight-line segments with slopes that are integral multiples of $\pm 20$ dB/decade. This fact, a simple consequence of the way in which the diagrams are defined, represents one of the principal practical advantages of the Bode diagram. The actual functions "depart" from the straight-line "asymptotes" only in the vicinity of changes in the slope of the asymptotic approximation. The ***breakpoint*** in slope occurs at the magnitude of the pole or zero, e.g., $1/T$ or $\omega_n$. For the second-order $j\omega$-Bodes the departure of the amplitude ratio at the breakpoint is $-|2\zeta|_{dB}$ for a pole or $|2\zeta|_{dB}$ for a zero. These and other properties of the first- and second-order

[10] A comprehensive set of $\xi$-Bode diagrams for first- and second-order elements appears in McRuer, "Unified Analysis of Linear Feedback Systems," ASD-TR-61-118, Wright-Patterson Air Force Base, Ohio, July 1961.

Table 2-4. Summary of the representations of simple transfer functions.

| Transfer Function G(s) | s - Plane Representation | σ - Bode Representation | jω - Bode Representation |
|---|---|---|---|
| K | | 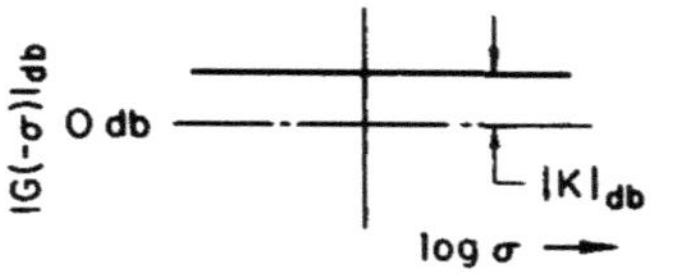 | 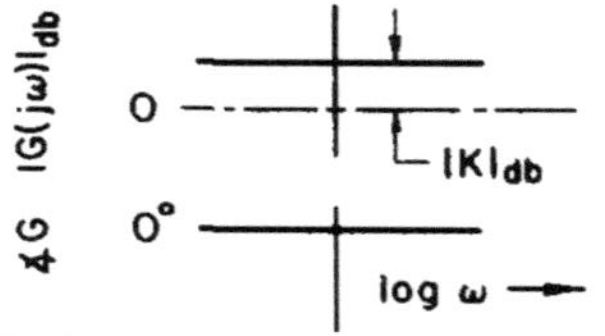 |
| $\frac{K}{s}$ | 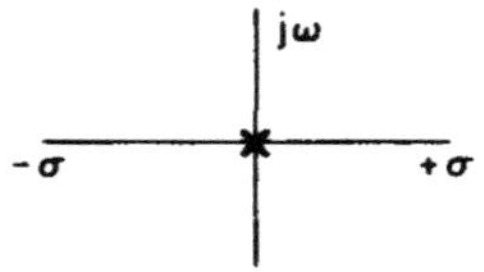 | 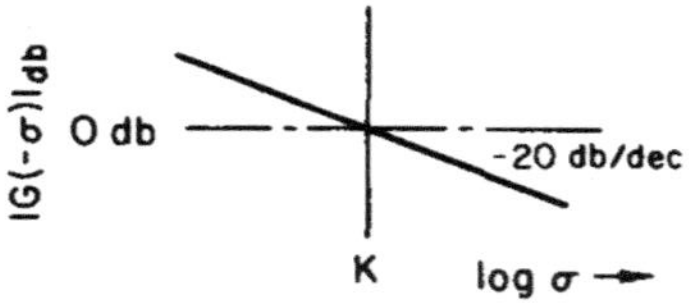 | 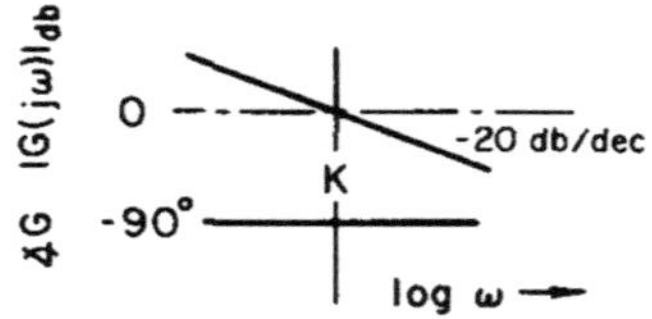 |
| $\frac{K}{s^2}$ |  | 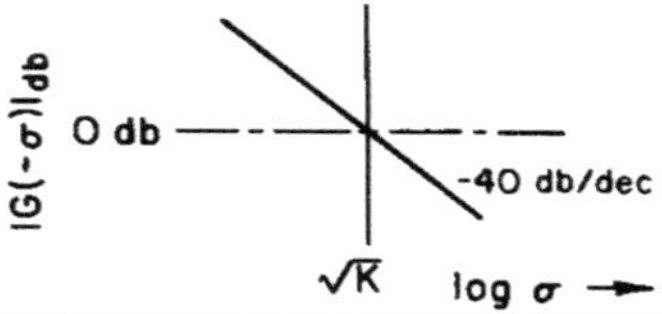 | 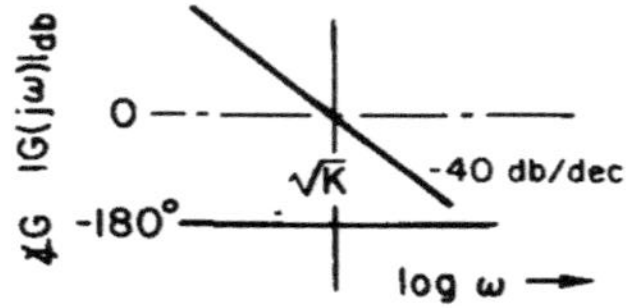 |
| Ks | 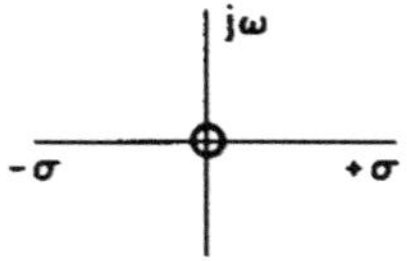 | 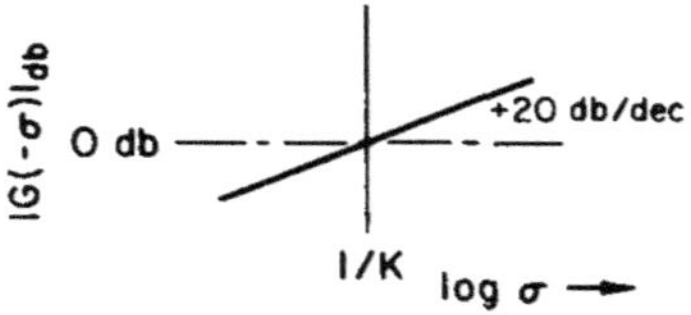 | 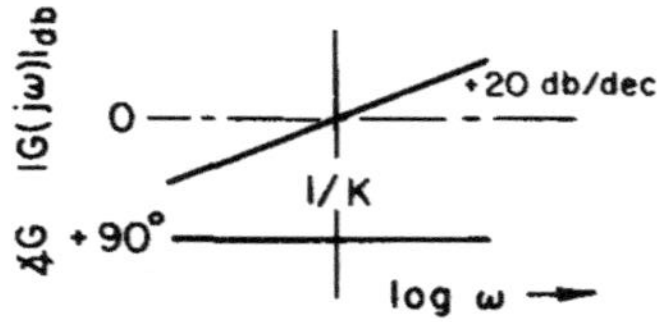 |
| $Ks^2$ | 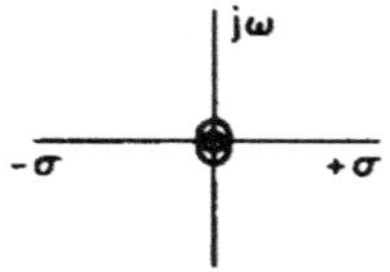 | 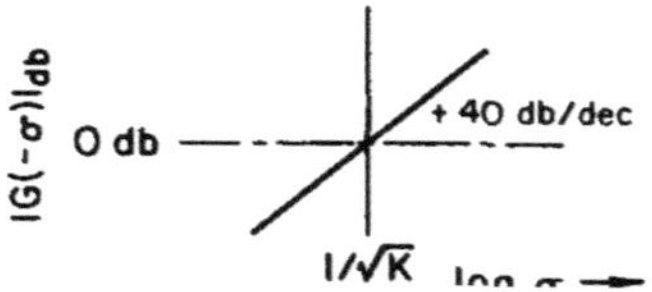 | 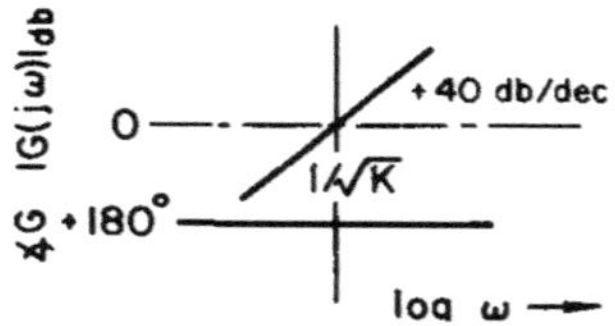 |

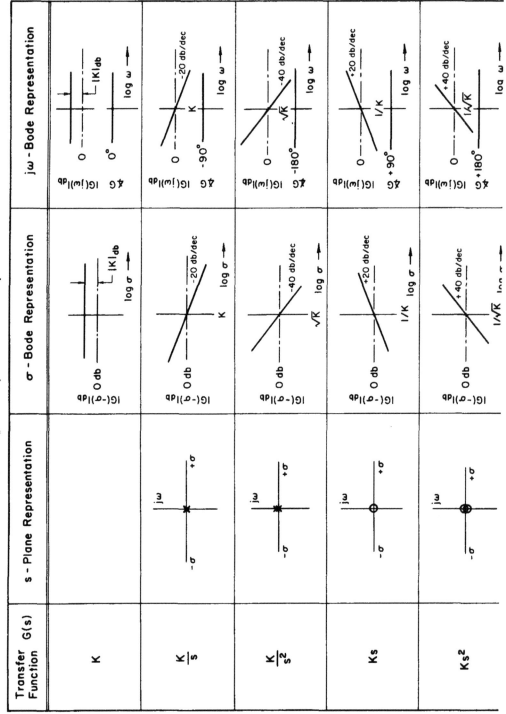

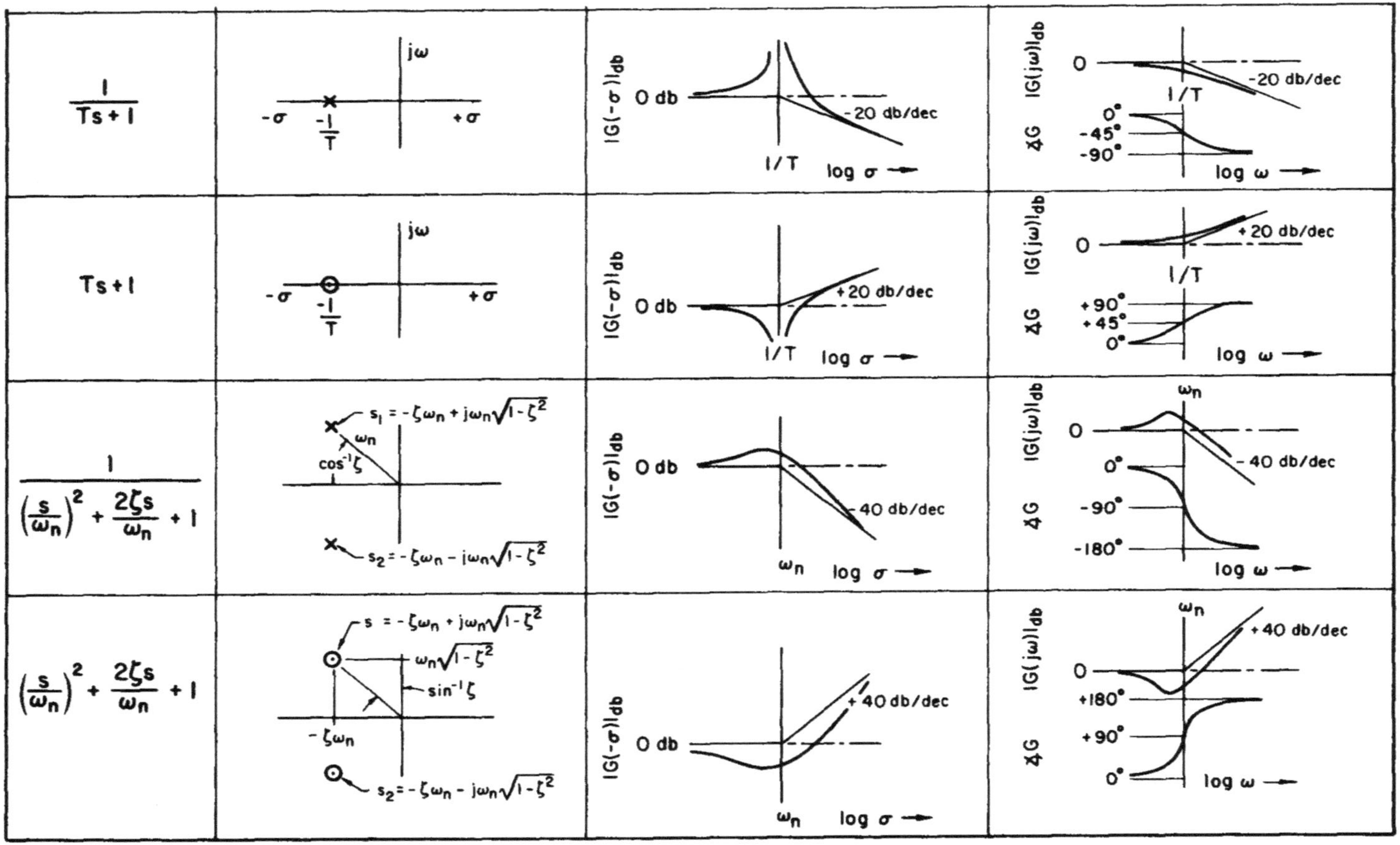

1/(Ts+1)
Ts+1
1/((s/ωn)² + 2ζs/ωn + 1)
(s/ωn)² + 2ζs/ωn + 1
jω
-σ
+σ
-1/T
s1 = -ζωn + jωn√(1-ζ²)
s2 = -ζωn - jωn√(1-ζ²)
s = -ζωn + jωn√(1-ζ²)
ωn
cos⁻¹ζ
sin⁻¹ζ
ωn√(1-ζ²)
-ζωn
|G(-σ)|db
0 db
1/T
log σ
-20 db/dec
+20 db/dec
-40 db/dec
+40 db/dec
|G(jω)|db
∡G
log ω
0°
-45°
-90°
+90°
+45°
-180°
+180°

| Transfer Function G(s) | s - Plane Representation | $\sigma$ - Bode Representation | $j\omega$ - Bode Representation |
|---|---|---|---|
| $\frac{1}{-Ts+1}$ | 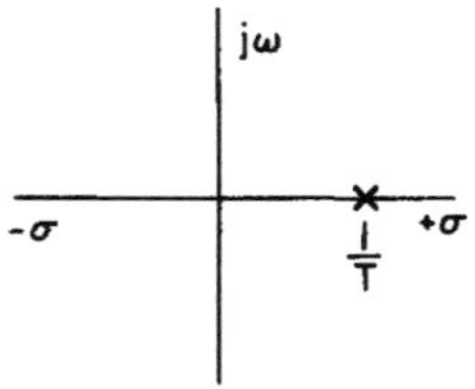  | 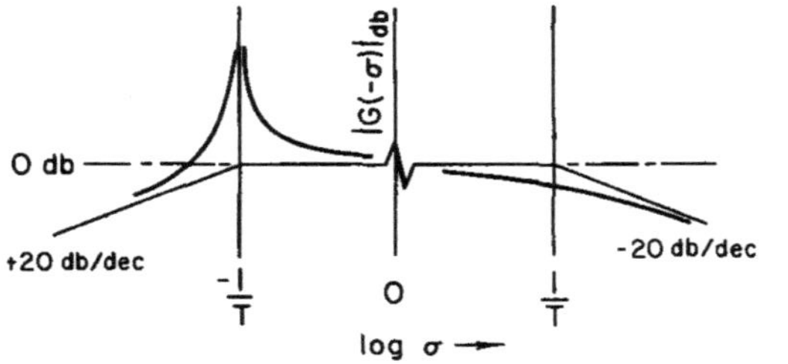  | 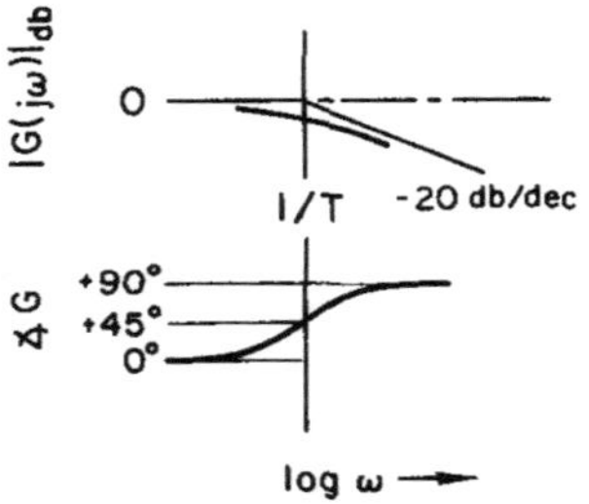  |
| $-Ts+1$ | 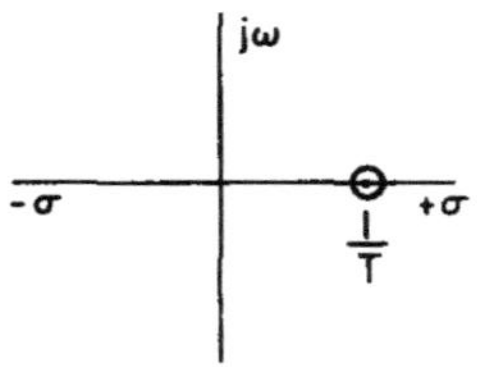  | 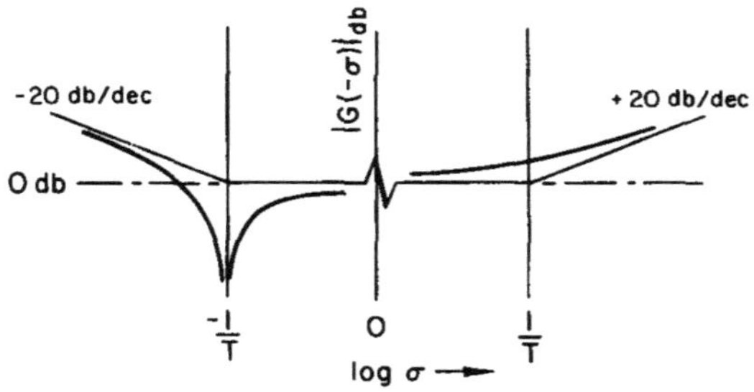  | 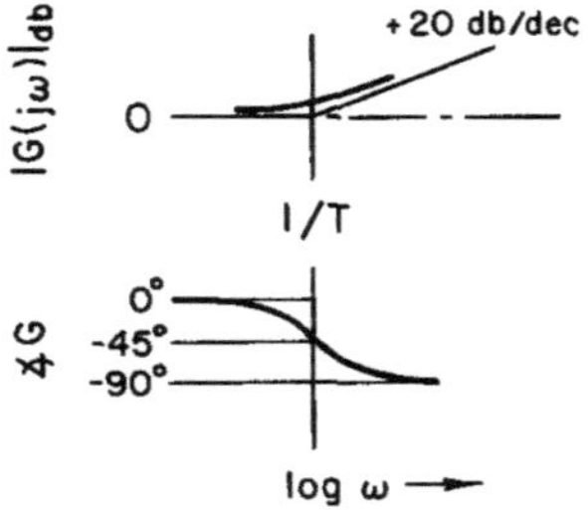  |

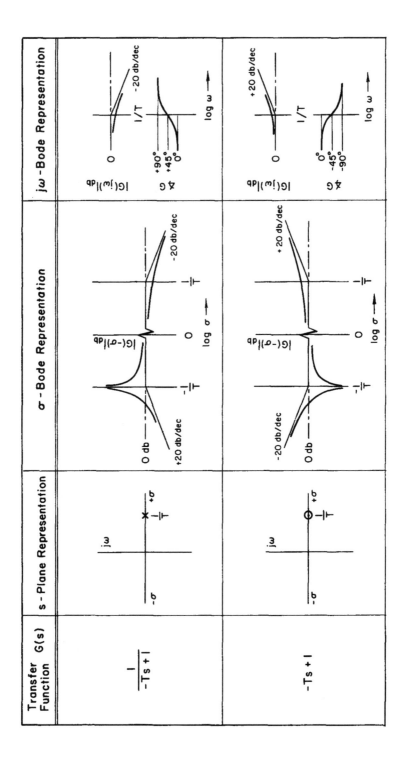

$$\frac{1}{\left(\frac{s}{\omega_n}\right)^2 - \frac{2\zeta s}{\omega_n} + 1}$$

$$\left(\frac{s}{\omega_n}\right)^2 - \frac{2\zeta s}{\omega_n} + 1$$

$$e^{-\tau s} \doteq \underbrace{\frac{(-\tau/2)\,s + 1}{(\tau/2)\,s + 1}}_{\text{First Order Pade' Approximation}}$$

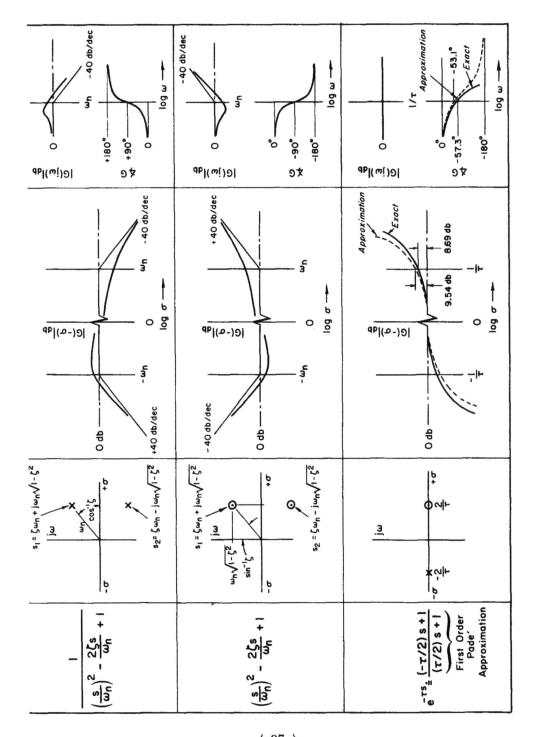

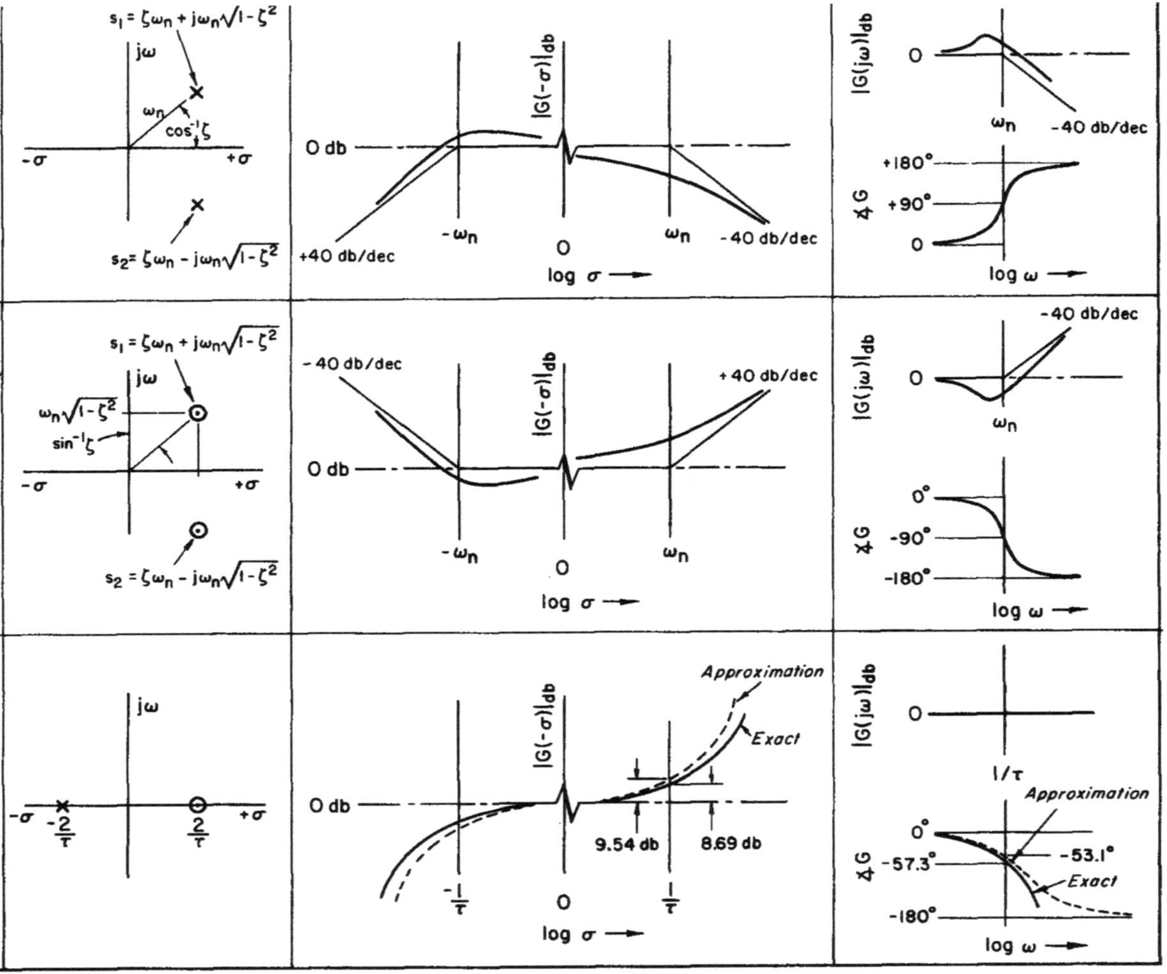

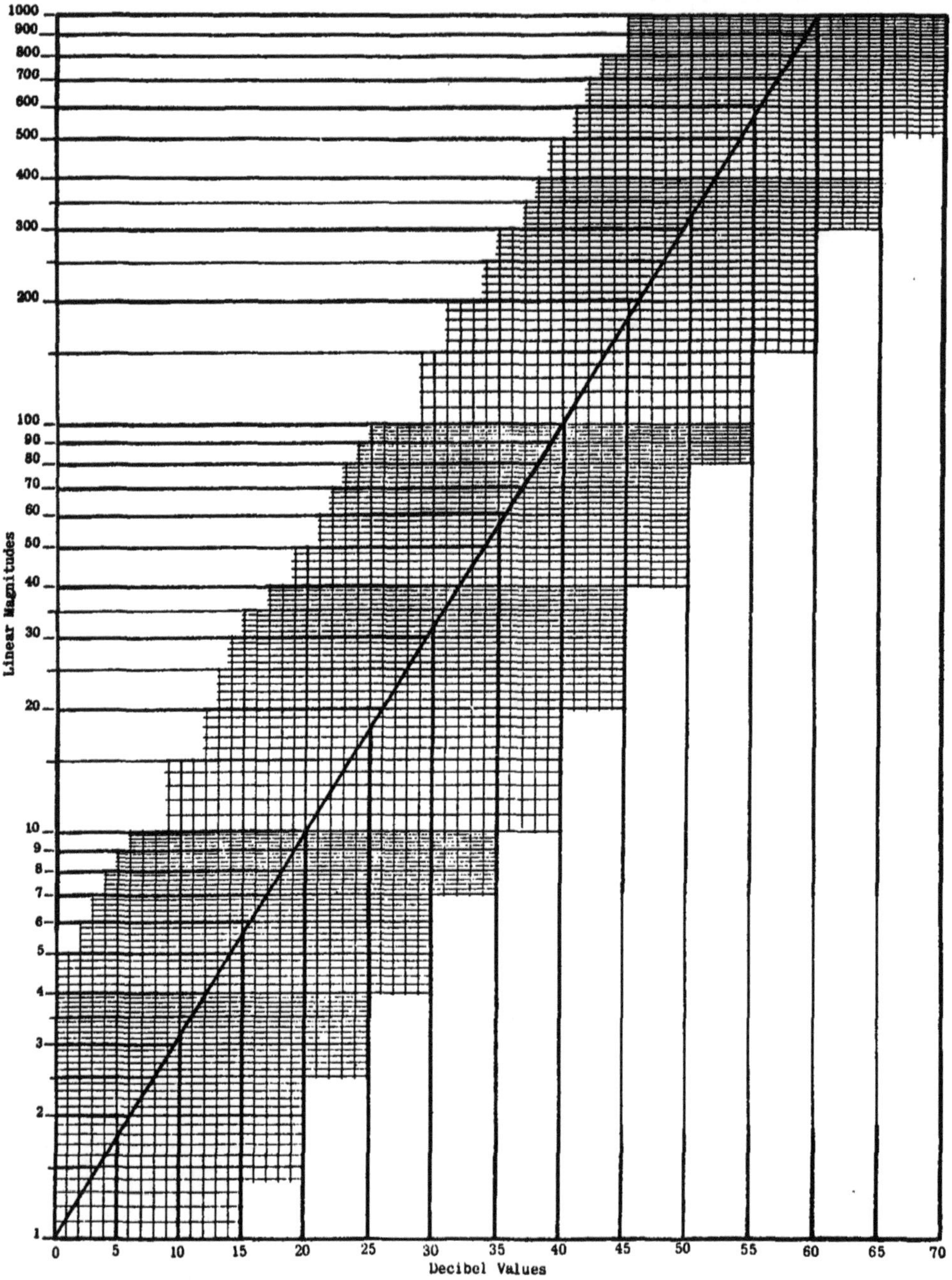


Fig. 2-8. Linear magnitudes vs. decibel values. Note: Log $1/x = -\text{Log } x$. Therefore, to obtain $x$ in dB, when $0 < x < 1$, find $1/x$, read dB from graph and prefix by negative sign, e.g., $|0.02| = 1/50 = -34$ dB.

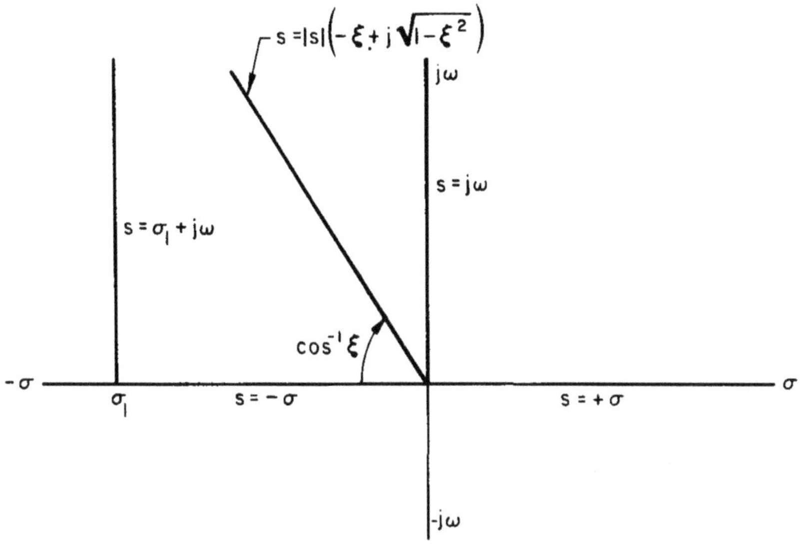


**Fig. 2-9. The s-plane pathways for a generalized Bode diagram.**

amplitude ratio components are shown in Fig. 2-10. Also at the breakpoint $(\omega_n)$, the phase slope, $d \measuredangle G/d\,(\ln \omega/\omega_n)$, is equal to $-1/\zeta$ $(-131.92/\zeta$ degree/decade) for a second-order pole and $+1/\zeta$ for a second-order zero. This provides the basis for phase asymptotes, as illustrated in Fig. 2-11, which are almost as convenient as the amplitude ratio asymptotes. As indicated by the slope formula above, and the $j\omega$-Bode summaries in Table 2-4, the direction of the phase shift depends on the sign of $\zeta$. In both the table and the figures described here the abscissa is a log scale but, since semilog graph paper is customarily used, the quantities along the scale are expressed in linear units.

For $\sigma$-Bodes, illustrations similar to those of Figs. 2-10 and 2-11 are given in Fig. 2-12. Only the amplitude ratios are shown because the phase is always zero for second-order factors and shifts abruptly from zero to 180° at $\sigma = 1/T$ for first-order factors. Items 10 and 14 in Table 2-4 indicate the lack of symmetry between the $\sigma$-Bode plots in the left and right half-planes. This has to be taken into account when right half-plane poles or zeros are present. In an actual problem the presentation with suppressed zero and both right and left half-plane $|G(-\sigma)|$ shown separately is awkward. It is usually more convenient to flip the right half-plane $|G(-\sigma)|_{\text{dB}}$ plot, so that it plots along the same axes as the left half-plane $\sigma$-Bode and the $j\omega$-Bode.

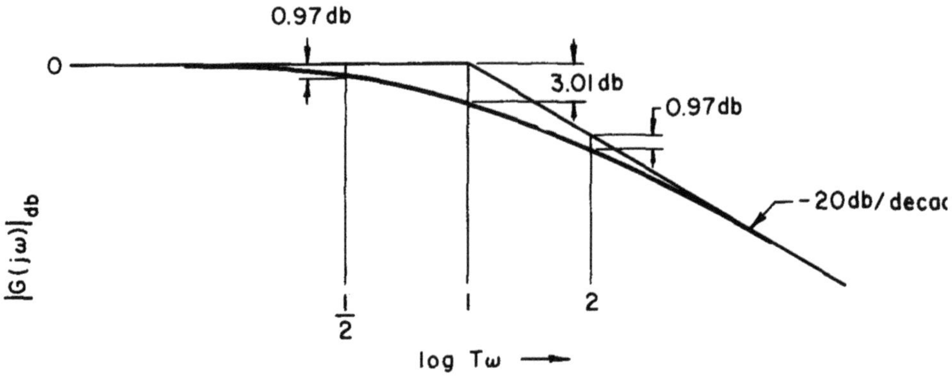


*(a) Amplitude Ratio for 1st Order Lag,* $G = \dfrac{1}{Tj\omega + 1}$

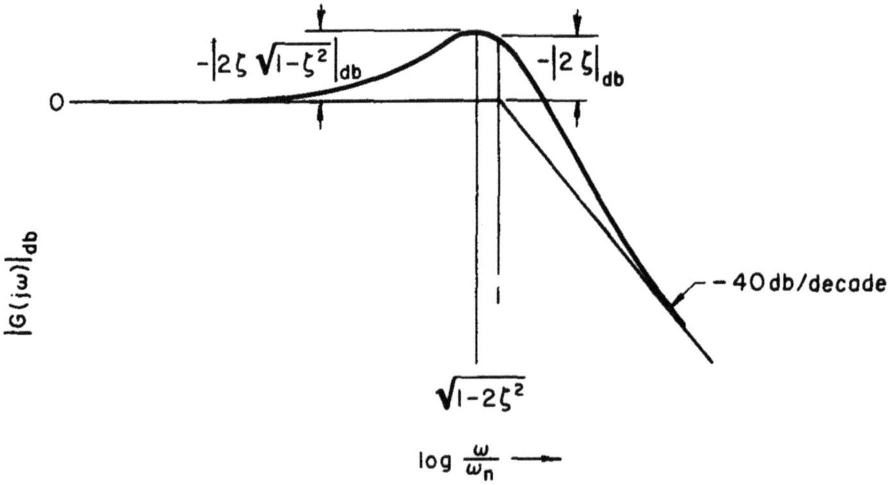


*(b) Amplitude Ratio for 2nd Order Lag,* $G = \dfrac{1}{\left(\dfrac{j\omega}{\omega_n}\right)^2 + 2\zeta\dfrac{j\omega}{\omega_n} + 1}$

**Fig. 2-10. Properties of the $j\omega$-Bode amplitude ratio for first- and second-order lag.**

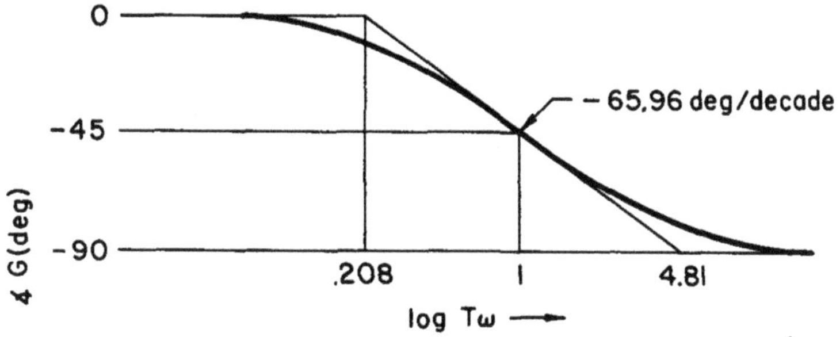


*(a) Phase Asymptotes for 1st Order Lag,* $G = \dfrac{1}{Tj\omega + 1}$, $T \geq 0$

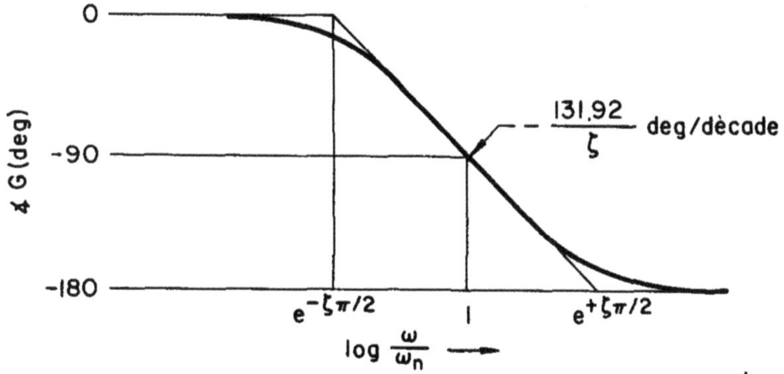


*(b) Phase Asymptotes for 2nd Order Lag,* $G = \dfrac{1}{\left(\dfrac{j\omega}{\omega_n}\right)^2 + 2\zeta \dfrac{j\omega}{\omega_n} + 1}$, $\zeta \geq 0$

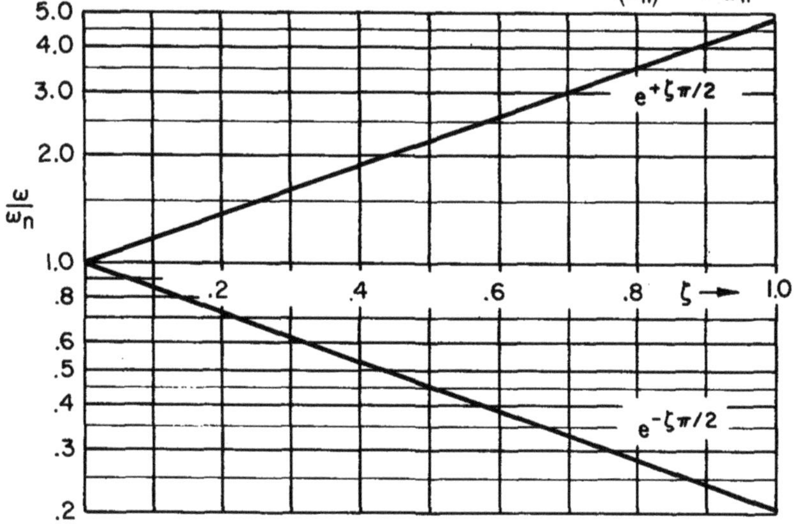


*(c) Breakpoints for 2nd Order Phase Asymptotes*

**Fig. 2-11.** Asymptotes of the $j\omega$-Bode phase for first- and second-order lag.

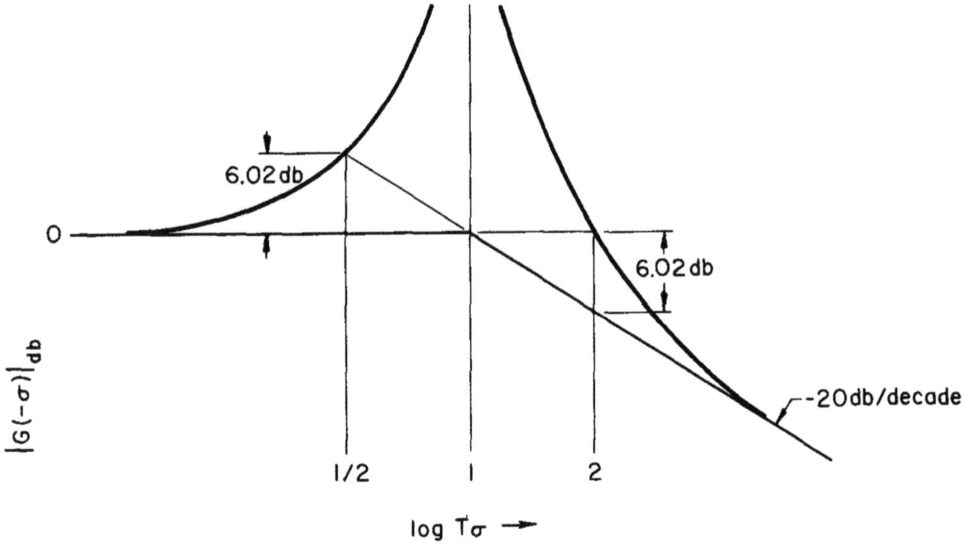


*(a) Amplitude Ratio for 1st Order Lag,* $G(-\sigma)=\dfrac{1}{-T\sigma+1}$, $T \geq 0$

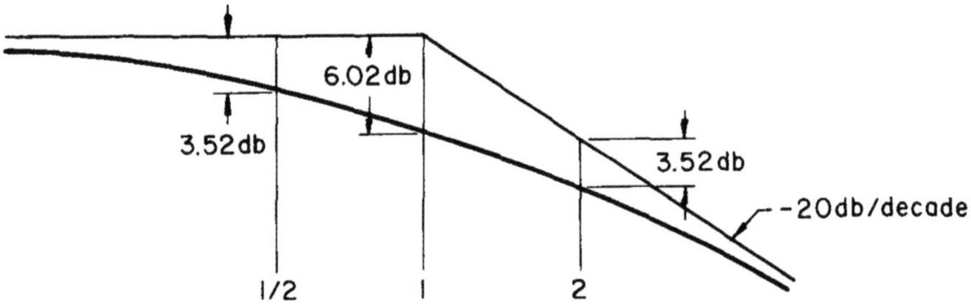


*(b) Amplitude Ratio for 1st Order Lag,* $G(-\sigma)=\dfrac{1}{-T\sigma+1}$, $T \leq 0$

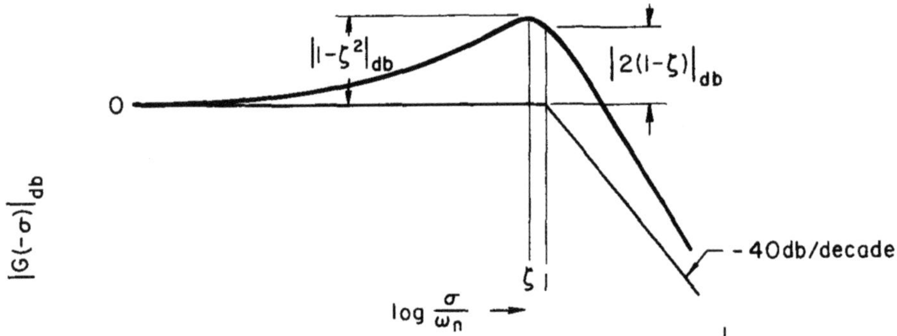


*(c) Amplitude Ratio for 2nd Order Lag,* $G(-\sigma)=\dfrac{1}{\left(\dfrac{-\sigma}{\omega_n}\right)^2-\dfrac{2\zeta\sigma}{\omega_n}+1}$

Fig. 2-12. Properties of the $\sigma$-Bode amplitude ratio for first- and second-order lag.

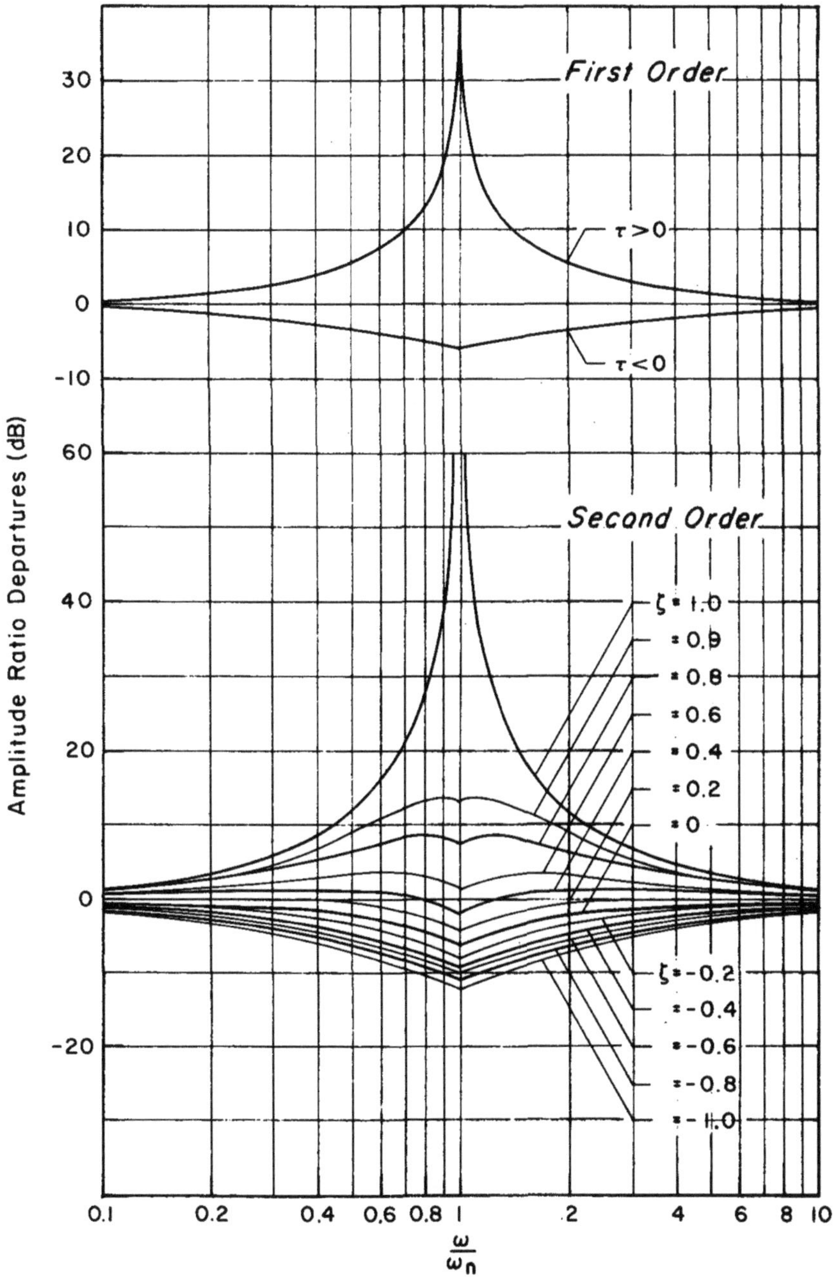


Fig. 2-13. Amplitude ratio departures for σ-Bode, first- and second-order poles.

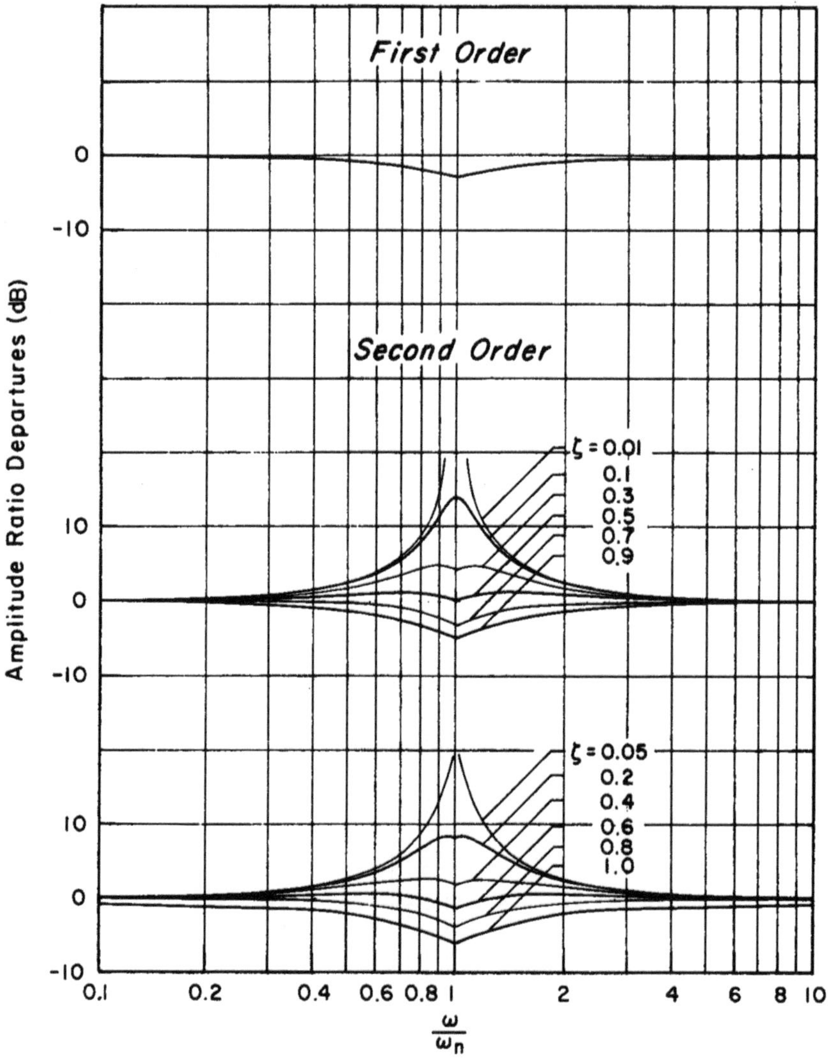


Fig. 2-14. Amplitude ratio departures for $j\omega$-Bode, first- and second-order poles.

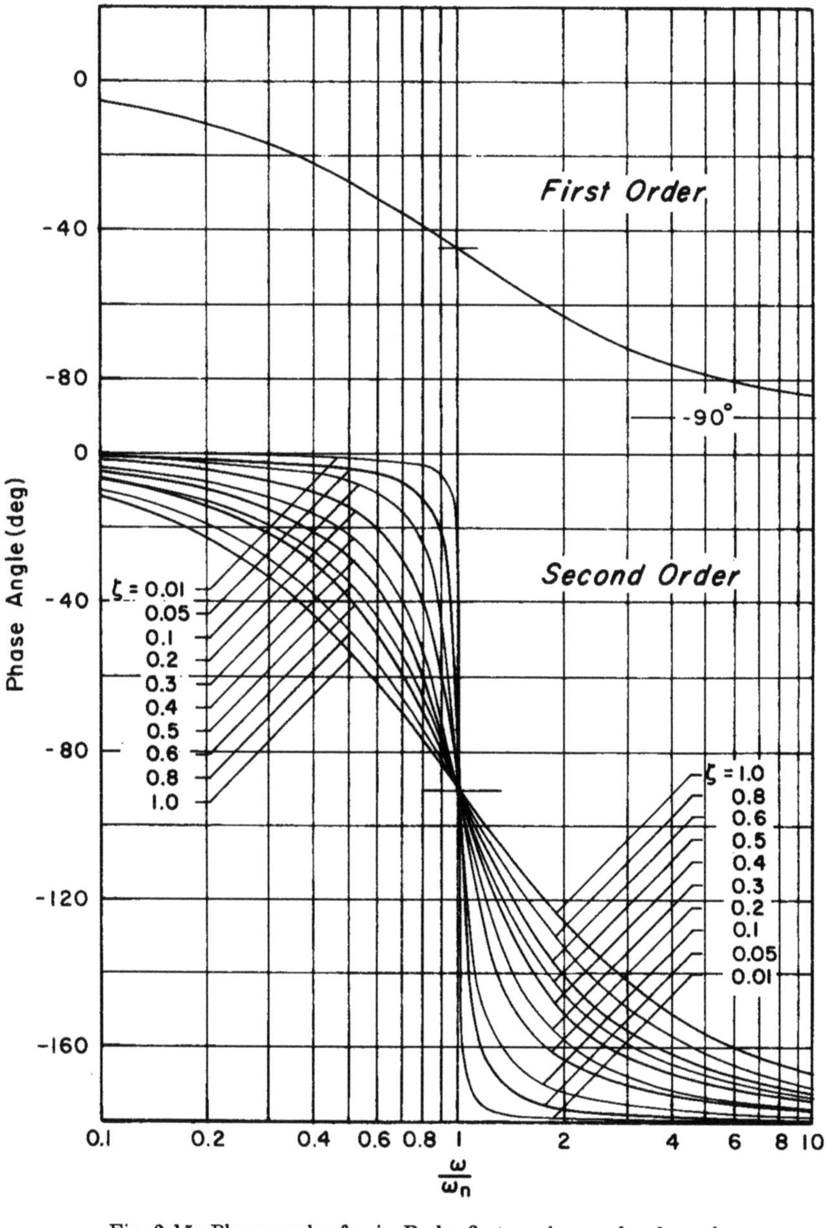


Fig. 2-15. Phase angles for $j\omega$-Bode, first- and second-order poles.

Charts that accurately show the "departures" for both the Bode and sigma diagrams, as well as the actual $\measuredangle G$ curves for the Bode diagrams of first-order and second-order factors, are given in Figs. 2-13, 2-14, and 2-15.

## 2-10. The Combining of Transfer Functions

A great advantage is enjoyed by the transfer function representation when the output of one of the elements or subsystems is the input to another; then the simple subsystem transfer function models are easily combined into the transfer function of the whole system. Thus, as already shown, the transfer function of the cascaded elements is simply the product of the transfer function of the individual elements.

It should be clear that the multiplication of concatenated transfer functions, illustrated in Fig. 2-16, is carried out on the pole-zero diagram by simply superposing the poles and zeros of the component transfer functions. On the $j\omega$- and $\sigma$-Bode diagrams the individual transfer function factors are represented by a quantity proportional to the logarithm of the magnitude and an angle. If one recalls that complex numbers are multiplied together by multiplying their magnitudes and adding their angles, it can be appreciated that the multiplication of transfer functions is carried out by the addition of the logarithmic magnitude curves together with the addition of the phase angles.

Since the Laplace transformation is a linear operation, we can justify the use of summers and differentials in the *block diagram* which show how the transfer functions of system elements may be combined. The summer, differential, and takeoff point are graphically illustrated in Fig. 2-17.

With the rules for combining the transfer functions of cascaded elements and the use of the symbols for the summer, differential, and takeoff point, it is possible to redraw block diagrams in a variety of ways by means of *block diagram algebra.*[11] The algorithms of block diagram algebra are justified by showing that two different configurations represent the same transformed equation. For example, the system of Fig. 2-2 could be represented in block diagram form by the configuration displayed in Fig. 2-18.

The most important algorithm of the block diagram algebra of feedback systems is the series of identities displayed in Figs. 2-19 and 2-20. Figure 2-20 shows the special case in which the feedback transfer function, $G_2$, is unity. The use of the identities of Figs. 2-19 and 2-20 allows the reduction of the diagram of Fig. 2-18 to the equivalent form shown in Fig. 2-7.

Figure 2-21 shows the general block diagram of a feedback control

[11] T. D. Graybeal, "Block Diagram Network Transformation," *Electrical Engineering*, 70, No. 11, 985–990 (Nov. 1951); D. T. McRuer, ed., "Methods of Analysis and Synthesis of Piloted Aircraft Flight Control Systems," Bureau of Aeronautics Rept. AE-61-4-I, 1952.

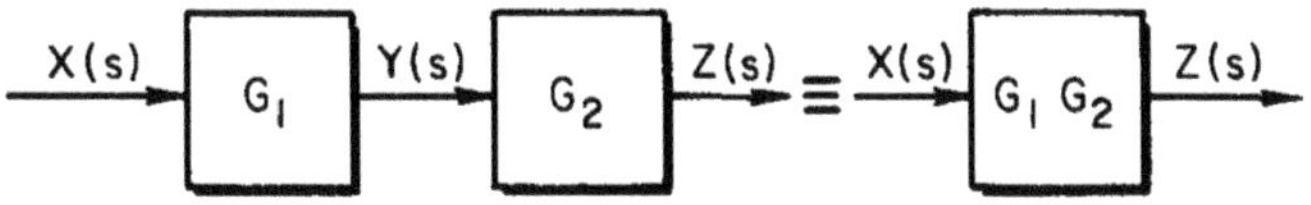


Fig. 2-16. The combining of the transfer functions of cascaded elements.

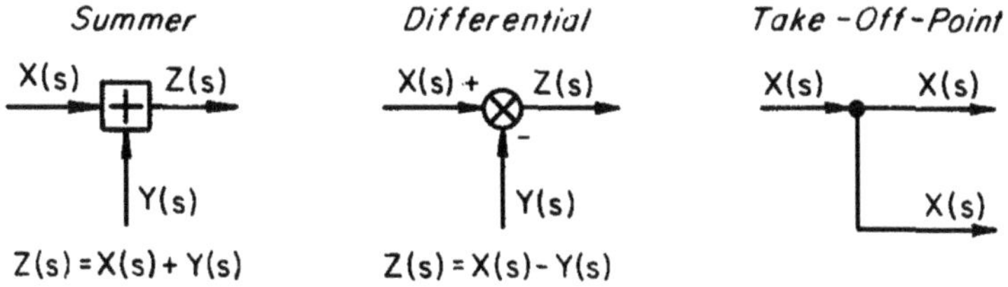


Fig. 2-17. Block diagram representations of the summer, differential, and takeoff point.

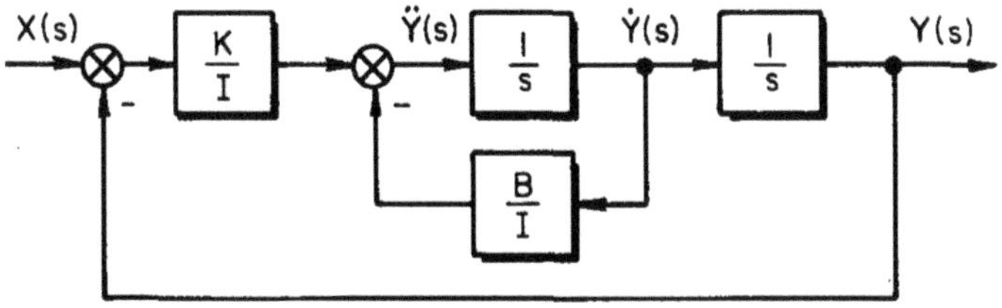


Fig. 2-18. Mathematical block diagram of the spring-mass-damper system.

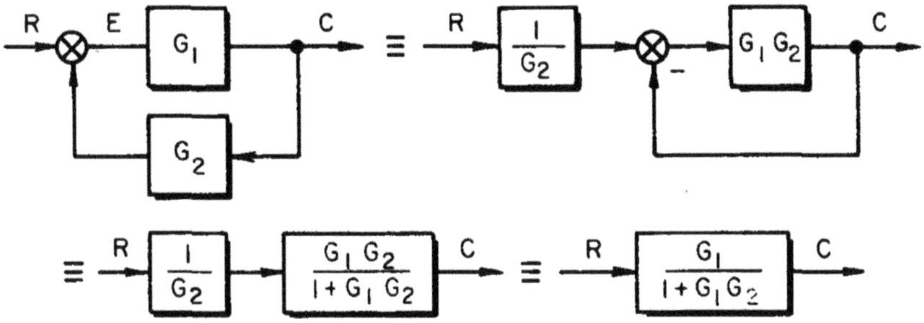


Fig. 2-19. Feedback system block diagram identities.

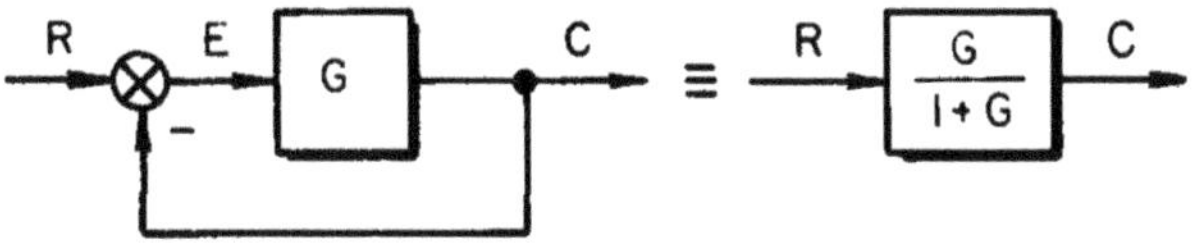


Fig. 2-20. Reduction of the unity feedback system block diagram.

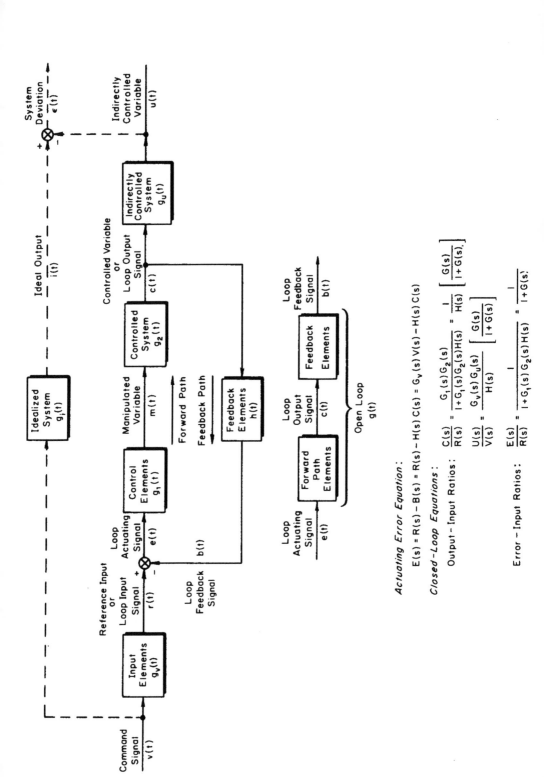

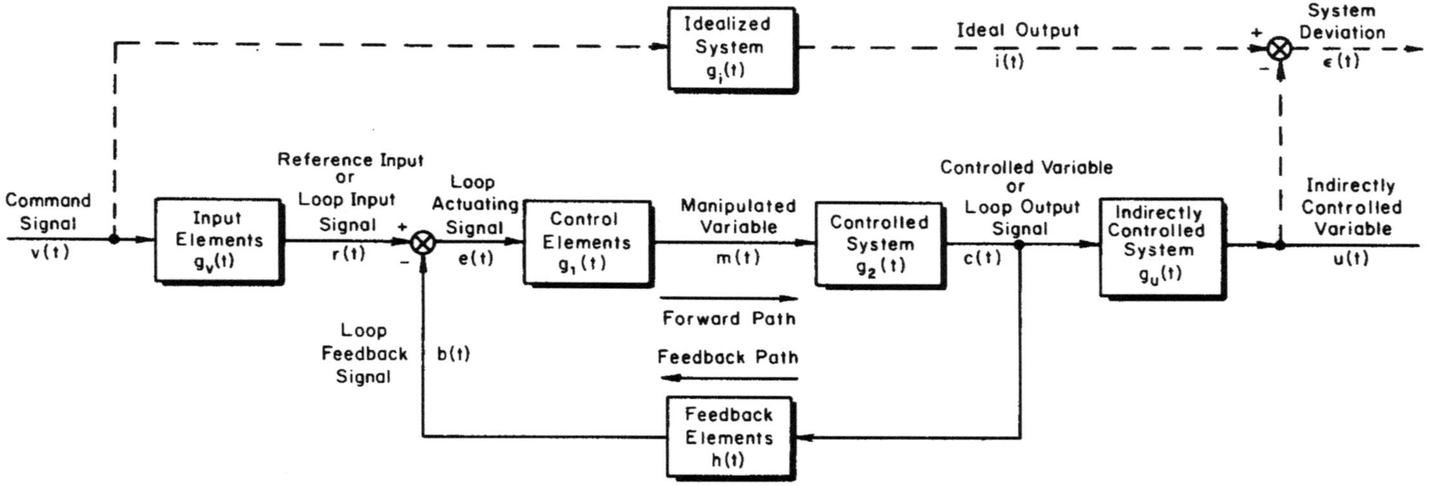


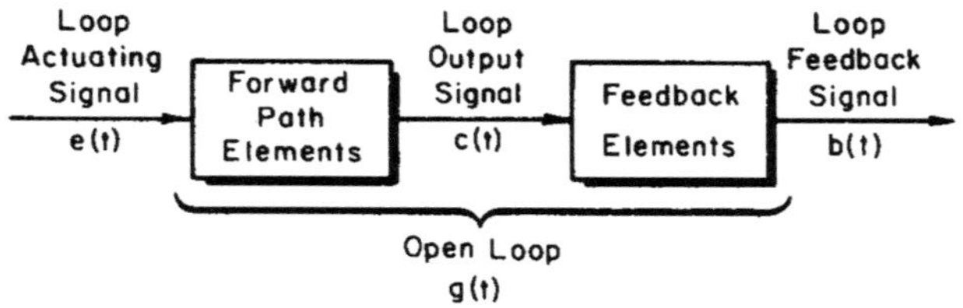


*Actuating Error Equation*:

$$E(s) = R(s) - B(s) = R(s) - H(s)\,C(s) = G_v(s)\,V(s) - H(s)\,C(s)$$

*Closed-Loop Equations*:

Output-Input Ratios: $$\frac{C(s)}{R(s)} = \frac{G_1(s)\,G_2(s)}{1 + G_1(s)\,G_2(s)\,H(s)} = \frac{1}{H(s)}\left[\frac{G(s)}{1+G(s)}\right]$$

$$\frac{U(s)}{V(s)} = \frac{G_v(s)\,G_u(s)}{H(s)}\left[\frac{G(s)}{1+G(s)}\right]$$

Error-Input Ratios: $$\frac{E(s)}{R(s)} = \frac{1}{1 + G_1(s)\,G_2(s)\,H(s)} = \frac{1}{1+G(s)}$$

system and the terms used to describe the several parts and signals. The algebraic derivation of the closed-loop transfer functions is also shown.

The transfer functions of the several blocks in the forward path, $G_1$, and the feedback path, $G_2$, are often known or are easily found in *factored form*, but both the error-input and the output-input transfer functions involve a denominator that appears as $1 + G$. Thus the problem inherent in linear feedback system analysis is to find the factors of $1 + G$ and other information about the closed-loop system such as the modal response coefficients, when given the open-loop characteristics, $G$. The means to accomplish this are discussed in the next chapter.

CHAPTER 3

# FEEDBACK SYSTEM ANALYSIS

## 3-1. Introduction

The early development of automatic flight controls evolved substantially independently of the use of any mathematics. By 1947, however, it was widely recognized that the dynamic problems of vehicle control could not be mastered by cut-and-try techniques or engineers' rules of thumb. To fill this need, the elaborate and extensive theory of linear feedback systems was further developed and then applied to an increasingly wide range of flight control problems. There was a dramatic interplay between theory and practice where, in many cases, aircraft and missiles provided both the inspiration for the theoretical developments and the examples of practical application.

In the intervening years, a large number of both introductory and advanced texts have been written on the subject of feedback control systems. It is assumed that the reader will already have acquainted himself with the contents of one or more of these texts.[1] In order to allow a connected account of our subject, however, it is necessary to review briefly some of the mathematical and physical bases on which it rests. This chapter continues in sequence from the last, which discussed the characterization of physical systems and system elements by means of mathematical models. The sections that follow contain condensed expositions of those elements of linear feedback control system theory needed to understand the later discussions of flight control systems.

All of these topics are involved with *system analysis* which, for linear constant-coefficient systems typified by the single-loop system shown in Fig. 3-1, consists of five essential steps:

1. Delineation of nominal open-loop system characteristics. This ordinarily starts with the differential equations which describe the nominal controlled element and one or more of the controller possibilities. This stage is concluded when one or more nominal open-loop transfer functions, $G(s)$, are available, in factored form, for further analysis.

[1] See n. 1, Chapter 1.

2. Determination of nominal closed-loop transfer functions, $G_{cr}(s)$ and $G_{er}(s)$, from the open-loop transfer function, $G(s)$.
3. Calculation of nominal closed-loop system responses for pertinent inputs.
4. Determination of the changes in $G(s)$ resulting from the expected variations in the controller and controlled element characteristics.
5. Consideration of the effects of open-loop system variations on closed-loop behavior.

The topics considered in the next two sections are concerned with Step 2 for single-loop systems. The similar analysis of multiloop systems is taken up in Section 3-5. Response calculations (Step 3) have already been treated in Chapter 2 and receive further attention in Chapter 10. Techniques for accomplishing Step 5 are discussed in Section 3-6 for both single-loop and multiloop systems. The discussion of Steps 1 and 4 is for the most part deferred to subsequent chapters.

The first step, the delineation of the open-loop characteristics in terms of a transfer function, $G(s)$, is relatively easy for linear time-invariant systems because transform methods can be used to convert the system differential equations to algebraic equations.

The equations for the single-loop system are as follows.

open-loop transfer function:

$$G(s) = \frac{C(s)}{E(s)} = \kappa \frac{s^n + a_1 s^{n-1} + a_2 s^{n-2} + \cdots a_n}{s^{m+n} + b_1 s^{m+n-1} + b_2 s^{m+n-2} + \cdots b_{m+n}} = \kappa \frac{\sum_{j=0}^{n} a_j s^{n-j}}{\sum_{i=0}^{m+n} b_i s^{m+n-i}} \tag{3-1a}$$

$$= \kappa \frac{\alpha(s)}{\beta(s)} = \kappa \frac{\prod_{j=1}^{n} (s + z_j)}{\prod_{i=1}^{m+n} (s + p_i)} \tag{3-1b}$$

$$= \frac{KA(s)}{s^k B(s)} = \frac{K \prod_{j=1}^{n} (s/z_j + 1)}{s^k \prod_{i=1}^{m+n-k} (s/p_i + 1)} \tag{3-1c}$$

output/input transfer function:

$$G_{cr}(s) = \frac{C(s)}{R(s)} = \frac{G(s)}{1 + G(s)} = \frac{\kappa\alpha(s)}{\beta(s) + \kappa\alpha(s)} = \frac{\kappa \prod_{j=1}^{n} (s + z_j)}{(1 + \kappa\delta_m^0) \prod_{i=1}^{m+n} (s + q_i)}$$

$$\delta_j^i = \begin{cases} 0, & i \neq j \\ 1, & i = j \end{cases} \tag{3-2}$$

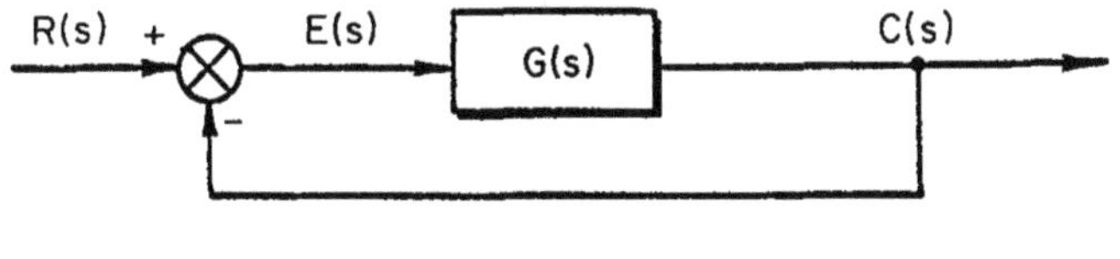


Fig. 3-1. Single-loop linear feedback system.

error/input transfer function:

$$G_{\mathrm{er}}(s) = \frac{E(s)}{R(s)} = \frac{1}{1 + G(s)} = \frac{\beta(s)}{\beta(s) + \kappa\alpha(s)} = \frac{\prod_{i=1}^{m+n}(s + p_i)}{(1 + \kappa\delta_m^0)\prod_{i=1}^{m+n}(s + q_i)} \tag{3-3}$$

This conversion permits the intermediate steps in an analysis sequence (e.g., the reduction of simultaneous equations using Cramer's rule and transfer function development, manipulation, and combination) to be carried out using algebraic forms. Most such forms either are, or can be approximated by, rational polynomials. Thus the delineation of open-loop transfer functions as indicated above is basically elementary, with the possible exception of polynomial factoring (which is also a part of Step 2).

The second step in the analysis sequence—given $G(s)$, find $G_{\mathrm{cr}}(s)$ and/or $G_{\mathrm{er}}(s)$—is the central problem of feedback system analysis. This requires only the solution for the roots of

$$1 + G(s) = 0 \tag{3-4}$$

Trivial as it may seem, a great deal of effort has been devoted to finding methods for performing this operation that are effective and at the same time promote insight.

The linear feedback analysis problem is usually presented in one of two artificially separated ways, referred to as the "root locus" and the "frequency response." The intent here is to show the interrelationships between the methods and to present a "unified" technique that is more efficient and flexible than either method used by itself.[2] By way of introduction we shall briefly review the features of the root locus method both in two dimensions (wherein gain is a parameter along the plot) and in three dimensions (where gain is a dimension),[3] and then discuss the use of the two principal logarithmic plots. Only then shall we emphasize the connections between the methods and point to the advantages of their coordinated use.

## 3-2. Conventional and Three-Dimensional Root Locus

Equation 3-5 expresses the fact that in a system with feedback the poles of the closed-loop transfer function or roots of the closed-loop

[2] The two sections following contain an abbreviated account of only those parts of a more complete report that are required for subsequent developments: D. T. McRuer, "Unified Analysis of Linear Feedback Systems," ASD-TR-61-118, Wright-Patterson Air Force Base, Ohio, July 1961.

[3] W. R. Evans, "Graphical Analysis of Control Systems," *Trans. AIEE*, 67, Part 1, 547–551 (1948); "Control System Synthesis by Root-Locus Methods," *Trans. AIEE*, 69, Part I, 66–69 (1950); *Control System Dynamics*, McGraw-Hill, New York, 1954.

characteristic equation are located at those values of the complex variable, $s$, where

$$G(s) = -1 \tag{3-5}$$

Because the function $G(s)$ is itself a complex number, we can write in place of Eq. 3-5,

$$|G(s)|\, e^{j\measuredangle G(s)} = e^{j(2k+1)\pi}; \qquad k = 0, \pm 1, \pm 2, \cdots \tag{3-6}$$

which requires the simultaneous satisfaction of

$$\measuredangle G(s) = (2k+1)\pi; \qquad k = 0, \pm 1, \pm 2, \ldots \tag{3-7}$$

and

$$|G(s)| = 1 \tag{3-8}$$

Each of the factors of $G(s)$ can itself be interpreted as a complex number, so that

$$G(s) = \kappa \frac{\prod_{h=1}^{n} (s + z_h)}{\prod_{i=1}^{m+n} (s + p_i)} = \kappa \left[\frac{\prod_{h=1}^{n} r_{N_h}}{\prod_{i=1}^{m+n} r_{D_i}}\right] \exp\left(\sum_{h=1}^{n} \varphi_{N_h} - \sum_{i=1}^{m+n} \varphi_{D_i}\right) \tag{3-9}$$

where $r_{N_h} = |s + z_h|$, $r_{D_i} = |s + p_i|$, $\varphi_{N_h} = \measuredangle(s + z_h)$, and $\varphi_{D_i} = \measuredangle(s + p_i)$. Then the two conditions expressed by Eqs. 3-7 and 3-8 can be rewritten,

$$\text{criterion angle:} \quad \left(\sum_{h=1}^{n} \varphi_{N_h} - \sum_{i=1}^{m+n} \varphi_{D_i}\right) = \begin{cases} (2k+1)\pi; & \kappa > 0 \\ 2k\pi; & \kappa < 0 \end{cases} \tag{3-10}$$

$$\text{magnitude:} \quad |\kappa| \left[\frac{\prod_{h=1}^{n} r_{N_h}}{\prod_{i=1}^{m+n} r_{D_i}}\right] = 1 \tag{3-11}$$

or finally, as is often more convenient, Eq. 3-11 can be expressed as

$$20 \log_{10} \kappa + \sum_{h=1}^{n} 20 \log_{10} r_{N_h} - \sum_{i=1}^{m+n} 20 \log_{10} r_{D_i} = 20 \log_{10} 1 = 0 \text{ dB} \tag{3-12}$$

which in shorthand form becomes

$$|\kappa|_{\text{dB}} + \sum_{h=1}^{n} (r_{N_h})_{\text{dB}} - \sum_{i=1}^{m+n} (r_{D_i})_{\text{dB}} = 0$$

Figure 3-2 shows how at any point on the $s$-plane the individual factors of the open-loop function can be graphically represented as vectors. Note from the geometry of the diagrams that, when the factors are represented by vectors drawn *from* the singularities *to* the test point, their angles may be measured *counterclockwise* from a horizontal line through the test point.

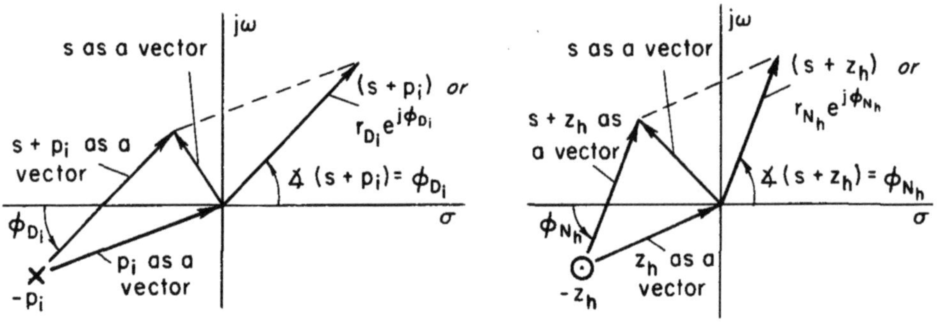


Fig. 3-2. Graphical representation of open-loop function factors.

The vectors from the singularity to all possible points in the $s$ plane can be represented by isomagnitude and isoargument (phase) curves.[4] Isomagnitude curves for first-order factors are constructed as loci in the $s$ plane for which $|(s + p_i)^{-1}|$ or $|s + z_h|$, or some measures thereof, are constant. If the logarithm is taken as the base for the magnitude measure, the plots for poles and zeros show useful symmetrical forms, and multiplication operations involving more than one singularity become additions. There is also a convenient tie-in later with other analysis techniques. Consequently, the isomagnitude plots for first-order factors are constructed for constant values of $20 \log_{10} |1/(s + p_i)| = -(r_{D_i})_{\text{dB}}$ or $20 \log_{10} |s + z_h| = (r_{N_h})_{\cdot\text{dB}}$ The result is a series of concentric circles, as shown in Fig. 3-3. If log magnitude is considered as a dimension measured along an axis perpendicular to the plane of the paper, then the isomagnitude loci shown would occur at heights appropriate to the magnitude dimension. The isomagnitude loci are thus contour maps of a surface that rises to a point for a pole or sinks to a point for a zero.

The isoargument curves for the first-order factors are simply straight lines, issuing radially from the origin which is located at the pole or zero location in the $s$-plane. These are also shown in Fig. 3-3.

When more than one pole or zero is present, the open-loop function can be represented as a contour map by adding logarithmic magnitudes of the several poles and zeros so as to represent the left-hand side of Eq. 3-12. Points of equal magnitude throughout the $s$ plane are then joined to make the contours. The isoargument lines are found similarly by adding the angle contribution of each pole and zero so as to represent the left-hand side of Eq. 3-10 and then joining points for which these are equal.

[4] Y. Chu, "Synthesis of Feedback Control System by Phase-Angle Loci," *Trans. AIEE*, 71, Part 2, 330–339 (1952).

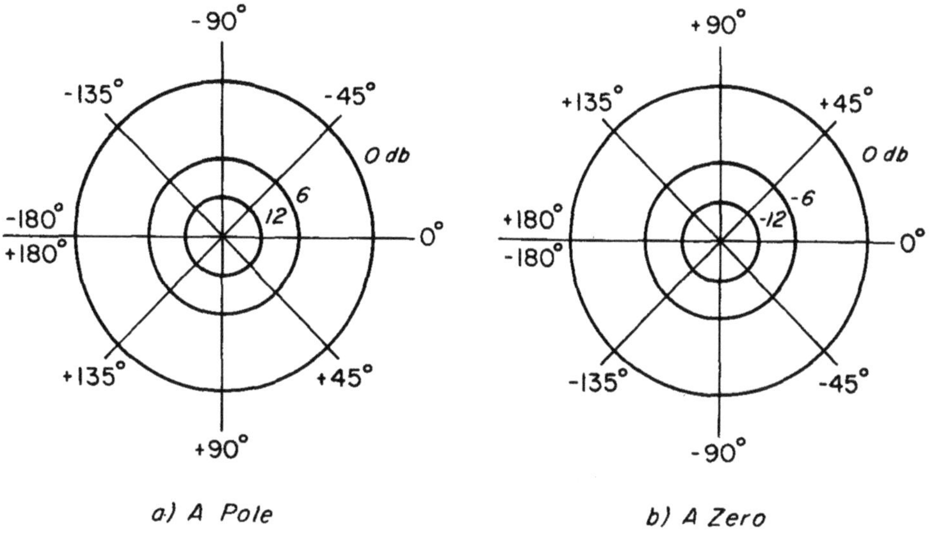


Fig. 3-3. Contour maps and isoargument curves for a pole and a zero.

The isoargument lines are always orthogonal to the contour lines, so that they are along the gradient. Several examples of such maps are presented later in this section.

In the root locus method the roots of the characteristic equation are found with semigraphical techniques based on the vector representation of the factors in $G(s)$. An attempt to determine a root of $1 + G(s) = 0$ is started by choosing a trial value of $s$ (a point on the $s$ plane) and imagining the vectors drawn to this point from each of the open-loop poles and zeros. Figure 3-4 illustrates the angle measurement convention in root locus construction. When a trial point is discovered that satisfies the phase condition of Eq. 3-10, a possible root location is identified. In principle, this procedure, repeated a sufficient number of times, delineates the *locus* of all possible closed-loop roots.

The second step in the method is to find the points on the locus corresponding to the satisfaction of the magnitude condition, i.e., Eq. 3-11, for particular values assigned to $\kappa$.[5]

Although the root locus can be found using the repetitive process outlined above, one of the most attractive features of the method is

[5] While both of these steps could be carried out with a protractor, dividers, and a slide rule, a device for mechanically adding angles and logarithmic magnitudes has been developed especially for the purpose. This is the "Spirule," copyright 1959 by North American Aviation.

$$G(s) = \frac{\kappa(s + z_1)}{s(s + p_2)(s + p_3)}$$

$$\sum_h \phi_{N_h} - \sum_i \phi_{D_i} = \phi_{N_1} - \left(\phi_{D_1} + \phi_{D_2} + \phi_{D_3}\right)$$

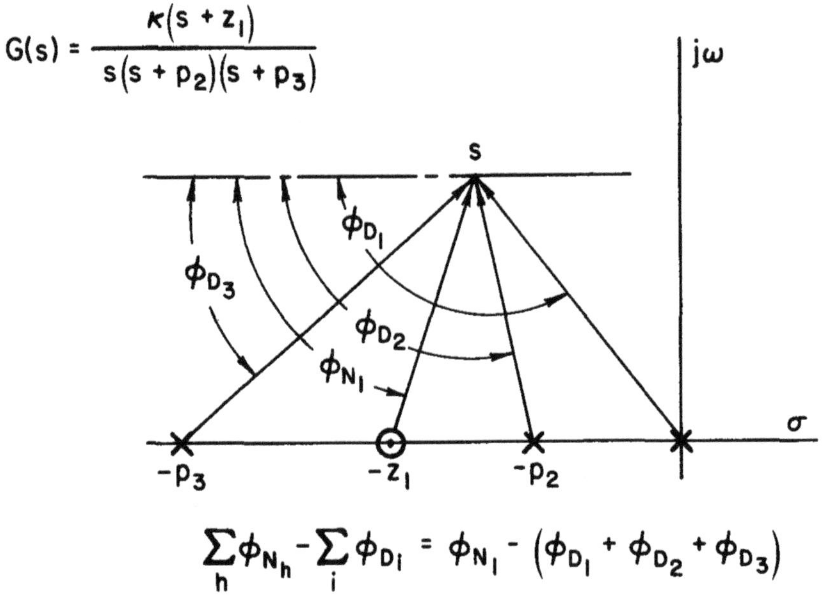

Fig. 3-4. Angle measurement convention in root locus construction.

that large segments of the locus can be found by applying root locus construction rules, which do not require the search for satisfactory trial points. The rules, which apply when $G(s)$ is a ratio of rational polynomials, are recapitulated, without proof, as follows:[6]

1. The total *number of separate branches* of the locus is equal to the total number of open-loop poles, $m + n$.

2. The root locus is *symmetrical* about the real axis ($\omega = 0$).

3. The branches of the locus *originate* at the open-loop poles.

4. The branches of the locus *terminate* at the open-loop zeros or at infinity: (a) $n$ (the total number of zeros) branches of the locus terminate on the open-loop zeros, and (b) $m$ branches approach points at infinity and are asymptotic to straight lines that originate at a point on the real

[6] D. T. McRuer, "Unified Analysis of Linear Feedback Systems," ASD-TR-61-118, Wright-Patterson Air Force Base, Ohio, July 1961. Also Evans, "Graphical Analysis of Control Systems," *Trans AIEE*, 67, Part 1, 547–551 (1948); "Control System Synthesis by Root Locus Methods," *Trans. AIEE*, 69, Part I, 66–69 (1950); F. M. Reza, "Some Mathematical Properties of Root Loci for Control System Design," *Trans. AIEE*, 75, Part 1, 103–108 (1956); H. Lass, "A Note on the Root-Locus Method," *Proc. IRE*, 44, 693 (May 1956).

axis, commonly called the "center of gravity," given by

$$\sigma_{c.g.} = -\frac{\sum_{i=1}^{m+n} p_i - \sum_{j=1}^{n} z_j}{m} \tag{3-13}$$

and which make angles with the real axis of

$$\varphi_m = \left.\begin{matrix} \dfrac{(2k+1)\pi}{m}; & \kappa > 0 \\ \text{or} & \\ \dfrac{2k\pi}{m}; & \kappa < 0 \end{matrix}\right\} k = 0, \pm 1, \pm 2, \cdots \tag{3-14}$$

5. The *root locus on the real axis* lies along alternate segments connecting real poles and zeros (or the point at infinity). When $\kappa$ is positive (negative), the locus exists in the intervals in which there is an odd (even) total number of poles and zeros to the right of the interval.

6. The *tangents* to the locus at *departure* from a pole or *arrival* at a zero are given by:

$$\text{departure angle} = \begin{pmatrix}\text{net angle from} \\ \text{other poles and zeros}\end{pmatrix} - (\text{criterion angle})$$

$$\text{arrival angle} = -\begin{pmatrix}\text{net angle from} \\ \text{other poles and zeros}\end{pmatrix} + (\text{criterion angle})$$

7. The *breakaway* of the locus from the real axis (or its rendezvous thereon) is located where the net change in angle caused by a small vertical displacement of the trial point is zero. These points correspond to the real roots of the equation

$$\frac{dG(s)}{ds} = 0 \tag{3-15}$$

8. At *junction points* (where the roots coalesce) the tangents to the branches of the locus are equally spaced over $2\pi$ rad $= 360$ degrees.

9. The *direction* in which the locus moves for increasing $|\kappa|$, at the point $-q_i$ on the locus, is shown by the orientation of the sensitivity vector:

$$S_\kappa^i = \frac{1}{[\partial G/\partial s]_{s=-q_i}} = \left[\sum_{j=1}^{m+n} \frac{1}{p_j - q_i} - \sum_{j=1}^{n} \frac{1}{z_j - q_i}\right]^{-1} \tag{3-16}$$

Note that this rule does not work to find the direction of the locus at breakaway and rendezvous points. On the other hand, in an algebraic form, it may be effectively employed to find the location of breakaway or rendezvous points when the poles and zeros are on the real axis.

10. When $m$ (the excess of poles over zeros) $\geqq 2$, the sum of the roots

is a constant equal to the sum of the open-loop poles. Then branches tending to the left must be "balanced" by branches tending to the right. When $m = 1$, the sum of the roots is the same constant added to $-\kappa$.

11. The locus crosses the imaginary axis when $1 + G(s) = 0$ has pure imaginary roots. This corresponds to the neutral stability condition indicated by the Routh-Hurwitz criterion or a similar test for stability, and is indicated by the vanishing of the imaginary part of the inverse open-loop function evaluated with $\sigma = 0$, i.e.,

$$\text{Im}\left[\frac{\beta(j\omega)}{\alpha(j\omega)}\right]_{\sigma=0} = 0 = \text{Im}\,[\beta(j\omega)\alpha^*(j\omega)]_{\sigma=0} \tag{3-17}$$

12. When a complete set of closed-loop roots is available for some value of $\kappa$, these roots may be used in the same fashion as open-loop poles for the purpose of continuing the plot of the locus. As a consequence, for example, the result of Rule 9 may be obtained by using Rule 6 in connection with any complete set of closed-loop roots.

13. The *product* of the negatives of the roots is equal to the sum of the product of the negatives of the open-loop poles, $b_{m+n}$, and $\kappa$ times the product of the negatives of the open-loop zeros, $\kappa a_n$, i.e.,

$$\prod_{i=1}^{m+n} q_i = b_{m+n} + \kappa a_n; \qquad m \geqq 1 \tag{3-18}$$

The application of the rules for the construction of root loci can now be illustrated with the aid of several simple examples. These are chosen so as to clarify the introduction to the unified method.

Example 1. First-order system

The open-loop function is $G(s) = \kappa/s$ and the only root of $1 + G(s) = 0$ is almost trivially found to be $s = -\kappa$. The closed-loop function is

$$\frac{C}{R} = \frac{G(s)}{1 + G(s)} = \frac{\kappa}{(s + \kappa)}$$

and, if necessary, the weighting function or indicial response could be found by the inverse Laplace transformation using Table 2-2. Suppose,

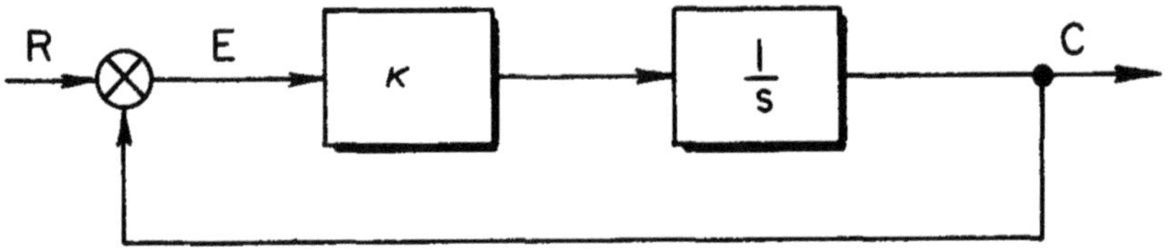


Fig. 3-5. Negative feedback around an integrator.

however, that the algebra were not so easy and we wanted to apply the techniques discussed so far. The contour map of the function $1/s$ has already been presented as Fig. 3-3(a). The pole-zero plot is simply a pole at the origin, and the use of Rules 1 to 5 tells us that the root locus shown in Fig. 3-6 has only one branch that lies along the negative real axis between the origin and the point at infinity. Rules 6 to 9 might be invoked but are not necessary. Rule 10 indicates the algebraic result already derived, and indeed Eq. 3-11 may be used to obtain the same result. The closed-loop roots are located along the locus at a radial distance from the origin equal to the gain constant $\kappa$. The closed-loop roots are marked along the locus for several values of the gain. A comparison of Fig. 3-6 with Fig. 3-3(a) shows that the locus of roots, marked with the position of the closed-loop roots for selected values of the gain constant, $\kappa$, is the same thing as the criterion angle contour ($-\pi$ in this case) marked with the intersections of the appropriate magnitude contours.

1. There is one branch of the locus.
2. The locus is symmetrical about the real axis.
3. The locus originates on the open-loop pole.
4. The locus terminates on a point at infinity and is asymptotic to a line that originates at $\sigma = 0$ and makes an angle with the real axis of $\pi$ rad.
5. The root locus exists in the interval between the pole and the point at infinity where there is one (an odd number) pole to the right.

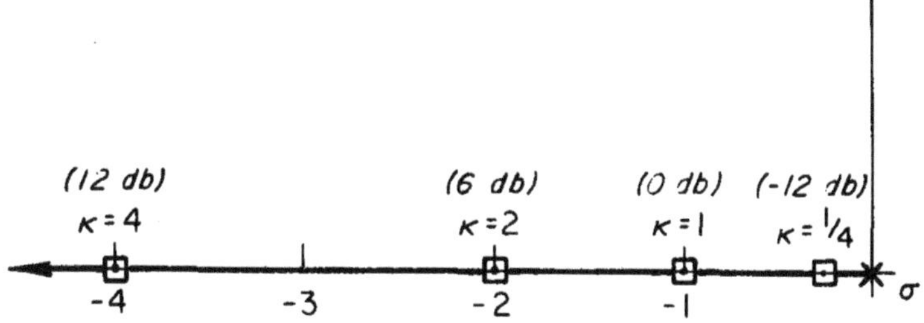


Fig. 3-6. Root locus for first-order system, $G(s) = \kappa/s$.

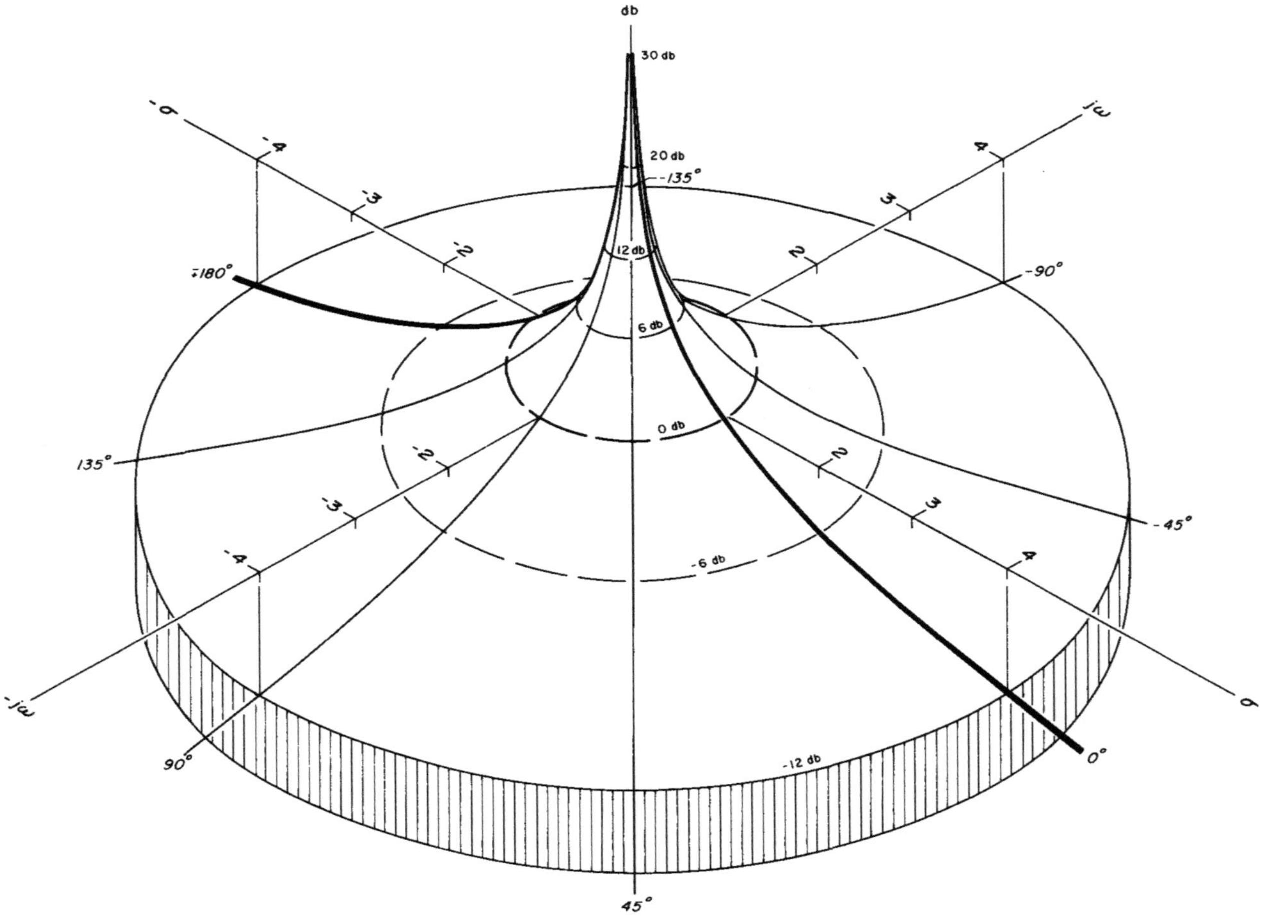


Fig. 3-7. The isometric view of the surface of $G(s) = 1/s$.

In Fig. 3-6 gain is a parameter along the locus. Alternatively, an additional dimension, coming out of the plane of the paper, can be introduced to show the variation of the closed-loop root with gain. This leads to the three-dimensional surface shown in isometric view in Fig. 3-7. The surface is, of course, identical to that represented by the contour map of Fig. 3-3(a). When the gain is very small, i.e., $\kappa \to 0$, the addition of the logarithmic magnitudes would remove the zero dB reference plane to an infinite distance, and only the top of the mountain (which stretches up to infinity) would protrude above the waterline. The closed-loop root would be at the location of the open-loop pole, i.e., $s = 0$. As the gain is increased, the mountain is shoved up like a volcano emerging from the ocean. The closed-loop root is always found at the intersection of the waterline and the criterion angle contour. For example, with $\kappa = 1$, the waterline for the function $\kappa/s$ would be in the reference plane (zero dB) With $\kappa = 2$, so that $20 \log_{10} \kappa = +6$ dB, the waterline for $\kappa/s$ would be at the level marked $-6$ dB for the function $1/s$ itself. The various positions of the closed-loop root, as the surface is raised or lowered with respect to the reference plane by raising or lowering the gain, $\kappa$, are illustrated for several values of $\kappa$ in Fig. 3-7. From this figure it can be readily appreciated that an increase in the gain, which raises the surface with respect to the reference plane, is entirely equivalent to lowering the reference plane with respect to the surface. We shall ordinarily take the latter view for convenience.

Example 2. Unit-numerator second-order system

Consider now the feedback system illustrated in Fig. 3-8. The closed-loop function is given by the expression

$$\frac{C}{R} = \frac{G(s)}{1 + G(s)} = \frac{\kappa}{(s^2 + ps + \kappa)}$$

and the closed-loop roots may be found from the quadratic formula

$$s_{1,2} = \frac{-p}{2}\left[1 \pm \sqrt{1 - \frac{4\kappa}{p^2}}\right]$$

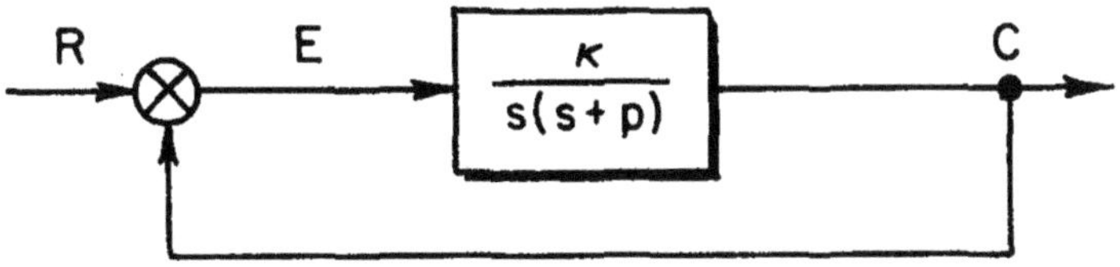


Fig. 3-8. A second-order servomechanism.

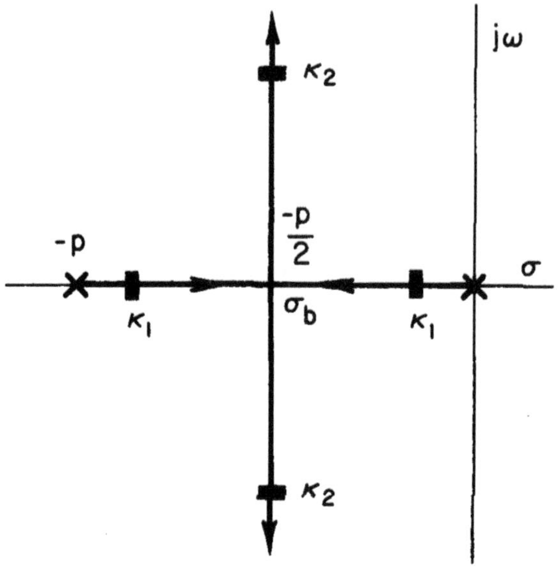


Fig. 3-9. Root locus for unit numerator second-order system $G(s) = \kappa/[s(s + p)]$.

As long as the gain, $\kappa$, is less than $p^2/4$, the roots are real, while, when $\kappa > p^2/4$, the roots are complex conjugates.

The root locus for the function $\kappa/s(s + p)$ is presented in Fig. 3-9 for $\kappa > 0$. The application of Rules 1-3, 5, and 6 is straightforward. The breakaway point coincides with the origin of the high gain asymptotes which make angles of $+90$ and $+270$ degrees with the real axis, so that Rules 4 and 7 yield the same result for this particular problem. In fact, the high gain asymptotes are the locus in this case. The breakaway condition is an example of the junction point (Rule 7), and it is seen that the tangents of both the coalescing and the departing locus branches are equally spaced over 360 degrees. It is also easy to see that, at the breakaway point, Rule 9 leads to

$$S_\kappa^b = \frac{1}{(\partial G/\partial s)_{s=-p/2}} = \left[\frac{1}{p - (p/2)} + \frac{1}{0 - (p/2)}\right]^{-1} \to \infty$$

This is an example of the blowup of the sensitivity vector at multiorder poles, a topic that will be discussed more fully later.

The application of Rule 10 (for $m \geqq 2$) to find closed-loop roots which are compatible in that they have the same open-loop gain is extremely easy for this example because the sum of the roots is $-p$ and there are only two roots present. The selection of one then immediately specifies the other (e.g., see points labeled $\kappa_1$ and $\kappa_2$ in Fig. 3-9). Specific values for the gains must be found using Eq. 3-11, although one can, of course, solve the characteristic equation at various values of gain in this elementary example. Also, the gain at breakaway will be $\kappa = p^2/4$, found, as already noted, by the condition for two equal roots.

The contour map for the function $1/s(s + p)$, with $p = 1$, is presented in Fig. 3-10. The locus of roots is identified with the $-180$-degree angle contour which lies between the poles and leaves the real axis at the breakaway point, $\sigma_b = -\frac{1}{2}$. The locus approaches the points at infinity along the two asymptotes. Points on the locus corresponding to closed-loop roots for particular values of the gain, $\kappa$, are indicated by the intersections of the magnitude contours with the locus. The values illustrated (+20 dB, +12 dB, +6 dB, 0 dB, −6 dB, −12 dB) correspond, in decibel measure, to the inverse of the gain, i.e., $\kappa = \frac{1}{10}, \frac{1}{4}, \frac{1}{2}, 1, 2, 4$.

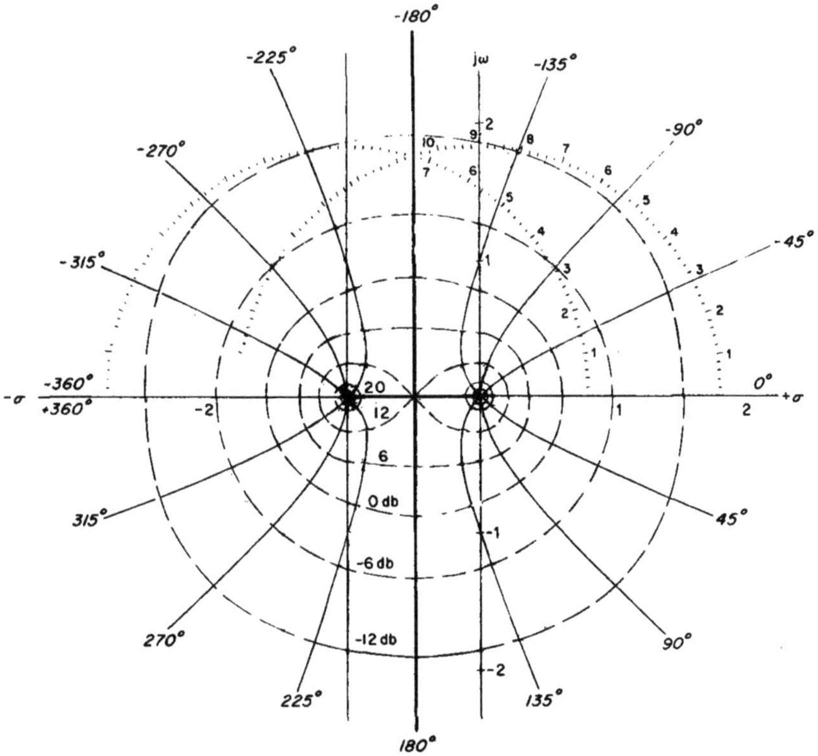


Fig. 3-10. Contour map and isoargument curves for $G(s) = 1/[s(s + 1)]$.

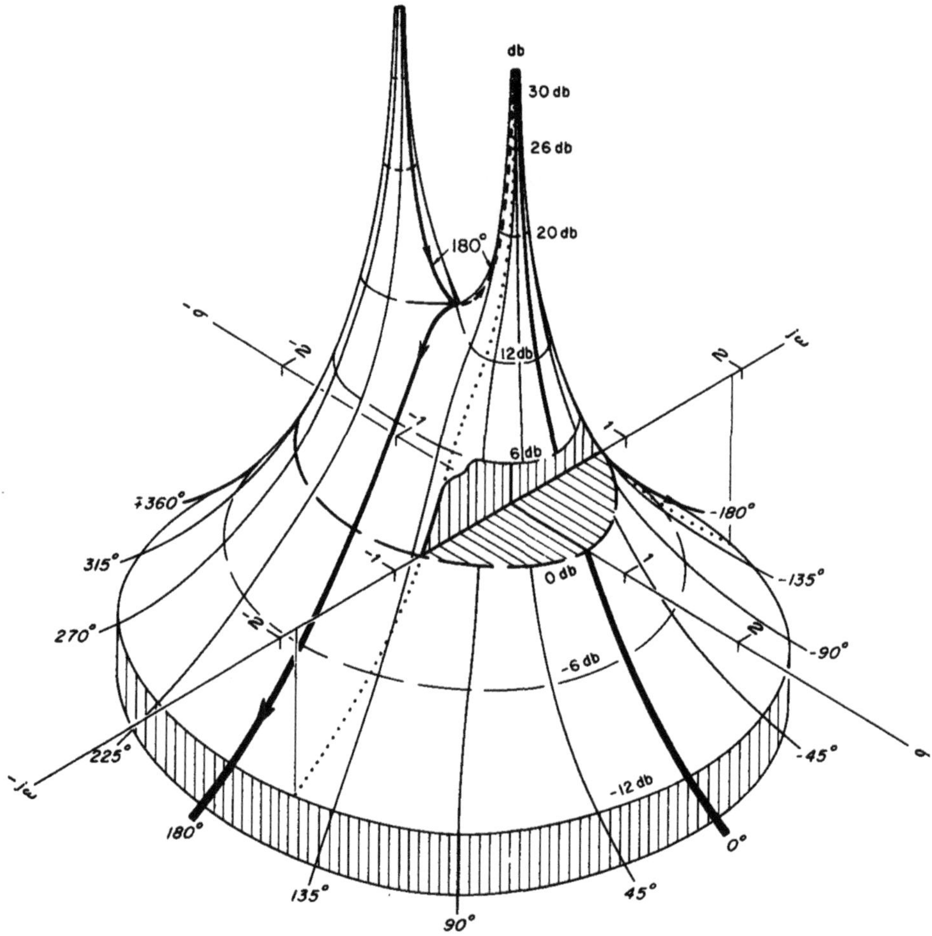


Fig. 3-11. Isometric view of the surface of $G(s) = 1/[s(s + 1)]$.

The isometric view of the surface corresponding to the transfer function $1/s(s + 1)$ is presented in Fig. 3-11. Several contours of constant angle and constant magnitude are shown on the surface. Remembering that the increase in gain from zero is analogous to starting the reference plane at infinity and moving it *down*, we can see that the closed-loop roots at the intersection of the $\pm 180$-degree angle contour and the waterline move downhill from the poles toward each other, coalesce at the saddle point, and then split apart and continue downhill.

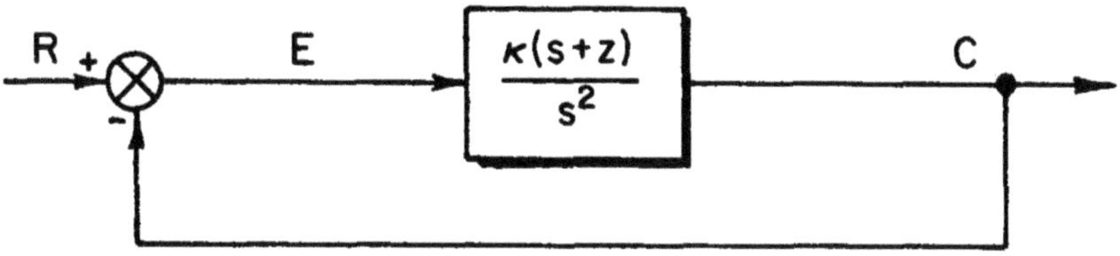


Fig. 3-12. A second-order feedback control with lead.

Example 3. Second-order system with a zero

Figure 3-12 shows another second-order system feedback controller which, in this case, contains a first-order lead. The closed-loop characteristic equation is given by

$$s^2 + \kappa s + \kappa z = 0$$

which has the roots

$$s_{1,2} = -\frac{\kappa}{2}\left[1 \pm \sqrt{1 - \frac{4z}{\kappa}}\right]$$

These will be complex for $0 < \kappa \leqq 4z$ and real for $\kappa > 4z$.

From Rules 4 and 5 the real axis to the left of the zero is seen to be on the root locus, with one of the closed-loop poles terminating on the zero as the other proceeds to infinity along the $-180$-degree (negative real axis) high gain asymptote. Rule 7, i.e.,

$$\frac{dG(s)}{ds} = -\frac{\kappa(s + 2z)}{s^3} = 0$$

indicates a rendezvous of two complex roots at $s = -2z$. The roots start their journey into the complex plane along pathways tangent to the imaginary axis, by Rule 6 or 8. As shown in Fig. 3-13, the pathway is a circle, centered at the zero $s = -z$. This can be shown analytically by considering the equation for the root locus as developed below.

The statement $G(s) = -1$ is equivalent to

$$\frac{\beta(s)}{\alpha(s)} = -\kappa \tag{3-19}$$

Because $\beta(s)/\alpha(s)$ is complex, Eq. 3-18 can be rewritten as

$$\text{Re}\left[\frac{\beta(s)}{\alpha(s)}\right] + j\,\text{Im}\left[\frac{\beta(s)}{\alpha(s)}\right] = -\kappa \tag{3-20}$$

Since $\kappa$ is real, the imaginary term must vanish on the locus of roots.

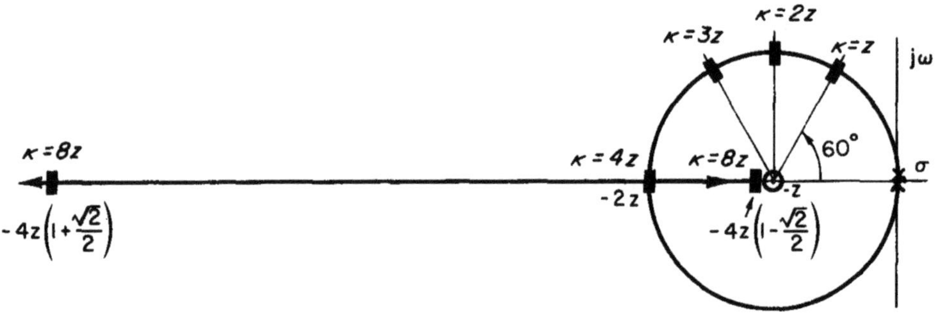


Fig. 3-13. Root locus for second-order system with lead, $G(s) = \kappa(s + z)/s^2$.

Consequently, the root locus criteria statements then become

$$\operatorname{Im}\left[\frac{\beta(s)}{\alpha(s)}\right] = 0 = \operatorname{Im}\,[\beta(s)\alpha^*(s)] \tag{3-21}$$

$$\operatorname{Re}\left[\frac{\beta(s)}{\alpha(s)}\right] = -\kappa \tag{3-22}$$

These correspond to Eqs. 3-10 and 3-11. Equation 3-21, in particular, can be very useful when equations for the root locus are desired.[7] Thus, applying Eq. 3-21 to the case at hand,

$$\begin{aligned}\operatorname{Im}\,[\beta(s)\alpha^*(s)] &= \operatorname{Im}\,[s^2(s+z)^*]\\ &= \operatorname{Im}\,[(\sigma+j\omega)^2(\sigma-j\omega+z)]\\ &= \operatorname{Im}\,[\sigma^2(\sigma+z)+\omega^2(\sigma-z)] + j\omega(\sigma^2+2\sigma z+\omega^2)\\ &= \omega[(\sigma+z)^2+\omega^2-z^2] = 0\end{aligned}$$

or

$$\omega = 0 \qquad \text{and} \qquad (\sigma+z)^2+\omega^2 = z^2$$

The first equation ($\omega = 0$) is that of the locus along the real axis, while the second is the circle, centered at $\sigma = -z$, with a radius $z$.

Figure 3-14 is the contour map for the function $G(s) = (s + z)/s^2$, with $z = 1$. It represents the *addition* of the logarithmic magnitudes and the angle contributions of *two* of the poles of Fig. 3-3(a) located at the origin and one of the zeros of Fig. 3-3(b) located at $s = -1$. Again, of course, the locus of roots is traced by the $\pm 180$-degree angle contour.

[7] V. C. M. Yeh, "Synthesis of Feedback Control Systems by Gain-Contour and Root-Contour Methods," *Trans. AIEE*, 75, Part 2, 85–95 (1956); H. Banerjee and T. J. Higgins, "Root Locus Delineations for Higher-Order Servomechanisms," Proc. National Electronic Conf., 13, 520–536 (1957).

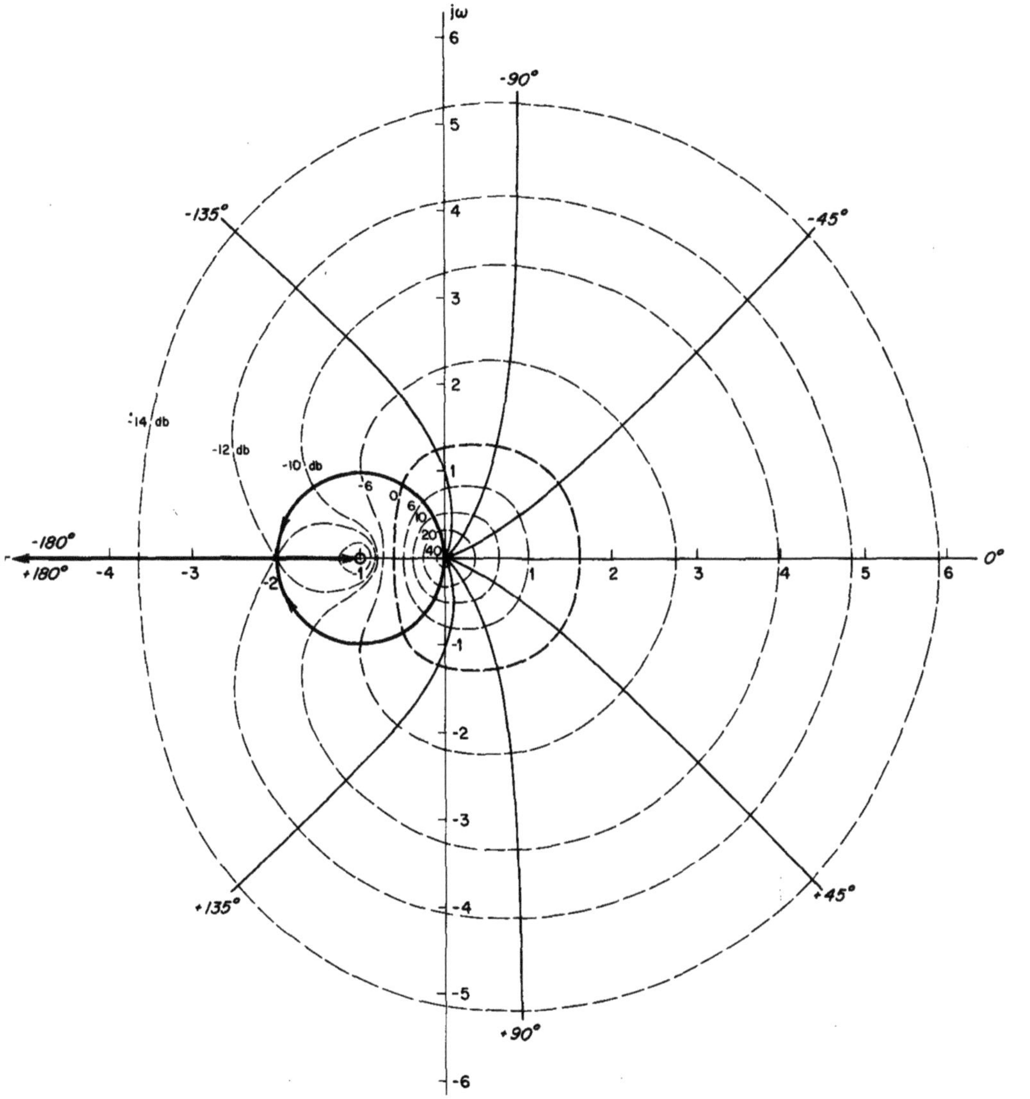


Fig. 3-14. Contour map and isoargument curves for $G(s) = (s + 1)/s^2$.

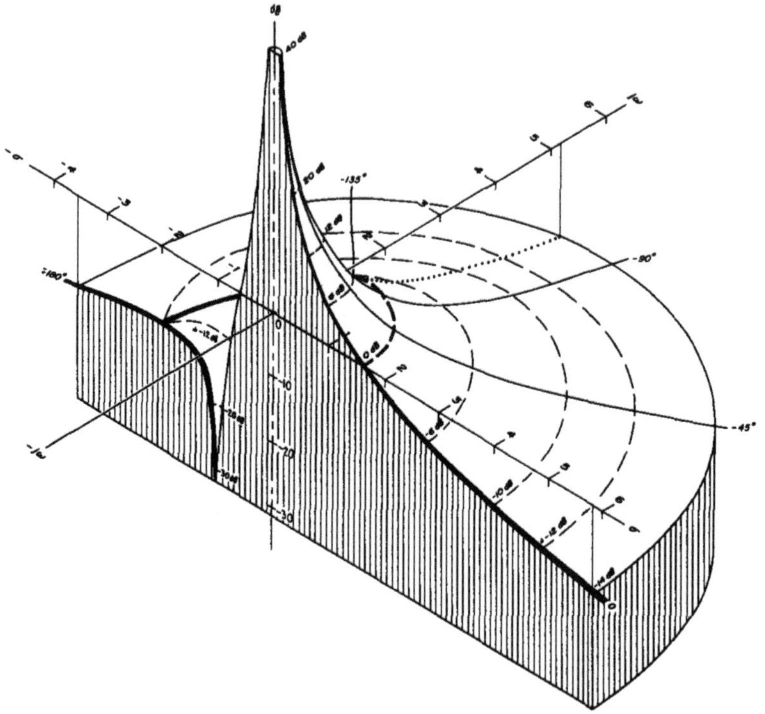

Fig. 3-15. Conventional root locus and isometric view of the surface $G(s) = (s + 1)/s^2$.

Figure 3-15 is a cutaway isometric view of the surface represented by the contour map of Fig. 3-14. The view shown emphasizes the real axis, the cavity due to the zero and the coincident twin peaks due to the poles at the origin. A conventional root locus is also given in Fig. 3-15 to show better the circular portion of the locus that is largely behind the hill on the isometric view. On these plots the change in sensitivity of the roots to gain changes is apparent. Starting with the open-loop poles at the origin with $\kappa = 0$, there is initially very little shift in the roots with gain, i.e., the slope of the peak is very steep and a large vertical movement produces but a small horizontal displacement. By the time a gain of $|\kappa|_{\text{dB}} = 0$ is reached, the locus has only progressed along an arc of 60 degrees in the plan view [see Fig. 3-15(a)]. From this point a mere factor of four (12 dB) moves the roots all the rest of the way to the rendezvous point where they coalesce. Then a very small additional increment in gain separates them very rapidly as the one travels along the asymptote to the point at infinity, while the other closed-loop root is driven toward the location of the zero. The region of high sensitivity is readily recognized in the isometric view because the surface is relatively flat there. Note again how the locus appears to run "downhill," approaching, in this case, the point at infinity and the zero.

Example 4. Unit-numerator third-order system

Figure 3-16 shows the block diagram of a third-order servomechanism.

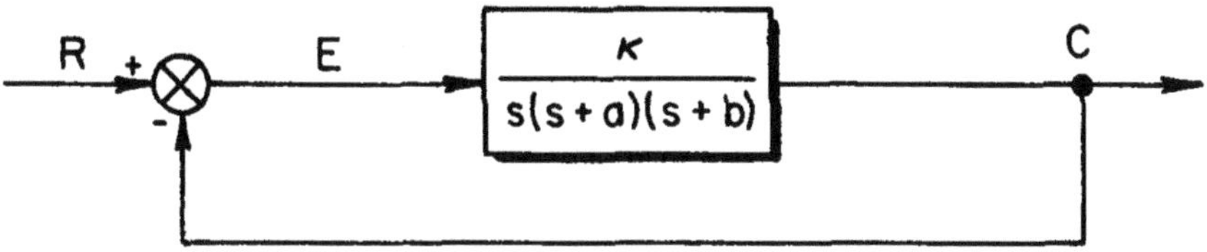


Fig. 3-16. Block diagram of a third-order servomechanism.

The closed-loop function in this case is

$$\frac{C}{R} = \frac{\kappa}{s^3 + (a + b)s^2 + abs + \kappa}$$

While, in principle, it is still possible to solve the characteristic equation algebraically, the job is now a little more difficult and, conversely, the advantages of the root locus method are now more prominent. In the case of higher order systems, the algebraic factoring of the characteristic function can only be done with repetitive numerical trials, and the semi-graphical methods really come into their own.

The root locus for the system of Fig. 3-16 is shown in Fig. 3-17. Rules 1, 3, and 5 are used to find the branches of the locus on the real axis. Rule 4 is invoked to determine the origin of the asymptotes and the angles that they make with the real axis. Either Rule 7 or Rule 9 may be used to determine the breakaway point,

$$\sigma_b = -\frac{1}{3}\left[(a + b) - \sqrt{a^2 - ab + b^2}\right]; \qquad \kappa > 0$$

Rule 8 says that here the four branches are at right angles to one another; Rule 11 shows where the locus crosses the imaginary axis, i.e., $\omega^2 = ab$.

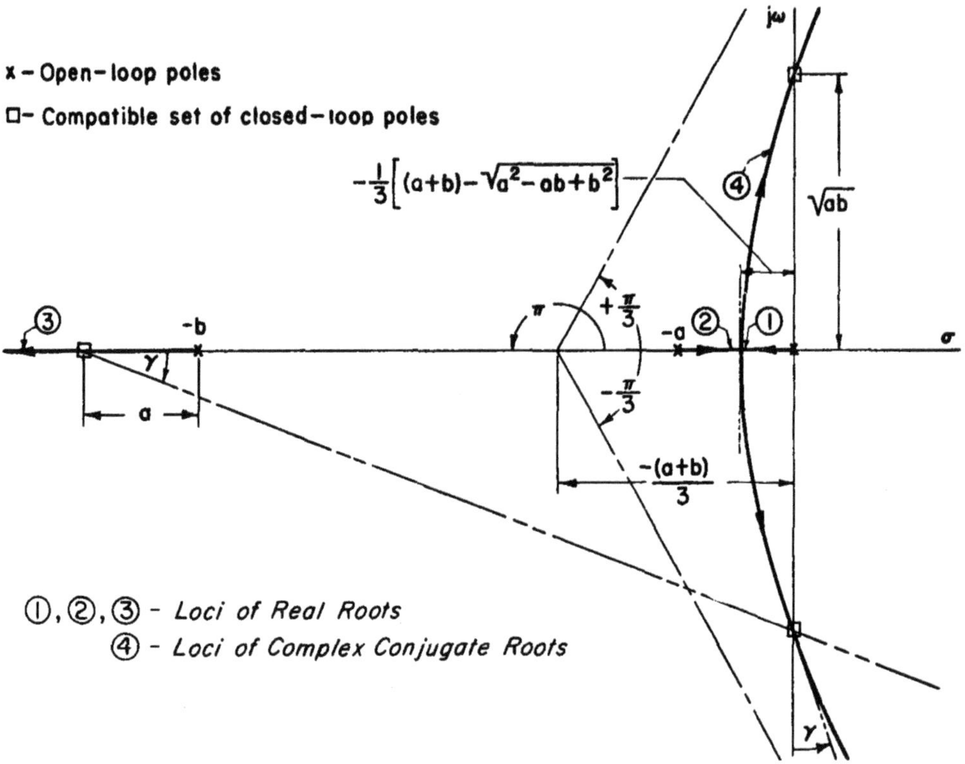


Fig. 3-17. Root locus for unit-numerator third-order system $G(s) = \kappa/s(s + a)(s + b)$.

[In this case, the Routh-Hurwitz stability criterion requires $ab(a + b) > \kappa$ for stability, so that the value of the gain at this point is also established.] Rule 10 can be used to determine the position of the third root when the gain is just sufficient to produce neutral oscillatory stability, and Rule 12 enables the angle, $\gamma$, to be found readily. To obtain the equation for the locus, Eq. 3-20 is applied:

$$\begin{aligned}\text{Im}\left[\frac{\beta(s)}{\alpha(s)}\right] &= \text{Im}\,[(j\omega + \sigma)(j\omega + \sigma + a)(j\omega + \sigma + b)] \\ &= \omega\{-\omega^2 + [3\sigma^2 + 2(a + b)\sigma + ab]\} \\ &= 0\end{aligned}$$

So the locus is given by

$$\omega = 0$$

$$\omega^2 = 3\sigma^2 + 2(a + b)\sigma + ab$$

After a few manipulations, the latter equation can be put into the conventional form for a hyperbola:

$$\left[\frac{\sigma + \frac{1}{3}(a + b)}{\frac{1}{3}\sqrt{a^2 - ab + b^2}}\right]^2 - \left[\frac{\omega}{(\sqrt{3}/3)\sqrt{a^2 - ab + b^2}}\right]^2 = 1$$

For higher order systems the locus equations become exceedingly complex, so that they are not too helpful in either the construction process or as an aid to the more analytically inclined.

With the choice $a = 1, b = 5$, the contour map of the open-loop function is shown in Fig. 3-18. The locus of roots, of course, lies along the angle contour lines corresponding to $\pm 180$ and $\pm 540$ degrees, and crosses the imaginary axis when the gain is just a little less than 30 dB. The actual gain for the onset of instability is $\kappa = 5(1 + 5) = 30 \doteq 29.5$ dB.

Figure 3-19 is an isometric view of the surface represented by the contour map of Fig. 3-18. It is particularly clear here that, as the gain is increased, the closed-loop roots do not move evenly along the locus represented in two dimensions. Instead, the closed-loop roots move a large distance along the locus for a small increment in gain, where the surface is flat; where the surface is steep, the roots move hardly at all for comparatively big increments in gain. Both this and similar observations for the last example are associated with the concept of "sensitivity," to which we shall return later.

Example 5. Generalization of the simple examples 1–3

As a final example of the root locus, the three simple systems described previously will be generalized to systems that have poles at locations other than the origin (for Example 1, a first-order lead is also added). Loci for all of these systems are given in Fig. 3-20. For these it is particularly worth remarking that the addition of the same real part to, or subtraction from, all the open-loop poles and zeros does not alter the *geometry* of the locus; it merely moves the whole locus right or left with respect to the imaginary axis.

As a final comment on the root locus method, it should be noted that the technique is useful not only to find the closed-loop roots of feedback systems but also to factor any polynomial, $\beta(s) + \kappa\alpha(s)$; $\alpha(s)$ and $\beta(s)$ do not have to represent respectively the numerator and denominator of the transfer function of a physical system, nor does $\kappa$ have to be a "gain." If $\alpha(s)$ and $\beta(s)$ are any polynomials whose factors are known,

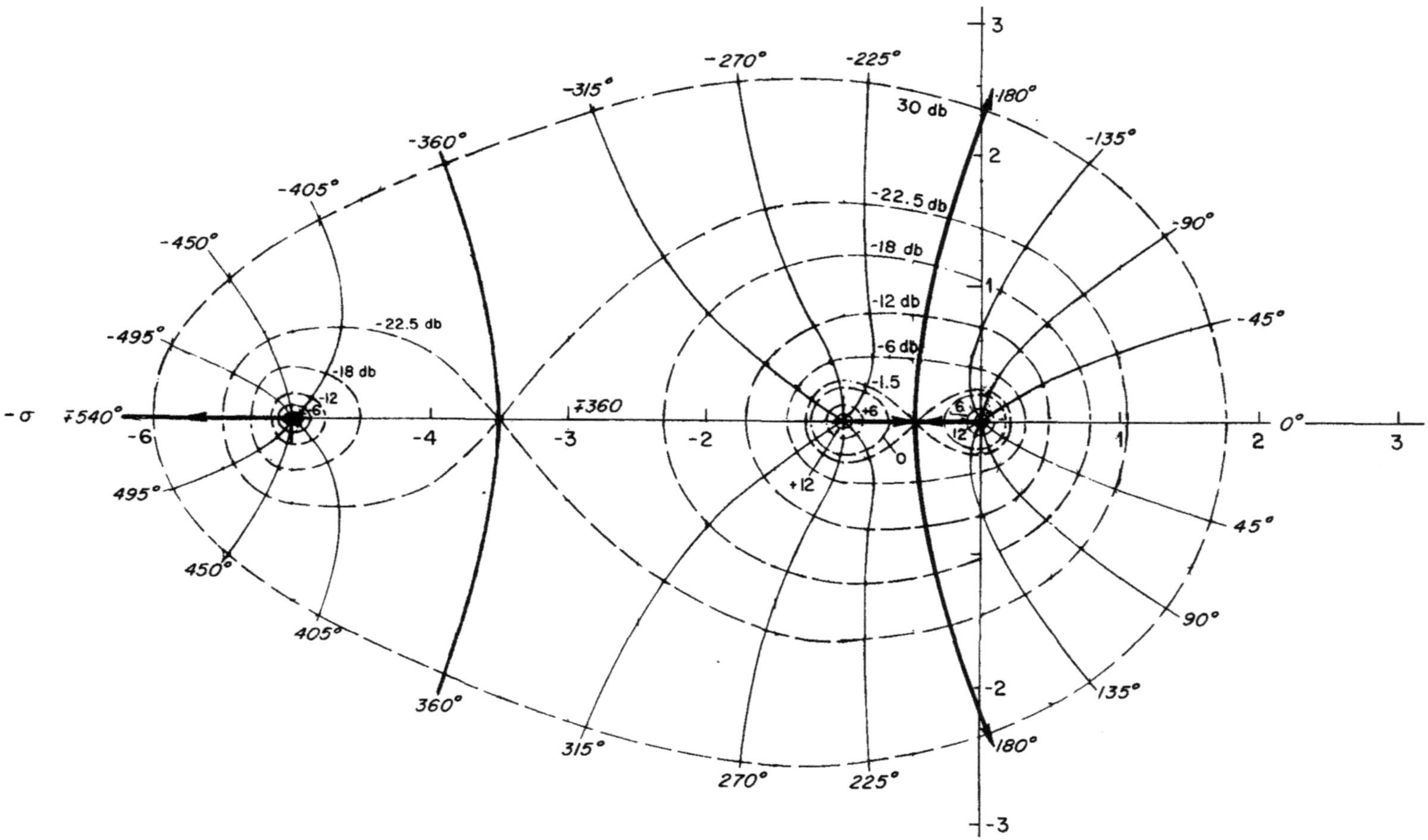


Fig. 3-18. Contour map and isoargument curves for $G(s) = \kappa/[s(s+1)(s+5)]$.

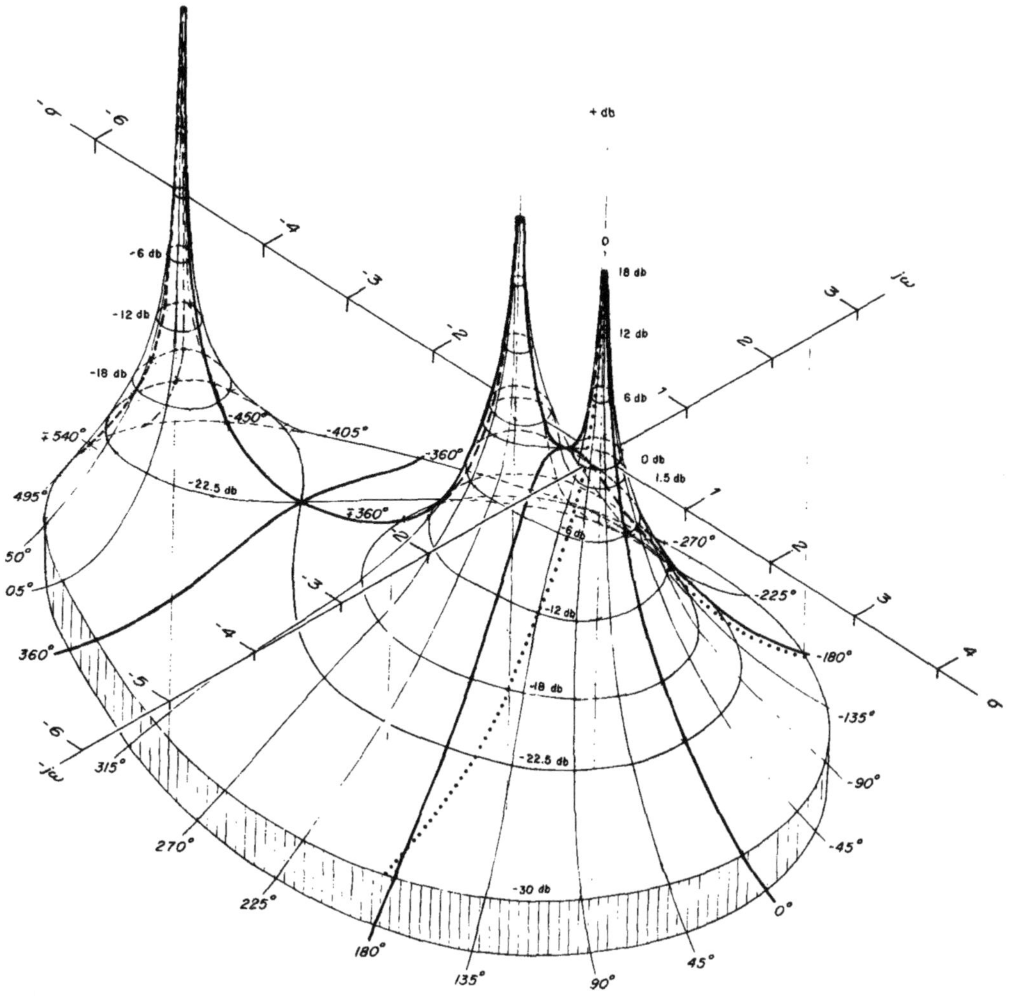


Fig. 3-19. Isometric view of the surface for $G(s) = \kappa/[s(s + 1)(s + 5)]$.

and $\kappa$ is any parameter (which appears only linearly in the complete polynomials) whose influence we wish to trace, we can put the problem into root locus form, i.e., write:

$$\beta(s) + \kappa\alpha(s) = 0$$

$$\frac{\kappa\alpha(s)}{\beta(s)} = -1 \tag{3-23}$$

$$P_c = \frac{p + Kz}{K+1}$$

O — OPEN-LOOP ZEROS

X — OPEN-LOOP POLES

□ — CLOSED-LOOP POLES

(a) System with single pole & zero $G(s) = \dfrac{K(s+z)}{(s+p)}$

For $K < \left(\dfrac{p_1 + p_2}{2}\right)^2 - p_1 p_2$ (real roots)

$$\sigma_{1,2} = -\frac{p_1 + p_2}{2} \pm \sqrt{\left(\frac{p_1 + p_2}{2}\right)^2 - p_1 p_2 - K}$$

For $K > \left(\dfrac{p_1 + p_2}{2}\right)^2 - p_1 p_2$ (complex roots)

$$\omega_c^2 = P_{c_1} P_{c_2} = p_1 p_2 + K$$

$$2\zeta_c \omega_c = P_{c_1} + P_{c_2} = p_1 + p_2$$

(b) System with two poles $G(s) = \dfrac{K}{(s+p_1)(s+p_2)}$

For $K < 4(z-p)$, (complex roots)

$$2\zeta_c \omega_c = P_{c_1} + P_{c_2} = 2p + K$$

$$\omega_c^2 = P_{c_1} P_{c_2} = p^2 + Kz$$

$$\omega_c = \sqrt{p^2 + Kz}$$

For $K > 4(z-p)$, (real roots)

$$\sigma_{1,2} = -\frac{(2p+K)}{2} \pm \sqrt{\left(\frac{2p+K}{2}\right)^2 - (p^2 + Kz)}$$

(c) System with lead & second order pole $G(s) = \dfrac{K(s+z)}{(s+p)^2}$ ; $z > p$

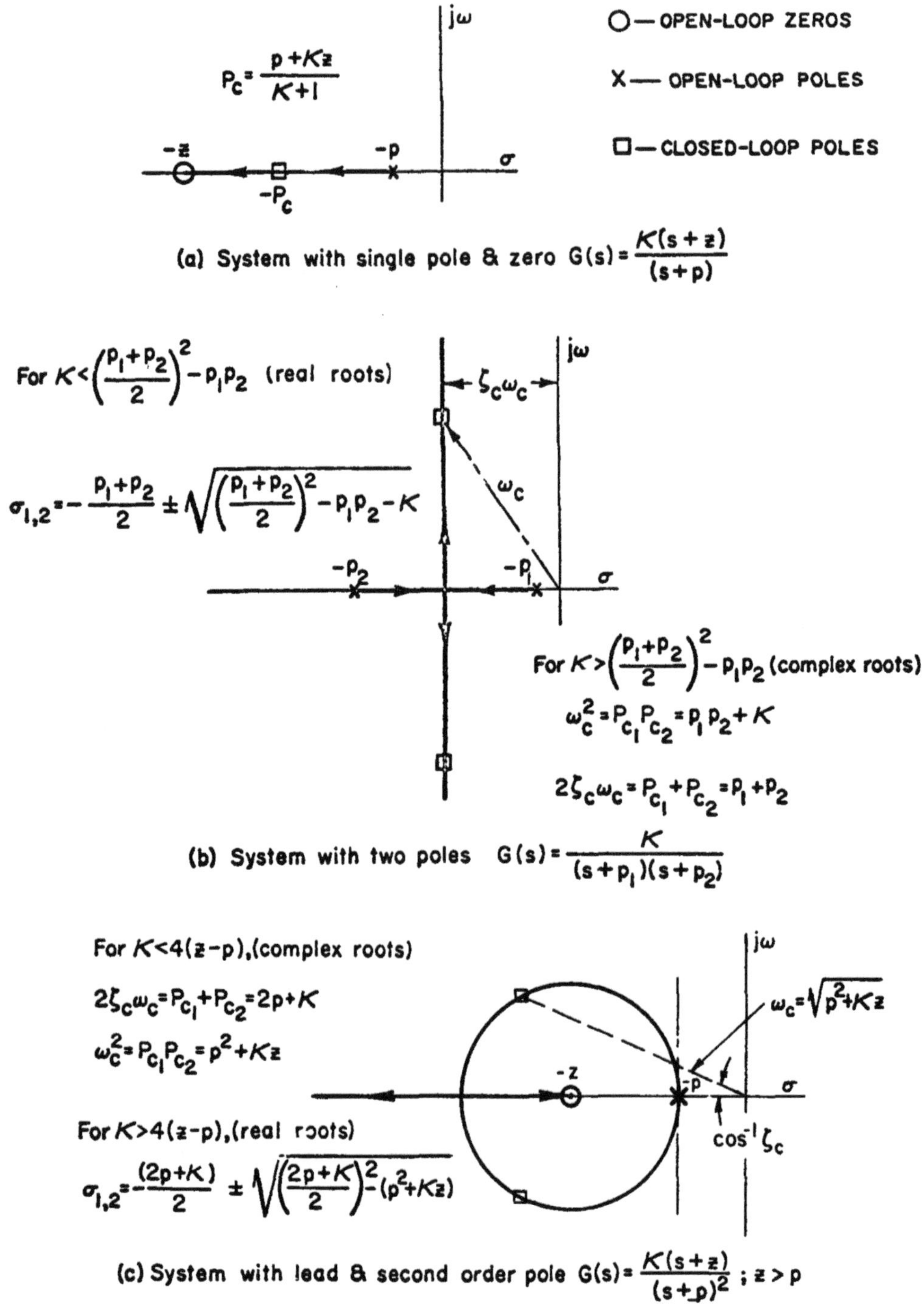

Fig. 3-20. Root loci for simple systems.

## 3-3. Bode Root Locus and Generalized Bode Diagram

We have had occasion to note that, for a given value of $s$, $G(s)$ is a complex number that can be expressed in terms of a modulus or magnitude and an argument or phase angle. The variable $s$ itself is complex $(\sigma + j\omega)$ so that a plot of $G(s)$ as a continuous function of $s$ would, in general, require four dimensions or two plots each of three dimensions, e.g., $|G(s)|$ vs. $s$ and $\measuredangle\, G(s)$ vs. $s$. The contour maps presented in the last section are plan views of such plots, whereas the isometric views directly show $|G(s)|_{\mathrm{dB}}$ vs. $s$. Unfortunately, such constructions are impractically difficult, even for moderately uncomplicated cases. Consequently, the usual practice is to construct only the root locus plan view with compatible sets of gain contour intersections marked on each branch; even this construction, which contains only the most essential information, can be tedious without automatic computation.

To obtain more complete information without undue labor, the graphical requirements can be reduced to one plot of three or two plots of two dimensions if the real and imaginary parts of $s$, $\sigma$, and $\omega$, are taken to be linearly connected. That is to say, we should find it relatively easy to plot $|G|$ and $\measuredangle\, G$ vs. $s$, as $s$ takes all values along a straight line in the $s$ plane. The simplest and most practical forms of such plots correspond to $s = \pm\sigma$; $\omega \equiv 0$ and $s = \pm j\omega$; $\sigma \equiv 0$. Sometimes, a plot with $s = \sigma_i \pm j\omega$; $\sigma_i \equiv$ a nonzero constant is also used and, finally, there is a version where $s = (\xi + j\sqrt{1 - \xi^2})u$, $\xi \equiv$ a constant. In Bode plot form these are the $\sigma$-Bode, $j\omega$-Bode, shifted Bode, and $\xi$-Bode diagrams, all of which have been described in Section 2.9.

In the analysis of closed-loop systems our objective is to find the roots of $1 + G(s) = 0$. This may be accomplished in either of two ways by means of the logarithmic Bode plots. The first, and more direct, procedure is to find the conditions under which $G(s) = -1$. The second procedure, often combined with the first, involves two steps: (a) development of a graphical representation of the closed-loop function and (b) decomposition of this closed-loop form into its zeros and poles. The poles of the closed-loop function are the roots of the characteristic equation. Both operations can be clarified by solving the same set of simple examples used to illustrate the root locus method.

Example 1. First-order system

Consider the system of Fig. 3-5 with the open-loop function $G(s) = K/s$. The surface of Fig. 3-7 is symmetric around the origin. A section in a vertical plane containing the real axis would appear as in Fig. 3-21, as would a section in a vertical plane containing the axis of imaginaries.

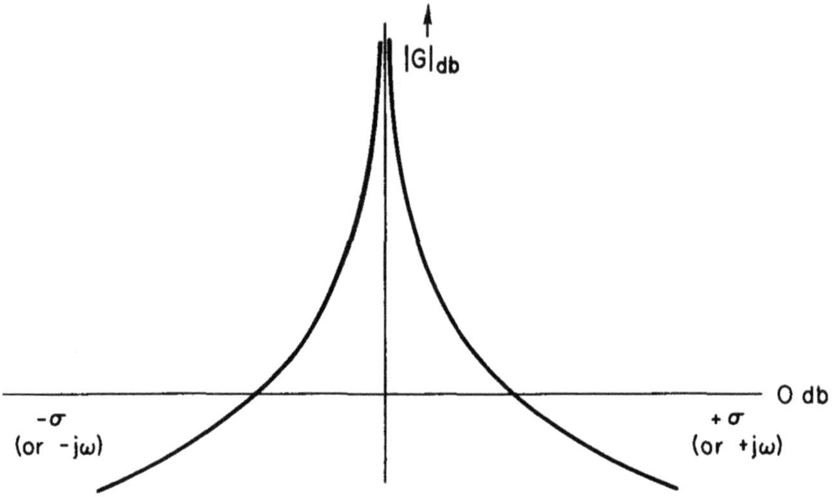


Fig. 3-21. A section containing the real axis or the axis of imaginaries.

When the abscissas are distorted so that 20 $\log_{10}|G(-\sigma)|$ is plotted against $\log_{10}|+\sigma|$, the branch of the sigma diagram illustrated in Table 2-4, Item 2, is the result. It often happens, as is the case here, that only the branch corresponding to the section containing the negative real axis, where $s = -\sigma$ so $G(s) = G(-\sigma)$, is of any interest. Because the poles and zeros of the transfer functions with which we are concerned are always real or occur in complex conjugate pairs, the $j\omega$-Bode diagram is symmetrical about the origin. The plot for negative frequencies is the reflection of the plot for positive frequencies. Therefore, it is customary to superpose the logarithmic plot of $G(s)$; $s = -\sigma$ on the logarithmic plot of $G(s)$; $s = +j\omega$. This is done in Fig. 3-22, where it is seen that not only do $|G(-\sigma|$ and $|G(j\omega)|$ have the same asymptotes, but, in this case, are themselves *identical.* In fact, the more general plot of $|G(s)|_{\text{dB}}$ vs. log $|s|$ will also coincide with the two special cases for $s = -\sigma$ and $s = j\omega$.

In the plot of Fig. 3-22 the effect of *adding* the $\sigma$-Bode or $j\omega$-Bode representation of the constant, $K$, would be to shift the whole plot up or down an amount 20 $\log_{10} K$ with respect to the reference zero-dB line, This, of course, has the same appearance as moving the zero-dB line. We can therefore identify positions of the zero-dB line with values of the constant. When the line is high, the gain constant, $K$, is small; when the zero-dB line is set low, it corresponds to high gain. When the plot is made on semilogarithmic paper, the transfer function, $K/s$, is plotted by making the line with $-20$ dB/decade slope intersect the zero-dB line where the value of the independent variable on the logarithmic scale is

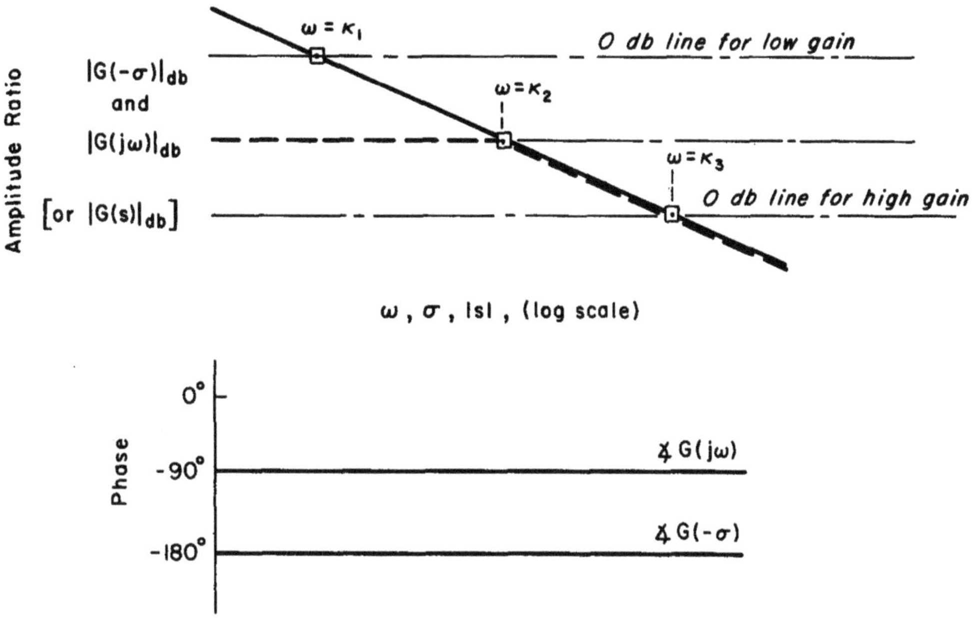


Fig. 3-22. $\sigma$-Bode and $j\omega$-Bode diagrams; $G(s) = K/s$.

numerically equal to the value of $K$. Note that, in this example, the phase angle of the $j\omega$-Bode plot is always $-90$ degrees, while the phase angle of the $G(-\sigma)$ plot is always $-180$ degrees. That this should be so has already been made evident in the contour map of Fig. 3-3(a) and the isometric view of Fig. 3-7.

While it is somewhat like using an elephant gun to kill a flea, we can use the plot of Fig. 3-22 to demonstrate the *two* methods of determining the closed-loop root(s) from the logarithmic plots. Using the direct method, it is seen that the condition $G(s) = -1$ is satisfied where the zero-dB line intersects the $|G(-\sigma)|$ plot and $\measuredangle\, G(-\sigma) = -180$ degrees. Since, in this case, the angle criterion is satisfied over the whole range of the plot, each intersection of a zero-dB line with the sigma diagram represents a closed-loop factor $(s + K)$ or, alternatively, a closed-loop root, $s = -K$. This result, of course, is precisely the same as the ones already obtained by means of algebra and the root locus method. The $\sigma$-Bode is seen, in fact, to be a root locus plot wherein the negative of the root is given as a function of gain. This is the simplest example of the so-called Bode root locus.

The decomposition method of using the logarithmic plot can be illustrated as follows:

$$G_{cr} = \frac{G(s)}{1 + G(s)} \begin{cases} \doteq 1; & G(s) \gg 1 \\ = 1; & s = 0 \end{cases}$$

$$G_{cr} = \frac{G(s)}{1 + G(s)} \doteq G(s); \quad G(s) \ll 1$$

Therefore, the asymptotes of the closed-loop function lie along the zero-dB (magnitude = 1) line when $G(s)$ is large, and along the same asymptote as the open-loop function when $G(s)$ is small. Starting from the low frequency end, where $G(s) \gg 1$, and the high frequency end, where $G(s) \ll 1$, these asymptotes can be projected toward each other. They intersect at $\omega = K$. This, then, is the asymptotic magnitude diagram of the closed-loop function $G_{cr}$. It is shown by the dashed lines in Fig. 3-22 for the gain $K_2$. This closed-loop function may be recognized as a first-order lag, Item 6 of Table 2-4, with an inverse time constant $1/T = K$. The pole of the closed-loop function, $s = -1/T = -K$, is the root of the characteristic equation, and we have discovered for the fourth time the dependence of this root on $K$.

Example 2. Unit-numerator second-order system

Figure 3-23 shows the $\sigma$-Bode and $j\omega$-Bode diagrams for the system of Fig. 3-8 with $p = 1$. Several possible positions of the zero-dB line corresponding to both low and relatively high values of the gain are marked on the diagram. Closed-loop roots on the real axis are indicated by the intersections of the zero-dB line with the sigma diagram, where $\measuredangle\, G(-\sigma) = -180$ degrees. Only the portion of the sigma diagram between $\sigma = 0$ and $\sigma = 1$ is therefore of any interest. The diagram shows two separate real roots (marked with squares) for a low value of gain. As the gain is increased, the indicated positions of these roots move toward each other and coalesce at the local minimum of the sigma diagram. As the gain is further increased, no real roots are indicated. In fact, as we know, the roots have become complex.

The second procedure is still applicable, however. The low frequency and high frequency asymptotes can be projected so as to meet at a point where there is a local change in the slope of the asymptotic approximation of from zero to $-40$ dB/decade.

The figure shows this construction in dashed lines. This closed-loop asymptotic approximation is recognized as the representation of a second-order transfer function with damping ratio $\zeta < 1$ and an undamped natural frequency, $\omega_{CL}$, indicated by the break or corner frequency. Then, since the sum of the roots is the sum of the poles, $2\zeta_{CL}\omega_{CL} = 1$, and the closed-loop damping ratio will be $\zeta_{CL} = \frac{1}{2}\omega_{CL}$.

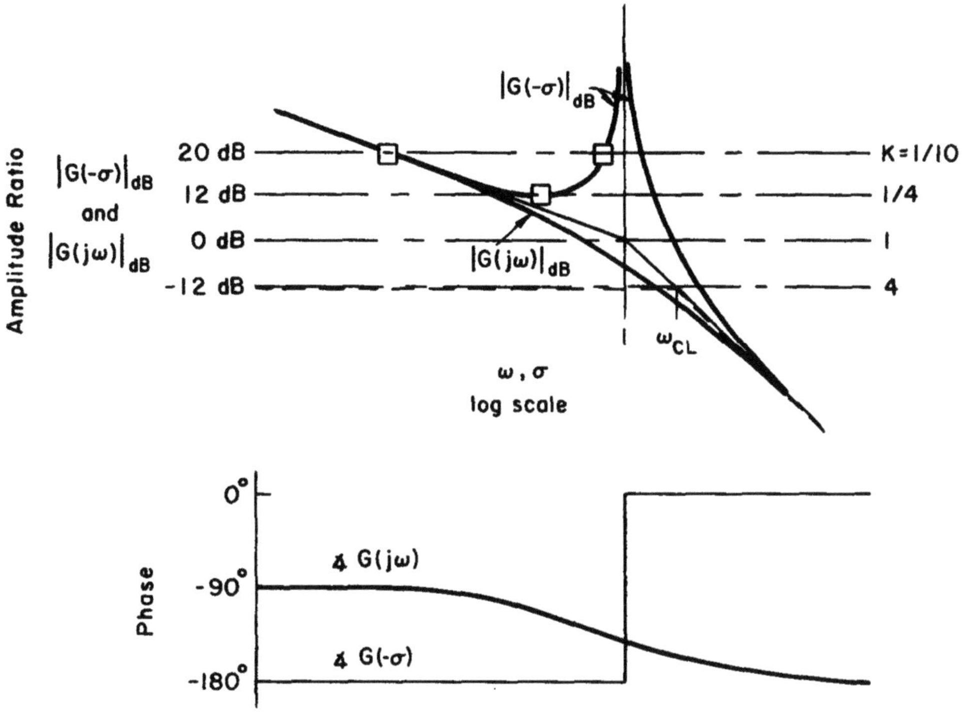


Fig. 3-23. $\sigma$-Bode and $j\omega$-Bode diagrams; $G(s) = K/s(s + 1)$.

A complete root locus, in which the magnitudes of the roots appear vs. gain in Bode diagram coordinates, is the result of the combined procedures described above. This is shown in Fig. 3-24, for the case where the pole, $-p$, is kept general. The conventional root locus of Fig. 3-9 is also given for comparison and correlation.

As in the first example, the locus of real roots coincides with that portion of the $\sigma$-Bode diagram for which $\measuredangle G(-\sigma) = -180$ degrees. These correspond to the real axis roots on the conventional root locus, with Branch 1 moving toward Branch 2 and meeting at the breakaway point, $\sigma_b$. On this part of the Bode root locus the abscissa scale used is log $\sigma$, with the closed-loop characteristics at a particular gain, $K_a$, being read as the negative of the roots, $\sigma_1$ and $\sigma_2$, i.e., as they appear in the closed-loop factors, $(s + \sigma_1)(s + \sigma_2)$.

For gains larger than that for breakaway, there are two complex conjugate roots, which have a damping ratio, $\zeta_{CL}$, and a magnitude, given by the undamped natural frequency, $\omega_{CL}$. In this elementary

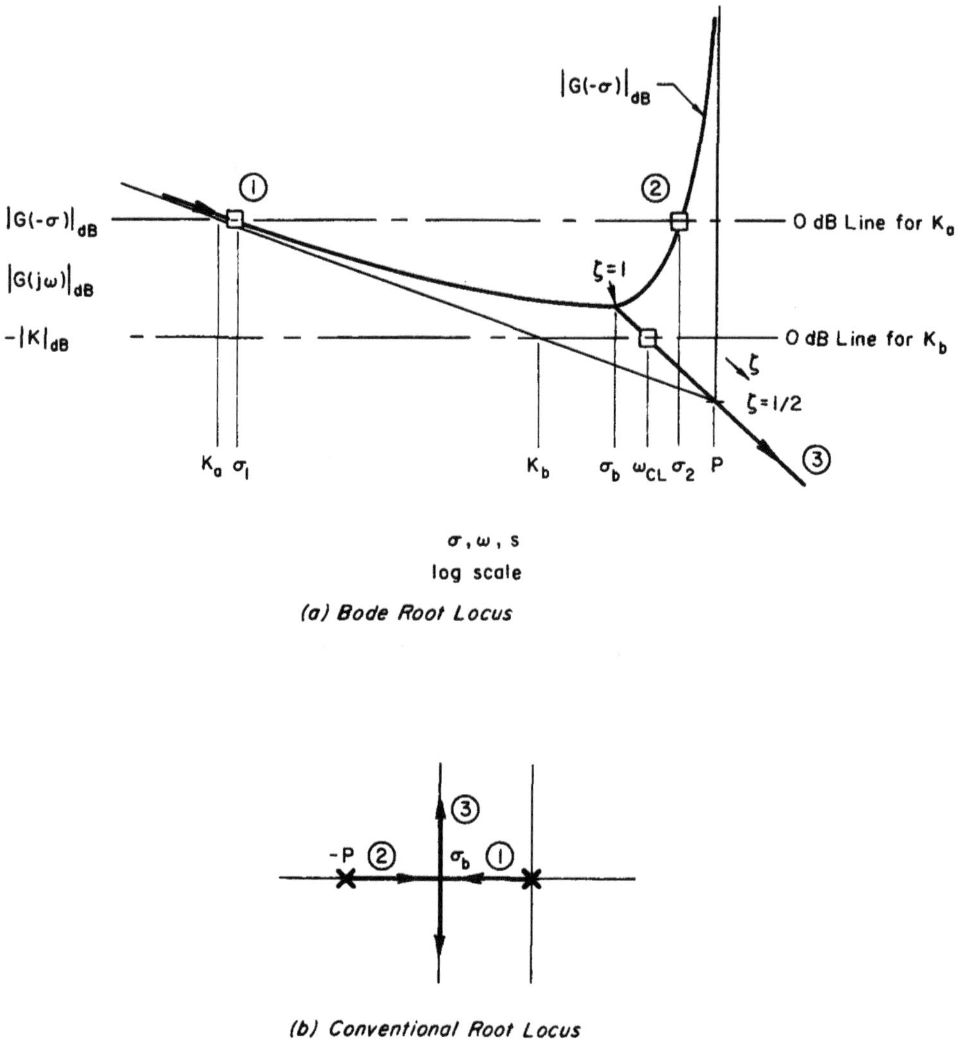


Fig. 3-24. Bode and conventional root locus for $G(s) = K/s[(s/p) + 1]$.

system the locus of $\omega_{\text{CL}}$ vs. gain coincides with the high frequency asymptote of $|G(j\omega)|_{\text{dB}}$ and its extrapolation back to the breakaway point. This branch labeled, 3, corresponds to the similarly labeled branch on the conventional root locus. For a given gain, $K_b$ (ordinate), the closed-loop undamped natural frequency, $\omega_{\text{CL}}$, is read as the abscissa (log $|s|$) of Branch 3. The damping ratio, on the other hand, cannot be considered as a dimension; instead, it must be treated as a parameter and noted along the plot. At breakaway it is, of course, identically 1, and it decreases as gain increases. For the current example it is just $\zeta_{CL} = p/2\omega_{CL}$, so it is readily obtained.

In more complicated problems than this one, the closed-loop complex roots are seldom obtained as simply as illustrated here. For these, recourse to supplementary techniques is often required. One such technique is the shifted Bode diagram, which can be readily illustrated by this example. If, in the numerical version of the example, the variable $s$ is changed to $s' = s + (\frac{1}{2})$, then the open-loop function becomes

$$G(s') = \frac{\kappa}{(s' + \frac{1}{2})(s' - \frac{1}{2})} = \frac{-K}{(2s' + 1)(-2s' + 1)}; \qquad K = 4\kappa$$

This shift in $s$ to $s'$ corresponds to a shift in the origin of one-half unit to the left. The addition of the logarithmic representations of the $j\omega$-Bode magnitude diagrams of the individual factors yields the second-order magnitude characteristic illustrated in Fig. 3-25. The negative sign of the Bode gain, $-K$, is taken into account by making the phase angle $-180$ degrees at low frequency. The phase angle contributions of the other transfer function factors exactly cancel, so that it is readily appreciated that the angle criterion is satisfied over the whole range of $\omega$. This, of course, is to be expected, since Fig. 3-25 is a representation of the section through the transfer function surface of Figs. 3-10 and 3-11 at $\sigma = -\frac{1}{2}$ where the angle is always $-180$ degrees. As for the magnitude criterion, the figure shows that there is no intersection of the zero-dB line and the magnitude plot for $K \leqq 1$, but as soon as $K$ exceeds 1, there is an intersection. The intersection moves along the actual magnitude curve (with the departures applied to the asymptotes) toward the right as the gain, $K$, is increased. A typical intersection is shown by the square. The frequency, $|\omega'|$, at which this intersection occurs is the *damped* frequency of the closed-loop root. For the case in which $K = 4$, the intersection is at $\omega = \sqrt{3}/2$ and the closed-loop factor can be written in terms of the real part, $\alpha$, and imaginary part, $\beta$, as $(s + \alpha)^2 + \beta^2 = [s + (\frac{1}{2})]^2 + (\sqrt{3}/2)^2 = 0$.

Example 3. Second-order system with a zero

The $\sigma$-Bode and $j\omega$-Bode diagrams for the system of Fig. 3-12 are presented in Fig. 3-26. The rendezvous point on the real axis is again indicated by the extremum of the $|G(-\sigma)|$ plot, where $\measuredangle G(-\sigma) = -180$ degrees. Also, the near cancellation of the zero and the closed-loop pole at high gain is shown by the closed-loop asymptotic plot which is constructed proceeding from both ends. The zero of the open-loop function is

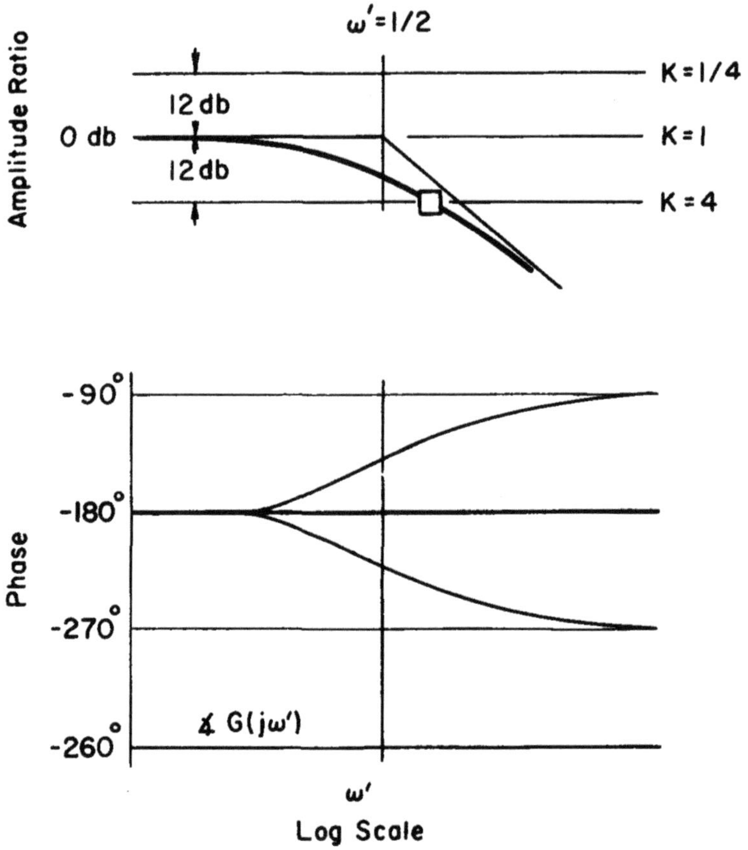


Fig. 3-25. Shifted $j\omega$-Bode diagram.

also a zero of the closed-loop function, while the roots are poles of the closed-loop function. The asymptotic approximation therefore has appropriate breakpoint corners at each of the magnitudes corresponding to these singularities.

At gains less than that for breakaway the roots are complex (i.e., along the circular part of the locus in Fig. 3-13). The magnitudes of these roots are readily found by decomposition, as exemplified in Fig. 3-27 for two values of gain, $K_1$ and $K_2$. At very small values of $\omega$ the amplitude ratio, $|G(j\omega)|$, is very large compared with unity, so that the low frequency closed-loop asymptote coincides with the open-loop, zero-dB line. This low frequency asymptote of $|G_{cr}|_{dB}$ either runs into the $|G(j\omega)|_{dB}$ asymptotes, e.g., at $\omega_{CL_1}$ for $K_1$, or reaches the magnitude of the zero, e.g., at $z$ for

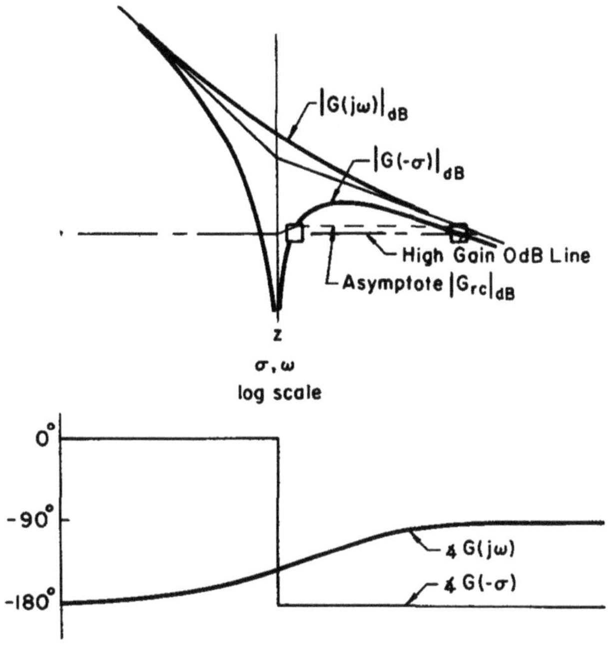


Fig. 3-26. $\sigma$-Bode and $j\omega$-Bode diagrams for $G(s) = K[(s/z) + 1]/s^2$.

$K_2$, before intersecting the asymptote $|G(j\omega)|_{\text{dB}}$ plot. In either event there is a change in slope. This is $-40$ dB/decade for the $K_1$ case, corresponding to the closed-loop undamped natural frequency $\omega_{\text{CL}_1}$, and $+20$ dB/decade for the $K_2$ example, associated with the appearance of the zero. The asymptotic $|G_{cr}|_{\text{dB}}$ plot for the low gain case has its final breakpoint at the zero, whereas the last slope change for the higher gain example occurs at the closed-loop undamped natural frequency $\omega_{\text{CL}_2}$. Finally, the damping ratio is found from the sum of the roots relationship, which in this case is

$$2\zeta_{\text{CL}}\omega_{\text{CL}} = \kappa = \frac{K}{z}$$

$$\zeta_{\text{CL}} = \frac{\kappa}{2\omega_{\text{CL}}} = \frac{K}{2z\omega_{\text{CL}}}$$

Gathering the above data together into a common presentation results in the Bode root locus of Fig. 3-28. The branches shown thereon correspond to the similarly numbered ones on the conventional root locus that is also shown. The closed-loop natural frequency is presented along Branch 1

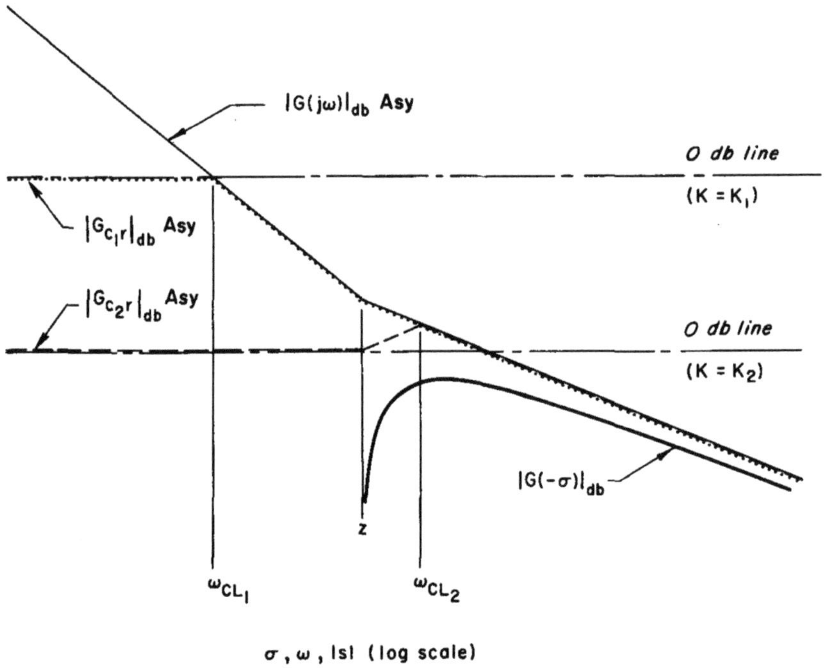


Fig. 3-27. Determination of the magnitude of complex roots by decomposition.

of the Bode root locus and is read on the logarithmic abscissa scale as $|s|$. On this branch $\zeta_{CL}$ is a parameter. At breakaway, where the gain is $|4z^2|_{dB}$, Branch 1 rendezvouses with its mirror image (with which it coincides on the Bode root locus); the two branches then depart in opposite directions along the real axis as Branches 2 and 3.

Example 4. Unit-numerator third-order system

When the $\sigma$-Bode and $j\omega$-Bode diagrams for the system of Fig. 3-16 are constructed, they appear as in Fig. 3-29. The saddle point at which the real roots coalesce and then break away, becoming complex conjugates, is recognized as the local minimum of the $|G(-\sigma)|$ diagram where $\measuredangle\, G(-\sigma) = -180$ degrees. The diagram also shows that at the value of gain at which this occurs, the third real root has hardly moved from the open-loop pole. The point at which the locus of roots crosses the axis of imaginaries is recognized as the frequency where $\measuredangle\, G(j\omega)$ is $-180$ degrees and the gain required is read off by inspection. (This corresponds to the simultaneous satisfaction of the angle and magnitude criteria in the plane containing the axis of imaginaries.)

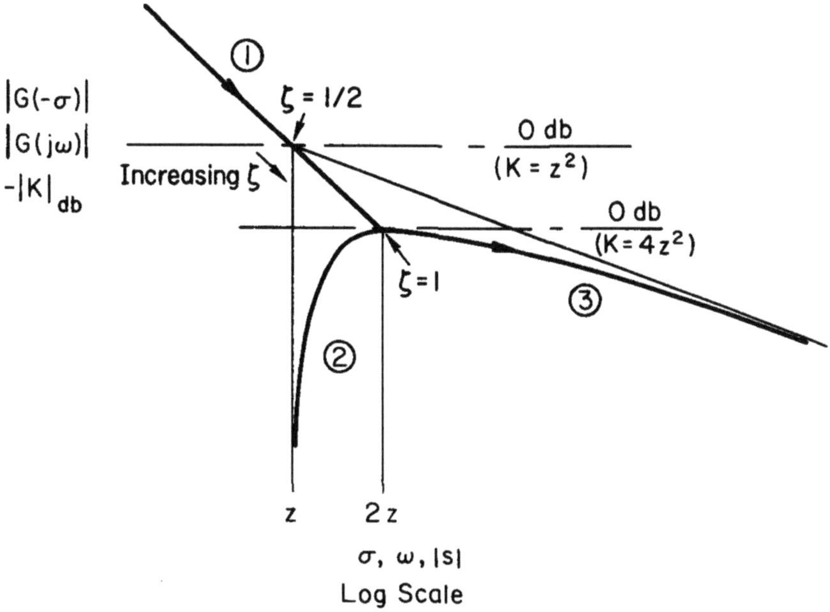


*(a) Bode Root Locus*

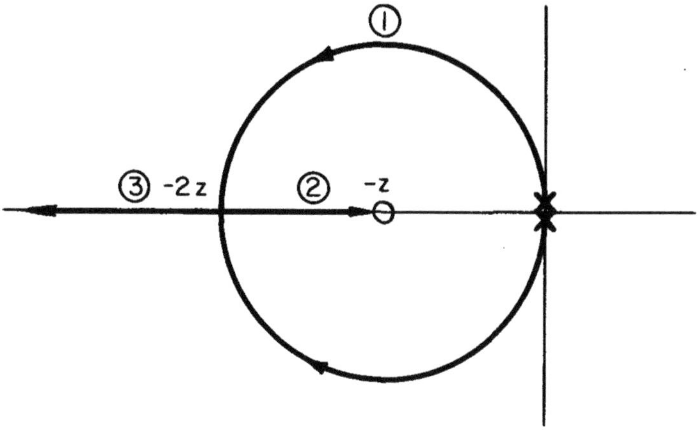


*(b) Conventional Root Locus*

Fig. 3-28. Bode and conventional root locus for $G(s) = K[(s/z) + 1]/s^2$.

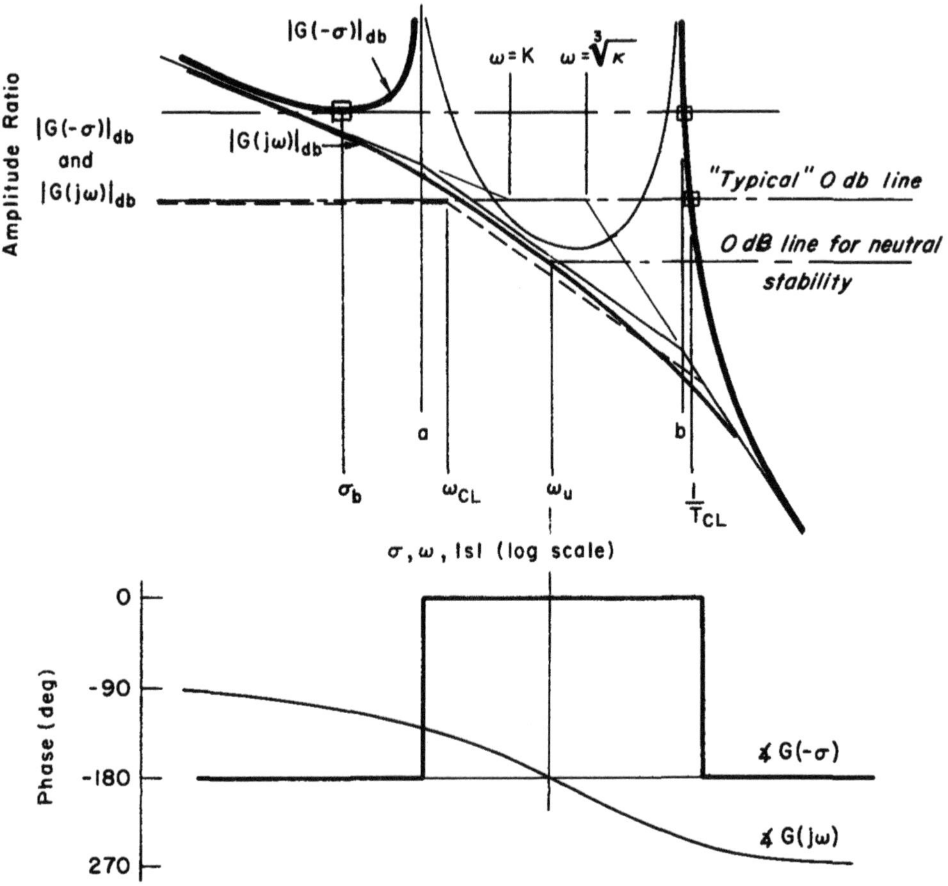


Fig. 3-29. $\sigma$-Bode and $j\omega$-Bode diagrams: $G(s) = K/s[(s/a) + 1][(s/b) + 1]$.

The $\sigma$-Bode and $j\omega$-Bode plots show without effort the breakaway points and the condition for marginal stability. These important points are usually only tediously determined by algebraic or root locus methods.

The location of the closed-loop roots for a typical value of the gain may again be determined by decomposition of the closed-loop function. The low frequency asymptote of the closed-loop function lies along the zero-dB line, and the high frequency asymptote lies along the high frequency asymptote of the open-loop function. At the frequency at which the typical zero-dB line intersects the $|G(-\sigma)|_{\text{dB}}$ curve, where $\measuredangle\, G(-\sigma) = -180$ degrees, there is a closed-loop real pole, with magnitude $1/T_{\text{CL}}$. The negative slope of the asymptotic approximation therefore changes here by −20 dB/decade. With the use of this change (decrease in slope), the asymptotic approximation can be continued back from the high frequency so as to intersect the low frequency asymptote. This intersection gives the undamped frequency of the closed-loop quadratic factor, $\omega_{CL}$. The damping ratio then may be determined from the sum of the roots formula:

$$\zeta_{CL} = \frac{1}{2\omega_{CL}}\left(a + b - \frac{1}{T_{CL}}\right)$$

Again, the closed-loop roots for a given value of gain are completely determined by simple operations on the logarithmic plots without the necessity of repetitive trial and error manipulations. When several such are combined, the Bode root locus of Fig. 3-30 results (here $a = 1$ and $b = 5$). The branches are again numbered to correspond with the conventional root locus shown in Fig. 3-17.

Now that the several examples have been worked out using both conventional and $G(s)$ logarithmic methods to find root locus plots for the closed-loop system, it should be apparent that, as a practical matter, the feedback analysis problem can be solved in either way. Each method does, however, present some difficulties when used alone. For instance,

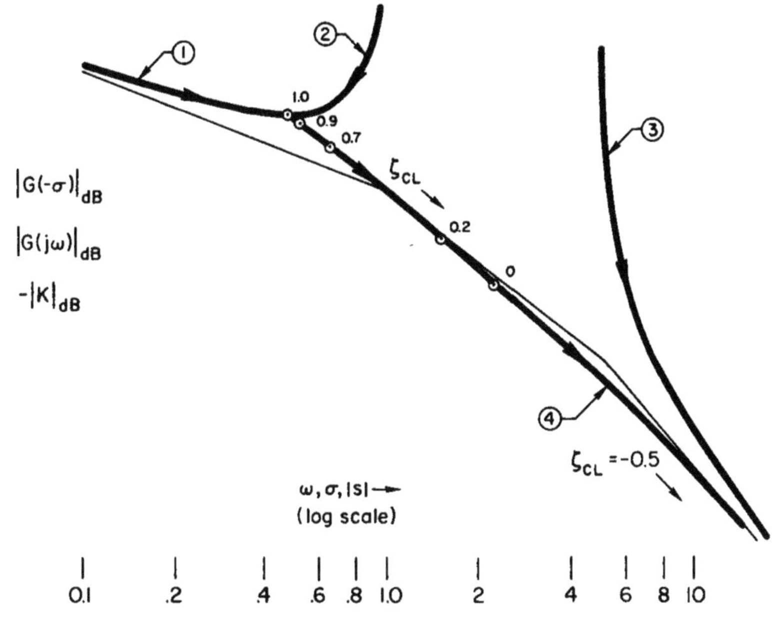


Fig. 3-30. Bode root locus for $G(s) = K/s(s + 1)[(s/5) + 1]$.

the calculation of breakaway points or of all the roots compatible with a given gain is tedious in conventional root locus; the determination of closed-loop quadratic factors using the logarithmic methods can be equally tiresome if decomposition is not completely applicable and auxiliary shifted Bodes must be constructed. Fortunately, the awkward or difficult aspect of one technique is usually a strong point of the other, so that the methods tend to be highly supplementary. Consequently, for many practical problems an intermix of techniques often provides the most effective and efficient solution. Since the best combination depends on the specifics of a given problem, an example provides the simplest way to illustrate some of the possibilities for joint use.

Example 5

Consider an open-loop transfer function given by

$$G(s) = \frac{4\{(s/7.5)^2 + [2(0.1)s/7.5] + 1\}}{s\{(s/10)^2 + [2(0.1)s/10] + 1\}\{(s/15)^2 + [2(0.1)s/15] + 1\}}$$

$$= \frac{1600[s^2 + 2(0.1)(7.5)s + (7.5)^2]}{s[s^2 + 2(0.1)(10)s + (10)^2][s^2 + 2(0.1)(15)s + (15)^2]}$$

The first step in the solution for the closed-loop roots is the construction of the asymptotic $|G(s)|_{\text{dB}}$ plot (Fig. 3-31). The departures from the asymptote of $|G(-\sigma)|_{\text{dB}}$ are then added, in the immediate region of crossover, to establish the value of the one real root, $\sigma = -4.4$. Portions of the closed-loop asymptotic plot for $|G_{\text{cr}}(s)|_{\text{dB}}$ can then be found, including:

1. The low frequency asymptote, $a$–$b$, which extends from zero to 4.4.
2. A mid-frequency asymptote with a −20 dB/decade slope, $b$–$c$, going from 4.4 to the frequency (7.5) of the complex zeros.
3. Another mid-frequency asymptote, starting at $c$ and having a slope of +20 dB/decade, reflecting the +40 dB/decade increment due to the complex zeros.
4. The high frequency asymptote, extending back from $f$ with a slope of −60 dB/decade.

At this stage only one closed-loop factor is known and only one asymptote (of the two required) is available for each of the two complex pairs remaining to be found. To find the missing intermediate asymptote, by far the simplest procedure is to construct one branch of the root locus and solve for the root when $\kappa = 1600$ ($K = 4$). This is shown in Fig. 3-32, where the high frequency closed-loop factor is found to be $s^2 - 2(0.123)(16.6)s + (16.6)^2$. The value of 16.6 for the undamped natural

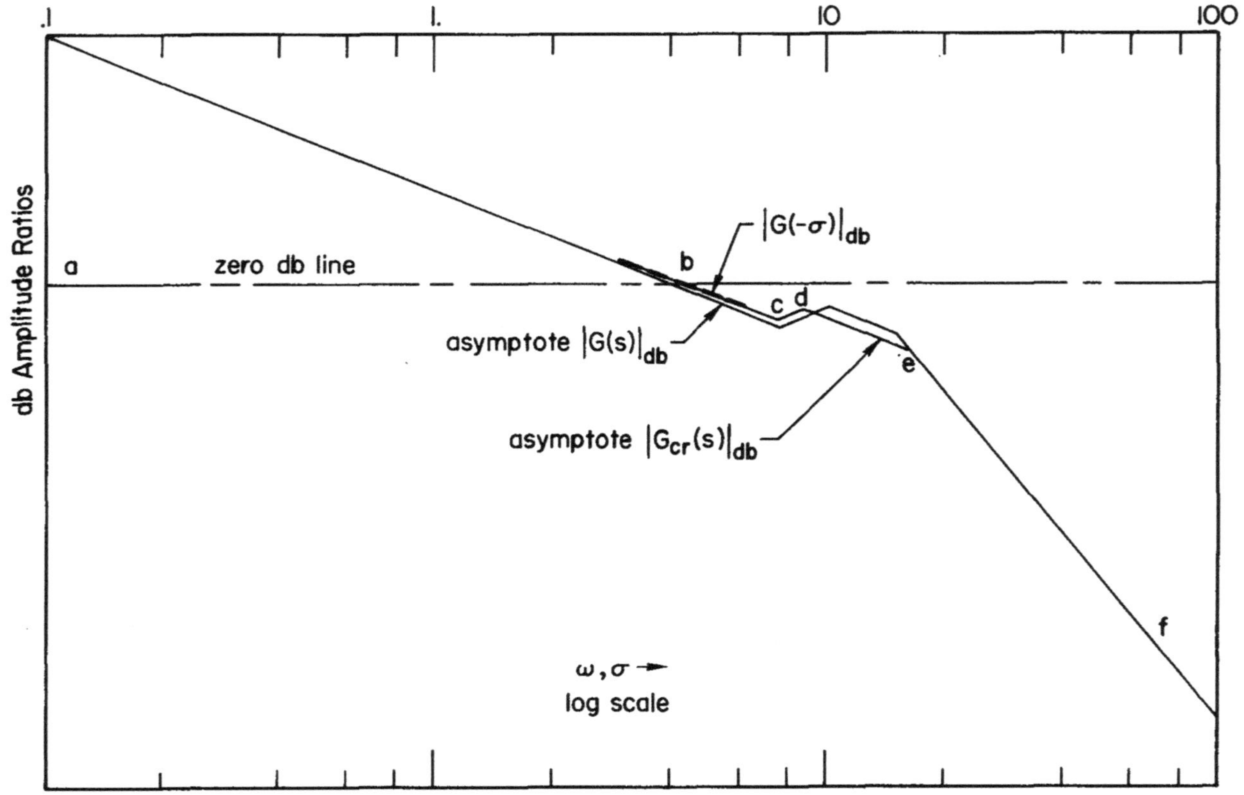


Fig. 3-31. Solution for closed-loop roots for $G(s) = \dfrac{4\{(s/7.5)^2 + [2(0.1)s/7.5] + 1\}}{s\{(s/10)^2 + [2(0.1)s/10] + 1\}\{(s/15)^2 + [2(0.1)s/15] + 1\}}$.

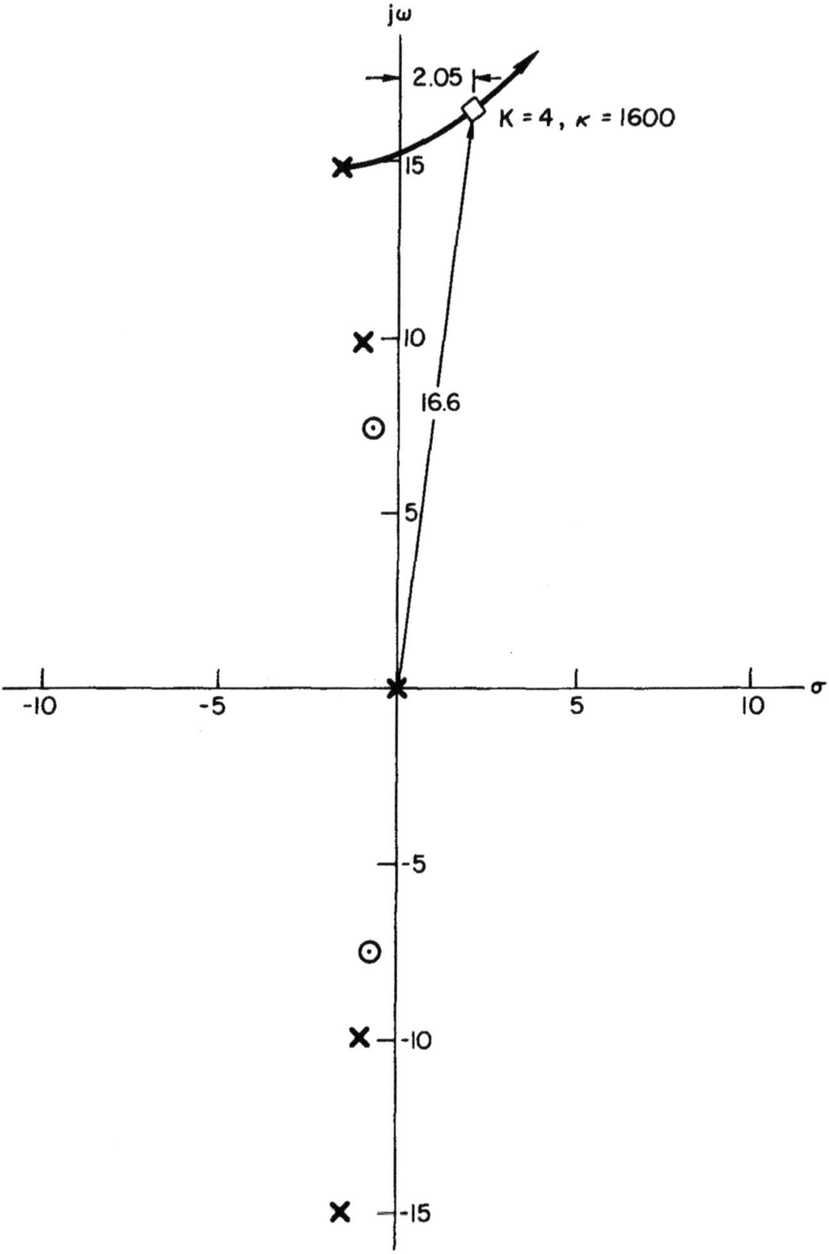


Fig. 3-32. High frequency portion of root locus for

$$G(s) = \frac{1600[s^2 + 2(0.1)(7.5)s + (7.5)^2]}{s[s^2 + 2(0.1)(10)s + (10)^2][s^2 + 2(0.1)(15)s + (15)^2]}.$$

frequency establishes the point $e$ on the high frequency asymptote of Fig. 3-31. The final intermediate asymptote is then constructed through $e$ with a slope of $-20$ dB/decade. Its intersection at $d$ with the $+20$ dB/decade asymptote from $c$ determines the value of the final undamped natural frequency, 8.8. This undamped natural frequency can also be found using the product of the roots relationship (Rule 13, Eq. 3-18). The fact that all the roots must sum to $-5$ by Rule 10 is used to determine a value for the last remaining damping ratio. Thus the final result for the closed-loop transfer function is

$$G_{cr}(s) = \frac{\{(s/7.5)^2 + [2(0.1)s/7.5] + 1\}}{[(s/4.4) + 1]\{(s/8.8)^2 + [2(0.267)s/8.8] + 1\} \times \{(s/16.6)^2 - [2(0.123)s/16.6] + 1\}}$$

The closed-loop transfer function obtained using this integrated graphical procedure compares favorably with the more precise version:

$$G_{cr}(s) = \frac{\{(s/7.5)^2 + [2(0.1)s/7.5] + 1\}}{[s/4.32 + 1]\{s/8.81)^2 + [2(0.262)s/8.81] + 1\} \times \{(s/16.39)^2 - [2(0.120)s/16.39] + 1\}}$$

obtained by factoring the characteristic equation. The numerical differences are, of course, a result solely of the graphical processes involved and are not fundamental.

This last example nicely illustrates the efficiency gained by adopting an eclectic viewpoint in which all available feedback system analysis methods are used in concert. It can hardly have escaped the reader that the ties between the conventional root locus and the $G(s)$ logarithmic plots are extremely close. These have been emphasized by the contour maps and the isometric views whose sections would amount to magnitude $G(s)$ logarithmic plots, except for the distortions accompanying the use of logarithmic abscissa scales. Such "distortion" is only a matter of convenience to achieve the very valuable asymptotic and symmetric properties exhibited by the $G(s)$ logarithmic plots; it does not constitute anything fundamental.

The common bonds revealed by the various forms of the open-loop transfer function have their origin in potential theory. First $G(s)$ may be expressed as

$$G(s) = G(\sigma + j\omega) = |G(\sigma, \omega)|\, e^{j\varphi(\sigma,\omega)} = U(\sigma, \omega) + jV(\sigma, \omega) \qquad (3\text{-}24)$$

where

$$|G(\sigma, \omega)|^2 = U^2(\sigma, \omega) + V^2(\sigma, \omega)$$

$$\varphi(\sigma, \omega) = \tan^{-1} \frac{V(\sigma, \omega)}{U(\sigma, \omega)}$$

and $\ln G(s)$ is given by

$$\ln G(s) = \ln |G(\sigma, \omega)| + j\varphi(\sigma, \omega) \tag{3-25}$$

Here $G(s)$ is an analytic function for all values of $s$ except those that correspond to poles and zeros. Consequently, $G(s)$ and $\ln G(s)$, or their real and imaginary components in Eqs. 3-24 and 3-25, will obey Laplace's equation in the two variables $\sigma$ and $\omega$ in all regions of the $s$-plane devoid of singularities. Thus,

$$\nabla^2 U(\sigma, \omega) = \frac{\partial^2}{\partial \sigma^2} U(\sigma, \omega) + \frac{\partial^2}{\partial \omega^2} U(\sigma, \omega) = 0$$

$$\nabla^2 V(\sigma, \omega) = 0$$

$$\nabla^2 \ln |G(\sigma, \omega)| = 0$$

$$\nabla^2 \varphi(\sigma, \omega) = 0$$

While $\ln |G(\sigma, \omega)|$ obeys Laplace's equation, $|G(\sigma, \omega)|$ and $e^{j\varphi(\sigma,\omega)}$ do not. This points out the theoretical basis for the use of logarithmic magnitudes in the isometric plots. The fact that Laplace's equation also describes a wide variety of physical phenomena suggests that physical analogies be used to help delineate the connections between the various forms of transfer function representations.[8]

As one such analogy, consider the $s$ plane as an infinite sheet of uniformly conducting resistive material. The open-loop poles and zeros can then be represented as point sources and sinks of current, placed at the pole and zero locations, with strength proportional to the order of the pole or zero. At any point $s$ on the sheet the potential $\Phi(s)$, measured with respect to the potential $\Phi(s_0)$ existing at some reference point $s_0$, will be

$$\Phi(s) \propto \ln \frac{G(s)}{G(s_0)} \tag{3-26}$$

Similarly, the current, $i$, flowing across a path between the two points will be

$$i \propto \varphi(s) - \varphi(s_0) \tag{3-27}$$

[8] V. C. M. Yeh, "Synthesis of Feedback Control Systems by Gain-Contour and Root-Contour Methods," *Trans. AIEE*, 75, Part 2, 85–95 (1956); P. J. Daniell, *Analogy Between the Interdependence of Phase-Shift and Gain in a Network and the Interdependence of Potential and Current Flow in a Conducting Sheet*, Ref. B39, Ministry of Supply Servo Library, 1942; A. R. Boothroyd and J. H. Westcott, "The Application of the Electrolytic Tank to Servo-mechanism Design," in *Automatic and Manual Control*, A. Tustin, ed., Butterworths, London, 1952, pp. 87–103; M. W. Fossier and H. A. Rosen, "A Field-Mapping Method for Analysis and Synthesis of Linear Closed-Loop Systems; "*J. IAS*, 20, 205–209 (Mar. 1953). H. S. Tsien, *Engineering Cybernetics*, McGraw-Hill, New York, 1954, pp. 46–58.

so that the lines of constant current in the $s$ plane correspond to lines of constant phase.

By applying this analogy to the elementary third-order system, the three poles would be represented as unit sources of current located at $s = 0$, $-a$, and $-b$, as shown in Fig. 3-33. This figure also shows the results that would be obtained by measuring the potential along the $\sigma$ and $j\omega$ axes, and along the line $s = (\xi + j\sqrt{1 - \xi^2})\mu$. From Eq. 3-26 it is apparent that these potential functions are proportional to the open-loop logarithmic amplitude ratio plots for $s = \sigma, j\omega$, and $(-\xi + j\sqrt{1 - \xi^2})\,|s|$, except that the abscissa is in linear rather than logarithmic units. They correspond directly to sections through the surface of Fig. 3-19. Lines of constant potential form the constant gain lines on the contour map of Fig. 3-18; the isoargument curves, or lines of constant phase, correspond to constant current flow lines.[9]

## 3-4. Simplified System Characteristics and Literal Approximate Factors

In almost all flight control problems the open-loop transfer functions are of very high order, with $m + n$ seldom less than 4, more often of the order of 10 and occasionally as large as 20 or 30. The techniques described in the previous sections still apply and, in fact, are in everyday use in the analytical design of flight control systems. But inevitably the price of complexity in analyses is a reduction in the physical appreciation of the essential nature of a problem, and an accompanying diminished insight into potential solutions. Fortunately, there are two counters available. These are the concept of the simplified or equivalent system and that of literal approximate factors. Both concepts can be converted into practical reality by the application of the feedback analysis techniques summarized above.

A simplified system is, in essence, a lower order approximation to a higher order system that is valid for specifiable conditions. Literal approximate factors are approximate expressions for transfer function poles and zeros in terms of the basic parameters of the system, defined only as symbols rather than as numbers. Their most important application in flight control is for aircraft transfer functions, for which the approximate factors are expressions that relate the poles and zeros with the stability derivatives and inertial properties of the vehicle. Typical examples of

[9] Similar analogies can be made with fluid dynamic, gravitational, magnetostatic, elastic, or electrostatic potential problems. For two-dimensional irrotational flow of an incompressible fluid, for example, the poles and zeros are again sources and sinks, $\ln |G(\sigma, \omega)|$ is the potential function and $\varphi(\sigma, \omega)$ is the stream function. Phase loci become streamlines, the root locus is the one-half streamline.

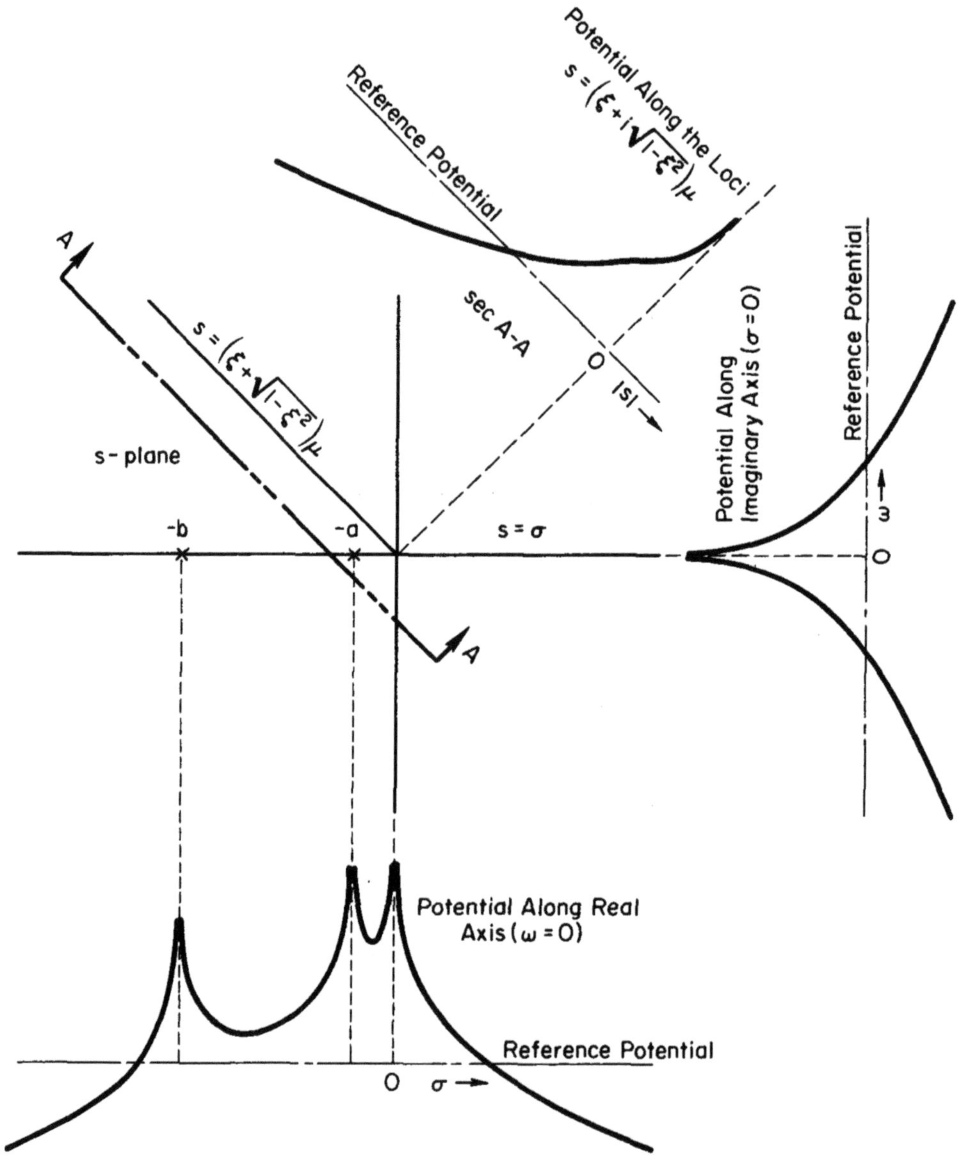


Fig. 3-33. Potential surveys along various lines in the $s$ plane.

simplified systems and approximate factors will be described below to introduce the key notions; many other examples will follow throughout this book.

*Simplified or Equivalent Systems*

As long ago as 1939 the feedback systems analysis pioneer, H. W. Bode, considered two central concepts in his mathematical definition of feedback. The first was the idea of a loop transmission, or return of output to input measured by a "return difference"; the second was "sensitivity," or the effective reduction of open-loop system variations when seen in a closed-loop context.

On the surface, these two concepts might be considered as a cause-effect pair, but both are equally fundamental in either an analytical or physical sense. In fact, Bode's "mathematical definition of feedback" [10] had the two entities the inverse of one another; subsequent writers have redefined "sensitivity" so that, as a physical measure, it is identical to the "return difference." The point of bringing this up is not to give a history of feedback system definitions, but instead to focus attention on sensitivity as a fundamental concept inseparable from feedback systems. In a gross sense, if there is no reduction in the effects of open-loop system variations on closed-loop behavior, there is no feedback worthy of mention; whereas large feedback can reduce such effects to negligible levels. This limiting case of large feedback leads directly to the equivalent system concept, for, if changes in certain open-loop parameters have no appreciable effect on closed-loop behavior, the open-loop system can be replaced by simpler, albeit approximate, descriptions that yield substantially the same closed-loop results.

To make these remarks more concrete, consider the elementary feedback system shown in Fig. 3-34.

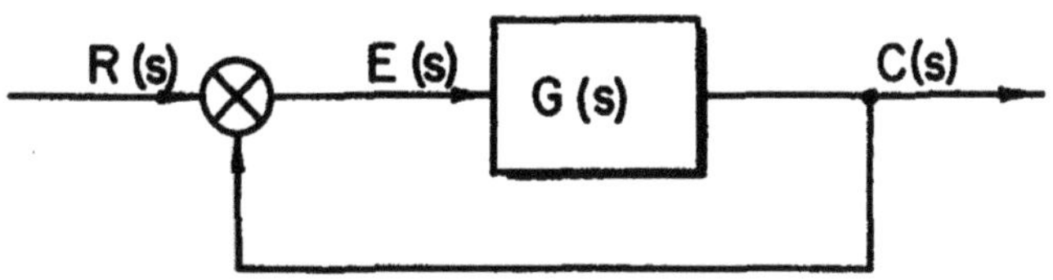


Fig. 3-34. Elementary feedback system.

[10] H. W. Bode, *Network Analysis and Feedback Amplifier Design*, Van Nostrand, New York, 1945.

The closed-loop transfer function, $G_{cr}(s)$, is given in terms of the open-loop transfer function by

$$G_{cr}(s) = \frac{C(s)}{R(s)} = \frac{G(s)}{1 + G(s)} \tag{3-28}$$

The classical sensitivity function, $S_G^{G_{cr}}(s)$, which measures the relative effects of open- and closed-loop changes, is [where $G_{cr}(s)$ and $G(s)$ are analytic]:

$$S_G^{G_{cr}}(s) = \left\{\frac{[dG_{cr}(s)/G_{cr}(s)]}{dG/G}\right\} = \frac{1}{1 + G(s)} \tag{3-29}$$

It provides a comparison between the relative change in closed-loop characteristics and the causative relative change in open-loop characteristics. Now, when feedback is really operable, sensitivity, from the discussion in the paragraph above, must be very small, i.e., the percentage change in closed-loop characteristics must be much less than the percentage change in open-loop characteristic. This implies that $G(s)$ is very large, which also implies that

$$G_{cr}(s) \doteq 1, \qquad |G(s)| \gg 1 \tag{3-30}$$

*When this condition applies, the closed-loop $G_{cr}(s)$ is insensitive to the* precise form of $G(s)$ in the region of $s$ where $|G(s)| \gg 1$, so that the actual $G(s)$ could conceivably be replaced by a simpler form in this region. On the other hand, when $|G(s)|$ is not large relative to one, the sensitivity approaches one, $G_{cr}(s)$ approaches $G(s)$, and such a replacement cannot be made. All of these esoteric remarks amount to the obvious, i.e., that

$$G_{cr}(s)\begin{cases} \doteq 1 & \text{when } s \text{ is such that } |G(s)| \gg 1 \\ = \dfrac{G(s)}{1 + G(s)} & \text{when } s \text{ is such that } |G(s)| = O(1) \\ \doteq G(s) & \text{when } s \text{ is such that } |G(s)| \ll 1 \end{cases} \tag{3-31}$$

Viewed another way, the relationships of Eq. 3-30 essentially partition the $s$ domain into regions where $G(s)$ needs to be specified with little accuracy ($|G(s)| \gg 1$), or fairly good accuracy [$|G(s)| = O(1)$ or less] if the closed-loop $G(s)$ is to be known to a reasonable approximation.

This function of feedback can be used to reduce the analytical complexity of practical problems. The basic idea is to replace the actual controlled-element transfer function (or equation) with a far simpler approximate transfer function (or set of equations) that would yield approximately the same closed-loop results. The simpler or equivalent system would then be suitable for use in many calculations.

To illustrate the procedure for equivalent system evolution, consider the concrete example of a high performance pitch attitude autopilot.

As will be more thoroughly described in later chapters, the vehicle transfer function and a sensor-controller transfer function might be described by the following open-loop transfer function:

$$G(s) = K\underbrace{\left\{\frac{(T_{\theta_1}s + 1)(T_{\theta_2}s + 1)}{\underbrace{[(s/\omega_p)^2 + (2\zeta_p s/\omega_p) + 1]}_{\text{phugoid}}\underbrace{[(s/\omega_{sp})^2 + (2\zeta_{sp}s/\omega_{sp}) + 1]}_{\text{short period}}}\right\}}_{\text{vehicle dynamics}}$$

$$\underbrace{\times\left\{\frac{\overbrace{(T_E s + 1)}^{\text{equalization}}}{\underbrace{[(s/\omega_\sigma)^2 + 2\zeta_\sigma s/\omega_\sigma + 1]}_{\text{sensor and servo actuator}}}\right\}}_{\text{controller}} \qquad (3\text{-}32)$$

Open-loop $G(j\omega)$ and $G(-\sigma)$ Bode diagrams of this system are shown in Fig. 3-35, with values for equalization, $1/T_E$, and open-loop gain set within ranges compatible with a good system. The complete open-loop system amounts to a controlled element (vehicle plus controller) equation of order six. By the direct decomposition technique of unified servo analysis, the closed-loop transfer function will be

$$G_{cr}(s) = \frac{K}{1 + K}\,\frac{(T_{\theta_1}s + 1)(T_{\theta_2}s + 1)(T_E s + 1)}{(T'_{\theta_1}s + 1)(T'_{\theta_2}s + 1)(T'_E s + 1)(T_c s + 1) \times [(s/\omega'_\sigma)^2 + (2\zeta'_\sigma s/\omega'_\sigma) + 1]} \qquad (3\text{-}33)$$

For the condition shown, the closed-loop denominator factor $(T'_{\theta_1}s + 1)$ is very close to the numerator factor $(T_{\theta_1}s + 1)$.

If $K$ is much greater than one, the amplitude ratio in the entire frequency region to the left of about $1/T_{\theta_2}$ will be much greater than one (and, in particular, $|G(j\omega_p)| \gg 1$). This suggests that the open-loop system could be approximated by the transfer function

$$G_1(s) \doteq \frac{KT_{\theta_1}\omega_p^2(T_{\theta_2}s + 1)(T_E s + 1)}{s[(s/\omega_{sp})^2 + (2\zeta_{sp}s/\omega_{sp}) + 1][(s/\omega_\sigma)^2 + (2\zeta_\sigma s/\omega_\sigma) + 1]} \qquad (3\text{-}34)$$

The closed-loop transfer function formed from this equivalent system, $|G_1/(1 + G_1)|$, will be almost identical to the exact closed-loop transfer function given by Eq. 3-33. The major difference will be that the closed-loop d-c gain is 1 instead of $K/(1 + K)$, and the nearly canceling dipole pair, $(T_{\theta_1}s + 1)/(T'_{\theta_1}s + 1)$, will not appear at all. When $|K|_{\text{dB}}$ is 20 dB or so these effects are trivial. The equivalent system is of order 5, which is one step in the right direction.

A further step can be taken by noting that the complete closed-loop system denominator factors $(T'_{\theta_2}s + 1)$ and $(T'_E s + 1)$ are not far removed

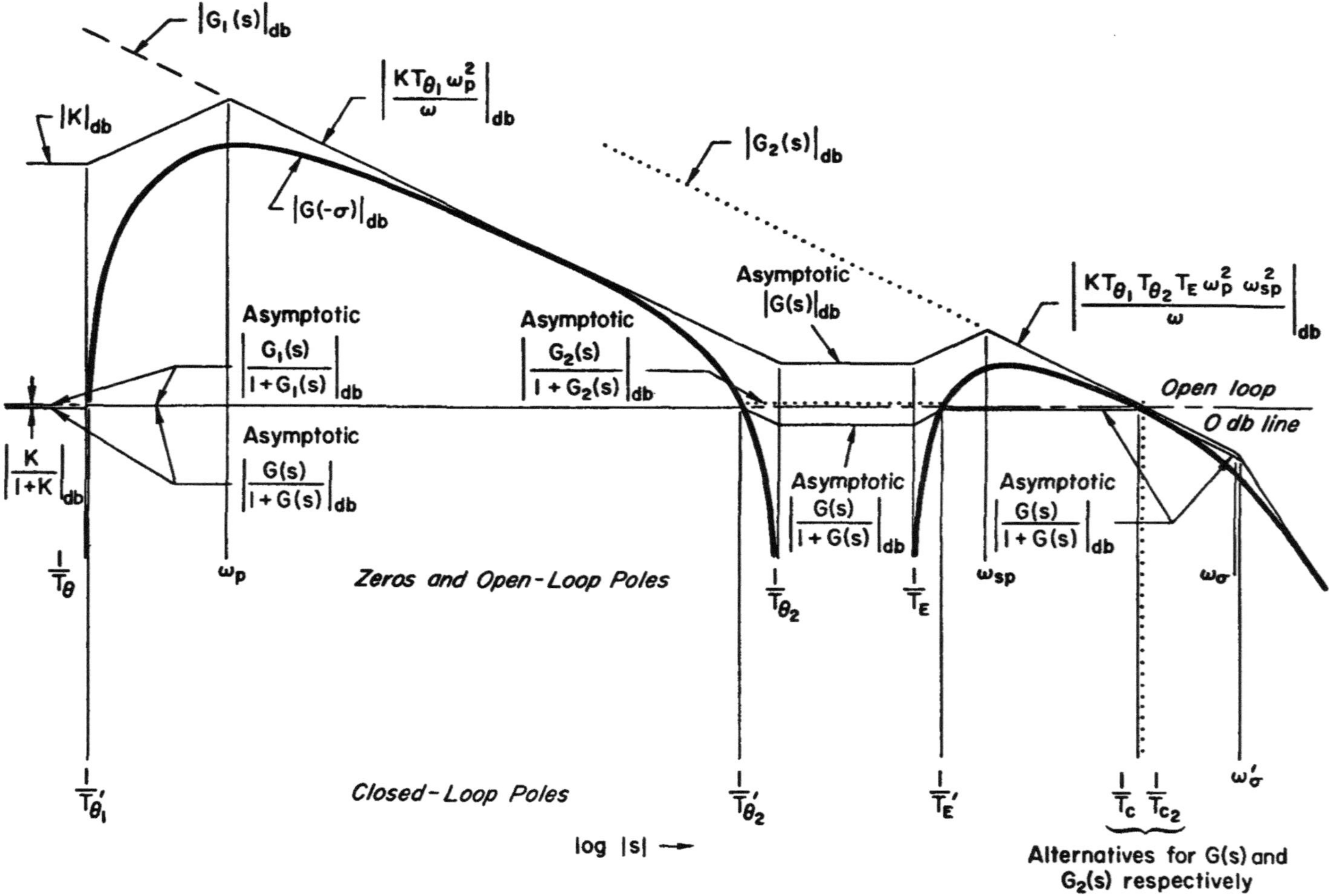


Fig. 3-35. Open- and closed-loop asymptotic plots, illustrating the development of equivalent systems.

from the numerator factors $(T_{\theta_2}s + 1)$ and $(T_E s + 1)$. Their proximity makes them act as dipole effects in the closed-loop system response, i.e., the modal response coefficients for the modes corresponding to $T'_{\theta_2}$ and $T'_E$ are relatively small because of the proximity of these time constants to $T_{\theta_2}$ and $T_E$, respectively. Thus these modes in the closed-loop system response will be minor if not negligible. To the extent that they can be ignored, a further equivalent system can be defined. This will be

$$G_2(s) = \frac{KT_{\theta_1}T_{\theta_2}T_E\omega_p^2\omega_{sp}^2}{s[(s/\omega_\sigma)^2 + (2\zeta_\sigma s/\omega_\sigma) + 1]} \tag{3-35}$$

Equation 3-35 is a striking reduction in dimensions to three. The resulting closed-loop transfer function will be

$$\frac{G_2}{1 + G_2} \doteq \frac{1}{(T_{c_1}s + 1)[(s/\omega'_{\sigma_1})^2 + (2\zeta'_{\sigma_1}s/\omega'_{\sigma_1}) \quad 1]} \tag{3-36}$$

Except for the absence of the two dipoles and the very low frequency effects, this closed-loop transfer function differs from that of Eq. 3-33 primarily in that $T_{c_1}$ will be slightly greater than $T_c$, and $\omega'_{\sigma_1}$ slightly less than $\omega'_\sigma$. However, the range of validity based on the use of Eq. 3-34 will be considerably narrower than that based on the higher order equivalent system approximation given by Eq. 3-35.

For some problems the equivalent systems given by Eqs. 3-34 and 3-35 can be further simplified. For low-pass inputs the major gross effect of the highest frequency modes, assuming that they are well beyond the crossover frequency, is an initial time delay. This can be approximated by noting that all high frequency leads and lags having breakpoints beyond a given frequency affect the characteristics below that frequency primarily in the phase shift. Thus, if the actual system's high frequency characteristics are

$$G_{\text{hf}}(s) = \frac{\prod_i^n (T_i s + 1)}{\prod_i^m (\tau_i s + 1)} \tag{3-37}$$

the contributions of these terms to the amplitude ratio and phase at frequencies, $\omega$, much less than $1/T_i$ and $1/\tau_i$ are

$$|G_{\text{hf}}(j\omega)| = \left| \frac{\prod_i^n \sqrt{(T_i\omega)^2 + 1}}{\prod_i^m \sqrt{(\tau_i\omega)^2 + 1}} \right| \doteq 1$$

$$\measuredangle G_{\text{hf}}(j\omega) = \sum_i^n \tan^{-1} T_i\omega - \sum_i^m \tan^{-1} \tau_i\omega$$

$$\doteq \sum_i^n T_i\omega - \sum_i^m \tau_i\omega = \omega\left[\sum_i^n T_i - \sum_i^m \tau_i\right] \tag{3-38}$$

These are the same contributions as those of a pure time delay, $e^{-j\omega\tau}$, where $\tau = \sum T_i - \sum \tau_i$. That is, for low frequency equivalence the delay $\tau$ is just the sum of the high frequency leads minus the sum of the high frequency lags. For the illustrative case, the servo is the only term for which this type of approximation might apply. In this case $\tau$ would be $-2\zeta_\sigma/\omega_\sigma$, so that the equivalent systems, for low-pass inputs, could be further reduced to

$$G_1(s) \doteq \frac{(KT_{\theta_1}\omega_p^2)(T_{\theta_2}s + 1)(T_E s + 1)}{s[(s/\omega_{sp})^2 + (2\zeta_{sp}s/\omega_{sp}) + 1]} e^{-(2\zeta_\sigma/\omega_\sigma)s} \tag{3-39}$$

and

$$G_2(s) \doteq \frac{(KT_{\theta_1}T_{\theta_2}T_E\omega_p^2\omega_{sp}^2)}{s} e^{-(2\zeta_\sigma/\omega_\sigma)s} \tag{3-40}$$

Once this point has been reached, one further step obtains an even lower order set of systems, formed by neglecting the minor high frequency effects entirely. This will often lead to better approximations than the fairly major shift from the first to the second equivalent systems in the first place. In other words, the $G_1(s)$ shown in Eq. 3-39 without the $e^{-(2\zeta_\sigma s/\omega_\sigma)}$ term will ordinarily be a better approximation to reality (and of no higher degree) than the $G_2(s)$ shown in Eq. 3-35. Indeed, a form

$$G_1(s) \doteq \frac{(KT_{\theta_1}\omega_p^2)(T_{\theta_2}s + 1)(T_E s + 1)}{s[(s/\omega_{sp})^2 + (2\zeta_{sp}s/\omega_{sp} + 1)]}$$

is ordinarily a very good approximation for problems of pitch attitude control.

*Literal Approximate Factors*

Transfer functions are totally specified, with the exception of a multiplying constant, by their poles and zeros. The poles and zeros are functions of the system's constants and are therefore directly affected by changes in any of the constants. Because of the intimate relationships between poles and zeros and the system constants, an extremely important prerequisite to a rational system design is an understanding of the effects of changes in the physical system configuration (as reflected by the system constants) on the transfer function poles and zeros.

Unfortunately, in complex systems the transfer functions are made up of ratios of higher order rational polynomials that are difficult to factor in general and meaningful terms. In airframe transfer functions, for example, the polynomials involved are largely of third or fourth order, having coefficients that are complicated functions of the stability derivatives. The usual approach to the determination of the effect on aircraft motions of varying the airframe configuration requires numerical values for the derivatives; this is ordinarily a time-consuming and irksome computational task. To alleviate this situation we should like to have some relatively simple, albeit approximate, expressions for the poles and zeros in terms of literal aircraft stability derivatives. But the derivation of approximate factors depends very strongly on the relative magnitudes and signs of the various polynomial coefficients, making approximations difficult to determine and the degree of approximation difficult to assess. One way out of this seeming quandary is the use of servo analysis methods where, in general:

1. An algebraic equation can be manipulated to have the form of a feedback system with:
   a. the open-loop transfer function given by $G(s)$,
   b. the closed-loop transfer function given by $G(s)/[1 + G(s)]$,
   c. the characteristic equation $1 + G(s) = 0$, identical to the original algebraic equation. Then
2. all of the servo-type methods for solving for the roots of $1 + G(s) = 0$ from a knowledge of $G(s)$ are applicable to the determination of factors for the algebraic equation; and, in particular,
3. a Bode plot representation, where wide regions of a graphical representation of $G(s)$ can be approximated by asymptotes having simplified equations, can be invaluable. The asymptotic and other properties of the Bode plot make the regions where simple approximate solutions apply quite clear. Further, since exact solutions are possible using the plot, the degree of error involved in a particular case is readily determined.

Although a quadratic is trivially simple, it will serve to illustrate the details of this "equivalent servo" technique for approximate factoring. Many more complex examples appear elsewhere.[11]

The second degree equation

$$s^2 + 2\zeta\omega_n s + \omega_n^2 = 0$$

can, of course, be factored exactly, i.e.,

$$(s + \zeta\omega_n + \omega_n\sqrt{\zeta^2 - 1})(s + \zeta\omega_n - \omega_n\sqrt{\zeta^2 - 1}) = 0$$

When dealing with transfer functions containing second degree terms with $\zeta < 1$, the unfactored form is usually suitable as is, while the factored form is called for when $\zeta > 1$. When $\zeta^2 \gg 1$, the factors become

[11] I. L. Ashkenas and D. T. McRuer, "Approximate Airframe Transfer Functions and Application to Single Sensor Control Systems," WADC-TR-58-82, June 1958.

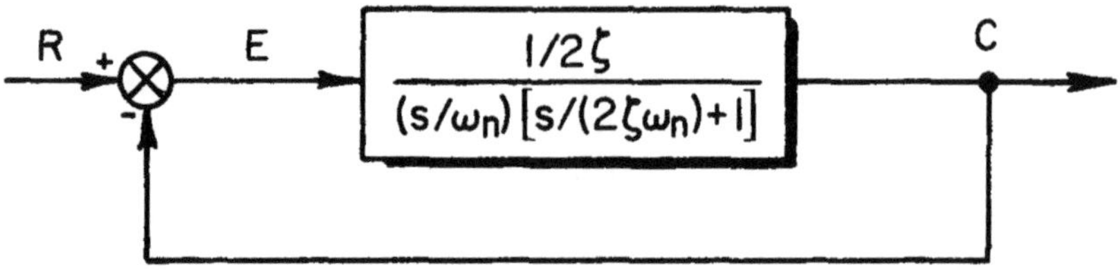


Fig. 3-36. Closed-loop system with closed-loop transfer function $[(s/\omega_n)^2 + (2\zeta s/\omega_n) + 1]^{-1}$.

approximately

$$\left[s + \zeta\omega_n + \zeta\omega_n\left(1 - \frac{1}{2\zeta^2} + \cdots\right)\right]\left[s + \zeta\omega_n - \zeta\omega_n\left(1 - \frac{1}{2\zeta^2} + \cdots\right)\right]$$

$$\doteq \left[s + \left(2\zeta\omega_n - \frac{\omega_n}{2\zeta}\right)\right]\left(s + \frac{\omega_n}{2\zeta}\right)$$

which, as $\zeta$ becomes larger still, approaches $(s + 2\zeta\omega_n)s$. The degree of adequacy of these various approximations can be seen readily by considering the second degree equation as the characteristic equation of the closed-loop system shown in Fig. 3-36.

Because only the real roots will be of interest here, only the $\sigma$-Bode need be used. This is shown in Fig. 3-37, with the abscissa normalized to $\sigma/\omega_n$. For $\zeta > 0$, the phase, although not shown, will be $-180°$ over the range $0 \leqq \sigma/\omega_n \leqq 2\zeta$. Relating this plot to the previous approximations, one can readily see that any value of the "gain," $1/2\zeta$, which results in an intersection of the zero-dB line with the amplitude ratio curve when it is near the low frequency asymptote, will give a root magnitude that is approximately

$$\frac{1}{2\zeta}\frac{\omega_n}{\sigma_1} \doteq 1 \qquad \text{or} \qquad \sigma_1 = \frac{\omega_n}{2\zeta}$$

Also, the magnitude of the larger root can be seen to approach $\sigma_2/\omega_n = 2\zeta$ as $\zeta$ becomes very large (gain very small). Since the system total damping, $2\zeta\omega_n$, is constant, the sum of the two real closed-loop roots must always be $-2\zeta\omega_n$; therefore the magnitude of the second root will be given by

$$\sigma_2 \doteq 2\zeta\omega_n - \frac{\omega_n}{2\zeta}$$

as long as an intersection of the zero-dB line and the amplitude ratio occurs near the low frequency asymptote. The error involved in the

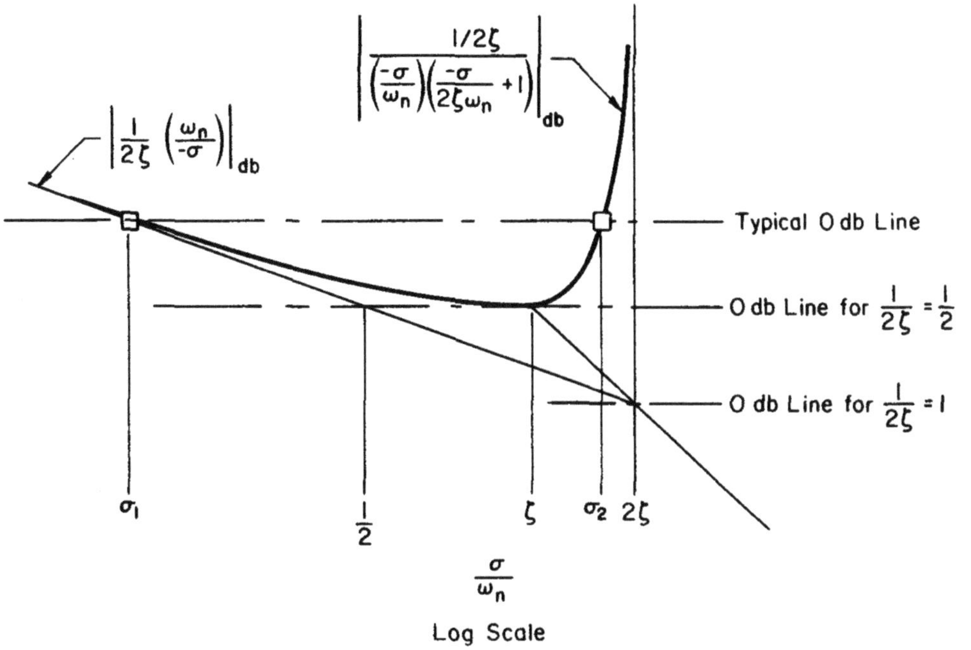


**Fig. 3-37.** A $\sigma$-Bode diagram for
$G(s/\omega_n) = (1/2\zeta)/\{(s/\omega_n)[(1/2\zeta)(s/\omega_n) + 1]\}$.

approximate roots becomes increasingly larger as the intersections of the amplitude ratio and the zero-dB line depart further from the low frequency asymptote. The error will be a maximum at open-loop gains where $\zeta = 1$, i.e., at the breakaway condition. The exact root magnitudes in this case are both $\omega$, while reliance on the approximations would give $\sigma_1 = \omega/2$ and $\sigma_2 = 3\omega/2$. In all cases it should be noted that the errors involved in the use of the approximate roots can always be determined readily by noting the departure of the actual amplitude ratio from the asymptotic plot at the point of intersection.

### 3-5. Analysis of Multiloop Vehicular Control Systems

It often happens in the design of feedback control systems for vehicles that the system cannot be conveniently represented by single-loop block diagrams. Until comparatively recently, this made the pencil-and-paper analysis of vehicular control systems either oversimplified or entirely too difficult, and the most usual approach to the design of multiloop systems was repetitive analysis using the analog computer as a tool. Complete dependence on an analog computer, however, does have deficiencies. First, because of the dominance of only particular modes in the time histories, modes that might be of great importance under slightly changed conditions may be effectively suppressed. Then, not only will these modes tend to be overlooked but their variation with the governing parameters will be difficult to evaluate; accordingly, some understanding of the overall performance of the system is lost. Second, elements described only in frequency response terms such as subsystem and human pilot describing functions cannot be used directly in computer operations. Third, and perhaps most important, all problems solved by the computer can provide only specific results which, in the absence of a suitable theory, cannot be readily and effectively extrapolated to different conditions or generalized. Thus gross trends and grand simplifications are harder to come by, insight is constricted, and initiative is stifled, as always happens when only a single approach to a problem is used.

To surmount these deficiencies, we should like to have a multiloop analysis technique with the following properties:

1. A formulation that clearly displays vehicle-alone and controller-alone characteristics in conventional and well-understood terms.
2. Analytical operations that can be performed using the most efficient graphical techniques of servoanalysis, so as to enhance the transfer of skill and intuition from the simpler single-loop situations.
3. Sequences and procedures that are highly responsive to physical insights and intuition, so as to lead to "good" systems with a minimum of iteration.
4. A presentation of results that is supplementary as well as equivalent to the results obtained using an analog computer.

The elements of such an analytical technique are presented below for a fairly simple case. Detailed developments for more complex systems are given elsewhere,[12] and specific examples that elucidate the use of the method are deferred to later chapters.

A generalized notation for vehicle and controller transfer characteristic quantities is introduced and used throughout this development. A compact matrix formation is appropriate for multiloop problems, and this is employed from the outset; but, to make the developments easier to follow, explicit equations in matrix form are used concurrently with the more compact matrix generalizations. This makes possible an inductive approach, wherein the matrix equations are both a shorthand for the equations of relatively simple systems and, viewed more broadly, the appropriate equations for far more complex systems.

[12] D. T. McRuer, I. L. Ashkenas, and H. R. Pass, "Analysis of Multiloop Vehicular Control Systems," ASD-TDR-62-1014, Wright-Patterson Air Force Base, Ohio, Mar. 1964.

*Development of Closed-Loop Transfer Functions for Multiloop Systems*

A multiloop vehicular control system that is relatively simple, yet complex enough for our present purposes, is shown in the block diagram of Fig. 3-38. It consists of a vehicle and control equipment comprising sensing, equalizing, and actuating elements. The vehicle has three independent degrees of freedom and is subject to control forces and moments applied by two control deflections and an external disturbance. The control deflections are functions of a command input and feedbacks from two of the three degrees of freedom.

The Laplace-transformed linearized equations of motion of the vehicle can be written in matrix form as

$$\begin{bmatrix} a_{11}(s) & a_{12}(s) & a_{13}(s) \\ a_{21}(s) & a_{22}(s) & a_{23}(s) \\ a_{31}(s) & a_{32}(s) & a_{33}(s) \end{bmatrix} \begin{bmatrix} X_1(s) \\ X_2(s) \\ X_3(s) \end{bmatrix} = \begin{bmatrix} b_{11}(s) & b_{12}(s) \\ b_{21}(s) & b_{22}(s) \\ b_{31}(s) & b_{32}(s) \end{bmatrix} \begin{bmatrix} \delta_1(s) \\ \delta_2(s) \end{bmatrix} + \begin{bmatrix} e_{11}(s) & e_{12}(s) \\ e_{21}(s) & e_{22}(s) \\ e_{31}(s) & e_{32}(s) \end{bmatrix} \begin{bmatrix} \eta_1(s) \\ \eta_2(s) \end{bmatrix} \tag{3-41}$$

or, more compactly,

$$[A(s)][X(s)] = [B(s)][\delta(s)] + [E(s)][\eta(s)] \tag{3-42}$$

The $a_{ij}$, $b_{ij}$, and $e_{ij}$ are, in general, functions of $s$ and the vehicle characteristics (stability derivatives); but to go along with the economy of notation of matrix methods, the functional dependence on $s$ of these quantities and the various transfer functions is not indicated in the remainder of this section.

The vehicle transfer functions for control or disturbance inputs are found from Eq. 3-41 by Cramer's rule. Typical examples are:

$$\frac{X_1}{\delta_1} = \frac{N_{\delta_1}^{x_1}}{\Delta} = \frac{\begin{vmatrix} b_{11} & a_{12} & a_{13} \\ b_{21} & a_{22} & a_{23} \\ b_{31} & a_{32} & a_{33} \end{vmatrix}}{\begin{vmatrix} a_{11} & a_{12} & a_{13} \\ a_{21} & a_{22} & a_{23} \\ a_{31} & a_{32} & a_{33} \end{vmatrix}} \tag{3-43}$$

$$\frac{X_2}{\delta_1} = \frac{N_{\delta_1}^{x_2}}{\Delta} = \frac{1}{\Delta} \begin{vmatrix} a_{11} & b_{11} & a_{13} \\ a_{21} & b_{21} & a_{23} \\ a_{31} & b_{31} & a_{33} \end{vmatrix} \tag{3-44}$$

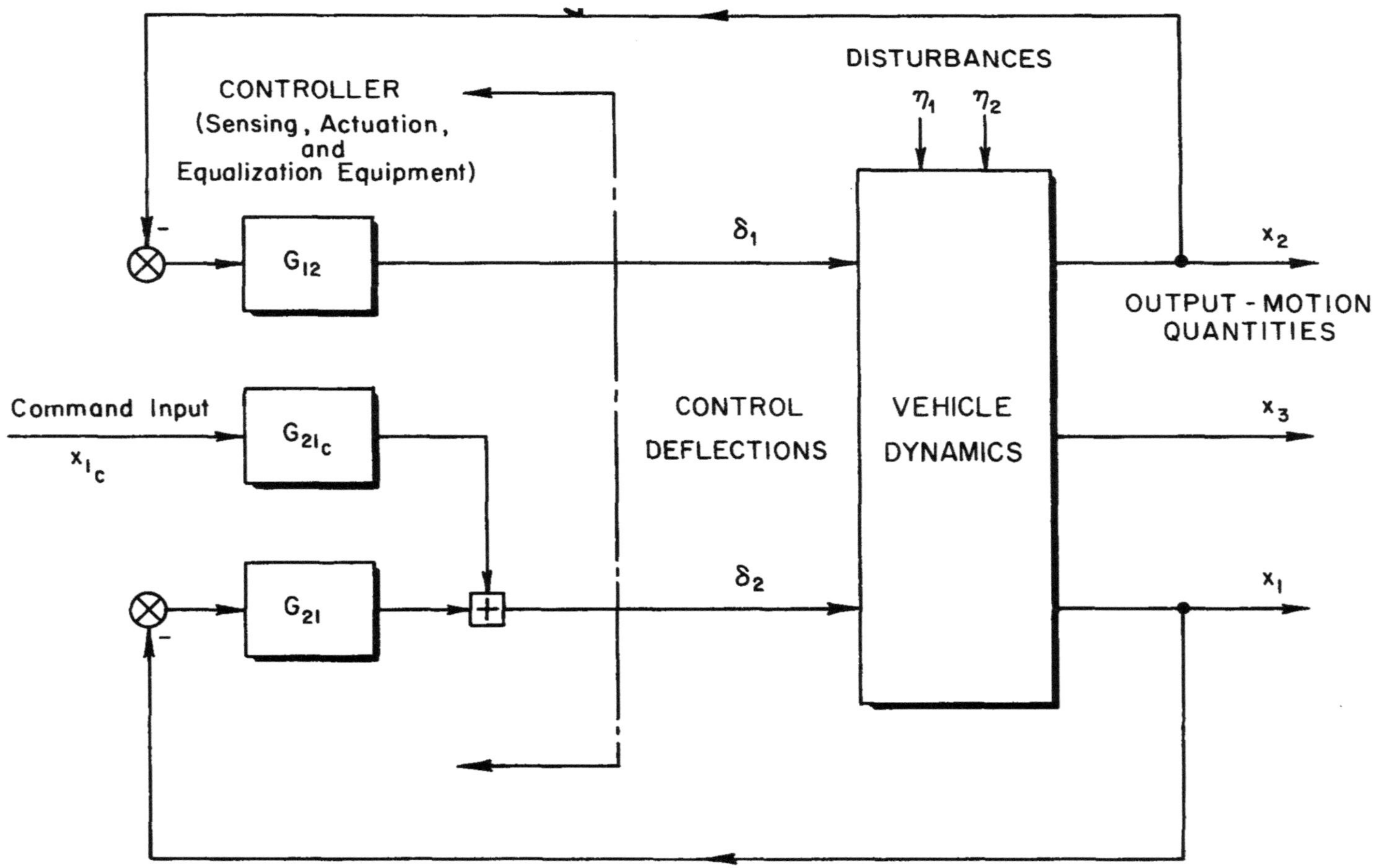


Fig. 3-38. Elementary multiloop vehicular control system.

In these transfer functions $\Delta$ is the determinant of the coefficients of the left-hand side (characteristic determinant of the vehicle),

$$\Delta = \begin{vmatrix} a_{11} & a_{12} & a_{13} \\ a_{21} & a_{22} & a_{23} \\ a_{31} & a_{32} & a_{33} \end{vmatrix} \tag{4-45}$$

and the numerator, $N_{\delta_j}^{x_i}$, is obtained by replacing the column of $x_i$ coefficients in $\Delta$ by the column of $\delta_j$ coefficients from the right-hand side of Eq. 3-41.

From the block diagram the controller characteristics are seen to be

$$\begin{bmatrix} \delta_1(s) \\ \delta_2(s) \end{bmatrix} = \begin{bmatrix} 0 & 0 \\ G_{21_c} & 0 \end{bmatrix} \begin{bmatrix} X_{1_c} \\ X_{2_c} \end{bmatrix} - \begin{bmatrix} 0 & G_{12} & 0 \\ G_{21} & 0 & 0 \end{bmatrix} \begin{bmatrix} X_1 \\ X_2 \\ X_3 \end{bmatrix} \tag{3-46}$$

or

$$[\delta] = [G_c][X_c] - [G][X] \tag{3-47}$$

where $[G_c]$ will be called the command matrix and $[G]$ the feedback matrix. The subscript convention used to identify the components of $G$ is that the first number identifies the controller output (control surface), the second number the controller input (sensed motion quantity). Equations 3-42 and 3-47 can be depicted as the deceptively simple matrix block diagram shown in Fig. 3-39.

The substitution of Eq. 3-47 into Eq. 3-42 gives

$$\begin{aligned} [A][X] &= [B][\delta] + [E][\eta] \\ &= [B]\{[G_c][X_c] - [G][X]\} + [E][\eta] \end{aligned}$$

Collecting like terms, we obtain

$$\{[A] + [B][G]\}[X] = [B][G_c][X_c] + [E][\eta] \tag{3-48}$$

and, after premultiplying by the inverse of $\{[A] + [B][G]\}$, the explicit expression for $[X]$ becomes

$$[X] = \{[A] + [B][G]\}^{-1}\{[B][G_c][X_c] + [E][\eta]\} \tag{3-49}$$

Equation 3-49 is the formal matrix solution for the transform of the outputs of the closed-loop system. It is not restricted to the equations of the example but is, in fact, applicable to systems with larger or smaller matrices.

The determinant of the coefficients of the matrix expression

$$\{[A] + [B][G]\}$$

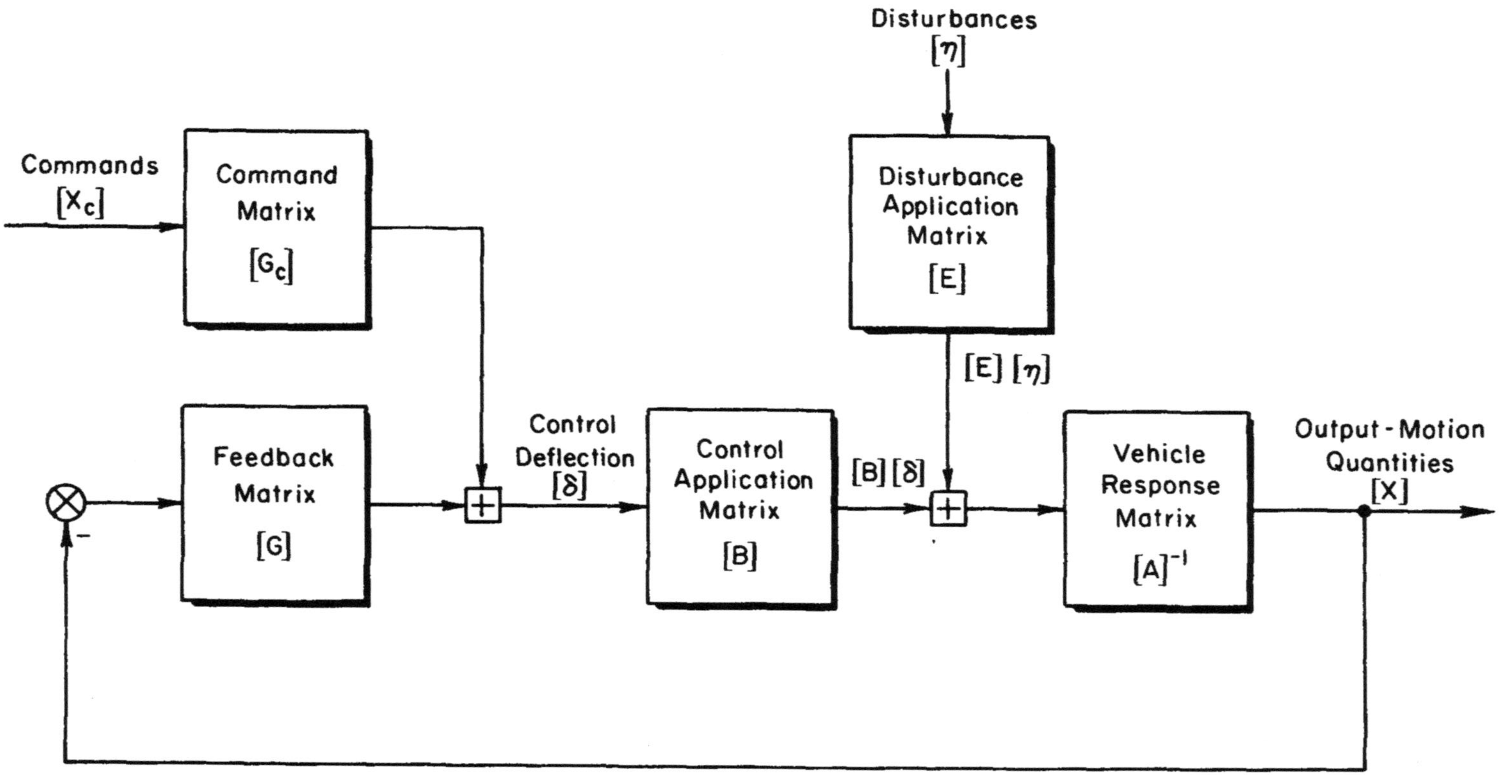


Fig. 3-39. Matrix block diagram for general multiloop control system.

will become the characteristic function of the closed-loop system, i.e.,

$$\Delta_{sys} = \det \{[A] + [B][G]\} \tag{3-50}$$

This determinant can be expanded in such a way as to retain explicitly the vehicle-alone characteristics, which is a powerful advantage. Also, the resulting expressions can be made amenable to the conventional servo-analysis techniques described in the previous section. Thus, in the case of the simplified system of Eqs. 3-41 and 3-46,

$$\begin{aligned}\Delta_{sys} &= \det \{[A] + [B][G]\} \\ &= \begin{vmatrix} a_{11} + b_{12}G_{21} & a_{12} + b_{11}G_{12} & a_{13} \\ a_{21} + b_{22}G_{21} & a_{22} + b_{21}G_{12} & a_{23} \\ a_{31} + b_{32}G_{21} & a_{32} + b_{31}G_{12} & a_{33} \end{vmatrix} \\ &= \Delta + G_{12}N_{\delta_1}^{x_2} + G_{21}N_{\delta_2}^{x_1} + G_{12}G_{21}N_{\delta_1\delta_2}^{x_2x_1} \end{aligned} \tag{3-51}$$

where $\Delta$ and $N_{\delta_j}^{x_i}$ have the same significance as in Eqs. 3-43, 3-44, and 3-45, while terms of the form $N_{\delta_1\delta_2}^{x_ix_k}$, called coupling numerators, are found by replacing both the $i^{th}$ and $k^{th}$ columns of the determinant of the left-hand coefficients in Eq. 3-41 by the columns of $\delta_1$ and $\delta_2$ coefficients, respectively. The awkward, but descriptive, symbol with two subscripts and two superscripts is intended to suggest this replacement. For example,

$$N_{\delta_1\delta_2}^{x_2x_1} = \begin{vmatrix} b_{12} & b_{11} & a_{13} \\ b_{22} & b_{21} & a_{23} \\ b_{32} & b_{31} & a_{33} \end{vmatrix} \tag{3-52}$$

The coupling numerator $N_{\delta_1\delta_2}^{x_ix_k}$ has no meaning when $i = k$ and is arbitrarily defined to be equal to zero. The properties of determinants can also be used to show:

$$N_{\delta_1\delta_1}^{x_ix_k} = N_{\delta_2\delta_2}^{x_ix_k} = 0 \tag{3-53}$$

$$N_{\delta_1\delta_2}^{x_ix_k} = -N_{\delta_2\delta_1}^{x_ix_k} = N_{\delta_2\delta_1}^{x_kx_i} \tag{3-54}$$

$$N_{\delta_1\delta_2}^{x_ix_k} = \frac{1}{\Delta}(N_{\delta_1}^{x_i}N_{\delta_2}^{x_k} - N_{\delta_2}^{x_i}N_{\delta_1}^{x_k}) \tag{3-55}$$

For the most general case of two control deflections fed by each of the three degrees of freedom, the system characteristic determinant can be shown to have the form[13]

$$\Delta_{sys} = \Delta + \sum_{i=1}^{3}\sum_{j=1}^{2} G_{ji}N_{\delta_j}^{x_i} + \sum_{i=1}^{3}\sum_{k=1}^{3} G_{1i}G_{2k}N_{\delta_1\delta_2}^{x_ix_k} \tag{3-56}$$

[13] *Ibid.*

While the system characteristic function, $\Delta_{sys}$, is the denominator for all closed-loop transfer functions, regardless of the command or disturbance input, the numerator of a closed-loop transfer function will depend on the particular command or disturbance.

The inverse matrix, $\{[A] + [B][G]\}^{-1}$, which appears in Eq. 3-49, can be expressed in terms of the basic matrix by the standard form[14]

$$\{[A] + [B][G]\}^{-1} = \frac{\begin{bmatrix} \Delta_{11} & \Delta_{21} & \Delta_{31} \\ \Delta_{12} & \Delta_{22} & \Delta_{32} \\ \Delta_{13} & \Delta_{23} & \Delta_{33} \end{bmatrix}}{\Delta_{sys}} \tag{3-57}$$

where the numerator is the transpose of the matrix of the cofactors, and the denominator is the determinant, of the basic matrix, $[A] + [B][G]$. Thus, for the example, $\Delta_{sys} = \det\{[A] + [B][G]\}$ (Eq. 3-50), and

$$\Delta_{11} = \begin{vmatrix} a_{22} + b_{21}G_{12} & a_{23} \\ a_{32} + b_{31}G_{12} & a_{33} \end{vmatrix} \tag{3-58}$$

is the cofactor of $(a_{11} + b_{12}G_{21})$ in the determinant of Eq. 3-51.

For command inputs the other matrix involved in Eq. 3-49 is $[B][G_c][X_c]$. For the case at hand, the matrix multiplication yields

$$\begin{aligned} \frac{X_1}{X_{1_c}} &= \frac{\Delta_{11}b_{12}G_{21_c} + \Delta_{21}b_{22}G_{21_c} + \Delta_{31}b_{32}G_{21_c}}{\Delta_{sys}} \\ &= \frac{G_{21_c}(b_{12}\Delta_{11} + b_{22}\Delta_{21} + b_{32}\Delta_{31})}{\Delta_{sys}} \\ &= \frac{\begin{vmatrix} b_{12}G_{21_c} & a_{12} + b_{11}G_{12} & a_{13} \\ b_{22}G_{21_c} & a_{22} + b_{21}G_{12} & a_{23} \\ b_{32}G_{21_c} & a_{32} + b_{31}G_{12} & a_{33} \end{vmatrix}}{\begin{vmatrix} a_{11} + b_{12}G_{21} & a_{12} + b_{11}G_{12} & a_{13} \\ a_{21} + b_{22}G_{21} & a_{22} + b_{21}G_{12} & a_{23} \\ a_{31} + b_{32}G_{21} & a_{32} + b_{31}G_{12} & a_{33} \end{vmatrix}} \end{aligned} \tag{3-59}$$

This same result is obtained, more directly but with less general carryover to more complex situations, by the application of Cramer's rule to Eq. 3-48 expanded to include the example matrix elements. It is apparent in Eq. 3-59 that the cofactors, such as $\Delta_{11}$, which appear in the numerator are *identical* to the terms that would appear multiplied by $G_{21}$ in the expansion

[14] For example, L. A. Pipes, *Matrix Methods for Engineering*, Prentice-Hall, Englewood Cliffs, N.J., 1963.

of the denominator. Thus the expansion of the closed-loop transfer function numerator can be inscribed by analogy, and the complete closed-loop transfer function for a command input, $x_{1_c}$, becomes

$$\frac{X_1}{X_{1_c}} = \frac{G_{21_c}(N^{x_1}_{\delta_2} + G_{12}N^{x_2 x_1}_{\delta_1 \delta_2})}{\Delta + G_{12}N^{x_2}_{\delta_1} + G_{21}N^{x_1}_{\delta_2} + G_{12}G_{21}N^{x_2 x_1}_{\delta_1 \delta_2}} \tag{3-60}$$

A similar development would yield the transfer function relating the response, $x_1$, to the disturbance, $\eta_1$:

$$\frac{X_1}{\eta_1} = \frac{N^{x_1}_{\eta_1} + G_{12}N^{x_1 x_2}_{\eta_1 \delta_1} + [G_{21}N^{x_1 x_1}_{\eta_1 \delta_2}]}{\Delta + G_{12}N^{x_2}_{\delta_1} + G_{21}N^{x_1}_{\delta_2} + G_{12}G_{21}N^{x_1 x_2}_{\delta_2 \delta_1}} \tag{3-61}$$

Here, of course, the disturbance does *not* go through the block $G_{21_c}$ to get into the doubly closed-loop system. Therefore, the leading numerator term is not multiplied by $G_{21_c}$ in this case. Note also that the term in square brackets is identically zero.

The pattern that is evident in Eqs. 3-60 and 3-61 can be described by the following rules:[15]

1. The effective denominator is equal to
   a. the open-loop denominator,
   b. plus the sum of all the feedback transfer functions, each one multiplied by the appropriate numerator,
   c. plus the sum of all the feedback transfer functions taken two at a time, each pair multiplied by the appropriate coupling numerator.
2. The effective numerator is equal to
   a. the open-loop numerator,
   b. plus the sum of all the feedback transfer functions, each one multiplied by the appropriate coupling numerator.
3. The responses to a command input are obtained from the matrix equation

$$[X] = \begin{bmatrix} \text{matrix of effective} \\ \text{transfer functions} \end{bmatrix} [G_c][X_c] \tag{3-62}$$

   In the common situation in which a command is only fed to one control, Eq. 3-61 reduces to

$$\frac{X_i}{X_{j_c}} = \frac{[(\delta_k/x_{j_c}) \text{ transfer function}][\text{effective } (x_i/\delta_k) \text{ numerator}]}{(\text{effective denominator})}$$

In Rule 1(b) the "appropriate" numerator is the one with the same output/input pair as the feedback, e.g., the product $G_{21}N^{x_1}_{\delta_2}$ is appropriate

[15] These rules are due to R. L. Stapleford. See Appendix C of ASD-TDR-62-1094, "Analysis of Multiloop Vehicular Control Systems," Mar. 1964.

to the specific example used earlier, while in Rule 1(c) the appropriate coupling numerator has the same two output/input pairs as the feedbacks, e.g., $G_{12}G_{21}N^{x_1x_2}_{\delta_2\delta_1}$. Recall that $N^{x_ix_j}_{\delta_k\delta_1}$ is zero (from the properties of determinants) if $i = j$ or $k = 1$. In general cases with many feedbacks, several of the coupling numerators will be zero.

In Rule 2 the appropriate coupling numerator is the one for the output/input pair of the original numerator, as well as for the feedback output/input pair. For example, the feedback $x_2 \rightarrow \delta_1$ modifies the $x_i/\delta_j$ numerator by adding to it the term $G_{12}N^{x_ix_2}_{\delta_j\delta_1}$, and modifies the $x_i/\eta_j$ numerator by adding to it the term $G_{12}N^{x_ix_2}_{\eta_j\delta_1}$.

Although the above discussion and the earlier derivations are adequate for determining the command response with feedbacks to two independent controls or the disturbance response with feedbacks to one control, a complete generalization (for a three degree of freedom system) requires the introduction of a second type of coupling numerator. In a type-two coupling numerator, three columns of the open-loop characteristic determinant are replaced by the appropriate control or disturbance coefficients. A type-two coupling numerator is zero if any two of the outputs or any two of the inputs are identical, i.e.,

$$N^{x_ix_ix_j}_{\delta_k\delta_l\delta_m} = 0 \tag{3-63}$$

$$N^{x_ix_jx_k}_{\delta_l\delta_l\delta_m} = 0 \tag{3-64}$$

The rules for forming the effective denominator and numerator are generalized by adding:

1(d). plus the sum of all the feedback transfer functions taken three at a time, each combination multiplied by the appropriate type-two coupling numerator,

2(c). plus the sum of all the feedback transfer functions taken two at a time, each pair multiplied by the appropriate type-two coupling numerator.

It often happens that it is necessary or desirable to feed back a quantity that is a *linear* combination of terms in the variables which appear in the equations of motion. Two alternatives are available. An additional, *not* linearly independent equation can be inscribed, together with the equations of motion; developments similar to the previous ones can be carried out, expanding the determinants which are now larger. On the other hand, all the terms that are required can be developed by adopting a special definition for the word "numerator":

$$\text{numerator} \equiv \Delta\ (\text{transfer function})$$

This definition does not, as we shall see later, exclude the possibility that a "numerator" can include a "denominator." For convenience the words will be used from here on as they are defined above, and the use of quotation marks will not be continued. Then, for example, if

$$x_4 = ax_1 + bx_2 + cx_3 \tag{3-65}$$

$$N^{x_4}_{\delta_1} = aN^{x_1}_{\delta_1} + bN^{x_2}_{\delta_1} + cN^{x_3}_{\delta_1} \tag{3-66}$$

and

$$N^{x_4x_1}_{\delta_2\delta_1} = [aN^{x_1x_1}_{\delta_2\delta_1}] + bN^{x_2x_1}_{\delta_2\delta_1} + cN^{x_3x_1}_{\delta_2\delta_1} \tag{3-67}$$

etc. The term in square brackets is again identically zero. Otherwise, it is worth noting that the replacements indicated by the right column of subscripts and superscripts are the same *throughout* the equation for the coupling numerator.

*Factoring the Transfer Function*

Returning now to the case of the simple example, we find that the transfer function of Eq. 3-60 contains terms that are readily recognized as vehicle transfer function numerators and denominators. Dividing the numerator and denominator by $\Delta$, $X_1/X_{1_c}$ can be expressed in terms of vehicle-alone transfer functions.

$$\frac{X_1}{X_{1_c}} = \frac{G_{21}\mathrm{X}_{1_{\delta_2}}[1 + G_{12}(N^{x_2x_1}_{\delta_1\delta_2}/N^{x_1}_{\delta_2})]}{1 + G_{12}\mathrm{X}_{2_{\delta_1}} + \{G_{21}\mathrm{X}_{1_{\delta_2}}[1 + G_{12}(N^{x_2x_1}_{\delta_1\delta_2}/N^{x_1}_{\delta_2})]\}} \tag{3-68}$$

where

$$\mathrm{X}_{1_{\delta_2}} = \frac{X_1(s)}{\delta_2(s)}$$

$$\mathrm{X}_{2_{\delta_1}} = \frac{X_2(s)}{\delta_1(s)}$$

Since the term in braces in the denominator of Eq. 3-68 is identical to the numerator, it makes it easy to recognize that the open-loop function $N(s)/D(s)$, corresponding to the closed-loop function $N(s)/[D(s) + N(s)]$, is

$$\frac{X_1}{X_{1_\epsilon}} = \frac{G_{21}\mathrm{X}_{1_{\delta_2}}[1 + G_{12}(N^{x_2x_1}_{\delta_1\delta_2}/N^{x_1}_{\delta_2})]}{1 + G_{12}\mathrm{X}_{2_{\delta_1}}}$$

$$= G_{21}\mathrm{X}_{1_{\delta_2}}\left[\frac{1 + G_{12}(N^{x_2x_1}_{\delta_1\delta_2}/N^{x_1}_{\delta_2})}{1 + G_{12}\mathrm{X}_{2_{\delta_1}}}\right] \tag{3-69}$$

Equation 3-69 can be represented in block diagram form in a particularly instructive way. Figure 3-41 shows that the effect of feeding back $x_2 \rightarrow \delta_1$ is to modify the $\mathrm{X}_{1_{\delta_2}}$ transfer function of the vehicle so that, when the

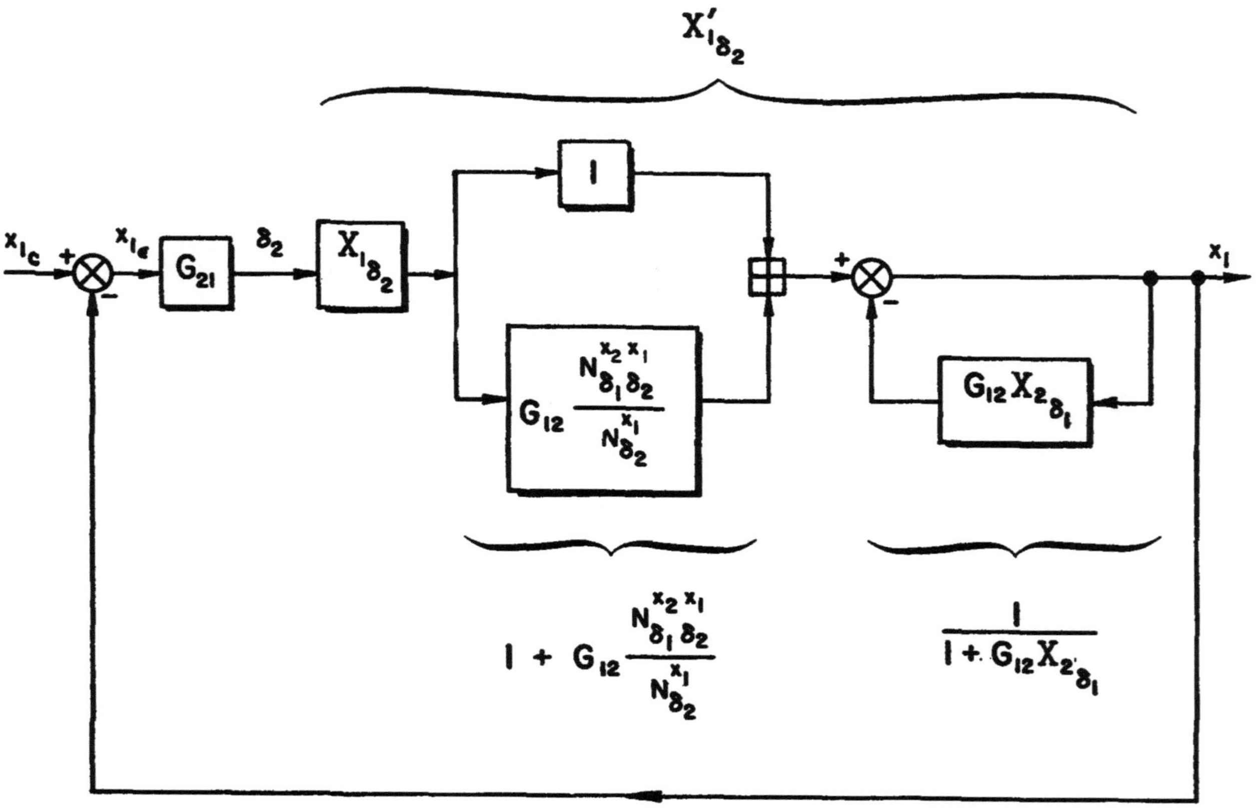


Fig. 3-40. Equivalent block diagram for the system: $x_{1_c} \to \delta_2$; $x_2 \to \delta_1$.

$x_1 \to \delta_2$ feedback loop is opened, the "effective" vehicle transfer function is $X'_{1_{\delta_2}}$ and the open-loop function for the second loop closure is $G_{21}X'_{1_{\delta_2}}$. In this case, both the poles and zeros of $X_{1_{\delta_2}}$ have been modified.

The primed notation $X'_{1_{\delta_2}}$ is used to indicate merely that one loop has been closed, and, by itself, it is not intended to specify the particular loop closure that is involved. The meaning of the primed notation in terms of which loops have been closed, therefore, depends on the local context. Later, the prime is similarly added to individual transfer function terms to indicate the number of prior loop closures.

Note here that since, typically, $s$ appears raised to higher powers in the denominators of $G_{12}X_{2_{\delta_1}}$ and $G_{12}N^{x_2x_1}_{\delta_1\delta_2}/N^{x_1}_{\delta_2}$ than in the numerators,

$$\lim_{s\to\infty}(G_{12}X_{2_{\delta_1}}) \doteq 0 \tag{3-70}$$

and

$$\lim_{s\to\infty}\left(G_{12}\frac{N^{x_2x_1}_{\delta_1\delta_2}}{N^{x_1}_{\delta_1}}\right) \doteq 0 \tag{3-71}$$

so that the so-called root locus or high frequency gain of $X'_{1_{\delta_2}}$ is identical to the one for $X_{1_{\delta_2}}$.

We may also remark, in passing, that Fig. 3-40 shows particularly clearly that the modification made by the first loop closure, $x_2 \to \delta_1$, might be made in such a way as to compensate or "equalize" the open-loop transfer function, $G_{21}X_{1_{\delta_2}}$. This is a thought to which we shall return later.

Now, in order to make an analysis of the effects of closing the second loop, $x_1 \to \delta_2$, it is first necessary to know the factors of $X'_{1_{\delta_2}}$, and finding the *factors* of $X'_{1_{\delta_2}}$ involves finding the factors of

$$1 + G_{12}X_{2_{\delta_1}} \qquad \text{given the factors of} \qquad G_{12}X_{2_{\delta_1}}$$

and of

$$1 + G_{12}\frac{N^{x_2x_1}_{\delta_1\delta_2}}{N^{x_1}_{\delta_2}} \qquad \text{given the factors of} \qquad G_{12}\frac{N^{x_2x_1}_{\delta_1\delta_2}}{N^{x_1}_{\delta_1}}$$

Of course, these operations are relatively easily accomplished by conventional servo analysis techniques. They correspond to finding the closed-loop roots of the characteristic equations for the feedback systems of Fig. 3-41. Note that the roots of the system of Fig. 3-41(a) are the poles of $X'_{1_{\delta_2}}$, while the roots of the system of Fig. 3-41(b) become the zeros of $X'_{1_{\delta_2}}$.

$$X'_{1_{\delta_2}} = \frac{N^{x_1}_{\delta_2} + G_{12}N^{x_2x_1}_{\delta_1\delta_2}}{\Delta + G_{12}N^{x_2}_{\delta_1}} = \frac{\text{closed-loop pole factors of closure 2}}{\text{closed-loop pole factors of closure 1}} \tag{3-72}$$

If the characteristics of the vehicle are fixed, $G_{12}$ is the only variable in both relationships. Thus, choosing $G_{12}$ appropriate to either Closure 1

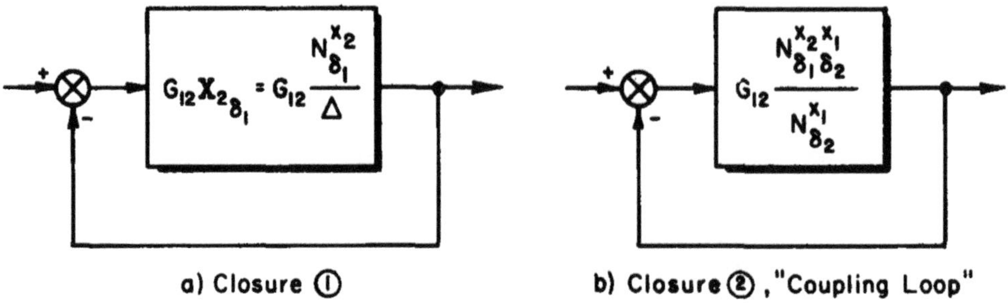


Fig. 3-41. The $x_2$ loop closures involved in the system: $x_{1_c} \to \delta_2$; $x_2 \to \delta_1$.

or Closure 2 completely determines the characteristics of the other closure.

The two loop closures considered here are further related, in typical applications to aircraft, in that the high frequency open- and closed-loop asymptotes of Systems 1 and 2 are often nearly identical; therefore the root locus gains are nearly the same. This observation is easy to appreciate because typically the terms of highest degree in $s$ in the expansion of $\Delta$ come from the product of main diagonal terms $(a_{11}a_{22}a_{33})$, while the control effectiveness terms $(b_{ij})$ are usually constants. Then, at high frequencies, denoted as $|s|$ large,

$$N^{x_1}_{\delta_2}\Big]_{|s|\text{ large}} \to b_{12}a_{22}a_{33}\Big]_{|s|\text{ large}} \qquad \text{(see Eq. 3-43)} \tag{3-73}$$

$$N^{x_2x_1}_{\delta_1\delta_2}\Big]_{|s|\text{ large}} \to (b_{12}b_{21} - b_{11}b_{22})a_{33}\Big]_{|s|\text{ large}} \qquad \text{(see Eq. 3-51)} \tag{3-74}$$

$$\left.\frac{N^{x_2x_1}_{\delta_1\delta_2}}{N^{x_1}_{\delta_2}}\right]_{|s|\text{ large}} \to \left.\frac{b_{21}}{a_{22}} - \frac{b_{11}b_{22}}{b_{12}a_{22}}\right]_{|s|\text{ large}} \tag{3-75}$$

Similarly (Eq. 3-44),

$$X_{2_{\delta_1}}\Big]_{|s|\text{ large}} \to \left.\frac{a_{11}b_{21}a_{33}}{a_{11}a_{22}a_{33}}\right]_{|s|\text{ large}} = \left.\frac{b_{21}}{a_{22}}\right]_{|s|\text{ large}} \tag{3-76}$$

Quite typically, for aircraft controls, $b_{21} \gg b_{11}b_{22}/b_{12}$, so that in general for large $|s|$,

$$\frac{N^{x_2x_1}_{\delta_1\delta_2}}{N^{x_1}_{\delta_2}} = \frac{b_{21}}{a_{22}} - \frac{b_{11}b_{22}}{a_{22}b_{12}} \doteq \frac{b_{21}}{a_{22}} \doteq X_{2_{\delta_1}} \tag{3-77}$$

It is helpful to keep this fact in mind while actually making the analysis of the simultaneous closures for Systems 1 and 2.

When $X'_{1_{\delta_2}}$ has been found in factored form, the closed-loop system characteristics can readily be determined by a conventional analysis of the system of Fig. 3-42.

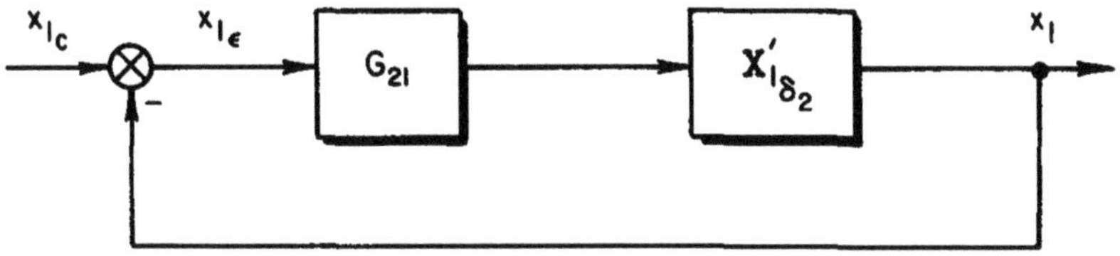


Fig. 3-42. Command loop closure.

In summary, the steps involved in the analysis of the system in Fig. 3-38 are:

1. The control channels are divided into two categories, *inner* and *outer* loops, reflecting the closure sequence. The $x_2$ (or $\delta_1$) loop, closed first, is thus the inner loop; whereas the $x_1$ (or $\delta_2$) loop, being closed second, is the outer loop.
2. The inner loop, $G_{12}X_{2_{\delta_1}}$ (Closure 1, Fig. 3-40), is closed with tentatively selected equalization and gains, and the closed inner loop roots are found. These roots become the vehicle's *poles* for the outer loop closure.
3. The same gain and equalization selected above are used; i.e., the same $G_{12}$, the coupling loop (Closure 2, Fig. 3-40) is closed. The closed-loop roots resulting from this closure become the vehicle's *zeros* for the outer loop closure.
4. The outer loop is closed in a conventional manner around the modified outer loop vehicle transfer function.
5. Possible repetitions of Steps 2 through 4 with different equalizations and gains may be required if the result of Step 4 is not satisfactory.

**3-6. Sensitivity of Closed-Loop Roots to System Parameter Variations**

We have thus far treated the system analysis problem for nominal characteristics, including the determination of open- and closed-loop transfer functions and time responses. We turn now to consideration of off-nominal behavior resulting from variations in the system parameters from their assigned values. The many ramifications of this subject are commonly encompassed by the name "sensitivity," referring generally to the change of some aspect of system behavior due to some change in the system elements. Because the sensitivity measures developed here are directly related to the modal response coefficients needed in response calculations, sensitivity and response factors become inseparable. Therefore, while this section is on sensitivity, it also implicitly covers the calculation of modal response coefficients.

Sensitivity considerations have long been of dominant interest in vehicle dynamic and control system design activities. As noted in Section 3.4, the sensitivity of a feedback system was originally one of the bases for the mathematical definition of feedback. This was natural, since feedback systems possess the "fundamental physical property that the effects of variations in the forward loop, whether they are taken as changes in $G(s)$ or as departures from strict linearity, or as freedom from extraneous noise, are reduced by the factor $1/(1 + G)$ in comparison with the effects which would be observed in a system without feedback."[16] The $1/(1 + G)$ factor is the classical sensitivity function, $S_G^{G_{cr}}(s)$, which compares the relative change in closed-loop characteristics to the causative relative change in open-loop characteristics, i.e.,

$$S_G^{G_{cr}}(s) = \left[\frac{dG_{cr}(s)/G_{cr}(s)}{dG(s)/G(s)}\right] = \frac{1}{1 + G} \tag{3-78}$$

Thus the measure $1/(1 + G)$ indicates directly the reduction in the sensitivity of the system to many of the influences that would otherwise tend to alter its performance.

Although classical sensitivity has been a popular subject in control topics ever since Bode's original work, we shall depart from the classical view to emphasize newer concepts of sensitivity. These concepts are more pertinent here because they result in measures that are directly related to the pole-zero description of system characteristics.[17] The basic measures, called here "gain," "open-loop pole," "open-loop zero," and "open-loop parameter" sensitivities were evolved to determine changes in the position of closed-loop poles due to shifts or changes in open-loop gain, poles, zeros, or other parameters. In the simplest terms these sensitivities connect open-loop differential variations with closed-loop differential shifts.

The sensitivity factors can play a significant role in synthesis, as well

[16] H. W. Bode, *Network Analysis and Feedback Amplifier Design*, Van Nostrand, New York, 1945, p. 44.

[17] K. Mitchell, "Estimation of the Effect of a Parameter Change on the Roots of Stability Equations," *Aeron. Quart.*, 1, 5, pp. 39–58 (May 1949). Ordway B. Gates, Jr., and C. H. Woodling, "A Method for Estimating Variations in the Roots of the Lateral-Stability Quartic due to Changes in Mass and Aerodynamic Parameters of an Airplane," NACA TN 3134, Jan. 1954; A. Papoulis, "Displacement of the Zeros of the Impedance $Z(p)$ Due to Incremental Variations in the Network Elements," *Proc. IRE*, 43, 1, pp. 79–82 (1955); R. Y. Huang, "The Sensitivity of the Poles of Linear Closed Loop Systems," *Trans. AIEE*, 77, Part 2, 182–186 (1958); F. F. Kuo, *Pole-Zero Sensitivity in Network Functions*, Ph.D. dissertation, University of Illinois, 1958; H. Ur, "Root Locus Properties and Sensitivity Relations in Control Systems," *IRE Trans.*, AC-5, No. 1, pp. 57–65 (Jan. 1960); D. T. McRuer, and R. L. Stapleford, "Sensitivity and Modal Response for Single-Loop and Multiloop Systems," ASD-TDR-62-812, Jan. 1963; I. M. Horowitz, *Synthesis of Feedback Systems*, Academic, New York, 1963.

as analysis, activities. In analysis, as already remarked, design calculations are inherently nominal because the system assumed in an analytical study can never precisely match the actual physical system. An assessment of the effects of system uncertainties and/or the implications of parameter tolerances thus becomes a matter of great concern to the designer. Sensitivity concepts provide the direct means to make such assessments. For synthesis the use of sensitivity is more subtle. Practical control systems usually require complex dynamical descriptions and are expected to meet a variety of diverse and often conflicting criteria. Because of this intrinsic complexity, the interrelationships between the system parameters and the diverse system performance measures are always involved and often obscure. Consequently, system synthesis is invariably accomplished by iterative analysis. In such iterations, knowledge of the effects of parameter variations is of great value in determining appropriate modifications. The guidance provided by sensitivity factors can be used to suggest both improvement in nominal system characteristics and the reduction of system sensitivity to parameter variations.

Our development sequence will first consider single-loop systems with first-order, closed-loop poles. The notions here are elementary and easily traced, as befits an introduction to the subject. This will be followed by a more general development that subsumes all forms of root sensitivities for open-loop, single-loop, and multiloop systems.

*Sensitivities for Single-Loop Systems*

The single-loop system, open-loop transfer function has the form

$$\begin{aligned} G(s) &= \kappa \frac{\alpha(s)}{\beta(s)} = \kappa \frac{\prod_{j=1}^{n} (s + z_j)}{\prod_{j=1}^{m+n} (s + p_j)} \\ &= G(s, \kappa, z_j, p_j) \end{aligned} \tag{3-79}$$

The effects of small variations in the parameters of $G(s)$ on the closed-loop roots can be found by starting with the total derivative of the closed-loop characteristic function, $1 + G$,

$$\begin{aligned} d(1 + G) &= dG \\ &= \frac{\partial G}{\partial s} ds + \frac{\partial G}{\partial \kappa} d\kappa + \sum_{j=1}^{n} \frac{\partial G}{\partial z_j} dz_j + \sum_{j=1}^{m+n} \frac{\partial G}{\partial p_j} dp_j \end{aligned} \tag{3-80}$$

Now because the closed-loop poles, $\lambda_i$, are defined by the equation

$$[1 + G(s)]_{s=\lambda_i} = 0 \tag{3-81}$$

the total differential of Eq. 3-80 is zero at the position of a closed-loop pole $s = \lambda_i$. Setting $dG = 0$ and $s = \lambda_i$ in Eq. 3-80, and rearranging terms, we find that

$$d\lambda_i = \left[\frac{-1}{\partial G/\partial s}\right]_{s=\lambda_i}\left\{\frac{\partial G}{\partial \kappa}\,d\kappa + \sum_{j=1}^{n}\frac{\partial G}{\partial z_j}\,dz_j + \sum_{j=1}^{m+n}\frac{\partial G}{\partial p_j}\,dp_j\right\}_{s=\lambda_i} \tag{3-82}$$

The variation, $d\lambda_i$, in the closed-loop pole can be expressed in another way by noting that the closed-loop roots depend only on the open-loop gain, poles, and zeros. Functionally,

$$\lambda_i = \lambda_i(\kappa, z_j, p_j) \tag{3-83}$$

The total derivative $d\lambda_i$ is then

$$\begin{aligned} d\lambda_i &= \kappa\frac{\partial \lambda_i}{\partial \kappa}\frac{d\kappa}{\kappa} + \sum_{j=1}^{n}\frac{\partial \lambda_i}{\partial z_j}\,dz_j + \sum_{j=1}^{m+n}\frac{\partial \lambda_i}{\partial p_j}\,dp_j \\ &= S_\kappa^i\frac{d\kappa}{\kappa} + \sum_{j=1}^{n}S_{z_j}^i\,dz_j + \sum_{j=1}^{m+n}S_{p_j}^i\,dp_j \end{aligned} \tag{3-84}$$

As indicated in the last form of Eq. 3-84, the factors $\kappa(\partial\lambda_i/\partial\kappa)$, $\partial\lambda_i/\partial z_j$, and $\partial\lambda_i/\partial p_j$ are accorded a special symbol, $S$. These are the first-order sensitivity factors. The subscript and superscript notation indicates that a differential increment in the open-loop parameter (defined by the subscript) results in a differential increment of the $i^{\text{th}}$ closed-loop root (denoted in the superscript) which is equal to the sensitivity factor times the open-loop parametric variation.

By equating like coefficients in Eqs. 3-82 and 3-84,

$$\begin{aligned} S_\kappa^i &\equiv \frac{\partial\lambda_i}{\partial\kappa/\kappa} = -\left[\frac{\partial G/\partial\kappa}{\partial G/\partial s}\right]_{s=\lambda_i}\kappa \\ S_{z_j}^i &\equiv \frac{\partial\lambda_i}{\partial z_j} = -\left[\frac{\partial G/\partial z_j}{\partial G/\partial s}\right]_{s=\lambda_i} \\ S_{p_j}^i &\equiv \frac{\partial\lambda_i}{\partial p_j} = -\left[\frac{\partial G/\partial p_j}{\partial G/\partial s}\right]_{s=\lambda_i} \end{aligned} \tag{3-85}$$

Note that the gain sensitivity is based on a fractional (percentage) change in $\kappa$, while the pole and zero sensitivities are based on absolute shifts of $p_j$ and $z_j$. These definitions were adopted here in order to provide some simplifications in relationships.

In terms of the open-loop transfer function form of Eq. 3-79, the gain sensitivity factor, remembering that $G(\lambda_i) = -1$, is,

$$S_\kappa^i = \left[-\frac{1}{\partial G/\partial s}\right]_{s=\lambda_i}\left[\frac{\partial G}{\partial \kappa}\right]_{s=\lambda_i}\kappa$$

$$= \left[-\frac{1}{\partial G/\partial s}\right]_{s=\lambda_i}[G(s)]_{s=\lambda_i}$$

$$= \left[\frac{1}{\partial G/\partial s}\right]_{s=\lambda_i} \tag{3-86}$$

Using Eq. 3-79, we obtain

$$S_\kappa^i = -\left[\frac{\kappa\alpha(s)}{\kappa[d\alpha(s)/ds] + [d\beta(s)/ds]}\right]_{s=\lambda} \tag{3-87}$$

or, since $\kappa\alpha(\lambda_i) = -\beta(\lambda_i)$,

$$S_\kappa^i = \left[\frac{\beta(s)}{\kappa[d\alpha(s)/ds] + [d\beta(s)/ds]}\right]_{s=\lambda_i} \tag{3-88}$$

With the factored form of Eq. 3-79,

$$\frac{\partial G}{\partial s} = G(s)\left[\frac{1}{\alpha(s)}\frac{\partial \alpha}{\partial s} - \frac{1}{\beta(s)}\frac{\partial \beta}{\partial s}\right]$$

$$= G(s)\left[\frac{1}{\alpha}\sum_{j=1}^{n}\prod_{\substack{k=1\\k\neq j}}^{n}(s+z_k) - \frac{1}{\beta}\sum_{j=1}^{m+n}\prod_{\substack{k=1\\k\neq j}}^{m+n}(s+p_k)\right]$$

$$= G(s)\left[\sum_{j=1}^{n}\frac{1}{(s+z_j)} - \sum_{j=1}^{m+n}\frac{1}{(s+p_j)}\right]$$

so that

$$S_\kappa^i = \left[\frac{1}{\partial G/\partial s}\right]_{s=\lambda_i} = \left[\sum_{j=1}^{m+n}\frac{1}{\lambda_i+p_j} - \sum_{j=1}^{n}\frac{1}{\lambda_i+z_j}\right]^{-1} \tag{3-89}$$

With the use of Eqs. 3-79, 3-85, and 3-86, the zero sensitivity becomes

$$S_{z_j}^i = -\left[\frac{\partial G/\partial z_j}{\partial G/\partial s}\right]_{s=\lambda_i} = \left[\frac{1}{\partial G/\partial s}\right]_{s=\lambda_i}\left[-\frac{\partial G}{\partial z_j}\right]_{s=\lambda_i}$$

$$= S_\kappa^i\left[\frac{-\kappa\prod_{k=1,k\neq j}^{n}(s+z_k)}{\prod_{k=1}^{m+n}(s+p_k)}\right]_{s=\lambda_i}\left[\frac{(s+z_j)}{(s+z_j)}\right]_{s=\lambda}$$

$$= S_\kappa^i[-G(\lambda_i)]\left(\frac{1}{s+z_j}\right)_{s=\lambda_i}$$

$$= \frac{S_\kappa^i}{\lambda_i+z_j} \tag{3-90}$$

By a similar development, the pole sensitivity is found to be

$$S_{p_j}^i = -\frac{S_\kappa^i}{\lambda_i + p_j} \tag{3-91}$$

Further examination and interpretation reveals six interesting and useful properties of the sensitivity factors:

1. The gain sensitivity is a measure of the slope along a conventional root locus.
2. The gain sensitivity is a factor in each of the sensitivity terms. Thus Eq. 3-84 becomes

$$d\lambda_i = S_\kappa^i\left(\frac{d\kappa}{\kappa} + \sum_{j=1}^{n}\frac{dz_j}{\lambda_i + z_j} - \sum_{j=1}^{m+n}\frac{dp_j}{\lambda_i + p_j}\right) \tag{3-92}$$

3. The gain sensitivity is the negative of the modal response coefficient for the root $\lambda_i$. This will be shown for the general case of $N^{\text{th}}$ order roots in the next article but will be taken for granted here. Thus,

$$S_\kappa^i = -Q_i \tag{3-93}$$

This equality is very useful, since all of the properties previously derived for the modal response coefficients are applicable to the gain sensitivity. With the use of this correspondence, other formulas for the gain sensitivity can be developed.

$$\begin{aligned} S_\kappa^i &= -[(s - \lambda_i)G_{\text{cr}}(s)]_{s=\lambda_i} \\ &= -\left[\frac{(s - \lambda_i)\kappa \prod_{j=1}^{n}(s + z_j)}{(1 + \kappa\delta_m^0)\prod_{j=1}^{m+n}(s + q_j)}\right]_{s=\lambda_i}, \qquad \delta_m^0 = \begin{matrix} 1, \ m = 0 \\ 0 \text{ otherwise} \end{matrix} \end{aligned}$$

Remembering that $\lambda_i = -q_i$, and that $\kappa\alpha(\lambda_i) = -\beta(\lambda_i)$, we obtain

$$\begin{aligned} S_\kappa^i &= \frac{-\kappa\prod_{j=1}^{n}(\lambda_i + z_j)}{(1 + \kappa\delta_m^0)\prod_{j=1, j\neq i}^{m+n}(\lambda_i + q_j)} \\ &= \frac{\prod_{j=1}^{m+n}(\lambda_i + p_j)}{(1 + \kappa\delta_m^0)\prod_{j=1, j\neq i}^{m+n}(\lambda_i + q_j)} \end{aligned} \tag{3-94}$$

4. Some combinations of gain sensitivities or modal response ratios are simple functions of the open-loop transfer function polynomial coefficients and gains. These are conveniently developed by matching coefficients in

expressions for the closed-loop transfer function and in the partial fraction expansion.[18] Among the relationships are the following for $m \geqq 1$:

$$\sum_{i=1}^{m+n} Q_i = -\sum_{i=1}^{m+n} S_\kappa^i = \kappa\delta_m^1 \tag{3-95}$$

$$\sum_{i=1}^{m+n} Q_i q_i = -\sum_{i=1}^{m+n} S_\kappa^i q_i = \kappa(b_1 - a_1 + \kappa)\,\delta_m^1 - \kappa\delta_m^2 \tag{3-96}$$

$$\sum_{i=1}^{m+n} \frac{Q_i}{q_i} = -\sum_{i=1}^{m+n} \frac{S_\kappa^i}{q_i} = \begin{cases} 1, & k \geqq 1 \\ \dfrac{K}{1+K}, & k = 0 \\ 0, & k \leqq -1 \end{cases} \tag{3-97}$$

5. The sum of all the zero and pole sensitivities for each closed-loop pole must equal $-1$. This follows directly from Eqs. 3-89, 3-90, and 3-91. Thus,

$$\sum_{j=1}^{n} S_{z_j}^i + \sum_{j=1}^{m+n} S_{p_j}^i = -1 \tag{3-98}$$

This result is easily explained by recalling that, on a root-locus plot, if all the open-loop zeros and poles are moved the same amount, all the closed-loop poles will also be moved by that amount. The minus sign is used because $z_j$ and $p_j$ are the negatives of the open-loop zeros and poles.

6. The gain sensitivities are the residues, evaluated at the closed-loop poles, of the classical sensitivity $S_G^{G_{cr}}$. This is a direct consequence of the relationship between the modal response coefficients and gain sensitivities.

$$\begin{aligned} S_G^{G_{cr}} = G_{er} &= 1 - \sum_{i=1}^{m+n} \frac{Q_i}{s + q_i} \\ &= 1 + \sum_{i=1}^{m+n} \frac{S_\kappa^i}{s + q_i} \end{aligned} \qquad m \geqq 1 \tag{3-99}$$

*Evaluation Procedures for Sensitivities*

The evaluation of single-loop gain sensitivities, or modal response coefficients, may be accomplished in several ways.[19] The fundamental bases are the formulas given by Eqs. 3-86, 3-87, 3-88, 3-89, or 3-94. All of these are appropriate for direct computation which is probably the most common means employed for sensitivity calculation.

Some of the gain sensitivity equations are readily interpreted graphically in terms of vectors in the complex plane. Equation 3-89, for example,

[18] D. T. McRuer, and R. L. Stapleford, "Sensitivity and Modal Response for Single-Loop and Multiloop Systems," ASD-TDR-62-812, Jan. 1963.

[19] *Ibid.*

$$G = \frac{\kappa(s+z)}{s^2}$$

$$S_\kappa^1 = \frac{-\kappa(\lambda_1+z)}{(\lambda_1+q_2)} = -Q_1$$

$$S_\kappa^2 = \frac{-\kappa(\lambda_2+z)}{(\lambda_2+q_1)} = -Q_2$$

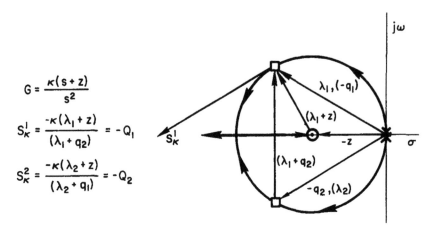

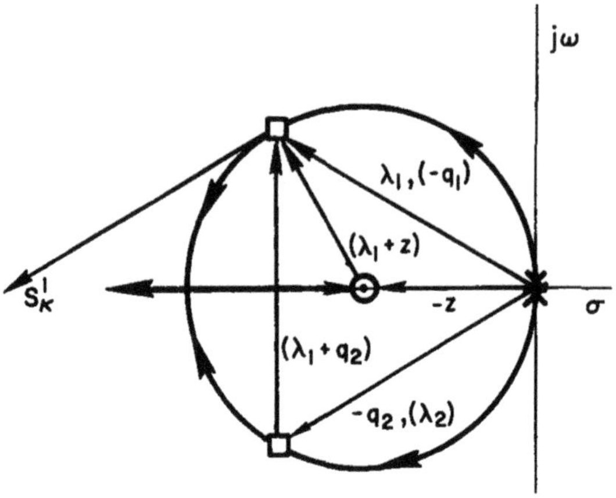


Fig. 3-43. Illustration of gain sensitivity and modal response coefficient calculations, using vectors from the complete root locus.

instructs us to draw, at any point, $\lambda_i$, on the locus of roots, vectors directed toward the zeros and away from the poles of the open-loop function, inversely proportional in magnitude to the distance from $\lambda_i$ to the singularity in question. The sum of these vectors shows the direction of the locus with increasing gain. A second graphical method using vectors in the complex plane is based on conventional root locus constructions. The basis is Eq. 3-94 which involves a ratio of vector products. With a complete set of closed-loop roots plotted on the locus, this calculation can be quickly evaluated, as exemplified in Fig. 3-43. The operations involved can be accomplished particularly rapidly when the logarithmic spiral features of the "Spirule"[20] are used as an aid. This is the method most often given for the graphical evaluation of the modal response coefficients. Its disdavantage for sensitivity calculations is that it requires a complete set of compatible closed-loop roots plotted on the locus.

Beside these vector forms, sensitivities can be obtained from any of the common plots used in the unified servo-analysis techniques treated earlier in this chapter.[21] Most of the methods depend on the interpretation of Eq. 3-86 in terms of slopes from the particular plot available (e.g., $j\omega$-Bode, conventional root locus, etc.).

When the complete set of conventional and Bode root locus plots are available, graphical methods using increments in the gain and closed-loop

[20] See n. 5.

[21] McRuer and Stapleford, "Sensitivity and Modal Response for Single-Loop and Multiloop Systems," ASD-TDR-62-812, Jan. 1963.

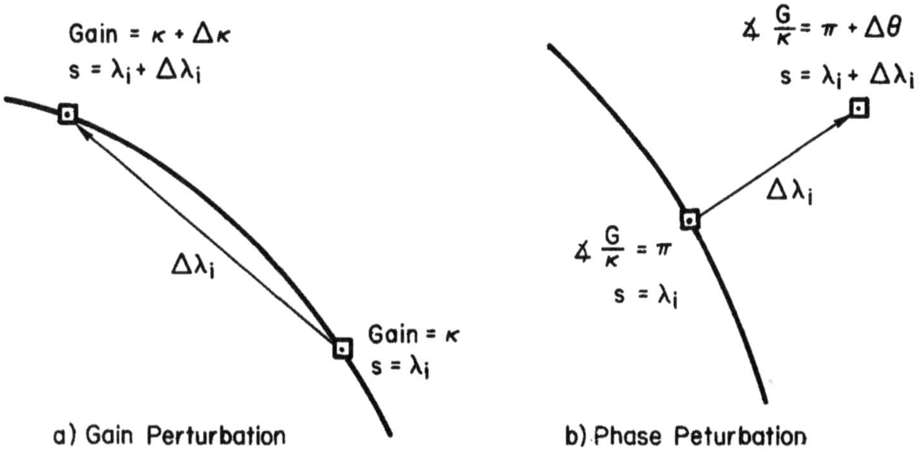


Fig. 3-44. Root-locus perturbation methods for gain sensitivity.

poles offer the simplest way to obtain a rapid appreciation of sensitivity considerations. The direction of the gain sensitivity is most easily determined from the conventional root locus, as it is simply along the tangent to the locus. A close approximation to the magnitude can be obtained, using the location of the closed-loop pole for two slightly different gains such as the segment of a locus shown in Fig. 3-44. The gain perturbation method requires the locations of the closed-loop pole for two slightly different gains, such as the segment of a locus shown in Fig. 3-44(a). With the use of finite increments as approximations to differential changes, Eq. 3-85 gives

$$S_\kappa^i = \frac{d\lambda_i}{d\kappa/\kappa} \doteq \frac{\kappa(\Delta\lambda_i)}{\Delta\kappa} \tag{3-100}$$

Equation 3-100 can also be used if the change in the closed-loop pole is obtained by some other technique such as from root decomposition.

The phase perturbation method is based on considering $\kappa$ to be a complex quantity. The normal root locus is then a graph of the closed-loop pole locations for $\kappa$ real. Now consider a small perturbation in the phase of $\kappa$. The closed-loop pole must then be perturbed a small distance normal to the conventional root locus, and the phase perturbation of $(G/\kappa)$ must be minus that of $\kappa$ [see Fig. 3-44(b)]. Consequently, for perturbations normal to the root locus, $\Delta\kappa/\kappa = -j\,\Delta\theta$, and Eq. 3-100 becomes

$$S_\kappa^i = \frac{d\lambda_i}{d\kappa/\kappa} \doteq j\frac{\Delta\lambda_i}{\Delta\theta} \tag{3-101}$$

where $\Delta\theta$ is the phase change in radians of $G/\kappa$.

For the Bode root locus a more natural sensitivity factor is the proportional change in the negative of the closed-loop root, i.e., $dq_i/q_i$ instead of $d\lambda_i$. If we consider gain sensitivity alone,

$$dq_i = -d\lambda_i = -S^i_\kappa \frac{d\kappa}{\kappa}$$

or

$$\frac{dq_i}{q_i} = -\frac{S^i_\kappa}{q_i}\frac{d\kappa}{\kappa}$$

$$= M^i_\kappa \frac{d\kappa}{\kappa} \tag{3-102}$$

where $M^i_\kappa$ is the sensitivity factor relating proportional changes in the negative of the closed-loop pole, $q_i$, and the gain. Because $dx/x = d \ln x$, the proportional sensitivity, $M^i_\kappa$, will be

$$M^i_\kappa = -\frac{S^i_\kappa}{q_i} = \frac{dq_i/q_i}{d\kappa/\kappa}$$

$$= \frac{d \ln q_i}{d \ln \kappa} \tag{3-103}$$

The proportional sensitivity can thus be estimated from the Bode root locus by considering small increments about the nominal gain. For real roots this will simply be the slope of the $\sigma$-Bode plot at $q_i$. As often as not, this would be expressed in decades/dB. To convert to a unit-less form of $M^i_\kappa$, we note that

$$M^i_\kappa]_{\text{decades/dB}} = \frac{d(\log_{10} q_i)}{d(20 \log_{10} \kappa)} = \frac{\log_{10} e(d \ln q_i)}{20 \log_{10} e(d \ln \kappa)}$$

$$= \tfrac{1}{20} M^i_\kappa \tag{3-104}$$

When the roots are complex, the most convenient quantities to measure along the Bode root locus are increments in the closed-loop damping ratio, $\Delta\zeta$, and logarithmic increments in the closed-loop natural frequency, $\Delta \ln \omega_n$, and open-loop gain $\Delta \ln \kappa$. These can readily be related to the proportional sensitivity. The negative of the closed-loop root in the upper half of the $s$ plane is

$$q_i = \zeta\omega_n - j\sqrt{1 - \zeta^2}\,\omega_n$$

so that

$$dq_i = \frac{\partial q_i}{\partial \zeta} d\zeta + \frac{\partial q_i}{\partial \omega_n} d\omega_n$$

$$= \frac{j[\zeta\omega_n - j\sqrt{1-\zeta^2}\,\omega_n]}{\sqrt{1-\zeta^2}} d\zeta + (\zeta - j\sqrt{1-\zeta^2})\, d\omega_n$$

$$= q_i\left[\frac{j}{\sqrt{1-\zeta^2}} d\zeta + \frac{d\omega_n}{\omega_n}\right]$$

The proportional sensitivity, $M_\kappa^i$, will then be

$$M_\kappa^i = \frac{d \ln q_i}{d \ln \kappa} = \frac{dq_i/q_i}{d \ln \kappa}$$

$$= \frac{(d\omega_n/\omega_n) + (j/\sqrt{1-\zeta^2})\, d\zeta}{d \ln \kappa}$$

$$= \frac{d \ln \omega_n}{d \ln \kappa} + \frac{1}{\sqrt{1-\zeta^2}} \frac{d\zeta}{d \ln \kappa} \tag{3-105}$$

The first term is converted from dB/decades by multiplying by 20, just as with first-order terms. The imaginary part will usually be measured as

$$\frac{\Delta\zeta}{\Delta(20 \log_{10} \kappa)} = \frac{\Delta\zeta}{20 \log_{10} e(\Delta \ln \kappa)}$$

so that

$$\frac{d\zeta}{d \ln \kappa} = 20 \log_{10} e \frac{d\zeta}{d(20 \log_{10} \kappa)}$$

$$= 8.6859 \frac{d\zeta}{d(20 \log_{10} \kappa)} \tag{3-106}$$

The gain sensitivity in these terms will then be

$$S_\kappa^i = \lambda_i M_\kappa^i = \lambda_i\left[\frac{20\, d(\log_{10} \omega_n)}{d(20 \log_{10} \kappa)} + \frac{j8.6859}{\sqrt{1-\zeta^2}} \frac{d\zeta}{d(20 \log_{10} \kappa)}\right] \tag{3-107}$$

The pole and zero sensitivities are easily obtained once the gain sensitivities have been found. A geometric appreciation is gained by considering them as vectors in the $s$ plane, as in Fig. 3-45. Note that the zero sensitivity vector is in a direction tending to pull the locus more toward the zero, whereas the pole sensitivity vector would tend to push the locus away from the pole.

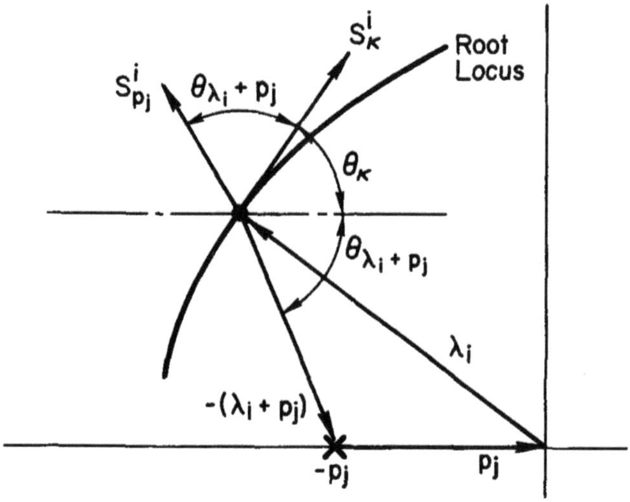

$$S^i_{p_j} = \frac{-S^i_\kappa}{\lambda_i + p_j} = \frac{|S^i_\kappa| \, \measuredangle \theta_\kappa}{|\lambda_i + p_j| \, \measuredangle -(\theta_{\lambda_i + p_j})} = \left| \frac{S^i_\kappa}{\lambda_i + p_j} \right| \measuredangle (\theta_\kappa + \theta_{\lambda_i + p_j})$$

*a) Pole Sensitivity*



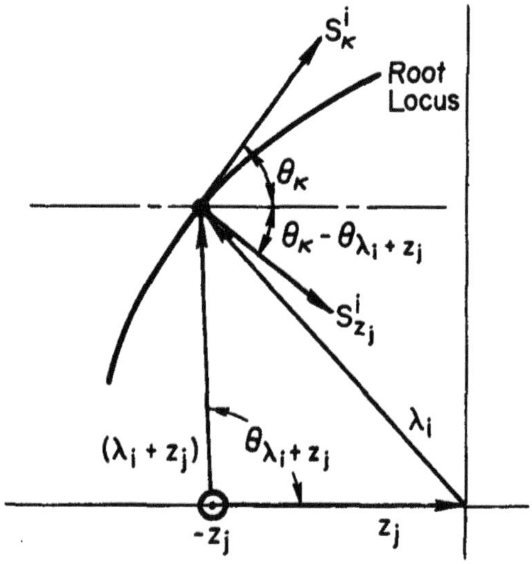

$$S^i_{z_j} = \frac{S^i_\kappa}{\lambda_i + z_j} = \frac{|S^i_\kappa| \, \measuredangle \theta_\kappa}{|\lambda_i + z_j| \, \measuredangle \theta_{\lambda_i + z_j}} = \left| \frac{S^i_\kappa}{\lambda_i + z_j} \right| \measuredangle (\theta_\kappa - \theta_{\lambda_i + z_j})$$

*b) Zero Sensitivity*

Fig. 3-45. Geometric illustrations of pole and zero sensitivity vectors.

**Example 1. Unit-numerator third-order system**

The analysis of the system

$$G(s) = \frac{\kappa}{s(s+1)(s+5)}$$

has been extensively treated in Examples 4 of Sections 3-2 and 3-3. A complete system survey, comprising block diagram, $G(j\omega)$ and $G(-\sigma)$ Bode plots, and Bode and conventional root locus plots is given in Fig. 3-46. In this example we shall compute the gain sensitivities in several ways for illustration and as an indication of relative accuracies.

The gain is set so that a pair of complex closed-loop poles with damping ratio of $\sqrt{2}/2$ exists. For this situation the important parameters are: $\kappa = 31\sqrt{26} - 156 = 2.070$, $q_1 = -\lambda_1 = \sqrt{26} = 5.099$, $q_2 = -\lambda_2 = \frac{1}{2}(6 - \sqrt{26})(1 - j) = 0.450(1 - j)$, and $q_3 = -\lambda_3 = \frac{1}{2}(6 - \sqrt{26})(1 + j) = 0.450(1 + j)$.

*1. Direct Calculation.* Using Eq. 3-87 for the numerator and denominator derivatives method, we find that

$$S_\kappa^i = \left(\frac{-\kappa\alpha}{\kappa(\partial\alpha/\partial s) + (\partial\beta/\partial s)}\right)_{s=\lambda_i}$$

where

$$\alpha = 1$$

$$\frac{\partial\alpha}{\partial s} = 0$$

$$\beta = s^3 + 6s^2 + 5s$$

$$\frac{\partial\beta}{\partial s} = 3s^2 + 12s + 5$$

Therefore,

$$S_\kappa^i = \frac{-\kappa}{3\lambda_i^2 + \kappa 2\lambda_i + 5}$$

Using the value for $\lambda_2$, we obtain

$$\begin{aligned} S_\kappa^2 &= \frac{-3276 + 701\sqrt{26} + j(8112 - 987\sqrt{26})}{6290} \\ &= 0.0474 + j0.490 \\ &= 0.492 \measuredangle\ 84.47 \text{ degrees} \end{aligned}$$

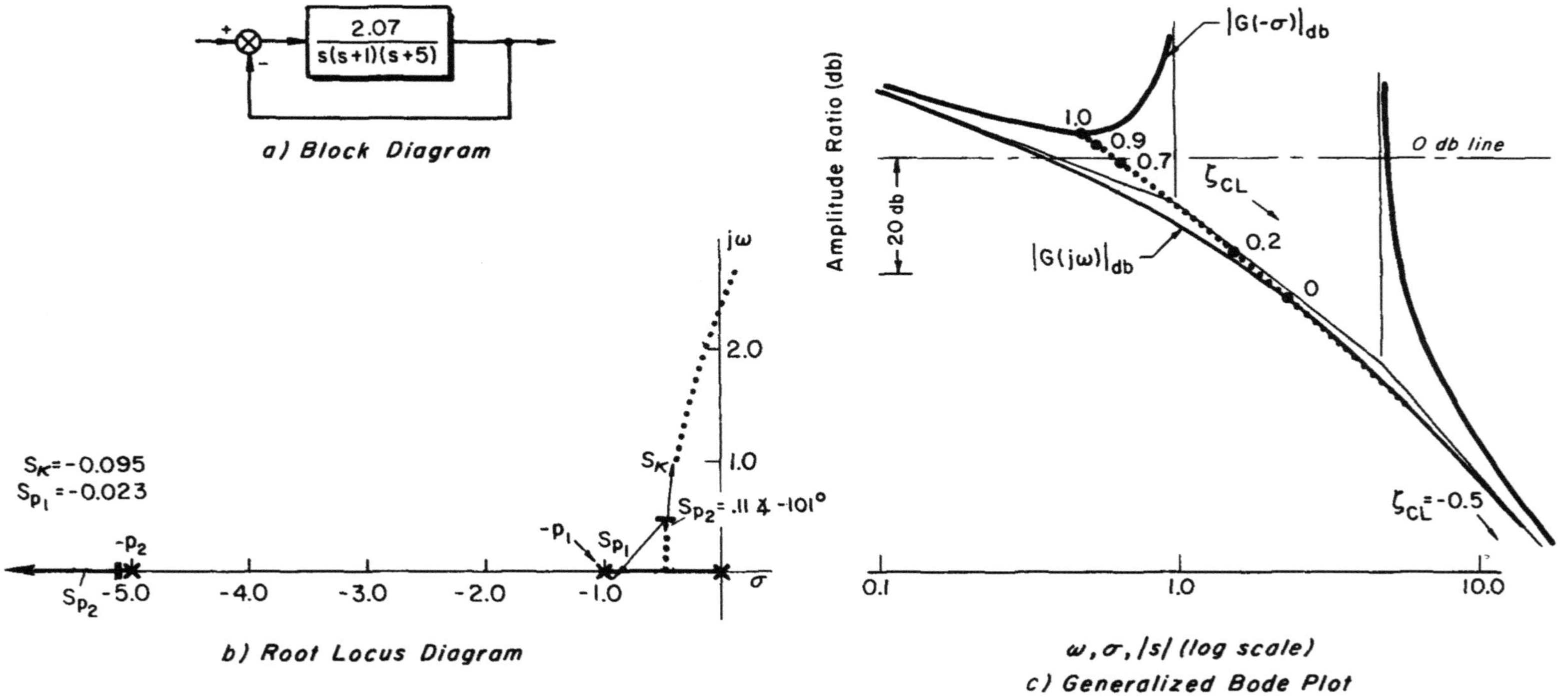


Fig. 3-46. System survey for unit-numerator third-order system.

Because $\lambda_3$ is the complex conjugate of $\lambda_2$, $S_\kappa^3$ is the complex conjugate of $S_\kappa^2$ or

$$S_\kappa^3 = \frac{-3276 + 701\sqrt{26} - j(8112 - 987\sqrt{26})}{6290}$$

$$= 0.0474 - j0.490$$

$$= 0.492 \measuredangle -84.47 \text{ degrees}$$

Using the value of $\lambda_1$ in the equation for $S_\kappa^i$, we obtain

$$S_\kappa^1 = \frac{3276 - 701\sqrt{26}}{3145} = -0.0949$$

The modal response coefficients are, of course, the negatives of these sensitivities.

As a check, recall that, if the number of system poles is greater than the number of zeros by two or more ($m \geqq 2$), then the sum of the modal response coefficients or gain sensitivities is zero. These conditions are met for this example, and we can see that the values do sum to zero.

The second method of direct calculation, called the summation of terms method, uses Eq. 3-89, i.e.,

$$S_\kappa^i = \frac{1}{\sum_{j=1}^{m+n} 1/(\lambda_i + p_j) - \sum_{j=1}^{n} 1/(\lambda_i + z_j)}$$

These computations may be performed in the following manner:

| $p_j$ | $(p_j + \lambda_2)$ | $\dfrac{1}{p_j + \lambda_2}$ |
|---|---|---|
| 0 | $-0.45049 + j0.45049$ | $-1.1099 - j1.1099$ |
| 1 | $0.54951 + j0.45049$ | $1.0883 - j0.8922$ |
| 5 | $4.54951 + j0.45049$ | $0.2177 - j0.0216$ |

$$\sum \left(\frac{1}{p_j + \lambda_2}\right) = 0.1961 - j2.0237$$

$$S_\kappa^2 = 0.0474 + j0.490 = 0.492 \measuredangle 84.47 \text{ degrees}$$

$$S_\kappa^3 = 0.0474 - j0.490 = 0.492 \measuredangle -84.47 \text{ degrees}$$

and

| $p_j$ | $(p_j + \lambda_1)$ | $\dfrac{1}{p_j + \lambda_1}$ |
|---|---|---|
| 0 | −5.09902 | −0.1961 |
| 1 | −4.09902 | −0.2440 |
| 5 | −0.09902 | −10.0990 |

$$\sum\left(\frac{1}{p_j + \lambda_1}\right) = -10.5391$$

$$S_\kappa^1 = -0.0949$$

The results naturally agree with the first method because both methods are exact.

The sensitivities to the poles at −1 and −5 are easily computed from Eq. 3-91. For the pole at −1,

$$S_{p_1}^1 = \frac{-S_\kappa^1}{\lambda_1 + p_1} = \frac{0.0949}{-4.10} = -0.0231$$

$$S_{p_1}^2 = \frac{-(0.0474 + j0.490)}{0.550 + j0.450} = -0.487 - j0.494$$

$$= 0.695 \measuredangle -135 \text{ degrees}$$

For the pole at −5,

$$S_{p_2}^1 = \frac{0.0949}{-0.0990} = -0.959$$

$$S_{p_2}^2 = \frac{-(0.0474 + j0.490)}{4.55 + j0.450} = -0.0208 - j0.1058$$

$$= 0.1077 \measuredangle -101 \text{ degrees}$$

2. *Vector Method.* The vector method uses Eq. 3-94. For this example, the equation reduces to

$$S_\kappa^1 = \frac{-\kappa}{(\lambda_1 + q_2)(\lambda_1 + q_3)}$$

$$S_\kappa^2 = \frac{-\kappa}{(\lambda_2 + q_1)(\lambda_2 + q_3)}$$

The following were measured with a Spirule:

$$\frac{1}{(\lambda_1 + q_2)(\lambda_1 + q_3)} = 0.0459$$

$$\frac{1}{(\lambda_2 + q_1)(\lambda_2 + q_3)} = 0.237 \measuredangle -95.5 \text{ degrees}$$

which gives

$$S_\kappa^1 \doteq -0.0952$$

$$S_\kappa^2 \doteq 0.490 \measuredangle\ 84.5 \text{ degrees}$$

These values are extremely close to the exact values. Both amplitude errors are less than 1 percent, and the angular error of $S_\kappa^2$ is less than 1 degree.

*3. Root-Locus Perturbation Methods.* To estimate the gain sensitivity from Eq. 3-100, a perturbed position on the locus of $-0.41 + j0.76$ was chosen. The measured gain at that point was 3.89. This gives an estimate of

$$S_\kappa^2 \doteq \frac{\kappa \Delta\lambda_2}{\Delta\kappa} = 2.07 \frac{[(-0.41 + j0.76) - (-0.45 + j0.45)]}{3.89 - 2.07}$$

$$= 0.045 + j0.353 = 0.356 \measuredangle\ 82.7 \text{ degrees}$$

which has a magnitude error of 28 percent and an angular error of 2 degrees. For $S_\kappa^1$ the point $-5.42$ was selected as the perturbed value of $\lambda$, for which the gain was 10.06. Then

$$S_\kappa^1 \doteq \frac{2.07[-5.42 - (-5.10)]}{10.06 - 2.07} = -0.083$$

Although the gain was increased to nearly five times its original value, this estimate is within 13 percent of the exact value.

To obtain estimates from Eq. 3-101, perturbations normal to the locus were considered. For $S_\kappa^2$ a perturbed $\lambda$ of $-0.64 + j0.47$ was selected, for which the measured phase change was $-24$ degrees. Then

$$S_\kappa^2 \doteq \frac{j\Delta\lambda_2}{\Delta\theta} = \frac{[(-0.64 + j0.47) - (-0.45 + j0.45)]}{-24/57.3}$$

$$= 0.0477 + j0.454 = 0.457 \measuredangle\ 84 \text{ degrees}$$

For this estimate the amplitude error is 7 percent and the angular error is less than 1 degree.

For $S_\kappa^1$ a perturbed $\lambda$ of $-5.10 + j0.10$ was chosen. The phase change was 47.5 degrees, so that

$$S_\kappa^1 \doteq \frac{j(j0.10)(57.3)}{47.5} = -0.121$$

This estimate is in error by 27 percent.

*4. Bode Root Locus Increments.* The pertinent slope on the Bode root locus for the complex root was measured as

$$\frac{\Delta 20 \log_{10} \kappa}{\Delta \log \omega_{\mathrm{CL}}} \doteq 39 \text{ dB/decade}$$

and a gain decrease of 3.8 dB increases $\zeta_{\mathrm{CL}}$ to 0.9 over the original 0.7, or 0.0527/dB. The proportional sensitivity (Eq. 3-105) is then

$$M_\kappa^2 = \frac{d \ln \omega_n}{d \ln \kappa} + \frac{j}{\sqrt{1-\zeta}} \frac{d\zeta}{d \ln \kappa}$$

$$\doteq \frac{20}{39} + \frac{j}{0.707}(8.686)\frac{0.2}{(-3.8)}$$

$$= 0.513 - j0.647$$

The conventional gain sensitivity is then

$$S_\kappa^2 = \lambda_2 M_\kappa^2 = (0.513 - j0.647)(-0.450)(1 - j)$$

$$= 0.0603 + j0.522$$

$$= 0.525 \measuredangle 83.4 \text{ degrees}$$

This is an error of about 7 percent in amplitude and 1 degree in angle.

To obtain $S_\kappa^1$, the slope of the high frequency branch of the $\sigma$-Bode is used. A slope measured from this plot is −830 dB/decade, although the root is too close to the open-loop pole to get an accurate value. Nevertheless,

$$M_\kappa^1 \doteq \tfrac{20}{830} = 0.0241$$

and

$$S_\kappa^1 = \lambda_1 M_\kappa^1 = (-5.1)(0.0241) = -0.123$$

This estimate is about 29 percent higher than the correct value.

*Limiting Behavior and Special Cases*

The magnitudes of gain sensitivities can cover the entire range of values from minus to plus infinity. Yet, intuitive notions of "sensitivity" as a general concept in closed-loop systems make part of this range appear unreasonable. One part of the problem is a direct consequence of the sensitivity definition, while another is associated with its nature as a first-order approximation. A better understanding of both facets can be gained by an examination of limiting cases.

In general, closed-loop poles are close to open-loop poles for low values of gain and proceed to either open-loop zeros or unbounded values as the open-loop gain becomes very large. Throughout this travel the gain sensitivity is (Eq. 3-89 repeated here)

$$S_\kappa^i = \frac{1}{\sum_{j=1}^{m+n} 1/(\lambda_i + p_j) - \sum_{j=1}^{n} 1/(\lambda_i + z_j)} \tag{3-108}$$

As $\kappa$ approaches zero, the closed-loop root, $\lambda_i$, approaches the open-loop pole from which it derives, i.e., $\lambda_i \to -p_i$. Then the term $1/(\lambda_i + p_i)$ in

Eq. 3-108 is dominant, so that

$$S^i_\kappa]_{\kappa\to 0} \to \frac{1}{1/(\lambda_i + p_i)} \to 0 \tag{3-109}$$

Similarly, as $\kappa$ becomes very large, $n$ of the closed-loop poles approach open-loop zeros. If the $i^{\text{th}}$ closed-loop pole is one of these, and it approaches the $j^{\text{th}}$ open-loop zero, so that $\lambda_i \to -z_j$, then

$$S^i_\kappa]_{\kappa\to\infty} \to \frac{-1}{1/(\lambda_i + z_j)} \to 0 \tag{3-110}$$

Finally, $m$ of the closed-loop poles have no zeros to approach and hence become very large relative to the $|p_j|$ and $|z_j|$. The sensitivity for these poles is

$$S^i_\kappa]_{\lambda_i \gg |z_j|, |p_j|} \to \frac{1}{\sum_{j=1}^{m} 1/\lambda_i} = \frac{\lambda_i}{m} \tag{3-111}$$

When the gain is sufficiently large for the open-loop zero dB line to intersect the high frequency asymptote, the open-loop transfer function is approximately

$$G(s) \doteq \frac{\kappa}{s^m}$$

so that $\lambda_i$ will be

$$\lambda_i \doteq \sqrt[m]{-\kappa} \tag{3-112}$$

Thus the sensitivity of the unbounded pole will be

$$S^i_\kappa]_{\lambda_i \gg |z_j|, |p_j|} \to \frac{\sqrt[m]{-\kappa}}{m} \tag{3-113}$$

Equation 3-113 indicates that the sensitivity increases as the $m^{\text{th}}$ root of $\kappa$ as the gain is increased, although for finite gains the sensitivity is always finite.

Another circumstance in which the sensitivity can become very large is revealed by Eq. 3-94. Here it is apparent that the gain sensitivity for a closed-loop pole becomes very large as that pole nears another closed-loop pole. Indeed, as $\lambda_i$ becomes equal to $\lambda_j$, indicating a branch point on the root locus, the gain sensitivity goes to infinity. This is to be expected, since the sensitivity factors defined thus far have not considered multiple-order, closed-loop roots. As long as the gain is finite, an infinite gain sensitivity always indicates multiple-order, closed-loop poles.

A special situation of considerable interest can occur when a closed-loop root lies between an open-loop pole and zero which are much closer to each other than to all other open-loop poles and zeros. This is the so-called dipole case. The sensitivity for the bounded closed-loop pole will be, approximately,

$$S_\kappa^i \doteq \frac{1}{1/(\lambda_i + p_i) - 1/(\lambda_i + z_i)}$$

$$\doteq \frac{(\lambda_i + z_i)(\lambda_i + p_i)}{z_i - p_i} \tag{3-114}$$

The maximum value of $S_\kappa^i$ will occur when $\lambda_i = -(z_i + p_i)/2$, for which $S_\kappa^i$ becomes

$$S_\kappa^i]_{\max} \doteq \tfrac{1}{4}(p_i - z_i) \tag{3-115}$$

*First-Order Root Sensitivities for a General Characteristic Equation*[22]

We turn now to the general case in which multiple-order, closed-loop roots are permitted, and in which the open-loop system parameters are not confined to gains, poles, and zeros. To accomplish the generalization, we shall repeat many of the steps involved in the special case with first-order roots only.

For linear constant-coefficient systems of the $n^{\text{th}}$ order, the system characteristic equation can be written as

$$F(\lambda_i, \boldsymbol{\alpha}) = 0, \qquad i = 1, 2, \ldots, n \tag{3-116}$$

where $\lambda_i$ is a root and $\boldsymbol{\alpha}$ is a vector with components $\alpha_j$ representing the set of system parameters. Each component, $\alpha_j$, is nominally constant but potentially variable. Open-loop poles, zeros, and gains are typical $\alpha_j$'s, although other system parameters are also included. If $\lambda_i$ is a first-order root, the total derivative of $F$ will be

$$dF = \frac{\partial F}{\partial \lambda_i} d\lambda_i + \sum_j \frac{\partial F}{\partial \alpha_j} d\alpha_j = 0 \tag{3-117}$$

When $\lambda_i$ is an $N^{\text{th}}$ order root, $F(s)$ has the form $(s - \lambda_i)^N F_1(s)$. Then the first $N - 1$ derivatives of $F$ with respect to $s$, when evaluated at $s = \lambda_1$, become zero. In this event the total differential given above must be expanded to incorporate the first nonzero higher order term. Equation 3-116 is then modified to

$$dF = \frac{1}{N!}\left(\frac{\partial^N F}{\partial \lambda_i{}^N}\right)(d\lambda_i)^N + \sum_j \frac{\partial F}{\partial \alpha_j} d\alpha_j = 0 \tag{3-118}$$

[22] This subsection is a condensed version of the paper: R. L. Stapleford and D. T. McRuer, "Sensitivity of Multiloop Flight Control System Roots to Open-Loop Parameter Variations," *AIAA J.*, 4, No. 9, 1655–1661 (Sept. 1966).

Equation 3-117 can be solved for $d\lambda_i$ to give

$$d\lambda_i = \left\{\frac{N!}{[(\partial^N F/\partial s^N)(s)]}\sum_j \frac{\partial F}{\partial \alpha_j}(s)\, d\alpha_j\right\}^{1/N}_{s=\lambda_i} \tag{3-119}$$

This relates differential changes, $d\alpha_j$, in the system parameters to the differential change in a system root. The expression can be written more compactly by defining parameter sensitivities as

$$S^i_{\alpha_j} = -N!\left\{\frac{[(\partial F/\partial \alpha_j)(s)]}{[(\partial^N F/\partial s^N)(s)]}\right\}_{s=\lambda_i} \tag{3-120}$$

The parameter sensitivity, $S$, is made specific to a particular system root by the superscript, $i$; while the pertinent system parameters, $\alpha_j$, is indicated by the subscript. While it then has the same form as the previous sensitivities, it is not yet totally equivalent, since the differential increment in the $i^{\text{th}}$ system root is given by

$$d\lambda_i = \left[\sum_j S^i_{\alpha_j}\, d\alpha_j\right]^{1/N} \tag{3-121}$$

The parameter sensitivities are partial derivatives, so that shifts in system roots predicted using the sensitivities are only exact in the limit as the parameter variations approach zero. The practical extent to which this linear first-order theory will apply cannot be defined generally, although experience and an intimate knowledge of system details usually provides an appreciation for the conditions under which the first-order results should be accurate or questionable. When in doubt, practical restrictions on ranges can be evolved by comparing the first-order results with auxiliary complete solutions or by developing second-order sensitivities.

The simplest special case using the above results is the single-loop system with the open-loop transfer function $G(s)$. Here $F(s) = 1 + G(s)$ and the components of $\boldsymbol{\alpha}$ are the gain, $\kappa$, the open-loop zeros, $-z_j$, and the open-loop poles, $-p_j$. Using the open-loop transfer function form given by Eq. 3-79, the relation $G(\lambda_i) = -1$ in Eq. 3-119, and the basic formulations of Eqs. 3-82 through 3-89, we obtain the result,

$$\begin{aligned} d\lambda_i &= \left\{\frac{N!}{[(\partial^N G/\partial s^N)(s)]_{s=\lambda_i}}\left[\frac{d\kappa}{\kappa} + \sum_{j=1}^{m}\frac{dz_j}{\lambda_i + z_j} - \sum_{j=1}^{m+n}\frac{dp_j}{\lambda_i + p_j}\right]\right\}^{1/N} \\ &= \left[S^i_\kappa \frac{d\kappa}{\kappa} + \sum_{j=1}^{m} S^i_{z_j}\, dz_j + \sum_{j=1}^{n} S^i_{p_j}\, dp_j\right]^{1/N} \end{aligned} \tag{3-122}$$

The comparison of Eq. 3-122 with its predecessor equivalent for first-order poles, Eq. 3-84, shows that key changes for the multiple-pole case are in the gain sensitivity and in the exponent $N$ on the closed-loop pole shift, $d\lambda_i$ (or, as written in Eq. 3-122, an exponent $1/N$ on the sum of the sensitivities). The pole and zero sensitivities are also changed as the gain sensitivity,

$$S_{\kappa}^{i} = \frac{N!}{[(\partial^N G/\partial s^N)(s)]_{s=\lambda_i}} \tag{3-123}$$

is a common factor. Equation 3-122 reflects the well-known characteristics of a branch point on a root locus plot. The incoming branches, which at their junction represent an $N^{\text{th}}$ order closed-loop root, are evenly spaced and separated from each other by $2\pi/N$. The departing branches are also separated from each other by $2\pi/N$ and are midway between the incoming branches.

Except for these isolated multiple-order root points, the root locus consists of branches characterizing the paths of the closed-loop roots as open-loop gain is changed. The gain sensitivity is tangent to these paths, whereas the zero and pole sensitivities will make angles with the tangents given by $\measuredangle\ (\lambda_i + z_j)^{-1}$ and $-\measuredangle\ (\lambda_i + p_j)^{-1}$, respectively.

Sensitivity factors for system parameters other than poles and zeros can also be determined. For instance, for the system parameter $\alpha_j$, where $G(s, \alpha_j) = KN(s, \alpha_j)/D(s, \alpha_j)$, the sensitivity factor is

$$S_{\alpha_j}^{i} = S_{\kappa}^{i}\left[\frac{1}{N}\frac{\partial N}{\partial \alpha_j} - \frac{1}{D}\frac{\partial D}{\partial \alpha_j}\right]_{s=\lambda_i} \tag{3-124}$$

*Connections between Sensitivity and Modal Response*

As noted previously, without proof, the gain sensitivities are simply related to the transient response characteristics of a unity feedback system via the modal response coefficients for the closed-loop roots. The partial fraction expansion of the system's weighting function will contain the terms

$$\mathcal{L}^{-1}\left[\frac{Q_{i_1}}{s-\lambda_i} + \frac{Q_{i_2}}{(s-\lambda_i)^2} + \cdots + \frac{Q_{i_N}}{(s-\lambda_i)^N}\right]$$
$$= e^{\lambda_i t}\left[Q_{i_1} + Q_{i_2}t + \cdots + \frac{Q_{i_N}t^{N-1}}{(N-1)!}\right] \tag{3-125}$$

as the contribution of an $N^{\text{th}}$ order root. The $N$ modal response coefficients, $Q_{i_k}$, $k = 1, 2, \ldots, N$, are evaluated by

$$Q_{i_k} = \frac{1}{(N-k)!}\left\{\frac{d^{N-k}}{ds^{N-k}}\left[\frac{(s-\lambda_i)^N G(s)}{1 + G(s)}\right]\right\}_{s=\lambda_i} \tag{3-126}$$

The gain sensitivity is the negative of the $N^{\text{th}}$ modal response coefficient, i.e.,

$$S_\kappa^i = -Q_{i_N} = -\left[\frac{(s-\lambda_i)^N G(s)}{1+G(s)}\right]_{s=\lambda_i} \tag{3-127}$$

To prove this relationship, it is convenient to introduce the variable $Q_{i_N}^*(s)$ which is defined by

$$Q_{i_N}^*(s) = \frac{(s-\lambda_i)^N G(s)}{1+G(s)} \tag{3-128}$$

where $Q_{i_N}^*(\lambda_i) = Q_{i_N}$. Rewriting Eq. 3-128 as

$$[1 + G(s)]Q_{i_N}^*(s) = (s-\lambda_i)^N G(s) \tag{3-129}$$

and differentiating with respect to $s$, we obtain

$$[1 + G(s)]\frac{dQ_{i_N}^*(s)}{ds} + Q_{i_N}^*(s)\frac{dG(s)}{ds} = N(s-\lambda_i)^{N-1}G(s) + (s-\lambda_i)^N\frac{dG(s)}{ds} \tag{3-130}$$

Evaluating Eq. 3-130 at $s = \lambda_i$, and noting that $G(\lambda_i) = -1$, we find that

$$Q_{i_N}\left[\frac{dG(s)}{ds}\right]_{s=\lambda_i} = 0 \qquad \text{if} \quad N > 1 \tag{3-131}$$

Therefore,

$$\left[\frac{dG(s)}{ds}\right]_{s=\lambda_i} = 0 \qquad \text{if} \quad N > 1 \tag{3-132}$$

The repeated differentiation of Eq. 3-130 shows that the first $N - 1$ derivatives of $G$ are zero at $s = \lambda_i$, i.e.,

$$\left[\frac{d^kG(s)}{ds^k}\right]_{s=\lambda_i} = 0, \qquad 1 \leq k \leq N-1 \tag{3-133}$$

and that

$$Q_{i_N} = \frac{-N!}{[d^NG(s)/ds^N]_{s=\lambda_i}} \tag{3-134}$$

Finally, comparison with Eq. 3-123 gives

$$Q_{i_N} = -S_\kappa^i \tag{3-135}$$

*Multiloop Sensitivity Ratios*

The multiloop situation and notation are straightforward extensions of those introduced above. Instead of one open-loop transfer function, $G$, there are $L$ open-loop transfer functions, $G_k$. Each of these has a gain, $\kappa_k$; $n_k$ zeros, $-z_{k_j}$; and $m_k + n_k$ poles, $-p_{k_j}$. The closed-loop system characteristic function $F$ is, in general, a summation of terms, each of which may be the product of several transfer functions, e.g., $F = 1 + G_1 + G_1G_2 + G_2G_3G_4 + \cdots$. Under these circumstances Eq. 3-119 becomes

$$d\lambda_i = \left[\frac{-N!}{(\partial^N F/\partial s^N)(s)} \sum_{k=1}^{L} \frac{\partial F}{\partial G_k}\left(\frac{\partial G_k}{\partial \kappa_k}\, d\kappa_k + \sum_{j=1}^{n_k} \frac{\partial G_k}{\partial z_{k_j}}\, dz_{k_j} + \sum_{j=1}^{m_k+n_k} \frac{\partial G_k}{\partial p_k}\, dp_{k_j}\right)\right]_{s=\lambda_i}^{1/N} \tag{3-136}$$

or in terms of the sensitivity factors

$$d\lambda_i = \left[\sum_{k=1}^{L}\left(S^i_{\kappa_k}\frac{d\kappa_k}{\kappa_k} + \sum_{j=1}^{n_k} S^i_{z_{k_j}}\, dz_{k_j} + \sum_{j=1}^{m_k+n_k} S^i_{p_{k_j}}\, dp_{k_j}\right)\right]^{1/N} \tag{3-137}$$

Thus the gain sensitivities are

$$S^i_{\kappa_k} = \left[\frac{-N!}{(\partial^N F/\partial s^N)(s)}\frac{\partial F}{\partial G_k}\frac{\partial G_k}{\partial \kappa_k}\kappa_k\right]_{s=\lambda_i} \tag{3-138}$$

The pole and zero sensitivities are given in terms of ratios involving the gain sensitivities by

$$\frac{S^i_{z_{k_j}}}{S^i_{\kappa_k}} = \left[\frac{\partial G_k/\partial z_{k_j}}{\kappa_k(\partial G_k/\partial \kappa_k)}\right]_{s=\lambda_i} = \left[\frac{\partial G_k/\partial z_{k_j}}{G_k}\right]_{s=\lambda_i} \tag{3-139}$$

$$\frac{S^i_{p_{k_j}}}{S^i_{\kappa_k}} = \left[\frac{\partial G_k/\partial p_{k_j}}{\kappa_k(\partial G_k/\partial \kappa_k)}\right]_{s=\lambda_i} = \left[\frac{\partial G_k/\partial p_{k_j}}{G_k}\right]_{s=\lambda_i} \tag{3-140}$$

These zero and pole sensitivity ratios can be considerably simplified. For example, for the $z_{k_j}$ term

$$\frac{\partial G_k}{\partial z_{k_j}} = \frac{G_k}{s + z_{k_j}} \tag{3-141}$$

Combining this with the sensitivity ratio, Eq. 3-139, gives

$$\frac{S^i_{z_{k_j}}}{S^i_{\kappa_k}} = \left(\frac{1}{s + z_{k_j}}\right)_{s=\lambda_i} = \frac{1}{\lambda_i + z_{k_j}} \tag{3-142}$$

This and several other sensitivity ratios are summarized in Table 3-1. Beside the ratios of simple pole and zero to gain sensitivities, similar ratios are also provided for quadratic elements that result when complex conjugate pairs of poles or zeros are joined. The basic forms for these quadratics are $s^2 + 2\zeta\omega s + \omega^2$ and $s^2 + 2as + (a^2 + b^2)$.

Table 3-1. Sensitivity ratios.

$$\frac{S^i_{\kappa_j}}{S^i_{\kappa_k}} = \left[\frac{G_j(\partial F/\partial G_j)}{G_k(\partial F/\partial G_k)}\right]_{s=\lambda_i}$$

$$S^i_{z_{kj}} = \frac{S^i_{\kappa_k}}{z_{k_j} + \lambda_i}$$

$$S^i_{p_{kj}} = \frac{-S^i_{\kappa_k}}{p_{k_j} + \lambda_i}$$

$$S^i_\omega = \frac{\pm 2(\omega + \zeta\lambda_i)S^i_\kappa}{\lambda_i^2 + 2\zeta\omega\lambda_i + \omega^2} {}^*$$

$$S^i_\zeta = \frac{\pm 2\omega\lambda_i S^i_\kappa}{\lambda_i^2 + 2\zeta\omega\lambda_i + \omega^2} {}^*$$

$$S^i_a = \frac{\pm 2(a + \lambda_i)S^i_\kappa}{\lambda_i^2 + 2a\lambda_i + a^2 + b^2} {}^*$$

$$S^i_b = \frac{\pm 2bS^i_\kappa}{\lambda_i^2 + 2a\lambda_i + a^2 + b^2} {}^*$$

* Use upper sign for zeros, lower sign for poles.

Also included in Table 3-1 is a key relationship between gain sensitivities for different transfer functions but relating to the same closed-loop root. The use of Eq. 3-138 for two different gains gives

$$\frac{S^i_{\kappa_j}}{S^i_{\kappa_k}} = \left[\frac{G_j(\partial F/\partial G_j)}{G_k(\partial F/\partial G_k)}\right]_{s=\lambda_i} \qquad (3\text{-}143)$$

Since $F$ is a linear function of $G_k$, i.e., each transfer function $G_k$ appears in $F$ only with an exponent of unity, the term $G_k(\partial F/\partial G_k)$ is simply the sum of all the terms of $F$ that include $G_k$. When evaluating sensitivity factors, Eq. 3-143 is of central importance, for all the gain sensitivities pertinent to a given closed-loop root can be calculated, using ratios, once a single-gain sensitivity is known. Since a multiloop analysis is usually done as a sequence of loop closures, the single-loop techniques can be applied to obtain the gain sensitivity for the last or outermost closure, thereby providing a starting point. The pole and zero sensitivity factors follow directly once the gain sensitivities are available.

The evaluation of the ratios of Eq. 3-143 can be quite involved. Considerable simplifications are sometimes possible by recalling that $F = 0$ at $s = \lambda_i$. Thus, if $F$ were given by

$$F = G_1 G_2 + G_1 G_3 G_4 + G_1 G_2 G_3 + G_3 G_4 \qquad (3\text{-}144)$$

then

$$\frac{S^i_{\kappa_2}}{S^i_{\kappa_1}} = \left(\frac{G_1G_2 + G_1G_2G_3}{G_1G_2 + G_1G_3G_4 + G_1G_2G_3}\right)_{s=\lambda_i} = -\left[\frac{G_1G_2(1 + G_3)}{G_3H_4}\right]_{s=\lambda_i} \tag{3-145}$$

$$\frac{S^i_{\kappa_3}}{S^i_{\kappa_1}} = \left(\frac{G_1G_3G_4 + G_1G_2G_3 + G_3G_4}{G_1G_2 + G_1G_3G_4 + G_1G_2G_3}\right)_{s=\lambda_i} = \left(\frac{G_1G_2}{G_3G_4}\right)_{s=\lambda_i} \tag{3-146}$$

$$\frac{S^i_{\kappa_4}}{S^i_{\kappa_1}} = \left(\frac{G_1G_3G_4 + G_3G_4}{G_1G_2 + G_1G_3G_4 + G_1G_2G_3}\right)_{s=\lambda_i} = -(1 + G_1)_{s=\lambda_i} \tag{3-147}$$

CHAPTER 4

# VEHICLE EQUATIONS OF MOTION

## 4-1. Introduction

With our background in feedback control and analysis methods established in the preceding chapters, we come now to the object of such control—the vehicle. We want to characterize the vehicle in a way that is especially instructive to the flight control system designer, rather than to the stability and control aerodynamicist or dynamic specialist. In order to do this, we deliberately emphasize the vehicle dynamic properties as a whole, not the sum of its component parts. For example, we avoid the fine-grain details of stability derivatives and their dependence on configuration layout, in favor of a rudimentary understanding of the origins of aerodynamic forces and moments. Also, we look for and identify simplifying but generally valid assumptions that eventually lead us to a direct appreciation of the important factors governing the vehicle's response characteristics. Such an understanding of the over-all aspects of the object to be controlled is an implicit requirement for effective and efficient flight control system design activities. It affords a basic understanding of the vehicle/control system interactions and of the flight controller possibilities which are most likely to succeed.

With this object and bias in mind we proceed, in this chapter, to establish the most generally useful (and used) sets of the basic equations of motion. The process is specifically designed to appeal to readers who are not necessarily conversant with formal advanced dynamic methods. Accordingly, we relate the developments to simple physical pictures of the phenomena involved, rather than only to compact mathematical formulation. All the assumptions required to get to the final set of equations are specifically identified, as are their simplifying effects. The factors involved in selecting an appropriate axis system, and the process of converting from one set of axes to another, are set forth. The effects of linearizing about an operating point or trim condition and the influences of the selected trim condition on the resulting linearized perturbed equations of motion are noted. Finally, the general origins of the aerodynamic forces that produce the usually important stability derivatives are discussed and illustrated. The overall treatment[1] presents an ensemble of

[1] See n. 4, Chapter 1.

selected descriptions, explanations, and formulations eclectically combined to illuminate, as brightly as possible, an area that is often incompletely understood.

Proceeding from this base, in Chapters 5 and 6, we establish the various transfer functions of interest, in factored form, for longitudinal and lateral control. Here some of the feedback notions and methods exposed in past chapters are used to illustrate the effects of derivative changes and to extract approximate transfer function factors in literal, rather than numeric, terms.[2] Such literal expressions show the direct connections between the transfer function poles and zeros as a function of the relative values of certain key derivatives and give valuable insight into the probable nature of the associated control problem. A feature of both Chapters 5 and 6 is the tabulation of all presently known literal approximate factors and the specific conditions under which they are expected to be reasonably valid. The tabulations cover not only conventional single-loop numerators, but also the coupling numerators required for the analysis of conventional multiloop flight control systems; they also cover vehicles ranging from entry gliders to VTOL machines.

With this preview of the ramifications issuing therefrom, we turn now to the formulation of the vehicle equations of motion.

## 4-2. Newton's Second Law and Reference Frames

In this section we lay the basic groundwork for the developments that follow, starting with some assumptions and definitions.

Assumption 1. The airframe is assumed to be a rigid body.

In a rigid body the distances between any specified points in the body are fixed, so that this assumption eliminates consideration of the forces acting between individual elements of mass. Consequently, the airframe motion can be described completely by a translation of the center of mass and by a rotation about this point.

Actual vehicles depart from the rigid body assumptions in two ways—they are composed of several major elements that are required to move relative to one another such as engines, rotors, or control devices; incidental elastic deformations of the structure do occur, as in wing bending caused by air loads. Some of the changes required in the description of the

[2] Such approximations go back to Bairstow, with more modern examples, of varying validity, appearing in the following: "Dynamics of the Airframe," Bureau of Aeronautics Rept. AE-61-4, Vol. 2, Sept. 1952; H. H. Thomas and S. Neumark, "Interim Note on Stability and Response Characteristics of Supersonic Aircraft," RAE TN Aero 2412, 1955; I. L. Ashkenas and D. T. McRuer, "Approximate Airframe Transfer Functions and Application to Single Sensor Control Systems," WADC-TR-58-82, June 1958; A. J. Ross, "The Lateral Oscillation of Slender Aircraft," RAE Rept. Aero 2666, June 1962.

aerodynamic forces due to such static deflection characteristics are illustrated later in this chapter (Section 4-9). Other changes, involving the dynamics of the structure,[3] which can greatly increase the degrees of freedom to be considered in the equations of motion, are beyond the scope of the present treatment. Since all motion is relative, a suitable frame of reference must be selected to describe airframe motion.

Assumption 2. The earth is assumed to be fixed in space.

The inertial frame of reference defined by this assumption, i.e., one that is fixed or moves at constant velocity relative to the earth, permits a description of vehicle motion which is accurate for relatively short-term guidance and control analysis purposes. It does have practical limitations when very long term navigation or extra-atmospheric operations are of interest.

With the above two assumptions as a basis, we have a reference frame in which Newton's laws are valid and a rigid body to which these laws may be applied. Consider that the aircraft has a linear momentum vector, $\mathbf{p}$, and an angular momentum vector, $\mathbf{H}$, each measured in the inertial coordinate frame. By Newton's second law the time rate of change of linear momentum equals the sum of all *externally* applied forces,

$$\frac{d\mathbf{p}}{dt} = \mathbf{F} \tag{4-1}$$

and the rate of change of the angular momentum is equal to the sum of all applied torques

$$\frac{d\mathbf{H}}{dt} = \mathbf{M} \tag{4-2}$$

These vector differential equations provide the starting point for a complete description of the rigid body motions of the vehicle.

In almost all aeronautical vehicles some part of the thrust forces is produced by the expenditure of the vehicle mass; the mass variation must be considered in determining the rate of change of linear momentum.[4]

[3] A good basic treatment of such considerations is given in R. L. Bisplinghoff, H. Ashley, and R. L. Halfman, *Aeroelasticity*, Addison-Wesley, Reading, Mass., 1955. Example equations of motion and resulting transfer functions for specific aircraft are limited features of this flight control literature; an exemplary cross section is contained in: B. F. Pearce, W. A. Johnson, and R. K. Siskind, "Analytical Study of Approximate Longitudinal Transfer Functions for a Flexible Airframe," ASD-TDR-62-279, June 1962.

[4] A. Sommerfield, *Mechanics*, Academic, New York, 1952; J. B. Rosser, R. R. Newton, and G. L. Gross, *Mathematical Theory of Rocket Flight*, McGraw-Hill, New York, 1947.

At time $t$ let the linear momentum be

$$\mathbf{p}_1 = m\mathbf{V} \tag{4-3}$$

where $m$ is the mass and $\mathbf{V}$ the velocity of the vehicle. Then, if at time $t + \Delta t$, $-\Delta m$ is the net mass that has left the vehicle with an effective exhaust velocity, $\mathbf{v}_e$, relative to the vehicle, the linear momentum will be

$$\mathbf{p}_2 = (m + \Delta m)(\mathbf{V} + \Delta\mathbf{V}) + (-\Delta m)(\mathbf{V} + \mathbf{v}_e) \tag{4-4}$$

Here $\mathbf{V} + \Delta\mathbf{V}$ and $m + \Delta m$ are the velocity and mass of the vehicle at time $t + \Delta t$, and $\mathbf{V} + \mathbf{v}_e$ is the effective velocity relative to inertial space of the mass increment exhausted. It should be noted in passing that the effective exhaust velocity depends on the exit area, the differential between exit and ambient pressures, and the exit velocity of the mass leaving the vehicle. The incremental change in momentum from time $t$ to time $t + \Delta t$ is then

$$\begin{aligned}\Delta\mathbf{p} = \mathbf{p}_2 - \mathbf{p}_1 &= (m + \Delta m)(\mathbf{V} + \Delta\mathbf{V}) - \Delta m(\mathbf{V} + \mathbf{v}_e) - m\mathbf{V} \\ &= m\Delta\mathbf{V} - \mathbf{v}_e\,\Delta m + \Delta m\,\Delta\mathbf{V}\end{aligned}$$

Dividing by $\Delta t$, and taking the limit as $\Delta t \to 0$, we find that the time rate of change of momentum becomes

$$\frac{d\mathbf{p}}{dt} = m\,\frac{d\mathbf{V}}{dt} - \mathbf{v}_e\,\frac{dm}{dt} = \mathbf{F}$$

or

$$m\,\frac{d\mathbf{V}}{dt} = \mathbf{F} + \mathbf{v}_e\,\frac{dm}{dt} = \mathbf{F} + \mathbf{T}_e \tag{4-5}$$

In Eq. 4-5 the thrust term, $\mathbf{T}_e$, is defined by $\mathbf{v}_e(dm/dt)$ and represents only that component of thrust due directly to the expulsion of vehicle mass. Thus the rate of change of linear momentum can be computed as if the *mass were constant* and the product of the change in mass per unit time and the relative velocity between the exhausted mass and the vehicle were an external force. Equation 4-5 directly follows from Eq. 4-1 when the thrust is developed by a momentum exchange other than one directly involving the vehicle mass, as in a propeller (in the present context, such a thrust would constitute an external force). Thus, if the thrust force includes exhaust products, Eq. 4-5 is correct in general for vehicles traveling at speeds small relative to the speed of light. Consequently, from this point on we shall consider that the thrust component, $\mathbf{T}_e$, is contained in the general applied force, $\mathbf{F}$.

If inertial space is now represented as a right-hand system of Cartesian axes, $\mathfrak{X}$, $\mathfrak{Y}$, $\mathfrak{Z}$, and the velocity vector, $\mathbf{V}$, and total applied force, $\mathbf{F}$, are resolved into their components $\mathfrak{U}$, $\mathfrak{V}$, and $\mathfrak{W}$ and $F_{\mathfrak{X}}$, $F_{\mathfrak{Y}}$, and $F_{\mathfrak{Z}}$ along

$\mathfrak{X}$, $\mathfrak{Y}$, and $\mathfrak{Z}$, respectively, then the vector equation, Eq. 4-5, can be written as the three scalar equations

$$m\frac{d\mathfrak{U}}{dt} = F_{\mathfrak{X}}$$

$$m\frac{d\mathfrak{V}}{dt} = F_{\mathfrak{Y}} \tag{4-6}$$

$$m\frac{d\mathfrak{W}}{dt} = F_{\mathfrak{Z}}$$

These equations would describe the motion of the airframe center of mass as seen by an observer in the $\mathfrak{X}$, $\mathfrak{Y}$, $\mathfrak{Z}$ frame. So far, the definition of $\mathfrak{X}$, $\mathfrak{Y}$, $\mathfrak{Z}$ is still quite arbitrary—it can have any orientation and can move at any constant velocity relative to the earth. Later we shall present a more specific definition.

The rotary equivalent of the linear momentum equation for the angular momentum, **H**, is, unfortunately, far more complicated. The rotary analog of the mass, $m$, is the moment of inertia, which is a dyad,[5] $\underline{\underline{\mathbf{I}}}$. The angular momentum is the vector dot product of the inertia dyad with the angular velocity $\mathbf{\Omega}$, i.e.,

$$\mathbf{H} = \underline{\underline{\mathbf{I}}} \cdot \mathbf{\Omega} \tag{4-7}$$

In the simplest case, in which the applied moment is about (or, as a vector, is along) a principal axis, $\xi$, of the rigid body, the angular velocity will also be about the same axis. Then the vector equation will reduce to the scalar equation

$$H = I_{\xi\xi}\Omega$$

where $I_{\xi\xi}$ is the moment of inertia about the principal axis considered.

In the general case there are up to nine angular momentum components which differ in detail, depending on the axis system selected as a reference frame for the equations of motion. Consider the derivative of Eq. 4-7,

$$\frac{d\mathbf{H}}{dt} = \underline{\underline{\mathbf{I}}} \cdot \frac{d\mathbf{\Omega}}{dt} + \frac{d\underline{\underline{\mathbf{I}}}}{dt} \cdot \mathbf{\Omega} = \mathbf{M} \tag{4-8}$$

If the inertias and angular velocities are computed in the space-fixed axis system, $\mathfrak{X}$, $\mathfrak{Y}$, $\mathfrak{Z}$, the moment of inertia about each axis will, in general vary continuously as the vehicle moves with respect to the axis origin. Even if the aircraft's motions are restricted to constant velocities so that the $\mathfrak{X}$, $\mathfrak{Y}$, $\mathfrak{Z}$ axes can be attached to the aircraft center of mass, the inertias will change as the craft rotates about the axes. Such variations in inertia

[5] See e.g., A. P. Wills, *Vector Analysis with an Introduction to Tensor Analysis*, Dover, New York, 1958.

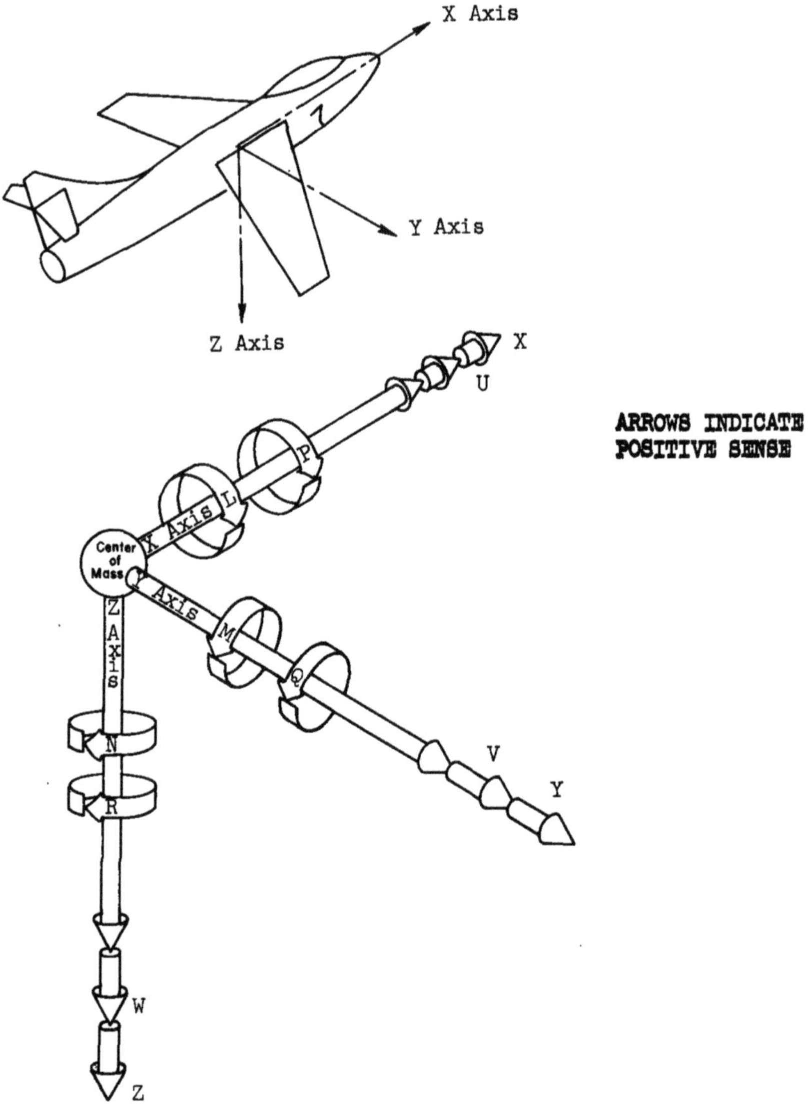


Fig. 4·1. Vehicle-fixed axis system and notation.

| | Velocities | Applied forces and moments | Distances |
|---|---|---|---|
| Forward | $U$ | $X$ | $x$ |
| Side | $V$ | $Y$ | $y$ |
| Vertical | $W$ | $Z$ | $z$ |
| Roll | $P$ | $L$ | |
| Pitch | $Q$ | $M$ | |
| Yaw | $R$ | $N$ | |

will contribute to the time rate of change of angular momentum via the $(d\underline{\underline{\mathbf{I}}}/dt) \cdot \mathbf{\Omega}$ term in Eq. 4-8. The resulting equations are therefore complicated by time-varying parameters, an extremely undesirable feature.

An attractive alternative is to select an axis system that is *fixed in the body*. In this kind of system the measured rotary inertial properties are constant, to the extent that the mass can also be considered constant; also, terms containing such quantities as $dI_{\xi\xi}/dt$ are always zero. A body-fixed axis system has another virtue: it is the natural frame of reference for most vehicle-borne observations and measurements of the vehicle's motions. For example, a pilot is fundamentally aware of rotary motions about the vehicle's center of gravity, rather than about a spaced-fixed point. He feels accelerating forces with respect to his own alignment with the body-fixed frame, not with respect to a space-fixed frame. Also, many flight instruments and sensors, especially those used for short-term control of the vehicle, are similarly constituted, i.e., in general they measure motions with respect to body-fixed axes.

Unfortunately, the above advantages are not obtained without some penalty. For instance, the very simple inertial force forms of Eq. 4-6 are replaced by more complex ones when the linear velocities are measured in a body-fixed axis system. Some of the applied forces, as well, are more simply and naturally expressed in the inertial reference frame. Also, instrumentation designed for long-term navigation and guidance is of necessity aligned to measure linear velocities and distances in space or earth-fixed coordinate systems, and the pilot himself is visually oriented to earth-fixed, as well as vehicle-fixed axes. These factors raise the possibility of a desirable intermixing of axis systems, e.g., an earth-fixed reference for forces and linear accelerations, and a body-fixed reference for moments and angular motions. Such dual systems are, in fact, needed for a complete treatment of the vehicle's motion. However, for flight-control applications the expansion of the left-hand sides of Eqs. 4-1 and 4-2 is usually made in a manner appropriate to a *body-fixed axis system with its origin on the vehicle's center of mass*. This axis system is shown in Fig. 4-1, together with the notation and sign conventions used to identify linear and angular velocities and applied forces and moments. The body-fixed $X$, $Y$, and $Z$ axes are oriented in the aircraft with $X$ forward, $Y$ out the right wing, and $Z$ out the bottom as shown.

### 4-3. Expansion of the Inertial Forces and Moments

The left-hand side of Eqs. 4-2 and 4-5 can readily be expanded formally, especially when vector or matrix algebra is used; however, this process affords little physical insight into the origins of the resulting terms. To better narrow the gap between simple mathematical routine and difficult physical interpretation, we shall carry out the development of the inertial forces and moments in two different ways. First, the developments will be related as much as possible to simple figures of the phenomena involved; then it will be repeated as an essentially mathematical exercise.

*Rectilinear Acceleration Components*

The total rectilinear acceleration is $d\mathbf{V}/dt$. The components of $\mathbf{V}$ are $U$, $V$, and $W$ along the $X$, $Y$, and $Z$ body-fixed axes. The translational acceleration components of $d\mathbf{V}/dt$ along these axes will contain not only the obvious components $dU/dt$, $dV/dt$, and $dW/dt$ but also centripetal acceleration components due to the rotation of the body-fixed axis system relative to inertial space. A simple two-dimensional example is helpful to develop an intuitive understanding of these statements. Figure 4-2 shows the plan view of an airplane at two slightly separated points along its flight path. At the left the airplane has the velocity components $U$ and $V$ along the $X$ and $Y$ axes, respectively; at the right, a small increment of time, $\Delta t$, later, the aircraft's velocity has changed in both magnitude and direction. The craft has rotated through an incremental angle,

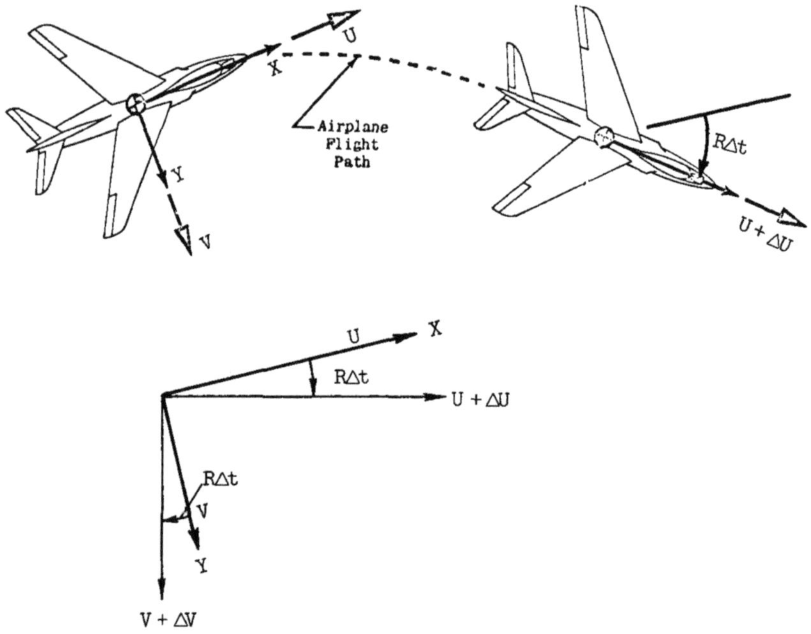


Fig. 4-2. An airplane in two-dimensional accelerated flight.

$R\,\Delta t$, and the linear velocity components are now $U + \Delta U$ and $V + \Delta V$.

When the acceleration is written as a time rate of change of the velocity, the components with respect to the *original* $X$ and $Y$ axes are

$$a_X = \lim_{\Delta t \to 0} \frac{(U + \Delta U)\cos R\,\Delta t - U - (V + \Delta V)\sin R\,\Delta t}{\Delta t}$$

$$= \dot{U} - VR \tag{4-9}$$

and

$$a_Y = \lim_{\Delta t \to 0} \frac{(V + \Delta V)\cos R\,\Delta t - V + (U + \Delta U)\sin R\,\Delta t}{\Delta t}$$

$$= \dot{V} + UR \tag{4-10}$$

where the dot above $U$ and $V$ indicates the time derivatives of these velocity components measured in body-fixed axes. The vector acceleration, $\mathbf{a}$, is the vector sum of the two components given in Eqs. 4-9 and 4-10. With $\mathbf{i}$, $\mathbf{j}$, and $\mathbf{k}$ taken as unit vectors along $X$, $Y$, and $Z$, the acceleration vector becomes

$$\begin{aligned} \mathbf{a} &= a_X\mathbf{i} + a_Y\mathbf{j} \\ &= [\dot{U}\mathbf{i} + \dot{V}\mathbf{j}] + [-VR\mathbf{i} + UR\mathbf{j}] \end{aligned} \tag{4-11}$$

The first term is just the time derivative of the velocity measured in the body-fixed coordinates. The second term can be identified, by recalling the rules for vector cross-multiplication,[6] as the cross-product of the angular and linear velocities; in this case $R\mathbf{k}$ and $\mathbf{V} = U\mathbf{i} + V\mathbf{j}$, respectively, i.e.,

$$\begin{aligned} R\mathbf{k} \times [U\mathbf{i} + V\mathbf{j}] &= UR(\mathbf{k} \times \mathbf{i}) + VR(\mathbf{k} \times \mathbf{j}) \\ &= UR\mathbf{j} - VR\mathbf{i} \end{aligned} \tag{4-12}$$

since $\mathbf{i} \times \mathbf{j} = \mathbf{k}$, $\mathbf{j} \times \mathbf{k} = \mathbf{i}$, $\mathbf{k} \times \mathbf{i} = \mathbf{j}$, $\mathbf{k} \times \mathbf{j} = -\mathbf{j} \times \mathbf{k}$, etc. If these results are now generalized to the three-dimensional case, the vector acceleration will be

$$\mathbf{a} = \frac{d\mathbf{V}}{dt} = \dot{\mathbf{V}} + \boldsymbol{\Omega} \times \mathbf{V} \tag{4-13}$$

Here $d\mathbf{V}/dt$ is used to denote the time derivative of velocity relative to inertial space axes, whereas the superior dot notation indicates the time derivative of velocity observed in the body-fixed coordinates.

Physically, the cross-product term arises from the centripetal accelerations along any given axis due to angular velocities about the remaining two axes. Thus, for rotations about the $Y$ axis, the pitching velocity, $Q$,

[6] *Ibid.*

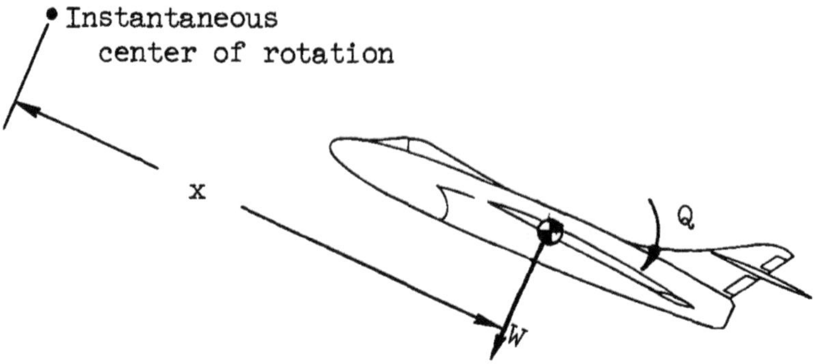


Fig. 4-3. Centripetal acceleration along $X$ due to pitching and plunging.

and the vertical velocity, $W$, can both be represented by simple angular motions about some instantaneous center in space defined by $x = W/Q$, as shown in Fig. 4-3. The centripetal acceleration component is then directed forward (positive) and is equal in magnitude to the product, $WQ$. The picture about the $Z$ axis is similar except that, following the right-hand rule, the sign of the side velocity is negative and the corresponding acceleration component is $-VR$ (Fig. 4-4).

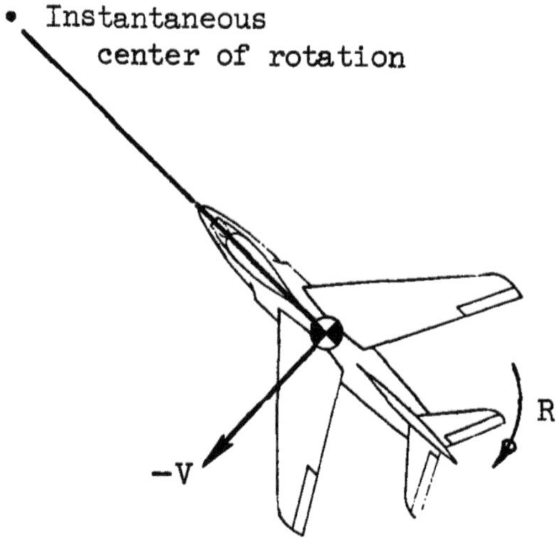


Fig. 4-4. Centripetal acceleration along $X$ due to yawing and sideslipping.

These constructions can be repeated for each axis in turn, with the similar and general result (including linear acceleration) that

$$a_1 = \dot{\mathsf{v}}_1 + \omega_2 \mathsf{v}_3 - \omega_3 \mathsf{v}_2 \tag{4-14}$$

where $\mathsf{v}$ and $\omega$ are linear and angular velocities, respectively, and the subscripts 1, 2, and 3 represent axes taken in right-hand progression (e.g., if 1 is $Y$, then 2 and 3 are $Z$ and $X$, respectively). Accordingly, the complete set of linear accelerations, obtained by specializing Eq. 4-14, is given by

$$\begin{aligned} a_x &= \dot{U} + QW - RV \\ a_y &= \dot{V} + RU - PW \\ a_z &= \dot{W} + PV - QU \end{aligned} \tag{4-15}$$

Precisely this same result can be derived formally using the vector equation, Eq. 4-13, with

$$\mathbf{V} = U\mathbf{i} + V\mathbf{j} + W\mathbf{k} \tag{4-16}$$

$$\mathbf{\Omega} = P\mathbf{i} + Q\mathbf{j} + R\mathbf{k} \tag{4-17}$$

Inserting these equations into Eq. 4-13, we obtain

$$\begin{aligned} \mathbf{a} = \frac{d\mathbf{V}}{dt} &= [\dot{U}\mathbf{i} + \dot{V}\mathbf{j} + \dot{W}\mathbf{k}] + \begin{vmatrix} \mathbf{i} & \mathbf{j} & \mathbf{k} \\ P & Q & R \\ U & V & W \end{vmatrix} \\ &= [\dot{U} + QW - RV]\mathbf{i} + [\dot{V} + RU - PW]\mathbf{j} + [\dot{W} + PV - QU]\mathbf{k} \end{aligned} \tag{4-18}$$

*Inertial Torque Components*

The total inertial torque about a given axis is due to both direct angular acceleration about that axis and to components arising from linear acceleration gradients resulting from combined rotations about all axes. To obtain a physical understanding of how these effects arise, we shall first investigate the dynamics of an infinitesimal element of mass, $dm$, as shown in Fig. 4-5.

Figure 4-5 shows the linear velocity components of the elemental mass due to the angular velocity components $P$, $Q$, and $R$. The accuracy of this representation can be verified by inspection or by multiplying the angular velocity and the radius (distance) from the axis system origin at

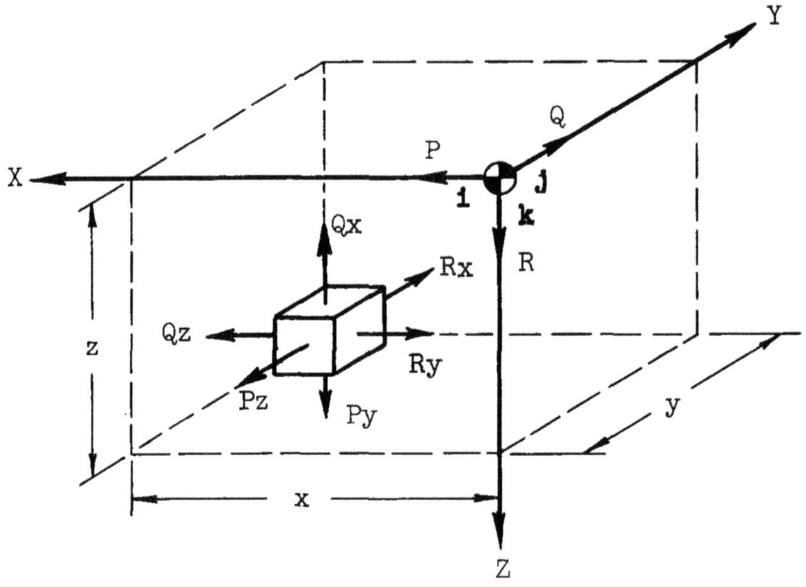


Fig. 4-5. Linear velocity components of an element of mass due to an angular velocity $\mathbf{\Omega}$ having components $P$, $Q$, and $R$.

the vehicle center of mass to the elemental mass. That is,

$$\mathbf{v}_{dm} = \mathbf{\Omega} \times \boldsymbol{\rho} = [P\mathbf{i} + Q\mathbf{j} + R\mathbf{k}] \times [x\mathbf{i} + y\mathbf{j} + z\mathbf{k}]$$

$$= \begin{vmatrix} \mathbf{i} & \mathbf{j} & \mathbf{k} \\ P & Q & R \\ x & y & z \end{vmatrix}$$

$$= \mathbf{i}[Qz - Ry] + \mathbf{j}[Rx - Pz] + \mathbf{k}[Py - Qx] \qquad (4\text{-}19)$$

Associated with each of these linear velocities is a linear momentum which is simply the velocities multiplied by $dm$. The moment of momentum (i.e., angular momentum) is obtained by summing the moments of these linear momenta about each axis. For example, about the $X$ axis the angular momentum component will arise from the $\mathbf{j}$ and $\mathbf{k}$ components multiplied, respectively, by the $z$ and $y$ lever arms (see Fig. 4-5),

$$dH_X = [-z(Rx - Pz) + y(Py - Qx)]\, dm \qquad (4\text{-}20)$$

or, upon expanding terms,

$$dH_X = y(yP)\, dm + z(zP)\, dm - z(xR)\, dm - y(xQ)\, dm$$

The complete set of equations obtained in this way is

$$
\begin{aligned}
dH_X &= (y^2 + z^2)\,P\,dm - zx\,R\,dm - yx\,Q\,dm \\
dH_Y &= (z^2 + x^2)\,Q\,dm - xy\,P\,dm - yz\,R\,dm \\
dH_Z &= (x^2 + y^2)\,R\,dm - zx\,P\,dm - zy\,Q\,dm
\end{aligned}
\tag{4-21}
$$

Again, these can be derived by inspection of Fig. 4-5 or by using the vector relationship for the angular momentum, i.e.,

$$
\begin{aligned}
d\mathbf{H} &= \boldsymbol{\rho} \times d\mathbf{p} = \boldsymbol{\rho} \times \mathbf{v}_{dm}\,(dm) \\
&= dm \begin{vmatrix} \mathbf{i} & \mathbf{j} & \mathbf{k} \\ x & y & z \\ [Qz - Ry] & [Rx - Pz] & [Py - Qx] \end{vmatrix}
\end{aligned}
\tag{4-22}
$$

For a finite rigid body the components of the moment of momentum will be the integrals of Eq. 4-21 over the entire mass of the vehicle:

$$
\begin{aligned}
H_X &= P \int (y^2 + z^2)\,dm - Q \int xy\,dm - R \int xz\,dm \\
H_Y &= Q \int (z^2 + x^2)\,dm - R \int yz\,dm - P \int yx\,dm \\
H_Z &= R \int (x^2 + y^2)\,dm - P \int zx\,dm - Q \int zy\,dm
\end{aligned}
\tag{4-23}
$$

Notice now that the integral $\int (y^2 + z^2)\,dm$ is, by definition, the moment of inertia, $I_{xx}$, of the entire mass of the airplane about the $X$ axis. Similarly, the integral $\int xy\,dm$ is defined as the product of inertia, $I_{xy}$. The remaining integrals in Eq. 4-23 are similarly defined, and the equations may be rewritten as

$$
\begin{aligned}
H_X &= PI_{xx} - QI_{xy} - RI_{xz} \\
H_Y &= QI_{yy} - RI_{yz} - PI_{xy} \\
H_Z &= RI_{zz} - PI_{xz} - QI_{yz}
\end{aligned}
\tag{4-24}
$$

where $I_{yz} = I_{zy}$, from the form of the integrals.

An alternative, still satisfactory, approach to the derivation of Eq. 4-24 is to consider the aircraft at the outset as a rigid body having principal axis moments of inertia, $I_{x_0}$, $I_{y_0}$, and $I_{z_0}$. Because the body-fixed $X$, $Y$, $Z$ axes do not necessarily coincide with the principal axes, we require an appreciation of the angular momentum components due to product of inertia terms. This can be obtained by reference to Fig. 4-6 which shows the $X$, $Z$ coordinates rotated with respect to the principal axes of inertia. Resolving the angular velocity, with components $P$ and $R$ in body-fixed

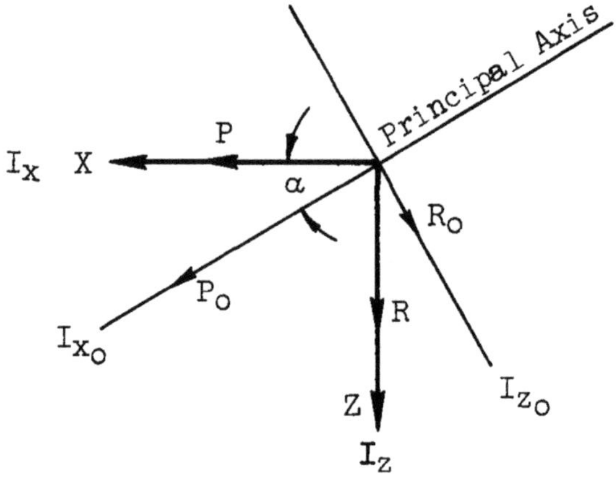


Fig. 4-6. Principal and body-fixed axes.

axes, along the principal axes, we find that

$$\begin{aligned} P_0 &= P \cos \alpha + R \sin \alpha \\ R_0 &= R \cos \alpha - P \sin \alpha \end{aligned} \tag{4-25}$$

Accordingly, the total angular momenta about the principal axes become

$$\begin{aligned} H_{X_0} &= P_0 I_{x_0} = I_{x_0}(P \cos \alpha + R \sin \alpha) \\ H_{Z_0} &= R_0 I_{z_0} = I_{z_0}(R \cos \alpha - P \sin \alpha) \end{aligned} \tag{4-26}$$

By resolving these back into the $X$, $Z$ axes,

$$H_X = H_{X_0} \cos \alpha - H_{Z_0} \sin \alpha$$

$$H_Z = H_{Z_0} \cos \alpha + H_{X_0} \sin \alpha$$

and, by substituting Eq. 4-26 and collecting terms,

$$H_X = P(I_{x_0} \cos^2 \alpha + I_{z_0} \sin^2 \alpha) + R\left(\frac{(I_{x_0} - I_{z_0})}{2} \sin 2\alpha\right)$$

$$H_Z = P\left(\frac{(I_{x_0} - I_{z_0})}{2} \sin 2\alpha\right) + R(I_{z_0} \cos^2 \alpha + I_{x_0} \sin^2 \alpha)$$

The terms in parentheses are moments and products of inertia in the $X$, $Z$ axis system, i.e.,

$$I_{xx} = I_{x_0} \cos^2 \alpha + I_{z_0} \sin^2 \alpha$$
$$I_{zz} = I_{z_0} \cos^2 \alpha + I_{x_0} \sin^2 \alpha \qquad (4\text{-}27)$$
$$I_{xz} = \frac{(I_{z_0} - I_{x_0})}{2} \sin 2\alpha$$

Consequently,

$$H_X = PI_{xx} - RI_{xz}$$
$$H_Z = RI_{zz} - PI_{xz} \qquad (4\text{-}28)$$

Repetition of the above process for the other two planes results finally in

$$H_X = PI_{xx} - QI_{xy} - RI_{xz}$$
$$H_Y = QI_{yy} - RI_{yz} - PI_{yx} \qquad (4\text{-}29)$$
$$H_Z = RI_{zz} - PI_{zx} - QI_{zy}$$

which is identical to Eq. 4-24.

Upon turning to calculation of the time rate of change of angular momentum, it is apparent that this will be simplified if the moments and products of inertia do not vary with time. Formally, this requires the following assumption.

Assumption 3. The mass and mass distribution of the vehicle are assumed to be constant.

Actually, there may be considerable differences in mass and its distribution throughout a mission as fuel is burned, stores expended, etc. The assumption is, nevertheless, ordinarily reasonable because the rates of change are relatively small and may be safely neglected for the time periods covered by most analyses.

With this assumption it is easy to see (Eq. 4-29) that angular accelerations $\dot{P}$, $\dot{Q}$, and $\dot{R}$ will lead directly to the contributions

$$\dot{H}_X = \dot{P}I_{xx} - \dot{Q}I_{xy} - \dot{R}I_{xz}$$
$$\dot{H}_Y = \dot{Q}I_{yy} - \dot{R}I_{yz} - \dot{P}I_{xy} \qquad (4\text{-}30)$$
$$\dot{H}_Z = \dot{R}I_{zz} - \dot{P}I_{xz} - \dot{Q}I_{yz}$$

These amount to the time derivatives of the angular momentum components as expressed in body-fixed coordinates. They do not constitute the total time rate of change of the angular momentum because, even with zero angular acceleration, steady rotations about a given axis change the *direction* of the angular momentum vectors about the remaining axes. For instance, from Fig. 4-7 it is clear that a steady angular velocity, $Q$,

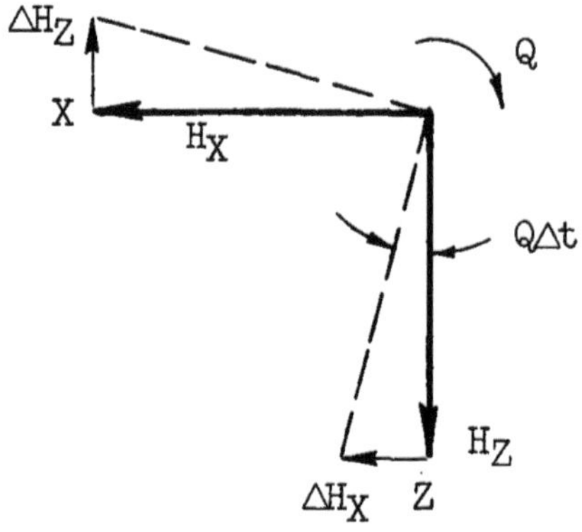


Fig. 4-7. Angular momentum change due to steady rotation.

about the $Y$ axis produces a change of angular momentum that is directed along the $X$ axis, i.e., $\Delta H_X = H_Z Q\,\Delta t$. This rotation will add a component of $dH_X/dt$ equal to $H_Z Q$ to the $X$ component and, similarly, a term $-H_X Q$ to the $Z$ component of $d\mathbf{H}/dt$. Analogous constructions in the remaining planes, when added to the above and to the time variation of the angular momentum relative to the body-fixed axes, yield the components of the total time rate of change of angular momentum:

$$\frac{dH_X}{dt} = \dot{H}_X + QH_Z - RH_Y$$

$$\frac{dH_Y}{dt} = \dot{H}_Y + RH_X - PH_Z \tag{4-31}$$

$$\frac{dH_Z}{dt} = \dot{H}_Z + PH_Y - QH_X$$

After substituting the angular momentum components of Eq. 4-24, the total time rate of change of momentum about a typical axis ($X$) will be

$$\begin{aligned}\frac{dH_X}{dt} &= \dot{H}_X + QH_Z - RH_Y \\ &= \underbrace{(\dot{P}I_{xx} - \dot{Q}I_{xy} - \dot{R}I_{xz})}_{\dot{H}_X} + \underbrace{Q(RI_{zz} - PI_{xz} - QI_{yz})}_{QH_Z} - \underbrace{R(QI_{yy} - RI_{yz} - PI_{xz})}_{RH_Y} \\ &= \dot{P}I_{xx} + QR(I_{zz} - I_{yy}) - (PQ + \dot{R})I_{xz} + (PR - \dot{Q})I_{xy} - (Q^2 - R^2)I_{yz}\end{aligned} \tag{4-32}$$

By considering the simplified drawing in Fig. 4-7 and such appropriate momentum change components as $QH_Z$, the physical origin of even such obscure terms as $-Q^2I_{yz}$ is readily seen.

When all of the components are inserted in Eq. 4-31, and the notation on the inertias $I_{xx}$, $I_{yy}$, and $I_{zz}$ is simplified to $I_x$, $I_y$, and $I_z$, the result is

$$\frac{dH_X}{dt} = \dot{P}I_x + QR(I_z - I_y) - (PQ + \dot{R})I_{xz} + (PR - \dot{Q})I_{xy} - (Q^2 - R^2)I_{yz}$$

$$\frac{dH_Y}{dt} = \dot{Q}I_y + PR(I_x - I_z) - (QR + \dot{P})I_{xy} + (PQ - \dot{R})I_{yz} - (R^2 - P^2)I_{xz}$$

$$\frac{dH_Z}{dt} = \dot{R}I_z + PQ(I_y - I_x) - (PR + \dot{Q})I_{yz} + (QR - \dot{P})I_{xz} - (P^2 - Q^2)I_{xy} \tag{4-33}$$

This completes the simplified development of the inertial torque components.

For a formal development of the inertial moments we note that the form of Eq. 4-13, and the steps leading to it, are actually suitable for any vector whose components are measured in the body-fixed $X$, $Y$, $Z$ axis system. Thus, substituting **H** for **V** in Eq. 4-13, the time rate of change of the angular momentum will be

$$\frac{d\mathbf{H}}{dt} = \dot{\mathbf{H}} + \mathbf{\Omega} \times \mathbf{H} \tag{4-34}$$

This equation can be used directly once a vector expression for the angular momentum is obtained.

The moments and products of inertia in Eq. 4-24 are the rotational analog of the mass in Eq. 4-5. There are, however, nine components or six magnitudes, since $I_{ij} = I_{ji}$, required to specify the rotary inertia properties. For use in vector algebra operations, this inertia is written as the dyad $\underline{\underline{\mathbf{I}}}$:

$$\begin{aligned} \underline{\underline{\mathbf{I}}} = I_{xx}\mathbf{ii} &- I_{xy}\mathbf{ij} - I_{xz}\mathbf{ik} \\ - I_{xy}\mathbf{ji} &+ I_{yy}\mathbf{jj} - I_{yz}\mathbf{jk} \\ - I_{xz}\mathbf{ki} &- I_{yz}\mathbf{kj} + I_{zz}\mathbf{kk} \end{aligned} \tag{4-35}$$

In forming the dot product between a dyad and a vector, the unit vectors in the vector quantity are dotted into the closer of the unit vector pairs shown in the dyad, i.e.,

$$\begin{aligned} (\mathbf{ij}) \cdot \mathbf{j} &= \mathbf{i} \\ (\mathbf{ki}) \cdot \mathbf{i} &= \mathbf{k} \\ (\mathbf{ij}) \cdot \mathbf{i} &= 0, \text{ etc.} \end{aligned} \tag{4-36}$$

With this convention the angular momentum can now be written directly as

$$\begin{aligned}\mathbf{H} = \underline{\underline{\mathbf{I}}} \cdot \mathbf{\Omega} &= H_X\mathbf{i} + H_Y\mathbf{j} + H_Z\mathbf{k} \\ &= [PI_{xx} - QI_{xy} - RI_{xz}]\mathbf{i} \\ &\quad + [-PI_{xy} + QI_{yy} - RI_{yz}]\mathbf{j} \\ &\quad + [-PI_{xz} - QI_{yz} + RI_{zz}]\mathbf{k}\end{aligned} \tag{4-37}$$

This result is the same as that given in Eq. 4-29 and, when used with Eq. 4-34, produces the Eq. 4-33 values of the components of $d\mathbf{H}/dt$ along $X$, $Y$, and $Z$. There is no instructional advantage to be gained in carrying out the algebraic operation (which is analogous to that of Eq. 4-18 for the rate of change of linear momentum).

*Recapitulation of Inertial Forces and Simplification of Inertial Moments*

The linear acceleration and moment equations expressed in quantities referred to body-fixed axes are given in Eqs. 4-15 and 4-33, respectively. The linear acceleration set will be recapped here just as is, but the rotary equations are overcomplicated and can be simplified by adopting an additional assumption.

Assumption 4. The $XZ$ plane is assumed to be a plane of symmetry.

Assumption 4 is a very good approximation for most airborne vehicles. When it applies, we see from Fig. 4-8 that there is both a positive and a negative value of $y$ for each value of $x$ and $z$; consequently, $I_{yz} = \int yz\,dm = 0$ and $I_{xy} = \int xy\,dm = 0$. The expanded form of the equations of motion referred to body-fixed axes can then be written as

$$\begin{aligned}\textstyle\sum X &= m(\dot{U} + QW - RV) \\ \textstyle\sum Y &= m(\dot{V} + RU - PW) \\ \textstyle\sum Z &= m(\dot{W} + PV - QU) \\ \textstyle\sum L &= \dot{P}I_x - \dot{R}I_{xz} + QR(I_z - I_y) - PQI_{xz} \\ \textstyle\sum M &= \dot{Q}I_y + PR(I_x - I_z) - R^2I_{xz} + P^2I_{xz} \\ \textstyle\sum N &= \dot{R}I_z - \dot{P}I_{xz} + PQ(I_y - I_x) + QRI_{xz}\end{aligned} \tag{4-38}$$

## 4-4. Expansion of the Gravity Force

*Components of the Gravity Force*

Among the forces and moments acting externally on the vehicle, those due to gravity are always present. By neglecting gravity gradient considerations which are only important when all other external forces are

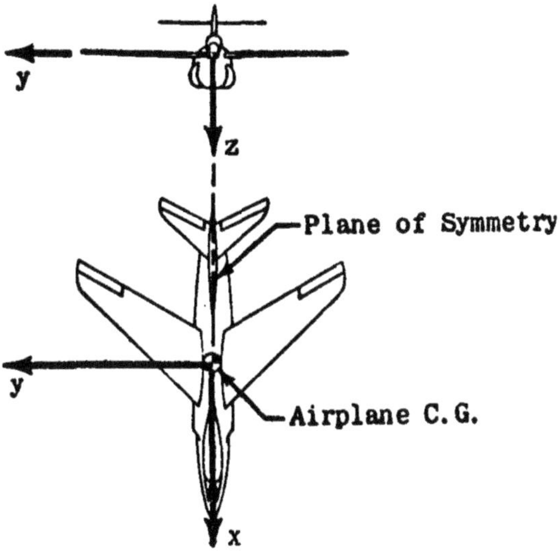


Fig. 4-8. Airframe plane of symmetry.

essentially nonexistent (e.g., in extra-atmospheric flight), the gravity force can be considered to act at the vehicle's center of gravity. Since the center of gravity coincides with the center of mass, the gravity force produces no external moments about that point. Thus, for our body-fixed axes, gravity can contribute components only to the summation of the external forces.

To determine these components, consider the alignment of the gravity vector with respect to the $X$, $Y$, $Z$ axes as shown in Fig. 4-9. Here the problem is essentially that of defining the relative orientation of a line (the $\mathbf{g}$ vector direction) with a rectangular coordinate reference frame. Rotations (in azimuth) of the frame about the $\mathbf{g}$ vector have no effect on the relative orientation of the $\mathbf{g}$ vector/reference frame, so that only the two angles $\Theta$ and $\Phi$ are needed to describe the situation physically. Direct resolution of the gravity-induced force, $m\mathbf{g}$, results in

$$\begin{aligned} \Delta X &= mg \sin(-\Theta) = -mg \sin \Theta \\ \Delta Y &= mg \cos(-\Theta) \sin \Phi = mg \cos \Theta \sin \Phi \\ \Delta Z &= mg \cos(-\Theta) \cos \Phi = mg \cos \Theta \cos \Phi \end{aligned} \tag{4-39}$$

Unfortunately, the angles $\Theta$ and $\Phi$ are not in general simply the integrals of $Q$ and $P$, respectively, so that we have, in effect, introduced

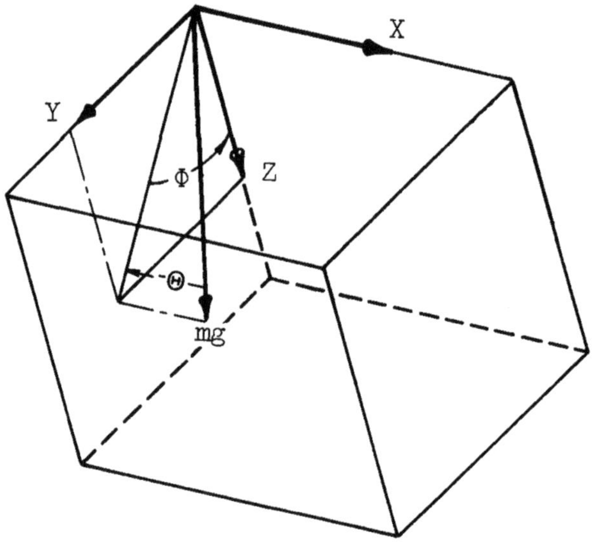


Fig. 4-9. Orientation of gravity vector with $X$, $Y$, $Z$ body-fixed axis system.

two new motion quantities. We must, therefore, connect the angular velocities $P$, $Q$, and $R$ with $\Theta$ and $\Phi$ and their derivatives. The details of this will depend on whether the gravitational vertical as seen in the vehicle is assumed to be fixed or rotating relative to inertial space. The first situation is generally an excellent approximation whenever the vehicle speed is small relative to orbital velocities, whereas the second is more appropriate for very high speed flight. Both situations will be discussed below in separate articles. Interspersed between these two treatments will be a short résumé of axis transformation considerations which are needed here for the first time in this book.

*Angular Velocity of XYZ Relative to the Gravitational Vertical*

To express the angular orientation and velocity of the $XYZ$ system with respect to the gravity vector requires the introduction of the angular velocity of the axes *about* the **g** vector. This is the azimuth rate, $\dot{\Psi}$. With this angular velocity added to $\dot{\Phi}$ and $\dot{\Theta}$, the resolutions of interest are shown in Fig. 4-10. Note that the $\dot{\Psi}$ vector is *not* normal to either $\dot{\Theta}$ or $\dot{\Phi}$ but that its projection in the $Y$, $Z$ plane is. Then, if we recognize that $\cos(-\Theta) = \cos\Theta$ and $\sin(-\Theta) = -\sin\Theta$, and assuming that *the* **g** *vector does not rotate relative to inertial space*, the following relations are apparent

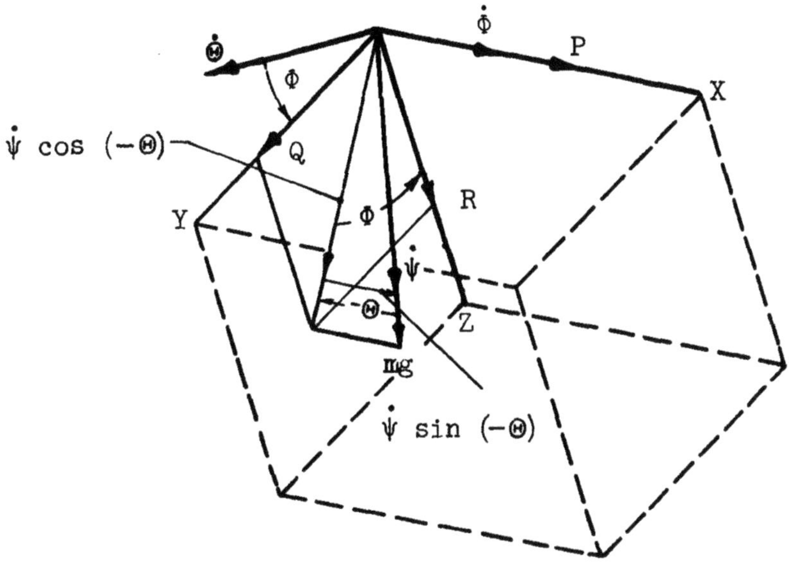


**Fig. 4-10. Angular orientation and velocities of g relative to *XYZ*.**

by direct resolution:

$$P = \dot{\Phi} - \dot{\Psi} \sin \Theta$$

$$Q = \dot{\Theta} \cos \Phi + \dot{\Psi} \cos \Theta \sin \Phi \qquad (4\text{-}40)$$

$$R = -\dot{\Theta} \sin \Phi + \dot{\Psi} \cos \Theta \cos \Phi$$

The converse case is also directly apparent from Fig. 4-10, i.e.,

$$\dot{\Phi} - \dot{\Psi} \sin \Theta = P$$

$$\dot{\Theta} = Q \cos \Phi - R \sin \Phi$$

$$\dot{\Psi} \cos \Theta = R \cos \Phi + Q \sin \Phi$$

which reduces to

$$\dot{\Phi} = P + Q \tan \Theta \sin \Phi + R \tan \Theta \cos \Phi$$

$$\dot{\Theta} = Q \cos \Phi - R \sin \Phi \qquad (4\text{-}41)$$

$$\dot{\Psi} = R\left(\frac{\cos \Phi}{\cos \Theta}\right) + Q\left(\frac{\sin \Phi}{\cos \Theta}\right)$$

These relationships supply the required connections between Θ and Φ and $P$, $Q$, and $R$.

*Vector Resolution and Axis Transformations*

When two or more frames of reference are needed to define physical relationships, it becomes necessary to orient these reference frames one to another and to establish the transformation relations needed to define a given vector quantity in terms of components in the several frames. Perhaps the simplest way to accomplish these requirements is to define a matrix of direction cosines that relates unit vectors in one axis system to those in another. The determination of the elements in such matrices can be accomplished efficiently by a combination of inspection and matrix multiplication procedures. These will be described below.

In Fig. 4-11 the unit vectors **a**′,**b**′, and **c**′ of the right-handed orthogonal coordinate system are found in terms of unit vectors **a**, **b**, and **c** of another. The **a**′, **b**′, **c**′ system is obtained from **a**, **b**, **c** by a rotation about **c** through the angle $\xi$. As a directional cosine array this transformation becomes

| | **a** | **b** | **c** |
|---|---|---|---|
| **a**′ | $\cos \xi$ | $\sin \xi$ | 0 |
| **b**′ | $-\sin \xi$ | $\cos \xi$ | 0 |
| **c**′ | 0 | 0 | 1 |

(4-42)

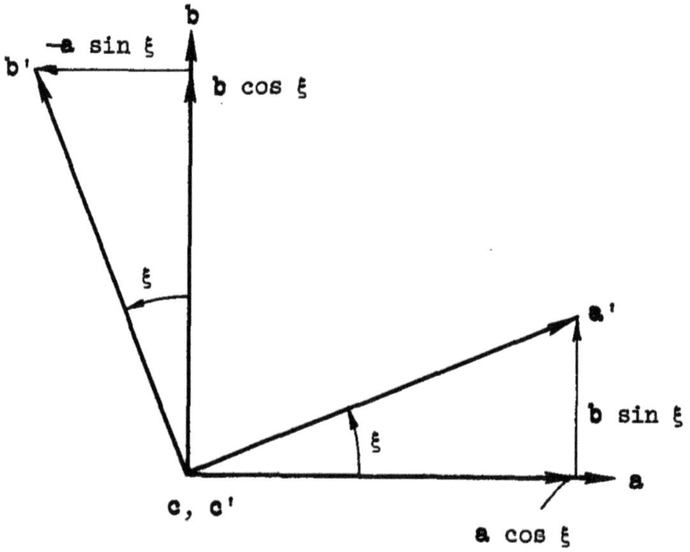


Fig. 4-11. Resolution of vectors.
$\mathbf{a}' = \mathbf{a} \cos \xi + \mathbf{b} \sin \xi$;
$\mathbf{b}' = -\mathbf{a} \sin \xi + \mathbf{b} \cos \xi$;
$\mathbf{c}' = \mathbf{c}$.

This array may be read either left to right or down. For instance, reading down in the **a** column,

$$\mathbf{a} = \mathbf{a}' \cos \xi - \mathbf{b}' \sin \xi + 0\mathbf{c}'$$

whereas reading **b**′ from left to right,

$$\mathbf{b}' = -\mathbf{a} \sin \xi + \mathbf{b} \cos \xi + 0\mathbf{c}$$

The array can also be interpreted as a matrix and its inverse; these correspond, respectively, to the array being read left to right and the array being read down.[7] The matrices will be

$$[\xi] = \begin{bmatrix} \cos \xi & \sin \xi & 0 \\ -\sin \xi & \cos \xi & 0 \\ 0 & 0 & 1 \end{bmatrix} \tag{4-43}$$

which is appropriate when going from **a**′, **b**′, **c**′ to **a**, **b**, **c**, i.e.,

$$\begin{bmatrix} \mathbf{a}' \\ \mathbf{b}' \\ \mathbf{c}' \end{bmatrix} = [\xi] \begin{bmatrix} \mathbf{a} \\ \mathbf{b} \\ \mathbf{c} \end{bmatrix} \tag{4-44}$$

and the inverse,

$$[\xi^{-1}] = \begin{bmatrix} \cos \xi & -\sin \xi & 0 \\ \sin \xi & \cos \xi & 0 \\ 0 & 0 & 1 \end{bmatrix} \tag{4-45}$$

which applies when going from **a**, **b**, **c** to **a**′, **b**′, **c**′, i.e.,

$$\begin{bmatrix} \mathbf{a} \\ \mathbf{b} \\ \mathbf{c} \end{bmatrix} = [\xi^{-1}] \begin{bmatrix} \mathbf{a}' \\ \mathbf{b}' \\ \mathbf{c}' \end{bmatrix} \tag{4-46}$$

This rotation and resolution can also be symbolized conveniently by a "transformation" box, as shown in Fig. 4-12.[8] The interconnections therein

[7] Such substitution of rows for columns in a matrix in general produces a *transposed* matrix. In the present case the product of the matrix and its transpose is a *unit* matrix that makes the transpose also an inverse. Aside from this, perhaps extraneous, explanation, the gist of the matrix operations needed for axis transformations can be followed directly in this presentation. If additional background is required, the reader may refer to any good text on matrix operations, e.g., A. C. Aitken, *Determinants and Matrices*, Interscience, New York, 1956; E. A. Guillemin, *The Mathematics of Circuit Analysis*, Wiley, New York, 1949; L. A. Pipes, *Matrix Methods for Engineering*, Prentice-Hall, Englewood Cliffs, N.J., 1963.

[8] This convenient rubric for vector resolution is due to Robert W. Bond.

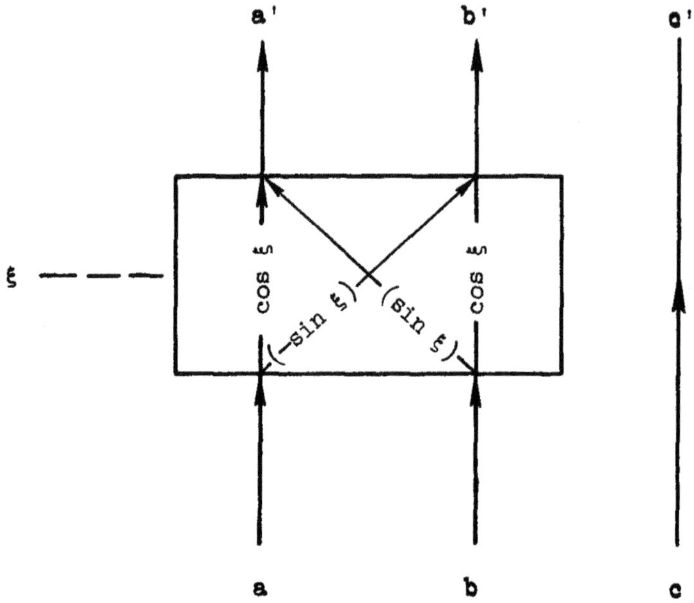


Fig. 4-12. Transformation box for single rotation.

indicate the multiplication and additions required in the resolutions. Although the arrows indicate progression from **a**, **b**, **c** to **a**′, **b**′, **c**′, the transformation works equally well in both directions.

Equation 4-42 exhibits certain general properties that, once recognized, make it possible to write out the appropriate array for any single rotation by inspection.

1. The main diagonal terms are always either the cosine of the angle of rotation or one.
2. The one is always associated with the axis about which rotation occurs.
3. The remaining elements in the row and column containing the one are all zeros.
4. The remaining places in the matrix contain the sine of the angle of rotation and are always symmetrically placed relative to the cosine terms.
5. In the direct right-handed rotation the negative sine always appears in the row above the one (this is to be interpreted as the third row if the one is in the first).

Any set of axes can be obtained from any other by a sequence of three rotations. For each rotation a matrix or transformation box corresponding to the array described above applies. The total array for the three rotations is simply obtained by multiplication of the three matrices representing the individual rotations.[9]

The most common transformation set in vehicle dynamics is that between an axis system which incorporates the **g** vector as one axis and the body-fixed $X$, $Y$, $Z$ axes. The actual rotations involved, following the usual order, are indicated in Fig. 4-13. This shows the unit vectors **l**, **m**, and **n** in the vehicle-centered, gravity-directed set, and **i**, **j**, and **k** in the vehicle body axes. The matrices for each of the three rotations shown can be written by inspection, in terms of the "Euler angles," $\Psi$, $\Theta$, $\Phi$, as

$$[\Psi] = \begin{bmatrix} \cos\Psi & \sin\Psi & 0 \\ -\sin\Psi & \cos\Psi & 0 \\ 0 & 0 & 1 \end{bmatrix}$$

$$[\Theta] = \begin{bmatrix} \cos\Theta & 0 & -\sin\Theta \\ 0 & 1 & 0 \\ \sin\Theta & 0 & \cos\Theta \end{bmatrix} \tag{4-47}$$

$$[\Phi] = \begin{bmatrix} 1 & 0 & 0 \\ 0 & \cos\Phi & \sin\Phi \\ 0 & -\sin\Phi & \cos\Phi \end{bmatrix}$$

These rotations are also shown in the transformation boxes of Fig. 4-14.

Carrying out the matrix multiplications in the order appropriate to the rotations, i.e.,

$$[E] = [\Phi][\Theta][\Psi] \tag{4-48}$$

we find the complete direction cosine array. That is,

| | **l** | **m** | **n** |
|---|---|---|---|
| **i** | $\cos\Psi\cos\Theta$ | $\sin\Psi\cos\Theta$ | $-\sin\Theta$ |
| **j** | $\cos\Psi\sin\Theta\sin\Phi$ $-\sin\Psi\cos\Phi$ | $\sin\Psi\sin\Theta\sin\Phi$ $+\cos\Psi\cos\Phi$ | $\cos\Theta\sin\Phi$ |
| **k** | $\cos\Psi\sin\Theta\cos\Phi$ $+\sin\Psi\sin\Phi$ | $\sin\Psi\sin\Theta\cos\Phi$ $-\cos\Psi\sin\Phi$ | $\cos\Theta\cos\Phi$ |

(4-49)

[9] A matrix product has the form

$$\begin{bmatrix} a_{11} & a_{12} \\ a_{21} & a_{22} \end{bmatrix} \begin{bmatrix} b_{11} & b_{12} \\ b_{21} & b_{22} \end{bmatrix} = \begin{bmatrix} a_{11}b_{11} + a_{12}b_{21} & a_{11}b_{12} + a_{12}b_{22} \\ a_{21}b_{11} + a_{22}b_{21} & a_{21}b_{12} + a_{22}b_{22} \end{bmatrix}$$

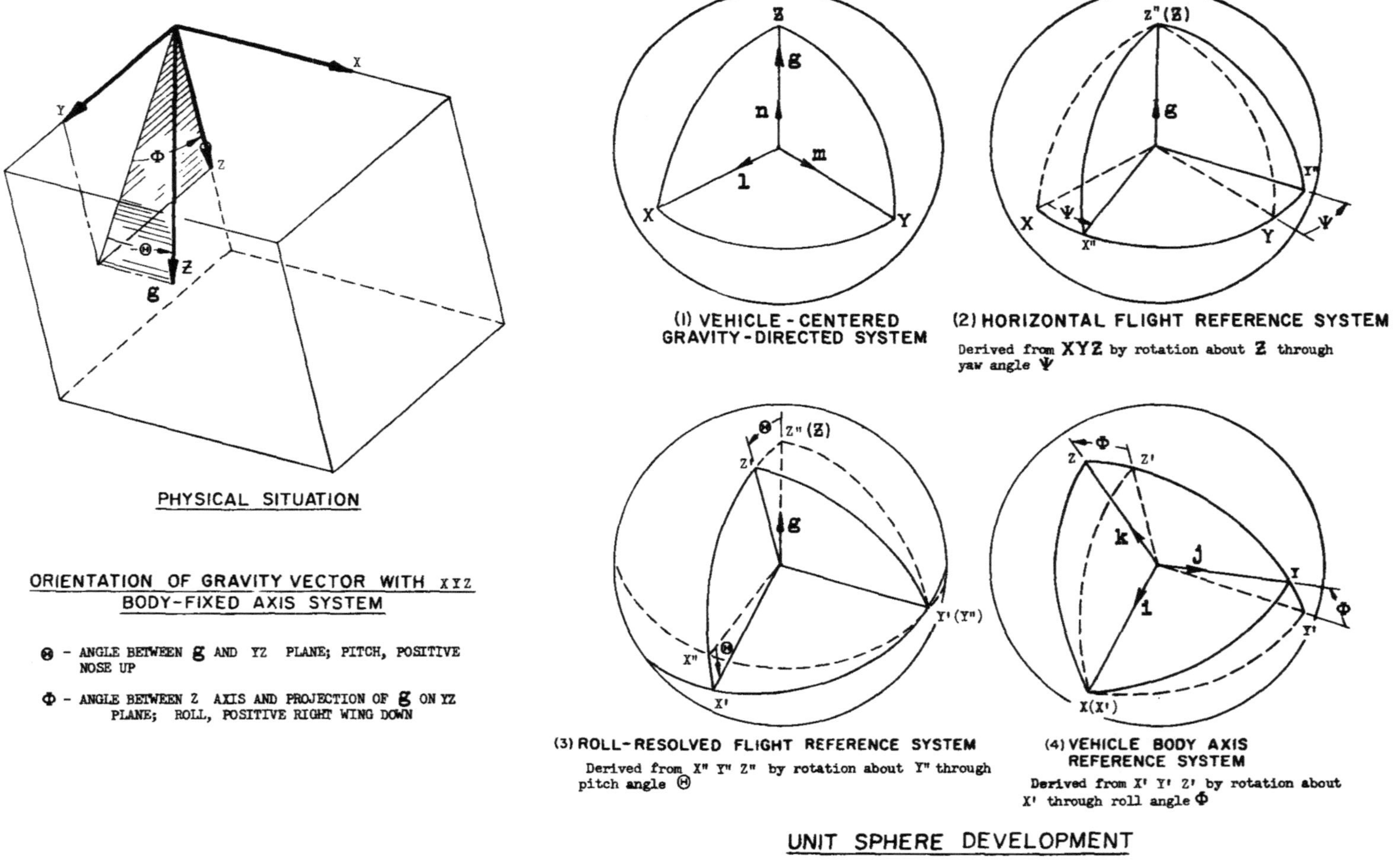


Fig. 4-13. Relationship between vehicle-centered gravity-directed and vehicle-fixed axis systems.

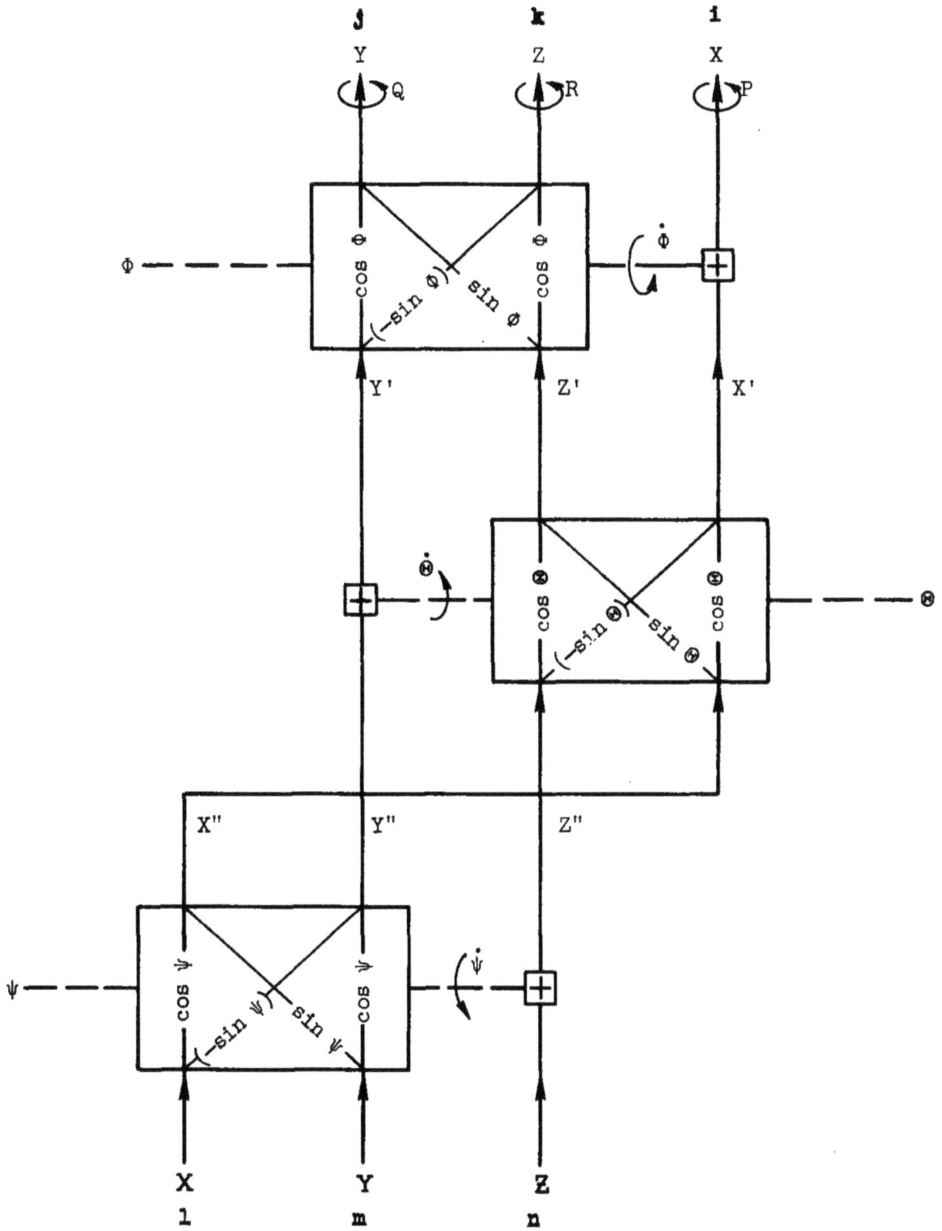


Fig. 4-14. Transformation boxes for relationships between vehicle-centered gravity-directed and vehicle-fixed axis systems.

Just as in the simple single-rotation example, the array can be read from left to right or down. As matrices these correspond to

$$[E]\begin{bmatrix}\mathbf{l}\\ \mathbf{m}\\ \mathbf{n}\end{bmatrix} = \begin{bmatrix}\mathbf{i}\\ \mathbf{j}\\ \mathbf{k}\end{bmatrix} \tag{4-50}$$

the array being read left to right, or

$$[E^{-1}]\begin{bmatrix}\mathbf{i}\\ \mathbf{j}\\ \mathbf{k}\end{bmatrix} = \begin{bmatrix}\mathbf{l}\\ \mathbf{m}\\ \mathbf{n}\end{bmatrix} \tag{4-51}$$

the array being read down.

In general, any pair of coordinate systems can be associated by a sequence of three rotations such as those exemplified above. There are, however, six possible sets of rotation angles corresponding to the six possible ways of getting from one axis system to the other. Each of these results in a different directional cosine array. The forms of all are similar in that, of the nine terms involved, one contains a single angle, four involve two angles, and the remaining four contain all three angles. While in principle any of the possible rotational sequences can be used, there is in practice ordinarily one to be preferred. In the present case the rotational sequence and angles were defined to result in the least complicated resolution of the **g** vector into the body-fixed $X$, $Y$, $Z$ system. This is given by

$$\mathbf{g} = g\mathbf{n} = g[-\sin\Theta\mathbf{i} + \cos\Theta\sin\Phi\mathbf{j} + \cos\Theta\cos\Phi\mathbf{k}] \tag{4-52}$$

The basic simplicity of the components shown is insured by taking the first rotation about the gravity vector through an azimuth angle, $\Psi$. As already remarked, rotations of the body-fixed axes about the **g** vector can have no effect on the relative orientation of the **g** vector and the $X$, $Y$, $Z$ frame. The two remaining rotations can conceivably be accomplished in two different ways without complicating the **g** vector description. The sequence and angles selected are the more natural cyclic sequence, and also have the advantage of being those measured by a typically oriented vertical gyro. A two degree of freedom, gravity-erected vertical gyro oriented with its outer gimbal bearing axis along $X$ measures $\Theta$ and $\Phi$ on its inner and outer gimbals, respectively.

The angular velocities are readily found using the transformation boxes of Fig. 4-14. Thus the angular velocities $P$, $Q$, and $R$ about $X$, $Y$, and $Z$, respectively, can be found by tracking the angular velocities associated with the transformation boxes $\dot{\Psi}$, $\dot{\Theta}$, and $\dot{\Phi}$ through the intervening transformations, with the result

$$\begin{aligned} P &= \dot{\Phi} - \dot{\Psi} \sin \Theta \\ Q &= \dot{\Theta} \cos \Phi + \dot{\Psi} \cos \Theta \sin \Phi \\ R &= \dot{\Psi} \cos \Theta \cos \Phi - \dot{\Theta} \sin \Phi \end{aligned} \tag{4-53}$$

This is the same result as that previously derived on a more physical basis in the discussion leading to Eq. 4-40. Although the axis transformations can be run in either direction through the transformation boxes, the angular velocities go only one way. Consequently, the equations (Eq. 4-41) for $\dot{\Psi}$, $\dot{\Theta}$, and $\dot{\Phi}$ in terms of $P$, $Q$, and $R$ are most conveniently obtained via the physical setup in Fig. 4-10 or by solution of Eq. 4-53.

*Angular Velocity of the Gravitational Vertical Relative to Inertial Space*

If the aircraft travels at a very high speed, the angular velocity relative to inertial space of the gravitational vertical is no longer approximately zero. Consequently, the right-hand side of Eq. 4-53 must have additional terms added that take into account the rotation of the X, Y, Z axes as the aircraft moves about the earth. To take this effect into account, inertial space will now be defined as an $\mathfrak{X}$, $\mathfrak{Y}$, $\mathfrak{Z}$ axis system, with its origin at the center of the earth and its axes nonrotating relative to the fixed stars. For any given flight path this system is oriented so that the $\mathfrak{X}$, $\mathfrak{Z}$ plane contains the equilibrium flight velocity and gravity vectors. Then, as shown in Fig. 4-15, the vehicle-centered, gravity-directed system, X, Y, Z, is oriented so that its X, Z plane also contains the equilibrium flight vector. Then the angular velocity of X, Y, Z relative to inertial space is $\dot{\nu}$, or

$$\dot{\nu} = \frac{\mathsf{U}}{\mathsf{r}} \tag{4-54}$$

Thus the angular velocity of X, Y, Z is

$$\mathbf{\Omega}_{\mathrm{X,Y,Z}} = -\dot{\nu}\mathbf{m} \tag{4-55}$$

or, referenced to measurements in the body-fixed frame via Eq. 4-49,

$$\begin{aligned} \mathbf{\Omega}_{\mathrm{X,Y,Z}} = -\dot{\nu}[&(\sin \Psi \cos \Theta)\mathbf{i} + (\sin \Psi \sin \Theta \sin \Phi + \cos \Psi \cos \Phi)\mathbf{j} \\ &+ (\sin \Psi \sin \Theta \cos \Phi - \cos \Psi \sin \Phi)\mathbf{k}] \end{aligned}$$

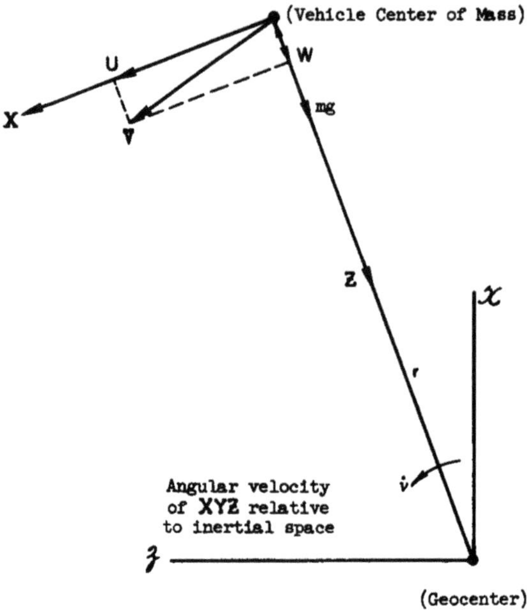


| AXES | UNIT VECTORS | DESCRIPTION |
|---|---|---|
| $\mathcal{X}\mathcal{Y}\mathcal{Z}$ | $\ell m n$ | Inertial system with origin at geocenter. $\mathcal{X}\mathcal{Z}$ plane coincident with equilibrium flight velocity vector. |
| XYZ | lmn | Rotating system with origin at vehicle center of mass. Z directed toward geocenter. XZ plane coincident with equilibrium flight velocity vector. |

Fig. 4-15. Definition of inertial space and radially directed rotating axis system.

Consequently, the components of the angular velocity relative to inertial space become

$$P = \dot{\Phi} - \Psi' \sin \Theta \qquad \frac{-U}{r}(\sin \Psi' \cos \Theta)$$

$$Q = \dot{\Theta} \cos \Phi + \Psi' \cos \Theta \sin \Phi \qquad \frac{-U}{r}(\sin \Psi' \sin \Theta \sin \Phi + \cos \Psi' \cos \Phi)$$

$$R = \Psi' \cos \Theta \cos \Phi - \dot{\Theta} \sin \Phi \qquad \frac{-U}{r}(\sin \Psi' \sin \Theta \cos \Phi - \cos \Psi' \sin \Phi)$$

$$\Omega \text{ relative to X, Y, Z} \longleftrightarrow \Omega_{X,Y,Z} \text{ relative to } \mathfrak{X}, \mathfrak{Y}, \mathfrak{Z} \tag{4-56}$$

## 4-5. Linearization of the Inertial and Gravitational Components

Both the inertial forces of Eq. 4-38 and the gravitational forces of Eq. 4-39 are proportional to the vehicle mass. It is therefore convenient to combine these terms into components that represent the accelerations which would be *sensed by instruments* located with their input axes coincident with the body-fixed $X$, $Y$, and $Z$ axes. When this is done and the remaining external aerodynamic and propulsive forces and moments are represented by their generalized symbols, the six degree of freedom, rigid body equations of motion become

$$
\begin{aligned}
ma_{x_{c.g.}} &= m[\dot{U} + QW - RV + g \sin \Theta] = X \\
ma_{y_{c.g.}} &= m[\dot{V} + RU - PW - g \cos \Theta \sin \Phi] = Y \\
ma_{z_{c.g.}} &= m[\dot{W} + PV - QU - g \cos \Theta \cos \Phi] = Z \\
\dot{P}I_x - \dot{R}I_{xz} &+ QR(I_z - I_y) - PQI_{xz} = L \\
\dot{Q}I_y + PR(I_x - I_z) &- R^2 I_{xz} + P^2 I_{xz} = M \\
\dot{R}I_z - \dot{P}I_{xz} &+ PQ(I_y - I_x) + QRI_{xz} = N
\end{aligned}
\qquad (4\text{-}57)
$$

To these must be added the auxiliary relations of Eq. 4-56 which relate $\Phi$, $\Theta$, $\Psi$ to $P$, $Q$, and $R$.

These equations contain products of the dependent variables, some of which appear as transcendental functions; therefore they are in general nonlinear. To reduce them to tractable form, the total motion can be considered as composed of two parts: an average or mean motion that is representative of the operating point or trim conditions, and a dynamic motion that accounts for small perturbations about the mean motion.

The operating point equations are obtained by recognizing that zero translational and rotational accelerations are implicit in the concept of a trim condition. Then, denoting such a condition with zero subscript, the *trim equations* are given directly by

$$
\begin{aligned}
m[W_0 Q_0 - V_0 R_0 + g \sin \Theta_0] &= X_0 \\
m[U_0 R_0 - W_0 P_0 - g \cos \Theta_0 \sin \Phi_0] &= Y_0 \\
m[V_0 P_0 - U_0 Q_0 - g \cos \Theta_0 \cos \Phi_0] &= Z_0 \\
Q_0 R_0 (I_z - I_y) - P_0 Q_0 I_{xz} &= L_0 \\
P_0 R_0 (I_x - I_z) - R_0^2 I_{xz} + P_0^2 I_{xz} &= M_0 \\
P_0 Q_0 (I_y - I_x) + Q_0 R_0 I_{xz} &= N_0
\end{aligned}
\qquad (4\text{-}58)
$$

Because steady-state rolling, pitching, and yawing are possible as a trim condition, the operating point equations for $P_0$, $Q_0$, and $R_0$ will, in general, contain all the terms shown in Eq. 4-56, although each will have a zero subscript.

The perturbed motions are by definition those obtained by subtracting the trim motions from the total motion. Thus the perturbed equations of motion can be obtained by substituting $U = U_0 + dU$, $P = P_0 + dP$, $\Phi = \Phi_0 + d\Phi$, etc., into Eq. 4-57, expanding, and then *subtracting* the trim equation. A more straightforward process is to differentiate both sides of Eq. 4-57 to obtain the perturbed equations directly. To simplify the notation, we designate the perturbed motion quantities by their lower case equivalents (e.g., $dU = u$, etc.); also we make another assumption.

Assumption 5. The disturbances from the steady flight conditions are assumed to be small enough so that the sines and cosines of the disturbance angles are approximately the angles themselves and one, respectively, and so that the products and squares of the disturbance quantities are negligible in comparison with the quantities themselves.

Using this assumption, the perturbed equations of motion for the vehicle become

$$\begin{aligned}
&m[\dot{u} + W_0 q + Q_0 w - V_0 r - R_0 v + (g \cos \Theta_0)\theta] && = dX \\
&m[\dot{v} + U_0 r + R_0 u - W_0 p - P_0 w \\
&\qquad - (g \cos \Theta_0 \cos \Phi_0)\varphi + (g \sin \Theta_0 \sin \Phi_0)\theta] && = dY \\
&m[\dot{w} + V_0 p + P_0 v - U_0 q - Q_0 u \\
&\qquad + (g \cos \Theta_0 \sin \Phi_0)\varphi + (g \sin \Theta_0 \cos \Phi_0)\theta] && = dZ \\
&\dot{p} I_x - \dot{r} I_{xz} + (Q_0 r + R_0 q)(I_z - I_y) - (P_0 q + Q_0 p) I_{xz} && = dL \\
&\dot{q} I_y + (P_0 r + R_0 p)(I_x - I_z) - (2R_0 r - 2P_0 p) I_{xz} && = dM \\
&\dot{r} I_z - \dot{p} I_{xz} + (P_0 q + Q_0 p)(I_y - I_x) + (Q_0 r + R_0 q) I_{xz} && = dN
\end{aligned} \tag{4-59}$$

Perturbed equations are also required for the auxiliary relations, Eq. 4-56. The portions of these equations that represent the rotation of the vertical relative to inertial space are excessively complicated when presented in general and are seldom if ever used in their entirety. On the other hand, the angular velocity components that represent the rotation of the body-fixed $X$, $Y$, $Z$ axes relative to X, Y, Z are occasionally needed. These are presented below:

$$\begin{aligned}
p &= \dot{\varphi} - \dot{\psi} \sin \Theta_0 - \theta(\dot{\Psi}_0 \cos \Theta_0) \\
q &= \dot{\theta} \cos \Phi_0 - \theta(\dot{\Psi}_0 \sin \Theta_0 \sin \Phi_0) \\
&\qquad + \varphi(\dot{\Psi}_0 \cos \Theta_0 \cos \Phi_0 - \dot{\Theta}_0 \sin \Phi_0) + \dot{\psi} \cos \Theta_0 \sin \Phi_0 \\
r &= \dot{\psi} \cos \Theta_0 \cos \Phi_0 - \varphi(\dot{\Psi}_0 \cos \Theta_0 \sin \Phi_0 + \dot{\Theta}_0 \cos \Phi_0) \\
&\qquad - \dot{\theta} \sin \Phi_0 - \theta(\dot{\Psi}_0 \sin \Theta_0 \cos \Phi_0)
\end{aligned} \tag{4-60}$$

Also, in view of Assumption 5, the rotation array of Eq. 4-49 can be written in its most general linearized form as follows.

| | $\mathbf{l}$ | $\mathbf{m}$ | $\mathbf{n}$ |
|---|---|---|---|
| $\mathbf{i}$ | $\left\{\begin{matrix}\underline{[\cos\Psi_0\cos\Theta_0]}\\ -\psi(\sin\Psi_0\cos\Theta_0)\\ -\theta(\cos\Psi_0\sin\Theta_0)\end{matrix}\right\}$ | $\left\{\begin{matrix}\underline{[\sin\Psi_0\cos\Theta_0]}\\ +\psi(\cos\Psi_0\cos\Theta_0)\\ -\theta(\sin\Psi_0\sin\Theta_0)\end{matrix}\right\}$ | $\left\{\begin{matrix}\underline{[-\sin\Theta_0]}\\ -\theta(\cos\Theta_0)\end{matrix}\right\}$ |
| $\mathbf{j}$ | $\left\{\begin{matrix}\underline{[\cos\Psi_0\sin\Theta_0\sin\Phi_0-\sin\Psi_0\cos\Phi_0]}\\ -\psi(\sin\Psi_0\sin\Theta_0\sin\Phi_0+\cos\Psi_0\cos\Phi_0)\\ +\theta(\cos\Psi_0\cos\Theta_0\sin\Phi_0)\\ +\varphi(\cos\Psi_0\sin\Theta_0\cos\Phi_0+\sin\Psi_0\sin\Phi_0)\end{matrix}\right\}$ | $\left\{\begin{matrix}\underline{[\sin\Psi_0\sin\Theta_0\sin\Phi_0+\cos\Psi_0\cos\Phi_0]}\\ +\psi(\cos\Psi_0\sin\Theta_0\sin\Phi_0-\sin\Psi_0\cos\Phi_0)\\ +\theta(\sin\Psi_0\cos\Theta_0\sin\Phi_0)\\ +\varphi(\sin\Psi_0\sin\Theta_0\cos\Phi_0-\cos\Psi_0\sin\Phi_0)\end{matrix}\right\}$ | $\left\{\begin{matrix}\underline{[\cos\Theta_0\sin\Phi_0]}\\ -\theta(\sin\Theta_0\sin\Phi_0)\\ +\varphi(\cos\Theta_0\cos\Phi_0)\end{matrix}\right\}$ |
| $\mathbf{k}$ | $\left\{\begin{matrix}\underline{[\cos\Psi_0\sin\Theta_0\cos\Phi_0+\sin\Psi_0\sin\Phi_0]}\\ +\psi(-\sin\Psi_0\sin\Theta_0\cos\Phi_0+\cos\Psi_0\sin\Phi_0)\\ +\theta(\cos\Psi_0\cos\Theta_0\cos\Phi_0)\\ +\varphi(-\cos\Psi_0\sin\Theta_0\sin\Phi_0+\sin\Psi_0\cos\Phi_0)\end{matrix}\right\}$ | $\left\{\begin{matrix}\underline{[\sin\Psi_0\sin\Theta_0\cos\Phi_0-\cos\Psi_0\sin\Phi_0]}\\ +\psi(\cos\Psi_0\sin\Theta_0\cos\Phi_0+\sin\Psi_0\sin\Phi_0)\\ +\theta(\sin\Psi_0\cos\Theta_0\cos\Phi_0)\\ -\varphi(\sin\Psi_0\sin\Theta_0\sin\Phi_0+\cos\Psi_0\cos\Phi_0)\end{matrix}\right\}$ | $\left\{\begin{matrix}\underline{[\cos\Theta_0\cos\Phi_0]}\\ -\theta(\sin\Theta_0\cos\Phi_0)\\ -\varphi(\cos\Theta_0\sin\Phi_0)\end{matrix}\right\}$ |

$$= [E] - [E_0] + [e] \quad \textbf{(4-61)}$$

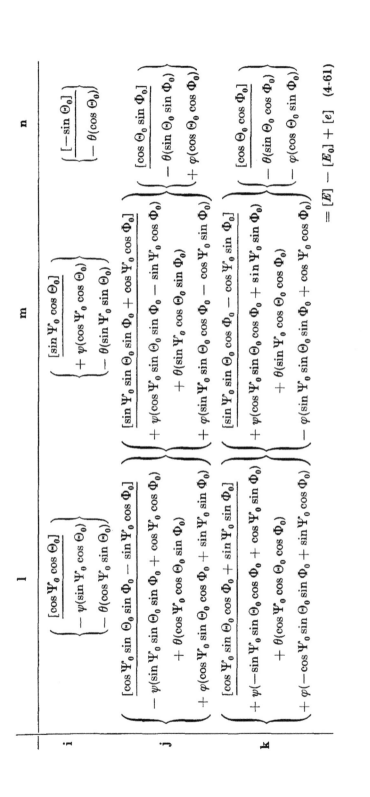

Here $[E_0]$ is the matrix made up of the operating point quantities alone (underlined, square-bracketed terms) and $[e]$ is the matrix made up of the perturbation quantities alone.

Assumption 5 limits the applicability of these equations to what are called "small perturbations." In return for these restrictions, the non-linearities are removed and sets of linear equations result. This permits an important simplification in the mathematical methods necessary to analyze aircraft motions. Also, while the linearizations are applicable in theory only to infinitesimal disturbances from trim, experience has shown that quite accurate results can be obtained for disturbances of much larger magnitude.

Although the equations are linear, they are also formidably complex because of the high degree of generality presumed for the trim conditions. Because of this complexity the equations are seldom used in this complete form. Instead, simpler cases that nevertheless reveal fundamental control and operating problems are more commonly used. For instance, a situation of great interest is that for steady, *straight* ($\dot{\Psi}_0 = \dot{\Theta}_0 = 0$), *symmetric* ($\Psi_0 = V_0 = 0$), *wings-level* ($\Phi_0 = 0$) flight. Under these circumstances, the trim, or operating point, conditions consist of $\mathsf{U}_0$, $U_0$, $W_0$, and $\Theta_0$, as connected by

$$\mathsf{U} = U \cos \Theta_0 + W \sin \Theta_0$$

$$\mathsf{U}_0 + \mathsf{u} = U_0 \cos \Theta_0 + W_0 \sin \Theta_0 + u \cos \Theta_0 + w \sin \Theta_0 - U_0 (\sin \Theta_0)\theta + W_0(\cos \Theta_0)\theta \qquad (4\text{-}62)$$

The stability characteristics, with or without automatic control for such conditions, are of major interest because, if they are unsatisfactory, the vehicle is useless. Then, neglecting changes in the distance to the geocenter, $\mathsf{r}$, the perturbed angular velocities, including the rotation of the vertical, become

$$\begin{aligned}
p &= \dot{\varphi} - \dot{\psi} \sin \Theta_0 && - \psi \frac{\mathsf{U}_0}{\mathsf{r}} \cos \Theta_0 \\
q &= \dot{\theta} && + \theta \left( \frac{U_0 \sin \Theta_0 - W_0 \cos \Theta_0}{\mathsf{r}} \right) - u \frac{\cos \Theta_0}{\mathsf{r}} - w \frac{\sin \Theta_0}{\mathsf{r}} \\
r &= \dot{\psi} \cos \Theta_0 && - \psi \frac{\mathsf{U}_0}{\mathsf{r}} \sin \Theta_0 + \varphi \frac{\mathsf{U}_0}{\mathsf{r}} \qquad (4\text{-}63) \\
Q_0 &= && \frac{-\mathsf{U}_0}{\mathsf{r}} = \frac{-1}{\mathsf{r}} (U_0 \cos \Theta_0 + W_0 \sin \Theta_0) \\
P_0 = R_0 &= && 0 \\
& && \text{rotation of vertical relative to } \mathfrak{X}, \mathfrak{Y}, \mathfrak{Z}
\end{aligned}$$

In Laplace transform style, $\varphi$, $\psi$, and $\theta$ are given by

$$\varphi = \frac{[s\cos\Theta_0 - (\mathsf{U}_0/\mathsf{r})\sin\Theta_0]p + [s\sin\Theta_0 + (\mathsf{U}_0/\mathsf{r})\cos\Theta_0]r}{\cos\Theta_0[s^2 + (\mathsf{U}_0^2/\mathsf{r}^2)]}$$

$$\psi = \frac{sr - (\mathsf{U}_0/\mathsf{r})p}{\cos\Theta_0[s^2 + (\mathsf{U}_0^2/\mathsf{r}^2)]} \tag{4-64}$$

$$\theta = \frac{q + [(u\cos\Theta_0)/\mathsf{r}] + [(w\sin\Theta_0)/\mathsf{r}]}{s + [(U_0\sin\Theta_0 - W_0\cos\Theta_0)/\mathsf{r}]}$$

Substituting Eqs. 4-63 and 4-64 into the Laplace-transformed perturbed equations (4-59) specialized for straight symmetric, wings-level flight, we get Eqs. 4-65 and 4-66.

$$\left(s - \frac{W_0}{\mathsf{r}}\cos\Theta_0\right)u - \frac{\mathsf{U}_0}{\mathsf{r}}\left(1 + \frac{W_0\sin\Theta_0}{\mathsf{U}_0}\right)w + \left[W_0 s + g\cos\Theta_0 + \frac{W_0}{\mathsf{r}}(U_0\sin\Theta_0 - W_0\cos\Theta_0)\right]\theta = \frac{dX}{m}$$

$$\frac{\mathsf{U}_0}{\mathsf{r}}\left(1 + \frac{U_0\cos\Theta_0}{\mathsf{U}_0}\right)u + \left(s + \frac{U_0\sin\Theta_0}{\mathsf{r}}\right)w - \left[U_0 s - \left(g - \frac{U_0^2}{\mathsf{r}}\right)\sin\Theta_0 - \frac{U_0 W_0}{\mathsf{r}}\cos\Theta_0\right]\theta = \frac{dZ}{m}$$

$$\frac{-\cos\Theta_0}{\mathsf{r}}su - \frac{\sin\Theta_0}{\mathsf{r}}sw + s\left(s + \frac{U_0\sin\Theta_0 - W_0\cos\Theta_0}{\mathsf{r}}\right)\theta = \frac{dM}{I_y} \tag{4-65}$$

$$sv - \left\{W_0 + \frac{g[s\cos\Theta_0 - (\mathsf{U}_0/\mathsf{r})\sin\Theta_0]}{s^2 + (\mathsf{U}_0^2/\mathsf{r}^2)}\right\}p + \left\{U_0 - \frac{g[s\sin\Theta_0 + (\mathsf{U}_0/\mathsf{r})\cos\Theta_0]}{s^2 + (\mathsf{U}_0^2/\mathsf{r}^2)}\right\}r = \frac{dY}{m}$$

$$0 + \left(s + \frac{I_{xz}}{I_x}\frac{\mathsf{U}_0}{\mathsf{r}}\right)p - \left[\frac{I_{xz}}{I_x}s + \left(\frac{I_z - I_y}{I_x}\right)\frac{\mathsf{U}_0}{\mathsf{r}}\right]r = \frac{dL}{I_x} \tag{4-66}$$

$$0 - \left[\frac{I_{xz}}{I_z}s + \left(\frac{I_y - I_x}{I_z}\right)\frac{\mathsf{U}_0}{\mathsf{r}}\right]p + \left(s - \frac{I_{xz}}{I_z}\frac{\mathsf{U}_0}{\mathsf{r}}\right)r = \frac{dN}{I_z}$$

Beside being considerably simplified over the formidable complete linearized set, these equations are seen to be separated into two groups of three each. In the first, Eqs. 4-65, the dependent motion variables are $u$, $w$, and $\theta$, and the motions described are thus confined to the $X$, $Z$

plane. These are generally referred to as the "longitudinal motions." The other set, Eqs. 4-66, are lateral/directional and consist of sideslipping, rolling, and yawing. Actually, as shown in Eq. 4-66, the rolling and yawing acceleration equations are not explicitly coupled with the sideslip equation, although implicitly, and in practice, a great deal of coupling can exist through the medium of the aerodynamic forces presently contained on the right-hand side. In fact, the lateral and longitudinal separation indicated above is, at this point, only a separation of the gravitational and inertial forces; the six degrees of freedom may actually be coupled through propulsive or aerodynamic forces and moments. The separation occurs largely because of the assumed trim conditions and is equally true whether or not effects from rotation of the vertical are included.

In passing we should note, in connection with Eqs. 4-65 and 4-66, that whereas the pitch *displacement*, $\theta$, is the angular motion quantity used in the longitudinal equations, the lateral equations utilize the angular *velocities* $p$ and $r$. This is only a matter of convenience and simplification; substituting for $p$ and $r$ (Eq. 4-63) would greatly expand Eq. 4-66 and Eq. 4-65 would suffer similarly if expressed in terms of $q$ rather than $\theta$. However, the choice of motion variables purely as a matter of simplifying the resulting equations can mask important physical effects and, further, may not be consistent with the sensory equipment commonly used to exercise feedback control.

To illustrate, consider the side acceleration equation in Eq. 4-66. It is apparent from the expression for $\varphi$ (Eq. 4-64) that an alternate form is

$$sv - W_0 p + U_0 r - g \cos \Theta_0 \varphi = \frac{dY}{m} \tag{4-67}$$

and that an added contribution to the $\varphi$ term appears when the Eq. 4-63 expression for $r$ is substituted. That is, doing so results in

$$sv - W_0 p + U_0\left(s\psi \cos \Theta_0 - \psi \frac{\mathrm{U}_0}{\mathrm{r}} \sin \Theta_0\right) - \left(\mathrm{g} \cos \Theta_0 - U_0 \frac{\mathrm{U}_0}{\mathrm{r}}\right)\varphi = \frac{dY}{m} \tag{4-68}$$

We see now that the net side acceleration from $\varphi$ perturbations is proportional to the difference between the gravitational attraction, $g \cos \Theta_0$, and the centrifugal forces due to high speed flight at a nearly constant value of r, $U_0(\mathrm{U}_0/\mathrm{r})$. What happens is that a constant value of $\varphi$, which is a gravity-directed angle, produces a body axis turn rate, $r$, proportional to $\varphi(\mathrm{U}_0/\mathrm{r})$ as illustrated in Fig. 4-16 for the simply visualized case of $\varphi = +90$ degrees. Note that the positive direction of $r$ is reversed (counterclockwise, rather than clockwise) because we are viewing the underside of the vehicle.

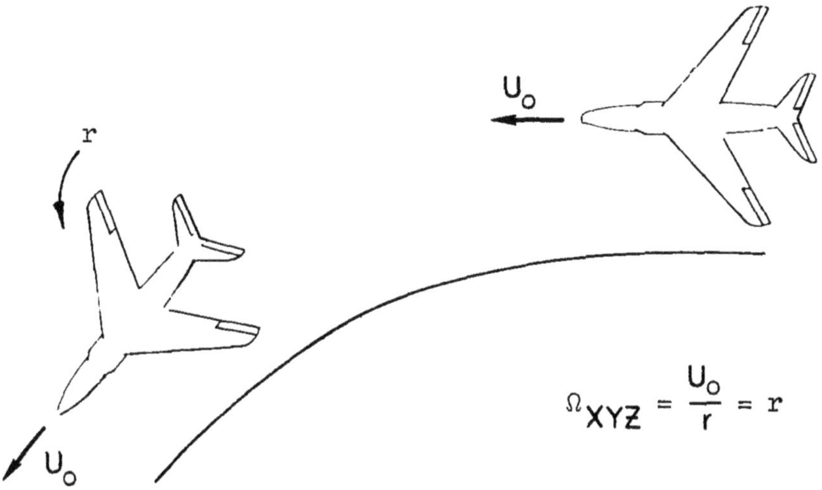


Fig. 4-16. Origins of body axis turn rate.

The similar-appearing "weightlessness" term, $\theta(g - U_0^2/r) \sin \Theta_0$, in the $Z$ acceleration equation of Eq. 4-65 is, correspondingly, eliminated by using body axis pitch rate, $q$, rather than $\theta$. That is, from the perturbed equations (4-59) specialized for the initial conditions of interest and the relations in Eqs. 4-63 and 4-64,

$$\frac{U_0}{r} u + sw - U_0 q + g \sin \Theta_0 \left\{ \frac{q + (u \cos \Theta_0/r) + (w \sin \Theta_0/r)}{s + [(U_0 \sin \Theta_0 - W_0 \cos \Theta_0)/r]} \right\} = \frac{dZ}{m} \tag{4-69}$$

where the bracketed term is just the expression for $\theta$ (Eq. 4-64).

If we consider now the appropriateness of Eqs. 4-65 and 4-66 for feedback control application, it is immediately clear that Eqs. 4-66 are directly usable in conjunction with body-oriented rate gyro sensors. Conversely, Eq. 4-65 is most appropriate for a gravity-erected gyro or platform. One final point worth mentioning is the appearance of the second-order denominator in the side force expression of Eq. 4-66. This oscillation, with a period $(2\pi r/U_0)$ equal to the time to circle the earth, is of special importance for inertial guidance and navigation systems.

## 4-6. Expansion of the Aerodynamic Forces and Moments

The aerodynamic forces are exerted on the vehicle by the surrounding atmosphere; they are present whenever there are any reactive forces between the air mass and the vehicle. In steady flight they may result from relative motion between the vehicle and the air mass or accelerated flows produced by the propulsion system. Although the specific forces depend on their peculiar origins, the *form* of the expressions that describe perturbations in these forces is not so particularly dependent on origin. For instance, while on vehicles designed to fly at very low speeds (e.g., helicopters and VTOL aircraft), the dominant forces and moments are produced by the accelerated flow surrounding the propulsion-lift system, the distortions of the accelerated flows (produced by motion disturbances from the trim condition) result in force and moment changes of similar character to those associated with "pure" aerodynamic flight. Consequently, the end results of the treatment here are pertinent to most kinds of air vehicles. However, whenever details of specific forces are needed in the ensuing discussion we shall consider pure aerodynamic flight of fixed-wing craft for simplicity.

It can be shown by dimensional analysis[10] that the forces acting on solids moving through fluids can be expressed in the form

$$F = C_F \tfrac{1}{2} \rho \mathsf{V}_a^2 S \tag{4-70}$$

where $C_F$ is a dimensionless coefficient, $\rho$ the density of the fluid, $\mathsf{V}_a$ the velocity of the solid relative to the fluid, and $S$ the characteristic area of the solid. Since a moment is the product of a force by a moment arm, the expression for a moment can also be written in a form similar to that of Eq. 4-70. The aerodynamic moments and forces acting on an airplane in flight (see Fig. 4-17) may then be written as

$$\begin{aligned}
X_a &= C_x(\tfrac{1}{2})\rho \mathsf{V}_a^2 S = \text{aerodynamic force along the } X \text{ axis} \\
Y_a &= C_y(\tfrac{1}{2})\rho \mathsf{V}_a^2 S = \text{aerodynamic force along the } Y \text{ axis} \\
Z_a &= C_z(\tfrac{1}{2})\rho \mathsf{V}_a^2 S = \text{aerodynamic force along the } Z \text{ axis} \\
L_a &= C_l(\tfrac{1}{2})\rho \mathsf{V}_a^2 Sb = \text{rolling moment} \\
M_a &= C_m(\tfrac{1}{2})\rho \mathsf{V}_a^2 Sc = \text{pitching moment} \\
N_a &= C_n(\tfrac{1}{2})\rho \mathsf{V}_a^2 Sb = \text{yawing moment}
\end{aligned} \tag{4-71}$$

where $S$ is the wing area; $c$ the mean aerodynamic chord,[11] the wing chord that has the average characteristics of all chords in the wing; and $b$ the wing span.

The total steady aerodynamic force on an aircraft is conventionally decomposed into lift and drag components. As shown in Fig 4-18, the

[10] C. B. Millikan, *Aerodynamics of the Airplane*, Wiley, New York, 1941.
[11] *Ibid.*

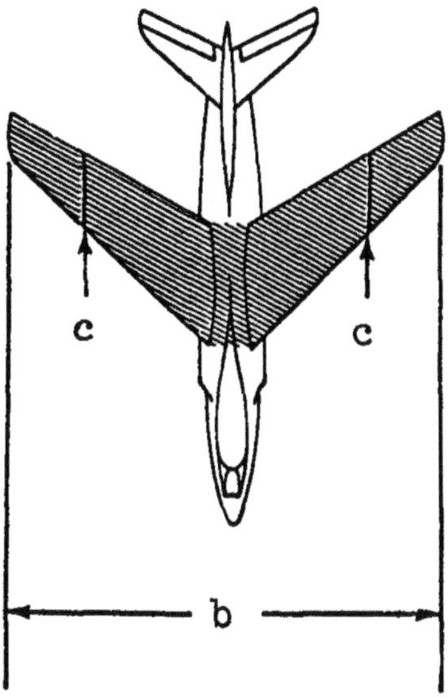


Fig. 4-17. $S$, $b$, and $c$ of wing. Shaded area $= S$.

lift acts normal to the flight path (i.e., to the relative wind) and the drag acts parallel to the flight path. Akin to Eq. 4-71, the lift and drag are

$$\begin{aligned} L &= C_L \tfrac{1}{2}\rho V_a^2 S = \text{lift} \\ D &= C_D \tfrac{1}{2}\rho V_a^2 S = \text{drag} \end{aligned} \tag{4-72}$$

Note that, to avoid confusion, the dimensionless coefficient in the lift equation is written with a capital "$L$," while a lower case "$l$" is used in the

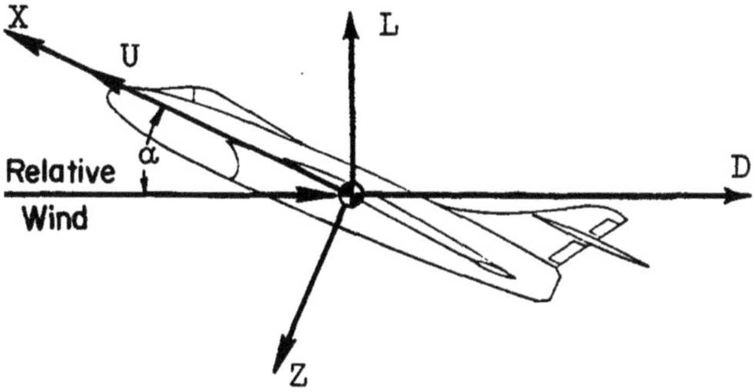


Fig. 4-18. Lift and drag acting on an airplane.

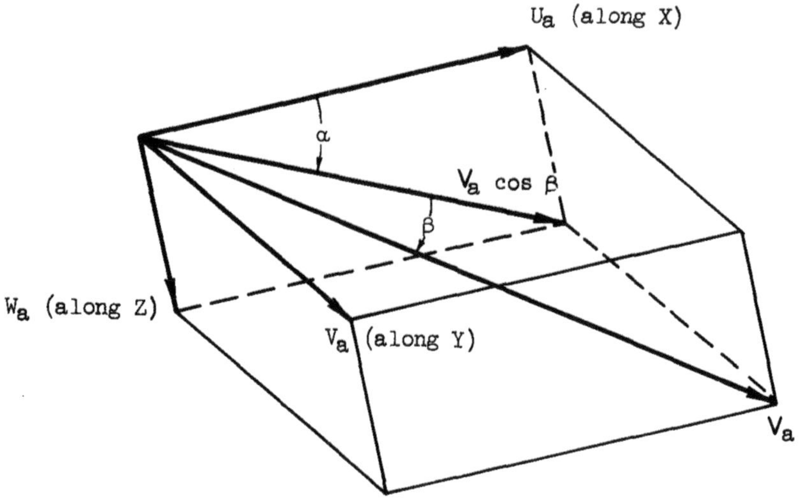


Fig. 4-19. Orientation of relative wind with body-fixed $X$, $Y$, $Z$ axes.

rolling moment coefficient. When resolved into body-fixed axes, the lift and drag become the $X$, $Y$, and $Z$ aerodynamic forces of Eq. 4-71.

The angles orienting lift and drag forces relative to the body-fixed axes are the angles of attack, $\alpha$, and of sideslip, $\beta$. These are shown in Fig. 4-19. Here, as previously, the subscript $a$ is used to indicate that the velocity and its components are *relative*, i.e., airframe relative to air mass. If the air mass velocity relative to inertial space is assumed to be constant, then the subscript can be removed. The velocity components of $\mathsf{V}_a$ along the body axes are

$$
\begin{aligned}
U_a &= \mathsf{V}_a \cos\beta \cos\alpha \\
V_a &= \mathsf{V}_a \sin\beta \\
W_a &= \mathsf{V}_a \cos\beta \sin\alpha
\end{aligned}
\tag{4-73}
$$

The angles of attack and sideslip are

$$
\begin{aligned}
\alpha &= \tan^{-1} \frac{W_a}{U_a} \\
\beta &= \sin^{-1} \frac{V_a}{\mathsf{V}_a}
\end{aligned}
\tag{4-74}
$$

and, as required by physical considerations, the latter is independent of the orientation of the $X$ axis in the plane of symmetry. Finally, the aerodynamic lift and drag force components along $X$, $Y$, and $Z$ are

$$\begin{aligned} X_a &= L \sin \alpha - D \cos \beta \cos \alpha \\ Y_a &= D \sin \beta \\ Z_a &= -L \cos \alpha - D \cos \beta \sin \alpha \end{aligned} \tag{4-75}$$

As indicated explicitly by the form of Eq. 4-70, the aerodynamic forces and moments are functions of air density and the relative velocity, and are therefore implicitly dependent on the altitude and the linear velocity components. Also, the dimensional coefficients $C_F$ are themselves dependent on the Reynolds and Mach numbers, angles of attack and sideslip, and linear and angular velocity components and their derivatives. If the aerodynamic forces are considered to be continuous functions of all these variables, each of the forces ($X_a$, $Y_a$, and $Z_a$) and the moments ($L_a$, $M_a$, and $N_a$) can be expressed in terms of the variables by expanding the forces in a Taylor series. These series have the form

$$F = F_0 + \left(\frac{\partial F}{\partial \Lambda_1}\right)_0 \lambda_1 + \left(\frac{\partial F}{\partial \Lambda_2}\right)_0 \lambda_2 + \left(\frac{\partial F}{\partial \Lambda_3}\right)_0 \lambda_3 + \cdots \tag{4-76}$$

where $\Lambda$ and $\lambda$ represent the variables, including their derivatives, and perturbations thereof, respectively, and the subscript zero indicates that the quantities are evaluated at the operating point or trim condition. In Eq. 4-76, terms of the order $(\partial^2 F/\partial\Lambda^2)_0(\lambda^2/2!)$ and all higher order terms have been omitted in accordance with Assumption 5. In spite of this, the number of potential contributing aerodynamic derivatives, $\partial F/\partial\Lambda_i$, is dishearteningly large; one way of eliminating some of them is to specialize the trim conditions as we have already done for the inertial and gravitational components.

### *Steady-State Trim Specialization*

To see how the trim situation can affect the aerodynamic derivatives, consider, as a pertinent example, the $X$ and $Z$ forces due to lift and drag as shown in Fig. 4-18. For an angle of sideslip, $\beta$, equal to zero, Eq. 4-75 becomes

$$\begin{aligned} X_a &= -D \cos \alpha + L \sin \alpha \\ Z_a &= -L \cos \alpha - D \sin \alpha \end{aligned}$$

or, in terms of the dimensionless coefficients of Eq. 4-71,

$$\begin{aligned} X_a &= \tfrac{1}{2}\rho V_a^2 S(-C_D \cos \alpha + C_L \sin \alpha) \\ Z_a &= \tfrac{1}{2}\rho V_a^2 S(-C_L \cos \alpha - C_D \sin \alpha) \end{aligned} \tag{4-77}$$

To simplify the example, both $C_L$ and $C_D$ are considered to be functions only of the angle of attack, $\alpha$, and the Mach number, $M$. Accordingly, the appropriate expansions of the perturbed forces, at constant density, involve derivatives with respect to $M$, $\alpha$, and $\mathsf{V}_a$, as follows:

$$\frac{\partial X_a}{\partial \Lambda} = \rho \mathsf{V}_a S \left\{ \frac{\partial \mathsf{V}_a}{\partial \Lambda} (-C_D \cos \alpha_0 + C_L \sin \alpha_0) + \frac{\mathsf{V}_a}{2} \frac{\partial M}{\partial \Lambda} \left( \frac{-\partial C_D}{\partial M} \cos \alpha_0 + \frac{\partial C_L}{\partial M} \sin \alpha_0 \right) + \frac{\mathsf{V}_a}{2} \frac{\partial \alpha}{\partial \Lambda} \left( \frac{-\partial C_D}{\partial \alpha} \cos \alpha_0 + \frac{\partial C_L}{\partial \alpha} \sin \alpha_0 + C_D \sin \alpha_0 + C_L \cos \alpha_0 \right) \right\} \tag{4-78}$$

$$\frac{\partial Z_a}{\partial \Lambda} = \rho \mathsf{V}_a S \left\{ \frac{\partial \mathsf{V}_a}{\partial \Lambda} (-C_L \cos \alpha_0 - C_D \sin \alpha_0) + \frac{\mathsf{V}_a}{2} \frac{\partial M}{\partial \Lambda} \left( \frac{-\partial C_L}{\partial M} \cos \alpha_0 - \frac{\partial C_D}{\partial M} \sin \alpha_0 \right) + \frac{\mathsf{V}_a}{2} \frac{\partial \alpha}{\partial \Lambda} \left( \frac{-\partial C_L}{\partial \alpha} \cos \alpha_0 - \frac{\partial C_D}{\partial \alpha} \sin \alpha_0 + C_L \sin \alpha_0 - C_D \cos \alpha_0 \right) \right\}$$

The various partial derivatives with respect to $\Lambda$ are to be taken holding the remaining primary motion variables constant. In this limited case the primary variables are simply $U_a$ and $W_a$, and from Eq. 4-73, for $\beta = 0$,

$$dU_a = d\mathsf{V}_a \cos \alpha_0 - \mathsf{V}_a \sin \alpha_0 \, d\alpha$$

$$dW_a = d\mathsf{V}_a \sin \alpha_0 + \mathsf{V}_a \cos \alpha_0 \, d\alpha$$

These can be solved to yield for $\Lambda = U_a$, $dW_a = 0$,

$$\frac{\partial \mathsf{V}_a}{\partial U_a} = \left( \frac{d\mathsf{V}_a}{dU_a} \right)_{W_a = \text{const}} = \cos \alpha_0$$

$$\frac{\partial \alpha}{\partial U_a} = \left( \frac{d\alpha}{dU_a} \right)_{W_a = \text{const}} = \frac{-\sin \alpha_0}{\mathsf{V}_a}$$

and, recognizing $M = \mathsf{V}_a/a$, where $a$ is the speed of sound, we obtain

$$\frac{\partial M}{\partial U_a} = \frac{\cos \alpha_0}{a}$$

Similarly, for $\Lambda = W_a$, $dU_a = 0$,

$$\frac{\partial V_a}{\partial W_a} = \sin \alpha_0$$
$$\frac{\partial \alpha}{\partial W_a} = \frac{\cos \alpha_0}{V_a} \tag{4-79}$$
$$\frac{\partial M}{\partial W_a} = \frac{\sin \alpha_0}{a}$$

Applying these values to Eq. 4-78, collecting terms, and simplifying the notation in the usual way, i.e., $\partial C_D/\partial \alpha \equiv C_{D_\alpha}$, etc., $(M/2)(\partial C_L/\partial M) \equiv C_{L_u}$, etc., we find that

$$\frac{\partial X_a}{\partial U_a} = \rho V_a S\{(-C_D - C_{D_u}) \cos^2 \alpha_0 + [\tfrac{1}{2}(C_{D_\alpha} - C_L) + C_L + C_{L_u}] \sin \alpha_0 \cos \alpha_0 - \tfrac{1}{2}(C_{L_\alpha} + C_D) \sin^2 \alpha_0\}$$

$$\frac{\partial X_a}{\partial W_a} = \rho V_a S\{\tfrac{1}{2}(C_L - C_{D_\alpha}) \cos^2 \alpha_0 + [\tfrac{1}{2}(C_{L_\alpha} + C_D) - C_D - C_{D_u}] \sin \alpha_0 \cos \alpha_0 + (C_L + C_{L_u}) \sin^2 \alpha_0\}$$

$$\frac{\partial Z_a}{\partial U_a} = \rho V_a S\{(-C_L - C_{L_u}) \cos^2 \alpha_0 + [\tfrac{1}{2}(C_{L_\alpha} + C_D) - C_D - C_{D_u}] \sin \alpha_0 \cos \alpha_0 - \tfrac{1}{2}(C_L - C_{D_\alpha}) \sin^2 \alpha_0\}$$

$$\frac{\partial Z_a}{\partial W_a} = \rho V_a S\{-\tfrac{1}{2}(C_{L_\alpha} + C_D) \cos^2 \alpha_0 - [\tfrac{1}{2}(C_{D_\alpha} - C_L) + C_L + C_{L_u}] \sin \alpha_0 \cos \alpha_0 - (C_D + C_{D_u}) \sin^2 \alpha_0\} \tag{4-80}$$

The recurring trigonometric terms are typical of the usual axis transformation relations (see Eq. 4-99) which, in effect, we have derived by this process for the quantities of interest. That is, the $\alpha_0 = 0$ derivatives or groups of them comprise the coefficients of the general trigonometric expansion for $\alpha_0 \neq 0$. It is clear that the choice of trim $\alpha_0$ cannot eliminate any of the derivatives but can only influence their assigned values and, for $\alpha_0 = 0$, considerably simplify their literal forms.

For the analogous situation in the $X$, $Y$ plane as sketched in Fig. 4-20, similar (lateral) lift and drag forces are acting, and expressions for $X$ and $Y$ corresponding to Eqs. 4-78 and 4-80 could easily be written. There is an important difference, however, in the values to be assigned the dimensionless coefficients, $C_L$ and $C_{D_\alpha}$. In the vertical ($X$, $Z$) plane, lift is necessary to sustain flight; since $C_{D_\alpha}$ is proportional to $C_L$, both parameters are usually positive and set by the trim condition. In the lateral ($X$, $Y$)

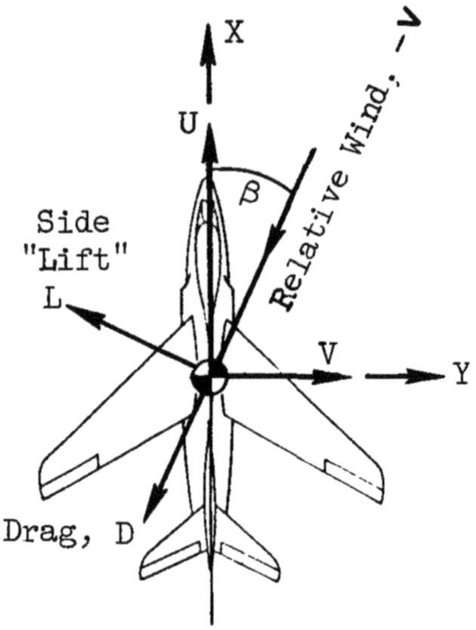


Fig. 4-20. Lateral lift and drag acting on an airplane.

plane, side lift is undesirable and ordinarily avoided, and it is unusual to set up trim conditions involving the steady sideslip, $\beta_0$. Thus the value of $\partial X/\partial V_a$, which can be inferred by substituting $\beta$ for $\alpha$ in the Eq. 4-80 expression for $\partial X/\partial W$, is usually zero. That is, for $\beta_0 = 0$ and the corresponding $C_L = \partial C_D/\partial \beta = 0$, $\partial X/\partial V_a = 0$. This really results because the $X$, $Z$ plane is not only a plane of inertial and geometric symmetry but usually also of aerodynamic trim symmetry. When this is true ($P_0 = R_0 = V_0 = 0$) then lateral perturbed motions, $p$, $r$, and $v$, do not produce $X$ and $Z$ forces or $M$ moments, as inferred by the foregoing discussion of $v$ effects. Thus we are led to Assumption 6.

Assumption 6. The steady lateral trim conditions are assumed to be $P_0 = R_0 = V_0 = \Phi_0 = 0$, and the longitudinal forces and moments *due to lateral perturbations about such trim conditions are assumed negligible.*

The assumed trim conditions are those prevailing for the great majority of all flying,[12] and are identical to those that lead to the "simplified" relationships of Eqs. 4-63 through 4-66. Although complete aerodynamic symmetry is not necessarily guaranteed by these conditions (as, for example, for propellers rotating in the same direction), the independence

[12] In Bairstow, *Applied Aerodynamics*, 1st edn., Longmans Green, London, 1920, turning and straight flight are considered, and minor differences are shown.

of longitudinal forces and moments from lateral perturbed motions is still assumed to hold with negligible error.

Assumption 6 has a corollary. With the aerodynamic trim forces essentially symmetrical about the $X$, $Z$ plane, there can be no appreciable lateral forces or moments induced by longitudinal perturbed motions ($u$, $w$, $q$). The net result therefore is that the perturbed longitudinal and lateral forces and moments are in general influenced, respectively, only by longitudinal and lateral perturbed motions, and of course by control inputs. This separation of the longitudinal and lateral aerodynamic forces and moments is completely parallel and analogous to the separation of the inertial and gravity terms shown in Eqs. 4-65 and 4-66. Accordingly, the complete equations are now, by virtue of Assumption 6, separated into longitudinal and lateral sets.

*Quasisteady Flow, Downwash and Sidewash*

Unfortunately, the sets are still potentially overcomplicated because of the implied inclusion in Eq. 4-76 of higher order terms due to body accelerations. Such effects arise because, when the vehicle is in accelerated motion, the air mass in its immediate surroundings must also accelerate in order to establish the *quasisteady flow*. The latter condition is one in which the aerodynamic forces and moments are dependent *only on the velocities* of the vehicle relative to the air mass. Generally speaking, the quasisteady forces and moments are of primary importance and the unsteady flow effects are usually secondary and negligible. However, the behavior of some modern high speed jet airplanes has exhibited marked discrepancies between the predicted damping and the observed flight test damping of high frequency oscillatory modes due, at least partly, to nonsteady flow effects. These effects become more important as the natural frequency is increased, as, for example, by a tight control system. Such possibilities should be kept in mind as perhaps requiring further investigation in specific cases. Nevertheless, because of its great applicability and essential simplification, we shall use Assumption 7.

Assumption 7. The flow is assumed to be quasisteady.

Because of Assumption 7 all derivatives with respect to the rates of change of velocities are omitted, with the exception of those with respect to $\dot{w}$ and $\dot{v}$, which are retained to account for the effect on the tail of the wing/body downwash and sidewash. This effect is present, as explained below, even when purely quasisteady considerations apply.

As an airplane wing travels through the air it leaves behind it an emanating downwash pattern, having a particular distribution and average value at the horizontal tail location. If the wing lift is suddenly increased by an abrupt change in $w$ (quasisteady assumption), the increased

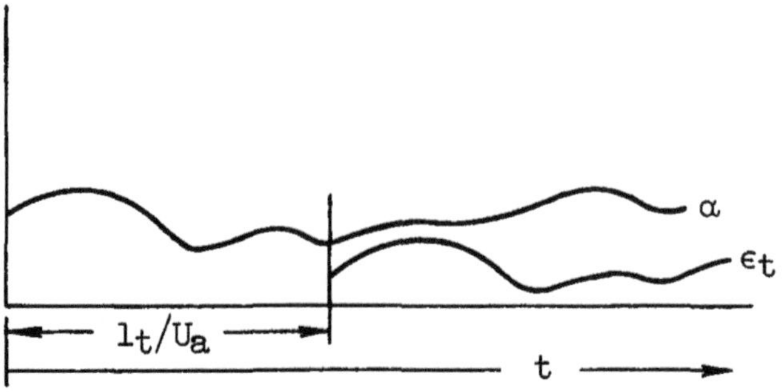


Fig. 4-21. Tail downwash delay following an angle of attack input.

downwash at the horizontal tail (or at the sweptback wing-tip for a tailless airplane), proportion to $\Delta w$, does not change immediately. The tail must actually arrive at the point in space where the wing lift was changed before it experiences the resulting downwash change; the time interval, $\Delta t$, for this to happen is just the tail length, $l_t$, divided by the forward speed, $U_a$. That is, for a given time history in $w_a$ and a resulting angle of attack variation with time, $\alpha(t)$, the corresponding downwash time history is given by $(\partial\epsilon_t/\partial\alpha)\alpha(t - l_t/U_a)$, as depicted in Fig. 4-21. The recognition that the Laplace transform of $f(t - \tau)$ is simply $F(s)e^{-\tau s}$ allows us to write the Laplace-transformed net tail angle of attack, $\alpha_t = \alpha - \epsilon_t$, as

$$\alpha_t(s) = \alpha(s)\left[1 - \frac{\partial\epsilon}{\partial\alpha}\, e^{-(l_t/U_a)s}\right]$$

The corresponding pitching moment is given by

$$C_M(s) = C_{M_{\alpha_t}}\alpha(s)\left[1 - \frac{\partial\epsilon}{\partial\alpha}\, e^{-(l_t/U_a)s}\right] \tag{4-81}$$

This most general representation of downwash lag effects can become quite important at high frequencies.[13] On a less accurate basis it can be translated into an approximate $\dot{\alpha}$ derivative by noting that, for the usual small values of $l_t/U_a$, $e^{-(l_t/U_a)s} \doteq 1 - (l_t/U_a)s$. Then,

$$C_M(s) \doteq C_{M_{\alpha_t}}\alpha(s)\left[1 - \frac{\partial\epsilon}{\partial\alpha}\left(1 - \frac{l_t}{U_a}\, s\right)\right]$$

$$\doteq C_{M_{\alpha_t}}\left[\alpha(s)\left(1 - \frac{\partial\epsilon}{\partial\alpha}\right) + \dot{\alpha}(s)\,\frac{l_t}{U_a}\,\frac{\partial\epsilon}{\partial\alpha}\right] \tag{4-82}$$

[13] I. C. Statler, "Dynamic Stability at High Speeds from Unsteady Flow Theory," *J. Aeron. Sci.*, 17, No. 4, pp. 232–242 (Apr. 1950).

and

$$C_{M\dot{\alpha}} \equiv \frac{\partial C_M}{\partial(\dot{\alpha}c/2U_a)} \doteq \frac{2l_t}{c}\frac{\partial \epsilon}{\partial \alpha} C_{M_{\alpha_t}}$$

Thus we have shown that aerodynamic partial derivatives with respect to $\dot{w} = U_a\dot{\alpha}$ can be included in the equations of motion on the basis of purely *quasisteady* considerations.

An analogous examination of the sidewash field due to a wing/body combination undergoing lateral velocity perturbations, and the resulting forces on the vertical fin, yields a similar conclusion with respect to $\dot{v}$ derivatives. Such derivatives are not usually as important as those caused by $\dot{w}$ motions and are sometimes neglected. For the time being we shall carry them along.

*Effects Caused by Nonuniform Atmosphere*

While both the aerodynamic and inertial plus gravity terms have now been suitably reduced, they are not yet on common ground. That is, the aerodynamic forces and moments are dependent on perturbations in the motions relative to the atmosphere, rather than to inertial space. To rectify this condition, we need to recognize the existence of atmospheric winds or gusts and to note that perturbed motions relative to the atmosphere are given by

$$\lambda_a = \lambda - \lambda_g \tag{4-83}$$

where now $\lambda$, being a general perturbed inertial motion quantity, is measurable in the body reference frame, and the subscript $g$ identifies the wind or gust component of the atmosphere. Substituting this relationship into Eq. 4-76 and suitably specializing $\lambda$ to those quantities of remaining interest, in view of the simplifying results of Assumptions 6 and 7, we can write the perturbed forces and moments as

$$dX = \frac{\partial X}{\partial U}(u - u_g) + \frac{\partial X}{\partial W}(w - w_g) + \frac{\partial X}{\partial \dot{W}}(\dot{w} - \dot{w}_g) + \frac{\partial X}{\partial Q}(q - q_g) + \sum \frac{\partial X}{\partial \delta}\delta$$

$$dZ = \frac{\partial Z}{\partial U}(u - u_g) + \frac{\partial Z}{\partial W}(w - w_g) + \frac{\partial Z}{\partial \dot{W}}(\dot{w} - \dot{w}_g) + \frac{\partial Z}{\partial Q}(q - q_g) + \sum \frac{\partial Z}{\partial \delta}\delta$$

$$dM = \frac{\partial M}{\partial U}(u - u_g) + \frac{\partial M}{\partial W}(w - w_g) + \frac{\partial M}{\partial \dot{W}}(\dot{w} - \dot{w}_g) + \frac{\partial M}{\partial Q}(q - q_g) + \sum \frac{\partial M}{\partial \delta}\delta \tag{4-84}$$

$$dY = \frac{\partial Y}{\partial V}(v - v_g) + \frac{\partial Y}{\partial \dot{V}}(\dot{v} - \dot{v}_g) + \frac{\partial Y}{\partial P}(p - p_g) + \frac{\partial Y}{\partial R}(r - r_g) + \sum \frac{\partial Y}{\partial \delta}\delta$$

$$dL = \frac{\partial L}{\partial V}(v - v_g) + \frac{\partial L}{\partial \dot{V}}(\dot{v} - \dot{v}_g) + \frac{\partial L}{\partial P}(p - p_g) + \frac{\partial L}{\partial R}(r - r_g) + \sum \frac{\partial L}{\partial \delta}\delta$$

$$dN = \frac{\partial N}{\partial V}(v - v_g) + \frac{\partial N}{\partial \dot{V}}(\dot{v} - \dot{v}_g) + \frac{\partial N}{\partial P}(p - p_g) + \frac{\partial N}{\partial R}(r - r_g) + \sum \frac{\partial N}{\partial \delta}\delta$$

where the summation of $\delta$ allows for more than one control input, e.g., elevator and/or throttle; aileron and/or rudder.

Motions of the atmosphere can be discrete or random; in any case they are usually characterized by only the three orthogonal components $u_g$, $v_g$, $w_g$. However, variations of these components in space and time can be considered to supply the additional *gust gradient* inputs shown in Eq. 4-84, $\dot{v}_g$, $\dot{w}_g$, $p_g$, $q_g$, $r_g$. Furthermore, for the long wavelengths characterizing atmospheric turbulence, the rotary gusts can be thought of as arising from the spatial distributions of linear gust velocities. For example, a rolling gust, $p_g$, is the result of a spanwise distribution of $w_g$, as shown in Fig. 4-22. Here the aircraft is considered to be encountering an effective vertical gust at its centerline plus an average spanwise gradient, $\partial w_g/\partial y$, negative in sign. Since a rolling velocity imparts a spanwise gradient in $w$ given by $py$, so that $\partial w/\partial y = p$, the effective $p_g$ is simply $\partial w_g/\partial y$. For a fore and aft distribution of $w_g$ as shown in Fig. 4-23, a similar decomposition of the gust produces $q_g = -\partial w_g/\partial x$; however, this can be better expressed in terms of the gust time variations felt by the aircraft. That is,

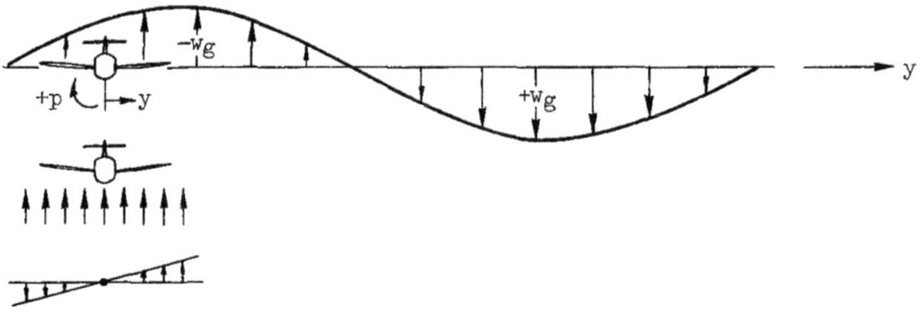


Fig. 4-22. Rolling gust, $p_g$.

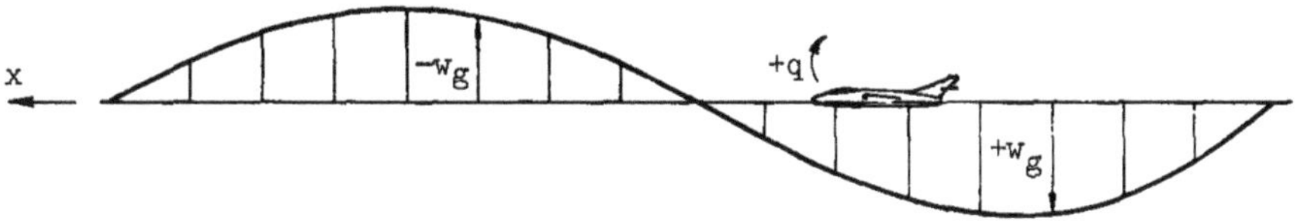


Fig. 4-23. Pitching gust, $q_g$.

differentiating numerator and denominator,[14]

$$q_g = \frac{-\partial w_g/\partial t}{\partial x/\partial t} = \frac{-\dot{w}_g}{U_0} = \frac{-sw_g}{U_0}$$

Similarly, the yawing gust due to a fore and aft distribution of side gust velocity, $v_g$, is given by $r_g = sv_g/U_0$.

These last expressions cannot be substituted directly into all parts of Eq. 4-84 because of the detailed way in which certain of the derivatives arise, as explained more fully later in this chapter. For instance, the rolling moment derivative $\partial L/\partial R$ stems primarily from the *spanwise* distribution of forward velocity due to yaw rate, which produces an asymmetric wing lift distribution. An effective $r_g$ resulting from fore and aft variations in the side gust velocity seen by an airplane traveling at speed $U_0$ would not produce a corresponding wing rolling moment. Similarly, the derivative $\partial X/\partial Q$, usually neglected, is theoretically the result of a vertical distribution of *horizontal* velocity due to pitch rate; therefore, an effective $q_g$ due to a fore and aft distribution (as above) does not produce forces proportional to $\partial X/\partial Q$.

In addition to the possibilities noted above, there are higher order gust input terms and derivatives that are sometimes considered.[15] Generally speaking, however, such terms are most appropriate to the detailed computation of random gust response spectra and are seldom considered necessary for the engineering solution of flight control problems.

In addition to the small scale atmospheric nonuniformities that produce winds and gusts in a given layer of air, there are also large scale nonuniformities with altitude. The most obvious of these is the variation in density, $\rho$, which through the basic Eq. 4-71 relations produces changes in

[14] W. H. Phillips and C. C. Kraft, Jr., "Theoretical Study of Some Methods for Increasing the Smoothness of Flight Through Rough Air," NACA TN 2416, July 1951; B. Etkin, "A Theory of the Response of Airplanes to Random Atmospheric Turbulence," University of Toronto, Inst. Aerophysics, UTIA Tech. Note 54, Nov. 1960.

[15] B. Etkin, "Theory of Flight of Airplanes in Isotropic Turbulence—Review and Extension," AGARD Rept. 372, 1961; J. M. Eggleston and W. H. Phillips, "The Lateral Response of Airplanes to Random Atmospheric Turbulence," NASA TR R-74, 1960.

the forces and moments with a change in altitude. The speed of sound, $a$, also varies in general, so that altitude displacement perturbations at a *given speed* can give rise to Mach number, $M \equiv V/a$, perturbations. If the dimensionless aerodynamic force and moment coefficients are Mach-number-dependent, as they are except for subsonic or hypersonic flight, then the resulting change in $M$ gives an additional *altitude-dependent* set of force and moment perturbations. Finally, there can be large scale changes in steady wind speed and direction with altitude (so-called wind shear) which will produce force and moment perturbations on a diving or climbing airplane. Notice that all of these effects are dependent, not on the vertical perturbed velocity measured along the body-fixed $Z$ axis, but rather on the perturbed altitude displacement and rate measured in earth-fixed coordinates.

Detailed investigations of the influence of altitude gradients[16] and the exact conditions under which they may be ignored indicate that these effects are insignificant for most flight situations and flight control problems. Consequently, a good assumption is as follows.

Assumption 8. Variations of atmospheric properties, such as density or speed of sound, are considered negligible for the small altitude perturbations of usual interest.

## 4-7. Expansion of the Direct Thrust Force

Many of the derivatives appearing in Eq. 4-84 will be the result not only of forces connected with the motion of the body through the atmosphere but also of forces arising from flows induced by the propulsion system. When such flows pass near or over portions of the vehicle, they can produce profound and difficult to predict effects on the derivatives which usually must be evaluated in model tests employing properly scaled jets or slipstreams. For configurations that inherently eliminate such slipstream interference (e.g., a subsonic jet centrally exhausting aft of the tail), the forces and moments associated with the direct thrust can still contribute significantly to various derivatives.

Among the contributions requiring general consideration are the forces produced on the inlet from the changed direction of the entering air mass; the moments caused by the angular velocity of a tube (engine plus tail-pipe) containing a moving air mass; and, finally, the forces and moments

[16] I. L. Ashkenas, "Effects of Atmospheric Gradients on Longitudinal Control," Section I of "Analysis of Several Handling Quality Topics Pertinent to Advanced Manned Aircraft," AFFDL-TR-67-2, 1967; J. K. Zbrozek, "Aircraft Behavior in a Vertical Gradient of Wind Velocity," RAE TN Aero 2810, 1962; S. Neumark, "Dynamic Longitudinal Stability in Level Flight, Including the Effects of Compressibility and Variations of Atmospheric Characteristics with Height," RAE Rept. Aero 2265, 1950.

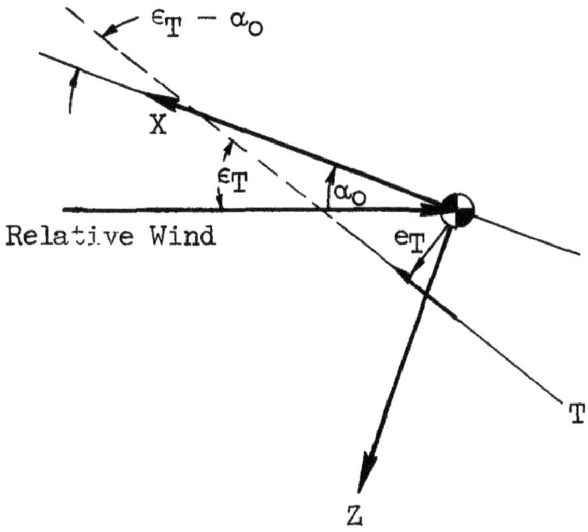


**Fig. 4-24. Thrust alignment notation.**

resulting from the thrust itself. All but the last of these contributions depend on detailed knowledge of the vehicle layout and powerplant performance characteristics; further consideration here is not warranted. However, an exposure of the direct thrust effects is pertinent, since it is simply accomplished, and the effects are always present and sometimes important. Consider Fig. 4-24, where the thrust line angle with the initial relative wind, $\epsilon_T$, is set by the trim condition and airplane geometry, and its angle with respect to the $X$ axis is *fixed* at $(\epsilon_T - \alpha_0) = \text{const.}$ Then, by direct resolution,

$$\begin{aligned} X_T &= T\cos(\epsilon_T - \alpha_0) \\ Z_T &= -T\sin(\epsilon_T - \alpha_0) \\ M_T &= Te_T \end{aligned} \tag{4-85}$$

where $e_T$, the thrust eccentricity (positive downward), is not a function of the axis orientation. With thrust a function only of density, control setting, and airplane relative speed (sometimes, but rarely, also of $\alpha_0$) the resulting perturbed forces are

$$\begin{aligned} dX_T &= \cos(\epsilon_T - \alpha_0)\left[\frac{\partial T}{\partial \mathsf{V}}\left(\frac{\partial \mathsf{V}}{\partial U}u + \frac{\partial \mathsf{V}}{\partial W}w\right) + \frac{\partial T}{\partial \delta_T}\delta_T\right] \\ dZ_T &= -\sin(\epsilon_T - \alpha_0)\left[\frac{\partial T}{\partial \mathsf{V}}\left(\frac{\partial \mathsf{V}}{\partial U}u + \frac{\partial \mathsf{V}}{\partial W}w\right) + \frac{\partial T}{\partial \delta_T}\delta_T\right] \end{aligned} \tag{4-86}$$

Applying Eq. 4-79 (dropping the $a$ subscript for simplicity) and expanding the trigonometric functions, we find that the partial derivatives are:

$$\begin{aligned}
\frac{\partial X_T}{\partial U} &= \frac{\partial T}{\partial V}(\cos \epsilon_T \cos^2 \alpha_0 + \sin \epsilon_T \sin \alpha_0 \cos \alpha_0) \\
\frac{\partial X_T}{\partial W} &= \frac{\partial T}{\partial V}(\cos \epsilon_T \sin \alpha_0 \cos \alpha_0 + \sin \epsilon_T \sin^2 \alpha_0) \\
\frac{\partial X_T}{\partial \delta_T} &= \frac{\partial T}{\partial \delta_T}(\cos \epsilon_T \cos \alpha_0 + \sin \epsilon_T \sin \alpha_0) \\
\frac{\partial Z_T}{\partial U} &= \frac{-\partial T}{\partial V}(\sin \epsilon_T \cos^2 \alpha_0 - \cos \epsilon_T \sin \alpha_0 \cos \alpha_0) \\
\frac{\partial Z_T}{\partial W} &= \frac{-\partial T}{\partial V}(\sin \epsilon_T \sin \alpha_0 \cos \alpha_0 - \cos \epsilon_T \sin^2 \alpha_0) \\
\frac{\partial Z_T}{\partial \delta_T} &= \frac{-\partial T}{\partial \delta_T}(\sin \epsilon_T \cos \alpha_0 - \cos \epsilon_T \sin \alpha_0)
\end{aligned} \tag{4-87}$$

When combined with Eqs. 4-80 these will provide composite derivatives such as $\partial X/\partial U = \partial X_a/\partial U + \partial X_T/\partial U$ for the force equation.

Unlike the perturbed forces, the perturbed moment is not obtained by simple differentiation of the Eq. 4-85 expression because of the influence of balancing aerodynamic trim moments. To show this, consider that at trim the total moment must be zero by definition, i.e., no resulting rotary acceleration and $P_0 = R_0 = 0$ (see Eq. 4-58) according to Assumption 6. Therefore, the thrust moment is balanced by an opposite and equal aerodynamic moment having the form of Eq. 4-71, i.e.,

$$M_0 = Te_T + \frac{\rho}{2} V^2 ScC_M = 0 \tag{4-88}$$

If we neglect aerodynamic $C_M$ variations with the Mach number as an additional effect to those of specific interest here, the moment change is

$$dM = e_T\left[\frac{\partial T}{\partial V}\left(\frac{\partial V}{\partial u} u + \frac{\partial V}{\partial w} w\right) + \frac{\partial T}{\partial \delta_T} \delta_T\right] + \rho V_0 ScC_M\left(\frac{\partial V}{\partial u} u + \frac{\partial V}{\partial w} w\right)$$

But by Eq. 4-88,

$$\rho V_0 ScC_M = \frac{-2Te_T}{V_0}$$

so that by applying Eq. 4-79,

$$dM = e_T\left[\left(\frac{\partial T}{\partial V} - \frac{2T_0}{V_0}\right)(u \cos \alpha_0 + w \sin \alpha_0) + \frac{\partial T}{\partial \delta_T} \delta_T\right] \tag{4-89}$$

We see that proper inclusion of the trim conditions has led to the additional $T_0/\mathsf{V}_0$ terms.

The foregoing is another good example of the importance of carefully considering the operating or trim point conditions when evaluating the force and moment derivatives. In the present case, neglecting such considerations would have led to possibly important errors in the moment perturbation.

## 4-8. Complete Linearized Equations of Motion

Now that the individual contributions have been evaluated, the remaining generally important terms can be assembled into complete sets of the linearized equations of motion.

Assumption 9. Effects associated with rotation of the vertical relative to inertial space will be assumed negligible; furthermore, the trim body-axis pitching velocity, $Q_0$, will be assumed zero.

The vertical rotation effects involve terms proportional to $\mathsf{U}_0/\mathsf{r}$, and $W_0/\mathsf{r}$ (see Eqs. 4-65 and 4-66) where $\mathsf{r}$ is approximately $2 \times 10^7$ ft. Even for orbital velocities, $\mathsf{U}_0 \doteq \sqrt{g\mathsf{r}}$, such terms are insignificant for the frequencies of usual concern in flight control problems. However, the weightlessness terms proportional to $U_0^2/\mathsf{r}$ (Eqs. 4-65 and 4-68) will not be negligible, except for speeds less than about 5000 ft/sec. For speeds greater than this we shall show in Section 4.9 how the weightlessness terms can be reincluded in the stability axis set of equations. In spite of this exception, the first part of Assumption 9 is generally valid for the great majority of airborne vehicular control situations of interest. The second part of Assumption 9, $Q_0 = 0$, is in keeping with the first part and the assumed $\dot{\Theta}_0 = 0$ corresponding to ***straight flight*** which leads to Eq. 4-63; that is, for $\mathsf{U}_0/\mathsf{r}$ negligible, Eq. 4-63 gives $Q_0 = 0$. In effect Assumption 9 confines our interest to operating points corresponding to ***straight flight over an effectively flat earth.***

Turning now to the process of assembling the pertinent components of the perturbed equations of motion, we divide the force equations by $m$ and the moment equations by the appropriate moment of inertia as in Eqs. 4-65 and 4-66, yielding terms similar to

$$\frac{1}{m}\frac{\partial X}{\partial U}u \qquad \text{and} \qquad \frac{1}{I_x}\frac{\partial L}{\partial R}r$$

If we replace these, respectively, by $X_u u$ and $L_r r$ and extend this convention to all the partial derivative terms the notation becomes considerably simplied. These quantities are called either "dimensional stability derivatives" or simply "stability derivatives"; they are understood to cover all effects of relative motion between the vehicle and the atmosphere, i.e., basic aerodynamic, as well as propulsion system effects, both induced and direct.

Using this notation and limiting the gust inputs to those discussed in connection with Eq. 4-84 (i.e., dropping the $X_q q_g$ and $L_r r_g$ terms) we obtain directly the right-hand side of the equations of motion. Since Eqs. 4-65 and 4-66 are no longer of interest in view of Assumption 9, the left-hand side is obtained most directly by Laplace-transforming Eq. 4-59 for the trim conditions of Assumptions 6 and 9, $P_0 = Q_0 = R_0 = V_0 = \Phi_0 = 0$. This procedure yields: longitudinal perturbed equations,

$$
\begin{aligned}
su + W_0 q + g\cos\Theta_0\theta &= X_u(u - u_g) + X_w(w - w_g) + X_{\dot{w}}s(w - w_g) \\
&\quad + X_q q \qquad + \sum X_\delta \delta \\
sw - U_0 q + g\sin\Theta_0\theta &= Z_u(u - u_g) + Z_w(w - w_g) + Z_{\dot{w}}s(w - w_g) \\
&\quad + Z_q\left(q + \frac{sw_g}{U_0}\right) + \sum Z_\delta \delta \qquad (4\text{-}90) \\
sq &= M_u(u - u_g) + M_w(w - w_g) + M_{\dot{w}}s(w - w_g) \\
&\quad + M_q\left(q + \frac{sw_g}{U_0}\right) + \sum M_\delta \delta
\end{aligned}
$$

and lateral perturbed equations,

$$
\begin{aligned}
&sv + U_0 r - W_0 p - g\cos\Theta_0\varphi \\
&\quad = Y_v(v - v_g) + Y_{\dot{v}}s(v - v_g) + Y_p(p - p_g) + Y_r\left(r - \frac{sv_g}{U_0}\right) + \sum Y_\delta \delta \\
&sp - \frac{I_{xz}}{I_x} sr \\
&\quad = L_v(v - v_g) + L_{\dot{v}}s(v - v_g) + L_p(p - p_g) + L_r r \qquad + \sum L_\delta \delta \\
&sr - \frac{I_{xz}}{I_z} sp \\
&\quad = N_v(v - v_g) + N_{\dot{v}}s(v - v_g) + N_p(p - p_g) + N_r\left(r - \frac{sv_g}{U_0}\right) + \sum N_\delta \delta
\end{aligned}
$$

(4-91)

The last two equations can be further simplified by substituting the expression for $sr$ obtained from the yawing moment equation into the roll equation (considering a single control input for simplicity), and, similarly, substituting the roll equation expression for $sp$ into the yawing equation:

$$sp = L_v(v - v_g) + L_{\dot{v}}s(v - v_g) + L_p(p - p_g) + L_r r + L_\delta \delta + \frac{I_{xz}}{I_x}\left\{N_v(v - v_g) + N_{\dot{v}}s(v - v_g) + N_p(p - p_g) + N_r\left(r - \frac{sv_g}{U_0}\right) + N_\delta \delta + \frac{I_{xz}}{I_z} sp\right\}$$

$$sr = N_v(v - v_g) + N_{\dot{v}}s(v - v_g) + N_p(p - p_g) + N_r\left(r - \frac{sv_g}{U_0}\right) + N_\delta \delta + \frac{I_{xz}}{I_z}\left\{L_v(v - v_g) + L_{\dot{v}}s(v - v_g) + L_r r + L_p(p - p_g) + L_\delta \delta + \frac{I_{xz}}{I_x} sr\right\}$$

Collecting like terms,[17] we define the "primed derivatives" for a general motion or input quantity designated by the subscript $i$,

$$L_i' \equiv \frac{L_i + (I_{xz}/I_x)N_i}{1 - (I_{xz}^2/I_x I_z)}; \qquad N_i' = \frac{N_i + (I_{xz}/I_z)L_i}{1 - (I_{xz}^2/I_x I_z)}$$

which implies, also,

$$L_i = L_i' - \frac{I_{xz}}{I_x} N_i'; \qquad N_i = N_i' - \frac{I_{xz}}{I_z} L_i' \tag{4-92}$$

Then the roll and yaw equations reduce to

$$\begin{aligned} sp &= L_v'(v - v_g) + L_{\dot{v}}'s(v - v_g) + L_p'(p - p_g) + L_r' r - (L_r')_g \frac{sv_g}{U_0} + \sum L_\delta' \delta \\ sr &= N_v'(v - v_g) + N_{\dot{v}}'s(v - v_g) + N_p'(p - p_g) + N_r' r - (N_r')_g \frac{sv_g}{U_0} + \sum N_\delta' \delta \end{aligned} \tag{4-93}$$

and we see that the use of primed derivatives eliminates the direct appearance of the product of inertia terms. The use of the $g$ subscript gust-gradient input term reflects the fact that there is no $L_r$ gust gradient term in Eq. 4-91, i.e., $(N_r')_g \equiv N_r/(1 - I_{xz}^2/I_x I_z)$, $(L_r')_g \equiv (I_{xz}/I_x)N_r/(1 - I_{xz}^2/I_x I_z)$.

[17] This procedure was apparently first reported by E. M. Frayn, "The Simplification of Lateral Response Calculations, when the Product of Inertia Is Not Negligible, by the Use of Modified Derivatives," Appendix D in K. Mitchell and E. M. Frayn, "Lateral Response Theory," ARC R & M 2297, 1952, pp. 35, 36. It was later rediscovered and emphasized in the United States by C. N. Tsu, "A Note about the Effects of Product of Inertia in Lateral Stability," *J. Inst. Aeron. Sci.*, 21, No. 7, 189 (July 1954).

The relationships between the gravity angles, $\theta$ and $\varphi$, and the motion parameters, which are needed to complete the foregoing sets of equations, are obtained by substituting the trim conditions of Assumptions 6 and 9 into Eq. 4-60; that is, for $\Phi_0 = P_0 = Q_0 = R_0 = 0$ and the corresponding $\dot{\Psi}_0 = \dot{\Theta}_0 = \dot{\Phi}_0 = 0$ (Eq. 4-41),

$$\begin{aligned} p &= \dot{\varphi} - \dot{\psi} \sin \Theta_0 \\ q &= \dot{\theta} \\ r &= \dot{\psi} \cos \Theta_0 \end{aligned} \tag{4-94}$$

The associated trim equations in Eq. 4-58 are

$$\begin{aligned} mg \sin \Theta_0 &= X_0 \\ -mg \cos \Theta_0 &= Z_0 \\ 0 &= M_0 \\ 0 &= Y_0 = L_0 = N_0 \end{aligned} \tag{4-95}$$

where $X_0$ and $Z_0$ are composed of the lift, drag, and thrust terms in Eqs. 4-77 and 4-85 and $M_0$ is given by Eq. 4-88.

*The Stability Axis System*

Thus far, considerable progress has been made in reducing the equations of motion to simple and analytically useful forms. A final simplification is immediately obtained by orienting the axis system to make $W_0 = 0$; that is, so that the $X$ axis in the steady state is pointed into the relative wind. Such an alignment results in a *stability axis* system which is initially inclined to the horizon at the flight path angle, $\gamma_0$ (since $\Theta_0 = \gamma_0 + \alpha_0$, and $\alpha_0 = \tan^{-1} W_0/U_0 = 0$).

It is important to recognize that the *initial alignment* does not alter the *body-fixed* nature of the axis system. All perturbed motions are still measured in a body-fixed frame; however, the alignment of the frame with respect to the body changes as a function of the operating or trim point condition. When the airframe is disturbed from the trim condition, the axes rotate with the airframe and do not change direction with respect to the airplane; consequently, the perturbed $X$ axis may or may not be parallel to the relative wind while the vehicle is in disturbed flight (see Fig. 4-25).

Using this axis system ($W_0 = 0$, $\Theta_0 = \gamma_0$) the Eq. 4-94 relationships, in Laplace transform notation, give

$$\begin{aligned} s\varphi &= p + r \tan \gamma_0 \\ s\theta &= q \\ s\psi &= \frac{r}{\cos \gamma_0} \end{aligned} \tag{4-96}$$

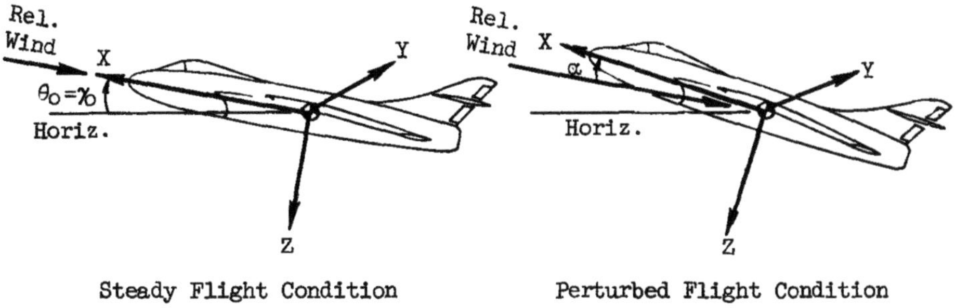


Fig. 4-25. Direction of stability axes with respect to the relative wind during the steady flight and disturbed flight conditions.

By incorporating this result, setting $W_0 = 0$, $\Theta_0 = \gamma_0$, considering one control input at a time (i.e., $\delta$ rather than $\sum \delta$), and collecting motion- and input-dependent terms on the left and right-hand sides, respectively, the resulting equations are: longitudinal set,

$$\begin{aligned}
&(s - X_u)u - (X_{\dot{w}}s + X_w)w + (-X_q s + g\cos\gamma_0)\theta \\
&\qquad = X_\delta \delta - [X_u u_g + (X_{\dot{w}}s + X_w)w_g] \\
&-Z_u u + (s - Z_{\dot{w}}s - Z_w)w + [(-U_0 - Z_q)s + g\sin\gamma_0]\theta \\
&\qquad = Z_\delta \delta - \left[Z_u u_g + (Z_{\dot{w}}s + Z_w)w_g - Z_q \frac{sw_g}{U_0}\right] \\
&-M_u u - (M_{\dot{w}}s + M_w)w + s(s - M_q)\theta \\
&\qquad = M_\delta \delta - \left[M_u u_g + (M_{\dot{w}}s + M_w)w_g - M_q \frac{sw_g}{U_0}\right]
\end{aligned} \tag{4-97}$$

and lateral set,

$$\begin{aligned}
&[s(1 - Y_{\dot{v}}) - Y_v]v - (Y_p s + g\cos\gamma_0)\frac{p}{s} + [(U_0 - Y_r)s - g\sin\gamma_0]\frac{r}{s} \\
&\qquad = Y_\delta \delta - \left[(Y_{\dot{v}}s + Y_v)v_g + Y_p p_g - Y_r \frac{sv_g}{U_0}\right] \\
&-(L'_{\dot{v}}s + L'_v)v + (s - L'_p)p - L'_r r \\
&\qquad = L'_\delta \delta - [(L'_{\dot{v}}s + L'_v)v_g + L'_p p_g] \\
&-(N'_{\dot{v}}s + N'_v)v - N'_p p + (s - N'_r)r \\
&\qquad = N'_\delta \delta - \left[(N'_{\dot{v}}s + N'_v)v_g + N'_p p_g - (N'_r)_g \frac{sv_g}{U_0}\right]
\end{aligned} \tag{4-98}$$

*Inclusion of Weightlessness Terms in Stability Axis Equations of Motion*

As mentioned earlier, the flat earth assumption, No. 9, limits the applicability of either the stability axis or body axis sets of equations to speeds less than about 5000 ft/sec. However, for the stability axis set, the specific assumption of $W_0 = 0$ allows a significant simplification without neglecting the weightlessness terms appearing in Eqs. 4-65 through 4-68. We still omit the generally negligible terms proportional to $\mathsf{U}_0/\mathsf{r}$ and $U_0/\mathsf{r}$, but now we take note of $W_0 = 0$, $\Theta_0 = \gamma_0$ and the simple relationships of Eq. 4-96. Then the only added (weightlessness) term in the $Z$ equation (4-65) is simply $(-U_0^2/\mathsf{r}) \sin \gamma_0 \theta$. For the side force equation we note that, for $W_0 = 0$, $\mathsf{U}_0 = U_0 \cos \gamma_0$ and, from Eq. 4-96, $s\psi \cos \Theta_0 = s\psi \cos \gamma_0 = r$; then Eq. 4-68 reduces to

$$sv + U_0 r - \left(g - \frac{U_0^2}{\mathsf{r}}\right) \cos \gamma_0 \varphi = \frac{dY}{m}$$

In effect, therefore, we simply replace $g$ by $g - U_0^2/\mathsf{r}$ to extend the applicability of Eqs. 4-97 and 4-98 to orbital speeds.

Further simplification of these equations is possible by dropping terms found to be negligible in specific instances, but, excepting this, these are the simplest of all generally applicable sets of the linearized equations of motion. They are especially convenient for computing and understanding the basic vehicle dynamic characteristics which are, of course, independent of the chosen axis system. However, as already mentioned, for situations in which the measurement or sensing axes differ markedly from the stability axes, motions computed in the latter system must be transformed before they can be compared with flight test results or used in closed-loop analyses. Under such circumstances the use of a body-fixed system aligned with the sensor axes may be preferable. The only difference in form between body axis and stability axis equations of motion for the *same set of assumptions* is the appearance in the former of the $W_0$ terms and the use of $\Theta_0 = \gamma_0 + \tan^{-1} W_0/U_0$, rather than $\gamma_0$ to orient the $\mathbf{g}$ vector. However, the stability derivatives used in each generally differ because of the axis rotation through $\alpha_0$.

*Stability Derivative Transformation Relationships*

The complete transformation from stability axis to body axis derivatives must consider resolution not only of the forces and moments but also of the perturbed motions and the changed inertias. Applying these considerations results in the following relationships between body axes, subscript $b$, and stability axes, no subscript. Notice that, for the lateral primed derivatives, the transformations are exactly the same as those for the basic (unprimed) dimensional derivatives $\partial L/\partial P$, $\partial N/\partial r$, etc.[18] For the unprimed derivatives, e.g., $L_p$, $N_r$, etc., the transformed values of $(\partial L/\partial P)_b$, $(\partial N/\partial r)_b$, etc., must be divided through by the transformed $(I_x)_b$, $(I_z)_b$ expressions, also given. This simplification accruing to the use of prime derivatives is due to the fact that they correctly account for *all* accelerations about a given axis from a given perturbation and can, accordingly, be treated as simple vectors. For the longitudinal case,

$$\begin{aligned}
(X_u)_b &= X_u \cos^2 \alpha_0 - (X_w + Z_u) \sin \alpha_0 \cos \alpha_0 + Z_w \sin^2 \alpha_0 \\
(X_w)_b &= X_w \cos^2 \alpha_0 + (X_u - Z_w) \sin \alpha_0 \cos \alpha_0 - Z_u \sin^2 \alpha_0 \\
(X_{\dot{w}})_b &= X_{\dot{w}} \cos^2 \alpha_0 - Z_{\dot{w}} \sin \alpha_0 \cos \alpha_0 \\
(X_{q;\delta})_b &= X_{q;\delta} \cos \alpha_0 - Z_{q;\delta} \sin \alpha_0 \\
(Z_u)_b &= Z_u \cos^2 \alpha_0 - (Z_w - X_u) \sin \alpha_0 \cos \alpha_0 - X_w \sin^2 \alpha_0 \\
(Z_w)_b &= Z_w \cos^2 \alpha_0 + (Z_u + X_w) \sin \alpha_0 \cos \alpha_0 + X_u \sin^2 \alpha_0 \\
(Z_{\dot{w}})_b &= Z_{\dot{w}} \cos^2 \alpha_0 + X_{\dot{w}} \sin \alpha_0 \cos \alpha_0 \qquad (4\text{-}99) \\
(Z_{q;\delta})_b &= Z_{q;\delta} \cos \alpha_0 + X_{q;\delta} \sin \alpha_0 \\
(M_u)_b &= M_u \cos \alpha_0 - M_w \sin \alpha_0 \\
(M_w)_b &= M_w \cos \alpha_0 + M_u \sin \alpha_0 \\
(M_{\dot{w}})_b &= M_{\dot{w}} \cos \alpha_0 \\
(M_{q;\delta})_b &= M_{q;\delta} \\
(I_y)_b &= I_y
\end{aligned}$$

Notice that the $X_u$, $X_w$, $Z_u$, and $Z_w$ expressions are equivalent to those in Eqs. 4-80 and 4-87. For the lateral case,

$$\begin{aligned}
(Y_{v;\delta})_b &= Y_{v;\delta} \\
(Y_{\dot{v}})_b &= Y_{\dot{v}} \\
(Y_p)_b &= Y_p \cos \alpha_0 - Y_r \sin \alpha_0 \\
(Y_r)_b &= Y_r \cos \alpha_0 + Y_p \sin \alpha_0 \\
(L'_{v;\delta})_b &= L'_{v;\delta} \cos \alpha_0 - N'_{v;\delta} \sin \alpha_0 \\
(L'_{\dot{v}})_b &= L'_{\dot{v}} \cos \alpha_0 - N'_{\dot{v}} \sin \alpha_0 \\
(L'_p)_b &= L'_p \cos^2 \alpha_0 - (L'_r + N'_p) \sin \alpha_0 \cos \alpha_0 + N'_r \sin^2 \alpha_0
\end{aligned}$$

[18] Edward Seckel, *Stability and Control of Airplanes and Helicopters*, Academic New York, 1964.

$$(L'_r)_b = L'_r \cos^2 \alpha_0 - (N'_r - L'_p) \sin \alpha_0 \cos \alpha_0 - N'_p \sin^2 \alpha_0 \qquad (4\text{-}100)$$

$$(N'_{v;\delta})_b = N'_{v;\delta} \cos \alpha_0 + L'_{v;\delta} \sin \alpha_0$$

$$(N'_{\dot{v}})_b = N'_{\dot{v}} \cos \alpha_0 + L'_{\dot{v}} \sin \alpha_0$$

$$(N'_p)_b = N'_p \cos^2 \alpha_0 - (N'_r - L'_p) \sin \alpha_0 \cos \alpha_0 - L'_r \sin^2 \alpha_0$$

$$(N'_r)_b = N'_r \cos^2 \alpha_0 + (L'_r + N'_p) \sin \alpha_0 \cos \alpha_0 + L'_p \sin^2 \alpha_0$$

$$(I_x)_b = I_x \cos^2 \alpha_0 + 2I_{xz} \sin \alpha_0 \cos \alpha_0 + I_z \sin^2 \alpha_0$$

$$(I_z)_b = I_z \cos^2 \alpha_0 - 2I_{xz} \sin \alpha_0 \cos \alpha_0 + I_x \sin^2 \alpha_0$$

$$(I_{xz})_b = (I_z - I_x) \sin \alpha_0 \cos \alpha_0 + I_{xz}(\cos^2 \alpha_0 - \sin^2 \alpha_0)$$

## 4-9. Description of the Dimensional and Nondimensional Stability Axis Derivatives

The adoption of Assumptions 6 and 7 has greatly reduced the number of stability derivatives appearing in the equations of motion. In this section each of the dimensional stability derivatives in Eqs. 4-90 and 4-91 is first given a brief physical interpretation, then expanded into a more basic form, and shown to be a function of what are called "basic nondimensional stability derivatives." Some discussion of these basic nondimensional stability derivatives is also given.[19] The longitudinal stability derivatives are treated first, and the lateral stability derivatives later. Equation 4-71, used in the discussion, is repeated here for reference, but the *a* subscript is dropped for simplicity, since the derivatives are not dependent on the specific origin of the relative motions between the vehicle and the atmosphere.

$$\begin{aligned} L &= C_L(\tfrac{1}{2})\rho V^2S & Z &= C_z(\tfrac{1}{2})\rho V^2S \\ D &= C_D(\tfrac{1}{2})\rho V^2S & L &= C_l(\tfrac{1}{2})\rho V^2Sb \\ X &= C_x(\tfrac{1}{2})\rho V^2S & M &= C_M(\tfrac{1}{2})\rho V^2Sc \\ Y &= C_y(\tfrac{1}{2})\rho V^2S & N &= C_n(\tfrac{1}{2})\rho V^2Sb \end{aligned} \qquad (4\text{-}101)$$

It can be readily appreciated that the direct conversion from the nondimensional aerodynamic coefficients to the corresponding forces and

[19] There is, of course, an extensive literature pertaining to the origin and estimation of nondimensional stability derivatives. Of particular note are "USAF Stability and Control Datcom," Air Force Flight Dynamics Lab., Wright-Patterson Air Force Base, Oct. 1960 (Rev. Nov. 1965); "Royal Aeronautical Society Data Sheets," Royal Aeronautical Society, London. Also, considerable bibliographic material is given in some of the texts already referenced, e.g., Seckel, *Stability and Control of Airplanes and Helicopters.*

moments requires a common axis system for both. That is, in this instance, the coefficients must be obtained in, or reduced to, a set applicable to stability axes. When other than stability axis systems are used for the equations of motion, the coefficients or the dimensional derivatives must be converted to, or set up in, that axis system (e.g., Eqs. 4-99 and 4-100).

Although none of the aerodynamic coefficients is guaranteed to behave linearly with any of the variables, in most cases, for small perturbations it is reasonably accurate to linearize the coefficients about the operating point as in Eq. 4-76. Some of the more significant nonlinearities are noted in the discussions to follow which are, at best, incomplete in this respect.

### *Dimensional and Nondimensional Forms*

The stability derivatives, dimensional and nondimensional, as we shall develop and use them, are of a particular, commonly-used form. Other forms appear in the literature relating to aircraft stability and control but little or no distinction in terminology is made among them; all are referred to as "stability derivatives," regardless of the particular form. For purposes of discussion and clarification, it is convenient to illustrate the four forms most generally found before selecting those to be given detailed treatment.

Example 1. Basic dimensional stability derivatives:

$$\frac{\partial L}{\partial P} \quad \text{or} \quad \frac{\partial L}{\partial p}$$

Example 2. Dimensional stability derivative parameters:

$$L_p = \frac{1}{I_x}\frac{\partial L}{\partial P} = \left(\frac{1}{I_x}\right)\left(\frac{\partial L}{\partial p}\right)$$

Example 3. Basic nondimensional stability derivatives:

$$C_{l_p} = \frac{\partial C_l}{\partial(pb/2U)} = \left(\frac{1}{qSb}\right)\left(\frac{\partial L}{\partial(pb/\partial U)}\right)$$

Example 4. Nondimensional stability derivative parameters:

$$l_p = \left(\frac{1}{4}\right)\left(\frac{b}{k_x}\right)^2 C_{l_p}; \qquad \text{where } k_x^2 = \frac{I_x}{m}$$

The equivalence between forms like $\partial L/\partial P$ and $\partial L/\partial p$ shown above comes about because we are normally dealing with trim conditions that eliminate accelerated steady motions, thereby making $dP/dp = 1$. Accordingly, we can express most derivatives in terms of either upper or lower case (perturbation) symbols to suit our convenience.

It may be seen from the examples that the dimensional forms (Examples 1 and 2) are concerned with *direct* forces and moments (or accelerations), and with motion velocities, whereas the nondimensional form (Examples 3 and 4) is concerned with force and moment *coefficients* and with nondimensional velocities (e.g., $pb/2U$ is the nondimensionalized form of the rolling velocity, $p$). It may also be seen that the basic stability derivatives (Examples 1 and 3) do not involve inertial quantities, whereas the stability derivative parameters (Examples 2 and 4) do. Conversion relations among these four general forms and also specific mathematical definitions of individual derivatives are given in Tables 4-1 and 4-4 at the end of this chapter.

It must be emphasized that the specific notation and the specific definitions used here are not necessarily employed by all writers on the subject. For example, Jones in a basic aerodynamic reference work[20] uses the notation $L_p$ to signify the partial derivative $\partial L/\partial p$, whereas most of the present-day writers use $L_p$ (or $L_{\dot{\varphi}}$) to represent the stability derivative parameter $(1/I_x)(\partial L/\partial p)$. On the other hand, almost everyone uses the same notation for such basic nondimensional stability derivatives as $C_{l_p}$, $C_{L_\alpha}$, and $C_{M_q}$.

Of the four forms listed above, the two of most practical importance are the basic nondimensional stability derivative (e.g., $C_{l_p}$) and the dimensional stability derivative parameter (e.g., $L_p$). The basic nondimensional form ($C_{l_p}$) is important because correlation between the aerodynamic characteristics of different airframes or the same airframe at different flight conditions is most easily attained with these stability derivatives; as a result, aerodynamic stability derivative data from wind tunnel tests, flight tests, and theoretical analyses are usually presented in the basic nondimensional form.[21] The dimensional stability derivative parameter form ($L_p$) is important because it leads directly to the numerical coefficients in the sets of simultaneous differential equations describing the real time dynamics of the airframe. Thus stability derivatives in this form are useful in determining the analytic transfer functions of the airframe and in setting up its mathematical model on an analog computer.

Accordingly, the discussion dealing with the evaluation of stability derivatives makes use of the *basic nondimensional stability derivative*

[20] B. Melvill Jones, "Dynamics of the Aeroplane," in W. F. Durand, ed., *Aerodynamic Theory*, Div. N., Vol. 5, Durand Reprinting Committee, Pasadena, Calif., 1943. Reprinted by Dover (Vols. 5 and 6 bound in one volume), New York, 1963.

[21] An important exception to this otherwise general state of affairs occurs for the low speed range of vehicles capable of hovering operation. For $V = 0$, as in hover, dynamic pressure $= (\rho/2)V^2$ is a poor measure of the aerodynamic forces on the vehicle which are more appropriately related to parameters indicative of slipstream (or jet efflux) velocities, e.g., R. L. Stapleford, J. Wolkovitch, R. E. Magdaleno, C. P. Shortwell, and W. A. Johnson, "An Analytical Study of V/STOL Handling Qualities in Hover and Transition," AFFDL-TR-65-73, May 1965.

*form* ($C_{l_p}$), and that dealing with airframe transfer functions makes use of the *dimensional stability derivative parameter form* ($L_p$).

*Perturbation Effects on the Total Velocity*

It should be noted that the quantity $\mathsf{V}^2$, which appears in Eq. 4-101, is the square of the total linear velocity. In the stability axis system the total linear velocity during the steady flight condition is equal to $U_0$, which is the velocity in the direction of the $X$ axis. When disturbed from steady flight, the airplane can have velocity components $U_0 + u$, $v$, and $w$ directed along the $X$, $Y$, and $Z$ axes, respectively. During disturbed flight the magnitude of the total linear velocity can be expressed as

$$\mathsf{V} = \sqrt{(U_0 + u)^2 + v^2 + w^2}$$

or

$$\mathsf{V} = \sqrt{U_0^2 + 2U_0 u + u^2 + v^2 + w^2} \tag{4-102}$$

In Assumption 5, $u$, $v$, and $w$ were assumed to be very small so that their products and squares could be neglected. Thus,

$$\mathsf{V} = \sqrt{U_0^2 + 2U_0 u}$$

$$\doteq U_0\left(1 + \frac{u}{U_0}\right) = U$$

Also, when one is very much greater than $u/U_0$, a very good approximation is $U \doteq U_0$. Therefore,

$$\mathsf{V} \doteq U \doteq U_0 \tag{4-103}$$

Thus, $\mathsf{V}$, $U_0$, and $U$ can be used somewhat interchangeably for stability axes and the conditions implicit in Assumption 5.

*Longitudinal Stability Derivatives*

The longitudinal force derivatives have already been treated for a general axis system (see Eq. 4-80); however, they are included and rederived in the present discussion for the sake of completeness.

EFFECT OF $u$, THE CHANGE IN FORWARD SPEED

As an airplane increases its forward speed, the lift, $L$; drag, $D$; and moment, $M$, change. Generally, but not always, each of these quantities increases. Also, the thrust component in the flight direction, $T_x$, changes.

$X_u$. Since drag acts along the negative axis, an increase in drag contributes a negative $X$ force; conversely, an increase in $T_x$ contributes a positive $X$ force. The change in the $X$ force due to a change in forward speed can be expressed mathematically in the form

$$\Delta X = \frac{\partial X}{\partial U} u = \frac{-\partial D}{\partial U} u + \frac{\partial T_x}{\partial U} u$$

$$X_u = \frac{1}{m}\frac{\partial X}{\partial U} = \frac{1}{m}\left(\frac{-\partial D}{\partial U} + \frac{\partial T_x}{\partial U}\right) \tag{4-104}$$

Using the drag equation from Eq. 4-101, substituting $U$ for V in accordance with Eq. 4-103, and performing the indicated differentiation, we obtain

$$X_u = \frac{-\rho S}{2m}\left(U^2 \frac{\partial C_D}{\partial U} + 2UC_D\right) + \frac{1}{m}\frac{\partial T_x}{\partial U}$$

and

$$X_u = \frac{-\rho SU}{m}\left(\frac{U}{2}\frac{\partial C_D}{\partial U} + C_D\right) + \frac{1}{m}\frac{\partial T_x}{\partial U} \tag{4-105}$$

The drag coefficient, $C_D$, is the equilibrium drag divided by $\rho U^2 S/2$. By definition it is always measured *along the direction of the relative wind*; hence the equilibrium drag coefficient is measured along the negative equilibrium $X$ axis in the stability axis system (Fig. 4-26) and is always positive in sign. In contrast, it should be pointed out that the derivative $X_u$ is at all times measured along the $X$ axis and is always negative in sign.

The equation for thrust can be written in a form similar to that of the lift and drag equations in Eq. 4-101:

$$T = \frac{\rho U^2 S}{2} C_T \tag{4-106}$$

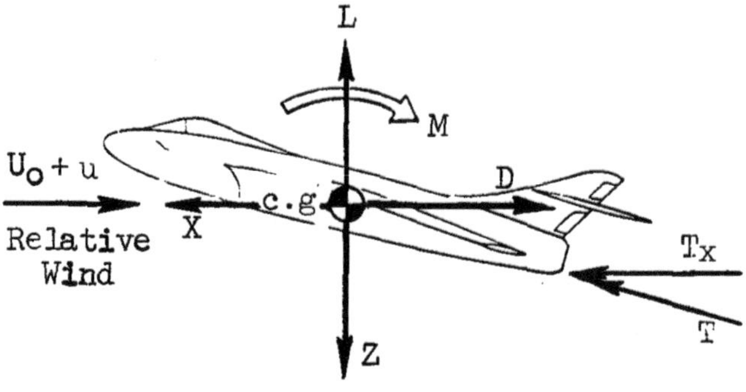


Fig. 4-26. Variation of lift, drag, and pitching moment with change in forward velocity.

However, this form is misleading, since the coefficient $C_T$ is basically *not* aerodynamic in nature. In fact, for jet engines the thrust, $T$, is more nearly constant than is $C_T$. Accordingly, the most generally applicable expression for the thrust contribution to $X_u$ is that given in Eq. 4-105, above, where the partial derivative $\partial T/\partial U$ is obtainable from powerplant performance estimates.

The change in drag coefficient with varying forward velocity for constant angle of attack and altitude, $\partial C_D/\partial U$, can arise from two sources: Mach number effects and aeroelastic effects. In most cases the latter are small and can be neglected. Changes due to Mach number are also small and negligible for low subsonic and high supersonic Mach numbers but become appreciable in the transonic region ($0.8 < M < 1.2$) where a large increase in drag occurs.

The appearance, in Eq. 4-105, of the nondimensional grouping $(U/2)(\partial C_D/\partial U)$ suggests a simplified notation:

$$\frac{U}{2}\frac{\partial C_\lambda}{\partial U} \equiv C_{\lambda_u} \tag{4-107}$$

where $C_\lambda$ is any basic nondimensional stability derivative. Accordingly,

$$X_u = \frac{-\rho SU}{m}(C_{D_u} + C_D) + \frac{1}{m}\frac{\partial T_x}{\partial U} \tag{4-108}$$

The direct thrust contributions to all other derivatives is generally negligible for conventional airplanes, except as it affects the equilibrium, operating point, conditions. Accordingly, its possible variation with other perturbations, and about or along other axes, is not considered in the formulation of the remaining derivative expressions.

$Z_u$. Since lift acts along the negative $Z$ axis, an increase in lift from a change in forward speed contributes a negative $Z$ force:

$$\begin{aligned} \Delta Z &= \frac{\partial Z}{\partial U}u = \frac{-\partial L}{\partial U}u \\ Z_u &= \frac{1}{m}\frac{\partial Z}{\partial U} = \frac{-1}{m}\frac{\partial L}{\partial U} \end{aligned} \tag{4-109}$$

By noting the similarity between Eqs. 4-109 and 4-104, and between the lift and drag equations of Eq. 4-101, we can immediately write

$$Z_u = \frac{-\rho SU}{m}(C_{L_u} + C_L) \tag{4-110}$$

The lift coefficient, $C_L$, is the equilibrium of trim lift divided by $\rho U^2 S/2$. For the trim conditions assumed (Assumptions 8 and 9) for the stability axis equations of motion, the trim lift is equal to the airplane weight times cos $\gamma_0$ minus the upward thrust component normal to the flight path. By definition, lift coefficient is always measured perpendicular to the relative wind and is positive upward (Fig. 4-26), and so the equilibrium lift coefficient is measured along the *negative* equilibrium $Z$ axis of the stability axis system. On the other hand, $Z_u$ is always measured along the *positive* $Z$ axis and is positive downward. Low values of $C_L$ are associated with low angles of attack and high speeds, whereas high values of $C_L$ are associated with high angles of attack and low speeds.

The derivative $C_{L_u}$ is the nondimensional change in lift coefficient with variations in forward velocity for constant angle of attack and altitude, $(U/2)(\partial C_L/\partial U)$; $C_{L_u}$ arises from Mach number and aeroelastic effects. The magnitude of the total $C_{L_u}$ can vary considerably and its sign can change, depending not only on the airframe geometry and its elastic properties but also on the Mach number and dynamic pressure at which it is flying. The magnitude of $C_{L_u}$ is negligibly small for low speed flight but it may reach a considerable value near the critical Mach number of the airframe.

$M_u$. The change in moment caused by a change in forward speed can be expressed as

$$\Delta M = \frac{\partial M}{\partial U} u$$
$$M_u = \frac{1}{I_y}\frac{\partial M}{\partial U} \tag{4-111}$$

The same mechanics used in the expansion of $X_u$ can be used to derive $M_u$.

$$M_u = \frac{\rho S U c}{I_y}(C_{M_u} + C_M) \tag{4-112}$$

Here, we must remember that $C_M$ represents only the aerodynamic portion of the total trimmed pitching moment and that the latter is, by Assumptions 6 and 9, zero. Thus, in general there will be a nonzero $C_M$ only in the presence of thrust asymmetry (Eq. 4-88).

The derivative $C_{M_u}$ is the nondimensional change in pitching moment coefficient with variation in forward velocity for the constant angle of attack and altitude, $(U/2)(\partial C_M/\partial U)$. The magnitude of $C_{M_u}$ can vary considerably and the sign can change, depending on such factors as the airframe's geometry and its elastic properties, and the Mach number and dynamic pressure at which it is flying. This derivative can arise from three sources: thrust or power effects, Mach number effects, and aeroelastic effects. The early treatment of $C_{M_u}$ was as a power effect arising from the propwash of propeller-driven aircraft. Today, because of the use of jet engines and the associated alleviation of power effects on dynamic stability, the $C_{M_u}$ from slipstream effects is small, except for "unconventional" VTOLs. (The direct thrust effects themselves were thoroughly discussed earlier in connection with Eq. 4-89.) On the other hand, the contributions to $C_{M_u}$ due to Mach number and aeroelastic effects have become increasingly important.

EFFECT OF $w$, THE CHANGE IN SPEED ALONG THE $Z$ AXIS.

In Fig. 4-27 the quantities $L_0$ and $D_0$ represent the lift and drag acting on the airplane during the steady flight condition. The lift and drag always act, respectively, normal and parallel to the relative wind. According to the definition of stability axes, the relative wind during the steady flight condition is parallel to the $X$ axis. Therefore, the only component of linear velocity during the steady flight condition is $U_0$; $L_0$ and $D_0$ are, respectively, perpendicular and parallel to the $X$ axis. When the airplane is disturbed from steady flight, so that it has a component of velocity, $w$, along the $Z$ axis as well as a forward velocity, $U_0$, the relative wind shifts to a new position, as shown in Fig. 4-27. This shift results in an increase in angle of attack, denoted by the perturbed angle $\alpha$. (The trim angle of attack, $\alpha_0$, is zero by definition.) The quantities $L$ and $D$ in Fig. 4-27 represent the lift and drag acting on the airplane during the disturbed flight condition, and they act normal and parallel to the relative wind. The relative wind acts in the direction opposite to the vector sum of $U_0$ and $w$.

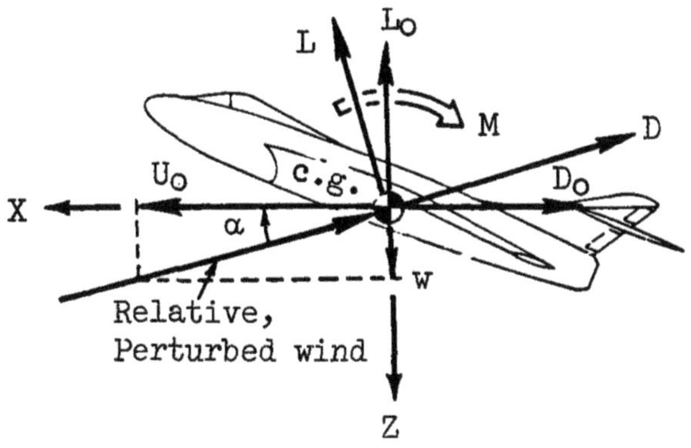


Fig. 4-27. Variation of lift and drag with change in $w$.

$X_w$. The perturbed $X$ forces due to $w$ are found by resolving $L$ and $D$ along the $X$ axis and taking the partial derivative with respect to $w$,

$$X = L \sin \alpha - D \cos \alpha$$

$$\frac{\partial X}{\partial w} = \frac{1}{U_0}\frac{\partial X}{\partial \alpha} = \frac{1}{U_0}\left(L \cos \alpha_0 + \frac{\partial L}{\partial \alpha} \sin \alpha_0 + D \sin \alpha_0 - \frac{\partial D}{\partial \alpha} \cos \alpha_0\right)$$

Neglecting perturbation products (Assumption 5), and recognizing $\alpha_0 = 0$, we find that

$$X_w \equiv \frac{1}{m}\frac{\partial X}{\partial w} = \frac{1}{mU_0}\left(L - \frac{\partial D}{\partial \alpha}\right) \tag{4-113}$$

Substitution of the values of lift and drag from Eq. 4-101 into Eq. 4-113 yields

$$\begin{aligned} X_w &= \frac{1}{mU_0}\left[\frac{\rho S U^2}{2} C_L - \frac{\partial}{\partial \alpha}\left(\frac{\rho S U^2}{2} C_D\right)\right] \\ X_w &= \frac{\rho S U^2}{2m}(C_L - C_{D_\alpha}) \end{aligned} \tag{4-114}$$

where $C_{D_\alpha} \equiv \partial C_D / \partial \alpha$.

The derivative $C_{D_\alpha}$ is the change in drag coefficient with varying angle of attack. When the angle of attack of an airframe increases from the equilibrium condition, the total drag increases, hence $C_{D_\alpha}$ is positive in sign. By far the largest contribution to $C_{D_\alpha}$ comes from the wing but there are small contributions from the horizontal tail and the fuselage. $C_{D_\alpha}$ is generally a nonlinear function of $\alpha$ but it can be considered piecewise linear for small perturbations.

$Z_w$. The change in the $Z$ force due to $w$ can be found by resolving the forces in Fig. 4-27 along the $Z$ axis and performing operations similar to those used in the derivation of Eq. 4-114:

$$Z_w = \frac{-\rho S U}{2m}(C_{L_\alpha} + C_D) \tag{4-115}$$

The derivative $C_{L_\alpha}$ is the change in lift coefficient with varying angle of attack; it is commonly known as the "lift curve slope." When the angle of attack of the airframe is increased, the lift force will increase more or less linearly until the wing stalls. The derivative $C_{L_\alpha}$ is therefore always positive in sign at angles of attack below the stall. The total airframe $C_{L_\alpha}$ is made up of contributions from the wing, the fuselage, and the horizontal tail. Ordinarily, the wing accounts for about 80 to 90 percent of the total $C_{L_\alpha}$, although it may account for less if the size of the fuselage is large in comparison with the size of the wing.

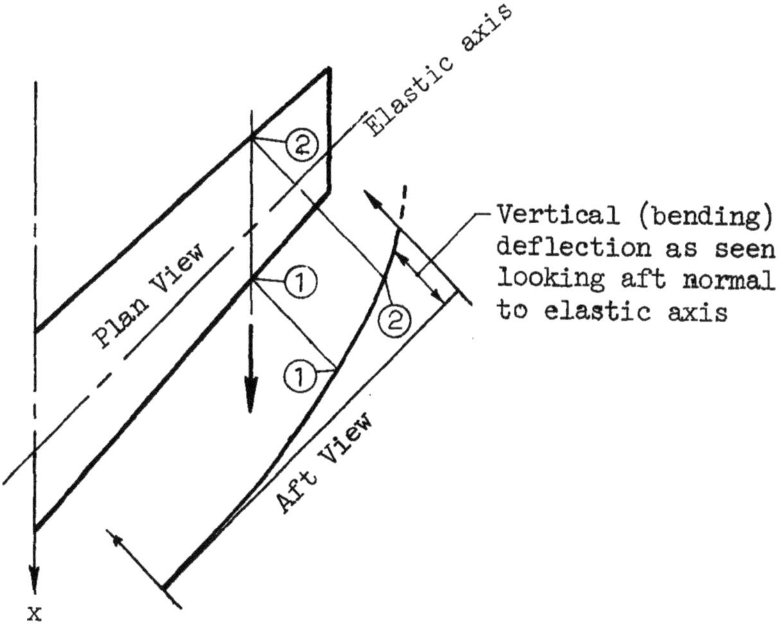


Fig. 4-28. Effect of wing bending on local angle of attack.

Aeroelastic distortion of the wing, under the incremental loads due to an angle of attack change, can alter its geometric twist to either increase or decrease the *net* change in lift and therefore the value of $C_{L_\alpha}$. Figure 4-28 illustrates how pure bending deflections of a sweptback wing cause a reduction in the local angle of attack (washout) of the tip region. Viewed along the flight direction, the trailing edge 2 moves up with respect to the leading edge 1, producing a negative increment in net angle of attack. On the other hand, airloads concentrated forward of the elastic axis tend to twist the wing sections to a positive increment in angle of attack.

$M_w$. The change in moment due to $w$ is most easily visualized by observing the components of the total lift and drag that act on the wing and the horizontal tail. Figure 4-29 shows these components; the subscripts $W$ and $t$ refer to the wing and the tail. A vertical velocity, $w$, causes a change in the angle of attack of both the wing and the horizontal tail, and consequently changes the lift and drag acting on these surfaces. The resulting moment can be found by summing the moments about the center of gravity caused by each of these forces to give the nondimensional coefficient $C_M$ (Eq. 4-101) as a function of $\alpha \equiv w/U_0$. In terms of this

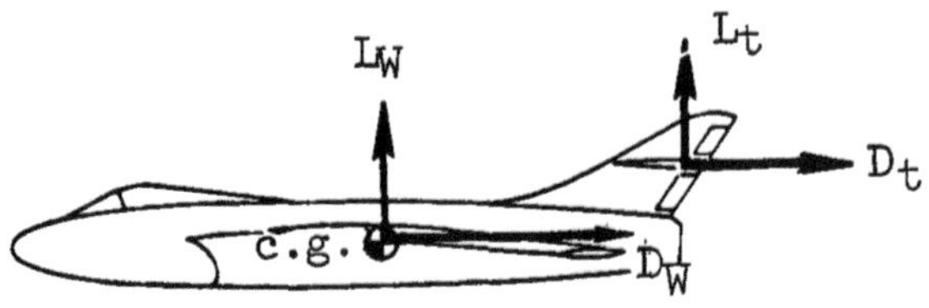


Fig. 4-29. Lift and drag acting on the wing and the horizontal tail.

coefficient

$$M = \frac{\rho U^2 S c}{2} C_M$$

$$\frac{\partial M}{\partial w} = \frac{1}{U_0}\frac{\partial M}{\partial \alpha} = \frac{\rho U S c}{2}\frac{\partial C_M}{\partial \alpha}$$

and

$$M_w \equiv \frac{1}{I_y}\frac{\partial M}{\partial w} = \frac{\rho S U c}{2 I_y} C_{M_\alpha} \tag{4-116}$$

where $C_{M_\alpha} \equiv \partial C_M / \partial \alpha$.

The derivative $C_{M_\alpha}$ is the change in pitching moment coefficient with varying angle of attack, and is commonly referred to as the "longitudinal static stability derivative." For a positive increment in the angle of attack, the increased lift on the horizontal tail causes a negative pitching moment about the center of gravity of the airframe. Simultaneously, the increased lift of the wing causes a positive or negative pitching moment, depending on the fore and aft location of the lift vector with respect to the center of gravity. These contributions, together with the pitching moment contribution of the fuselage, are combined to establish the derivative $C_{M_\alpha}$. Because of the distributed loads involved, $C_{M_\alpha}$ can be strongly influenced by aeroelastic distortions of the wing, the tail, and the fuselage. However, the major influence on the magnitude and sign of the total $C_{M_\alpha}$ for a particular airframe configuration is the center of gravity (c.g.) location. In fact, $C_{M_\alpha}$ is proportional to the distance between the center of gravity and the aerodynamic center (a.c.), the latter being the point about which the increment of lift due to a change in angle of attack effectively acts. If the center of gravity is on the aerodynamic center, $C_{M_\alpha}$ is zero; if it is ahead of the aerodynamic center, $C_{M_\alpha}$ is negative, and the airframe is said to possess static longitudinal stability; if it is aft of the aerodynamic center, $C_{M_\alpha}$ is positive, and the airframe is then statically unstable. The $C_{M_\alpha}$ or aerodynamic center can have important configuration-dependent nonlinear variations with $\alpha$ but can generally be considered piecewise linear over a limited range. Also, the aerodynamic center generally moves aft in going from subsonic to supersonic flight.

Another way of expressing static stability is in terms of the derivative $-\partial C_M/\partial C_L = -C_{M_\alpha}/C_{L_\alpha}$. This quantity, called "static margin," is identically equal to the $x$ distance from the aerodynamic center to the center of gravity, divided by the reference chord length,[22] i.e.,

$$\frac{\partial C_M}{\partial C_L} \equiv C_{M_{C_L}} = \frac{-x_{\text{a.c.}}}{c}$$

where $x$ is measured positive forward. A *positive* static margin corresponds to statically stable conditions and a *negative* $\partial C_M/\partial C_L$ and $C_{M_\alpha}$.

Static stability, characterized by either $C_{M_\alpha}$ or $\partial C_M/\partial C_L$, is perhaps the most important of the longitudinal derivatives and is a major factor in determining the response of the airframe to elevator motions and to gusts.

### Effect of $\dot{w}$, the Rate of Change of Speed Along the $Z$ Axis

In the earlier discussion of unsteady and quasisteady flows, the existence of a force due to $\dot{w}$ was explained on the basis of quasisteady flow considerations. It was pointed out that this rate of change of speed along the $Z$ axis results in an effective change of the angle of attack of the horizontal tail. This change in the angle of attack causes changes in the lift and drag acting on the horizontal tail. These are incremental forces and are represented by $\Delta L_t$ and $\Delta D_t$ in Fig. 4-30.

$X_{\dot{w}}$. The change in drag on the horizontal tail is the main contributor to the change in the $X$ force. But the drag on the horizontal tail is generally small in comparison with the total drag; the increment in the tail

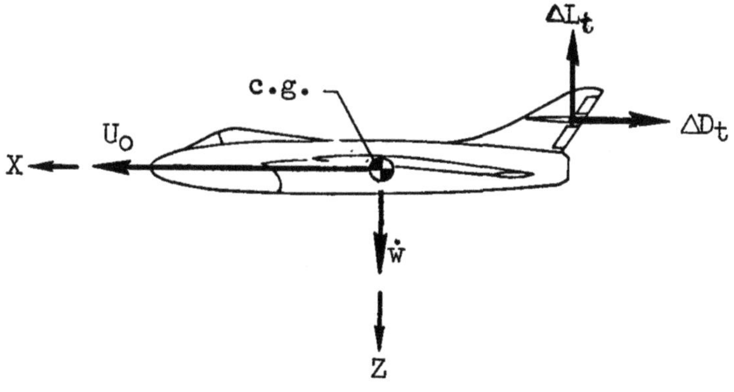


**Fig. 4-30. Lift and drag changes on the horizontal tail due to plunging acceleration.**

[22] e.g., Etkin, "Theory of Flight of Airplanes in Isotropic Turbulence—Review and Extension," AGARD Rept. 372, 1961.

drag due to $\dot{w}$ is even smaller. Therefore, $X_{\dot{w}}$ is considered zero in the first approximation.

$Z_{\dot{w}}$. The change in lift on the horizontal tail sometimes causes an important change in the total $Z$ force which, in terms of the total lift coefficient, can be written

$$\frac{\partial Z}{\partial \dot{w}} = \frac{1}{U}\frac{\partial Z}{\partial \dot{\alpha}} = \frac{-1}{U}\frac{\partial L}{\partial \dot{\alpha}} = \frac{-\rho S U}{2}\frac{\partial C_L}{\partial \dot{\alpha}}$$

$$Z_{\dot{w}} \equiv \frac{1}{m}\frac{\partial Z}{\partial \dot{w}} = \frac{-\rho S U}{2m}\frac{\partial C_L}{\partial \dot{\alpha}} \tag{4-117}$$

To form a nondimensional coefficient, Eq. 4-117 is multiplied and divided by $c/2U$:

$$Z_{\dot{w}} = \frac{-\rho S U}{2m}\frac{c}{2U}\frac{\partial C_L}{\partial(\dot{\alpha}c/2U)}$$

$$Z_{\dot{w}} = \frac{-\rho S c}{4m} C_{L\dot{\alpha}} \tag{4-118}$$

where $C_{L\dot{\alpha}} \equiv \partial C_L/\partial(\dot{\alpha}c/2U)$.

The derivative $C_{L\dot{\alpha}}$ arises essentially from two independent sources: an aerodynamic time lag effect, as explained in the discussion of quasi-steady flow, and various deadweight aeroelastic effects. For low speed flight, $C_{L\dot{\alpha}}$ arises mostly from the aerodynamic lag effect on the horizontal tail and its sign is positive. However, even tailless aircraft have $C_{L\dot{\alpha}}$'s because the wing must accelerate the air mass in its path as it accelerates (the apparent mass effect that goes beyond quasisteady flow assumption). For high speed flight the sign of $C_{L\dot{\alpha}}$ can be positive or negative, depending on aeroelastic effects such as wing twisting due to the deadweight moments of projecting nacelles and fuselage bending caused by the deadweight of the aft fuselage (see Fig. 4-31).

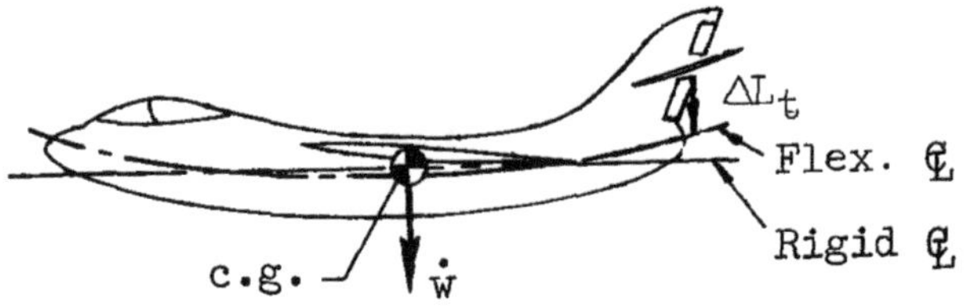


Fig. 4-31. The $\dot{w}$ effect on aeroelastic distortion due to fuselage flexibility.

The effect of $C_{L_{\dot{\alpha}}}$ on longitudinal dynamics is essentially the same as if the airframe's mass or inertia were changed in the equation relating the forces in the $Z$ direction. This effect is very small, and for this reason $C_{L_{\dot{\alpha}}}$ is often neglected in longitudinal dynamic analysis.

$M_{\dot{w}}$. This derivative can be expressed in terms of the total moment coefficient as

$$\frac{\partial M}{\partial \dot{w}} = \frac{1}{U_0}\frac{\partial M}{\partial \dot{\alpha}} = \frac{\rho S U c}{2}\frac{\partial C_M}{\partial \dot{\alpha}}$$

$$M_{\dot{w}} \equiv \frac{1}{I_y}\frac{\partial M}{\partial \dot{w}} = \frac{\rho S c^2}{4 I_y} C_{M_{\dot{\alpha}}} \tag{4-119}$$

where $C_{M_{\dot{\alpha}}} = \partial C_M / \partial(\dot{\alpha} c/2U)$.

The derivative $C_{M_{\dot{\alpha}}}$ is produced by the same aerodynamic and aeroelastic effects that produce $C_{L_{\dot{\alpha}}}$. However, whereas $C_{L_{\dot{\alpha}}}$ is usually negligible $C_{M_{\dot{\alpha}}}$ is relatively important in longitudinal dynamics because it does have a significant, if not powerful, effect on the damping of the short period mode. A negative value of $C_{M_{\dot{\alpha}}}$, which is its normal sign, increases the short period damping.

### Effect of $q$, the Pitching Velocity

$X_q$. In the light of the quasisteady flow assumption (Assumption 7), the major effect of the airplane's pitching about its center of gravity is to cause an increase in the angle of attack of the horizontal tail (see Fig. 4-32). As in the case of the effect of $\dot{w}$, the resulting drag increase is neglected in the first approximation and $X_q$ is normally set equal to zero.

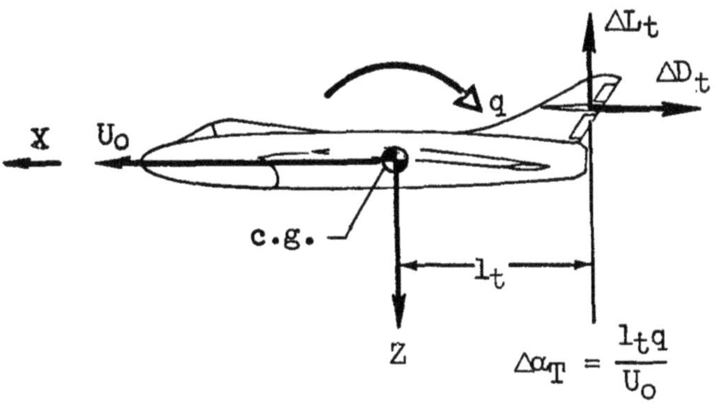


Fig. 4-32. Lift and drag changes on horizontal tail due to pitching velocity.

$Z_q$, $M_q$. The incremental lift produces a change both in the $Z$ force and in the pitching moment. The expressions for $Z_q$ and $M_q$ can be derived as follows:

$$\frac{\partial Z}{\partial q} = \frac{-\partial L}{\partial q} = \frac{-\rho S U^2}{2}\frac{\partial C_L}{\partial q} = \frac{-\rho S U^2}{2}\frac{c}{2U}\frac{\partial C_L}{\partial(qc/2U)}$$

$$Z_q \equiv \frac{1}{m}\frac{\partial Z}{\partial q} = \frac{-\rho S U c}{4m} C_{L_q} \tag{4-120}$$

where $C_{L_q} = \partial C_L/\partial(qc/2U)$, and

$$\frac{\partial M}{\partial q} = \frac{\rho S U^2 c}{2}\frac{\partial C_M}{\partial q} = \frac{\rho S U^2 c}{2}\frac{c}{2U}\frac{\partial C_M}{\partial(qc/2U)}$$

$$M_q \equiv \frac{1}{I_y}\frac{\partial M}{\partial q} = \frac{\rho S U c^2}{4 I_y} C_{M_q} \tag{4-121}$$

where $C_{M_q} \equiv \partial C_M/\partial(qc/2U)$.

The derivatives $C_{L_q}$ and $C_{M_q}$ are the change in lift and moment coefficients, respectively, with pitching velocity at a fixed angle of attack. As the airframe pitches about its center of gravity, the fore and aft angle of attack distribution changes and lift forces develop primarily on the horizontal tail and wing (see Fig. 4-33). These produce contributions to both derivatives, positive in sign for $C_{L_q}$ and negative for $C_{M_q}$.

There are also contributions because of various deadweight aeroelastic effects. Since the airframe is moving in a curved flight path because of its pitching, a centrifugal force is developed on all the components of the

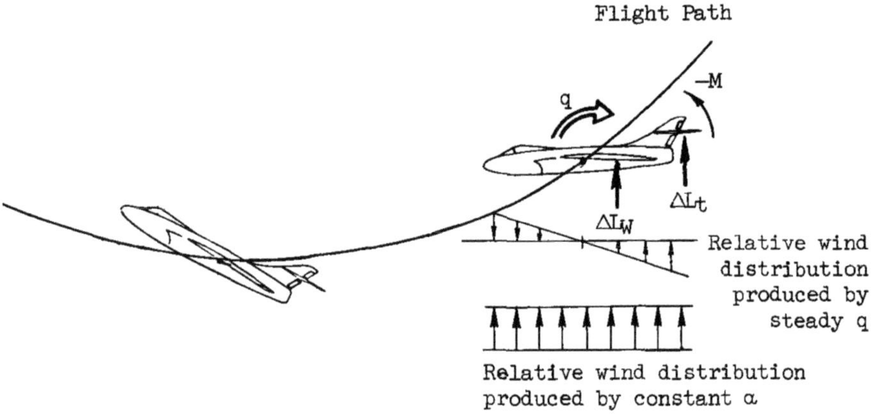


Fig. 4-33. Lift and moment coefficient change due to variation in pitching velocity.

airframe. This force can cause the wing to twist as a result of the dead-weight moment of overhanging nacelles, and can cause the horizontal tail angle of attack to change, as a result of fuselage bending due to the weight of the tail section. In low speed flight, both $C_{L_q}$ and $C_{M_q}$ occur mostly from the effect of the curved flight path on the horizontal tail; their signs are positive and negative, respectively. In high speed flight the signs of either can be positive or negative, depending on the nature of the aero-elastic effects.

As with $C_{L_{\dot{\alpha}}}$, the effect of $C_{L_q}$ on longitudinal stability is usually very small and it is ordinarily neglected in dynamic analyses. On the other hand, $C_{M_q}$ is very important in longitudinal dynamics because it contributes a large portion of the damping of the short period mode for conventional aircraft. As pointed out, this damping effect comes mostly from angle of attack changes at the horizontal tail, which are proportional to the tail length, $l_t$ (i.e., $\Delta\alpha_T = ql_t/U$). The tail length is also the lever arm converting the tail lift into the moment; therefore $C_{M_q}$ is proportional to $l_t^2$.

### Effects of $\delta_T$, the Change of Power Plant Control Setting (throttle)

An increase in power plant control setting yields an increase in thrust having the general orientation shown in Fig. 4-34. By simple resolution

$$X_{\delta_T} = \left(\frac{1}{m}\right)\left(\frac{\partial T}{\partial \delta_T}\right) \cos \alpha_T$$

$$Z_{\delta_T} = \left(\frac{-1}{m}\right)\left(\frac{\partial T}{\partial \delta_T}\right) \sin \alpha_T$$

$$M_{\delta_T} = \left(\frac{-z_T}{I_y}\right)\left(\frac{\partial T}{\partial \delta_T}\right) \cos \alpha_T$$

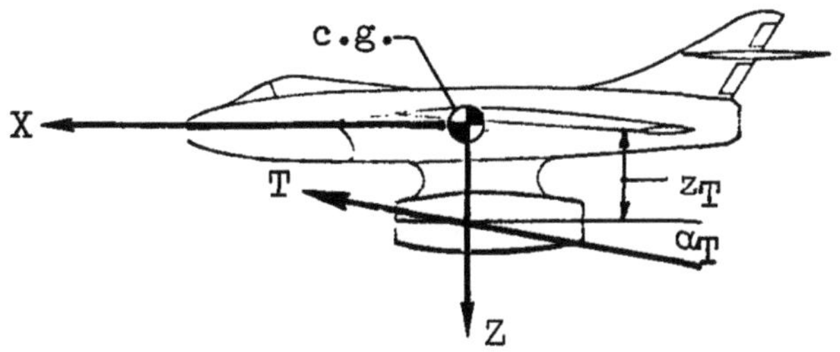


Fig. 4-34. Resolution of thrust into forces and moments.

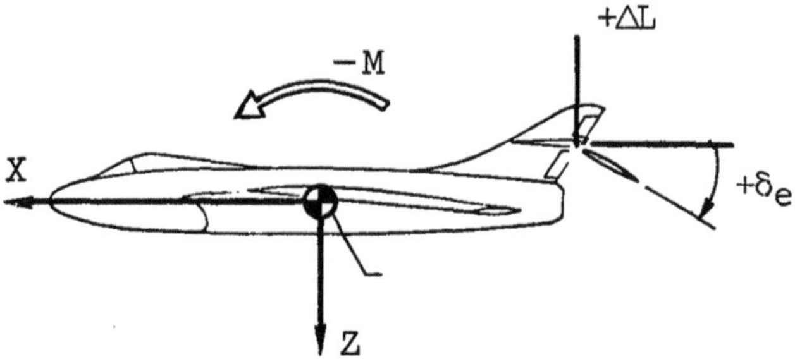


Fig. 4-35. Lift and moment changes due to surface deflection.

The further breakdown of these derivatives into conventional nondimensional coefficients is not warranted, as noted above in the discussion of $X_u$.

EFFECTS OF AERODYNAMIC CONTROL SURFACE DEFLECTIONS

Aerodynamic surfaces pertinent to longitudinal control include elevators, stabilizers, flaps, slats, dive brakes, etc. Where applicable, the positive direction is taken as that giving positive lift, as in the definition of positive $\alpha$ (see Fig. 4-35). Usually, the deflection of such surfaces, in addition to producing the sought-for primary effect (e.g., tail lift to produce a pitching moment for an elevator or drag to produce longitudinal deceleration for a dive brake), also produces secondary forces and moments. Generally, therefore, we must consider contributions to the $X$ and $Z$ forces and to the pitching moment, $M$, as follows:

$$
\begin{aligned}
X_\delta &\equiv \frac{1}{m}\frac{\partial X}{\partial \delta} = \frac{-\rho U^2 S}{2m} C_{D_\delta} \qquad \text{where } C_{D_\delta} \equiv \frac{\partial C_D}{\partial \delta} \\
Z_\delta &\equiv \frac{1}{m}\frac{\partial Z}{\partial \delta} = \frac{-\rho U^2 S}{2m} C_{L_\delta} \qquad \text{where } C_{L_\delta} \equiv \frac{\partial C_L}{\partial \delta} \qquad (4\text{-}122) \\
M_\delta &\equiv \frac{1}{I_y}\frac{\partial M}{\partial \delta} = \frac{\rho U^2 S c}{2I_y} C_{M_\delta} \qquad \text{where } C_{M_\delta} \equiv \frac{\partial C_M}{\partial \delta}
\end{aligned}
$$

The general form of Eq. 4-122 is specialized by suitable subscripts for particular control surfaces, e.g., $\delta_e$ for the elevator, $\delta_f$ for the flap, $\delta_B$ for the dive brakes, etc.

The elevator is, of course, the primary longitudinal control surface; its function is, through applied pitching moments, to control the angle of attack of the airframe in equilibrium and maneuvering flight. This function is usually considered to be the most important of all the control functions about the three axes, so that the elevator control effectiveness,

$C_{M_{\delta_e}}$ is of great importance in airframe design. Its design value is determined by the anticipated fore and aft center of gravity travel, and the maximum $C_L$ capability of the wing as influenced by high lift devices. In general, the larger the center of gravity range and the higher the maximum $C_L$, the larger the required value of $C_{M_{\delta_e}}$. Supersonic flight may impose additional requirements because of the attendant aft movement of the aerodynamic center. The sign of $C_{M_{\delta_e}}$ depends on the location of the elevator, fore or aft of the center of gravity; for aft locations (as in Fig. 4-35) and the elevator sign convention used here, $C_{M_{\delta_e}}$ is negative.

The elevator control effectiveness $C_{L_{\delta_e}}$ is always positive in accordance with the convention used here to define positive control deflection as a producer of positive lift (Fig. 4-35). On conventional aircraft with the horizontal tail mounted at an appreciable distance aft of the center of gravity, $C_{L_{\delta_e}}$ is usually very small and its effect is relatively unimportant, except for automatic control involving vertical acceleration feedback. The value of $C_{D_{\delta_e}}$ is invariably smaller than $C_{L_{\delta_e}}$ because of the usual variation of drag with lift and it is normally negligible. However, on tailless aircraft having small effective elevator lever arms, the values of $C_{L_{\delta_e}}$ and $C_{D_{\delta_e}}$ are relatively large with respect to the required $C_{M_{\delta_e}}$ and neither can be safely neglected. The sign of $C_{D_{\delta_e}}$ can be positive or negative, depending on the trim position of the elevator and the trim angle of attack.

### *Lateral Stability Derivatives*

#### EFFECT OF $v$, THE CHANGE IN SIDE VELOCITY

When an airplane is disturbed from steady flight so that it has a side velocity, $v$, a force along the $Y$ axis and moments about the $X$ and $Z$ axes are developed. The major forces caused by the side velocity are labeled $F_1$, $F_2$, $F_3$, and $F_4$ in Fig. 4-36, where $F_1$ arises from the change of the angle of attack of the vertical tail; $F_2$ is the side force acting on the fuselage; and $F_3$ and $F_4$ are differential lift forces acting on each semispan of the wing, due to its "effective dihedral."

$Y_v$. From Eq. 4-101, the side force equation has the form

$$Y = \frac{\rho U^2 S}{2} C_y$$

so that

$$\frac{\partial Y}{\partial v} = \frac{\rho U^2 S}{2} \frac{\partial C_y}{\partial v}$$

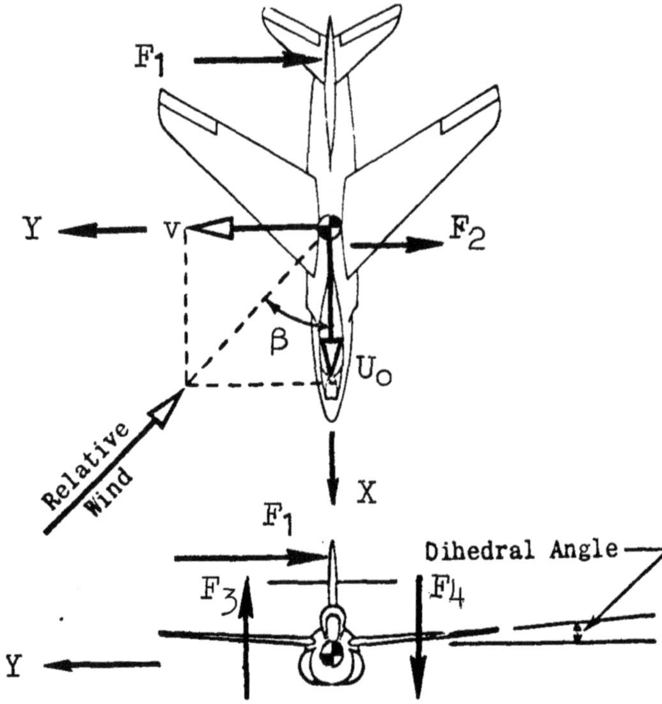


Fig. 4-36. Forces accompanying sideslipping motions.

and, in terms of $\beta = \tan^{-1} v/U_0 \doteq v/U_0$,

$$Y_v \equiv \frac{1}{m}\frac{\partial Y}{\partial v} = \frac{\rho U S}{2m} C_{y_\beta} \tag{4-123}$$

where $\partial C_y/\partial\beta = C_{y_\beta}$.

The major portion of $C_{y_\beta}$ normally comes from the vertical tail, with small contributions from the fuselage and wing. It is usually negative in sign for practical airframe configurations; i.e., the side force opposes the sideward motion. However, the forces on a slender fuselage can actually be in the aiding direction for high angles of attack.[23] These effects can apparently override the resisting tail forces to produce positive values of $C_{y_\beta}$ for certain (as yet rare) configurations with low aspect ratio wings that operate at high angles of attack.

[23] Bernard Spencer, Jr., and W. Pelham Phillips, "Transonic Aerodynamic Characteristics of a Series of Bodies Having Variations in Fineness Ratio and Cross-Sectional Ellipticity," NASA TN D-2622, Feb. 1965.

Small (or positive) values of $C_{y_\beta}$ are undesirable because the resulting small (or reversed) side forces make the detection of sideslip difficult; accordingly, coordination of banked turns becomes a piloting problem. Also, such values of $C_{y_\beta}$ contribute little to the damping of the dutch roll, whereas normal (negative) values of $C_{y_\beta}$ can contribute substantially to the total damping.

$L_v$. The rolling moment, $L$, about the $X$ axis is caused mainly by the components $F_3$ and $F_4$ which act normal to the wing, and by $F_1$ at the fin center of pressure which can be either above or below the $X$ axis. From Eq. 4-101,

$$L = \frac{\rho U^2 Sb}{2} C_l$$

and

$$\frac{\partial L}{\partial v} = \frac{\rho U^2 Sb}{2} \frac{\partial C_l}{\partial v} = \frac{\rho U Sb}{2} \frac{\partial C_l}{\partial \beta}$$

$$L_v \equiv \frac{1}{I_x} \frac{\partial L}{\partial v} = \frac{\rho U Sb}{2I_x} C_{l_\beta} \tag{4-124}$$

where $C_{l_\beta} = \partial C_l / \partial \beta$. Also,

$$L_\beta = U_0 L_v = \frac{\rho U^2 Sb}{2I_x} C_{l\beta}$$

The change in rolling moment coefficient with sideslip angle, $C_{l_\beta}$, is usually referred to as the "effective dihedral." This nomenclature is a holdover from earlier days when, in fact, the value of $C_{l_\beta}$ was governed largely by the geometric dihedral built into the wing (Fig. 4-36). For positive geometric dihedral $C_{l_\beta}$ is negative, and this leads to some confusion because *negative* values of $C_{l_\beta}$ are referred to as "*positive* dihedral." For modern configurations the wing contribution to $C_{l_\beta}$ is a function not only of geometric dihedral but, more so, of the sweep, aspect ratio, and angle of attack. In addition, the wing location on the fuselage, high or low, contributes negative or positive increments, respectively, to $C_{l_\beta}$; and the direct forces on the vertical tail contribute decreasingly negative increments as the fin moves down with respect to the $X$ axis with increasing trim angle of attack. While the general overriding importance of the wing contribution to $C_{l_\beta}$ cannot be denied, there may be flight conditions in which it is small relative to the vertical tail or wing/fuselage contribution.

Since $C_{l_\beta}$ is very important in lateral stability and control, it is imperative to consider it in the preliminary design of an airframe. It is involved in damping both the dutch roll mode and the spiral mode, and also in the maneuvering characteristics of an airframe, especially with regard to lateral control with the rudder alone near stall. To improve the dutch roll damping characteristics of an airframe, small negative values of

$C_{l_\beta}$ are sought but are difficult to obtain in general because of the influences noted above.

$N_v$. The yawing moment, $N$, due to a side velocity, $v$, is caused mainly by the force on the vertical tail, $F_1$. The forms of the stability derivatives $N_v$ and $N_\beta$ are similar to those of Eq. 4-124:

$$N_v = \frac{\rho U S b}{2I_z} C_{n_\beta}$$
$$N_\beta = \frac{\rho U^2 S b}{2I_z} C_{n_\beta} \tag{4-125}$$

The term $C_{n_\beta}$ is the change in yawing moment coefficient with variation in the sideslip angle. It is usually referred to as the static directional, or "weathercock," stability. The major portion of $C_{n_\beta}$ comes from the vertical tail area and lever arm, which stabilize the body of the airframe just as the tail feathers of an arrow stabilize the arrow shaft. The $C_{n_\beta}$ contribution due to the vertical tail is positive, signifying static directional stability, whereas the $C_{n_\beta}$ due to the body is negative, signifying static directional instability. There is also a contribution to $C_{n_\beta}$ from the wing, the value of which is usually positive but very small compared to the body and vertical tail contributions. Since both the major contributions depend essentially on the dimensions of the body, center of gravity variations which are limited by longitudinal considerations to a small fraction of the body length have little effect on the value of $C_{n_\beta}$. Increasing supersonic speeds generally have a deleterious effect on $C_{n_\beta}$ because the fin lift curve slope decreases, whereas the body moments remain about constant. Also, for high angles of attack, the fin may be immersed in the wing-body wake with consequent drastic reductions in $C_{n_\beta}$.

The derivative $C_{n_\beta}$ is very important in determining dynamic lateral stability and control characteristics. It primarily establishes the natural frequency of the dutch roll oscillatory mode of the airframe and is also a factor in determining the spiral stability characteristics. A high value of $C_{n_\beta}$ aids the pilot in effecting coordinated turns and prevents excessive sideslip and yawing motions in extreme flight maneuvers; however, in rough air, excessive $C_{n_\beta}$ magnifies the disturbances from side gusts. There are cases on record in which a reduction in the vertical tail resulted in improved overall (in rough and calm air) dynamic properties.

EFFECT OF $\dot{v}$, THE RATE OF CHANGE IN SIDE VELOCITY

The existence of quasisteady forces and moments due to $\dot{v}$ was explained in the discussion of unsteady flow as arising from sidewash lags that produce angle of attack variations at the vertical tail.

$Y_{\dot{v}}$. The change in side force with $\dot{v}$ in terms of the side force coefficient $C_y$, is

$$\frac{\partial Y}{\partial \dot{v}} = \frac{1}{U}\frac{\partial Y}{\partial \dot{\beta}} = \frac{\rho S U_0}{2}\frac{\partial C_y}{\partial \dot{\beta}}$$

To nondimensionalize, we multiply and divide by $b/2U$, whereby

$$Y_{\dot{v}} \equiv \frac{1}{m}\frac{\partial Y}{\partial \dot{v}} = \frac{\rho S b}{4m} C_{y_{\dot{\beta}}} \tag{4-126}$$

where $C_{y_{\dot{\beta}}} \equiv \partial C_y / \partial(\dot{\beta} b/2U)$.

$L_{\dot{v}}$. The change in rolling moment, $L$, in terms of the rolling moment coefficient, $C_l$, is

$$\frac{\partial L}{\partial \dot{v}} \equiv \frac{1}{U_0}\frac{\partial L}{\partial \dot{\beta}} = \frac{\rho S U_0 b}{2}\frac{\partial C_l}{\partial \dot{\beta}}$$

and nondimensionalizing, as above,

$$U_0 L_{\dot{v}} \equiv L_{\dot{\beta}} \equiv \frac{1}{I_x}\frac{\partial L}{\partial \dot{\beta}} = \frac{\rho S U_0 b^2}{4 I_x} C_{l_{\dot{\beta}}} \tag{4-127}$$

where $C_{l_{\dot{\beta}}} \equiv \partial C_l / \partial(\dot{\beta} b/2U)$.

$N_{\dot{v}}$. The corresponding yawing moment derivative is

$$U_0 N_{\dot{v}} \equiv N_{\dot{\beta}} \equiv \frac{1}{I_z}\frac{\partial N}{\partial \dot{\beta}} = \frac{\rho S U_0 b^2}{4 I_z} C_{n_{\dot{\beta}}} \tag{4-128}$$

where $C_{n_{\dot{\beta}}} \equiv \partial C_n / \partial(\dot{\beta} b/2U)$.

Generally speaking, not too much is known of the nondimensional aerodynamic derivatives appearing in the foregoing expressions. In fact, as concerns the usual formulation of the rigid body equations of motion, *all* of these derivatives are generally neglected. However, there are cases in which the observed dutch roll damping can be accounted for only by including the significant $C_{n_{\dot{\beta}}}$ effects, when $C_{n_{\dot{\beta}}}$ is the same order of magnitude as $C_{n_r}$. The difficulty is that there is no good way of estimating $C_{n_{\dot{\beta}}}$ or of knowing a priori for which configurations it may be important.

Aside from aerodynamic lag effects, $\dot{v}$ derivatives also arise as a result of aeroelastic effects. Figure 4-37 shows how the aft fuselage distortion, from lateral acceleration of the distributed mass, produces a vertical tail angle of attack and a concomitant aerodynamic side force. This side force, proportional to $\dot{v}$, reduces the airplane's resistance to lateral motion (i.e., its effective mass) but only by a negligible amount; therefore, even considering such aeroelastic effects, $Y_{\dot{v}}$ is negligible. Usually, the rolling moment contributed by the side force will also be negligible because of the

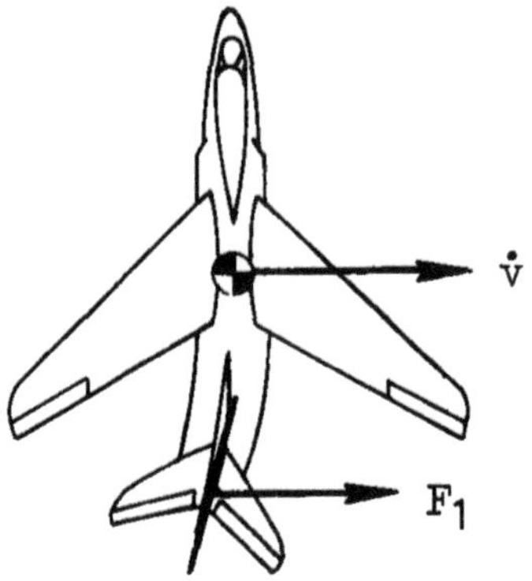


Fig. 4-37. Distortion effects due to $\dot{v}$.

small vertical moment arm involved. This leaves the yawing moment derivative, $N_{\dot{\beta}}$, as the most probable significant contribution of aeroelastic distortions and aerodynamic lag effects.

EFFECT OF $p$, THE CHANGE IN ROLLING VELOCITY

The rolling velocity, $p$, causes the various forces shown in Fig. 4-38.

$Y_p$. The change in the $Y$ force with $p$, illustrated as $F_1$ acting on the vertical tail in Fig. 4-38, is expressed as

$$\frac{\partial Y}{\partial p} = \frac{\rho U^2 S}{2} \frac{\partial C_y}{\partial p}$$

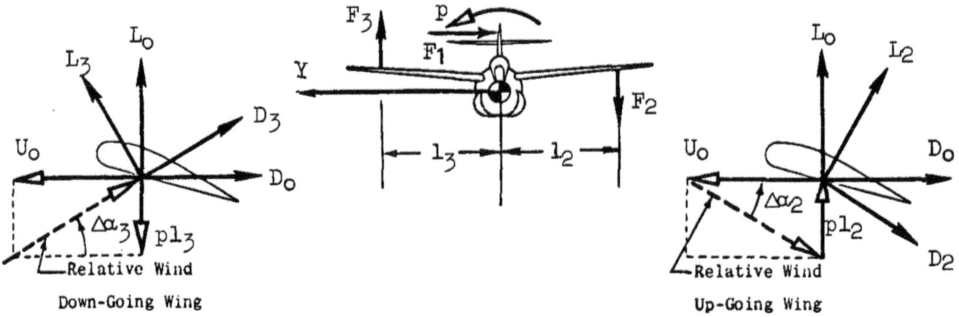


Fig. 4-38. Forces arising from roll rate perturbations, $p$.

To form a nondimensional coefficient, we multiply and divide by $b/2U$. Then,

$$Y_p \equiv \frac{1}{m}\frac{\partial Y}{\partial p} = \frac{\rho U^2 S}{2m}\frac{b}{2U}\frac{\partial C_y}{\partial(pb/2U)}$$
$$Y_p = \frac{\rho USb}{4m} C_{y_p} \tag{4-129}$$

where $C_{y_p} = \partial C_y/\partial(pb/2U)$.

The sign of $C_{y_p}$ can be positive or negative, depending on the vertical tail location with respect to the $X$ axis, as a function of angle of attack, and also depending on the sidewash from the wing. Since $C_{y_p}$ is of very little importance in lateral dynamics it is common practice to neglect this derivative in lateral dynamic calculations.

$L_p$. There are also incremental forces acting on the wing, illustrated as $F_2$ and $F_3$ in Fig. 4-38. The vertical velocity of the down-going wing at any station a distance $l_3$ from the $X$, $Z$ plane is $pl_3$. This vertical velocity increases the effective angle of attack at this station by an amount $\Delta\alpha_3$ (where $\Delta\alpha_3 = pl_3/U_0$). This increase in the angle of attack increases the lift and drag acting on the wing. The effective angle of attack of the up-going wing at a station a distance $l_2$ from the $X$, $Z$ plane is decreased by an amount $\Delta\alpha_2$ (where $\Delta\alpha_2 = pl_2/U_0$). This decrease in effective angle of attack decreases the lift and drag acting on the wing at this station. Usually, the change in drag force is a relatively negligible contribution to the change in rolling moment because of $p$, which can be expressed as

$$\frac{\partial L}{\partial p} = \frac{\rho U^2 Sb}{2}\frac{\partial C_l}{\partial p}$$

Multiplying and dividing by $b/2U$, we obtain

$$L_p \equiv \frac{1}{I_x}\frac{\partial L}{\partial p} = \frac{\rho USb^2}{4I_x} C_{l_p} \tag{4-130}$$

where $C_{l_p} \equiv \partial C_l/\partial(pb/2U)$.

The derivative $C_{l_p}$ is the change in rolling moment coefficient with the change in rolling velocity and is usually known as the roll damping derivative. It is composed of contributions, negative in sign, from the wing and the horizontal and vertical tails. However, unless the size of the tails is unusually large in comparison with the size of the wing, the major portion of the total $C_{l_p}$ comes from the wing; and the variations in $C_{l_p}$ with Mach number and $\alpha$ closely follow the variations in wing lift curve slope, $C_{L_\alpha}$.

Because $C_{l_p}$ essentially determines the damping in roll characteristics of the aircraft, it is quite important in lateral dynamics. Its value is more or less given by the wing planform geometry which is determined by other more important design criteria. The value of $C_{l_p}$ does directly affect the design of the ailerons, however, since $C_{l_p}$ in conjunction with $C_{l_{\delta_a}}$ establishes the airframe's maximum available rolling velocity; this is an important criterion of flying qualities.

Positive values of $C_{l_p}$ occur only when the wing, or portions thereof, are stalled. Flight situations involving stall or separated flow are generally avoided, except for demonstration purposes; they are seldom subjected to conventional dynamic analysis. Such situations usually lead to spinning motions which can only be successfully analyzed by including nonlinear terms and solving the equations of motion by numerical or analog methods.

$N_p$. In addition to the change in magnitude of the lift forces acting on each semispan of the wing, it may be seen from Fig. 4-38 that the lift forces acting on the down-going and up-going semispans are rotated forward and backward, respectively. The change in direction of these forces results in a negative yawing moment about the $Z$ axis.

Figure 4-38 represents the general case. However, for flight near the stall, the drag forces may become important and result in a yawing moment of opposite sign. The change in yawing moment because of $p$ can be written immediately by analogy to the Eq. 4-130 result:

$$N_p = \frac{\rho U S b^2}{4 I_z} C_{n_p} \tag{4-131}$$

where $C_{n_p} = \partial C_n / \partial(pb/2U)$.

The derivative $C_{n_p}$ is the change in yawing moment coefficient with varying rolling velocity. While it arises mainly from the wing, as discussed above, the vertical tail can also contribute (see Fig. 4-38). The contribution from the vertical tail can be either positive or negative, depending on the vertical tail geometry, the sidewash from the wing, and the equilibrium angle of attack of the airframe.

Because of its influence on dutch roll damping, $C_{n_p}$ is fairly important in lateral dynamics. It is usually negative in sign, and for most airframe configurations, the larger its negative value, the greater the reduction in dutch roll damping. Also, the more negative its value, the higher the sideslipping (uncoordinated) motions accompanying turn entry or exit. Therefore, positive values of $C_{n_p}$ are to be desired, although it is completely impractical to make this a design goal.

### Effect of $r$, the Change in Yawing Velocity

The yawing velocity, $r$, is illustrated in Fig. 4-39.

$Y_r$. A side force, $F_1$, is caused by a yawing velocity, $r$, which is mainly due to the fact that the effective angle of attack of the vertical tail is

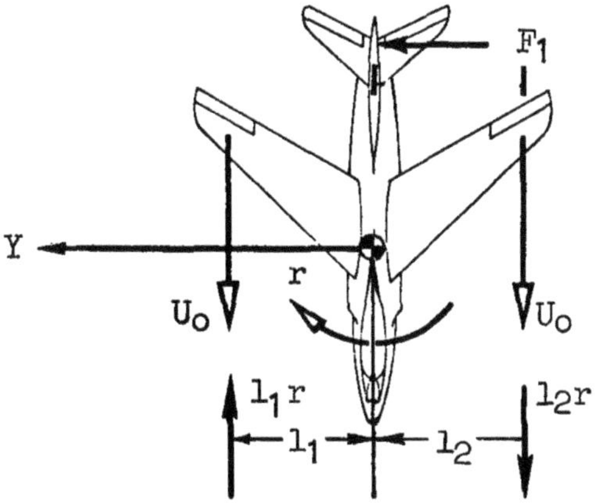


Fig. 4-39. Forces due to yaw rate, $r$.

increased. By analogy with $Y_p$, derived in Eq. 4-129, it is possible to write

$$Y_r \equiv \frac{1}{m}\frac{\partial Y}{\partial r} = \frac{\rho U S b}{4m} C_{y_r} \tag{4-132}$$

where $C_{y_r} = \partial C_y / \partial(rb/2U)$.

The derivative $C_{y_r}$, the change in side force coefficient with yawing velocity, is of little importance in lateral dynamics; it is common practice to neglect this derivative in lateral calculations.

$L_r$. As shown in Fig. 4-39, the forward speed of a station that is a distance $l_1$ from the $X$, $Z$ plane on the semispan of the wing is decreased an amount $l_1 r$, resulting in a decrease in lift at this station. Similarly, the forward speed of a station a distance $l_2$ normal to the $X$, $Z$ plane on the semispan of the wing is increased an amount $l_2 r$, resulting in an increase in lift at this section. The result of the changes in lift acting on each semispan is then a rolling moment, usually positive, whose variation with $r$ can be expressed by analogy with $L_p$ (Eq. 4-130) as:

$$L_r \equiv \frac{1}{I_x}\frac{\partial L_r}{\partial r} = \frac{\rho U S b^2}{4 I_x} C_{l_r} \tag{4-133}$$

where $C_{l_r} = \partial C_l / \partial(rb/2U)$.

The derivative $C_{l_r}$ is the change in rolling moment coefficient with the change in yawing velocity. In addition to the major wing contribution, discussed above, the vertical tail will also contribute to $C_{l_r}$ if it is located either above or below the $X$ axis. Its contribution can therefore be positive or negative, depending on the vertical tail geometry and the equilibrium angle of attack of the airframe. The sign of $C_{l_r}$ is usually dominated by the wing contribution which is positive and proportional to the trimmed lift coefficient.

Although $C_{l_r}$ is of secondary importance in lateral dynamics, it should not be neglected in lateral dynamic calculations. For a conventional airframe configuration, changes in $C_{l_r}$ of reasonable magnitude show only slight effects on the dutch roll damping characteristics. In the spiral mode, however, $C_{l_r}$ has a considerable effect. For stability in this mode it is desirable that the positive value of $C_{l_r}$ be as small as possible.

$N_r$. The side force, $F_1$, in Fig. 4-39 also causes a moment about the $Z$ axis, since the vertical tail is some distance aft of the center of gravity. This moment is usually negative and its variation with $r$ is, by analogy to $L_r$,

$$N_r = \frac{1}{I_z}\frac{\partial N}{\partial r} = \frac{\rho U S b^2}{4 I_z} C_{n_r} \tag{4-134}$$

where $C_{n_r} = \partial C_n / \partial (rb/2U)$.

The derivative $C_{n_r}$ is the change in yawing moment coefficient with the change of yawing velocity. Known as the "yaw damping derivative," it is made up of contributions, all of negative sign, from the wing, the fuselage, and, as discussed above, the vertical tail. The latter contribution is by far the largest, usually amounting to about 80 or 90 percent of the total $C_{n_r}$ of the airframe. Like the horizontal tail contribution to $C_{M_q}$, it is proportional to the square of the tail lever arm.

The derivative $C_{n_r}$ is very important in lateral dynamics because it is the main contributor to the damping of the dutch roll oscillatory mode and is important to the spiral mode. For each mode, large negative values of $C_{n_r}$ are desired.

### Effects of Control Surface Deflections

The conventional lateral control surfaces are rudder and aileron, shown in Figs. 4-40 and 4-41, respectively. The primary rudder function

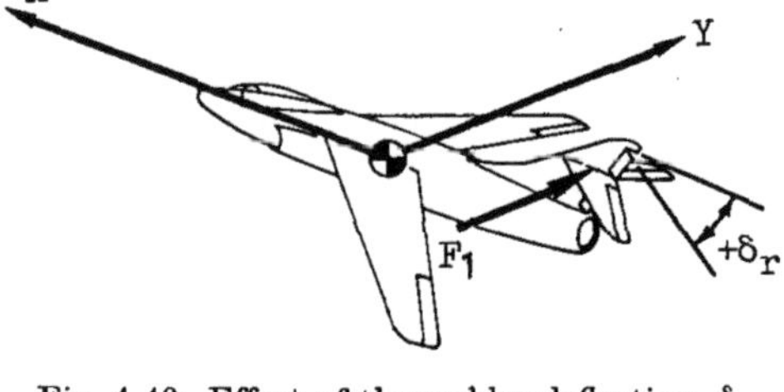


Fig. 4-40. Effect of the rudder deflection, $\delta_r$.

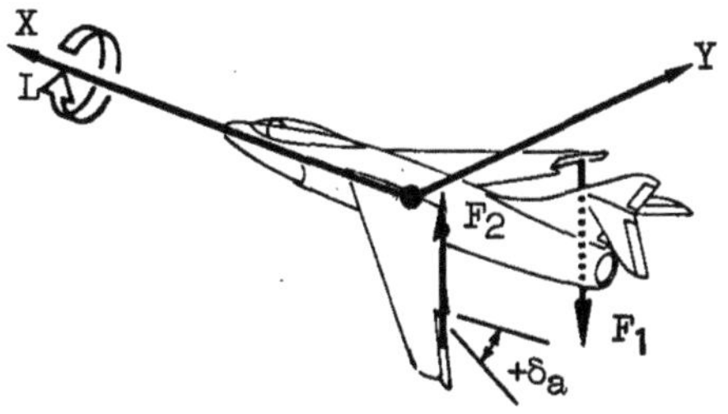


Fig. 4-41. Effect of the aileron deflection, $\delta_a$.

is the provision of controllable yawing moments; the primary aileron function is the generation of rolling moments. Positive rudder deflection is defined to produce positive side force, as in Fig. 4-40, and positive aileron deflection produces positive rolling moment, as in Fig. 4-41.

In addition to the direct (desired) moments, there are usually indirect (and undesirable) cross moments and side forces, so that in general either surface produces side forces and rolling and yawing moments. The lift, drag, and pitching moment effects of lateral/directional control deflections on longitudinal motion are generally ignored; however, there are special cases involving large differences between up-going and down-going ailerons (as in spoiler-type ailerons) in which significant changes in lift and pitching moment can occur. Such situations do not usually strongly influence the lateral/directional motions because the pilot, or autopilot, easily counters the effect with a small elevator deflection. Accordingly, for the general control surface deflection angle, $\delta$, we limit ourselves to the following important lateral/directional derivatives:

$$Y_\delta \equiv \frac{1}{m}\frac{\partial Y}{\partial \delta} = \frac{\rho U^2 S}{2m} C_{y_\delta} \qquad \text{where } C_{y_\delta} \equiv \frac{\partial C_y}{\partial \delta}$$

$$L_\delta \equiv \frac{1}{I_x}\frac{\partial L}{\partial \delta} = \frac{\rho U^2 S b}{2I_x} C_{l_\delta} \qquad \text{where } C_{l_\delta} \equiv \frac{\partial C_l}{\partial \delta}$$

$$N_\delta \equiv \frac{1}{I_z}\frac{\partial N}{\partial \delta} = \frac{\rho U^2 S b}{\partial I_z} C_{n_\delta} \qquad \text{where } C_{n_\delta} \equiv \frac{\partial C_n}{\partial \delta}$$

These quantities are specialized for the particular control surface in question by the addition of a suitable subscript to $\delta$, i.e., $\delta_r$ for the rudder, $\delta_a$ for the aileron.

The derivative $C_{y_{\delta_r}}$ is the change in side force coefficient with variation in rudder deflection. According to the sign convention adopted here, a positive rudder deflection gives a positive side force; hence the derivative $C_{y_{\delta_r}}$ is positive in sign. Its effects are relatively unimportant in lateral stability and control, except when considering lateral acceleration feedbacks to an autopilot.

The change in side force coefficient with aileron deflection is represented by $C_{y\delta_a}$. For most conventional airframe configurations, the magnitude of this derivative is essentially zero. However, for certain aircraft with highly swept wings of low aspect ratio or with inboard ailerons, the strong sidewash in the vicinity of the vertical tail caused by the asymmetrically deflected ailerons can produce finite values of either sign. Even then, however, the effects on lateral stability and control are usually negligibly small.

The derivative $C_{l_{\delta_r}}$ is the change in rolling moment coefficient with variation in rudder deflection. Because the rudder is usually located above the $X$ axis, a positive rudder deflection will create a positive rolling moment. Therefore, $C_{l_{\delta_r}}$ is usually positive in sign; however, it can be negative, depending on the particular airframe configuration and the angle of attack at which it is flying. The derivative $C_{l_{\delta_r}}$ is usually of only minor importance in the dynamic lateral control of conventional aircraft, and it is sometimes neglected.

The change in rolling moment coefficient with change in aileron deflection, $C_{l_{\delta_a}}$ is commonly referred to as the aileron effectiveness. According to the definition given above, left-aileron-down or right-aileron-up is a positive deflection. This produces a right-wing-down rolling moment that is positive; $C_{l_{\delta_a}}$ is therefore positive. As far as lateral dynamics are concerned, the derivative $C_{l_{\delta_a}}$ is the most important of the control surface derivatives. The aileron effectiveness in conjunction with the damping in roll ($C_{l_p}$) establishes the maximum available rate of roll of an airframe which is a very important consideration in fighter tactics at high speed. The aileron effectiveness is also very important in low speed flight during takeoffs and landings, where adequate lateral control is necessary to counteract asymmetric gusts tending to roll the aircraft.

The derivative $C_{n_{\delta_r}}$, the change in yawing moment coefficient with variation in rudder deflection, is usually referred to as the "rudder effectiveness" (or rudder power). When the rudder is deflected positively, i.e., to the left, a negative yawing moment is created on the airplane; hence the derivative $C_{n_{\delta_r}}$ is negative. The design value of $C_{n_{\delta_r}}$ is usually determined by considering such requirements as directional control for asymmetric power and crosswind takeoffs and landings, counteracting adverse yaw in rolling maneuvers, and spin recovery control.

Finally, $C_{n_{\delta_a}}$ is the change in yawing moment coefficient with change of aileron deflection. This derivative arises in part from the difference in drag due to the down-aileron compared to the drag of the up-aileron. Where such effects predominate, the sign of $C_{n_{\delta_a}}$ depends mainly on the rigging of the ailerons, their profile drag characteristics, and the angle of attack of the airframe. Aileron deflections can also produce side forces on the vertical tail, as discussed above in connection with $C_{y_{\delta_a}}$, and these can become important contributors to $C_{n_{\delta_a}}$. If negative, $C_{n_{\delta_a}}$ is called "adverse yaw" because it causes the airframe to yaw initially in a direction opposite to that desired by the pilot when he deflects the ailerons for a turn. If positive, it produces favorable or "proverse yaw" in the turning maneuver. Large values of either sign are undesirable for good lateral control qualities.

The longitudinal and lateral derivatives discussed above are listed in Tables 4-1 through 4-4. Appendix A contains dimensional derivative data for a variety of vehicles.

**Table 4-1. Longitudinal nondimensional stability derivatives (stability axis system).**[a]

| Basic nondimensional stability derivatives — Total airframe — Definitions | Unit | Theoretical aft horizontal tail contribution | Nondimensional stability derivative parameters |
|---|---|---|---|
| $C_D = \frac{\text{drag}}{qS}$ | $\frac{1}{1}$ | | |
| $C_{D_u} = \frac{U}{2}\frac{\partial C_D}{\partial U}$ | $\frac{1}{1}$ | | $x_u = (-C_D - C_{D_u})$ |
| $C_{D\alpha} = \frac{\partial C_D}{\partial \alpha}$ | $\frac{1}{\text{rad}}$ | | $x_w = \frac{1}{2}(C_L - C_{D_\alpha})$ |
| $C_{D\delta} = \frac{\partial C_D}{\partial \delta}$ | $\frac{1}{\text{rad}}$ | | $x_\delta = \frac{-1}{2} C_{D\delta}$ |
| $C_L = \frac{\text{lift}}{qS}$ | $\frac{1}{1}$ | | |
| $C_{L_u} = \frac{U}{2}\frac{\partial C_L}{\partial U}$ | $\frac{1}{1}$ | | $z_u = (-C_L - C_{L_u})$ |
| $C_{L\alpha} = \frac{\partial C_L}{\partial \alpha}$ | $\frac{1}{\text{rad}}$ | $C_{L\alpha_h}\frac{q_h}{q}\frac{S_h}{S}\left(1 - \frac{\partial \epsilon}{\partial \alpha}\right)$ | $z_w = \frac{1}{2}(-C_{L\alpha} - C_D)$ |
| $C_{L\dot{\alpha}} = \frac{\partial C_L}{\partial(\dot{\alpha}c/2U)}$ | $\frac{1}{\text{rad}}$ | $2C_{L\alpha_h}\frac{q_h}{q}\frac{S_h}{S}\frac{l_h}{c}\frac{\partial \epsilon}{\partial \alpha}$ | $z_{\dot{w}} = \frac{-1}{4} C_{L\dot{\alpha}}$ |
| $C_{L_q} = \frac{\partial C_L}{\partial(qc/2U)}$ | $\frac{1}{\text{rad}}$ | $2C_{L\alpha_h}\frac{q_h}{q}\frac{S_h}{S}\frac{l_h}{c}$ | $z_q = \frac{-1}{4} C_{L_q}$ |
| $C_{L\delta} = \frac{\partial C_L}{\partial \delta}$ | $\frac{1}{\text{rad}}$ | $C_{L\alpha_h}\frac{q_h}{q}\frac{S_h}{S}\frac{\partial \alpha_h}{\partial \delta}$ | $z_\delta = -\frac{1}{2} C_{L\delta}$ |
| $C_M = \frac{M}{qSc}$ | $\frac{1}{1}$ | | |
| $C_{M_u} = \frac{U}{2}\frac{\partial C_M}{\partial U}$ | $\frac{1}{1}$ | | $m_u = \frac{1}{2}\left(\frac{c}{k_y}\right)^2 (C_M + C_{M_u})$ |
| $C_{M\alpha} = \frac{\partial C_M}{\partial \alpha}$ | $\frac{1}{\text{rad}}$ | $-\frac{l_h}{c}[C_{L\alpha}]_h$ | $m_w = \frac{1}{2}\left(\frac{c}{k_y}\right)^2 C_{M\alpha}$ |
| $C_{M\dot{\alpha}} = \frac{\partial C_M}{\partial(\dot{\alpha}c/2U)}$ | $\frac{1}{\text{rad}}$ | $-\frac{l_h}{c}[C_{L\dot{\alpha}}]_h$ | $m_{\dot{w}} = \frac{1}{4}\left(\frac{c}{k_y}\right)^2 C_{M\dot{\alpha}}$ |
| $C_{M_q} = \frac{\partial C_M}{\partial(qc/2U)}$ | $\frac{1}{\text{rad}}$ | $-\frac{l_h}{c}[C_{L_q}]_h$ | $m_q = \frac{1}{4}\left(\frac{c}{k_y}\right)^2 C_{M_q}$ |
| $C_{M\delta} = \frac{\partial C_M}{\partial \delta}$ | $\frac{1}{\text{rad}}$ | $-\frac{l_\delta}{c} C_{L\delta}$ | $m_\delta = \frac{1}{2}\left(\frac{c}{k_y}\right)^2 C_{M\delta}$ |

[a] The symbol $q$, in addition to its normal use to designate pitching velocity, is also used in these tables to denote the dynamic pressure, $\rho U^2/2$, in accordance with long-established aeronautical practice. When particularized by the subscript $h$ (or $v$), it signifies the *local* dynamic pressure at the horizontal (or vertical) tail. The local flow angles relative to free stream conditions are denoted by $-\epsilon$ ($X$, $Z$ plane) and $-\sigma$ ($X$, $Y$ plane).

Table 4-2. Lateral nondimensional stability derivatives (stability or body axis systems).

| Basic nondimensional stability derivatives | | | Nondimensional stability derivative parameters |
|---|---|---|---|
| Total airframe | | Theoretical aft vertical tail contribution | |
| Definitions | Unit | | |
| $C_{y_\beta} = \frac{\partial C_y}{\partial \beta}$ | $\frac{1}{\text{rad}}$ | $-C_{L\alpha_v}\frac{q_v}{q}\frac{S_v}{S}\left(1 - \frac{\partial\sigma}{\partial\beta}\right)$ | $y_v = \frac{1}{2}C_{y_\beta}$ |
| $C_{y\dot{\beta}} = \frac{\partial C_y}{\partial(\dot{\beta}b/2U)}$ | $\frac{1}{\text{rad}}$ | Unknown | $y_{\dot{v}} = \frac{1}{4}C_{y_{\dot{\beta}}}$ |
| $C_{y_r} = \frac{\partial C_y}{\partial(rb/2U)}$ | $\frac{1}{\text{rad}}$ | $2C_{L\alpha_v}\frac{q_v}{q}\frac{S_v}{S}\frac{l_t}{b}\left(1 - \frac{U}{l_v}\frac{\partial\sigma}{\partial r}\right)$ | $y_r = \frac{1}{4}C_{y_r}$ |
| $C_{y_p} = \frac{\partial C_y}{\partial(pb/2U)}$ | $\frac{1}{\text{rad}}$ | $2C_{L\alpha_v}\frac{q_v}{q}\frac{S_v}{S}\frac{h_v}{b}\left(1 - \frac{U}{h_v}\frac{\partial\sigma}{\partial p}\right)$ | $y_p = \frac{1}{4}C_{y_p}$ |
| $C_{y_\delta} = \frac{\partial C_y}{\partial \delta}$ | $\frac{1}{\text{rad}}$ | $C_{L\alpha_v}\frac{q_v}{q}\frac{S_v}{S}\frac{\partial\alpha_v}{\partial\delta}$ | $y_\delta = \frac{1}{2}C_{y_\delta}$ |
| $C_{n_\beta} = \frac{\partial C_n}{\partial \beta}$ | $\frac{1}{\text{rad}}$ | $-\frac{l_v}{b}[C_{y_\beta}]_v$ | $n_v = \frac{1}{2}\left(\frac{b}{k_z}\right)^2 C_{n_\beta}$ |
| $C_{n\dot{\beta}} = \frac{\partial C_n}{\partial(\dot{\beta}b/2U)}$ | $\frac{1}{\text{rad}}$ | $-\frac{l_v}{b}[C_{y\dot{\beta}}]_v$ | $n_{\dot{v}} = \frac{1}{4}\left(\frac{b}{k_z}\right)^2 C_{n\dot{\beta}}$ |
| $C_{n_r} = \frac{\partial C_n}{\partial(rb/2U)}$ | $\frac{1}{\text{rad}}$ | $-\frac{l_v}{b}[C_{y_r}]_v$ | $n_r = \frac{1}{4}\left(\frac{b}{k_z}\right)^2 C_{n_r}$ |
| $C_{n_p} = \frac{\partial C_n}{\partial(pb/2U)}$ | $\frac{1}{\text{rad}}$ | $-\frac{l_v}{b}[C_{y_p}]_v$ | $n_p = \frac{1}{4}\left(\frac{b}{k_z}\right)^2 C_{n_p}$ |
| $C_{n_\delta} = \frac{\partial C_n}{\partial \delta}$ | $\frac{1}{\text{rad}}$ | $-\frac{l_{\delta_r}}{b}C_{y_\delta}$ | $n_\delta = \frac{1}{2}\left(\frac{b}{k_z}\right)^2 C_{n_\delta}$ |
| $C_{l_\beta} = \frac{\partial C_l}{\partial \beta}$ | $\frac{1}{\text{rad}}$ | $\frac{h_v}{b}[C_{y_\beta}]_v$ | $l_v = \frac{1}{2}\left(\frac{b}{k_x}\right)^2 C_{l_\beta}$ |
| $C_{l\dot{\beta}} = \frac{\partial C_l}{\partial(\dot{\beta}b/2U)}$ | $\frac{1}{\text{rad}}$ | $\frac{h_v}{b}[C_{y\dot{\beta}}]_v$ | $l_{\dot{v}} = \frac{1}{4}\left(\frac{b}{k_x}\right)^2 C_{l\dot{\beta}}$ |
| $C_{l_r} = \frac{\partial C_l}{\partial(rb/2U)}$ | $\frac{1}{\text{rad}}$ | $\frac{h_v}{b}[C_{y_r}]_v$ | $l_r = \frac{1}{4}\left(\frac{b}{k_x}\right)^2 C_l$ |
| $C_{l_p} = \frac{\partial C_l}{(pb/2U)}$ | $\frac{1}{\text{rad}}$ | $\frac{h_v}{b}[C_{y_p}]_v$ | $l_p = \frac{1}{4}\left(\frac{b}{k_x}\right)^2 C_{l_p}$ |
| $C_{l_\delta} = \frac{\partial C_l}{\partial \delta}$ | $\frac{1}{\text{rad}}$ | $\frac{h_{\delta_r}}{b}C_{y_\delta}$ | $l_\delta = \frac{1}{2}\left(\frac{b}{k_x}\right)^2 C_{l_\delta}$ |

Table 4-3. Longitudinal dimensional stability derivatives (stability axis system).

| Quantity | In terms of basic stability derivatives: Dimensional — Definitions | Dimensional — Unit | Nondimensional | In terms of nondimensional stability derivative parameters |
|---|---|---|---|---|
| $X_u$ | $\frac{1}{m}\frac{\partial X}{\partial u}$ | $\frac{1}{\text{sec}}$ | $\frac{\rho SU}{m}(-C_D - C_{D_u})^a$ | $\frac{1}{\tau}x_u{}^b$ |
| $X_w$ | $\frac{1}{m}\frac{\partial X}{\partial w}$ | $\frac{1}{\text{sec}}$ | $\frac{\rho SU}{2m}(C_L - C_{D_\alpha})$ | $\frac{1}{\tau}x_w$ |
| $X_\delta$ | $\frac{1}{m}\frac{\partial X}{\partial \delta}$ | $\frac{\text{ft}}{\text{rad-sec}^2}$ | $\frac{\rho SU^2}{2m}(-C_{D_\delta})$ | $\frac{U}{\tau}x_\delta$ |
| $Z_u$ | $\frac{1}{m}\frac{\partial Z}{\partial u}$ | $\frac{1}{\text{sec}}$ | $\frac{\rho SU}{m}(-C_L - C_{L_u})^c$ | $\frac{1}{\tau}z_u$ |
| $Z_w$ | $\frac{1}{m}\frac{\partial Z}{\partial w}$ | $\frac{1}{\text{sec}}$ | $\frac{\rho SU}{2m}(-C_{L_\alpha} - C_D)$ | $\frac{1}{\tau}z_w$ |
| $Z_{\dot{w}}$ | $\frac{1}{m}\frac{\partial Z}{\partial \dot{w}}$ | $\frac{1}{1}$ | $\frac{\rho Sc}{4m}(-C_{L_{\dot{\alpha}}})$ | $\frac{c}{\tau U}z_{\dot{w}}$ |
| $Z_q$ | $\frac{1}{m}\frac{\partial Z}{\partial q}$ | $\frac{\text{ft}}{\text{rad-sec}}$ | $\frac{\rho SUc}{4m}(-C_{L_q})$ | $\frac{c}{\tau}z_q$ |
| $Z_\delta$ | $\frac{1}{m}\frac{\partial Z}{\partial \delta}$ | $\frac{\text{ft}}{\text{rad-sec}^2}$ | $\frac{\rho SU^2}{2m}(-C_{L_\delta})$ | $\frac{U}{\tau}z_\delta$ |
| $M_u$ | $\frac{1}{I_y}\frac{\partial M}{\partial u}$ | $\frac{1}{\text{ft-sec}}$ | $\frac{\rho SUc}{I_y}(C_M + C_{M_u})$ | $\frac{2}{\tau c}m_u$ |
| $M_w$ | $\frac{1}{I_y}\frac{\partial M}{\partial w}$ | $\frac{1}{\text{ft-sec}}$ | $\frac{\rho SUc}{2I_y}C_{M_\alpha}$ | $\frac{1}{\tau c}m_w$ |
| $M_{\dot{w}}$ | $\frac{1}{I_y}\frac{\partial M}{\partial \dot{w}}$ | $\frac{1}{\text{ft}}$ | $\frac{\rho Sc^2}{4I_y}C_{M_{\dot{\alpha}}}$ | $\frac{1}{\tau U}m_{\dot{w}}$ |
| $M_q$ | $\frac{1}{I_y}\frac{\partial M}{\partial q}$ | $\frac{1}{\text{sec}}$ | $\frac{\rho SUc^2}{4I_y}C_{M_q}$ | $\frac{1}{\tau}m_q$ |
| $M_\delta$ | $\frac{1}{I_y}\frac{\partial M}{\partial \delta}$ | $\frac{1}{\text{rad-sec}^2}$ | $\frac{\rho SU^2c}{2I_y}C_{M_\delta}$ | $\frac{U}{\tau c}m_\delta$ |

[a] The thrust gradient terms are neglected here in the interests of symmetry and consistency.

[b] $\tau \equiv m/\rho US$ is the dimensionless time first proposed by H. Glauert, "A Non-dimensional Form of the Stability Equations of an Aeroplane," ARC R & M 1093, 1927.

[c] For $C_{L_u} = 0$, as in subsonic flight, and $C_L = W/(\rho U^2 S/2)$, as in trimmed flight for $\gamma_0 = 0$, $Z_u = -2g/U_0$.

**Table 4-4.** Lateral dimensional stability derivative parameters (stability or body axis systems).

| Quantity[a] | In terms of basic stability derivatives: Dimensional — Definitions | Dimensional — Unit | Nondimensional | In terms of nondimensional stability derivative parameters |
|---|---|---|---|---|
| $Y_v$ | $\frac{1}{mU}\frac{\partial Y}{\partial \beta}$ | $\frac{1}{\text{sec}}$ | $\frac{\rho SU}{2m}C_{y_\beta}$ | $\frac{1}{\tau}y_v$ |
| $Y_{\dot{v}}$ | $\frac{1}{mU}\frac{\partial Y}{\partial \dot{\beta}}$ | $\frac{1}{1}$ | $\frac{\rho Sb}{4m}C_{y_{\dot{\beta}}}$ | $\frac{b}{\tau U}y_{\dot{v}}$ |
| $Y_r^*$ | $\frac{1}{mU}\frac{\partial Y}{\partial r}$ | $\frac{1}{\text{rad}}$ | $\frac{\rho Sb}{4m}C_{y_r}$ | $\frac{b}{\tau U}y_r$ |
| $Y_p^*$ | $\frac{1}{mU}\frac{\partial Y}{\partial p}$ | $\frac{1}{\text{rad}}$ | $\frac{\rho Sb}{4m}C_{y_p}$ | $\frac{b}{\tau U}y_p$ |
| $Y_\delta^*$ | $\frac{1}{mU}\frac{\partial Y}{\partial \delta}$ | $\frac{1}{\text{rad-sec}}$ | $\frac{\rho SU}{2m}C_{y_\delta}$ | $\frac{1}{\tau}y_\delta$ |
| $N_\beta$ | $\frac{1}{I_z}\frac{\partial N}{\partial \beta}$ | $\frac{1}{\text{sec}^2}$ | $\frac{\rho SU^2b}{2I_z}C_{n_\beta}$ | $\frac{U}{\tau b}n_v$ |
| $N_{\dot{\beta}}$ | $\frac{1}{I_z}\frac{\partial N}{\partial \dot{\beta}}$ | $\frac{1}{\text{sec}}$ | $\frac{\rho SUb^2}{4I_z}C_{n_{\dot{\beta}}}$ | $\frac{1}{\tau}n_v$ |
| $N_r$ | $\frac{1}{I_z}\frac{\partial N}{\partial r}$ | $\frac{1}{\text{sec}}$ | $\frac{\rho SUb^2}{4I_z}C_{n_r}$ | $\frac{1}{\tau}n_r$ |
| $N_p$ | $\frac{1}{I_z}\frac{\partial N}{\partial p}$ | $\frac{1}{\text{sec}}$ | $\frac{\rho SUb^2}{4I_z}C_{n_p}$ | $\frac{1}{\tau}n_p$ |
| $N_\delta$ | $\frac{1}{I_z}\frac{\partial N}{\partial \delta}$ | $\frac{1}{\text{rad-sec}^2}$ | $\frac{\rho SU^2b}{2I_z}C_{n_\delta}$ | $\frac{U}{\tau b}n_\delta$ |
| $L_\beta$ | $\frac{1}{I_x}\frac{\partial L}{\partial \beta}$ | $\frac{1}{\text{sec}^2}$ | $\frac{\rho SU^2b}{2I_x}C_{l_\beta}$ | $\frac{U}{\tau b}l_v$ |
| $L_{\dot{\beta}}$ | $\frac{1}{I_x}\frac{\partial L}{\partial \dot{\beta}}$ | $\frac{1}{\text{sec}}$ | $\frac{\rho SUb^2}{4I_x}C_{l_{\dot{\beta}}}$ | $\frac{1}{\tau}l_{\dot{v}}$ |
| $L_r$ | $\frac{1}{I_x}\frac{\partial L}{\partial r}$ | $\frac{1}{\text{sec}}$ | $\frac{\rho SUb^2}{4I_x}C_{l_r}$ | $\frac{1}{\tau}l_r$ |
| $L_p$ | $\frac{1}{I_x}\frac{\partial L}{\partial p}$ | $\frac{1}{\text{sec}}$ | $\frac{\rho SUb^2}{4I_x}C_{l_p}$ | $\frac{1}{\tau}l_p$ |
| $L_\delta$ | $\frac{1}{I_x}\frac{\partial L}{\partial \delta}$ | $\frac{1}{\text{rad-sec}^2}$ | $\frac{\rho SU^2b}{2I_x}C_{l_\delta}$ | $\frac{U}{\tau b}l_\delta$ |

[a] The starred derivatives arise when $\beta$ rather than $v$ is used as the lateral motion parameter (see Chapter 6); in general, $Y_\lambda^* = Y_\lambda/U_0$.

CHAPTER 5

# *LONGITUDINAL DYNAMICS*

## 5-1. Introduction

The vehicle dynamic properties, defined in general by the equations of motion derived in the last chapter, are best specified for use in control system analysis by a series of transfer functions that relate output quantities (various airframe motions) to input variables (usually control motions or external disturbances). These transfer functions are readily obtained from the linearized Laplace-transformed airframe equations of motion as sets of ratios between transformed airframe output and input quantities or initial conditions. The ratios comprise numerators and denominators expressed as rational polynomials in the Laplace transform variable, $s$. The various polynomial coefficients are composed of combinations of stability derivatives and inertial gravitational quantities.

For maximum utility it is desirable to have the numerator and denominator polynomials in factored form. Each transfer function is then made up of a ratio of first- and second-order polynomial products and a gain. The gains, poles, and zeros thus obtained define the fundamental properties of dynamic elements and are essential in most servo analysis and synthesis methods and in response calculations. Even though most of the polynomials involved are of third degree or higher, numerical factorization is no problem, especially when digital computers are employed. However, unless a large number of cases are so computed, the specific connections between vehicle configuration (represented by stability derivatives, etc.) and transfer and response characteristics (represented by poles and zeros) are difficult to appreciate. Such an appreciation is important in the following.

1. Developing the insight required for the determination of airframe/automatic-control combinations that offer possible improvements in over-all system complexity.
2. Assessing the effects of configuration changes on aircraft response and on airframe/autopilot/pilot system characteristics.
3. Showing the detailed effects of particular stability derivatives (and their estimated accuracies) on the poles and zeros, and hence on aircraft and airframe/autopilot/pilot characteristics.
4. Obtaining stability derivatives from flight test data.

It is highly desirable, therefore, to express the locations of the poles and zeros directly in terms of the literal stability derivatives and inertial parameters. This can be accomplished directly by decomposing a fairly complete transfer function form, or indirectly by the use of fewer airframe degrees of freedom. In either case an approximation is required to arrive at reasonably compact and usable expressions that delineate dominant, as opposed to trivial, effects. Such effects can vary among vehicle types, so it is to be expected that literal approximate factors which apply to all vehicles for all flight conditions will be an exception, rather than the rule.

With this background the immediately succeeding articles in this chapter will: first, recapitulate the longitudinal equations of motion as normally used; present the polynomial forms of the most significant control-input transfer functions; and develop some appreciation for the transfer functions and responses with two specific numerical examples. Based on the physical insights provided by the examples, further simplified sets of the equations that apply to the individual modes of motion will be developed. Following this, the complete gust-input transfer functions in polynomial form are presented and our numerical examples are extended to cover these transfer functions.

The basic understanding accruing from the above process is further developed by considering the approximate literal expressions for the various transfer function factors. These approximate factors are shown to be related to the simplified equations of motion appropriate for each mode, and their implications as regards the direct influence of the dominant stability derivatives on the important poles and zeros are discussed. In the final article a discussion of the modal response ratios in literal form is presented.

## 5-2. Recapitulation and Further Simplification of the Longitudinal Equations of Motion

Equations 4-97 are somewhat more complicated than those generally used for transfer function computations; to simplify them further, we make the following two assumptions.

Assumption 10. It is assumed that $X_{\dot{w}} = X_q = Z_{\dot{w}} = Z_q = 0$.

Perhaps the best general evidence in justification of Assumption 10 is that the derivatives named in it rarely appear in the technical literature concerned with aircraft dynamics. The inference here is that, although individual investigators have evaluated the effects of these derivatives for a multitude of various airframe configurations, they have found them to be of only secondary importance. However, it must be remembered that, if any of these derivatives are of actual importance for a particular airframe, this assumption may produce somewhat erroneous quantitative results for that airframe.

In general, any stability derivative may be neglected if it is first determined that the term containing the given derivative is small in comparison with other terms in the same equations. For the derivatives in question, comparing the term $X_{\dot{w}}sw$ with $X_w w$ shows that, if the frequency range of interest extends as high as $|s| \doteq X_w/X_{\dot{w}}$, $X_{\dot{w}}$ can no longer be neglected a priori. Similarly, the upper frequency limit for the valid a priori neglect of $X_q$ is $|s| \doteq |g/X_q|$. As $Z_q$ occurs in the group $(Z_q + U_0)sw$, the appropriate criterion here is $|Z_q| \ll U_0$; by grouping $\dot{w}$ terms together as $(s - Z_{\dot{w}}s)w$ the criterion for neglecting $Z_{\dot{w}}$ is $|Z_{\dot{w}}| \ll 1$.

Assumption 11. In the steady flight condition, the flight path of the airplane is assumed to be horizontal, $\gamma_0 = 0$.

Assumption 11 is introduced solely to simplify the mechanics of the analysis. When the flight path of an airplane is initially inclined to the horizontal, $\gamma_0$ must of course be included in the transfer functions.

On the basis of these assumptions, the longitudinal equations of motion, referenced to *stability axes*, become

$$
\begin{aligned}
(s - X_u)u - X_w w + g\theta &= X_\delta\delta - X_u u_g - X_w w_g \\
-Z_u u + (s - Z_w)w - U_0 s\theta &= Z_\delta\delta - Z_u u_g - Z_w w_g \qquad (5\text{-}1) \\
-M_u u - (M_{\dot{w}}s + M_w)w + s(s - M_q)\theta &= M_\delta\delta - M_u u_g \\
&\quad - [(M_{\dot{w}} - M_q/U_0)s + M_w]w_g
\end{aligned}
$$

where (Eq. 4-96) $\qquad s\theta = q$

and $\begin{pmatrix}\text{inertial terms of } Z \\ \text{equation of Eq. 5-1}\end{pmatrix}$ $\quad a_z = \dot{w} - U_0 s\theta = -U_0(\dot{\gamma}) = -\ddot{h} \qquad (5\text{-}2)$

The auxiliary relationships of Eq. 5-2 are needed to convert the motion variables of Eq. 5-1 to the quantities sensed by flight instruments such as rate gyros, accelerometers, and altimeters.

## 5-3. Control-Input Transfer Functions

As fully discussed in Chapter 2, the transfer functions, for a given input, are obtained simply by solving the transformed simultaneous equations of motion for the output variable of interest with all other inputs considered to be zero. For example, using determinants, we can directly write the attitude to control input transfer function (i.e., neglecting gust inputs in Eq. 5-1) as

$$\frac{\theta(s)}{\delta(s)} = \frac{\begin{vmatrix} s - X_u & -X_w & X_\delta \\ -Z_u & s - Z_w & Z_\delta \\ -M_u & -(M_{\dot{w}}s + M_w) & M_\delta \end{vmatrix}}{\begin{vmatrix} s - X_u & -X_w & g \\ -Z_u & s - Z_w & -U_0 s \\ -M_u & -(M_{\dot{w}}s + M_w) & s(s - M_q) \end{vmatrix}} = \frac{N_\delta^\theta(s)}{\Delta(s)} \tag{5-3}$$

By expanding the determinants, the transfer function can be expressed as the ratio of a numerator polynomial in $s$ over a denominator polynomial. The denominator polynomial, $\Delta(s)$, is common to all the transfer functions and its factors determine the frequency and damping, or time constants, of the individual *modes* of motion. The numerator polynomials depend on the output quantity of interest. The general polynomial forms of the primary longitudinal transfer functions are given below, together with the most usual factored forms of the polynomial expressions.

$$\begin{aligned} \frac{\theta(s)}{\delta(s)} &= \frac{N_\delta^\theta(s)}{\Delta_{\text{long}}} = \frac{A_\theta s^2 + B_\theta s + C_\theta}{\Delta_{\text{long}}} \\ &= \frac{A_\theta(s + 1/T_{\theta_1})(s + 1/T_{\theta_2})}{\Delta_{\text{long}}} \end{aligned} \tag{5-4}$$

$$\begin{aligned} \frac{w(s)}{\delta(s)} &= \frac{N_\delta^w(s)}{\Delta_{\text{long}}} = \frac{A_w s^3 + B_w s^2 + C_w s + D_w}{\Delta_{\text{long}}} \\ &= \frac{A_w \begin{cases} (s + 1/T_{w_1})(s^2 + 2\zeta_w \omega_w s + \omega_w^2) \\ \text{or} \\ (s + 1/T_{w_1})(s + 1/T_{w_2})(s + 1/T_{w_3}) \end{cases}}{\Delta_{\text{long}}} \end{aligned} \tag{5-5}$$

$$\begin{aligned} \frac{u(s)}{\delta(s)} &= \frac{N_\delta^u(s)}{\Delta_{\text{long}}} = \frac{A_u s^3 + B_u s^2 + C_u s + D_u}{\Delta_{\text{long}}} \\ &= \frac{A_u \begin{cases} (s + 1/T_{u_1})(s + 1/T_{u_2})(s + 1/T_{u_3}) \\ \text{or} \\ (s + 1/T_{u_1})(s^2 + 2\zeta_u \omega_u s + \omega_u^2) \end{cases}}{\Delta_{\text{long}}} \end{aligned} \tag{5-6}$$

$$\begin{aligned} \frac{h(s)}{\delta(s)} &= \frac{N_\delta^h(s)}{\Delta_{\text{long}}} = \frac{A_h s^3 + B_h s^2 + C_h s + D_h}{s\,\Delta_{\text{long}}} \\ &= \frac{A_h \begin{cases} (s + 1/T_{h_1})(s + 1/T_{h_2})(s + 1/T_{h_3}) \\ \text{or} \\ (s + 1/T_{h_1})(s^2 + 2\zeta_h \omega_h s + \omega_h^2) \end{cases}}{s\,\Delta_{\text{long}}} \end{aligned} \tag{5-7}$$

$$\frac{a_z(s)}{\delta(s)} = \frac{-s^2 N_\delta^h(s)}{\Delta_{\text{long}}} \tag{5-8}$$

Table 5-1. Longitudinal control-input transfer function coefficients.

| | $A$ | $B$ | $C$ | $D$ | $E$ |
|---|---|---|---|---|---|
| $\Delta$ | $1$ | $-M_q - M_{\dot\alpha} - Z_w - X_u$ | $Z_w M_q - M_\alpha - X_w Z_u$ <br> $+ X_u(M_q + M_{\dot\alpha} + Z_w)$ | $-X_u(Z_w M_q - M_\alpha)$ <br> $+ Z_u(X_w M_q + gM_{\dot w})$ <br> $-M_u(X_\alpha - g)$ | $g(Z_u M_w - M_u Z_w)$ |
| $N_\delta^\theta$ | $M_\delta + Z_\delta M_{\dot w}$ | $X_\delta(Z_u M_{\dot w} + M_u)$ <br> $+ Z_\delta(M_w - X_u M_{\dot w})$ <br> $- M_\delta(X_u + Z_w)$ | $X_\delta(Z_u M_w - Z_w M_u)$ <br> $+ Z_\delta(M_u X_w - M_w X_u)$ <br> $+ M_\delta(Z_w X_u - X_w Z_u)$ | | |
| $N_\delta^w$ | $Z_\delta$ | $X_\delta Z_u - Z_\delta(X_u + M_q) + M_\delta U_0$ | $X_\delta(U_0 M_u - Z_u M_q)$ <br> $+ Z_\delta X_u M_q - U_0 M_\delta X_u$ | $g(Z_\delta M_u - M_\delta Z_u)$ | |
| $N_\delta^u$ | $X_\delta$ | $-X_\delta(Z_w + M_q + M_{\dot\alpha}) + Z_\delta X_w$ | $X_\delta(Z_w M_q - M_\alpha)$ <br> $- Z_\delta(X_w M_q + gM_{\dot w})$ <br> $+ M_\delta(X_\alpha - g)$ | $g(M_\delta Z_w - Z_\delta M_w)$ | |
| $sN_\delta^h$ | $-Z_\delta$ | $-X_\delta Z_u + Z_\delta(M_q + M_{\dot\alpha} + X_u)$ | $X_\delta Z_u(M_q + M_{\dot\alpha})$ <br> $- Z_\delta[X_u(M_q + M_{\dot\alpha}) - M_\alpha]$ <br> $- M_\delta Z_\alpha$ | $-X_\delta(Z_\alpha M_u - M_\alpha Z_u)$ <br> $+ Z_\delta[M_u(X_\alpha - g) - M_\alpha X_u]$ <br> $+ M_\delta[Z_\alpha X_u - Z_u(X_\alpha - g)]$ | |

where

$$\Delta_{\text{long}} = As^4 + Bs^3 + Cs^2 + Ds + E$$

$$= \begin{matrix} (s^2 + 2\zeta_p\omega_p s + \omega_p^2) & (s^2 + 2\zeta_{sp}\omega_{sp}s + \omega_{sp}^2) \\ \text{or} & \text{or} \\ (s + 1/T_{p_1})(s + 1/T_{p_2}) & (s + 1/T_{sp_1})(s + 1/T_{sp_2}) \end{matrix} \qquad (5\text{-}9)$$

The literal expressions for the various $A$, $B$, $C$, etc., coefficients of Eqs. 5-4 through 5-9 are given in Table 5-1 in terms of the stability derivatives. In many cases $\alpha$ derivatives ($w$ derivatives multiplied by $U_0$) are used to achieve conciseness.

## 5-4. Example Transfer Functions, Bode Forms, and Time Responses for a Conventional Airplane

To develop an appreciation for the usual responses and transfer function forms, we shall study some numerical examples. Accordingly, we consider first a conventional airplane having the following characteristics and the general arrangement shown in Fig. 5-1. The substitution of these

| | |
|---|---|
| Altitude (ft) | 20,000 |
| Weight (lb) | 30,500 |
| Mach number | 0.638 |
| True airspeed (ft/sec) | 660 |
| $X_u$ | −0.0097 |
| $X_w$ | 0.0016 |
| $X_{\delta_e}$ | 0.0 |
| $Z_u$ | −0.0955 |
| $Z_w$ | −1.430 |
| $Z_{\delta_e}$ | −69.8 |
| $M_u$ | 0.0 |
| $M_w$ | −0.0235 |
| $M_{\dot{w}}$ | −0.0013 |
| $M_q$ | −1.920 |
| $M_{\delta_e}$ | −26.10 |

data into the Table 5-1 forms and routine factoring of the resulting polynomials yields the following transfer functions (the denominator, shown only for $\theta$ is common to all the transfer functions as indicated):

$$\frac{\theta(s)}{-\delta_e(s)} = 4.85 \frac{[(s/0.0098) + 1][(s/1.371) + 1]}{\{[s^2/(0.0630)^2] + [2(0.0714)s/0.0630] + 1\} \times \{[s^2/(4.27)^2] + [2(0.493)s/4.27] + 1\}} \qquad (5\text{-}10)$$

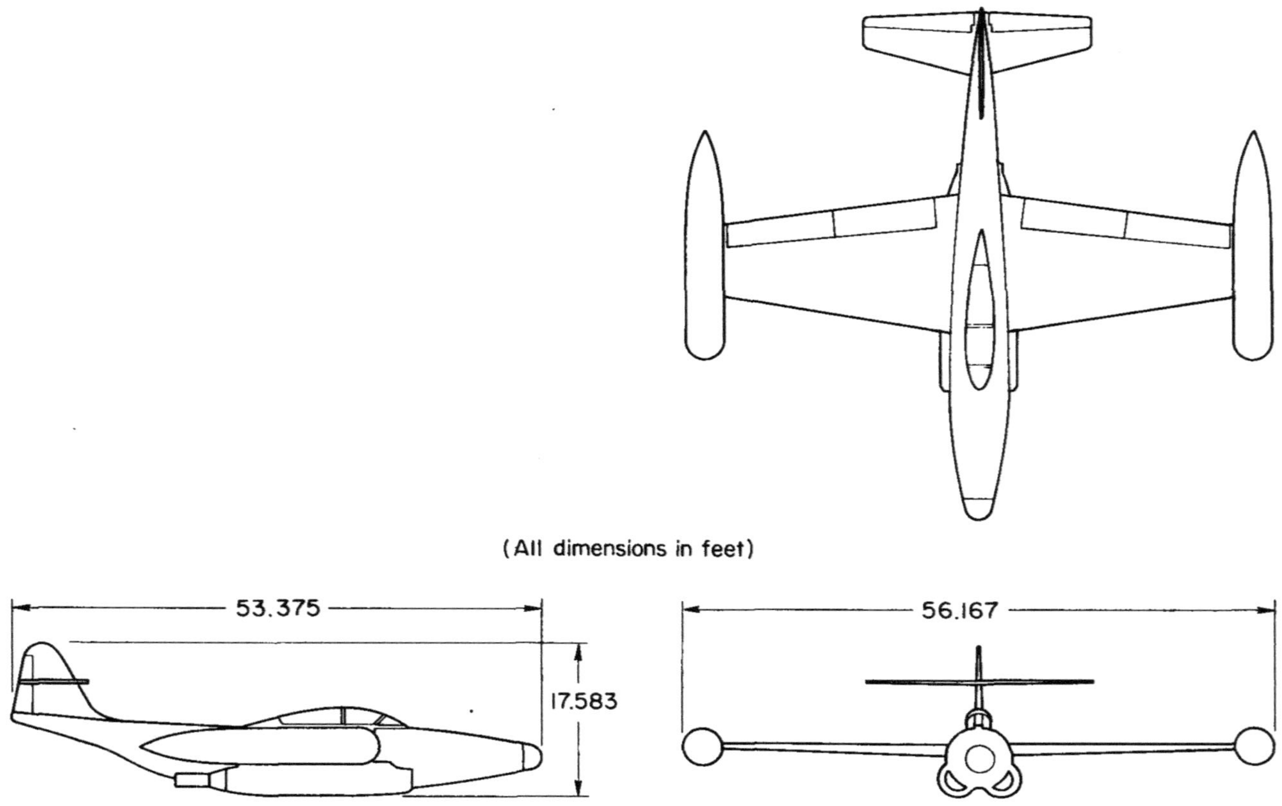


Fig. 5-1. Three-view of conventional airplane used for the numerical example.

$$\frac{\alpha(s)}{-\delta_e(s)} = \frac{-1}{U_0}\frac{w(s)}{\delta_e(s)}$$

$$= \frac{1110}{660}\frac{[(s/248.5) + 1]\{[s^2/(0.068)^2] + [2(0.0713)s/0.068] + 1\}}{\{\qquad\qquad\}\{\qquad\qquad\}} \tag{5-11}$$

$$\frac{u(s)}{-\delta_e(s)} = -15{,}920\frac{[(s/1.44) + 1][(s/-7251.4) + 1]}{\{\qquad\}\{\qquad\}} \tag{5-12}$$

$$\frac{a_z(s)}{-\delta_e(s)} = 2080\frac{s[(s/0.0064) + 1][(s/17.03) - 1][(s/19.81) + 1]}{\{\qquad\qquad\}\{\qquad\qquad\}} \tag{5-13}$$

Inspection of the roots of the denominator which, set to zero, is the characteristic equation (commonly called the "longitudinal stability quartic") shows that the free longitudinal motions consist of two oscillatory modes. One of these is a relatively well-damped, high frequency oscillation called the "short period," and the other is a lightly damped relatively low frequency oscillation called the "phugoid." The subscript notation employed in Eq. 5-9 reflects this nomenclature.

Figure 5-2 contains $j\omega$-Bode plots of the above transfer functions, including amplitude ratio and phase asymptotes. Among other things, the Bode diagrams are graphical representations of transforms of the weighting functions. In this view, they amount to response transforms for unit impulse inputs and can be used to draw conclusions concerning the appearance of the phugoid and short period modes in the transient response of the airplane. For example, Fig. 5-2(c) shows that the amplitude ratio, $|u/\delta_e|$, is much smaller at the natural frequency of the short period than at that of the phugoid. This indicates that much smaller airspeed changes occur in the short period transient mode than in the phugoid transient oscillation. Figure 5-2(b) shows that the quadratic in the numerator of the $\alpha/\delta_e$ transfer function very nearly cancels the denominator quadratic corresponding to the phugoid oscillation. Consequently, we expect almost no change in the angle of attack during the phugoid oscillation. A comparison between Figs. 5-2(a) and (c) shows that the values of the amplitude ratio, $|\theta/\delta_e|$, at the short period and phugoid frequencies are more nearly the same magnitude than those of $|u/\delta_e|$ at the same frequencies. This implies that, for the same inputs, the amplitudes of $\theta$ occurring in the characteristic modes are more similar than those of $u$. Finally, Fig. 5-2(d) shows that the vertical acceleration amplitudes at phugoid are only somewhat higher than those at the short period. However, considering the large differences in frequency, a given oscillatory vertical acceleration will involve much higher excursions in $h$ and $\dot{h}$ at phugoid than at short period frequency. This can also be visualized in

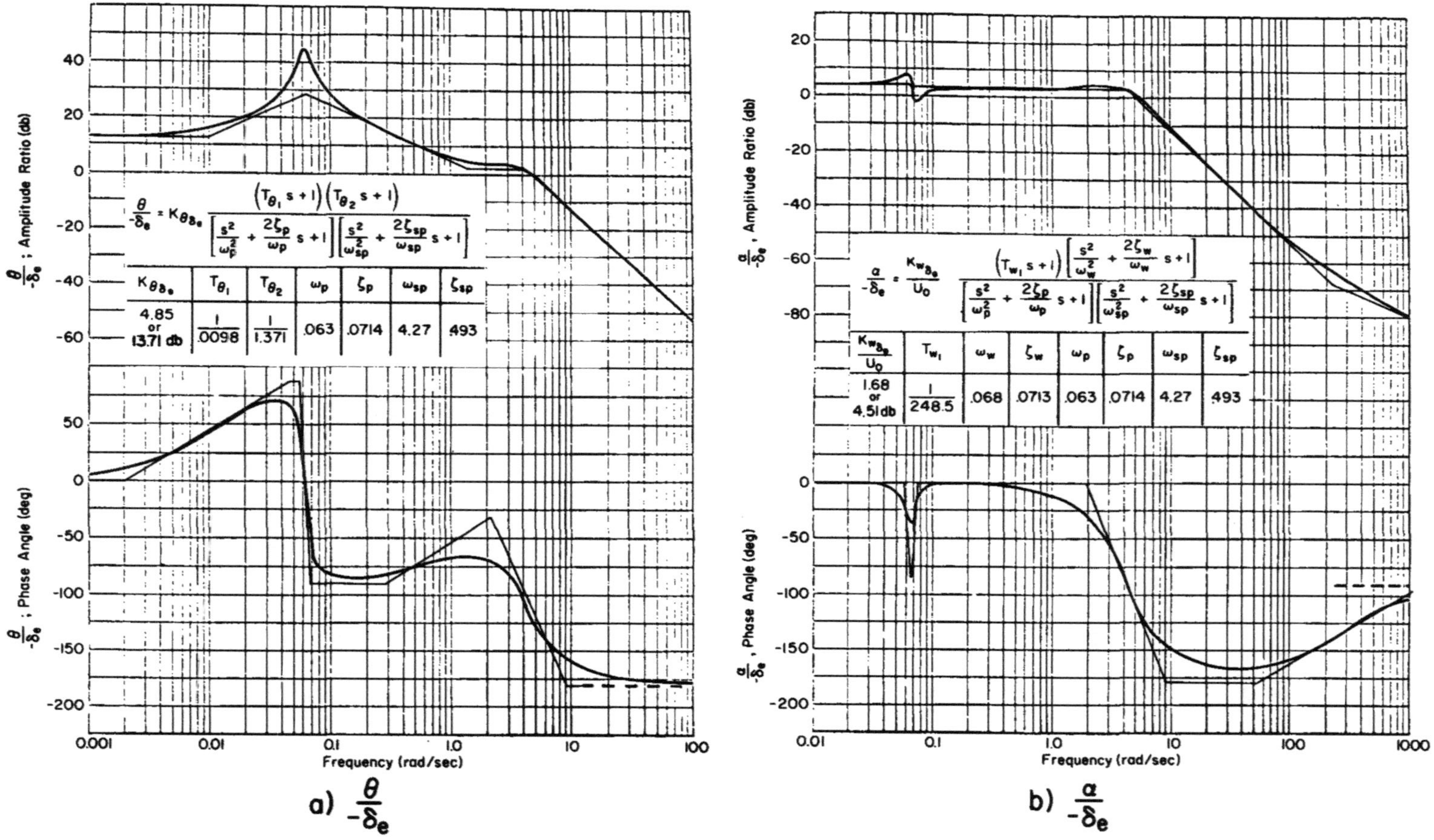


Fig. 5-2. Examples of control-input Bode plots for a conventional airplane.

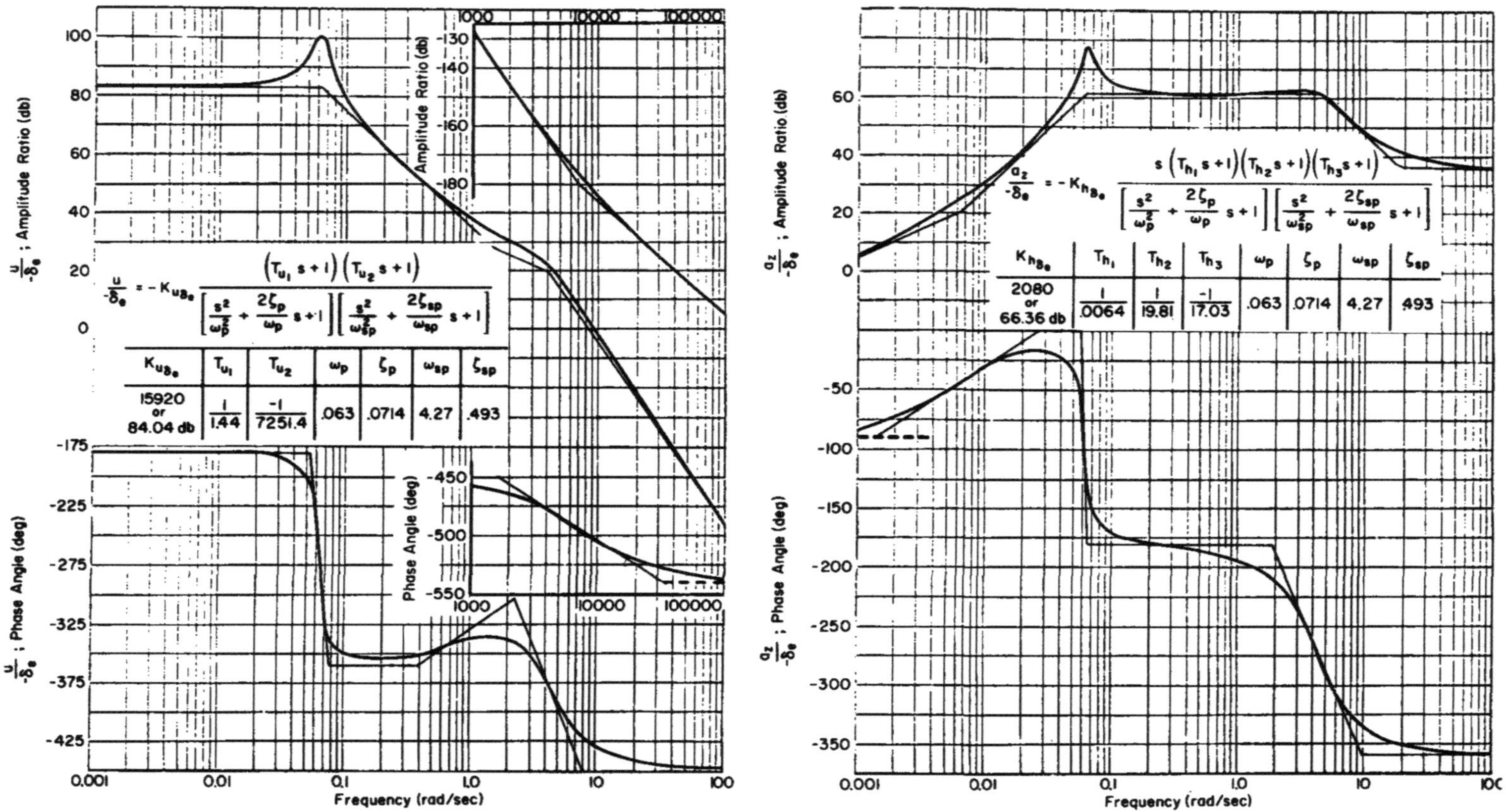


c) $\frac{u}{-\delta_e}$ d) $\frac{a_z}{-\delta_e}$

Fig. 5-2 (Continued)

Fig. 5-2(d) by considering that successive integrations of $a_z$ to obtain $\dot{h}$ and $h$ involve successive clockwise rotations of the $|a_z/\delta_e|$ Bode, each of 20 dB/decade. Such rotations progressively suppress the short period hump relative to the phugoid peak.

In summary, it appears from our study of Fig. 5-2 that only relatively small amplitudes of $u$ and $h$ occur in the short period mode and of $\alpha$ in the phugoid mode, whereas large amplitudes of $\theta$ can occur in both.

The response time history of the example airplane to an elevator pulse (the weighting functions themselves rather than their transforms) supports these conclusions. These are given in Fig. 5-3 using two time scales; that

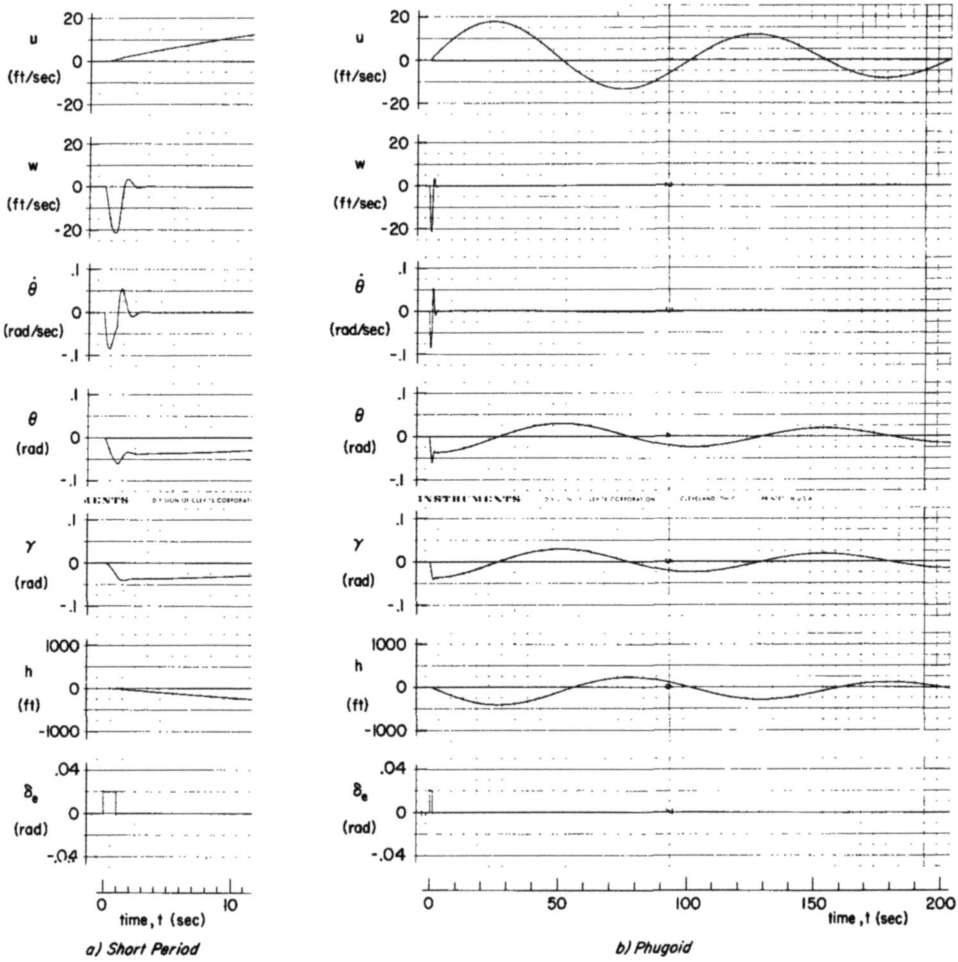


Fig. 5-3. Analog computer record of time history for pulse elevator deflection, conventional airplane.

of Fig. 5-3(a) emphasizes the short period, while that of Fig. 5-3(b) shows the phugoid best. Here we see that the maximum amplitudes of $u$ and $h$ are very much smaller in the short period than in the phugoid and that the maximum amplitude of $w$ ($\doteq U_0\alpha$) during the phugoid is very nearly zero. Further, the maximum amplitudes of $\theta$ in each mode are comparable in magnitude. All of these facts are in agreement with what we inferred from the Bode plots.

Another way of studying the relative motions is to draw the time vectors and force and moment polygons (Chapter 2) as we have done in Fig. 5-4. Note the relative magnitudes and phases of the component motions for each mode as shown by the time vectors and compare these with the appropriate time responses. The fore and aft perturbation from the undisturbed flight path ($\int u\,dt$) is negligible for the short period mode, but for the phugoid it is of the same order as the height perturbation. The time vector polygons indicate the relative importance of each term in the equation of motion for each mode; e.g., compare the $s^2\theta$ inertial pitch acceleration time vector for the phugoid and short period modes.

## 5-5. Two Degree of Freedom, Short Period Approximations

Using the foregoing observations of the detailed nature of the responses, we can construct a simplified set of equations applicable specifically to short period motions. This approximation involves setting the variation in forward velocity, $u$, equal to zero and deleting the first relation of Eq. 5-1. This reflects the previous statements that $u$ is of relatively small amplitude in the short period mode and that the two degrees of freedom, $\alpha$ (i.e., $w$) and $\theta$, are dominant; accordingly, with $u_g = 0$,

$$(s - Z_w)w - U_0 s\theta = Z_\delta \delta - Z_w w_g \tag{5-14}$$

$$-(sM_{\dot{w}} + M_w)w + (s - M_q)s\theta = M_\delta \delta - [(M_{\dot{w}} - M_q/U_0)s + M_w]w_g$$

Solving this system of equations with $w_g = 0$ leads to the transfer functions

$$\frac{\alpha(s)}{\delta(s)} = \frac{1}{U_0}\frac{w(s)}{\delta(s)}$$

$$= \frac{1}{U_0}\frac{Z_\delta s + (U_0 M_\delta - Z_\delta M_q)}{[s^2 - (U_0 M_{\dot{w}} + Z_w + M_q)s + (M_q Z_w - U_0 M_w)]} \tag{5-15}$$

$$\frac{\theta(s)}{\delta(s)} = \frac{(M_\delta + Z_\delta M_{\dot{w}})s + (Z_\delta M_w - M_\delta Z_w)}{s[s^2 - (U_0 M_{\dot{w}} + Z_w + M_q)s + (M_q Z_w - U_0 M_w)]} \tag{5-16}$$

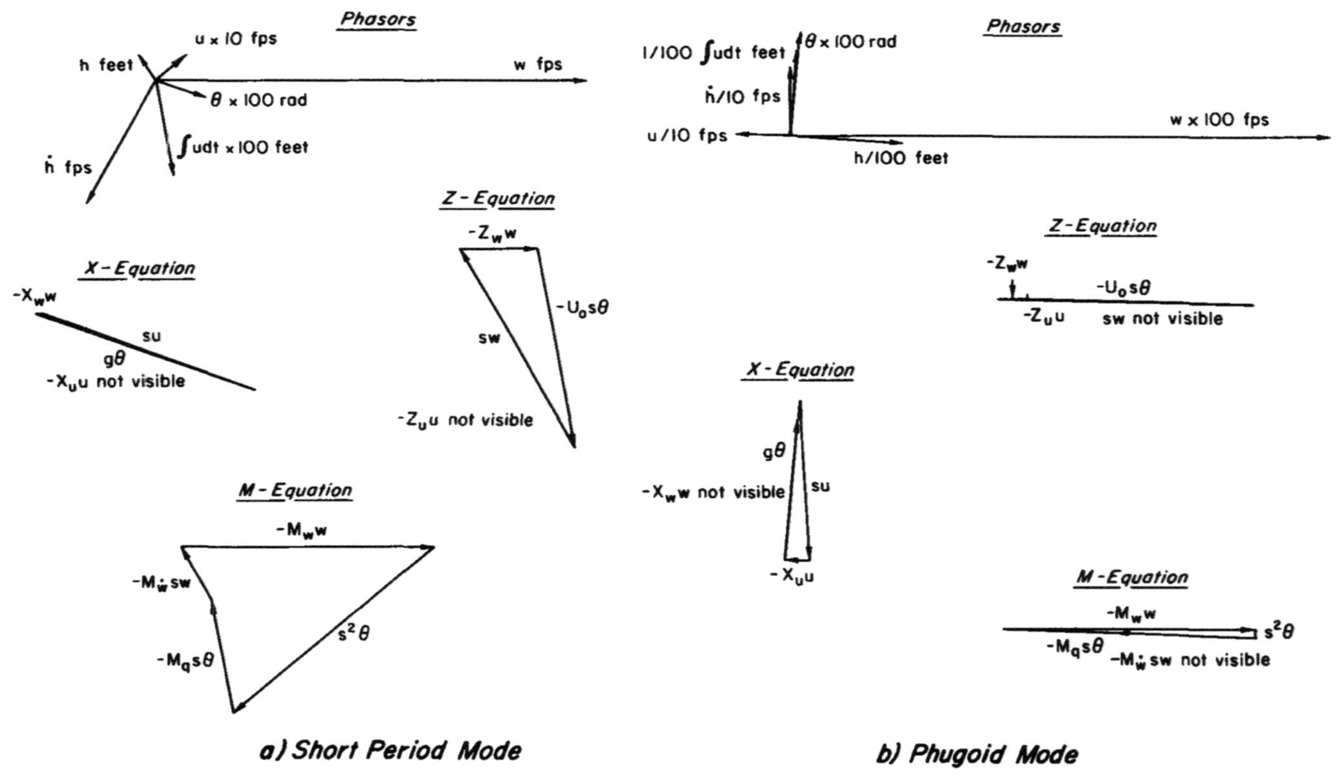


Fig. 5-4. Time vector diagrams for the example airplane.

The quadratic term in the denominator is of the form $s^2 + 2\zeta_{sp}\omega_{sp}s + \omega_{sp}^2$, with

$$\omega_{sp} = \sqrt{M_q Z_w - M_\alpha}$$
$$2\zeta_{sp}\omega_{sp} = -(Z_w + M_q + M_{\dot{\alpha}})$$

The evaluation of Eqs. 5-15 and 5-16 for the sample numerical values of the stability derivatives yields

$$\frac{\alpha(s)}{-\delta_e(s)} = 1.44 \frac{(s/248.5) + 1}{\{[s^2/(4.27)^2] + [2(0.493)s/4.27] + 1\}} \tag{5-18}$$

and

$$\frac{\theta(s)}{-\delta_e(s)} = 1.955 \frac{(s/1.371) + 1}{s\{\qquad\}} \tag{5-19}$$

These results are compared with Eqs. 5-10 and 5-11, obtained from the complete equations of motion, in Fig. 5-5. We see that for frequencies above the phugoid the two degree of freedom, short period transfer functions are very good approximations in both amplitude and phase. Furthermore, the two- and three degree of freedom time responses in $w$ and $\theta$ are in excellent agreement for times shorter than about 10 sec.

In summary, the two degree of freedom solution of the pitching moment and vertical force equations of motion is a suitable approximation to the short period mode. For typical flight conditions, the short period mode can be considered to consist of changes only in the angle of attack and in the angle of pitch; the short period motion occurs before there is any appreciable change in forward speed. We shall later examine the limiting conditions for which this typical behavior is still reasonably accurate.

## 5-6. Three Degree of Freedom Phugoid Approximations

Characteristically, the phugoid components of vehicle motions are very slow compared to the short period components. In an approximate sense the phugoid mode describes the long-term translatory motions of the vehicle center of mass, whereas the short period describes rotations about the center of mass. For practical purposes in the phugoid motion, the "dynamic" pitching, inertial, and damping moments are small compared to the "static" pitching moment changes with speed and angle of attack. Classically, with $M_u = 0$, the phugoid motion was conceived to involve fairly large oscillatory changes in forward speed, pitch attitude, and altitude, with approximately constant angle of attack; the static stability, $M_\alpha$, being sufficiently large to decouple the phugoid and short periods and to maintain $\alpha$ perturbations small. This is, of course, the case in the numerical example, where the $w$ time vector is very small indeed [Fig. 5-4(b)].

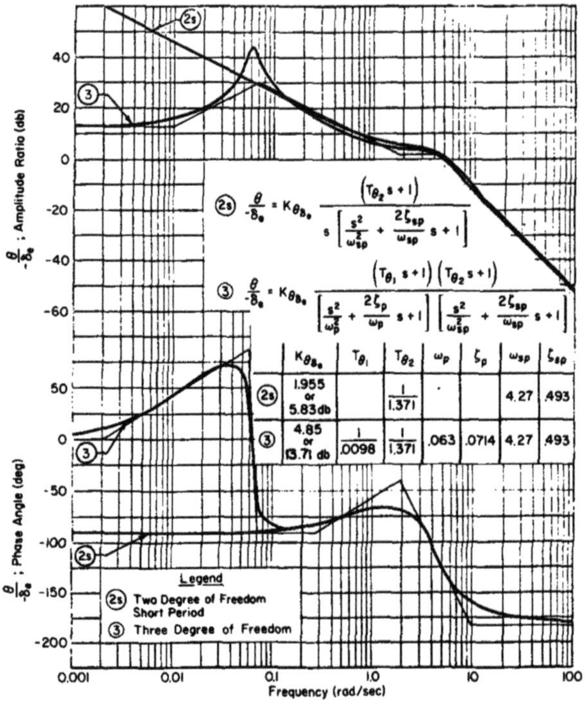


a) $\frac{\theta}{-\delta_e}$

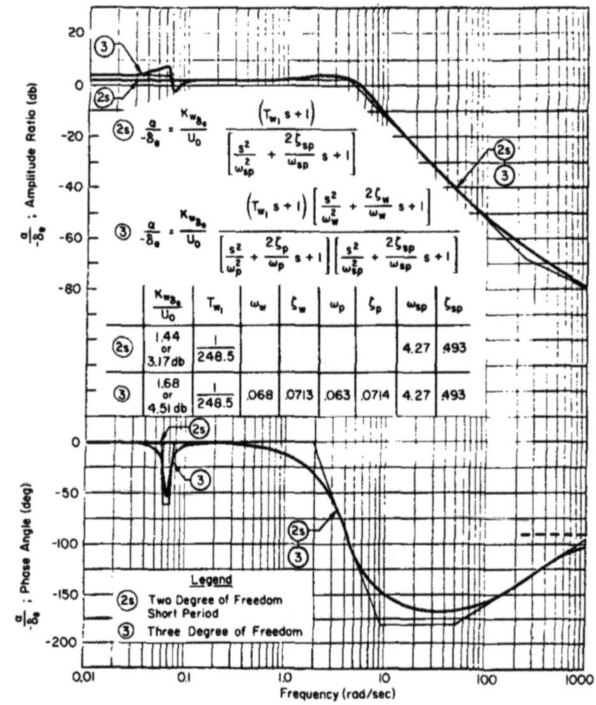


b) $\frac{\alpha}{-\delta_e}$

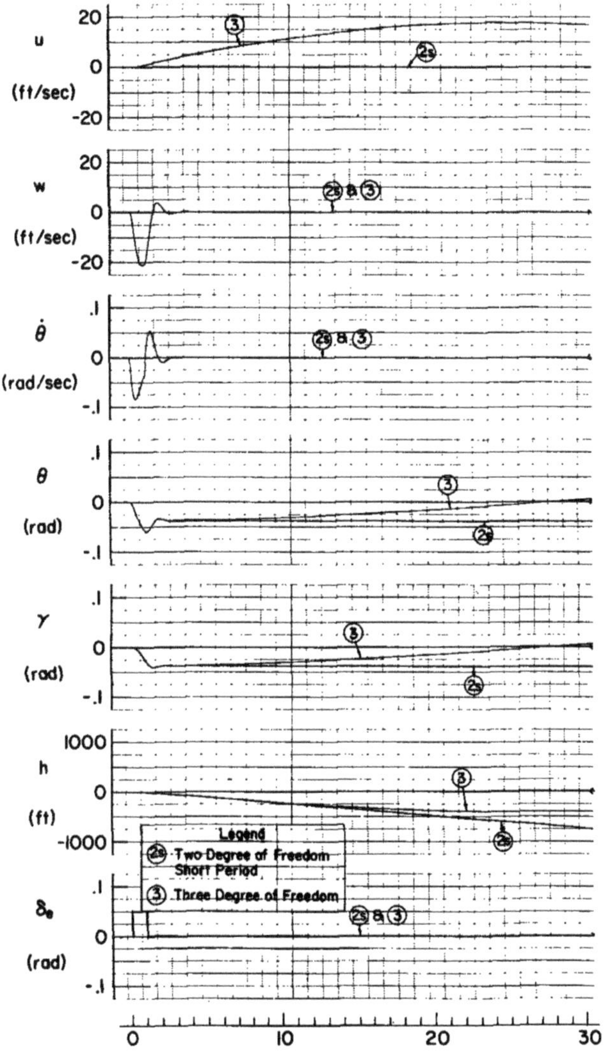


c) Time Responses to an Elevator Pulse
(Weighting Function)

Fig. 5-5. Comparisons of complete three degree of freedom and short period (two degree of freedom) control-input transfer functions and responses.

On modern craft $M_u$ is seldom zero, and the total static stability terms become $M_u u + M_w w$. When only these static moments are retained in the pitching moment equation, the three degree of freedom phugoid equations of motion become

$$\begin{aligned} (s - X_u)u - X_w w + g\theta &= X_\delta \delta - X_u u_g - X_w w_g \\ -Z_u u + (s - Z_w)w - U_0 s\theta &= Z_\delta \delta - Z_u u_g - Z_w w_g \\ -M_u u - M_w w &= M_\delta \delta - M_u u_g - M_w w_g \end{aligned} \quad (5\text{-}20)$$

With $u_g = w_g = 0$, these lead to the transfer functions

$$\frac{\theta(s)}{\delta(s)} = \frac{\begin{aligned} M_\delta s^2 + [M_u X_\delta + M_w Z_\delta - (X_u + Z_w)M_\delta]s \\ + [(Z_u M_w - Z_w M_u)X_\delta + (M_u X_w - M_w X_u)Z_\delta \\ + (Z_w X_u - X_w Z_u)M_\delta] \end{aligned}}{-M_\alpha \Delta_p} \quad (5\text{-}21)$$

$$\frac{u(s)}{\delta(s)} = \frac{[(X_\alpha - g)M_\delta - M_\alpha X_\delta]s + g(M_\delta Z_w - Z_\delta M_w)}{-M_\alpha \Delta_p} \quad (5\text{-}22)$$

$$\frac{w(s)}{\delta(s)} = \frac{U_0 M_\delta s^2 + (U_0 M_u X_\delta - U_0 X_u M_\delta)s + g(Z_\delta M_u - M_\delta Z_u)}{-M_\alpha \Delta_p} \quad (5\text{-}23)$$

where

$$\Delta_p = s^2 + 2\zeta_p \omega_p s + \omega_p^2$$

$$2\zeta_p \omega_p = -X_u + \frac{M_u(X_\alpha - g)}{M_\alpha}; \qquad \omega_p^2 = \frac{-g}{U_0}\left(Z_u - \frac{M_u}{M_w} Z_w\right) \quad (5\text{-}24)$$

By substitution of the sample numerical values, the following three degree of freedom phugoid approximate transfer functions result:

$$\frac{\theta(s)}{-\delta_e(s)} = 4.85 \frac{[(s/0.0098) + 1][(s/1.363) + 1]}{\{[s^2/(0.0683)^2] + [2(0.0710)s/0.0683] + 1\}} \quad (5.25)$$

$$\frac{u(s)}{-\delta_e(s)} = -15{,}920 \frac{[(s/1.415) + 1]}{\{\qquad\}} \quad (5\text{-}26)$$

$$\frac{\alpha(s)}{-\delta_e(s)} = \frac{-1}{U_0} \frac{w(s)}{\delta_e(s)} = \frac{1110}{660} \frac{\{[s^2/(0.0683)^2] + [2(0.0710)s/0.0683] + 1\}}{\{\qquad\}} \quad (5\text{-}27)$$

These transfer functions and the response to a pulse elevator input are compared with the complete three degree of freedom results in Fig. 5-6. The correspondence is very good for frequencies below about 1 rad/sec, roughly one-quarter of $\omega_{sp}$.

Because the example $M_u$ is zero, we return to Eq. 5-24 to obtain an appreciation for the general importance of $M_u$. We note that, for the normally positive values of $Z_w/M_w$ and $(X_\alpha - g)/M_\alpha$, the phugoid frequency and damping increase proportionally to $M_u$. Conversely, for sufficiently negative values of $M_u$, $\omega_p^2$ becomes negative and the phugoid mode is then characterized by two first orders, one convergent, the other a divergent "tuck," so called because, as speed increases, the airplane's nose has a tendency to tuck under (negative $M_u$).

When $M_u = 0$, the approximate factors for the classical case result. These are

$$2(\zeta\omega)_p = -X_u \tag{5-28}$$

$$\omega_p^2 = \frac{-g}{U_0} Z_u \tag{5-29}$$

The undamped natural frequency result can be further simplified by considering $C_{L_u} \doteq 0$, as is generally true for subsonic flight, and taking the trimmed lift equal to the weight so that $Z_u \doteq -2g/U_0$. Then,

$$\omega_p \doteq \sqrt{2}\,\frac{g}{U_0} \tag{5-30}$$

The period of this classical phugoid is $(\sqrt{2}\pi/g)U_0$, or about a fifth of the true airspeed in miles per hour. Based partly on these results, so-called two degree of freedom phugoid approximate equations have been used from time to time. These are

$$(s - X_u)u + g\theta = 0 \tag{5-31}$$

$$-Z_u u - U_0 s\theta = Z_\delta \delta \tag{5-32}$$

The characteristic equation of this set is

$$s^2 - X_u s - \frac{gZ_u}{U_0} = 0 \tag{5-33}$$

and the damping and undamped natural frequency are thereby the same as those given above (Eqs. 5-28, 5-29) for the classical ($M_u = 0$) case. Examination of the $X$ and $Z$ vector polygons in Fig. 5-4 indicates that the two degree of freedom set is reasonably good for this example. Unfortunately, however, some of the transfer functions (and thus some time responses) derived from these equations are very poor approximations to the complete three degree of freedom situation in the region of phugoid

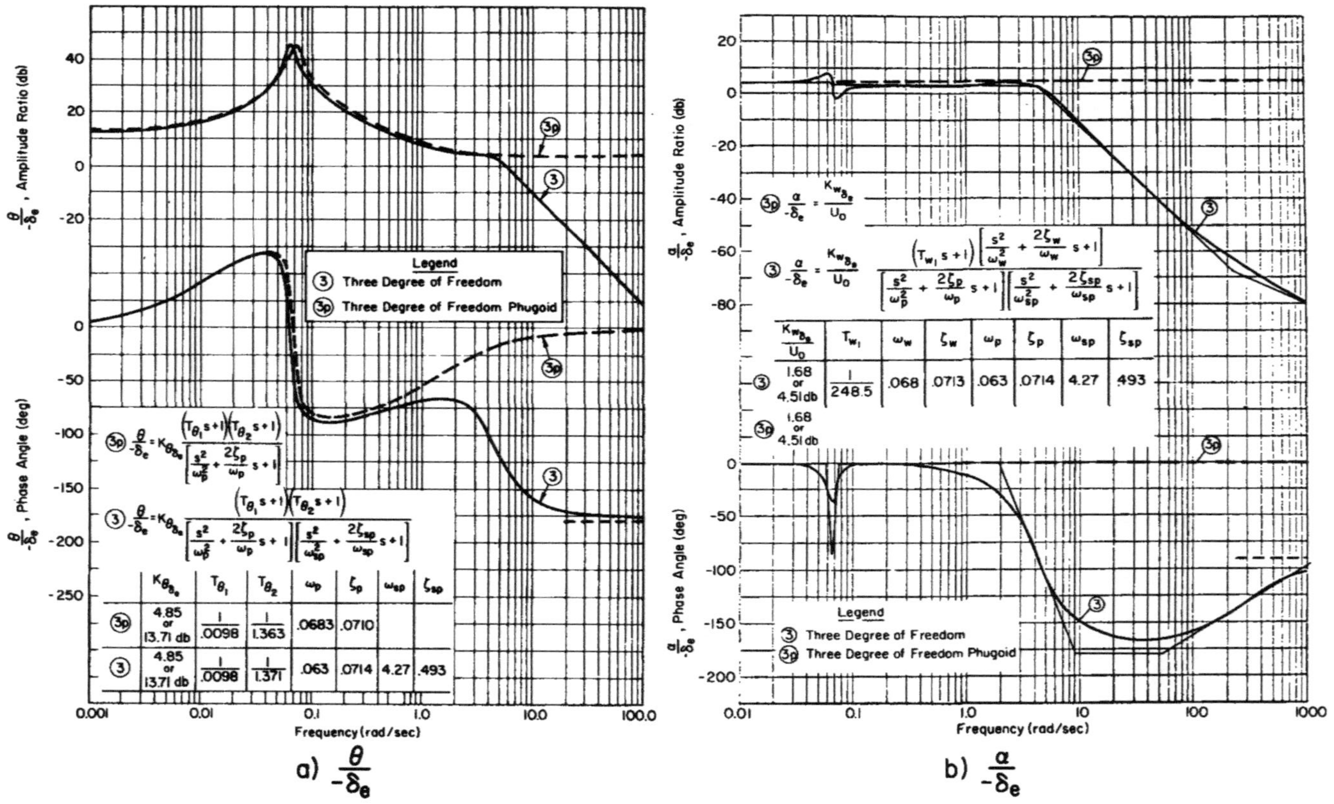


| | $K_{\theta_{\delta_e}}$ | $T_{\theta_1}$ | $T_{\theta_2}$ | $\omega_p$ | $\zeta_p$ | $\omega_{sp}$ | $\zeta_{sp}$ |
|---|---|---|---|---|---|---|---|
| 3p | 4.85 or 13.71 db | $\frac{1}{.0098}$ | $\frac{1}{1.363}$ | .0683 | .0710 | | |
| 3 | 4.85 or 13.71 db | $\frac{1}{.0098}$ | $\frac{1}{1.371}$ | .063 | .0714 | 4.27 | .493 |

| | $\frac{K_{w_{\delta_e}}}{U_0}$ | $T_{w_1}$ | $\omega_w$ | $\zeta_w$ | $\omega_p$ | $\zeta_p$ | $\omega_{sp}$ | $\zeta_{sp}$ |
|---|---|---|---|---|---|---|---|---|
| 3 | 1.68 or 4.51 db | $\frac{1}{248.5}$ | .068 | .0713 | .063 | .0714 | 4.27 | .493 |
| 3p | 1.68 or 4.51 db | | | | | | | |

Fig. 5-6. Comparison of complete and approximate phugoid (three degree of freedom) control-input transfer functions and responses.

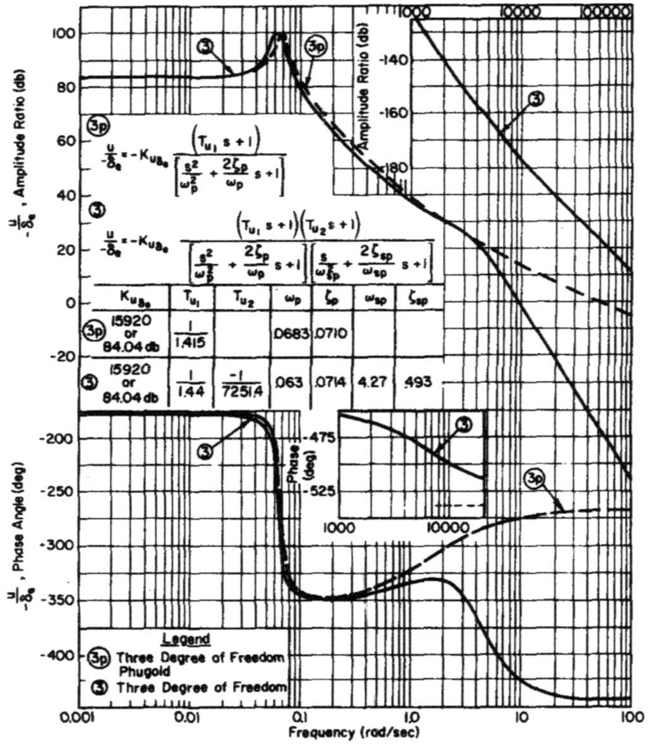

c) $\frac{u}{-\delta_e}$

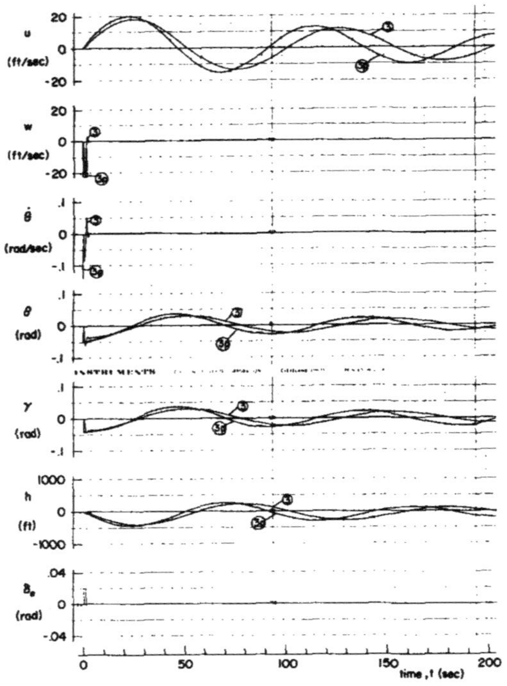

d) Time Response to an Elevator Pulse (Weighting Function)

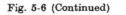

Fig. 5-6 (Continued)

frequencies; thus, the two degree of freedom set is oversimplified for most practical purposes.

## 5-7. Hovering Equations of Motion, Control-Input Transfer Functions, and Modal Responses

The preceding example has illustrated the phugoid and short period modes that characterize the longitudinal motion of "conventional" aircraft, that is, aircraft supported principally by fixed wings, flying at the relatively high speeds demanded by this form of sustentation. For vehicles such as helicopters and VTOLs operating at zero and very low forward speeds, the longitudinal motions display modes very different from the phugoid and short period modes. These differences arise because many of the usual derivatives disappear at hovering. It is pertinent, therefore, to delete such terms from the equations before considering a numerical example in detail.

For hovering vehicles the derivatives $M_w$, $M_{\dot{w}}$, and $X_w$ are usually negligible due to considerations of symmetry. This can easily be seen for certain very simple VTOL aircraft, e.g., a "stand-on" ducted fan.[1] With such vehicles perturbations in $w$ can only produce $Z$ forces because the configurations are symmetrical about the $Z$ axis. For more complex VTOLs such as the tilt-duct shown in Fig. 5-7, $M_w$ arises principally through flat-plate drag on the horizontal tail and fuselage for the usual center of gravity location near the duct axis which is vertical in hover. The $M_w$ resulting from these contributions is exceedingly small and may safely be neglected in calculating the frequencies, dampings, and time constants of the longitudinal modes. The contribution of the horizontal tail flat-plate drag to the $M_{\dot{w}}$ and $X_w$ derivatives is totally negligible for this configuration; in fact, for any reasonable hover arrangement, non-zero $X_w$ and/or $M_{\dot{w}}$ are very unlikely.

Significant $M_w$ effects can occur in hover in the following cases:

1. When the tail jet (or rotor) is of high disk loading and contributes significant lift in the unperturbed trimmed condition. The thrust of a high-disk-loading tail jet (or rotor) is virtually unaffected by $w$ perturbations and gives a negligible contribution to $Z_w$ (and hence $M_w$). However, the *main rotors* now have a moment arm about the center of gravity and thus induce some $M_w$.
2. When the tail disk loading (measured in the perturbed condition) is appreciably lower than the main disk loading. In this case the tail thrust

[1] J. P. Campbell, *Vertical Takeoff and Landing Aircraft*, Macmillan, New York, 1962.

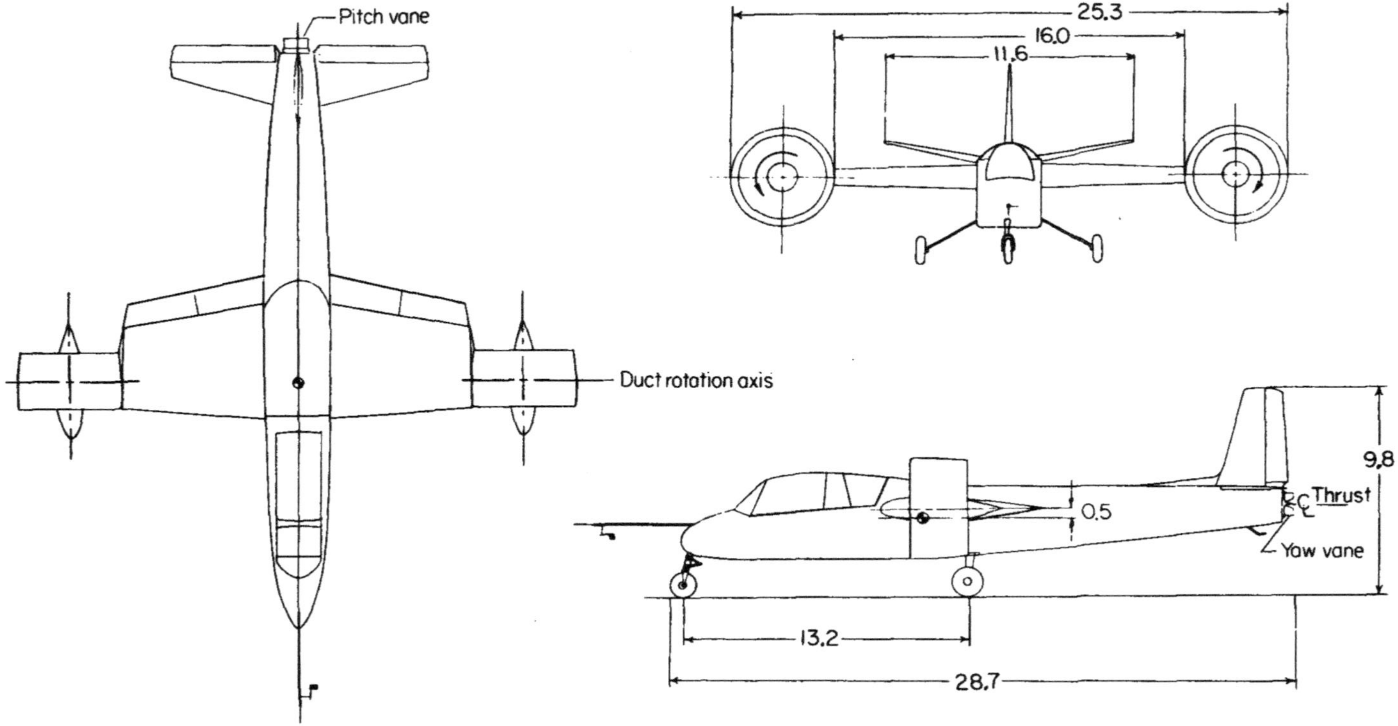


(All dimensions in feet)

**Fig. 5-7. Example of tilt-duct VTOL aircraft.**

is sensitive to perturbations in $w$ and it may give a significant contribution to $Z_w$ which induces some $M_w$ due to the long moment arm about the center of gravity.

Even in these cases nonzero $M_w$ usually does not change the character of the typical hovering modes. Their description can therefore proceed on the assumption for $U_0 = 0$, that, $M_{\dot{w}}$, $M_w$, and $X_w$ are negligible. The removal of these and gust-input terms from the equations of motion (Eq. 5-2) gives

$$\begin{aligned} (s - X_u)u + \qquad\qquad\qquad\qquad & g\theta = X_\delta \delta \\ -Z_u u + (s - Z_w)w \qquad\qquad & \quad\;\, = Z_\delta \delta \\ -M_u u + \qquad\qquad (s^2 - M_q s)& \theta = M_\delta \delta \end{aligned} \tag{5-34}$$

The corresponding control-input transfer functions in polynomial form are obtainable directly from Table 5-1 with $M_w$, $M_{\dot{w}}$, and $X_w$ terms deleted. Many of the numerator and denominator factors are directly separable, and canceling, because of the zero terms appearing on the left-hand side of Eq. 5-34. For example, for $X_\delta \doteq 0$, a common occurrence, the $w$, $\theta$, and $u$ transfer functions are most directly obtained by expanding the determinant forms (Eq. 5-3) as follows:

$$\frac{w(s)}{\delta(s)} = \frac{\begin{vmatrix} s - X_u & 0 & g \\ -Z_u & Z_\delta & 0 \\ -M_u & M_\delta & s^2 - M_q s \end{vmatrix}}{\begin{vmatrix} s - X_u & 0 & g \\ -Z_u & s - Z_w & 0 \\ -M_u & 0 & s^2 - M_q s \end{vmatrix}} = \frac{Z_\delta \cancel{\Delta_{\text{hover}}}}{(s - Z_w)\cancel{\Delta_{\text{hover}}}} \tag{5-35$^2$}$$

$$\frac{\theta(s)}{\delta(s)} = \frac{\begin{vmatrix} s - X_u & 0 & 0 \\ -Z_u & s - Z_w & Z_\delta \\ -M_u & 0 & M_\delta \end{vmatrix}}{(s - Z_w)\Delta_{\text{hover}}} = \frac{M_\delta(s - X_u)\cancel{(s - Z_w)}}{\cancel{(s - Z_w)}\Delta_{\text{hover}}} \tag{5-36}$$

$$\frac{u(s)}{\delta(s)} = \frac{\begin{vmatrix} 0 & 0 & g \\ Z_\delta & s - Z_w & 0 \\ M_\delta & 0 & s^2 - M_q s \end{vmatrix}}{(s - Z_w)\Delta_{\text{hover}}} = \frac{-gM_\delta\cancel{(s - Z_w)}}{\cancel{(s - Z_w)}\Delta_{\text{hover}}} \tag{5-37}$$

[2] Assuming that $|Z_u M_\delta| \ll |M_u Z_\delta|$ as usually appropriate to "throttle" (vertical thrust) control.

where

$$\Delta_{\text{hover}} = (s - X_u)(s^2 - M_q s) + gM_u \tag{5-38}$$

From these relationships we see that the $s - Z_w$ mode is associated only with $w$ perturbations and does not appear in either $\theta$ or $u$ motions. Since modal response ratios are independent of the input assumed (Chapter 2), this is a general conclusion that holds as well for $X_\delta \neq 0$. Furthermore, since for $U_0 = 0$, $\dot{h} = -w$ (Eq. 5-2), the mode is characterized by pure vertical translatory motions in altitude (heaving motions) with an aperiodic time constant of $-1/Z_w$, invariably positive. Conversely, the modes associated with the *longitudinal hovering cubic*,

$$\Delta_{\text{hover}} = s^3 - (X_u + M_q)s^2 + X_u M_q s + gM_u \tag{5-39}$$

do not usually involve $w$ or $h$ motions,[3] irrespective of whether there are three real or one real and two complex roots. That is, the longitudinal hovering cubic describes motions normally consisting of $u$ and $\theta$ perturbations only. The relative $u$ and $\theta$ motions can be most simply obtained from the $X$ equation of Eq. 5-34 as a *modal response ratio*:

$$\frac{\theta}{u} = \left[\frac{s - X_u}{-g}\right]_{s=-q_{\text{hover}}} \tag{5-40}$$

where $-q_{\text{hover}}$ represents the roots of the hovering cubic. For real roots, and resulting aperiodic motions, the ratio is real; for complex roots, and resulting oscillatory motions, the ratio is complex and must be characterized by an amplitude and phase.

## 5-8. Example Transfer Functions, Bode Forms, and Time Responses for a Hovering Vehicle

To illustrate the foregoing in more concrete terms, we now consider the VTOL airplane of Fig. 5-7 which has the following characteristics under hovering conditions:[4]

[3] The absence of $h$ motions depends on $Z_u$'s being negligibly small which is generally true in hover; however, the tilt-wing configuration examined in the following reference had $Z_u$ almost equal to $Z_w$: J. Wolkovitch and R. P. Walton, "VTOL and Helicopter Approximate Transfer Functions and Closed-Loop Handling Qualities," Systems Technology Tech. Rept. 128-1, Sept. 1963.

[4] *Ibid.*

| | |
|---|---|
| Weight (lb) | 3,100 |
| Pitch inertia (slug-ft$^2$) | 1,790 |
| Groundspeed, $U_0$ (ft/sec) | 0 |
| $X_u$ | $-0.137$ |
| $X_w$ | 0.0 |
| $X_{\delta_e}$ | 0.0 |
| $Z_u$ | 0.0 |
| $Z_w$ | $-0.137$ |
| $Z_q$ | 0.0 |
| $Z_{\delta_e}$ | $-1.08$ |
| $M_u$ | 0.0136 |
| $M_w$ | 0.0 |
| $M_{\dot{w}}$ | 0.0 |
| $M_q$ | $-0.0452$ |
| $M_{\delta_e}$ | $-1.0$ |

The resulting control-input transfer functions are

$$\frac{w}{-\delta_e} = \frac{\dot{h}}{\delta_e} = \frac{1.08}{(s + 0.137)} \tag{5-41}$$

$$\frac{\theta}{-\delta_e} = \frac{(s + 0.137)}{\Delta_{\text{hover}}} \tag{5-42}$$

$$\frac{u}{-\delta_e} = \frac{-32.2}{\Delta_{\text{hover}}} \tag{5-43}$$

where

$$\Delta_{\text{hover}} = \left(s + \frac{1}{T_{sp_2}}\right)[s^2 + 2\zeta_p\omega_p s + \omega_p^2]$$

$$= (s + 0.824)[s^2 - 2(0.439)(0.731)s + (0.731)^2] \tag{5-44}$$

Figure 5-8 contains the corresponding Bode diagrams and time responses to an impulsive elevator input. The Bodes are much simpler than those given for the conventional airplane in Fig. 5-2. The leading phase shown for $\theta$ and $u$ results from the negative damping of the hovering phugoid mode.[5]

The time responses show the divergent oscillatory character of the $u$ and $\theta$ motions and the first-order convergence of the $w$ motions. The relative amplitudes of $u$ and $\theta$ are clearly shown on the time vector diagrams of Fig. 5-9. Additionally, the time vectors for the phugoid mode,

[5] For such *nonminimum phase* situations it must be remembered that the stability criterion for the closed-loop system is no longer simply that the open-loop amplitude ratio be less than one for phase lags greater than 180° (see Chapter 7 for example closures under such circumstances).

Fig. 5-9(b), show that the instability is aggravated by a very small $|M_q|$ and a relatively large $M_u$. Such unstable oscillations are quite typical for hovering vehicles out of ground effect. The separate influences of these two derivatives can most easily be traced using the approximate literal transfer function factors considered subsequently.

## 5-9. Gust-Input Transfer Functions

The gust-input transfer functions are obtained from the equations of motion (Eq. 5-1), using the same procedures as for control inputs. Because the denominator remains the same regardless of the type of input, we need consider only gust-input numerators. These are given in polynomial form in Table 5-2, which follows the format established in Table 5-1. In addition, Table 5-3 contains the gust transfer function numerators for the truncated short period and phugoid equations of motion, Eqs. 5-14 and 5-20, respectively; the corresponding denominators are those given in Eqs. 5-16 and 5-23.

Figure 5-10 presents Bode plots of the $u$-gust transfer functions for the example of a conventional airplane. We see here that the primary response occurs at the phugoid frequency. Taking peak amplitudes as approximately proportional to the magnitudes of the phugoid modal response coefficients in the several degrees of freedom, we find that the magnitudes of the relative motions in decibels are given by

$$u:a_z:w:\theta = |18:-5:-23:-37|_{\text{dB}}$$

Converting $a_z$ to $\dot{h}$ (i.e., $|a_z(j\omega_p)/\omega_p| \doteq |\dot{h}(j\omega_p)|$) to correspond with the $u$ and $w$ velocities, $\theta$ to degrees as a more familiar measure, and, finally, considering linear amplitudes rather than decibels, we obtain

$$u:\dot{h}:w:\theta = (8:8.8:0.071)_{\text{fps}}:0.8°$$

These relative magnitudes are completely consistent with our already established concept of the phugoid motions, that is, they show that the phugoid responses to a $u$ gust are predominantly in $\dot{h}$ and $u$, secondarily in $\theta$, and hardly at all in $w$. Note, however, that the $u$ and $\theta$ responses drop off sharply with increasing frequency, whereas the $w$ and $a_z$ responses do not. Thus, in the short period frequency region, $u$ and $\theta$ are essentially nonexistent, while $w$ and $a_z$ are about the same as in the phugoid region. Furthermore, the $a_z$ response for frequencies beyond the short period remains constant, so that there is a distinct possibility of $u$ gust excitation of lightly damped structural modes which may actually exist in this region, but which are eliminated by the rigid body assumption.

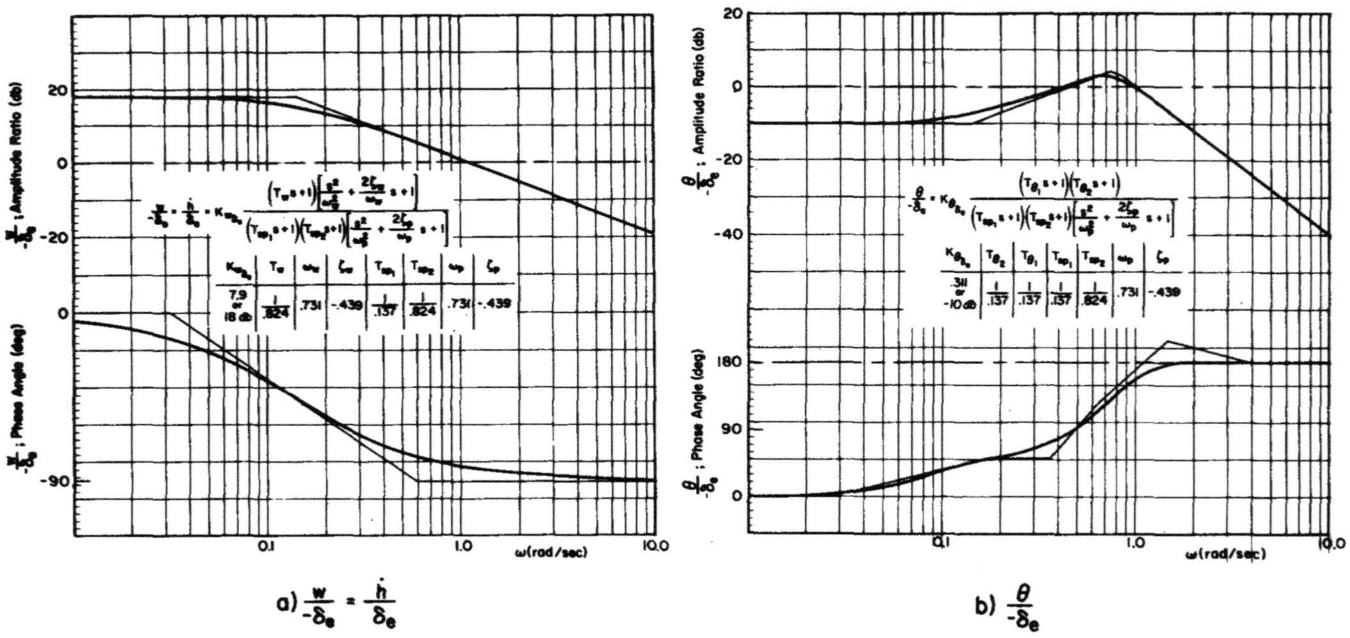


Fig. 5-8. Example control-input Bode plots and time responses for a hovering airplane.

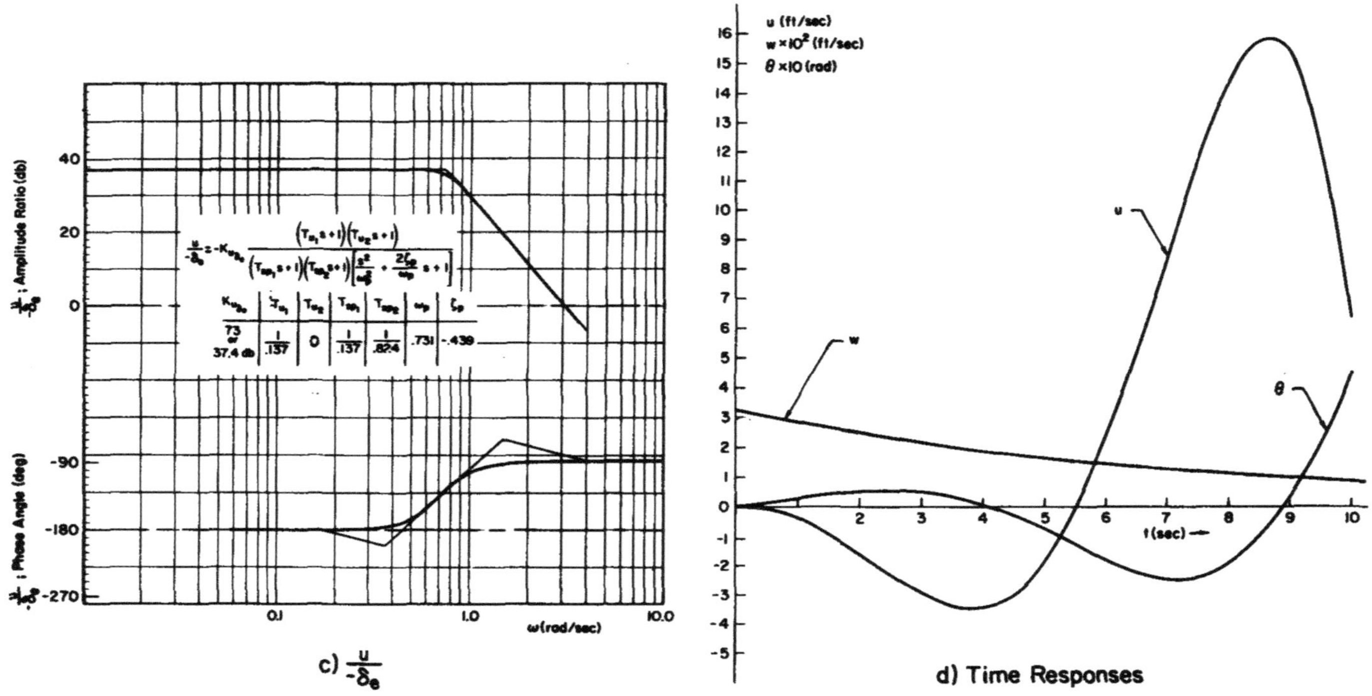


c) $\frac{u}{-\delta_e}$

d) Time Responses

Fig. 5-8 (Continued)

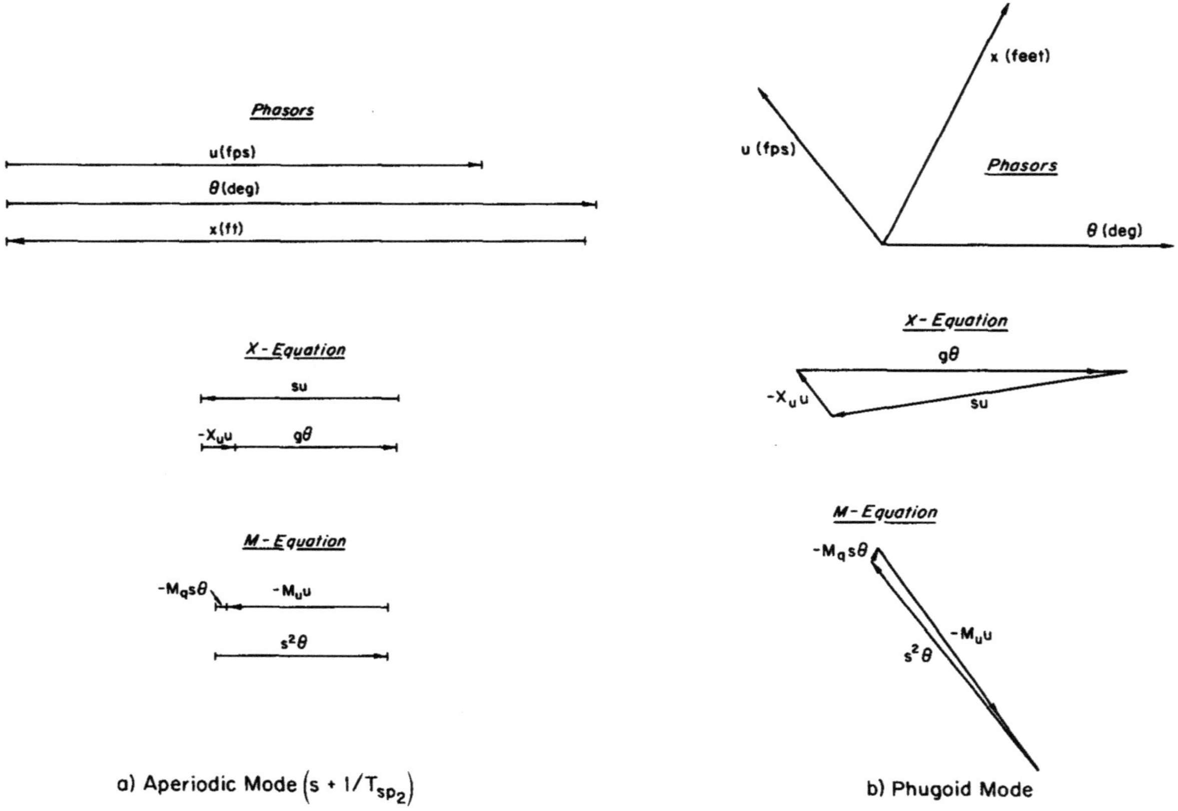


Fig. 5-9. Time vector diagram for the hovering airplane.

Table 5-2. Longitudinal gust-input transfer function numerator coefficients.

| | $A$ | $B$ | $C$ | $D$ |
|---|---|---|---|---|
| $\frac{N^{\theta}_{u_g}}{s}$ | $-(M_u + Z_u M_{\dot{w}})$ | $Z_w M_u - M_w Z_u$ | | |
| $\frac{N^{w}_{u_g}}{s^2}$ | $-Z_u$ | $Z_u M_q - U_0 M_u$ | | |
| $N^{u}_{u_g}$ | $-X_u$ | $X_u(Z_w + M_q + M_{\dot{\alpha}}) - Z_u X_w$ | $-X_u(Z_w M_q - M_\alpha) + Z_u(X_w M_q + gM_{\dot{w}})$ $-M_u(X_\alpha - g)$ | $g(Z_u M_w - M_u Z_w)$ |
| $N^{h}_{u_g}$ | $Z_u$ | $-Z_u(M_q + M_{\dot{\alpha}})$ | $-(Z_u M_\alpha - M_u Z_\alpha)$ | |
| $\frac{U_0 N^{\theta}_{w_g}}{s}$ | $M_q - M_{\dot{\alpha}}$ | $-M_\alpha - Z_w M_q - X_u(M_q - M_{\dot{\alpha}})$ | $X_u(M_\alpha + Z_w M_q) - X_w(U_0 M_u + Z_u M_q)$ | |
| $N^{w}_{w_g}$ | $-(Z_w - M_q + M_{\dot{\alpha}})$ | $X_u(Z_w - M_q + M_{\dot{\alpha}}) - X_w Z_u - M_\alpha + Z_w M_q$ | $X_u(M_\alpha - Z_w M_q) - X_w(U_0 M_u - Z_u M_q)$ $+ gZ_u\left(M_{\dot{w}} - \frac{M_q}{U_0}\right)$ | $g(Z_u M_w - M_u Z_w)$ |
| $\frac{U_0 N^{u}_{w_g}}{s}$ | $-X_\alpha$ | $2X_\alpha M_q + g(M_{\dot{\alpha}} - M_q)$ | $g(M_\alpha + Z_w M_q)$ | |
| $sN^{h}_{w_g}$ | $Z_w$ | $-2Z_w M_q + X_w Z_u - Z_w X_u$ | $-2M_q(X_w Z_u - Z_w X_u) - gZ_u\left(M_{\dot{w}} - \frac{M_q}{U_0}\right)$ | $-g(Z_u M_w - M_u Z_w)$ |

**Table 5-3.** Longitudinal gust-input transfer function numerator coefficients for the truncated short period and phugoid equations of motion.

| | Phugoid | | |
|---|---|---|---|
| | $A$ | $B$ | $C$ |
| $\dfrac{N^{\theta}_{u_g}}{s}$ | $-M_u$ | $Z_wM_u - M_wZ_u$ | |
| $\dfrac{N^{w}_{u_g}}{s^2}$ | $-U_0M_u$ | | |
| $N^{u}_{u_g}$ | $X_uM_\alpha - M_u(X_\alpha - g)$ | $g(Z_uM_w - M_uZ_w)$ | |
| $\dfrac{U_0N^{\theta}_{w_g}}{s}$ | $-M_\alpha$ | $(X_uM_\alpha - M_uX_\alpha)$ | |
| $N^{w}_{w_g}$ | $-M_\alpha$ | $(X_uM_\alpha - M_uX_\alpha)$ | $g(Z_uM_w - M_uZ_w)$ |
| $\dfrac{U_0N^{u}_{w_g}}{s}$ | $gM_\alpha$ | | |

| | Short period | |
|---|---|---|
| $U_0N^{\theta}_{w_g}$ | $(M_q - M_{\dot\alpha})$ | $-M_\alpha - Z_wM_q$ |
| $\dfrac{N^{w}_{w_g}}{s}$ | $-(Z_w - M_q + M_{\dot\alpha})$ | $-M_\alpha + Z_wM_q$ |

Finally, notice that the short period is almost exactly canceled by a corresponding numerator term in both the $u$ and $a_z$ transfer functions, indicating that the short period mode itself will scarcely be present in $u$ and $a_z$ responses to $u$ gusts.

The $w$ gust transfer functions given in Fig. 5-11 show that low frequency responses progressively decrease in importance relative to high frequency responses as we consider, in turn, the $u$, $\theta$, $w$, and $a_z$ outputs. The $u$ response is somewhat lower at the phugoid and higher at the short period than the corresponding $u$ response to $u_g$ (Fig. 5-10). Otherwise both plots are similar, the inference being that, regardless of input, $u$ perturbations are always small, except near phugoid frequency. The $\theta$ response at phugoid is about the same, but that at the short period is considerably higher than the similar responses obtained for a $u_g$ input. In this connection the similarity in form of the $w/u_g$ (Fig. 5-10) and $\theta/w_g$ transfer functions should be noted, i.e., $w$ responses to $u_g$ are almost identical, except for a scale factor, to $\theta$ responses to $w_g$. This similarity also extends, but not quite so markedly, to the $w/w_g$ transfer function where, due to the near cancellation of the phugoid, the response is almost completely flat out to the short period. The $a_z$ response occurs predominantly at the short period and beyond. Again, the possibility of obtaining structural $a_z$ responses in this region must be considered for an elastic airplane.

### 5-10. Coupling Numerators

Multiloop situations in which more than a single input to the vehicle is involved require consideration of vehicle-introduced coupling effects between loops. Such consideration is based on the use of coupling numerators. These are defined in Chapter 3, where the notation established is a direct indication of the operations required to compute the coupling numerator. For example, the coupling numerator, $N^{\theta\ u}_{\delta_e \delta_T}$, appropriate for simultaneous $\theta \rightarrow \delta_e$ and $u \rightarrow \delta_T$ feedbacks is computed by substituting the $\delta_e$ and $\delta_T$ input coefficients for the terms normally appearing in the $\theta$ and $u$ columns of the characteristic determinant. That is, from Eq. 5-1,

$$N^{\theta\ u}_{\delta_e \delta_T} = \begin{vmatrix} X_{\delta_T} & -X_w & X_{\delta_e} \\ Z_{\delta_T} & (s - Z_w) & Z_{\delta_e} \\ M_{\delta_T} & -(M_{\dot{w}}s + M_w) & M_{\delta_e} \end{vmatrix} \tag{5-45}$$

There can also be coupling effects between gust inputs and control inputs, and among more than two inputs, control, or disturbance (Chapter 3), so that the possible variations are quite numerous. However, the coupling numerators are always easily computed and factored because replacement of at least two columns that are generally functions of $s$ by two columns that are generally constants reduces the $s$ polynomial to second degree or lower. Accordingly, it is unnecessary to catalogue all possible input combinations. Consequently, only the more commonly used two-input, control-coupling numerators are given in Table 5-4.

Table 5-4 contains the polynomial coefficients in literal form, the corresponding factored form, and literal approximations to the factors. Since the polynomials are either first or second degree in $s$, factoring even in literal form is no problem and exact literal factors can easily be written. However, the approximate factors shown are considerably more compact than exact literal factors because usually unimportant terms have been eliminated. The assumptions involved in obtaining the approximate factors are readily apparent, almost by inspection, and are not explicitly listed. They are usually valid, however, for conventional elevator and throttle controls (i.e., $|X_{\delta_e}| \ll |X_{\delta_T}|$; $|M_{\delta_T}| \ll |M_{\delta_e}|$). For unconventional controls, e.g., cyclic and collective pitch on a helicopter, the approximate factors shown may be inappropriate; in such cases, the specifically important contributions to the exact polynomial coefficients should be used.

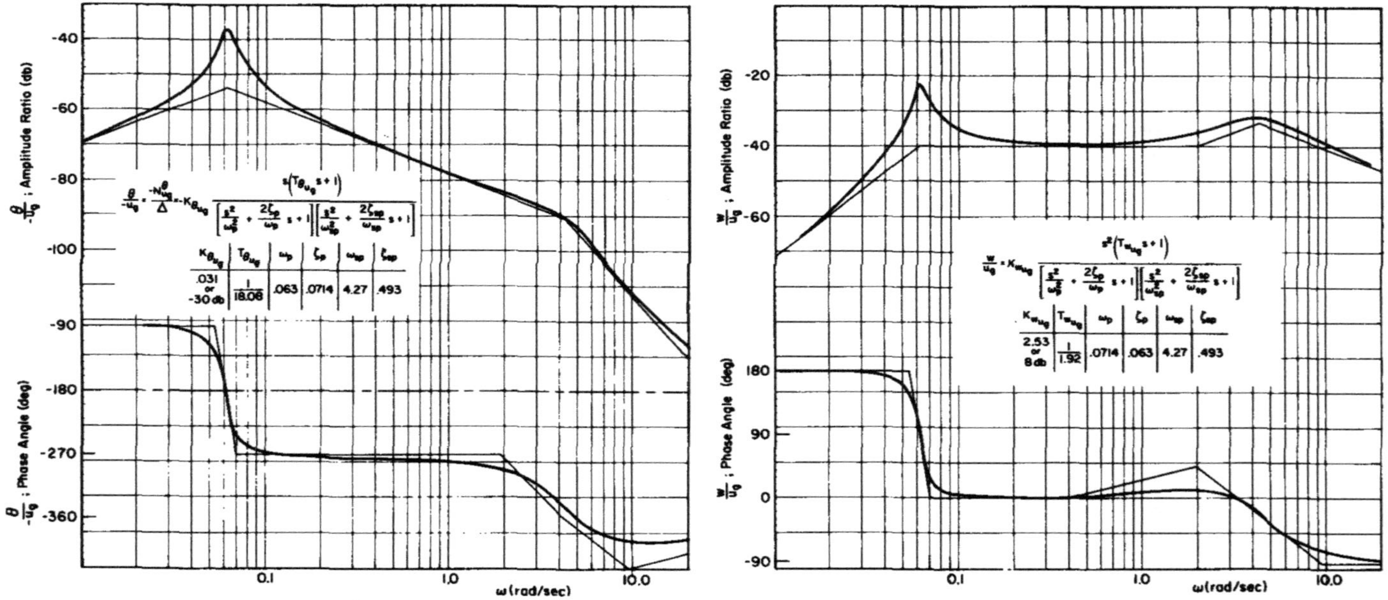


a) $\frac{\theta}{-u_g}$ b) $\frac{w}{u_g}$

Fig. 5-10. Example of $u$-gust Bode plots for a conventional airplane.

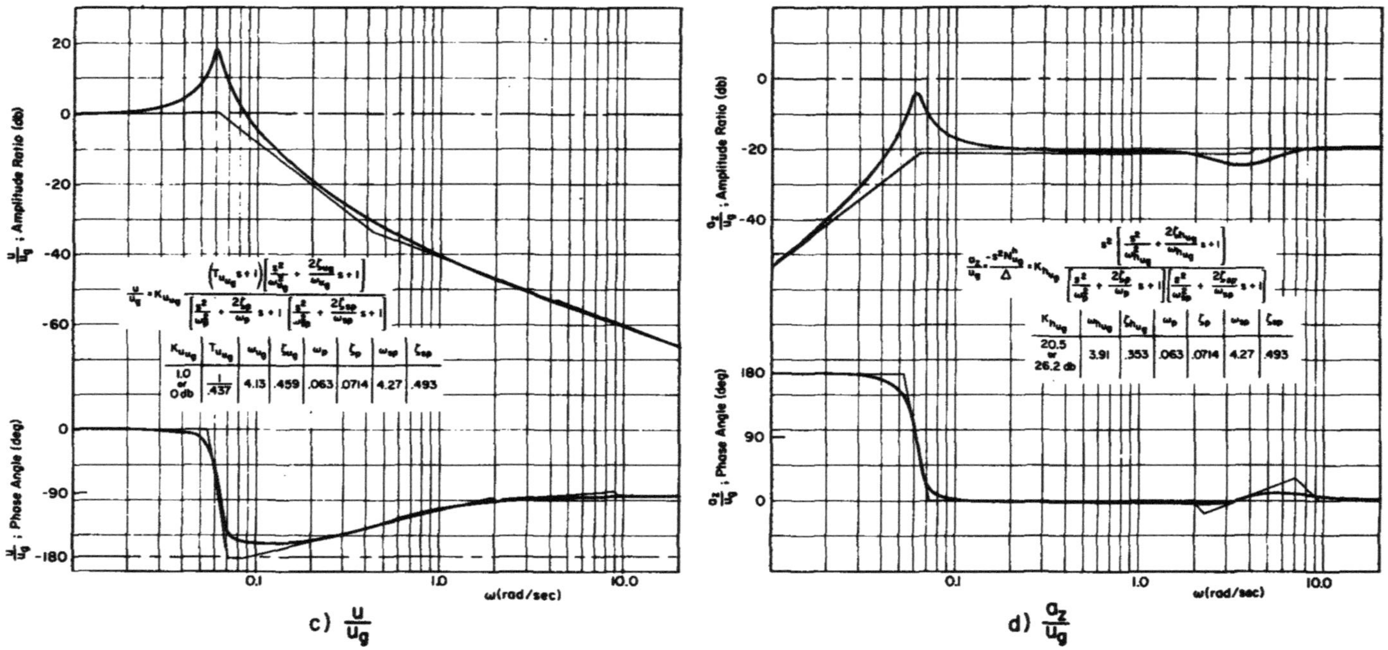


Fig. 5-10 (Continued)

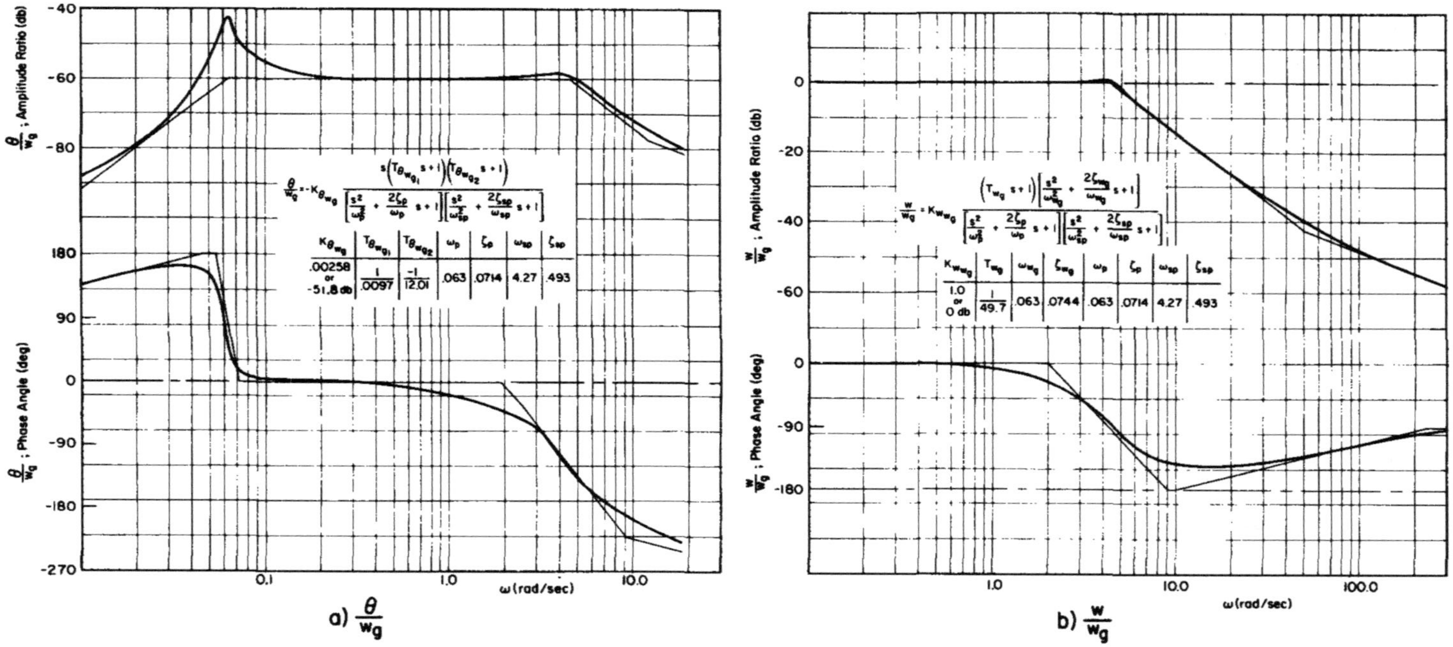


Fig. 5-11. Example of $w$-gust Bode plots for a conventional airplane.

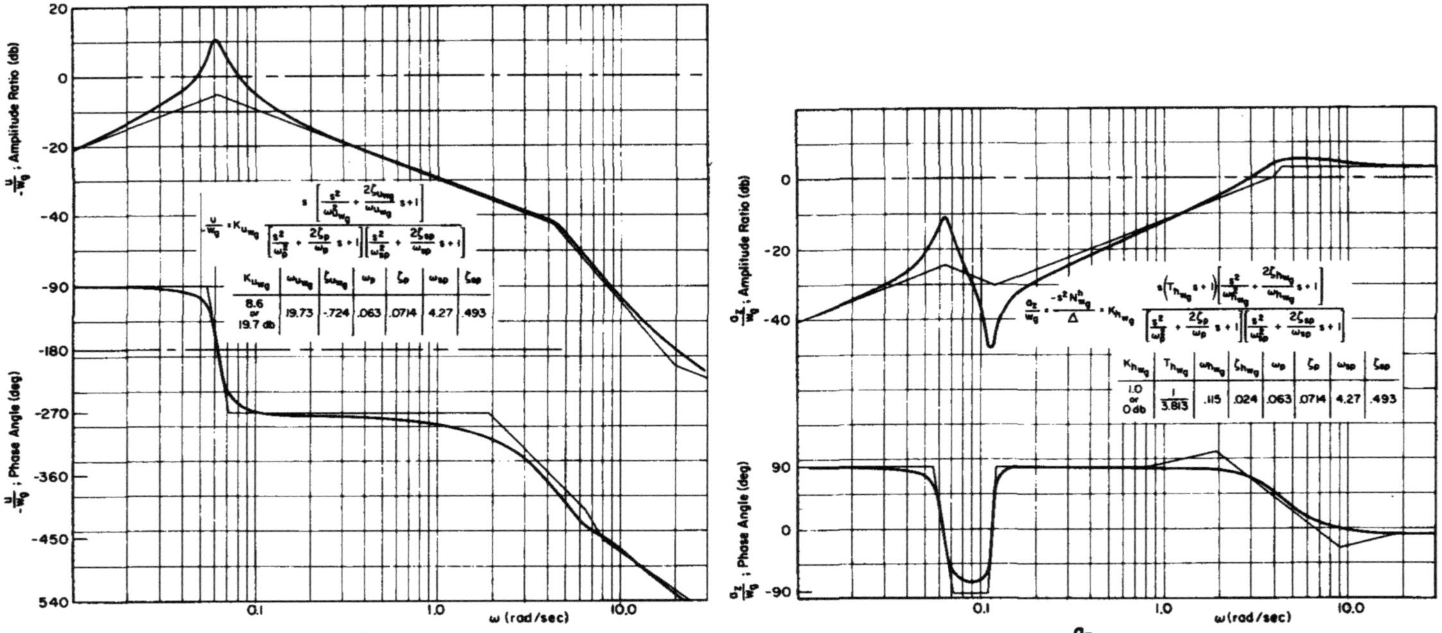


c) $\frac{u}{-w_g}$

d) $\frac{a_z}{w_g}$

Fig. 5-11 (Continued)

Table 5-4. Longitudinal coupling numerators for elevator and throttle inputs.

| Alternative numerators | $A$ | $B$ | $C$ |
|---|---|---|---|
| $N^{\theta\ u}_{\delta_e\delta_T}$<br>or<br>$-N^{u\ \theta}_{\delta_e\delta_T}$ | $M_{\delta_e}X_{\delta_T} - M_{\delta_T}X_{\delta_e}$<br>$+ M_{\dot{w}}(Z_{\delta_e}X_{\delta_T} - Z_{\delta_T}X_{\delta_e})$ | $M_w(Z_{\delta_e}X_{\delta_T} - Z_{\delta_T}X_{\delta_e})$<br>$+ Z_w(X_{\delta_e}M_{\delta_T} - X_{\delta_T}M_{\delta_e})$<br>$+ X_w(M_{\delta_e}Z_{\delta_T} - M_{\delta_T}Z_{\delta_e})$<br>$A_{\theta u}\left(s + \frac{1}{T_{\theta u}}\right)$<br>$A_{\theta u} \doteq X_{\delta_T}M_{\delta_e};$<br>$\frac{1}{T_{\theta u}} \doteq -Z_w + M_w\frac{Z_{\delta_e}}{M_{\delta_e}}$ | |
| $sN^{\theta}_{\delta_e\delta_T}\Big\|^{h}$<br>or<br>$-sN^{h\ \theta}_{\delta_e\delta_T}$<br>or<br>$U_0N^{\alpha\ \theta}_{\delta_e\delta_T}$<br>or<br>$-U_0N^{\theta\ \alpha}_{\delta_e\delta_T}$<br>or<br>$sN^{\alpha\ h}_{\delta_e\delta_T}$<br>or<br>$-sN^{h\ \alpha}_{\delta_e\delta_T}$ | $Z_{\delta_e}M_{\delta_T} - Z_{\delta_T}M_{\delta_e}$ | $X_u(M_{\delta_e}Z_{\delta_T} - M_{\delta_T}Z_{\delta_e})$<br>$+ Z_u(X_{\delta_e}M_{\delta_T} - X_{\delta_T}M_{\delta_e})$<br>$+ M_u(Z_{\delta_e}X_{\delta_T} - Z_{\delta_T}X_{\delta_e})$<br>$A_{\theta h}\left(s + \frac{1}{T_{\theta h}}\right)$<br>$A_{\theta h} \doteq -Z_{\delta_T}M_{\delta_e};$<br>$\frac{1}{T_{\theta h}} \doteq -X_u + \frac{X_{\delta_T}}{Z_{\delta_T}}\left(Z_u - M_u\frac{Z_{\delta_e}}{M_{\delta_e}}\right)$ | |

| | | | |
|---|---|---|---|
| $sN^{h\ u}_{\delta_e\delta_T}$ or $-sN^{u\ h}_{\delta_e\delta_T}$ | $X_{\delta_e}Z_{\delta_T} - X_{\delta_T}Z_{\delta_e}$ | $(X_{\delta_T}Z_{\delta_e} - Z_{\delta_T}X_{\delta_e})(M_q + M_{\dot{\alpha}})$ <br> $A_{hu}[s^2 + 2(\zeta\omega)_{hu}s + \omega_{hu}^2]$ <br> $A_{hu} \doteq -X_{\delta_T}Z_{\delta_e};$ <br> $2(\zeta\omega)_{hu} = -(M_q + M_{\dot{\alpha}});$ <br> $\omega_{hu}^2 \doteq -\left(M_\alpha - \frac{M_{\delta_e}}{Z_{\delta_e}} Z_\alpha\right)$ | $Z_\alpha(X_{\delta_e}M_{\delta_T} - X_{\delta_T}M_{\delta_e})$ <br> $+ (X_\alpha - g)(Z_{\delta_T}M_{\delta_e} - M_{\delta_T}Z_{\delta_e})$ <br> $+ M_\alpha(X_{\delta_T}Z_{\delta_e} - X_{\delta_e}Z_{\delta_T})$ |
| $U_0N^{\alpha\ u}_{\delta_e\delta_T}$ or $-U_0N^{u\ \alpha}_{\delta_e\delta_T}$ | $X_{\delta_T}Z_{\delta_e} - Z_{\delta_T}X_{\delta_e}$ | $U_0(X_{\delta_T}M_{\delta_e} - M_{\delta_T}X_{\delta_e})$ <br> $- M_q(X_{\delta_T}Z_{\delta_e} - Z_{\delta_T}X_{\delta_e})$ <br> $A_{\alpha u}\left(s + \frac{1}{T_{\alpha u_1}}\right)\left(s + \frac{1}{T_{\alpha u_2}}\right)$ <br> $A_{\alpha u} \doteq X_{\delta_T}Z_{\delta_e};$ <br> $\frac{1}{T_{\alpha u_1}} \doteq \frac{g}{U_0}\left(\frac{Z_{\delta_T}}{X_{\delta_T}} - \frac{Z_{\delta_e}}{M_{\delta_e}}\frac{M_{\delta_T}}{X_{\delta_T}}\right);$ <br> $\frac{1}{T_{\alpha u_2}} \doteq \frac{U_0M_{\delta_e}}{Z_{\delta_e}}$ | $g(M_{\delta_e}Z_{\delta_T} - Z_{\delta_e}M_{\delta_T})$ |

## 5-11. Approximate Factors

The literal approximate factors given in Table 5-4 for coupling numerators serve to reintroduce the subject of approximate factors in general. In addition to those given in Table 5-4 we have already implicitly derived some approximate factors in connection with the simplified short period, phugoid, and hover equations of motion (i.e., Eqs. 5-17, 5-24, 5-35, and 5-36). However, using such truncated equations does not always produce unambiguous and reasonably accurate approximate factors or adequately identify the conditions for which the approximations hold. A more direct attack, which meets these requirements, is simply to factor the literal polynomial expressions in Table 5-1 by first neglecting terms of minor significance and then applying an approximate factorization technique such as that described in Section 3-4. Of course, the relative importance of various derivatives or derivative groupings depends on the vehicle type and flight condition, so that there is no single set of literal approximate factors generally applicable to all situations. This is reflected in the profusion of approximate factors given in Tables 5-5 and 5-8.[6]

Table 5-5 presents the approximate factors for conventional airplanes, and these apply as well to VTOL machines and helicopters at cruise or high speed flight. Notice in particular that the approximate short period frequency and damping are identical to those of Eqs. 5-17, and the phugoid frequency and damping are quite similar to those of Eq. 5-24, except for the appearance of the $Z_w M_q$ term in the Table 5-5 expressions. Similarly, the numerator factors implicit in Eqs. 5-15, 5-16, and 5-21 to 5-23 have their counterpart in Table 5-5. For example, the numerator time constants in Eqs. 5-15 and 5-16 reduce to the same expressions as those for $1/T_{w_1}$ and $1/T_{\theta_2}$ in Table 5-5, when the Table 5-5 conditions of validity are invoked.

The various expressions in Table 5-5 tell us, almost at a glance, what derivatives are important for a given control situation. Furthermore, they also display the explicit connections between all the numerators and the denominators via the dominant derivatives. For example, changes in $1/T_{\theta_2}$ that involve changes in $Z_w$ are also accompanied by important changes in $\zeta_{sp}\omega_{sp}$, $1/T_{u_1}$, and $\omega_h^2$ (or $1/T_{h_2}T_{h_3}$).

The approximate factors can be simple and effective guides to the influences of speed, center of gravity, and configuration changes on the

[6] From I. L. Ashkenas and D. T. McRuer, "Approximate Airframe Transfer Functions and Application to Single Sensor Control Systems," WADC-TR-58-82, June 1958; J. Wolkovitch and R. P. Walton, "VTOL and Helicopter Approximate Transfer Functions and Closed-Loop Handling Qualities," Systems Technology Rept. 128-1, Sept. 1963; R. L. Stapleford, J. Wolkovitch, R. E. Magdaleno, C. P. Shortwell, and W. A. Johnson, "An Analytical Study of V/STOL Handling Qualities in Hover and Transition," AFFDL-TR-65-73, May 1965.

basic transfer functions and on the flight control system accommodations required for such changes. They are especially useful for the preliminary selection of appropriate flight control system feedbacks. For example, a very crude, albeit relatively accurate, way of considering the basic effects of a stability-augmenting flight control system is to regard its action primarily as one of directly augmenting various aerodynamic derivatives. From this point of view a pitch rate damper ($q \rightarrow \delta_e$) simply augments the derivatives $M_q$ and $Z_q$ (for $X_{\delta_e}$ negligible). Accordingly, the primary effect, related to the increased $-M_q$, is to improve the short period damping. Table 5-6 delineates the important derivatives and the qualitative consequences of their augmentation for conventional aircraft.

The conditions for which the Table 5-5 approximate factors are valid do not extend to those appropriate for VTOL aircraft and helicopters in hovering and transition flight. Such situations are treated in Table 5-8 which, because of the different vehicle types considered, contains a variety of possibly applicable expressions. Table 5-7 serves as a guide to Table 5-8 and contains additional information concerning the very simple transfer function forms appropriate for hover. Although the unconventional aircraft covered are all VTOL or helicopter types, we continue to call the primary control inputs "elevator" and "throttle." However, as noted in Table 5-7, "elevator" now means whatever device is used to produce a pitching moment (e.g., longitudinal cyclic pitch on a single-rotor helicopter), and "throttle" now means a thrust input regardless of direction (e.g., collective pitch on a helicopter that produces a vertical thrust).

The connections between the dominant derivatives and the resulting transfer function poles and zeros listed in Table 5-8 serve the general purpose of delineating potentially important feedbacks and furnishing direct insight into the basic vehicle dynamics and the effects of changes in configuration, speed, etc. In this regard notice particularly the expression for the tilt-duct hovering oscillation given in Table 5-8(a), and recall that the corresponding example of time vectors in Fig. 5-9(a) indicated strong $M_q$ and $M_u$ influences on the damping. We can examine such influences in detail by directly considering the approximate factor

$$2\zeta_p\omega_p \doteq -\sqrt[3]{gM_u} - \tfrac{2}{3}(X_u + M_q) \tag{5-46}$$

First we note that substituting the example values into Eq. 5-46 gives

$$\begin{aligned} 2\zeta_p\omega_p &\doteq -\sqrt[3]{32.2(0.0136)} + \tfrac{2}{3}(0.137 + 0.045) \\ &\doteq -0.759 + 0.121 \doteq -0.638 \end{aligned}$$

in good agreement with the exact value (Eq. 5-44) of $-0.642$. With our confidence in the approximate expression thus established, we can now fix the relative effects of the dominant derivatives on the total damping.

Table 5-5. Summary of conventional airplane longitudinal approximate factors.

| Factored forms | Approximate factors | Conditions of validity |
|---|---|---|
| $\Delta(s) = \underbrace{(s^2 + 2\zeta_p\omega_p s + \omega_p^2)}(s^2 + 2\zeta_{sp}\omega_{sp}s + \omega_{sp}^2)$ | $\omega_{sp}^2 \doteq M_q Z_w - M_\alpha$ | $\left\| \dfrac{X_w Z_u}{Z_w(X_u + Z_w)} \right\| \ll 1$ |
| | $2\zeta_{sp}\omega_{sp} \doteq -(Z_w + M_q + M_{\dot{\alpha}})$ | |
| or | $\omega_p^2 \text{ or } \dfrac{1}{T_{p_1}T_{p_2}} \doteq \dfrac{g(M_w Z_u - M_u Z_w)}{Z_w M_q - M_\alpha}$ | $\left\| \dfrac{-g}{U_0}\left(Z_u - \dfrac{M_u}{M_w} Z_w\right) \right\| < \left(\dfrac{Z_w}{2}\right)^2 \text{or} \left(\dfrac{M_q}{2}\right)^2$ |
| $\left(s + \dfrac{1}{T_{p_1}}\right)\left(s + \dfrac{1}{T_{p_2}}\right)$ | $2\zeta_p\omega_p \text{ or } \left(\dfrac{1}{T_{p_1}} + \dfrac{1}{T_{p_2}}\right) \doteq -X_u - \dfrac{M_u(X_\alpha - g)}{Z_w M_q - M_\alpha}$ | $\left\| M_{\dot{\alpha}}\left(\omega_p^2 + \dfrac{g}{U_0} Z_u\right) + M_q X_w Z_u \right\|$ $\ll \|-M_u(X_\alpha - g)\|$ |
| $N_{\delta_e}^{\theta}(s) = A_\theta\left(s + \dfrac{1}{T_{\theta_1}}\right)\left(s + \dfrac{1}{T_{\theta_2}}\right)$ | $A_\theta \doteq M_\delta$ | |
| | $\dfrac{1}{T_{\theta_1}} \doteq -X_u + \dfrac{X_w}{Z_w} Z_u \dfrac{[1 - (Z_\delta/M_\delta)(M_u/Z_u)]}{[1 - (Z_\delta/M_\delta)(M_w/Z_w)]}$ | $\|Z_\delta M_{\dot{w}}\| \ll \|M_\delta\|$ <br> $\|X_u\| \ll \|Z_w\|$ |
| | $\dfrac{1}{T_{\theta_2}} \doteq -Z_w + \dfrac{M_w}{M_\delta} Z_\delta$ | |
| $N_{\delta_e}^{w}(s) = A_w\left(s + \dfrac{1}{T_{w_1}}\right)(s^2 + 2\zeta_w\omega_w s + \omega_w^2)$ | $A_w \doteq Z_\delta$ | |
| | $\dfrac{1}{T_{w_1}} \doteq \dfrac{U_0 M_\delta}{Z_\delta}$ | $\|Z_\delta(M_q + X_u)\| \ll \|U_0 M_\delta\|$ |
| | $\omega_w^2 \doteq \dfrac{g}{U_0}\left(-Z_u + \dfrac{Z_\delta}{M_\delta} M_u\right)$ | |
| | $2\zeta_w\omega_w \doteq -X_u$ | |

| | | |
|---|---|---|
| $N^{u}_{\delta_e}(s) = A_u\left(s + \frac{1}{T_{u_1}}\right)\left(s + \frac{1}{T_{u_2}}\right)$ | $A_u \doteq Z_\delta X_w$<br>$\frac{1}{T_{u_1}} \doteq \frac{-g}{X_\alpha - g}\left(-Z_w + \frac{M_w}{M_\delta} Z_\delta\right)$<br>$\frac{1}{T_{u_2}} \doteq \frac{M_\delta(X_\alpha - g)}{Z_\delta X_w}$ | $\lvert Z_\delta(gM_{\dot{u}} + M_q X_w)\rvert \ll \lvert M_\delta(X_\alpha - g)\rvert$ |
| $sN^{h}_{\delta_e} = A_h\left(s + \frac{1}{T_{h_1}}\right)\underbrace{\left(s + \frac{1}{T_{h_2}}\right)\left(s + \frac{1}{T_{h_3}}\right)}$ Aft control<br>or<br>$(s^2 + 2\zeta_h\omega_h s + \omega_h^2)$ Forward control | $A_h \doteq -Z_\delta$<br>$\frac{1}{T_{h_1}} \doteq -X_u + (X_\alpha - g)\frac{Z_u}{Z_\alpha}\frac{[1 - (Z_\delta/M_\delta)(M_u/Z_u)]}{[1 - (Z_\delta/M_\delta)(M_\alpha/Z_\alpha)]}$<br>$\frac{1}{T_{h_2}} = \frac{-1}{T_{h_3}} \doteq \left(M_\alpha - \frac{M_\delta}{Z_\delta} Z_\alpha\right)^{1/2}$<br>$\omega_h^2 \doteq -M_\alpha + \frac{M_\delta}{Z_\delta} Z_\alpha$<br>$2\zeta_h\omega_h \doteq -\left(M_{\dot{\alpha}} + M_q + X_u + \frac{1}{T_{h_1}}\right)$ | $\lvert X_u(M_{\dot{\alpha}} + M_q)\rvert \ll \left\lvert M_\delta - \frac{M_\delta}{Z_\delta} Z_\alpha\right\rvert$ |

Notes: General assumption is $X_{\delta_e} = 0$.

Applicable also to VTOL airplanes in cruise or high speed configuration (i.e., tilt-wing speeds for which wing incidence is 45° greater than that at hover; tilt-duct speeds greater than 100 ft/sec; tilt-rotor at all speeds); and to single-rotor helicopters at speeds greater than 50 ft/sec.

$\gamma/\delta = sN_h(s)/U_0\,\Delta(s)$; $a_{z/\delta} = -s^2N_h(s)/\Delta(s)$.

Table 5-6. **Qualitative importance of normal airframe stability parameters to longitudinal elevator-input transfer function quantities**[a].

| Quantities affected | Derivative | | | | | | | Change in point of control application |
|---|---|---|---|---|---|---|---|---|
| | $M_u$ | $M_\alpha$ | $M_{\dot{\alpha}}$ | $M_q$ | $Z_u$ | $Z_\alpha$ | $X_u$ | $l_\delta$ |
| Denominator, $\Delta(s)$ | | | | | | | | |
| Short period undamped frequency, $\omega_{sp}$ | | xxx | | x | | x | | |
| Short period damping, $2\zeta_{sp}\omega_{sp}$ | | | xx | xx | | xx | | |
| Classical phugoid, $\omega_p = -\sqrt{2(g/U_0)}$, $2\zeta_p\omega_p \doteq -X_u$ | | | | | | | xxx | |
| Normal phugoid undamped frequency, $\omega_p \doteq [g(M_wZ_u - M_uZ_w)/(Z_wM_q - M_\alpha)]^{1/2}$ | xxx | x | | x | xx | x | | |
| "Tuck" and longitudinal subsidence, $1/T_{p_1}$ and $1/T_{p_2}$ | xxx | x | | x | xx | x | | |
| Normal or "tuck" damping, $2\zeta_p\omega_p$ or $(1/T_{p_1} + 1/T_{p_2})$ | xx | x | | x | | x | | |
| Pitch Numerator, $N_\theta(s)$ | | | | | | | | |
| Low frequency factor, $1/T_{\theta_1}$ | x | x | | | x | | | |
| High frequency factor, $1/T_{\theta_2}$ | | xx | | | | xxx | | xx |
| Airspeed Numerator, $N_u(s)$ | | | | | | | | |
| Low frequency factor, $1/T_{u_1}$ | | x | | | | xxx | | x |
| High frequency factor, $1/T_{u_2}$ | | | | | | | | xxx |
| Downward Velocity Numerator, $N_w(s)$ | | | | | | | | |
| First-order factor, $1/T_{w_1}$ | | | | | | | | xxx |
| Quadratic factor undamped frequency, $\omega_w$ | xx | | | | x | | | xx |
| Quadratic factor damping, $2\zeta_w\omega_w$ | | | | | | | | |
| Altitude (Flight Path, Acceleration, etc.) Numerator, $N_h(s)$ | | | | | | | | |
| First-order factor, $1/T_{h_1}$ | x | x | | | xx | xx | | x |
| Aft control quadratic, $1/T_{h_2}$ and $1/T_{h_3}$ | | x | | | | xxx | | xxx |
| Forward control quadratic undamped frequency, $\omega_h$ | | x | | | | xxx | | xxx |
| Forward control quadratic damping, $2\zeta_h\omega_h$ | x | x | xx | xx | x | x | | x |

Code: blank = little or no effect; x = moderate effect; xx = important effect; xxx = predominant effect.

[a] The only stability parameters considered in this table are those that (1) exist in the normal uncontrolled airframe or (2) can be augmented by relatively simple automatic control systems. Some of these ($M_\alpha$, $M_q$, and the point of control application) are also modified by fairly simple airframe geometry changes.

Table 5-7. Summary of the unconventional aircraft longitudinal approximate factors given in table 5-8.

| | | | Tilt-wing | | Helicopter | | | | Tilt-duct | | Tilt-rotor |
|---|---|---|---|---|---|---|---|---|---|---|---|
| | | | | | Single-rotor | | Tandem-rotor | | | | |
| | | | | | Hover | High speed | Hover | High speed | Hover | High speed | |
| | | | Low speed[a,c] | High speed[b,c] | $U_0 \doteq 0$ | $U_0 > 50$ fps | $U_0 \doteq 0$ | $U_0 > 40$ fps<br>$U_0 < 120$ fps | $U_0 \doteq 0$ | $U_0 > 100$ fps | All speeds |
| | Denominator | | √ | Table 5-5 | √ | Table 5-5 | √ | √ | √ | Table 5-5 | |
| Numerators | Elevator | $\theta$ | √ | √ | √ | √ | √ | √ | √ | √ | √ |
| | | $w$ | — | √ | Zero | √ | Zero | √ | — | √ | √ |
| | Pitching moment control | $u$ | √ | √ | √ | √ | √ | √ | √ | √ | √ |
| | | $\dot{h}$ | — | — | Zero | √ | Zero | √ | — | √ | — |
| | Throttle | $\theta$ | √ | √ | Zero | √ | Zero | √ | Zero | √ | √ |
| | | $w$ | * | — | * | √ | * | √ | * | — | — |
| | Thrust control | $u$ | — | — | Zero | √ | Zero | √ | Zero | — | — |
| | | $\dot{h}$ | * | — | * | — | * | √ | * | — | — |

"Zero" indicates that response is approximately zero.

Asterisk (*) indicates conditions for which numerator and denominator factors cancel to give, approximately, $\dot{h}/\delta_T \doteq -w/\delta_T \doteq -Z_{\delta_T}/(s - Z_w)$. In general, this requires close to zero response in $\theta$ and $u$, which can also occur for tilt-rotor and tilt-wing hover, provided that $M_{\delta_T}$ and $X_{\delta_T}$ are negligibly small.

[a] Including range of speed for which wing incidence is within 45° of wing incidence at hover.

[b] Wing incidence greater than 45° from incidence at hover.

[c] The separation of high and low speed factors by wing incidence is empirical.

Table 5-8. Approximate longitudinal factors for unconventional aircraft.

A. Denominator ($\Delta_{\text{long}}$) factors.

| First coeff. | Approximate factors | Conditions of validity | Applicable to: |
|---|---|---|---|
| | $\frac{1}{T_{sp_1}} \doteq -Z_w$ | $\lvert U_0 M_w \rvert \ll \lvert Z_w M_q \rvert$ | Tilt-wing |
| | $\frac{1}{T_{sp_2}} \doteq \sqrt[3]{gM_u} + \frac{-X_u - M_q}{3}$ | $\left\lvert \frac{gM_u}{M_q^3} \right\rvert > 1$ | Low speed[a,b] |
| | $2\zeta_p\omega_p \doteq -\sqrt[3]{gM_u} + \frac{2}{3}(-X_u - M_q)$ | $\left\lvert \frac{gM_u}{X_u^3} \right\rvert > 1$ | Tilt-duct |
| | $\omega_p^2 \doteq gM_u T_{sp_1}$ | | Hover |
| | $\frac{1}{T_{sp_1}} \doteq -Z_w$ | $\lvert M_q \rvert \gg \lvert X_u \rvert$ | |
| | $\frac{1}{T_{sp_2}} \doteq \frac{-3M_q}{4} + \sqrt{\frac{M_q^2}{16} - \frac{gM_u}{2M_q}}$ | | |
| | $2\zeta_p\omega_p \doteq -X_u - \frac{M_q}{4} - \sqrt{\frac{M_q^2}{16} - \frac{gM_u}{2M_q}}$ | $\left\lvert \frac{gM_u}{M_q^3} \right\rvert < 1$ | Helicopters: Hover |
| 1 | $\omega_p^2 \doteq \frac{gM_u}{(-3M_q/4) + \sqrt{(M_q^2/16) - (gM_u/2M_q)}}$ | If $\lvert X_u \rvert \gg \lvert M_q \rvert$ and $\lvert gM_u/X_u^3 \rvert < 1$, interchange $X_u$ and $M_q$ in equations | |

| | | | |
|---|---|---|---|
| | $\frac{1}{T_{sp_1}} \doteq -\left(\frac{M_q + Z_w + U_0 M_{\dot{w}}}{2}\right) + \sqrt{\left(\frac{M_q + Z_w + U_0 M_{\dot{w}}}{2}\right)^2 + U_0 M_w - M_q Z_w}$ <br> $\frac{-1}{T_{sp_2}} \doteq \omega_p \doteq -[g T_{sp_1}(M_w Z_u - M_u Z_w)]^{1/3}$ <br> $2\zeta_p \omega_p \doteq -X_u - \frac{1}{T_{sp_2}}$ | $\left\lvert\frac{1}{T_{sp_1}}\right\rvert > 5 \left\lvert\frac{1}{T_{sp_2}}\right\rvert$ <br> $\lvert gM_u \rvert \gg \lvert -X_u(M_q Z_w - U_0 M_w) - M_u U_0 X_w + M_q X_w Z_u + g Z_u M_{\dot{w}} \rvert$ <br> If $M_u$ is too small to satisfy this condition, use conventional airplane approximate factors (Table 5-5) | Helicopters: <br> 40 fps $< U_0 <$ 120 fps |
| | **B. $\theta$ Numerator ($N_\delta^\theta$) factors.** | | |
| $M_\delta$ | $\frac{1}{T_{\theta_1}} + \frac{1}{T_{\theta_2}} \doteq \frac{X_\delta}{M_\delta}(Z_u M_{\dot{w}} + M_u) + \frac{Z_\delta}{M_\delta}(M_w - X_u M_{\dot{w}}) - (X_u + Z_w)$ <br> $\frac{1}{T_{\theta_1}} \frac{1}{T_{\theta_2}} \doteq \frac{X_\delta}{M_\delta}(Z_u M_w - Z_w M_u) + \frac{Z_\delta}{M_\delta}(M_u X_w - M_w X_u) + (Z_w X_u - X_w Z_u)$ | $\lvert Z_\delta M_{\dot{w}} \rvert \ll \lvert M_\delta \rvert$ | Tilt-wing <br> All speeds, elevator, throttle <br> Tilt-duct <br> All speeds, elevator <br> $U_0 > 100$ fps, throttle <br> Tilt-rotor <br> All speeds, throttle |
| | $\frac{1}{T_{\theta_1}} \doteq \frac{X_\delta(M_u Z_w - Z_u M_w + Z_\delta(X_u M_w - M_u X_w) + M_\delta(Z_u X_w - X_u Z_w)}{Z_w M_\delta}$ <br> $\frac{1}{T_{\theta_2}} \doteq -Z_w$ | $\left\lvert\frac{1}{T_{\theta_1}}\right\rvert \ll \left\lvert\frac{1}{T_{\theta_2}}\right\rvert$, $\left\lvert\frac{M_w}{M_\delta} Z_\delta\right\rvert \ll \lvert Z_w \rvert$ | Helicopters <br> All speeds, elevator <br> Tilt-rotor <br> All speeds, elevator |

Table 5-8 (Continued)

B. $\theta$ Numerator ($N_\delta^\theta$) factors

| First coeff. | Approximate factors | Conditions of validity | Applicable to: |
|---|---|---|---|
| $M_\sigma$ | $\frac{1}{T_{\theta_1}} \doteq -X_u - \frac{M_u(Z_wX_\delta - X_wZ_\delta)}{M_wZ_\delta - Z_wM_\delta}$<br>$\frac{1}{T_{\theta_2}} \doteq -Z_w + \frac{M_wZ_\delta + M_uX_\delta}{M_\delta} \doteq -Z_w + \frac{M_w}{M_\delta}Z_\delta$ | $\left\|\frac{1}{T_{\theta_1}}\right\| \ll \|Z_w\|$<br>$\left\|\frac{1}{T_{\theta_1}}\right\| \ll \left\|\frac{1}{T_{\theta_2}}\right\|$<br>$M_\delta \neq 0$ | Single-rotor helicopter<br>$U_0 > 50$ fps, throttle |
| $Z_\delta M_{\dot{w}}$ | $\frac{1}{T_{\theta_1}} \doteq \frac{M_w}{M_{\dot{w}}}$<br>$\frac{1}{T_{\theta_2}} \doteq -X_u + M_u\frac{X_w}{M_w}$ | $M_{\dot{w}} \neq 0$<br>$\|M_\delta\| \ll \|Z_\delta M_{\dot{w}}\|$<br>$\|X_\delta\| \ll \|Z_\delta\|$<br>$U_0 \neq 0$ | Tandem-rotor helicopter<br>Throttle for<br>40 fps $< U_0 < 120$ fps |

C. $w$ Numerator ($N_\delta^w$) factors.

| First coeff. | Approximate factors | Conditions of validity | Applicable to: |
|---|---|---|---|
| | $\frac{1}{T_{w_1}} \doteq \frac{U_0M_\delta}{Z_\delta}$<br>$\omega_w^2 \doteq \frac{g}{U_0}\left(\frac{M_uZ_\delta}{M_\delta} - Z_u\right)$<br>$2\zeta_w\omega_w \doteq -X_u$ | $\|Z_\delta(M_q + X_u) - X_\delta Z_u\| \ll \|U_0M_\delta\|$<br>$\|X_\delta(U_0M_u - Z_uM_q)\| \ll \|X_u(Z_\delta M_q - U_0M_\delta)\|$ | Tilt wing<br>High speed,[c] elevator<br>Tilt duct<br>$U_0 > 100$ fps, elevator |
| $Z_\delta$ | $\frac{1}{T_{w_1}} \doteq \frac{U_0M_\delta}{Z_\delta}$<br>$\omega_w^2 \doteq \frac{g}{U_0}\left(\frac{M_uZ_\delta}{M_\delta} - Z_u\right)$<br>$2\zeta_w\omega_w \doteq \frac{1}{U_0M_\delta}[U_0(X_\delta M_u - X_uM_\delta) + M_q(X_uZ_\delta - X_\delta Z_u) - Z_\delta\omega_w^2]$ | $\left\|\frac{1}{T_w}\right\| \gg \|M_q\|$<br>$U_0 \neq 0$ (at $U_0 = 0$, $Z_\delta = 0$)<br>$\left\|\frac{1}{T_w}\right\| \gg \omega_w$<br>$\|U_0M_\delta\| \gg \|X_uZ_\delta\|$ | Helicopters: Elevator<br>Single-rotor,<br>$U_0 > 50$ fps<br>Tandem-rotor, 40 fps $< U_0 < 120$ fps<br>Tilt-rotor<br>All speeds, elevator |

| | | | |
|---|---|---|---|
| | $\frac{1}{T_{w_1}} \doteq -M_q$<br>$\frac{1}{T_{w_2}}\frac{1}{T_{w_3}}$ or $\omega_w^2 \doteq \frac{gM_u}{-M_q}$<br>$\frac{1}{T_{w_2}} + \frac{1}{T_{w_3}}$ or $2\zeta_w\omega_w \doteq -X_u - \frac{gM_u}{M_q^2}$ | $U_0 \neq 0 \left[\text{for } U_0 = 0, \frac{w(s)}{\delta(s)} = \frac{Z_\delta}{s - Z_w}\right]$<br>$\lvert X_u\rvert \ll \lvert 1/T_w\rvert$<br>$\lvert X_u\rvert \ll \lvert \omega_w\rvert$<br>$\lvert X_\delta\rvert \ll \lvert Z_\delta\rvert$<br>$\lvert U_0 M_\delta\rvert \ll \lvert Z_\delta M_q\rvert$ | Helicopters<br>For above speeds, throttle |
| D. $u$ Numerator ($N_\delta^u$) factors. | | | |
| $Z_\delta X_w$ | $\frac{1}{T_{u_1}} \doteq \frac{M_\delta(U_0X_w - g)}{Z_\delta X_w}$<br>$\frac{1}{T_{u_2}} \doteq \frac{g[Z_w - (Z_\delta/M_\delta)M_w]}{(U_0X_w - g)}$ | $\lvert Z_\delta(gM_\delta^u + M_qX_w\rvert \ll \lvert M_\delta(U_0X_w - g)\rvert$<br>$\left\lvert \frac{X_wZ_\delta}{X_\delta}\right\rvert \gg \left\lvert\frac{1}{T_{u_2}}\right\rvert \gg \left\lvert\frac{1}{T_{u_1}}\right\rvert$<br>For $U_0 \doteq 0, \frac{1}{T_{u_2}} = -Z_w, \frac{1}{T_{u_1}} \to \infty$ | Tilt-wing<br>All speeds, elevator<br>Tilt-duct<br>All speeds, elevator |
| $X_\delta$ | $\frac{1}{T_{u_1}} \doteq -Z_w$<br>$\omega_u^2 \doteq \frac{-g}{X_\delta}\left(M_\delta - Z_\delta\frac{M_w}{Z_w}\right)$<br>$2\zeta_u\omega_u \doteq -M_q$ | $\lvert X_\delta(M_q + Z_w)\rvert \gg \lvert U_0X_\delta M_{\dot{w}} - Z_\delta X_w\rvert$<br>$\lvert X_\delta Z_wM_q - gM_\delta\rvert \gg \lvert U_0(X_wM_\delta - X_\delta M_w)$<br>$- Z_\delta(gM_{\dot{w}} + X_wM_q)\rvert$ | Helicopters<br>All speeds, elevator<br>Tilt-rotor<br>All speeds, elevator |
| | $\frac{1}{T_{u_1}} \doteq -Z_w + X_w\frac{Z_\delta}{X_\delta}$<br>$\omega_u^2 \doteq \frac{-gM_\delta}{X_\delta}$<br>$2\zeta_u\omega_u \doteq -M_q$ | $X_\delta \neq 0$ (Note: $X_\delta = 0$ at $U_0 = 0$)<br>$\lvert X_wZ_\delta\rvert < \lvert Z_wX_\delta\rvert$ | Single-rotor helicopter<br>$U_0 > 50$ fps, throttle |

Table 5-8 (Continued)

D. $u$ Numerator ($N_\delta^u$) factors.

| | | | |
|---|---|---|---|
| $Z_\delta X_w$ | $\frac{1}{T_{u_1}}\frac{1}{T_{u_2}} \doteq \frac{-gM_w}{X_w}$<br>$\frac{1}{T_{u_1}} + \frac{1}{T_{u_2}} \doteq -M_q - \frac{gM_{\dot{w}}}{X_w}$ | $X_\delta = M_\delta = 0$<br>$U_0 \neq 0$ | Tandem-rotor helicopter<br>Throttle for 40 fps $< U_0 <$ 120 fps |

E. $\dot{h}$ Numerator ($sN_\delta^h$) factors.

| | | |
|---|---|---|
| $\frac{1}{T_{h_1}} \doteq -X_u + (U_0X_w - g)\,\frac{M_u - (M_\delta/Z_\delta)Z_u}{U_0(M_w - (M_\delta/Z_\delta)Z_w)}$<br>$-\frac{1}{T_{h_2}} \doteq \frac{1}{T_{h_3}} \doteq \sqrt{U_0\left(M_w - \frac{M_\delta}{Z_\delta}Z_w\right)}$ | $\left(\frac{1}{T_{h_2}}\right)^2 \gg \left\lvert\left(X_u - \frac{X_\delta Z_u}{Z_\delta}\right)(M_q + U_0M_{\dot{w}})\right\rvert$<br>$\left\lvert\frac{1}{T_{h_1}}\frac{1}{T_{h_2}}\frac{1}{T_{h_3}}\right\rvert \gg \left\lvert\frac{X_\delta}{Z_\delta}\,U_0(Z_wM_u - Z_uM_w)\right\rvert$<br>For $U_0 \doteq 0$, $\dot{h} \doteq -w$ | Tilt duct<br>$U_0 >$ 100 fps, elevator |

| | | | |
|---|---|---|---|
| $-Z_\delta$ | $\frac{1}{T_{h_1}} \doteq -\frac{gZ_u}{U_0Z_w} + \frac{Z_\delta}{M_\delta}\frac{gM_u}{U_0Z_w} + \frac{X_\delta}{M_\delta}\left(M_u - \frac{M_wZ_u}{Z_w}\right)$<br>$w_h^2 \doteq \frac{M_\delta U_0 Z_w}{Z_\delta}$<br>$2\zeta_h\omega_h \doteq -M_q - U_0M_{\dot{w}}$ | $\left\lvert\frac{1}{T_h}\right\rvert \ll \omega_h$<br>$\lvert M_wZ_\delta\rvert \ll \lvert Z_wM_\delta\rvert$<br>$\left\lvert\frac{1}{T_h}\omega_h^2\right\rvert \gg U_0\left\lvert(M_uX_w - M_wX_u) + \frac{M_\delta}{Z_\delta}(X_uZ_w - X_wZ_u)\right\rvert$ | Helicopters: Elevator<br>Single-rotor, $U_0 > 50$ fps<br>Tandem-rotor, 40 fps $< U_0 < 120$ fps |
| | $\frac{1}{T_{h_1}} \doteq -X_u - (g - U_0X_w)\frac{M_u}{U_0M_w}$<br>$\frac{1}{T_{h_2}} + \frac{1}{T_{h_3}}$ or $2\zeta_h\omega_h \doteq -M_q$<br>$\frac{1}{T_{h_2}}\frac{1}{T_{h_3}}$ or $\omega_h^2 \doteq -U_0M_w$ | $U_0 \neq 0\ \left[\text{for } U_0 = 0, \frac{\dot{h}(s)}{\delta(s)} = \frac{-Z_\delta}{s - Z_w}\right]$<br>$\lvert Z_u\rvert \ll \lvert U_0Z_w\rvert$<br>$\lvert M_u\rvert \ll \lvert U_0M_w\rvert$<br>$X_\delta = M_\delta = 0$ | Tandem-rotor helicopter Throttle for 40 fps $<$ $U_0 < 120$ fps |

[a] Including range of speed for which wing incidence is within 45° of wing incidence at hover.
[b] The separation of high and low speed factors by wing incidence is empirical.
[c] Wing incidence 45° greater than that at hover.

Applying Eq. 5-46, we see that, for the example values, the order of dominance is $M_u$, $X_u$, $M_q$ and that a 100 percent increase in $-M_q$ can be counteracted by about a 12 percent decrease in $M_u$. For large changes in $M_q$, Eq. 5-46 may not be appropriate because it requires as a condition of validity that $|gM_u| > |M_q^3|$. For example, if $-M_q$ is increased to 1.0, then $|gM_u| < |M_q^3|$ and the appropriate approximation, Table 5-8(a), is

$$2\zeta_p\omega_p \doteq -X_u - \frac{M_q}{4} - \sqrt{\frac{M_q^2}{16} - \frac{gM_u}{2M_q}} \tag{5-47}$$

## 5-12. Approximate Modal Response Ratios

The ways in which the various degrees of freedom enter into the total motion corresponding to a given dynamic mode can most simply be studied by considering the modal response ratios. We have already used these in the time vector diagrams for the specific examples of this chapter, but now we shall generalize to obtain the more interesting approximate literal relationships for conventional airplanes. As noted in Chapter 2, the modal response ratios can be expressed using any *one* of the $n$ cofactor sets (for $n$ degrees of freedom) of the characteristic determinant, i.e., denoting two of the $n$ degrees of freedom as $\alpha_i$, $\alpha_j$:

$$\left(\frac{\alpha_i}{\alpha_j}\right) = (-1)^{i-j}\left(\frac{\Delta_{ki}(s)}{\Delta_{kj}(s)}\right)_{s=-q} \tag{5-48}$$

where $k = 1, 2, \ldots, n$.

*Short Period*

Applying Eq. 5-48 to the short period modal responses, we first compute the ratio $u/w$ from the selected cofactors of the Eq. 5-3 denominator:

$$\left(\frac{u}{w}\right)_{sp} = \left(\frac{-\Delta_{11}}{\Delta_{12}}\right)_{s=-q_{sp}}$$

$$\left(\frac{u}{w}\right)_{sp} = -\frac{\left.\begin{vmatrix} s - Z_w & -U_0 s \\ -(M_{\dot{w}}s + M_w) & s(s - M_q) \end{vmatrix}\right|_{s=-q_{sp}}}{\left.\begin{vmatrix} -Z_u & -U_0 s \\ -M_u & s(s - M_q) \end{vmatrix}\right|_{s=-q_{sp}}}$$

$$= \left(\frac{s^2 - s(Z_w + M_q + M_{\dot{\alpha}}) - M_\alpha + M_q Z_w}{-Z_u(s - M_q) - U_0 M_u}\right)_{s=-\zeta_{sp}\omega_{sp}+j\omega_{sp}\sqrt{1-\zeta_{sp}^2}} \tag{5-49}$$

For the approximate short-period frequency and damping given by Eq. 5-17,

$$\omega_{sp}^2 \doteq -M_\alpha + M_q Z_w$$

$$2\zeta_{sp}\omega_{sp} \doteq -(Z_w + M_q + M_{\dot\alpha})$$

it is obvious that the numerator of Eq. 5-49 is identically zero; i.e.,

$$\left(\frac{u}{w}\right)_{sp} = 0 \tag{5-50}$$

This is the classical statement of the *conditions of validity* for the short period approximation. It is, in fact, taken as the starting point, based on flight test observations, for the development of the approximate *two degree of freedom, short period equation* given previously (Eq. 5-14).

From the characteristic determinant of the *short period* equation, we can now conveniently compute the ratio $\theta/w$ from either of the expressions:

$$\left(\frac{\theta}{w}\right)_{sp} \doteq \left(\frac{s - Z_w}{U_0 s}\right)_{s=-q_{sp}} \doteq \left[\frac{M_{\dot w} s + M_w}{s(s - M_q)}\right]_{s=-q_{sp}}$$

Using the first, and converting to angle of attack, $\alpha \equiv w/U_0$,

$$\left(\frac{\theta}{\alpha}\right)_{sp} \doteq \left(\frac{s - Z_w}{s}\right)_{s=-q_{sp}} \doteq \frac{-(\zeta\omega)_{sp} - Z_w + j\omega_{sp}\sqrt{1-\zeta_{sp}^2}}{-(\zeta\omega)_{sp} + j\omega_{sp}\sqrt{1-\zeta_{sp}^2}} \tag{5-51}$$

The magnitude of the modal response ratio, which is the square root of the squared sums of the real and imaginary parts, can be simplified by substituting the approximate expressions for $\omega_{sp}^2$ and $2(\zeta\omega)_{sp}$ (Eq. 5-17) to yield

$$\left|\frac{\theta}{\alpha}\right|_{sp} \doteq \left[1 + \frac{Z_w(-M_{\dot\alpha} - M_q)}{-M_\alpha + M_q Z_w}\right]^{1/2} \doteq \left[\frac{-M_\alpha - M_{\dot\alpha} Z_w}{-M_\alpha + M_q Z_w}\right]^{1/2} \tag{5-52}$$

The phase angle of the modal response ratio can be written as the difference between the numerator lead and the denominator lag contributions; these can be combined via the identity

$$\tan^{-1} x \pm \tan^{-1} y \equiv \tan^{-1}\frac{(x \pm y)}{(1 \mp xy)}$$

to give

$$\measuredangle\left(\frac{\theta}{\alpha}\right)_{sp} \doteq \tan^{-1}\frac{\sqrt{1-\zeta_{sp}^2}}{\zeta_{sp} + (\omega_{sp}/Z_w)} \tag{5-53}$$

For the usual $\zeta_{sp} \ll \omega_{sp}/Z_w$ this further reduces to

$$\measuredangle\left(\frac{\theta}{\alpha}\right)_{sp} \doteq \tan^{-1}\frac{Z_w}{\omega_{sp}}\sqrt{1-\zeta_{sp}^2} \tag{5-54}$$

In terms of the flight path angle, $\gamma \equiv \theta - \alpha$, Eq. 5-51 can be expanded to

$$\left(\frac{\gamma}{\alpha}\right)_{sp} \doteq \left(\frac{-Z_w}{s}\right)_{s=-q_{sp}} \doteq \frac{-Z_w}{-\zeta_{sp}\omega_{sp} + j\omega_{sp}\sqrt{1-\zeta_{sp}^2}} \tag{5-55}$$

whereby

$$\left|\frac{\gamma}{\alpha}\right|_{sp} \doteq \frac{-Z_w}{\omega_{sp}}$$
$$\measuredangle\left(\frac{\gamma}{\alpha}\right)_{sp} = -\tan^{-1}\frac{\omega_{sp}\sqrt{1-\zeta_{sp}^2}}{-\zeta_{sp}\omega_{sp}} = \tan^{-1}\frac{\sqrt{1-\zeta_{sp}^2}}{\zeta_{sp}} - 180^\circ \tag{5-56}$$

Notice that for small $\zeta_{sp}$, Eq. 5-55 becomes

$$-Z_w\alpha_{sp} \doteq j\omega_{sp}\gamma_{sp} = \dot{\gamma}_{sp}$$
$$-Z_\alpha\alpha_{sp} \doteq U_0\dot{\gamma}_{sp} = \dot{h}_{sp} \tag{5-57}$$

This is simply the $Z$ equation (5-1) with $u$ perturbations neglected.

Accordingly, typical short period motions for $-M_\alpha$ reasonably large involve $|\theta|$ approximately equal to but somewhat smaller than $|\alpha|$; $\theta$ lagging $\alpha$ (because $Z_w$ is invariably negative) by an angle whose tangent is given by $-Z_w\sqrt{1-\zeta_{sp}^2}/\omega_{sp}$; $|\gamma/\alpha|$ proportional to part of the same quantity, $-Z_w/\omega_{sp}$, and $\gamma$ lagging $\alpha$ by somewhat more than 90 degrees (the $\gamma$ vector lies in the third quadrant).

*Phugoid*

From the complete denominator of Eq. 5-3 the phugoid modal response ratio, $(u/w)_p$, is most simply evaluated in literal terms, using the expression

$$\left(\frac{u}{w}\right)_p = \left(\frac{-\Delta_{31}}{\Delta_{32}}\right)_{s=-q_p} = -\frac{\begin{vmatrix} -X_w & g \\ s - Z_w & -U_0 s \end{vmatrix}_{s=-q_p}}{\begin{vmatrix} s - X_u & g \\ -Z_u & -U_0 s \end{vmatrix}_{s=-q_p}}$$
$$= \left[\frac{s(X_\alpha - g) + gZ_w}{U_0\left(s^2 - X_u s - \dfrac{g}{U_0}Z_u\right)}\right]_{s=-q_p} \tag{5-58}$$

For $s = -\zeta_p\omega_p + j\omega_p\sqrt{1-\zeta_p^2}$,

$$\left(\frac{u}{w}\right)_p = \frac{(X_\alpha - g)(-\zeta_p\omega_p + j\omega_p\sqrt{1-\zeta_p^2}) + gZ_w}{U_0\left[2\zeta_p^2\omega_p^2 - \omega_p^2 - 2j\zeta_p\omega_p^2\sqrt{1-\zeta_p^2} - X_u(-\zeta_p\omega_p + j\omega_p\sqrt{1-\zeta_p^2}) - \dfrac{g}{U_0}Z_u\right]} \tag{5-59}$$

where (Table 5-5)

$$\omega_p^2 \doteq \frac{g(M_w Z_u - M_u Z_w)}{M_q Z_w - M_\alpha}; \qquad 2\zeta_p \omega_p \doteq -X_u - \frac{M_u(X_\alpha - g)}{M_q Z_w - M_\alpha}$$

Substitution of these relationships into Eq. 5-59 gives

$$\left(\frac{u}{w}\right)_p \doteq \frac{(X_\alpha - g)(-\zeta_p\omega_p + j\omega_p\sqrt{1-\zeta_p^2}) + gZ_w}{\dfrac{U_0}{M_q Z_w - M_\alpha}\left[M_u(X_\alpha - g)(-\zeta_p\omega_p + j\omega_p\sqrt{1-\zeta_p^2}) + gZ_w\left(M_u - \dfrac{M_q Z_u}{U_0}\right)\right]} \tag{5-60}$$

Considering that the $Z_w$ terms generally dominate both numerator and denominator, we find that

$$\left(\frac{u}{w}\right)_p \doteq \frac{-U_0 M_w + M_q Z_w}{U_0 M_u - M_q Z_u} \tag{5-61}$$

However, if we assume that $|M_q Z_u| \ll |U_0 M_u|$ and $|M_q Z_w| \ll |U_0 M_w|$ produce the same result for either Eq. 5-60 or 5-61, namely,

$$\left(\frac{u}{w}\right)_p = \frac{-M_w}{M_u} \qquad \text{or} \qquad (M_u u + M_w w)_p = 0 \tag{5-62}$$

The last equation corresponds to that assumed for the three degree of freedom phugoid equations of motion (Eq. 5-20), i.e., a balance of only the *static* moments. Accordingly, we can now specify more precisely, by Eq. 5-60, the conditions under which these equations are valid. For $M_u = 0$, Eq. 5-62 shows the classical result, $w = 0$; i.e., a constant angle of attack, $\alpha$, for the phugoid. The $w = 0$ result is, in fact, given exactly by Eq. 5-58 when the classical results for the phugoid undamped natural frequency and damping (Eqs. 5-28 and 5-29) are added.

Proceeding to another modal response ratio, using now the generally valid three degree of freedom phugoid equations of motion (Eq. 5-20), we compute

$$\left(\frac{\theta}{u}\right)_p = \left(\frac{\Delta_{23}}{\Delta_{21}}\right)_{s=-q_p} = \frac{\begin{vmatrix} s - X_u & -X_w \\ -M_u & -M_w \end{vmatrix}_{s=-q_p}}{\begin{vmatrix} -X_w & g \\ -M_w & 0 \end{vmatrix}_{s=-q_p}}$$

$$= -\left[\frac{s - X_u - (X_w M_u / M_w)}{g}\right]_{s=-\zeta_p\omega_p + j\omega_p\sqrt{1-\zeta_p^2}} \tag{5-63}$$

For the approximate factors of Eq. 5-24,

$$-X_u + \frac{X_w M_u}{M_w} = 2\zeta_p\omega_p + \frac{gM_u}{M_\alpha}$$

and

$$\left(\frac{-\theta}{u}\right)_p = \frac{\zeta_p\omega_p + (gM_u/M_\alpha) + j\omega_p\sqrt{1-\zeta_p^2}}{g} \tag{5-64}$$

The absolute magnitude, using the Eq. 5-24 approximations, is given by

$$\left|\frac{\theta}{u}\right|_p^2 = \frac{(g/U_0)(M_u/M_w)[-X_u + Z_w + (X_w M_u/M_w)] - (g/U_0)Z_u}{g^2} \tag{5-65}$$

Generally, $Z_w$ will be the dominant term of those in the parentheses; for such circumstances,

$$\left|\frac{\theta}{u}\right|_p^2 \doteq \frac{(-g/U_0)[Z_u - (M_u Z_w/M_w)]}{g^2} \doteq \frac{\omega_p^2}{g^2}$$
$$\left|\frac{\theta}{u}\right|_p \doteq \frac{\omega_p}{g} \qquad \text{or} \qquad |g\theta|_p \doteq |\dot{u}|_p \tag{5-66}$$

The phase angle from Eq. 5-64 is

$$\measuredangle\left(\frac{-\theta}{u}\right)_p = \tan^{-1}\frac{\omega_p\sqrt{1-\zeta_p^2}}{\zeta_p\omega_p + (gM_u/M_\alpha)} \tag{5-67}$$

and for the usual small values of $\zeta_p$ and $gM_u/\omega_p M_\alpha$ it approaches 90 degrees. Because there is also roughly a 90 degree rotation between $u$ and $\dot{u}$, $\theta$ is nearly aligned with $-u$; in view of this and Eq. 5-66,

$$(\dot{u} + g\theta)_p \doteq 0 \tag{5-68}$$

This result is clearly evident from Eq. 5-63 by considering the $s$ term to dominate the numerator.

For $M_u = 0$ we can multiply Eq. 5-68 by $U$ and substitute $\dot{h}$ for $U\theta$ (since $w \doteq 0$), whereby

$$U\dot{u} + g\dot{h} \doteq 0 \tag{5.69}$$

Now, by integrating,

$$\frac{U^2}{2} + gh \doteq \text{const}$$

which shows us that, very roughly, the sum of kinetic plus potential energies remains constant during the phugoid motion. This is again a "classical" result, first given by Lanchester. Another interesting result follows from Eq. 5-69 if we recognize (Eq. 5-30) that $g/U_0 \doteq \omega_p/\sqrt{2}$

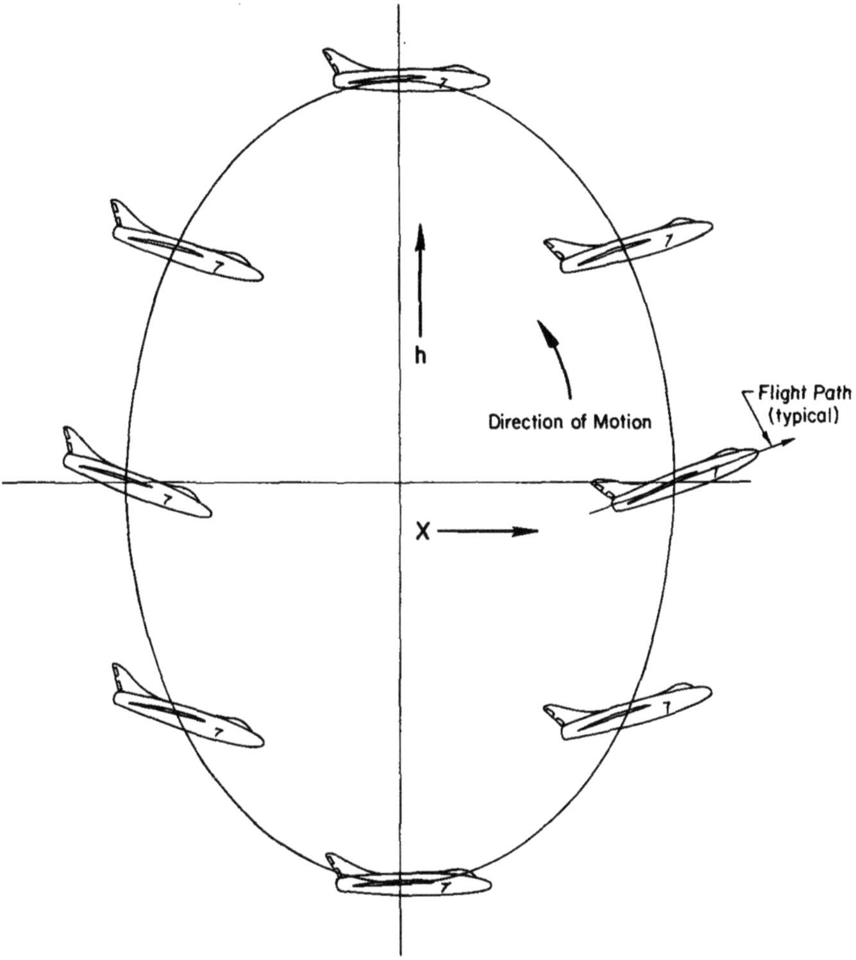


Fig. 5-12. Phugoid perturbations in space.

and that $\dot{h}_p \doteq j\omega_p h_p$; then

$$-\dot{u} \doteq \frac{g}{U_0}\, h \doteq \frac{\dot{h}}{j\sqrt{2}}$$

or

$$\dot{h} + j\sqrt{2}\, \dot{u} \doteq 0$$

Integrating twice and neglecting the first constant of integration (which leads to a time variation inappropriate to the assumed trim conditions), we obtain

$$h + j\sqrt{2}\, x \doteq \text{const}$$

or, considering absolute values,

$$h^2 + 2x^2 \doteq \text{const} \tag{5-70}$$

Equation 5-70 shows that, as seen by an observer flying in steady formation with the airplane, the phugoid motions have an elliptical pattern with an amplitude in $h$ about 1.4 times that in $x$. (See Fig. 5-12.) Furthermore, for a positive climb rate, $\dot{h}$, $\dot{u}$ is negative (Eq. 5-69) and $x$, 180 degrees out of phase with $\dot{u}$, is positive; the motion around the ellipse is therefore counter-clockwise, with $\dot{h}$ and $x$ being in phase. Finally, for positive $\dot{h}$ and negative $\dot{u}$, as above, $\theta$ is positive (Eq. 5-68), therefore in phase with $\dot{h}$ and $x$ and also aligned with the path (since $w \doteq 0$). Figure 5-12 is consistent with the time vectors shown in Fig. 5-4(b).

CHAPTER 6

# *LATERAL DYNAMICS*

## 6-1. Introduction

The treatment of the vehicle's lateral dynamic properties given in this chapter closely follows the form and content of the preceding chapter on longitudinal dynamics. The first article in this chapter recapitulates the lateral equations of motion as commonly used, presents the polynomial forms of the more important control-input transfer functions, and develops an appreciation for the transfer functions and responses with numerical examples. Based on the physical insights thus afforded, we then develop further-simplified sets of equations that apply to the individual modes of motion appropriate for conventional aircraft; this is followed by similar considerations of VTOL aircraft in hover. Then we present the complete gust-input transfer functions in polynomial form and extend our numerical examples to cover these. Approximate literal expressions for the various transfer function factors are given next; their relation to the simplified equations are established and their implications as concerns the direct influence of the dominant stability derivatives on the important poles and zeros are drawn. Finally, we develop some important modal response ratios in literal form and discuss their implications.

## 6-2. Recapitulation and Further Simplification of the Lateral Equations of Motion

Equations 4-98 are usually simplified according to the following assumptions.

Assumption 11. In the steady flight condition, the flight path of the airplane is assumed to be horizontal, $\gamma_0 = 0$.

This assumption has already been invoked in the simplification of the longitudinal equations (5-2) and is repeated here for convenience.

Assumption 12. It is assumed that $Y_{\dot{v}} = Y_p = Y_r = L'_{\dot{v}} = N'_{\dot{v}} = 0$.

Generally speaking, this is a good assumption for most configurations, especially when only control inputs are being considered. Notice, however, that for gust inputs the assumption eliminates completely the $p_g$ term in the $Y$ equation and requires $|Y_{\dot{v}}s| \ll Y_v$, in addition to the more readily evaluated $|Y_{\dot{v}}| \ll 1$. Also, in the $N$ equation, $N'_{r_g}/U_0$ is retained while $N'_{\dot{v}}$ is dropped, whereas they may both be of the same magnitude. In all cases the validity of the individual assumptions should be checked for the frequency range of interest, once the derivatives are known, by the process described in Chapter 5, Section 5-2, in connection with Assumption 10. Despite such reservations the approximations generally introduce only small errors and are in accord with common flight control practice.

With these assumptions the lateral equations of motion, referenced to *stability axes* (Eq. 4-98) and conventionally written in terms of $\beta \equiv v/U_0$ rather than $v$[1], become

$$
\begin{aligned}
&(s - Y_v)\beta && - \frac{g}{U_0}\frac{p}{s} && + r = Y^*_\delta \delta && - Y_v\beta_g \\
&-L'_\beta\beta + (s - L'_p)p && - L'_r r && = L'_\delta \delta - [L'_\beta + (L'_r)_g s]\,\beta_g - L'_p p_g \\
&-N'_\beta\beta && - N'_p p + (s - N'_r)r && = N'_\delta \delta - [N'_\beta + (N'_r)_g s]\beta_g - N'_p p_g
\end{aligned}
\tag{6-1}
$$

In terms of the unprimed derivatives, Equation 4-91 can also be written as

$$
\begin{aligned}
&(s - Y_v)\beta && - \frac{g}{U_0}\frac{p}{s} && + r = Y^*_\delta \delta && - Y_v\beta_g \\
&-L_\beta\beta && + (s - L_p)p - \left(\frac{I_{xz}}{I_x}s + L_r\right)r && = L_\delta \delta - (L_\beta + L_r s)\beta_g - L_p p_g \\
&-N_\beta\beta - \left(\frac{I_{xz}}{I_z}s + N_p\right)p && + (s - N_r)r && = N_\delta \delta - (N_\beta + N_r s)\beta_g - N_p p_g
\end{aligned}
\tag{6-2}
$$

The auxiliary relationships needed to convert the motion quantities to those usually sensed by flight instruments are (from Eq. 4-96 and the $Y$ inertial terms of Eq. 6-1):

$$
\begin{aligned}
p &= s\varphi \\
r &= s\psi \\
a_{y_{\mathrm{c.g.}}} &= U_0\dot{\beta} - g(p/s) + U_0 r = \dot{v} - g\varphi + U_0 r
\end{aligned}
\tag{6-3}
$$

## 6-3. Control-Input Transfer Functions

The control-input transfer functions are easily obtained from the equations of motion as previously discussed and illustrated. For example, the

[1] To do this, we define $Y^*_\delta \equiv Y_\delta/U_0$ and note that $L'_v v = L'_\beta\beta$, $N'_v v = N'_\beta\beta$.

$\beta$-output to $\delta$-input transfer function can be written directly from Eq. 6-2 (i.e., assuming no gust inputs) as

$$\frac{\beta(s)}{\delta(s)} = \frac{\begin{vmatrix} Y_\delta^* & \dfrac{-g}{U_0} & 1 \\ L_\delta & (s - L_p)s & -\left(\dfrac{I_{xz}}{I_x}s + L_r\right) \\ N_\delta & -\left(\dfrac{I_{xz}}{I_z}s + N_p\right)s & s - N_r \end{vmatrix}}{\begin{vmatrix} s - Y_v & \dfrac{-g}{U_0} & 1 \\ -L_\beta & (s - L_p)s & -\left(\dfrac{I_{xz}}{I_x}s + L_r\right) \\ -N_\beta & -\left(\dfrac{I_{xz}}{I_z}s + N_p\right)s & s - N_r \end{vmatrix}} \tag{6-4}$$

Expansion of the determinants gives the transfer function as a ratio of polynomials in $s$, where the denominator polynomial, $\Delta(s)$, is common to all transfer functions, and its factors determine the characteristic frequencies and dampings, or time constants, of the individual modes of motion. The numerator polynomials vary with the output quantity of interest. The polynomial forms of the primary lateral transfer functions are given below, together with their usual factored forms:

$$\begin{aligned} \frac{\beta(s)}{\delta(s)} = \frac{N_\delta^\beta(s)}{\Delta_{\text{lat}}} &= \frac{A_\beta s^3 + B_\beta s^2 + C_\beta s + D_\beta}{\Delta_{\text{lat}}} \\ &= \frac{A_\beta(s + 1/T_{\beta_1})(s + 1/T_{\beta_2})(s + 1/T_{\beta_3})}{\Delta_{\text{lat}}} \end{aligned} \tag{6-5}$$

$$\begin{aligned} \frac{\varphi(s)}{\delta(s)} = \frac{N_\delta^\varphi(s)}{\Delta_{\text{lat}}} &= \frac{A_\varphi s^2 + B_\varphi s + C_\varphi}{\Delta_{\text{lat}}} \\ &= \frac{A_\varphi(s^2 + 2\zeta_\varphi\omega_\varphi s + \omega_\varphi^2)}{\Delta_{\text{lat}}} \end{aligned} \tag{6-6}$$

$$\begin{aligned} \frac{r(s)}{\delta(s)} = \frac{N_\delta^r(s)}{\Delta_{\text{lat}}} &= \frac{A_r s^3 + B_r s^2 + C_r s + D_r}{\Delta_{\text{lat}}} \\ &= \frac{A_r\begin{cases} (s + 1/T_{r_1})(s^2 + 2\zeta_r\omega_r s + \omega_r^2) \\ \text{or} \\ (s + 1/T_{r_1})(s + 1/T_{r_2})(s + 1/T_{r_3}) \end{cases}}{\Delta_{\text{lat}}} \end{aligned} \tag{6-7}$$

$$\frac{a_{y_{c.g.}}(s)}{\delta(s)} = \frac{N_{\delta}^{a_{y_{c.g.}}}(s)}{\Delta_{\text{lat}}} = \frac{A_{a_y}s^4 + B_{a_y}s^3 + C_{a_y}s^2 + D_{a_y}s + E_{a_y}}{\Delta_{\text{lat}}}$$

$$= \frac{A_{a_y}(s + 1/T_{a_{y_1}})(s + 1/T_{a_{y_2}})(s + 1/T_{a_{y_3}})(s + 1/T_{a_{y_4}})}{\Delta_{\text{lat}}} \quad (6\text{-}8)$$

where

$$\Delta_{\text{lat}} = As^4 + Bs^3 + Cs^2 + Ds + E$$

$$= A\begin{Bmatrix}(s + 1/T_s)(s + 1/T_R) \\ \text{or} \\ (s^2 + 2\zeta_{SR}\omega_{SR}s + \omega_{SR}^2)\end{Bmatrix}(s^2 + 2\zeta_d\omega_d s + \omega_d^2) \quad (6\text{-}9)$$

An additional $a$ or $r$ subscript is often used on numerator quantities to specify aileron or rudder, respectively.

The literal expressions for the various $A$, $B$, $C$, etc., coefficients of Eqs. 6-5 through 6-9 are given in Table 6-1. There the coefficient expressions are in terms of unprimed derivatives but can easily be converted to prime derivatives by the simple expedient of eliminating all $I_{xz}$ terms and priming the $L$ and $N$ derivatives; the compactness thus achieved is self-evident. While either the primed or unprimed form may be used, as convenient, to obtain the polynomial numerator and denominator expressions and eventually the factors, the transfer function gain is properly computed only when the *same* derivative forms are used in both numerator and denominator. This follows directly from a consideration of the $A$ coefficients given in Table 6-1. For example, the high frequency gain of the $\beta$ transfer function is given by

$$\frac{\beta(s)}{\delta(s)} \rightarrow \frac{A_\beta s^3}{As^4} = \frac{A_\beta}{As} = \frac{Y_\delta^*[1 - (I_{xz}^2/I_xI_z)]}{[1 - (I_{xz}^2/I_xI_z)]s} = \frac{Y_\delta^*}{s} \quad (6\text{-}10)$$

If now the prime derivatives were used (i.e., $I_{xz} = 0$) for both numerator and denominator, the result would be the same. However, mixing the two forms would produce

$$\frac{\beta(s)}{\delta(s)} \rightarrow \frac{A_\beta}{A's} = \frac{Y_\delta^*[1 - (I_{xz}^2/I_xI_z)]}{s}$$

or

$$\frac{\beta(s)}{\delta(s)} \rightarrow \frac{A_\beta'}{As} = \frac{Y_\delta^*}{[1 - (I_{xz}^2/I_xI_z)]s}$$

which are both incorrect.

## 6-4. Example Transfer Functions, Bode Forms, and Time Responses for a Conventional Airplane

For the example conventional airplane of Chapter 5 (see Fig. 5-1) the lateral characteristics are given by

| | | | |
|---|---|---|---|
| Altitude (ft) | | | 20,000 |
| Weight (lb) | | | 30,500 |
| Mach number | | | 0.638 |
| True airspeed (ft/sec) | | | 660 |
| $Y_v$ | $-0.0829$ | $N_\beta$ | 3.55 |
| $Y_{\delta_a}$ | 0 | $N_p$ | $-0.0025$ |
| $Y^*_{\delta_r}$ | 0.0116 | $N_r$ | $-0.0957$ |
| $L_\beta$ | $-4.770$ | $N_{\delta_a}$ | $-0.615$ |
| $L_p$ | $-1.695$ | $N_{\delta_r}$ | $-1.383$ |
| $L_r$ | 0.1776 | $I_{xz}/I_x$ | 0.0663 |
| $L_{\delta_a}$ | 27.25 | $I_{xz}/I_z$ | 0.0370 |
| $L_{\delta_r}$ | 0.666 | | |

The corresponding primed derivatives are

| | | | |
|---|---|---|---|
| $L'_\beta$ | $-4.546$ | $N'_\beta$ | 3.382 |
| $L'_p$ | $-1.699$ | $N'_p$ | $-0.0654$ |
| $L'_r$ | 0.1717 | $N'_r$ | $-0.0893$ |
| $L'_{\delta_a}$ | 27.276 | $N'_{\delta_a}$ | 0.3952 |
| $L'_{\delta_r}$ | 0.5758 | $N'_{\delta_r}$ | $-1.362$ |

Substituting these data into the Table 6-1 forms and factoring the resulting polynomials yield the following transfer functions ($a_{y_{c.g.}}/\delta_a$ is not given because of its minor importance): for aileron inputs,

$$\frac{\beta(s)}{\delta_a(s)} = -14.38 \frac{[(s/0.0495) + 1][(s/-6.250) + 1]}{[(s/-0.001355) + 1][(s/1.777) + 1] \times \{[s^2/(1.8775)^2] + [2(0.0243)/1.8775]s + 1\}} \tag{6-11}$$

$$\frac{\varphi(s)}{\delta_a(s)} = -11070 \frac{\{[s^2/(1.859)^2] + [2(0.0047)/1.859]s + 1\}}{[\qquad][\qquad]\{\qquad\}} \tag{6-12}$$

$$\frac{r(s)}{\delta_a(s)} = -543.7 \frac{[(s/1.65) + 1]\{[s^2/(2.659)^2] + [2(-0.827)/2.659]s + 1\}}{[\qquad][\qquad]\{\qquad\}} \tag{6-13}$$

Table 6-1. Lateral control-input transfer function coefficients.

| | $A$ | $B$ | $C$ | $D$ | $E$ |
|---|---|---|---|---|---|
| $\Delta$ | $1 - \frac{I_{xz}^2}{I_x I_z}$ | $-Y_v\left(1 - \frac{I_{xz}^2}{I_x I_z}\right) - L_p - N_r - \frac{I_{xz}}{I_x} N_p - \frac{I_{xz}}{I_z} L_r$ | $N_\beta + L_p(Y + N_r) + N_p\left(\frac{I_{xz}}{I_x} Y_v - L_r\right) + Y_v\left(\frac{I_{xz}}{I_z} L_r + N_r\right) + \frac{I_{xz}}{I_z} L_\beta$ | $-N_\beta L_p + Y_v(N_p L_r - L_p N_r) + N_p L_\beta - \frac{g}{U_0}\left(L_\beta + \frac{I_{xz}}{I_x} N_\beta\right)$ | $\frac{g}{U_0}(L_\beta N_r - N_\beta L_r)$ |
| $N^\beta$ | $Y_\delta^*\left(1 - \frac{I_{xz}^2}{I_x I_z}\right)$ | $-Y_\delta^*\left[L_p + N_r + \frac{I_{xz}}{I_x} N_p + \frac{I_{xz}}{I_z} L_r\right] - \frac{I_{xz}}{I_z} L_\delta - N_\delta$ | $Y_\delta^*(L_p N_r - N_p L_r) + N_\delta L_p - L_\delta N_p + \frac{g}{U_0}\left(L_\delta + \frac{I_{xz}}{I_x} N_\beta\right)$ | $\frac{g}{U_0}(N_\delta L_r - L_\delta N_r)$ | |
| $N_\delta^\varphi$ | $L_\delta + \frac{I_{xz}}{I_x} N_\delta$ | $Y_\delta^*\left(L_\beta + \frac{I_{xz}}{I_x} N_\beta\right) - L_\delta(N_r + Y_v) + N_\delta\left(L_r - \frac{I_{xz}}{I_x} Y_v\right)$ | $Y_\delta^*(L_r N_\beta - L_\beta N_r) + L_\delta(Y_v N_r + N_\beta) - N_\delta(L_\beta + Y_v L_r)$ | | |

| | | | | | |
|---|---|---|---|---|---|
| $N_\delta^r$ | $N_\delta + \frac{I_{xz}}{I_z} L_\delta$ | $Y_\delta^*\left(N_\beta + \frac{I_{xz}}{I_z} L_\beta\right) + L_\delta\left(N_p - \frac{I_{xz}}{I_z} Y_v\right) - N_\delta(Y_v + L_p)$ | $Y_\delta^*(L_\beta N_p - N_\beta L_p) - L_\delta Y_v N_p + N_\delta Y_v L_p$ | $\frac{g}{U_0}(L_\delta N_\beta - N_\delta L_\beta)$ | |
| $N_\delta^{a_y\text{c.g.}}$ | $Y_\delta\left(1 - \frac{I_{xz}^2}{I_x I_z}\right)$ | $-Y_\delta\left(L_p + N_r + \frac{I_{xz}}{I_x} N_p + \frac{I_{xz}}{I_z} L_r\right)$ | $Y_\delta\left(N_\beta + L_p N_r - N_p L_r + \frac{I_{xz}}{I_z} L_\beta\right) - U_0 Y_v\left(\frac{I_{xz}}{I_z} L_\delta + N_\delta\right)$ | $Y_\delta\left[N_p L_\beta - \frac{g L_\beta}{U_0} - N_\beta\left(L_p + \frac{g I_{xz}}{U_0 I_x}\right)\right] + U_0 Y_v\left[L_\delta\left(\frac{g}{U_0} - N_p\right) + N_\delta\left(\frac{g I_{xz}}{I_x U_0} + L_p\right)\right]$ | $g[Y_\delta^*(L_\beta N_r - N_\beta L_r) + Y_v(N_\delta L_r - L_\delta N_r)]$ |

Note: To convert to primed derivatives, eliminate all $I_{xz}$ terms and substitute $L'$ and $N'$ for $L$ and $N$, respectively.

For rudder inputs,

$$\frac{\beta(s)}{\delta_r(s)} = 1.049 \frac{[(s/-0.00374) + 1][(s/1.75) + 1][(s/117.25) + 1]}{[(s/-0.001355) + 1][(s/1.777) + 1] \times \{[s^2/(1.8775)^2] + [2(0.0243)/1.8775]s + 1\}} \tag{6-14}$$

$$\frac{\varphi(s)}{\delta_r(s)} = 500 \frac{[(s/2.561) + 1][(s/-2.885) + 1]}{[\qquad\qquad][\qquad\qquad]\{\qquad\qquad\qquad\qquad\}} \tag{6-15}$$

$$\frac{r(s)}{\delta_r(s)} = 24.35 \frac{[(s/1.7767) + 1]\{[s^2/(0.293)^2] + [2(0.00563)/0.293]s + 1\}}{[\qquad\qquad][\qquad\qquad]\{\qquad\qquad\qquad\qquad\}} \tag{6-16}$$

$$\frac{a_{y_{c.g.}}(s)}{\delta_r(s)} = 7.00 \frac{[(s/0.000725) + 1][(s/1.705) + 1] \times [(s/2.551) + 1][(-s/2.468) + 1]}{[\qquad\qquad][\qquad\qquad]\{\qquad\qquad\qquad\qquad\}} \tag{6-17}$$

The roots of the denominator show that the lateral motions consist of three modes:

1. A relatively lightly damped oscillatory mode, called the "dutch roll."
2. A first-order divergent mode of relatively long time constant, called the "spiral" mode.
3. A first-order convergent mode of relatively short time constant, called the "roll subsidence" mode.

The subscript notation used in Eq. 6-9 and in subsequent tables and figures reflects the above nomenclature. The general aspects of these modes will be discussed in more detail later.

Figures 6-1 and 6-2 present the $j\omega$-Bode plots of the transfer functions, including amplitude and phase asymptotes. Using these as representative of response transforms to unit impulse inputs, as in Chapter 5, we can draw certain conclusions concerning the appearance of the various modes in the airplane's transient response.

Consider first the dutch roll mode. For all transfer functions but $\varphi/\delta_a$ the dutch roll peak is a dominant characteristic, indicating that this oscillation will be a major component of the weighting function responses. For the exception, $\varphi/\delta_a$, the quadratic numerator nearly cancels the dutch roll denominator. Because of this dipole effect, the modal response coefficient for the dutch roll mode in the $\varphi$ weighting function for the $\delta_a$ input will be small. If the cancellation were exact, the dutch roll would disappear entirely from this rolling response. Approximate values for the dutch roll modal response ratios can also be determined using the $j\omega$-Bode plots because $\zeta_d$ is small; therefore, a typical amplitude ratio evaluated at the dutch roll root becomes

$$|G[-(\zeta\omega)_d + j\omega_d\sqrt{(1-\zeta_d^2)}]| \doteq [G(j\omega_d)]$$

Just how well this works for the present example can be seen by comparing results from Figs. 6-1 and 6-2 with the actual modal response ratios. Thus, for rudder inputs (Fig. 6-1),

$$\begin{aligned}\beta:\varphi:r &\doteq [17:16:23]_{\mathrm{dB}}\\ &\doteq [0,-1,6]_{\mathrm{dB}}\\ &\doteq 1:0.89:2.0\end{aligned}$$

and for aileron inputs (Fig. 6-2),

$$\begin{aligned}\beta:\varphi:r &\doteq [14:18:20]_{\mathrm{dB}}\\ &\doteq [0:4:6]_{\mathrm{dB}}\\ &\doteq 1:1.6:2.0\end{aligned}$$

The exact model response ratio is

$$\beta:\varphi:r = 1:0.99:1.87$$

Making the comparison, we see that all the plots but $\varphi/\delta_a$ yield values that are quite comparable to the exact set. The difficulty with $\varphi/\delta_a$ again stems from the presence of the numerator quadratic. The amplitude ratio of this transfer function evaluated at the exact dutch roll root, $s_d$, differs considerably from the approximation based on $s_d \doteq j\omega_d$. Except for this effect, the dutch roll modal response ratios computed from amplitude frequency response plots evaluated at $\omega_d$ are quite insensitive to the damping ratio, $\zeta_d$, because its effect cancels when the response ratios are taken.

Turn now to the roll subsidence mode. For this, numerator terms in both $\beta$ and $r$ responses to rudder and $r$ response to aileron essentially cancel the roll subsidence mode denominator. These dipole effects imply that the roll subsidence mode will be small in the $\beta$ and $r$ rudder-input and the $r$ aileron-input weighting functions. The absence of a similar near-cancellation of the roll subsidence factor by a numerator term in the $\beta/\delta_a$ transfer function might, at first glance, be taken to imply that the roll subsidence mode component of the sideslip response to aileron would be substantial. Thus we might be tempted to conclude that the relative

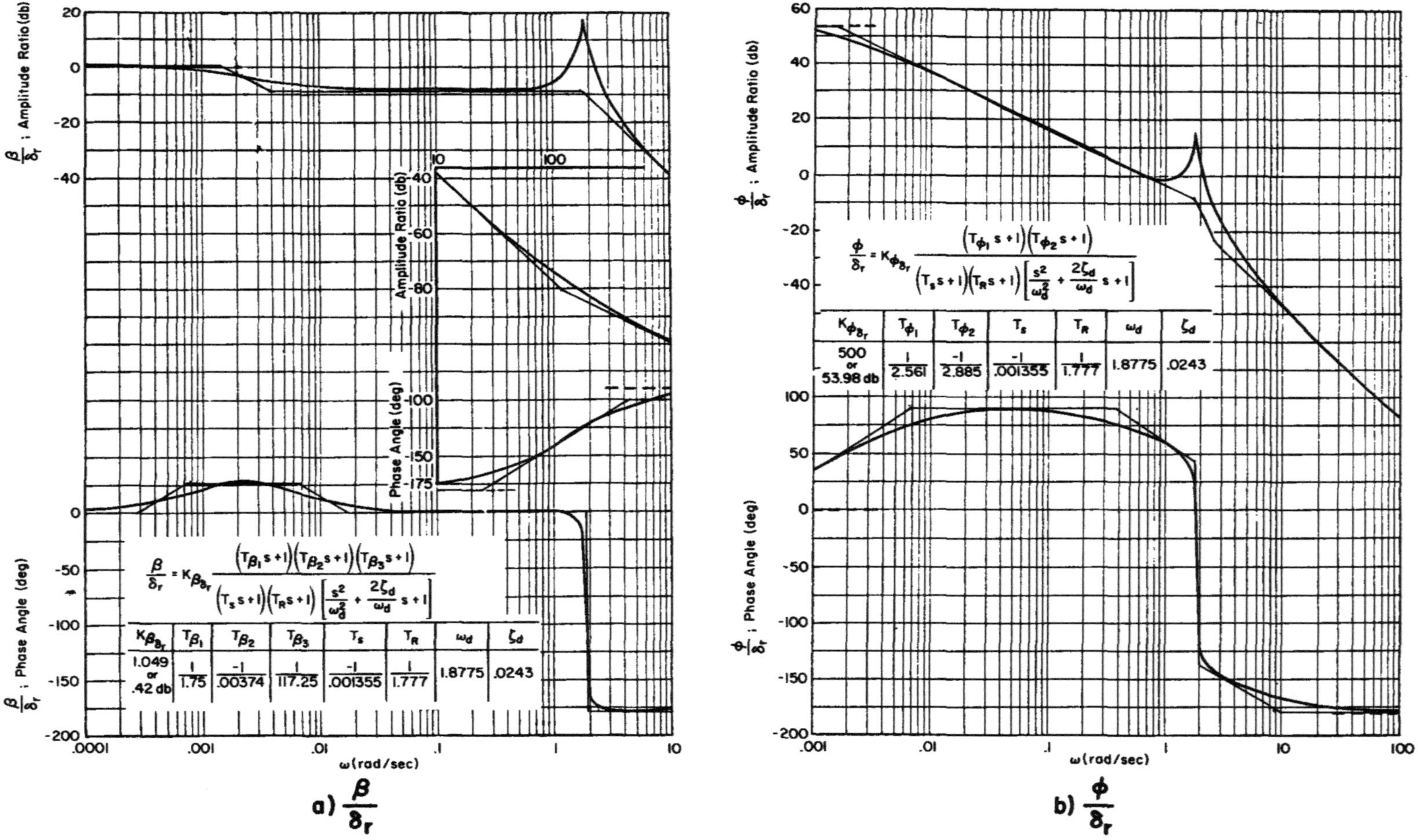


Fig. 6-1. Rudder-input Bode plots for a conventional airplane.

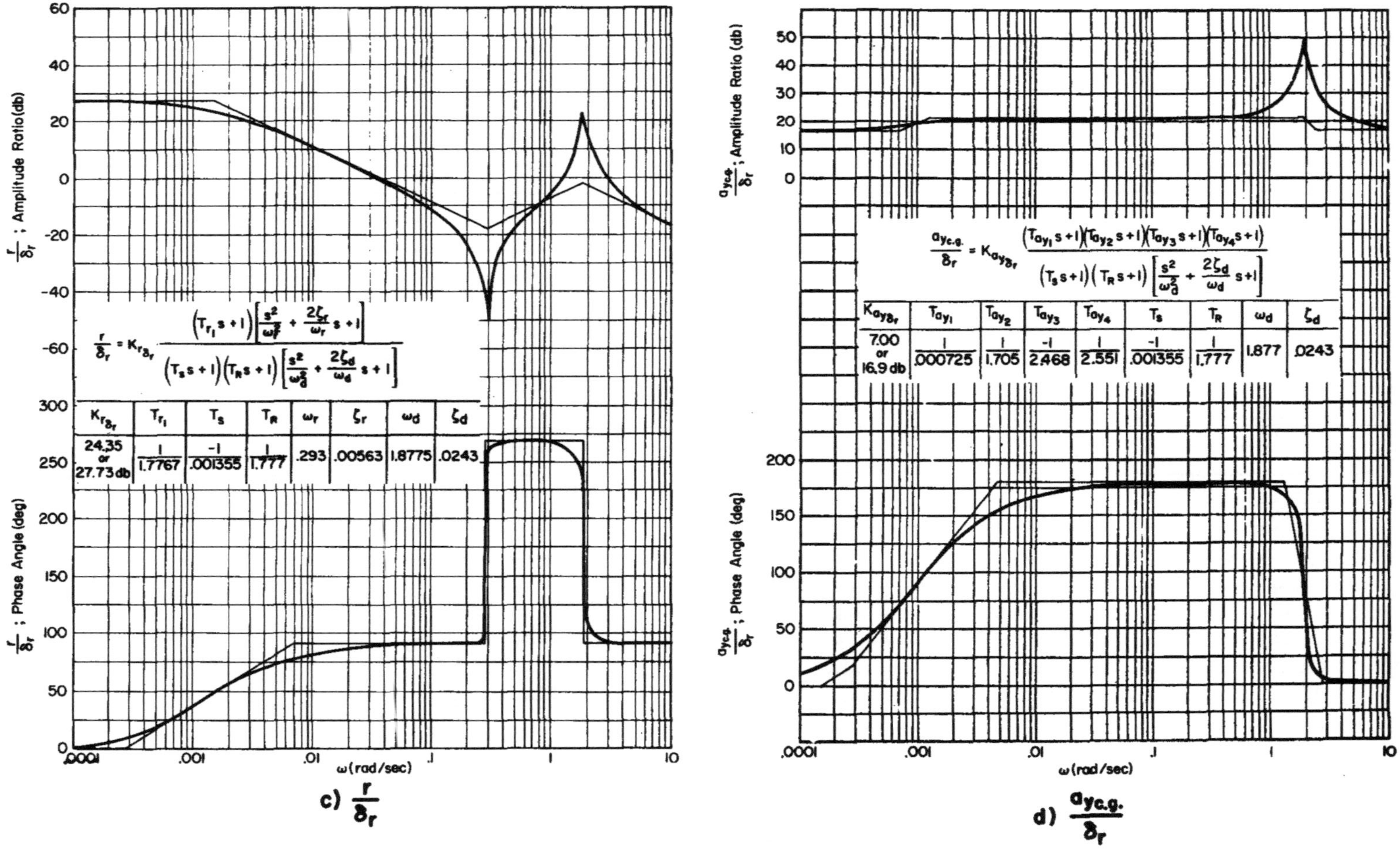


Fig. 6-1 (Continued)

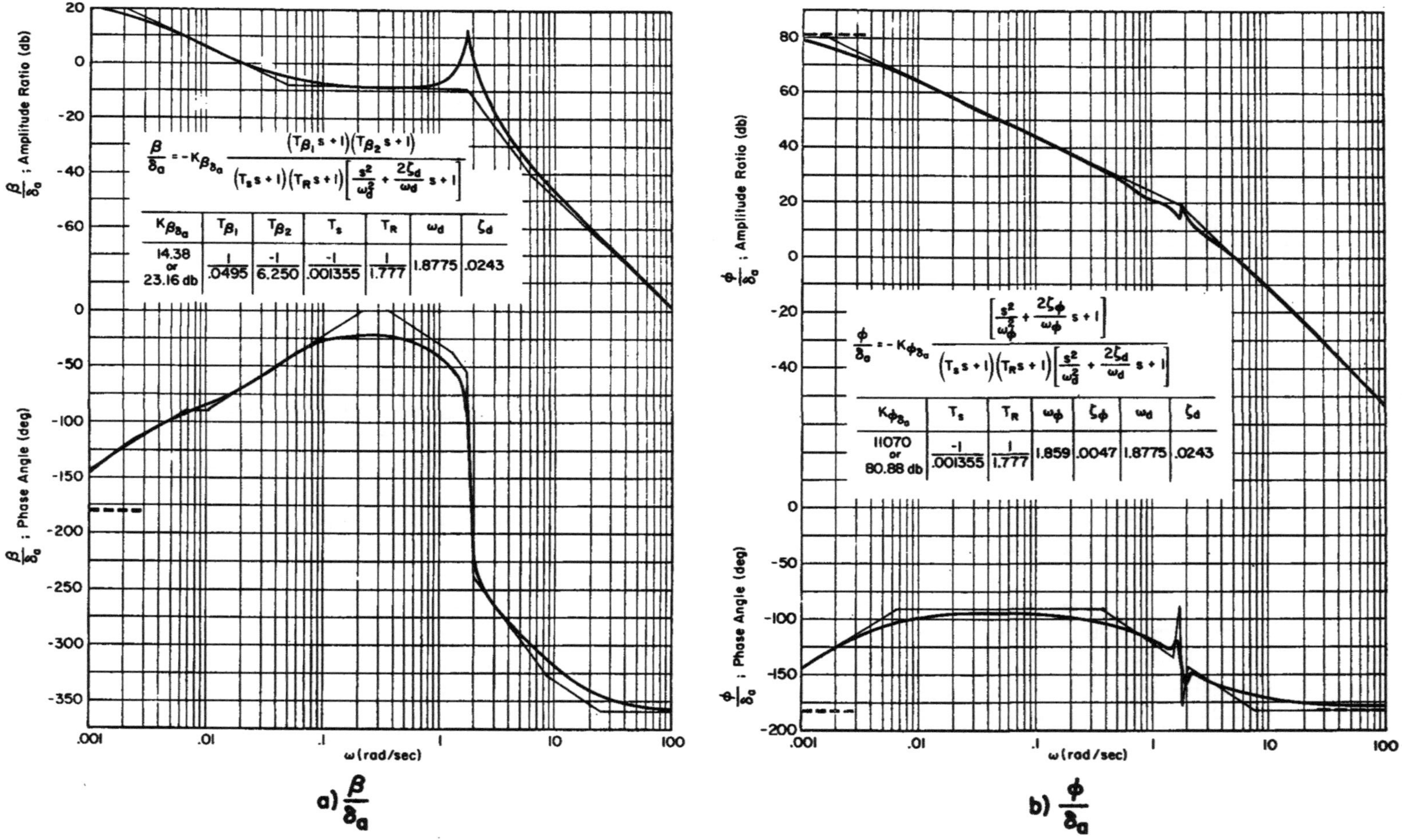


Fig. 6-2. Aileron-input Bode plots for a conventional airplane.

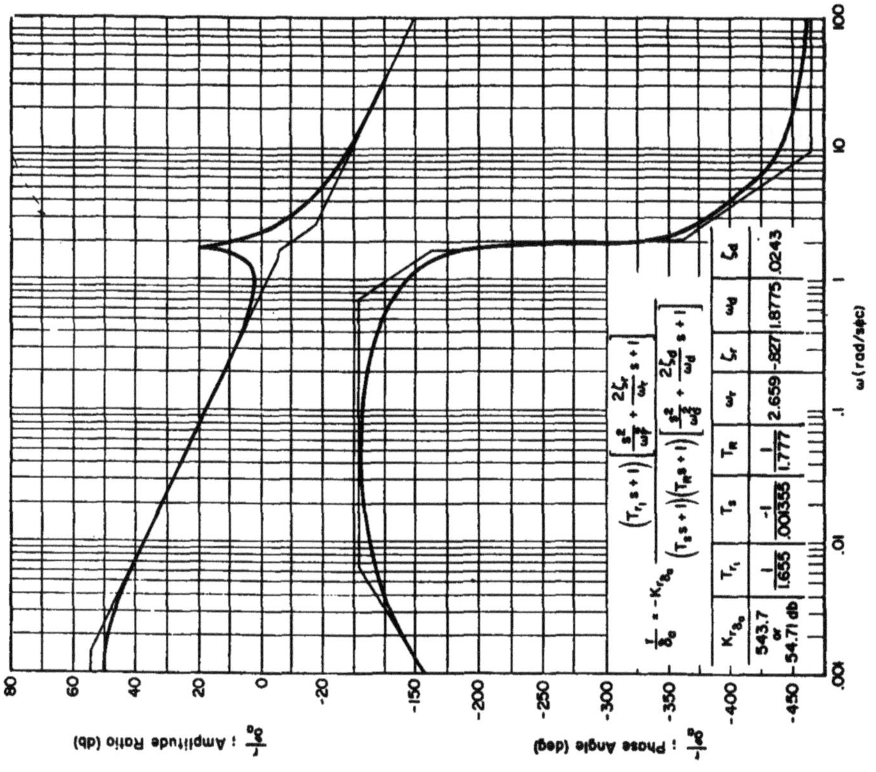


c) $\frac{r}{\delta_a}$

Fig. 6-2 (Continued)

magnitudes of the degrees of freedom occurring in a particular mode depend on whether the mode is excited by aileron or rudder. But we know from the modal response considerations presented in Section 2-5 that this cannot be so, that, in fact, the response ratios for a given mode are independent of the input. A second look at the $\beta/\delta_a$ and $\varphi/\delta_a$ transfer functions indicates a large difference in gains at $\omega = 1/T_R$ (when the dutch roll peaks are ignored) which results in a large $|\varphi/\beta|$ computed for aileron inputs, just as the near-cancellation of $\beta$ responses to rudder inputs also infers a large $|\varphi/\beta|$. All of these considerations imply that the roll subsidence is small in $r$ and $\beta$ weighting functions, and that this mode is then characterized almost entirely by rolling motions. This is, of course, the origin of its name. The roll subsidence modal response ratios evaluated from the $j\omega$-Bode plots will, in this case, be quite inaccurate if the plot itself is used because of the close proximity and dominance of the dutch roll peak. Good results can, however, be obtained if the asymptotic plot is used.

Finally, for the spiral mode the modal response ratios could be computed from the d-c gains if all the other poles and zeros were far enough removed. For the aileron responses (Fig. 6-2) this condition obtains, and the modal response ratios approximated by the gain ratios,

$$\begin{aligned} \beta:\varphi:r &\doteq (-14.38):(-11{,}070):(-543.7) \\ &\doteq 1:770:37.8 \end{aligned}$$

correspond favorably to the exact ratios,

$$\beta:\varphi:r = 1:756:36$$

The close proximity of a numerator factor to the spiral denominator term in the $\beta/\delta_r$ transfer functions makes the use of the gain ratios alone inaccurate for the rudder case; however, the $\varphi:r$ modal response ratio computed using the rudder-input d-c gains is satisfactory. The predominance of rolling and yawing in the spiral motion indicates that this mode, albeit unstable, is nevertheless nearly coordinated. Further, because of the relatively small sideslip, the $r:\varphi$ ratio is approximately $g/U_0$ (e.g., Eq. 6-3).

Figure 6-3 presents time histories of the responses to an aileron pulse and a rudder step which are in accord with the foregoing conclusions. Here we can see that the primary response to an aileron input is in roll angle; and that nearly constant angle is achieved in a short time, corresponding to a response time $3T_R$ of the roll subsidence mode. The spiral mode divergence is not discernible due to its extremely large time constant, 738 sec. The dutch roll excited by the aileron appears to be mostly in $\beta$ and $r$, but the much larger scale used for the $\varphi$ trace tends to mask its magnitude relative to $\beta$ and $r$. The magnitudes of these three motions in the dutch roll mode are more easily seen in the time histories for a rudder step. These also show a relatively small average roll rate response, indicative of the reduced excitation of the roll subsidence mode by the rudder as compared to the aileron.

Figure 6-4 presents the time-vector diagrams which directly indicate the relative magnitude and phasing of the free motions (phasors) and show the relative importance of each term in the equations of motion (polygons) for each mode. Again, we see that $\beta$ is almost nonexistent in the spiral mode which therefore consists of coordinated banking and turning (i.e., no sideslipping) motions. For the roll subsidence, $\varphi$ is the dominant motion parameter; for the dutch roll, all motions are the same order of magnitude. The small amount of lateral motion in the dutch roll is apparent in the approximate cancellation of the $r$ and $s\beta$ side force components, i.e.,

$$y = \frac{U_0}{s}(\beta + \psi) = \frac{U_0}{s^2}(s\beta + r) \doteq 0$$

## 6-5. Two Degree of Freedom Dutch Roll Approximations

Because the $\varphi$ oscillations occurring in the dutch roll mode can be just as large as the $\beta$ and $\psi$ motions, it is clear that they cannot safely be neglected in formulating a generally applicable set of simplified dutch roll equations. Nevertheless, by analogy with the longitudinal short period, this is commonly done in the interests of obtaining simple physical insight. If we assume, therefore, that the $\varphi$ motions are negligible, then the sum of the rolling moments must be zero at all times and the roll equation is eliminated along with $\varphi$ perturbations. Thus the equations of motion (6-1), neglecting gust inputs, reduce to

$$\begin{aligned} (s - Y_v)\beta + \qquad\qquad r &= Y_\delta^*\delta \\ -N'_\beta\beta + (s - N'_r)r &= N'_\delta\delta \end{aligned} \tag{6-18}$$

The corresponding transfer functions are given by

$$\begin{aligned} \frac{\beta(s)}{\delta(s)} &= \frac{Y_\delta^*[s - N'_r - (N'_\delta/Y_\delta^*)]}{\Delta_{d_2}} \\ \frac{r(s)}{\delta(s)} &= \frac{N'_\delta[s - Y_v + (Y_\delta^*/N'_\delta)N'_\beta]}{\Delta_{d_2}} \end{aligned} \tag{6-19}$$

where $\Delta_{d_2} = s^2 + (-Y_v - N'_r)s + N'_\beta + Y_vN'_r$. There is, of course, no $\varphi$ transfer function because of the assumptions made.

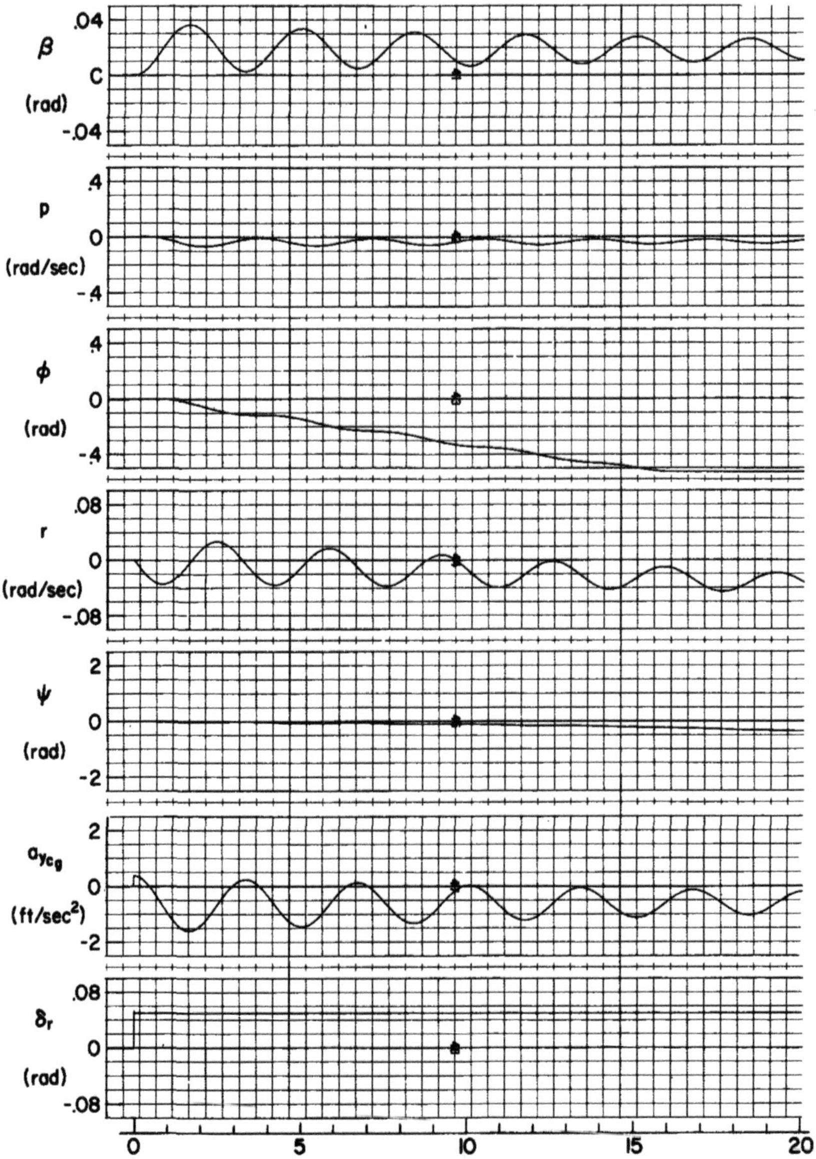


**a) Rudder**

Fig. 6-3. Time history of lateral motions to rudder and aileron inputs.

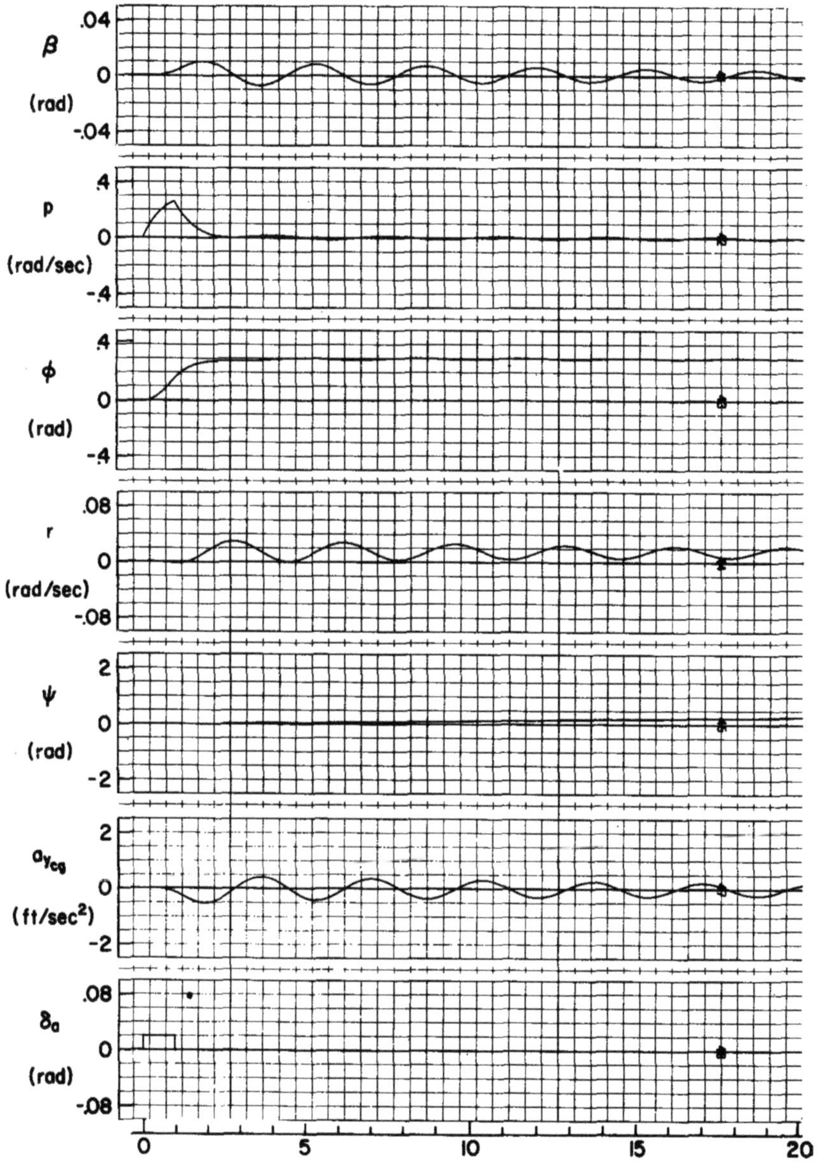


**b) Aileron**

Fig. 6-3 (Continued)

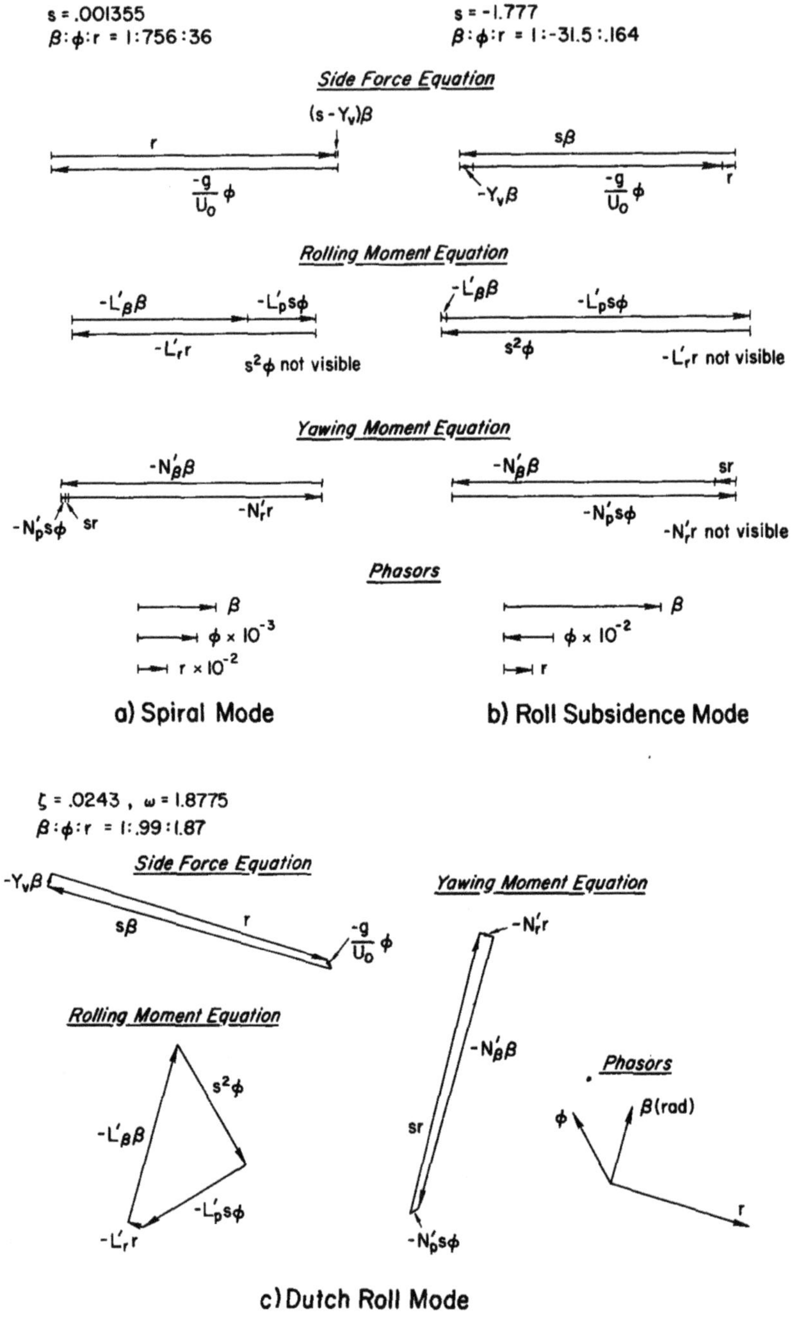


Fig. 6-4. Time-vector diagrams for a conventional airplane.

For the example airplane derivatives, the rudder and aileron numerical transfer functions are

$$\begin{aligned}
\frac{\beta(s)}{\delta_r(s)} &= 0.402\,\frac{[(s/117.6)+1]}{\{[s^2/(1.84)^2]+[2(0.0469)/1.84]s+1\}}\\
\frac{r(s)}{\delta_r(s)} &= -0.0218\,\frac{[(s/0.0541)+1]}{\{\qquad\qquad\}}\\
\frac{\beta(s)}{\delta_a(s)} &= -0.1166\,\frac{1}{\{\qquad\qquad\}}\\
\frac{r(s)}{\delta_a(s)} &= 0.00965\,\frac{[(s/0.0829)+1]}{\{\qquad\qquad\}}
\end{aligned} \tag{6-20}$$

Comparing the dutch roll characteristics, we see that the frequency is fairly close to the exact value (1.8775) but that the damping is about twice as high (i.e., should be 0.0243). Also, the direct Bode amplitude comparisons given in Fig. 6-5 show that, although the rudder transfer functions are fairly well matched, the two degree of freedom aileron approximations depart considerably from the complete three degree of freedom cases.

## 6-6. Three Degree of Freedom Dutch Roll Approximations

We can improve on our approximation to the dutch roll mode by considering that several of the smaller terms in the time-vector diagrams of Fig. 6-4(c) are negligible. Prime candidates are the gravity terms, $g\varphi/U_0$, the rolling acceleration due to the rate of yaw, $L_r'r$, and the yawing acceleration due to the rate of roll, $N_p'p$. With these simplifications the approximate three degree of freedom set, considering only control inputs, becomes

$$\begin{aligned}
(s-Y_v)\beta \qquad\qquad\qquad &+ r = Y_\delta^*\delta\\
-L_\beta'\beta + s(s-L_p')\varphi \qquad &= L_\delta'\delta\\
-N_\beta'\beta \qquad\qquad + (s &- N_r')r = N_\delta'\delta
\end{aligned} \tag{6-21}$$

For this set of equations the denominator determinant is

$$\Delta_{d_3} = s(s-L_p')[s^2 + (-Y_v - N_r')s + (N_\beta' + Y_vN_r')] \tag{6-22}$$

Here the free $s$ corresponds to the spiral mode, the $(s - L_p')$ factor to the roll subsidence, and the quadratic to the dutch roll. The dutch roll approximation is, of course, the same as that for the two degree of freedom set. Likewise, the $\beta$ and $r$ transfer functions are unchanged. The advantage

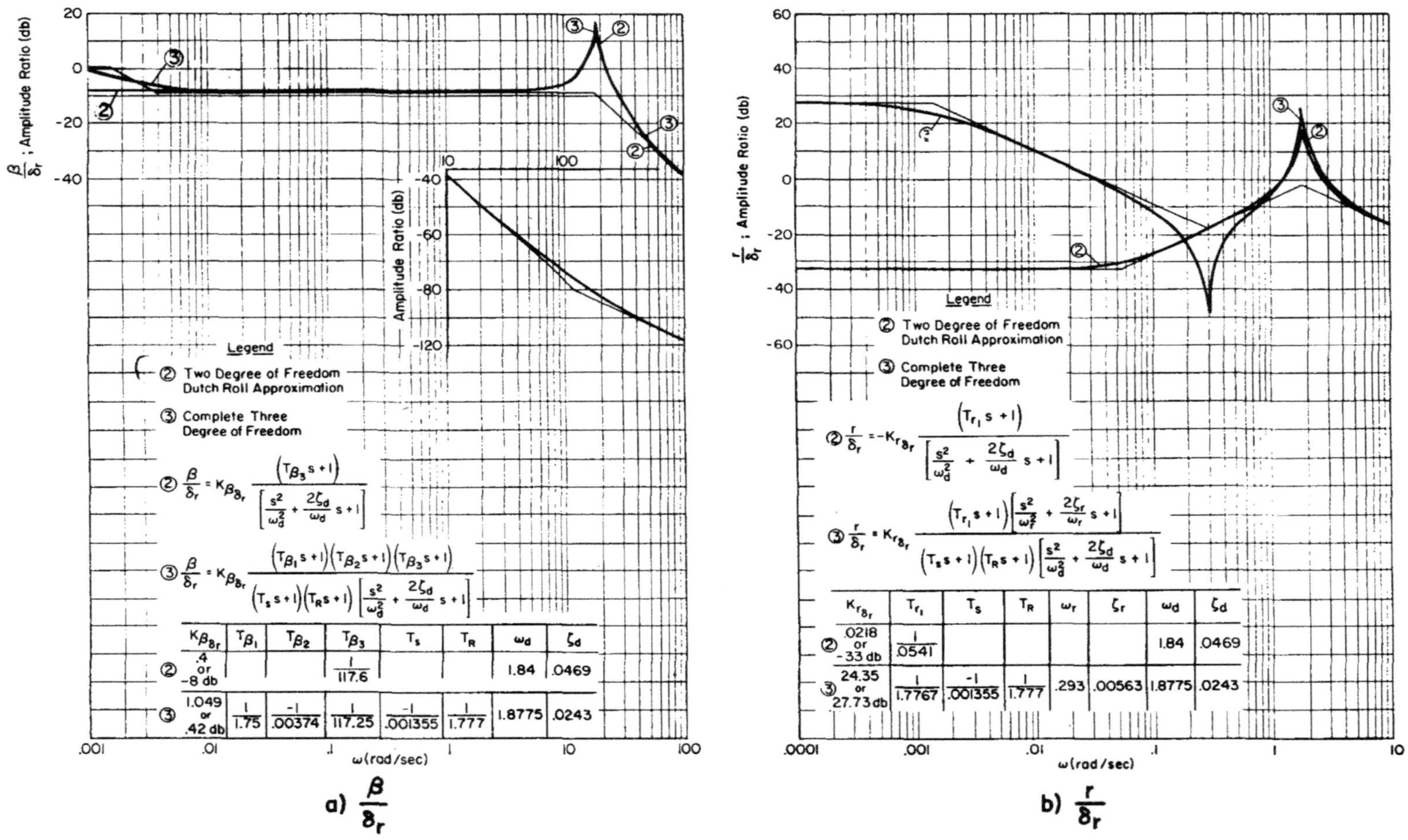


Fig. 6-5. Comparisons between complete three and two degree of freedom dutch roll approximate Bode amplitudes.

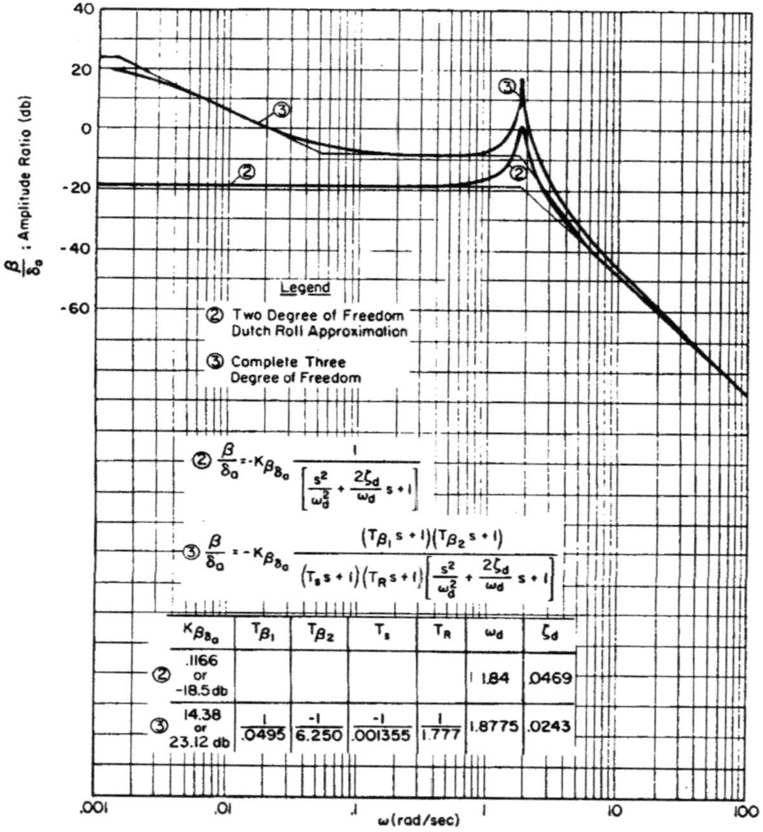


② $\frac{\beta}{\delta_a} = -K_{\beta_{\delta_a}} \frac{1}{\left[\frac{s^2}{\omega_d^2} + \frac{2\zeta_d}{\omega_d} s + 1\right]}$

③ $\frac{\beta}{\delta_a} = -K_{\beta_{\delta_a}} \frac{(T_{\beta_1} s + 1)(T_{\beta_2} s + 1)}{(T_s s + 1)(T_R s + 1)\left[\frac{s^2}{\omega_d^2} + \frac{2\zeta_d}{\omega_d} s + 1\right]}$

| | $K_{\beta_{\delta_a}}$ | $T_{\beta_1}$ | $T_{\beta_2}$ | $T_s$ | $T_R$ | $\omega_d$ | $\zeta_d$ |
|---|---|---|---|---|---|---|---|
| ② | .1166 or -18.5 db | | | | | 1.84 | .0469 |
| ③ | 14.38 or 23.12 db | $\frac{1}{.0495}$ | $\frac{-1}{6.250}$ | $\frac{-1}{.001355}$ | $\frac{1}{1.777}$ | 1.8775 | .0243 |

c) $\frac{\beta}{\delta_a}$

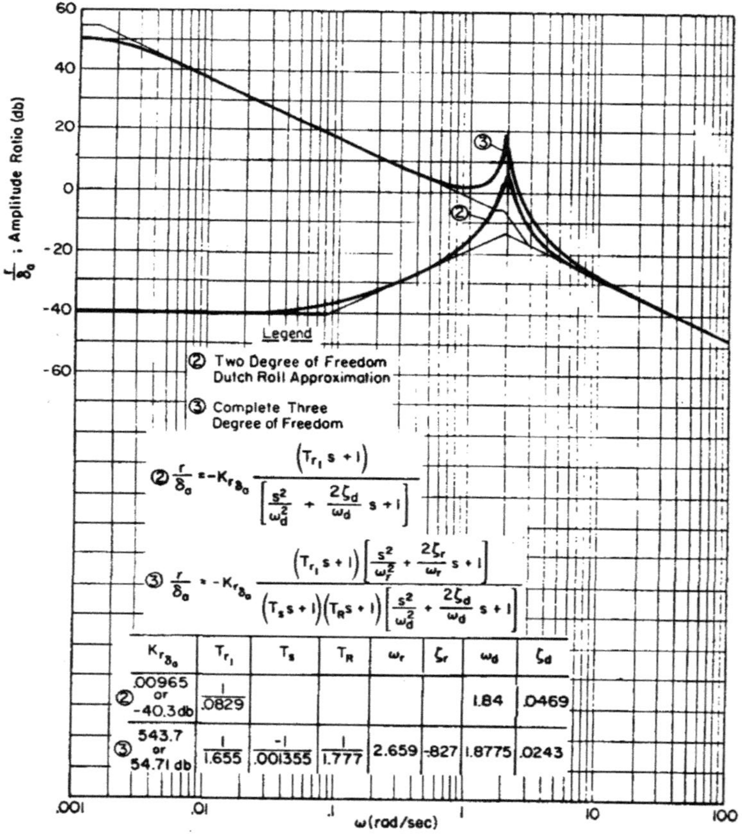


② $\frac{r}{\delta_a} = -K_{r_{\delta_a}} \frac{(T_{r_1} s + 1)}{\left[\frac{s^2}{\omega_d^2} + \frac{2\zeta_d}{\omega_d} s + 1\right]}$

③ $\frac{r}{\delta_a} = -K_{r_{\delta_a}} \frac{(T_{r_1} s + 1)\left[\frac{s^2}{\omega_r^2} + \frac{2\zeta_r}{\omega_r} s + 1\right]}{(T_s s + 1)(T_R s + 1)\left[\frac{s^2}{\omega_d^2} + \frac{2\zeta_d}{\omega_d} s + 1\right]}$

| | $K_{r_{\delta_a}}$ | $T_{r_1}$ | $T_s$ | $T_R$ | $\omega_r$ | $\zeta_r$ | $\omega_d$ | $\zeta_d$ |
|---|---|---|---|---|---|---|---|---|
| ② | .00965 or -40.3 db | $\frac{1}{.0829}$ | | | | | 1.84 | .0469 |
| ③ | 543.7 or 54.71 db | $\frac{1}{1.655}$ | $\frac{-1}{.001355}$ | $\frac{1}{1.777}$ | 2.659 | .827 | 1.8775 | .0243 |

d) $\frac{r}{\delta_a}$

Fig. 6-5 (Continued)

of the three degree of freedom dutch roll approximation is to obtain $\varphi$ transfer functions. Thus Eq. 6-21 yields the following $\varphi$ numerator:

$$N_\delta^\varphi = L'_\delta\left[s^2 + \left(-Y_v - N'_r + \frac{Y_\delta^*}{L'_\delta} L'_\beta\right)s + N'_\beta\left(1 - \frac{L'_\beta N'_\delta}{L'_\delta N'_\beta}\right) + Y_v N'_r - \frac{Y_\delta^*}{L'_\delta} L'_\beta N'_r\right] \quad (6\text{-}23)$$

The corresponding numerical transfer function for aileron input to the airplane is

$$\frac{\varphi(s)}{\delta_a(s)} = 15.71 \frac{\{[s^2/(1.82)^2] + [2(0.0474)/1.82]s + 1\}}{s[(s/1.699) + 1]\{\qquad\}} \quad (6\text{-}24)$$

Figure 6-6 compares the three degree of freedom dutch roll approximation Bode plot with the complete set. This comparison shows that the $\varphi/\delta_a$ transfer function, based on the approximate equations, is quite close to that found using the complete three degree of freedom set.

## 6-7. Three Degree of Freedom Spiral and Roll Subsidence Approximations

Approximate equations of motion appropriate to the spiral and roll subsidence modes are obtained quite simply from the observation that for both modes the $\beta$ motions are relatively small, as shown by the phasors of Fig. 6-4(a) and (b); and that $(s - Y_v)\beta$ for the spiral mode is negligible with respect to the remaining side force terms [see Fig. 6-4(a)]. Accordingly, we neglect this term to obtain the approximate equations of motion given by

$$\begin{aligned} -(g/U_0)\varphi \qquad\qquad + r &= Y_\delta^*\delta \\ -L'_\beta\beta + s(s - L'_p)\varphi \qquad - L'_r r &= L'_\delta\delta \\ -N'_\beta\beta \qquad - N'_p s\varphi + (s - N'_r)r &= N'_\delta\delta \end{aligned} \quad (6\text{-}25)$$

These lead to the transfer functions:

$$\frac{\beta(s)}{\delta(s)} = \frac{N_\delta^\beta}{N'_\beta\Delta_{SR}}$$

where $N_\delta^\beta$ is the complete form given in Table 6-1,

$$\frac{\varphi(s)}{\delta(s)} = \frac{Y_\delta^* L'_\beta\{s - N'_r + (N'_\beta/L'_\beta)\, L'_r + 1/Y_\delta^*[-N'_\delta + (N'_\beta/L'_\beta)\, L'_\delta]\}}{N'_\beta\Delta_{SR}}$$

$$\frac{r(s)}{\delta(s)} = \frac{Y_\delta^*\{s^2 + [-L'_p + (L'_\beta/N'_\beta)\, N'_p]s + (g/Y_\delta)L'_\delta - ([L'_\beta/N'_\beta)\, N'_\delta]\}}{\Delta_{SR}} \quad (6\text{-}26)$$

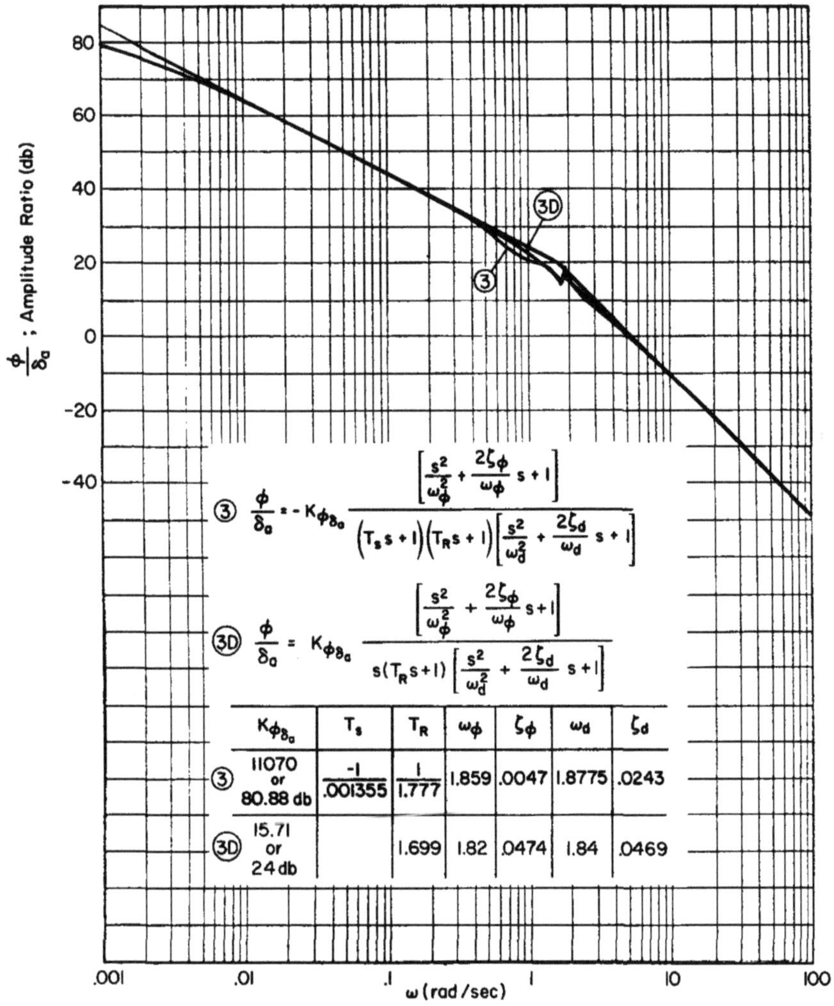


$$③\quad \frac{\phi}{\delta_a} = -K_{\phi\delta_a}\frac{\left[\frac{s^2}{\omega_\phi^2} + \frac{2\zeta_\phi}{\omega_\phi}s + 1\right]}{(T_s s + 1)(T_R s + 1)\left[\frac{s^2}{\omega_d^2} + \frac{2\zeta_d}{\omega_d}s + 1\right]}$$

$$③D\quad \frac{\phi}{\delta_a} = K_{\phi\delta_a}\frac{\left[\frac{s^2}{\omega_\phi^2} + \frac{2\zeta_\phi}{\omega_\phi}s + 1\right]}{s(T_R s + 1)\left[\frac{s^2}{\omega_d^2} + \frac{2\zeta_d}{\omega_d}s + 1\right]}$$

| | $K_{\phi\delta_a}$ | $T_s$ | $T_R$ | $\omega_\phi$ | $\zeta_\phi$ | $\omega_d$ | $\zeta_d$ |
|---|---|---|---|---|---|---|---|
| ③ | 11070 or 80.88 db | $\frac{-1}{.001355}$ | $\frac{1}{1.777}$ | 1.859 | .0047 | 1.8775 | .0243 |
| ③D | 15.71 or 24 db | | 1.699 | 1.82 | .0474 | 1.84 | .0469 |

Fig. 6-6. Comparison between complete three degree of freedom and dutch roll approximation three degree of freedom Bode amplitude ratios for $\varphi/\delta_a$.

where

$$\Delta_{SR} = s^2 + \left[-L_p' + \frac{L_\beta'}{N_\beta'}\left(N_p' - \frac{g}{U_0}\right)\right]s + \frac{g}{U_0}\left(\frac{L_\beta'}{N_\beta'}N_r' - L_r'\right)$$

For the example airplane, these transfer functions are

$$\frac{\beta(s)}{\delta_r(s)} = 1.049\,\frac{[(s/-0.00374)+1][(s/1.75)+1][(s/117.25)+1]}{[(s/-0.00135)+1][(s/1.852)+1]}$$

$$\frac{\varphi(s)}{\delta_r(s)} = 498\,\frac{[(s/80.4)+1]}{[\qquad][\qquad]}$$

$$\frac{r(s)}{\delta_r(s)} = 24.35\,\frac{[(s/-1.57)+1][(s/3.36)+1]}{[\qquad][\qquad]}$$

$$\frac{\beta(s)}{\delta_a(s)} = -14.38\,\frac{[(s/0.0495)+1][(s/-6.250)+1]}{[\qquad][\qquad]} \tag{6-27}$$

$$\frac{\varphi(s)}{\delta_a(s)} = -11080\,\frac{1}{[\qquad][\qquad]}$$

$$\frac{r(s)}{\delta_a(s)} = -543.7\,\frac{1}{[\qquad][\qquad]}$$

Directly comparing these results with the complete transfer functions given in Eqs. 6-11 to 6-17, we see first that the approximate values of the spiral and roll subsidence inverse time constants are quite close to the exact values (i.e., $-0.00135$ and 1.852 vs. $-0.001355$ and 1.777), that the low frequency gains for both sets are essentially the same, and that the $\beta$ numerators for both sets are identical. However, the $\varphi$ and $r$ numerator approximate factors are quite different from the exact factors. For the $\varphi$ numerators these differences occur at frequencies greater than that corresponding to the roll subsidence mode, as might be expected. For the $r$ numerators the breakpoint differences occur at frequencies below the roll subsidence, and it is quite clear that the approximate Bodes will not closely resemble the complete Bodes at the lower frequencies in the vicinity of the spiral root.

The spiral and roll subsidence approximations may often be further simplified by recognizing that, for most conventional airplanes, the value of $g/U_0[(L_\beta'/N_\beta')N_r' - L_r')]$ is small relative to the other terms appearing in $\Delta_{SR}$. Then, if $Y_{\delta_a}^* \doteq 0$,

$$\frac{\varphi(s)}{\delta_a(s)} = \frac{L_{\delta_a}'[1 - (N_{\delta_a}'/L_{\delta_a}')(L_\beta'/N_\beta')]}{s\{s - L_p' + (L_\beta'/N_\beta')[N_p' - (g/U_0)]\}} \tag{6-28}$$

This can be contrasted with the classical single degree of freedom roll approximation obtained by considering only the rolling equation of motion, and neglecting $\beta$ and $r$ motions.

$$\frac{\varphi(s)}{\delta_a(s)} = \frac{L'_{\delta_a}}{s(s - L'_p)} \tag{6-29}$$

We see that both the gain and the roll time constant are affected by three degree of freedom considerations, but that the temporal nature of the motion is not, i.e., the response to a step aileron input is an exponential increase from zero to a steady rolling velocity in either case.

### 6-8. Commentary on Approximate Equations of Motion

In this article we shall examine both the discrepant aspects of the approximations and their positive attributes. In this process we shall first take up the various approximations separately and then consider them as a whole.

*Two Degree of Freedom Dutch Roll*

Early versions of the simplified equations of motion for the dutch roll mode were very similar to Eq. 6-18 but used the unprimed derivatives in the yawing acceleration equation. A principal virtue of using the primed derivatives shows up in considering the denominator, $\Delta_{d_2}$, and its factors. The fact that $\omega_d^2 \doteq N'_\beta + Y_v N'_r$, as implied by Eq. 6-19, rather than $\omega_d^2 \doteq N_\beta + Y_v N_r$, is an important distinction. It shows that the effective directional stiffness is a function not only of the weathercock stability, $N_\beta$, but also of the dihedral, $L_\beta$, and the product of inertia, $I_{xz}$, i.e.,

$$N'_\beta \equiv \frac{N_\beta + (I_{xz}/I_z)L_\beta}{1 - (I_{xz}^2/I_x I_z)}$$

Thus, even though $N_\beta + Y_v N_r \rightarrow 0$, it is still possible to have positive stiffness and a finite dutch roll frequency if $I_{xz}$ and $L_\beta$ are of the same sign (i.e., for normally negative $L_\beta$ and nose-up inclination of the principal axis of inertia to the direction of flight). Such situations are not uncommon for high angle of attack conditions, where a negative $N_\beta$, e.g., due to fin immersion in the wing/body wake, can be stabilized by overriding negative values of $I_{xz}$ and $L_\beta$. Of course, unfavorable $N_\beta$ characteristics can also occur because of either Mach and/or aeroelastic effects at low angles of attack corresponding to high speed operation. The possibility of $I_{xz}L_\beta$ overriding $N_\beta$ in such situations is small because of the low angle of attack and correspondingly low $I_{xz}$ and $L_\beta$ generally involved.

The dutch roll damping implied by Eq. 6-19, $2\zeta_d\omega_d \doteq -Y_v - N'_r$, is almost always positive because usually both $Y_v$ and $N'_r$ are negative. There are rare cases, as noted in Chapter 4, in which $Y_v$ can become positive; it is conceivable that the same high $\alpha$ conditions could "blanket" the fin and greatly reduce $-N'_r$. Except for such uncommon possibilities, the implication of the approximation is that dutch roll damping will always be positive. Divergent oscillations have been observed in practice so that these cannot be explained by this set of simplified equations.

The possible motions for the two degree of freedom simplified equations must therefore be either a damped (or overdamped) oscillation or a subsidence/divergence combination. Both forms have been observed, but the lightly damped oscillation is the more usual characteristic.

*Three Degree of Freedom Dutch Roll*

This approximation is intended primarily to allow the computation of a bank angle transfer function or response to complement the two degree of freedom approximation results for yawing velocity and sideslip. It satisfies this requirement but is otherwise undistinguished.

*Spiral and Roll Subsidence*

Until recently, the roll subsidence and spiral modes were usually thought of as unconnected and independent. Physically, the roll subsidence was associated primarily with the rolling behavior of the aircraft which is described largely by the time lag in attaining a nearly steady-state rolling velocity after a step application of the ailerons. This lag is a result principally of the combination of the roll rate damping moment and the roll moment of inertia, so that the mode is conventionally considered to be essentially single degree of freedom. On the other hand, the spiral mode has long been recognized generally as involving at least two degrees of freedom: yawing and rolling. Since the describing time constant for this mode involves a very small root of the lateral characteristic equation, the approximation for the root (i.e., $E/D$ of $\Delta_{\text{lat}}$) has been well known for many years.

As can be appreciated from these remarks, no particular reason existed for further exploration of the underlying physics of the roll subsidence and spiral modes at a time when most aircraft possessed three degree of freedom rolling motions that approximated single degree of freedom characteristics. However, modern craft with high effective dihedral, low roll damping, etc., occasionally exhibit a long period lateral oscillation; also, rolling velocity calculations based on $L_{\delta_a}$, rather than $L'_{\delta_a}(\omega_\varphi/\omega_d)^2$, have led to later surprises. Early contacts with these phenomena called attention to the problem of obtaining a more adequate physical understanding of the spiral and roll subsidence modes. Equation 6-25 brings the two modes into their present closely interconnected context.

The history related above has certain overtones of repetition because the earliest partially successful gliders and unmanned aircraft possessed considerable "lateral stability." In the terms used here, the roll subsidence and spiral were both stable modes and fairly close to, or had actually reached, the coupled state. Model aircraft also use the same principles to obtain as much overall lateral dynamic stability as possible. It was not until the Wrights correctly deduced that a "neutral lateral stability" was desirable for lateral control that the roll subsidence became essentially single degree of freedom and the spiral simply a minor headache to the pilot in IFR conditions.

By considering the terms combined in $\Delta_{SR}(s)$, it can be shown that the time response characteristics for the lateral roll subsidence/spiral combination can take on almost any second-order form. The solitary exception on a normal winged aircraft is two divergent first orders. All other forms, i.e., positively or negatively damped oscillation, two subsidences, or a subsidence/divergence pair, have actually occurred in practice.

*Combined Considerations*

Preferably, we should like approximate relationships that could be connected in some logical fashion to give a fairly accurate picture of the complete three degree of freedom situation. As already noted, in terms of the denominator the two three degree of freedom approximations to the dutch roll and spiral/roll-subsidence modes offer a combined set of denominator dynamic characteristics in general quite representative of the complete situation. The major denominator deficiency is the dutch roll damping. This can be alleviated by considering that the damping for the spiral/roll subsidence mode is more correct than that for the two degree of freedom dutch roll, and by recalling that the $s^3$ coefficient of the complete denominator is the sum of all the damping terms. Then, a better estimate for the dutch roll damping will be

$$
\begin{aligned}
2(\zeta\omega)_d &\doteq (-\mathrm{Y}_v - \mathrm{N}'_r - \mathrm{L}'_p) - \left[-\mathrm{L}'_p + \frac{\mathrm{L}'_\beta}{\mathrm{N}'_\beta}\left(\mathrm{N}'_p - \frac{g}{U_o}\right)\right] \\
&\doteq (-\mathrm{Y}_v - \mathrm{N}'_r) - \frac{\mathrm{L}'_\beta}{\mathrm{N}'_\beta}\left(\mathrm{N}'_p - \frac{g}{U_o}\right)
\end{aligned}
\tag{6-30}
$$

For normally negative $L'_\beta$ and positive $N'_\beta$, negative values of $N'_p$ and/or small values of $U_0$ will result in a negative damping contribution due to the added term. This can be large enough to overpower the two degree of freedom damping provided by $(-Y_v - N'_r)$ and thereby to produce a divergent oscillation.

Unfortunately, as observed earlier, the various examples of approximate numerators show a very uneven correspondence with the exact numerators. In all cases, however, we can say that, for the spiral/roll subsidence, the asymptotic low frequency gain/dynamics closely approximate those for the complete transfer function; also, either of the dutch roll approximations approaches the correct asymptotic high frequency gain/dynamics. As for the numerator *factors*, those that are reasonably correct numerically for the example airplane are limited to the following.

1. Two degree of freedom dutch roll:
   $\beta/\delta$, $r/\delta$—good agreement at high frequencies
2. Three degree of freedom dutch roll:
   $\varphi/\delta_a$—identical to exact complete numerator if $Y^*_{\delta_a} = 0$
   $\beta/\delta$, $r/\delta$, $\varphi/\delta_r$—good agreement at high frequencies[2]
3. Three degree of freedom spiral/roll subsidence:
   $\varphi/\delta_a$—good except in immediate region of dutch roll
   $\beta/\delta_a$, $\beta/\delta_r$—identical to exact complete numerator

This is in contrast somewhat to the situation for the longitudinal approximations, where the numerator and denominator factors are in better agreement with the exact factors. It appears that the use of simplified modal equations of motion cannot in general yield good approximate numerator factors and that another approach to approximate factors is required to supplement the approximate equation technique. Such an approach was outlined in Section 3-4 and has been applied extensively to the problem of obtaining approximate factors.[3]

## 6-9. Hovering Equations, Control-Input Transfer Functions, and Time Responses

In their simplest practical form the lateral, small perturbation equations of motion in hover are

$$\begin{aligned}(s - Y_v)v \qquad\qquad - g\varphi \qquad\qquad &= Y_\delta \delta \\ -L_v v + s(s - L_p)\varphi \qquad\qquad &= L_\delta \delta \\ (s - N_r)r &= N_\delta \delta\end{aligned} \tag{6-31}$$

Equation 6-31 assumes $U_0 = N_p = L_r = I_{xz} = N_v = Y_p = Y_r = 0$ and applies fairly well to any hovering vehicle without a tail rotor or with a tail rotor of high disk loading. It is not completely valid for typical single-rotor helicopters because the tail rotor, being of low disk loading, is sensitive to local sideslip velocity perturbations and hence generates, e.g., $N_v$ and possibly $L_r$ and $N_p$. Notice that it is completely analogous

[2] Not given in text; computed factors of example case are $[(s/-2.67) + 1][(s/2.75) + 1]$.
[3] See I. L. Ashkenas and D. T. McRuer, "Approximate Airframe Transfer Functions and Application to Single Sensor Control Systems," WADC TR 58-82, June 1958, Appendix, pp. 191–210.

to the longitudinal hover equations (5-34) with $\varphi$ replacing $\theta$, $v$ replacing $u$, and $r$ replacing $w$; the $Y$, $N$, $L$ derivatives replace $X$, $Z$, and $M$ derivatives, respectively. The resulting transfer functions and modal responses are correspondingly similar in form to those given previously for the longitudinal motions, i.e.:

$$\frac{r(s)}{\delta(s)} = \frac{N_\delta}{s - N_r} \tag{6-32}$$

$$\frac{\varphi(s)}{\delta(s)} = \frac{L_\delta[s - Y_v + (Y_\delta/L_\delta)L_v]}{\Delta_{\text{hover}}} \tag{6-33}$$

$$\frac{v(s)}{\delta(s)} = \frac{Y_\delta[s^2 - L_p s + g(L_\delta/Y_\delta)]}{\Delta_{\text{hover}}} \tag{6-34}$$

where

$$\Delta_{\text{hover}} = s^3 + (-Y_v - L_p)s^2 + Y_v L_p s - gL_v \tag{6-35}$$

From these relationships we see that the $s - N_r$ mode is associated only with yaw rate perturbations, $r$, and does not appear in either $\varphi$ or $v$ motions. Conversely, there are no $r$ motions in the modes associated with the lateral *hovering cubic* (Eq. 6-35) which usually factors into a negative real root describing a stable mode[4], plus a complex pair associated with a lightly damped, or unstable, oscillatory mode.

The similarity, in form, of these and the longitudinal equations is matched by similarities in the derivatives. For most hovering vehicles[4] $X_u = Y_v$ and $I_x L_v = -I_y M_u$. However, the equations of motion are not numerically identical because usually $I_x \neq I_y$; hence $L_v \neq -M_u$ even if $I_x L_v = -I_y M_u$. Similarly, $M_q \neq L_p$, although often $I_y M_q = I_x L_p$. Despite this, the analogy between longitudinal and lateral equations is sufficiently complete to suggest that the root locations and modal response coefficients may be very similar.

As an example of this we again choose the configuration shown in Fig. 5-7 and consider small perturbations from a near-hover condition ($U_0 =$ 1.0 ft/sec) to detect any possibly significant effects of not being exactly at hover. As will be shown, the effects of small $U_0$ are trivial; $Y_p = Y_r = Y_{\delta_a} = L_r = N_v = N_p = N_{\delta_a}$ are zero, but $I_{xz} \neq 0$, and this produces some coupling of the yawing mode $(s - N_r)$ with the roll and sideslip modes. Because the example configuration has no tail rotor, it has essentially zero $N_v$, $L_r$, and $N_p$. The assumed characteristics are as follows.

[4] J. Wolkovitch and R. P. Walton, "VTOL and Helicopter Approximate Transfer Functions and Closed-Loop Handling Qualities," Systems Technology Tech. Rept. 128-1, Sept. 1963.

| | | | |
|---|---|---|---|
| Weight (lb) | | 3100 | |
| Roll inertia (slug-ft$^2$) | | 1990 | |
| Yaw inertia (slug-ft$^2$) | | 3450 | |
| Groundspeed, $U_0$ (ft/sec) | | 1 | |
| $Y_p$ | 0 | $L_p$ | −0.271 |
| $Y_r$ | 0 | $L_r$ | 0 |
| $Y_v$ | −0.14 | $L_v$ | −0.0122 |
| $Y_{\delta_a}$ | 0 | $L_{\delta_a}$ | 0.69 |
| $Y_{\delta_r}$ | 1.017 | $L_{\delta_r}$ | −0.185 |
| $N_p$ | 0 | $I_{xz}/I_x$ | −0.1246 |
| $N_r$ | −0.656 | $I_{xz}/I_z$ | −0.07188 |
| $N_v$ | 0 | | |
| $N_{\delta_a}$ | 0 | | |
| $N_{\delta_r}$ | −0.539 | | |

The corresponding primed derivatives are

| | | | |
|---|---|---|---|
| $L'_p$ | −0.273 | $N'_p$ | 0.0197 |
| $L'_r$ | 0.0825 | $N'_r$ | −0.662 |
| $L'_v$ | −0.0123 | $N'_v$ | 0.0008885 |
| $L'_{\delta_a}$ | 0.696 | $N'_{\delta_a}$ | −0.0500 |
| $L'_{\delta_r}$ | −0.119 | $N'_{\delta_r}$ | −0.531 |

For the above assumed characteristics, the control-input transfer functions are: for yaw control, $\delta_r$,

$$\begin{aligned}
\frac{\varphi}{\delta_r} &= \frac{-0.119(s + 0.267)(s + 1.008)}{\Delta_{\text{lat}}} \\
\frac{r}{\delta_r} &= \frac{-0.531(s + 0.886)[s^2 - 2(0.348)(0.674)s + (0.674)^2]}{\Delta_{\text{lat}}} \\
\frac{v}{\delta_r} &= \frac{1.017(s - 1.75)(s + 0.994)(s + 2.22)}{\Delta_{\text{lat}}}
\end{aligned} \tag{6-36}$$

for roll control, $\delta_a$,

$$\begin{aligned}
\frac{\varphi}{\delta_a} &= \frac{0.696(s + 0.14)(s + 0.6564)}{\Delta_{\text{lat}}} \\
\frac{r}{\delta_a} &= \frac{-0.0500s^2(s + 0.14)}{\Delta_{\text{lat}}} \\
\frac{v}{\delta_a} &= \frac{0.0500(s + 0.657)(s + 447.0)}{\Delta_{\text{lat}}}
\end{aligned} \tag{6-37}$$

where $\Delta_{\text{lat}} = (s + 0.653)(s + 0.888)[s^2 - 2(0.347)(0.669)s + (0.669)^2]$.

Bode diagrams of these transfer functions and time responses for impulsive $\delta_a$ and $\delta_r$ inputs are given in Figs. 6-7 and 6-8. The diagrams, in general, show the approximate cancellations of the hovering cubic in the $r$ transfer function and of the $(s - N_r)$ subsidence mode in the $\varphi$ and $v$ transfer functions, as anticipated by the approximate Eqs. 6-32 through 6-34. The one exception is the $r/\delta_a$ Bode which is radically different in appearance from $r/\delta_r$. This difference is traceable to the product of inertia influence which "recreates" many of the derivatives, e.g., $L_r'$, $N_v'$, $N_p'$, considered zero in their unprimed state, and to the size of these derivatives relative to $N_{\delta_a}'$, the normally important yaw-rate-input excitation term.

The time responses are consistent with the foregoing differences and similarities. In particular, the yaw rate first-order subsidence can be clearly seen in the rudder-input response, and the negatively damped oscillation associated with the hovering cubic is clearly visible in the $\varphi$ and $v$ traces. However, the first-order portion of the cubic cannot easily be separated from the initial portions of these latter responses because its time constant is near the period of the second-order oscillation. This mode "subsides" almost completely in about 3 sec, so that the residual trace thereafter is almost entirely due to the second order.

The time-vector diagrams of Fig. 6-9 enhance considerably our understanding of the physical nature of the modes of free motion and the significance of each derivative. For example, the yawing acceleration time vectors for the spiral mode, Fig. 6-9(a), show that the approximation $s = N_r$ for the root is very accurate. The yawing moment induced by roll acceleration, $I_{xz}s^2\varphi$, is not visible on the same scale, so for this mode the sideslip and roll equations of motion are superfluous.

The time vectors for the remaining modes present quite a different appearance. Considering both Figs. 6-9(b) and 6-9(c), we see that the balance of yawing accelerations in free motion includes an appreciable contribution from $I_{xz}$. However, the time vectors representing the rolling and side accelerations show negligible contributions due to yawing velocity or yawing acceleration. Hence these modes can be calculated using roll and side force equations of motion only.

Thus, despite the nonzero $I_{xz}$, free yawing motions are essentially uncoupled, and the roll and sideslip motions are similar to the hovering forward displacement and pitching motions described in Chapter 5. In this regard, the comparison of the $X$ force and pitching moment time vectors of Fig. 5-9 with the side force and rolling moment vectors of Fig. 6-9 shows that the principal difference between longitudinal and lateral motions is caused by the greater damping in roll, $I_xL_p$, rather than that in pitch, $I_yM_q$. (The difference in the inertias is secondary, $I_x = 1990$ vs. $I_y = 1790$ slug-ft$^2$.) The increased $L_ps\varphi$ (compared to $M_qs\theta$) vector tends

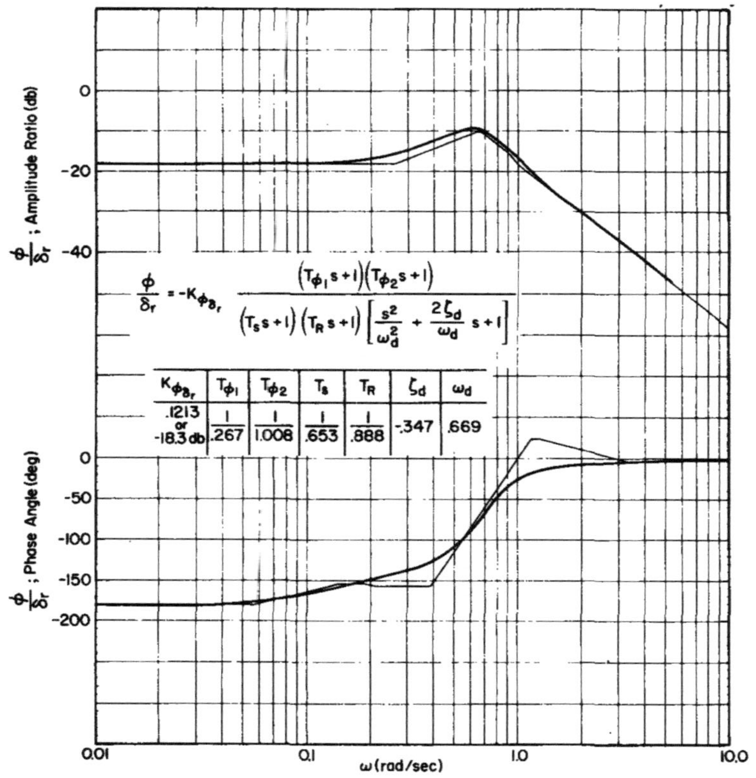


a) $\frac{\phi}{\delta_r}$

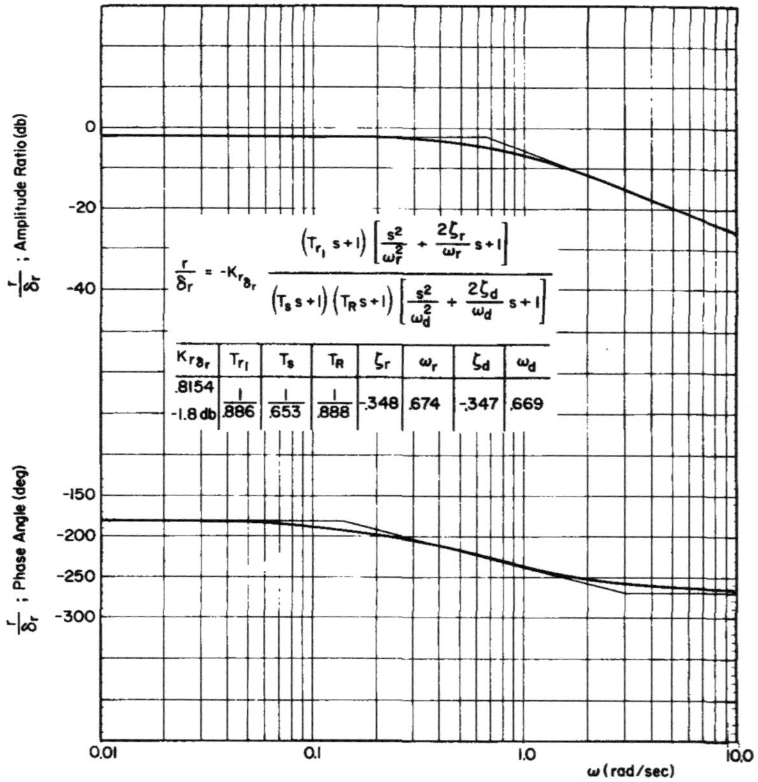


b) $\frac{r}{\delta_r}$

Fig. 6-7. Example of rudder control-input Bode plots and time responses for a hovering airplane.

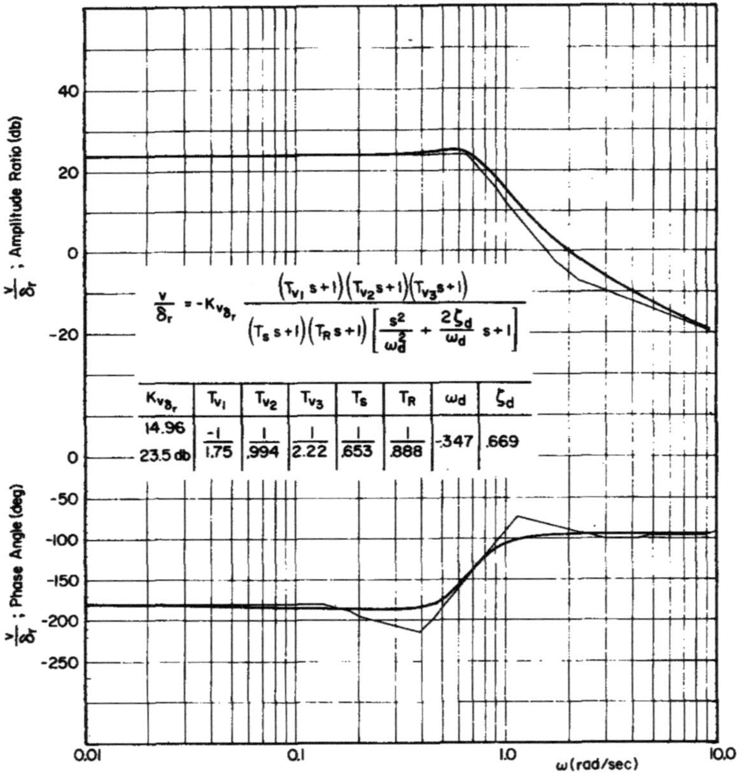


$$\frac{v}{\delta_r} = -K_{v_{\delta_r}} \frac{(T_{v_1}s+1)(T_{v_2}s+1)(T_{v_3}s+1)}{(T_s s+1)(T_R s+1)\left[\frac{s^2}{\omega_d^2} + \frac{2\zeta_d}{\omega_d}s+1\right]}$$

| $K_{v_{\delta_r}}$ | $T_{v_1}$ | $T_{v_2}$ | $T_{v_3}$ | $T_s$ | $T_R$ | $\omega_d$ | $\zeta_d$ |
|---|---|---|---|---|---|---|---|
| 14.96<br>23.5 db | $\frac{-1}{1.75}$ | $\frac{1}{.994}$ | $\frac{1}{2.22}$ | $\frac{1}{.653}$ | $\frac{1}{.888}$ | .347 | .669 |

c) $\frac{v}{\delta_r}$

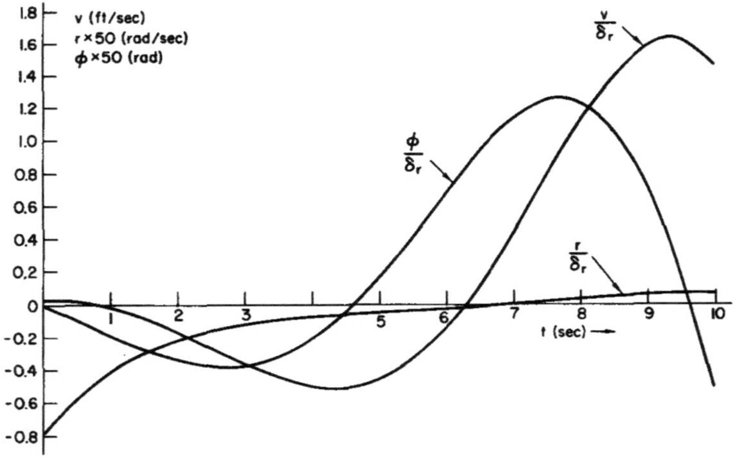


d) Time Responses

Fig. 6-7 (Continued)

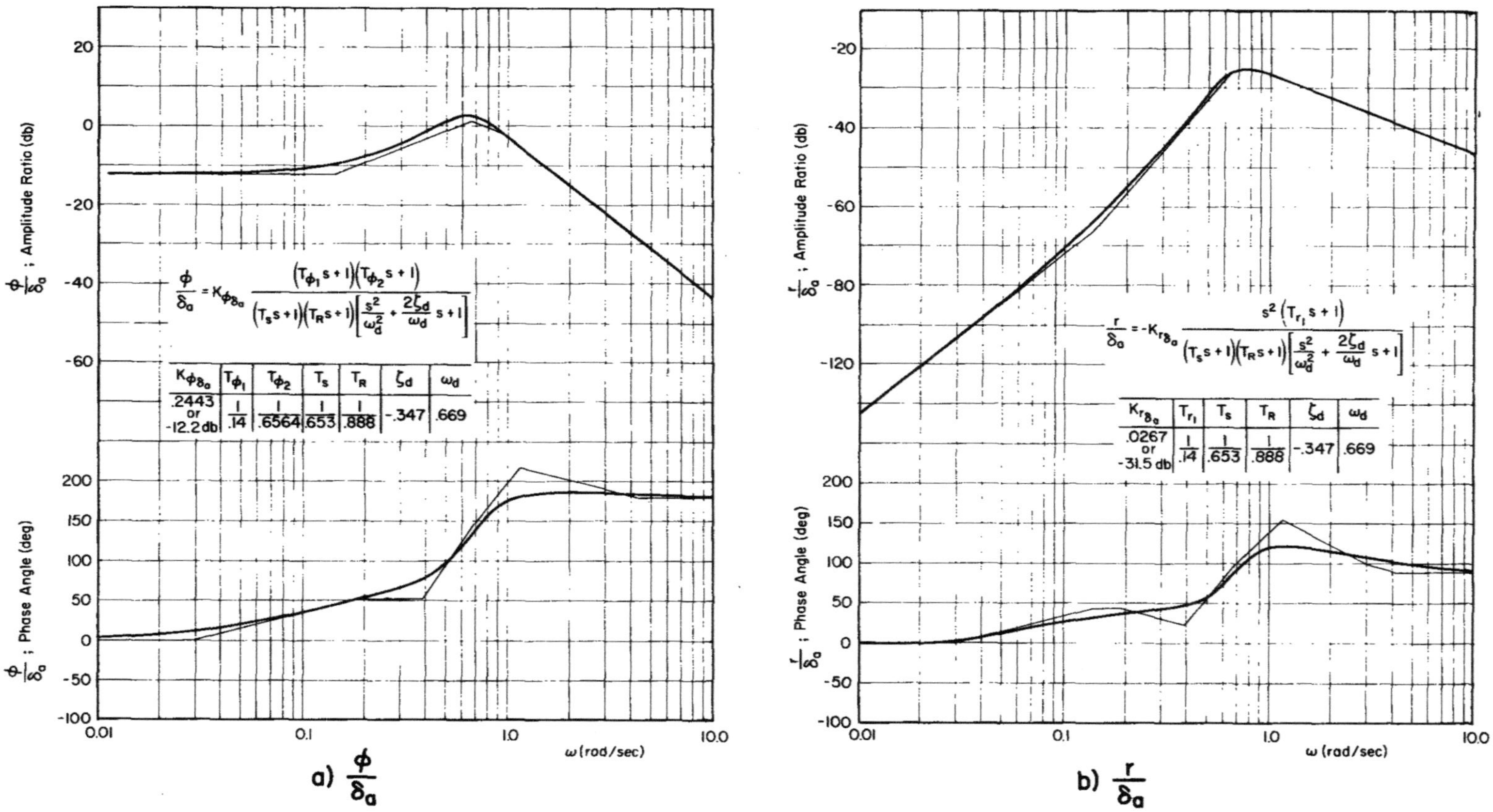


Fig. 6-8. Example of aileron control-input Bode plots and time responses for a hovering airplane.

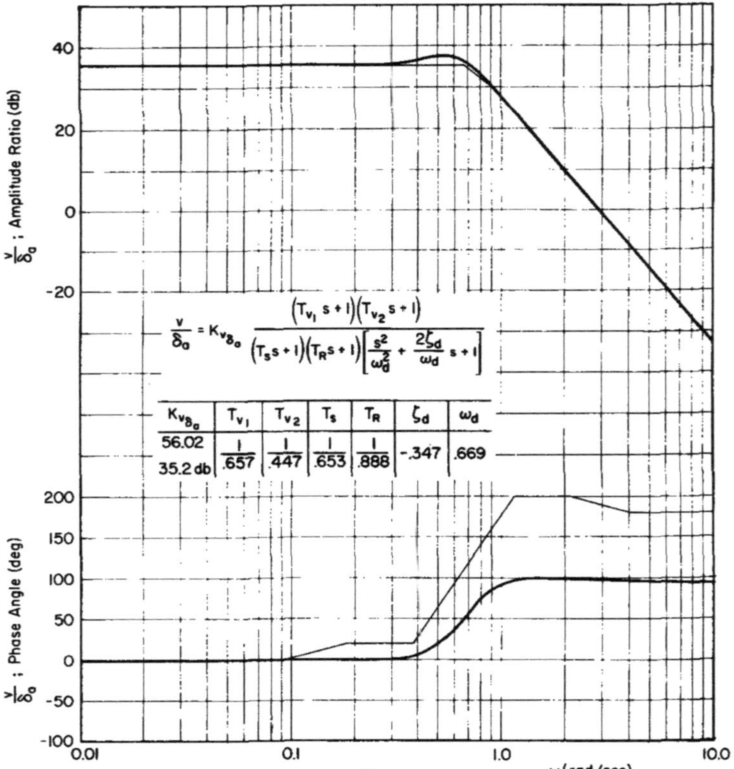


$$\frac{v}{\delta_a} = K_{v_{\delta_a}} \frac{(T_{v_1} s + 1)(T_{v_2} s + 1)}{(T_s s + 1)(T_R s + 1)\left[\frac{s^2}{\omega_d^2} + \frac{2\zeta_d}{\omega_d} s + 1\right]}$$

| $K_{v_{\delta_a}}$ | $T_{v_1}$ | $T_{v_2}$ | $T_s$ | $T_R$ | $\zeta_d$ | $\omega_d$ |
|---|---|---|---|---|---|---|
| 56.02<br>35.2 db | $\frac{1}{.657}$ | $\frac{1}{.447}$ | $\frac{1}{.653}$ | $\frac{1}{.888}$ | -.347 | .669 |

c) $\frac{v}{\delta_a}$

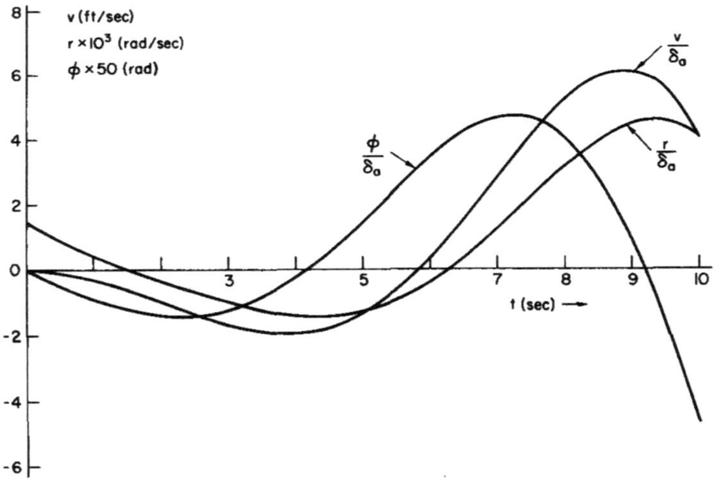


d) Time Responses

Fig. 6-8 (Continued)

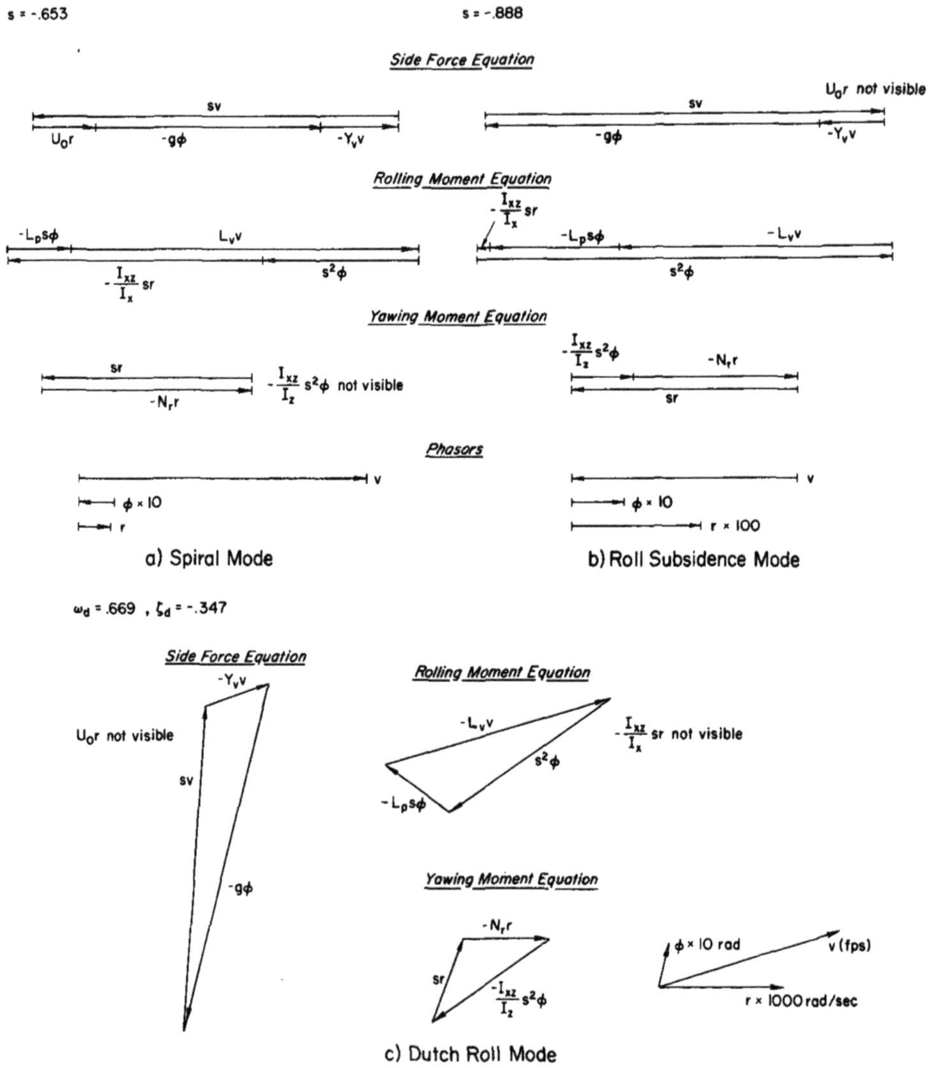


**Fig. 6-9. Time-vector diagrams for a hovering vehicle.**

to distort the appropriate time-vector polygon, so as to make the damping angle, $\sin^{-1} \zeta_d$, less negative than its longitudinal counterpart, $\sin^{-1} \zeta_p$.

Criteria for the neglect of $I_{xz}$ in the hovering modes can be found by substituting the missing $I_{xz}$ terms into Eq. 6-31. The characteristic equation then becomes

$$(s - N_r)[(s - Y_v)(s^2 - L_p s) - gL_v] - \left(\frac{I_{xz}^2}{I_x I_z}\right)s^4 + Y_v\left(\frac{I_{xz}^2}{I_x I_z}\right)s^3 = 0 \tag{6-38}$$

and the conditions for the hovering cubic roots to be hardly changed by $I_{xz}$ are

$$\begin{gathered} \frac{I_{xz}^2}{I_x I_z} \ll 1 \\ \left|\frac{Y_v I_{xz}^2}{I_x I_z}\right| \ll |Y_v + L_p + N_r| \end{gathered} \tag{6-39}$$

The unimportance of a small nonzero forward speed on the dynamics is shown by the small size of the $U_0 r$ time vector which is invisible in the roll subsidence and dutch roll vector polygons; and, although visible in the spiral mode, it appears in the summation of side forces that are not needed to describe the mode, as noted earlier.

### 6-10. Gust-Input Transfer Functions

The gust-input transfer function numerators (the denominator is independent of input and has already been fully considered) are obtained in their most general polynomial form from Eq. 6-2, and are given in Table 6-2. Here, in order to allow direct conversion to the primed notation (Eq. 6-1) by conventional rules (i.e., $I_{xz} = 0$, $L \rightarrow L'$, $N \rightarrow N'$), we have included an $L_r s\beta_g$ input term not shown in Eq. 6-2; further, we have distinguished $\beta$-gust-gradient-input derivatives with a $g$ subscript so that potential differences between response and input derivatives, as in $(N_r')_g$ of Eq. 6-1, can be identified.

Figure 6-10 presents Bode plots of the $p$ gust transfer functions for the example conventional airplane; it shows quite clearly that the major response is in $\varphi$. In fact, considering the roll rate response, $p$ (which corresponds to rotating the $\varphi$ amplitude Bode plot counterclockwise 20 dB/decade about the asymptotic zero dB intersection, as indicated), we see that it is exactly equal to the $p_g$ input for low frequencies and is sharply attenuated beyond frequencies corresponding to the roll subsidence mode. Because the dutch roll mode in $\varphi$ almost cancels exactly, the dutch roll modal response coefficient for $\varphi$ will be small, resulting in little $\varphi$ dutch roll motion. For frequencies below dutch roll, the $\beta$ motions

Table 6-2. Lateral gust-input transfer function numerators.

| | $A$ | $B$ | $C$ | $D$ |
|---|---|---|---|---|
| $N^{\beta}_{v_g}$ | $N_p + \frac{I_{xz}}{I_z} L_p$ | $\frac{g}{U_0}\left(-L_p - \frac{I_{xz}}{I_x} N_p\right)$ | $\frac{g}{U_0}(L_pN_r - N_pL_r)$ | |
| $N^{\varphi}_{v_g}$ | $-L_p - \frac{I_{xz}}{I_x} N_p$ | $L_pN_r - N_pL_r + Y_v\left(L_p + \frac{I_{xz}}{I_x} N_p\right)$ | $L_\beta N_p - N_\beta L_p$<br>$+ Y_v(N_pL_r - L_pN_r)$ | |
| $N^{r}_{v_g}$ | $-N_p - \frac{I_{xz}}{I_z} L_p$ | $Y_v\left(N_p + \frac{I_{xz}}{I_z} L_p\right)$ | $0$ | $\frac{g}{U_0}(L_\beta N_p - N_\beta L_p)$ |
| $N^{\beta}_{\beta_g}$ | $-Y_v\left(1 - \frac{I_{xz}^2}{I_xI_z}\right) + (N_r)_g$<br>$+ \frac{I_{xz}}{I_z}(L_r)_g$ | $N_\beta + \frac{I_{xz}}{I_z} L_\beta - (N_r)_g\left(L_p + \frac{I_{xz}}{I_x}\frac{g}{U_0}\right)$<br>$+ Y_v\left(L_p + \frac{I_{xz}}{I_x} N_p + N_r\right.$<br>$\left.+ \frac{I_{xz}}{I_z} L_r\right) + (L_r)_g\left(N_p - \frac{g}{U_0}\right)$ | $-Y_v(L_pN_r - L_rN_p)$<br>$-\frac{g}{U_0}\left(L_\beta + \frac{I_{xz}}{I_x}N_\beta\right) + L_\beta N_p$<br>$- N_\beta L_p - \frac{g}{U_0}[(N_r)_gL_r - (L_r)_gN_r]$ | $\frac{g}{U_0}(L_\beta N_r - N_\beta L_r)$ |
| $\frac{N^{\varphi}_{\beta_g}}{s}$ | $-(L_r)_g - \frac{I_{xz}}{I_x}(N_r)_g$ | $Y_v\left[(L_r)_g + \frac{I_{xz}}{I_x}(N_r)_g\right] + N_r(L_r)_g$<br>$- L_r(N_r)_g - L_\beta - \frac{I_{xz}}{I_x} N_\beta$ | $Y_v[L_r(N_r)_g - N_r(L_r)_g]$<br>$+ L_\beta[N_r + (N_r)_g] - N_\beta[L_r + (L_r)_g]$ | |
| $\frac{N^{r}_{\beta_g}}{s}$ | $-(N_r)_g - \frac{I_{xz}}{I_z}(L_r)_g$ | $Y_v\left[(N_r)_g + \frac{I_{xz}}{I_z}(L_r)_g\right] + L_p(N_r)_g$<br>$-N_p(L_r)_g - N_\beta - \frac{I_{xz}}{I_z} L_\beta$ | $N_\beta L_p - L_\beta N_p$<br>$- Y_v[L_p(N_r)_g - N_p(L_r)_g]$ | $\frac{g}{U_0}[L_\beta(N_r)_g - N_\beta(L_r)_g]$ |

Note: To convert to primed derivatives, eliminate all $I_{xz}$ terms and substitute $L'$ and $N'$ for $L$ and $N$, respectively.

are quite small and the $r$ amplitudes are roughly $g/U_0$ times the $\varphi$ amplitudes. All of these observations are in agreement with our previous ideas regarding the relative motions appearing in the various modes.

The $\beta$ gust Bode plots of Fig. 6-11 assume zero gust gradient terms [i.e., $(N_r)_g = (L_r)_g = 0$ (see Table 6-2)] because such terms are inconsistent with the step gusts assumed for the time histories to be subsequently considered (Fig. 6-12). At dutch roll frequencies the relative amplitudes between $\beta$, $\varphi$, and $r$ are substantially the same as those for the rudder-input situations. If the peak amplitudes are used to estimate the modal response ratios, the result is

$$\beta:\varphi:r \doteq [26:26:31]_{\text{dB}}$$
$$\doteq 1:1:1.78$$

This is, of course, very close to the exact values, 1:0.99:1.87.

For low frequencies, Fig. 6-11 shows that $\beta \doteq \beta_g$, so that $\beta - \beta_g \doteq 0$; this means that the sideslip *relative to the air mass is zero* (see Eq. 4-83). Thus the aerodynamic side forces are zero and the banking and turning motions for the spiral mode are coordinated, as we expect them to be, and as reflected in the $g/U_0$ ($-26.3$ dB) relationship between $\varphi$ and $r$ at the spiral inverse time constant. However, whereas for other inputs (e.g., $\delta$ or $p_g$) the relationship $r \doteq (g/U_0)\varphi$ holds as well for frequencies below the (open- or closed-loop) spiral mode, for $\beta_g$ inputs it does not. For example, for a $\beta_g$ ramp input, $\dot{\beta}_g =$ constant, the resulting steady motions, applying the final value theorem, are $\dot{\beta} = \dot{\beta}_g$, $r = 0$, $\varphi = (U_0/g)\dot{\beta}$.

Figures 6-10 and 6-11 show that, for a given input, all motions (except $\beta/\beta_g$ discussed above) follow roughly the same pattern with frequency; that is, responses to $p_g$ excitation are primarily low frequency in nature, whereas those due to $\beta_g$ excitation occur primarily near dutch roll. Although not shown in Fig. 6-11, the lateral acceleration ($a_y$) response at high frequencies would be quite large. For example, using the basic Eq. 6-3 relationship,

$$a_{y_{\text{c.g.}}} = U_0(\dot{\beta} + r) - g\varphi$$

The high frequency asymptote is given simply by

$$\left.\frac{a_{y_{\text{c.g.}}}}{\beta_g}\right]_{\text{hf}} \rightarrow \frac{U_0(A_\beta s^4 + A_r s^4) - gA_\varphi s^3}{[1 - (I_{xz}^2/I_x I_z)]s^4}$$

where the $A$ terms are those in Table 6-2. Accordingly,

$$\left.\frac{a_{y_{\text{c.g.}}}}{\beta_g}\right]_{\text{hf}} \rightarrow \frac{U_0(A_\beta + A_r)}{[1 - (I_{xz}^2/I_x I_z)]} = -U_0 Y_v \tag{6-40}$$

Notice that this asymptote is constant (i.e., not a function of $s$) and therefore does not attenuate with frequency. This situation is analogous to

Fig. 6-10. Bode plots for rolling gust $(p_g)$ input.

**a) $\dfrac{\beta}{p_g}$**

$$\frac{\beta}{p_g} = -K_{\beta p_g} \frac{(T_{\beta_1}s+1)(T_{\beta_2}s+1)}{(T_s s+1)(T_R s+1)\left[\frac{s^2}{\omega_d^2} + \frac{2\zeta_d}{\omega_d}s+1\right]}$$

| $K_{\beta p_g}$ | $T_{\beta_1}$ | $T_{\beta_2}$ | $T_s$ | $T_R$ | $\zeta_d$ | $\omega_d$ |
|---|---|---|---|---|---|---|
| .9363 or -.57 dB | $\frac{-1}{1.357}$ | $\frac{-1}{.0896}$ | $\frac{-1}{.00135}$ | $\frac{-1}{1.78}$ | .0243 | 1.88 |

**b) $\dfrac{\phi}{p_g}$**

$$\frac{\phi}{p_g} = -K_{\phi p_g} \frac{\left[\frac{s^2}{\omega_\phi^2} + \frac{2\zeta_\phi}{\omega_\phi}s+1\right]}{(T_s s+1)(T_R s+1)\left[\frac{s^2}{\omega_d^2} + \frac{2\zeta_d}{\omega_d}s+1\right]}$$

| $K_{\phi p_g}$ | $T_s$ | $T_R$ | $\zeta_\phi$ | $\omega_\phi$ | $\zeta_d$ | $\omega_d$ |
|---|---|---|---|---|---|---|
| 713.5 or 57.1 dB | $\frac{-1}{.00135}$ | $\frac{-1}{1.78}$ | .0474 | 1.888 | .0243 | 1.88 |

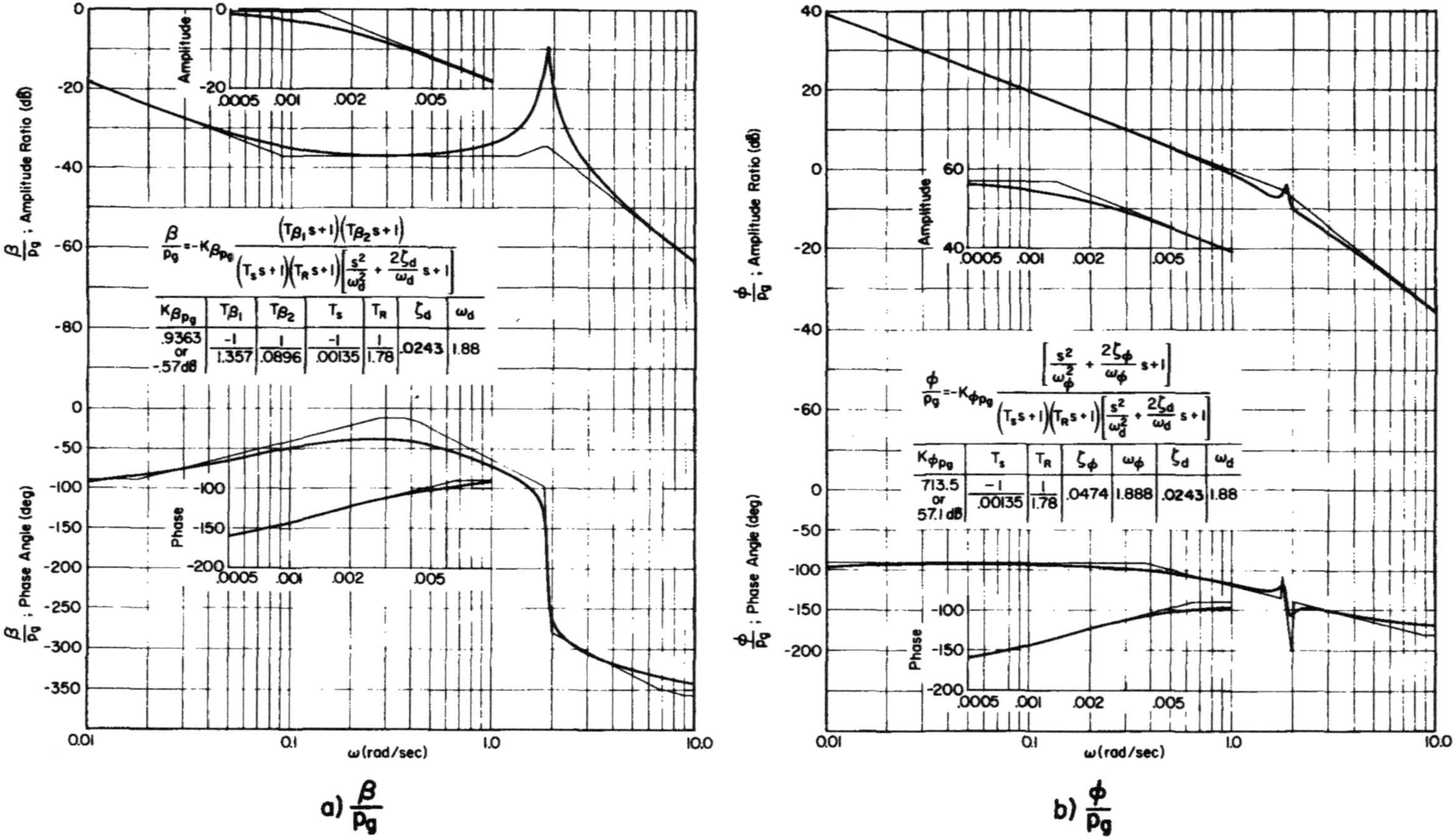

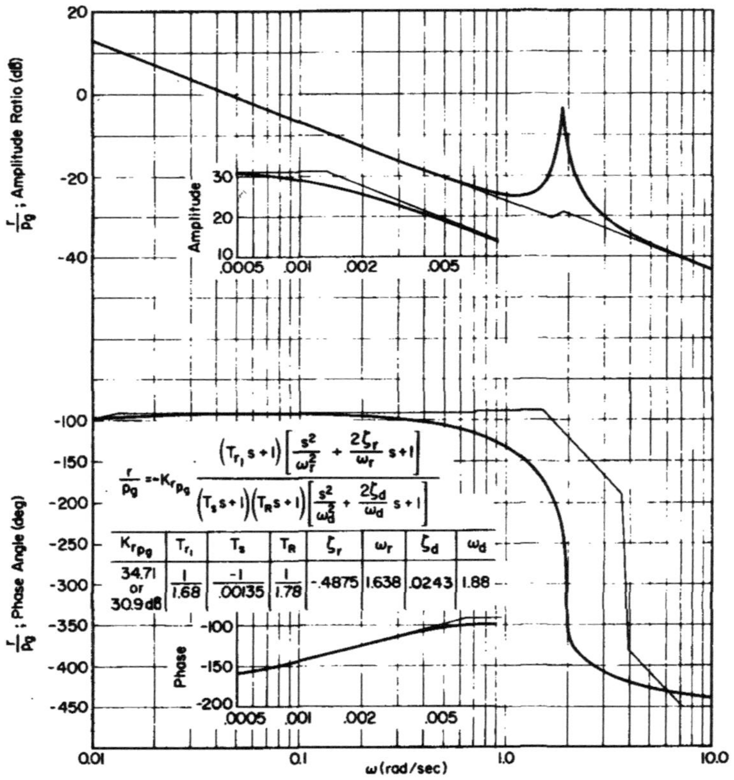


c) $\frac{r}{p_g}$

Fig. 6-10 (Continued)

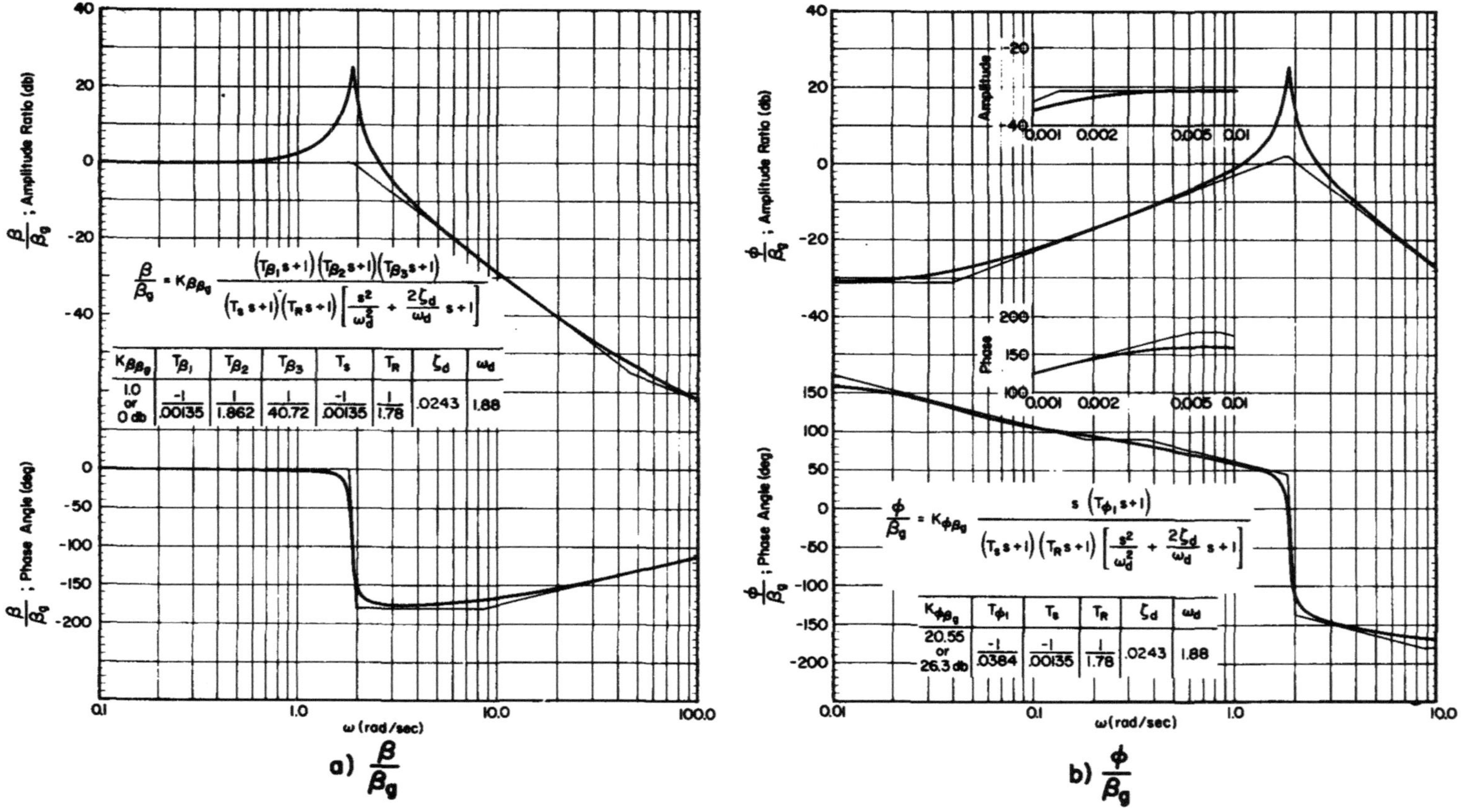


Fig. 6-11. Bode plots for side gust ($\beta_g$) input.

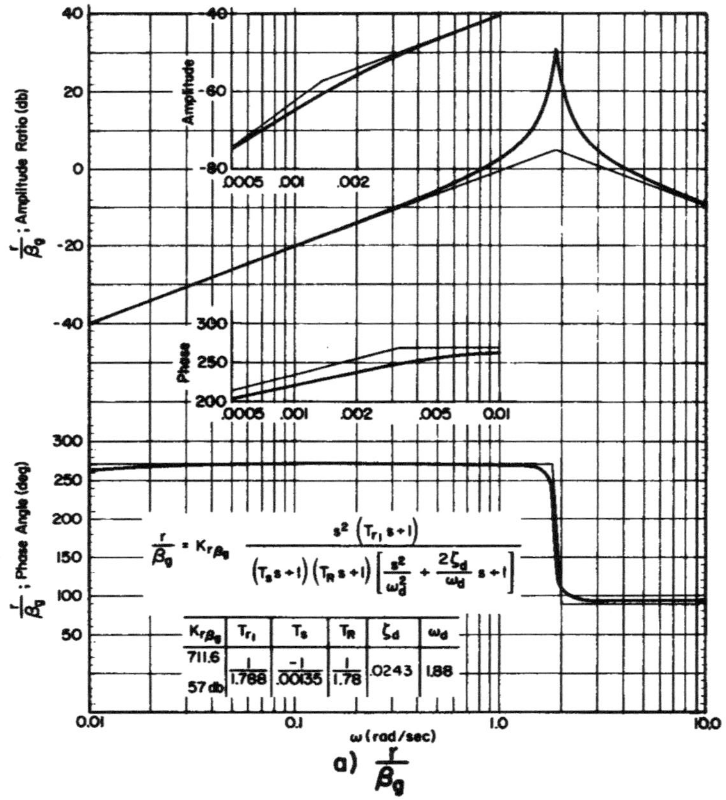


Fig. 6-11 (Continued)

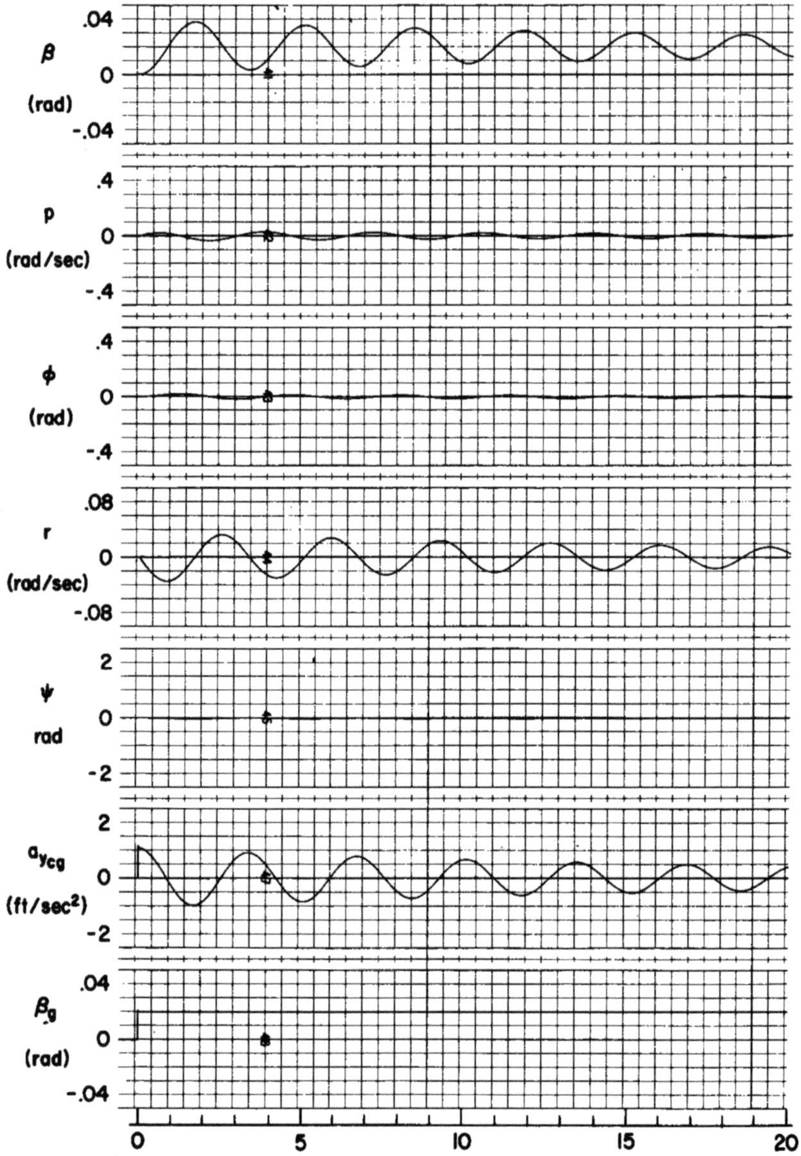


a) $\beta_g$ Step

Fig. 6-12. Responses of a conventional airplane to gust inputs.

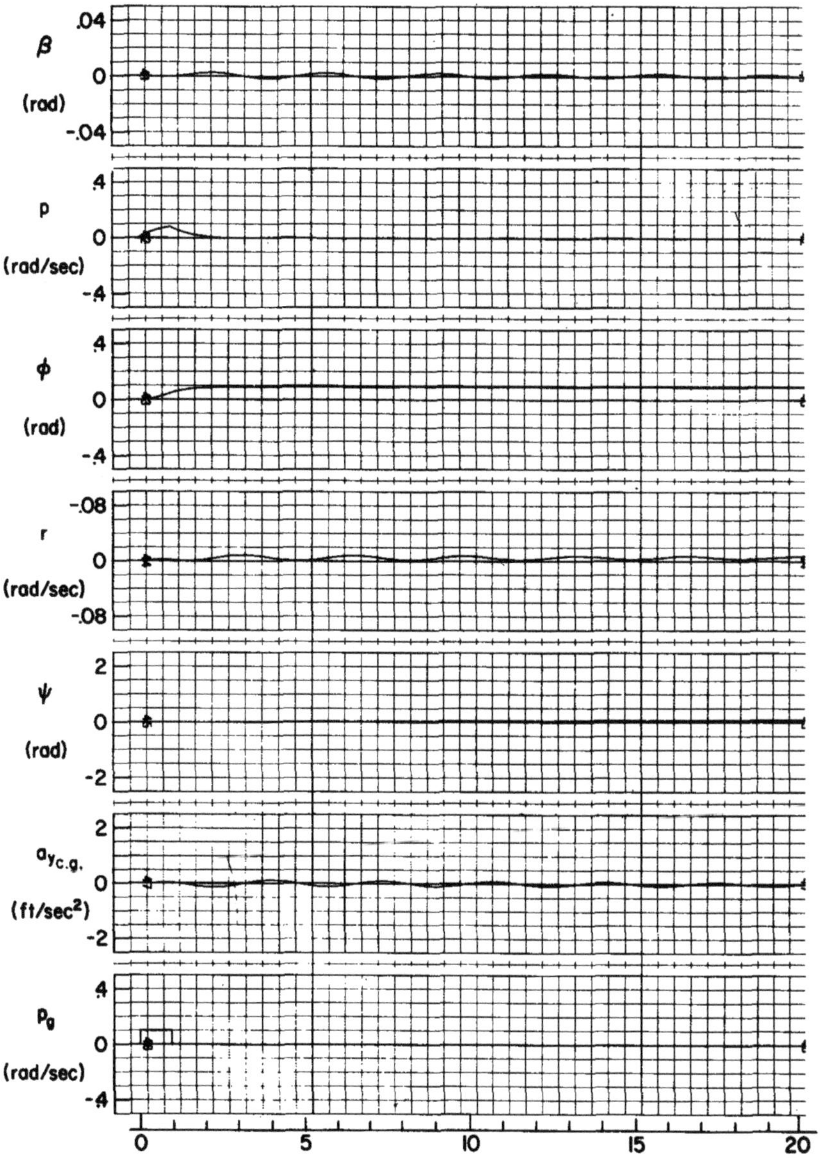


**b) $p_g$ Pulse**

Fig. 6-12 (Continued)

that noted with respect to the $a_z$ high frequency asymptote in Chapter 5 and the same comment applies; that is, the possibility of excitation and coupling with high frequency structural modes must be considered in any system involving $a_y$ feedbacks.

Figure 6-12 presents time histories of the response to a $\beta_g$ step and to a $p_g$ pulse (rather than a step to keep the bank angle trace within bounds). The motions shown are consistent with those deduced above from the Bode plots.

### 6-11. Coupling Numerators

The general uses of, and appreciation for, the longitudinal coupling numerators given in Chapter 5 apply equally well to the lateral coupling numerators:

1. They are required for analysis of multiloop situations involving more than a single input to the vehicle.
2. They are simply obtained by replacing motion derivatives in the characteristic determinant by the appropriate feedback control-input derivatives.
3. They can be a result of coupling between gust- and control-inputs and, among more than two inputs, control or disturbance.
4. They are always of order $s^2$ or lower and are easily computed; therefore, it is unnecessary to catalogue all possible combinations.

The most commonly used aileron and rudder coupling numerators are given in Table 6-3, which contains the polynomial coefficients in literal primed-derivative terms, the corresponding factored forms, and literal approximations to the factors. The latter are based on neglecting terms usually unimportant for conventional airplanes (terms neglected are obvious by comparison with the complete polynomial coefficients). For unconventional configurations the approximate factors shown may not be appropriate, in which case the specifically important contributions to the exact polynomial coefficients can be used.

### 6-12. Approximate Factors

We have already derived some approximate factors in association with the various sets of simplified equations. However we have also noted that, in general, the numerator factors thus obtained are not necessarily good approximations to the complete situation. For these we can use a direct approach that involves factoring the literal polynomial expressions in Table 6-1 by partitioning the polynomial into a form $N(s) + KD(s)$

which can be factored by servo analysis methods (see Section 3-4). This approach yields relatively accurate factors for specific "conditions of validity."

Naturally, the relative significance of given derivatives or groups thereof depends on the vehicle configuration and/or flight condition; therefore, as in the longitudinal case, there is no single set of literal approximate factors that applies to all situations. This results in the large collection of approximate factors given in Tables 6-4 and 6-7, which are taken from the same references used for the corresponding longitudinal tables.

Table 6-4 contains the approximate factors for conventional airplanes, and these apply also to single-rotor helicopters at speeds over 50 ft/sec. Notice that the denominator factors are essentially identical to those obtainable (Eqs. 6-22 and 6-26) from the three degree of freedom dutch roll and spiral/roll subsidence approximate equations of motion. Some of the numerator factors are expressed in terms of unprimed derivatives and some in terms of primed derivatives, depending on which results in the more compact form. As noted earlier, this may lead to confusion regarding the proper value of the transfer function gain; this is, however, easily resolvable according to the rules given on p. 356.

The various expressions in Table 6-4 indicate directly the derivatives of primary importance for a given dynamic mode and the connections between denominator and numerator factors. These connections are specifically spelled out for conventional aircraft in Table 6-5 which can also be used to indicate the gross effects of artificial stability derivative augmentation via motion feedback to the control surfaces.

In general, the conditions for which the Table 6-4 approximate factors apply are inappropriate for hovering and transition flight of helicopters and VTOL aircraft. The latter situations are treated in Table 6-7, to which Table 6-6 serves as a guide. In Table 6-7 we continue to refer to aileron and rudder controls; as noted in Table 6-6, these are to be taken to be whatever devices are used to produce rolling and yawing moments, respectively (e.g., main rotor lateral cyclic pitch and tail rotor collective pitch for a single-rotor helicopter).

We can check some of the VTOL approximate factors against our sample hovering vehicle. Doing so, we find, for instance, from Table 6-7(A) and (B) that the approximate expressions yield

$$\frac{\varphi(s)}{\delta_a(s)} = \frac{0.696(s + 0.14)(s + 0.662)}{(s + 0.662)(s + 0.873)[s^2 - 2(0.347)(0.662) + (0.662)^2]} \qquad (6\text{-}41)$$

which is in excellent agreement with Eq. 6-37. Because of this correspondence, we can use the approximate factor expressions to answer a variety of questions. For example, to determine what changes will stabilize the

Table 6-3. Lateral coupling numerators for aileron and rudder inputs.

| | A | B | C |
|---|---|---|---|
| $N^{\varphi\, r}_{\delta_a\delta_r}$ | $L'_{\delta_a}N'_{\delta_r} - N'_{\delta_a}L'_{\delta_r}$ | $-Y_v(L'_{\delta_a}N'_{\delta_a}L'_{\delta_r})$ | |
| | | $+L'_\beta(Y^*_{\delta_a}N'_{\delta_r} - N'_{\delta_a}Y^*_{\delta_r})$ | |
| or | | $-N'_\beta(Y^*_{\delta_a}L'_{\delta_r} - L'_{\delta_a}Y^*_{\delta_r})$ | |
| $-N^{r\, \varphi}_{\delta_a\delta_r}$ | | $A_{\varphi r}\left(s + \dfrac{1}{T_{\varphi r}}\right)$ | |
| | | $A_{\varphi r} \doteq L'_{\delta_a}N'_{\delta_r};\ \dfrac{1}{T_{\varphi r}} \doteq -Y_v + \dfrac{Y^*_{\delta_r}}{N'_{\delta_r}}\left(N'_\beta - \dfrac{N'_{\delta_a}}{L'_{\delta_a}}L'_\beta\right)$ | |
| $N^{\beta\, \varphi}_{\delta_a\delta_r}$ | $L'_{\delta_a}Y^*_{\delta_r} - Y^*_{\delta_a}L'_{\delta_r}$ | $-N'_r(L'_{\delta_a}Y^*_{\delta_r} - Y^*_{\delta_a}L'_{\delta_a})$ | |
| | | $+ L'_r(N'_{\delta_a}Y^*_{\delta_r} - Y^*_{\delta_a}N'_{\delta_r})$ | |
| or | | $+ (N'_{\delta_a}L'_{\delta_r} - L'_{\delta_a}N'_{\delta_r})$ | |
| $-N^{\varphi\, \beta}_{\delta_a\delta_r}$ | | $A_{\varphi\beta}\left(s + \dfrac{1}{T_{\varphi\beta}}\right)$ | |
| | | $A_{\varphi\beta} \doteq L'_{\delta_a}Y^*_{\delta_r};\quad \dfrac{1}{T_{\varphi\beta}} \doteq -N'_r - \dfrac{N'_{\delta_r}}{Y^*_{\delta_r}}$ | |

$$N^{\varphi\, a_y}_{\delta_a \delta_r}$$

$$U_0(L'_{\delta_a} Y^*_{\delta_r} - Y^*_{\delta_a} L'_{\delta_r} + x_a(L'_{\delta_a} N'_{\delta_r} - N'_{\delta_a} L'_{\delta_r})$$

$$-U_0 N'_r(L'_{\delta_a} Y^*_{\delta_r} - Y^*_{\delta_a} L'_{\delta_r}) + U_0 L'_r(N'_{\delta_a} Y^*_{\delta_r} - Y^*_{\delta_a} N'_{\delta_r}) + x_a\begin{bmatrix} -Y_v(L'_{\delta_a} N'_{\delta_r} - N'_{\delta_a} L'_{\delta_r}) \\ + L'_\beta(Y^*_{\delta_a} N'_{\delta_r} - N'_{\delta_a} Y^*_{\delta_r}) \\ - N'_\beta(Y^*_{\delta_a} L'_{\delta_r} - L'_{\delta_a} Y^*_{\delta_r}) \end{bmatrix}$$

$$U_0\begin{bmatrix} -Y_a(L'_{\delta_a} N'_{\delta_r} - N'_{\delta_a} L'_{\delta_r}) \\ + L'_\beta(Y'_{\delta_a} N'_{\delta_r} - N'_{\delta_a} Y^*_{\delta_r}) \\ -N'_\beta(Y^*_{\delta_a} L'_{\delta_r} - L'_{\delta_a} Y^*_{\delta_r}) \end{bmatrix}$$

or

where the $a_y$ sensor is located $x_a$ ft ahead of the center of gravity

$$-N^{a_y \varphi}_{\delta_a \delta_r} \qquad A_{\varphi y}[s^2 + 2(\zeta_\varphi \omega_\varphi)_y s + (\omega_\varphi)^2_y]$$

$$x_{a_0} \equiv -\frac{U_0 Y^*_{\delta_r}}{N'_{\delta_r}} \sim \text{center of rotation}$$

$$x_a \equiv x_{a_0} + \Delta x_a$$

$$A_{\varphi y} \doteq L'_{\delta_a}(U_0 Y^*_{\delta_r} + x_a N'_{\delta_r}) \doteq -L'_{\delta_a} U_0 Y^*_{\delta_r} \frac{\Delta x_a}{x_{a_0}}$$

$$(\omega_\varphi)^2_y \doteq \frac{U_0}{\Delta x_a N'_{\delta_r}}\left[Y^*_{\delta_r}\left(N'_\beta - \frac{N'_{\delta_a}}{L'_{\delta_a}} L'_\beta\right) - Y_v N'_{\delta_r}\right] \doteq -\frac{U_0 Y_v}{\Delta x_a}$$

$$2(\zeta_\varphi \omega_\varphi)_y \doteq \frac{x_a}{\Delta x_a N'_{\delta_r}}\left[Y^*_{\delta_r}\left(N'_\beta - \frac{N'_{\delta_a}}{L'_{\delta_a}} L'_\beta\right) - Y_v N'_{\delta_r}\right] - \frac{U_0 Y^*_{\delta_r}}{\Delta x_a N'_{\delta_r}}\left(N'_r - \frac{N'_{\delta_a}}{L'_{\delta_a}} L'_r\right) \doteq -\frac{x_a}{\Delta x_a} Y_v$$

Table 6-4. Summary of conventional airplane lateral approximate factors.
A. Denominator.[a]

| Factored forms | Approximate factors | Conditions of validity |
| --- | --- | --- |
| $\Delta(s) = \underbrace{\left(s + \frac{1}{T_s}\right)\left(s + \frac{1}{T_R}\right)}_{\text{or}}(s^2 + 2\zeta_d\omega_d s + \omega_d)$ | $\omega_d^2 \doteq N'_\beta$ | $\left[\begin{array}{r} Y_r = Y_p = Y_{\dot v} = L_{\dot v} = N_{\dot v} \doteq 0 \\ \lvert Y_v N'_r - N'_p L'_r \rvert \ll \lvert N'_\beta \rvert \\ \left\lvert \frac{g}{U_0} \sin \gamma_0 \right\rvert \ll \lvert L'_p \rvert \\ \lvert Y_v L'_r \rvert \ll \lvert L'_\beta \rvert \end{array}\right]$ |
| | $2(\zeta\omega)_d \doteq -(Y_v + N'_r) - \frac{L'_\beta}{N'_\beta}\left(N'_p - \frac{g}{U_0}\right)$ | $\left\lvert \frac{L_p}{N'_\beta}(Y_v + N'_r) \right\rvert < 0.3$ |
| | $\frac{1}{T_R} \doteq - L'_p + \frac{L'_\beta}{N'}\left(N'_p - \frac{g}{U_0}\right)$ | $\left\lvert \frac{L'_\beta}{L'_p N'_\beta}\left(N'_p - \frac{g}{U_0}\right) \right\rvert < 0.4$ |
| | $\frac{1}{T_s} \doteq T_R \frac{g}{U_0}\left(\frac{L'_\beta}{N'_\beta} N'_r - L'_r\right)$ | $\left\lvert \frac{L'_r N'_p}{L'_p N'_r} \right\rvert \ll 1, \quad \left\lvert \frac{1}{T_s} \right\rvert \ll \left\lvert \frac{1}{T_R} \right\rvert$ |
| ↓ $(s^2 + 2\zeta_{SR}\omega_{SR}s + \omega_{SR}^2)$ | $\omega_{SR}^2 \doteq \frac{g/U_0[(L'_\beta N'_r/N'_\beta) - L'_r]}{1 + (L'_p/N'_\beta)(Y_v + N'_r)}$ | Same conditions applicable. |
| | $2(\zeta\omega)_{SR} \doteq \frac{-L'_p + (L'_\beta/N'_\beta)[N'_p - (g/U_0)]}{1 + (L'_p/N'_\beta)(Y_v + N'_r)}$ | $\left\lvert \frac{(g/U_0)(L'_\beta N'_r - N'_\beta L'_r)}{N'^2_\beta[1 + (L'_p/N'_\beta)(Y_v + N'_r)]^2} \right\rvert \ll 1$ |
| | $2(\zeta\omega)_d \doteq -(Y_v + N)'_r - \left\{\frac{L'^2_p(Y_v + N'_r) + L'_\beta[N'_p - g/U_0]}{N'_\beta[1 + (L'_p/N'_\beta)(Y_v + N'_r)]}\right\}$ | $\left\lvert \frac{-(Y_v + L'_p + N'_r)[-L'_p N'_\beta + L'_\beta(N'_p - g/U_0)]}{N'^2_\beta[1 + (L'_p/N'_\beta)(Y_v + N'_r)]^2} \right\rvert < 0.25$ |
| | $\omega_d^2 \doteq N'_\beta\left[1 + \frac{L'_p}{N'_\beta}(Y_v + N'_r)\right] - \omega_{SR}^2 - 4(\zeta\omega)_{SR}(\zeta\omega)_d$ | $0.005 < \frac{\{-L'_p + (L'_\beta/N'_\beta)[N'_p - (g/U_0)]\}^2}{4(g/U_0)[(L'_\beta N'_r/N'_\beta) - L'_r] \times [1 + (L'_p/N'_\beta)(Y_v + N'_r)]} < 1$ |

B. Aileron numerators.

| | | |
|---|---|---|
| $N_\varphi(s) = A_\varphi(s^2 + 2\zeta_\varphi\omega_\varphi s + \omega_\varphi^2)$ | $A_\varphi - L'_{\delta_a}$<br>$\omega_\varphi^2 \doteq N'_\beta\left(1 - \frac{N'_{\delta_a}L'_\beta}{L'_{\delta_a}N'_\beta}\right)$<br>$2\zeta_\varphi\omega_\varphi \doteq -(Y_v + N'_r) + \frac{N'_{\delta_a}}{L'_{\delta_a}} L'_r$ | $Y_{\delta_a} \doteq 0$<br>$\lvert Y_v(L'_{\delta_a}N'_r - L'_rN'_{\delta_a})\rvert \ll \lvert L'_{\delta_a}N'_\beta - L'_\beta N'_{\delta_a}\rvert$ |
| $N_r(s) = A_r\left(s + \frac{1}{T_{r_1}}\right)(s^2 + 2\zeta_r\omega_r s + \omega_r^2)$ | $A_r \doteq \frac{I_{xz}}{I_z} L_{\delta_a}$<br>$\frac{1}{T_{r_1}} \doteq \omega_r \doteq \left(\frac{I_z}{I_{xz}}\frac{g}{U_0} N_\beta\right)^{1/3}$<br>$2\zeta_r\omega_r \doteq \frac{I_z}{I_x} N_p - Y_v - \frac{1}{T_{r_1}}$ | $N_{\delta_a} = Y_{\delta_a} \doteq 0$<br>$B - AC^{1/3} \ll C^{2/3}$<br>$B = -\frac{I_z}{I_{xz}} Y_v N_p$<br>$A = \frac{I_z}{I_{xz}} N_p - Y_v$<br>$C = \frac{I_z}{I_{xz}}\frac{g}{U_0} N_\beta$ |
| $N_\beta(s) = A_\beta\left(s + \frac{1}{T_{\beta_1}}\right)\left(s + \frac{1}{T_{\beta_2}}\right)$ | $A_\beta \doteq -N'_{\delta_a}$<br>$\frac{1}{T_{\beta_1}} \doteq \left(\frac{g}{U_0}\right)\frac{L'_r - (L'_{\delta_a}/N'_{\delta_a})N'_r}{L'_p - (L'_{\delta_a}/N'_{\delta_a})[N'_p - (g/U_0)]}$<br>$\frac{1}{T_{\beta_2}} \doteq -L'_p + \frac{L'_{\delta_a}}{N'_{\delta_a}}\left(N'_p - \frac{g}{U_0}\right)$ | $Y^*_{\delta_a} \doteq 0$<br>$\left\lvert\frac{g}{U_0}\left(L'_r - \frac{L'_{\delta_a}}{N'_{\delta_a}} N'_r\right)\right\rvert \ll \left[L'_p - \frac{L'_{\delta_a}}{N'_{\delta_a}}\left(N'_p - \frac{g}{U_0}\right)\right]^2$ |

Table 6-4 (Continued)
C. Rudder numerators.

| Factored forms | Approximate factors | Conditions of validity |
|---|---|---|
| $N_\varphi(s) = A_\varphi\underbrace{(s^2 + 2\zeta_\varphi\omega_\varphi s + \omega_\varphi^2)}$ | $A_\varphi = L'_{\delta_r}$ | |
| or | $\omega_\varphi^2$ or $\frac{1}{T_{\varphi_1}T_{\varphi_2}} \doteq N'_\beta\left[1 - \frac{L'_\beta N'_{\delta_r}}{N'_\beta L'_{\delta_r}} + \frac{Y^*_{\delta_r}L'_r}{L'_{\delta_r}}\left(1 - \frac{L_\beta N'_r}{N'_\beta L'_r}\right)\right]$ | $\left\lvert Y_v\left(N'_r - \frac{L'_r N'_{\delta_r}}{L'_{\delta_r}}\right)\right\rvert \ll \omega_d^2$ |
| $\left(s + \frac{1}{T_{\varphi_1}}\right)\left(s + \frac{1}{T_{\varphi_2}}\right)$ | $2(\zeta\omega)_\varphi$ or $\frac{1}{T_{\varphi_1}} + \frac{1}{T_{\varphi_2}} = -(Y_v + N'_r) + \frac{Y^*_{\delta_r}L'_\beta + N'_{\delta_r}L'_r}{L'_{\delta_r}}$ | |
| $N_r(s) = A_r\left(s + \frac{1}{T_{r_1}}\right)\underbrace{(s^2 + 2\zeta_r\omega_r s + \omega_r^2)}$ | $A_r \doteq N_{\delta_r}$ | $L_{\delta_r} \doteq 0$ |
| | $\left.\begin{array}{c} \frac{1}{T_{r_1}} = \left(-\frac{g}{U_0}L_\beta\right)^{1/3} \\ \omega_r = \left\lvert\frac{1}{T_{r_1}}\right\rvert \\ 2\zeta_r\omega_r = -Y_v - L_p - \frac{1}{T_{r_1}} \end{array}\right\}$ | $\left\lvert\frac{g}{U_0}L_\beta\right\rvert \gg \lvert L_p^3\rvert, \quad \left\lvert\frac{I_{xz}L_\beta}{I_z N_\beta}\right\rvert \ll 1$ |
| or | $\frac{1}{T_{r_1}} \doteq -L_p$ | $\left\lvert\frac{Y^*_{\delta_r}L_\beta I_{xz}}{N_{\delta_r}L_p I_z}\right\rvert \ll 1, \quad \left\lvert\frac{L_\beta N_p}{N_\beta L_p}\right\rvert \ll 1$ |
| $\left(s + \frac{1}{T_{r_2}}\right)\left(s + \frac{1}{T_{r_3}}\right)$ | $\left.\begin{array}{c} \frac{1}{T_{r_2}} \doteq -\left(Y_v - \frac{Y^*_{\delta_r}}{N_{\delta_r}}N_\beta\right) \\ \frac{1}{T_{r_3}} \doteq -\frac{gL_\beta}{U_0L_p[Y_v - (Y^*_{\delta_r}/N_{\delta_r})N_\beta]} \end{array}\right\}$ | $\left\lvert\frac{g}{U_0}\frac{L_\beta}{L_p}\right\rvert < \left\lvert\frac{[Y_v - (Y^*_{\delta_r}/N_{\delta_r})N_\beta]^2}{4}\right\rvert$ |
| | $\left.\begin{array}{c} \omega_r^2 \doteq \frac{g}{U_0}\frac{L_\beta}{L_p} \\ 2\zeta_r\omega_r \doteq -\left(Y_v - \frac{Y^*_{\delta_r}N_\beta}{N_{\delta_r}} - \frac{g}{U_0}\frac{L_\beta}{L_p^2}\right) \end{array}\right\}$ | $\left\lvert\frac{[Y_v - (Y^*_{\delta_r}/N_{\delta_r})N_\beta]^2}{4}\right\rvert < \left\lvert\frac{g}{U_0}\frac{L_\beta}{L_p}\right\rvert < \left\lvert\frac{L_p^2}{10}\right\rvert$ |

| | | |
|---|---|---|
| $N_\beta(s) = A_\beta\left(s + \frac{1}{T_{\beta_1}}\right)\left(s + \frac{1}{T_{\beta_2}}\right)\left(s + \frac{1}{T_{\beta_3}}\right)$ | $A_\beta = Y^*_{\delta_r}$ | |
| | $\frac{1}{T_{\beta_1}} \doteq \left(\frac{g}{U_0}\right)\frac{L'_r - (L'_{\delta_r}/N'_{\delta_r})N'_r}{L'_p - [L'_{\delta_r}/N'_{\delta_r})(N'_p - (g/U_0)]}$ | $\lvert L_p + N'_r\rvert \ll \left\lvert\frac{N'_{\delta_r}}{Y^*_{\delta_r}}\right\rvert$ |
| | $\frac{1}{T_{\beta_2}} \doteq -L'_p + \frac{L'_{\delta_r}}{N'_{\delta_r}}\left(N'_p - \frac{g}{U_0}\right)$ | $\lvert L'_pN'_r - N'_pL'_r\rvert \ll \left\lvert\frac{N'_{\delta_r}}{Y^*_{\delta_r}}\left[L'_p - \frac{L'_{\delta_r}}{N'_{\delta_r}}\left(N'_p - \frac{g}{U_0}\right)\right]\right\rvert$ |
| | $\frac{1}{T_{\beta_3}} \doteq -\frac{N'_{\delta_r}}{Y^*_{\delta_r}}$ | $\left\lvert\frac{1}{T_{\beta_1}}\left(\frac{1}{T_{\beta_2}} + \frac{1}{T_{\beta_3}}\right)\right\rvert \ll \left\lvert\frac{1}{T_{\beta_2}T_{\beta_3}}\right\rvert$ |
| General: | | |
| | $A_{a_y} = Y_{\delta_r} + x_aN'_{\delta_r}$ | $N'_{\delta_r}L'_p \doteq N_{\delta_r}L_p$ |
| | $\frac{1}{T_{a_{y_1}}} \doteq \frac{g}{U_0}\left[\frac{L_r}{L_p} - \frac{N_r(Y_vL_{\delta_r} - Y^*_{\delta_r}L_\beta)}{L_p(Y_vN_{\delta_r} - Y^*_{\delta_r}N_\beta)}\right]$ | $\lvert L_{\delta_r}N_p\rvert < \lvert Y^*_{\delta_r}N'_p\rvert$ |
| $N_{a_y}(s) = A_{a_y}\left(s + \frac{1}{T_{a_{y_1}}}\right)\left(s + \frac{1}{T_{a_{y_2}}}\right)$ | $\frac{1}{T_{a_{y_2}}} \doteq -L_p$ | $\left\lvert\frac{L_\beta N_p}{N_\beta L_p}\right\rvert \ll 1$ |
| $\times \underbrace{(s^2 + 2\zeta_{a_y}\omega_{a_y}s + \omega^2_{a_y})}$ or | $\omega^2_{a_y}$ or $\frac{1}{T_{a_{y_3}}T_{a_{y_4}}} \doteq \frac{U_0}{\Delta x_aN'_{\delta_r}}(Y^*_{\delta_r}N_\beta - Y_vN_{\delta_r})$ | $\lvert N_{\delta_r}L_p\rvert \gg \lvert Y^*_{\delta_r}N'_\beta - Y_vN'_{\delta_r}\rvert$ |
| $\left(s + \frac{1}{T_{a_{y_3}}}\right)\left(s + \frac{1}{T_{a_{y_4}}}\right)$ | $2\zeta_{a_y}\omega_{a_y}$ or $\frac{1}{T_{a_{y_3}}} + \frac{1}{T_{a_{y_4}}} \doteq -\frac{Y_{\delta_r}}{\Delta x_a(N'_{\delta_r})^2}$ | $\left\lvert\frac{L_pY^*_{\delta_r}}{N_{\delta_r}}\right\rvert \ll 1$ |
| where $a_y$ sensor is located $x_a$ feet ahead of the center of gravity | $\times (Y^*_{\delta_r}N_\beta - Y_vN_{\delta_r})$ | $\lvert L_{\delta_r}N_p\rvert \ll \lvert N_{\delta_r}L_p\rvert$ |
| At the center of rotation: | $B_{a_y} \doteq Y_{\delta_r}\left(-N'_r + Y_v - \frac{Y_{\delta_r}}{N'_{\delta_r}}N'_v\right)$ | $\frac{g}{U_0}$ and $\lvert N_p\rvert \ll \lvert L_p\rvert$ |
| $N_{a_y}(s) = B_{a_y}\left(s + \frac{1}{T_{a_{y_1}}}\right)\left(s + \frac{1}{T_{a_{y_2}}}\right)\left(s + \frac{1}{T_{a_{y_3}}}\right)$ | $\frac{1}{T_{a_{y_3}}} \doteq \frac{-U_0(Y_vN'_{\delta_r} - Y_{\delta_r}N'_v)}{Y_{\delta_r}\left(-N'_r + Y_v - \frac{Y_{\delta_r}}{N'_{\delta_r}}N'_v\right)}$ | $Y_vN'_{\delta_r} - Y^*_{\delta_r}N'_\beta \doteq Y_vN_{\delta_r} - Y^*_{\delta_r}N_\beta$ |

[a] Applicable also to single-rotor helicopters at high speed ($U_0 > 50$ fps).

Table 6-5. Qualitative importance of normal airframe stability parameters to lateral, aileron and rudder input, transfer function quantities.[a]

| Derivative | $L_\beta$ | $L_p$ | $L_r$ | $N_\beta$ and $Y_v$ | $N_p$ | $N_r$ | Change in fore and aft sensor location | Change in $I_{xz}$ | Change in point of control application |
|---|---|---|---|---|---|---|---|---|---|
| 1. Denominator, $\Delta(s)$ | | | | | | | | | |
| a. Spiral mode, $1/T_s$ | xxx | | xx | xx | | xx | | | |
| b. Roll subsidence mode, $1/T_R$ | x | xxx | | x | x | | | | |
| c. Dutch roll undamped frequency, $\omega_d$ | xx | | | xxx | | | | xx | |
| d. Dutch roll damping, $2\zeta_d\omega_d$ | xx | | | xx | x | xxx | | xx | |
| e. Lateral phugoid undamped frequency, $\omega_{SR}$ | xxx | | x | xxx | | xx | | | |
| f. Lateral phugoid damping, $2\zeta_{SR}\omega_{SR}$ | xx | xxx | | xx | x | | | | |
| 2. Roll (Aileron input) numerator, $N^{\varphi}_{\delta_a}(s)$ | | | | | | | | | |
| a. Undamped frequency, $\omega_{\varphi_a}$ | xx | | | xxx | | | | | x to xxx [as reflected in $C_{n_{\delta(a)}}/C_{l_{\delta(a)}}$] |
| b. Damping, $2\zeta_{\varphi_a}\omega_{\varphi_a}$ | | | x | xx | | xxx | | | x |
| 3. Roll (rudder input) numerator, $N^{\varphi}_{\delta_r}(s)$ | | | | | | | | | |
| a. Undamped frequency, $\omega_{\varphi_r}$ or product of first orders, $1/T_{\varphi_1}T_{\varphi_2}$ | x | | x | xx | | x | | | x to xxx |
| b. Damping, $2\zeta_{\varphi_r}\omega_{\varphi_r}$ or sum of first orders, $1/T_{\varphi_1} + 1/T_{\varphi_2}$ | x | | x | | | xx | | | x to xxx |

| | | | | | | | | | |
|---|---|---|---|---|---|---|---|---|---|
| **4.** Yawing Velocity (Aileron Input) Numerator, $N^{r}_{\delta_a}(s)$ | | | | | | | | | |
| a. First-order factor, $1/T_{r_a}$ undamped frequency of quadratic factor, $\omega_{r_a}$ | | | | xx | | | | xx | |
| b. Quadratic factor damping, $2\zeta_{r_a}\omega_{r_a}$ | | | | xx | xx | | | x | xx to xxx |
| **5.** Yawing Velocity (Rudder Input) Numerator, $N^{r}_{\delta_r}(s)$ | | | | | | | | | |
| a. First factor, $1/T_{r_1}$ | | xxx | | | | | | | |
| b. Second factor, $1/T_{r_2}$ | | | | xx | | | | | xx to xxx |
| c. Third factor, $1/T_{r_3}$ | xx | xx | | xx | | | | | x |
| d. Quadratic factor frequency, $\omega_r$ | xx | xx | | | | | | | |
| e. Quadratic factor damping, $2\zeta_r\omega_r$ | x | x | | x | | | | | x |
| **6.** Sideslip Numerator, $N^{\beta}_{\delta}$ | | | | | | | | | |
| a. First factor, $1/T_{\beta_1}$ | | xx | xx | | x | x | | | x |
| b. Second factor, $1/T_{\beta_2}$ | | xxx | | | x | | | | x |
| c. Third factor, $1/T_{\beta_3}$ (rudder input) | | | | | | | | | xxx |
| **7.** Side Acceleration (Rudder Input) Numerator, $N^{a_y}_{\delta_r}(s)$ | | | | | | | | | |
| a. First factor, $1/T_{a_{y_1}}$ | x | x | x | x | | x | | | x |
| b. Second factor, $1/T_{a_{y_2}}$ | | xxx | | | | | | | |
| c. Third factor, $1/T_{a_{y_3}}$ | | | | xx | | | xxx | x | xx |
| d. Fourth factor, $1/T_{a_{y_4}}$ | | | | xx | | | xxx | x | xx |
| e. Quadratic factor frequency, $\omega_{a_y}$ | | | | xx | | | xxx | x | xx |
| f. Quadratic factor damping, $2\zeta_{a_y}\omega_{a_y}$ | | | | xx | | | xxx | x | xx |

Code: Blank—Little or no effect; x—moderate effect; xx—important effect; xxx—predominant effect.

[a] The only stability parameters considered in this table are those that (1) exist in the normal uncontrolled airframe, and (2) can be augmented by relatively simple automatic control systems.

Table 6-6. Summary of the unconventional aircraft lateral approximate factors given in table 6-7.

| | | Tilt-wing | | Helicopter | | | | Tilt-duct | | Tilt-rotor |
|---|---|---|---|---|---|---|---|---|---|---|
| | | | | Single-rotor | | Tandem-rotor | | | | |
| | | Low speed[a,c] | High speed[b,c] | Hover ($U_0 \doteq 0$) | High speed ($U_0 > 50$ ft/sec) | Hover ($U_0 \doteq 0$) | High speed ($U_0 > 40$ ft/sec) ($U_0 < 120$ ft/sec) | Hover ($U_0 \doteq 0$) | High speed ($U_0 > 100$ ft/sec) | All speeds |
| Denominator | | — | — | — | Table 6-4 | √ | √ | √ | — | — |
| Numerators: Aileron | $\varphi$ | √ | √ | √ | √ | √ | √ | √ | √ | √ |
| Numerators: (rolling moment control) | $r$ | — | — | √ | √ | √ | √ | — | — | — |
| | $v$ | — | — | — | — | — | — | √ | √ | — |
| Numerators: Rudder | $\varphi$ | √ | √ | √ | √ | √ | √ | √ | √ | √ |
| Numerators: (yawing moment control) | $r$ | — | — | √ | √ | √ | √ | * | — | — |
| | $v$ | — | — | — | √ | √ | √ | — | √ | — |

Note: Asterisk (*) indicates condition for which numerator and denominator factors cancel to give, approximately, $r/\delta_r \doteq N_{\delta_r}/(s - N_r)$.

[a] Includes range of speed for which wing incidence is within 45° of wing incidence at hover.

[b] Wing incidence greater than 45° from incidence at hover.

[c] The separation of high and low speed factors by wing incidence is empirical.

Table 6-7. Approximate lateral factors for unconventional aircraft.
A. Denominator ($\Delta_{lat}$) factors.

| First coeff. | Approximate factors | Conditions of validity | Applicable to: |
|---|---|---|---|
| 1 | $\frac{1}{T_s} \doteq -N'_r$<br>$\frac{1}{T_R} \doteq -\sqrt[3]{gL'_v} + \left(\frac{-L_p - Y_v}{3}\right)$<br>$2\zeta_d\omega_d \doteq \sqrt[3]{gL'_v} + \frac{2}{3}(-L'_p - Y_v)$<br>$\omega_d^2 \doteq (gL'_v)^{2/3}\left[1 + \frac{-L'_p - Y_v}{3\sqrt[3]{gL'_v}}\right]$ | $\lvert U_0N'_v\rvert \ll \lvert N'_rL'_p\rvert$<br>$\lvert L'_p\rvert > \lvert Y_v\rvert$<br>$\left\lvert\frac{gL'_v}{L'^3_p}\right\rvert > 1$ | Tilt-duct—hover |
| 1 | $\frac{1}{T_s} \doteq -N'_r\left[1 + \frac{U_0}{g}\left(N'_p - \frac{N'_vL'_p}{L'_v}\right) - \frac{L'_rN'_v}{N'_rL'_v}\right]$<br>$\frac{1}{T_R} \doteq -L'_p - \frac{gL'_v}{L'^2_p}$<br>$2\zeta_d\omega_d \doteq -Y_v + \frac{gL_v}{L'^2_p}$<br>In hover, a better approximation is<br>$2\zeta_d\omega_d \doteq -Y_v - \frac{L'_p}{4} - \sqrt{\frac{L'^2_p}{16} + \frac{gL'_v}{2L'_p}}$<br>$\omega_d^2 \doteq \frac{-gL'_v + U_0(L'_vN'_p - N'_vL'_p)}{-L'_p - (gL'_v/L'^2_p)}$ | | Tandem-rotor helicopters—all speeds |

Table 6-7 (Continued)

B. $\varphi$-Numerator ($N_{\delta}^{\varphi}$) factors.

| First coeff. | Approximate factors | Conditions of validity | Applicable to: |
|---|---|---|---|
| $L'_{\delta}$ | $\left.\begin{array}{l}\frac{1}{T_{\varphi_1}}+\frac{1}{T_{\varphi_2}}\\ \text{or } 2\zeta_{\varphi}\omega_{\varphi}\end{array}\right\} = -(N'_r + Y_v) + \frac{N'_{\delta}}{L'_{\delta}} L'_r + \frac{Y_{\delta}}{L'_{\delta}} L'_v$ <br> $\left.\begin{array}{l}\frac{1}{T_{\varphi_1}}\frac{1}{T_{\varphi_2}}\\ \text{or } \omega_{\varphi}^2\end{array}\right\} = \frac{-N'_{\delta}}{L'_{\delta}}(L'_v U_0 + Y_v L'_r) + (Y_v N'_r + U_0 N'_v) + \frac{Y_v}{L'_{\delta}}(L'_r N'_v - L'_v N'_r)$ | None; this is *exact* expression. | All configurations and speed, and specifically to: <br> Tilt-wing and tilt-rotor—aileron and rudder <br> Tilt-duct—rudder <br> Tandem-rotor helicopter—rudder at hover |
| $L'_{\delta_r}$ | $2\zeta_{\varphi}\omega_{\varphi} \doteq -Y_v - N'_r + \frac{N'_{\delta_r}}{L'_{\delta_r}} L'_r$ | $\lvert Y_{\delta_r} L'_v \rvert \ll \lvert N'_{\delta_r} L'_r + L'_{\delta_r}(Y_v + N'_r) \rvert$ | Single-rotor helicopter—rudder at all speeds |
| | $\omega_{\varphi}^2 \doteq U_0 N'_v + \frac{Y_{\delta_r}}{L'_{\delta_r}} L'_v N'_r - \frac{N'_{\delta_r}}{L'_{\delta_r}} U_0 L'_v$ | $2\zeta_{\varphi}\omega_{\varphi}$ is given by small difference between $(-Y_v - N'_r)$ and $(N_{\delta_r}/L_{\delta_r})L'_r$; hence, accuracy is low for small $\zeta$. | Tandem-rotor helicopter—rudder at high speed |
| $L'_{\delta_a}$ | $2\zeta_{\varphi}\omega_{\varphi} \doteq -N'_r - Y_v$ <br> $\omega_{\varphi}^2 \doteq U_0 N'_v + Y_v N'_r - \frac{N'_{\delta_a}}{L'_{\delta_a}} L'_v U_0$ | $Y_{\delta_a} = 0$ | Tilt-duct—aileron |
| $L'_{\delta_a}$ | $2\zeta_{\varphi}\omega_{\varphi} \doteq -N'_r - Y_v$ <br> $\omega_{\varphi}^2 \doteq U_0 N'_v$ | $\lvert Y_{\delta_a} L'_r \rvert \ll \lvert U_0 L'_{\delta_a} \rvert$ <br> $(-Y_v - N'_r)^2 < \lvert 4 U_0 N'_v \rvert$ | Single-rotor helicopter—aileron |

| | | | |
|---|---|---|---|
| $L_{\delta_a}$ | $\left.\begin{matrix}\dfrac{1}{T_{\varphi_1}T_{\varphi_2}} \\ \text{or } \omega_\varphi^2\end{matrix}\right\} \doteq U_0 N_v + Y_v N_r$<br>$\left.\begin{matrix}\dfrac{1}{T_{\varphi_1}} + \dfrac{1}{T_{\varphi_2}} \\ \text{or } 2\zeta_\varphi\omega_\varphi\end{matrix}\right\} \doteq -Y_v - N_r$ | $\|N_{\delta_a} I_{xz}\| \ll \|L_{\delta_a} I_x\|$<br>$\left\| \dfrac{Y_{\delta_a}}{L_{\delta_a}}\left(L_v + \dfrac{I_{xz}}{I_x} N_v\right) + \dfrac{N_{\delta_a}}{L_{\delta_a}}\left(L_r - \dfrac{I_{xz}}{I_x} Y_v\right)\right\| \ll \|N_r + Y_v\|$<br>$\|Y_{\delta_a}(L_r N_v - L_v N_r) - N_{\delta_a}(U_0 L_v + Y_v L_r)\| \ll \|L_{\delta_a}(U_0 N_v + Y_v N_r)\|$<br>For $U_0 \doteq 0$, use $\dfrac{1}{T_{\varphi_1}} \doteq -N_r$,<br>$\dfrac{1}{T_{\varphi_2}} \doteq -Y_v + \left(\dfrac{Y_{\delta_a}}{L_{\delta_a}}\right) L_v$ | Tandem-rotor helicopter—aileron |
| | C. $r$-Numerator ($N_\delta^r$) factors. | | |
| $N_{\delta_r}$ | $\dfrac{1}{T_r} \doteq -L_p'$<br>$\omega_r^2 \doteq \dfrac{-g}{L_p'}\left(-L_v' + \dfrac{L_{\delta_r}'}{N_{\delta_r}'} N_v'\right)$<br>For single-rotor,<br>$2\zeta_r\omega_r \doteq 0$<br>For tandem-rotor,<br>$2\zeta_r\omega_r \doteq -Y_v + \dfrac{\omega_r^2}{L_p'}$ | $\|L_p'\| \gg \left\|\dfrac{L_{\delta_r}'}{N_{\delta_r}'} N_p' + \dfrac{Y_{\delta_r}}{N_{\delta_r}'} N_v'\right\|$<br>$Y_{\delta_r} \doteq 0$<br>$\left\|\dfrac{1}{T_r}\right\| \gg \omega_r$<br>$\left\|\dfrac{1}{T_r}\right\| \gg \|2\zeta_r\omega_r\|$ | Single- and tandem-rotor helicopter—rudder |

Table 6-7 (Continued)

C. $r$-Numerator $(N_\delta^r)$ factors (continued).

| First Coeff. | Approximate factors | Conditions of validity | Applicable to: |
|---|---|---|---|
| $Y_{\delta_a}N'_v + L'_{\delta_a}N'_p$ | $2\zeta_r\omega_r \doteq \dfrac{Y_{\delta_a}(L'_vN'_p - N'_vL'_p) - L'_{\delta_a}Y_vN'_p}{L'_{\delta_a}N'_p + Y_{\delta_a}N'_v}$ <br> $\omega_r^2 \doteq \dfrac{gL'_{\delta_a}N'_v}{L'_{\delta_a}N'_p + Y_{\delta_a}N'_v}$ | Expressions are exact for $N'_{\delta_a} = 0$. $N'_{\delta_a} \neq 0$ introduces third root and changes first coefficient to $N'_{\delta_a}$. | Single-rotor helicopter—aileron |
| $N'_{\delta_a}$ | $\dfrac{1}{T_r} \doteq \left[g\left(-L_v + N'_v\dfrac{L'_{\delta_a}}{N'_{\delta_a}}\right)\right]^{1/3}$ <br> $\omega_r \doteq \lvert 1/T_r \rvert$ <br> $\zeta_r \doteq 0.5$ for $1/T_r < 0$ <br> $\zeta_r \doteq -0.5$ for $1/T_r > 0$ | $\dfrac{L'_p}{N'_p} \doteq \dfrac{L'_{\delta_a}}{N'_{\delta_a}}$ | Tandem-rotor helicopter—aileron |

D. $v$-Numerator $(N_\delta^v)$ factors.

| First Coeff. | Approximate factors | Conditions of validity | Applicable to: |
|---|---|---|---|
| $Y_{\delta_r}$ | $\dfrac{1}{T_{v_1}} \doteq 0$ <br> $\dfrac{1}{T_{v_2}} + \dfrac{1}{T_{v_3}} \doteq -L'_p - \dfrac{U_0N'_{\delta r}}{Y_{\delta r}}$ <br> $\dfrac{1}{T_{v_2}}\dfrac{1}{T_{v_3}} \doteq \left[L'_pN'_r - L'_rN'_p - \dfrac{L'_{\delta_r}}{Y_{\delta_r}} \times (U_0N'_p - g) + \dfrac{N'_{\delta_r}}{Y_{\delta_r}}U_0L'_p\right]$ | $U_0 \neq 0$ <br> $\left\lvert\dfrac{1}{T_{v_1}}\right\rvert \ll \dfrac{1}{T_{v_2}}$ <br> $\left\lvert\dfrac{1}{T_{v_1}}\right\rvert \ll \dfrac{1}{T_{v_3}}$ | Single-rotor helicopter—rudder at high speed |

| | | | |
|---|---|---|---|
| $-U_0N'_{\delta_r}$ | $\frac{1}{T_{v_1}} + \frac{1}{T_{v_2}} \doteq -\left[L'_p + \frac{L'_{\delta_r}}{N'_{\delta_r}}\left(\frac{g}{U_0} - N'_v\right)\right]$ <br> $\frac{1}{T_{v_1}}\frac{1}{T_{v_2}} \doteq -\left[\frac{g}{U_0}\left(L'_r - \frac{L'_{\delta_r}}{N'_{\delta_r}}N'_r\right)\right]$ | $Y_{\delta_r} = 0$, $U_0 \neq 0$ (but valid for $U_0 \doteq 0$) <br> At $U_0 = 0$, numerator becomes first order: $g[L'_{\delta_r}(s - N_r) + L'_rN'_{\delta_r}]$. | Tandem-rotor helicopter—rudder |
| $Y_{\delta_r}$ | $\frac{1}{T_{v_3}} \doteq -L'_p - N'_r - \frac{U_0N'_{\delta_r}}{Y_{\delta_r}}$ <br> $\left.\begin{matrix}\frac{1}{T_{v_1}}\frac{1}{T_{v_2}} \\ \text{or } \omega_v^2\end{matrix}\right\} \doteq \frac{gT_{v_3}}{Y_{\delta_r}}(-L'_{\delta_r}N'_r + L'_rN'_{\delta_r})$ <br> $\left.\begin{matrix}\frac{1}{T_{v_1}} + \frac{1}{T_{v_2}} \\ \text{or } 2\zeta_v\omega_v\end{matrix}\right\} \doteq T_{v_3}\left[(L'_pN'_r - L'_rN'_p) + \frac{U_0}{Y_{\delta_r}}(L'_pN'_{\delta_r} - L'_{\delta_r}N'_p) - \omega_v^2\right]$ | $\left\|\frac{1}{T_{v_3}}\right\| \gg \left\|\frac{1}{T_{v_1}}\right\|$ <br> $\left\|\frac{1}{T_{v_3}}\right\| \gg \left\|\frac{1}{T_{v_2}}\right\|$ | Tilt-duct—rudder at high speed |
| $-U_0N'_{\delta_a}$ | $\left.\begin{matrix}\frac{1}{T_{v_1}} + \frac{1}{T_{v_2}} \\ \text{or } 2\zeta_v\omega_v\end{matrix}\right\} \doteq \frac{L'_{\delta_a}}{N'_{\delta_a}}\left(N'_p - \frac{g}{U_0}\right)$ <br> $\left.\begin{matrix}\frac{1}{T_{v_1}}\frac{1}{T_{v_2}} \\ \text{or } \omega_v^2\end{matrix}\right\} \doteq \frac{g}{U_0}\left(\frac{L'_{\delta_a}}{N'_{\delta_a}}N'_r - L'_r\right)$ | $Y_{\delta_a} = 0$; $N'_{\delta_a} \neq 0$ <br> For hover, if $N'_{\delta_a} = 0$, first coefficient $= gL'_{\delta_a}$, $1/T_v = -N'_r$. | Tilt-duct—aileron |

negatively damped dutch roll mode in the above denominator, one should make

$$2\zeta_d\omega_d \doteq \sqrt[3]{gL'_v} + \tfrac{2}{3}(-L'_p - Y_v) \tag{6-42}$$

greater than zero. It takes about a 2.7-fold increase in $-L'_p - Y_v$ as opposed to a reduction by a factor of about $\frac{1}{20}$ on $-L'_v$ to accomplish such a change; these relative effects are consistent with the interpretation of the time-vector polygons of Fig. 6-9(c). For the stable system that results for such changes, the final value theorem applies (Chapter 2), but now a step aileron input, instead of producing a steady-state roll rate, as for a conventional airplane (with small $1/T_s$), results in a steady-state bank *angle* given approximately by (Eq. 6-33):

$$\lim \varphi(t)_{t\to\infty} \doteq \frac{L_{\delta_a}\delta_a(-Y_v)}{-gL_v} \tag{6-43}$$

## 6-13. Approximate Modal Response Ratios

We have already studied the modal responses of selected configuration examples by means of the time-vector diagrams. Now, on the basis of the approximate factors and equations of motion, we can generalize to show the most interesting literal relationships for conventional airplanes.

*Spiral and Roll Subsidence*

From the three degree of freedom spiral and roll subsidence equation (6-25), and first considering the side forces, we can say that both spiral and roll subsidence (free) motions involve an approximate balance between gravitational forces ($g\varphi$) and centripetal forces ($U_0r$), and that side accelerations are produced only by sideslip, i.e., $(s - Y_v)\beta = 0$. Unfortunately, the latter condition tells us little about the values of $\beta$ relative to the remaining motion quantities. To obtain information on this point we can evaluate the modal response ratio $\beta/r$ separately for the spiral and roll subsidence modes, using the simplified equations.

We choose $\beta/r = \Delta_{21}/\Delta_{23}$ (see Eq. 5-48) for the *spiral* evaluation, whereby

$$\left(\frac{\beta}{r}\right)_S \doteq \frac{\begin{vmatrix} \dfrac{-g}{U_0} & 1 \\ -N'_p s & s - N'_r \end{vmatrix}_{s=-1/T_s}}{\begin{vmatrix} 0 & \dfrac{-g}{U_0} \\ -N'_\beta & -N'_p s \end{vmatrix}_{s=-1/T_s}} \tag{6-44}$$

$$\left(\frac{\beta}{r}\right)_S \doteq \left[\frac{s[N'_p - (g/U_0)] + (g/U_0)N'_r}{(-g/U_0)N'_\beta}\right]$$

for

$$s = \frac{-1}{T_s} = -T_R \frac{g}{U_0}\left(N'_r \frac{L'_\beta}{N'_\beta} - L'_r\right)$$

Substituting the values of $s$, and collecting terms, we obtain

$$\left(\frac{\beta}{r}\right)_S \doteq \frac{-N'_r}{N'_\beta}\left[1 - T_R \frac{L'_\beta}{N'_\beta}\left(N'_p - \frac{g}{U_0}\right)\right] - \frac{L'_r T_R}{N'_\beta}\left(N'_p - \frac{g}{U_0}\right)$$

$$\doteq \frac{-N'_r T_R}{N'_\beta}\left[\frac{1}{T_R} - \frac{L'_\beta}{N'_\beta}\left(N'_p - \frac{g}{U_0}\right) + \frac{L'_r}{N'_r}\left(N'_p - \frac{g}{U_0}\right)\right]$$

and then by substituting for $1/T_R$ (from Table 6-4),

$$\left(\frac{\beta}{r}\right)_S \doteq \frac{-N'_r}{N'_\beta} \frac{\{-L'_p + (L'_r/N'_r)[N'_p - (g/U_0)]\}}{\{-L'_p + (L'_\beta/N'_\beta)[N'_p - (g/U_0)]\}} \doteq \frac{-N'_r}{N'_\beta} \tag{6-45}$$

The final approximation follows because of the usual magnitude of the $(N'_p - g/U_0)$ term relative to $L'_p$. Note that it accurately predicts the balance of yawing moment contribution shown in Fig. 6-4(a).

For the values appropriate to conventional airplane flight, $N'_r/N'_\beta$ is invariably small. The spiral mode is therefore usually characterized by banking and turning motions that produce little side acceleration which, therefore, are approximately coordinated. In the case of an unstable spiral $(1/T_s < 0)$, the uncontrolled bank angle and turn rate gradually increase (diverge); as the bank angle increases, the vertical component of lift is reduced so that the airplane's descent rate increases. Thus the motion through space consists of a tightening spiral dive, from which the mode takes its name.

For the roll subsidence mode, we evaluate $\beta/r = \Delta_{11}/\Delta_{13}$:

$$\left(\frac{\beta}{r}\right)_R = \frac{\begin{vmatrix} s(s - L'_p) & -L'_r \\ -N'_p s & s - N'_r \end{vmatrix}}{\begin{vmatrix} -L'_\beta & s(s - L'_p) \\ -N'_\beta & -N'_p s \end{vmatrix}} = \left[\frac{(s - L'_p)(s - N'_r) - L'_r N'_p}{N'_\beta(s - L'_p) + L'_\beta N'_p}\right]_{s=-1/T_R}$$

Recognizing that $|L'_r N'_p / L'_p N'_r| \ll 1$ and dividing through by $(s - L'_p)$, we find that

$$\left(\frac{\beta}{r}\right)_R \doteq \frac{s - N'_r}{N'_\beta\{1 + [L'_\beta N'_p / N'_\beta(s - L'_p)]\}}$$

where

$$s = \frac{-1}{T_R} \doteq L'_p - \frac{L'_\beta}{N'_\beta}\left(N'_p - \frac{g}{U_0}\right)$$

$$\left(\frac{\beta}{r}\right)_R \doteq \frac{-[N'_p - (g/U_0)]}{(g/U_0)N'_\beta}\left[L'_p - N'_r - \frac{L'_\beta}{N'_\beta}\left(N'_p - \frac{g}{U_0}\right)\right] \tag{6-46}$$

When $(N'_p - g/U_0)$ is identically equal to zero, $(\beta/r)_R$ is also identically zero (Eq. 6-46). Then, to the extent that $L'_r$ contributions are generally negligible, the single degree of freedom dynamics apply regardless of the value of $g/U_0$. This result is also indicated because, for $(N'_p - g/U_0) = 0$, there is no difference between $-L'_p$ and $1/T_R$. Nevertheless, even for these ideally suitable circumstances the single degree of freedom approximation is not adequate to describe accurately the *magnitude* of the rolling response that can be affected by $\beta$ motions induced by aileron yaw (e.g., Eqs. 6-28 and 6-29).

*Dutch Roll*

The dutch roll motions, being oscillatory, are basically much more complex than the first-order spiral and roll subsidence modes. Nevertheless, as we shall see, there are pertinent generalizations that can be made. To derive these, we examine the modal response ratios, $\varphi/\beta$ and $\psi/\beta$, as obtained from the complete characteristic determinant of Eq. 6-1:

$$\left(\frac{\varphi}{\beta}\right)_d = \frac{-\Delta_{32}}{\Delta_{31}} = \frac{-\begin{vmatrix} s - Y_v & 1 \\ -L'_\beta & -L'_r \end{vmatrix}}{\begin{vmatrix} \dfrac{-g}{U_0} & 1 \\ s(s - L'_p) & -L'_r \end{vmatrix}}$$

$$= \frac{-L'_r\{s - [Y_v + (L'_\beta/L'_r)]\}}{s^2 - L'_p s - (g/U_0)L'_r}$$

where $s = -\zeta_d\omega_d + j\omega_d\sqrt{1 - \zeta_d^2}$. Noting that, usually, $|Y_v L'_r| \ll |L'_\beta|$ and $|(g/U_0)L'_r| \ll \omega_d^2$, and substituting the expression for $s$ but dropping the subscript $d$ for convenience, we obtain

$$\frac{\varphi}{\beta} \doteq \frac{L'_\beta + \zeta\omega L'_r - j\omega L'_r\sqrt{1 - \zeta^2}}{2\zeta^2\omega^2 - \omega^2 + L'_p\zeta\omega - j\omega\sqrt{1 - \zeta^2}\,(L'_p + 2\zeta\omega)} \tag{6-47}$$

The magnitude is given by

$$\left|\frac{\varphi}{\beta}\right|^2 = \frac{(L'_\beta + \zeta\omega L'_r)^2 + \omega^2 L'^2_r(1 - \zeta^2)}{[\omega^2(1 - 2\zeta^2) - L'_p\zeta\omega]^2 + \omega^2(1 - \zeta^2)(L'_p + 2\zeta\omega)^2}$$

$$= \frac{L'^2_\beta + \omega^2 L'^2_r + 2\zeta\omega L'_r L'_\beta}{\omega^4 + \omega^2 L'^2_p + 2\zeta\omega^3 L'_p} \tag{6-48}$$

and for negligible $L'_r$ terms, $\omega_d^2 \doteq N'_\beta$ and $-L'_p \doteq 1/T_R$.

$$\left|\frac{\varphi}{\beta}\right| \doteq \left|\frac{L'_\beta}{N'_\beta}\right| \frac{1}{\sqrt{1 + (1/\omega_d^2 T_R^2)(1 - 2\zeta_d \omega_d T_R)}} \qquad (6\text{-}49)$$

The $\zeta_d$ term is usually negligible and is often discarded.

The phase angle (Eq. 6-47) is given by

$$\measuredangle \frac{\varphi}{\beta} = -\tan^{-1} \frac{\omega L_r \sqrt{1 - \zeta^2}}{L'_\beta + \zeta \omega L'_r} + \tan^{-1} \frac{\omega \sqrt{1 - \zeta^2}\,(L'_p + 2\zeta\omega)}{\omega^2(2\zeta^2 - 1) + L'_p \zeta \omega}$$

which, using the identity $\tan^{-1} x - \tan^{-1} y = \tan^{-1} (x - y)/(1 + xy)$, reduces to

$$\measuredangle \frac{\varphi}{\beta} = \tan^{-1} \frac{\omega \sqrt{1 - \zeta^2}\,(L'_p L'_\beta + 2\zeta\omega L'_\beta + \omega^2 L'_r)}{\omega^2(2\zeta^2 - 1)L'_\beta + \zeta\omega^3 L'_r + \zeta\omega L'_p L'_\beta + \omega^2 L'_r L'_p} \qquad (6\text{-}50)$$

For $\omega^2 \doteq N'_\beta$ and $\zeta \ll 1$, and dividing numerator and denominator by $\omega N'_\beta = \omega^3$, we obtain

$$\measuredangle \frac{\varphi}{\beta} \doteq \tan^{-1} \frac{(L'_\beta/N'_\beta)(L'_p + 2\zeta_d\omega_d) + L'_r}{(L'_\beta/N'_\beta)(\zeta_d L'_p - \omega_d) + L'_r[\zeta_d + (L'_p/\omega_d)]} \qquad (6\text{-}51)$$

Assuming that $L'_r$ *terms are negligible* and $2\zeta_d\omega_d \ll -L'_p \doteq 1/T_R$, we find that

$$\measuredangle \frac{\varphi}{\beta} \doteq \tan^{-1} \frac{1}{\zeta_d + \omega_d T_R} \qquad (6\text{-}52)$$

For $L'_\beta/N'_\beta \to 0$, so that only $L'_r$ terms are left,

$$\measuredangle \frac{\varphi}{\beta} \doteq \tan^{-1} \frac{1}{\zeta_d - (1/\omega_d T_R)}$$

But, from Eq. 6-48 $|\varphi/\beta|^2 = L'^2_r/(\omega_d^2 + L'^2_p)$ is very small, so that effectively $\varphi \to 0$ and the phasing of $\varphi/\beta$ is unimportant. In general, therefore, when significant $\varphi$ motions do occur, Eqs. 6-49 and 6-52 are pertinent.

These relationships show that the predominant effect on the magnitude of the rolling motions is the ratio $L'_\beta/N'_\beta$, with the roll damping of increasing importance in reducing the motions as $1/\omega_d^2 T_R^2 \doteq L'^2_p/N'_\beta$ approaches and exceeds unity. The phasing between the $\varphi$ and $\beta$ motions is influenced primarily by the value of $\omega_d T_R$; i.e., from Eq. 6-52 for small $\zeta_d$,

$$\measuredangle \frac{\varphi}{\beta} \doteq \tan^{-1} \frac{1}{\omega_d T_R} \doteq \tan^{-1}\left(\frac{-L'_p}{\sqrt{N'_\beta}}\right)$$

and the bank angle therefore leads sideslip by less than 90 degrees for $N'_\beta > 0$. The leading phase relationship results because, in the dutch roll mode, $\beta$ is of opposite sign to the heading change, $\psi$, as we shall see below. Thus, referenced to the yawing oscillations, $\varphi$ *lags* $\psi$ by from approximately 90 to 180 degrees as the roll damping decreases from large to small values.

To evaluate $\psi/\beta$ as simply as possible, we express the side force equation (Eq. 6-1) as

$$\frac{r}{\beta} - \frac{g}{U_0}\frac{\varphi}{\beta} + (s - Y_v) = 0$$

so that

$$\left(\frac{r}{\beta}\right)_d = -s_d + Y_v + \frac{g}{U_0}\left(\frac{\varphi}{\beta}\right)_d \tag{6-53}$$

where

$$s_d = -\zeta_d\omega_d + j\omega_d\sqrt{1 - \zeta_d^2}$$

Generally, $Y_v$ is very small with respect to $\omega_d$; for reasonable flight speeds, $U_0$, so is $(g/U_0)(\varphi/\beta)$; accordingly,

$$\left(\frac{r}{\beta}\right)_d \doteq -s_d \qquad \text{or} \qquad \left(\frac{r}{s\beta}\right)_d = -1$$

Recognizing that (for $\gamma_0 = 0$) $\psi = r/s$, we get finally

$$\left(\frac{\psi}{\beta}\right)_d \doteq -1 \tag{6-54}$$

We see, therefore, that the usual heading and sideslip motions in dutch roll are consistent with those in relatively flat yawing oscillations. Implying from this that the rolling motions are not of primary importance, we obtain the two degree of freedom approximate dutch roll equations of motion (Eq. 6-18) presented earlier.

CHAPTER 7

# *ELEMENTARY LONGITUDINAL FEEDBACK CONTROL*

A most powerful approach to obtain an appreciation for the effects of automatic control on the motions of an aircraft is to consider closed-loop systems formed by direct feedback of aircraft motion quantities to the controls. Such systems are idealizations since, in fact, the controls cannot be moved without lag, and instruments cannot sense and reproduce the motion quantities instantaneously and in a pure form. Nevertheless, consideration of these idealized systems shows the ultimate performance approachable by some practical system or, by way of contrast, reveals directions in which it would be unprofitable to proceed.

The prototype for all the systems to be discussed is the single-loop flight controller shown in Fig. 7-1. This illustrates the direct feedback of a generalized aircraft motion quantity, $\xi$, to a control deflection, $\delta$. The reader will recognize that the transfer function that belongs in the controlled element block may be any one of the several developed in Chapters 5 or 6 which relate the aircraft motion quantities to control deflections. Table 7-1 lists the most promising possibilities in connection with longitudinal motions. Control using all of the output quantities listed will be discussed below using elevator as the actuation quantity.

In many situations it will be instructive to consider that the controller is simply a gain, but in other cases it will be desirable to provide for lead, or lag and lead equalization. Yet, in each instance, only a single output motion variable will be of interest or concern.

## 7-1. Feedback of Pitch Angle and Pitch Rate to the Elevator

Historically, the earliest automatic pilots comprised a vertical gyroscope and an associated actuator that deflected the elevator in such a way as to oppose departures from the reference or commanded pitch attitude. In modern terms this would be described as negative feedback of the pitch attitude, $\theta$, to the elevator control deflection, $\delta_e$. The appropriate controlled element transfer function is therefore $\theta(s)/\delta_e(s)$,

$$\frac{\theta(s)}{\delta_e(s)} = \frac{N^{\theta}_{\delta_e}}{\Delta_{\text{long}}} = \frac{A_\theta[s + (1/T_{\theta_1})][s + (1/T_{\theta_2})]}{[s^2 + 2\zeta_p\omega_p s + \omega_p^2][s^2 + 2\zeta_{sp}\omega_{sp}s + \omega_{sp}^2]} \tag{7-1}$$

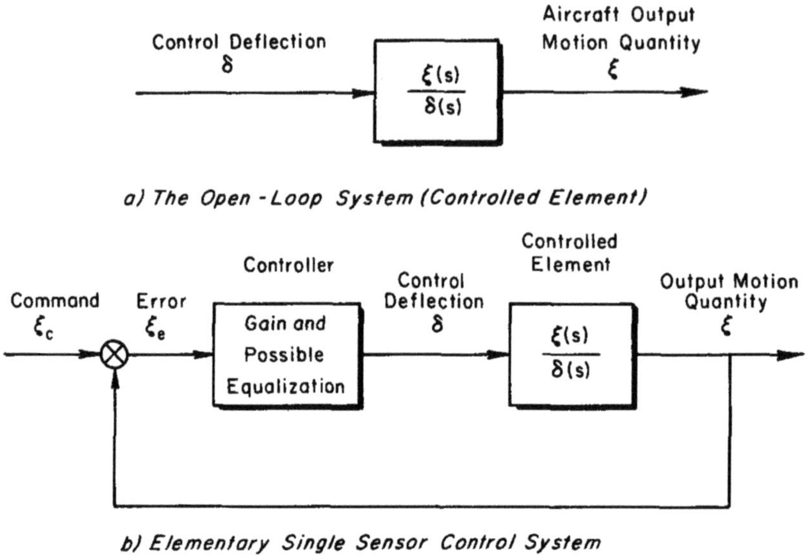


Fig. 7-1. Feedback of aircraft motion quantities.

In what follows, several variants of this controlled element transfer function will be considered. First, we assume that the airplane pitch attitude characteristics are stable, with good short period damping, large separation between short period and phugoid frequencies, etc. Then we explore a variety of less favorable characteristics, including insufficient short period damping, longitudinal divergences of one kind or another, nonzero position error, etc.

*Conventional Pitch Attitude Control*

Figure 7-2 shows the block diagram of the $\theta$ to $\delta_e$ ($\theta \rightarrow \delta_e$) feedback control system with a pure gain controller. The figure also contains a system survey using Bode and conventional root loci to show the location

Table 7-1. Longitudinal motion airframe output and actuating quantities.

| Output quantities | Actuating quantities |
|---|---|
| $\theta$, pitch angle | $\delta_e$, elevator deflection |
| $q$, pitch rate | $\delta_f$, flap deflection |
| $u$, forward velocity | $\delta_T$, engine power control deflection |
| $\alpha$, angle of attack, $w/U_0$ | |
| $a_x$, longitudinal acceleration | |
| $a_z$, normal acceleration, $a_{z_{c.g.}} - x_a\dot{q}$ | |
| $h$, altitude | |

of the closed-loop roots as a function of the closed-loop system gain. The dynamics of the uncontrolled vehicle represented by the open-loop characteristics illustrated in Fig. 7-2 are typical of a well-behaved aircraft in cruising flight at moderate altitudes. The main features of these characteristics are the wide separation between the short period and phugoid breakpoints, in both amplitude ratio and frequency, and the relatively heavy damping of the short period mode.

From the system surveys it can be appreciated that, at moderate gain, the modified (closed-loop) phugoid roots are driven into close proximity to the zeros, while the short period roots move to a higher frequency and lower damping ratio. This is not undesirable, provided that the open-loop short period damping is not initially already marginal. The key point is that the phugoid damping increase is obtained at the expense of the short period. In fact, the total system damping is unchanged by the feedback of terms that, when considered as creating or augmenting stability derivatives, do not affect the coefficient of the second term in the characteristic equation. Such feedbacks, which do not augment $X_u$, $Z_w$, $M_{\dot{\alpha}}$, $M_q$, etc., can only redistribute the open-loop damping among the closed-loop modes. This is a specific example in somewhat different terms of a general algebraic rule previously given (in Chapter 3) as Rule 10 for the root locus method.

The closed-loop asymptotic amplitude Bode constructed for moderate gain shows, by its nearly flat properties in the vicinity of the modified phugoid, that this mode will be suppressed almost completely in the pitch attitude response to $\theta_c$ commands. Also, as a consequence of the heavily damped phugoid, any fluctuations of the other longitudinal degrees of freedom such as speed and altitude will exhibit well-damped long period characteristics. This will be true even if the uncontrolled aircraft has negative phugoid damping. For these reasons the feedback of pitch angle alone to the elevator has been, and will continue to be, a successful control technique in many aircraft.

### *Attitude Control for Small Static/Short Period Gain*

In the feedback system described above, the open-loop gain does not become infinite at zero frequency; therefore, the closed-loop frequency response has an amplitude ratio slightly less than one at low frequencies. This corresponds to a small steady-state position error in response to step commands which is not serious for the condition described in Fig. 7-2. However, the situation is likely to be aggravated at low speed, as in landing approach, or at very high altitude. Here the static to short period gain ratio, $1/\omega_p^2 T_{\theta_1} T_{\theta_2}$, is likely to be small and indeed may well be less than one. When this occurs the long-term response of the closed-loop system to commands is very poor. Figure 7-3 presents the $\theta \rightarrow \delta_e$ system

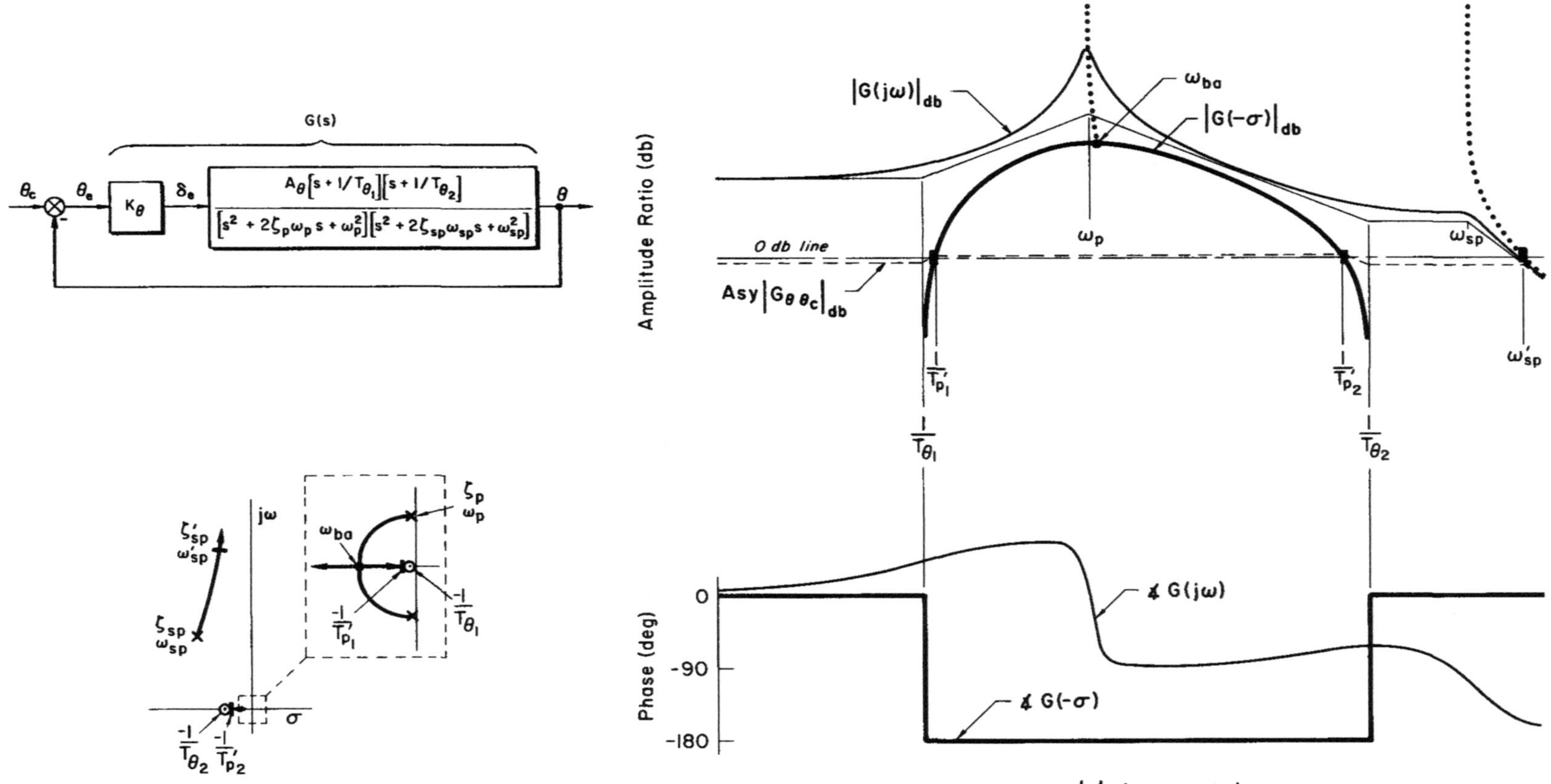


*b) Root Locus*

*a) Bode G(jω), G(−σ) Diagrams and Bode Root Locus*

Fig. 7-2. System survey of pitch attitude ($\theta \rightarrow \delta_e$) control system for well-behaved aircraft.

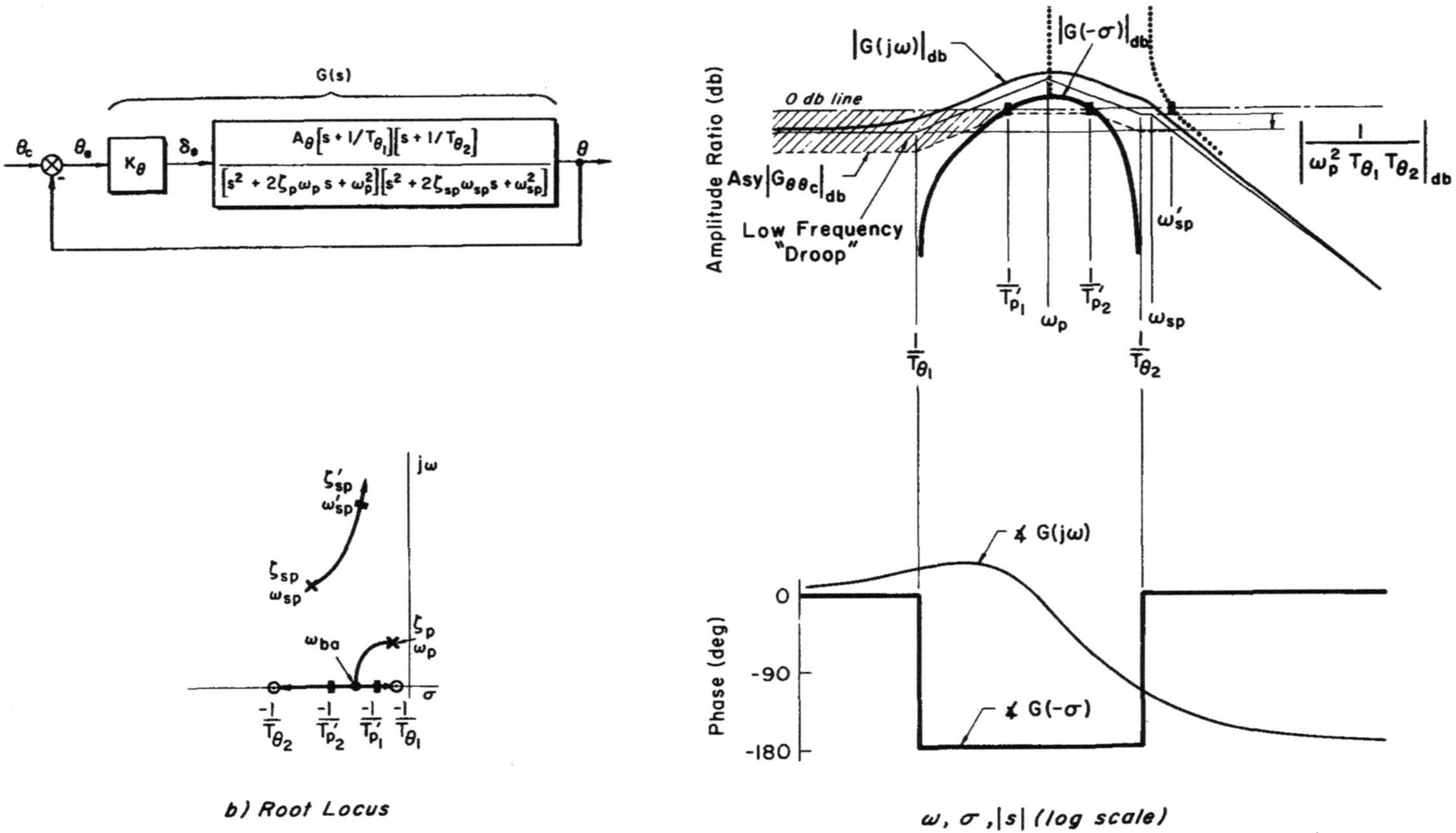


*b) Root Locus*

*a) Bode $G(j\omega)$, $G(-\sigma)$ Diagrams and Bode Root Locus*

Fig. 7-3. System survey of pitch attitude ($\theta \to \delta_e$) control system illustrating a low frequency "droop."

survey plots for such an airplane in the approach configuration. The typical closed-loop zero dB line shown is for a gain that is about as large as possible without seriously degrading the closed-loop, short period damping. Yet, as indicated graphically by the closed-loop asymptotes, there will be a very low frequency lead/lag and a d-c gain less than unity in the closed-loop pitch attitude transfer function. In the indicial response of $\theta$ to a step $\theta_c$, these correspond to a very long time constant mode and to a steady-state position error. The closed-loop asymptotic plot, when considered as a frequency response, also exhibits this low frequency "droop" as a region, indicated by the cross-hatched area, where the output/input amplitude ratio is less than unity.

To define more precisely the aircraft characteristics that govern the static to short period gain ratio, $1/\omega_p^2 T_{\theta_1} T_{\theta_2}$, we can express it in terms of the approximate factors

$$\frac{1}{\omega_p^2 T_{\theta_1} T_{\theta_2}} \doteq \frac{[1-(Z_w M_q/M_\alpha)][Z_w X_u - X_w Z_u + (Z_{\delta_e}/M_{\delta_e})(X_w M_u - X_u M_w)]}{-(g/U_0)[Z_u - (M_u Z_w/M_w)]} \tag{7-2}$$

For the simplified but quite normal conditions where $|Z_w M_q/M_\alpha| \ll 1$, $|Z_{\delta_e} M_w/Z_w M_{\delta_e}| \ll 1$, and the $M_u$ terms are negligibly small, the static to short period gain ratio becomes

$$\begin{aligned}\frac{1}{\omega_p^2 T_{\theta_1} T_{\theta_2}} &\doteq \frac{Z_w X_u - X_w Z_u}{-(g/U_0) Z_u} \\ &\doteq \frac{U_0 Z_w}{g Z_u}\left[-X_u + \left(X_w - \frac{g}{U_0}\right)\frac{Z_u}{Z_w} + \frac{g}{U_0}\frac{Z_u}{Z_w}\right] \\ &\doteq \left[1 + \frac{U_0 Z_w}{g Z_u}\left(\frac{1}{T_{h_1}}\right)\right]\end{aligned} \tag{7-3}$$

The magnitude of $1/\omega_p^2 T_{\theta_1} T_{\theta_2}$ will be unity when $1/T_{h_1} = 0$ and less than unity as $1/T_{h_1}$ decreases to negative values. It can easily be shown that $1/T_{h_1}$ is given by

$$\frac{1}{T_{h_1}} \doteq \frac{1}{m}\left(\frac{dD}{dU} - \frac{\partial T}{\partial U}\right) \tag{7-4}$$

Thus a static to short period gain of 1.0 occurs approximately at performance reversal, $m/T_{h_1} \doteq dD/dU - \partial T/\partial U = 0$. This is also usually close to the condition for minimum drag, i.e., $dD/dU = 0$. Because near-minimum drag flight is sometimes desirable from a performance standpoint, flight conditions near the performance reversal are not uncommon.

For automatic pilot systems that are intended to follow commands, such as systems with attitude-hold features, the deficiency in low frequency gain can be made up with a form of integral control. A "pitch integrator" is added in parallel to the straight-through gain of the controller. This configuration is shown in Fig. 7-4. The transfer function of the controller is now $K_\theta + K_{\bar{\theta}}/s$ and an integration and higher frequency lead, $(K_\theta/s)(s + K_{\bar{\theta}}/K_\theta)$, are cascaded with the open-loop function representing the dynamics of the airplane. As indicated by the amplitude ratio asymptotes for the compensated system, the lead time constant represented by the ratio $K_\theta/K_{\bar{\theta}}$ is chosen so that its breakpoint, $K_{\bar{\theta}}/K_\theta$, is greater than $\omega_p$, thus making the low frequency amplitude ratio in the region of $\omega_p$ as large as feasible. This effectively eliminates the "droop" shown in Fig. 7-3. The addition of the $K_{\bar{\theta}}$ feedback introduces a fifth root to the characteristic equation without changing the sum of the roots. The total system damping therefore remains constant and, since the added root is a low frequency subsidence, the effect of the integral feedback is to detract from the phugoid damping. This is, of course, offset by the $K_\theta$ feedback that damps the phugoid by taking damping away from the short period.

*Attitude Control with Deficient Short Period Damping*

Somewhat by contradistinction to the two cases just considered, Fig. 7-5 is a system survey of $\theta \rightarrow \delta_e$ feedback for open-loop dynamics appropriate to an interceptor at supersonic speed and high altitude. Here the damping of the longitudinal short period motion is weak, and the feedback causes it to deteriorate rapidly with increases in the controller gain. The suppression of the phugoid, therefore, may be considered to exact too high a price with regard to the short period. To alleviate the short period damping deficiency, a pitch rate signal can be fed back, as in a pitch damper, to produce an elevator deflection and a corresponding pitching moment proportional to pitching velocity, $q$, i.e.,

$$\delta_e = -K_q q \qquad \text{or} \qquad M_{\delta_e}\delta_e = -K_q M_{\delta_e} q \tag{7-5}$$

This feedback then has the same effect as an increase in magnitude of the stability derivative $M_q$. As seen from the short period approximate factor

$$2(\zeta\omega)_{sp} \doteq -(Z_w + M_q + M_{\dot{\alpha}}) \tag{7-6}$$

an increase in $-M_q$ directly increases the short period damping.

This same effect is illustrated in the system survey in Fig. 7-6. These plots show that, at a comparatively low value of the loop gain, the short period motion can easily be overdamped. The phugoid motion is hardly altered by the action of this relatively low gain feedback, although it can

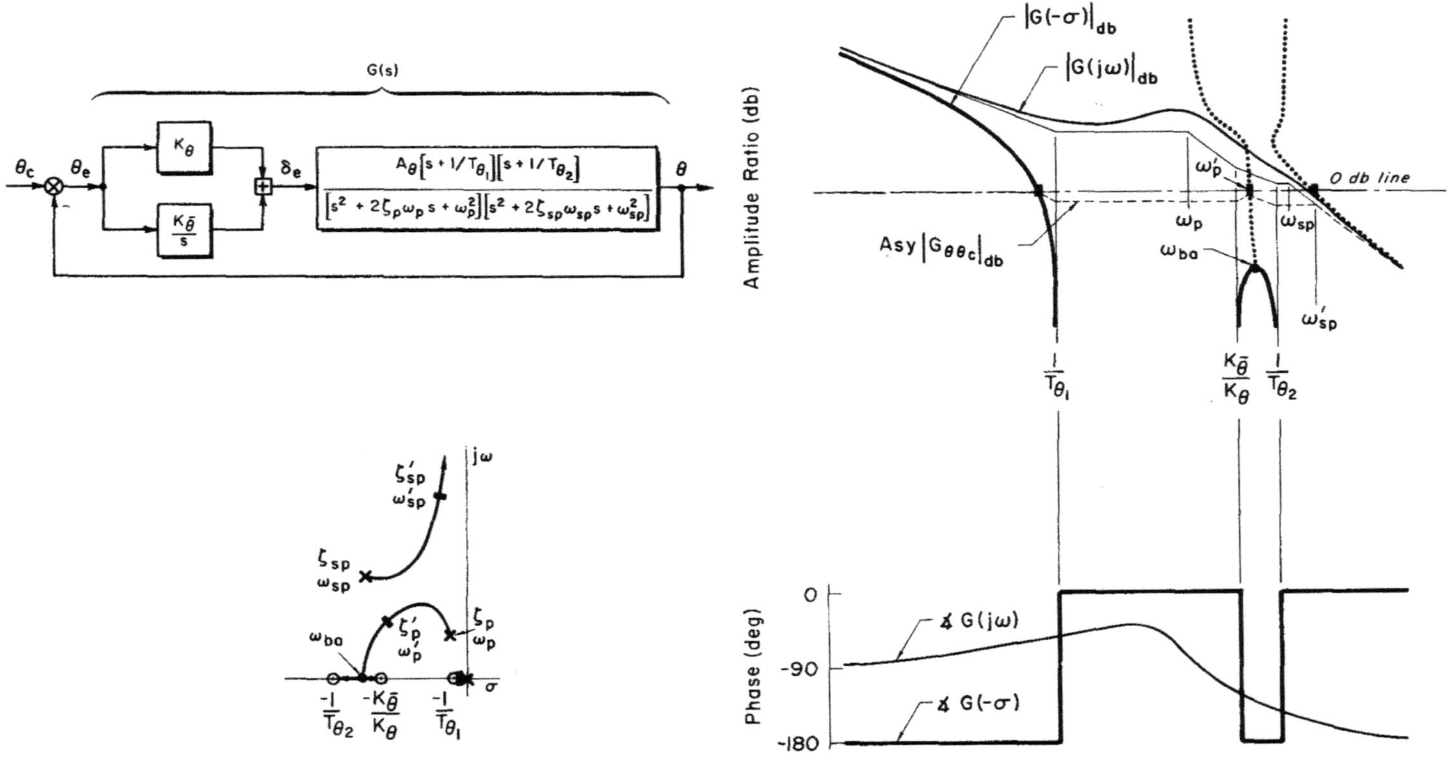


*b) Root Locus*

*ω, σ, |s| (log scale)*

*a) Bode G(jω), G(-σ) Diagrams and Bode Root Locus*

Fig. 7-4. System survey of pitch attitude ($\theta \rightarrow \delta_e$) control system with integrator.

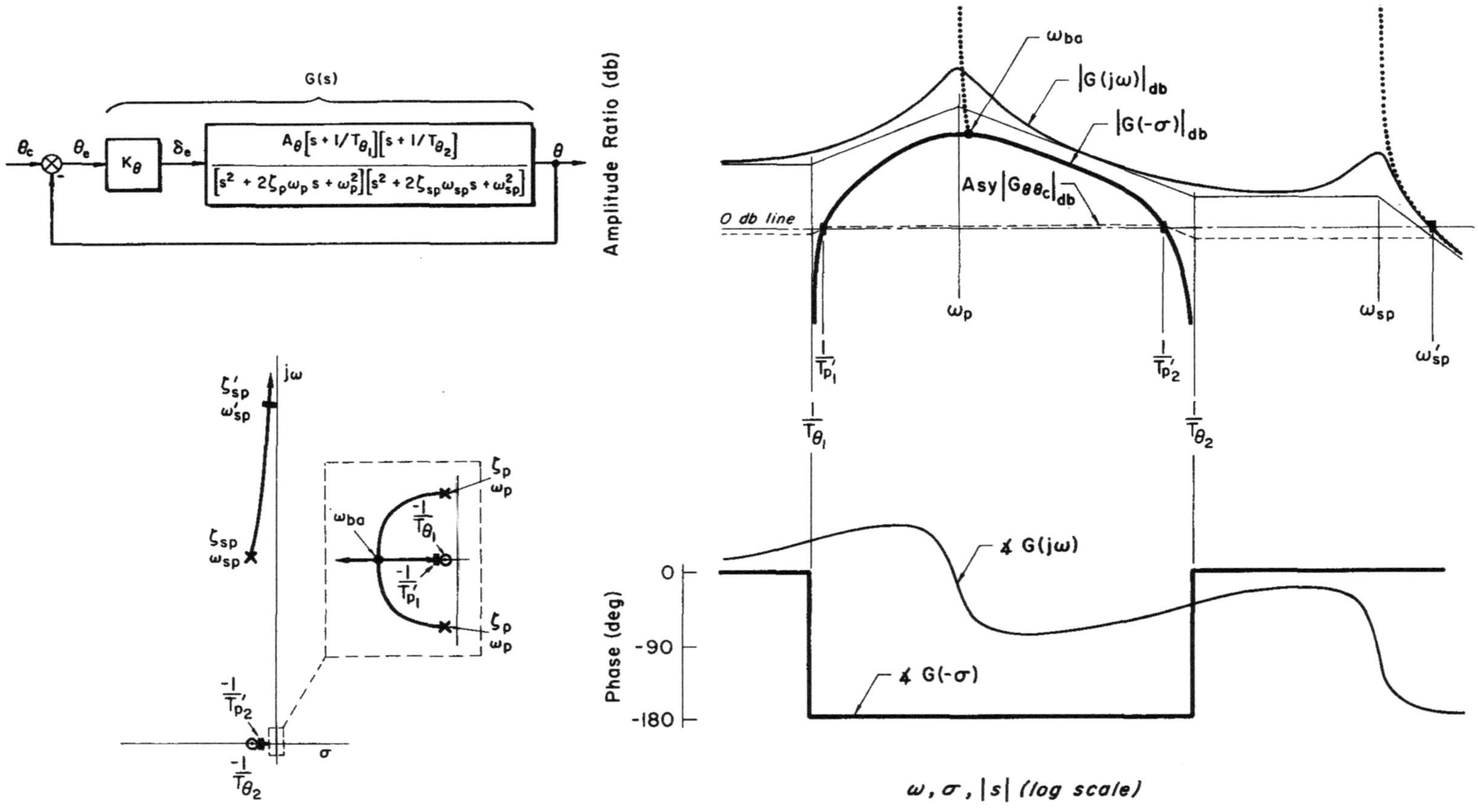


*b) Root Locus*

*a) Bode G(jω), G(−σ) Diagrams and Bode Root Locus*

Fig. 7-5. System survey of pitch attitude ($\theta \rightarrow \delta_e$) control system for low short period damping case.

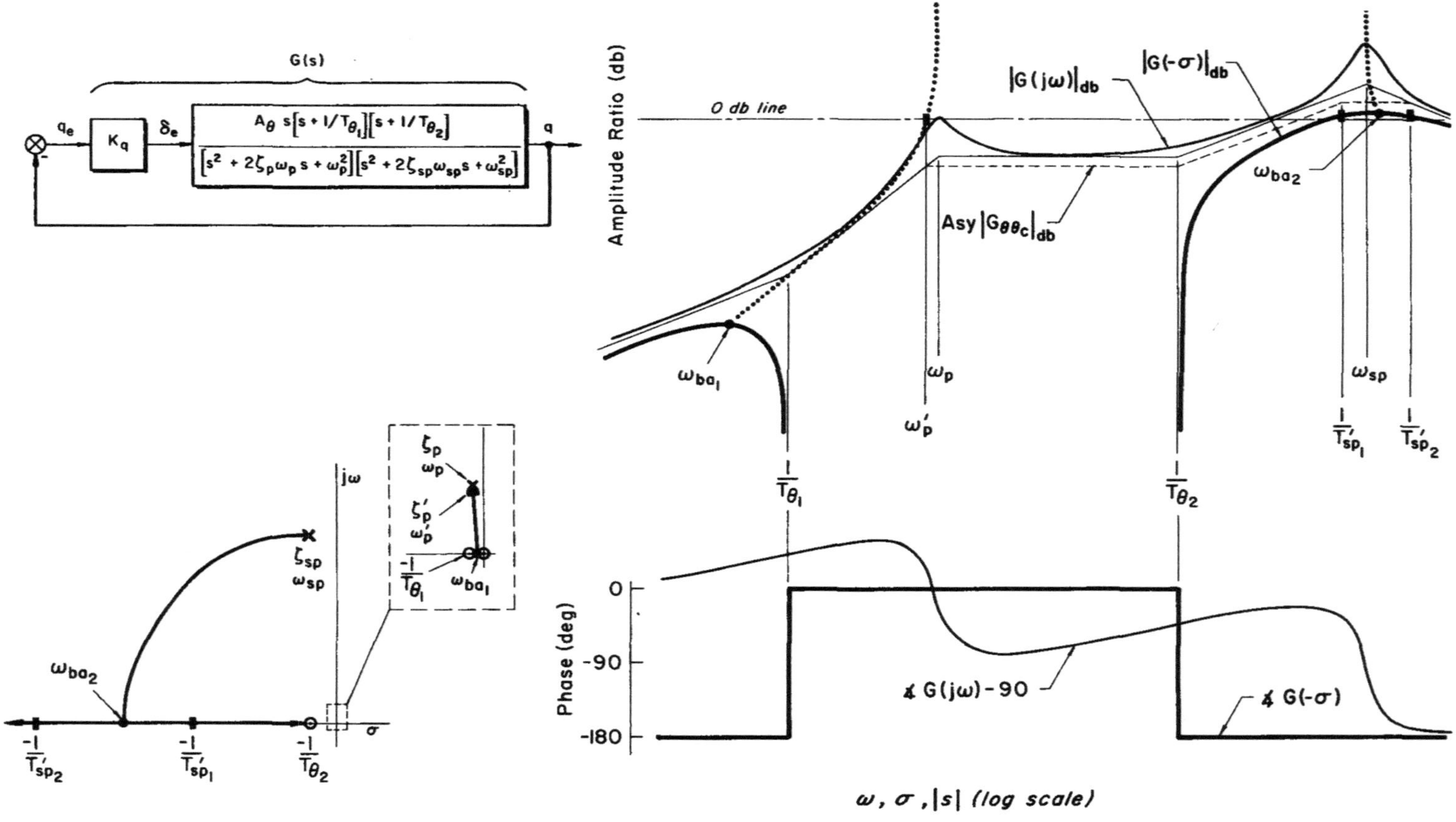


*b) Root Locus*

*a) Bode $G(j\omega)$, $G(-\sigma)$ Diagrams and Bode Root Locus*

Fig. 7-6. System survey of pitching velocity ($q \rightarrow \delta_e$) control system for low short period damping case.

eventually be overdamped, with an attendant reduction in frequency, for sufficiently high gain.

Figure 7-7 illustrates the combined effect of pitch rate and pitch angle feedback for the preceding high speed, high altitude case. Now, at only moderately high controller gains, the phugoid is heavily damped and suppressed for attitude commands and the short period motion is also heavily damped. This technique is so efficacious that it has become nearly universal, and most modern automatic pilots feed back pitch angle and pitch rate signals to the elevator. The technique is applicable not only to aircraft with typical dynamics, such as the ones that have been illustrated, but also in cases in which the longitudinal motions of the vehicle alone (open-loop) are unstable.

*Pitch Attitude Control of Longitudinal Divergences*

As we have seen in Chapter 5, the phugoid oscillation may become a convergence and a divergence, especially in the high subsonic flight regime. The unstable divergence is sometimes called the "tuck" mode because it usually manifests itself as a slow increase in speed and nose-down pitch attitude. "Tuck" is, in essence, a static instability due to an $M_u$ that is sufficiently negative to make

$$M_w Z_u - M_u Z_w < 0 \tag{7-7}$$

The result is a pole on the positive real axis.

Figure 7-8 presents a system survey for the feedback of pitch angle to elevator when the aircraft exhibits a tuck mode. The $\sigma$-Bode diagram shows that at a very moderate gain, corresponding to an open-loop d-c gain of 1.0 (zero dB), the closed-loop root crosses from the right half to the left half-plane, and the closed-loop system becomes stabilized. The value of pitch angle feedback is evident here. Pitch rate feedback would serve, as before, to damp the short period mode.

Figure 7-9 shows the effect of pitch angle feedback to the elevator in connection with another form of longitudinal instability. In this case the short period oscillation has become a convergence and divergence associated with the inequality

$$M_q Z_w - M_\alpha < 0 \tag{7-8}$$

due to a sufficiently positive value of $M_\alpha$. In all practical cases of this kind, the inequality of Eq. 7-7 also applies. One possible solution for this condition is a pitch attitude control system. It may be seen in Fig. 7-9 that the feedback of pitch angle and pitch angle rate can stabilize the violent instability and that, here again, the value of this particular feedback is strikingly illustrated.

Still in reference to pitch angle feedback, consider the case of a hovering VTOL aircraft or helicopter. The transfer function relating pitch angle to

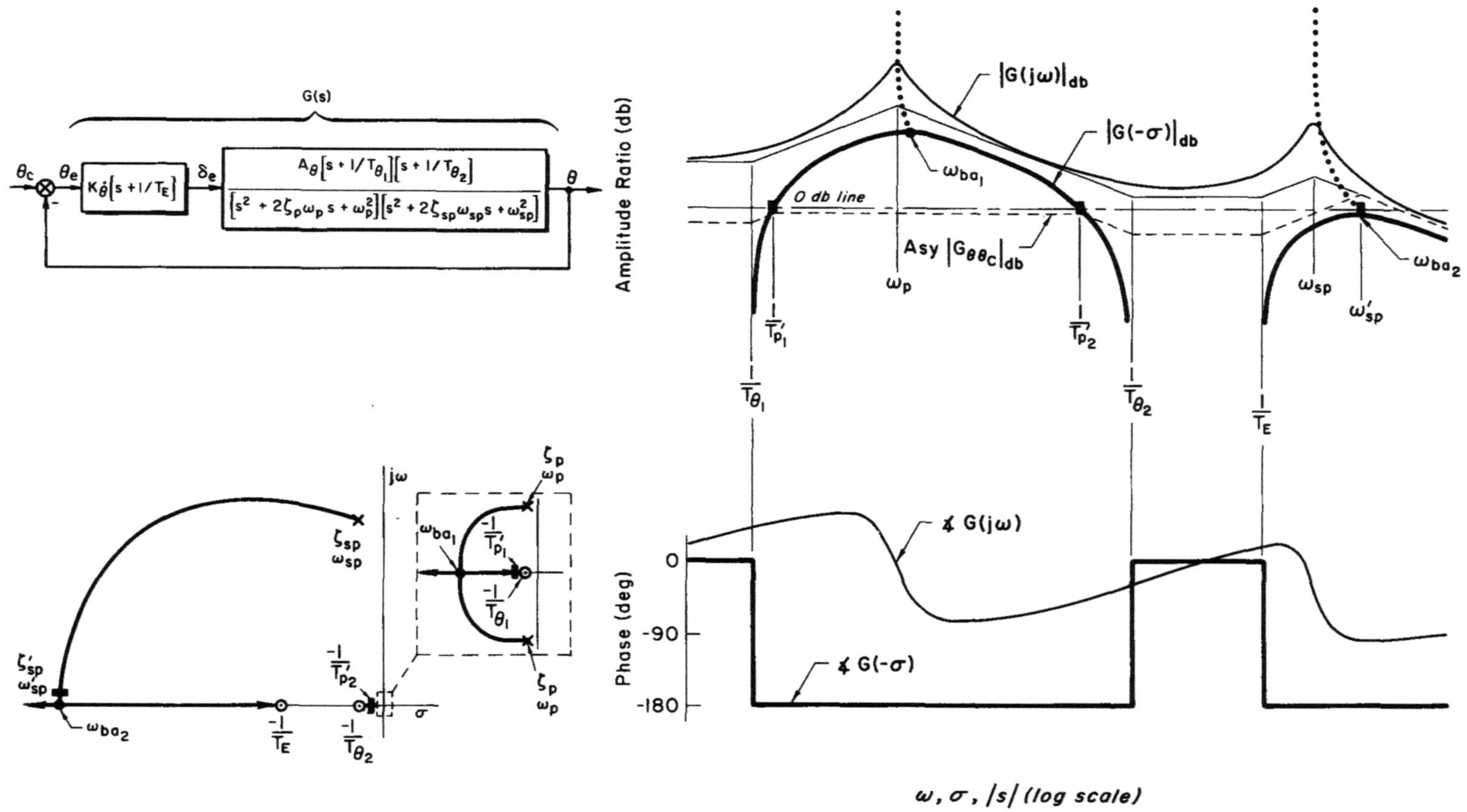


*b) Root Locus*

*a) Bode $G(j\omega)$, $G(-\sigma)$ Diagrams and Bode Root Locus*

Fig. 7-7. System survey of pitch attitude and rate ($\theta, \dot\theta \rightarrow \delta_e$) control system for low short period damping case.

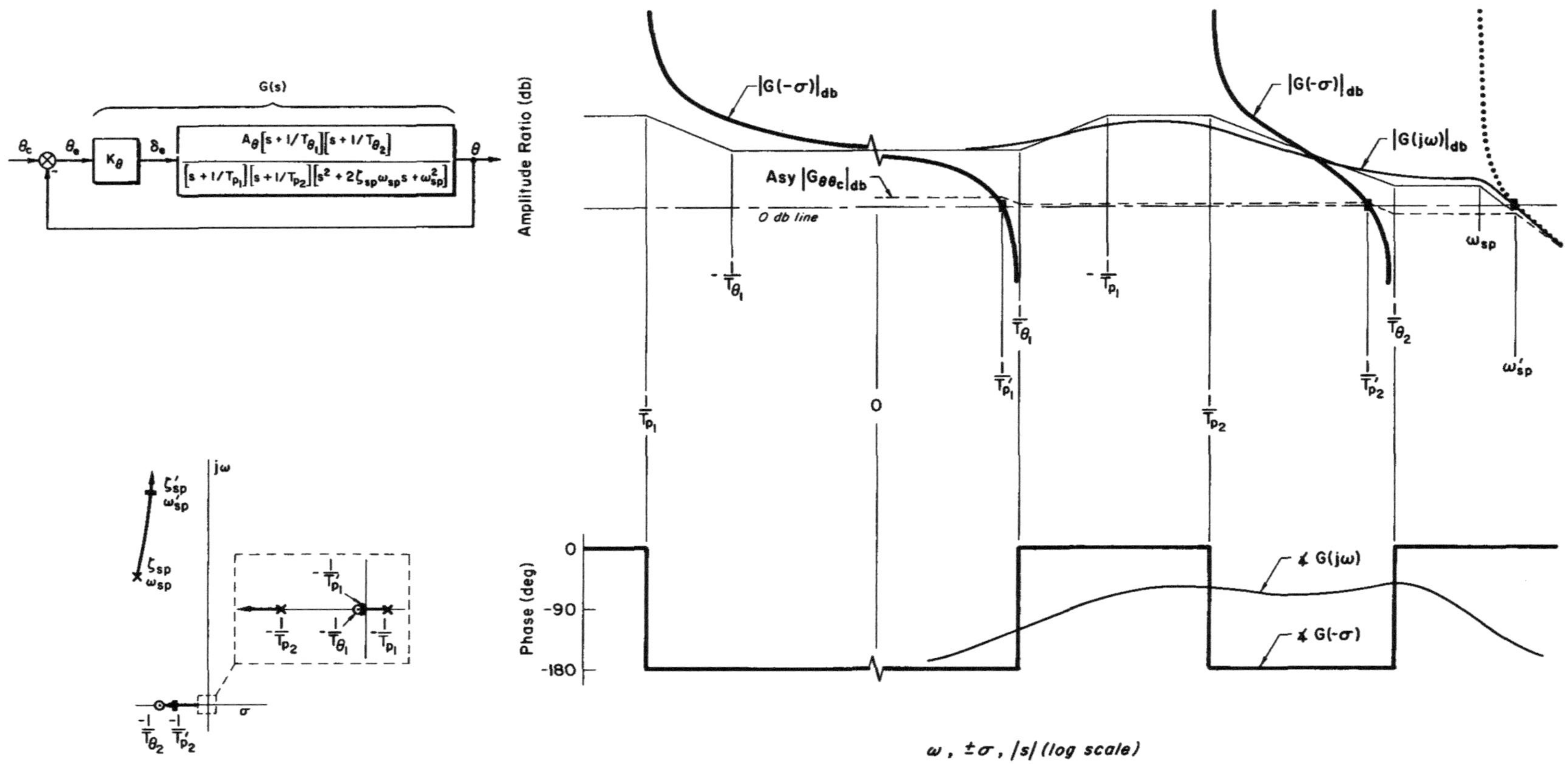


*b) Root Locus*

*a) Bode $G(j\omega)$, $G(-\sigma)$ Diagrams and Bode Root Locus*

Fig. 7-8. System survey of pitch attitude ($\theta \rightarrow \delta_e$) control system for aircraft with "tuck" mode.

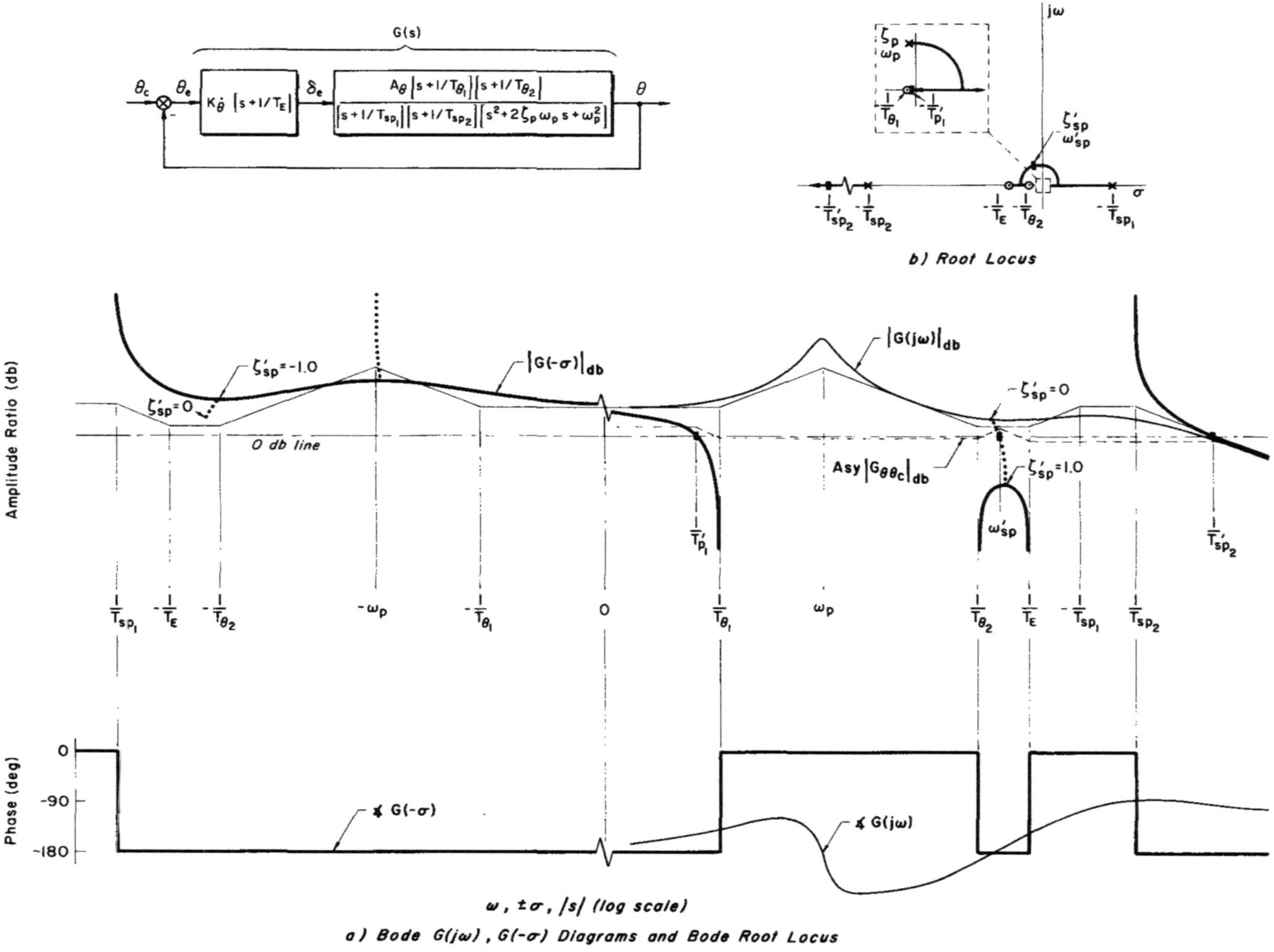


Fig. 7-9. System survey of pitch attitude and rate $(\theta, \dot{\theta} \rightarrow \delta_e)$ control system for aircraft with short period instability.

longitudinal control is given by

$$\frac{\theta(s)}{\delta_e(s)} = \frac{M_{\delta_e}[s + (1/T_\theta)]}{[s + (1/T_{sp})][s^2 + 2\zeta_p\omega_p s + \omega_p^2]} \tag{7-9}$$

This is quite different, in both form and kind, from the corresponding $\theta/\delta$ for a typical aircraft in forward flight. Not only does the transfer function have a third-order denominator, but typically the low frequency oscillation is unstable ($\zeta_p < 0$). The system survey of Fig. 7-10 shows that when the feedback loop is closed at moderate gain, the divergent oscillation is made stable. Of course this possibility depends on $1/T_{sp}$ being sufficiently large. If it is not, pitch rate feedback should also be employed. This necessity is real for the majority of helicopters and VTOL machines, and Fig. 7-11 shows the effect of both pitch angle and pitch rate feedback. For simplicity the lead is set equal to $1/T_{sp}$, although this will not be true in general. Once again the benefits of these feedbacks, in combination, are apparent.

All of the systems described in this article are conditionally stable. That is, they require a minimum value of gain for stability to be attained. In actual designs this property demands careful consideration of nonlinearities (such as limiting), which tend to reduce gain from the nominal, small perturbation, linearized values.

*Generalized Pitch Attitude Control and Nature of Gain Adjustments*

Accumulating the experience gained from all the cases treated above indicates that a general pitch control law with rate, proportional, and integral terms would be adequate for all conceivable pitch command control and attitude stabilization situations. This would have the form

$$\delta_e = K_{\dot\theta}\dot\theta_e + K_\theta\theta_e + K_{\bar\theta}\int \theta_e\,dt \tag{7-10}$$

or, in transform style,

$$\frac{\delta_e}{\theta_e} = \frac{K_{\dot\theta}}{s}\left(s^2 + \frac{K_\theta}{K_{\dot\theta}}s + \frac{K_{\bar\theta}}{K_{\dot\theta}}\right)$$

$$= \frac{K_{\dot\theta}}{s}\left(s + \frac{1}{T_{E_1}}\right)\left(s + \frac{1}{T_{E_2}}\right)$$

In most systems in which pitch command, $\theta_c$, is often steplike in nature the rate term, shown here as $K_{\dot\theta}\dot\theta_e$, is actually mechanized as $K_{\dot\theta}\dot\theta$ to avoid bad effects from a pulsed elevator signal at the initiation of the command. This is, however, a detail that need not concern us further. Although the control law form of Eq. 7-10 is generally satisfactory, the gains must be made consonant with the control requirements imposed by the airplane characteristics. Since the latter vary with changes in the conditions of

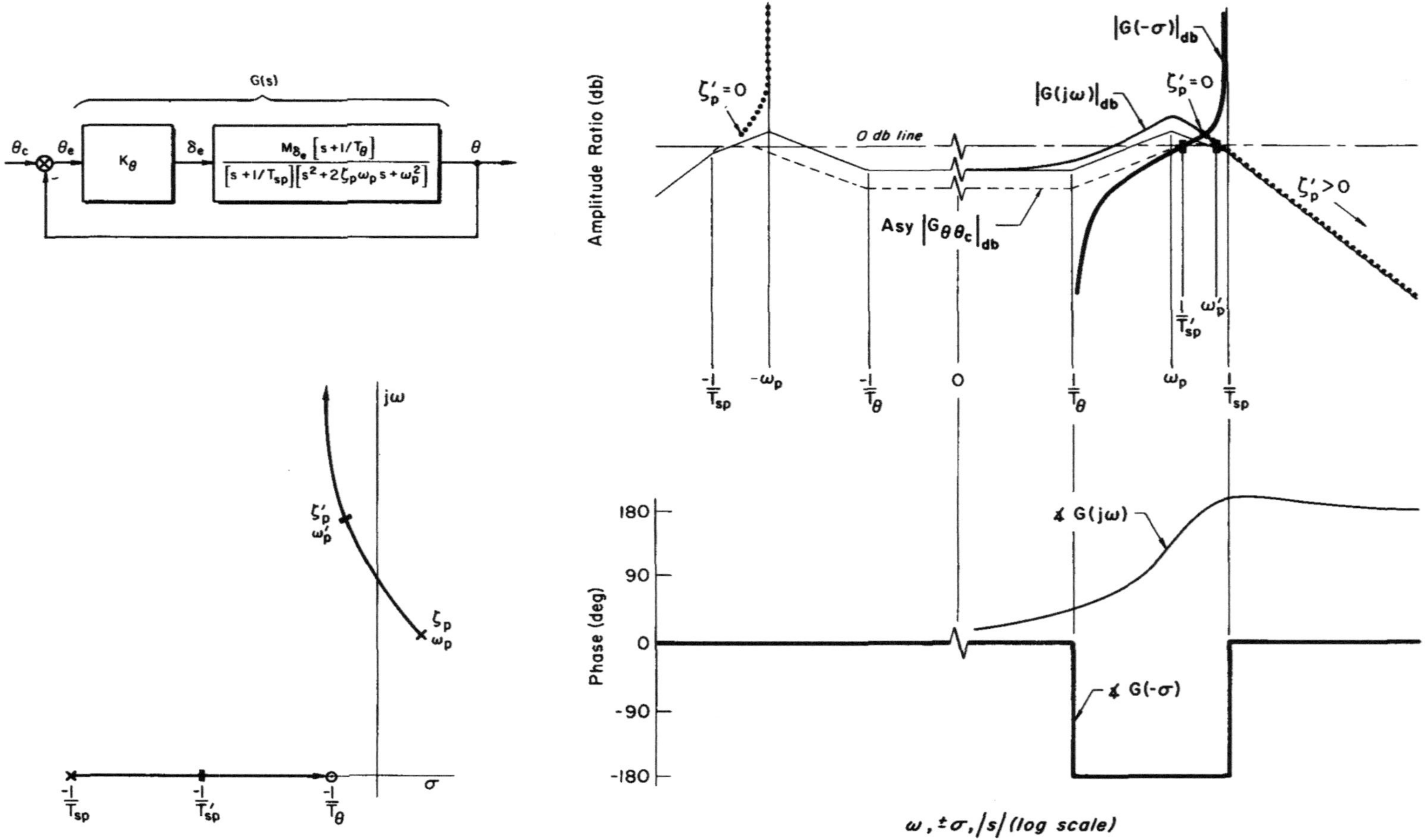


*b) Root Locus*

*a) Bode $G(j\omega)$, $G(-\sigma)$ Diagrams and Bode Root Locus*

Fig. 7-10. System survey of pitch attitude ($\theta \to \delta_e$) control system for hovering VTOL aircraft or helicopter.

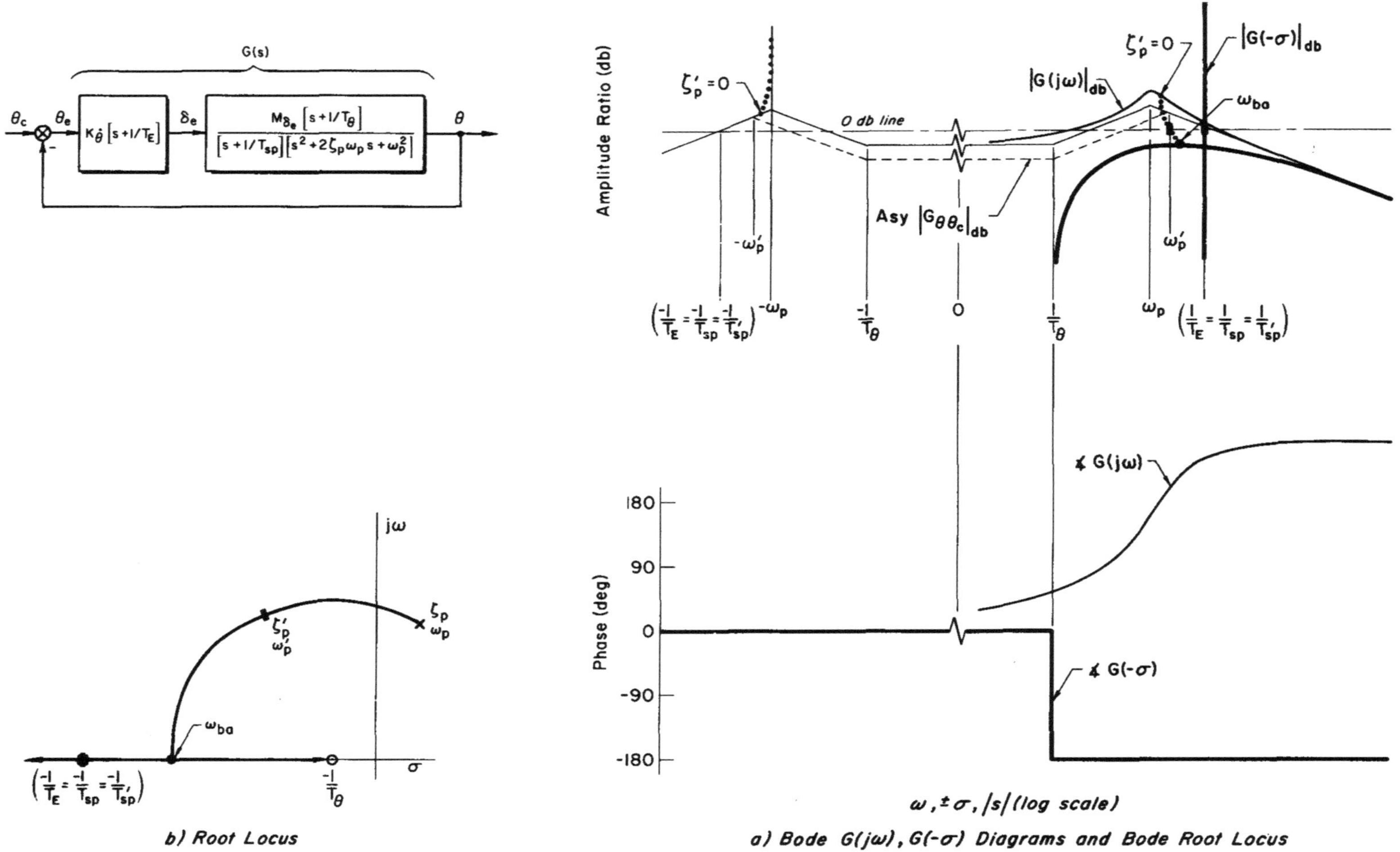


Fig. 7-11. System survey of pitch attitude and rate $(\theta, \dot\theta \rightarrow \delta_e)$ control system for hovering VTOL aircraft or helicopter.

flight, the gains must sometimes also be modified to accommodate these changes. The characteristics of the variations actually employed depend on the specific variations of key stability derivatives, the functional mechanization of the gain-changing devices, and the specified closed-loop performance envelope. However, some insight into the nature of appropriate gain variations can be obtained on an elementary basis by considering the pitch attitude control system sketched in Fig. 7-12. This shows the controller time constants appropriately oriented relative to each other and to the aircraft breakpoints, and indicates a desirable zero-dB line location.

When the controller time constants are well separated as shown, they may be expressed in terms of the controller gains as

$$\frac{1}{T_{E_1}} \doteq \frac{K_{\bar{\theta}}}{K_\theta}, \qquad \frac{1}{T_{E_2}} \doteq \frac{K_\theta}{K_{\dot{\theta}}} \tag{7-11}$$

If the relative locations illustrated can be maintained over the flight

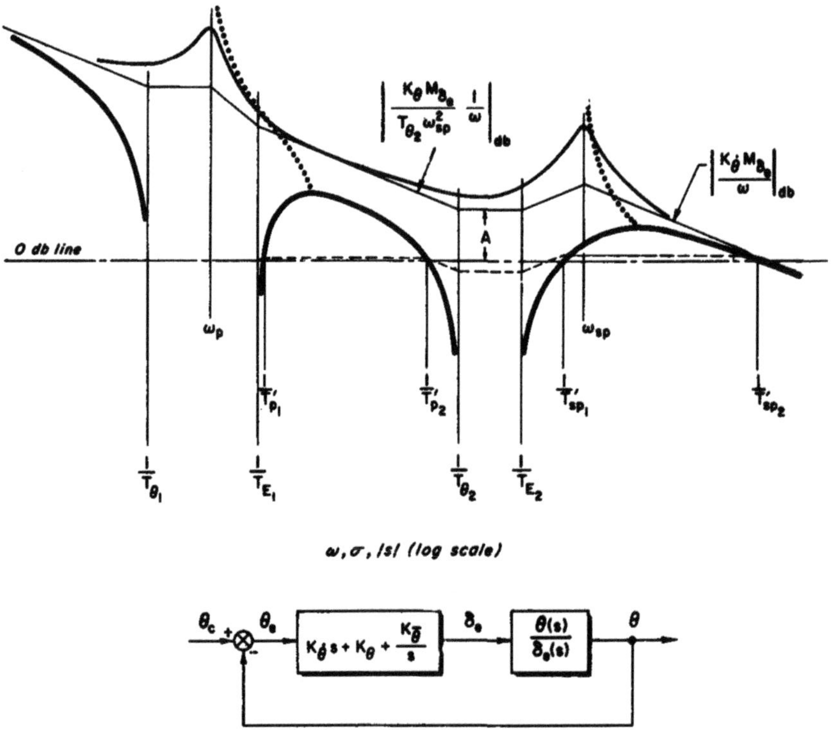


Fig. 7-12. Pitch attitude control with $\delta_e = K_{\dot{\theta}}\dot{\theta}_e + K_\theta\theta_e + K_{\bar{\theta}}\int \theta_e\,dt$.

envelope, the pitch attitude control will generally be excellent, with both short period and phugoid extremely well damped and with good disturbance suppression.

The region where feedback is least effective occurs in the flat stretch between $1/T_{\theta_2}$ and $1/T_{E_2}$. Ideally, the amplitude ratio in this region should be held at, or above, some minimum level to assure positive control and disturbance suppression and to avoid dynamic "droops" due to the dipole pairs at $1/T'_{p_2}$ and $1/T'_{sp_1}$. The amplitude ratio of the asymptote will be given here approximately by[1]

$$\begin{aligned} A &\doteq \frac{K_\theta M_{\delta_e}}{\omega_{sp}^2} \\ &\doteq K_\theta\left(\frac{C_{m_{\delta_e}}}{-C_{m_\alpha}}\right) \end{aligned} \tag{7-12}$$

The term $A$ can be kept approximately constant by varying $K_\theta$ so as to offset the variations in $C_{m_{\delta_e}}/C_{m_\alpha}$. With a fixed center of gravity location this ratio varies only slightly with the Mach number for subsonic and supersonic flight, although it is different in the two regimes. The ratio fluctuates in the transonic region and is, of course, in general a direct function of center of gravity location. All of these factors enter into determining the required change in $K_\theta$ with flight condition to maintain $A$ at or near a desired level.

The equalization time constants should be kept approximately in the same positions relative to the key airplane breakpoints. For instance, $1/T_{E_2}$ should remain somewhere between $1/T_{\theta_2}$ and $\omega_{sp}$. Thus, as indicated below, its basic variation with flight environmental parameters should be proportional to a function intermediate between $\rho U_0$ and $\rho^{1/2}U_0$, i.e.,

$$\frac{1}{T_{\theta_2}} \propto \rho U_0 C_{L_\alpha} < \frac{1}{T_{E_2}} < \omega_{sp} \propto \rho^{1/2} U_0 \sqrt{-C_{m_\alpha}} \tag{7-13}$$

For the simplest condition in which a fixed $K_\theta$ is feasible, appropriate adjustment of $1/T_{E_2}$ could be obtained by varying $K_{\dot\theta}$ inversely with $\rho^{1/2}U_0$; then the closed-loop, short period time constant, $T'_{sp_2}$ will vary as

$$\frac{1}{T'_{sp_2}} \doteq K_{\dot\theta} M_{\delta_e} \propto \rho^{1/2} U_0 \tag{7-14}$$

In theory, at least, a variation of $1/T_{E_1}$ with $\rho U_0$ or $\rho^{1/2}U_0$ can keep this breakpoint properly located relative to $\omega_p$ and $1/T_{\theta_2}$. It will be

[1] Notice that, because of the negative sign on $M_{\delta_e}$, $K_\theta$ must also be negative. In other words, a positive $\theta_c$ requires an up (negative) elevator deflection.

recalled, however, that the addition of integral control was only required at conditions near minimum drag. These are quite specific, so that generally one or more fixed gain settings can be used instead of a continuous variation.

In concluding this discussion of single-loop attitude control, two further points should be mentioned. The first is a deficiency, one of the few associated with attitude as a feedback. The tight high-gain attitude control system described immediately above will tend to hold the pitch angle constant in the presence of disturbances. As a consequence, when the aircraft is subjected to gusts this rigidity in attitude prohibits any weathercocking tendency to nose into the wind and thereby reduce accelerations. It also tends to make the angle of attack change coincide with the gust, considered as an equivalent angle of attack. The net results are somewhat increased structural loads and linear accelerations due to gusts over those that would be present in the uncontrolled aircraft with the same short period damping. The second point relates to the use of attitude as an inner loop for subsequent outer loop controls. Although multiloop systems are not discussed until Chapter 11, it is pertinent to remark here that pitch attitude modifies the effective controlled element characteristics in such a way as to enhance outer loops involving either altitude or speed control operations.

## 7-2. Feedback of Speed Error to the Elevator

Consider next the feedback of speed error to the elevator. The appropriate controlled element transfer function (with $X_{\delta_e} \doteq 0$) is

$$\frac{u(s)}{\delta_e(s)} = \frac{N^u_{\delta_e}}{\Delta_{\text{long}}} = \frac{A_u[s + (1/T_{u_1})][s + (1/T_{u_2})]}{[s^2 + 2\zeta_p\omega_p s + \omega_p^2][s^2 + 2\zeta_{sp}\omega_{sp}s + \omega_{sp}^2]} \tag{7-15}$$

The zero, $1/T_{u_2}$, can be in either the right or left half-plane, depending on elevator location aft or forward and $C_L/C_{D_\alpha}$ larger or smaller than one.[2] In any event it is typically very remote from the origin, so that for most practical purposes it can be neglected.

Figure 7-13 presents the system survey for negative feedback of a speed error. This control may be seen to have a powerful effect on the phugoid undamped natural frequency and also to increase the phugoid damping. The modified phugoid roots move up rapidly and further into the left half-plane, and comparatively large damping ratios for the phugoid mode can be achieved before the short period mode has been much altered.

[2] Irving L., Ashkenas, and Duane T. McRuer, "Approximate Transfer Functions and Application to Single Sensor Control Systems," WADC-TR-58-82, June 1958.

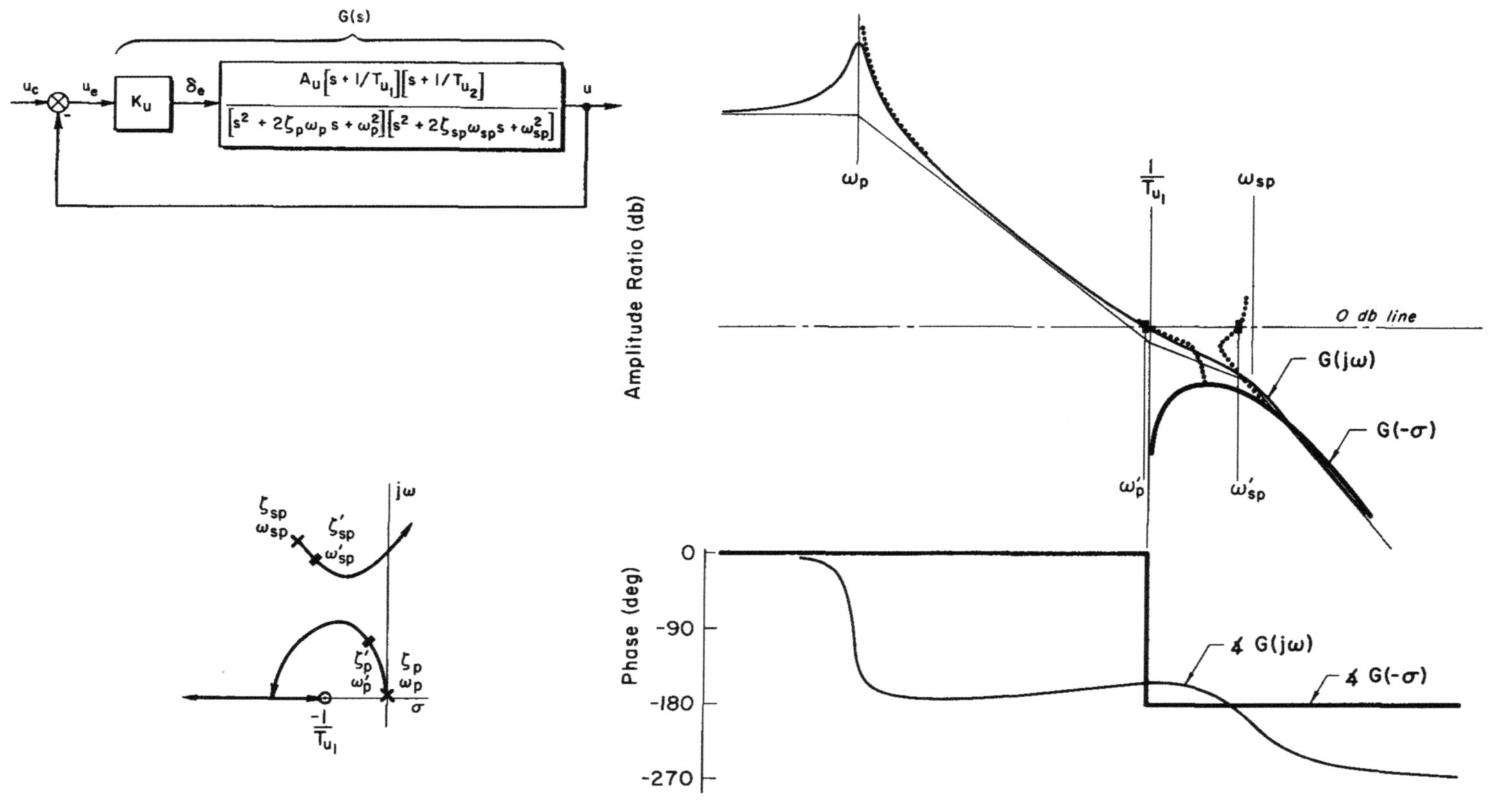


*b) Root Locus*

*a) Bode $G(j\omega)$, $G(-\sigma)$ Diagrams and Bode Root Locus*

Fig. 7-13. System survey of airspeed ($u \rightarrow \delta_e$) control system for stable aircraft.

The changes to both the phugoid frequency and damping illustrated in Fig. 7-13 are in accord with the approximate factors, i.e.,

$$\omega_p^2 \doteq \frac{g[M_w Z_u - Z_w M_{u_{\text{aug}}}]}{\omega_{sp}^2}$$

$$2(\zeta\omega)_p \doteq -X_u - \frac{(X_\alpha - g)}{\omega_{sp}^2} M_{u_{\text{aug}}} \tag{7-16}$$

where $M_{u_{\text{aug}}} = M_u - K_u M_{\delta_e}$.[3] The same beneficial effects on the phugoid occur when a tuck is present; $M_u$ augmentation is often used to counter such tendencies.

Although the closed-loop phugoid damping is somewhat improved by the feedback of speed error alone, it is materially increased if $\dot{u}$ as well is fed back. This action creates a new stability derivative, $M_{\dot{u}}$; its effect on phugoid characteristics can be examined with the aid of the phugoid approximation equations. When an $M_{\dot{u}}su$ pitching acceleration is added to these (Eq. 5-20), the characteristic equation becomes

$$\Delta_p = \begin{vmatrix} (s - X_u) & -X_w & g \\ -Z_u & (s - Z_w) & -U_0 s \\ -(M_{\dot{u}}s + M_u) & -M_w & 0 \end{vmatrix} = 0 \tag{7-17}$$

By the use of the primed notation to indicate that a loop has been closed, the characteristic equation for the modified phugoid becomes

$$s^2 + 2\zeta_p'\omega_p's + \omega_p'^2 = s^2 - \frac{X_u - [M_u(X_\alpha - g)/M_\alpha] - (gM_{\dot{u}}Z_w/M_\alpha)}{1 + [M_{\dot{u}}(X_\alpha - g)/M_\alpha]}s$$

$$- \frac{(g/U_0)[Z_u - (M_u Z_w/M_w)]}{1 + [M_{\dot{u}}(X_\alpha - g)/M_\alpha]} \tag{7-18}$$

While $M_{\dot{u}}$ is present in the denominator for both the damping and the undamped natural frequency, the quantity in which it appears is generally small relative to one. On the other hand, the $gM_{\dot{u}}Z_w/M_\alpha$ addition to the $s$ term in Eq. 7-18 can constitute a major modification to the damping. These conclusions hold, of course, only for the relatively small values of gain for which the approximate equations of motion are still applicable. The system survey of Fig. 7-14 shows that substantially the same effects are present for much larger values of gain.

To increase the damping, $M_{\dot{u}}$ should be positive, as should $M_u$ to increase the undamped natural frequency. The effect of the control is

[3] Notice that, because of the negative sign on $M_{\delta_e}$, $K_u$ must be positive to augment $M_u$. In other words, a positive speed error (speed less than the set speed) requires a down (positive) elevator which tends to restore the command speed.

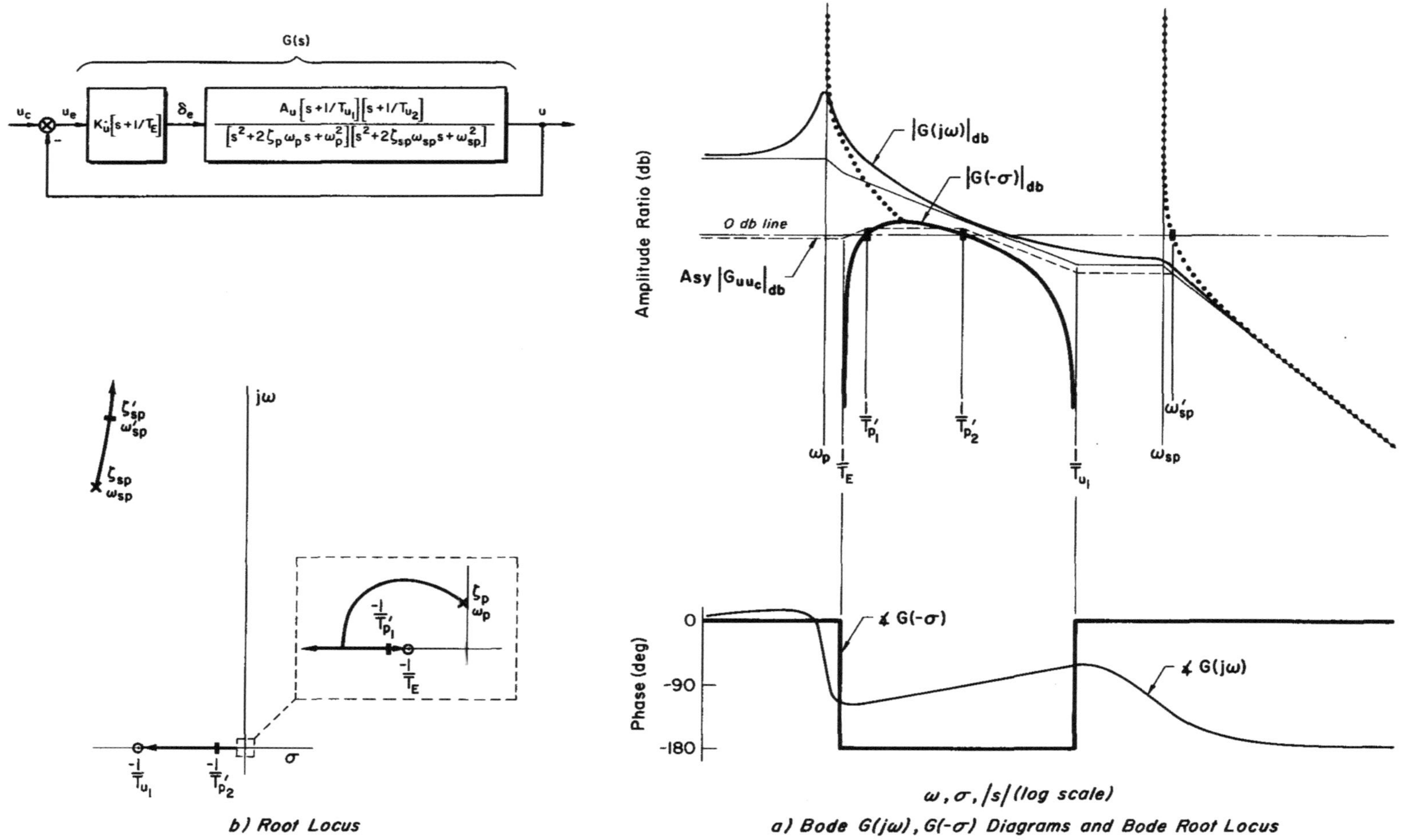


Fig. 7-14. System survey of airspeed and airspeed rate ($u, \dot{u} \rightarrow \delta_e$) control system.

thus to apply a positive pitching moment whenever the speed is greater than the set speed and is increasing. Speed is thus controlled at the expense of attitude changes. This property can cause difficulties if the pitching moments used to control speed are too great. The British experience cited in Chapter 1 is just such a case, in which the atmospheric turbulence components in the direction of flight were sensed by the speed error instrument which actuated the elevator in such a way that large and disconcerting changes in the pitch angle of the aircraft occurred.

An explanation of this phenomenon was discovered by computing transient responses,[4] but it can likewise be appreciated by considering the ratio of the airspeed to attitude numerators for gust excitation. These are readily computed from the simplified phugoid equations with $M_{\dot{u}}$ added:

$$\begin{aligned} (s - X_u)u \quad - X_w w \quad + g\theta &= -X_u u_g \\ -Z_u u + (s - Z_w)w - U_0 s\theta &= -Z_u u_g \\ -(M_{\dot{u}}s + M_u)u \quad - M_w w \quad &= -(M_{\dot{u}}s + M_u)u_g \end{aligned} \tag{7-19}$$

The appropriate numerator ratio is

$$\frac{N^u_{u_g}}{N^\theta_{u_g}} = \frac{-g}{s} + \frac{X_\alpha(M_{\dot{u}}s + M_u) - M_\alpha X_u}{(M_{\dot{u}}s + M_u)(s - Z_w) + Z_u M_w} \tag{7-20}$$

At frequencies in the neighborhood of either the open- or closed-loop phugoid, i.e., either $M_u$ and $M_{\dot{u}}$ approximately zero or $M_u$ and $M_{\dot{u}}$ dominant, respectively, this ratio of numerators is very nearly the first term in Eq. 7-20 alone. That is,

$$\frac{N^u_{u_g}}{N^\theta_{u_g}} \doteq \frac{-g}{s} \tag{7-21}$$

or

$$su \doteq -g\theta, \qquad \text{implying } \frac{du}{dt} \doteq -g\theta \tag{7-22}$$

as far as the phugoid motion components are concerned. (Note that this result is consistent with the phugoid approximate modal response ratios developed in Chapter 5.) The increase in phugoid undamped natural frequency and damping brought about by the $u$, $\dot{u} \rightarrow \delta_e$ feedbacks causes a corresponding increase in the $du/dt$ and attitude responses to gusts. That is, Eq. 7-22 indicates that making the speed response more rapid must incidentally result in larger values of $\theta$.

[4] S. Neumark, "The Disturbed Longitudinal Motion of an Uncontrolled Aircraft and of an Aircraft with Automatic Control," ARC R & M 2078, His Majesty's Stationery Office, London, Jan. 1943.

To avoid the phenomenon described here, an attitude loop is also ordinarily used when speed control is desired. This loop has, among other things, the effect of favorably modifying the ratio of closed-loop $u$ and $\theta$ gust numerators.

### 7-3. Feedback of Angle of Attack to the Elevator

Angle of attack and its derivative are, in principle, very desirable feedbacks to the elevator. Considered as stability derivative augmentation, $\dot{\alpha} \rightarrow \delta_e$ and $\alpha \rightarrow \delta_e$ increase the magnitude of $M_{\dot{\alpha}}$ and $M_\alpha = U_0 M_w$ directly which will increase the damping ratio and undamped natural frequency, respectively. These features are indicated by the short period approximate factors

$$\begin{aligned} 2(\zeta\omega)_{sp} &= -(Z_w + M_q + M_{\dot{\alpha}_{\text{aug}}}) \\ \omega_{sp}^2 &= Z_w M_q - M_{\alpha_{\text{aug}}} \end{aligned} \tag{7-23}$$

The phugoid approximate factors do not contain an $M_{\dot{\alpha}}$ term, and $M_\alpha$ enters only in conjunction with $M_u$, so that there will be little effect of these feedbacks on the phugoid when $M_u \doteq 0$.

The appropriate transfer function for the controlled element is

$$\frac{\alpha(s)}{\delta_e(s)} = \frac{N_{\delta_e}^w}{U_0\Delta_{\text{long}}} = \frac{A_w[s + (1/T_{w_1})][s^2 + 2\zeta_w\omega_w s + \omega_w^2]}{U_0[s^2 + 2\zeta_p\omega_p s + \omega_p^2][s^2 + 2\zeta_{sp}\omega_{sp}s + \omega_{sp}^2]} \tag{7-24}$$

and Fig. 7-15 illustrates the system survey for the feedback of angle of attack alone to the elevator. The close proximity of the complex zeros to the phugoid poles implies very little angle of attack change in the phugoid mode. This, of course, is an inference already well established above and in Chapter 5. Under the condition of closed-loop operation, the angle of attack motion in the phugoid mode will be suppressed to an even greater extent than in the open-loop case, but otherwise this feedback does not appreciably alter the phugoid characteristics. The short-period roots, on the other hand, are greatly affected by the feedback. They are seen to move rapidly to a very large frequency, just as would be expected from the approximate factors. In the idealized case extremely high gains can give very heavy damping to the modified short period. However, this is difficult to achieve in the practical case because of servo and sensor lag effects at the modified short period frequencies, and because the high gains would tend to drive the servo to its limits for all but the smallest inputs or disturbance.

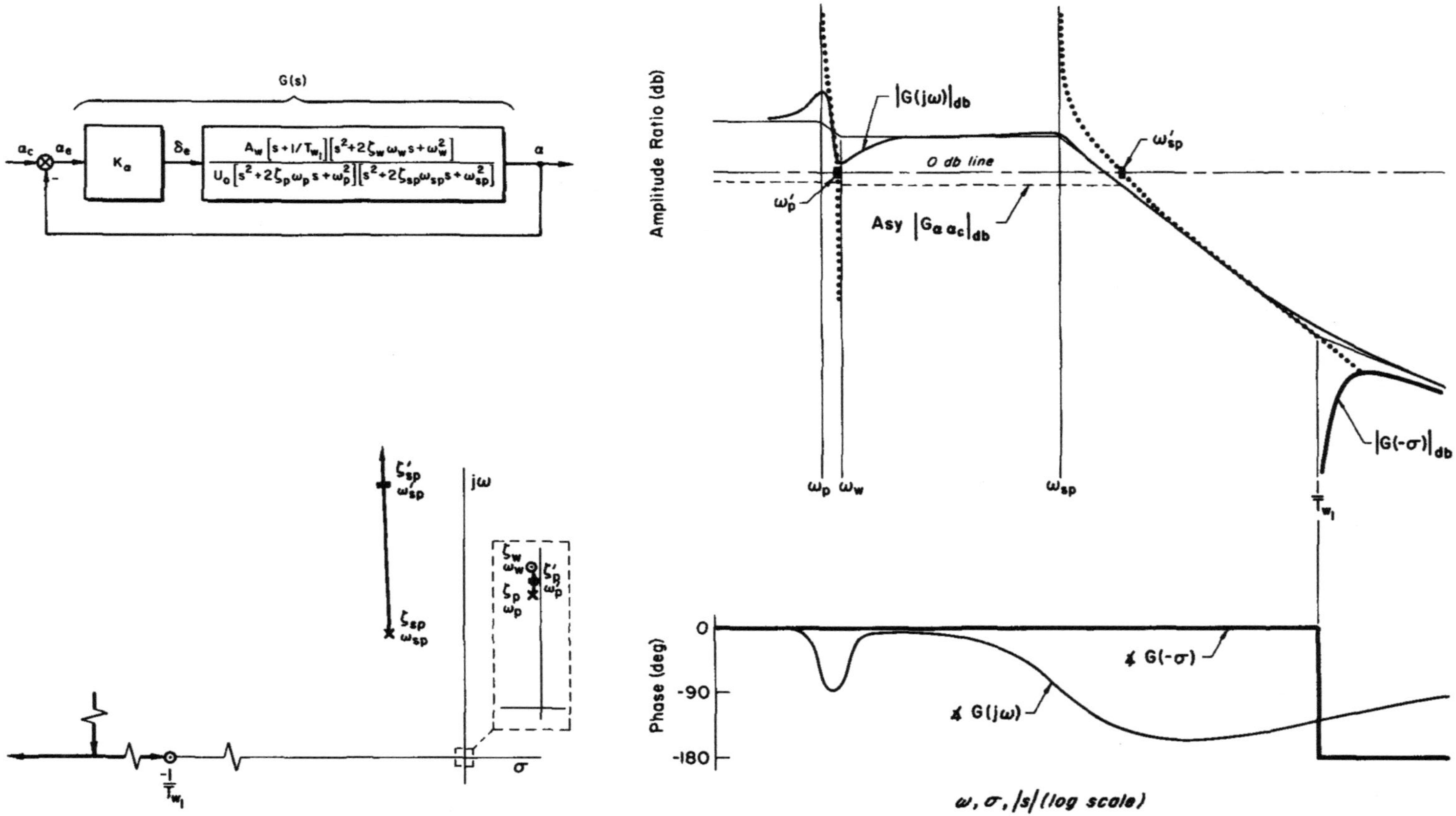


*b) Root Locus*

*a) Bode $G(j\omega)$, $G(-\sigma)$ Diagrams and Bode Root Locus*

Fig. 7-15. System survey of angle of attack ($\alpha \rightarrow \delta_e$) control system.

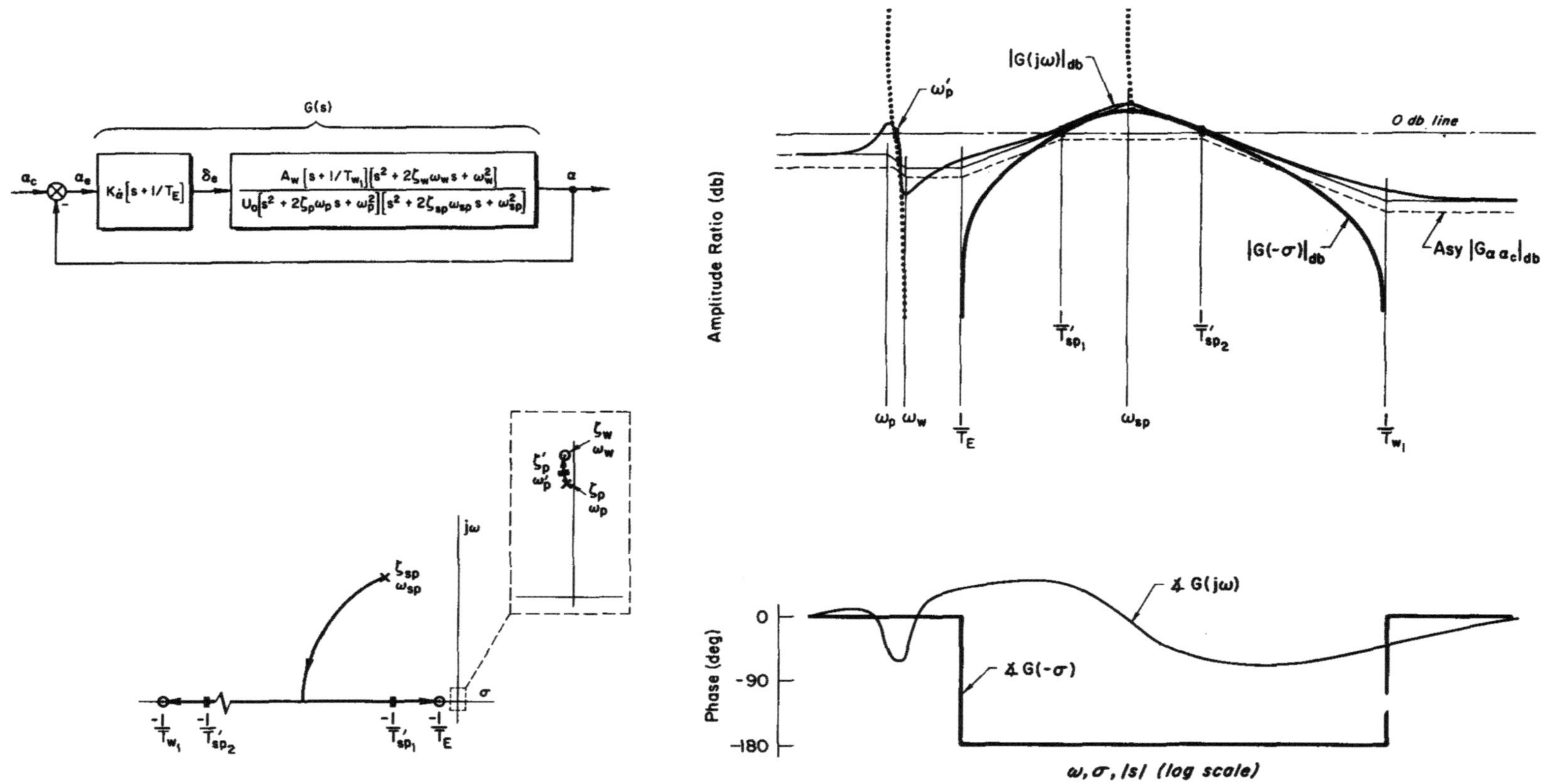


*b) Root Locus*

*a) Bode G(jω), G(−σ) Diagrams and Bode Root Locus*

Fig. 7-16. System survey of angle of attack attitude and rate $(\alpha, \dot{\alpha} \rightarrow \delta_e)$ control system.

To improve the short period damping in practice, lead equalization is added, with the result shown by the system survey in Fig. 7-16. Here again the effect of the lead at moderate gains is just as would be predicted by the approximate factors.

The angle of attack system, when tightly closed, is similar to the pitch attitude plus rate systems described in Section 7.1. Thus both systems stabilize attitude and damp short period, and the nature of gain compensation requirements is similar. However, the reference for orientation stabilization is quite different, being the **g** vector (or horizon) for pitch attitude and the relative wind for angle of attack. Consequently, a substantial difference occurs when vertical gust disturbances are encountered. The angle of attack system rotates the craft into the new relative wind direction, thereby tending to maintain the angle of attack and load factor constant. In these same circumstances pitch attitude control, as already mentioned, will resist any tendencies to rotate the craft relative to inertial space. Also, the pitch attitude system has a very beneficial effect on phugoid stability, whereas the angle of attack system, as noted above, has essentially none.

## 7-4. Feedback of Normal Acceleration to the Elevator

The favorable effects on the short period characteristics of $\alpha, \dot{\alpha} \rightarrow \delta_e$ are often difficult to achieve with practical controls because of sensing problems. However, an alternative control system, using a normal accelerometer as the basic sensor, has many similar features. A major component in the normal acceleration signal is proportional to angle of attack. For instance, for $\gamma_0 = 0$, and considering only short period characteristics, the acceleration at the center of gravity is

$$a_{z_{\text{c.g.}}} = \dot{w} - U_0 q \doteq Z_\alpha \alpha + Z_{\delta_e} \delta_e \tag{7-25}$$

Actually, it is seldom possible, or even desirable, to measure normal acceleration at the aircraft's center of gravity under all operational loading distributions. A case of more interest is the feedback of normal acceleration measured in the plane of symmetry at some distance, $x_a$, from the center of gravity ($x_a$ is positive forward). The normal acceleration at this point is

$$a_z = a_{z_{\text{c.g.}}} - x_a \dot{q} \doteq Z_\alpha \alpha + Z_{\delta_e} \delta_e - x_a \dot{q} \tag{7-26}$$

The frequency range over which the normal acceleration and angle of attack transfer functions are similar can be compared directly, using the short period approximations. The three transfer functions of interest are

$a_{z_{c.g.}}/\delta_e$, $a_z/\delta_e$, and $\alpha/\delta_e$. These are

$$\frac{a_{z_{c.g.}}}{\delta_e} = \frac{N_{\delta_e}^{a_{z_{c.g.}}}}{\Delta_{sp}} = \frac{Z_{\delta_e}\{s^2 - (M_{\dot{\alpha}} + M_q)s - [M_\alpha - (M_{\delta_e}/Z_{\delta_e})Z_\alpha]\}}{\Delta_{sp}}$$

$$= \frac{Z_{\delta_e}\overbrace{[s + (1/T_{h_2})][s + (1/T_{h_3})]}^{(s^2 + 2\zeta_h\omega_h s + \omega_h^2)}}{s^2 + 2\zeta_{sp}\omega_{sp}s + \omega_{sp}^2} \tag{7-27}$$

$$\frac{a_z}{\delta_e} = \frac{N_{\delta_e}^{a_z}}{\Delta_{sp}} = \frac{\begin{array}{l}[Z_{\delta_e} - x_a(M_{\delta_e} + Z_{\delta_e}M_{\dot{w}})]s^2 - [Z_{\delta_e}(M_{\dot{\alpha}} + M_q) \\ \quad + x_a(Z_{\delta_e}M_w - M_{\delta_e}Z_w)]s + M_{\delta_e}[Z_\alpha - (Z_{\delta_e}/M_{\delta_e})M_\alpha]\end{array}}{\Delta_{sp}}$$

$$= \frac{[Z_{\delta_e} - x_a(M_{\delta_e} + Z_{\delta_e}M_{\dot{w}})]\overbrace{[s + (1/T_{a_{z_2}})][s + (1/T_{a_{z_3}})]}^{(s^2 + 2\zeta_{a_z}\omega_{a_z}s + \omega_{a_z}^2)}}{s^2 + 2\zeta_{sp}\omega_{sp}s + \omega_{sp}^2} \tag{7-28}$$

$$\frac{\alpha}{\delta_e} = \frac{N_{\delta_e}^{\alpha}}{\Delta_{sp}} = \frac{(Z_{\delta_e}/U_0)\{s + [(U_0 M_{\delta_e}/Z_{\delta_e}) - M_q]\}}{\Delta_{sp}}$$

$$= \left(\frac{Z_{\delta_e}}{U_0}\right)\frac{[s + (1/T_\alpha)]}{s^2 + 2\zeta_{sp}\omega_{sp}s + \omega_{sp}^2} \tag{7-29}$$

The break frequencies corresponding to the numerators of Eqs. 7-27 and 7-29 ($a_{z_{c.g.}}/\delta_e$ and $\alpha/\delta_e$) are generally considerably greater than the short period undamped natural frequency. This will also be true for the $a_z/\delta_e$ numerator breakpoints for any accelerometer location suitable for control purposes. Below these breakpoints the acceleration transfer functions become

$$\frac{a_z}{\delta_e} \doteq \frac{M_{\delta_e}Z_\alpha - M_\alpha Z_{\delta_e}}{s^2 + 2\zeta_{sp}\omega_{sp}s + \omega_{sp}^2}, \qquad |s| < \left|\frac{1}{T_{h_2}}\right|, \left|\frac{1}{T_{h_3}}\right|; \left|\frac{1}{T_{a_{z_2}}}\right|, \left|\frac{1}{T_{a_{z_3}}}\right| \text{ or } \omega_{a_z} \tag{7-30}$$

Similarly, for frequencies less than $1/T_\alpha$ and the usual inequality, $Z_{\delta_e}M_q/M_{\delta_e}U_0 \ll 1$, the angle of attack transfer function will be approximately

$$\frac{\alpha}{\delta_e} \doteq \frac{M_{\delta_e}}{s^2 + 2\zeta_{sp}\omega_{sp}s + \omega_{sp}^2}, \qquad |s| < \left|\frac{1}{T_\alpha}\right| \tag{7-31}$$

It is apparent, therefore, that for the frequency band defined by the minimum of the several numerator breakpoints, the acceleration and angle of attack transfer functions are directly related by a proportionality factor, $a_z/\alpha \doteq Z_\alpha - (Z_{\delta_e}/M_{\delta_e})M_\alpha$; this can also be recognized from the approximate $\theta/\delta_e$ factors as equivalent to $-U_0(1/T_{\theta_2})$.

To determine the frequency range over which the above simple proportionality applies, and to consider means to make and keep this frequency range reasonably large (and thereby retain the desirable features of $\alpha$ feedback), we must now consider the magnitude of the smallest numerator breakpoint. Of those that require consideration (Eqs. 7-30 and 7-31), $1/T_\alpha$ is easily eliminated as being generally the largest. That is,

$$\frac{1}{T_\alpha} \doteq \frac{U_0 M_{\delta_e}}{Z_{\delta_e}} \doteq \frac{-cU_0}{k_y^2}\frac{C_{m_{\delta_e}}}{C_{L_{\delta_e}}} \doteq \frac{U_0 l_{\delta_e}}{k_y^2} \tag{7-32}$$

which is of the order of, and varies directly with, airspeed and takes the sign of $l_{\delta_e}$ (positive for aft control). Next, consider the numerator terms for the acceleration at the center of gravity. When the point of control application is aft ($M_{\delta_e} < 0$), the constant term in the quadratic is negative, since in general $|M_{\delta_e}Z_\alpha/Z_{\delta_e}| > M_\alpha$. The numerator then consists of two real roots that are nearly equal but of opposite sign (the negative root is somewhat smaller in magnitude than the positive one because their sum, which is $-M_{\dot\alpha} - M_q$, is invariably positive). The magnitude of these breakpoints is approximately

$$\frac{1}{T_{h_3}} \doteq \frac{-1}{T_{h_2}} \doteq \sqrt{M_\alpha - (M_{\delta_e}/Z_{\delta_e})Z_\alpha} \tag{7-33}$$

When the longitudinal control is forward, the quadratic form does not factor into real roots and its undamped natural frequency becomes

$$\omega_h \doteq \sqrt{-M_\alpha + (M_{\delta_e}/Z_{\delta_e})Z_\alpha} \tag{7-34}$$

For either the forward or aft elevator the numerator breakpoint is greater than $\omega_{sp}$, although seldom by a large factor. This relatively close proximity has a profound effect on control systems using the center-of-gravity-mounted accelerometer. As illustrated in the system survey of Fig. 7-17, the $a_z$ feedback increases the short period undamped natural frequency, but the right half-plane zero, $-1/T_{h_2}$, pulls the short period roots into the right half-plane. As is apparent from the Bode root locus, much of the movement of the short period roots occurs with but a slight change in the gain. For instance, only a tiny increment in gain is needed to drive the short period from a value in which $\omega'_{sp}$ is about equal to $1/T_{h_3}$ to the negative real axis rendezvous and thence unstable.

To extend the frequency range over which the $a_z/\alpha$ correspondence holds and, specifically, to alleviate the deleterious effects associated with the nonminimum phase numerator of the center-of-gravity-located accelerometer, other locations are desirable. From Eq. 7-28 it is apparent that,

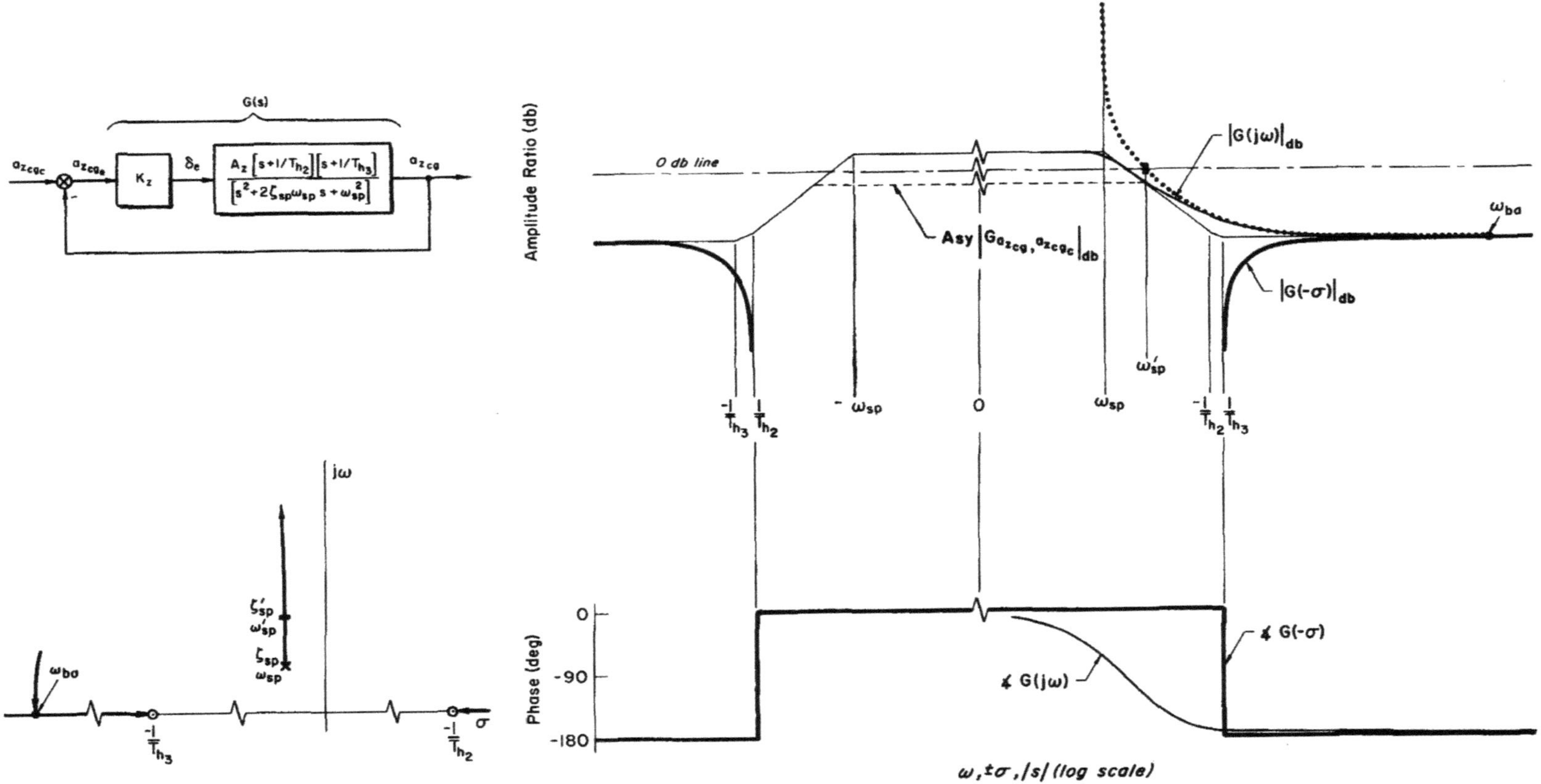


*b) Root Locus*

*a) Bode G(jω), G(−σ) Diagrams and Bode Root Locus*

Fig. 7-17. System survey of normal acceleration ($a_{z_{c.g.}} \rightarrow \delta_e$) control system for short period approximation.

including location effects,

$$\frac{1}{T_{a_{z_3}}} \doteq \frac{-1}{T_{a_{z_2}}} \doteq \sqrt{[M_\alpha - (M_{\delta_e}/Z_{\delta_e})Z_\alpha]/\{1 - x_a[(M_{\delta_e}/Z_{\delta_e}) + M_{\dot{w}}]\}} \quad (7\text{-}35)$$

We see now that the numerator roots can be increased in magnitude by locating the accelerometer so that the $x_a$ term subtracts from one. For an aft elevator control ($M_{\delta_e}/Z_{\delta_e} > 0$) the appropriate $x_a$ is positive, implying an accelerometer location ahead of the center of gravity. Up to a point, the change in location simply moves the numerator breakpoints farther out and away from the short period breakpoint, thereby permitting a higher gain without instability.

When the term involving $x_a$ is greater than one, the numerator becomes a quadratic with an undamped natural frequency:

$$\omega_{a_z} \doteq \sqrt{[-M_\alpha + (M_{\delta_e}/Z_{\delta_e})Z_\alpha]/\{1 - x_a[(M_{\delta_e}/Z_{\delta_e}) + M_{\dot{w}}]\}} \quad (7\text{-}36)$$

With a further increase in $x_a$, $\omega_{a_z}$ is decreased and the numerator breakpoint starts back toward the short period.

The detailed nature of the numerator root variations with the accelerometer-position, control-location parameter, $x_a M_{\delta_e}/Z_{\delta_e}$, can be explored by considering the general $a_z$ numerator as an equivalent servo system that has $x_a M_{\delta_e}/Z_{\delta_e}$ as a gain. From Eqs. 7-27 and 7-28, the numerator is given by

$$\begin{aligned} N_{\delta_e}^{a_z}(s) &= N_{\delta_e}^{a_{z\,\mathrm{c.g.}}}(s) - x_a M_{\delta_e} s\left[\left(1 + \frac{Z_{\delta_e} M_{\dot{w}}}{M_{\delta_e}}\right)s - Z_w + \frac{Z_{\delta_e}}{M_{\delta_e}} M_w\right] \\ &= N_{\delta_e}^{a_{z\,\mathrm{c.g.}}}(s)\left[1 + \frac{-x_a M_{\delta_e} s\{[1 + (Z_{\delta_e} M_{\dot{w}}/M_{\delta_e})]s - Z_w + (Z_{\delta_e}/M_{\delta_e})M_w\}}{N_{\delta_e}^{a_{z\,\mathrm{c.g.}}}(s)}\right] \end{aligned} \quad (7\text{-}37)$$

If we consider that usually $|Z_{\delta_e} M_{\dot{w}}/M_{\delta_e}| \ll 1$ and $|(Z_{\delta_e}/M_{\delta_e})(M_w/Z_w)| \ll 1$, then the zeros of $N_{\delta_e}^{a_z}(s)$ will be the zeros of $(1 + G)$, where $G$ is given by

$$G \doteq \frac{-x_a M_{\delta_e}}{Z_{\delta_e}} \frac{s(s - Z_w)}{[s^2 - (M_\alpha + M_q)s + (M_{\delta_e}/Z_{\delta_e})Z_\alpha]} \quad (7\text{-}38)$$

or, with an aft control point,

$$G \doteq \left(\frac{-x_a M_{\delta_e}}{Z_{\delta_e}}\right) \frac{s(s - Z_w)}{[s + (1/T_{h_2})][s + (1/T_{h_3})]} \quad (7\text{-}39)$$

Furthermore, $|1/T_{h_{2,3}}|/|-Z_w| \doteq (M_{\delta_e}/Z_{\delta_e})(U_0/Z_w)$ will invariably be larger than one.

A survey of this equivalent servo system is shown in Fig. 7-18. At low gains the accelerometer zeros (the poles of the equivalent servo system)

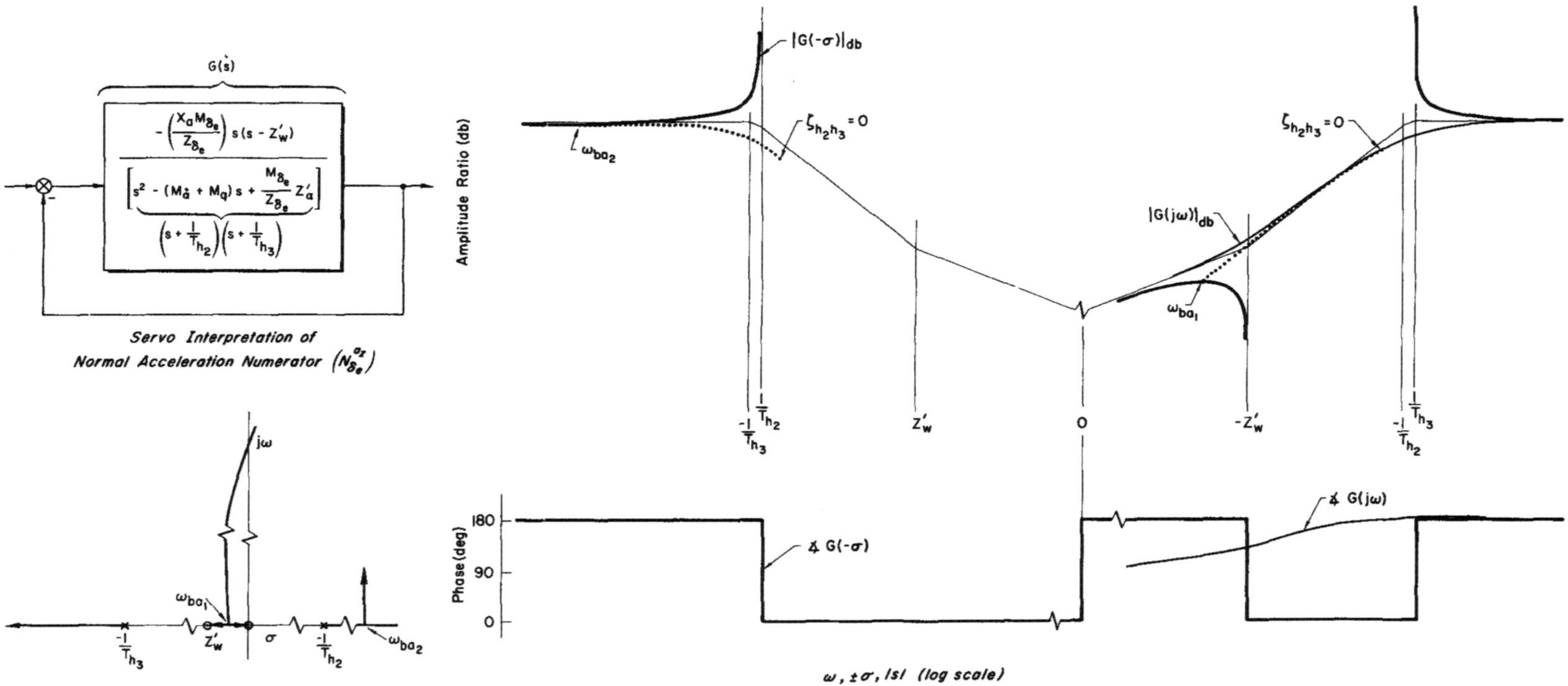


b) Root Locus

a) Bode $G(j\omega)$, $G(-\sigma)$ Diagrams and Bode Root Locus

Fig. 7-18. System survey showing accelerometer location $(x_a)$ effects on normal acceleration numerator $(N^{a_z}_{\delta_e})$.

are near those for the center of gravity location. As the gain is increased, both the minimum and nonminimum phase zeros increase in magnitude until $|x_a M_{\delta_e}/Z_{\delta_e}| = 1$. At this point the minimum phase zero emanating from $-1/T_{h_3}$ has gone to $-\infty$, but the nonminimum phase zero (from $-1/T_{h_2}$) is still finite. For values of gain just slightly greater than one, the minimum phase zero has gone through $-\infty$ and has become a very large positive number; as the gain is slightly increased further, this zero and the nonminimum phase zero rendezvous to form a high frequency quadratic pair.[5] The undamped natural frequency of this pair initially decreases rapidly as the gain is further increased, and then more gradually as the zeros proceed down the +40 dB/decade Bode root locus asymptote between $-Z_w$ and $-1/T_{h_2}$.

From the features just described, the most appropriate location for the accelerometer, in the sense that the numerator breakpoints are as far removed from $\omega_{sp}$ as possible, corresponds to locations at which $|x_a M_{\delta_e}/Z_{\delta_e}|$ is near one. When this is exactly the case, the coefficient of the $s^2$ term in the $a_z/\delta_e$ numerator is zero and the transfer function becomes

$$\frac{a_z}{\delta_e} \doteq \frac{-Z_{\delta_e}(M_{\dot\alpha} + M_q - Z_w)\{s - [(M_{\delta_e} Z_\alpha)/Z_{\delta_e}(M_{\dot\alpha} + M_q - Z_w)]\}}{\Delta_{sp}} \qquad (7\text{-}40)$$

The location $x_a = Z_{\delta_e}/M_{\delta_e}$ corresponds to an "instantaneous center of rotation" for which the center of pressure of the aerodynamic load due to deflecting the control surface is a "center of percussion." At the instantaneous center of rotation, a step function elevator input will produce an initial vertical acceleration due to $Z_{\delta_e}\delta_e$ which is just balanced by that due to the pitching acceleration, $x_a\dot{q}$. (Most readers will recall that the trademark on a baseball bat is at the center of percussion, and that the impulse on the bat when the ball is hit at this point does not result in translational forces on the batter's hands, i.e., the handle is the center of rotation.) This particular location is, of course, ideal for accelerometer control systems. In practice, however, the effective center of rotation shifts because of center of gravity and effective control arm changes so that a location in its general neighborhood is the best that can normally be expected.

In all of the above discussion the accelerometer system has been considered as a replacement for systems involving the angle of attack. The analogy can be carried further; for instance, an $\dot{a}_z$ signal can be used to increase the short period damping, effectively augmenting $M_{\dot\alpha}$. The major difference, then, between a good $a_z$ system and an $\alpha$ system, other

[5] This rendezvous point can occur in either the right half or left half-plane, depending on the sign of $(M_{\dot\alpha} + M_q - Z_w)$. For an aft elevator and $(M_{\dot\alpha} + M_q - Z_w) > 0$, the rendezvous point is in the left half-plane; for $(M_{\dot\alpha} + M_q - Z_w) < 0$, it is in the right half-plane.

than those dwelt on at length above, is in the controller gain variation required for any reasonable control. This is generally quite extreme for accelerometer systems intended to perform angle of attack stabilization functions. More specifically, in an angle of attack controller for which a pure gain might be suitable, the $a_z$ gain would have to vary inversely as $\rho U_0^2$.

For control in the range of phugoid frequencies, the properties of a normal acceleration control are considerably different from those of an angle of attack system. This can perhaps best be appreciated by comparing the system survey of Fig. 7-19, which shows an $a_z \to \delta_e$ system, with the angle of attack control for the same airplane illustrated in Fig. 7-15. Unlike the angle of attack system, the normal acceleration control can have a significant effect on the phugoid. In this case the major effect is a reduction in the phugoid undamped natural frequency. However, for flight conditions that exhibit a performance reversal, i.e., $1/T_{h_1}$ negative, this zero will appear in the right half-plane and a normal acceleration system without washout will tend to drive the phugoid unstable.

## 7-5. Feedback of Altitude to the Elevator

The altitude is an important and very physical motion quantity that must often be controlled accurately. In the linearized equations for small perturbations from straight, level, and horizontal flight, the altitude is proportional to the double integral of the normal acceleration at the center of gravity,

$$h = U_0 \int \gamma \, dt = -U_0 \iint a_{z_{c.g.}} \, dt \, dt \tag{7-41}$$

and the altitude to elevator transfer function is given by

$$\frac{h(s)}{\delta_e(s)} = \frac{N_{\delta_e}^h}{\Delta_{\text{long}}} = \frac{A_h \overbrace{[s + (1/T_{h_1})][s + (1/T_{h_2})]}^{s^2 + 2\zeta_h\omega_h s + \omega_h^2 \text{ or }}[s + (1/T_{h_3})]}{s[s^2 + 2\zeta_p\omega_p s + \omega_p^2][s^2 + 2\zeta_{sp}\omega_{sp} s + \omega_{sp}^2]} \tag{7-42}$$

Just as in the $a_{z_{c.g.}}$ acceleration system discussed in the last article, the numerator quadratic or two first order terms correspond to forward and aft elevator control, respectively. Because of the free $s$, the feedback of altitude by itself drives the modified phugoid roots into the right half-plane at very low values of gain (see Fig. 7-20). Consequently, some form of equalization is required to make an effective altitude control. An inner attitude loop to damp the phugoid is one such possibility. Since this

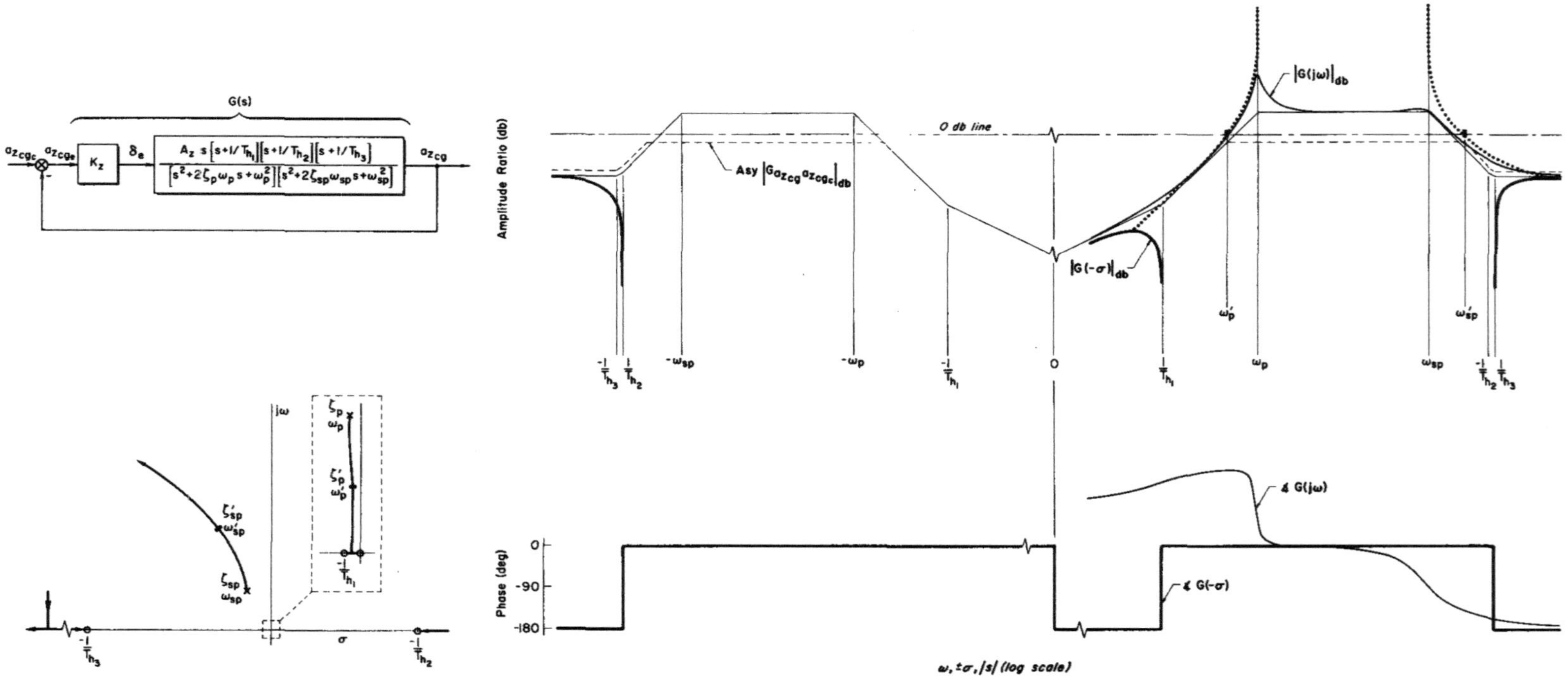


b) Root Locus

a) Bode G(jω), G(−σ) Diagrams and Bode Root Locus

Fig. 7-19. System survey of normal acceleration ($a_{z_{c.g.}} \to \delta_e$) control system.

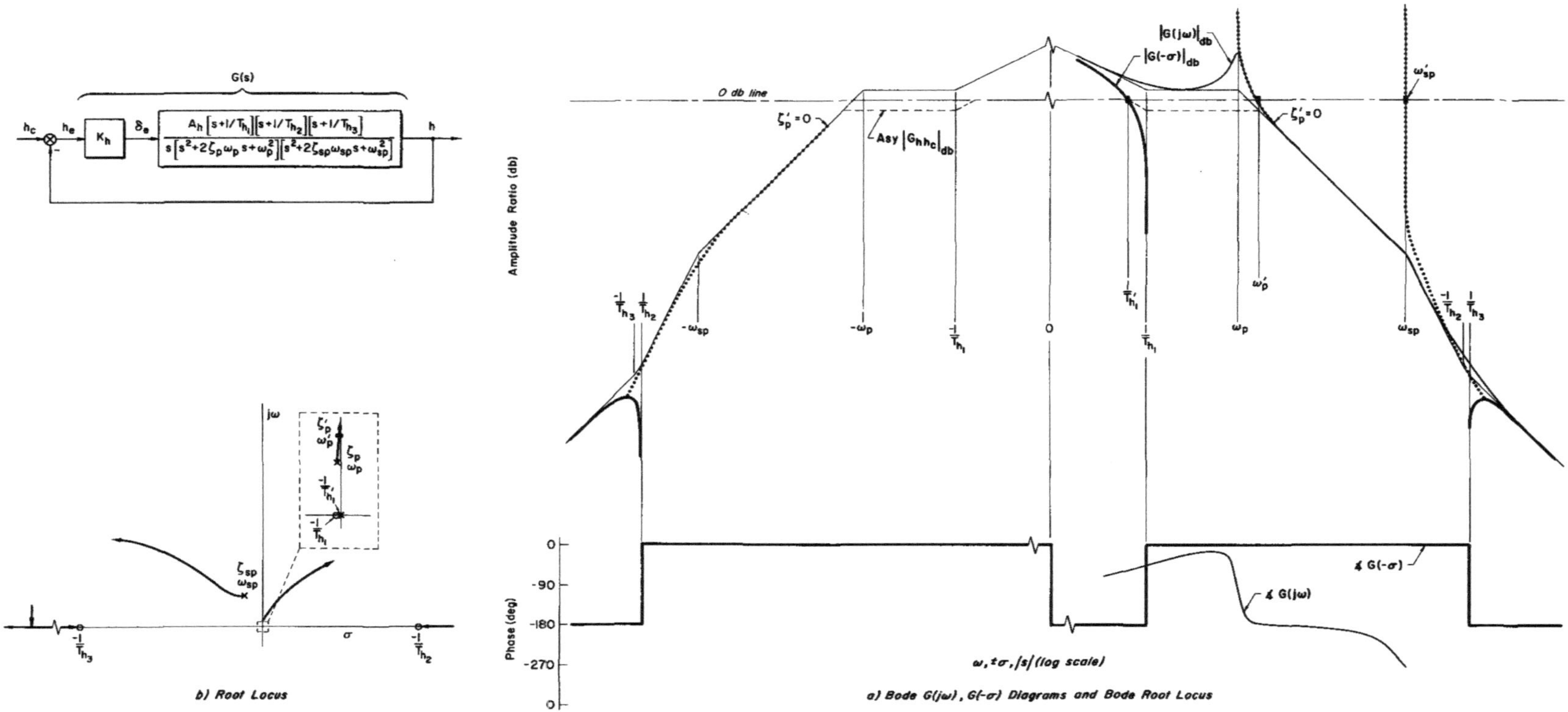


Fig. 7-20. System survey of altitude ($h \rightarrow \delta_e$) control system.

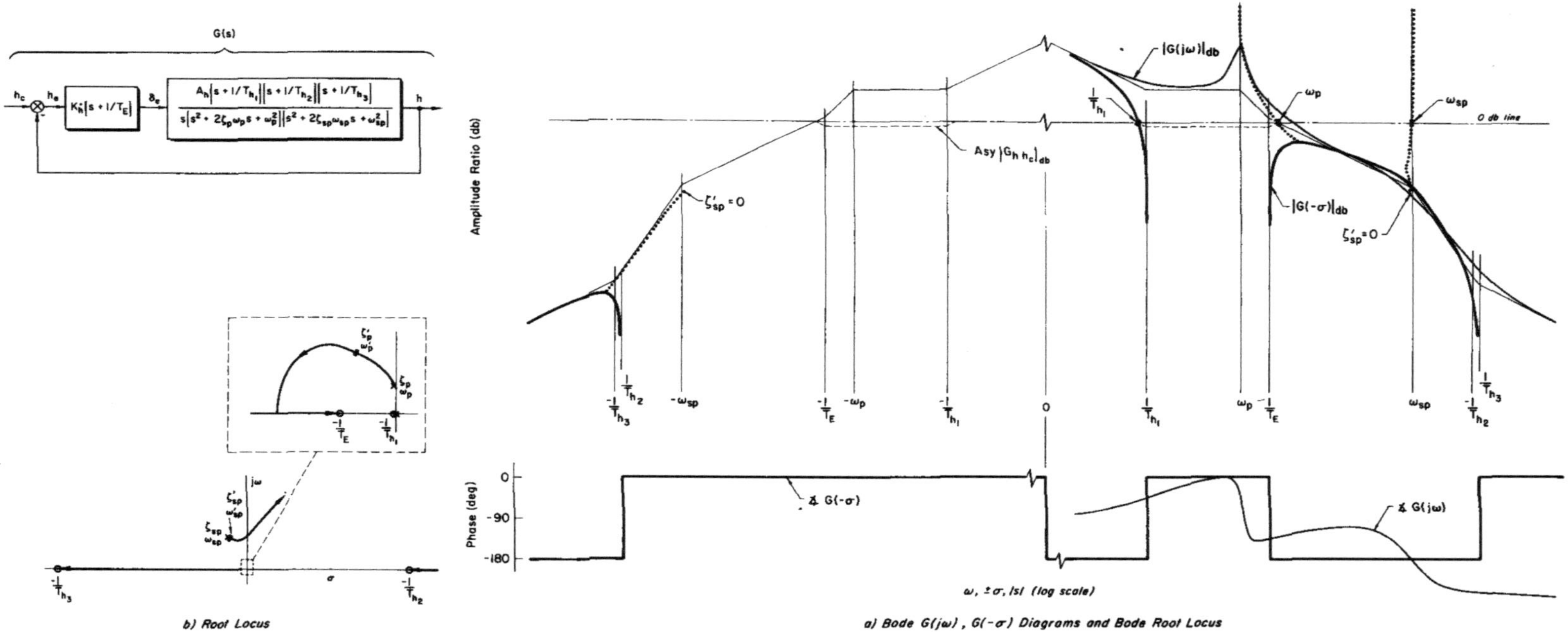


b) Root Locus

a) Bode G(jω), G(−σ) Diagrams and Bode Root Locus

Fig. 7-21. System survey of altitude and altitude rate $(h, \dot{h} \rightarrow \delta_e)$ control system.

constitutes a multiloop system, its consideration is deferred until Chapter 11. We confine our attention here to single-loop or effective single-loop controllers, so that the appropriate equalization is a lead on the altitude error. Practically, this might be accomplished by rate equalization in series or by the use of an altitude rate or rate of climb signal. In any event, the inverse of the lead time constant, $1/T_E$, is adjusted to lie between the phugoid and short period breakpoints. A system survey indicating the efficacy of this technique is shown in Fig. 7-21.

A major problem with altitude control is encountered for performance reversal situations in which $1/T_{h_1}$ becomes negative. In terms of drag and thrust derivatives, $1/T_{h_1}$ is approximated by

$$\frac{1}{T_{h_1}} \doteq \frac{1}{m}\left(\frac{dD}{dU} - \frac{\partial T}{\partial U}\right) \tag{7-43}$$

It will change sign whenever the airplane is on the reverse of the thrust-required vs. speed curve. This is a common situation on very low speed approaches (e.g., carrier approaches), steep climbs, and other flight conditions in which flight at near-minimum drag is desirable. When this performance reversal occurs, an altitude loop will drive the pole at the origin into the right half-plane toward $1/T_{h_1}$. The result is a divergent instability at any value of closed-loop gain. In principle, the performance reversal point could be detected and the sign of the gain changed to avoid divergence, but this would also change the root locus departure from the phugoid roots by 180 degrees, thereby tending to make the phugoid unstable at very low values of gain. Consequently, even in principle such a change is impractical. The conclusion to be drawn from these considerations is that altitude control using the elevator alone cannot be achieved for flight conditions in which the performance reversal exists. An additional control deflection (other than the elevator) must be added.

A similar conclusion on the same basis can be drawn for rate of climb systems. However, there are many situations in which a *specific* value of rate of climb is not the basic requirement; rather, it is desired that the nearly best rate of climb as a function of altitude be maintained. Under these conditions the use of airspeed-like feedbacks to control either indicated airspeed or Mach number to values approximating those for best climb is most appropriate. Such feedback systems do not suffer from performance reversal problems and, further, they do not require command scheduling consistent with available flight performance as a function of altitude.

CHAPTER 8

# ELEMENTARY LATERAL FEEDBACK CONTROL

The usefulness of studying feedback loops closed around the various transfer functions of aircraft is exactly the same in the case of lateral motions as it is for longitudinal. Idealized systems still serve to show the ultimate performance that practical systems can approach or, on the other hand, tend to show which feedbacks are unlikely to prove useful.

A general block diagram for the cases of lateral motion quantity feedback is identical to the one for longitudinal motion quantity feedback (see Fig. 7-1). Again it will be sufficient, in most instances, to consider that the controller is simply a gain, although in some cases simple forms of compensation or equalization may be provided to improve the performance of the feedback control loop.

The main lateral motion output quantities that have been controlled directly and the actuating quantities for the lateral motions of aircraft

Table 8-1. Lateral motion airframe output and actuating quantities.

| Output quantities | Actuating quantities |
|---|---|
| $\varphi$, bank angle | $\delta_a$, aileron deflection |
| $p$, rolling velocity | $\delta_r$, rudder deflection |
| $r$, yawing velocity | |
| $\psi$, heading angle | |
| $\beta$, sideslip angle, $v/U_0$ | |
| $a_y$, side acceleration, $a_{y_{c.g.}} + x_a\dot{r}$ | |

are summarized in Table 8-1. Because there are so many possible combinations, only the more useful and/or instructive of these will be described in this chapter.

## 8-1. Feedback of Bank Angle and Rolling Velocity to the Ailerons

The first successful aircraft automatic pilots employed the same vertical gyroscope that sensed the pitching motions to detect departures from a wings-level attitude and to cause the ailerons to move so as to oppose the bank angle. This action can be thought of as negative feedback of the bank angle motion variable to the aileron actuating quantity. The appropriate transfer function for the controlled element is

$$\frac{\varphi(s)}{\delta_a(s)} = \frac{N^{\varphi}_{\delta_a}}{\Delta_{\text{lat}}} = \frac{A_\varphi[s^2 + 2\zeta_\varphi\omega_\varphi s + \omega_\varphi^2]}{(s + (1/T_s))(s + (1/T_R))[s^2 + 2\zeta_d\omega_d s + \omega_d^2]} \tag{8-1}$$

Typically, in cruising flight the spiral time constant is very large and the mode itself can be either neutrally stable, or a slow convergence, or a slight divergence. Both the neutral and the unstable conditions are undesirable for unattended operation, and the slow convergence is little better. Consequently, one purpose of an automatic pilot is to impose a higher degree of spiral stability. With roll attitude control this is achieved by the creation of static stability in roll. Directly associated with the improvement in spiral mode stability with this kind of system is the provision of bank angle stability and a tendency to maintain roll attitude orientation in the presence of aircraft disturbances. Finally, the bank angle system permits the imposition of roll commands on the aircraft. In the discussion below, several aspects of roll attitude and rolling velocity control will be considered for a variety of aircraft configurations that correspond to several degrees of control difficulty.

*Conventional Roll Attitude Control*

Figure 8-1 displays the feedback system analysis for the bank angle to aileron closure ($\varphi \rightarrow \delta_a$) for an aircraft with good rolling characteristics. The $\sigma$-Bode diagram (shown for both positive and negative values) indicates that, at a gain greater than $|K_\varphi K_{\varphi_{\delta_a}}| = 1$, the spiral motion is made stable. For the vehicle dynamics presumed here this loop may, in fact, be closed with much higher gains without the danger of instability.

Although the dutch roll mode in this airplane is weakly damped, the close proximity of the numerator zeros to the dutch roll poles implies that there will be very little excitation of this mode in response to aileron inputs. (However, the dutch roll mode will still be excited by rudder or gust inputs.) As far as the regulation of bank angle and the ability to follow bank angle commands are concerned, this system would be eminently satisfactory when the loop is closed with the typical zero dB line location shown.

Control of the bank angle of a hovering VTOL aircraft, or helicopter, by means of feedback to the rolling moment control is the analog of pitch angle control for this type of vehicle (see Section 7-1, Eq. 7-8, and Fig. 7-10). The equations of motion and transfer functions have the same form and, except for changes in the numerical values of the stability derivatives or transfer function factors, they are appropriate for the description of

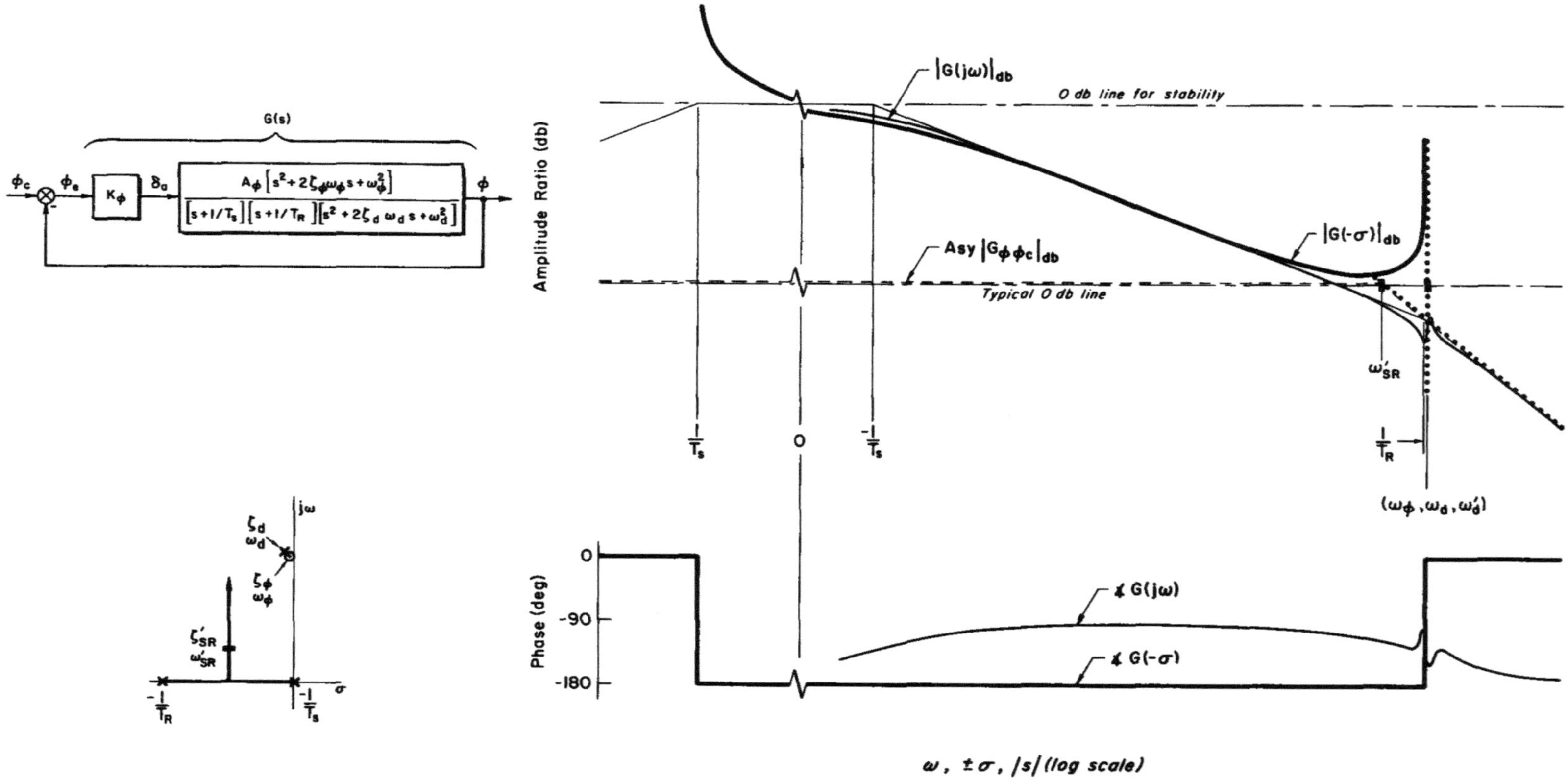


*b) Root Locus*

*a) Bode $G(j\omega)$, $G(-\sigma)$ Diagrams and Bode Root Locus*

Fig. 8-1. System survey of roll attitude ($\varphi \rightarrow \delta_a$) control system for aircraft with good rolling characteristics.

either longitudinal or lateral motions at hover and at very low forward speeds. Figure 7-10, therefore, can be considered to be equally as good an analysis of the bank angle to rolling moment control of a hovering helicopter as it is of pitch angle to pitching moment control.

*Nature of Gain Adjustments for Conventional Roll Attitude Control*

To maintain these satisfactory closed-loop characteristics over a broad regime of flight conditions may require changes in the controller gain to accommodate changes in the conditions of flight. Some appreciation for the nature of appropriate gain variations can be obtained by considering roll control of a much simplified airplane. If we assume that $1/T_R \gg 1/T_s$, and that the dutch roll mode is exactly canceled by the numerator quadratic, then the pure-gain roll control system has an open-loop transfer function given by

$$\frac{\varphi}{\varphi_e} = \frac{K_\varphi L'_{\delta_a}}{s[s + (1/T_R)]} \tag{8-2}$$

The open- and closed-loop characteristics of this system are shown in Fig. 8-2.

The closed-loop roots for this system are defined by

$$2\zeta'_{SR}\omega'_{SR} = \frac{1}{T_R} \tag{8-3}$$

$$\omega'^2_{SR} = K_\varphi L'_{\delta_a} \tag{8-4}$$

The closed-loop damping ratio is therefore

$$\zeta'_{SR} = \frac{1}{2T_R\sqrt{K_\varphi L'_{\delta_a}}} \tag{8-5}$$

The controller gain required to obtain a specific closed-loop damping ratio (or, equivalently, a specific phase margin) is then

$$K_\varphi = \frac{1}{L'_{\delta_a}(2\zeta'_{SR}T_R)^2} \doteq \frac{L'^2_p}{4\zeta'^2_{SR}L'_{\delta_a}}$$

$$\doteq \frac{\rho S b^3}{32\zeta'^2_{SR}I_x}\frac{(C_{l_p})^2}{C_{l_{\delta_a}}} \tag{8-6}$$

The ratio $(C_{l_p})^2/C_{l_{\delta_a}}$ is reasonably constant at subsonic conditions but is a function of Mach number in the transonic and supersonic regimes. To the extent that $(C_{l_p})^2/C_{l_{\delta_a}}$ is approximately constant, an appropriate variation for $K_\varphi$ to maintain constant closed-loop damping ratio and phase margin would be

$$K_\varphi \propto \rho \tag{8-7}$$

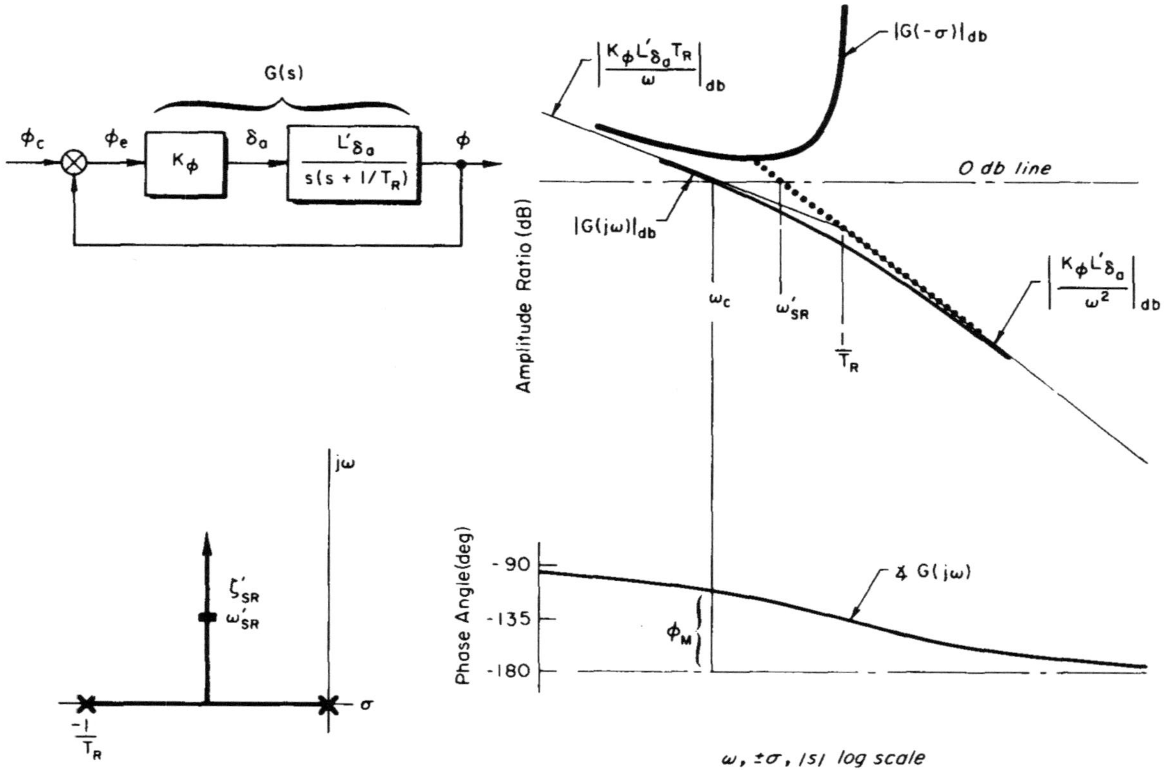


a) Root Locus

b) Bode $G(j\omega)$, $G(-\sigma)$ Diagrams and Bode Root Locus

Fig. 8-2. Roll attitude control for simplified airframe ($1/T_R \gg 1/T_s$; $\omega_\varphi \doteq \omega_d$).

With constant phase margin, it is clear from Fig. 8-2 that

$$\omega_c \propto \frac{1}{T_R} \doteq -L'_p \tag{8-8}$$

or that the variation of the crossover frequency with flight environmental parameters will be

$$\omega_c \propto \rho U_0 \tag{8-9}$$

If the desire is to keep the closed-loop undamped natural frequency, $\omega'_{SR}$, constant, a somewhat more drastic gain variation is required. From Eq. 8-4 it follows that $K_\varphi$ must vary inversely with $L'_{\delta_a}$, or in terms of flight environmental parameters

$$K_\varphi \propto \frac{1}{\rho U_0^2} \tag{8-10}$$

Controller gain variation between the two extremes of Eqs. 8-7 and 8-10 results in a compromise between constant closed-loop damping ratio, $\zeta'_{SR}$, and constant undamped natural frequency, $\omega'_{SR}$.

*Roll Attitude Control with Roll/Yaw Coupling*

One facet of the good lateral control behavior exhibited by the aircraft considered above is the near absence of the dutch roll mode in rolling motions. In fact, the significant roll dynamics could have been treated almost as well using the three degree of freedom spiral roll subsidence approximations in Chapter 6, in which neither numerator nor denominator quadratics appear. When the quadratic numerator in the $\varphi/\delta_a$ transfer function does not approximately cancel the dutch roll denominator, the magnitude of the dutch roll component in rolling motions becomes more significant and may in certain conditions lead to serious control problems. In the context of the simplified case, this change amounts to the multiplication of the simplified transfer function by a ratio of quadratics.

The possible relative orientations of the quadratic pole/zero can best be developed by considering the simplified approximate factors

$$\begin{aligned} \frac{\omega_\varphi}{\omega_d} &\doteq \left(1 - \frac{N'_{\delta_a} L'_\beta}{L'_{\delta_a} N'_\beta}\right)^{\frac{1}{2}} \\ 2\zeta_\varphi \omega_\varphi &\doteq -(Y_v + N'_r) \\ 2\zeta_d \omega_d &\doteq 2\zeta_\varphi \omega_\varphi - \frac{L'_\beta}{N'_\beta}\left(N'_p - \frac{g}{U_0}\right) \end{aligned} \tag{8-11}$$

Great efforts are ordinarily made in the basic airframe design to keep $L'_\beta/N'_\beta$ negative. To the extent that these succeed, $\omega_\varphi/\omega_d$ will be less than one for adverse aileron yaw ($N'_{\delta_a}/L'_{\delta_a} < 0$) and greater than one for so-called favorable aileron yaw ($N'_{\delta_a}/L'_{\delta_a} > 0$). The difference in the damping terms depends mainly on the sign of ($N'_p - g/U_0$). On most aircraft each sign occurs somewhere in the flight regime, so that the relative magnitudes of $\zeta_\varphi\omega_\varphi$ and $\zeta_d\omega_d$ alternate. In general, however, the magnitude of $\zeta_\varphi\omega_\varphi$ will be roughly that of $\zeta_d\omega_d$, except for those cases in which $\zeta_d$ becomes negative, since $\zeta_\varphi$ is invariably positive. Finally, the relative location of $1/T_R$ and $\omega_d$ affects the roll attitude system closure characteristics. Thus the following matrix of possible conditions is of interest:

| Case | $\omega_\varphi/\omega_d$ | $1/T_R\omega_d$ |
|---|---|---|
| 1 | $<1$ | $>1$ |
| 2 | $>1$ | $>1$ |
| 3 | $<1$ | $<1$ |
| 4 | $>1$ | $<1$ |

These various cases are illustrated in the $G(j\omega)$ Bode plots of Fig. 8-3. These plots are drawn with $\zeta_\varphi$ and $\zeta_d$ equal, and use phase asymptotes to emphasize the phase dip due to the quadratic pair.

Examination of Fig. 8-3 reveals that Cases 2 and 4 may become unstable for values of gain in which the zero dB line/amplitude ratio intersection is in the region of the sharp dip in phase. The phase dip must, of course, take the phase angle to values greater than 180 degrees for such an instability to occur. The magnitude of the dip depends on the values of $\zeta_\varphi$, $\zeta_d$, and the $\omega_\varphi/\omega_d$ ratio; in many circumstances the total maximum phase dip is not sufficient to reach $-180$ degrees. For example, in aircraft with large values of dutch roll damping, both $\zeta_\varphi$ and $\zeta_d$ are fairly large and the phase dip is therefore small. Such craft are seldom affected by an unstable condition of this nature. The surest way to avoid the phase dip of Cases 2 and 4 is, of course, to have $\omega_\varphi < \omega_d$, corresponding to Cases 1 and 3. These latter cases, for the minimum-phase airframe conditions shown in Fig. 8-3 and the ideal sensor and servo characteristics assumed here, cannot become unstable at any value of gain.

Root locus diagrams for the four cases are shown in Fig. 8-4. In all cases the closed-loop spiral and roll subsidence roots proceed toward each other, couple, and break off from the real axis. This behavior is similar to that of the stable aircraft situation, being precisely the same for low gains. A divergent spiral does not change this picture appreciably, although there is a minimum value of gain required for stability. Also, the direct correlation between the phase dip and the entry and subsequent exit of the closed-loop dutch roll mode from the right half-plane is apparent.

The behavior of the spiral/roll subsidence coupled pair after breakoff, and that of the modified dutch roll roots, is most interesting and varied.

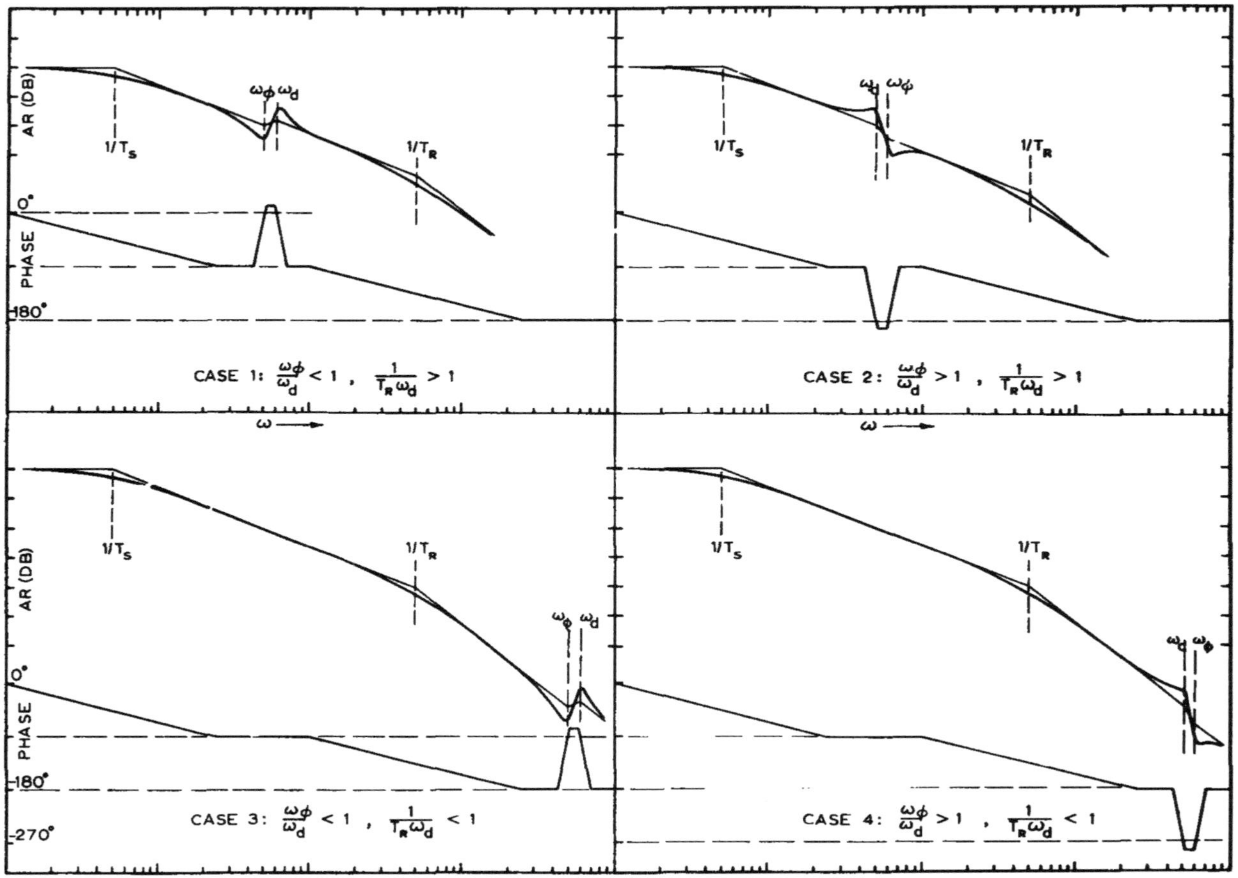


Fig. 8-3. Generic $G(j\omega)$ Bode diagrams of $K_\varphi(\varphi/\delta_a)$ for roll control system cases.

Case 1. $\omega_\phi/\omega_d < 1$ ; $1/T_R\omega_d > 1$

Case 2. $\omega_\phi/\omega_d > 1$ ; $1/T_R\omega_d > 1$

Case 3. $\omega_\phi/\omega_d < 1$ ; $1/T_R\omega_d < 1$

Case 4. $\omega_\phi/\omega_d > 1$ ; $1/T_R\omega_d < 1$

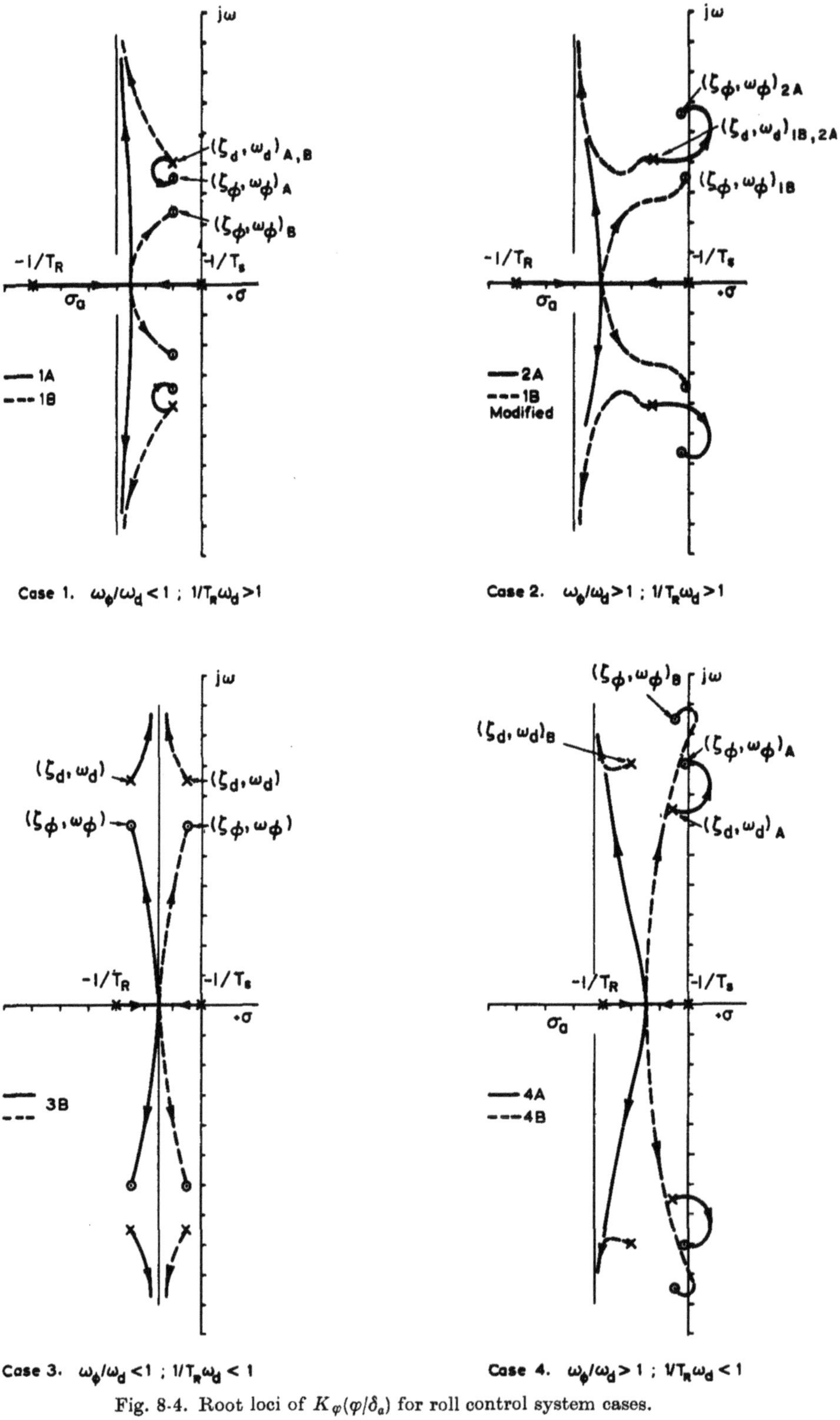

Fig. 8-4. Root loci of $K_\varphi(\varphi/\delta_a)$ for roll control system cases.

One of these pairs must go to the roll numerator at high gain, while the other proceeds toward the high gain asymptotes. Which pair goes where is the basis for further classification—in Subcase A the dutch roll poles go into the roll numerator pair, while the coupled spiral/roll-subsidence mode goes to the asymptote; Subcase B is the opposite of Subcase A. Whether a particular pole/zero configuration fits into a specific category depends primarily on the relative location of the root, especially that of the dutch roll and of the roll numerator. When $\zeta_\varphi\omega_\varphi \doteq \zeta_d\omega_d$ and $\omega_\varphi/\omega_d$ is near unity (Cases 1A and 4A), the infinite gain dutch roll characteristics are those of the roll numerator. As $\omega_\varphi/\omega_d$ decreases (Cases 1B and 3) or $\zeta_\varphi\omega_\varphi$ and $\zeta_d\omega_d$ become separated in value (Cases 1B modified and 4B), the coupled spiral/roll-subsidence roots drive into the zeros. In either case the high gain characteristic of the closed-loop transfer function, $\varphi/\varphi_c$, is basically similar, since one or the other of the denominator pairs will approximately cancel the zeros. The net result for the high gain values of $G/(1 + G)$ will then be

$$\left.\frac{G}{1 + G}\right]_{K_{\varphi_{\text{high}}}} \doteq \frac{\omega_{\text{CL}}^2}{s^2 + 2\zeta_{\text{CL}}\omega_{\text{CL}}s + \omega_{\text{CL}}^2} \tag{8-12}$$

The value of $\zeta_{\text{CL}}\omega_{\text{CL}}$ for high gain approaches the negative of the high gain root locus asymptote, i.e.,

$$\zeta_{\text{CL}}\omega_{\text{CL}} \doteq -\sigma_{\text{c.g.}} = (\zeta_d\omega_d - \zeta_\varphi\omega_\varphi) + \frac{1}{2}\left(\frac{1}{T_R} + \frac{1}{T_s}\right) \tag{8-13}$$

Of all the varied airframe forms considered in Figs. 8-3 and 8-4, a "preferred," or most desirable, condition will usually be Case 1A. This would be nearly ideal when $\zeta_\varphi\omega_\varphi$ is large enough to satisfy dutch roll damping requirements, for then the roll loop can be made high gain and will have a response to commands that is essentially second order in character.

In a practical control with actuator and sensor lags, the unstable conditions due to the phase dip cannot ordinarily be overcome by raising the gain to force the dutch roll poles into the $\omega_\varphi$ zeros. Instead, more elaborate means of compensation must be sought. These typically involve the use of the rudder either to provide damping for the dutch roll or the use of a crossfeed from the aileron to alter the apparent aileron yawing moment characteristics. The analysis of this important example of a separate loop closure as a means of compensation is deferred until Chapter 11.

### *Roll Attitude Control with Deficient Roll Damping*

In many practical situations a roll control system based on only roll angle feedback is inadequate from the standpoint of tightness of control in response to disturbances and commands. This deficiency can be traced back, fundamentally, to the fact that, with pure roll attitude feedback, the open-loop and closed-loop total effective dampings are the same. To improve this situation the degree of the open-loop transfer function numerator must be made no less than one below that of the denominator. Then, when the quantity $1 + G$ is formed and divided through by the coefficient of the highest degree $s$ term, the second highest degree $s$ term (which represents the total system damping or the sum of all the roots) will be a function of open-loop gain. The damping factors of the various closed-loop modes can then be considerably increased, with attendant reduction in response time and increase in system tightness.

To accomplish this increase in relative numerator/denominator degree in the $\varphi \rightarrow \delta_a$ roll control system, the controller transfer function is changed from $K_\varphi$ to $K_\varphi(1 + T_\varphi s)$. From the mechanization standpoint this introduction of lead equalization can be obtained in several ways, all adding lags at high frequency. These lags will be ignored here because they are no more important to the discussion at hand than sensor or servo lags. Also, the lead can be introduced in either the feedback or forward loops, although it is usually inserted in the feedback path. In this way the closed-loop system characteristics at high gain can be made to approach $\varphi/\varphi_c \doteq 1/(T_\varphi s + 1)$. To the extent that high gain can be maintained throughout the flight regime, this provides constant closed-loop system dynamics in response to commands. Although such considerations are important in connection with $\varphi_c$ inputs, the major interest in what follows is on the effective vehicle dynamics that are independent of input, i.e., on the closed-loop modes defined by $1 + G(s) = 0$.

In most respects the significant changes in lateral characteristics due to the addition of lead in the controller, i.e.,

$$\delta_a = (K_{\dot\varphi} s + K_\varphi)\varphi_\epsilon = K_{\dot\varphi}\left(s + \frac{1}{T_\varphi}\right)\varphi_\epsilon$$

can be treated using the three degree of freedom spiral/roll subsidence approximation. The open-loop transfer function will then be

$$G = \frac{K_{\dot\varphi} L'_{\delta_a}(\omega_\varphi/\omega_d)^2[s + (1/T_\varphi)]}{[s + (1/T_s)][s + (1/T_R)]} \tag{8-14}$$

Since the roll subsidence is always greater than the spiral root, the simplified system characteristics will be strongly dependent on the relative location of $1/T_\varphi$ and $1/T_R$. The two possible cases (excluding the trivial one when $1/T_\varphi = 1/T_R$) are shown in Figs. 8-5 and 8-6. When $1/T_\varphi > 1/T_R$ (Fig. 8-5) the low gain closed-loop roots are similar to those for the simple roll case (which is a limiting condition corresponding to

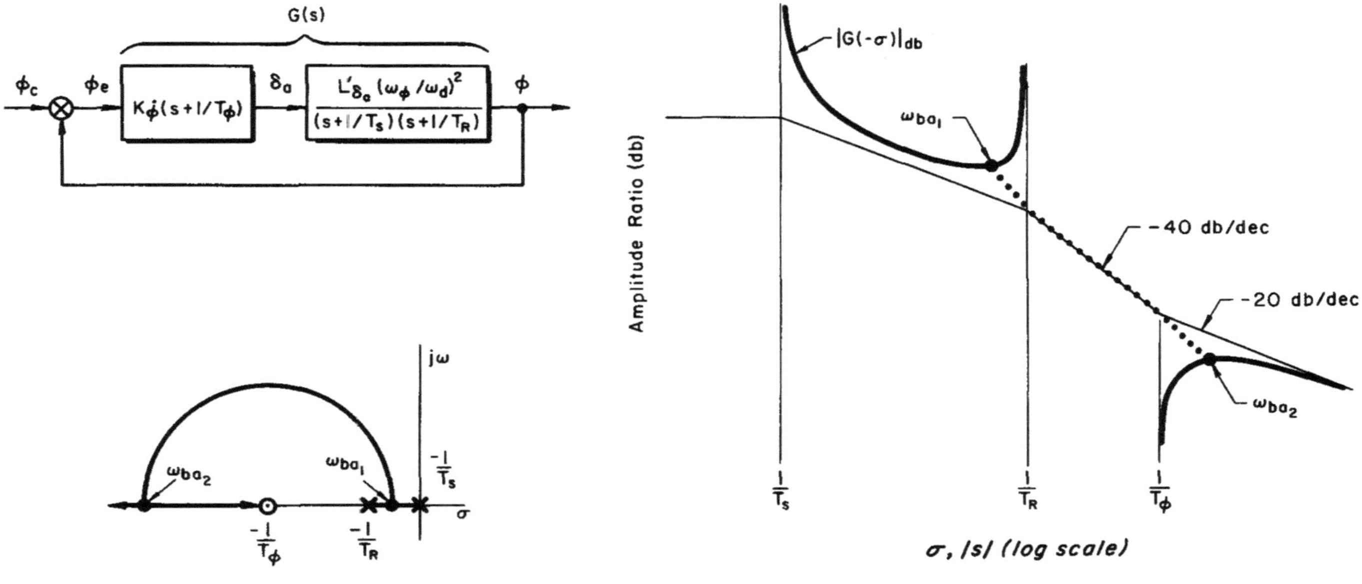


*b) Root Locus*

*a) Bode G(-σ) Diagram and Bode Root Locus*

**Fig. 8-5.** System survey of roll attitude and roll rate ($\varphi$, $\dot{\varphi} \rightarrow \delta_a$) control system for $1/T_s < 1/T_R < 1/T_\varphi$.

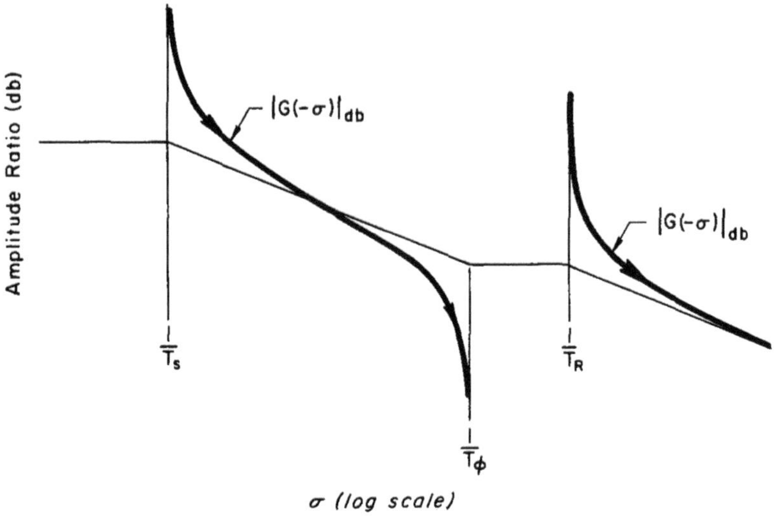


Fig. 8-6. Bode $G(-\sigma)$ diagram of roll attitude and roll rate ($\varphi, \dot{\varphi} \rightarrow \delta_a$) control system for $1/T_s < 1/T_\varphi < 1/T_R$.

$1/T_\varphi \rightarrow \infty$). With an increase in gain the modified spiral and roll subsidence approach one another until the zero dB line reaches the breakaway point, $\omega_{ba_1}$. At higher gains the roots become complex. At still higher gains the presence of $1/T_\varphi$ becomes a more emphatic factor in the character of the system. For the zero dB line at the rendezvous point, $\omega_{ba_2}$, or lower (higher gains), the two oscillatory roots have returned to the real axis—one goes into ($-1/T_\varphi$), while the other increases (negatively) with increased gain. The complex branch from $\omega_{ba_1}$ to $\omega_{ba_2}$ is, of course, a perfect circle, centered at $-1/T_\varphi$, on the conventional root locus.

In Fig. 8-6 the presence of the zero, $-1/T_\varphi$, between the $-1/T_s$ and $-1/T_R$ poles leads to a system that exhibits considerable differences from the $\varphi \rightarrow \delta_a$ case. The behavior of the closed-loop roots as gain is increased is straightforward: the modified spiral goes to the lead term, $1/T_\varphi$, and the modified roll subsidence approaches the high frequency asymptote.

Different as the plots of Figs. 8-5 and 8-6 appear, they indicate very similar system behavior at low values of gain. At very high gains the results shown are also similar. Only at intermediate gains is the behavior of the two basically different.

*Roll Velocity Feedback* (*Roll Damper*)

One of the simplest stability augmentation systems imaginable is the feedback of rolling velocity to aileron ($p \rightarrow \delta_a$). This system is seldom used as a command controller, i.e., there is no $p_c$ command. Instead, its function is to augment the roll damping derivative $L'_p$. This may be desirable for one or more of several reasons. One was discussed in the last article, in which the $\dot{\varphi} \rightarrow \delta_a$ component of the $\varphi$, $\dot{\varphi} \rightarrow \delta_a$ system was used to improve the response of a roll command system. Other reasons will be described below, using the equivalent stability derivative approach and approximate factors.

When the aileron deflection is made proportional to rolling velocity, i.e.,

$$\delta_a = -K_p p \tag{8-15}$$

the effect on the roots of the closed-loop system is the same as those that would be caused by changing the stability derivative $L'_p$ to $L'_{p_{\text{aug}}}$ where

$$L'_{p_{\text{aug}}} = L'_p - K_p L'_{\delta_a} \tag{8-16}$$

if $Y^*_{\delta_a}$ and $N'_{\delta_a}$ are neglected. The direct effect of this action is to increase the roll subsidence break frequency, expressed by the approximate factor

$$\frac{1}{T'_R} \doteq -L'_{p_{\text{aug}}} + \frac{L'_\beta}{N'_\beta}\left(N'_p - \frac{g}{U_0}\right) \tag{8-17}$$

This may be desirable to improve the handling qualities in manual control or, as treated above, to allow a tighter roll attitude loop closure in automatic control.

An increase in $1/T'_R$ can also improve the aircraft response to rolling gusts. For instance, by using the simplified airplane equations with $1/T_s \doteq 0$, the rolling response to a rolling velocity gust, $p_g$, can be described approximately by the transfer function

$$\frac{\varphi}{p_g} \doteq \frac{-[L'_p - (L'_\beta N'_p/N'_\beta)]}{s[s + (1/T'_R)]} \tag{8-18}$$

In this relationship the low frequency gain is changed from approximately unity $(-L'_p T_R)$ for the airplane-alone condition to $L'_p/L'_{p_{\text{aug}}}$ with augmentation.

Both the ratio of rolling and sideslipping amplitudes in the dutch roll mode and the roll response to a $\beta_g$ gust input are measured by the ratio of $\varphi$ and $\beta$ modal response coefficients for the dutch roll mode. As developed in Chapter 6, this is given for the airplane-alone by the approximate expression

$$\left|\frac{\varphi}{\beta}\right|_d \doteq \left|\frac{L'_\beta}{N'_\beta}\right| \frac{1}{\sqrt{1 + L'^2_p/N'_\beta}} \tag{8-19}$$

if the dutch roll damping ratio is fairly small. Taking account of the augmentation by replacing $L'_p$ with $L'_{p_{\text{aug}}}$ in Eq. 8-19, we find that $|\varphi/\beta|_d$

can be reduced substantially by a roll damper if $|L'_{p_{aug}}|/\sqrt{N'_\beta} > 1$; otherwise, the effect of augmentation on $|\varphi/\beta|_d$ will be minor.

Finally, the use of a roll damper to improve effective airframe numerator characteristics for outer loop equalization must not be overlooked, although it is beyond the scope of this chapter.

## 8-2. Feedback of Other Quantities to the Ailerons

Bank angle and rolling velocity are the primary feedbacks in single-loop roll control systems. They are generally effective and are almost always present in automatic flight control systems. In addition to $\varphi$ and/or $p$, other quantities have from time to time been used as feedbacks to the ailerons. The most common are lateral deviation and heading, as might be obtained from a localizer or other lateral guidance device and a directional gyro. Neither of these is generally suitable for direct feedback to the ailerons without extensive equalization. This is normally supplied by virtue of inner attitude loops, so that the feedback of these quantities, in practice, constitutes a multiple-loop situation, and thus is a subject for Chapter 11.

Quantities other than bank angle and rolling velocity have been used occasionally in single-loop aileron control systems. These include feedbacks of yawing velocity ($r \to \delta_a$), sideslip ($\beta \to \delta_a$), and lateral acceleration ($a_y \to \delta_a$). On particular craft each of these has been made to serve some useful purpose, but their success has invariably been dependent on the existence of favorable vehicle characteristics that are by no means universally present. The favorable characteristics are exhibited in the transfer function numerators and, as already remarked in Chapter 6, such numerators as $N^r_{\delta_a}(s)$ are mavericks of the most independent sort, in that their basic characteristics differ not only between different aircraft but even within a single vehicle in different flight regimes. Consequently, a thorough discussion of such single-loop feedbacks would have to cover a very large number of mostly unsuitable conditions. To avoid this, we shall confine our attention here to the low gain behavior of such loops, thereby permitting us to use approximate factors and equivalent stability derivatives as the basis for the discussion. Most of the interesting features and deficiencies of $\beta$, $r$, and $a_y$ feedbacks to aileron can be treated in this way in fairly general terms.

*Feedback of Sideslip to the Ailerons*

The feedback of sideslip angle to the ailerons modifies the stability derivatives $L'_\beta$ and $N'_\beta$ as follows:

$$\begin{aligned} L'_{\beta_{aug}} &= L'_\beta - K_\beta L'_{\delta_a} \\ N'_{\beta_{aug}} &= N'_\beta - K_\beta N'_{\delta_a} \end{aligned} \tag{8-20}$$

For most aircraft with reasonably large directional weathercock stability, the principal effect will be on $L'_{\beta_{\text{aug}}}$; in this sense the feedback of sideslip angle is analogous to changing the dihedral of the airplane. A very modest amount of this feedback is sufficient to stabilize an unstable spiral motion; that is, with $K_\beta$ positive, the magnitude of $L'_{\beta_{\text{aug}}}$ can easily be increased such that

$$L'_{\beta_{\text{aug}}}N'_r - L'_rN'_\beta > 0 \tag{8-21}$$

Unfortunately, at the same time this feedback may significantly decrease the damping of the dutch roll mode whenever $(N'_p - g/U_0)$ is negative, as shown by the approximate factor

$$2\zeta'_d\omega'_d \doteq -(Y_v + N'_r) - \frac{L'_{\beta_{\text{aug}}}}{N'_\beta}\left(N'_p - \frac{g}{U_0}\right) \tag{8-22}$$

Sideslip-to-aileron feedbacks have also been proposed as a means to decrease the effective dihedral and thereby the dutch roll modal response ratio $|\varphi/\beta|_d$. This requires a negative $K_\beta$, and all of the effects noted above are simply reversed. To achieve significant reduction in the effective $|L'_\beta|$ over a reasonable range of inputs, the gain, $K_\beta$, and the aileron authority must be fairly large. Further, the degree of spiral instability must be limited. This is automatically accomplished if the $\beta/\delta_a$ zeros are such that one of them is negative and slightly larger in magnitude than the spiral, thereby providing a zero for the spiral to approach. Unfortunately, this numerator has a basic dependence on such variable stability derivatives as $N'_{\delta_a}$ and $(N'_p - g/U_0)$, so that the zero locations can shift drastically with flight condition.

We can conclude from the above that $\beta \to \delta_a$ systems can provide some good features but are usually accompanied by deleterious side effects. Since the disadvantages often outweigh the advantages, such systems are seldom used, except in special circumstances.

*Feedback of Side Acceleration to the Ailerons*

The feedback of side acceleration, $a_{y_{\text{c.g.}}}$, to the ailerons is often practically equivalent to the feedback of side velocity. This follows from the relationship

$$a_{y_{\text{c.g.}}} = Y_{\delta_a}\delta_a + Y_{\delta_r}\delta_r + Y_pp + Y_rr + Y_vv \tag{8-23}$$

where the terms $Y_{\delta_a}$, $Y_p$, and $Y_r$ are negligible so that

$$a_{y_{\text{c.g.}}} \doteq Y_{\delta_r}\delta_r + Y_vv \tag{8-24}$$

Then, overlooking for the moment the possible effect of rudder deflections, we find that only the very much simplified formula remains:

$$v = \frac{a_{y_{\text{c.g.}}}}{Y_v} \tag{8-25}$$

Consequently, as a single-loop control the lateral accelerometer will share most of the characteristics of $\beta = v/U_0 \to \delta_a$ systems (similar to the correspondence between $a_z$ and $\alpha$ in the longitudinal control).

In some automatic pilots the connection between side acceleration and side velocity has been used as the basis for a coordination control, in which the ailerons are driven so as to null a side acceleration signal. While this may seem contrary to the customary use of the rudder as the sole coordination control, it has been made to work as part of a multiloop automatic flight control system on aircraft with characteristics favorable to this type of coordination device.

*Feedback of Yawing Velocity to the Ailerons*

An $r \to \delta_a$ feedback system corresponds to the alteration of the airplane stability derivative $L'_r$. This can be an effective means of stabilizing the spiral mode by making $L'_{r_{\text{aug}}}$ sufficiently negative ($K_r$ positive) that

$$L'_\beta N'_{r_{\text{aug}}} - L'_{r_{\text{aug}}} N'_\beta > 0 \tag{8-26}$$

If the aileron yaw is favorable ($N'_{\delta_a}$ positive), then this feedback will also tend to increase the magnitude of $N'_{r_{\text{aug}}}$. This will make the left side of the inequality still larger and will incidentally improve the dutch roll damping. On the other hand, for the more common adverse yaw ($N'_{\delta_a}$ negative) the $r \to \delta_a$ system gain cannot be made very large without making the modified dutch roll oscillation unstable.

There have been some proposed automatic pilot configurations that used the properties $r \to \delta_a$ feedback to advantage. These are especially effective when a tilted rate gyro can be satisfactorily employed. This will sense a combination of rolling and yawing velocity; when this signal is used to actuate the ailerons, the damping in roll and spiral stability are simultaneously improved. For more detailed studies of this type of system, the airplane angular velocity to aileron transfer function, $\omega/\delta_a$, should be used. In fact, this is generally necessary in practice, for the angular velocity is sensed by a rate gyro fixed to the vehicle. Because the gyro input axis does not vary with flight condition, while the vehicle's stability axes do, the lateral angular velocity sensed will not always coincide with yawing velocity.

## 8-3. Feedback of Heading Angle to the Rudder

The earliest automatic pilots employed a feedback of heading angle to the rudder as a means of steering the airplane. For small perturbations about straight, level, and horizontal flight the appropriate transfer function for the study of this as an elementary feedback control system is

$$\frac{\psi(s)}{\delta_r(s)} = \frac{N^r_{\delta_r}}{s\Delta_{\text{lat}}} = \frac{A_r[s + (1/T_r)][s^2 + 2\zeta_r\omega_r s + \omega_r^2]}{s[s + (1/T_s)][s + (1/T_R)][s^2 + 2\zeta_d\omega_d s + \omega_d^2]} \tag{8-27}$$

Figure 8-7 presents a system survey of this system for a typical case. This survey indicates that the feedback of heading to rudder stabilizes the divergent spiral motion by forcing it to combine with the pole at the origin. At a comparatively high value of the feedback gain, the resulting low frequency oscillation is made stable, while at the same time the damping of the dutch roll mode is only slightly altered. Consequently, this type of system can be a very satisfactory directional stabilizer for aircraft with heavy dutch roll damping.

In the absence of actuator lags no danger of instability at high frequencies exists, although the dutch roll undamped natural frequency becomes much greater as the feedback gain is increased. However, for the typical location shown for the complex zeros of the transfer function, the dominant closed-loop heading mode ($\zeta_s'$, $\omega_s'$) can never become very satisfactorily damped because the $\zeta_r$, $\omega_r$ zeros are themselves rather close to the imaginary axis. In fact, it is not at all uncommon for these zeros to be in the right half-plane during some phase of flight. (This will be treated in more detail in the discussion of the $r \rightarrow \delta_r$ system given below.) To avoid the light or negative closed-loop damping accompanying a high gain closure, the gain is generally made fairly low. The closed-loop bandwidth of the system is, therefore, also bound to remain low. Some improvement in these deficiencies can be obtained by a combination of bank angle and heading angle feedback to the rudder, and a tilted directional gyroscope can provide a suitable signal. The first British automatic pilot was mechanized in this way.[1]

## 8-4. Feedback of Yawing Velocity to the Rudder

In a number of modern airplane automatic pilots the rudder axis, instead of being controlled by a direction-sensing instrument, is actuated by a rate gyro that senses yawing velocity. This is particularly valuable in those aircraft that need dutch roll damping augmentation. Assuming that the yawing velocity sensor's input axis coincides approximately with the aircraft's stability axes,[2] the transfer function relating yaw rate to

[1] F. W. Meredith and P. A. Cooke, "Aeroplane Stability and the Automatic Pilot," J. Roy. Aeron. Soc., 41, No. 318, 415–436 (June 1937).

[2] In practice the $\omega/\delta_r$ transfer function should be used to take into account sensor tilt angles that invariably exist. Alternatively, the transformation between instrument axes and body-fixed axes can be treated as a multiloop problem in which both $r$ and $p$ are fed back.

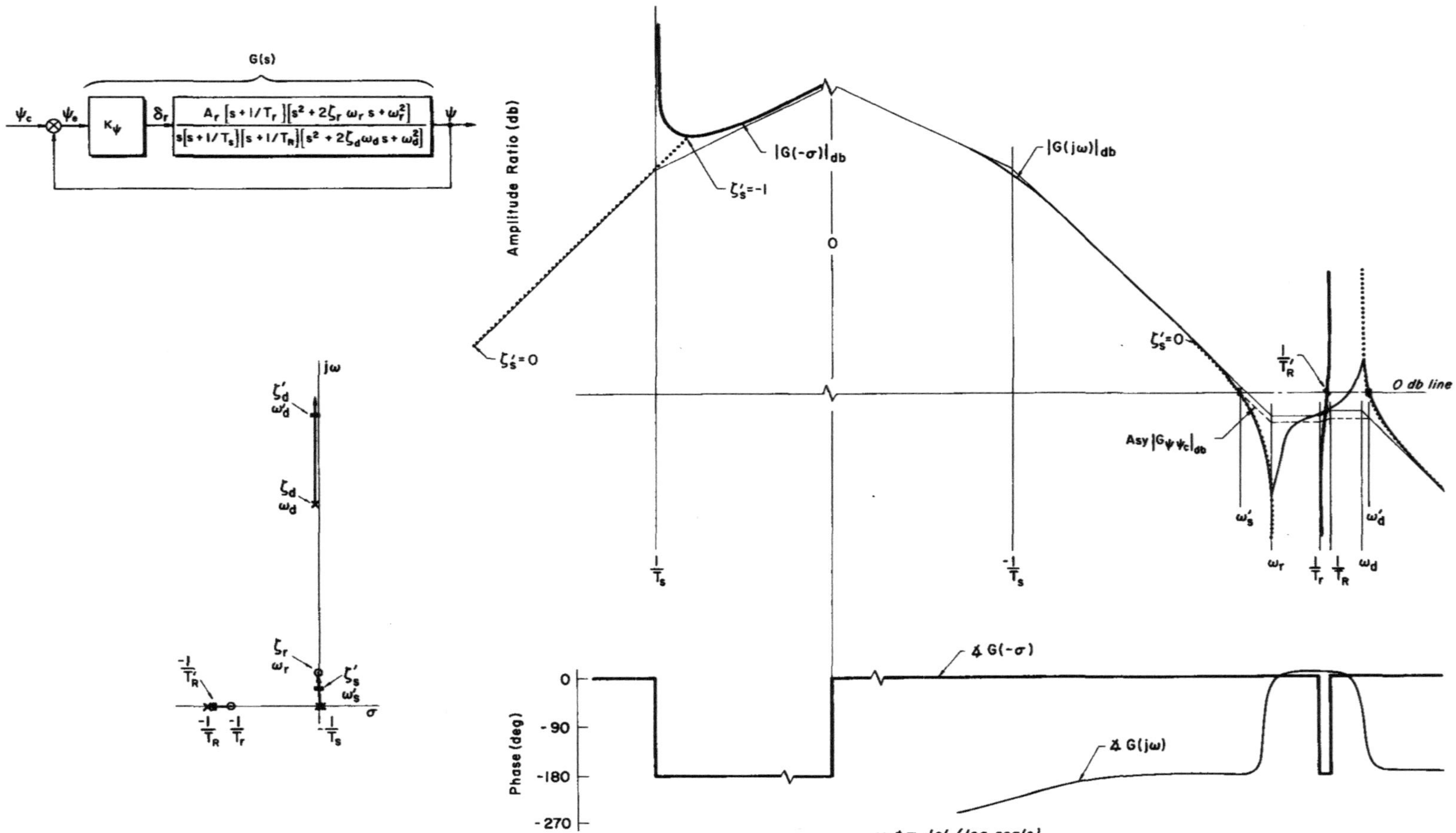


*a) Bode $G(j\omega)$, $G(-\sigma)$ Diagrams and Bode Root Locus*

Fig. 8-7. System survey of yaw attitude ($\psi \rightarrow \delta_r$) control system.

rudder has the same numerator and denominator factors as that given in Eq. 8-27 relating heading to rudder deflection, except for the free $s$, i.e.,

$$\frac{r(s)}{\delta_r(s)} = \frac{N^r_{\delta_r}}{\Delta_{\text{lat}}} = \frac{A_r[s + (1/T_r)][s^2 + 2\zeta_r\omega_r s + \omega_r^2]}{[s + (1/T_s)][s + (1/T_R)][s^2 + 2\zeta_d\omega_d s + \omega_d^2]} \tag{8-28}$$

*Spiral and Dutch Roll Stabilization with* $\omega_r/\omega_d \ll 1$

Figure 8-8 presents the system analysis for an $r \rightarrow \delta_r$ feedback around the transfer function of Eq. 8-28. The aircraft characteristics are typical for a high aspect ratio, straight wing aircraft at cruising speed and altitude.

The particular dynamics of this airplane and flight condition illustrate the powerful effect of yaw rate feedback on the lateral motions of the airplane. At a very low value of the feedback gain the spiral divergence is made stable, and at higher gains the very weakly damped dutch roll motion is made heavily damped. In fact, as the figure shows, in this idealized system it is not difficult to provide nearly critical damping for the closed-loop dutch roll roots. If $1/T_r$ and $1/T_R$ were somewhat closer together, as commonly occurs, all the roots could be made negative and real.

*Washed-Out Yawing Velocity to the Rudder*

In turning flight a straight-through feedback opposes the turn, requiring an increment of rudder or aileron into the turn to compensate for the action of the feedback system. Any necessity for supplying this increment of control has been found to be unacceptable on highly maneuverable craft. A *washout* is therefore usually installed in the feedback loop. This device has the property of having no output at d-c, so that the feedback no longer opposes a steady turn. (In some so-called single-axis automatic pilots that consist of a yaw rate feedback to the rudder, it is precisely the fact that they *do* oppose the almost steady turn of a developing spiral instability which is considered valuable. These devices are designed, therefore, without the washout feature.)

The transfer function of a washout has the form

$$\frac{e_o(s)}{e_i(s)} = \frac{s}{s + (1/T_{wo})} \tag{8-29}$$

The step function response of a washout is shown in Fig. 8-9. The device gets its name from the fact that the step function response decays or is washed out.

A system survey of the washed-out yaw rate to rudder system is shown in Fig. 8-10. To minimize the opposition of steady turns, it is desirable to choose the inverse washout time constant, $1/T_{wo}$, as large as can be without detracting too much from the dutch roll damping. The larger the

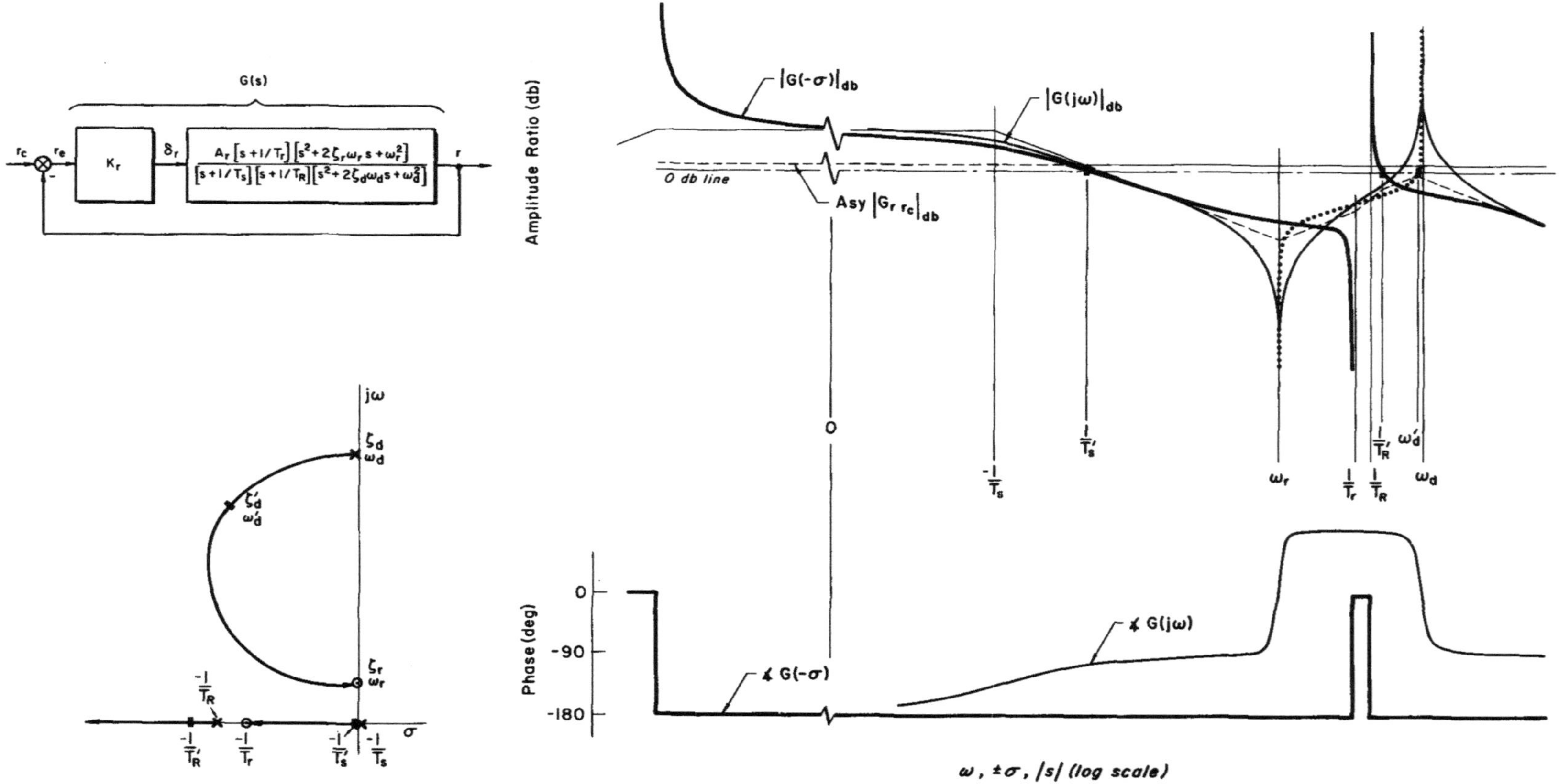


*b) Root Locus*

*a) Bode G(jω), G(−σ) Diagrams and Bode Root Locus*

Fig. 8-8. System survey of yaw rate ($r \rightarrow \delta_r$) control system.

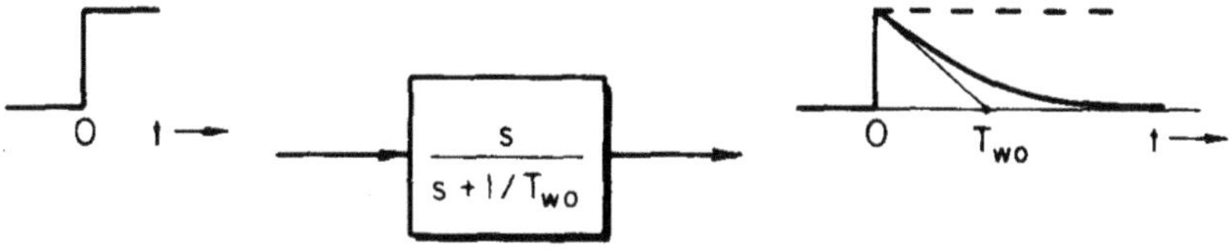


Fig. 8-9. Step function response of a washout.

inverse washout time constant becomes, the smaller the maximum obtainable value of the damping ratio of the modified dutch roll mode. This fact is illustrated by the sensitivity vectors on the root loci of the system survey shown in Fig. 8-11 which also serves as the basis of discussion in the next article.

*Spiral and Dutch Roll Stabilization with Large $\omega_r/\omega_d$*

A less favorable configuration of the open-loop poles and zeros than those illustrated in Figs. 8-8 and 8-10 is easily possible. This is illustrated in Fig. 8-11 which is drawn for the dynamics of a jet interceptor operating at high lift coefficient. In this case the quadratic zeros ($\zeta_r$, $\omega_r$) are much closer to the dutch roll poles and the feedback of yaw rate to the rudder is of strictly limited effectiveness in damping the dutch roll. Further, the sensitivity vector $S_{\omega_r}$ clearly indicates that still closer $\omega_r$, $\omega_d$ spacings will be even more undesirable.

Although the $N^r_{\delta_r}$ numerator is another maverick, it is instructive to consider two limiting cases:

dihedral effect dominant, $|(g/U_0)L'_\beta| \gg |L'^3_p|$,

$$\frac{1}{T_r} \doteq \omega_r \doteq \left(-\frac{g}{U_0}L'_\beta\right)^{1/3}$$
$$2\zeta_r\omega_r \doteq -(Y_v + L'_p) - \left(-\frac{g}{U_0}L'_\beta\right)^{1/3} \tag{8-30}$$

roll damping dominant, $|(gL'_\beta/U_0L'^3_p) + (Y\delta_r N'_\beta/U_0N'_{\delta_r}L'_p)| \ll 1$

$$\frac{1}{T_r} \doteq -L'_p$$
$$\omega_r^2 \doteq \frac{g}{U_0}\frac{L'_\beta}{L'_p} \tag{8-31}$$
$$2\zeta_r\omega_r \doteq -\left(Y_v - \frac{Y_{\delta_r}N'_\beta}{U_0N'_{\delta_r}} - \frac{g}{U_0}\frac{L'_\beta}{L'^2_p}\right)$$

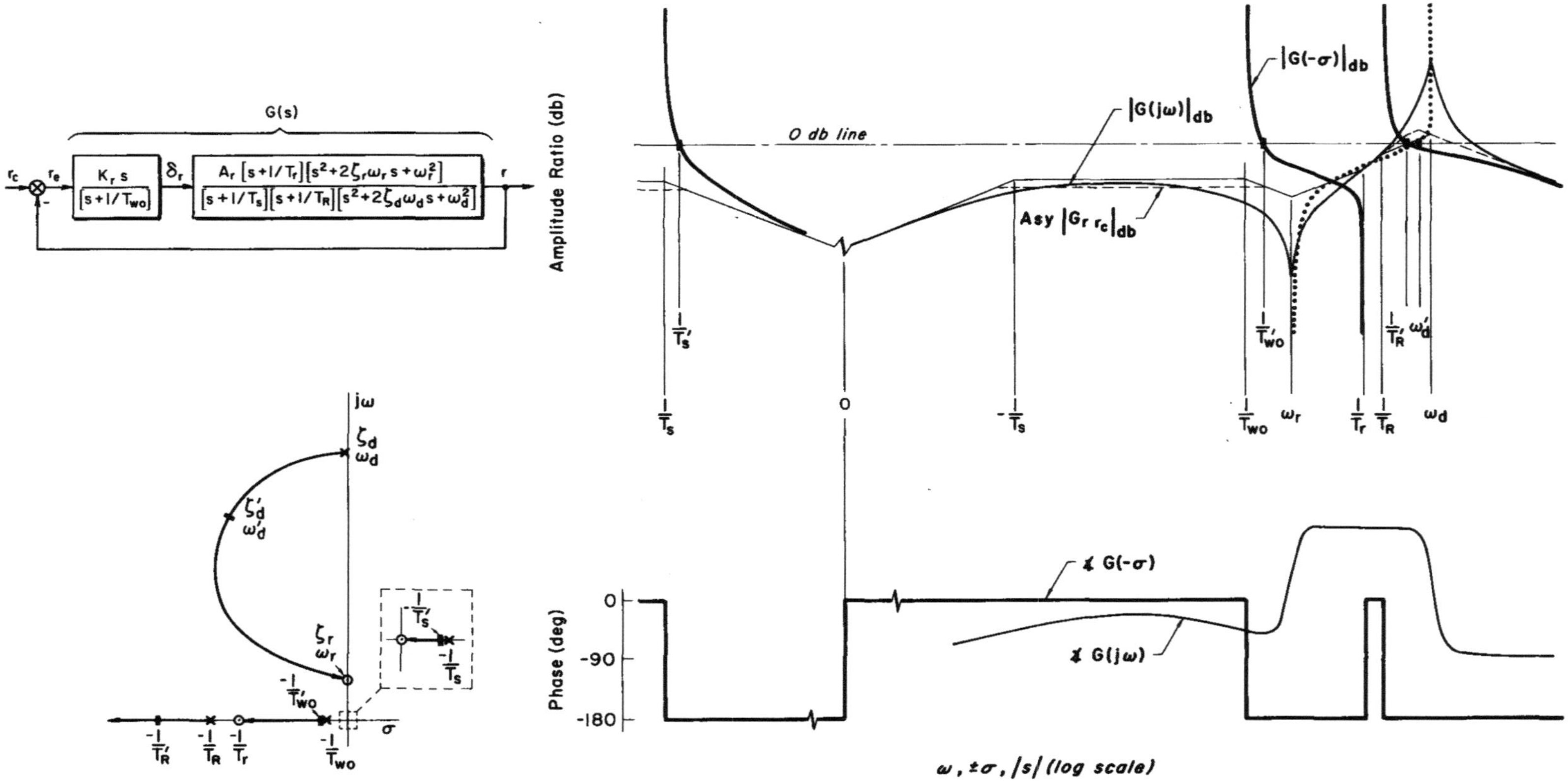


*b) Root Locus*

*a) Bode $G(j\omega)$, $G(-\sigma)$ Diagrams and Bode Root Locus*

Fig. 8-10. System survey of "washed-out" yaw rate ($r \rightarrow \delta_r$) control system for aircraft with small $\omega_r/\omega_d$.

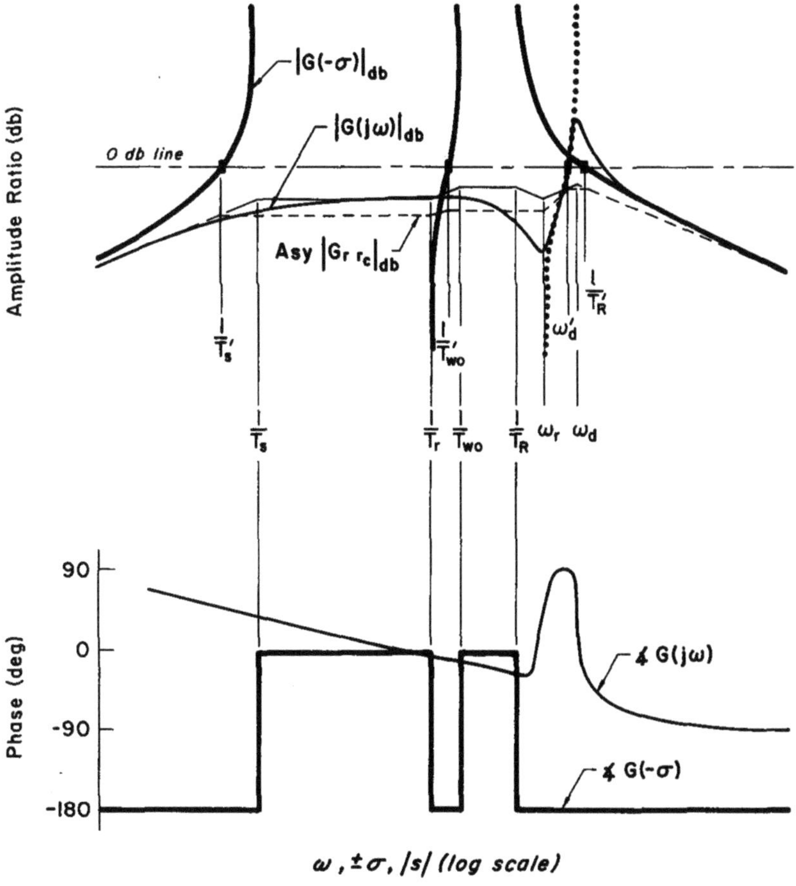


*a) Bode G(jω), G(-σ) Diagrams and Bode Root Locus*

Fig. 8-11. System survey of "washed-out" yaw rate ($r \rightarrow \delta_r$) control system for aircraft with large $\omega_r/\omega_d$.

If we presume that $L'_\beta$ is negative, $2\zeta_r\omega_r$ can easily become negative for either case. Thus the complex zeros are not only typically lightly damped but are likely to be disadvantageously located in the right half-plane. This can happen for all the systems illustrated in Figs. 8-8, 8-10, and 8-11.

The difficulties in improving the dutch roll damping occasioned by $\omega_r$ approaching $\omega_d$ can be alleviated by increasing $L'_p$. This is obvious in Eq. 8-31 but not at all apparent for Eq. 8-30. In the latter equation a sufficient increase in $L_p$ causes the inequality $|(g/U_0)L'_\beta| \gg |L'^3_p|$ to be violated, so that the approximate factors no longer apply and are replaced by the second set. The actual decrease of $\omega_r/\omega_d$ is ordinarily accomplished with the aid of a roller damper.

*Nature of Gain Adjustment for Dutch Roll Damping Augmentation*

Just as with the other systems considered, the nature of appropriate gain adjustments with flight environmental parameters for the yaw damper depends on the desired closed-loop characteristics. If dutch roll damping is to be constant, then

$$K_r N'_{\delta_r} \doteq -N'_{r_{\text{aug}}} \doteq \text{const} \tag{8-32}$$

or

$$K_r \propto \frac{1}{N'_{\delta_r}} \propto \frac{1}{\rho U_0^2} \tag{8-33}$$

On the other hand, if the closed-loop dutch roll damping ratio is to be nearly invariant, then the desired relationship will be

$$\zeta'_d \doteq \frac{-N'_{r_{\text{aug}}}}{2\sqrt{N'_\beta}} \doteq \frac{K_r N'_{\delta_r}}{2\sqrt{N'_\beta}} \doteq \text{const} \tag{8-34}$$

This implies that the variation of gain with flight parameters should be

$$K_r \propto \frac{\sqrt{N'_\beta}}{N'_{\delta_r}} \propto \frac{1}{\rho^{1/2} U_0} \tag{8-35}$$

Again, as with all the gain compensation considerations discussed in this and the previous chapter, the actual nature of the compensation will depend on the specifics of vehicle dynamics, closed-loop dynamic performance envelope specifications, functional mechanization of the gain-changing devices, etc.

## 8-5. Feedback of Sideslip to the Rudder

The feedback of sideslip angle or sideslipping velocity to the rudder is practically equivalent to the augmentation of the directional stability derivative $N'_\beta$ (or $N'_v$). It serves in a very useful way to minimize the sideslip angle (uncoordination).

The transfer function that relates sideslip angle to rudder deflection has the form

$$\frac{\beta(s)}{\delta_r(s)} = \frac{N^v_{\delta_r}}{U_0\Delta_{\text{lat}}} = \frac{A_\beta[s + (1/T_{\beta_1})][s + (1/T_{\beta_2})][s + (1/T_{\beta_3})]}{[s + (1/T_s)][s + (1/T_R)][s^2 + 2\zeta_d\omega_d s + \omega_d^2]} \tag{8-36}$$

Figure 8-12 shows the system survey for negative feedback of the $\beta \rightarrow \delta_r$ system. This has the effect of destabilizing the aperiodic spiral motion and of moving the dutch roll roots to a higher frequency. This, of course, is the expected result of an increase in directional stability.

To improve the dutch roll damping, a $\dot{\beta}$ component can be added to the controller, i.e.,

$$\begin{aligned}\delta_r &= -(K_\beta\beta + K_{\dot{\beta}}s\beta)\\ &= -K_{\dot{\beta}}\left(s + \frac{1}{T_E}\right)\beta\end{aligned} \tag{8-37}$$

A survey of this system is given in Fig. 8-13. As indicated, this type of system has excellent characteristics as a single-loop control for the improvement of the dutch roll and the minimization of sideslip. Appropriate gain adjustments with flight environmental parameters are not extreme, being similar to those for $\alpha \rightarrow \delta_e$ systems. The system is also an excellent inner loop for roll attitude and other outer loop controllers. Its primary deficiency is a practical one in instrumenting an adequate sideslip sensor. This is, to some extent, alleviated by the possibility of a lateral acceleration system.

### 8-6. Feedback of Lateral Acceleration to the Rudder

Some of the favorable features of $\beta \rightarrow \delta_r$ or $\beta, \dot{\beta} \rightarrow \delta_r$ systems can be obtained by substituting a properly located lateral accelerometer for the sideslip sensor. The similarity between the two systems can be seen conceptually from the side acceleration equation and the expression for side acceleration at a general location,

$$a_{y_{\text{c.g.}}} \doteq Y_v v + Y_{\delta_r}\delta_r \tag{8-38}$$

$$\begin{aligned}a_y &= a_{y_{\text{c.g.}}} + x_a\dot{r}\\ &\doteq Y_v v + Y_{\delta_r}\delta_r + x_a\dot{r}\end{aligned} \tag{8-39}$$

These equations are the lateral analogs to the longitudinal $a_z$ discussed at length in Chapter 7, and the same kind of center of rotation and center of percussion relationships apply. In this case the center of rotation corresponds to that accelerometer location distance, $x_a$, for which the $x_a\dot{r}$ component will tend to offset the $Y_{\delta_r}\delta_r$ component, thereby leaving

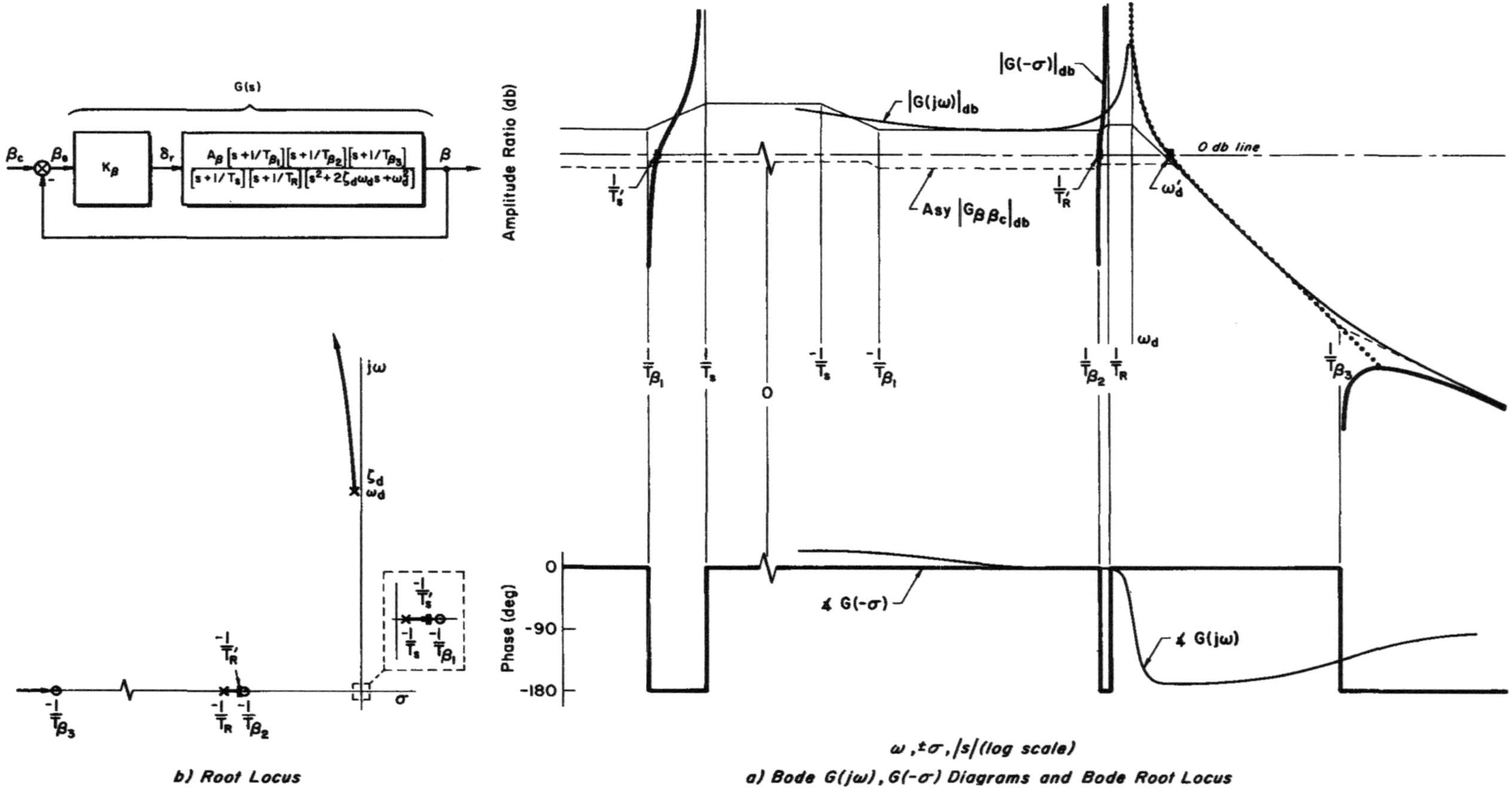


Fig. 8-12. System survey of sideslip attitude ($\beta \rightarrow \delta_r$) control system.

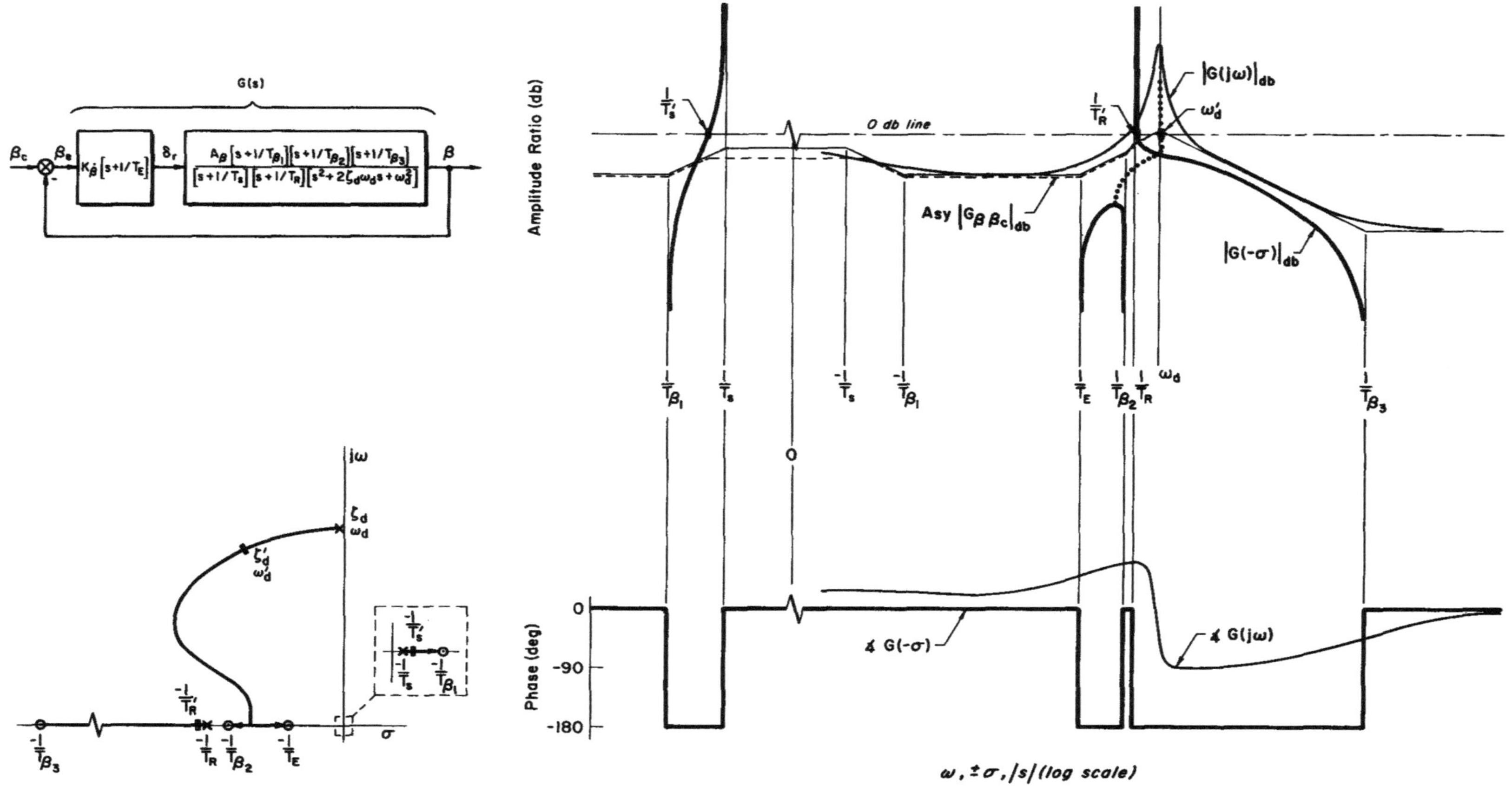


*b) Root Locus*

*a) Bode $G(j\omega)$, $G(-\sigma)$ Diagrams and Bode Root Locus*

Fig. 8-13. System survey of sideslip attitude and rate $(\beta, \dot\beta \rightarrow \delta_r)$ control system.

$Y_v v$ (or $Y_\beta \beta$) as the major part of the acceleration measured. The relationships can be developed in a fashion similar to the longitudinal case or, alternatively, directly by finding the instantaneous center of rotation. To do this, consider the initial yawing acceleration due to a rudder step input, $\delta_{r_0}/s$. This will be

$$\begin{aligned}\dot{r}(t)_{t\to 0} &= \lim_{s\to\infty} \left\{ s\left[\frac{r(s)}{\delta_r(s)}\right]\frac{\delta_{r_0}}{s}\right\} \\ &= N'_{\delta_r}\delta_{r_0}\end{aligned} \tag{8-40}$$

Consequently, if

$$Y_{\delta_r}\delta_{r_0} + x_a N'_{\delta_r}\delta_{r_0} = 0 \tag{8-41}$$

or

$$x_a = \frac{-Y_{\delta_r}}{N_{\delta_r}}$$

the initial part of the lateral acceleration response to a step rudder input will be proportional to the sideslip response to the rudder deflection.

In terms of nondimensional derivatives the ideal $x_a$ becomes

$$\begin{aligned}x_a &\doteq \frac{-I_z}{mb}\frac{C_{y_{\delta_r}}}{C_{n_{\delta_r}}} \\ &\doteq \frac{-k_z^2 C_{y_{\delta_r}}}{b[(-l_{\delta_r}/b)]C_{y_{\delta_r}}} = \frac{k_z^2}{l_{\delta_r}}\end{aligned} \tag{8-42}$$

The lateral acceleration numerator is given by

$$\begin{aligned}N_{\delta_r}^{a_y} = {} & (Y_{\delta_r} + x_a N'_{\delta_r})s^4 + [-Y_{\delta_r}(L'_p + N'_r) \\ & + x_a(Y_{\delta_r}N'_v - N'_{\delta_r}Y_v + L'_{\delta_r}N'_p - N'_{\delta_r}L'_p)]s^3 \\ & + [Y_{\delta_r}(L'_pN'_r - L'_rN'_p + N'_\beta) - N'_{\delta_r}Y_\beta \\ & + x_a(Y_{\delta_r}L'_vN'_p - Y_{\delta_r}L'_pN'_v - L'_{\delta_r}Y_vN'_p + N'_{\delta_r}Y_vL'_p)]s^2 \\ & + [Y_{\delta_r}(L'_\beta N'_p - L'_pN'_\beta - gL'_v) + N'_{\delta_r}Y_\beta L'_p - L'_{\delta_r}Y_\beta N'_p \\ & + gL'_{\delta_r}Y_v + x_a g(L'_{\delta_r}N'_v - N'_{\delta_r}L'_v)]s \\ & + g[Y_{\delta_r}(L'_vN'_r - L'_rN'_v) + N'_{\delta_r}Y_vL'_r - L'_{\delta_r}Y_vN'_r]\end{aligned} \tag{8-43}$$

From this it is seen that placing the accelerometer at the instantaneous center of rotation will make the leading coefficient zero, just as in the analogous longitudinal case. For the accelerometer at the instantaneous center of rotation, the numerator then becomes

$$N_{\delta_r}^{a_y} \doteq Y_{\delta_r}\left(-N_r' + Y_v - \frac{Y_{\delta_r}}{N_{\delta_r}'}N_v'\right)\left(s + \frac{1}{T_{a_{y_1}}}\right)\left(s + \frac{1}{T_{a_{y_2}}}\right)\left(s + \frac{1}{T_{a_{y_3}}}\right) \tag{8-44}$$

The approximate factors for this numerator are (Table 6-4)

$$\frac{1}{T_{a_{y_1}}} \doteq \frac{g}{U_0}\left[\frac{L_r'}{L_p'} - \frac{N_r'}{L_p'}\frac{(Y_vL_{\delta_r}' - Y_{\delta_r}L_v')}{(Y_vN_{\delta_r}' - Y_{\delta_r}N_v')}\right]$$

$$\frac{1}{T_{a_{y_2}}} \doteq -L_p'; \qquad \frac{1}{T_{a_{y_3}}} \doteq \frac{-U_0(Y_vN_{\delta_r}' - Y_{\delta_r}N_v')}{Y_{\delta_r}[-N_r' + Y_v - (Y_{\delta_r}/N_{\delta_r}')N_v']} \tag{8-45}$$

The sign of $1/T_{a_{y_3}}$ will generally depend only on the location of the rudder with respect to the center of gravity, and will be negative for aft and positive for forward locations. The breakpoint corresponding to $1/T_{a_{y_3}}$ provides a convenient upper limit of approximate correspondence between $a_y$ and $\beta$. At frequencies below this the two are almost directly proportional, i.e.,

$$\frac{a_y}{\delta_r} \doteq Y_\beta\frac{\beta}{\delta_r}, \qquad |s| < \frac{1}{T_{a_{y_3}}} \tag{8-46}$$

Above this frequency the rudder deflection will destroy the correspondence between side acceleration and sideslip angle or sideslipping velocity.

Figures 8-14 and 8-15 show system surveys of $a_y \to \delta_r$ and $a_y$, $\dot{a}_y \to \delta_r$ control systems, respectively, for a situation in which the accelerometer is near, but behind, the center of rotation. As expected, the dynamics of these systems are quite similar to those of the corresponding sideslip controllers illustrated in Figs. 8-12 and 8-13. A major difference, however, between these two general types of systems is not revealed explicitly in the system surveys. This is the matter of necessary gain variation to keep the closed-loop system dynamics within specified bounds. Again, the differences between the lateral $a_y$ and $\beta$ systems are similar to those between the longitudinal $a_z$ and $\alpha$ controllers. In complete analogy to the latter systems, the controller gain, $K_y$, must vary as $1/\rho U_0^2$ if it is to simulate a constant-gain sideslip controller.

Although rudder control systems using $a_y$, $\dot{a}_y \to \delta_r$ are not as common as washed-out yaw dampers, they have been successful in past applications. This type of system is particularly advantageous for gun-firing aircraft which require two-control (aileron and elevator) lead/pursuit maneuvers without sideslip, and rocket-firing craft for which sideslip minimization is desirable to simplify the fire control equipment design.

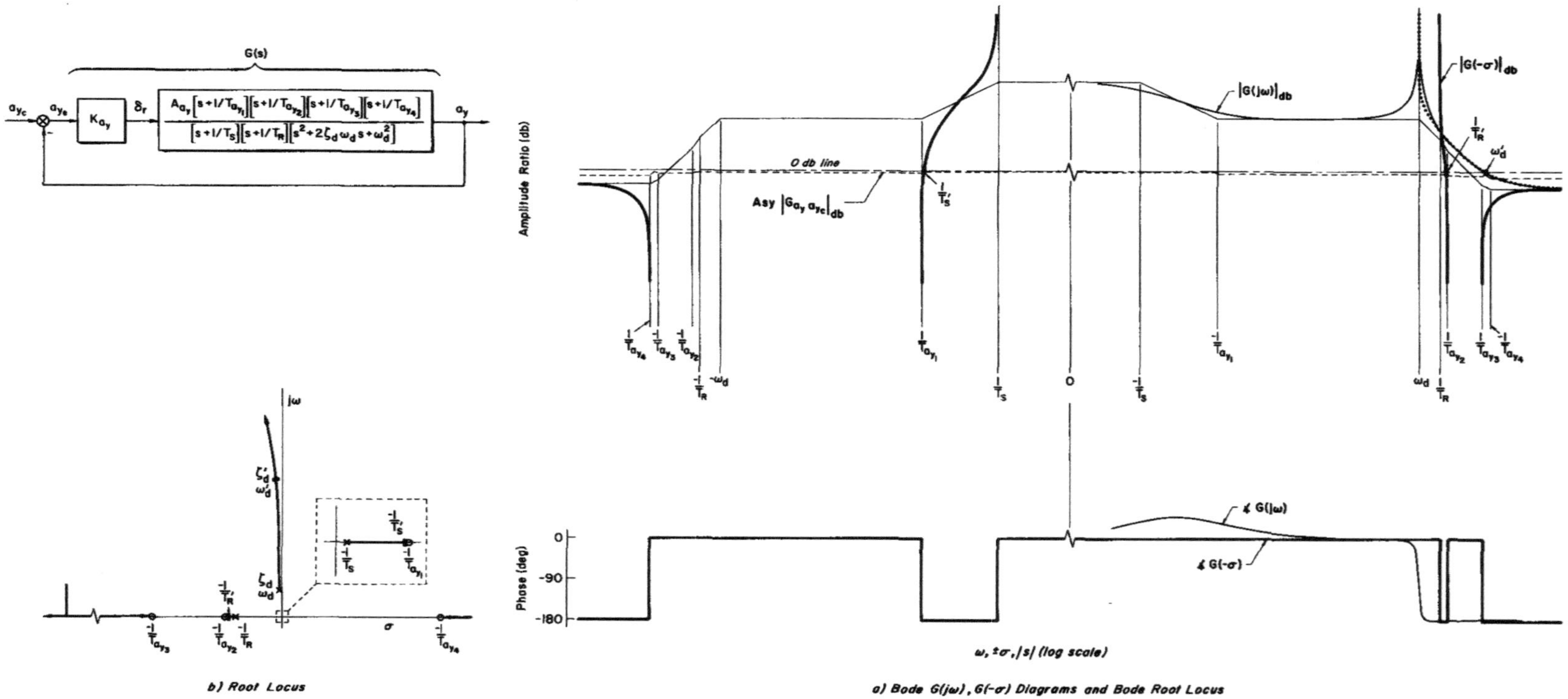


Fig. 8-14. System survey of side acceleration ($a_y \rightarrow \delta_r$) control system.

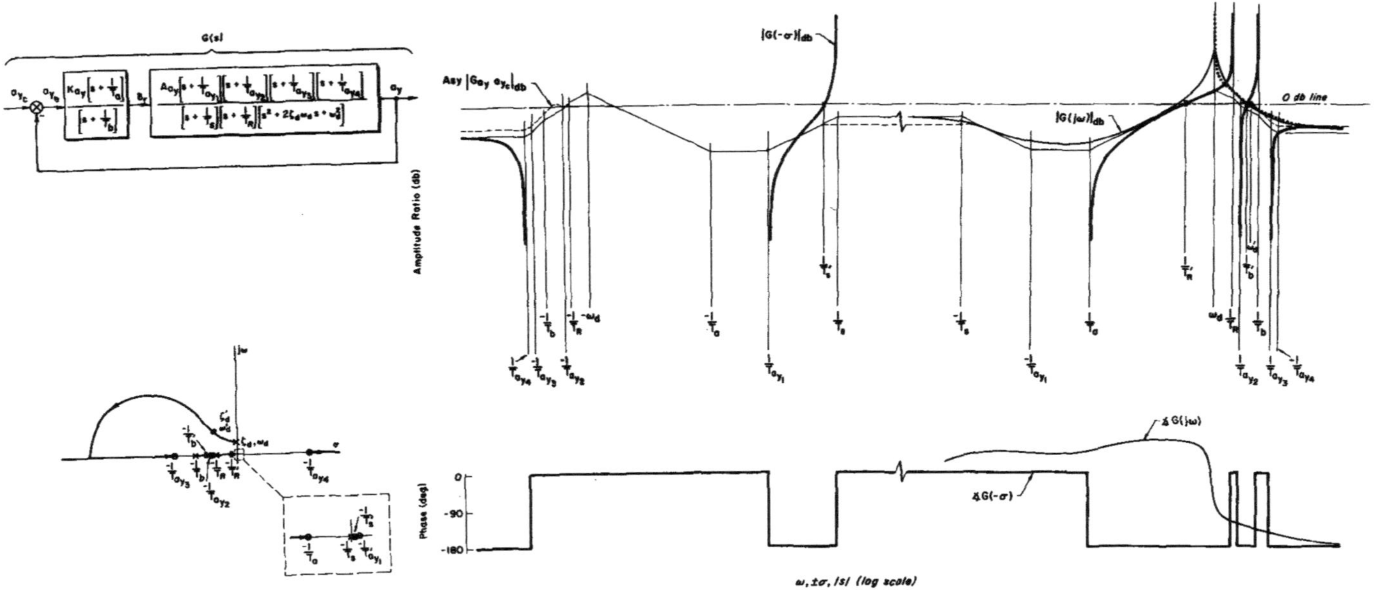


b) Root Locus

a) Bode $G(j\omega)$, $G(-\sigma)$ Diagrams and Bode Root Locus

Fig. 8-15. System survey of side acceleration ($a_y \rightarrow \delta_r$) control system with lead/lag equalization.

It also shares the excellent properties of the $\beta$, $\dot{\beta} \rightarrow \delta_r$ system as an inner loop in more complex multiloop systems.

Early models of the B-52 bomber also depended on a side acceleration feedback to the rudder to damp the dutch roll, but the principle of operation of the damper was completely different from the one discussed above. In the all-mechanical B-52 damper the side acceleration was sensed in the fin (where it consisted almost entirely of an $x_a\dot{r}$ signal) and was pseudointegrated (lagged) so as to produce a signal roughly proportional to yawing velocity at the dutch roll frequency. This latter signal was then employed to deflect the rudder.

CHAPTER 9

# *REQUIREMENTS, SPECIFICATIONS, AND TESTING*

## 9-1. Introduction: The System Design Process

In Chapter 1 there was occasion to suggest the importance of feedback to the solution of the problem of control and guidance of aeronautical vehicles. In particular, its roles in making the vehicle amenable to following guidance commands and in suppressing the effects of disturbances were emphasized. Since then the text has exposed the mathematical and physical principles necessary to the solution of the deterministic *analysis* problem, i.e., given the (deterministic) input and the mathematical description of the system, find the outputs or errors. This, however, by no means represents a solution to the design or *synthesis* problem, i.e., given the inputs and the desired outputs or tolerable errors, find the system.

The final aim of system design is to integrate components into a functional system that performs its assigned tasks "satisfactorily." The design process leading to this end can be broken into several phases that are more or less chronological, yet are extensively interrelated and interconnected. A typical set of such phases might include the following.

1. Establishment of System Purpose and Overall System Requirements. At the design stage, system purpose can be equated with the mission phase or task definitions. Requirements are partially derivable from the functions needed to be performed to accomplish these mission phases (operational requirements), and less directly from the characteristics of the interconnected components and the environment in which they operate (implied requirements).

2. Determination of Unalterable Element, Command, and Disturbance Environment Characteristics. Typically, the characteristics of some component parts of the system are not easily changed by choice of the system designer. In aeronautical control and guidance such relatively "unalterable" elements often include the vehicle itself and possibly the control surface actuators and some of the motion quantity sensors. The "structure" of the commands and disturbances is even less subject to the choice

of the designer but is instead a direct consequence of the mission or task and the environment (which may, however, be open to some interpretation).

While the controlled element is thus usually considered to be unalterable for the duration of a design calculation, this viewpoint should not be allowed to prohibit vehicle modifications that may improve the system. Similarly, sensing and actuating elements are at least "quasi-alterable" in practice because they can be changed by selection of a different item of the same general class. Nevertheless, for the delineation of competing system possibilities (next item), such elements are generally considered as having fixed characteristics.

3. Evolution of Competing Feasible Systems (Determination of Basic Functional Block Diagrams). Usually, the requirements can be met in more than one way, e.g., with different sensed motion quantities and equalization elements which are completely alterable within the limits of physical realizability and practicality. Then it is possible to evolve competing systems that become candidates for selection on the basis of certain desirable properties.

4. Competing System Assessment, System Selection. The competing systems can be compared on a very large number of bases which can be divided into two categories: design *quantities* and design *qualities*. Design quantities include the dynamic performance (relative stability, accuracy, speed of response or bandwidth, etc.) and the physical characteristics (weight, volume, power or energy consumption, etc.). Design qualities include safety, operational capability, reliability, maintainability, cost, etc. An optimum system is one that has some "best" combination of all these features.

5. Detailed Study of the Selected System. Once a best system has been selected, it is still necessary to validate it for all nominal and abnormal operating conditions. The components that do not yet exist as hardware must be specified, designed, fabricated and tested as components. As many of these as possible should be assembled in a series of system simulations which culminate in flight tests of the complete system in its actual operating environment. At each stage of the testing process the assumptions that were made in previous phases of the design should be checked for validity. If actual conditions violate the assumptions, a new iteration of the design should be begun at the point at which the incorrect assumption was made.

The above steps in the orderly process of evolving, or synthesizing, a system that satisfactorily meets all its objectives are governed initially by a set of *functional requirements* stemming (see Item 1 above) from operational needs and system integration implications. These latter, implied, requirements depend on the final system selected (see Item 4) which is then completely specified functionally by the total set of (functional) requirements. The converting of these requirements on each element of the system into usable hardware generally requires a specification that includes not only the functional requirements but also applicable specifications relating to hardware design and fabrication practice, including extreme operating environmental factors.

In the sections that follow, we present the concepts on which the requirements for automatic flight control systems are based. This process begins with a definition of the mission, and becomes increasingly intricate and detailed as the flight control system is examined in its more intimate details. Accordingly, after the discussion of mission-centered and operational requirements, we turn to a consideration of the more obscure requirements implied by component or system design specifics. Since flight control systems are largely feedback controllers, many of the implied requirements and comparisons are exposed, in the next article, by considering generic feedback system properties. A final article pertinent to the establishment of functional requirements establishes the bases for compromise in selecting system bandwidth. After covering the first or requirements phase in the above-described manner, the chapter concludes with a brief section on specification, simulation, and flight testing as the methods by which it is ultimately demonstrated that the automatic flight control system satisfies the requirements.

Some important aspects of the design process are not discussed in this chapter. For instance, the characteristics of the unalterable element, the aircraft itself, have been treated in Chapters 4–6, and the subject of analytic descriptions of commands and disturbances, as well as an introduction to system dynamic performance assessment, is deferred to Chapter 10. This scrambling of the order of presentation is, in part, a reflection of the interrelationships between the subjects, and is used partly to permit an orderly development of the mathematical background for the several topics.

## 9-2. Mission Phases and Operational Requirements

In any aeronautical system design the requirements for subsystems evolve in a pyramidal fashion, and become more numerous and detailed as definition of the actual equipment is approached. The apex of the pyramid is the mission purpose and definition (see Fig. 9-1). Immediately below this central point are three blocks involving considerations that interact strongly in the earliest preliminary design stages: mission phases, vehicle operating point profile, and guidance possibilities. When the mission is realistic, one or more feasible vehicle operating-point profiles are joined with one or more guidance possibilities to enable the overall system to perform through the constituent phases of the mission.

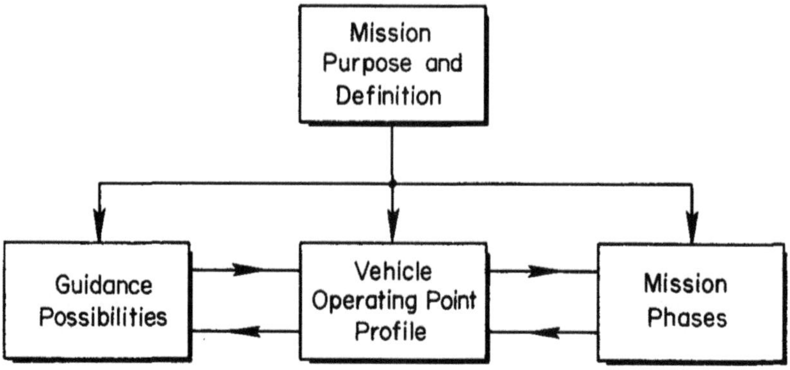


Fig. 9-1. First step in flight control requirements evolution.

Vehicle operating-point profiles and guidance possibilities can ordinarily be expressed in concrete numerical terms. The mission phase structure, however, is basically open to choices. For flight control design purposes, the mission phase categories should be selected so that the quantities required to define flight control activities are determined once the phase is identified. A mission phase structure based on maneuvers is ideal for this purpose. In Table 9-1, for example, a number of aerospace

Table 9-1. V/STOL aircraft mission-mission phase matrix.

| Mission phase \ Mission | Transport | Rescue | ASW | Close support (attack) | Interdiction (strike) |
|---|---|---|---|---|---|
| Short field or deck takeoff or launch | x | x | x | x | x |
| Ferry or cruise | x | x | x | x | x |
| Approach, transition, hover | * | * | * | * | * |
| Land at forward site | * | * | | * | * |
| Takeoff, transition from hover | | x | x | | x |
| Takeoff from forward site | x | x | | x | x |
| Loiter | (In traffic) | x | x | x | |
| Low altitude penetration | | | | | x |
| Weapon delivery | | | x | x | x |
| Approach transition and land at short field or recover on deck | * | * | * | * | * |

Note: x denotes applicable nonlanding segment; * denotes approach transition, hover, and landing segments.

missions pertinent to V/STOL aircraft are broken down into mission phases that indicate the sequences required for their accomplishment.

At this point the flight control system commands and the basic flight control references are, in principle, determined, since these must be such as to allow the vehicle to accomplish each mission phase. This usually amounts to the specification of at least the outer loops required, as is illustrated in Table 9-2. Two other elements of the flight control problem also enter at this stage. These are those forcing functions and vehicle dynamic properties that depend on the environment (as defined by the various mission phases in a gross way) and on the operating-point profile. Such a dependence is illustrated in Fig. 9-2 and Table 9-2.

In Fig. 9-2 we see that decisions about guidance possibilities, operating-point profiles, and mission phases have defined the direct inputs to the flight control system (dotted) block by establishing the command structure and references, kinematics, forcing functions, and controlled element properties. All but the last of these constitute "mission-centered" requirements; and all result in given starting points for subsequent dynamic requirements evolution. These subsequent requirements, which arise from "component-centered" and "system interaction" considerations are *implied* by the flight control/guidance system structure and its elements. For example, considering Fig. 1-6, it is evident that dynamic requirements are imposed on the flight control system by virtue of its operation as an important subsystem of the overall system—specifically, that it operate properly during all phases of the mission. Based on this general requirement and the interactions and unalterable characteristics defined by the specific system block diagram, a series of possible specifications for desirable flight control system properties can be expressed in servo terms (e.g., as dominant features of Bode or root plots, time response properties, etc.). Recognition of this fact and its implications on the overall system provide the starting point for considering this particular system interaction.

## 9-3. An Approach to Implied Requirements for System Design

Unfortunately, any gross transfer function requirements derived from considerations of the mission-centered requirements and system interaction structure are not necessarily complete or even consistent with acceptable performance in other necessary phases of the mission, such as takeoff, response to nonguidance commands, etc. In many systems these other mission phases, in which the guidance loop is not tightly closed, impose stringent requirements on the flight control system over and above those required for proper operation of the overall flight control and guidance system. In such cases an appreciation of the interaction that

Table 9-2. Mission phase-forcing function structure.

| Mission phase | Guidance | | Disturbance | |
|---|---|---|---|---|
| | External | Internal | External | Internal |
| Short field or vertical deck takeoff or launch | Radio, ground radar, and data link | Airborne radar, inertial, doppler | Turbulence, wind shear, intrusions, traffic | |
| Ferry or cruise | Ground radar, loran or Decca, etc. | Airborne radar, inertial, dead reckoning | Turbulence | |
| Approach, transition, hover | Radio, ground radar, and data link | ↑ | ↑ | |
| Land at forward site | Radio, ground radar, and data link | ↑ | ↑ | Configuration change, inadvertent stores release, crew motions |
| Takeoff, transition from hover at forward site | Radio, ground radar, data link | Doppler, inertial | Turbulence, wind shear, intrusions, traffic | |
| Loiter | Radio, ground radar, data link, loran, or Decca, etc. | ↓ | ↓ | |
| Low altitude penetration | Radio, ground radar, and data link | Radar, doppler, inertial | Turbulence, wind shear, terrain, obstacles, intrusions | |
| Weapon delivery | Radio, ground radar, and data link | | ↓ | Stores release |
| Approach, transition, and land at short field or recover on deck | Radio, ground radar, data link, laser | Doppler, inertial | Turbulence, intrusions, traffic, obstacles | Configuration change, inadvertent stores release |

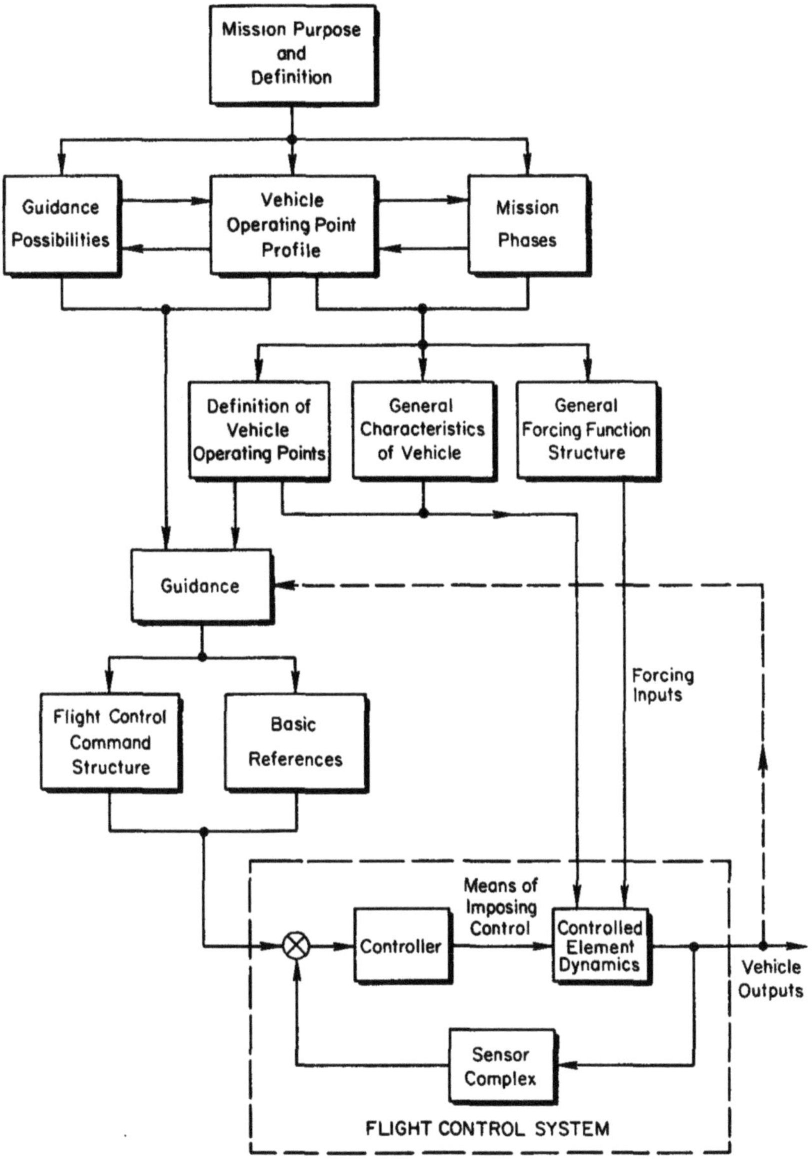


Fig. 9-2. Evolution of flight control requirements.

exists between the vehicle and the controller requires a detailed examination of the flight control loop itself. Specifically, one must consider the requirements imposed on the flight control equipment by various types of controlled element (vehicle) characteristics, flight control system command inputs, and external disturbances; and conversely, requirements implied on the controlled element by various possible flight control equipments.

This leads us to delay consideration of system interaction effects until the component-centered quantities, which involve the primary elements of the flight control loop, the controller and the vehicle, have been established. Definition and a thorough understanding of the general dynamic characteristics and operating features of the vehicle readily lead to identification of the physical quantities that might be controlled and to the means through which control can readily be imposed; these now provide a more solid basis for considering system interactions. The entire set of considerations, including the establishment of mission-centered requirements, discussed in the preceding article, is summarized in the annotated outline of Table 9-3. When all the outlined considerations, quantities and possibilities are thoroughly understood, a detailed set of functional requirements may be set up for the flight control system. For example, by assuming the vehicle characteristics to be unalterable and that desirable sensors fall into a relatively narrow class, the flight control system interactions, for a specific case, change from general to likely possibilities. The functional requirements may then be established on the basis of alternatives that are limited largely to those possible at the detailed mechanization level. This procedure is the one most commonly adopted in flight control system design. On the other hand, if the vehicle characteristics are considered subject to modification, system interactions between vehicle and controller again assume the status of possibilities, and competition between alternatives can proceed on a somewhat broader basis.

In either event, the process of evolving a *final* set of functional requirements is inevitably complicated and involves iterative consideration of the factors and possibilities listed in Table 9-3. When properly performed it demands ruthless objectivity coupled with ingenuity and a flair for selecting among competing possibilities that *one system configuration* which is most favorably constituted to accomplish given purposes. In principle, a complete evolution of functional requirements is the result of an optimization process that, reduced to its logical constitutents, requires:

1. The establishment of an overall criterion that measures the degree to which systems accomplish their given purposes. The overall criterion depends upon a weighted intermix of subcriteria. Subcriteria are measures of those quantities contributing to system

Table 9-3. Steps and considerations involved in the evolution of flight control system requirements.

1. Mission-Centered Requirements (Mission Phase and Vehicle Operating-Point Profile)
   Establish:
   - a. Flight control system commands
   - b. Command modes and sequences
   - c. Basic flight control references

   Define:
   - a. Disturbance, or unwanted input, environment to be encountered throughout the operating-point profile
   - b. General characteristics of the controlled element
2. Component-Centered Quantities
   - a. Vehicle-centered quantities, including
     - Vehicle steady-state maneuvering characteristics
     - Vehicle steady-state asymmetrical performance conditions
     - Vehicle output quantities (vehicle motions)
     - Vehicle input quantities (means of imposing control)
     - Vehicle dynamic (transfer) characteristics:
       - Linear, constant-coefficient
       - Nonlinear and/or time-varying linear
   - b. Controller-centered quantities, including
     - Controller input possibilities (ability to sense vehicle output quantities)
     - Controller output possibilities (actuation of vehicle input quantities)
3. System Interaction Aspects
   - a. Overall Flight Control and Guidance System
     Interactions establish possible requirements imposed on the flight control system by operation within the overall system. Logically the guidance-control dichotomy has both static and dynamic interactions:
     - Operating Point Control: The fundamental role of guidance is to establish the desired vehicle operating-point profiles in some suitable physical form. Thus both the physical flight control command structure and basic references derive from guidance; the role of operating-point control is to translate guidance desires into vehicle mass center motions.
     - Limiting: Practical guidance equipment may impose limits upon allowable vehicle departures or rates thereof from operating-point conditions.
     - Dynamic Interactions: In homing systems, guidance and control are inseparable dynamically, so that proper dynamic operation of the total overall system imposes demands upon the transfer characteristics of the flight control system.
   - b. Flight Control System
     System interactions between vehicle and controller for various controller input and output possibilities establish bases for requirements selection and/or evolution. Interactions must consider:
     - Sensing and equalization possibilities
     - Relationships between loop dynamics, accuracy, stability, and response
     - Partition of required system dynamic characteristics between vehicle and controller:
       - Vehicle dynamic requirements
       - Controller dynamic requirements

success (e.g., performance, pilot and/or vehicle safety, weight, energy, reliability, schedule, cost, etc.).

2. The establishment of competing configurations.
3. The assessment of the competing configurations in terms of the sub-criteria and then the overall criterion.

The most fundamental element in the optimization procedure is, of course, the establishment of possible alternative configurations, since criteria become academic and assessment meaningless unless there is a choice to be made. It is at the configuration level that requirements implied by a particular system configuration meet in a contest with those that stem from physical limitations, the result being the functional requirements for that particular system. Assessment across competing systems then leads to a final system choice and a final set of requirements.

### 9-4. General Feedback Control System Considerations in Flight Control

The complete solution of flight control problems normally requires a suitable intermix of open- and closed-loop systems. By their very nature open-loop systems are calibrated; therefore their application is restricted largely to situations in which the inputs acting and desired control responses are fairly well known a priori. In flight control practice this required foreknowledge largely confines the role of open-loop elements to partial solution of operating-point control problems. Examples include throttle and flap setting, trim adjustment simultaneous with the release of stores, near-impulsive velocity corrections, etc. Even for these examples low response feedback loops may exist or the open-loop elements may operate in conjunction with closed-loop vernier-like supplementary control systems. In any event, pure open-loop control actions depend only on the response characteristics of the controller and vehicle in series, so that dynamic interactions are either practically nonexistent or relatively simple and straightforward. Consequently, we shall here say no more about open-loop control serving by itself as a complete flight system.

The vast majority of flight control systems are multi-input, multi-output, multimode devices with many coupled degrees of freedom. Functionally and operationally, they behave as multiloop feedback control mechanisms in which dynamic interaction between elements plays the central role in overall system dynamic performance. In this and other ways flight control systems share all the generic qualities of servomechanisms and regulators. Adopting a point of view in which flight control systems are considered as members of a broad class of feedback control systems is especially useful in the evolution of flight control system dynamic requirements; this view is the basis for the remainder of the discussion of this chapter.

In order to provide concreteness in the subsequent development, we shall use the system shown in Fig. 9-3 as a prototype. Although some flight control systems may require a more complex block diagram representation, the system shown is suitable for a wide variety known as "single-sensor-loop" control systems. The system types classified under this title have the distinguishing characteristic that single-loop block diagrams composed of analytically simple transfer function blocks provide suitable descriptions of their functional operation. On the other hand, the forward-loop transfer function may actually include the closed-loop transfer functions of the inner loops in a multiloop structure. Specifically, in Fig. 9-3 the actuator loop has not yet been replaced by its closed-loop equivalent, although this step shall be taken subsequently. Also, the feedback transfer function may be a combined representation of two or more sensors. As will be seen later, nearly every flight control system, including some with two different control inputs, can be represented with profit to the analyst by single-sensor-loop block diagrams.

A few remarks about the physical realities represented in Fig. 9-3 are now in order. Consider first the ideal system block, $H(s)$. This is, of course, not physical in that $H(s)$ represents the desired transfer function between output and command. [In subsequent developments $H(s)$ may represent: an ideal system between input, $R(s)$, and output, $C(s)$; input, $R(s)$, and indirectly controlled output, $Q(s)$; or command, $V(s)$, and indirectly controlled output, $Q(s)$; instead of between $C(s)$ and $V(s)$ as shown on the diagram. The distinction will always be clear from the context.] Turn now to the disturbances which enter the system in four different locations. Disturbances may occur physically in yet other places, but essentially all of these can be lumped into those shown. Of the two internal disturbances, $\eta_e(s)$ and $\eta_a(s)$, the second is by far the more important. Normally, the noise, or unwanted input, near the error point is minimized by using high quality sensing instruments and by the careful design of signal circuits, so that input signal to noise ratios are extremely high. Further, a large proportion of the error point disturbance is subsequently rejected in such elements as phase-sensitive demodulators incorporated in the forward-loop amplification and equalization block. On the other hand, the actuator error disturbance, $\eta_a(s)$, while often small in absolute magnitude, is almost invariably followed by large amplification factors within the actuator loop. These are required in practice to minimize the effects of actuator load disturbances and closed-loop actuator dynamics on the overall system. These high gains coupled with the internal disturbance, $\eta_a(s)$, and limiting within the actuator loop, can result in very deleterious consequences. The actuator load disturbances may arise physically from several sources inherent in the actual actuator installation. In addition they may include external disturbances imposed upon the control

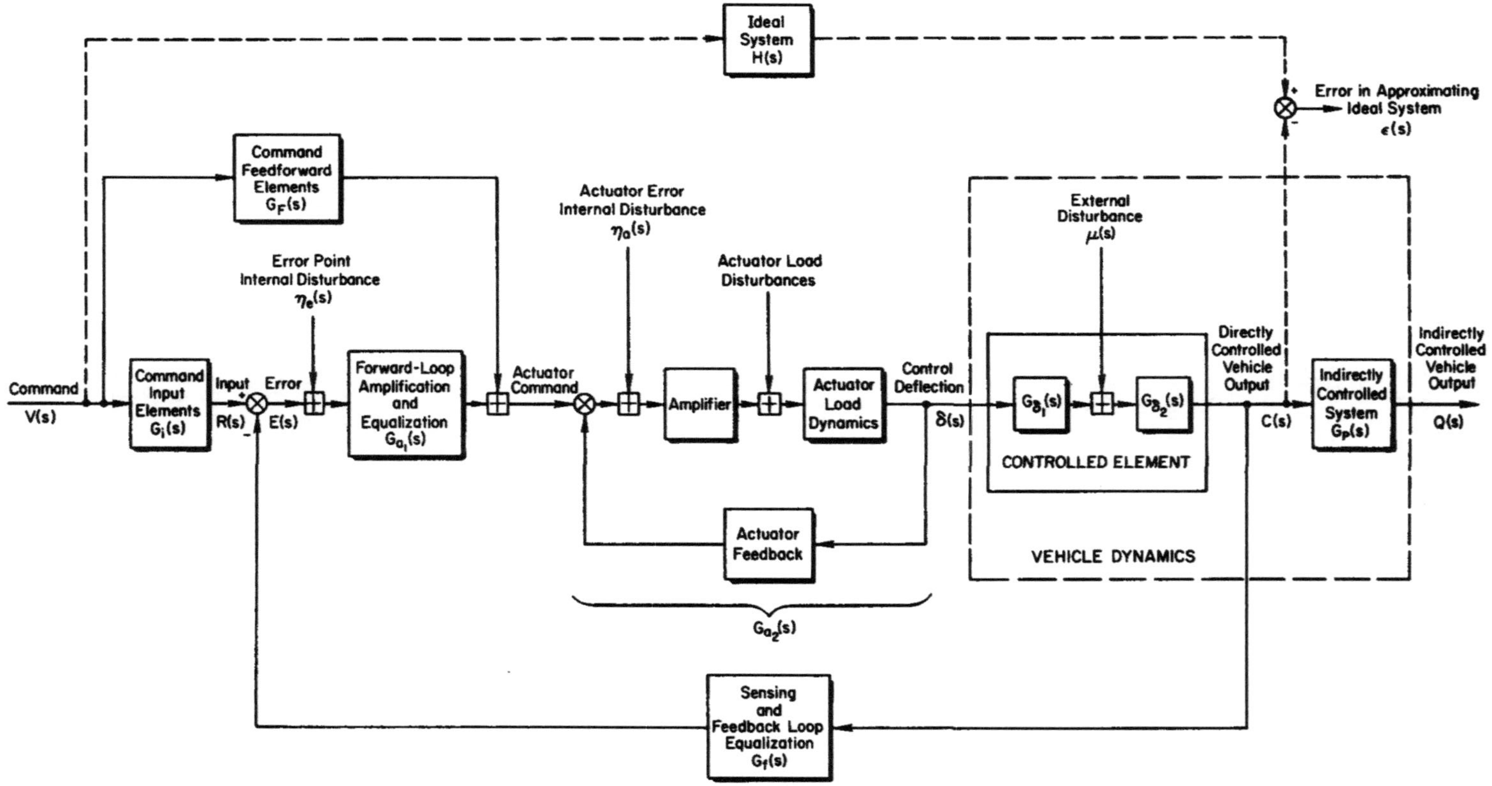


Fig. 9-3. Prototype "single-sensor-loop" flight control system.

surfaces; e.g., for aerodynamic controls, a feedback path (not shown) also exists between the vehicle dynamics and the actuator load disturbance point.

As already mentioned, the actuator loop will ordinarily be closed with gains sufficiently high nearly to eliminate the effects of actuator load disturbances on the control deflection. In these circumstances the block diagram of Fig. 9-3 reduces to the one shown in Fig. 9-4. Further, if the internal and external disturbances are neglected, the even simpler diagram of Fig. 9-5 appears. Finally, for those circumstances in which command feedforward and command input elements are not present, the overall block diagram takes on the especially simple form shown in Fig 9-6. All of these block diagrams shall be used to illustrate particular points in the following discussion of flight control system functions.

The flight control system's basic overall function in accomplishing flight control purposes implies a control system that provides:

Stability (either general or at specified times)
Desired responses to specified inputs
Suppression of the effects of undesired inputs

To provide these features, the flight control system is called upon to perform certain specific functions. Since the functions are interrelated, lines of demarcation between them are difficult to draw distinctly. One breakdown of specific functions follows:

1. Provide vehicle stability and controllability.
2. Reduce "effective" order of vehicle dynamics.
3. Adjust "effective" vehicle dynamic response.
4. Provide specified command-response relationships.
5. Reduce effects of unwanted inputs and disturbances.
6. Suppress effects of vehicle and component variations and uncertainties.
7. Improve linearity.
8. Modify or eliminate vehicle cross-coupling effects.

Each of these specific functions and the implications drawn from their satisfaction is discussed in the following paragraphs.

1. *Provide vehicle stability and controllability.* Most aeronautical and aero space vehicles that operate over a wide range of flight conditions inevitably encounter open- or closed-loop stability difficulties in one or more flight regimes. Because of large variations in the types and magnitudes of the forces acting, sophisticated vehicle configuration "tailoring,"

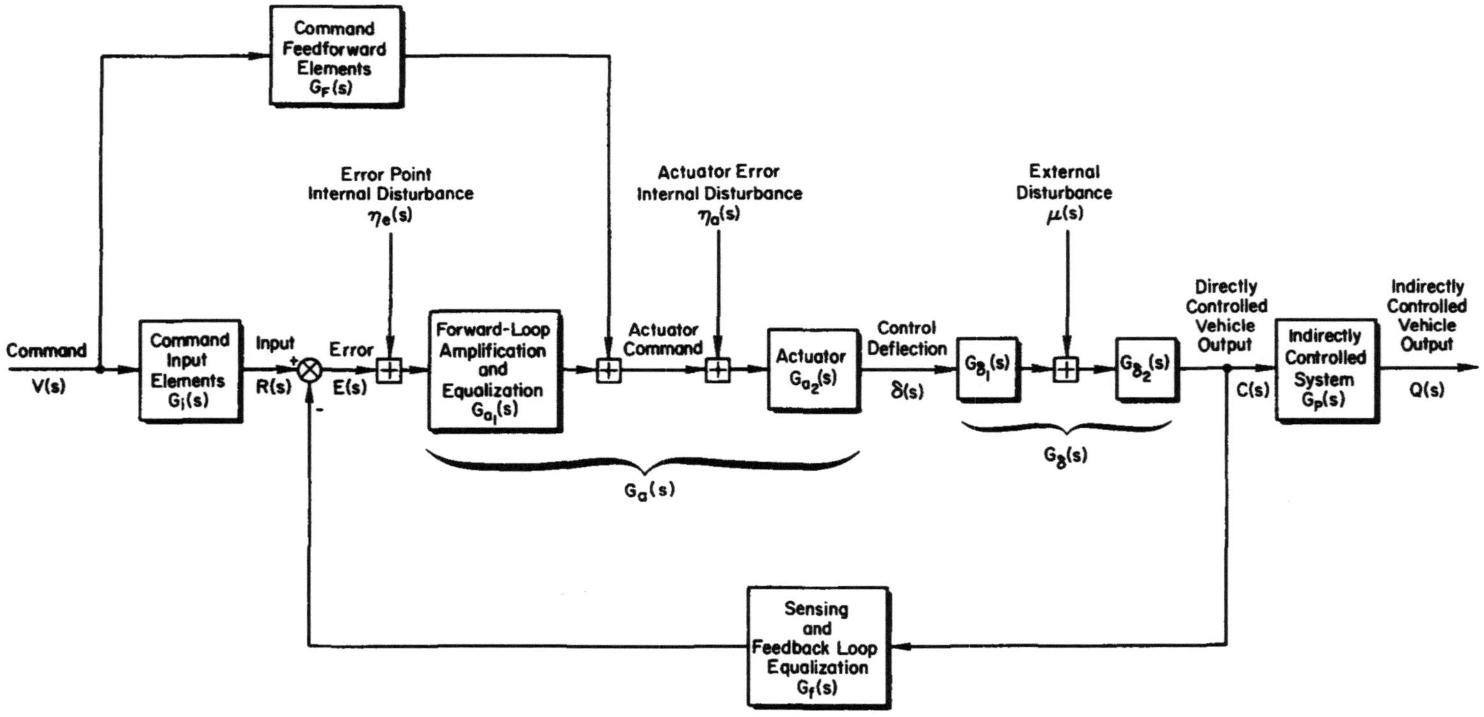


Open-Loop Transfer Function $G(s) = G_a(s)G_\delta(s)G_f(s)$

Fig. 9-4. Simplification of "single-sensor-loop" flight control system.

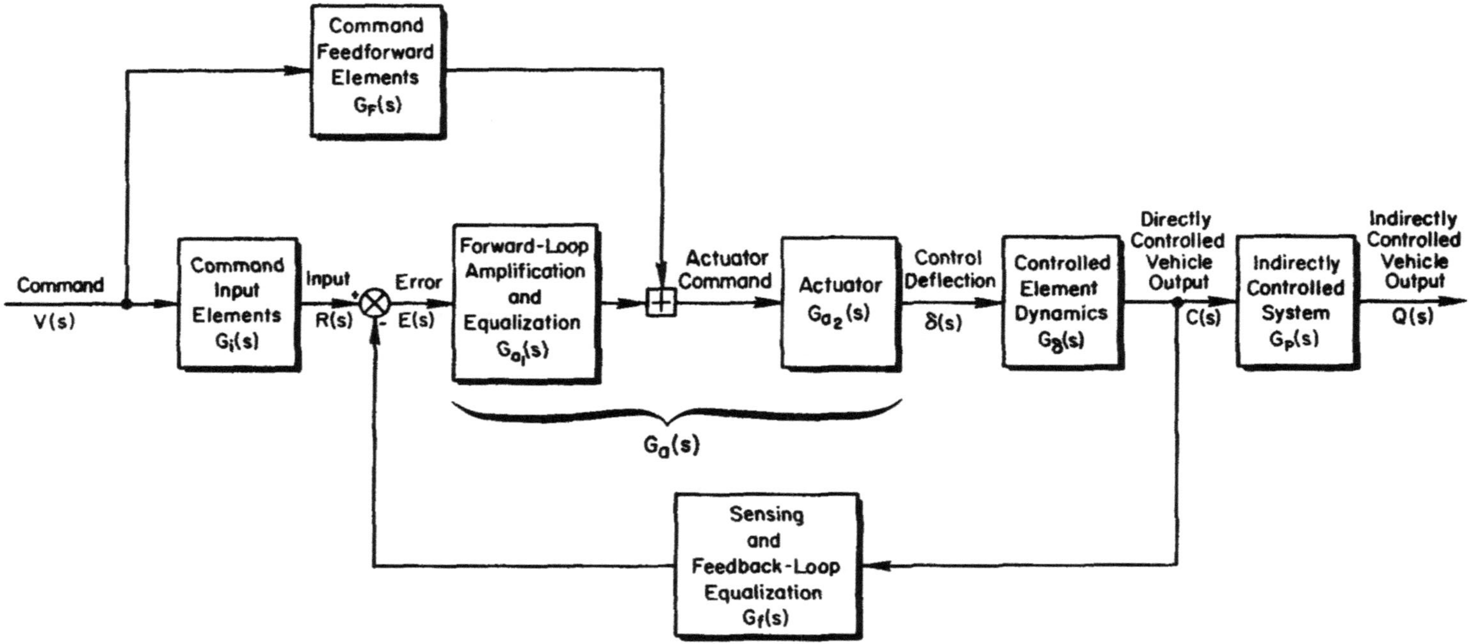


Open-Loop Transfer Function $G(s) = G_{a_1}(s)\,G_{a_2}(s)\,G_\delta(s)\,G_f(s)$

Fig. 9-5. Prototype "single-sensor-loop" flight control system without external and internal disturbances.

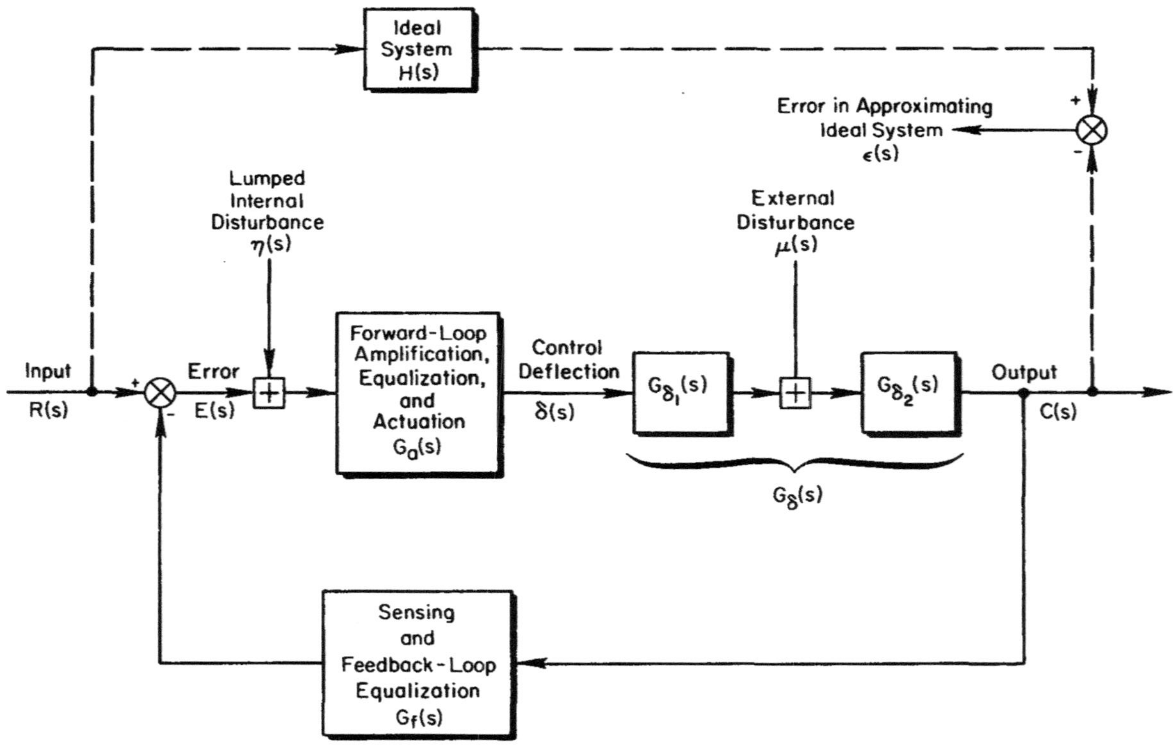


Open-Loop Transfer Function $G(s) = G_a(s)G_\delta(s)G_f(s)$

**Fig. 9-6.** Prototype "single-sensor-loop" flight control system without feed forward and command input elements.

alone, is seldom sufficient to insure stability over the entire performance envelope. Consequently, a prime flight control system function is ordinarily the augmentation or creation of vehicle dynamic stability. In terms of Fig. 9-6, a vehicle having unstable dynamic characteristics will possess a transfer function $G_\delta(s)$ containing poles in the right half of the $s$-plane (RHP). With a flight control system added, the "effective" vehicle characteristics are modified from

$$\frac{C(s)}{\delta(s)} = G_\delta(s) \text{ (unstable, i.e., poles in RHP)} \tag{9-1}$$

to

$$\begin{aligned}\frac{C(s)}{R(s)} &= \frac{G_a(s)G_\delta(s)}{1 + G_a(s)G_f(s)G_\delta(s)} \\ &= \frac{1}{G_f(s)}\left[\frac{G(s)}{1 + G(s)}\right]\end{aligned} \tag{9-2}$$

With proper design, the poles of the effective vehicle transfer function, given by the zeros of $G_f(s)[1 + G(s)]$, lie in the left half of the $s$-plane, thereby providing stability for the "effective" vehicle.

In some instances the attainable bandwidth for the Eq. 9-2 closed-loop is limited to unacceptably low values by unstable, or poorly placed, zeros in the vehicle, $G_\delta(s)$, numerator. In order, then, to provide adequate controllability, as well as basic stability, the flight control system must modify these vehicle zeros to permit an acceptable bandwidth to be obtained. Such modification, when needed, establishes requirements for auxiliary feedbacks, e.g., those that decouple the vehicle's motions. (See Item 8.)

2. *Reduce effective order of vehicle dynamics.* A desirable consequence of feedback is to reduce the effective degree of the system transfer function within important frequency regions. For instance, the denominator of the open-loop transfer function, $G(s)$, in Fig. 9-6 is ordinarily of high degree; the denominator of the closed-loop characteristics given by $G(s)/[1 + G(s)]$ will be of this same degree. However, the fact that $|G(j\omega)|$ can be made much greater than unity over wide frequency ranges sets up regions in which either the closed-loop poles are nearly cancelled by zeros or are otherwise removed from these regions to considerably higher or lower frequencies. So, while, theoretically, the denominators of $G(s)$ and $G(s)/[1 + G(s)]$ are of the same degree, practically, the degree of $G(s)/[1 + G(s)]$, in a frequency range of interest, can be effectively lowered from that of $G(s)$. Expressed another way, the closed-loop transfer function

$$\left|\frac{G(j\omega)}{1 + G(j\omega)}\right| \doteq 1, \qquad \text{when } |G(j\omega)| \gg 1 \tag{9-3}$$

so all the modes of $G(j\omega)$ in the frequency regions where $|G(j\omega)| \gg 1$ have been effectively removed. An example of the reduction of the effective order of vehicle dynamics has already been presented in Chapter 3 (Section 3-4).

This reduction in effective order is one of the most profound and significant features of feedback systems. One of its more important consequences is an improvement in performance potential. This is demonstrated in Fig. 9-7 which shows minimum values of two figures of merit vs. the system order, $n$. These results are obtained with unit numerator systems, i.e., systems having output/input transfer functions of the form

$$G_{cr}(s) = \frac{1}{s^n + a_1 s^{n-1} + \cdots + a_i s^{n-i} + \cdots + 1} \tag{9-4}$$

subjected to unit step inputs. In each case, the system coefficients (the $a_i$'s) were adjusted to values for which

$$\frac{\partial(\text{figure of merit})}{\partial a_i} = 0 \tag{9-5}$$

so that the figures of merit shown are those for optimum systems, using a zero value of the total differential as the definition of optimum.[1] From Fig. 9-7 it is apparent that a high order system which is reduced to act like a second-order system has the potential, if properly adjusted to give lower (i.e., better) figures of merit than a similar system reduced only to the point at which it acts like a fourth-order system. However, the process of reduction usually requires the insertion of equalization and therefore the consequent addition of numerator terms to the unit numerator output/input transfer function, $G_{cr}(s)$. Such considerations must be included in any specific attempts to improve a system's performance by reducing its effective order.

3. *Adjust "effective" vehicle dynamic response.* This function is related to the one directly above as the second step in a two-step process intended to improve the effective vehicle dynamic response for desired inputs. Reduction in the effective system order stems directly from setting up the condition $|G(j\omega)| \gg 1$ over some desired frequency range; adjustment of effective vehicle dynamic response follows in the selection of the restricted frequency region where $|G(j\omega)| = O(1)$, and in the tailoring of $G(j\omega)$ in this region. Normally, the closed-loop system has three regions of interest,

[1] The validity of particular figures of merit is beside the point for the present discussion, although of extreme interest if a figure of merit is used as a performance criterion. For a detailed discussion of the latter point, see J. Wolkovitch, R. Magdaleno, D. McRuer, D. Graham, and J. McDonnell, "Performance Criteria for Linear Constant-Coefficient Systems with Deterministic Inputs," ASD-TR-61-501, Feb. 1962.

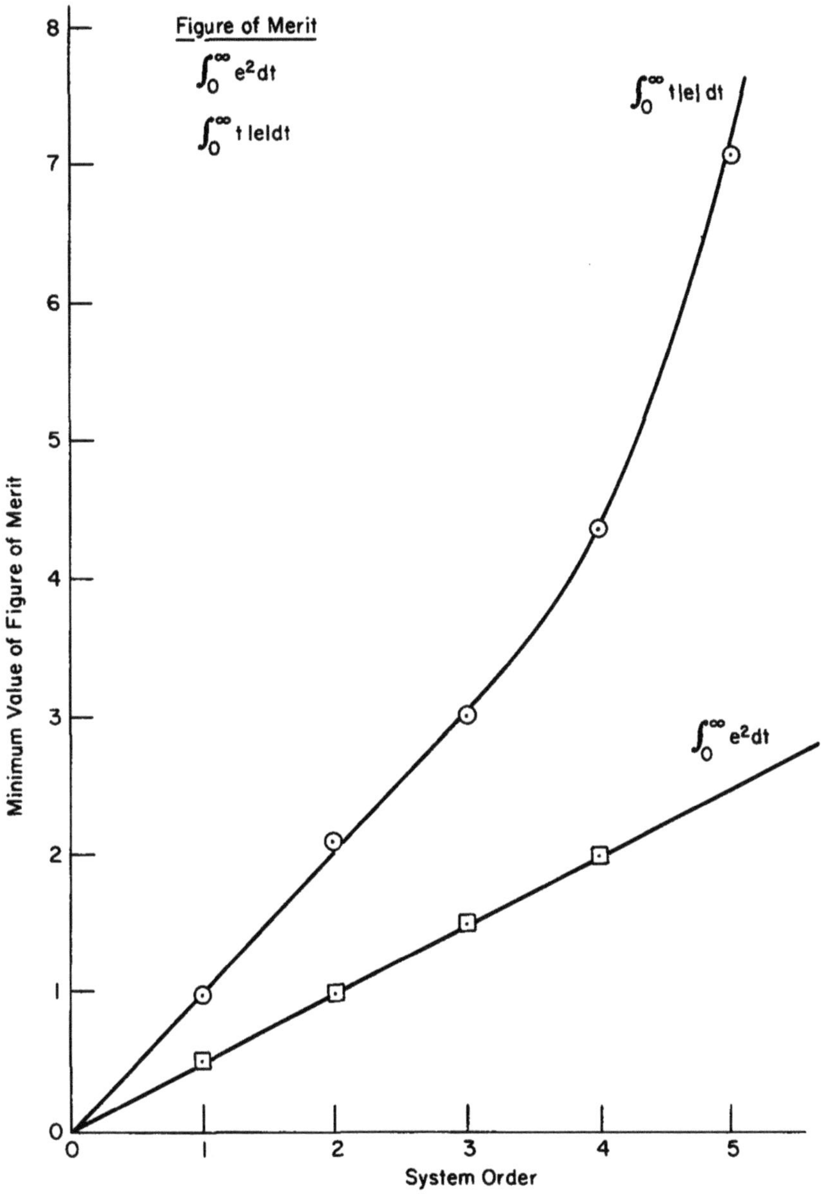


Fig. 9-7. Variation of minimum value for figure of merit with system order.

defined by

$$|G(j\omega)| \gg 1, \qquad \text{over which} \left|\frac{G(j\omega)}{1 + G(j\omega)}\right| \doteq 1$$

$$|G(j\omega)| \ll 1, \qquad \text{over which} \left|\frac{G(j\omega)}{1 + G(j\omega)}\right| \doteq |G(j\omega)| \tag{9-6}$$

$$|G(j\omega)| = O(1)$$

The form of $|G(j\omega)/[1 + G(j\omega)]|$ in this last region defines the dominant modes of the closed-loop system dynamic response. In most cases, $G(j\omega)/[1 + G(j\omega)]$ in the region where $|G(j\omega)|$ is of the order of unity can be approximated by a first-, second-, or third-order system and one, two, or three modes will be dominant in the response.

4. *Provide specified command-response relationships.* It is often desired that the flight control system provide some specified relationship between command and response. Referring to Fig. 9-5, it can be seen that the closed-loop transfer function in response to command inputs has the form:

$$G_{cv}(s) = \frac{G_{a_2}(s)G_\delta(s)[G_i(s)G_{a_1}(s) + G_F(s)]}{1 + G(s)} \tag{9-7}$$

When the magnitude of the open-loop transfer function, $|G(j\omega)|$, is much larger than unity, the relationship between the command and response reduces to

$$G_{cv}(j\omega) \doteq \frac{1}{G_f(j\omega)}\left[G_i(j\omega) + \frac{G_F(j\omega)}{G_{a_1}(j\omega)}\right] \tag{9-8}$$

The designer has complete control over the transfer functions $G_i(s)$ and $G_F(s)$ and some control over $G_f(s)$ and $G_{a_1}(s)$. The latter two are, of course, involved in the open-loop transfer function, $G(s)$, so that they cannot be arbitrarily adjusted without affecting the basic constraint $|G(j\omega)| \gg 1$. By proper selection of the form of the transfer function combination shown in Eq. 9-8, the designer can provide almost any specified command-response relationship, at least over the frequency regions where $|G(j\omega)| \gg 1$. For manual control purposes these transfer functions can be adjusted to provide a vehicle response to piloted command that approximates the best vehicle in the sense of pilot goals. Among other things, the effective vehicle dynamics as seen by the pilot can be made substantially invariant with flight conditions. However, this may not be altogether desirable, depending on the resulting indirectly controlled responses (see Fig. 9-5).

If we assume a desired command-response relationship, $H(s)$, for the single-sensor-loop control system shown in Fig. 9-6, this implies that the system output and command be related, over a limited range of frequency, by

$$C(j\omega)_{\text{desired}} = H(j\omega)R(j\omega) \tag{9-9}$$

The actual output for this system, assuming no disturbances ($\mu = \eta = 0$), is

$$C(j\omega) = \frac{G(j\omega)}{G_f(j\omega)[1 + G(j\omega)]} R(j\omega) \tag{9-10}$$

and the "error," $\epsilon$, which measures the degree to which the actual system approximates the desired system is

$$\epsilon(j\omega) = C(j\omega)_{\text{desired}} - C(j\omega) = \left\{H(j\omega) - \frac{G(j\omega)}{G_f(j\omega)[1 + G(j\omega)]}\right\} R(j\omega) \tag{9-11}$$

This error could be reduced to zero by designing the open-loop transfer function to be

$$G(j\omega) = \frac{G_f(j\omega)H(j\omega)}{1 - G_f(j\omega)H(j\omega)} \tag{9-12}$$

Altering the system to match this formula would, in principle, implicitly satisfy the flight control function of providing good response to specified inputs. Unfortunately, Eq. 9-12 is only of qualitative value in flight control applications because disturbances, ignored in its formulation, impose additional requirements on $G(j\omega)$, as next discussed; and because a system so formulated is essentially calibrated, so that its performance is highly sensitive to controller or controlled element tolerances and uncertainties.

5. *Reduce effects of unwanted inputs and disturbances.* The primary regulator function of a flight control system is to reduce the effects of external disturbances acting on the vehicle and internal disturbances acting on the flight controller itself. For all but those disturbances acting at the input, this function is again accomplished by the feedback aspects of the system. In Fig. 9-4, for example, the system output with no command input is given by

$$C(s) = \frac{1}{1 + G(s)} [G_{\delta_2}(s)\mu(s) + G_a(s)G_\delta(s)\eta_a(s) + G_a(s)G_\delta(s)\eta_e(s)] \tag{9-13}$$

To reduce these disturbance effects again requires $|G(j\omega)|$ much greater than one; then the output is approximately

$$C(j\omega) \doteq \frac{\mu(j\omega)}{G_a(j\omega)G_{\delta_1}(j\omega)G_f(j\omega)} + \frac{\eta_a(j\omega)}{G_{a_1}(j\omega)G_f(j\omega)} + \frac{\eta_e(j\omega)}{G_f(j\omega)} \tag{9-14}$$

Ordinarily, the transfer functions $G_a(j\omega)G_{\delta_1}(j\omega)G_f(j\omega)$ and $G_{a_1}(j\omega)G_f(j\omega)$ can be made large compared with unity over the frequency regions for which the external disturbances, $\mu(j\omega)$, and the actuator error internal disturbances, $\eta_a(j\omega)$, are important. Unfortunately, it is seldom possible to specify independently the feedback transfer function, $G_f(j\omega)$, such that a similar reduction in output due to the internal disturbance, $\eta_e(j\omega)$, also occurs. Even if this could be done, the desired input, $R(j\omega)$, would have the same closed-loop transfer function as $\eta_e(j\omega)$, so that no effective reduction in signal to noise would accrue by closing the loop. These facts are reflected in the emphasis placed on procuring high quality instruments for flight control systems.

6. *Suppress effects of vehicle and component variations and uncertainties.* One of the principal differences between flight control and other feedback systems is the enormous variation possible in the controlled-element (vehicle dynamics) over the entire operating-point profile. Consequently, an extremely important function of an adequate flight control system is to maintain the effective closed-loop dynamics more or less constant. If we consider Fig. 9-6, a fractional change in $G_{cr}(s)$, denoted as $dG_{cr}(s)/G_{cr}(s)$, in terms of fractional changes in the elements within the loop, is given by

$$\frac{dG_{cr}(s)}{G_{cr}(s)} = \frac{1}{1 + G(s)}\left[\frac{dG_a(s)}{G_a(s)} + \frac{dG_\delta(s)}{G_\delta(s)} - G(s)\frac{dG_f(s)}{G_f(s)}\right] \tag{9-15}$$

Again considering the open-loop transfer function magnitude $|G(j\omega)| \gg 1$, we find that the fractional change in overall closed-loop characteristics reduces to

$$\frac{dG_{cr}(j\omega}{G_{cr}(j\omega)} \doteq \frac{1}{G(j\omega)}\left[\frac{dG_a(j\omega)}{G_a(j\omega)} + \frac{dG_\delta(j\omega)}{G_\delta(j\omega)}\right] - \frac{dG_f(j\omega)}{G_f(j\omega)} \tag{9-16}$$

The beneficial consequences of large values for the open-loop transfer function in lowering the effects of forward-loop controller and vehicle variations and uncertainties on the overall transfer function are apparent from this equation. Equally worth noting is the complete absence of any beneficial effect on uncertainties in the feedback path—another reason for emphasizing very high quality instruments.

7. *Improve linearity.* The elements of flight control systems often contain nonlinearities, either inadvertently or by design. With the exception of command and other limiting functions, it is ordinarily desirable to suppress the dynamic effects of these nonlinearities so that more or less proportional cause-effect relationships exist among the various system inputs and outputs. Although there are several ways of accomplishing this "approximate linearization," the most common method is also another attribute of feedback. In Fig. 9-8, for example, when the

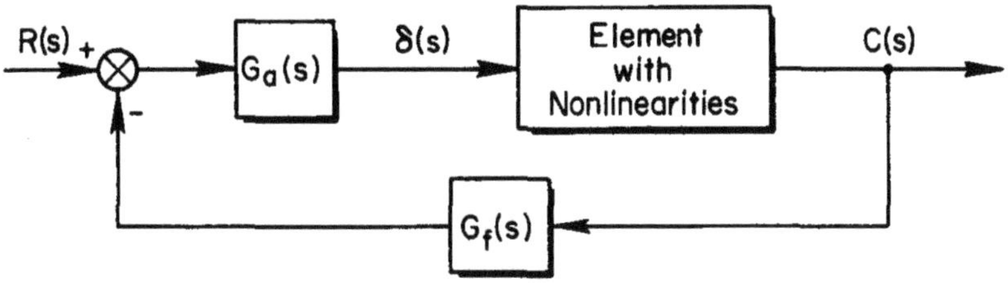


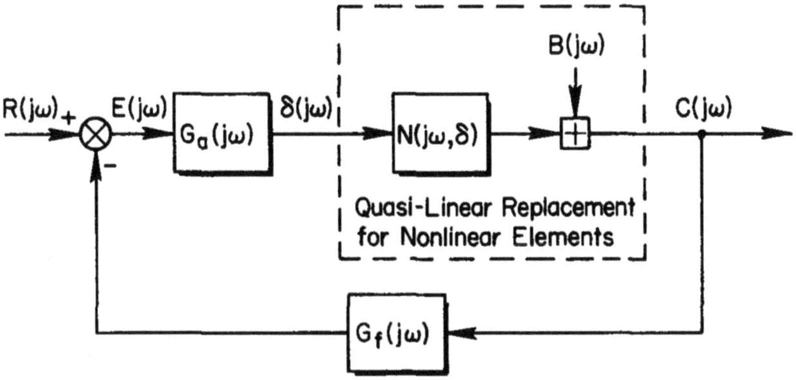


Fig. 9-8. Replacement of nonlinear elements with describing function and remnant.

element containing nonlinearities is replaced by a describing function, $N(j\omega;\,\delta)$, and a remnant, $B(j\omega)$,[2] the system output becomes

$$C(j\omega) = \frac{1}{G_f(j\omega)}\left[\frac{G_a(j\omega)G_f(j\omega)N(j\omega;\,\delta)}{1 + G_a(j\omega)G_f(j\omega)N(j\omega;\,\delta)}\right](Rj\omega) + \left[\frac{1}{1 + G_a(j\omega)G_f(j\omega)N(j\omega;\,\delta)}\right]B(j\omega) \quad (9\text{-}17)$$

This substitution of a quasilinear representation for a nonlinear element can be accomplished precisely, in principle, when the input signal is a periodic or stationary-random function of time. For other inputs a quasilinear representation is sometimes useful heuristically, although caution must be exercised to avoid carrying linear system notions too far.

[2] See D. Graham, and D. McRuer, *Analysis of Nonlinear Control Systems*, Wiley, New York, 1961.

When the input signal to the nonlinearity is confined within certain amplitude limits, the open-loop describing function can be made much larger than unity over some specified range of frequencies. For such conditions the output reduces to

$$C(j\omega) \doteq \frac{1}{G_f(j\omega)} R(j\omega) + \frac{1}{G_a(j\omega)G_f(j\omega)N(j\omega;\delta)} B(j\omega)$$

Because the open-loop describing function, $G_aG_fN$, is so large over this frequency range, the remnant term will be much smaller than without feedback, and the output becomes, approximately,

$$C(j\omega) \doteq \frac{1}{G_f(j\omega)} R(j\omega) \tag{9-18}$$

To the extent that these concepts apply, the input and output bear an approximately linear relationship that is independent of the nonlinearity. Outside either the amplitude (of the nonlinearity's input) or frequency region, where the open-loop describing function is much larger than unity, this relationship will, of course, no longer be valid.

Another illustrative analysis of the linearizing properties of feedback has been given by West for the nondynamic case.[3] Suppose, considering Fig. 9-8(a), that the element with nonlinearities is such that $C$ is the function of $\delta$ illustrated in Fig. 9-9. Call this function $g(\delta)$. Then, from the

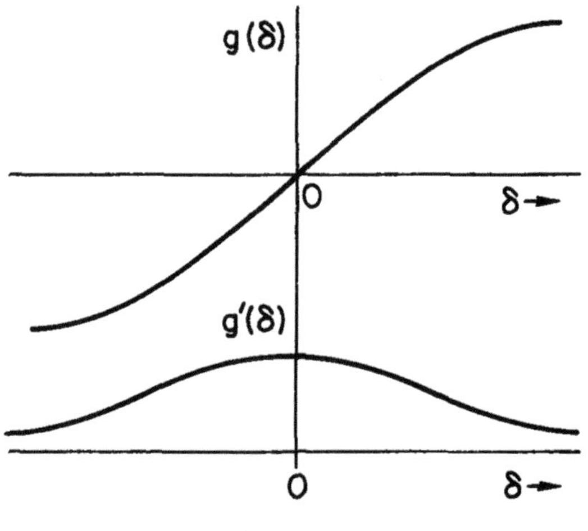


Fig. 9-9. A nonlinear function and its derivative with respect to the input to the nonlinearity.

[3] J. C. West, *Analytical Techniques for Nonlinear Control Systems*, English Universities Press, London, 1960.

block diagram,

$$\delta = G_a(R - G_f C) \tag{9-19}$$

$$C = g(\delta) = g(G_a R - G_a G_f C) \tag{9-20}$$

Now

$$\frac{dC}{dR} = \frac{dg}{d\delta}\frac{d\delta}{dR} = g'(\delta)\frac{d\delta}{dR} \tag{9-21}$$

Differentiating Eq. 9-19 and substituting, we obtain

$$\frac{dC}{dR} = g'(\delta)G_a\left(1 - G_a\frac{dC}{dR}\right) \tag{9-22}$$

or

$$\frac{dC}{dR} = \frac{G_a g'(\delta)}{1 + G_a g'(\delta) G_f} \tag{9-23}$$

When the open-loop function $G_a G_f g'(\delta)$ is much larger than unity, this becomes

$$\frac{dC}{dR} \doteq \frac{1}{G_f}; \qquad \text{for } G_f G_a g'(\delta) \gg 1$$

Again, this illustrates the fact that the action of the feedback is to linearize the nonlinearity as long as the loop gain is high.

8. *Modify or eliminate vehicle cross-coupling effects.* Both aeronautical and aerospace vehicles often have several coupled degrees of freedom. Some of these couplings are useful, while others can cause a great deal of difficulty. High gain feedback loops are used to swamp many of these cross-couplings by suppressing the influence of the vehicle itself. However, it is occasionally desirable to cancel out cross-couplings by generating almost exactly opposing forces and moments via the flight control system. For instance, with vehicle equations of motion given by

$$\begin{bmatrix} a_{11}(s) & a_{12}(s) & a_{13}(s) \\ a_{21}(s) & a_{22}(s) & a_{23}(s) \\ a_{31}(s) & a_{32}(s) & a_{33}(s) \end{bmatrix} \begin{bmatrix} \beta \\ \varphi \\ r \end{bmatrix} = \begin{bmatrix} 0 \\ 0 \\ N_\delta \delta \end{bmatrix} \tag{9-24}$$

the $a_{32}(s)$ off-diagonal term could be removed from the effective vehicle equations of motion by letting

$$\delta = \frac{a_{32}(s)\varphi}{N_\delta} \tag{9-25}$$

Attempts to decouple vehicle modes of motion in this direct additive (or subtractive) fashion do not have as widespread application in flight control as one might expect because of the drastic variation of vehicle characteristics throughout the flight envelope. However, there are instances in which deliberate crossfeeds of the type illustrated above can be used with great success.

To summarize the foregoing, most of the specific functions of flight control systems are accomplished by the "swamping" action provided by feedback over limited regions of frequency by making $|G(j\omega)| \gg 1$. In addition, the specific flight control functions involving suppression of external disturbances, minimization of effects of system component variation, linearization, etc., are all obtained in an approximate way using the same conditions, i.e., $|G(j\omega)| \gg 1$ over the frequency ranges in which these effects are important. Consequently, a set of limiting, ideal specifications on the overall flight control system would include:

1. $G(j\omega) \gg 1$
2. $\dfrac{G(j\omega)}{G_f(j\omega)[1 + G(j\omega)]} \doteq H(j\omega)$

} Over the frequency range for which the desired $H(j\omega)$ is to be closely approximated and/or over the frequencies at which $\mu(j\omega)$ and $R(j\omega)$ have substantial frequency components.

3. $\dfrac{G(j\omega)}{G_f(j\omega)[1 + G(j\omega)]} \ll 1$

} Outside the frequency region in which the major frequency components of the input and/or external disturbances occur.

The first and second conditions reflect the regulation (suppression of external disturbances) and servomechanism (development of an output that is a specified function of a command) aspects of the problem. The third condition minimizes the effects of internal disturbances to the extent possible without detracting seriously from the system's action as a servo or regulator. It is possible, depending upon the frequency content of the command signals and the external and internal disturbances, for the several conditions to be in conflict. In flight control systems, however, a reasonable compromise is ordinarily attainable.

The internal disturbances, $\eta(j\omega)$, grouped at the error point in Fig. 9-6 are not substantially reduced by feedback action (although the third condition above reduces their effect outside the frequency region where feedback operation is dominant). Therefore, these should be reduced to a minimum relative to the input, especially over the frequency range in which $H$ is to be closely approximated. In practice, this reduction is accomplished primarily by detailed mechanizational considerations.

For control systems as a general class, techniques to synthesize mathematically an approximation to the conditions enumerated above range from elementary repetitive analyses to sophisticated exercises employing constraints and resulting in optimal systems (approximations to $H$)

which meet specified criteria.[4] In general the more sophisticated synthesis techniques are best suited to problems in which the characteristics of the commands and disturbances are known (e.g., power-spectral and cross-spectral densities); in which considerable latitude is allowable in the form of the open-loop transfer function; and in which the controlled-element characteristics are no more than moderately variable. Unfortunately, none of these conditions is present in most flight control problems to any appreciable extent. Consequently, the emphasis in practical synthesis activities is heavily centered on an attempt to provide suitable control and stabilization for a wide range of vehicle dynamics in the presence of a large variety of command and disturbance signals with a controller that can operate satisfactorily in several modes. For this kind of problem, intelligent repetitive analysis is usually the most suitable synthesis procedure.

Regardless of the detailed procedure used, the basic question as to desirable forms for $H$ and $G_f H$ must somehow be resolved. Certain very general aspects of this question deserve discussion here. As noted, the functions of flight control systems involving regulation, or suppression of external disturbances, require an open-loop transfer function, $G$, which is large over the disturbance bandwidth. Similarly, the desire to cut off internal disturbances outside the command frequency band of interest requires $G$ to be small at higher frequencies. The desired, feedback-modified, closed-loop transfer function, $G_f H$, which is approximated by the actual closed-loop transfer function, $G/(1 + G)$, can thus be seen to be basically that of a low pass filter.[5]

The three regions of interest, defined in terms of $|G(j\omega)|$, which shed light on the desired low pass form for $G_f H$ are

$$\begin{aligned} |G(j\omega)| &\gg 1, \qquad \frac{G}{1+G} \doteq 1 \\ |G(j\omega)| &\ll 1, \qquad \frac{G}{1+G} \doteq G \\ |G(j\omega)| &= 0(1), \qquad \frac{G}{1+G} \end{aligned} \tag{9-26}$$

[4] See, e.g., A. Bryson, "Applications of Optimal Control Theory in Aerospace Engineering," *J. Spacecraft and Rockets*, 4, No. 5, 545–553 May 1967; E. G. Rynaski and R. F. Whitbeck, "The Theory and Application of Linear Optimal Control," AFFDL-TR-65-28, Jan. 1966; Arthur E. Bryson, Jr., and Yu-Chi Ho, *Applied Optimal Control*, Blaisdell, Waltham, Mass., 1969.

[5] $G/(1 + G)$ does not necessarily possess a low pass nature all the way to zero frequency when only the vehicle short period modes are to be controlled. Even in these cases, however, the desire is still to have a low-pass, closed-loop system within and above the short period frequency regions.

Since $G$ itself is, in general, a low pass transfer function above some frequency, the first two relations satisfy the spirit of the desirable conditions enumerated above. The crux of the matter occurs, of course, in the crossover region about frequencies where $|G(j\omega)| \doteq 1$. The behavior of $G/(1 + G)$ in this region defines the dominant modes of the closed-loop response, or their frequency domain correlates of bandwidth, peak magnification, etc. Consequently, the major part of the system closed-loop dynamic behavior is implicit in what goes on when $|G(j\omega)| = O(1)$; $G$ and $G/(1 + G)$ are so closely interconnected in this region that both ordinarily require consideration. At this point, gross general considerations must cease, and we must become more specific about system and element details.

## 9-5. Bases for Compromise in Selecting Crossover Region

If we consider Fig. 9-6, it is an easy task to find the closed-loop transfer function $G(s)/[1 + G(s)]$ once $G(s)$ is given (the subject of Chapter 3). The dominant modes of the closed-loop system are also easy to find in these circumstances. On the other hand, it is often difficult to specify what the characteristic frequencies of the dominant modes *should* be. In terms of closed-loop frequency response, this last statement is equivalent to noting that the bandwidth is difficult to specify with any degree of precision. (Bandwidth is that frequency at which $|G(j\omega)/[1 + G(j\omega)]|_{\text{dB}}$ is down $-3$ dB from the low frequency value, which is usually zero dB when the closed-loop system is low pass.) Bandwidth is closely related to the crossover frequency [at which $|G(j\omega)| = 1$] and is also a crude measure of the undamped natural frequencies and/or inverse time constants of the dominant modes. So, even though the latter quantities are actually of principal interest, the following discussion shall use all of these terms loosely as if they were more or less interchangeable.

The root of the difficulty in deciding upon a closed-loop bandwidth requirement stems from incomplete knowledge of the inputs and disturbances. Consequently, uncertainty exists about just how far in frequency the inequality $|G(j\omega)| \gg 1$ should hold in order to attain good performance in response to commands and suppression of disturbances. On the other hand, the total frequency range over which $|G(j\omega)|$ can, or should, be made large relative to unity is restricted by considerations of stability, equalization economy, actuator-loop characteristics, and suppression of internal disturbances appearing at the input. Each of these factors is discussed in relatively general terms below to indicate the broad bases for compromise involved in bandwidth selection.

*Stability Considerations*

Perhaps the biggest factor involved in bandwidth selection is the general form of the open-loop transfer function $G(s)$. For example, in a single-sensor-loop system, $G(s)$ is made up of two primary components, the controlled-element transfer function, $G_\delta(s) = G_{\delta_1}(s)G_{\delta_2}(s)$, and the controller transfer function, $G_a(s)G_f(s)$. The product $G_a(s)G_f(s)G_\delta(s)$ must be such that a stable closed-loop system is possible. When the form of $G_a(s)G_f(s)$ is fairly limited for reasons of equalization economy (see below), and a relatively unalterable form exists for $G_\delta(s)$, crossover can occur only in particular frequency regions if stability is to be assured. If well-damped responses to inputs and disturbances are also necessary, these restricted frequency regions become even further limited.

The simplest case in which a bandwidth specification and stability interact occurs when the vehicle transfer function is unstable. A single-sensor-loop control system is then conditionally stable, and some minimum gain (and hence some minimum bandwidth) is necessary if stability is to be achieved.

For systems that contain a stable vehicle as an element, or for loops that do not require a stable closure (e.g., an inner loop of a multiaxis flight control system), a theoretical minimum bandwidth does not exist. However, the regions in which gain crossovers consistent with stability and reasonable transient response can occur can be defined within narrow limits. For a system to have neutrally stable oscillations (i.e., closed-loop poles on the imaginary axis of the $s$ plane), the open-loop transfer function must satisfy

$$|G(j\omega)| = 1, \quad \measuredangle G(j\omega) = -\pi, \qquad G(0) \text{ positive} \tag{9-27}$$

for some value of $\omega$. Most of the transfer functions encountered in flight control applications can be adjusted to avoid either neutrally stable or unstable dominant modes simply by making

$$\begin{aligned} |G(j\omega)| &< 1 \qquad \text{when } \measuredangle G(j\omega) = -\pi \\ \measuredangle G(j\omega) &> -\pi \qquad \text{when } |G(j\omega)| = 1 \end{aligned} \tag{9-28}$$

These are the common conditions of positive gain margin and phase margin, respectively. These simple statements always apply for minimum-phase transfer functions. Equivalent simplified conditions may be delineated for any given transfer function form by starting with the Nyquist criterion, defining new gain and phase margins and modifying Eq. 9-28, etc., as appropriate for the particular system. For a given system the derived gain margin and phase margin conditions define a maximum bandwidth (crossover frequency).

Compatibility with requirements for stable closure can also be inferred from the general slope characteristics of the open-loop amplitude ratio, $|G(j\omega)|$, in the crossover region. This follows from the well-known relationships existing between amplitude ratio and phase. The phase angle of a minimum phase transfer function, at a frequency $\omega_c$, in terms of the slope of the amplitude ratio is

$$\varphi(\omega_c) = \frac{\pi}{40}\left\{\frac{d\,|G(\omega)|}{d[\ln(\omega/\omega_c)]}\right\}_{\omega=\omega_c} + \frac{1}{20\pi} \times \int_{-\infty}^{\infty}\left\{\frac{d\,|G(\omega)|}{d[\ln(\omega/\omega_c)]} - \left[\frac{d\,|G(\omega)|}{d[\ln(\omega/\omega_c)]}\right]_{\omega=\omega_c}\right\} \ln\coth\frac{|\ln(\omega/\omega_c)|}{2}\, d\left(\ln\frac{\omega}{\omega_c}\right) \tag{9-29}$$

where the slopes, $dG/d[\ln \omega/\omega_c]$ are expressed in dB/decade.[6] The $\ln\coth|\ln(\omega/\omega_c)|/2$ term in the integral, shown in Fig. 9-10, applies a large weighting to slope change in the immediate vicinity of $\omega_c$. Consequently, the phase at $\omega_c$ is affected primarily by the local dB amplitude ratio slope (the first term in the expression) and local changes in this slope (the integral term). If the dB amplitude ratio slope is essentially constant over a wide region about $\omega_c$, the expression reduces to the first term only. For

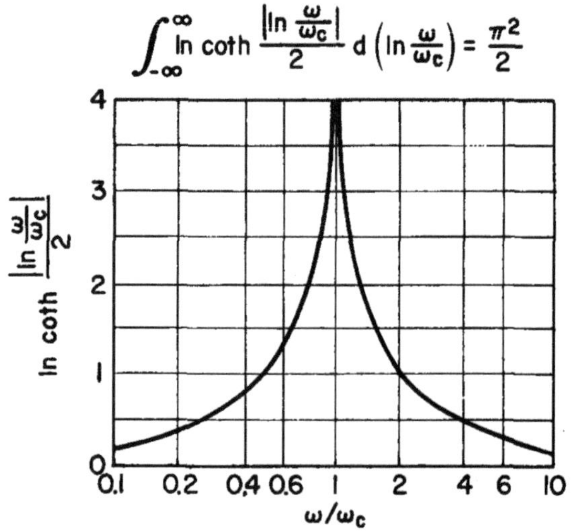


Fig. 9-10. Weighting function in phase integral.

[6] H. W. Bode, *Network Analysis and Feedback Amplifier Design*, Van Nostrand, New York, 1945.

low pass, open-loop transfer functions the amplitude ratio slope at gain crossover is negative, so a positive phase margin can exist only when $d\,|G(j\omega)|_{\mathrm{dB}}/d(\ln \omega/\omega_c)$ in the immediate region is less (numerically) than $-40$ dB/decade, and the local changes in slope are moderate. Consequently, the available crossover regions for most transfer functions are confined to areas in which the local amplitude ratio slope fulfills these conditions and the possible bandwidths are restricted accordingly.

When a reasonable margin of stability and/or good transient response is to be provided, the possible gain crossover frequencies are still more limited. For instance, when phase margin is viewed as a measure of stability margin alone (ignoring for the moment its interpretation as a measure of response and other qualities), a minimum phase margin specification of 35 to 40 degrees results in both a maximum possible bandwidth specification (for a given system) and in relatively restricted regions of possible crossover. In terms of the implications of Eq. 9-29 such phase margins can be interpreted as limiting the crossover frequency to regions in which the amplitude ratio slope is approximately $-20$ dB/decade. These limitations on crossover regions may be quite gross, but the main point of the present argument is that they do exist.

*Equalization Economy*

In its narrowest sense, series equalization is the process of modifying signals derived from sensors and presenting these modified signals to actuation devices. Physically, the equalization elements form the connecting links between sensors and actuators; they establish the form of the dominant low frequency cause/effect relationships between vehicle output motions sensed and control actions imposed upon the vehicle. Equalization thus reflects directly into the possible forms for $G_f(s)$ and $G_a(s)$, at least in the sense that the product $G_aG_f$ is the means by which a control action, proportional to some function of vehicle output quantities, is applied to the vehicle. Practical considerations impose limitations on the types of operational functions derivable from a given sensor. Examples are: equipment complexity; the ratio maximum-signal:minimum-increment-of-control (dynamic range); noise amplification, especially as it affects saturation in subsequent elements; etc. Ordinarily, such factors restrict equalization generation (per sensor) to no more than integral, proportional, and rate signals. Thus, if the single-sensor-loop control system of Fig. 9-6 should contain only one sensor, the low frequency form of $G_a(s)G_f(s)$ is fairly well restricted to

$$G_a(s)G_f(s) \doteq \frac{K_{\bar{c}}}{s} + K_c + K_{\dot{c}}s \tag{9-30}$$

Sensor and actuator dynamics, as well as high frequency lags from the equalization elements, will be present in more precise mathematical descriptions of $G_a G_f$, but the above form serves to establish an upper limit on total equalization.

The primary role of equalization is to modify the transfer function $G(s)$ in a fashion calculated to provide good closed-loop dominant modes. As the means to this end, equalization elements may be employed to raise low frequency $|G(j\omega)|$ levels to approximate the conditions $|G(j\omega)| \gg 1$ over a desired frequency band and also to modify the form of $G(s)$ in the crossover region such that good dominant closed-loop modes result. Therefore, *equalization can be an important factor in the crossover and* bandwidth limitations imposed by stability considerations.

None of these comments are news to anyone even remotely aware of elementary servo theory, but the point of the discussion—that the total form of equalization available is often narrowly limited by practical considerations—still needs to be emphasized.

*Actuator-Loop Characteristics*

The actuation element, so blithely dismissed thus far as an inner loop, must often contend with a wide variety of nonlinearities and disturbances, as well as bewildering dynamic effects occasionally inherent in the elements themselves. Because of the difficulty encountered in coping with the physical realities imposed upon the actuation loop, many experienced flight control designers steadfastly believe that a sound solution of the actuation problem is the secret of an ultimately successful design.

In general terms, the suppression of load disturbances (Fig. 9-3) within the actuator loop demands that the open-loop amplitude ratio for this loop be large over a wide range of frequencies. Further, the closed-loop dynamics of the actuator loop are generally a predominant factor in setting the outer-loop crossover to some value consistent with overall flight control system stability. On the basis of both of these considerations, the bandwidth of the actuator loop should be as large as possible.

These generalities can be given more substance by considering that the open-loop actuator transfer function can often be approximated by

$$G(s) = \frac{Ke^{-\tau s}}{s} \tag{9-31}$$

where $\tau$ is an equivalent, pure time delay (Chapter 3) which accounts for all high frequency "additional" dynamics; the basic $K/s$ form corresponds to the $-20$ dB/decade rule noted above.[7] The properties of the

[7] We should note that this open-loop system form is an excellent approximation for the dominant characteristics of a great many "good" systems. Consequently, its properties are of more general interest than in this local context.

closed loop actuator transfer function,

$$G_a(s) = \frac{G(s)}{1 + G(s)} = \frac{Ke^{-\tau s}}{s + Ke^{-\tau s}} \tag{9-32}$$

are readily mapped as a function of gain, $K$, in the $j\omega$ and $\sigma$-Bode diagrams and in the Bode root locus of Fig. 9-11. The corresponding conventional root locus is also given in this figure.

Figure 9-12 shows the relationship between closed-loop damping, $\zeta_{\text{CL}}$, and phase margin,

$$\varphi_M = \measuredangle G(j\omega_c) - (-\pi) = -\left(\frac{\pi}{2} + \tau\omega_c\right) + \pi = \frac{\pi}{2} - \tau\omega_c \tag{9-33}$$

Note also that $\omega_c = K$ for this system.

If, as a measure of performance, we consider the actuator system's response to a spectral input, the mean-squared error is given by[8]

$$\overline{e^2} = \frac{1}{2\pi}\int_0^\infty \Phi_{ee}(\omega)\,d\omega = \frac{1}{2\pi}\int_0^\infty |G_{\text{er}}|^2\,\Phi_{\eta\eta}\,d\omega \tag{9-34}$$

For the rectangular power spectral density input specified in Fig. 9-13,

$$\overline{e^2} = \frac{1}{2\pi}\int_0^{\omega_i} |G_{\text{er}}|^2\,\Phi_{ii}\,d\omega \qquad \text{or} \qquad \frac{\overline{e^2}}{\sigma^2} = \frac{1}{\omega_i}\int_0^{\omega_i} |G_{\text{er}}|^2\,d\omega \tag{9-35}$$

Substituting the closed-loop, input-error characteristics, we get

$$|G_{\text{er}}(j\omega)| = \left|\frac{j\omega}{j\omega + \omega_c e^{-j\omega\tau}}\right| = \frac{\omega}{\sqrt{\omega^2 - 2\omega\omega_c \sin\omega\tau + \omega_c^2}} \tag{9-36}$$

$$\frac{\overline{e^2}}{\sigma^2} = \frac{1}{\omega_i}\int_0^{\omega_i} \frac{\omega^2\,d\omega}{\omega^2 - 2\omega\omega_c \sin\omega\tau + \omega_c^2} \tag{9-37}$$

This function is plotted in Fig. 9-14 and can be approximated for $\sin\omega\tau \doteq \omega\tau$ and $\omega_i\tau \ll 1$ by

$$\begin{aligned}\frac{\overline{e^2}}{\sigma^2} &\doteq \frac{1}{\omega_i}\int_0^{\omega_i} \frac{\omega^2\,d\omega}{\omega^2(1 - 2\omega_c\tau) + \omega_c^2} \\ &= \frac{1}{(2\omega_c\tau - 1)}\left[1 - \frac{\omega_c/\omega_i}{\sqrt{2\omega_c\tau - 1}}\tanh^{-1}\frac{\omega_i}{\omega_c}\sqrt{2\omega_c\tau - 1}\right]\end{aligned} \tag{9-38}$$

[8] See Chapter 10.

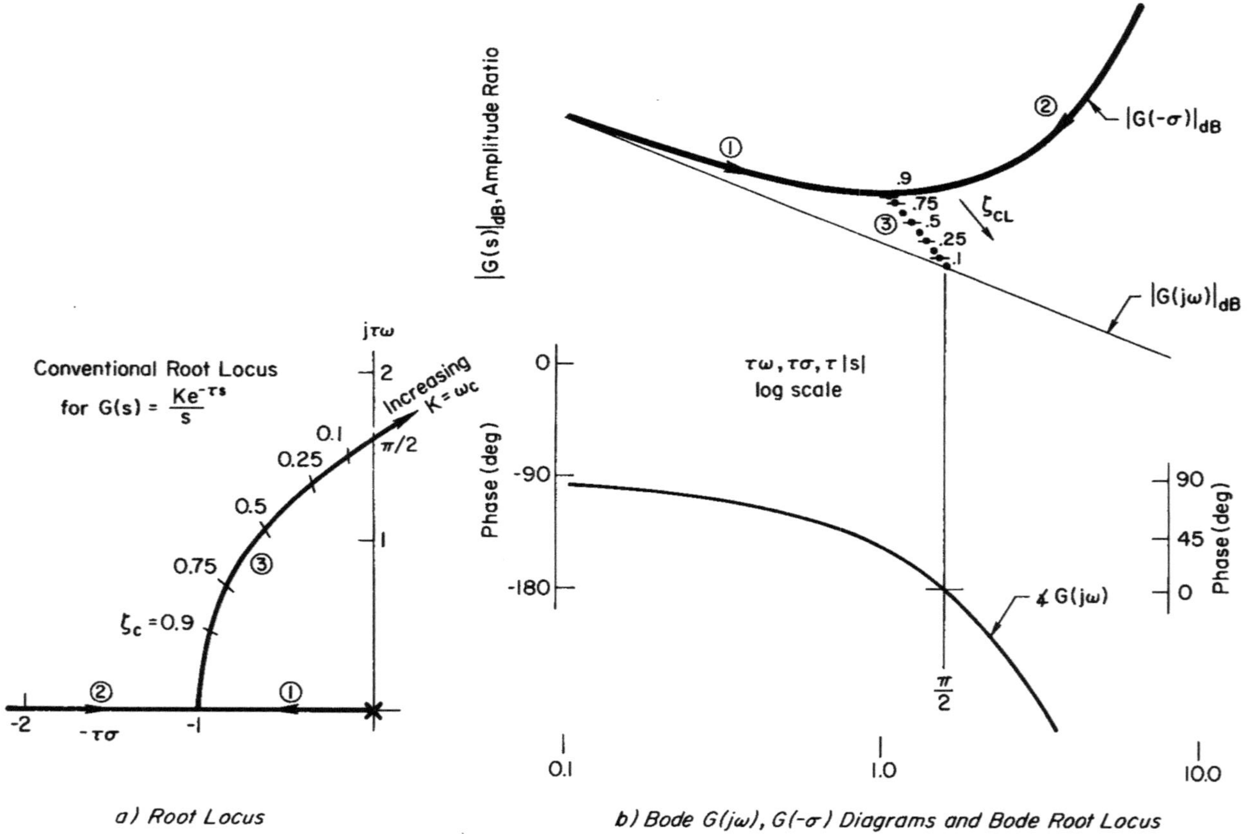


Fig. 9-11. System survey for $G(s) = Ke^{-\tau s}/s$.

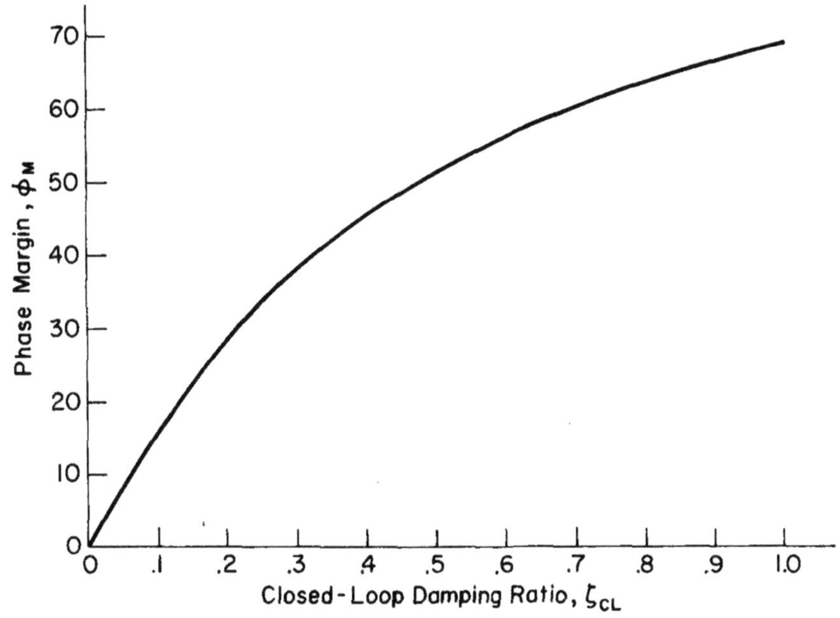


Fig. 9-12. The phase-margin/closed-loop damping ratio relationship for $G(s) = Ke^{-\tau s}/s$.

If the $\tanh^{-1}$ is expanded, and carried only to the second term,

$$\frac{\overline{e^2}}{\sigma^2} \doteq \frac{1}{3}\left(\frac{\omega_i}{\omega_c}\right)^2 \tag{9-39}$$

This extremely simple law fits the exact values from Fig. 9-14 with reasonable accuracy for $\omega_i/\omega_c$ less than about 0.4. It shows, as does the Fig. 9-14 plot, that increasing the system bandwidth (or crossover frequency, $\omega_c$) beyond the largest frequency in the input ($\omega_i$) has a directly beneficial effect in reducing system errors. Thus, if the frequency content of the actuator command and actuator load disturbance signals shown in Fig. 9-13 is largely composed of frequency components considerably less than $\omega_c$, the system would be satisfactory in both its servo and regulator functions. In this sense the actuation system would have very high integrity indeed.

However, if we assume that the flight control system *internal* disturbances presented at the input of the actuator loop, $\eta_a$, can be represented

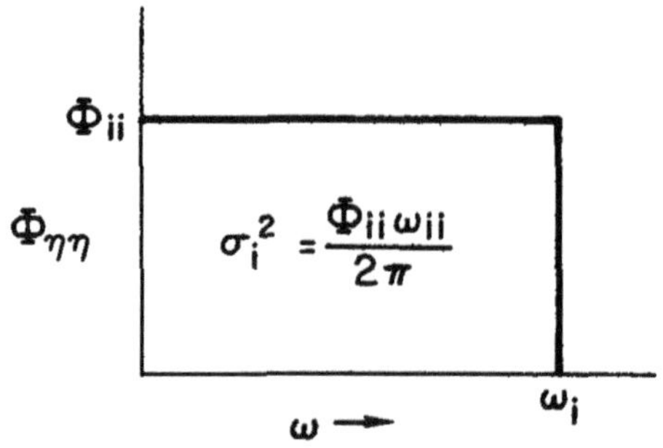


Fig. 9-13. Input Spectrum.

by a constant power spectral density, $\Phi_{\eta\eta}$, over the actuator loop bandwidth (i.e., $\omega_i = \omega_c$), then increasing actuator bandwidth is actually deleterious (see Fig. 9-14 for $\omega_i/\omega_c = 1.0$). This is perhaps easier to see by considering, as a gross approximation, that the *closed-loop* actuator dynamics have a form similar to that of the ideal low pass filter characteristics shown in Fig. 9-15. A system possessing these characteristics would pass all signals without attenuation up to a frequency of $\omega_b$; the phase angle between an input sine wave and the steady-state sinusoidal output at the same frequency would vary linearly with frequency from zero to a maximum lag of $\varphi_b$ at $\omega_b$. These properties are, of course, unrealizable in a nonanticipatory device, as is made obvious in Fig. 9-15 by the initiation of the output response prior to the onset of the step input. Neither the frequency response nor the transient step response, however, is very far from approximating the characteristics of practical servomechanisms.

For this ideal low pass actuator, the mean-squared output due to the internal disturbance input will be given by (refer to Eq. 9-34)

$$\overline{\delta^2} = \frac{1}{2\pi}\int_0^\infty \Phi_{\delta\delta}(\omega)\, d\omega$$

$$= \frac{\Phi_{\eta\eta}}{2\pi}\int_0^{\omega_b} d\omega = \frac{\Phi_{\eta\eta}\omega_b}{2\pi} \qquad (9\text{-}40)$$

Equation 9-40 indicates that this undesired mean-squared output increases directly with the bandwidth, $\omega_b$. Since the signals within the actuator loop are directly related to the output, these signals will also increase more or less proportionately to the actuator closed-loop bandwidth. However, saturation invariably occurs in one or more of the actuation loop elements, so that an increase in the general signal levels within the

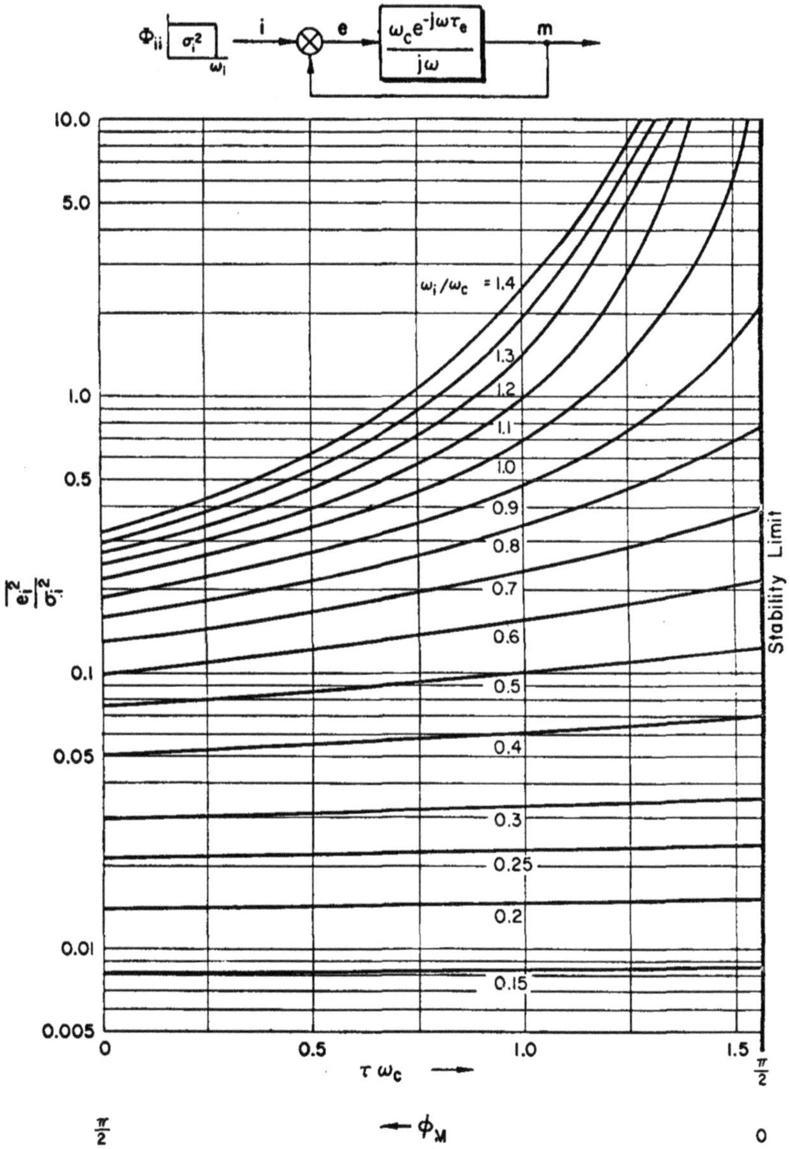


Fig. 9-14. Mean-squared error as function of $\tau\omega_c$ for $G(s) = Ke^{-\tau s}/s$.

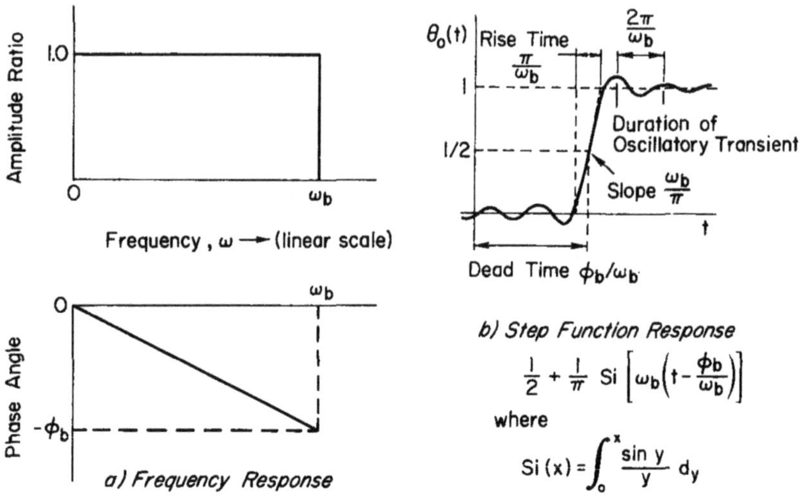


Fig. 9-15. Response of the ideal low pass filter.

loop will eventually unavoidably result in more saturation occurrences per unit time. In the limit, of course, this can become bad enough almost to entirely remove any connection between the actuator output and its command input.

We have remarked previously that the effects of unwanted signals appearing at the input of the overall system are seldom deleterious in a good design, despite the existence, in principle, of the effects mentioned here. The significant differences in this regard between the actuator and outer loops are that actuator loop bandwidth is an order of magnitude or more higher than those of the outer loops; and the relative magnitude of command and disturbance signals occurring at the actuator loop inputs are much more comparable.

Another physical effect directly related to the foregoing discussion is the increase, with actuator bandwidth, in the power and energy required by the actuator to perform its totally useless task in following the unwanted disturbances.

Other practical points militating against indefinite extension of actuator loop bandwidth are the economic, weight, size, and reliability costs paid for extremely high dynamic performance devices operating at high power levels; and the almost inevitable troubles encountered due to higher order dynamic modes. The latter may stem from either the vehicle dynamics or the actuator load dynamics.

All of the above discussion is intended to lead to the conclusion that actuator loop bandwidth should be as low as possible, consistent with suppression of actuator loop disturbances, effective linearization of nonlinearities, and outer loop stability and dynamic performance.

*Effects of Noise at the Input*

As noted above, the argument for restricted actuator loop bandwidth can be made in principle for the outer loop. To the extent that this argument applies, one conceptual basis for bandwidth selection as a tradeoff is apparent. The competition involved is that between command or input signals, $R(s)$, and internal disturbances, $\eta(s)$ (Fig. 9-6); the tradeoff would be made between response to commands and suppression of external disturbances on the one hand, and the reduction of errors due to the internal disturbances on the other. This basis of compromise is seldom of much consequence in flight control systems, since the scaling of desired signals relative to unwanted signals acting at the input can usually be made very large with careful design.

*Summary of Compromises in Selecting Crossover Region*

With equalization restricted to the reasonable forms noted, all of the discussion indicates that the bandwidth, $\omega_b$, of the closed-loop transfer function, $G(s)/[1 + G(s)]$, is often not a very independent parameter. Instead, it is ordinarily restricted in value by vehicle or actuator characteristics that may be relatively unalterable. This feature can possibly be avoided by the introduction of extensive equalization and by efforts to extend the bandwidth of sensors and actuators. But the latter can be accomplished only by some degradation in performance, an increase in overall cost and unreliability, and an education in the almost inevitable troubles encountered with higher frequency dynamic effects. Thus the general conclusions of this subsection may be summarized as

1. $|G(j\omega)/[1 + G(j\omega)]| \doteq 1 \quad \text{for} \quad \omega < \omega_b$
2. $|G(j\omega)/[1 + G(j\omega)]| \ll 1 \quad \text{for} \quad \omega \gg \omega_b$
3. The bandwidth $\omega_b$ must be great enough to *control* and/or stabilize the vehicle modes for which automatic control is desired, and, in any event, cannot be made significantly larger without exchanging the increased bandwidth for trouble.
4. Ranges of $\omega_b$ are essentially fixed, by open-loop transfer function and stability considerations, at values that result in good closed-loop system stability.

## 9-6. Specifications and Testing

The reader will recall from the discussion at the beginning of this chapter that we proposed to defer consideration of the methods for determining the command and disturbance environment characteristics, and the methods for assessing the dynamic performance, until Chapter 10. We turn instead to a very brief description of the methods employed in the detailed study of a selected system.

Consider a case in which several competing flight control systems have been evolved on a block diagram basis, and the performance and design qualities of these systems have been assessed. One of them is then selected as having the best combination of desirable qualities and is used to establish the detailed functional requirements. Further study of the selected system, evolved from the preliminary design process, usually involves a series of increasingly realistic tests that attempt to thoroughly analyze the system in both normal and abnormal operating conditions. The purpose of all this testing, insofar as possible, is to insure that, when the equipment is put into service, it will not have any surprises left for its designers and builders. Surprises occur when some mode or condition of operation has been overlooked or when the assumptions used in constructing the mathematical models of the system elements are invalid.

Very likely, the first model of the system will have been based on linearized descriptions of the system elements, including the perturbation equations of motion of the airframe written with respect to straight and level flight. The first checks to be made concern the effects of small parasitic nonlinearities in the elements of the system. This may be done on an individual and approximate basis by analytical means, but anything approaching a comprehensive study of the effect of multiple nonlinearities in the elements of the system is, of necessity, accomplished through *simulation*. Even before any hardware is built, estimates may be made of the characteristics of, e.g., amplifier saturation, sensor thresholds, control system friction, and mechanical hysteresis; these may be "mechanized" on a computer, together with the other elements of the system, to evaluate performance in the presence of the system element nonlinearities.

It is next necessary to consider the performance of the system for motions that are no longer small, and possibly also to relax the assumption of the airframe as a rigid body. Some facets of the subject of large motions and elastic modes are covered in the reports and papers listed in the bibliography at the end of this book. These references, however, may represent only an introduction to some of the difficulties of a specific system, and, again, the performance of the system in high climb or dive angles, at large bank angles, or in rapid rolls is usually studied by means of simulation. For this purpose the six degree of freedom equations with nonlinear aerodynamic terms as functions of several variables are mechanized on a computer and often several elastic modes are also included in the simulation.

If, at this point, the system and its elements still show satisfactory performance, consideration may be given to the construction of prototype hardware. In general, the equipment may evolve through several stages such as the breadboard model, the flightworthy prototype, the preproduction model, and finally the production model. At each stage the pieces of equipment are subjected to tests as components and are assembled for tests as a system. It may very well happen, however, that not all the elements are in the same state of evolution during any particular system test. Thus, for example, a breadboard control amplifier might be combined with a preproduction actuator, and so on.

Each component by itself, and the system as a whole, may be required to comply with general specifications or standards, in addition to meeting functional requirements. The general specification governing automatic flight control systems for piloted, naval aircraft is: "Control and Stabilization Systems—Automatic, Piloted Aircraft, General Specification for, MIL-C-18244." This specification makes reference to many other specifications and standards. Some of these are listed in Table 9-4. The roughly corresponding specification for U.S. Air Force aircraft flight control systems is: "Fight Control Systems—Design, Installation, and Test of, Piloted Aircraft, General Specification for, MIL-F-9490."

Many of the applicable specifications prescribe specific tests such as the radio noise interference tests, and the environmental tests, including vibration, altitude, sand and dust, humidity, high and low temperature, and salt spray. Otherwise, each component and the system as a whole, insofar as is practical, are subjected to tests for the purpose of determining the frequency response, loading effects, linearity, effects of saturation and saturation levels, switching transients and noise characteristics. Further, the automatic flight control system is usually tested in closed-loop operation in conjunction with an aircraft control system functional test stand and a computer simulation of the vehicle equations of motion. The ground equipment required for environmental tests and the closed-loop tests with the control system functional test stand is illustrated in Fig. 9-16.

A typical test stand comprises a steel frame on which are mounted all the important elements of the actual aircraft control system. It includes the complete control surface actuating system, pilot's seat, cockpit controls, and artificial feel devices. Stick forces that would appear in response to motion, such as those which might be produced by a bobweight, are applied by a force servo driven by the computer. Any automatic control equipment to be tested is installed on the control test stand in a manner representing, as closely as possible, the installation in the actual aircraft. Simulated aerodynamic loads are applied to the control surfaces by means of mechanical or hydraulic springs and dampers.

Some elements of the pilot's display are also often included to simulate the stimuli to which the pilot responds in flight. Cockpit instrument

Table 9-4. Specifications and standards applicable to the design, installation, and operation of automatic control and stabilization systems in naval aircraft.

| | |
|---|---|
| MIL-C-18244 | Control and Stabilization Systems: Automatic, Piloted Aircraft, General Specification for |
| JAN-I-225 | Interference Measurements, Radio, Methods of, 150 Kilocycles to 20 Megacycles (For Components and Complete Assemblies) |
| JAN-T-781 | Terminal; Cable, Steel (For Swaging) |
| MIL-F-3541 | Fittings, Lubrication |
| MIL-S-3950 | Switches, Toggle |
| MIL-E-4682 | Electron Tubes and Transistors, Choice and Application of |
| MIL-W-5088 | Wiring, Aircraft, Installation of |
| MIL-E-5272 | Environmental Testing, Aeronautical and Associated Equipment, General Specification for |
| MIL-E-5400 | Electronic Equipment, Aircraft, General Specification for |
| MIL-H-5440 | Hydraulic System; Aircraft Type 1 and 2, Installation and Data Requirements for |
| MIL-I-6115 | Instrument Systems, Pitot Tube and Flush Static Port Operated, Installation of |
| MIL-I-6181 | Interference Control Requirements, Aircraft Equipment |
| MIL-L-6880 | Lubrication of Aircraft, General Specification for |
| MIL-E-7080 | Electrical Equipment, Piloted Aircraft Installation and Selection of, General Specification for |
| MIL-M-7793 | Meter, Time Totalizing |
| MI2-M-7969 | Motors, Alternating Current, 400-Cycle, 115/200 Volt System, Aircraft, General Specification for |
| MIL-A-8064 | Actuators and Actuating Systems, Aircraft, Electro-Mechanical, General Requirements for |
| MIL-H-8501 | Helicopter Flying Qualities, Requirements for |
| MIL-S-8512 | Support Equipment Aeronautical, Special, General, Specification for Design of |
| MIL-M-8609 | Motors, Direct Current, 28-Volt System, Aircraft, General Specification for Class A and B |
| MIL-D-8706 | Data, Design; Contract Requirement for Aircraft |
| MIL-F-8785 | Flying Qualities of Piloted Airplanes |
| MIL-D-18300 | Design Data Requirements for Contracts Covering Airborne Electronic Equipment |
| MIL-N-18307 | Nomenclature and Nameplates for Aeronautical Electronic and Associated Equipment |
| MIL-E-19600 | Electronic Modules, General Aircraft Requirements for |
| MIL-R-22256 | Reliability Requirements for Design of Electronic Equipment or Systems |
| MIL-R-23094 | Reliability Assurance for Production Acceptance of Avionic Equipment, General Specification for |
| MIL-STD-203 | Cockpit Controls; Location and Actuation of For Fixed Wing Aircraft |
| MIL-STD-704 | Electric Power, Aircraft, Characteristics and Utilization of |
| MS15001 | Fittings, Lubrication (Hydraulic) Surface Check, 1/4-28 Taper Threads, Steel, Type 1 |
| MS15002 | Fittings, Lubrication (Hydraulic) Surface Check, Straight Threads, Steel, Type 2 |

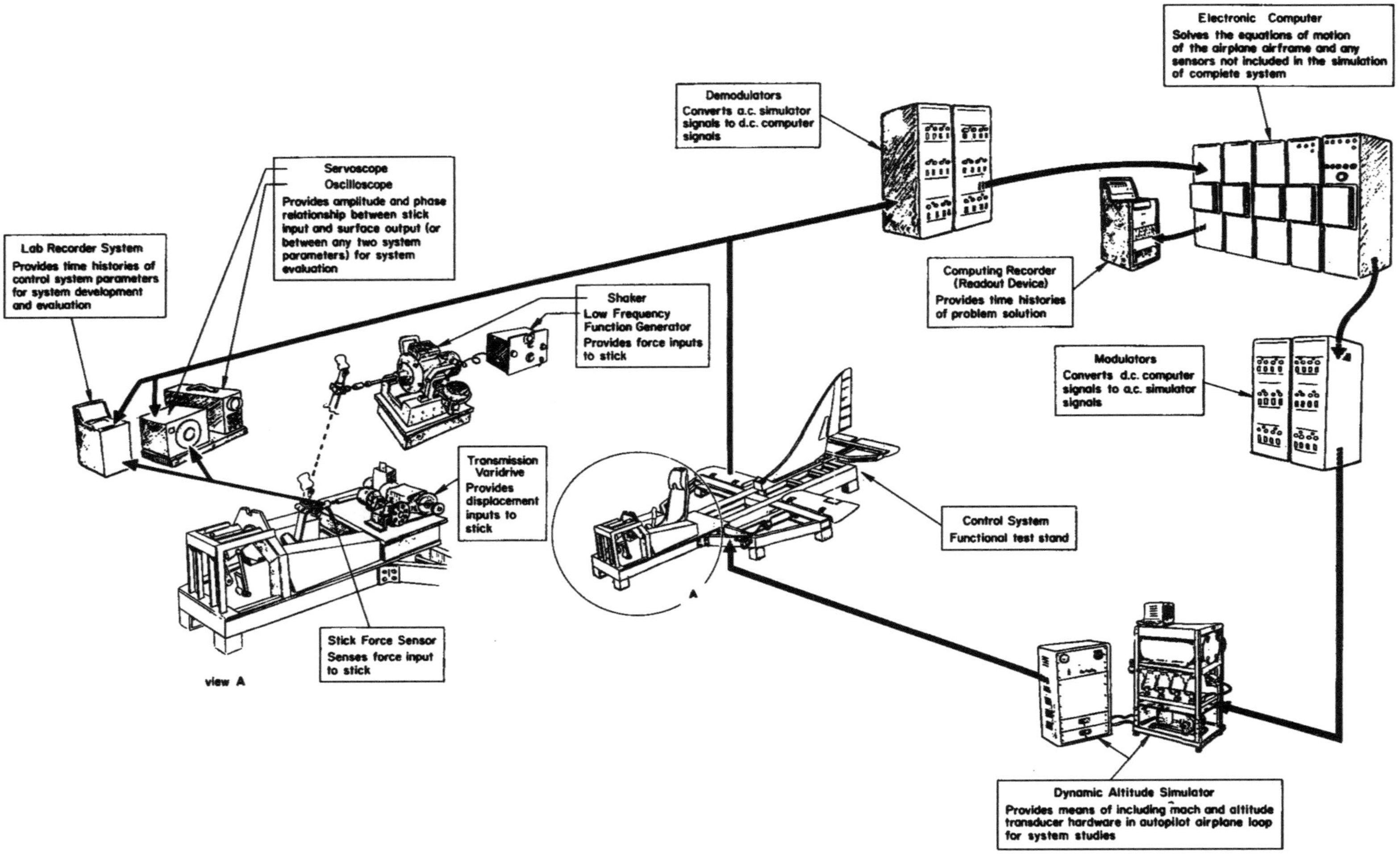
Electronic Computer
Solves the equations of motion of the airplane airframe and any sensors not included in the simulation of complete system
Demodulators
Converts a.c. simulator signals to d.c. computer signals
Servoscope
Oscilloscope
Provides amplitude and phase relationship between stick input and surface output (or between any two system parameters) for system evaluation
Lab Recorder System
Provides time histories of control system parameters for system development and evaluation
Shaker
Low Frequency
Function Generator
Provides force inputs to stick
Computing Recorder
(Readout Device)
Provides time histories of problem solution
Modulators
Converts d.c. computer signals to a.c. simulator signals
Transmission
Varidrive
Provides displacement inputs to stick
Control System
Functional test stand
A
Stick Force Sensor
Senses force input to stick
view A
Dynamic Altitude Simulator
Provides means of including mach and altitude transducer hardware in autopilot airplane loop for system studies

Fig. 9-16. Flight control functional test stand arrangement.

presentations that are commonly provided include the artificial horizon, altimeter, airspeed and Mach meters, the heading indicator, possibly the turn and slip indicator, and localizer and glide slope needles. In certain applications fire control or other weapon delivery displays may also be provided.

Airframe dynamics are simulated on the computer which can be analog, hybrid, or digital. For analog simulation, the computer inputs are voltages proportional to control surface deflections and its outputs can be voltages proportional to any or all of the airframe output quantities (motions). These voltages are then used to drive the pilot's displays, the simulated force producers, as simulated sensor inputs to the controller and for recording the responses. Modulators, demodulators, converters, and scale changing devices are used where necessary to change the form and level of electric signals.

The use of the test stand permits additional and more realistic tests to be conducted, especially for those modes of operation in which the human pilot plays an important role. Since the actual operating conditions of the equipment are also much more closely simulated than was the case when each component of the system was represented only by a mathematical model, the test stand provides information that might otherwise have to be obtained in flight. This serves the desirable purpose of cutting down the number of flight test hours required to tune up and test the automatic flight control equipment.

Another important application of the control test stand is in the investigation of the results of possible component failures. A systematic program of inducing specific failures such as mechanical malfunctions and open and shorted electrical circuits is conducted. These tests may be made both with and without the human pilot, since it is important to observe whether or not the pilot can detect the failure and can make a successful correction that does not result in dangerous motions of the airframe. In cases in which there may be a question regarding the structural safety of the airplane, test stand investigations are mandatory because of the danger involved in determining the effects of component failures in flight.

Of course, the test stand described above does not simulate the motion cues to which the pilot will be subject in flight. (Moving-base simulators are currently largely employed for research purposes but are not yet commonly used in flight control system development.) In fixed-base simulation, the tests should therefore be confined to cases in which the motion cues can be considered to be unimportant or for which fixed-base results are conservative.

At the conclusion of the test-stand program the prototype equipment is installed in the airplane for ground tests. Some of the test-stand results should be checked to assure that the actual installation in the aircraft has not altered their nature. It may even be desirable to perform some closed-loop tests with the computer representing the motion sensors and the dynamics of the vehicle but with everything else being the actual equipment. An essential feature of the first tests of the flight control system in the actual vehicle, however, is the establishment of inspection test procedures for the production system. Inspection test procedures are required to insure that malfunctioning components are not installed in the airplane. In addition, many automatic flight control systems require individual adjustments after installation in the airplane to compensate for component and airplane tolerances. Procedures for accomplishing this must be developed and written in such a way as to be readily understandable by typical mechanics and technicians. Normally, the procedure must be developed by cut and try, and the occasion of the prototype installation is the first chance to do this.

Final evaluation of the operating characteristics of an automatic flight control system is, of course, made by means of flight tests. The magnitude of the flight test program depends on the complexity of the system being tested and on the amount of ground testing that has preceded the flight tests. If a thorough flight simulation program has been carried out on the control test stand, the flight tests, with luck, may consist of very little more than verification of the results previously obtained on the ground. On the other hand, it may be found that considerable development work has to be carried out during the flight test phase. This will be the case if, e.g., the previously neglected motion cues for the pilot assume importance or if previously neglected higher structural modes cause closed-loop instability. It will almost certainly occur *if great care is not taken at all phases of the design to forecast and to take into account the unexpected and uncertain.*

In any case, many of the same aircraft input and output quantities that were recorded during the ground tests should be recorded in flight. In fact, insofar as possible, some of the same tests should be repeated so as to reveal any discrepancies between the performance in the ground tests and the performance in flight. A number of additional quantities that may facilitate the analysis of system operation in the event of unexpected modes of operation should also be recorded.

It is then desirable to explore the conditions that may not have been examined in the ground tests. These might be simulated tactical situations, spins, or large scale maneuvers such as the Immelman turn sometimes used in bomb delivery. (It may have been difficult or impossible to simulate these.) Since these new test conditions may very well reveal new modes of automatic flight control system performance, some redesign or at least the adjustment of system parameters may well be called for at this juncture.

Following the successful completion of flight tests, the system design is frozen, the production design of the components is completed, detailed component test specifications are finished, and the system test specification (described above) is revised to accommodate the results of the flight test program. The *design* process is then, in principle, complete. Actually, it will usually happen that the system design engineer may be called upon to resolve difficulties in manufacturing or in the in-service use of the automatic flight control system. The resolution of some of these difficulties may even involve new changes to component or system designs.

CHAPTER 10

# INPUTS AND SYSTEM PERFORMANCE ASSESSMENT

## 10-1. Introduction

The elementary feedback control system concepts reviewed in the last chapter provide a basis for the view that flight control system performance requirements, in the large, serve to define a system that will follow desired inputs, reject internal disturbances, and suppress external disturbances. There remain the inevitable questions of how well the "following," "rejecting," and "suppressing" needs to be done; and what the tradeoffs may be between the design qualities of reliability, weight, power demands, cost, etc., and the dynamic performance quantities of following, rejecting, and suppressing. Further, in flight control systems the controlled element (vehicle) is not entirely unalterable, so that the consequences of possible interchanges of function between controller and vehicle are also subject to compromise. In optimization efforts, interactions between all of these considerations occur, but only the dynamic performance quantities are at first involved in the satisfaction of dynamic requirements.

The three key dynamic performance attributes in any system are stability, response, and accuracy. There are many definitions of stability, most of them coined to satisfy requirements of generality and mathematical rigor. For our purposes a qualitative operational definition will suffice, i.e., "if a small temporary input applied to the system in equilibrium results only in a temporary change in the output or response, the system is stable." For constant coefficient linear systems the stability according to this definition amounts to a specification on pole location; in particular, denying all of the right half-plane and the axis of imaginaries, except the origin, to the poles of stable systems. Poles at the origin are permitted in principle by appropriately defining the output or response quantity. Because stability for constant coefficient linear systems is specified by pole location, and because the methods of analysis given heretofore give precise pole locations, we have bypassed discussion of conventional stability criteria in this book. Stability is, of course, readily determined by several analytical and graphical techniques without determining the precise location of the poles. These are fundamental and well known tests,[1] so that no further mention is needed here.

Once stability, relative to some operating point, is attained in a constant coefficient linear system, a unique stable equilibrium state exists insofar as the system dynamic excursions about that operating point are concerned. The stable equilibrium state establishes a datum from which perturbed responses may be computed and accuracy assessed. Because the equilibrium is unique, accuracy and response are closely related, with accuracy being assessed by an error response related directly to the difference between input and output. If the equilibrium were not unique, as might occur in a nonlinear system, the accuracy might be completely independent of the input and output response. But, for linear systems, accuracy may be considered to be just another response quantity. In view of this, response considerations are central to flight control system performance assessment. Flight control systems are subjected to both deterministic and random inputs, so that responses to both categories are of interest and will be considered. To make this possible, several simplified and idealized inputs will be introduced in the first part of this chapter. These can be used with the system mathematical model to determine the output responses that serve as the bases for system dynamic's assessments. In this connection we should remark that, if all the details are considered, flight control systems are so complex and of such high order that the possibility of obtaining a simple and adequate system representation might seem remote. Yet a simple analytical form is almost an essential for preliminary design purposes if physical understanding is to be gained rapidly and if repetitious, ad hoc procedures are to be minimized. Fortunately, the feedbacks inherent in flight control systems usually result in a large range of frequencies over which pertinent open-loop transfer functions have magnitudes much greater than unity. As a consequence simple systems can usually be derived that serve as close approximations to the actual systems. The development of simple low order systems to approximate complex high order systems has already been illustrated with an example in Chapter 3 (Section 3-4).

In a fundamental sense a control system, to be useful, must contend successfully with all the inputs imposed upon it. A major step in the design process is, accordingly, the identification of the type, form, and general character of the input environment, followed by a detailed

[1] The more prominent are the Routh and Hurwitz tests and the Nyquist stability criterion. For the Routh-Hurwitz criteria see, e.g., D. Graham and D. McRuer, *Analysis of Nonlinear Control Systems*, Wiley, New York, 1961, pp. 457–460. For a comprehensive treatment of conventional and several generalized versions of the Nyquist criterion, see D. T. McRuer, ed., "Methods of Analysis and Synthesis of Piloted Aircraft Flight Control Systems," BuAer Rept. AE 61-4I, Mar. 1952, pp. 3-2 to 3-10.

assessment of the more critical input-system response combinations. The critical combinations are especially important in the selection of bandwidth (the general region of crossover) and the character of the response.

Because a flight control system has many different command tasks that must ordinarily be performed over a large range of environmental conditions and mission phases, a variety of inputs must be considered. These may be classified, according to their point of entry into the system block diagram, in three general categories: command point, internal, and external. The command point inputs comprise all the flight control system command signals generated by the guidance loop or otherwise inserted directly. Some of these inputs are desired instructions for control while others are unwanted consequences of nature or the result of simplifications in the guidance system mechanization. The internal inputs are, in general, all unwanted disturbances acting primarily on the controller. They arise either from design compromise or from the operation of nature's side effects. The external inputs are forces or moments, induced by the vehicle or the external environment, which act as disturbances upon the vehicle. Suppression of the effects of these forces and moments constitutes the action of the flight control system as a regulator.

A fair cross section of flight control system inputs is presented in Table 10-1. Here another classification is introduced in tabulating the inputs as a function of their analytical form. The first five entries (steps, pulses, cutoff ramps; initial conditions or impulses; gust function; power series; and periodic or repetitive functions) are all members of the deterministic class. They can all be expressed as functions of time and used to obtain system responses as functions of time under specific circumstances. The last entry, random or nondeterministic inputs, contains two somewhat different types within the random class. The first type are time functions that are either random by their nature or so complicated that a deterministic description is impractical. The second kind are generalizations of a wide range of inputs each of which, for a specific case, could be treated as deterministic. For some purposes it is more useful to consider these signals as an ensemble and to describe the whole class of inputs by a single random composite. Both types of random inputs have characteristics that can be expressed only in statistical terms, with probability distributions replacing a precise specification of the input as a function of time.

The random inputs based upon generalization of an ensemble of time functions that can also be treated as deterministic, in specific particular cases, might seem, at first glance, to be of little value, since their use in response calculations can only give a smoothed-over, average view of reality. Two important points in their favor should, therefore, be noted. The first advantage can be seen directly by taking a tack opposite to the one already mentioned. A smoothed-over, average answer can be

Table 10-1. Flight control system inputs; application point and source.

| Input form | Command point input | | Internal disturbances | External disturbances | |
|---|---|---|---|---|---|
| | Desired inputs | Unwanted inputs | | Vehicle induced | Environment |
| Steps, pulses cutoff ramps | Operating point changes (altitude, speed, etc.) | Errors in operating point changes | Mode switching; power supply variations with load switching | Thrust eccentricities; c.g. shifts; vehicle asymmetries; stores release | Wind; wind shear |
| Initial conditions or impulses | | Automatic control engagement | | Tipoff; mechanical shocks | Nonsimultaneous release of restraints; particle impact; shock and blast waves; exit-entry (incl. water) |
| Gust function | | | | | Discrete Gust |
| Power series | Target motion; terrain variation; programmed operating point changes | Higher order target motions; errors in operating point reference drifts | Unbalanced component drifts | Variable burning rate; c.g. shift | Wind shear; wind gradient |
| Periodic or repetitive | Some terrain variations | | | Dynamic unbalance; gun firing; vibration; flutter; reference coordinate rotation | Ambient fields (orbital motion only) |
| Random | Generalization of target motions; generalization of terrain variations; generalization of step-pulse sequences | Generalization of higher order target maneuvers; geometry-target noise (scintillation, beam reflection, etc.); guidance system internal noise; generalization of errors in step-pulse sequences | Power-supply fluctuations; sensor noise; local flow changes on control surfaces; electrical noise | Vibration; buffeting; acoustic noise; random variations in burning rates | Random variations in ambient fields; gusts and turbulence |

significant, especially when classes of systems are being compared. The second advantage is less obvious. It derives from the fact that a statistical view, in practice, utilizes power spectra and correlation concepts. These, in turn, provide bases for the partition of the frequency domain into regions in which either the desired or the unwanted signals are dominant, and for system design procedures that make use of the distinction. No such basis exists for distinguishing between the two when the inputs are considered as deterministic, for then the desired signal and any errors (unwanted signal) are inseparable and indistinguishable. For these reasons, several deterministic input entries in Table 10-1 also have their random counter-parts.

In succeeding sections of this chapter the more critical input types given in Table 10-1 will be summarized and, in some cases, idealized into special "equivalent" forms that depend on only one or two parameters. The initial discussion deals with the deterministic forms and with the response of linear systems to deterministic inputs. This proceeds quickly, since no new mathematical techniques are required beyond those already exposed in Chapter 2.

The third section of this chapter is devoted to a presentation of the methods of describing random functions of time. This requires some new mathematics and the development of the connections between the statistical descriptors and behavior in the time domain. It is followed by the discussion, in detail, of the analytical description of some special random functions of time useful for the representation of flight control system inputs, and further by a discussion of the properties of the very practically important class of random processes with Gaussian amplitude distributions.

Methods for calculating the response of linear systems to random inputs are treated next, and the chapter concludes with two simple examples of the application of these methods.

## 10-2. Response to Deterministic Inputs

Deterministic inputs listed in Table 10-1 include steps, pulses, impulses, cutoff ramps, power series, periodic functions, and a special function used to characterize a discrete gust. The type of time variation and the time domain and Laplace transform descriptions for the simpler of these functions are summarized in Table 10-2. Most of them are familiar.

The special function used to approximate a discrete gust, shown in Table 10-2, is made up of a step and a cosine wave, both cut off after one period of the cosine function (Fig. 10-1). The gust function is defined as

$$x(t) = \begin{cases} \dfrac{k}{T}\left(1 - \cos\dfrac{2\pi t}{T}\right); & 0 \leqq t \leqq T \\ 0; & t > T \end{cases} \tag{10-1}$$

Table 10-2. Summary of simple transient deterministic input characteristics.

| INPUT FORM | TIME CHARACTERISTICS | Time Domain | |
|---|---|---|---|
| | | Exact | Approximate |
| Impulse | 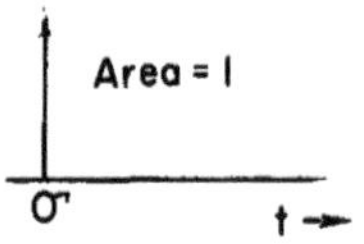 | $\delta(t)$ | — |
| Step | 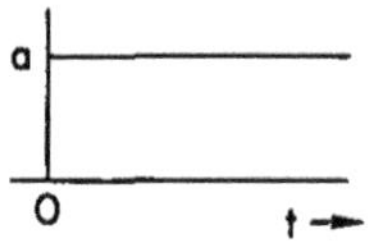 | $au(t)$ | — |
| Square pulse | 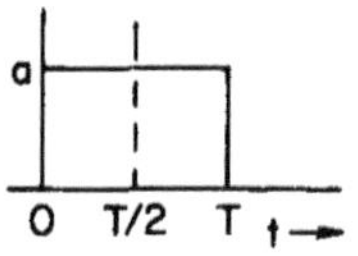 | $a\left[u(t) - u(t-T)\right]$ $u(t-T) = \begin{cases} 0, t < T \\ 1, t > T \end{cases}$ | $aT\delta\left(t - \frac{T}{2}\right)$ |
| Cutoff ramp | 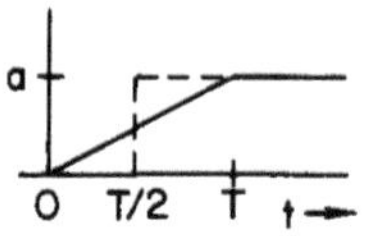 | $\frac{a}{T}\left[tu(t) - (t-T)u(t-T)\right]$ | $au\left(t - \frac{T}{2}\right)$ |
| Exponential | 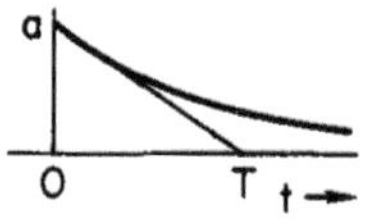 | $au(t)e^{-t/T}$ | — |
| Gust function | 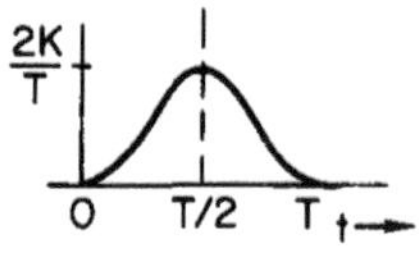 | $\frac{k}{T}\left(1 - \cos\frac{2\pi t}{T}\right)$ $0 \leq t \leq T$ | $k\delta\left(t - \frac{T}{2}\right)$ |

| ANALYTICAL DESCRIPTIONS | | | |
|---|---|---|---|
| Laplace Transform | | Energy Spectrum | |
| Exact | Approximate | Exact | Approximate |
| 1 | — | 1 | — |
| $\frac{a}{s}$ | — | $\frac{a^2}{\omega^2}$ | — |
| $\frac{a}{s}\left(1 - e^{-Ts}\right)$ | $aTe^{-Ts/2}$ | $\frac{a^2}{\omega^2}\, 2(1 - \cos T\omega)$ | $(aT)^2$ |
| $\frac{a}{Ts^2}\left(1 - e^{-Ts}\right)$ | $\frac{a}{s}\, e^{-Ts/2}$ | $\frac{a^2}{T^2\omega^4}\, 2(1 - \cos T\omega)$ | $\frac{a^2}{\omega^2}$ |
| $\frac{a}{s + 1/T}$ | — | $\frac{(aT)^2}{1 + T^2\omega^2}$ | — |
| $\frac{k(2\pi)^2}{T^3}\, \frac{(1 - e^{Ts})}{s\left[s^2 + \left(\frac{2\pi}{T}\right)^2\right]}$ | $ke^{-Ts/2}$ | $\left(\frac{2\pi}{T}\right)^4 \left(\frac{k}{T}\right)^2 \left\{\frac{2(1 - \cos T\omega)}{\omega^2\left[\omega^4 - 2\left(\frac{2\pi}{T}\right)^2\omega^2 + \left(\frac{2\pi}{T}\right)^4\right]}\right\}$ | $k^2$ |

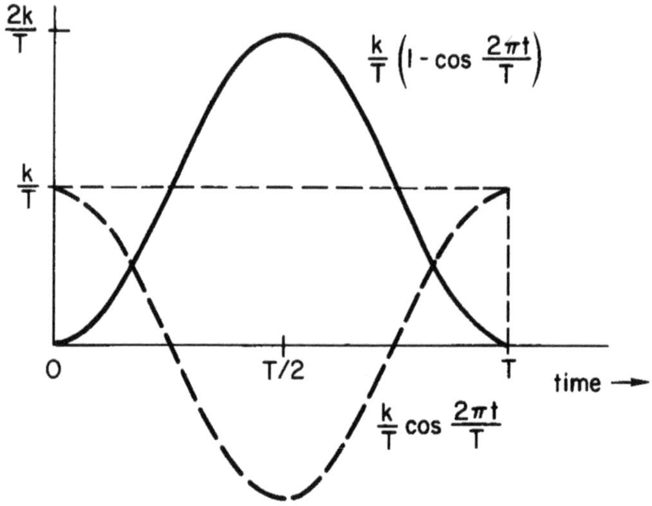


Fig. 10-1. Unit area gust function.

This form of gust description is intended to account for a gust gradient with a gradual buildup. The proportionality constant, $k$, can be selected to adjust the magnitude of the gust to any desired value. For aeronautical applications the gust length, $T$, is ordinarily taken to be equivalent to 25 chord lengths, so that

$$T \doteq \frac{25c}{U_0} \tag{10-2}$$

As a Laplace transform, the discrete gust input becomes

$$\begin{aligned} x(s) &= \frac{k}{T}\left[\frac{1}{s} - \frac{s}{s^2 + (2\pi/T)^2}\right](1 - e^{-Ts}) \\ &= \frac{(2\pi)^2 k}{T^3} \frac{(1 - e^{-Ts})}{s[s^2 + (2\pi/T)^2]} \end{aligned} \tag{10-3}$$

It will be noted that the table gives approximations for square pulses, cutoff ramps, and the gust function in terms of delayed impulses and steps. These approximations follow from reasoning that will be illustrated for the pulse. The Laplace transform for the square pulse function is

$$\begin{aligned} \mathcal{L}[\text{square pulse}] &= \frac{a}{s}(1 - e^{-Ts}) \\ &= \frac{a}{s}\left\{1 - 1\left[-Ts + \frac{(Ts)^2}{2} - \frac{(Ts)^3}{3!} + \cdots\right]\right\} \end{aligned} \tag{10-4}$$

If the system's dominant modes are characterized by inverse time constants and undamped natural frequencies which are small when compared with $1/T$, then the higher order terms in the series for $e^{-Ts}$ will make only minor contributions to all but the very first portion of the system's time response. So, carrying the exponential series only to terms in $s^2$:

$$\mathcal{L}[\text{square pulse}] \doteq \frac{a}{s}\left\{Ts\left[1 - \frac{(Ts)}{2}\right]\right\}$$
$$\doteq aTe^{-Ts/2} \tag{10-5}$$

Thus the square pulse can be replaced, under the conditions noted, by an impulse having strength equal to the area under the pulse and applied at a time $t = T/2$.[2] System response calculations for the square pulse, cutoff ramp, and gust function inputs can thus be approximated simply by moving the time origin for impulse and step responses back from zero to $-T/2$ sec. The approximations to output responses obtained in this way will range from very good when $t \gg T/2$ to very poor for $t < T/2$ (where the approximate responses are zero). The adequacy of the approximation for time values between $T/2$ and $t \gg T/2$ depends, as noted, upon the relative values of the system dominant mode characteristic and $T$. When the characteristic frequencies (inverse time constants and/or undamped natural frequencies) of the dominant modes are five or more times greater than $1/T$ the mid-range approximation ($t > T$) for the response is ordinarily quite satisfactory. The approximations improve as $t$ becomes very large relative to $T/2$, regardless of the relationships between dominant mode characteristics frequencies and $1/T$. The fact that the initial parts of the responses to the approximating input forms are very poor approximations to actual system outputs is of no great consequence. Instead of using the approximating input forms, the responses in the time period from $t = 0$ to $T$ can be found exactly by using a step for the square pulse, a ramp function for the cutoff ramp input, and the initial portion of a continuous $1 - \cos 2\pi t/T$ wave for the gust function.

To the extent that the approximations given above hold for the square pulse, cutoff ramp, and gust function inputs, systems subjected to the inputs given in Table 10-2 can be assessed by their step and impulse responses, i.e., by their indicial responses and weighting functions. Because the indicial response is the time integral of the weighting function, these two responses are so closely related that only one need be used in most dynamic response assessment procedures. The weighting function is superior when emphasis is placed upon the response to impulses, short

[2] For a more detailed justification of this approximation see, e.g., J. L. Bower and P. M. Schultheiss, *Introduction to the Design of Servomechanisms*, Wiley, New York, 1958.

pulses, and initial conditions. The indicial response is most suitable for response assessments for steps, cutoff ramps, and long pulses. When the short time ($0 < t < T$) response to cutoff ramps is also of interest, a pure ramp can be added to the test-input inventory. However, because the ramp response is just the integral of the indicial response it is, again, seldom necessary to consider ramp responses in detail apart from the closely related weighting function or indicial response. The key inputs from Table 10-2 are, therefore, impulses and steps, sometimes delayed, with the ramp being useful occasionally. In terms of system response the corresponding quantities are related, for a general input-output set $x(t)$ and $y(t)$, by a transfer function $W_{yx}$, the weighting function (system response to unit impulse input) is

$$w_{yx}(t) = \mathcal{L}^{-1}[W_{yx}(s)] = \mathcal{L}^{-1}\left[\frac{Y(s)}{X(s)}\right] \tag{10-6}$$

the indicial response or indicial admittance (system response to unit step input) is

$$I_{yx}(t) = \int_0^t w_{yx}(\lambda)\, d\lambda = \mathcal{L}^{-1}\left[\frac{W_{yx}(s)}{s}\right] \tag{10-7}$$

and the ramp response (system response to unit ramp input) is

$$V_{yx}(t) = \int_0^t I_{yx}(\lambda)\, d\lambda = \mathcal{L}^{-1}\left[\frac{W_{yx}(s)}{s^2}\right] \tag{10-8}$$

In the case of initial conditions or of impulse functions applied so as to represent portions of the initial conditions, it is impossible to distinguish these from inputs, once the proper substitutions have been made.[3] For this reason it is unnecessary to discuss these inputs separately.

The next of the general deterministic inputs to be considered is the power series in time. As should be evident from Table 10-2, this type of input is a convenient catchall for a wide variety of physical effects, when it is desired to examine particular responses rather than an ensemble. Power series are also convenient for the description of the average effects of random input time functions that have stationary statistics about a time-varying mean. The random fluctuations about the mean can then be treated separately as a stationary process.

When represented as a power series in time, an input is described by a series:

$$x(t) = a_0 + a_1 t + a_2 t^2 + \cdots + a_N t^N = \sum_{n=0}^{N} a_n t^n, \qquad 0 \leq t \leq T \tag{10-9}$$

[3] J. A. Aseltine, *Transform Method in Linear System Analysis*, McGraw-Hill, New York, 1958, pp. 29–30.

The derivatives of $x(t)$ are then:

$$\frac{dx}{dt} = a_1 + 2a_2t + 3a_3t^2 + \cdots + Na_Nt^{N-1} = \sum_{n=0}^{N} na_nt^{n-1}$$

$$\frac{d^2x}{dt^2} = 2a_2 + 6a_3t + 12a_4t^2 + \cdots + N(N-1)a_Nt^{N-2} = \sum_{n=0}^{N} n(n-1)a_nt^{n-2}$$

$$\vdots$$

$$\frac{d^ix}{dt^i} = \sum_{n=0}^{N} n(n-1)(n-2)\cdots[n-(i-1)]a_nt^{n-i} = \sum_{n=i}^{N} \binom{n}{i} i!\, a_nt^{n-i} \tag{10-10}$$

The constants, $a_n$, may be derived from theoretical considerations or possibly from physical limits (e.g., target maneuvers with maximum load factors), or may simply be the results of curve fits to empirical data (e.g., a representation of a short segment of terrain). In this case a Maclaurin's series equivalent of Eq. 10-9 is useful:

$$x(t) = x(0) + t\frac{dx}{dt}\bigg|_{t=0} + \frac{t^2}{2}\frac{d^2x}{dt^2}\bigg|_{t=0} + \frac{t^3}{3!}\frac{d^3x}{dt^3}\bigg|_{t=0} + \cdots + \frac{t^N}{N!}\frac{d^Nx}{dt^N}\bigg|_{t=0} + \cdots \tag{10-11}$$

For most inputs represented as power series in time, this approximating series is valid only for very restricted intervals. For convenience, the interval can be taken to start at $t = 0$. Because the primary concern in response calculations for power series inputs centers on the system steady-state behavior, the system transient resulting from this selection of a distinct time origin is ordinarily of secondary interest compared with the steady-state response. When the transient is of significance, other input forms such as steps and ramps are still more convenient for assessment purposes. Consequently, only the steady-state facet of the system response shall be discussed here.

When a generalized input, $x(t)$, is inserted into a linear system having a transfer function $W_{yx}(s)$, the output $y(t)$ will be given by the inverse transform of

$$Y(s) = W_{yx}(s)X(s) \tag{10-12}$$

As shown in Chapter 2 (pp. 68–69), the inverse transform of the steady-state response is

$$y(t) \doteq W_{yx}(0)x(t) + W'_{yx}(0)\dot{x}(t) + \frac{1}{2!}W''_{yx}(0)\ddot{x}(t) + \cdots + \frac{1}{n!}W^n_{yx}(0)\frac{d^nx}{dt^n} + \cdots$$

$$= C_0x(t) + C_1\dot{x}(t) + C_2\ddot{x}(t) + \cdots + C_n\frac{d^nx}{dt^n} + \cdots \tag{10-13}$$

The primes denote differentiation with respect to $s$ in the first expression; and the general output response coefficients $C_n$ are used to replace $d^n W_{yx}(s)/n!\ ds^n|_{s=0}$ in the second expression. The discussion in Chapter 2 that follows Eq. 2-42 is also applicable here. The reader will recall that a convenient method of calculating the $C_n$ in terms of the polynomial coefficients of the system transfer function was presented there. It may also be recalled (from Section 2-6) that the output response coefficients, $C_n$, can be interpreted as time-weighted moments of the weighting function, $w_{yx}(t) = \mathcal{L}^{-1}[W_{yx}(s)]$, i.e., that

$$C_n = \frac{1}{n!}\frac{d^n W_{yx}(s)}{ds^n}\bigg|_{s=0} = \frac{(-1)^n}{n!}\int_0^\infty \tau^n W_{yx}(\tau)\, d\tau \tag{10-14}$$

The final category of deterministic inputs from Table 10-2 to be discussed is the periodic variety. Strictly speaking, a periodic function is one that recurs or repeats itself, with some period $T$, over all time, i.e.,

$$x(t) = x(t \pm T), \qquad -\infty < t < \infty \tag{10-15}$$

Because the time origin is indefinite and the time span infinite, such functions do not generate transient terms in the system output response. They are, however, suitable representations for a large class of flight control system inputs that are recurring in nature and can be thought of as having no distinct time origin. The natural way to describe such functions analytically is with the familiar Fourier series which provides an expansion of $x(t)$ in terms of trigonometric functions. If a function $x(t)$ is periodic with period $T$, as in Eq. 10-15, then the Fourier series expansion will be

$$x(t) = a_0 + \sum_{n=1}^{\infty}(a_n \cos \omega_n t + b_n \sin \omega_n t) \tag{10-16}$$

where

$$\omega_n = n\omega_0 = \frac{2\pi n}{T}; \qquad n = 1, 2, 3, \ldots$$

$$a_0 = \frac{1}{T}\int_{-T/2}^{T/2} x(t)\, dt$$

$$a_n = \frac{2}{T}\int_{-T/2}^{T/2} x(t) \cos \omega_n t\, dt$$

$$b_n = \frac{2}{T}\int_{-T/2}^{T/2} x(t) \sin \omega_n t\, dt$$

Alternative forms are

$$x(t) = \sum_{n=0}^{\infty} c_n \cos (\omega_n t - \psi_n) \tag{10-17}$$

where

$$c_n = \sqrt{a_n^2 + b_n^2}, \qquad \psi_n = \tan^{-1}\frac{b_n}{a_n}$$

or

$$a_n = c_n \cos \psi_n, \quad b_n = c_n \sin \psi_n, \qquad a_0 = c_0, \quad \psi_0 = 0$$

and

$$x(t) = \sum_{n=-\infty}^{\infty} \alpha_n e^{j\omega_n t} \tag{10-18}$$

where

$$\alpha_n = \frac{a_n - jb_n}{2}, \quad \alpha_{-n} = \frac{a_n + jb_n}{2}, \qquad \alpha_0 = a_0$$

The response to a periodic input involves the system frequency response transfer function, $W_{yx}(j\omega)$, where:

$$\begin{aligned} W_{yx}(j\omega) &= [W_{yx}(s)]_{s=j\omega} = |W_{yx}(j\omega)|\, e^{j\measuredangle W_{yx}(j\omega)} \\ &= \mathrm{Re}\,[W_{yx}(j\omega)] + j\,\mathrm{Im}\,[W_{yx}(j\omega)] \end{aligned} \tag{10-19}$$

The system output $y(t)$ for an input $x(t)$ expressed as a Fourier series in terms of Eq. 10-16 is then:

$$y(t) = \sum_{n=0}^{\infty} [(a_n R_n + b_n I_n)\cos \omega_n t + (b_n R_n - a_n I_n)\sin \omega_n t] \tag{10-20}$$

where

$$R_n = \mathrm{Re}\,[W_{yx}(jn\omega_0)] \qquad \text{and} \qquad I_n = \mathrm{Im}\,[W_{yx}(jn\omega_0)]$$

When $x(t)$ is given by Eq. 10-17, the corresponding output expression becomes

$$y(t) = \sum_{n=0}^{\infty} |W_{yx}(jn\omega_0)|\, c_n \cos [\omega_n t - \psi_n + \measuredangle W_{yx}(jn\omega_0)] \tag{10-21}$$

Fourier series for a number of the most interesting periodic functions or waveforms are shown in Table 10-3. Also shown there are the probability density function, autocorrelation function, and power spectral density for each of the waveforms. These latter descriptions are most appropriate for random functions, the subject to which we now turn.

## 10-3. The Description of Random Processes[4]

[4] For more detailed discussions of the description of random time functions see, e.g., H. M. James, N. B. Nichols, and R. S. Phillips, *Theory of Servomechanisms*, McGraw-Hill, New York, 1947, Chapter 6; W. R. Bennett, "Methods of Solving Noise Problems," *Proc. IRE*, 44, 609–638 (1956); J. H. Laning, Jr., and R. H. Battin, *Random Processes in Automatic Control*, McGraw-Hill, New York, 1956; W. B. Davenport, Jr., and W. L. Root, *An Introduction to the Theory of Random Signals and Noise*, McGraw-Hill, New York, 1958; V. V. Solodovnikov, *Introduction to the Statistical Dynamics of Automatic Control Systems*, Dover, New York, 1960.

Table 10-3. Description of simple periodic functions.

| TIME FUNCTION | FOURIER SERIES |
|---|---|
| Alternating Impulses 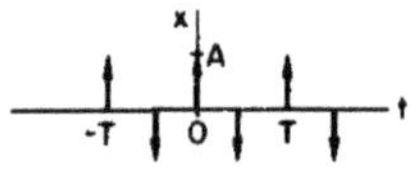  | $x(t) = \frac{4A}{T} \sum_{n=1,3,5,\dots}^{\infty} \cos(n\omega_0 t)\ ;\ \omega_0 = \frac{2\pi}{T}$ <br> $= \frac{4A}{T}\left[\cos(\omega_0 t) + \cos(3\omega_0 t) + \cos(5\omega_0 t) + \dots\right]$ <br> Note: This series does not converge |
| Impulses   | $x(t) = \frac{2A}{T}\left[\frac{1}{2} + \sum_{n=1,3,5,\dots}^{\infty} \cos(n\omega_0 t)\right]\ ;\ \omega_0 = \frac{2\pi}{T}$ <br> $= \frac{2A}{T}\left[\frac{1}{2} + \cos(\omega_0 t) + \cos(2\omega_0 t) + \dots\right]$ <br> Note: This series does not converge |
| Alternating Pulses 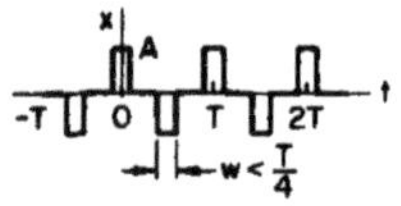  | $x(t) = \frac{4A}{\pi} \sum_{n=1,3,5,\dots}^{\infty} \frac{1}{n}\sin\left(n\pi \frac{w}{T}\right)\cos(n\omega_0 t)\ ;\ \omega_0 = \frac{2\pi}{T}$ <br> $= \frac{4A}{\pi}\left[\sin\left(\pi \frac{w}{T}\right)\cos(\omega_0 t) + \frac{1}{3}\sin\left(3\pi \frac{w}{T}\right)\cos(3\omega_0 t) + \frac{1}{5}\sin\left(5\pi \frac{w}{T}\right)\cos(5\omega_0 t) + \dots\right]$ |
| Pulses   | $x(t) = \frac{2A}{\pi}\left[\frac{w}{T}\frac{\pi}{2} + \sum_{n=1,2,3,\dots}^{\infty} \frac{1}{n}\sin\left(n\pi \frac{w}{T}\right)\cos(n\omega_0 t)\right]\ ;\ \omega_0 = \frac{2\pi}{T}$ <br> $= \frac{2A}{\pi}\left[\frac{w}{T}\frac{\pi}{2} + \sin\left(\pi \frac{w}{T}\right)\cos(\omega_0 t) + \frac{1}{2}\sin\left(2\pi \frac{w}{T}\right)\cos(2\omega_0 t) + \frac{1}{3}\sin\left(3\pi \frac{w}{T}\right)\cos(3\omega_0 t) + \dots\right]$ |
| Square Wave 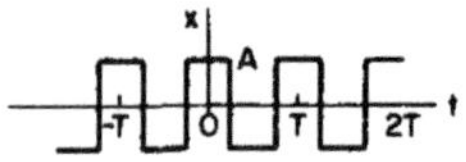  | $x(t) = \frac{4A}{\pi} \sum_{n=1,3,5,\dots}^{\infty} \frac{(-1)^{[(n-1)/2]}}{n}\cos(n\omega_0 t)\ ;\ \omega_0 = \frac{2\pi}{T}$ <br> $= \frac{4A}{\pi}\left[\cos(\omega_0 t) - \frac{1}{3}\cos(3\omega_0 t) + \frac{1}{5}\cos(5\omega_0 t) + \dots\right]$ |

| PROBABILITY DENSITY FUNCTION | AUTOCORRELATION FUNCTION | POWER SPECTRAL DENSITY |
|---|---|---|
| 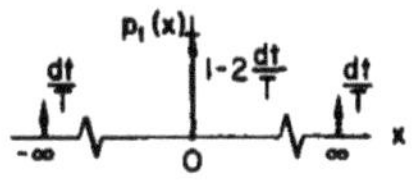  | 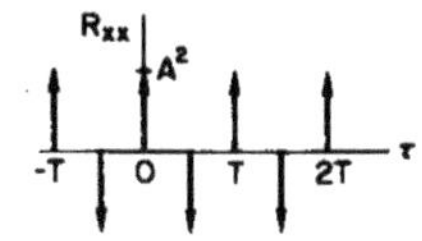  | 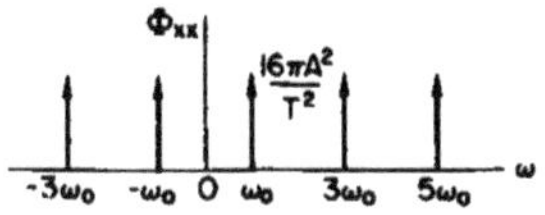  |
| 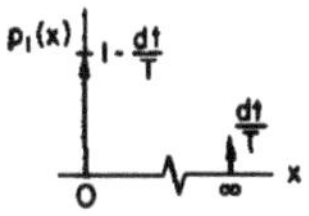  | 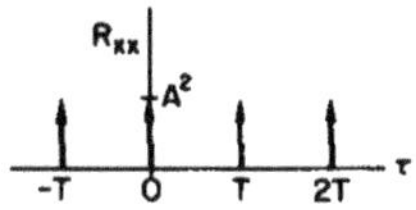  | 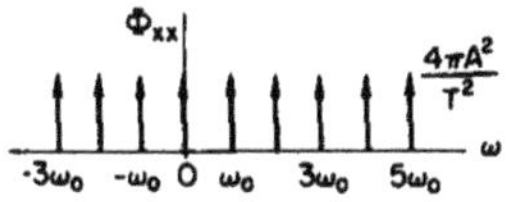  |
| 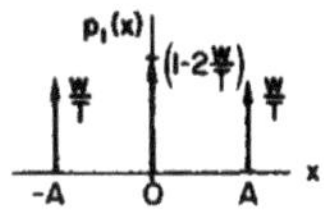  | 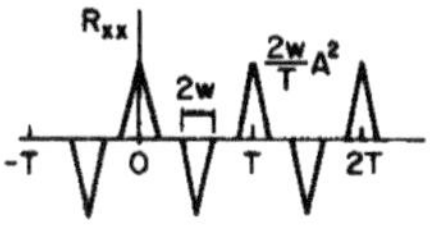  | $\Phi_{xx}(\omega) = \frac{16\pi A^2 w^2}{T^2} \frac{\sin^2\left(\frac{w}{2}\omega\right)}{\left(\frac{w}{2}\omega\right)^2} \sum_{n=1,3,5,\ldots}^{\infty} \left[\delta(n\omega_0-\omega)+\delta(n\omega_0+\omega)\right]$ 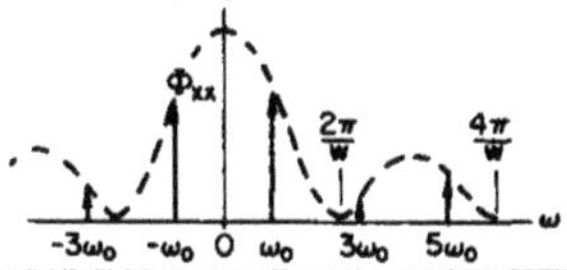  |
| 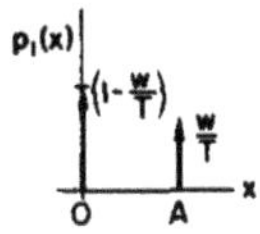  | 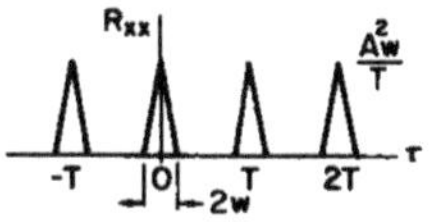  | $\Phi_{xx}(\omega) = \frac{4\pi A^2 w^2}{T^2} \frac{\sin^2\left(\frac{w}{2}\omega\right)}{\left(\frac{w}{2}\omega\right)^2} \sum_{n=0,1,2,\ldots}^{\infty} \left[\delta(n\omega_0-\omega)+\delta(n\omega_0+\omega)\right]$ 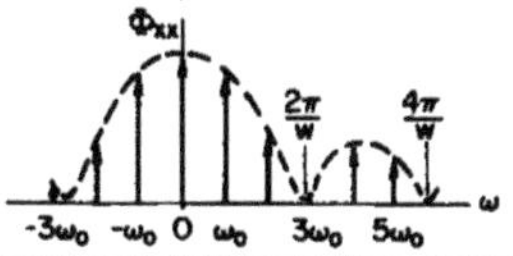  |
| 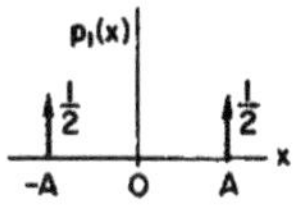  | 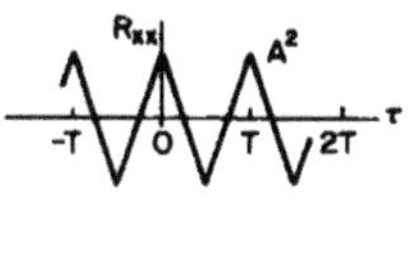  | $\Phi_{xx}(\omega) = \frac{64\pi A^2}{\omega^2 T^2} \sum_{n=1,3,5,\ldots}^{\infty} \left[\delta(n\omega_0-\omega)+\delta(n\omega_0+\omega)\right]$ 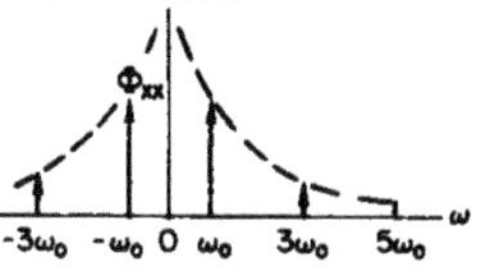  |

**Table 10-3 (Continued)**

| TIME FUNCTION | FOURIER SERIES |
|---|---|
| Sawtooth Wave 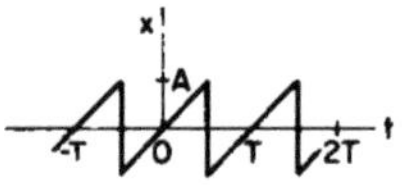 | $x(t) = \frac{2A}{\pi}\sum_{n=1,2,3,\ldots}^{\infty}\frac{(-1)^{n+1}}{n}\sin(n\omega_0 t)\;;\;\omega_0 = \frac{2\pi}{T}$ <br> $= \frac{2A}{\pi}\left[\sin(\omega_0 t) - \frac{1}{2}\sin(2\omega_0 t) + \frac{1}{3}\sin(3\omega_0 t) - \ldots\right]$ |
| Triangular Wave 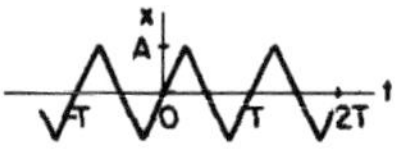 | $x(t) = \frac{8A}{\pi^2}\sum_{n=1,3,5,\ldots}^{\infty}\frac{(-1)^{[(n-1)/2]}}{n^2}\sin(n\omega_0 t)\;;\;\omega_0 = \frac{2\pi}{T}$ <br> $= \frac{8A}{\pi^2}\left[\sin(\omega_0 t) - \frac{1}{9}\sin(3\omega_0 t) + \frac{1}{25}\sin(5\omega_0 t) - \ldots\right]$ |
| Rectified Triangular (or Sawtooth) Wave 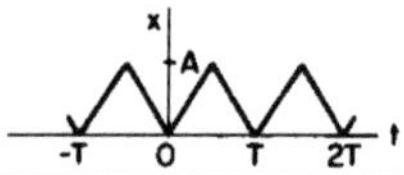 | $x(t) = \frac{4A}{\pi^2}\left[\frac{\pi^2}{8} - \sum_{n=1,3,5,\ldots}^{\infty}\frac{1}{n^2}\cos(n\omega_0 t)\right]\;;\;\omega_0 = \frac{2\pi}{T}$ <br> $= \frac{4A}{\pi^2}\left[\frac{\pi^2}{8} - \cos(\omega_0 t) - \frac{1}{9}\cos(3\omega_0 t) - \frac{1}{25}\cos(5\omega_0 t) - \ldots\right]$ |
| Sinusoid <br> $x(t) = A\sin(\omega_0 t)$ 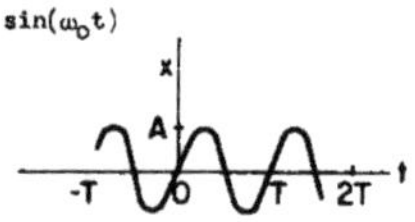 | $x(t) = A\sin(\omega_0 t)\;;\;\omega_0 = \frac{2\pi}{T}$ |
| Rectified Sinusoid <br> $x(t) = A\left\|\sin\left(\frac{\omega_0}{2}t\right)\right\|$ 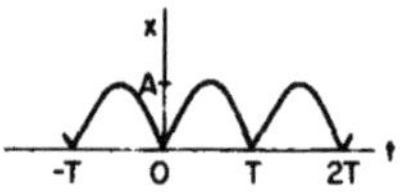 | $x(t) = \frac{4A}{\pi}\left[\frac{1}{2} - \sum_{n=1,2,3,\ldots}^{\infty}\frac{1}{4n^2-1}\cos(n\omega_0 t)\right]\;;\;\omega_0 = \frac{2\pi}{T}$ <br> $= \frac{4A}{\pi}\left[\frac{1}{2} - \frac{1}{3}\cos(\omega_0 t) - \frac{1}{15}\cos(2\omega_0 t) - \frac{1}{35}\cos(3\omega_0 t) - \ldots\right]$ |

| PROBABILITY DENSITY FUNCTION | AUTOCORRELATION FUNCTION | POWER SPECTRAL DENSITY |
|---|---|---|
| $p_1(x) = \begin{cases} \frac{1}{2A} ; & -A < x < +A \\ 0 ; & \text{elsewhere} \end{cases}$ 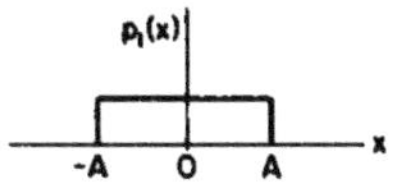  | $R_{xx}(\tau) = \frac{2A^2}{\pi^2} \sum_{n=1,2,3,\ldots}^{\infty} \frac{1}{n^2} \cos(n\omega_0\tau)$ 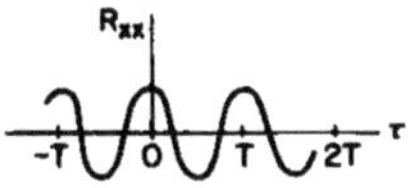  | $\Phi_{xx}(\omega) = \frac{16\pi A^2}{\omega^2 T^2} \sum_{n=1,2,3,\ldots}^{\infty} [\delta(n\omega_0 - \omega) + \delta(n\omega_0 + \omega)]$ 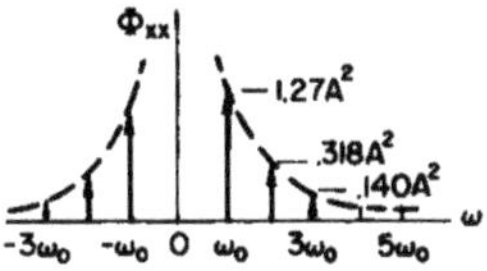  |
| $p_1(x) = \begin{cases} \frac{1}{2A} ; & -A < x < A \\ 0 ; & \text{elsewhere} \end{cases}$ 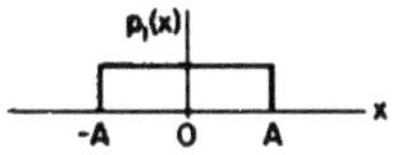  | $R_{xx}(\tau) = \frac{32A^2}{\pi^4} \sum_{n=1,3,5,\ldots}^{\infty} \frac{1}{n^4} \cos(n\omega_0\tau)$ 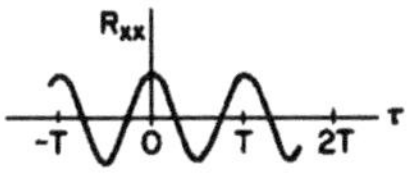  | $\Phi_{xx}(\omega) = \frac{1{,}024\pi A^2}{\omega^4 T^4} \sum_{n=1,3,5,\ldots}^{\infty} [\delta(n\omega_0 - \omega) + \delta(n\omega_0 + \omega)]$ 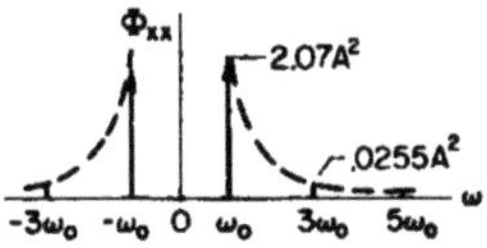  |
| $p_1(x) = \begin{cases} \frac{1}{A} ; & 0 < x < A \\ 0 ; & \text{elsewhere} \end{cases}$ 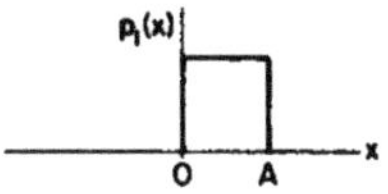  | $R_{xx}(\tau) = \frac{8A^2}{\pi^4} \left[ \frac{\pi^4}{32} + \sum_{n=1,3,5,\ldots}^{\infty} \frac{1}{n^4} \cos(n\omega_0\tau) \right]$ 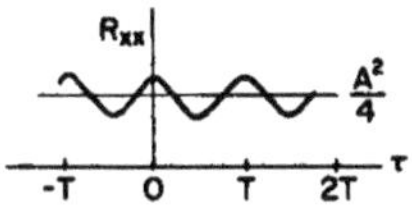  | $\Phi_{xx}(\omega) = \frac{\pi A^2}{4} \delta(\omega) + \frac{256\pi A^2}{\omega^4 T^4} \sum_{n=1,3,5,\ldots}^{\infty} [\delta(n\omega_0 - \omega) + \delta(n\omega_0 + \omega)]$ 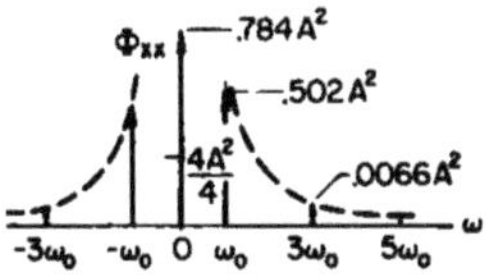  |
| $p_1(x) = \begin{cases} \frac{1}{\pi\sqrt{A^2 - x^2}} ; & -A < x < +A \\ 0 ; & \text{elsewhere} \end{cases}$ 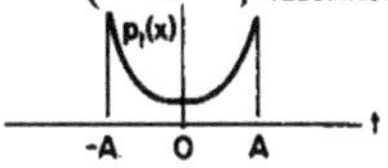  | $R_{xx}(\tau) = \frac{A^2}{2} \cos(\omega_0\tau)$ 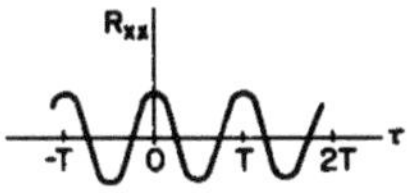  | $\Phi_{xx}(\omega) = \pi A^2 [\delta(\omega_0 - \omega) + \delta(\omega_0 + \omega)]$ 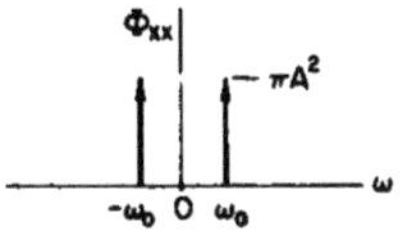  |
| $p_1(x) = \begin{cases} \frac{2}{\pi\sqrt{A^2 - x^2}} ; & 0 < x < +A \\ 0 ; & \text{elsewhere} \end{cases}$ 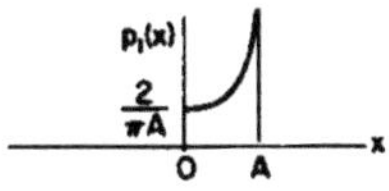  | $R_{xx}(\tau) = \frac{2A^2}{\pi^2} \left[ 1 + 4 \sum_{n=1,2,3,\ldots}^{\infty} \frac{1}{4n^2 - 1} \cos(n\omega_0\tau) \right]$ 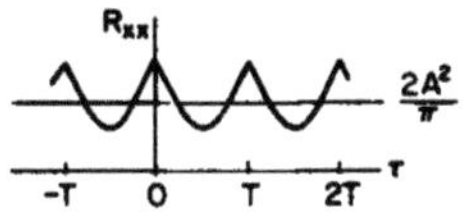  | $\Phi_{xx}(\omega) = \frac{16A^2}{\pi} \delta(\omega) + \frac{16A^2}{(\omega^2 T^2 - \pi^2)^2} \sum_{n=1,2,3,\ldots}^{\infty} [\delta(n\omega_0 - \omega) + \delta(n\omega_0 + \omega)]$ 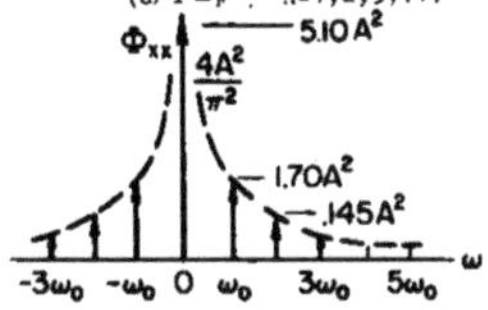  |

While, in practice, the design of linear feedback systems is often carried out considering only the response to simple, deterministic inputs, it is also often necessary or desirable to determine the response of the system to signals whose characteristics are specified only in statistical (probabilistic) terms. As remarked in the introduction to the chapter, the use of statistical properties for the description of signals may be for any one of several reasons. The signal may be random by its nature; or so complicated that a deterministic description is impractical; or perhaps a range of typical signals can best be expressed as a single random composite. In all of these cases, analysis can reveal a smoothed-over (average) picture of the true state of affairs if the actual signals are considered to be known solely by their statistics.

If we consider, in the abstract, an experiment whose outcome is determined by chance, we may associate with each basic possible outcome (called a sample point, $\zeta_i$) a *probability* of its occurrence. Outcomes or functions of outcomes are called random variables, $X(\zeta)$. (See Appendix B.) A real valued function of both the sample point, $\zeta$, and time, $t$, is called a *random*, or stochastic, *process.*

$$X = X(\zeta, t) \tag{10-22}$$

This definition is broad enough to include functions that we do not ordinarily think of as "random." Take, for example, a sine wave whose phase angle is determined by chance:

$$X(\varphi, t) = \sin(\omega t + \varphi) \tag{10-23}$$

This is, of course a deterministic function of time, once the phase angle is chosen. Actually, a random process may be in one of four categories:

1. A family of random time functions ($\zeta$ and $t$ variable)
2. A single time function ($\zeta$ fixed, $t$ variable)
3. A random variable ($\zeta$ variable, $t$ fixed)
4. A constant number ($\zeta$ fixed, $t$ fixed)

Ordinarily, in considering random processes, the first of these numbered categories will be the one of the most interest, but functions from the other categories may well be used for illustrative purposes.

The members of a family of random processes that all arise in the same way are called an ensemble. If $X_n(t)$ is considered to be a typical member of such an ensemble, where we now omit to note the functional dependence on the sample point, then $X_n(t)$ itself could possibly be thought of as a deterministic function of time. An analytical expression, e.g., some series, could then be found that approximates $X_n(t)$ arbitrarily closely over a given time period, $-T \leqq t \leqq T$. For another random signal, $X_m(t)$, which arises from the same physical source, an analytical expression derived on the same basis as that used in characterizing $X_n(t)$ might be entirely different, even if $X_m(t)$ were simply a section of $X_n(t)$ taken for a time period other than $-T \leqq t \leqq T$. For this reason any attempt to describe a truly random input as a meaningful and typical function of time is doomed at the outset. Attention must therefore be given to the ensemble of functions, $X_1(t)$, $X_2(t)$, . . . , $X_n(t)$, . . . , which stem from common causes, and where any one $X_i(t)$ is no more or less typical of the ensemble than any other member. The behavior of the ensemble, $X(t)$, must then be described in terms of *averages* of one sort or another. Average views are substituted for a precise knowledge of the nature of the signal variation with time. These remarks are the intuitive essence of what is meant by random signals and their analytical treatment.

The most fundamental averages are probability density functions. Consider Fig. 10-2. Illustrated there are a number of members of a family of random functions of time that comprise an ensemble of random processes. At any given time, $t_1$, the probability that one of the functions $X(t)$ will be greater than $x$, and smaller than $x + \Delta x$, can be computed in a direct arithmetic fashion. For an ensemble with $N$ members,

$$\Pr\,[x < X(t) < x + \Delta x \text{ at } t = t_1] = \frac{\begin{Bmatrix}\text{no. of functions } X_1(t_1), X_2(t_1), \ldots\ X_N(t_1) \\ \text{that have values between } x \text{ and } x + \Delta x\end{Bmatrix}}{N} \qquad (10\text{-}24)$$

When the amplitude interval $\Delta x$ is very small this probability will be roughly proportional to $\Delta x$, so that

$$\Pr\,[x < X(t) < x + \Delta x, t_1] \doteq p_1(x, t_1)_N\,\Delta x \qquad (10\text{-}25)$$

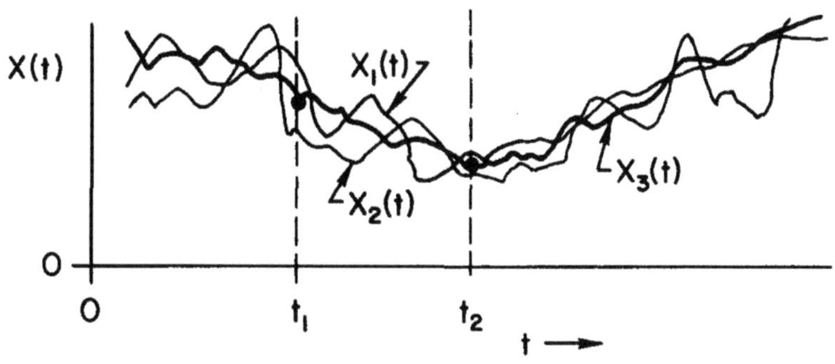


Fig. 10-2. An ensemble of random processes.

where $p_1(x, t_1)_N$ is the *first probability density function* for a finite ensemble. It will depend, in general, upon the total number of functions in the ensemble, the value of $x$, and the time $t_1$ at which the measurements are made. To be of any practical use as a means to describe the average amplitude characteristic of a typical signal, the number of members in the ensemble must be sufficiently large so that $p_1(x, t_1)_N$ is not materially changed when the ensemble is made even larger. This factor can be taken into account, and the "roughly proportional" restriction can be removed if the first probability density function is defined as

$$p_1(x, t_1) = \lim_{\substack{N \to \infty \\ \Delta x \to 0}} \frac{\text{number of values lying between } x \text{ and } x + \Delta x}{N(\Delta x)} \tag{10-26}$$

When the probability that $X(t)$ lies within a range from $x_1$ to $x_2$, at time $t_1$, is desired, the density function is integrated over the range. In general,

$$\Pr [x_1 < X(t) < x_2, t_1] = \int_{x_1}^{x_2} p_1(x, t_1)\, dx \tag{10-27}$$

If the range from $x_1$ to $x_2$ is extended from $-\infty$ to $+\infty$, the probability that $X(t)$ lies within the range becomes a certainty.

$$\Pr [-\infty < X(t) < +\infty, t_1] = \int_{-\infty}^{+\infty} p_1(x, t_1)\, dx = 1 \tag{10-28}$$

When it is desired to find only the probability that $X(t)$ is less than $x$ at time $t_1$,

$$\Pr [X(t) < x, t_1] = \int_{-\infty}^{x} p_1(x, t_1)\, dx = P_1(x, t_1) \tag{10-29}$$

where $P_1(x, t)$ is the *first probability distribution function*. It is an alternative measure of the amplitude characteristics of a typical signal $X(t)$, and has the properties:

$$\begin{aligned} P_1(-\infty, t_1) &= 0 \\ P_1(\infty, t_1) &= 1 \\ \frac{dP_1(x, t_1)}{dx} &= p_1(x, t_1) \end{aligned} \tag{10-30}$$

To generalize the calculation of averages, suppose that the average value of some arbitrary function, $f[X(t_1)]$, of the random signal $X(t)$ is desired. The average of $f[X(t_1)]$ can be computed in the usual way by first finding the sum of a large number of observed values for $f[X(t_1)]$, and then dividing by the total number of values. The fraction of the values of $X(t_1)$ which lie between $x$ and $x + dx$ is $p_1(x, t_1)\, dx$; and the value of $f[X(t_1)]$ when $X(t_1)$ lies in this range is just $f[x]$. Consequently, the fraction of values of $f[X(t_1)]$ which correspond to values of $X(t)$ falling between $x$ and $x + dx$ is just $f[x]p_1(x, t_1)\, dx$. The average value of $f[X(t_1)]$ can then be found by summing these fractions over the entire range of values for $X$, or:

$$\text{average of } f[X(t_1)] = E[f(X)] = \int_{-\infty}^{\infty} f(x)p_1(x, t_1)\,dx \qquad (10\text{-}31)$$

The notation $E[\ ]$ means "expected value" or average. The averages in which $f[X(t_1)]$ is $X^n(t_1)$ are particularly important. These are the *moments* of the distribution and have the form:

$$m_n(t_1) = E[X^n(t_1)] = \int_{-\infty}^{\infty} x^n p_1(x, t_1)\,dx \qquad (10\text{-}32)$$

The three lowest order moments are: zero-order moment,

$$m_0 = 1 \qquad (10\text{-}33)$$

first moment,

$$m_1(t_1) = \int_{-\infty}^{\infty} x p_1(x, t_1)\,dx = E[X(t_1)] \qquad (10\text{-}34)$$

second moment,

$$m_2(t_1) = \int_{-\infty}^{\infty} x^2 p_1(x, t_1)\,dx = E[X^2(t_1)] \qquad (10\text{-}35)$$

The first and second moments are interpreted as the arithmetic mean (usually called simply the mean) and the mean square values of the typical random function $X(t)$ at time $t_1$.

If the mean value is subtracted from all values of $X_i(t_1)$ in the ensemble, the moments derived from the results become *central moments:*

$$\mu_n(t_1) = E[X(t_1) - m_1(t_1)]^n = \int_{-\infty}^{\infty} [x - m_1(t_1)]^n p_1(x, t_1)\,dx \qquad (10\text{-}36)$$

The first central moment, $\mu_1(t_1)$, is zero. The second central moment is

$$\begin{aligned} \mu_2(t_1) &= \int_{-\infty}^{\infty} [x - m_1(t_1)]^2\, p_1(x, t_1)\,dx \\ &= m_2(t_1) - [m_1(t_1)]^2 \qquad (10\text{-}37) \\ &= E[X^2(t_1)] - \{E[X(t_1)]\}^2 \end{aligned}$$

$\mu_2(t_1)$ is the *variance,* $[\sigma(t_1)]^2$. It is a measure of the average alternating fluctuating power in the signal. In terms of the density function, $\mu_2(t_1)$ is a measure of its width about the mean. Similarly, the third moment, $\mu_3(t_1)$, gives an indication of the skew of the density function. The square root of the variance is the *standard deviation,* $\sigma$, of the distribution for the ensemble $X_1(t_1)$, $X_2(t_1)$, . . . , $X_n(t_1)$, . . . . When the mean is zero, the variance and the mean-square are identical.

In general, not only the first probability density and the distribution function but also the expected or mean value, the variance, and other statistics of an ensemble of stochastic processes may depend on the time of observation. This is suggested in Fig. 10-2, in which clearly both the average value and average fluctuating component of the signals are smaller at $t_2$ than at $t_1$.

The first probability density function, together with the moments and other averages derived from it, provide information about the expected values of $X(t_1)$ that will occur on the average and about the probability that various magnitudes may occur. However, these quantities give no information concerning the time scale in which the values might occur. To rectify this situation, a *second probability density function* must be defined. The second probability density function, $p_2(x_1, t_1; x_2, t_2)$, when multiplied by $dx_1\, dx_2$ is the probability that $X(t)$ will be within the bounds $x_1$ and $x_1 + dx_1$ at time $t_1$, and that this same $X(t)$ will be between $x_2$ and $x_2 + dx_2$ at time $t_2$, i.e.,

$$\Pr[x_1 < X(t_1) < x_1 + dx_1; x_2 < X(t_2) < x_2 + dx_2] = p_2(x_1, t_1; x_2, t_2)\, dx_1\, dx_2 \quad (10\text{-}38)$$

The concept is illustrated in Fig. 10-3. The second probability density function provides the means to find average values such as average $[X(t_1)X(t_2)]$.

$$\text{average}\,[X(t_1)X(t_2)] = E[X(t_1)X(t_2)] = \int_{-\infty}^{\infty}\int_{-\infty}^{\infty} x_1 x_2 p_2(x_1, t_1; x_2, t_2)\, dx_1 dx_2 \quad (10\text{-}39)$$

This average will be a function of $t_1$ and $t_2$. When $t_1 = t_2$, the second probability density, $p_2(x_1, t_1; x_2, t_1)$, is just a product of the first probability

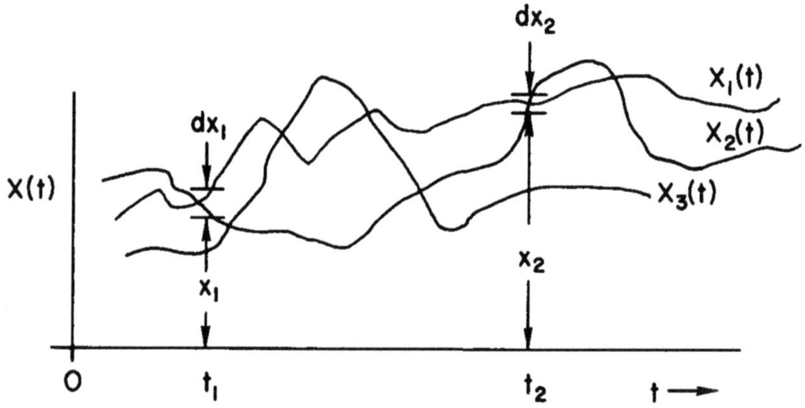


Fig. 10-3. Illustration for second probability density function.

density functions, $p_1(x_1, t_1)p_1(x_2, t_1)$. The expected value $E[X_1X_2]$ then becomes simply the mean square value or second moment, $m_2(t_1)$.

Higher order probability density functions are also defined, following the pattern established above, as the fraction of the ensemble members that have values which lie within given ranges at respectively given times. To completely describe the random process, *all* of the probability density functions must be defined. Stated in another way, the process is defined in successively greater detail the higher the order of the known probability density functions. Each lower order density function can be derived from the highest order function by a succession of special cases, so that the degree of detailed knowledge about the complete process is incorporated in the highest order density function which is available.

In the above discussion, as we have noted, the density functions and averages derived therefrom all may be functions of the times of observation. In many applications, however, the underlying mechanism that generates the random function does not vary with time or can be considered to be time invariant for the time periods of interest in response calculations. The time signals actually measured as physical manifestations of the underlying process may not reflect this time invariance property directly because of the kinematics of the measurement situation, but transformations can often be made to find time signals (related to the actual signals measured) that do. When these circumstances apply, the probability density functions will not depend upon the observation times themselves but rather only upon time differences. For the first probability density function, for example, the same function would be obtained regardless of the time when measurements were taken. Then $p_1(x, t_1)$ would be equal to $p_1(x, t_2)$, so that the time dependence would no longer be present. The first probability density function would then be simply $p_1(x)$; the second probability density function becomes $p_2(x_1, x_2, \tau)$, where $\tau = t_2 - t_1$, etc. When all the statistics describing the ensemble of time functions are not dependent upon the absolute times of observation, the random processes are *strictly stationary*. As a practical matter, however, it is usually impossible to determine if this is the case. If it can be shown, for example, that the first and second probability density functions are not dependent upon the time of observation, the random process is said to be stationary of order two. It is often only shown that the process is stationary of order one and that the autocorrelation function (defined below) depends not on $t_1$ and $t_2$ but only on $\tau = t_2 - t_1$. We then say that the process is *wide sense* or *weakly stationary*.

When the statistics of the ensemble of functions, $X(t)$, are independent of the absolute time, an alternative possibility for finding averages exists. Instead of averaging in the conventional arithmetic way across an entire ensemble of functions at fixed times, one could consider just one member of the ensemble and perform averages over all time. Such averages as

$$\overline{X(t)} = \lim_{T\to\infty} \frac{1}{2T} \int_{-T}^{+T} X(t)\,dt = \text{the mean value} \tag{10-40}$$

$$\overline{X^2(t)} = \lim_{T\to\infty} \frac{1}{2T} \int_{-T}^{+T} X^2(t)\,dt = \text{the mean square value} \tag{10-41}$$

$$R_{xx}(\tau) = \overline{X^*(t)X(t+\tau)} = \lim_{T\to\infty} \frac{1}{2T} \int_{-T}^{+T} X^*(t)X(t+\tau)\,dt \tag{10-42}$$

would then be the time-average equivalent of the ensemble averages $E[X(t)] = m_1$, $E[X^2(t)] = m_2$, and $E[X^*(t)X(t+\tau)]$, respectively. The last time average here is called the autocorrelation function.[5]

As a practical matter stationarity for the time averages requires only that averaging time be sufficiently long so that the results obtained would be substantially unchanged if it were made still longer. The practical equivalent to the averaging time for ensemble averages would be that the number of signals observed be sufficiently great so that the results are substantially unchanged when even more signals are added to the ensemble. (This, however, does not assure stationarity for the ensemble.)

The most extensive applications of the random input characteristics, discussed above, occur when time averages and ensemble averages are equal, i.e., when the random process is *ergodic*. Whether a given physical process is ergodic or not is very difficult to show, in general. Consequently, when processes are stationary (a necessary but not sufficient condition for ergodicity) and the lower averages (e.g., mean, mean square, etc.) appear to be equal, ergodic properties are often assigned to the physical situation by hypothesis. Clearly, the process may be ergodic for some statistical parameters and not for others.[6] With the hypothesis of ergodicity, time averages are used interchangeably with ensemble averages, and the statistical (probability density functions) and moment properties can be determined from either averages performed on an ensemble or time averages performed on a single member of the ensemble. This is an enormous practical advantage when data must be obtained from physical measurement. The assumption of ergodicity is also helpful from the theoretical

[5] The asterisk notation in the autocorrelation function indicates that $X^*(t)$ is the conjugate of $X(t)$. Since $X(t)$ is a real function, this may seem superfluous. However, $X(t)$ might be expanded in the complex form of a Fourier series, and the conjugate would then be required when working with the series representation.

[6] Conditions for the ergodicity of the more commonly interesting statistical parameters are summarized in A. Papoulis, *Probability, Random Variables, and Stochastic Processes*, McGraw-Hill, New York, 1965.

standpoint, since it simplifies analytical work tremendously. However, theoretical treatments of nonstationary random phenomena, while complicated, are now well advanced; so the major practical importance of ergodicity currently resides in the simplifications introduced in measurement. In any event, attention from this point on will be confined to random input descriptions which, as a practical matter, can be assumed to possess ergodic properties for the time periods of interest.

When the time functions are ergodic:

$$\begin{aligned}
\overline{X(t)} &= E[X(t)] = \int_{-\infty}^{\infty} x p_1(x)\,dx = m_1 \\
\overline{X^2(t)} &= E[X^2(t)] = \int_{-\infty}^{\infty} x^2 p_1(x)\,dx = m_2 \\
&\vdots \\
\overline{X^n(t)} &= E[X^n(t)] = \int_{-\infty}^{\infty} x^n p_1(x)\,dx = m_n
\end{aligned} \tag{10-43}$$

and

$$\begin{aligned}
\overline{X^*(t)X(t+\tau)} &= E[X^*(t)X(t+\tau)] \\
&= \int_{-\infty}^{\infty}\int_{-\infty}^{\infty} x_1 x_2 p_2(x_1, x_2, \tau)\,dx_1\,dx_2 = R_{xx}(\tau)
\end{aligned} \tag{10-44}$$

Higher order averages could also be added to this list, but these are seldom used in practical calculations. With the functions defined thus far, it is possible to compute both mean and mean square values and, for many cases of practical interest, the first probability density function may also be available. Thus a good indication of the response characteristics, on the average, can be found, and the probability that certain values of various signals will occur can be assessed.

From this point on, we shall designate a function of time that may be either a random process or a deterministic periodic signal by a lower case letter.

For a stable, constant-coefficient, linear system with a weighting function $w(t)$, the autocorrelation function of the output response $y(t)$ to an input $x(t)$ having an autocorrelation function $R_{xx}(\tau)$ is:[7]

$$R_{yy}(\tau) = \int_{-\infty}^{\infty}\int_{-\infty}^{\infty} w(\lambda)w(\gamma)R_{xx}(\tau + \lambda - \gamma)\,d\gamma\,d\lambda \tag{10-45}$$

Thus the output signal autocorrelation can be computed from the input autocorrelation from what amounts to two convolutions with the system weighting function. Because the mean-square value of a stationary time

[7] James, Nichols, and Phillips, *Theory of Servomechanisms.*

function is given by its autocorrelation function at $\tau = 0$, the mean square output will be

$$\overline{y^2(t)} = R_{yy}(0) = \int_{-\infty}^{\infty}\int_{-\infty}^{\infty} w(\lambda)w(\gamma)R_{xx}(\lambda - \gamma)\,d\gamma\,d\lambda \qquad (10\text{-}46)$$

Integrals of this nature are often awkward to work with, whereas their frequency domain equivalent is considerably simpler. To proceed into the frequency domain requires the definition of *power spectral density functions*. For either periodic or stationary random time functions, this is here defined as four times the Fourier cosine transform of the autocorrelation function:

$$\Phi_{xx}(\omega) = 4\int_0^{\infty} R_{xx}(\tau)\cos\omega\tau\,d\tau \qquad (10\text{-}47)$$

The autocorrelation function is also a Fourier cosine transform of the power spectral density, i.e.,

$$R_{xx}(\tau) = \int_0^{\infty}\Phi_{xx}(f)\cos 2\pi f\tau\,df = \frac{1}{2\pi}\int_0^{\infty}\Phi_{xx}(\omega)\cos\omega\tau\,d\omega \qquad (10\text{-}48)$$

The mean square value of $x(t)$ is given in terms of the power spectral density by

$$\overline{x^2(t)} = \frac{1}{2}\int_{-\infty}^{\infty}\Phi_{xx}(f)\,df = \int_0^{\infty}\Phi_{xx}(f)\,df$$
$$= \frac{1}{2\pi}\int_0^{\infty}\Phi_{xx}(\omega)\,d\omega = R_{xx}(0) \qquad (10\text{-}49)$$

The factor of four in Eq. 10-47 arises from the desire to have the mean square value be the integral of the power spectral density over all positive frequencies. This usage agrees with James, Nichols, and Phillips and with Rice.[8] Others use a factor of two, which is compatible with having $\overline{x^2(t)}$ be the result of integrating $\Phi_{xx}(f)$ over both positive and negative frequencies.

Analogously to Eq. 10-44 we can define a cross-correlation function, $R_{xy}(\tau)$, between two variables $x(t)$ and $y(t)$ as a species of average:

$$\overline{x^*(t)y(t+\tau)} = E[x^*(t)y(t+\tau)] = R_{xy}(\tau)$$
$$= \lim_{T\to\infty}\frac{1}{2T}\int_{-T}^{T} x^*(t)y(t+\tau)\,dt \qquad (10\text{-}50)$$

If $x(t)$ is the input and $y(t)$ the output of a linear, constant coefficient

[8] S. O. Rice, "Mathematical Analysis of Random Noise," originally published in the *Bell Systems Tech. J.*, Vols. 23 and 24, reprinted in N. Wax, ed., *Selected Papers on Noise and Stochastic Processes*, Dover, New York, 1954.

device such that

$$y(t) = \int_{-\infty}^{\infty} w(\tau)x(t-\tau)\,d\tau = \int_{-\infty}^{\infty} w(t-\tau)x(\tau)\,d\tau \tag{10-51}$$

then

$$R_{xy}(\tau) = \int_{-\infty}^{\infty} w(u)R_{xx}(\tau-u)\,du = R_{yx}(-\tau) \tag{10-52}$$

The cross-spectral density is again proportional to the Fourier transform of the cross-correlation function

$$\Phi_{xy}(j\omega) = 2\int_{-\infty}^{\infty} R_{xy}(\tau)e^{-j\omega\tau}\,d\tau \tag{10-53}$$

and the cross-correlation function can be found by inverse Fourier transforming the cross-spectral density.

$$R_{xy}(\tau) = \frac{1}{4\pi}\int_{-\infty}^{\infty} \Phi_{xy}(j\omega)e^{j\omega\tau}\,d\omega \tag{10-54}$$

Finally, if $W(j\omega)$ is the frequency response transfer function,

$$\Phi_{xy}(j\omega) = W(j\omega)\Phi_{xx}(\omega) \tag{10-55}$$

An instructive way in which to approximate a stationary random process is as a sum of a large number of sinusoids comprising the Fourier series approximation to an arbitrary function in the interval $-T/2$ to $T/2$. (See Eq. 10-18.)

$$x(t) = \sum_{n=-\infty}^{\infty} \alpha_n e^{j\omega_n t} \tag{10-56}$$

From Eq. 10-42 the autocorrelation function will be

$$\begin{aligned} R_{xx}(\tau) &= \lim_{T\to\infty} \frac{1}{2T}\int_{-T}^{T} x^*(t)x(t+\tau)\,dt = \overline{x^*(t)x(t+\tau)} \\ &= \lim_{T\to\infty} \frac{1}{2T}\int_{-T}^{T}\left\{\sum_{-\infty}^{\infty}\alpha_n^* e^{-j\omega_n t}\sum_{-\infty}^{\infty}\alpha_k e^{j\omega_k(t+\tau)}\right\}dt \\ &= \sum_{n=-\infty}^{\infty}\sum_{k=-\infty}^{\infty}\alpha_n^*\alpha_k e^{j\omega_k\tau}\left\{\lim_{T\to\infty}\frac{1}{2T}\int_{-T}^{T} e^{j(\omega_k-\omega_n)t}\right\}dt \end{aligned} \tag{10-57}$$

The quantity in brackets is

$$\begin{aligned} &= \lim_{T\to\infty}\frac{1}{2T}\left[\frac{e^{j(\omega_k-\omega_n)t}}{j(\omega_k-\omega_n)}\right]_{-T}^{+T} = 0; \qquad \omega_k \neq \omega_n \\ &= \lim_{T\to\infty}\frac{1}{2T}\int_{-T}^{T} e^{j\omega_n t}e^{-j\omega_n t}\,dt = 1; \qquad \omega_k = \omega_n \end{aligned} \tag{10-58}$$

Therefore,

$$R_{xx}(\tau) = \sum_{k=-\infty}^{\infty} |\alpha_k|^2 e^{j\omega_k \tau} \tag{10-59}$$

But since

$$|\alpha_n|^2 = \frac{a_n - jb_n}{2} \cdot \frac{a_n + jb_n}{2} = \frac{c_n^2}{4} \quad \text{and} \quad \frac{e^{j\omega_k\tau} + e^{-j\omega_k\tau}}{2} = \cos \omega_k \tau$$

then

$$R_{xx}(\tau) = c_0^2 + \sum_{n=1}^{\infty} \frac{c_n^2}{2} \cos \omega_n \tau \tag{10-60}$$

Consider now the power spectral density of the function $x(t)$. Substituting the expression on the right-hand side of Eq. 10-60 in Eq. 10-47, we get

$$\Phi_{xx}(\omega) = 4c_0^2 \int_0^\infty \cos \omega\tau \, d\tau + 4 \sum_{n=1}^{\infty} \frac{c_n^2}{2} \int_0^\infty \cos \omega_n \tau \cos \omega\tau \, d\tau \tag{10-61}$$

The evaluation of this integral is somewhat circuitous. The Fourier transform of the delta function is identically unity, so that the inverse Fourier transform of unity must be the delta function, i.e.,

$$\int_{-\infty}^{\infty} \delta(t) e^{-j\omega t} \, dt = 1, \quad \text{then} \ \frac{1}{2\pi} \int_{-\infty}^{\infty} (1) e^{j\omega t} \, d\omega = \delta(t) \tag{10-62}$$

Consequently,

$$\int_{-\infty}^{\infty} (e^{j\omega_n t}) e^{-j\omega t} \, dt = \int_{-\infty}^{\infty} e^{-j(\omega - \omega_n)t} \, dt = 2\pi\delta(\omega - \omega_n) \tag{10-63}$$

From the trigonometric identity,

$$\cos \omega_n \tau \cos \omega\tau = \frac{1}{2} [\cos \tau(\omega_n - \omega) + \cos \tau(\omega_n + \omega)] \tag{10-64}$$

both the cosine and the delta function are even functions, so that

$$\begin{aligned}
\int_0^\infty \cos \omega_n \tau \cos \omega\tau \, d\tau
&= \frac{1}{4} \int_{-\infty}^{\infty} \cos \tau(\omega_n - \omega) \, d\tau + \frac{1}{4} \int_{-\infty}^{\infty} \cos \tau(\omega_n + \omega) \, d\tau \\
&= \frac{1}{4} \int_{-\infty}^{\infty} \frac{e^{j(\omega_n - \omega)\tau} + e^{-j(\omega_n - \omega)\tau}}{2} \, d\tau + \frac{1}{4} \int_{-\infty}^{\infty} \frac{e^{j(\omega_n + \omega)\tau} + e^{-j(\omega_n + \omega)\tau}}{2} \, d\tau \\
&= \frac{\pi}{4} [\delta(\omega_n - \omega) + \delta(\omega - \omega_n) + \delta(\omega_n + \omega) + \delta(-\omega_n - \omega)] \\
&= \frac{\pi}{2} [\delta(\omega - \omega_n) + \delta(\omega + \omega_n)]
\end{aligned} \tag{10-65}$$

Finally, therefore, the power spectral density of the function $x(t)$ is

$$\Phi_{xx}(\omega) = 4\pi c_0^2\,\delta(\omega) + \pi \sum_{n=1}^{\infty} c_n^2[\delta(\omega - \omega_n) + \delta(\omega + \omega_n)] \qquad (10\text{-}66)$$

This function may thus be seen to represent the distribution of the harmonic components in $x(t)$ with each component here proportional to the square of the amplitude coefficient in the original Fourier series.

If each wave in the original Fourier series were a voltage, the power, at that frequency, dissipated in a unit resistance would be proportional to the square of the voltage—hence the name *power* spectral density.

The first probability density functions, autocorrelation functions, and power spectral density functions for a number of periodic waveforms have already been presented in Table 10-3.

We next consider the probabilistic description of some particular random processes.

## 10-4. Analytical Description and Catalog of Special Random Processes

From the discussion above, it should be apparent that the random inputs listed in Table 10-1 must be characterized as autocorrelations or power spectral densities if even minimal information (mean-square values) is to be obtained about system response to these inputs. If more response information is required, at least some of the lower order probability density functions are required. Even if these are known, however, the probability density functions for the system output are difficult to find, unless the amplitudes of the random signals are characterized by a Gaussian or normal distribution. (See Appendix B.)

Table 10-4 shows a number of the more interesting random processes that may be used for describing automatic flight control commands and disturbances. In most cases a typical segment of the random process is shown, together with its autocorrelation function and power spectral density.

A random binary transmission is the signal generated by abstracting the result of a coin tossing experiment whose outcome may change at intervals of $T$ sec. Calling heads $+1$ and tails $-1$, so that short sequences in which the signal is unchanged from toss to toss could occur, but that long ones are unlikely, and then shifting the signal along the time axis an amount, $\epsilon$, where $\epsilon$ is equally likely to take any value in the interval $O$, $T$, the result would be as illustrated in Item 1 in Table 10-4.

*Boxcar sequences* are generated by taking positive or negative steps whose size is determined in some random way at intervals whose statistics are also prescribed. There are at least three interesting cases.

**Table 10-4. Description of random signals.**

| TYPICAL ENSEMBLE MEMBER DEFINITION | EXPECTATIONS | |
|---|---|---|
| | MEAN VALUE | MEAN-SQUARE VALUE |
| 1. Random Binary Transmission 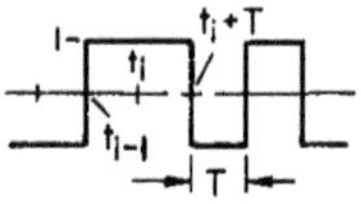 $f(t) = \pm 1;\ t_i < t < (t_i + T)$ depending on the flip of a fair coin at $t_i$. (Ensemble members randomly phased — uniform distribution in $0, T$) | 0 | 1 |
| 2. Semi-random Boxcar Sequence 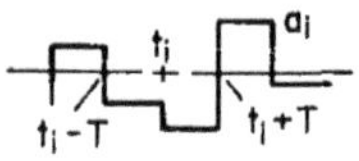 $f(t) = a_i;\ t_i < t < (t_i + T)$ where the time characteristics are as in item 1 and $a_i$ is a random independent variable with zero mean-value. | 0 | $\widetilde{a^2}$ |
| 3. Boxcar Sequence — Uniform Distribution of Time Intervals 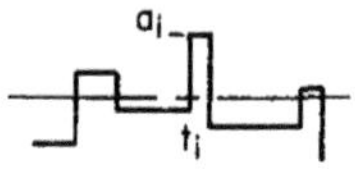 $f(t) = a_i;\ t_i < t < t_{i+1}$ where $a_i$ is defined as in item 2 and $t_i$ is a random uniformly distributed variable. | 0 | $\widetilde{a^2}$ |
| 4. Shot Noise (Special Case) 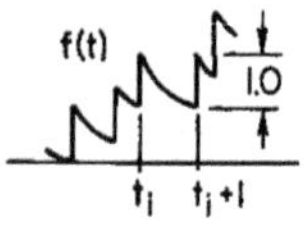 $f(t) = \sum_i h(t - t_i);\ h(t) = e^{-\alpha t}$ where $t_i \triangleq$ poisson distributed points, that is $P_{t_i}(n, T) = \frac{(\lambda T)^n}{n!} e^{-\lambda T}$; $1/\lambda$ = average interval between $t_i$ and $t_{i+1}$; $T$ = some time interval; $n$ = the number of $t_i$ in $T$ | $\sqrt{2}\left(\frac{\lambda}{\alpha}\right)$ | $2\left(\frac{\lambda}{\alpha}\right)^2 + \frac{1}{2}\left(\frac{\lambda}{\alpha}\right)$ |
| 5. Semi-Random Telegraph Signal 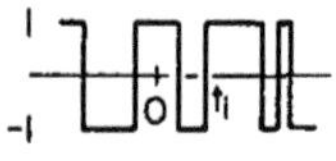 $f(t) = \begin{cases} 1; & t = 0 \\ \pm 1; & \text{depending on number of Poisson distributed zero crossings (see item 4)} \end{cases}$ (origin specified) | $e^{-2\lambda t}$ | 1 |
| | Not Stationary! | |
| 6. Boxcar Sequence — Poisson Time Intervals 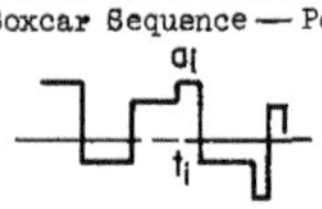 $f(t) = a_i;\ t_i < t < t_{i+1}$ where $t_i$ is defined in item 4 and $a_i$ is a random independent variable with zero mean-value. | 0 | $\widetilde{a^2}$ |

| AUTOCORRELATION FUNCTION | POWER SPECTRAL DENSITY | TRANSIENT ANALOG |
|---|---|---|
| 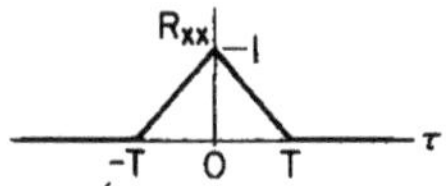<br>$R_{xx}(\tau) = \begin{cases} (1-\lvert\tau\rvert/T)\,; & \lvert\tau\rvert < T \\ 0\,; & \text{elsewhere} \end{cases}$ | 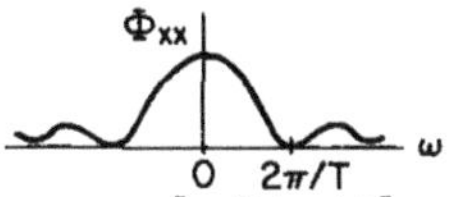<br>$\Phi_{xx}(\omega) = 2T\left[\frac{\sin^2(\omega T/2)}{(\omega T/2)^2}\right]$ | 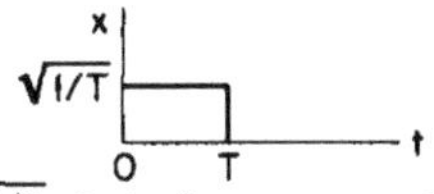<br>$x(t) = \begin{cases} \sqrt{1/T}\,; & 0 < t < T \\ 0\,; & \text{elsewhere} \end{cases}$ $X(s) = \sqrt{1/T}\,\frac{1-e^{-sT}}{s}$ |
| 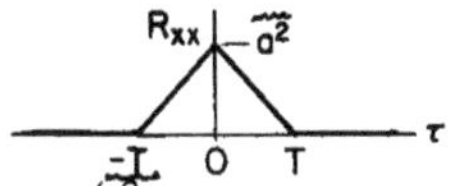<br>$R_{xx}(\tau) = \begin{cases} \overline{a^2}(1-\lvert\tau\rvert/T)\,; & \lvert\tau\rvert < T \\ 0\,; & \text{elsewhere} \end{cases}$ | 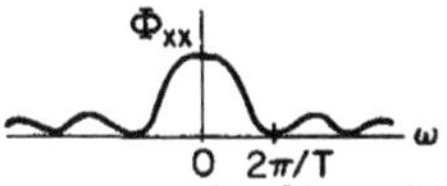<br>$\Phi_{xx}(\omega) = 2T\overline{a^2}\left[\frac{\sin^2(\omega T/2)}{(\omega T/2)^2}\right]$ | 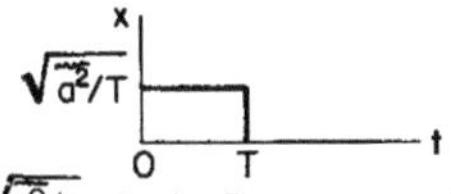<br>$x(t) = \begin{cases} \sqrt{\overline{a^2}/T}\,; & 0 < t < T \\ 0\,; & \text{elsewhere} \end{cases}$ $X(s) = \sqrt{\overline{a^2}/T}\,\frac{1-e^{-sT}}{s}$ |
| 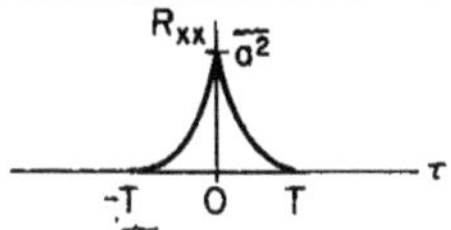<br>$R_{xx}(\tau) = \begin{cases} \overline{a^2}(1-\lvert\tau\rvert/T)^2\,; & \lvert\tau\rvert < T \\ 0\,; & \text{elsewhere} \end{cases}$ | 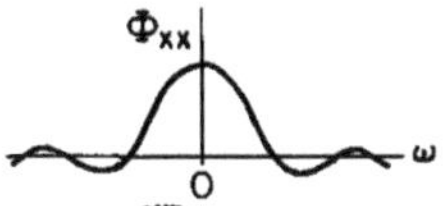<br>$\Phi_{xx}(\omega) = \frac{8\overline{a^2}T}{(\omega T)^2}\left[1-\frac{\sin(\omega T)}{(\omega T)}\right]$ | 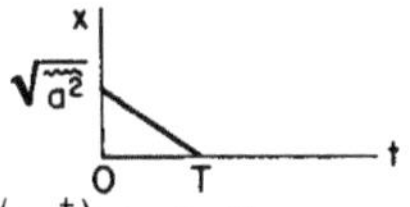<br>$x(t) = \begin{cases} \sqrt{\overline{a^2}}\left(1-\frac{t}{T}\right); & 0 < t < T \\ 0\,; & \text{elsewhere} \end{cases}$ $X(s) = \sqrt{\overline{a^2}}\left[\frac{Ts-1}{Ts^2}\right]\left[1-e^{-sT}\right]$ |
| 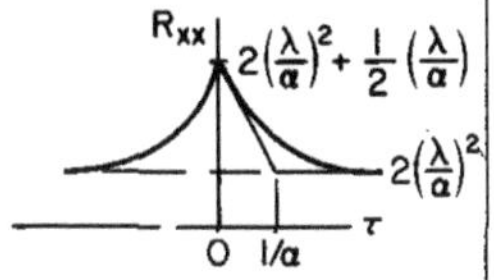<br>$R_{xx}(\tau) = 2\left(\frac{\lambda}{\alpha}\right)^2 + \frac{1}{2}\left(\frac{\lambda}{\alpha}\right)e^{-\alpha\lvert\tau\rvert}$ | 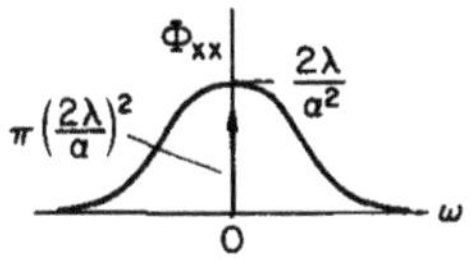<br>$\Phi_{xx}(\omega) = \pi\left(\frac{2\lambda}{\alpha}\right)^2\delta(\omega) + \frac{2\lambda}{\alpha}\left(\frac{\alpha}{\omega^2+\alpha^2}\right)$ | 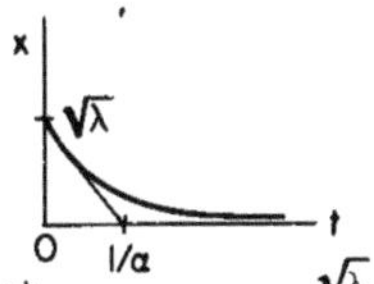<br>$x(t) = \sqrt{\lambda}e^{-\alpha t}$ $X(s) = \frac{\sqrt{\lambda}}{s+\alpha}$<br>Note: The transient analog cannot and need not model impulsive features of the power spectral density |
| 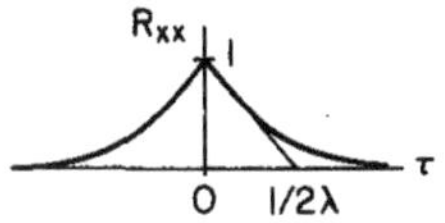<br>$R_{xx}(\tau) = e^{-2\lambda\lvert\tau\rvert}$ | 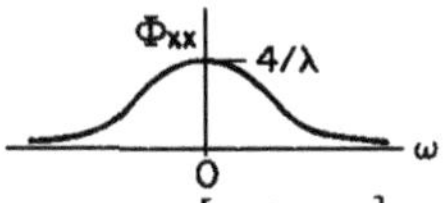<br>$\Phi_{xx}(\omega) = 4\left[\frac{2\lambda}{\omega^2+(2\lambda)^2}\right]$ | 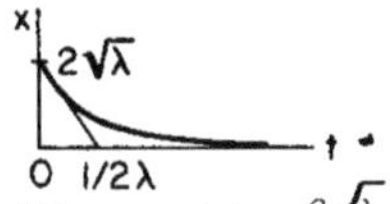<br>$x(t) = 2\sqrt{\lambda}e^{-2\lambda t}$ $X(s) = \frac{2\sqrt{\lambda}}{s+2\lambda}$ |
| 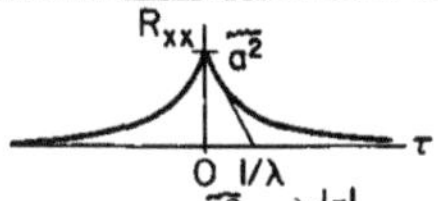<br>$R_{xx}(\tau) = \overline{a^2}\,e^{-\lambda\lvert\tau\rvert}$ | 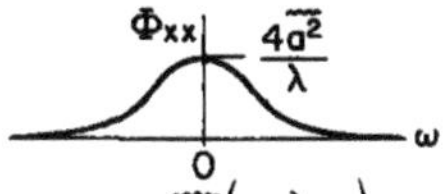<br>$\Phi_{xx} = 4\overline{a^2}\left(\frac{\lambda}{\omega^2+\lambda^2}\right)$ | 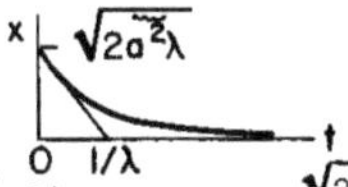<br>$x(t) = \sqrt{2\overline{a^2}\lambda}\,e^{-\lambda t}$ $X(s) = \frac{\sqrt{2\overline{a^2}\lambda}}{s+\lambda}$ |

Table 10-4 (Continued)

| TYPICAL ENSEMBLE MEMBER DEFINITION | EXPECTATIONS | |
|---|---|---|
| | MEAN VALUE | MEAN-SQUARE VALUE |
| 7. Sum of Random Sine Waves<br>$f(t) = A_o + \sum_{t=1}^{N} A_k \cos(\omega_k t + \varphi_k);$<br><br>$A_k$ arbitrary<br>$\omega_k$ noncommensurate<br>$\varphi_k$ random, independent and uniformly distributed in $0, 2\pi$ | $A_o$ | $A_0^2 + \sum_{k=1}^{N} \frac{A_k^2}{2}$ |
| 8. Sum of Sine Waves of Equal Amplitude and Frequency<br>$f(t) = \sum_{k=1}^{N} A \cos(\omega_o t + \varphi_k);$ $\varphi_k$ defined as in item 7<br>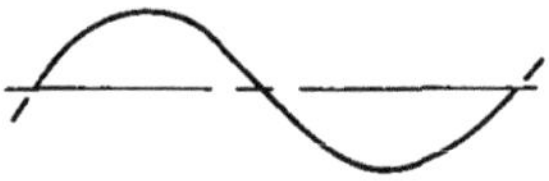<br>$= A_1 \sin(\omega_o t + \theta);$ where<br>$A_1 = A\left[N + 2 \sum_{j=k}^{N} \cos(\varphi_j - \varphi_k)\right]^{1/2}$<br>$\theta = \tan^{-1}\left(-\frac{\cos \varphi_1 + \cos \varphi_2 + \cdots}{\sin \varphi_1 + \sin \varphi_2 + \cdots}\right)$ | $E[f(t)] = 0$<br>$\overline{f}(t) = 0$<br>Not Ergodic! | $E[f^2(t)] = \frac{NA^2}{2}$<br>$\overline{f^2}(t) = \frac{A_1^2}{2}$ |
| 9. Gaussian White Noise<br> | 0 | Infinity |
| 10. Gaussian "Colored" Noise — First-Order Spectrum<br>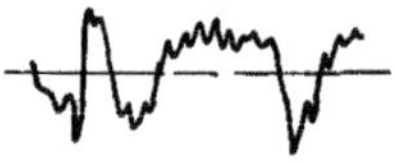 | 0 | $\widetilde{a^2}$ |
| 11. Gaussian "Colored" Noise — Second-Order Spectrum<br>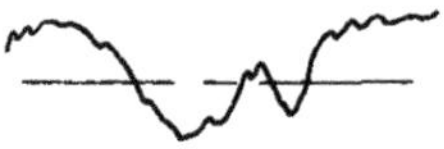 | 0 | $\widetilde{a^2}$ |
| 12. Typical Error Signal<br>(Specified as the difference between two first-order spectra, see power spectral density column)<br>$\frac{\widetilde{a^2}}{\alpha} = \frac{\widetilde{b^2}}{\beta}$ | 0 | $\widetilde{a^2} - \widetilde{b^2}$ |
| 13. Typical Error Signal<br>(Specified as the difference between two rectangular spectra)<br>$\frac{\widetilde{a^2}}{\alpha} = \frac{\widetilde{b^2}}{\beta}$ | 0 | $\widetilde{a^2} - \widetilde{b^2}$ |

| AUTOCORRELATION FUNCTION | POWER SPECTRAL DENSITY | TRANSIENT ANALOG |
|---|---|---|
| 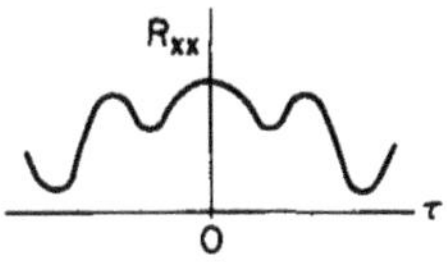 $R_{xx}(\tau) = A_0^2 + \sum_{k=1}^{N} \frac{A_k^2}{2}\cos(\omega_k\tau)$ | 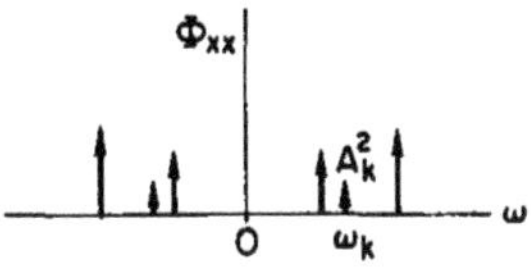 $\Phi_{xx}(\omega) = \pi \sum_{k=-N}^{N} A_k^2\,\delta(\omega-\omega_k)$ | See note for item 4 |
| 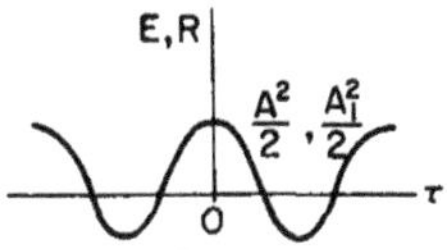 $E = \frac{A^2}{2}\cos(\omega_0\tau)$ <br> $R(\tau) = \frac{A_1^2}{2}\cos(\omega_0\tau)$ | 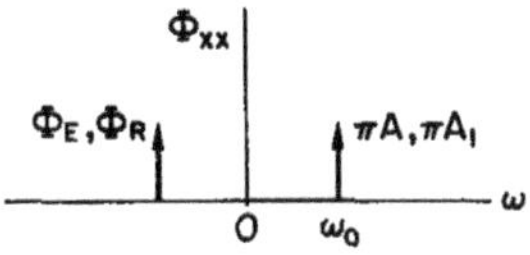 $\Phi_E(\omega) = \pi A\delta(\omega - \lvert\omega_0\rvert)$ <br> $\Phi_R(\omega) = \pi A_1^2\delta(\omega - \lvert\omega_0\rvert)$ | See note for item 4 |
| 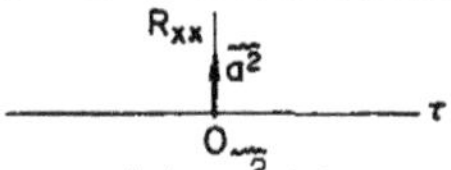 $R_{xx}(\tau) = \tilde{a}^2\delta(\tau)$ | 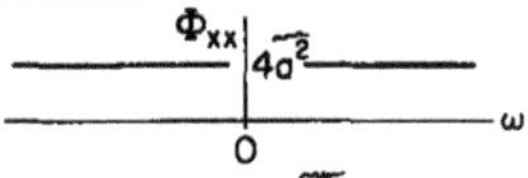 $\Phi_{xx}(\omega) = 4\tilde{a}^2$ | 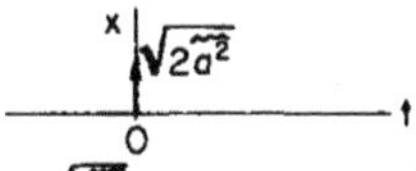 $x(t) = \sqrt{2\tilde{a}^2}\,\delta(t) \quad X(s) = \sqrt{2\tilde{a}^2}$ |
| 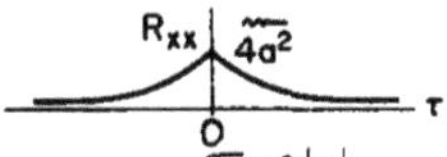 $R_{xx}(\tau) = \tilde{a}^2 e^{-\beta\lvert\tau\rvert}$ | 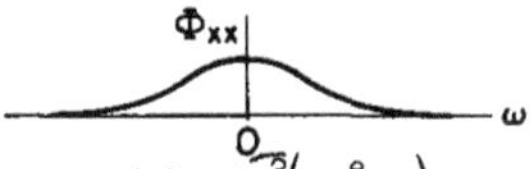 $\Phi_{xx}(\omega) = 4\tilde{a}^2\left(\frac{\beta}{\omega^2+\beta^2}\right)$ | 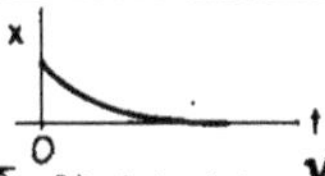 $x(t) = \sqrt{2\beta\tilde{a}^2}\,e^{-\beta t}u(t); X(s) = \frac{\sqrt{2\beta\tilde{a}^2}}{s+\beta}$ |
| 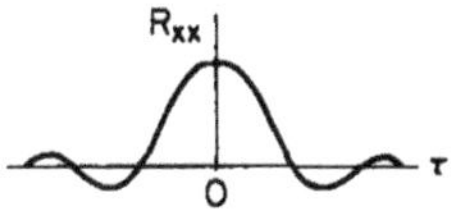 $R_{xx}(\tau) = \tilde{a}^2 e^{-\zeta\omega_1\lvert\tau\rvert}\cos\sqrt{1-\zeta^2}\,\omega_1\tau$ | 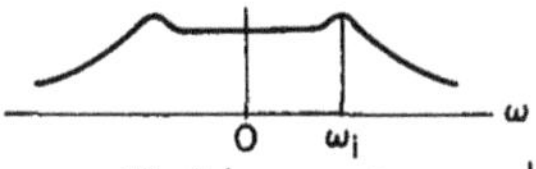 $\Phi_{xx}(\omega) = 4\tilde{a}^2\,\frac{\zeta}{\omega_1}\left\lvert\frac{1}{\left(\frac{j\omega}{\omega_1}\right)^2 + 2\zeta\left(\frac{j\omega}{\omega_1}\right) + 1}\right\rvert^2$ | $x(t) = \sqrt{\frac{2\tilde{a}^2\zeta\omega_1}{1-\zeta^2}}\,e^{-\zeta\omega_1 t}\sin\sqrt{1-\zeta^2}\,\omega_1 t$ 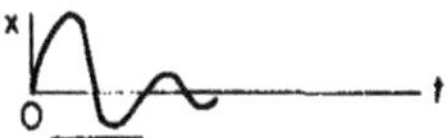 $X(s) = \frac{\sqrt{2\tilde{a}^2\zeta\omega_1}}{\omega_1}\left[\frac{1}{\left(\frac{s}{\omega_1}\right)^2 + 2\zeta\left(\frac{s}{\omega_1}\right) + 1}\right]$ |
| 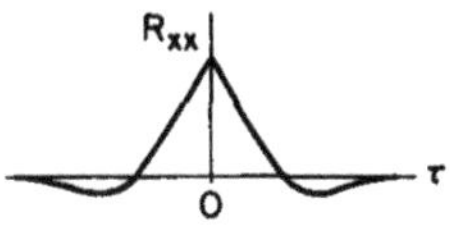 $R_{xx}(\tau) = \tilde{a}^2 e^{-\alpha\tau} - \tilde{b}^2 e^{-\beta\tau}$ | 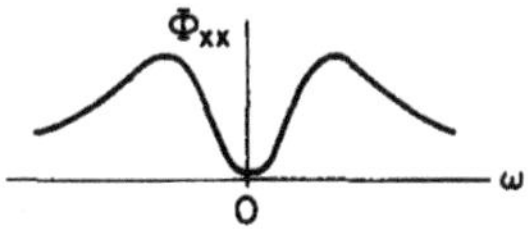 $\Phi_{xx} = 4\tilde{a}^2\,\frac{\alpha}{(\omega^2+\alpha^2)} - 4\tilde{b}^2\,\frac{\beta}{(\omega^2+\beta^2)}$ | — |
| 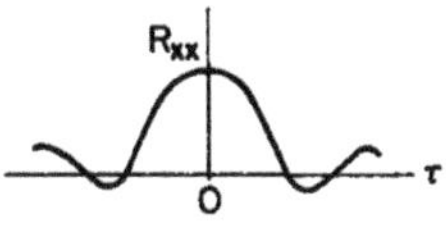 $R_{xx}(\tau) = \tilde{a}^2\,\frac{(\sin\alpha\tau - \sin\beta\tau)}{\alpha\tau}$ | 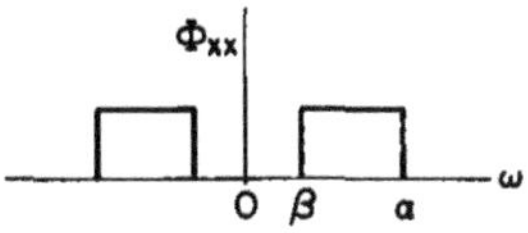 $\Phi_{xx}(\omega) = \frac{2\pi\tilde{a}^2}{\alpha}: \ \beta < \lvert\omega\rvert < \alpha$ | — |

There is a boxcar sequence, Item 2 in the table, in which the axis crossings are always $T$ sec apart, as in the random binary transmission, but in which the magnitude of the signal at any time is a random variable. If, however, the mean square value here were one, this signal would have an autocorrelation and power spectral density *identical* to the random binary transmission. This serves to emphasize the fact that the autocorrelation function and power spectral density are not specific to a particular time function.

Another type of boxcar sequence is one with random amplitudes and in which the intervals during which the signal is constant have a uniform distribution of lengths in the interval $O$, $T$. See Item 3 in the table.

The third boxcar sequence of interest, discussed below, depends for its description on a time function that comprises a sequence of impulses, each of weight (area), $q_i$, occurring randomly in time at a constant average rate. In Appendix B it is shown that the probability of finding a given number $s = 0, 1, 2, \ldots n$ impulses in a given interval of length, $T$, is then governed by the Poisson distribution. Such a sequence of impulses is, therefore, called a sequence of *generalized Poisson impulses*. It is also shown in Appendix B that the probability of finding an impulse in an interval more than $t$ and less than $t + dt$ sec after its predecessor is governed by the exponential distribution. The mean interval between impulses, $T_{av} = 1/\beta$, where $\beta$ is the density of the impulses along the time axis. The expression for the generalized Poisson impulse sequence is

$$z(t) = \sum_i q_i \delta(t - t_i) \tag{10-67}$$

Here the $t_i$ are understood to represent the (random) times of occurrence of the impulses. The manner of choosing the weights, $q_i$, remains to be specified. Figure 10-4 shows a typical generalized Poisson impulse sequence in which $q_i$ is a random variable that may take either positive or negative values.

If the generalized Poisson impulse sequence is specialized in such a way that each $q_i = +1$, i.e., each impulse is a positive unit impulse occurring randomly in time at a constant average rate, and this signal is passed through a linear, constant coefficient filter with a weighting function, $h(t)$, the result is a random process known as "shot noise." Shot noise, from our point of view, is primarily of historical interest.[9] The shot noise process is illustrated as Item 4 in Table 10-4 for the particular case in which $h(t) = e^{-\alpha t}$.

A time function called the semirandom telegraph signal is generated by switching from $+1$ to $-1$ at times $t_i$ which are determined in the same

[9] Rice, "Mathematical Analysis of Random Noise."

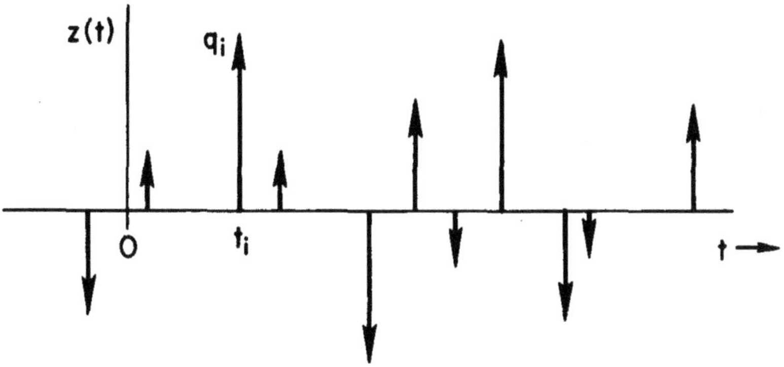


Fig. 10-4. Generalized Poisson impulse sequence.

way as the times of occurrence of the Poisson impulses. It is illustrated as Item 5 in the table.

Next, Item 6, take a boxcar sequence in which the occurrence of the steps in time is governed by the Poisson distribution. In fact, the function we wish to consider is simply the integral of the generalized Poisson impulse sequence, Eq. 10-67. Note interestingly, that, except for constants, the autocorrelation function and power spectral density for this function are exactly the same as the ones for the quite different semirandom telegraph signal. The exponential autocorrelation function and the corresponding power spectral density are typical of many physical random processes. See also Item 10 in the table. In this case, if the amplitudes of the steps in the boxcar sequence were chosen in such a way that the amplitude of the original function were normally (Gaussian) distributed, it would be quite impossible to distinguish between this process and the one represented by Item 10 on the basis of the first probability density function and the autocorrelation function or its transform, the power spectral density. Higher order density functions would be required to define each function in more detail, in order to be able to tell the difference.

Sums of sine waves are very important special functions with random like properties. This is because in computing and in experimental work it is often convenient to have a segment of a "random process," that can be repeated, in which the mean square value and power spectral density can easily be adjusted to meet particular requirements. The probability distribution for the amplitude of a sum of a comparatively small number of sine waves closely approaches a Gaussian distribution. Figure 10-5 shows the first probability density function for 1 sine wave and for sums

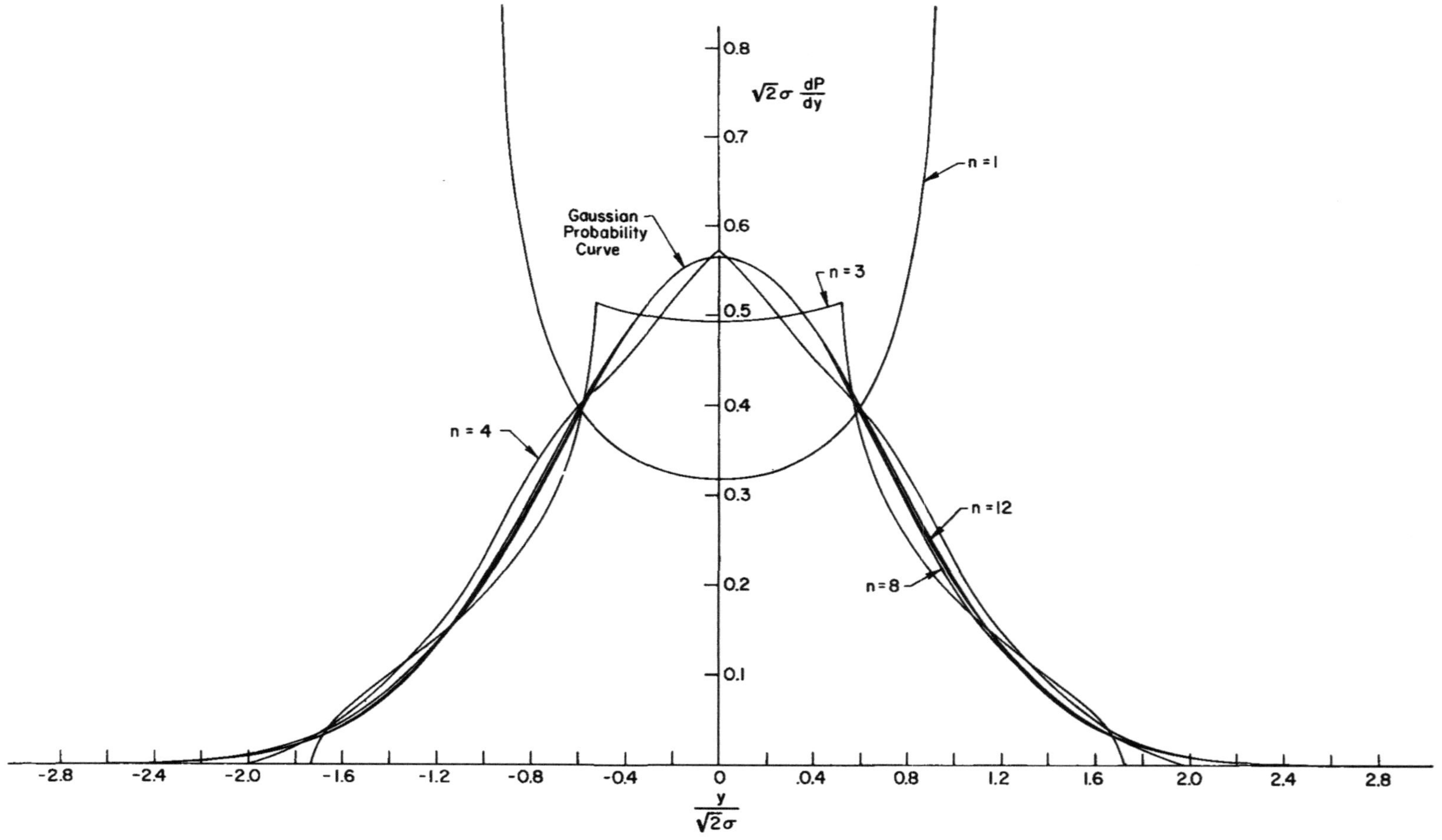


Fig. 10-5. Plot of the probability density distributions of the sum of $n$ sine waves of equal amplitude.

of 3, 4, 8, and 12 sine waves of equal amplitude. In the case in which the sine waves have unequal amplitudes, the density function does not approximate the density function of the Gaussian distribution as well for the same number of sine waves.[10] It is still amazing, however, how few sine waves need to be added together to yield a very good approximation of a random function of time with a Gaussian amplitude distribution. This is an illustration of the operation of the central limit theorem of statistics.

For analytical and experimental work, it is Item 7 in Table 10-4 that is most important. Item 8 is interesting because of its statistical properties, but, of course, as a function of time it is merely a (periodic) sine wave with an initially randomly chosen phase angle.

The next five items in Table 10-4 are perhaps somewhat out of place, since they are defined primarily in terms of the properties of the power spectral density function, rather than as time functions.

"White" noise, Item 9, is a signal whose power spectral density as a function of frequency is a constant. (The name is derived from the fact that white light is a composite of all the colors with their different wavelengths or frequencies.) Our principal interest in white noise is in connection with the signal at the output of a linear filter whose input may be an approximation to white noise. (The nature of the approximation lies in the fact that white noise, with a power spectral density flat to infinite frequency, is a physically unrealizable signal, but if the bandwidth of the noise is very large compared to the bandwidth of the filter, the approximation is accurate.)

When white noise is passed through a first-order filter, i.e., one with a transfer function $1/(Ts + 1)$, the result is a stochastic process that is typical of a very large number of physical phenomena (see Item 10). Among other things, with a suitable choice of the time constant, $T$, it may be used as a model for atmospheric turbulence or as a model of terrain elevation as seen from an airplane in rectilinear flight. White noise through a first-order filter is, therefore, a signal of particular interest and importance in flight control system engineering. This is especially true if the signal has a Gaussian amplitude distribution. (See the next section of this chapter.)

White noise through an underdamped second-order filter, Item 11, is a signal with a predominant band of frequencies that may give it some properties similar to a periodic signal. In this case, the filter has a transfer function $1/[(s/\omega_n)^2 + (2\zeta/\omega_n)s + 1]$. As the damping ratio, $\zeta$, of the filter approaches zero, the output of the filter approaches a purely periodic sinusoidal wave with a noisy bias.

[10] W. R. Bennett, "Distribution of the Sum of Randomly Phased Components," *Quarterly of Applied Mathematics*, Vol. 5, No. 4, pp. 385–393 (Jan. 1948).

A typical error signal spectrum, Item 12, may be simply the result of taking the difference between two spectra of the type illustrated as Item 10.

Finally, the rectangular spectrum, Item 13, is of interest particularly in connection with experimental or computer work in which the signal is a sum of sine waves. Actually, Items 10 to 13 may be representative special cases of Item 7, since the amplitudes of the constituent sine waves can be adjusted so as to produce any desired spectral shape. (See Eq. 10-61.)

## 10-5. Properties of Random Processes with Gaussian Amplitude Distributions

The most important stationary random functions, both theoretically and practically, are those that can be described in terms of Gaussian or normal amplitude distributions. Fortunately, these are also among the simplest distributions to handle analytically.

Recall (or see Appendix B) that the first probability density and distribution functions for signals that have Gaussian statistics are

$$p_1(x) = \frac{1}{\sqrt{2\pi}\,\sigma} e^{-(x-m_1)^2/2\sigma^2} \tag{10-68}$$

and

$$P_1(x) = \frac{1}{2}\left[1 + \operatorname{erf}\frac{(x - m_1)}{\sqrt{2}\,\sigma}\right] \tag{10-69}$$

where

$$\operatorname{erf} z = \frac{2}{\sqrt{\pi}} \int_0^z e^{-\lambda^2}\, d\lambda$$

It may be noted that the distributions are specified completely when the mean, $m_1$, and standard derivation, $\sigma$, are known. In linear systems, response calculations for the mean, or steady-state, value can be separated from those for the fluctuating portions of the response, so it is possible to consider the fluctuations about $m_1$ alone. Under these circumstances the distributions about the mean are pertinent, $m_1$ can be taken equal to zero without loss of generality, and the variance, $\sigma^2$, becomes simply the mean-square value of the signal.

When a signal within the system may be described by a Gaussian distribution, information about the probabilities of the signal's having various values, being within certain ranges, etc., is readily obtained from Eqs. 10-27 and 10-68 and tables of the distribution,[11] once the standard

[11] See, e.g., A. Hald, *Statistical Tables and Formulas*, Wiley, New York, 1952; M. M. Abramowitz and I. Stegun, eds., *Handbook of Mathematical Functions*, National Bureau of Standards Applied Mathematics Series 55, U.S. Government Printing Office, Washington, D.C., 1964.

deviation, $\sigma$, is known. The simplest probabilities of all to obtain are those given by the first distribution function directly. These take on especially simple forms when low probability values are considered, since then the error function can be approximated by

$$\operatorname{erf} z \doteq 1 - \frac{e^{-z^2}}{z\sqrt{\pi}}; \qquad z \gg 1$$

Thus, when it is desired to find the probability that a Gaussian distribution signal is a factor $k$ times greater than the standard deviation,

$$\Pr[x(t) > k\sigma] = 1 - P_1(k\sigma) \doteq \frac{e^{-k^2/2}}{\sqrt{2\pi}\,k} \tag{10-70}$$

when $k \gg 1$. (See also Table B-4, Appendix B.)

One of the most interesting properties of time signals that have Gaussian distributions is the fact that for some purposes they can be thought of as a sum of sinusoids. The time signal $x(t)$, as expressed in a Fourier series with zero mean,

$$x(t) = \sum_{k=1}^{N} A_k \cos(\omega_k t + \varphi_k) \tag{10-71}$$

will have a probability density function as $N \to \infty$ which approaches a Gaussian distribution function if the frequencies are given by $\omega_k = k\,\Delta\omega$, and the phase angles $\varphi_k$ are random and distributed uniformly (i.e., as in the rectangular distribution of Table B-1, Appendix B) over the range from zero to $2\pi$. (Recall Fig. 10-5.) The constants $A_k$ must be fixed at values to match the power spectral density of the random process. The power spectral density of the function described by Eq. 10-71 is

$$\Phi_{xx}(\omega) = \pi \sum_{k=1}^{N} A_k^2[\delta(\omega - \omega_k) + \delta(\omega + \omega_k)] \tag{10-72}$$

This expression represents a series of spectral lines at frequencies $\omega = \pm\omega_k$ and with weights $\pi A_k^2$. In response calculations the same answers will be obtained using Eq. 10-72 and an ordinary, continuous power spectral density function $\Phi(\omega)$ if $\pi A_k^2 = \Phi(\omega_k)$ and if $N$ is made sufficiently large.

A consequence of surpassing importance follows almost intuitively from the use of a sum of sinusoids to represent a Gaussian random signal. In a stable, constant coefficient, linear system the steady-state response to a sinusoidal input is itself sinusoidal, although the amplitude and phase are, in general, modified. When many sinusoids are used in the input, they all appear in the output, again with different amplitudes and phases. Now, if the input phases of the sinusoids are randomly distributed, the phases in the output waves will still be randomly distributed, even though phase shifts occur for each of the sinusoids. In the limit, then, as the

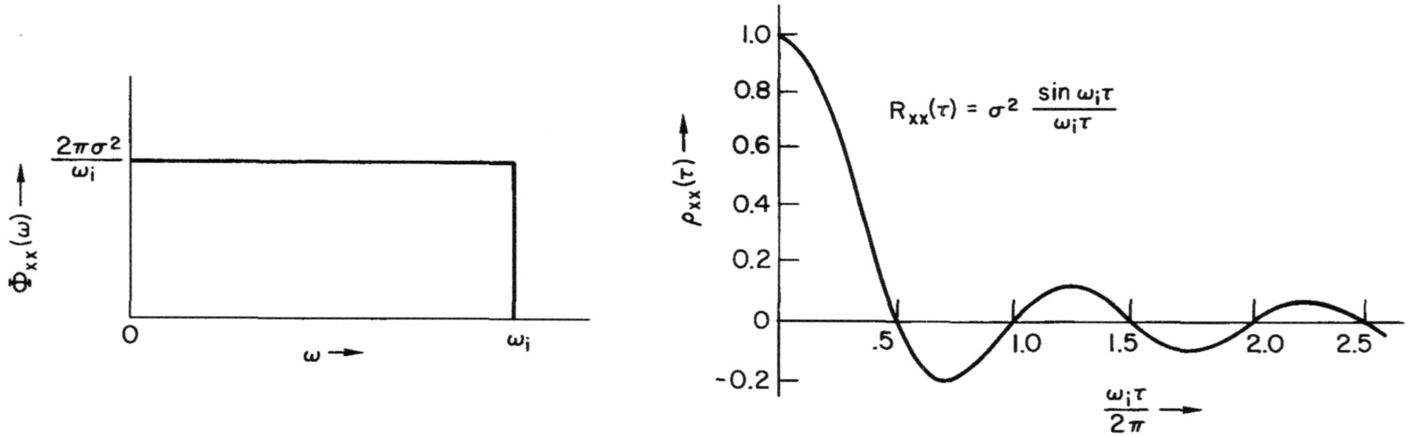


Power Spectral Density

Autocorrelation Function

Average Axis Crossing per second, $N_0 = \frac{\omega_i}{\sqrt{3}\pi} = 0.1837\,\omega_i$

Average Number of Maxima per second, $N_m = \frac{1}{2\pi}\sqrt{\frac{3}{5}}\,\omega_i = 0.1232\,\omega_i$

Average Number of Maxima > x, $N_L(x) = \frac{\sigma}{\sqrt{6}\pi}\, p_1(x)\omega_i = 0.230\,\sigma p_1(x)\omega_i$

Fig. 10-6. Characteristics of band-limited, flat spectrum, Gaussian-distributed signal.

number of sinusoids in the input is increased to come closer and closer to a normal distribution, the distribution of the output will also approach a normal distribution. The variance of the output will, in general, be different from that of the input because of the change in the amplitudes of the individual waves. Thus, *in a time-invariant, linear system, Gaussian inputs will produce Gaussian outputs*, and the computation of the mean-square output will give complete information about the first probability density and distribution functions of the output signal.

Beside the interesting and valuable characteristics noted above, Rice[12] has presented a number of additional properties of Gaussian random processes that are useful in obtaining a grasp of the time history characteristics of Gaussian-distributed time signals. These are summarized in Table 10-5.

A useful idealization of a Gaussian random signal has the band-limited power spectral density shown in Fig. 10-6. This is, of course, a special case of Item 13, Table 10-4. The corresponding autocorrelation function is also illustrated, along with a tabulation of the characteristics given in Table 10-5.

Table 10-5. Some properties of a Gaussian random process computed from the autocorrelation function or the power spectral density.

| Property | From the autocorrelation function | From the power spectral density |
|---|---|---|
| Mean square | $\overline{x^2} = R_{xx}(0) = \sigma^2$ | $\frac{1}{2\pi}\int_0^\infty \Phi_{xx}(\omega)\,d\omega$ |
| Axis crossings per second | $N_0 = \frac{1}{\pi}\left[\frac{-R''_{xx}(0)}{R_{xx}(0)}\right]^{1/2}$ (See note.) | $\frac{1}{\pi}\left[\frac{\int_0^\infty \omega^2\Phi_{xx}(\omega)\,d\omega}{\int_0^\infty \Phi_{xx}(\omega)\,d\omega}\right]^{1/2}$ |
| Maxima per second | $N_m = \frac{1}{2\pi}\left[\frac{-R^{\text{IV}}(0)}{R''(0)}\right]^{1/2}$ | $\frac{1}{2\pi}\left[\frac{\int_0^\infty \omega^4\Phi_{xx}(\omega)\,d\omega}{\int_0^\infty \omega^2\Phi_{xx}(\omega)\,d\omega}\right]^{1/2}$ |
| Positive crossings of level $x$, or number of maxima per second greater than $x$; $x > 2\sigma$ (exceedences) | $N_L(x) = \frac{N_0}{2}e^{-x^2/2\sigma^2}$ (See above for $N_0$ and $\sigma$.) | $N_L(x) = \frac{N_0}{2}e^{-x^2/2\sigma^2}$ (See above for $N_0$ and $\sigma$.) |

Note: $R''_{xx}(0) = [d^2R(\tau)/d\tau^2]_{\tau=0}$ etc.

[12] Rice, "Mathematical Analysis of Random Noise."

## 10-6. Response of Linear Systems to Random Inputs

The effect of a linear filter, or feedback system, operating on a particular power spectral density, $\Phi_{xx}(\omega)$, may be found as follows. Let the system have a weighting function, $w(t)$, such that the output, $y(t)$, in response to an input, $x(t)$, is given by the convolution integral:

$$y(t) = \int_{-\infty}^{\infty} w(\tau_1)x(t - \tau_1)\, d\tau_1 = \int_{-\infty}^{\infty} w(t - \tau_1)x(\tau_1)\, d\tau_1 \qquad (10\text{-}73)$$

With the use of Eq. 10-42 the autocorrelation function of the output can then be written in terms of the input as follows:

$$R_{yy}(\tau) = \lim_{T\to\infty} \frac{1}{2T}\int_{-T}^{T} dt \Bigg[\underbrace{\int_{-\infty}^{\infty} w(\tau_1)x(t - \tau_1)\, d\tau_1}_{y(t)} \underbrace{\int_{-\infty}^{\infty} w(\tau_2)x(t + \tau - \tau_2)\, d\tau_2}_{y(t+\tau)}\Bigg] \qquad (10\text{-}74)$$

Interchanging the order of the integrations, we find that

$$R_{yy}(\tau) = \lim_{T\to\infty} \int_{-\infty}^{\infty} d\tau_1 \int_{-\infty}^{\infty} d\tau_2 w(\tau_1)w(\tau_2)\left[\frac{1}{2T}\int_{-T}^{T} dt x(t - \tau_1)x(t + \tau - \tau_2)\right] \qquad (10\text{-}75)$$

but, in the limit, the quantity in square brackets may be seen to be the autocorrelation function, $R_{xx}(\tau_1 + \tau - \tau_2)$. Therefore, in the limit,

$$R_{yy}(\tau) = \int_{-\infty}^{\infty} d\tau_1 \int_{-\infty}^{\infty} d\tau_2 w(\tau_1)w(\tau_2)R_{xx}(\tau_1 + \tau - \tau_2) \qquad (10\text{-}76)$$

Because the autocorrelation function is even, the Fourier transform of Eq. 10-76 is

$$\begin{aligned}\Phi_{yy}(\omega) &= 2\int_{-\infty}^{\infty} R_{yy}(\tau)e^{-j\omega\tau}\, d\tau = 4\int_{0}^{\infty} R_{yy}(\tau)\cos\omega\tau\, d\tau \\ &= 2\int_{-\infty}^{\infty} d\tau \int_{-\infty}^{\infty} d\tau_1 \int_{-\infty}^{\infty} d\tau_2 e^{-j\omega(\tau+\tau_1-\tau_2)} \\ &\qquad e^{j\omega\tau_1}e^{-j\omega\tau_2}R_{xx}(\tau + \tau_1 - \tau_2)w(\tau_1)w(\tau_2)\end{aligned} \qquad (10\text{-}77)$$

Upon a change of the variable of integration $\tau$ to $(\tau + \tau_1 - \tau_2)$, the volume integral breaks up into the product of three independent integrals which may be recognized as representing the Fourier transforms of $w(t)$, its conjugate, and the Fourier transform of $R_{xx}(\tau)$. Therefore, we have the important relationship in the transform domain:

$$\Phi_{yy}(\omega) = W(j\omega)W(-j\omega)\Phi_{xx}(\omega) = |W(j\omega)|^2\,\Phi_{xx}(\omega) \qquad (10\text{-}78)$$

where

$$W(j\omega) = \int_{-\infty}^{\infty} w(t)\, e^{-j\omega t}\, dt$$

and, of course, the mean-squared value of the signal of interest is found from the formula:

$$\overline{y^2} = \frac{1}{2\pi}\int_0^\infty \Phi_{yy}(\omega)\, d\omega = \frac{1}{2\pi}\int_0^\infty |W(j\omega)|^2\, \Phi_{xx}(\omega)\, d\omega = R_{yy}(0) \qquad (10\text{-}79)$$

Integrals of this type, evaluated for systems through the seventh-order, appeared first in James, Nichols, and Phillips.[13] The tables have since been modified in form and extended through tenth-order systems, with corrections found for the $I_7$ expression.[14] These are duplicated in a condensed form in Table 10-6. The higher order literal expressions are lengthy, and the integrals can be expressed more compactly in terms of Hurwitz determinants.[15] The actual tabulated forms are for the integral

$$\begin{aligned} I_n = \int_0^\infty y^2(t)\, dt &= \lim_{\sigma\to 0}\frac{1}{2\pi j}\int_{c-j\infty}^{c+j\infty} Y(s)\,Y(\sigma - s)\, ds \\ &= \frac{1}{2\pi j}\int_{-j\infty}^{j\infty} Y(s)\,Y(-s)\, ds \\ &= \frac{1}{2\pi j}\int_{-j\infty}^{j\infty} \frac{d(s)}{a(s)}\frac{d(-s)}{a(-s)}\, ds \end{aligned} \qquad (10\text{-}80)$$

This applies only for stable systems. Note in comparing Eq. 10-79 with Eq. 10-80 that

$$\begin{aligned} \overline{y^2} &= \frac{1}{2\pi}\int_0^\infty \Phi_{yy}(\omega)\, d\omega = \frac{1}{2\pi}\int_{-\infty}^\infty \frac{\Phi_{yy}(\omega)}{2}\, d\omega \\ &= \frac{1}{2\pi j}\int_{-j\infty}^{j\infty} Y(s)\,Y(-s)\, ds = \frac{1}{2\pi}\int_{-\infty}^\infty Y(j\omega)\,Y(-j\omega)\, d\omega \end{aligned}$$

or

$$Y(j\omega)\,Y(-j\omega) = \frac{\Phi_{yy}}{2}$$

The factor of two is a consequence of the definition of the power spectral density such that $\overline{y^2}$ is given as an integral power density over only real frequencies. It is a point to be remembered when using Table 10-6.

An immediate application of Eq. 10-79 is the computation of mean-squared amplitude for an output quantity in response to an input spectrum, $\Phi_{xx}(\omega)$. This is accomplished by evaluating Eq. 10-79 for $\Phi_{yy}(\omega)$

[13] James, Nichols, and Phillips, *Theory of Servomechanisms*, pp. 369–370.

[14] R. C. Booten, Jr., Max V. Mathews, and W. W. Seifert, "Nonlinear Servomechanisms with Random Inputs," Rept. 70, Dynamic Analysis and Control Lab., Massachusetts Institute of Technology, Aug. 20, 1953, pp. 38–42.

[15] G. C. Newton, Jr., L. A. Gould, and J. F. Kaiser, *Analytical Design of Linear Feedback Controls*, Wiley, New York, 1957, pp. 366–381.

Table 10-6. Abbreviated table of integrals.[a]

This is a table of integrals of the type

$$\overline{y^2} = \int_0^\infty y(t)^2\,dt = \frac{1}{2\pi j}\int_{-j\infty}^{j\infty} Y(s)Y(-s)$$

where

$$Y(s) = \frac{d_0 s^{n-1} + d_1 s^{n-2} + \cdots + d_{n-1}}{a_0 s^n + a_1 s^{n-1} + \cdots + a_n}$$

and the roots of $a_0 s^n + a_1 s^{n-1} + \cdots + a_n$ are all in the left half-plane.

$$I_1 = \frac{d_0^2}{2a_0 a_1}$$

$$I_2 = \frac{d_0^2 + (a_0/a_2)\,d_1^2}{2a_0 a_1}$$

$$I_3 = \frac{a_2 d_0^2 + a_0(d_1^2 - 2d_0 d_2) + (a_0 a_1/a_3)\,d_2^2}{2a_0(a_1 a_2 - a_0 a_3)}$$

$$I_4 = \frac{d_0(a_2 a_3 - a_1 a_4) + a_0 a_3(d_1^2 - 2d_0 d_2) + a_0 a_1(d_2^2 - 2d_1 d_3) + (a_0/a_4)\,d_3^2(a_1 a_2 - a_0 a_3)}{2a_0(a_1 a_2 a_3 - a_0 a_3^2 - a_1^2 a_4)}$$

[a] Adopted from the table in Booten, Mathews, and Seifert, "Nonlinear Servomechanisms with Random Inputs," Rept. 70, Dynamic Analysis and Control Lab., Massachusetts Institute of Technology, Cambridge, Mass., Aug. 20, 1953, pp. 38–42.

to obtain $\overline{y^2}$, or the root mean square (rms), $\sigma$. This technique allows computation of a variety of performance indices. Examples include flight path deviation about the ideal profile in terrain following or the localizer course in an automatic approach. Design considerations such as root mean square control surface deflection and deflection rate can be studied as well.

Another application of Eq. 10-79 is in the modeling of a command input spectral density. If $\Phi_{nn}(\omega)$ is a Gaussian white noise source, then a shaping filter, $G_f(j\omega)$, can be fitted to yield the input spectrum, $\Phi_{xx}(\omega)$. The input spectrum is then given by

$$\Phi_{xx}(\omega) = |G_f(j\omega)|^2\,\Phi_{nn}(\omega) \tag{10-81}$$

Since $\Phi_{nn}(\omega)$ is a constant, $|G_f(j\omega)|^2$ must have the shape of the desired input spectrum.[16]

[16] Actual disturbance and command data for automatic flight control system analyses are summarized in J. E. Hart, L. A. Adkins, and L. L. Lacau, "Stochastic Disturbance Data for Flight Control System Analysis," ASD-TDR-62-347, Sept. 1962; D. H. Weir, "Compilation and Analysis of Flight Control System Command Inputs," AFFDL-TR-65-119, Nov. 1965.

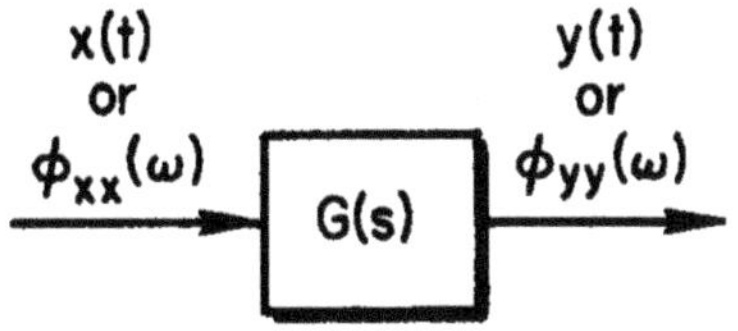


Fig. 10-7. A simple transfer characteristic.

A more general case is the one in which the command input is the sum of several random signals. For $N$ input components and one output, the transfer function characteristic of Fig. 10-7 becomes the one shown in Fig. 10-8. The inputs are assumed to have zero means but are not necessarily uncorrelated. Denote the cross-spectral density between the $i^{\text{th}}$ and $k^{\text{th}}$ inputs as $\Phi_{ki}(\omega)$. Defining $G_k^*(j\omega)$ as the complex conjugate of $G_k(j\omega)$, yields the following equation for the spectrum of the sum:

$$\Phi_{xx}(\omega) = \sum_{i=1}^{N} \sum_{k=1}^{N} G_i(j\omega) G_k^*(j\omega) \Phi_{ki}(\omega) \tag{10-82}$$

When the $N$ inputs are uncorrelated, Eq. 10-82 reduces to

$$\Phi_{xx}(\omega) = \sum_{k=1}^{N} |G_k(j\omega)|^2 \, \Phi_{kk}(\omega) \tag{10-83}$$

If the inputs are summed prior to entering a single transfer block, the situation would be the one illustrated in Fig. 10-9. In this case Eq. 10-83 simplifies to

$$\Phi_{xx}(\omega) = |G_1(j\omega)|^2 \sum_{k=1}^{N} \Phi_{kk}(\omega) \tag{10-84}$$

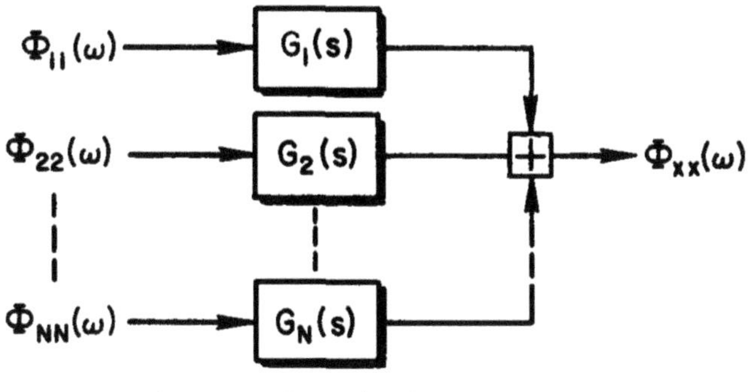


Fig. 10-8. A multiple-input system.

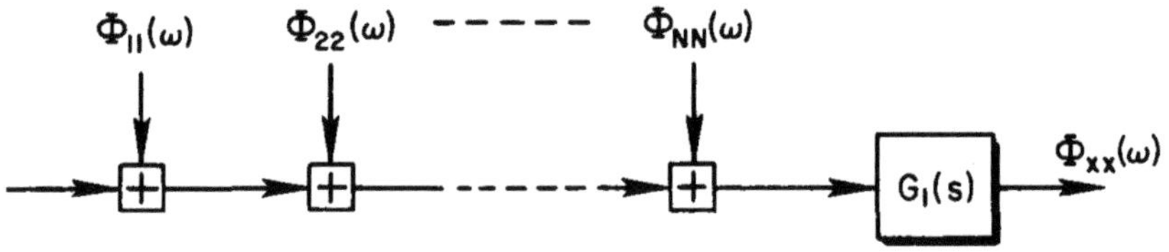


Fig. 10-9. Summation of random inputs.

Many random disturbance and command input data are well approximated by the Gaussian probability density function.[17] With single or multiple Gaussian random inputs and a linear transfer characteristic (see Fig. 10-7), the output will also be Gaussian-distributed, so that evaluation of the mean square value from the output power spectral density is only the beginning of the interesting calculations that can be made. (See the previous section.)

Example 1. Following error in first-order servomechanism

Given a rectangular band limited input spectrum $\Phi_{rr}(\omega)$ as the input to a first-order servomechanism, find the dependence of the mean square error in following this input on the cutoff frequency of the input, $\omega_i$, and the inverse time constant (crossover frequency) of the closed-loop system, $\omega_c$ (Fig. 10-10).

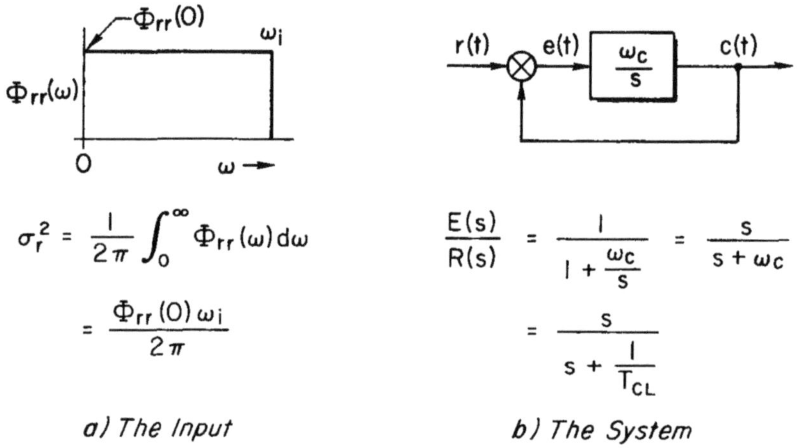


Fig. 10-10. A band limited input spectrum and first-order servomechanism.

[17] *Ibid.*

From Eq. 10-77, we obtain

$$\overline{e^2} = \frac{1}{2\pi}\int_0^\infty \left|\frac{E(j\omega)}{R(j\omega)}\right|^2 \Phi_{rr}(\omega)\,d\omega$$

$$= \frac{\Phi_{rr}(0)}{2\pi}\int_0^{\omega_i}\left[\frac{1}{1+(\omega_c/j\omega)}\right]\left[\frac{1}{1-(\omega_c/j\omega)}\right]d\omega$$

$$= \frac{\Phi_{rr}(0)}{2\pi}\int_0^{\omega_i}\frac{\omega^2}{\omega^2+\omega_c^2}\,d\omega$$

$$= \frac{\Phi_{rr}(0)}{2\pi}\,\omega_i\left[1-\frac{\omega_c}{\omega_i}\tan^{-1}\left(\frac{\omega_i}{\omega_c}\right)\right]$$

The continued fraction[18] for the inverse tangent is

$$\tan^{-1} x = \cfrac{x}{1+\cfrac{x^2}{3+\cfrac{4x^2}{5+\cfrac{9x^2}{7+\cdots}}}}$$

Taking the third convergent as an approximation, we find that

$$1-\frac{1}{x}\tan^{-1} x \doteq 1-\frac{1}{x}\left[\cfrac{x}{1+\cfrac{x^2}{3+(4x^2/5)}}\right]$$

$$= \frac{x^2}{3+(9x^2/5)}$$

Finally,

$$\frac{\overline{e^2}}{\sigma_r^2} \doteq \frac{(\omega_i/\omega_c)^2}{3+\frac{9}{5}(\omega_i/\omega_c)^2} \doteq \frac{1}{3}\left(\frac{\omega_i}{\omega_c}\right)^2; \qquad \omega_i \ll \omega_c$$

The mean square error in following the input is roughly proportional to the mean square input, $\sigma_r^2$, to the square of the input bandwidth, $\omega_i^2$, and is inversely proportional to the square of the crossover frequency, $\omega_c^2$.

Example 2. Optimum fixed-form adjustment of first-order servomechanism with signal and noise inputs

Given a first-order servomechanism with command signal and noise inputs, we seek the best adjustment of the crossover frequency, $\omega_c$, so as to minimize the error, $\epsilon(t)$, in following the desired signal $r(t)$ (see

[18] C. D. Olds, *Continued Fractions*, Random House, New York, 1963.

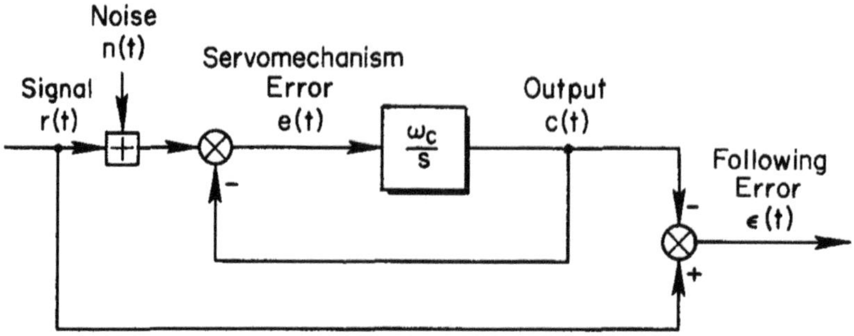


*a) The System*

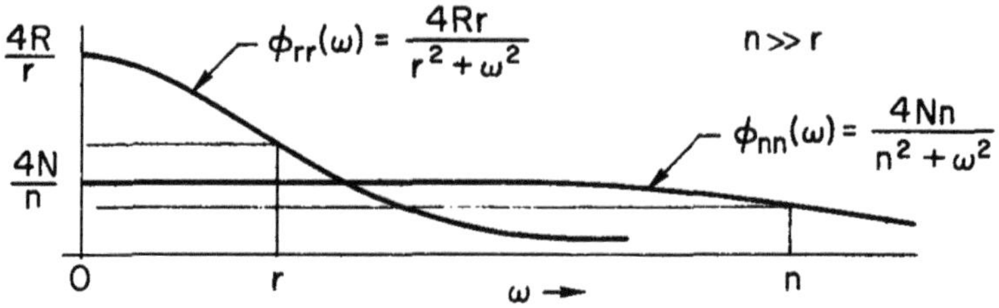


*b) The Signal and Noise Inputs*

**Fig. 10-11. Minimizing the following error in the presence of noise.**

Fig. 10-11). Note that the injection of the noise signal might be in the feedback path (measurement noise) or at the servomechanism error point (amplifier noise) without altering this problem.

The error in following is given by the expression,

$$\epsilon = r - c$$

Then, with a substitution for $C(s)$, the Laplace transform is

$$\epsilon(s) = R(s) - [R(s) + N(s)]\frac{\omega_c}{s + \omega_c}$$

or

$$\epsilon(s) = R(s)\frac{s}{s + \omega_c} - N(s)\frac{\omega_c}{s + \omega_c} = E_r + E_n$$

Assuming that there is no correlation between signal and noise, we get

$$\Phi_{\epsilon\epsilon}(\omega) = \frac{4Rr\omega^2}{|(r+\omega)(\omega_c+\omega)|^2} + \frac{4Nn\omega_c^2}{|(n+\omega)(\omega_c+\omega)|^2}$$

$$= \Phi_{\epsilon\epsilon_r}(\omega) + \Phi_{\epsilon\epsilon_n}(\omega)$$

Then the mean-squared error will be

$$\overline{\epsilon^2} = \overline{\epsilon_r^2} + \overline{\epsilon_n^2}$$

$$= \frac{1}{2\pi}\int_0^\infty \Phi_{\epsilon\epsilon_r}(\omega)\, d\omega + \frac{1}{2\pi}\int_0^\infty \Phi_{\epsilon\epsilon_n}(\omega)\, d\omega$$

$$= \frac{1}{2\pi j}\int_{-j\infty}^{j\infty} E_r(s)E_r(-s)\, ds + \frac{1}{2\pi j}\int_{-j\infty}^{j\infty} E_n(s)E_n(-s)\, ds$$

where

$$E_r(s) = \frac{\sqrt{2Rr}}{(s+r)}\,\frac{s}{(s+\omega_c)}$$

$$E_n(s) = \frac{\sqrt{2Rr}}{(s+n)}\,\frac{\omega_c}{(s+\omega_c)}$$

Both of the following error components are given by a specialized form of the integral $I_2$ from Table 10-6, i.e.,

$$I_2 = \frac{d_0^2 + (a_0/a_2)d_1^2}{2a_0a_1}$$

For the following error in response to the signal,

$$d(s) = d_0 s + d_1 = \sqrt{2Rr}$$

$$d_0 = \sqrt{2Rr}, \qquad d_1 = 0$$

$$a(s) = a_0 s^2 + a_1 s + a_2 = s^2 + (r + \omega_c)s + r\omega_c$$

$$a_0 = 1, \quad a_1 = r + \omega_c, \quad a_2 = r\omega_c$$

so that

$$\overline{\epsilon_r^2} = \frac{Rr}{(r+\omega_c)}$$

Similarly, for the following error in response to the noise,

$$d_0 = 0, \quad d_1 = \sqrt{2Nn}\,\omega_c$$

$$a_0 = 1, \quad a_1 = n + \omega_c, \quad a_2 = n\omega_c$$

and

$$\overline{\epsilon_n^2} = \frac{(1/n\omega_c)(2Nn)\omega_c^2}{2(n + \omega_c)}$$

$$= \frac{N\omega_c}{n + \omega_c}$$

Combining the contributions to the mean square following error, we obtain

$$\overline{\epsilon^2} = \overline{\epsilon_r^2} + \overline{\epsilon_n^2} = \frac{Rr}{r + \omega_c} + \frac{N\omega_c}{n + \omega_c}$$

To find the crossover frequency, $\omega_c$, which minimizes the total mean square error, we differentiate the expression for $\overline{\epsilon^2}$ with respect to $\omega_c$, set the result equal to zero, and try to solve for $\omega_c$.

$$\frac{d\overline{\epsilon^2}}{d\omega_c} = -\frac{Rr}{(r + \omega_c)^2} + \frac{N}{(n + \omega_c)} - \frac{N\omega_c}{(n + \omega_c)^2} = 0$$

or

$$\frac{Nn}{(n + \omega_c)^2} = \frac{Rr}{(r + \omega_c)^2} = \frac{N/n}{[1 + (\omega_c/n)]^2} = \frac{R/r}{[1 - (\omega_c/r)]^2}$$

In the practical case in which $\omega_c \ll n$,

$$\frac{N}{n} \doteq \frac{R}{r}\frac{1}{[1 + (\omega_c/r)]^2}$$

or

$$\left(\frac{\omega_c}{r}\right)^2 + \frac{2\omega_c}{r} + 1 \doteq \frac{R/r}{N/n}$$

Therefore,

$$\frac{\omega_c}{r} \doteq \sqrt{\frac{Rn}{Nr}} - 1$$

If, now, the "signal to noise" is large (i.e., $R/N \gg 1$) and the signal to noise bandwidth ratio is small (i.e., $n \gg r$), then

$$\frac{\omega_c}{r} \doteq \sqrt{\frac{Rn}{Nr}}$$

In 1951 J. R. Dutilh[19] conjectured that, in general, the half power frequency of the servomechanism (crossover frequency) should be adjusted to the frequency, in which the command signal power is equal to

[19] Comment in A. Tustin, ed., *Automatic and Manual Control*, Butterworths, London, 1951, p. 156.

the noise power. In this problem, we could set the expressions for the two spectra equal and solve for the frequency, $\omega$.

$$\frac{4Rr}{r^2 + \omega^2} = \frac{4Nn}{n^2 + \omega'}$$

or

$$\frac{R/r}{1 + (\omega^2/r^2)} = \frac{N/n}{1 + (\omega^2/n^2)}$$

or again if $\omega \ll n$

$$1 + \frac{\omega^2}{r^2} = \frac{Rn}{Nr}$$

and

$$\frac{\omega}{r} \doteq \sqrt{\frac{Rn}{Nr} - 1}$$

Under similar assumptions of signal to noise power and bandwidth ratios as above, this will become

$$\frac{\omega}{r} \doteq \sqrt{\frac{Rn}{Nr}}$$

## 10-7. Computer Methods

In many practical cases the response of vehicle control systems to random inputs and disturbances will be determined by automatic computation. If the calculations are to be performed on a computer anyway, there may be no advantage to working with the spectral descriptions. Computer methods of evaluating the mean square, and other statistics, of signals in control systems, including cases that often arise in guidance problems in which the system is time-varying, are discussed in Laning and Battin.[20] Two particularly powerful and useful methods are the *transient analog*,[21] and the *adjoint system*.[22] The methods are closely related, with the transient analog being a special feature of the adjoint technique. Here, however, they will be treated separately, for simplicity.

[20] Laning and Battin, *Random Processes in Automatic Control.*

[21] T. R. Benedict and V. C. Rideout, "Error Determination for Optimum Predicting Filters," *Proc. National Electronics Conference,* Vol. 13, 875–887 (1957); B. Etkin, "A Simple Method for the Analogue Computation of the Mean-Square Response of Airplanes to Atmospheric Turbulence," *J. Aerospace Sciences,* 28, No. 10, 825–826 (Oct. 1961); R. Magdaleno and J. Wolkovitch, "Performance Criteria for Linear Constant-Coefficient Systems with Random Inputs," ASD-TDR-62-470, Jan. 1963.

[22] Laning and Battin, *Random Processes in Automatic Control.*

For a system with transfer function, $W(s)$, excited by a random process $x(t)$, the mean-squared output is, of course,

$$\overline{y^2} = \lim_{T\to\infty} \frac{1}{2T}\int_{-T}^{T} y^2(t)\,dt = \frac{1}{2\pi}\int_0^{\infty} |W(j\omega)|^2 \Phi_{xx}(\omega)\,d\omega \qquad (10\text{-}85)$$

The idea of the transient analog of a power spectral density, $\Phi_{xx}(\omega)$, is based on finding an input time function, $x_t(t)$, such that the system output response, $y_t$, has an integral square value that is the same as $\overline{y^2}$. That is,

$$\int_0^{\infty} y_t^2\,dt = \overline{y^2} \qquad (10\text{-}86)$$

where the transient response $y_t$,

$$y_t = \int_0^t w(\tau)x_t(t-\tau)\,d\tau = \int_0^t w(t-\tau)x_t(\tau)\,d\tau; \qquad t \geqq 0 \quad (10\text{-}87)$$

To find the relationship between $x_t$ and $\Phi_{xx}$ required to make this so, we shall resort to the real multiplication, real integration, and final value theorems of the Laplace transformation (see Table 2-1). Using the real multiplication theorem, we find that the transform of $y_t^2$ will be

$$\begin{aligned}\mathcal{L}[y_t^2] &= \frac{1}{2\pi j}\int_{c-j\infty}^{c+j\infty} Y_t(s-\lambda)Y_t(\lambda)\,d\lambda \\ &= \frac{1}{2\pi j}\int_{c-j\infty}^{c+j\infty} W(s-\lambda)X_t(s-\lambda)W(\lambda)X_t(\lambda)\,d\lambda \qquad (10\text{-}88)\end{aligned}$$

From the real integration theorem,

$$\begin{aligned}\mathcal{L}\left[\int_0^t y_t^2\,dt\right] &= \frac{1}{s}\mathcal{L}[y_t^2] \\ &= \frac{1}{s}\frac{1}{2\pi j}\int_{c-j\infty}^{c+j\infty} W(s-\lambda)X_t(s-\lambda)W(\lambda)X_t(\lambda)\,d\lambda \qquad (10\text{-}89)\end{aligned}$$

The final value theorem is subject to restrictions on $W(s)$ and $X_t(s)$ which insure that the integral square output has a finite value. As a practical matter this requires the system to be stable and $X_t(s)$ to be analytic on and to the right of the imaginary axis, thereby allowing $c$ to be zero in the contour integral. The final value theorem can now be applied:

$$\begin{aligned}\int_0^{\infty} y_t^2\,dt &= \lim_{t\to\infty}\int_0^t y_t^2\,dt = \lim_{s\to 0} s\mathcal{L}\left[\int_0^t y_t^2\,dt\right] \\ &= \lim_{s\to 0} s\,\frac{1}{2\pi j s}\int_{-j\infty}^{j\infty} W(s-\lambda)W(\lambda)X_t(s-\lambda)X_t(\lambda)\,d\lambda \\ &= \frac{1}{2\pi j}\int_{-j\infty}^{j\infty} W(-\lambda)W(\lambda)X_t(-\lambda)X_t(\lambda)\,d\lambda \qquad (10\text{-}90)\end{aligned}$$

Changing the variable $\lambda$ to $j\omega$, we obtain

$$\int_0^\infty y_t^2\,dt = \frac{1}{2\pi}\int_{-\infty}^{\infty} W(-j\omega)W(j\omega)X_t(-j\omega)X_t(j\omega)\,d\omega$$

$$= \frac{1}{2\pi}\int_{-\infty}^{\infty} |W(j\omega)|^2\,|X_t(j\omega)|^2\,d\omega \qquad (10\text{-}91)$$

Comparison of Eqs. 10-85 and 10-91 reveals that the two integrals will be the same, provided that the transient analog input, $x_t(t)$, is chosen so that

$$X_t(j\omega)X_t(-j\omega) = \tfrac{1}{2}\Phi_{xx}(\omega) \qquad (10\text{-}92)$$

Thus the transient analog, $x_t(t)$, appropriate to the power spectral density $\Phi_{xx}(\omega)$, must have an *energy spectral density* that is the same function of frequency (except for the units) as half the power spectral density of the random input.

The *unit* delta function is the transient analog of a white noise spectrum with two units of power per radian per second. If the random input of interest is considered to be the result of passing a unit white noise spectrum through a linear filter with a Fourier transfer function, $G_f(j\omega)$ (see Fig. 10-12 and recall Eq. 10-81), then the transient analog of the random input of interest is the output of the same filter excited by an impulse with weight $1/\sqrt{2}$.

For constant coefficient situations the transient analog technique is all that is required for the computation of integral-square transient responses, and thus the analogous mean square response. When time variations become important, however, we need a different approach. This is provided by the adjoint techniques described below.

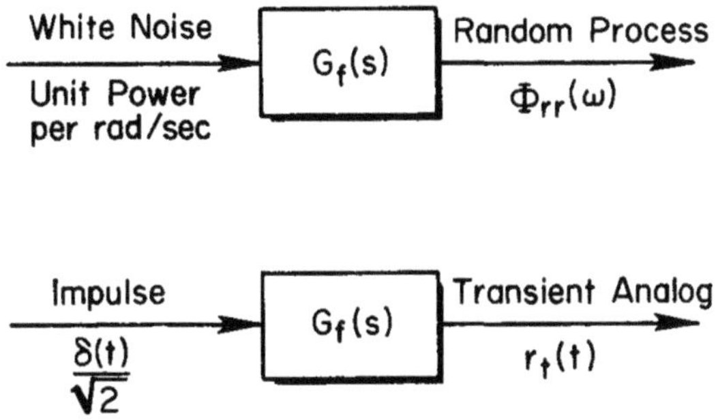


Fig. 10-12. The transient analog of a random process formed by passing white noise through a filter.

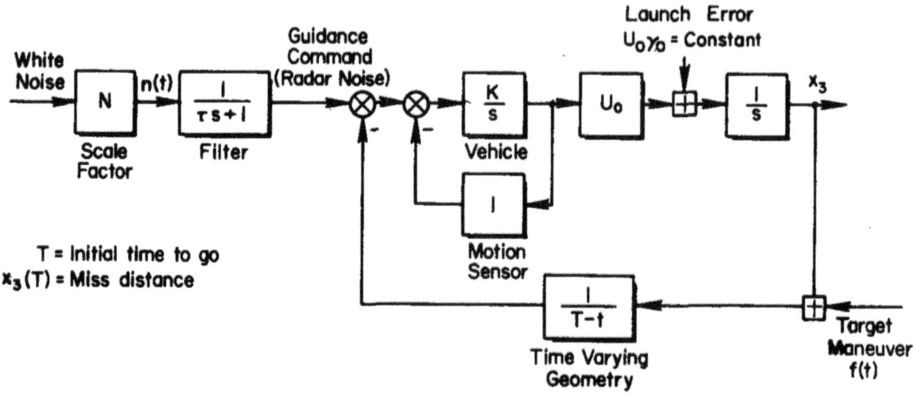


Fig. 10-13. Simplified block diagram for guidance of a homing missile in the vertical plane.

To make the discussion concrete, consider the problem of determining the effect of noise in the radar guidance system on the miss distance of a homing missile. A highly simplified block diagram of the automatic flight control and guidance system for control in a vertical plane is presented in Fig. 10-13. (Cf. Fig. 1-5 in Chapter 1.) The block diagram represents a system that is linear but with time-varying coefficients. Even this very simple system is relatively intractable, and the usual recourse is to simulation. Figure 10-14 represents a simulation diagram for the system,[23] drawn in terms of components that are called

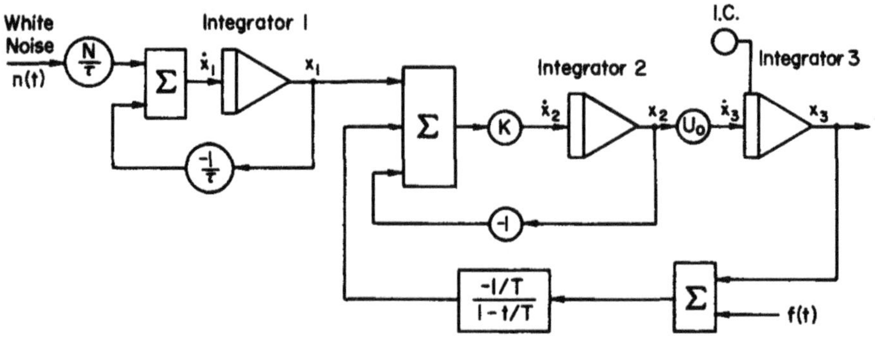


Fig. 10-14. Simulation diagram; homing missile.

[23] C. L. Johnson, *Analog Computer Techniques*, McGraw-Hill, New York, 1956; G. A. Korn and T. M. Korn, *Electronic Analog and Hybrid Computers*, McGraw-Hill, New York, 1964; S. Fifer, *Analogue Computation*, McGraw-Hill, New York, 1961, 4 vols; A. E. Rogers and T. W. Connolly, *Analog Computation in Engineering Design*, McGraw-Hill, New York, 1960.

"integrators," "summers," and "coefficient potentiometers." (We have neglected the usual sign change in integrators and summers.) From either the block diagram or the simulation diagram and the definitions of the operation of the several components, we may write the differential equations of the system. Considering only the noise input, $n(t)$, for the moment for simplicity, we find that

$$\begin{aligned} \dot{x}_1 &= -\frac{1}{\tau}x_1 + \frac{N}{\tau}n(t) \\ \dot{x}_2 &= +Kx_1 - Kx_2 - \frac{K}{T}\left[\frac{1}{1-(t/T)}\right]x_3 \\ \dot{x}_3 &= U_0 x_2 \end{aligned} \tag{10-93}$$

The equations for $x_1$, $x_2$, and $x_3$, sometimes called "state variables," may also be written in matrix form:

$$\begin{bmatrix} \dot{x}_1 \\ \dot{x}_2 \\ \dot{x}_3 \end{bmatrix} = \begin{bmatrix} -\frac{1}{\tau} & 0 & 0 \\ +K & -K & -\frac{K}{T}\left(\frac{1}{1-(t/T)}\right) \\ 0 & +U_0 & 0 \end{bmatrix} \begin{bmatrix} x_1 \\ x_2 \\ x_3 \end{bmatrix} + \begin{bmatrix} \frac{N}{\tau} \\ 0 \\ 0 \end{bmatrix} n(t) \tag{10-94}$$

In general, equations such as Eq. 10-94 may be written in the form:

$$\dot{\mathbf{x}}(t) = \mathbf{A}(t)\mathbf{x}(t) + \mathbf{B}(t)\mathbf{u}(t) \tag{10-95}$$

where $\mathbf{x}(t)$ is the $n \times 1$ matrix of system outputs, $\mathbf{u}(t)$ is the $m \times 1$ matrix of system inputs, $\mathbf{A}(t)$ is the square $n \times n$ matrix of system coefficients, and $\mathbf{B}(t)$ is the $n \times m$ matrix of input coefficients.

Note that, in the concrete example of Fig. 10-14 and Eq. 10-93, the impulse responses of the system to a unit impulse $n(t) = \delta(t - z)$ applied at the first summer are the same things as the initial condition responses of the system if $x_1(z) = N/\tau$ and all other initial conditions (in this case $x_2$ and $x_3$) are zero. This is, of course, because with no other inputs or initial conditions, the integration of the impulse by Integrator 1 produces a step, or equivalently, an initial condition on the output when the problem begins to run at $t = z$. In general, the impulse responses of the system are the same as the initial condition responses for the initial condition on the same integrator to which the impulse would have been applied.

Consider next what it might take to give an answer to the original question, i.e., the effect of noise on the miss distance. Suppose that the guidance system noise could be represented by a white noise power spectral density, so that a suitably scaled impulse would be its transient analog. The impulse response, $w(t, z)$ of a linear, time-varying system, however, is a function of two time variables:

1. The time, $z$, at which the impulse is applied.
2. The time, $t$, at which the response is measured.

By way of contrast, the impulse response of a linear, constant coefficient system is a function *only* of the difference in time, $\tau = t - z$, between the application of the impulse and the time at which the response is measured.

Similar to the transient analog, the *ensemble* mean square response[24] at a time $t$, $E[y^2(t)]$, of a time-varying system responding to a white noise input, $x$, with unit power per radian per second applied at Input Terminal 1 continuously from $-\infty$ to $t$, is derived below:

$$E[y^2(t)] = E\left[\int_{-\infty}^{t} x(z)w_{y^1}(t, z)\,dz\right]^2 \tag{10-96}$$

where $w_{y^1}(t, z)$ is the response in $y$ at time, $t$, to a unit impulse applied at time, $z$, to Input Terminal (integrator) 1.

Expanding Eq. 10-96 and moving the expected value operator inside the integrations, we find that

$$\begin{aligned} E[y^2(t)] &= E\left[\int_{-\infty}^{t}\int_{-\infty}^{t} w_{y^1}(t, z)w_{y^1}(t, \lambda)x(z)x(\lambda)\,dz\,d\lambda\right] \\ &= \int_{-\infty}^{t}\int_{-\infty}^{t} w_{y^1}(t, z)w_{y^1}(t, \lambda)E[x(z)x(\lambda)]\,dz\,d\lambda \\ &= \int_{-\infty}^{t}\int_{-\infty}^{t} w_{y^1}(t, z)w_{y^1}(t, \lambda)R_{xx}(z - \lambda)\,dz\,d\lambda \end{aligned} \tag{10-97}$$

Since $x$ is unit white noise,

$$R_{xx}(z - \lambda) = \tfrac{1}{2}\delta(z - \lambda) \tag{10-98}$$

Inserting Eq. 10-98 into 10-97, we obtain

$$E[y^2(t)] = \frac{1}{2}\int_{-\infty}^{t} [w_{y^1}(t, z)]^2\,dz \tag{10-99}$$

Suppose that we wished to evaluate the ensemble mean-squared response using simulation and Eq. 10-96. The expression $w_{y^1}(t, z)$ might be obtained from direct measurements. If a unit impulse is applied at time, $z$, to Input Terminal 1, the simulation yields $w_{y^1}(t, z)$ as a function of $t$. The variable of integration in Eq. 10-96, however, is $z$. It would, therefore, seem necessary to record a number of impulse responses for various values of $z$, cross plot the results, and integrate the kernel of the equation

[24] The ensemble average is the value of a statistical parameter at time, $t$, averaged over a large number of trials with random initial conditions and noise but all having the same statistical properties. In time-varying problems, of course, ensemble averages are not equal to time averages.

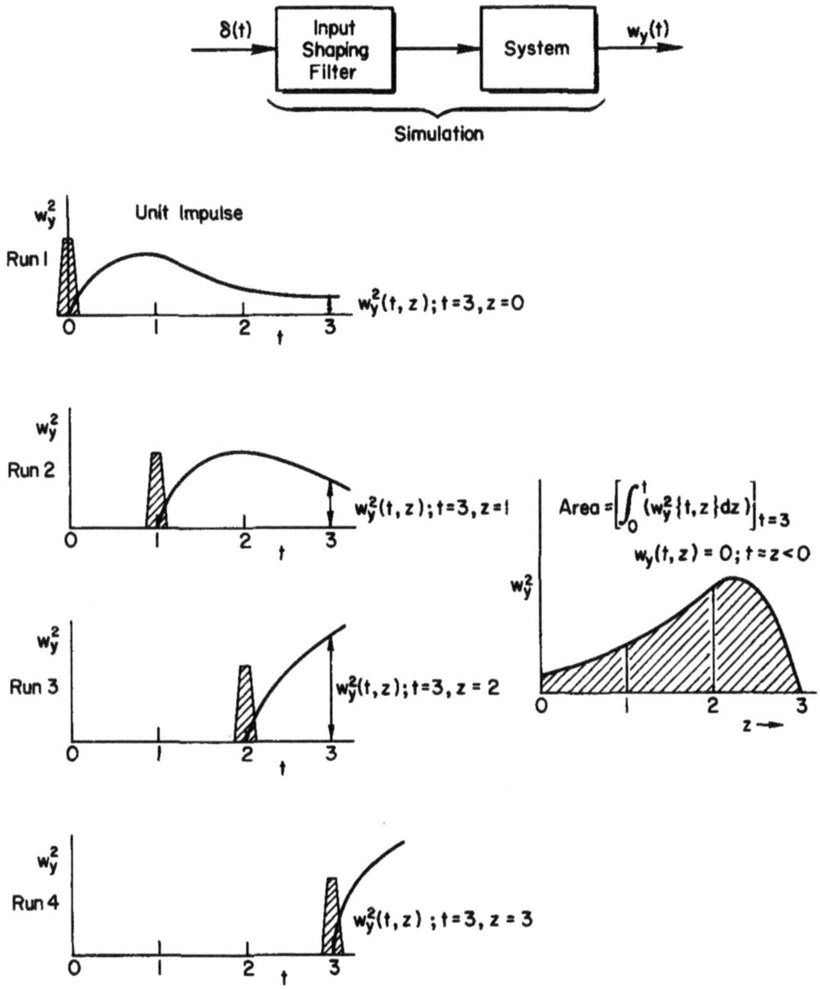


Fig. 10-15. Steps required to calculate mean-squared response by the direct (nonadjoint) method.

numerically or graphically. This procedure is illustrated in Fig. 10-15. In this example several simulator trials would be required in order to evaluate $E[y^2(t)]_{t=3}$, and the whole process would have to be repeated for any other values of $t$ of interest. Although it is entirely feasible to do this, it is tedious and uneconomical. Fortunately, it is also unnecessary.

The system simulation we should like to use would produce $w_{y1}(t, z)$ as a continuous function of $z$. The response would then be evaluated according to the formula (Eq. 10-96). The operations of squaring and integrating the impulse response could be mechanized in the simulation. Then only one run would be required for each value $t = T$ which might be of interest. Is there such a system? The answer is: yes. It is called the modified adjoint system.[25]

Heuristically, the basic idea of the adjoint system can be introduced by considering the principle of reciprocity for linear systems. A familiar example of the reciprocity principle occurs in the theory of structures, in which a concentrated load applied at Point A produces a deflection at Point B equal to that which would be produced at Point A by the same load applied at Point B. The nature of the load influences are characterized by structural influence coefficients, analogous to weighting functions.

How does this apply to the present context, i.e., how do we obtain the same response, $w(t, z)$, in two different ways? We already know one way: a unit impulse applied to the system at time $z$ produces an output at time $t$ given by $w(t, z)$. Now how do we get an output $w(t, z)$ at time $z$ by applying an input impulse at time $t$? The direct answer is obvious enough: measure $w(t, z)$ at the original input terminal, apply the impulse at the original output terminal, and run time *backwards*. This last requirement is awkward, to say the least, on analog computers! Consequently, to exploit the reciprocity idea, it is necessary to develop a simulator diagram that is related to the original simulator circuit by a simple set of rules to be given below. This new simulator setup is that of the adjoint system. When it is achieved, then the adjoint output will produce a time history $w_y$ vs. $z$. When squared, this produces $w_y^2$ vs. $z$, identical to that illustrated in Fig. 10-15. Then, in terms of this example, integration of the squared time history from zero to three yields $[\int_0^t w^2(t, z)\, dz]_{t=3}$ in a single run.

In order to understand *why* the modified adjoint system may operate so as to produce the desired results, it is necessary to take a modest excursion via matrix calculus.

Take a system of linear, time-varying coefficient, differential equations of the form of Eq. 10-95, and a closely related system of equations:

$$\dot{\boldsymbol{\alpha}}(t) = -\mathbf{A}^{\mathrm{T}}(t)\boldsymbol{\alpha}(t) \tag{10-100}$$

Here the matrix $\mathbf{A}^{\mathrm{T}}(t)$ is the transpose of $\mathbf{A}(t)$, formed by interchanging the rows and columns of $\mathbf{A}(t)$. If Eq. 10-95 is the original equation, then the Eq. 10-100 homogeneous equation is said to be the *adjoint* equation. This simply means that they are formed from the original set of homogeneous equations $\dot{\mathbf{x}} = \mathbf{A}(t)\mathbf{x}(t)$ according to the rule which specifies that the

[25] Laning and Battin, *Random Processes in Automatic Control;* G. L. Teper and R. L. Stapleford, "Adjoint Computer Techniques for a Homing Missile System," STI TM 241-5, Systems Technology, Hawthorne, Calif., Jan. 1963; P. M. DeRusso, R. J. Roy, and C. M. Close, *State Variables for Engineers*, Wiley, New York, 1965, pp. 383–388; Fifer, *Analogue Computation*, pp. 1052–1085; Rogers and Connolly, *Analog Computation in Engineering Design*, pp. 233–246.

coefficient matrix of the adjoint equations, $\mathbf{A}^T$, is the negative transpose of the coefficient matrix of the original homogeneous equations, $\mathbf{A}$.

We form the product $\boldsymbol{\alpha}^T(t)\mathbf{x}(t)$ and take the derivative with respect to time,

$$\frac{d}{dt}[\boldsymbol{\alpha}^T(t)\mathbf{x}(t)] = \dot{\boldsymbol{\alpha}}^T(t)\mathbf{x}(t) + \boldsymbol{\alpha}^T(t)\dot{\mathbf{x}}(t) \tag{10-101}$$

but from Eq. 10-100,

$$\dot{\boldsymbol{\alpha}}^T(t) = -[\mathbf{A}^T(t)\boldsymbol{\alpha}(t)]^T = -\boldsymbol{\alpha}^T(t)[\mathbf{A}^T(t)]^T = -\boldsymbol{\alpha}^T(t)\mathbf{A}(t) \tag{10-102}$$

Substituting this result and the expression on the right-hand side of the first of Eq. 10-95 in Eq. 10-101, we obtain

$$\begin{aligned}\frac{d}{dt}[\boldsymbol{\alpha}^T(t)\mathbf{x}(t)] &= -\boldsymbol{\alpha}^T(t)\mathbf{A}(t)\mathbf{x}(t) + \boldsymbol{\alpha}^T(t)[\mathbf{A}(t)\mathbf{x}(t) + \mathbf{B}(t)\mathbf{u}(t)] \\ &= \boldsymbol{\alpha}^T(t)\mathbf{B}(t)\mathbf{u}(t)\end{aligned} \tag{10-103}$$

Integration of Eq. 10-103 from $t_1$ to $t_2$ yields

$$\boldsymbol{\alpha}^T(t_2)\mathbf{x}(t_2) = \boldsymbol{\alpha}(t_1)\mathbf{x}(t_1) + \int_{t_1}^{t_2}\boldsymbol{\alpha}^T(t)\mathbf{B}(t)\mathbf{u}(t)\,dt \tag{10-104}$$

It is ordinarily possible to define the starting time $t_1$ such that $\mathbf{x}(t_1) \equiv 0$, thereby simplifying Eq. 10-104 somewhat. Also, the boundary conditions on $\boldsymbol{\alpha}(t)$ are arbitrary, so that when we are interested in one particular output, $x_i(t_2)$, all the components of $\boldsymbol{\alpha}(t_2)$ except $\alpha_i(t_2)$ can be set equal to zero, i.e.,

$$\alpha_j(t_2) = \begin{cases} 0 & i \neq j \\ 1 & i = j \end{cases} \tag{10-105}$$

Eq. 10-104 can now be written as

$$x_i(t_2) = \int_{t_1}^{t_2}\boldsymbol{\alpha}^T(t)\mathbf{B}(t)\mathbf{u}(t)\,dt \tag{10-106}$$

With $\boldsymbol{\alpha}(t_2)$ fixed, everything is known in Eq. 10-106, except $\boldsymbol{\alpha}(t)$ which is the output of the adjoint system. To make the *boundary* conditions equal to initial conditions, we substitute a new time variable $t^* = t_2 - t_1$ and rewrite Eqs. 10-100 and 10-106 in terms of this new independent variable. Then they become

$$\frac{d}{dt^*}[\boldsymbol{\alpha}(t_2 - t^*)] = \mathbf{A}^T(t_2 - t^*)\boldsymbol{\alpha}(t_2 - t^*) \tag{10-107}$$

$$\begin{aligned}x_i(t_2) &= -\int_{t_2-t_1}^{0}\boldsymbol{\alpha}^T(t_2 - t^*)\mathbf{B}(t_2 - t^*)\mathbf{u}(t_2 - t^*)\,dt^* \\ &= \int_0^{t_2-t_1}\boldsymbol{\alpha}^T(t_2 - t^*)\mathbf{B}(t_2 - t^*)\mathbf{u}(t_2 - t^*)\,dt^*\end{aligned} \tag{10-108}$$

where

$$(\alpha_j)_{t^*=0} = \begin{cases} 0 & i \neq j \\ 1 & i = j \end{cases} \tag{10-109}$$

Equations 10-107 and 10-108 are the equations of the modified adjoint system (modified by the change in the time variable). Their solution after the imposition of the initial conditions of Eq. 10-109 will yield $\boldsymbol{\alpha}^{\mathrm{T}}(t_2 - t^*)$. All of the rules for the formulation of the adjoint computer diagram can be derived from Eqs. 10-107 and 10-108.

The coefficient matrix, $\mathbf{A}^{\mathrm{T}}$, in the modified adjoint system is the transpose (rows and columns interchanged) of the coefficient matrix in the original system, Eq. 10-96, and with time, $t^*$, in the modified adjoint system running backwards, the coefficients are started at their values appropriate to $t = t_2$, i.e., $t^* = 0$. Thus if Integrator $i$ in the original system produces $x_i$ from $\dot{x}_i$ and feeds $n$ integrators via coefficients $a_{ij}(t)$, $j = 1, 2, \ldots, n$, in the modified adjoint system, the inputs to Integrator $i$ come from $j = 1, 2, \ldots, n$ integrators via coefficients $a_{ji}(t_2 - t^*)$. This is illustrated in Fig. 10-16. Similar considerations apply to summers when they are necessary and to the coefficient potentiometers for the elements of the **B** matrix. (Actually summers become merely takeoff points and takeoff points become summers in the simulation of the modified adjoint system.)

These facts lead to the following simple rules for simulating the modified adjoint system, starting with a simulation diagram of the original system.[26]

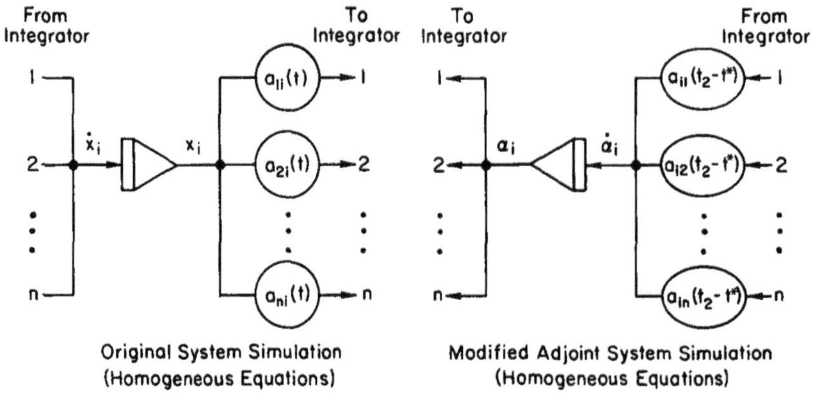


Fig. 10-16. Interchange of inputs and outputs on integrators and coefficient potentiometers.

[26] T. S. Durand and G. L. Teper, "An Analysis of Terminal Flight Path Control in Carrier Landings," TR 137-1, Systems Technology, Hawthorne, Calif., June 1964.

1. The outputs of each summer or integrator in the system analog become the inputs to that summer or integrator in the modified adjoint analog; the inputs in the system analog become the outputs in the modified adjoint analog.
2. The input and output are exchanged on all coefficient potentiometers.
3. Multipliers used to represent time-varying coefficients are replaced by multipliers representing the same coefficient as a function of the new variable $t^* = t_2 - t$, i.e., time-varying coefficients are started at their final values and run backward toward their initial values.
4. A unit impulse is put into the one integrator or summer from which the output of interest was taken in the system analog. Since, however, the integral of a unit impulse is a unit step, instead of attempting to generate an actual impulse, a step or initial condition can be put on the output of the integrator, or summer, but the step or initial condition should not exist at the output of the integrator until the problem begins to run.
5. The modified adjoint impulse response functions are then measured in the modified adjoint analog, at the points at which the corresponding inputs in the system analog were introduced.
6. Components necessary to square, scale, and integrate the impulse response functions are added as needed. (See Fig. 10-17.)

Step (initial condition) or ramp responses of the original system may be obtained by once or twice integrating the impulse responses of the modified adjoint system at the terminals at which the steps or ramps would have been introduced. Similarly, the response to an arbitrary input may be obtained by multiplying the impulse response at the terminal in the modified adjoint system corresponding to the point of introduction of the arbitrary function in the original system and then integrating the result. This represents the computer mechanization of Eq. 10-108. The operations are shown symbolically in Fig. 10-17.

It is necessary to point out that it is not usually possible to associate physical quantities with all of the signals in the simulation of the modified adjoint system. This very likely leads to problems in the scaling of the simulation that are only overcome by trial and error.

Now, by way of illustration, the drawing of the simulation diagram will be carried out explicitly in connection with the simple example of the guidance of a homing missile. The simulation diagram of the original system has already been presented as Fig. 10-14. By identifying $t_2 = T$ and carrying out the instructions of the numbered rules above the simulation diagram of Fig. 10-18 results. Reversing the connections on the integrators and coefficient potentiometers is particularly straightforward, as is

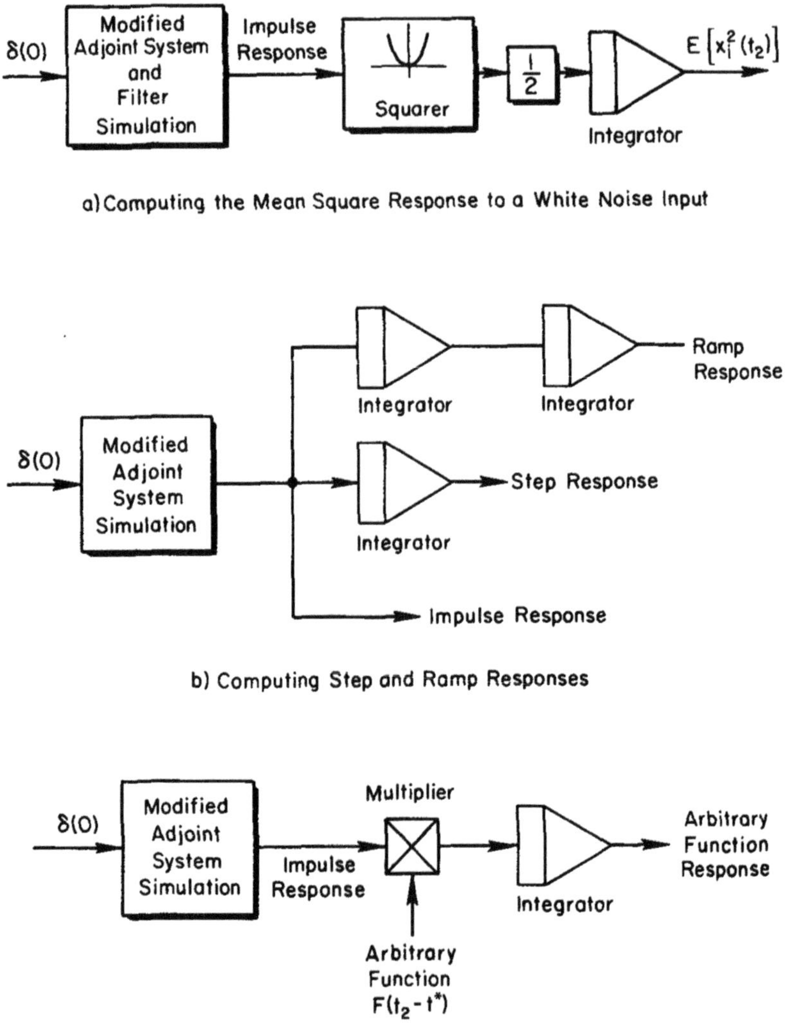


a) Computing the Mean Square Response to a White Noise Input

b) Computing Step and Ramp Responses

c) Computing an Arbitrary Function Response

Fig. 10-17. Responses of interest can be generated from the impulse responses of the modified adjoint system simulation.

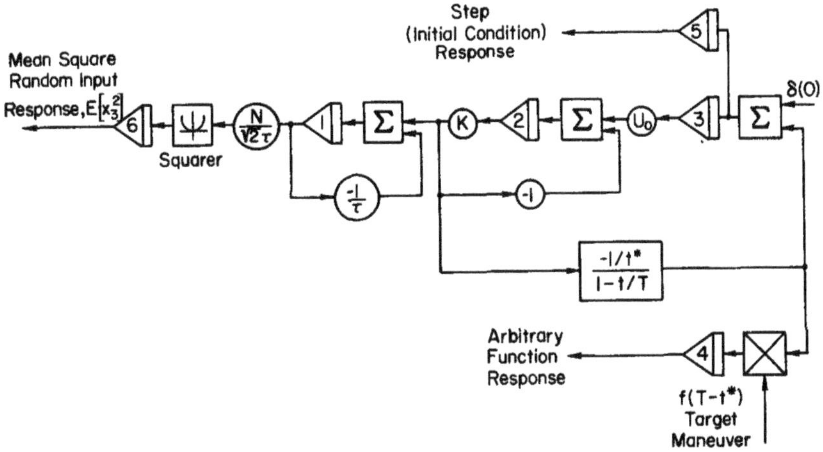


Fig. 10-18. Modified adjoint system simulation derived from the simulation diagram of the original system.

making the summers takeoff points and the takeoff points summers. Taking the outputs from the points for the inputs in the original system and providing the impulse input at the original output point completes the diagram for the modified adjoint system. Notice that the factor of $\sqrt{2}$ is inserted as a modifier for $N/\tau$. This permits the multiplier and Integrators 4 and 5 to be added to compute the miss distance responses to the target maneuver and launch error without any modifiers at all.

# CHAPTER 11

# *MULTILOOP FLIGHT CONTROL SYSTEMS*

According to one system of Chinese philosophy, the cycles of affairs in the circumscribed area of the terrestrial globe can be accounted for by the alternating influence of the counterpoised forces *yin* and *yang*. The development of the material so far presented here has perhaps afforded the occasion to observe such cycles and indeed cycles on the cycles. We now wish to complete the last circumvolution by demonstrating the mathematical synthesis of multiloop automatic flight control to meet requirements. In this chapter, we complete the system and circumscribe our subject. We limit the subject by bringing the discussion to a close, and we draw the boundary around the subject in such a way as to cover the largest number of points; that is, to make the best use of the material that has gone before.

## 11-1. Introduction

The history of aircraft flight control system development, as with other histories, may be considered in terms of stimulus and response. The stimuli have been flight control aims or problems; these have led to responses in the form of system configurations to satisfy the goals or to remedy the basic problems. In the course of this challenge-response evolution, there has been competition between possible system configurations, with the inevitable result that some configurations have emerged as far more feasible and desirable than others. The factors that enter into the competition may be many and varied, and perhaps change with time. But in any given era, the best systems exhibit their superiority in such qualities as the following:

1. Simplicity of mechanization.
2. Economy of equalization.
3. Commonality of elements and settings for different operational modes.
4. Simplicity of gain compensation.
5. Versatility across vehicles.
6. Lack of response to unwanted inputs.
7. Lack of susceptibility of the sensors to unwanted inputs.
8. Lack of sensitivity to controller tolerances and airframe configuration changes.

9. Lack of sensitivity to controlled element uncertainties and parasitic nonlinearities.
10. Inherent reliability and maintainability.

In actuality, of course, few flight control systems have been formally competitive, one with another. Rather, the competition has been similar to historical evolution. But it is important to recognize the results of this history: *the root problems of flight control and the successful system structures that solve these problems*. From the systems viewpoint, the systems structures are defined by particular combinations of feedbacks. These are what we call here "essential feedbacks." They constitute those quantities needed for feedback control purposes to solve, in a practical fashion, the several basic flight control problems.

Since, in fact, almost every conceivable airframe motion quantity has been used as a feedback at some time or other, the essential feedbacks constitute in themselves a record of historical successes. The record is extended as new flight control problems appear, or also by revolutionary advances in sensor technology or computational capacity in the flight controller. Accordingly, the technical content of this chapter begins with a discussion of essential feedbacks for both the longitudinal and lateral-directional motions. These essential feedbacks are then summarized in symbolic form in convenient tables to which the reader may refer in the course of the development of the longitudinal and lateral-directional multiloop examples that comprise most of the remaining material.

A final consequence of the essential feedback discussion is a set of system configurations. The configurations are incomplete, however, because gains and equalization are not specified; and the overall synthesis would be incomplete in any case because dynamic responses and other measures of performance are still, at this stage, undetermined. To complete the synthesis, a sequence of iterative analyses is needed, ending with final system characteristics and performance estimates. Such procedures are conventionally carried out using a combination of analytical, computational, and simulation procedures. The analysis and much of the computation uses the theory and techniques developed in Chapter 3, in general, and particularly the theory of Section 3-5.

## 11-2. Essential Feedbacks[1]

The essential feedbacks which we shall summarize below derive from one or both of two basic flight control system purposes:

[1] This section is adapted from R. L. Stapleford, D. T. McRuer, L. G. Hofmann, and G. L. Teper, "A Practical Optimization Design Procedure for Stability Augmentation Systems," AFFDL-TR-70-11, Wright-Patterson Air Force Base, Ohio, Oct. 1970.

1. To establish and maintain certain specified equilibrium states of vehicle motion.
2. To remedy aircraft handling quality (piloted control) deficiencies.

The establishment and maintenance of an equilibrium state of motion requires an outer control loop pertinent to the vehicle motion quantity defining that state. But more often than not such outer loops, closed about unmodified aircraft dynamics, do not result in stable, well-damped, rapidly responding systems. Instead, equalization, provided by inner loops, may be required. The inner loops needed to make possible stable, well-damped, rapidly responding outer loop system responses are often similar or identical to the ones needed to alleviate flying quality deficiencies. Consequently, many of the feedbacks needed to accomplish the first purpose listed above are also useful for the second. This will become more evident as we consider specific cases.

In the following, we shall develop and summarize sets of essential feedbacks that are needed to solve the more fundamental longitudinal and lateral flight control problems. The development, which will separately consider longitudinal and lateral situations, comprises a number of steps. The first step is a listing of the basic operational modes and/or states of equilibrium that may be desired. These are, in almost all cases, operational modes that are common to both manual and automatic control. The only notable exceptions involve highly maneuvering flight which is conducted almost exclusively under manual control. For each operational function and mode, the vehicle dynamic parameters, key metrics, and stability derivatives, which together define the key features of the effective controlled element, are then listed in a table. When the key metrics, or certain combinations thereof, have values outside specifiable ranges, a control system using only the outer loop feedback will suffer from possible deficiencies. Correction is accomplished by using control crossfeeds or feedbacks from other variables. These are listed in the table as methods of modifying the effective controlled element and represent potentially useful additional feedbacks.

From the list of potential feedback possibilities, a set of systems may be constructed. Each such system amounts to a compatible set of essential feedbacks. To some extent these systems, even though they all are practical in that they satisfy operational requirements and, in almost all cases, have been reduced to practice, may still be considered competitive. Thus another table is subsequently presented in which each system is appraised with respect to equalization and compensation requirements and with respect to other practical design problems. These tables amount to verbal compromise summaries for the basic essential feedbacks. The detailed background for the story presented in summary form here is a result of single-loop analyses such as the ones exemplified in Chapters 7 and 8.

*Essential Longitudinal Feedbacks*

Among possible requirements for the automatic control of the longitudinal motions of aircraft are the provision of facilities for the command and maintenance of

Pitch attitude
Indicated airspeed
Mach number
Rate of climb or descent
Glide path

As discussed in Section 9-2, each of these control requirements specifies an outer loop feedback or operating mode of control. For our purposes, however, it will be more convenient to subsume these operational modes under different classifications which we shall call "operational functions." These are more pertinent when considering inner loop characteristics.

The key operational functions for longitudinal control are

Pitch attitude control
Vertical gust regulation
Path (deviation from prescribed datum) control, frontside operation[2]
Path control, backside operation[3]
Speed control
Horizontal gust regulation

For each of these operational functions, Table 11-1 lists the effective dynamic parameters, key metrics, and potentially useful feedbacks. The individual entries are briefly discussed below.

Pitch attitude control is a most important function, both in its own right and as a possible inner loop for path and/or speed control. Pitch attitude, $\theta$, is the essential feedback directly specified by the operational function. The aircraft dynamic parameters central to the establishment of an adequate pitch angle control loop are usually the short period damping

[2] These terms refer to the airplane's curves of thrust and drag vs. velocity and more specifically to the total derivative, $dX/dU$. Frontside operation is when $dX/dU < 0$. Backside operation is when $dX/dU > 0$. Since the $x$-component of the thrust may often be taken to be constant with changes in speed, the distinction is roughly the same as would be indicated by the change in sign of the slope of drag vs. speed curve. See Eqs. 4-103 and 4-104.

[3] *Ibid.*

Table 11-1. Longitudinal operational functions governing augmentation requirements.

| Operational functions | Effective dynamic parameters | Key metrics | Methods of modifying |
|---|---|---|---|
| 1. Pitch attitude control ($\theta_c \rightarrow \delta_e$) | $\zeta_{sp}, \omega_{sp}, T_{\theta_2}$ | $2\zeta_{sp}\omega_{sp} \doteq -(M_q + M_{\dot{\alpha}} + Z_w)$ | $q \rightarrow \delta_e, \dot{\alpha} \rightarrow \delta_e, \dot{a}_z \rightarrow \delta_e$ |
| | | $\omega_{sp}^2 \doteq M_q Z_w - M_\alpha$ | $\theta \rightarrow \delta_e, \alpha \rightarrow \delta_e, a_z \rightarrow \delta_e$ |
| | | $\dfrac{1}{T_{\theta_2}} \doteq -Z_w + M_w \dfrac{Z_{\delta_e}}{M_{\delta_e}}$ | $\delta_e \rightarrow \delta_f, \alpha \rightarrow \delta_f, a_z \rightarrow \delta_f$ |
| 2. Vertical gust regulation | $\zeta_{sp}, \omega_{sp}, T_{\theta_2}, Z_w, M_w$ | Open-loop | |
| | | $\dfrac{\theta}{w_g} \doteq \dfrac{-M_w}{s^2 + 2\zeta_{sp}\omega_{sp}s + \omega_{sp}^2}$ | $q \rightarrow \delta_e, \dot{\alpha} \rightarrow \delta_e, \dot{a}_z \rightarrow \delta_e$ |
| | | | $\alpha \rightarrow \delta_e, a_z \rightarrow \delta_e$ |
| | | $\dfrac{\dot{h}}{w_g} \doteq \dfrac{Z_w(s - M_q - M_{\dot{\alpha}})}{s^2 + 2\zeta_{sp}\omega_{sp}s + \omega_{sp}^2}$ | $\alpha \rightarrow \delta_f, a_z \rightarrow \delta_f$ |
| | | $\theta \rightarrow \delta_e$ loop closed | |
| | | $\left(\dfrac{\dot{h}}{w_g}\right)_{\theta \rightarrow \delta_e} \doteq \dfrac{-1}{T_{\theta_2}s + 1}$ | $\alpha \rightarrow \delta_f, a_z \rightarrow \delta_f$ |
| 3. Speed control ($u_c \rightarrow \delta_e$) | $\zeta_p, \omega_p$ (or $1/T_{p_1}, 1/T_{p_2}$) | $2\zeta_p\omega_p$ or $\left(\dfrac{1}{T_{p_1}} + \dfrac{1}{T_{p_2}}\right) \doteq -X_u - \dfrac{M_u(X_\alpha - g)}{Z_w M_q - M_\alpha}$ | $\theta \rightarrow \delta_e, u \rightarrow \delta_e$ |
| | | $\omega_p^2$ or $\dfrac{1}{T_{p_1}T_{p_2}} \doteq \dfrac{g(M_w Z_u - M_u Z_w)}{Z_w M_q - M_\alpha}$ | $\dot{h} \rightarrow \delta_e$ |

| | | | |
|---|---|---|---|
| 4. Path control, front side ($h_c \to \delta_e$) | Above plus $T_{h_2}$, $T_{h_3}$ | Above plus $\frac{1}{T_{h_2}} \doteq -\frac{1}{T_{h_3}} \doteq \left[M_\alpha - Z_\alpha \frac{M_{\delta_e}}{Z_{\delta_e}}\right]^{1/2}$ <br> $\tau_h \doteq \left[\frac{12 Z_{\delta_e}}{-Z_\alpha M_{\delta_e}}\right]^{1/2}$ | $\delta_e \to \delta_f$ |
| | | $\frac{\gamma}{\theta} \doteq \frac{T_{h_1}s + 1}{(T_{\theta_1}s + 1)(T_{\theta_2}s + 1)}$ ; $\frac{1}{T_{\theta_1}} \doteq -X_u$ | See below |
| 5. Path control on the backside of the drag curve ($h_c \to \delta_e$) | $T_{h_1}$ | $\frac{1}{T_{h_1}} \doteq -X_u$ <br> $+ \frac{(X_\alpha - g)(M_u Z_{\delta_e} - Z_u M_{\delta_e}) + X_{\delta_e}(M_\alpha Z_u - Z_\alpha M_u)}{M_\alpha Z_{\delta_e} - M_{\delta_e} Z_\alpha}$ | $u \to \delta_T$, $\alpha \to \delta_T$ <br> $\delta_e \to \delta_T$ |
| 6. Horizontal gust regulation | $T_{h_1}$ | | Same as above |

ratio, $\zeta_{sp}$, frequency, $\omega_{sp}$, and the time constant of the pitch/elevator numerator, $T_{\theta_2}$ (see Section 7-1). In some circumstances, such as, e.g., backside operation, the phugoid characteristics can also become important as quantities that limit the speed, accuracy, or damping of the closed-loop response. Pertinent feedbacks that might be used to modify the short period (and phugoid) characteristics are listed beside their appropriate key metric. Notice that, if a direct lift control (DLC), $\delta_f$, is available, either an elevator to DLC crossfeed or one of the indicated feedbacks can be used to modify the pitch zero, $1/T_{\theta_2}$.

The second operational function, vertical gust regulation, is closely associated with pitch attitude control. The key metrics are the same as for pitch attitude control, with the addition of the stability derivatives $Z_w$ and $M_w$, in their roles as gain factors acting on gust inputs. With only elevator control, beneficial potential feedbacks can do little more than increase the short period damping and frequency. Altitude disturbances because of gusts can be reduced by washing out any $\theta$ feedback, or by using a normal acceleration, $a_z$, or angle of attack, $\alpha$, feedback as the primary means to increase the short period undamped natural frequency. Similarly, pitching motions can be reduced at the expense of increased altitude disturbances. If DLC, $\delta_f$, is available, both the pitch and altitude gust responses can, in principle, be attenuated. However, there are instances when altitude control even with an inner attitude loop is unsatisfactory because of the low frequency "droop" associated with $\gamma/\theta$ response characteristics. Unless such characteristics ($1/T_{\theta_1}$, $1/T_{h_1}$) are modified by additional throttle feedbacks, *elevator* control of altitude is unsatisfactory and use of the throttle to directly control altitude is indicated, as for backside operation (see below).

Speed control (that is, true speed, TAS, indicated airspeed, IAS, or Mach number, $M$, control) considered as an operational function is useful for air traffic control procedures, for cruise control, and for precise control during landing approach. It also represents a potential problem when a divergent phugoid or tuck mode is present. These latter forms of behavior bring the phugoid characteristics into consideration as key metrics, although they are ordinarily important in this regard only at transonic speeds, where large positive values of $M_u$ are possible. The conventional fixes use attitude or airspeed feedback to the elevator.

Path control is an important operational mode for normal flying, and it can become critical for such precision tasks as landing approach, terrain following, and in-flight refueling. The essential feedback is the deviation, $h$, from the datum line defined by the task. (Here $h$ is used as the symbol for deviation because "altitude" from a datum line, as defined by $h = U_0 \int \gamma \, dt$, is the most common deviation of interest.) This feedback, however, by itself, is not sufficient to achieve a well-damped response of the basic path mode, as may be recalled from the discussion in Section 7-5. To provide path damping, either an $\dot{h}$ feedback or an inner pitch attitude control loop is required. For the latter case, the key metrics and feedbacks are exactly as described previously for pitch attitude control. For large, short-coupled aircraft, i.e., ones with a short tail length, the altitude elevator numerator time constants, $T_{h_2}$ and $T_{h_3}$, can become important as can also the effective deadtime for altitude response, $\tau_h$. Methods of modifying any of these parameters rely primarily on DLC, $\delta_f$, as a device that may be used with crossfeeds to alter the effective value of the normal force due to elevator deflection, $Z_{\delta_e}$.

For path control when the aircraft is operating on the backside of the drag curve, the key parameter is $T_{h_1}$ (see Section 7-5). Because of the backside operation, $T_{h_1}$ is negative. When an altitude loop is closed, this fact causes a divergence (instability) that is apparent in both altitude and airspeed responses. The solution is to make $T_{h_1}$ positive. Because this is an elevator transfer function numerator quantity, another control is required (see Section 3-5). The most effective "other" control is the throttle. This is indicated in the right-hand column of Table 11-1 in line with other considerations of path control on the backside. An alternative is to use throttle control for the outer, altitude loop, thereby eliminating the $1/T_{h_1}$-caused instability.

Regulation against horizontal gusts is particularly important during critical path control situations, such as landing. Again, $T_{h_1}$, the altitude to elevator very low frequency numerator time constant, is the most important parameter because it is the limiting time constant in airspeed response when the altitude loop is tightly closed. To improve the airspeed time response, the effective $1/T_{h_1}$ can be changed with either of the throttle feedbacks noted or also by elevator to throttle crossfeed.

Table 11-1 shows a number of potentially useful feedbacks and crossfeeds. For each of these, the primary functions performed, the equalization requirements, and practical design problems are listed in Table 11-2. For clarity, further elaboration of some of the Table 11-2 entries is given below.

Gain variations indicated among the equalization requirements are based on maintaining a constant crossover frequency for each individual feedback. These indicated requirements may be overly severe, and smaller gain variations would normally be adequate. The listed gain variations should be considered merely as indications of the relative amount of gain changing between competing possibilities.

Feedback of structural modes may be an important practical design problem for any of the feedbacks. However, the relative magnitude of the problem varies greatly, depending on the aircraft motion variable being fed back. The problem is most severe when the feedback is normal

Table 11-2. Longitudinal competing systems.

| Feedback or crossfeed | Primary functions performed | Equalization requirements | Practical design problems |
|---|---|---|---|
| $q \rightarrow \delta_e$ | 1. Increase short period damping.<br>2. Reduce pitch response to gusts. | Gain, $K_q$, should vary as $1/M_{\delta_e}$ i.e.,<br>$K_q \propto \dfrac{I_y}{\rho U_0^2 C_{m\delta_e}}$ | Gain adjustment with flight condition. |
| $\theta \rightarrow \delta_e$ | 1. Increase short period damping and frequency.<br>2. Reduce pitch response to gusts.<br>3. Increase phugoid damping.<br>4. Stabilize tuck mode. | 1. Gain should vary as $1/M_{\delta_e}$.<br>2. Lead/lag element desirable. | 1. Gain adjustment with flight condition.<br>2. Increased $h$ and $a_z$ response to vertical gusts. |
| $a_z \rightarrow \delta_e$ | 1. Increase short period damping and frequency.<br>2. Reduce $h$ and $a_z$ response to gusts. | 1. Gain, $K_{a_z}$, should vary as $1/Z_\alpha M_{\delta_e}$, i.e.,<br>$K_{a_z} \propto \dfrac{mI_y}{\rho^2 U_0^4 C_{L_\alpha} C_{m\delta_e}}$<br>2. Lead/lag element desirable. | 1. Severe gain adjustment with flight condition.<br>2. Sensor location adequate for all flight conditions.<br>3. Structural mode feedback.<br>4. Increased $\theta$ response to vertical gusts. |
| $\alpha \rightarrow \delta_e$ | 1. Increase short period damping and frequency.<br>2. Reduce $h$ and $a_z$ response to vertical gusts. | 1. Gain should vary as $1/M_{\delta_e}$.<br>2. Lead/lag element desirable. | 1. Gain adjustment with flight condition.<br>2. Sensor instrumentation.<br>a. Determination of operating point.<br>b. Errors due to aerodynamic interference.<br>c. Elaborate sensor complex required to suppress gust inputs. |

| | | | |
|---|---|---|---|
| $u \rightarrow \delta_e$ | Stabilize tuck mode. | | Sensor instrumentation (see above). |
| $\alpha \rightarrow \delta_T$ | Prevent altitude instability. | | Sensor instrumentation (see above). |
| $u \rightarrow \delta_T$ | Prevent altitude instability. | | Sensor instrumentation (see above). |
| $a_z \rightarrow \delta_f$ | Reduce $h$ and $a_z$ response to gusts. | 1. Gain, $K_{a_z}$, should vary as $1/Z_w Z_{\delta_f}$, i.e., $K_{a_z} \propto \dfrac{m^2}{\rho^2 U_0^3 C_{L\alpha} C_{L\delta_f}}$<br>2. Crossfeed, $\delta_f \rightarrow \delta_e$, probably desirable to adjust effective $M_{\delta_f}/Z_{\delta_f}$. | 1. Severe gain adjustment with flight condition.<br>2. Sensor location adequate for all flight conditions.<br>3. Structural mode feedback.<br>4. Probable drag penalty due to direct lift control surface. |
| $\delta_e \rightarrow \delta_f$ | Reduce effective $T_{\theta_2}$. | 1. Gain, $K_{\delta_e}$, should vary as $M_{\delta_e}/M_w Z_{\delta_f}$, i.e., $K_{\delta_e} \propto \dfrac{m C_{m\delta_e}}{\rho U_0 C_{m\alpha} C_{L\delta f}}$<br>2. Washout may be desirable. | 1. Gain adjustment with flight condition.<br>2. Probable drag penalty due to direct lift control surface. |
| $\delta_e \rightarrow \delta_T$ | Prevent altitude instability. | Low pass filter desirable. | Will not work if stick force/knot gets too low. |

acceleration because:

1. The $a_z/\delta$ rigid body transfer function is flat at high frequency, i.e., there is no attenuation of higher frequency modes, except the attenuation inherent in the modes themselves.
2. The sensor location is restricted by rigid body considerations.

In contrast, for example, a pitch angle to elevator feedback is highly attenuated at higher frequencies, even if lead/lag equalization is used; and the gyro location can be selected to minimize structural mode feedback. Consequently, Table 11-2 lists structural mode feedback as a practical design problem only for the normal acceleration feedbacks although it should be considered in all cases.

The use of airspeed or angle of attack feedbacks present several problems in sensor instrumentation. One is setting the operating point or desired steady-state condition, e.g., setting the reference or command speed in an airspeed feedback. Another problem with any aerodynamic sensor is the errors that may be introduced by variations in the local flow conditions. The flow in the vicinity of the sensor can change significantly with the Mach number and angle of attack. The third problem relates to the basic sensitivity of the sensor to gust disturbances. In modern systems the gust inputs may possibly be suppressed by using an elaborate sensor complex, instead of a single aerodynamic sensor. For example, the sensitivity of an airspeed sensor to horizontal gusts may be reduced by adding, with appropriate gains and equalization, fore-and-aft acceleration and pitch feedbacks.

At this point, combinations of feedbacks and crossfeeds can be selected from Table 11-2 to meet the requirements implied by Table 11-1. The outer loops are, of course, already specified by an operational mode selection. Consequently, for simplicity we shall focus on the inner loop system configurations. These are inner loops for automatic control or for manual control, wherein the pilot performs the outer loop function to accomplish a particular operational mode. In the latter case the inner loops taken together are commonly called "augmentation systems"; since that nomenclature is less awkward than "inner loops," we shall use it henceforth.

For an aircraft without DLC, the resultant multiloop augmenter system might include:

$$q \rightarrow \delta_e \quad a_z \text{ or } \alpha \rightarrow \delta_e \quad u \rightarrow \delta_e$$

A block diagram of this type of system, using $a_z$ in preference to $\alpha$, is shown in Fig. 11-1.

For an aircraft with DLC, the system might include (see Fig. 11-2):

$$q \rightarrow \delta_e \quad a_z \rightarrow \delta_e \text{ and } \delta_f \quad \delta_e \rightarrow \delta_f \quad u \rightarrow \delta_e$$

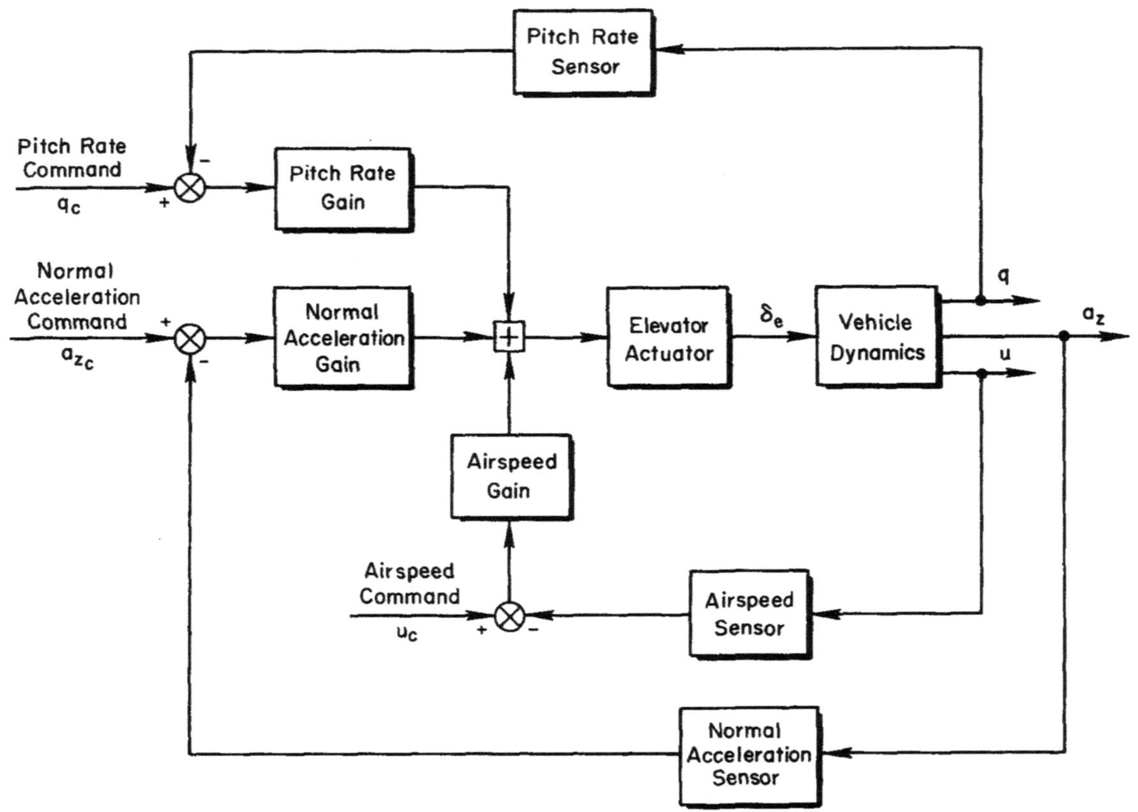


Fig. 11-1. Longitudinal augmenter mechanization without DLC.

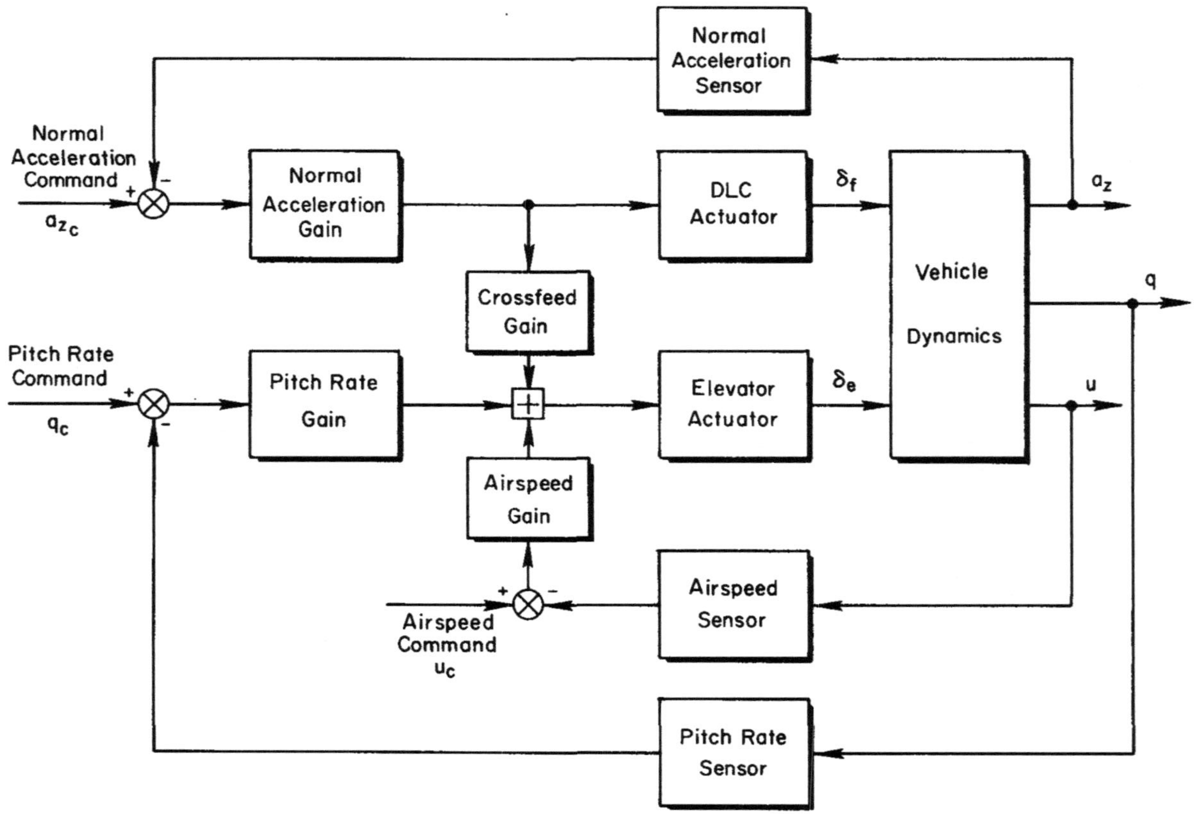


Fig. 11-2. Longitudinal augmenter mechanization with DLC.

In either case, the following would be added for operation on the backside of the drag curve:

$$u,\ \alpha,\ \text{or}\ \delta_e \rightarrow \delta_T$$

*Essential Lateral-Directional Feedbacks*

The automatic flight control system lateral-directional modes ordinarily require the provision of facilities for command or selection and maintenance of:

Roll attitude
Heading
Turns
Instrument low-approach system (ILS) localizer course
Visual omini-range (VOR) radial course
Tracks derived from inertial navigation system information (or other "off course" computer)

Again we shall substitute operational functions as a generalization of the flight control system outer loop modes. These will include:

Roll attitude control
Roll attitude regulation
Rudder-induced rolling
Turn coordination
Path control
Inertial cross-coupling suppression

For each of these operational functions, Table 11-3 lists the effective dynamic parameters, key metrics, and potential modifying feedback paths. Appropriately related dynamic parameters, metrics, and modifying feedbacks are shown in the table. The progression from one operational function to the next is somewhat more complex than is the case for the longitudinal essential feedbacks, in that there is more interaction between the several functions. A partially sequential buildup, however, is possible.

Those operational functions that serve directly to define an equilibrium state of vehicle motion provide some essential feedbacks at the outset. In particular the need for roll attitude control and regulation and for path control implies bank angle and lateral flight path or heading as necessary feedback quantities. Other essential feedbacks are the ones needed to make feasible control systems having these outer loops. They comprise the inner loops or augmenter feedbacks.

The first such function is roll attitude control. This is of great importance,

Table 11-3. Lateral/directional operational functions governing augmentation requirements.

| Operational functions | Effective dynamic parameters | Key metrics | Feedback or crossfeed |
|---|---|---|---|
| 1. Roll attitude control ($\varphi_c \rightarrow \delta_a$) | $T_R$ | $\frac{1}{T_R} \doteq -L'_p + \left(N'_p - \frac{g}{U_0}\right)\frac{L'_\beta}{N'_\beta}$ | $p \rightarrow \delta_a$ |
| | $p_{\max}$ | $p_{\max} = \left(\frac{\omega_\varphi}{\omega_d}\right)^2 T_R L'_{\delta_a}\delta_{a_{\max}}$ | $\delta_a \rightarrow \delta_r$ |
| | $\omega_\varphi/\omega_d$ | $\left(\frac{\omega_\varphi}{\omega_d}\right)^2 \doteq 1 - \frac{N'_{\delta_a}L_\beta}{L'_{\delta_a}N'_\beta}$ | $\delta_a \rightarrow \delta_r$ |
| | $\zeta_\varphi\omega_\varphi - \zeta_d\omega_d$ | $\zeta_\varphi\omega_\varphi - \zeta_d\omega_d \doteq \frac{L'_\beta}{2N'_\beta}\left(N'_p - \frac{g}{U_0}\right)$ | $p \rightarrow \delta_r$, $\delta_a \rightarrow \delta_r$ |
| | $\zeta_d\omega_d$ | $\zeta_d\omega_d \doteq -\frac{1}{2}\left[N'_r + Y_v + \frac{L'_\beta}{N'_\beta}\left(N'_p - \frac{g}{U_0}\right)\right]$ | $r \rightarrow \delta_r$, $\dot{\beta} \rightarrow \delta_r$, $\dot{a}_y \rightarrow \delta_r$ |
| 2. Roll attitude regulation (gust and yaw input disturbances) | $\left\|\frac{\varphi}{\beta}\right\|_d, \left\|\frac{\dot{p}}{\beta}\right\|_d$ | $\left\|\frac{\varphi}{\beta}\right\|_d \doteq \left\|\frac{L'_\beta}{N'_\beta}\right\| \frac{1}{\sqrt{1 + [(1 - 2\zeta_d\omega_d T_R)/(\omega_d T_R)^2]}}$ ; $\left\|\frac{\dot{p}}{\beta}\right\|_d \doteq N'_\beta \left\|\frac{\varphi}{\beta}\right\|_d$ | $\beta \rightarrow \delta_r$, $a_y \rightarrow \delta_r$<br>$\beta \rightarrow \delta_a$, $a_y \rightarrow \delta_a$<br>$p \rightarrow \delta_a$ |
| | $\omega_d$ | $\omega_d \doteq \sqrt{N'_\beta}$ | $\beta \rightarrow \delta_r$, $a_y \rightarrow \delta_r$ |
| | $\zeta_d\omega_d$ | As above<br>Roll response to random side gust<br>$\frac{\sigma_\varphi}{\sigma_{\beta_g}} \doteq \frac{\omega_g}{2\zeta_d\omega_d}\frac{\|\varphi/\beta\|_d}{[1 + (\omega_g/\omega_d)^2]}$ where $\omega_g$ is gust break frequency<br>Roll response to step side gust<br>$\frac{\varphi}{\beta_g} \doteq \left\|\frac{\varphi}{\beta}\right\|_d \left[\frac{\omega_d T_R}{\sqrt{1 + (\omega_d T_R)^2}} e^{-t/T_R} + e^{-\zeta_d\omega_d t}\sin(\omega_d t - \tan^{-1}\omega_d T_R)\right]$ | $r \rightarrow \delta_r$, $\dot{\beta} \rightarrow \delta_r$, $\dot{a}_y \rightarrow \delta_r$ |

| | | | |
|---|---|---|---|
| 3. Rudder-induced rolling maneuver | Same as above | Same as above | Same as above |
| 4. Turn coordination | | Rudder deflection to coordinate<br>$\delta_r \doteq \dfrac{1}{N'_{\delta_r}}\left[-N'_{\delta_a}\delta_a + \left(\dfrac{g}{U_0} - N'_p\right)p\right]$ | $\delta_a \rightarrow \delta_r$<br>$p \rightarrow \delta_r$ |
| 5. Path control<br>$(\lambda_c \rightarrow \delta_a)$ or $(\psi_c \rightarrow \delta_a)$ | | $\dfrac{N^{\lambda}_{\delta_a}}{N^{\varphi}_{\delta_a}} \doteq \dfrac{g}{U_0 s} + \dfrac{Y_v}{s}\dfrac{N^{\beta}_{\delta_a}}{N^{\varphi}_{\delta_a}};\ \dfrac{N^{\psi}_{\delta_a}}{N^{\varphi}_{\delta_a}} \doteq \dfrac{g}{U_0 s} - \dfrac{(s - Y_v)}{s}\dfrac{N^{\beta}_{\delta_a}}{N^{\varphi}_{\delta_a}}$<br>$\dfrac{N^{\beta}_{\delta_a}}{N^{\varphi}_{\delta_a}} \doteq$<br>$\dfrac{-(N'_{\delta_a}/L'_{\delta_a})s^2 + \{(N'_{\delta_a}/L'_{\delta_a})L'_p - [N' - (g/U_0)]\}s + (g/U_0)[(N'_{\delta_a}/L'_{\delta_a})L'_r - N'_r]}{s^2 + 2\zeta_{\varphi}\omega_{\varphi}s + \omega^2_{\varphi}}$ | $\delta_a \rightarrow \delta_r$<br>$p \rightarrow \delta_r$<br>$\beta \rightarrow \delta_r$<br>$a_y \rightarrow \delta_r$ |
| 6. Inertial cross coupling; pitch/yaw divergence at critical roll rate; roll rate buildup from pitch input | | $p_{\text{crit}} = \sqrt{N'_{\beta}}$<br>$\dfrac{p(s)}{p_0} \doteq 1 + \dfrac{L'_{\beta}[(I_y - I_x)/I_z]\dot{\theta}(s)}{(s - L'_p)(s^2 - N'_r s + N'_{\beta})}$<br>$\left.\dfrac{p(t)}{p_0}\right]_{t\rightarrow\infty} \doteq \left\{1 + \dfrac{L'_{\beta}}{N'_{\beta}L'_p}\left(\dfrac{I_y - I_x}{I_z}\right)[\dot{\theta}(t)]_{t\rightarrow\infty}\right\}^{-1}$ | $\beta \rightarrow \delta_r$<br>$a_y \rightarrow \delta_r$<br>$r \rightarrow \delta_r$<br>$p \rightarrow \delta_a$<br>$\beta \rightarrow \delta_r$<br>$a_y \rightarrow \delta_r$ |

since banking to rotate the lift vector is the primary means that may be used to change the horizontal flight path of the airplane. In an airplane with idealized rolling characteristics (see Fig. 8-2), the roll response dynamic parameters are the roll subsidence time constant, $T_R$, and the maximum steady-state roll rate, $p_{max}$, due to the aileron. The latter parameter is a basic limitation on the performance of the aircraft; it can be modified only by configuration changes or, sometimes, by aileron to rudder crossfeeds. The roll subsidence time constant, on the other hand, is subject to direct modification using a feedback of rolling velocity to the ailerons.

In order actually to obtain the inherently available idealized roll control properties, the dutch roll characteristics must often be modified. In many aircraft this is a requirement because ordinary aileron-induced rolling maneuvers excite excessive dutch roll oscillations, sometimes to the point of instability of the closed-loop roll attitude control system (see Fig. 8-4). We have already noted that, as long as the dutch roll of the aircraft alone is stable, this tendency to instability under aileron control can be eliminated by making $\omega_\varphi/\omega_d \leqq 1$ or by increasing the dutch roll damping. The value of $\omega_\varphi/\omega_d$ is most simply adjusted to an optimum value near 1.0 by modifying the effective yawing moment due to aileron deflection, $N'_{\delta_a}$, to reduce thereby the amount of dutch roll excitation. When $\omega_\varphi$ exactly equals $\omega_d$, dutch roll excitation is minimized, except for the pole/zero difference, $\zeta_\varphi\omega_\varphi - \zeta_d\omega_d$. The reduction of this difference to zero by e.g., roll rate to rudder feedback, would then completely eliminate dutch roll excitation from aileron inputs. Increase of the dutch roll damping also reduces the dutch roll excitation, and has the additional merit of reducing dutch roll settling time and improving the roll response to side gusts. This will be described further below.

Several metrics relating to the function of roll attitude regulation and yaw roll coupling are listed next. With most aircraft that have high effective dihedral, any yawing moment, which may be because of a gust, engine out, or asymmetric store release, is converted via the dihedral effect into a rolling moment. When the yaw roll coupling metrics are out of bounds, the major fixes are to increase the directional stability, the dutch roll damping, and the roll damping. The metrics shown first in connection with roll attitude regulation are the same as those for the next item: rudder-induced rolling. Because low values of the metrics are desirable for disturbance regulation and higher values are desirable for deliberate rudder-induced rolls, a compromise must often be made between these factors.

When the roll attitude control and dutch roll properties are properly adjusted, the airplane can then be made to bank effectively. However, to bank is not necessarily to turn, at least not immediately. So, the function of turn coordination now enters our considerations. The conventional definition of a coordinated turn is that the sideslip, $\beta$, is zero. To the extent that the dutch roll mode is excited in turning maneuvers (e.g., at entry to, or exit from, the turn) the sideslip cannot be zero. Dutch roll can be excited in aileron-only maneuvers by $N'_{\delta_a}$ which is also the primary quantity making $\omega_\varphi$ different from $\omega_d$, or by $N'_p$ which principally affects the difference in roll numerator and dutch roll denominator dampings. The modification of these stability derivatives by aileron to rudder crossfeed or by roll rate to rudder feedback provides an effective means of reducing transient yawing during rolling maneuvers. An alternative or additional solution is to reduce the effects of these dutch roll exciting terms by increasing the directional stability.

Turn coordination factors are also important for path control using aileron. Turn coordination difficulties make the lateral path angle over bank angle transfer function, $\gamma/\varphi$, or the closely related transfer function $\psi/\varphi$ depart from the ideal $g/U_0 s$ relationship. Consequently, the outer loop bandwidth attainable in closing a heading or path control loop is reduced, and there is an associated degradation in precision of control. Fortunately, the correction of turn coordination during turn entry and exit is tantamount to solution of these path control problems.

Finally, large maneuvers require us to consider inertial cross-coupling suppression as an operational function.[4] There are two important inertial cross-coupling problems. One is a pitch/yaw divergence, which occurs at a critical roll rate (approximated here for the situation where $\omega_d < \omega_{sp}$ and $\zeta_d \doteq 0$). A function of the control system can be to increase the critical roll rate beyond that which will ever be used in the operation of the aircraft (e.g., by increasing $N'_\beta$), or additionally or alternatively to increase the damping. The second inertial cross-coupling problem is roll rate buildup. This occurs if the aircraft is pitched nose-down during a rolling maneuver. The buildup can be quite severe for swept-wing aircraft with high effective dihedral, low directional stability, and low roll damping. Increased directional stiffness and roll damping are the most effective solutions to this problem. Both of these solutions may be implemented with automatic feedback control.

From the above discussion, we see that the most important functions of the augmentation (inner loop) control systems are to provide any needed improvements in roll damping, directional stability, and dutch roll damping, and to reduce the excitation of the dutch roll in turning maneuvers. There are, as shown by the Feedback or crossfeed column in Table 11-3, a number of sets of possible systems that might be used to perform these functions. These sets are considered as competing systems in Table 11-4, where equalization and compensation requirements and

[4] Consideration of the coupled lateral-longitudinal dynamics involved in inertial cross-coupling problems is beyond the scope of this book. For references to complete treatments, see the bibliography at the end of the text.

Table 11-4. Lateral/directional competing systems.

| Primary controller functions | Feedbacks | Equalization requirements | Practical problems |
|---|---|---|---|
| 1. Improve roll response.<br>2. Reduce roll sensitivity to gusts and other inputs.<br>3. Reduce $\omega_r/\omega_d$. | $p \to \delta_a$ | 1. Gain inversely proportional to $L'_{\delta_a}$, i.e., $K_p \propto \frac{I_x}{\rho U^2 C_{l_{\delta_r}}}$<br>2. Use $p$ command system so that feedback does not oppose pilot inputs and reduce $p_{max}$. | 1. Gain adjustment to compensate for changes in dynamic pressure, Mach no. effects on $C_{l_{\delta_r}}$, and loading effects on $I_x$.<br>2. Effects of flaps, slats, and other aerodynamic devices on $L'_{\delta_a}$.<br>3. Failure may result in unstable $r \to \delta_r$ loop. |
| 1. Increase directional stability.<br>2. Increase dutch roll damping.<br>3. Reduce inertial cross coupling.<br>4. Improve turn coordination. | $\beta \to \delta_r$ | 1. Gain inversely proportional to $N'_{\delta_r}$, i.e., $K_\beta \propto \frac{I_z}{\rho U^2 C_{n_{\delta_r}}}$<br>2. Lead/lag element. | 1. Instrumenting $\beta$ sensor (type and location).<br>2. Gain adjustment to compensate for changes in dynamic pressure and Mach no. effects on $C_{n_{\delta_r}}$.<br>3. Lead/lag equalization on sensor possibly sensitive to gusty inputs. |
| | $\beta \to \delta_r$<br>$r \to \delta_r$ | 1. Both gains inversely proportional to $N'_{\delta_r}$ (see above).<br>2. Washout in $r \to \delta_r$ feedback. | 1. Instrumenting $\beta$ sensor.<br>2. Gain adjustments.<br>3. Unless $\omega_r/\omega_d \ll 1$, $r \to \delta_r$ does not greatly increase dutch roll damping (may be unstable if $p \to \delta_a$ fails).<br>4. Location of washout time constant; washout effect may be deleterious in spin recovery.<br>5. Rate gyro inclination effects. |

| | | | |
|---|---|---|---|
| | $a_y \rightarrow \delta_r$ | 1. Gain inversely proportional to $Y_\beta N'_{\delta_r}$, i.e., $K_{a_y} \propto \dfrac{mI_z}{\rho^2 U^4 C_{y_\beta} C_{n\delta_r}}$ <br> 2. Lead/lag element. | 1. Severe range requirements on gain adjustments. <br> 2. Finding a sensor location adequate for all flight conditions. <br> 3. Gain limited by high frequency modes (actuator and structural). <br> 4. Location of lead/lag time constants. |
| | $a_y \rightarrow \delta_r$ <br> $r \rightarrow \delta_r$ | 1. $a_y$ gain inversely proportional to $Y^\beta N'_{\delta_r}$ (see above). <br> 2. $r$ gain inversely proportional to $N'_{\delta_r}$. <br> 3. Washout in $r \rightarrow \delta_r$ feedback. | 1. Severe range requirements on gain adjustments. <br> 2. Finding an accelerometer location adequate for all flight conditions. <br> 3. $a_y$ gain limited by high frequency modes but not as much as equalized $a_y \rightarrow \delta_r$. <br> 4. Unless $\omega_r/\omega_d \ll 1$, $r \rightarrow \delta_r$ does not greatly increase dutch roll damping (may be unstable if $p \rightarrow \delta_a$ fails). <br> 5. Location of washout time constant; washout effect may be deleterious in spin recovery. <br> 6. Rate gyro inclination effects. |
| 1. Provide precise turn coordination. <br> 2. Eliminate lateral/directional coupling. | $p \rightarrow \delta_r$ | 1. Lead/lag element. <br> 2. Gain and lead time constant complex functions of several parameters. | 1. Programming gain and lead time constant for various flight configurations and conditions. <br> 2. May be deleterious near stall and in spins. |
| | $\delta_a \rightarrow \delta_r$ | 1. Lead/lag or lag/lead element. <br> 2. Gain and time constants complex functions of several parameters. | 1. Programming gain and time constants for various flight configurations and conditions. <br> 2. May significantly increase pilot rudder input required for deliberate sideslip maneuvers. |

practical problems are listed as compromise factors.

For the functions of improving roll damping and reducing roll response time, there is no competition, since roll rate to aileron feedback is the only practical solution. The gain adjustment requirement given in Table 11-4 is that necessary to provide a constant increment in roll damping. However, in many cases fixed gain systems have been satisfactory and the changes in the roll damping increment across the spectrum of flight conditions have been acceptable. Perhaps the most significant practical problem with the $p \rightarrow \delta_a$ feedback is that failure of this loop might result in an instability in an $r \rightarrow \delta_r$ loop, unless $\omega_r/\omega_d$ is much less than unity with the roll damper off (see Fig. 8-11). To avoid this eventuality, if a yaw rate to rudder loop is used, this could require a fail-operational augmentation capability of the roll damper.

Four competing directional control systems involving different combinations of rudder feedbacks are shown in Table 11-4. The first is feedback of sideslip with lead/lag equalization. The gain adjustment indicated in the table is the one required to maintain a constant increment in directional stability and dutch roll damping. By far the most significant problem with this system is that of obtaining an adequate sensor or sensor complex.

In the second system, yaw rate feedback has been substituted for the lead/lag equalization in the first system. This eliminates the problem of lead equalization on a possibly noisy sensor, but it introduces a variety of problems associated with the yaw rate to rudder feedback. These primarily come into play with relatively large $\omega_r/\omega_d$ (near unity), rate gyro inclination relative to stability axes and the use of a washout to prevent the rudder from opposing a steady turn (see Section 8-4). Even with washout, high gain on the yaw rate feedback is detrimental to turn coordination, as the feedback tries to keep the airplane from turning (yawing).

In the other two systems lateral acceleration has been substituted for sideslip. This substitution gets around the problem of the sideslip sensor, but unfortunately it introduces several others. For one thing, the gain adjustments required for an acceleration feedback are more severe than for the systems with a sideslip sensor. Also, in this case, it is necessary to locate the accelerometer fairly close to the airplane center of rotation for rudder inputs. Because the center of rotation can shift appreciably with changes in Mach number and vehicle loading, a serious practical problem may consist in finding an accelerometer location that is adequate for all flight and loading conditions (see Section 8-6).

Another problem closely related to the one of sensor location is the gain limitation imposed by high frequency modes such as those associated with actuator and structural dynamics. As illustrated in Fig. 11-3, this gain limitation difficulty becomes more severe as the accelerometer is moved

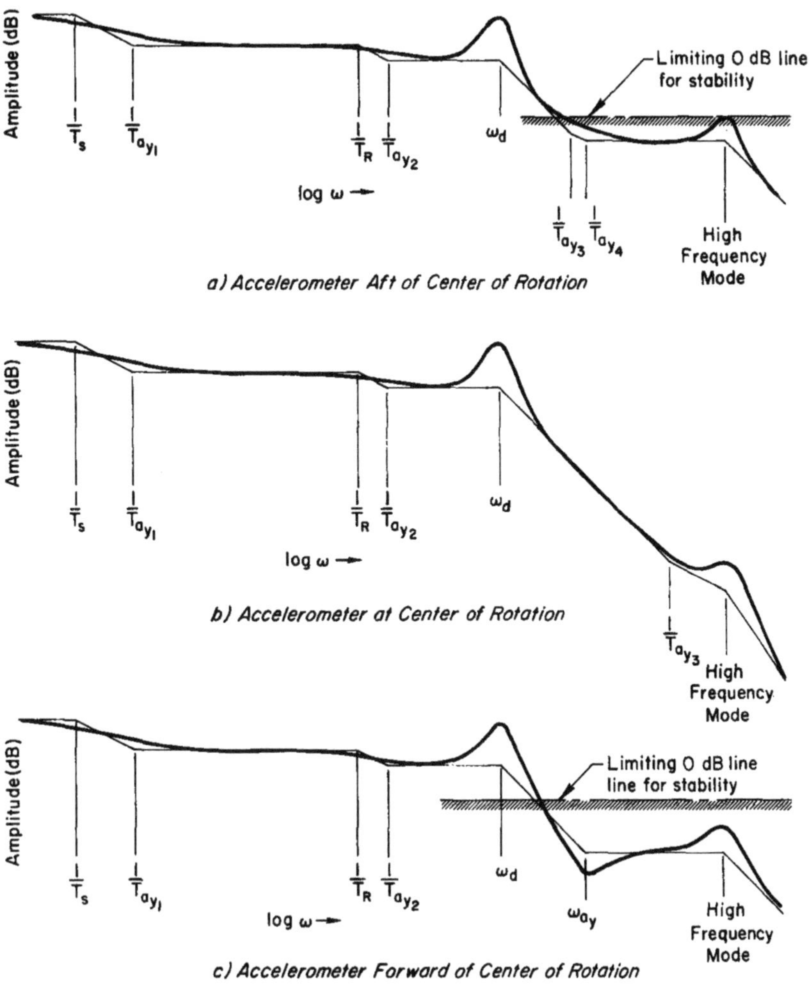


Fig. 11-3. Typical $a_y \rightarrow \delta_r$ Bode diagram with a high frequency mode.

further from the center of rotation. The ideal accelerometer location is at the center of rotation for rudder inputs. An open loop $j\omega$-Bode plot for this case is shown in Fig. 11-3(b). The corresponding diagram for the case in which the accelerometer is aft of the center of rotation appears in Fig. 11-3(a) which corresponds to Fig. 8-14 when a high frequency mode is added. Figure 11-3(c) shows the open loop $j\omega$-Bode diagram for location of the accelerometer forward of the center of rotation for rudder inputs. For both cases in which the accelerometer is displaced from the center of rotation, there is a very restricted upper limit on the gain, although the

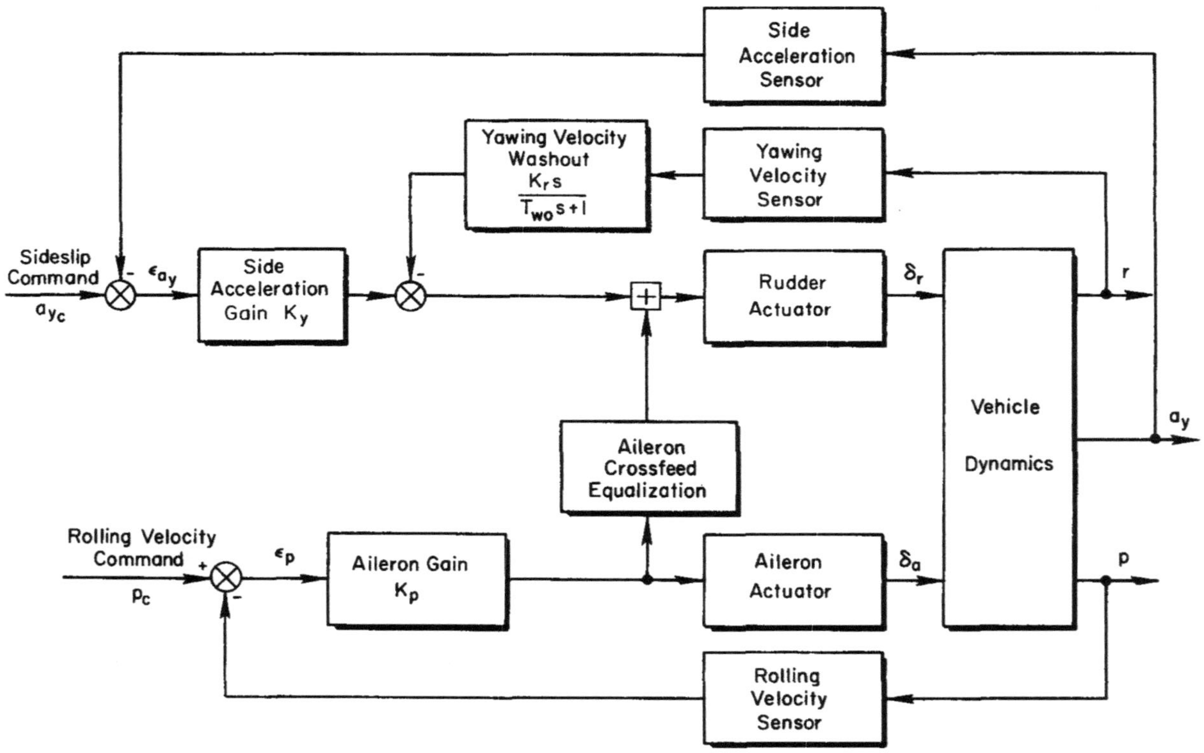


Fig. 11-4. A basic mechanization of essential feedbacks; acceleration plus yaw rate.

reasons for the limit are somewhat different. For the aft accelerometer, it is simply a matter of the conventional gain margin's disappearance. For the forward accelerometer, it is a question of a phase dip analogous to the $\omega_\varphi/\omega_d$ effect (see Section 8-1). The gain limitation problems noted here are, of course, accentuated if lead/lag equalization is used on the accelerometer feedback.

While increases in directional stability will tend to improve turn coordination, the resulting coordination may not be sufficiently precise for such mission phases as intercept, landing, and tactical support. In these cases, the systems described above may be supplemented with a special, precise turn coordination device. Two possibilities are shown in Table 11-4: roll rate to rudder feedback and aileron to rudder crossfeed. The basic purpose of these devices is to augment both of the derivatives $N'_{\delta_a}$ and $N'_p$. A pure gain $p \rightarrow \delta_r$ feedback will augment $N'_p$; but to augment $N'_{\delta_a}$, we attempt to reproduce $\delta_a$ by putting lead with a time constant near $T_R$ in the $p \rightarrow \delta_r$ feedback. With the $\delta_a \rightarrow \delta_r$ system, a pure gain augments $N'_{\delta_a}$, and we attempt to reproduce $p$ by putting $\delta_a$ through a suitable lag (again, with time constant near $T_R$). The net results are lag/lead equalization in the $\delta_a \rightarrow \delta_r$ crossfeed or lead/lag in a $p \rightarrow \delta_r$ feedback. Either system will greatly improve the turn coordination, although the $\delta_a \rightarrow \delta_r$ crossfeed will generally be slightly superior to the $p \rightarrow \delta_r$ feedback because it is easier to reproduce $p$ from $\delta_a$ than vice versa. With either system, however, the primary practical problem is in programming the gain and equalization time constants for the various flight configurations and conditions. The crossfeed can also present another problem, noted in Table 11-4, in that it may require the pilot to use additional rudder input to sideslip the airplane deliberately.

The block diagrams of Figs. 11-4 and 11-5 are typical versions of augmentation systems that incorporate the essential features described here. The primary difference between these two mechanizations is in the dutch roll damping features.

These remarks conclude the discussion of essential or inner loop feedbacks. We turn now to the exposition of an example of a more complete feedback control system including some aspects of outer loop feedback control.

## 11-3. Longitudinal Approach Control System

The longitudinal multiloop example is constructed in connection with an approach coupler for automatically following the instrument low approach system (ILS) glide path down to a very low altitude above the runway. In order to accomplish automatic landing approach, the aircraft must become an element in a feedback control system. The principal

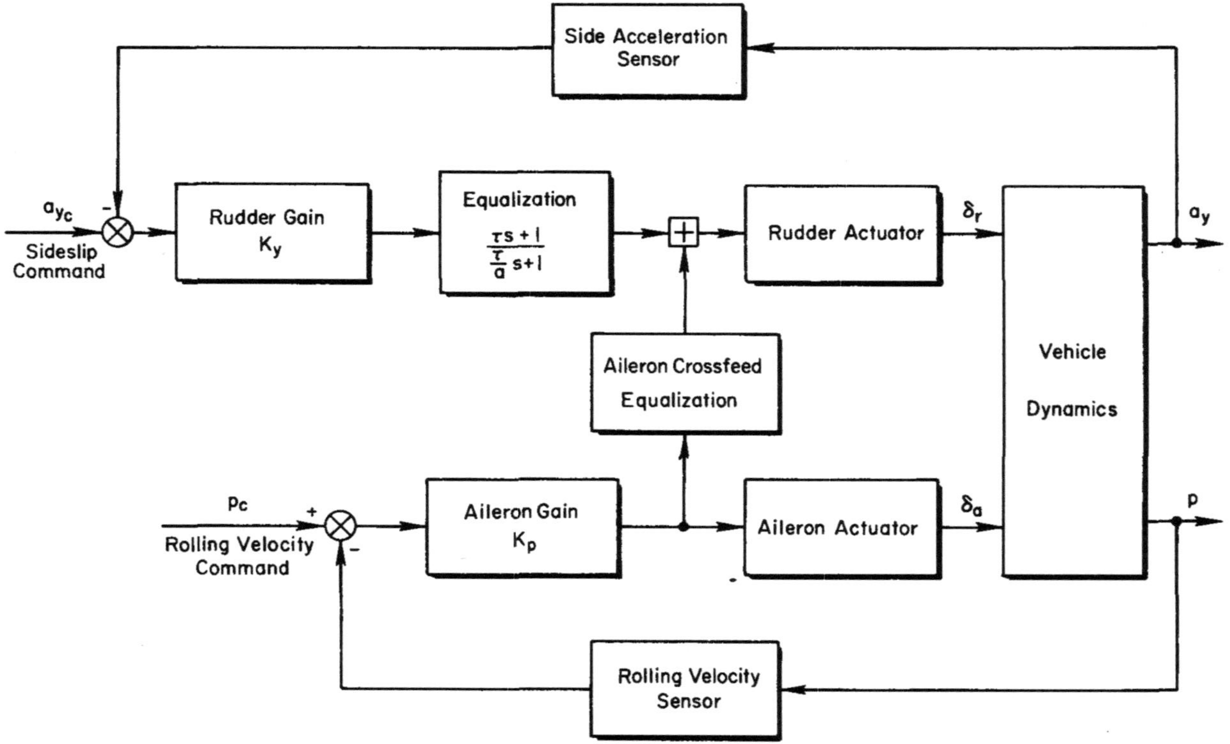


Fig. 11-5. A basic mechanization of essential feedbacks; equalized acceleration.

features of the system structure are the feedbacks themselves, their equalization, and their combinations to create control commands. For systems that demonstrate uniform, reliable, high quality approach performance, the possible feedback structures are very limited. They derive primarily from guidance, control, and regulation demands, and secondarily from dynamic response characteristics that may be desired by the pilot.

Stated verbally, the primary guidance and control requirements for low approach systems are:

1. To establish and maintain the aircraft on a specified spatial pathway or beam [e.g., the instrument low approach system (ILS) glide path].
2. To reduce flight path errors to zero in a stable, well-damped, and rapidly responding manner.
3. To establish an equilibrium flight condition.
4. To limit the speed or angle of attack excursions from this established equilibrium flight condition.

The regulation requirements are similar, i.e.,

1. To maintain the established flight path in the presence of disturbances such as gusts, winds, and wind shears.
2. To provide a degree of short term attitude stability in the presence of disturbances.

These requirements relate in the main to the relatively low frequency path modes of the aircraft control system. In essence, they define outer control loops, involving those vehicle motion quantities that define the desired equilibrium state of motion. As we have seen, more often than not, such outer loops, by themselves, are unusable. Equalization of either a series or a parallel nature is needed. Parallel equalization is the most common form, and it is achieved by the use of inner loops that feed back such quantities as attitude, angular velocity, and sometimes linear acceleration. These inner loops dominate the high frequency characteristics of the aircraft/controller system. They correspond to the augmentation systems discussed in the last section.

In order to obtain a better appreciation of just what feedbacks the verbal requirement statements presented above actually imply, we shall consider in detail the determination of feedback structures for a longitudinal controller for the particular operational function of glide path control. Then we shall show the results of an analysis of the system that has been evolved. The analysis is actually presented in three parts; namely, in terms of the closed loop system analysis, the transient response characteristics, and the steady-state system response to random disturbances.

*Development of the Feedback Structure*

A concrete example of a multiloop longitudinal approach control system is shown in Fig. 11-6. This diagram emphasizes the guidance, or outer loop, elements in the forward path; in particular, it emphasizes the ground to air transmission of guidance data, together with the airborne reception and decoding of these data to produce control system commands.

The fundamental error signal in the system is the displacement error from the radio beam, $d_e$. This signal is equal to the difference between the aircraft's actual position, $d$, and a beam command signal, $d_i$, where both of these are measured relative to a nominal beam datum. When the guidance is provided by a conventional ILS glide slope, $d_i$ is the beam centerline, and is thus identically equal to zero; for variable (curved) path systems, on the other hand, $d_i$ is a path command from a nominal or initial path.

The error signal, $d_e$, is not available as a physical quantity uncontaminated by noise. At the command point the beam noise, $n_b$, is added; this sum is then converted to an angle seen in the aircraft via division by the range from the transmitter (i.e., without noise, $d_e = R\lambda$); finally, the addition of electromagnetic disturbances and the further noise, $n_r$, gives the actual glide slope signal at the input of the aircraft receiver. Typically, $n_b$ is used to represent those unwanted inputs that are approximately stationary when represented as lengths (i.e., range-independent noise), whereas $n_r$ is made up of noises that are approximately stationary when represented as angles (range-dependent noise). Range-dependent noise includes the effective receiver noise, which tends to have a constant root mean square value at the input of the receiver and thus is larger relative to a given error signal ($d_e$) at the longer ranges. An example of range-independent noise is main beam multipath transmissions. These are caused by fixed structures, by changes in ground reflection coefficients due to stratified wet and dry layers in the ground ("fixed" for a particular approach), and by other deviations of the ground plane from the characteristics of an ideal reflecting surface. Both $n_b$ and $n_r$ can be considered and treated as random processes.

The electromagnetic disturbances indicated in Fig. 11-6 can come from a lead aircraft casting an electromagnetic shadow on a following aircraft on the same approach beam pathway, from multipath transmissions because of overflying aircraft (either directly or from side lobes), and so forth. These disturbances are, in general, complex discrete inputs, rather than random phenomena.

The receiver (as well as filtering, gain changing, and other operations)

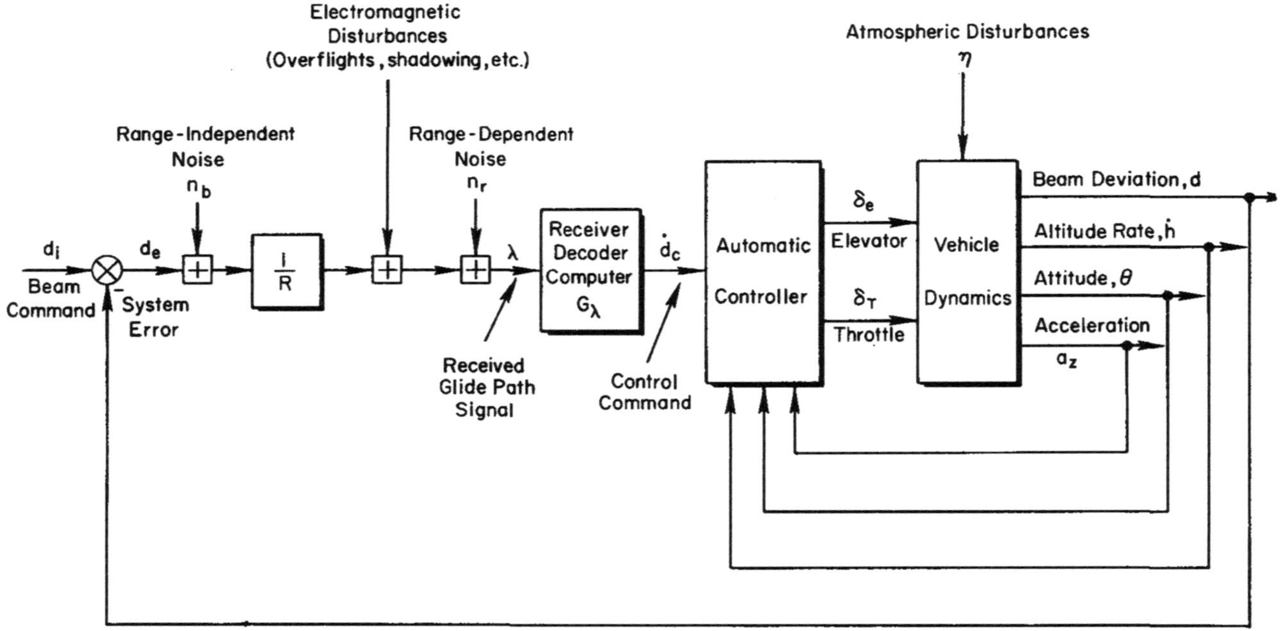


Fig. 11-6. Longitudinal control system for vertical plane approach.

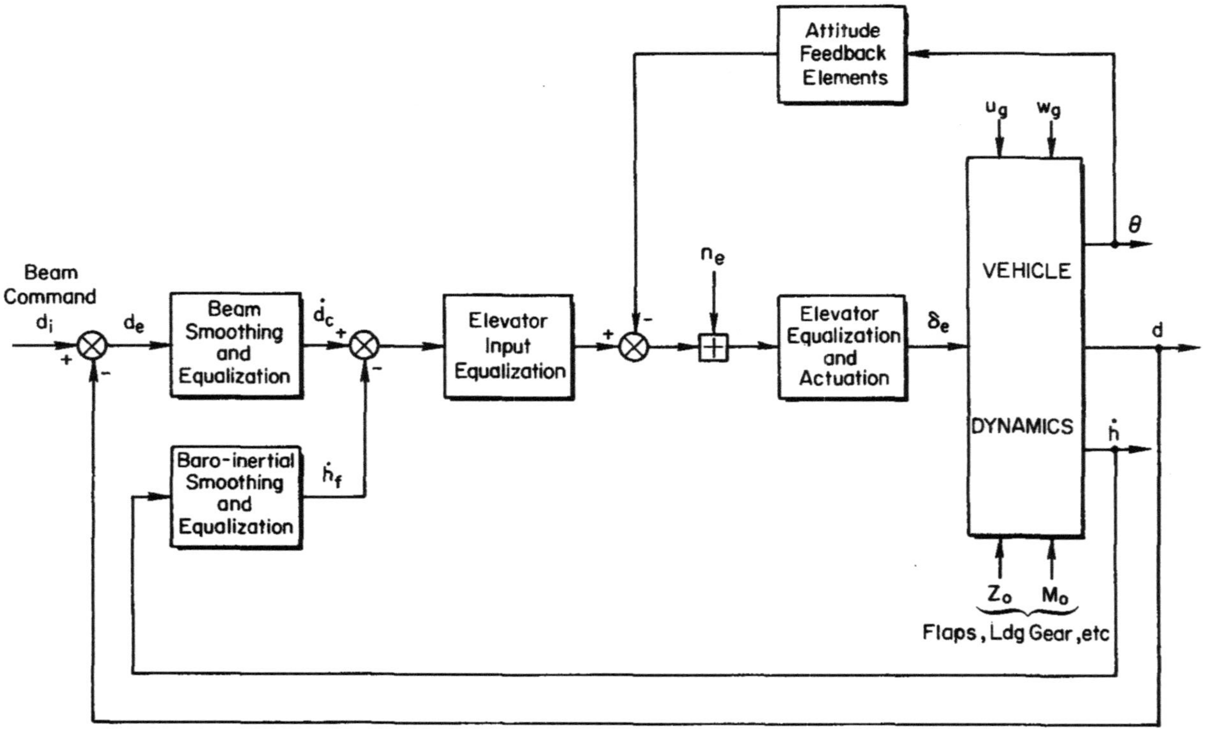


Fig. 11-7. Block diagram of longitudinal control for approach and landing.

is represented in the transfer characteristic $G_\lambda$ which has the output $\dot{d}_c$. This is the effective command to the aircraft/flight control system. The latter comprises the airframe and its inner-loop controllers.

Just as Fig. 11-6 emphasizes the forward path of the outer loop, Fig. 11-7 provides a more detailed breakdown of the inner loops of the control system. The control system feedbacks shown here are actually based on an interpretation of the requirements yet to come. However, matters are made more concrete and comprehensible if the end result (the system) is given first and then justified by the subsequent discussion. This is the procedure that is to be adopted here. The most outer loop subsystem is lumped into the "beam smoothing and equalization" block, and the output of this block, $\dot{d}_c$, has a filtered altitude rate signal, $\dot{h}_f$, subtracted from it so as to create an error in the rate of sink (or climb).

Note that the $\dot{h}_f$ signal, here supposed to be derived from a "complementary filter" combining barometric and inertial elements, comes from a different path than the deviation from the beam, $d_e$.[5] This is done in order to obtain an $\dot{h}$ signal (which is similar to a $\dot{d}$ signal) without actually deriving a rate from the noisy beam deviation signal. Here the composite signal is derived from a barometric rate of climb and accelerometer sensors, as shown in Fig. 11-8. For simplicity the higher frequency lags inherent in the sensors are neglected. The actual rate of climb is $\dot{h}$; the various $n$'s are unwanted signals and noises. The composite signal, $\dot{h}_f$, is determined

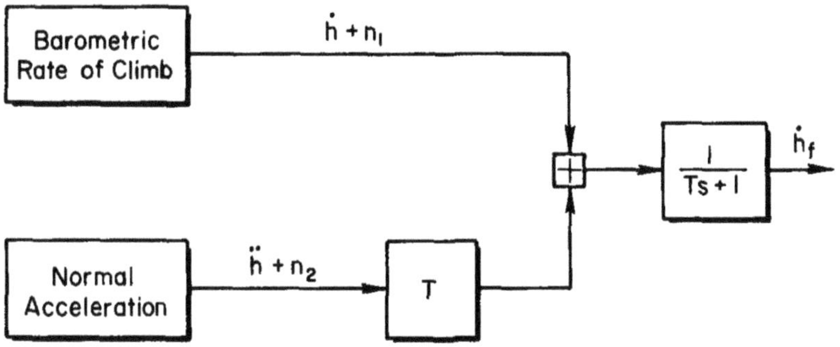


Fig. 11-8. Derivation of rate of climb signal by means of complementary filtering.

[5] The process of combining two signals, one good at low frequencies—in this case the barometric rate of climb signal—and the other good at high frequencies—in this case the accelerometer signal—so as to produce a composite signal more useful than either alone, is known as complementary filtering. The process has many applications in automatic flight control.

by manipulation of transformed quantities as follows:

$$\mathcal{L}[h_f] = sh_f = \frac{1}{Ts + 1}[(sh + n_1) + T(s^2h + n_2)] = sh + \frac{n_1 + Tn_2}{Ts + 1} \tag{11-1}$$

In the time domain, the major component of the signal output is then just $\dot{h}$, while the noises are heavily attenuated by the lag filter. The limitation on this technique is in the magnitude of the time constant, $T$. Since it appears as a gain on the accelerometer noise, $n_2$, the time constant can only be large if the accelerometer noise is small in the frequencies of interest. This is ordinarily the case.

In the remainder of the control system block diagram of Fig. 11-7, both elevator and throttle controls may be involved, but, for the sake of simplicity, only the elevator control is shown and the throttle control is not explicitly indicated. The controller equation for the elevator control is given by

$$\delta_e = G_{d_i} d_i - (G_d d + G_\theta \theta) \tag{11-2}$$

where each $G$ is a shorthand notation for the product of all the transfer functions in the blocks between the subscript variable and the elevator. For example, $G_{d_i}$ is the product of the transfer functions for the blocks labeled: Beam Smoothing and Equalization, Elevator Input Equalization, and Elevator Equalization and Actuation—or, briefly, everything between $d_i$ and $\delta_e$. Using this notation, we can write the complete closed-loop approach system equations according to the rules of Section 3.5.

$$d = \frac{G_{d_i} N^d_{\delta_e} d_i + G_{n_e} N^d_{\delta_e} n_e + \sum_\eta (N^d_\eta + G_\theta N^{d\,\theta}_{\eta\,\delta_e})\eta}{s\Delta + sG_\theta N^\theta_{\delta_e} + G_d N^d_{\delta_e}} \tag{11-3}$$

$$\theta = \frac{sG_{d_i} N^\theta_{\delta_e} d_i + sG_{n_e} N^\theta_{\delta_e} n_e + \sum_\eta (sN^\theta_\eta + G_d N^{\theta\,d}_{\eta\,\delta_e})\eta}{s\Delta + sG_\theta N^\theta_{\delta_e} + G_d N^d_{\delta_e}} \tag{11-4}$$

$$u = \frac{sG_{d_i} N^u_{\delta_e} d_i + sG_{n_e} N^u_{\delta_e} n_e + \sum_\eta (sN^u_\eta + G_d N^{ud}_{\eta\delta_e} + sG_\theta N^{u\theta}_{\eta\delta_e})_\eta}{s\Delta + sG_\theta N^\theta_{\delta_e} + G_d N^d_{\delta_e}} \tag{11-5}$$

where $\eta$ is $u_g$, $w_g$, $Z_0$, $M_0$; $\Delta$ the airframe-alone longitudinal characteristic function; $N^\theta_{\delta_e}$, $N^d_{\delta_e}$, etc. the airframe-alone transfer function numerators; $N^{u\theta}_{\eta\delta_e}$, $N^{\theta d}_{\eta\delta_e}$, etc. the airframe-alone coupling numerators; and $n_e$ the lumped noise effectively acting in the elevator channel.

Equations 11-3 to 11-5 combine the controller equations with those of the vehicle which are characterized by the transfer function numerators, coupling numerators, and the characteristic function. These functions characterizing the aircraft are summarized in Table 11-5 which for completeness also includes some of the "conventional" functions given in Chapter 5. Notice that the trim and atmospheric disturbances are denoted by a general disturbance input, $\eta$, and that the noises may be lumped into an equivalent noise $n_e$. Equations for other aircraft motion quantities, such as $w$, can be obtained from Eq. 11-5 simply by replacing the $u$ superscripts in the numerators by the new variable.

Finally, notice that the deviation from the beam, $d$, becomes equal to the perturbed altitude, $h$, when the steady-state glide path, $\Theta_0$, is zero. In that case the characteristic function, numerators, and coupling numerators reduce to the ones already given in Chapter 5.

With the general control system now described, we can turn to definitions of the necessary general forms of the controller transfer functions. To do this, we return to the verbal requirements given in the introduction to this section. Many of these qualitative requirements can readily be translated into required forms for the feedbacks by considering certain steady-state aspects of the system equations. Other required feedbacks can be deduced from simple stability and response arguments. We shall consider the steady-state features first.

*Guidance Requirements*

The most fundamental guidance requirement is that the aircraft establish itself on the beam when the system is engaged. In other words, the deviation from the beam, $d$, must ultimately become zero when the system "input" is an initial condition, $d(0+)$.

The response transform of an $n^{\text{th}}$-order system with the characteristic function

$$\Delta_{\text{sys}} = s^n + a_1 s^{n-1} + \cdots + a_{n-1}s + a_n \tag{11-6}$$

to an initial condition of position, $d(0+)$, is readily shown to be

$$d(s) = \frac{(s^{n-1} + a_1 s^{n-2} + \cdots + a_{n-1})d(0+)}{s^n + a_1 s^{n-1} + \cdots + a_{n-1}s + a_n} \tag{11-7}$$

Assuming that the final value theorem holds, we get

$$\begin{aligned} \lim_{t\to\infty} d(t) &= \lim_{s\to 0} sd(s) \\ &= \lim_{s\to 0} s\,\frac{(s^{n-1} + a_1 s^{n-2} + \cdots + a_{n-1})a(0+)}{s^n + a_1 s^{n-1} + \cdots + a_{n-1}s + a_n} \end{aligned} \tag{11-8}$$

This limit will approach zero only when $a_n \neq 0$. Consequently, the characteristic function must contain a constant term for the system to acquire the beam. The characteristic function, as we have already seen, is

$$\Delta_{\text{sys}} = s\Delta + sG_\theta N_{\delta_e}^{\theta} + G_d N_{\delta_e}^{d} \tag{11-9}$$

**Table 11-5.** Longitudinal equations, transfer functions, and coupling numerators for the low approach example.

---

Equations of Motion:

$$\begin{bmatrix} s - X_u & -X_w & g\cos\Theta_0 \\ -Z_u & s - Z_w & -U_0 s + g\sin\Theta_0 \\ -M_u & -(M_{\dot w}s + M_w) & s(s - M_q) \end{bmatrix}\begin{bmatrix} u \\ w \\ \theta \end{bmatrix} = \begin{bmatrix} X_{\delta_e} & X_{\delta_T} & -X_\eta \\ Z_{\delta_e} & Z_{\delta_T} & -Z_\eta \\ M_{\delta_e} & M_{\delta_T} & -M_\eta \end{bmatrix}\begin{bmatrix} \delta_e \\ \delta_T \\ \eta \end{bmatrix}$$

$$\dot d = -w + U_0\theta$$

$$\dot h = -U_0\sin\Theta_0 - w\cos\Theta_0 + u\sin\Theta_0 + U_0\cos\Theta_0\theta$$

Characteristic Function:[a]

$$\Delta = (A_\Delta s^4 + B_\Delta s^3 + C_\Delta s^2 + D_\Delta s + E_\Delta)$$

$$A_\Delta = 1$$

$$B_\Delta = -(M_q + Z_w + U_0 M_{\dot w} + X_u)$$

$$C_\Delta = M_q Z_w - U_0 M_w + X_u(M_q + Z_w + U_0 M_{\dot w}) - X_w Z_u + gM_{\dot w}\sin\Theta_0$$

$$D_\Delta = -X_u(Z_w M_q - U_0 M_w) - M_u U_0 X_w + M_q Z_u X_w + g\cos\Theta_0(M_u + Z_u M_{\dot w}) + g\sin\Theta_0(M_w - X_u M_{\dot w})$$

$$E_\Delta = g\cos\Theta_0(M_w Z_u - M_u Z_w) + g\sin\Theta_0(M_u X_w - M_w X_u)$$

Numerators

$$N^{\dot d}_\delta = A^{\dot d}_\delta s^3 + B^{\dot d}_\delta s^2 + C^{\dot d}_\delta s + D^{\dot d}_\delta$$

$$A^{\dot d}_\delta = -Z_\delta$$

$$B^{\dot d}_\delta = Z_\delta(U_0 M_{\dot w} + M_q + X_u) - X_\delta Z_u$$

$$C^{\dot d}_\delta = U_0[M_{\dot w}(X_\delta Z_u - Z_\delta X_u) + (Z_\delta M_w - Z_w M_\delta)] + M_q(Z_u X_\delta - X_u Z_\delta) + gM_\delta\sin\theta$$

$$D^{\dot d}_\delta = U_0[X_\delta(Z_u M_w - M_u Z_w) - Z_\delta(M_w X_u - X_w M_u) - M_\delta(X_w Z_u - Z_w X_u)] + g\cos\theta_0(Z_u M_\delta - M_u Z_\delta) + g\sin\theta_0(M_u X_\delta - X_u M_\delta)$$

$$N^{\dot d}_\eta = A^{\dot d}_\eta s^3 + B^{\dot d}_\eta s^2 + C^{\dot d}_\eta s + D^{\dot d}_\eta$$

$$A^{\dot d}_\eta = +Z_\eta$$

$$B^{\dot d}_\eta = -Z_\eta(U_0 M_{\dot w} + M_q + X_u) + X\ Z_u$$

$$C^{\dot d}_\eta = U_0[M_{\dot w}(Z_\eta X_u - X_\eta Z_u) + (M_\eta Z_w - Z_\eta M_w)] + M_q(X_u Z_\eta - Z_u X_\eta) - gM_\eta\sin\theta_0$$

$$D^{\dot d}_\eta = U_0[X_\eta(M_u Z_w - Z_u M_w) + Z_\eta(M_w X_u - X_w M_u) + M_\eta(X_w Z_u - Z_w X_u)] + g\cos\theta_0(Z_\eta M_u - Z_u M_\eta) + g\sin\theta_0(X_u M_\eta - M_u X_\eta)$$

---

Table 11-5 (Continued)

$$N^\theta_\delta = A^\theta_\delta s^2 + B^\theta_\delta s + C^\theta_\delta$$

$$A^\theta_\delta = Z_\delta M_{\dot w} + M_\delta$$

$$B^\theta_\delta = X_\delta(M_{\dot w}Z_u + M_u) + Z_\delta(M_w - M_{\dot w}X_u) - M_\delta(Z_w + X_u)$$

$$C^\theta_\delta = X_\delta(M_wZ_u - M_uZ_w) + Z_\delta(M_uX_w - M_wX_u) + M_\delta(Z_wX_u - X_wZ_u)$$

$$N^\theta_\eta = A^\theta_\eta s^2 + B^\theta_\eta s + C^\theta_\eta$$

$$A^\theta_\eta = -(Z_\eta M_{\dot w} + M_\eta)$$

$$B^\theta_\eta = +[-X_\eta(M_{\dot w}Z_u + M_u) - Z_\eta(M_w - M_{\dot w}X_u) + M_\eta(Z_w + X_u)]$$

$$C^\theta_\eta = -X_\eta(M_wZ_u - M_uZ_w) - Z_\eta(M_uX_w - M_wX_u) - M_\eta(Z_wX_u - X_wZ_u)$$

Coupling Numerators:

$$N^{\dot d\theta}_{\eta\delta} = A^{\dot d\theta}_{\eta\delta} s + B^{\dot d\theta}_{\eta\delta}$$

$$A^{\dot d\theta}_{\eta\delta} = Z_\eta M_\delta - M_\eta Z_\delta$$

$$B^{\dot d\theta}_{\eta\delta} = -X_u(Z_\eta M_\delta - M_\eta Z_\delta) + Z_u(X_\eta M_\delta - M_\eta X_\delta) - M_u(X_\eta Z_\delta - Z_\eta X_\delta)$$

$$N^{u\dot d}_{\eta\delta} = A^{u\dot d}_{\eta\delta} s^2 + B^{u\dot d}_{\eta\delta} s + C^{u\dot d}_{\eta\delta}$$

$$A^{u\dot d}_{\eta\delta} = X_\eta Z_\delta - Z_\eta X_\delta$$

$$B^{u\dot d}_{\eta\delta} = (X_\delta X_\eta - Z_\delta X_\eta)(M_q + U_0 M_{\dot w})$$

$$C^{u\dot d}_{\eta\delta} = U_0[X_\eta(Z_w M_\delta - M_w Z_\delta) + X_w(M_\eta Z_\delta - Z_\eta M_\delta) + X_\delta(M_w Z_\eta - Z_w M_\eta)] + g[(X_\delta M_\eta - M_\delta X_\eta)\sin\theta_0 - (Z_\delta M_\eta - M_\delta Z_\eta)\cos\theta_0]$$

$$N^{u\theta}_{\eta\delta} = A^{u\theta}_{\eta\delta} s + B^{u\theta}_{\eta\delta}$$

$$A^{u\theta}_{\eta\delta} = X_\delta(M_\eta + Z_\eta M_{\dot w}) - X_\eta(M_\delta + Z_\delta M_{\dot w})$$

$$B^{u\theta}_{\eta\delta} = X_\eta(Z_w M_\delta - M_w Z_\delta) + X_w(M_\eta Z_\delta - Z_\eta M_\delta) + X_\delta(M_w Z_\eta - Z_w M_\eta)$$

---

[a] The deviation transfer function numerators are defined in terms of rate (rather than position), in order to avoid the confusion that can arise when transfer function 'numerators' have a free $s$ denominator. Thus the transfer function for $\dot d/\delta_e$ is defined to be $N^{\dot d}_{\delta_e}/\Delta_{long}$ and the transfer function for $d/\delta_e$ is $N^{\dot d}_{\delta_e}/s\Delta_{long}$.

As the Laplace transform variable, $s$, approaches zero, $s\Delta$, $sN^{\theta}_{\delta_e}$, and $N^{d}_{\delta_e}$ approach $sE_{\Delta}$, $sC^{\theta}_{\delta_e}$, and $D^{d}_{\delta}$, respectively (see Table 11-5). So, as $s$ approaches zero in Eq. 11-9,

$$\lim_{s\to 0} \Delta_{\text{sys}} = \lim_{s\to 0} \{s[E_{\Delta} + G_{\theta}(s)C^{\theta}_{\delta_e}] + G_d(s)D^{d}_{\delta_e}\} \tag{11-10}$$

Assuming (realistically) that $G_{\theta}(0)$ is not of lower degree in $s$ than a constant or zero, we find that the right-hand side of Eq. 11-8 will be zero if $G_d(s)$ contains a proportional ($K_d$) or lower degree term in $s$ (this will insure that the denominator of Eq. 11-8 is a constant or lower degree in $s$). Thus all of this justifies the intuitively obvious requirement for a proportional feedback of path deviation, in order to achieve static stability relative to the beam.

An associated guidance requirement may be that the system follow guidance commands, $d_i$. These might arise from a two segment beam or possibly from even a higher order curvature command in more advanced systems.

In following guidance commands, $d_i$, the error $d_e$ is given by

$$d_e = d_i - d = d_i\left(1 - \frac{d}{d_i}\right) = \frac{s\Delta + sG_{\theta}N^{\theta}_{\delta_e} + N^{\dot{d}}_{\delta_e}(G_d - G_{d_i})}{s\Delta + sG_{\theta}N^{\theta}_{\delta_e} + G_d N^{\dot{d}}_{\delta_e}} d_i \tag{11-11}$$

If $G_d = G_{d_i} + Y_d$, the expression for $d_e$ may be simplified to

$$d_e = \frac{s\Delta + sG_{\theta}N^{\theta}_{\delta_e} + Y_d N^{\dot{d}}_{\delta_e}}{s\Delta + sG_{\theta}N^{\theta}_{\delta_e} + G_d N^{\dot{d}}_{\delta_e}} d_i \tag{11-12}$$

If then the commanded path were given by a power series in time, i.e.,

$$d_i(t) = d_1 + d_2 t + d_3 t^2 + \cdots \tag{11-13}$$

the Laplace transform of the command would be

$$d_i(s) = \frac{d_1}{s} + \frac{d_2}{s^2} + \frac{2d_3}{s^3} + \cdots \tag{11-14}$$

and the term of lowest degree in $s$ would be $(n - 1)!\, d_n/s^n$. Using the final value theorem, we obtain

$$d_{e_{\text{steady-state}}} = \lim_{s\to 0}\left[\frac{s\Delta + sG_{\theta}N^{\theta}_{\delta_e} + Y_d N^{\dot{d}}_{\delta_e}}{s\Delta + sG_{\theta}N^{\theta}_{\delta_e} + G_d N^{\dot{d}}_{\delta_e}}\right]\frac{(n-1)!\, d_n}{s^{n-1}} \tag{11-15}$$

For the system to follow a command path of $n^{\text{th}}$ degree with no steady state error, the numerator of the bracketed expression in Eq. 11-15 must contain a free $s^n$.

From Fig. 11-9 it is apparent that a system that is stabilized on the first segment of a two-segment glide path system must follow a ramp function in $d$ without steady-state error if it is successfully to make the

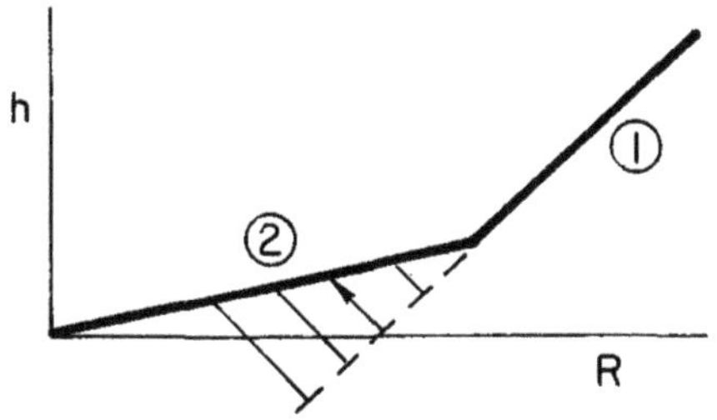


Fig. 11·9. Two-segment glide path.

transition from paths 1 and 2. So a free $s^2$ is needed in the bracketed portion of Eq. 11-15. Using the same limiting properties as in the beam acquisition case, we find that

$$\lim_{s\to 0}[\quad] = \lim_{s\to 0}\left[\frac{s[E_\Delta + G_\theta(s)C^\theta_{\delta_e}] + Y_d(s)D^d_{\delta_e}}{G_d(s)D^d_{\delta_e}}\right] \tag{11-16}$$

Typically, $Y_d$ is either zero or contains a single free $s$, so that the total numerator in Eq. 11-16 has a net free $s$. Then, to provide the second numerator free $s$ (needed to satisfy the steady-state requirement), the path deviation/elevator transfer function must have an integral term, $K_{\bar{d}}/s$. This is also obvious intuitively from an examination of Fig. 11-7. There it is plain that a steady-state signal must be developed at the $\dot{d}_c$ point to offset, in this case, a change in the steady-state output of the baro-inertial smoothing and equalization block.

*Regulation Requirements*

The path deviation response to an external disturbance $\eta$ is given by

$$d_\eta = \frac{N^d_\eta + G_\theta N^{d\theta}_{\eta\delta_e}}{\Delta_{sys}}\eta \tag{11-17}$$

For our purposes here, $\eta$ may be a $w_g$ or $u_g$ wind disturbance or a $Z_0$ or $M_0$ trim change. Any of these can have a constant component, $\eta(s) = \eta_1/s + \cdots$, for example. For the constant component to have no long-term effect requires

$$d_{\eta_{ss}} = \lim_{s\to 0} s\left(\frac{N^d_\eta + G_\theta N^{d\theta}_{\eta\delta_e}}{\Delta_{sys}}\right)\eta(s)$$

$$= \eta_1 \lim_{s\to 0}(\quad) = 0 \tag{11-18}$$

This will be zero when there is at least one net free $s$ in the numerator of the bracketed expression.

*Case 1:* $\eta_1 = w_g$. In terms of the Table 11-5 identities,

$$d_{w_{g_{ss}}} = w_g \lim_{s\to 0} \left(\frac{D^d_{w_g} + G_\theta B^{d\ \theta}_{w_g\delta_e}}{G_d D^d_{\delta_e}}\right) \tag{11-19}$$

Because the numerator is a constant as $s \to 0$, the free $s$ needed to make the whole expression approach zero must come from the denominator. Consequently, a $K_d/s$ component is needed in the $G_d$ control path. A similar argument applies for trim changes resulting in $Z_0$ and $M_0$ lift and pitching accelerations applied to the vehicle. Note also that any shear component in $w_g$ will result in a steady-state error, even with an integral controller.

*Case 2:* $\eta_1 = u_g$. Again, in terms of the Table 11-5 coefficients and noting that $D^d_{u_g}$ and $B^{d\ \theta}_{u_g\delta}$ are identically zero,

$$d_{u_{g_{ss}}} = u_g \lim_{s\to 0} s\left(\frac{C^d_{u_g} + G_\theta A^{d\ \theta}_{u_g\delta_e}}{G_d D^d_{\delta_e}}\right) \tag{11-20}$$

A free $s$ occurs naturally in the numerator of this expression, so that for $d_{u_{g_{ss}}}$ to be zero requires only that $\Delta_{\rm sys}(0)$ be a constant or of lower degree in $s$, which means that $G_d(s)]_{s\to 0} \to K_d$ is satisfactory. However, head or tail winds invariably have a shear component, so that a $K_{\bar{d}}/s$ is desirable in order to help counter these effects.

These elementary considerations indicate that the most undesirable disturbance inputs are wind shears. In principle, the worst is a shear normal to the flight path, for this will cause a steady-state error in path following. In practice, however, all shears occurring near the terminal condition (landing) give cause for concern because their effects are countered primarily by the integral control. This is inherently slow in action, as will be appreciated better with the aid of the numerical example of the next article. The promise of zero steady-state error is, accordingly, more academic than real, in that it may only be redeemed over a fairly long period of time.

Additional steady-state requirements on regulation against disturbances may be derived when attention is confined to short periods of time. This can be done by considering the two degree of freedom short period characteristics of the aircraft, instead of the complete three degree of freedom equations. Any final values found using the short period equations apply only for time intervals that are large compared with the system's settling time but not so long as to be comparable with phugoid periods. Thus the short period approximation may be valuable for considering some mid-frequency response properties.

Using the short period approximation (see Table 11-6), we obtain

$$d_\eta = \left(\frac{Z_\eta\{s^2 - (M_q + M_{\dot{\alpha}})s - [M_\alpha - (M_\eta/Z_\eta)Z_\alpha]\} + G_\theta(M_{\delta_e}Z_\eta - Z_{\delta_e}M_\eta)}{\Delta_{\text{sys}}}\right)\eta \tag{11-21}$$

This equation is most pertinent for a $w$-gust disturbance. When $\eta_1 = w_g$,

$$d_{w_{g_{ss}}} = w_g \lim_{s\to 0}\left\{\frac{Z_w s[s - (M_q + M_{\dot{\alpha}})] + G_\theta(M_{\delta_e}Z_w - Z_{\delta_e}M_w)}{\Delta_{\text{sys}}}\right\} \tag{11-22}$$

Also, $\Delta_{\text{sys}}(0)$ will be a constant if $K_d$ is the lowest frequency feedback or if the time span considered is relatively short such that a $K_{\bar{d}}/s$ control will have little effect. Then, to obtain a net free $s$ in the numerator, $G_\theta$ must contain one. Thus we can establish a desire for pitching velocity, rather than attitude feedback, to improve the gust regulation at mid-frequencies. This permits the drift from the beam caused by a normal gust to be reduced by virtue of the aircraft's weathercocking tendency. This would not be possible if the aircraft maintained its pitch attitude rigidly.

From the above steady-state (long time, as well as short time) considerations, we have shown that the minimum forms of the feedbacks

Table 11-6. Short period equations and transfer functions.

Equations of Motion:

$$[s - Z_w]sd + Z_\alpha\theta = Z_{\delta_e}\delta_e + Z_\eta\eta$$

$$(M_{\dot{w}}s + M_w)sd + [s^2 - (M_q + M_{\dot{\alpha}})s - M_\alpha]\theta = M_{\delta_e}\delta_e - M_\eta\eta$$

Characteristic Function:

$$\Delta = s^2\{s^2 - (M_q + M_{\dot{\alpha}} + Z_w)s - (M_\alpha - Z_wM_q)\}$$

Numerator[a]

$$N_\delta^d = -Z_\delta\left[s^2 - (M_q + M_{\dot{\alpha}})s - \left(M_\alpha - \frac{M_\delta}{Z_\delta}\right)Z_\alpha\right]$$

$$N_\eta^d = Z_\eta\left[s^2 - (M_q + M_{\dot{\alpha}})s - \left(M_\alpha - \frac{M_\eta}{Z_\eta}\right)Z_\alpha\right]$$

$$N_\delta^\theta = s\{[M_\delta + Z_\delta M_{\dot{w}}]s - (M_\delta Z_w - Z_\delta M_w)\}$$

$$N_\eta^\theta = -s\{[M_\eta + Z_\eta M_{\dot{w}}]s - (M_\eta Z_w - Z_\eta M_w)\}$$

Coupling Numerators:

$$N_{\eta\delta_e}^{d\theta} = M_{\delta_e}Z_\eta - Z_{\delta_e}M_\eta$$

Modal Response Ratio:

$$\left[\frac{U_0\theta}{d}\right]_{s_i} = \left[\frac{1}{-Z_w}(s - Z_w)\right]_{s_i}$$

[a] For these short period equations it is possible to define $N_\delta^d$ without having to resort to numerators that are ratios of polynomials.

needed to satisfy reasonable steady-state guidance, control, and regulation requirements are

$$G_d = \frac{K_{\dot{d}}}{s} + K_d \tag{11-23}$$

$$G_\theta = K_{\dot{\theta}} s \tag{11-24}$$

We shall now turn to a short discussion of the higher frequency feedbacks. The simplest of these is the attitude transfer function, $G_\theta$. When the short term attitude regulation requirement is considered—it implies attitude stiffening at short period frequencies so as to provide a craft that is stable and solid in attitude—the $G_\theta$ of Eq. 11-24 is modified to a simple washout. This retains the free $s$ in the numerator of $G_\theta$, with its favorable consequences for regulating against wind disturbances, while it still provides an attitude feedback at short period frequencies. The washout time constant $1/T_{wo}$ must, of course, be such that $1/T_{wo}$ is less than $\omega_{sp}$. Further, taking higher frequency effects into account, a pitch rate feedback $K_{\dot{\theta}}s$ would be desirable to provide greater short period attitude damping and attitude loop bandwidth. As will be seen later, the latter permits improved deviation loop gain margins and, thereby, a greater outer loop bandwidth. Consequently, the desired general form for the pitch attitude feedback is

$$G_\theta = \frac{K_\theta s}{s + (1/T_{wo})} + K_{\dot{\theta}} s \tag{11-25}$$

The remaining requirement for the general form of the feedback controller dynamics can be derived by considering the damping of the path mode. In general, as has already been pointed out, this requires a term such as $\dot{d}$, or its near equivalents at low frequency, $\dot{h}$ or $\theta$. (Recall that $\dot{h} \doteq U_0\theta$ if $\dot{w} \doteq 0$.) With the attitude feedback washed out at low frequencies, $K_{\dot{d}}$ or $K_{\dot{h}}$ feedbacks are the only remaining possibilities for the improvement of the path damping. These are nearly equivalent, in practice, so that the path deviation feedback transfer function can be simply represented as

$$G_d = \frac{K_{\dot{d}}}{s} + K_d + K_{\dot{d}} s \tag{11-26}$$

A general summary of the feedbacks, their purposes and the qualitative requirements are displayed in Table 11-7.

*Detailed System Analysis*

As a concrete example we shall now consider the controller forms given by Eqs. 11-25 and 11-26 as applied to a DC-8 aircraft in landing approach.

Table 11-7. Purposes and qualitative requirements for longitudinal feedbacks.

| Feedback terms | Functions | Requirements | Remarks |
|---|---|---|---|
| $K_\theta, K_{\dot\theta}, T_{w_0}$ | Short period attitude stiffness | $\theta \to \delta_e$ in short period frequency range<br>$(1/T_{w_0} < \omega_{sp})$ | |
| | Short period damping; path loop bandwidth extension capability | $\dot\theta \to \delta_e$ in short period frequency range | Windproofing and attitude stiffening conflict |
| | Short term $w_g$ windproofing | $s[s - (M_q + M_{\dot{z}})] + G_\theta M_{\delta_e}$<br>Small over $w_g$ frequencies | |
| $K_{\bar{d}}$ | Higher order path following<br>Trim<br>Windproofing ($w_g$ step) | $K_d/K_{\bar{d}}$ sets trim response time | |
| $K_d$ | Path acquisition and stiffness<br>Windproofing ($u_g$ step, $w_g$ pulse) | $K_d$ sets dominant path mode frequency; made as large as possible consistent with stability and limiting | Altitude control bandwidth with $d$, $\theta \to \delta_e$ is limited by $1/T_{\theta_2}$; can be increased significantly by $d$, $\dot d$, or $\dot h \to \delta_f$. Therefore $\delta_f$ can be big help in approach and flare precision. |
| $K_{\dot d}$ or $K_{\dot h}$ | Path damping | Sets path mode damping ratio<br>$0.64 < \zeta_{\text{path mode}} < 1$[a] | $\dot d$ suffers from beam noise; $\dot h$ requires a bias to offset steady-state sink rate. |
| or $K_\theta$ | | $\theta \to \delta_e$ in long period frequency range | Conflicts with windproofing. |

[a] Broader limits, e.g., $0.2 \leqq \zeta_{\text{path mode}} < 2.0$, were recommended by Special Committee SC-79 of the Radio Technical Commission for Aeronautics. See "Standard Performance Criteria for Autopilot/Coupler Equipment," Paper 31-63/DO-118, Radio Technical Commission for Aeronautics, Washington, D.C., March 14, 1963. These limits reflect, to some extent, the path damping ratio attainable without beam compensation for range, as well as an emphasis on automatic approach rather than automatic approach and landing.

Table 11-8. DC-8 parameters for landing approach configuration.

| Geometry and inertial properties | | Stability derivatives | |
|---|---|---|---|
| $h$ (ft) | 0 | $X_u$ (1/sec) | −0.0373 |
| $M$ (−) | 0.204 | $X_w$ (1/sec) | 0.136 |
| $V_{T_0}$ (ft/sec) | 228. | $X_{\delta_e}$ (ft/sec²/rad) | 0 |
| $\gamma_0$ (degree) | −2.8° | $Z_u$ (1/sec) | −0.283 |
| $q$ (lb/ft²) | 61.8 | $Z_w$ (1/sec) | −0.750 |
| $S$ (ft²) | 2,758. | $Z_{\delta_e}$ (ft/sec²/rad) | −9.25 |
| $b$ (ft) | 142.4 | $M_u$ (1/sec-ft) | 0 |
| $c$ (ft) | 22.16 | $M_w$ (1/sec-ft) | −0.00461 |
| $W$ (lb) | 180,000 | $M_{\dot{w}}$ (1/ft) | −0.00085 |
| $m$ (slugs) | 5,580 | $M_q$ (1/sec) | −0.594 |
| $I_x$ (slug-ft²) | $3.8 \times 10^6$ | $M_{\delta_e}$ (1/sec²) | −0.923 |
| $x_{cg}$ (% c) | 25.2 | $M_\alpha$ (1/sec²) | −1.05 |
| $\delta_{f_0}$ (degree) | 50 | $M_{\dot{\alpha}}$ (1/sec) | −0.1936 |

The dynamics of the aircraft are defined by the landing approach configuration parameters given in Table 11-8. The aircraft transfer function characteristics for control inputs are shown in the $j\omega$-Bode plots of Fig. 11-10. The notation in the numerical transfer functions shown on these plots is a shorthand in which $(1/T)$ represents a first-degree factor $(s + 1/T)$ and $[\zeta, \omega_n]$ represents a second-degree factor $[s^2 + 2\zeta\omega_n s + \omega_n^2]$. For example, in Fig. 11-10, $1/T_{\theta_1} = 0.101$, $1/T_{\theta_2} = 0.646$, $\zeta_{sp} = 0.626$ and $\omega_{sp} = 1.231$. This same notation is used in the complete compilation of transfer functions and coupling numerators given in Table 11-9.

The control equation, based on the requirements of the previous article, is

$$\delta_e = \left[\frac{K_{\bar{d}} + K_d s}{s(T_f s + 1)} d + K_{\dot{h}} \dot{h} + \frac{K_\theta T_{wo} s}{T_{wo} s + 1} \theta + K_{\dot{\theta}} \dot{\theta}\right] \frac{1}{(T_a s + 1)} \tag{11-27}$$

A block diagram for the system is shown in Fig. 11-11. The beam deviation signal is filtered by a low pass filter with time constant $T_f$ (taken as 0.5 sec) and then put through parallel integrator and proportional channels. Path damping, pitch rate, and washed-out attitude feedbacks are as already described. The actuator characteristics will be approximated by a first-order lag with $1/T_a$ equal to 15 sec$^{-1}$. Although grossly oversimplified, this should be an adequate approximation to the actuator properties for the low frequency range of primary interest here.

For simplicity we can replace $\dot{h}$ by $\dot{d}$ in Eq. 11-27 with only very small errors in considering perturbed conditions (the steady-state component of the $K_{\dot{h}} \dot{h}$ signal will be offset by part of the integrator's output and then the difference between $\dot{h} - \dot{h}_{ss}$ and $\dot{d}$ will be relatively minor). This substitution reduces by one the number of loop closures that need to be

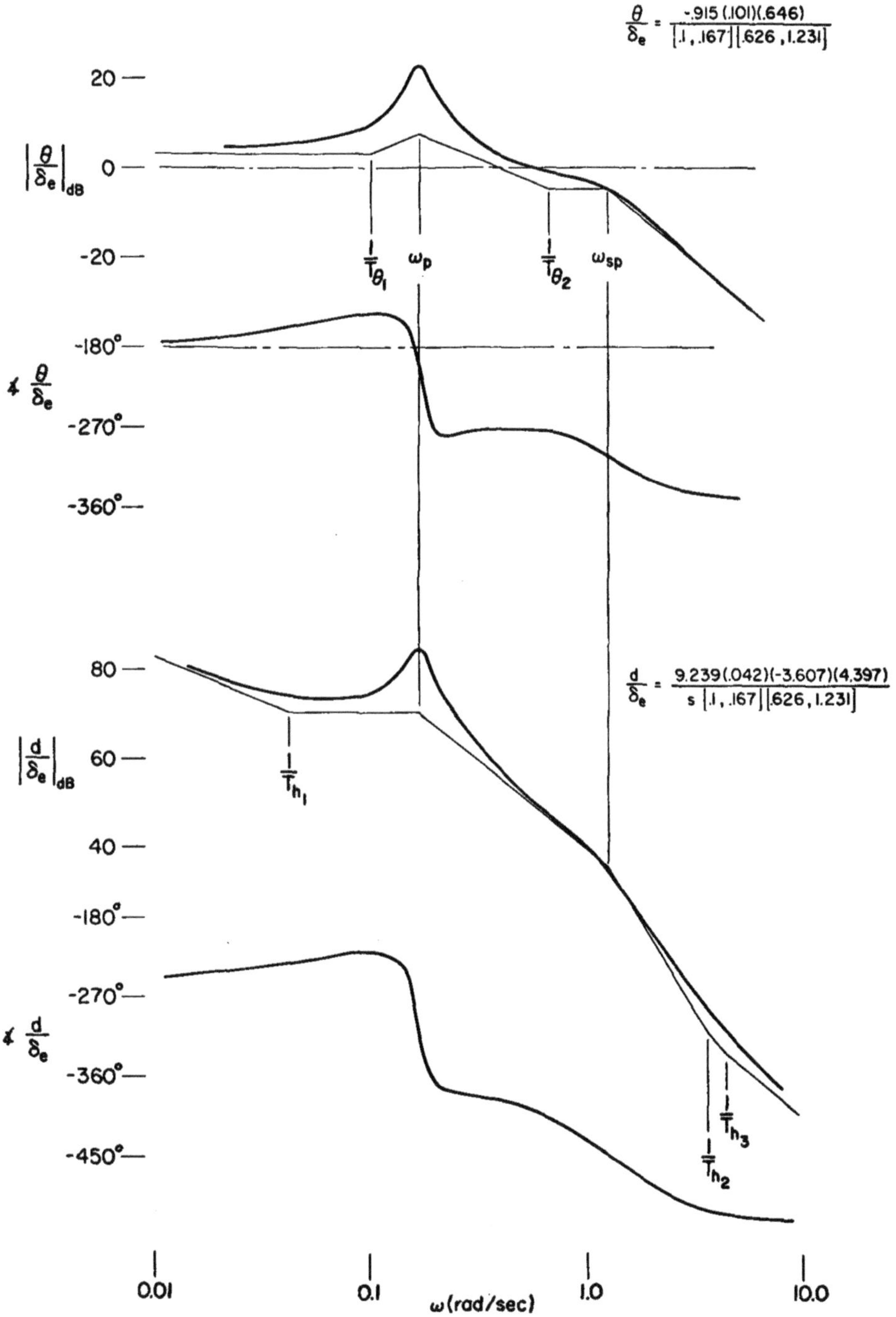


Fig. 11-10. Bode plots of elevator input transfer functions for the airframe alone.

Table 11-9. Longitudinal transfer functions for the DC-8 in a landing approach configuration.

Denominator:

$$\Delta\ [0.10,\ 0.167][0.626,\ 1.231]$$

Numerators:

$\delta_e$ control input

$$N^{u}_{\delta_e} = -1.258(4.03)(-4.082)$$
$$N^{w}_{\delta_e} = -9.25(23.34)[0.107,\ 0.198]$$
$$N^{\theta}_{\delta_e} = -0.9151(0.101)(0.646)$$
$$N^{\dot h}_{\delta_e} = 9.239(0.042)(-3.607)(4.397)$$
$$N^{\dot d}_{\delta_e} = 9.25(0.035)(-3.606)(4.396)$$

$u_g$ gust input

$$N^{u}_{u_g} = 0.0373(1.543)[0.599,\ 0.857]$$
$$N^{w}_{u_g} = 0.283(0)(0)(0.594)$$
$$N^{\theta}_{u_g} = -0.0002406(0)(5.424)$$
$$N^{\dot h}_{u_g} = -0.2845(0.007)[0.386,\ 1.027]$$
$$N^{\dot d}_{u_g} = -0.283[0.384,\ 1.025]$$

$w_g$ gust input

$$N^{u}_{w_g} = -0.1360(0)[0.486, 0.795]$$
$$N^{w}_{w_g} = 0.3498(4.379)[0.118,\ 0.166]$$
$$N^{\theta}_{w_g} = -0.001755(0)(-1.475) \times (-0.00464)$$
$$N^{\dot h}_{w_g} = -0.7425(1.243)[0.081,\ 0.214]$$
$$N^{\dot d}_{w_g} = -0.75(1.234)[0.091,\ 0.214]$$

Coupling Numerators:

$$N^{\theta\ u}_{\delta_e u_g} = -0.03413(1.751)$$
$$N^{\theta\ w}_{\delta_e u_g} = -0.2612(0)$$
$$N^{u\ w}_{\delta_e u_g} = 0.345(1.102)(22.243)$$
$$N^{\theta\ u}_{\delta_e w_g} = 0.1223(0)$$
$$N^{\theta\ w}_{\delta_e w_g} = -0.01624(0.092)(39.955)$$
$$N^{u\ w}_{\delta_e w_g} = -1.258(-0.696)(23.626)$$

analyzed. The control equation may then be represented as

$$\delta_e = \left[\frac{K_{\bar d} + K_d s}{s(T_f s + 1)}\, d + K_{\dot d}\, \dot d + \frac{K_\theta T_{wo} s}{(T_{wo} s + 1)}\, \theta + K_{\dot\theta}\dot\theta\right]\frac{1}{(T_a s + 1)}$$

$$= \left\{\frac{K_{\dot d}[s^3 + (1/T_f)s^2 + (K_d/K_{\dot d})s + (K_{\bar d}/K_{\dot d})]}{s(s + 1/T_f)}\, d + \frac{K_{\dot\theta} s[s + (K_\theta/K_{\dot\theta} + 1/T_{wo})]}{(s + 1/T_{wo})}\, \theta\right\}\frac{1}{(T_a s + 1)} \tag{11-28}$$

$$= \left\{\frac{K_{\dot d}(s + 1/T_{d_1})(s + 1/T_{d_2})(s + 1/T_{d_3})}{s(s + 1/T_f)}\, d + \frac{K_{\dot\theta} s(s + 1/T_E)}{(s + 1/T_{wo})}\, \theta\right\}\frac{1}{(T_a s + 1)}$$

The lead equalization in attitude, $1/T_E$, as may be seen in the development above, is

$$\frac{1}{T_E} = \frac{K_\theta}{K_{\dot\theta}} + \frac{1}{T_{wo}} \tag{11-29}$$

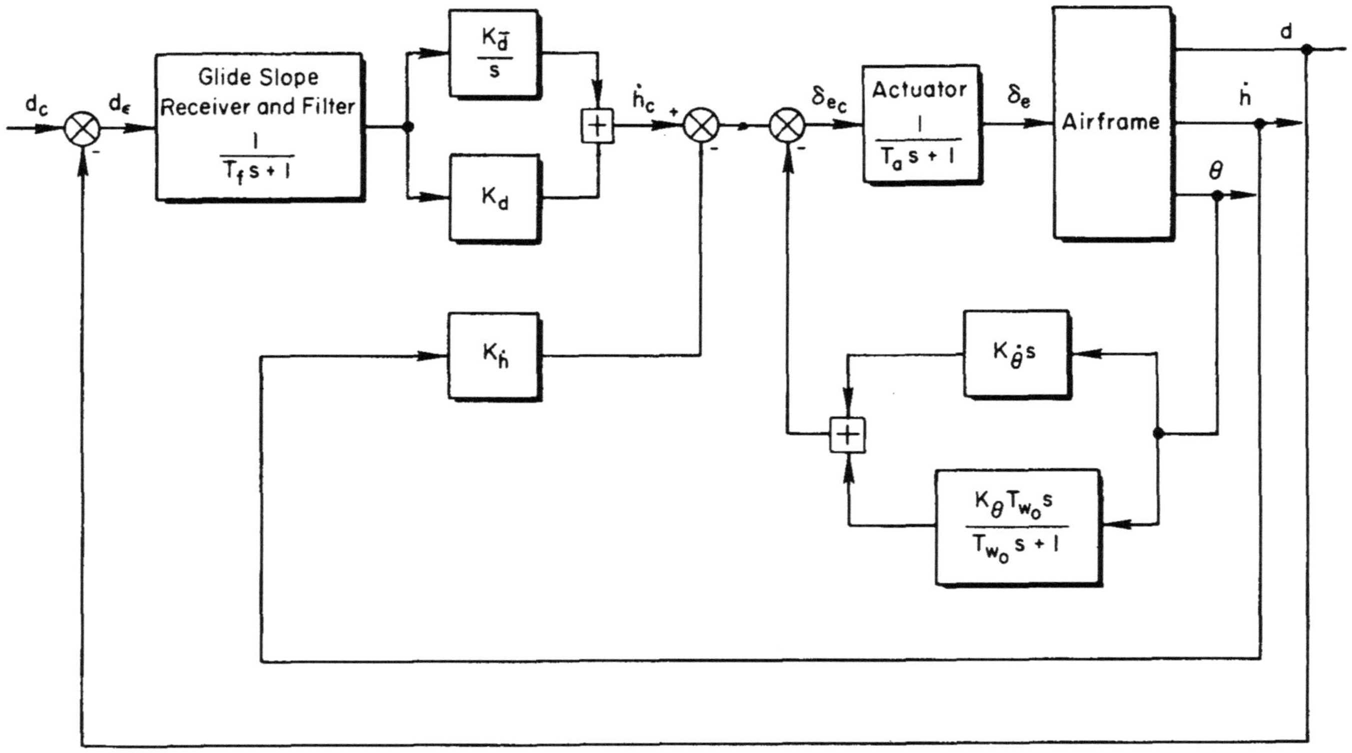


Fig. 11-11. Block diagram of advanced system.

The path deviation numerator time constants are not simply related to the feedback and filter gains, but for conventional values they may be factored approximately as

$$\frac{1}{T_{d_1}} \doteq \frac{K_{\bar{d}}}{K_d}, \qquad \frac{1}{T_{d_2}} \doteq \frac{K_d}{K_{\bar{d}}}, \qquad \frac{1}{T_{d_3}} \doteq \frac{1}{T_f} \tag{11-30}$$

With the system defined qualitatively, we shall now turn to its quantitative description. First, the closed-loop dynamic characteristics will be considered in terms of $j\omega$-Bode root locus plots for the equalization as adjusted. Then, typical time histories of the system will be discussed for command and disturbance inputs.

The attitude loop closure analysis is presented in Fig. 11-12. Here the attitude washout is made somewhat less than the short period undamped natural frequency (i.e., $1/T_{wo} = 0.7$ sec$^{-1}$, $\omega_{sp} = 1.231$ sec$^{-1}$) so as to assure nearly pure gain attitude feedback at the short period frequency. If, in addition, the washout inverse time constant is placed near the singularity at $1/T_{\theta_2}$, the closed-loop root, $1/T'_{wo}$, starting at the washout inverse time constant at low gain and proceeding to $1/T_{\theta_2}$ as gain is increased, will form a dipole pair with the $1/T_{\theta_2}$ singularity. This will effectively remove any $1/T'_{wo}$ mode from the short term $\theta$ response to a $\theta$ command or a similar input. Then the second-order, closed-loop short period mode alone will be dominant in the attitude response.

Location of the lead equalization breakpoint, $1/T_E$, is the primary means available for adjusting the modified short period damping ratio. If this breakpoint is placed at a frequency somewhat greater than $\omega_{sp}$ (at 1.7 sec$^{-1}$ in Fig. 11-12), a long stretch of $-20$ dB per decade slope will be established between $1/T_E$ and the actuator breakpoint at $1/T_a$. Gain crossover anywhere in this stretch will result in excellent damping of the short period mode, while it still permits reasonably large values of the (open-loop) amplitude ratio at mid-frequencies.

The Bode root locus on Fig. 11-12 also shows that the phugoid and short period damping ratios and the short period undamped natural frequency are all increased with an increase in gain, whereas the phugoid undamped natural frequency is decreased. This decrease of phugoid frequency with gain is a result of the washout of $\theta$ at low frequency.

Because there is a wide range of available gains compatible with good closed-loop attitude characteristics, the precise gain that is chosen will have relatively wide tolerances. A nominal value of $K_\theta = -2$ sec (as shown in Fig. 11-12) will be used. This is compatible with the following:

1. Excellent closed-loop short period damping and undamped natural frequency, i.e., $\zeta'_{sp} = 0.778$, $\omega'_{sp} = 2.312$ sec$^{-1}$.
2. Good phugoid damping, $\zeta'_p = 0.409$.

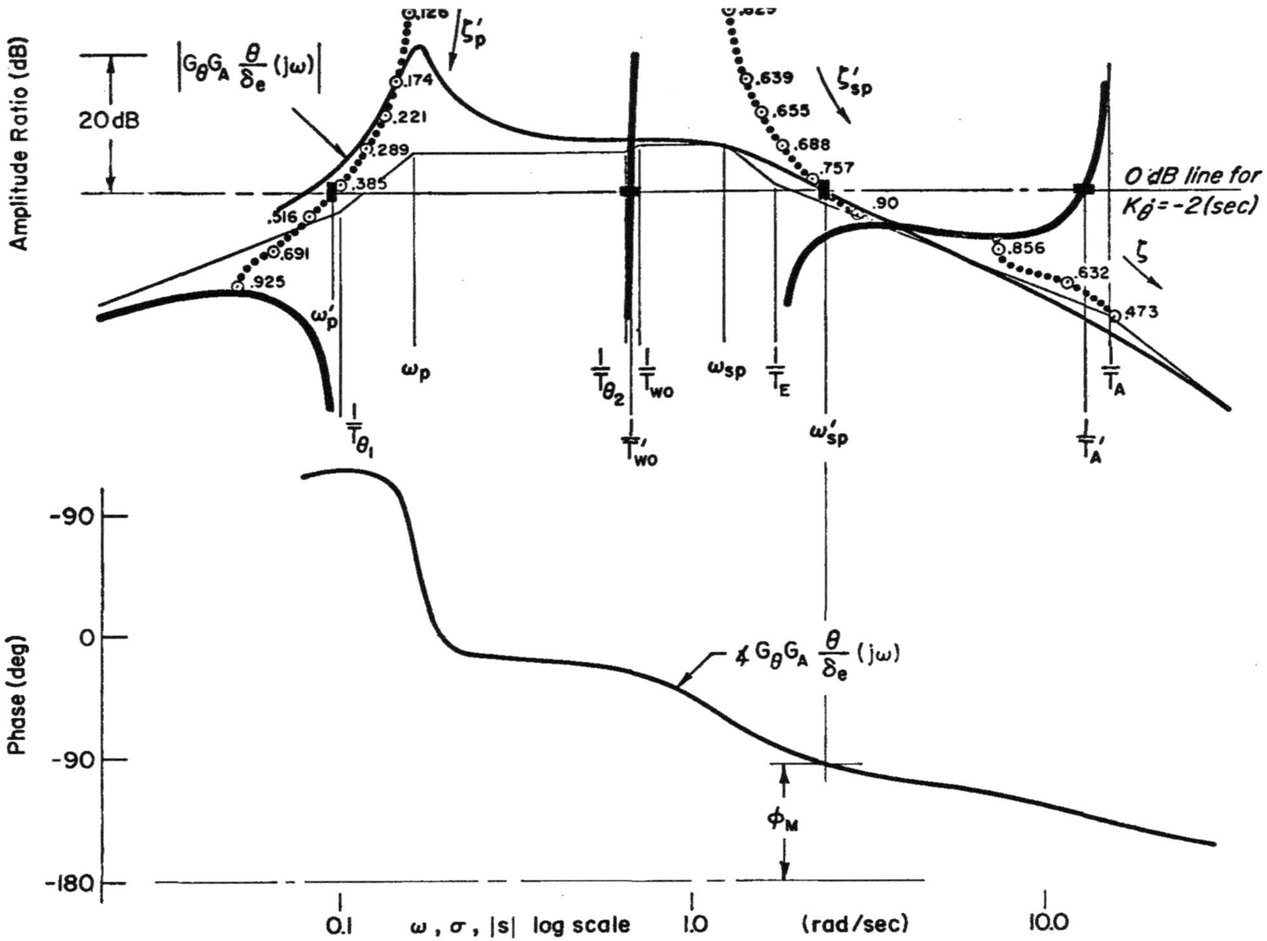


Fig. 11-12. Bode root locus for attitude loop closure.

3. Near cancellation of $1/T'_{wo}$ and $1/T_{\theta_2}$.
4. Good phase margin, $\varphi_M = 88$ degrees, with consequent insensitivity to parasitic nonlinearities and margin for additional high frequency system lags not considered explicitly in this analysis.
5. Excellent proportional gain sensitivity (recall Eq. 3-103) in connection with both phugoid and short period dominant modes:

$$M_{K\theta}^{sp'} \doteq +0.425 + 0.42j = 0.598\underline{/+44.66^\circ}$$

$$S_{K\theta}^{sp'} \doteq 1.36\underline{/+186^\circ}$$

$$M_{K\theta}^{p'} \doteq -0.37 + 0.254j = 0.449\underline{/+145.5}$$

$$S_{K_\theta}^{p'} \doteq 0.0438\underline{/-99^\circ}$$

6. Near-maximum available path deviation loop bandwidth (this is, it will turn out, primarily dependent on $\zeta'_{sp}$ and $\omega'_{sp}$).[6]

Consider now the beam deviation closure. Here, as given by Eq. 11-28, the equalization has both a rate and an integral term, in addition to the glide-slope-beam noise filter. The poles of the glide slope deviation to elevator transfer function are, of course, modified by the closure of the $\theta$ to $\delta_e$ loop (i.e., to $\zeta'_p$, $\omega'_p$; $\zeta'_{sp}$, $\omega'_{sp}$; $1/T'_{wo}$; $1/T'_A$), while the numerator ($1/T_{h_1}$, $1/T_{h_2}$, $1/T_{h_3}$) is still the same as the numerator of the airframe alone. At very low frequencies, the total open-loop system comprising beam deviation controller and airframe modified by the pitch attitude closure, appears as a double integrator [$G(s) = K/s^2$] because of the integral term in the deviation controller. The rest of the equalization is needed to adjust the amplitude ratio to approximate a $-20$ dB per decade slope in the desired region of crossover. While the $d$ equalization has three time constants available, $1/T_{d_3}$ is intrinsically close to the beam noise filter breakpoint, so that the best that can be achieved with it is to create a dipole pair near $1/T_f$ which yields a small phase lead in the region of the filter breakpoint. The principal remaining adjustments are then $1/T_{d_1}$ and $1/T_{d_2}$. When these breakpoints are positioned relative to the $\omega'_p$ breakpoint as shown in Fig. 11-13, the low frequency amplitude ratio attainable with a given zero dB line location is made nearly as large as possible and a very long stretch of approximately $-20$ dB per decade slope is achieved in the higher frequency amplitude ratio plot, starting at a frequency somewhat greater than $1/T_{d_1}$. Note, however, that the placement of $1/T_{d_1}$ and

[6] The reader is reminded that the primes used here, for example, on $\zeta'_{sp}$, $\omega'_{sp}$, and $1/T'_{wo}$, signify parameters that have been *modified* by a *single* loop closure. The particular loop that has been closed is not indicated by the notation and must be inferred from the context. Later we shall have occasion to use *double* primes which, of course, indicate that the parameters have been modified by two loop closures. The process may be continued if necessary.

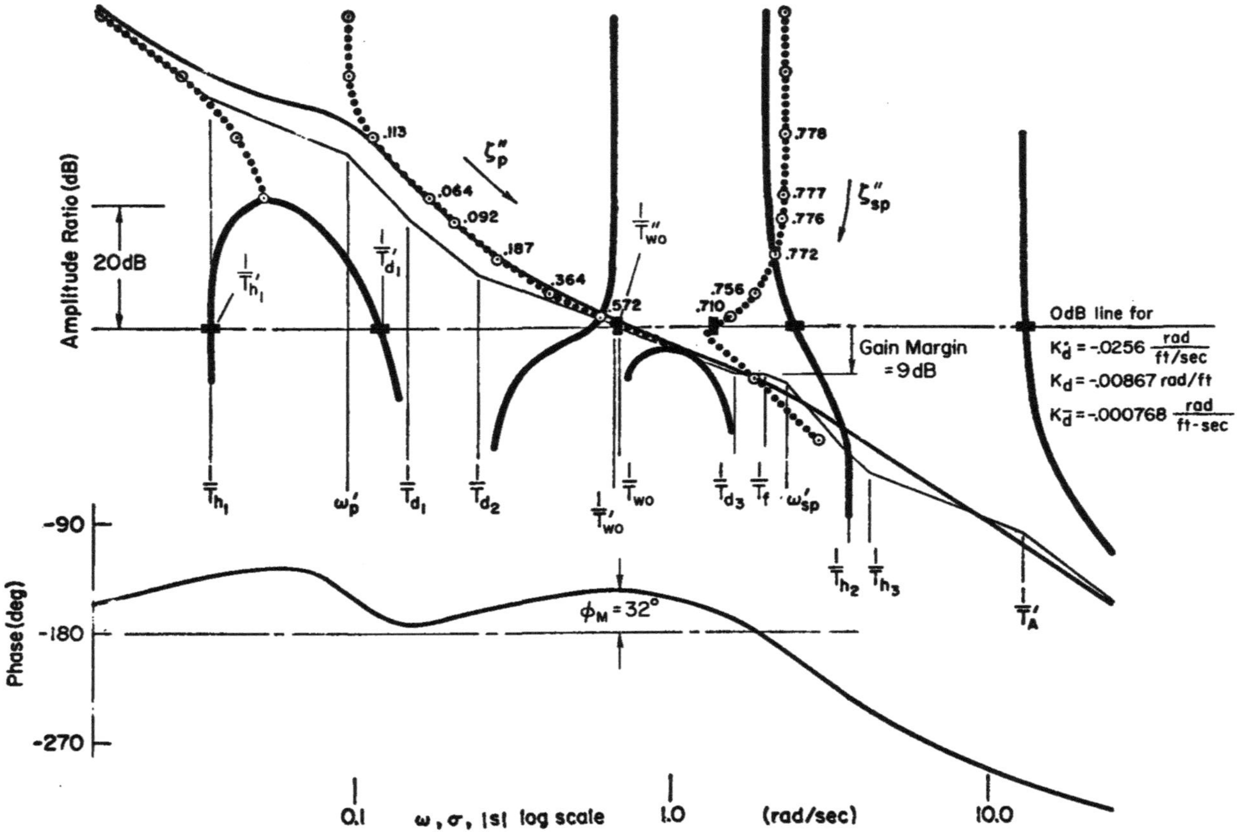


Fig. 11-13. Bode root locus for beam deviation loop closure.

$1/T_{d_2}$ should be such that the phase in the low to medium frequency region is not permitted to exceed −180 degrees. This is important to avoid a low frequency, low amplitude oscillation because of any system threshold or hysteresis characteristics. Either of these will result in reduced gain at very low amplitudes which could set the conditions for a *limit cycle* at a frequency at which the phase angle reaches −180 degrees.[7]

The selection of the gain, $K_d$, and the final loop closure is based on a compromise considering the following factors:

1. A desire to have as high a bandwidth system (maximum $\omega_c$) as is possible, while retaining a well-damped, approximately second order, dominant response in $d$ to $d_c$. With the gain selected, the $\zeta_p''$, $\omega_p''$ second-order mode is dominant with the $1/T_{wo}''$ first-order mode present but having a relatively small modal response coefficient (inversely proportional to the $\sigma$-Bode plot slope at $1/T_{wo}''$).
2. Near-minimum sensitivity to variations in gain and in airframe characteristics. Note that the $\sigma$-Bode slope at $1/T_{h_1}'$ and $1/T_{d_1}'$ is large. The proportional sensitivities of the closed-loop dominant path mode ($\zeta_p''$, $\omega_p''$) and short period mode ($\zeta_{sp}''$, $\omega_{sp}''$) are then:

$$M_{K_\alpha}^{p''} \doteq +1.18 + 0.645j = 1.345\underline{/+28.64^\circ}$$

$$S_{K_{\dot{d}}}^{p''} \doteq 0.95\underline{/+160^\circ}$$

$$M_{K_\alpha}^{sp''} \doteq -0.77 - 0.437j = 0.885\underline{/-150.425^\circ}$$

$$S_{K_{\dot{d}}}^{sp''} \doteq 1.265\underline{/-18^\circ}$$

3. Adequate phase margin, $\varphi_M = 32^\circ$; and a large gain margin of 9 dB.

The complete closed-loop transfer functions are tabulated in the shorthand notation in Table 11-10.

*Transient Response Characteristics*

For the system analyzed in the previous article, typical transient responses for a deviation command input, $d_c$, and step $u_g$ and $w_g$ disturbances are shown in Figs. 11-14, 11-15, and 11-16.

The $d$ response to a $d_c$ step command is, as anticipated, dominated by the second-order path mode. The slight, persistent overshoot present is

[7] A limit cycle oscillation is a form of behavior that may occur only in nonlinear, not in linear systems. See, for example, D. Graham and D. McRuer, *Analysis of Nonlinear Control Systems*, Wiley, New York, 1961.

Table 11-10. Closed-loop transfer functions for longitudinal approach control system.

$$\frac{d}{u_g} = \frac{-0.283(0)(0)(0.174)(2.0)(12.918)[0.767,\ 2.215]}{(0.036)(0.123)(0.582)(2.462)(13.232)[0.657,\ 0.699][0.673,\ 1.428]}$$

$$\frac{d}{w_g} = \frac{-0.75(0)(2.0)(13.75)[0.464,\ 0.103][0.936,\ 2.018]}{(0.036)(0.123)(0.582)(2.462)(13.232)[0.657,\ 0.699][0.673,\ 1.428]}$$

$$\frac{d}{d_{\text{command}}} = \frac{-2.406(0.035)(0.089)(0.7)(-3.606)(4.396)}{(0.036)(0.123)(0.582)(2.462)(13.232)[0.657,\ 0.699][0.673,\ 1.428]}$$

$$\frac{u}{u_g} = \frac{-0.0373(0.136)(1.596)(2.777)(13.261)[0.5,\ 0.276][0.58,\ 1.918]}{(0.036)(0.123)(0.582)(2.462)(13.232)[0.657,\ 0.699][0.673,\ 1.428]}$$

$$\frac{u}{w_g} = \frac{-0.136(0.153)(0.215)(2.409)(13.262)[-0.082,\ 1.023][0.872,\ 1.492]}{(0.036)(0.123)(0.582)(2.462)(13.232)[0.657,\ 0.699][0.673,\ 1.428]}$$

$$\frac{u}{d_{\text{command}}} = \frac{0.3272(0)(0.089)(0.7)(-4.082)(4.03)}{0.036)(0.123)(0.582)(2.462)(13.232)[0.657,\ 0.699][0.673,\ 1.428]}$$

primarily because of the lead at $1/T_{d_2}$. With the gains selected, this response is, in general, entirely satisfactory. The short period properties show up primarily in the $w$ response and elevator traces. The attitude response closely parallels the altitude rate, as can be seen by comparing the $\dot{h}$ and $\theta$ traces. Finally, a small speed deviation occurs with a dominant time constant, given by the closed-loop mode characterized by the inverse time constant, $1/T'_{h_1}$, which approaches $1/T_{h_1}$. This is more graphically demonstrated in connection with the speed responses to $u_g$ and $w_g$ step disturbances which appear next in Figs. 11-15 and 11-16, respectively. These speed response transients reflect the absence of a speed control loop. The basic time constant of the speed deviation is limited by $1/T_{h_1}$. In terms of approximate factors, this is given by

$$\frac{1}{T_{h_1}} \doteq -X_u + (X_\alpha - g)\frac{Z_u}{Z_\alpha}\frac{(1 - Z_{\delta_e}M_u/M_{\delta_e}Z_u)}{(1 - Z_{\delta_e}M_\alpha/M_{\delta_e}Z_\alpha)} \tag{11-31}$$

When a speed-control system is considered as a stability augmenter, an $\alpha$ or $u$ to $\delta_T$ feedback will modify $X_u$ or $X_\alpha$; thus it can be used to change $1/T_{h_1}$. Then the deviation loop closure will result in a larger value of $1/T'_{h_1}$, and there will be a concomitant improvement in the closed-loop system speed response.

### *Steady-State Performance Characteristics with Random Gust Disturbances*

The average performance of the system when subjected to random excitation by gusts could be computed using the techniques described in

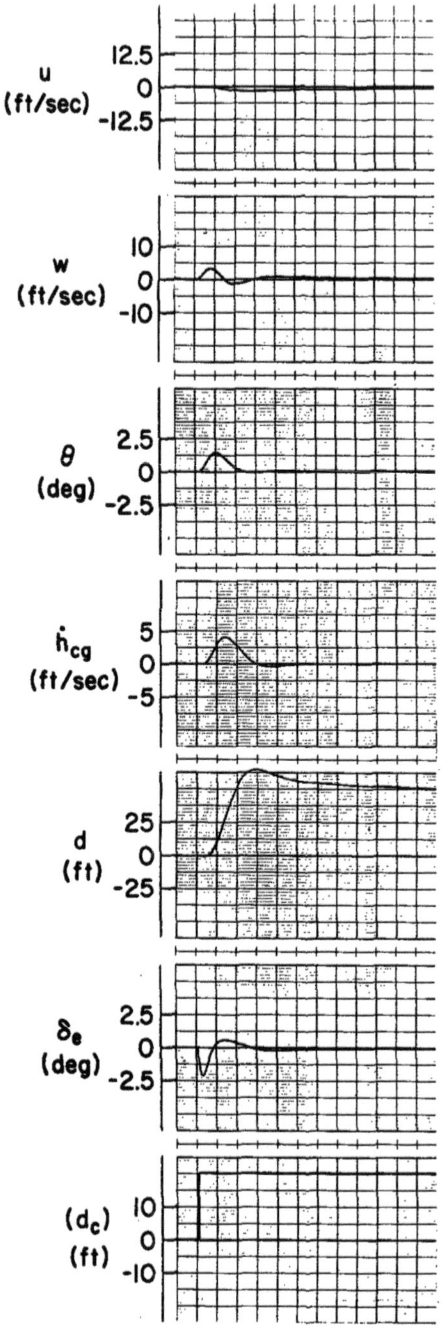


Fig. 11-14. Transient responses for a step deviation command, $d_c$.

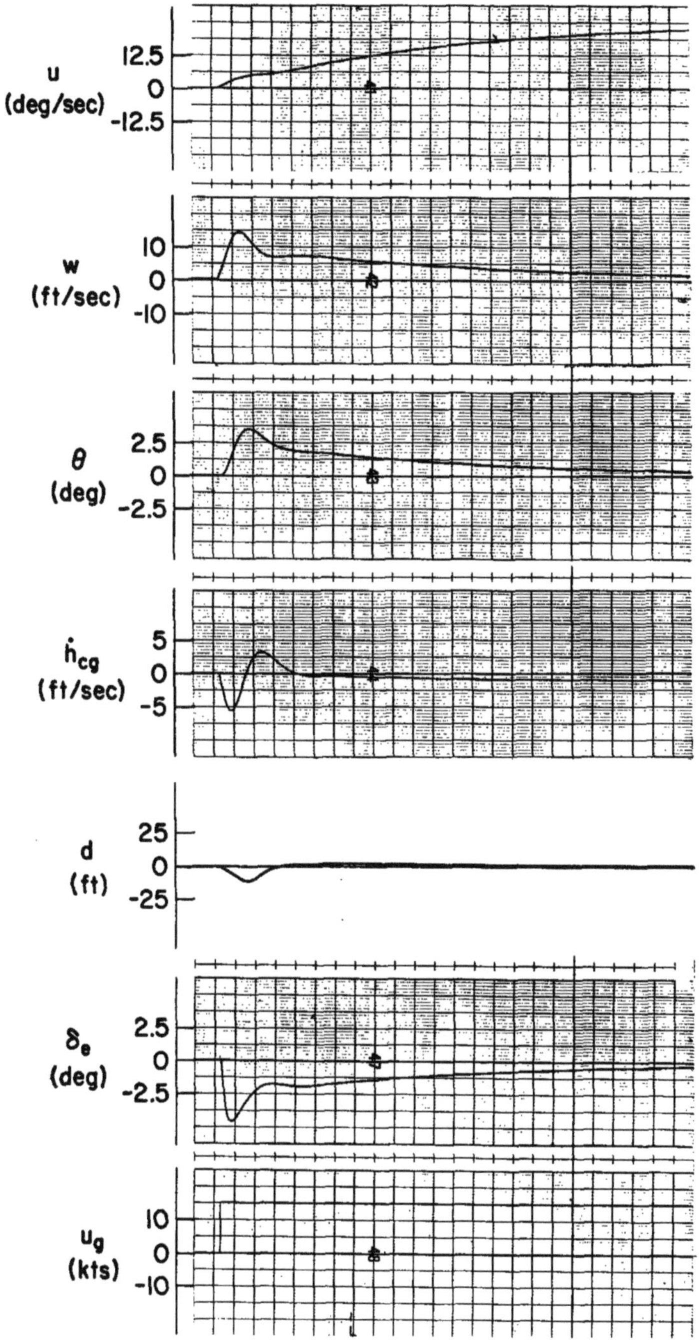


Fig. 11-15. Transient responses for a step $u_g$ gust input.

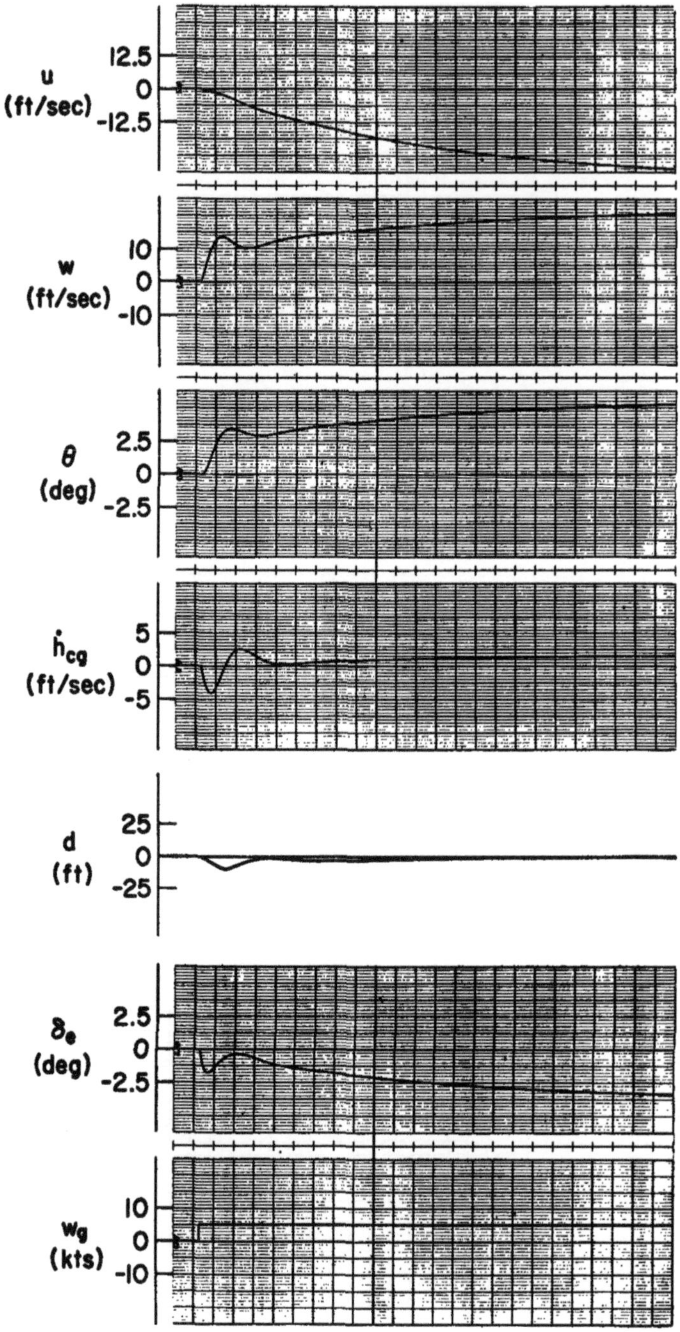


Fig. 11-16. Transient responses for a step $w_g$ gust input.

Section 10-6. The fundamental formulas are

$$\sigma_d^2]_{w_g} = \frac{1}{2\pi}\int_0^\infty \left|\left(\frac{d}{w_g}\right)_{\text{CL}}\right|^2 \Phi_{w_g}(\omega)\, d\omega \tag{11-32}$$

$$\sigma_d^2]_{u_g} = \frac{1}{2\pi}\int_0^\infty \left|\left(\frac{d}{u_g}\right)_{\text{CL}}\right|^2 \Phi_{u_g}(\omega)\, d\omega \tag{11-33}$$

$$\sigma_u^2]_{w_g} = \frac{1}{2\pi}\int_0^\infty \left|\left(\frac{u}{w_g}\right)_{\text{CL}}\right|^2 \Phi_{w_g}(\omega)\, d\omega \tag{11-34}$$

$$\sigma_u^2]_{u_g} = \frac{1}{2\pi}\int_0^\infty \left|\left(\frac{u}{u_g}\right)_{\text{CL}}\right|^2 \Phi_{u_g}(\omega)\, d\omega \tag{11-35}$$

The closed-loop transfer functions for substitution in these equations are already available from Table 11-9. In order to complete the required information, representative forms for the gust spectra, such as $\Phi_{w_g}(\omega)$, are required.

Spectral descriptions of gust characteristics have been intensively studied for the past two decades.[8] Although by no means universally accepted, considerable gust data from numerous sources have been integrated to produce a gust model appropriate for design purposes.[9] The basic form of this model assumes that the three velocity components $u_g$, $v_g$, and $w_g$ are mutually uncorrelated, so that analysis can be carried out using the three components of the gust model separately. While this assumption is justified for clear air turbulence at high altitude, it is not strictly correct at low altitudes because of prevailing anisotropy in the boundary layer. Nevertheless, at low altitudes the cross-correlations between the gust components are weak and are therefore disregarded in the model. The model further assumes a Gaussian distribution of gust velocities and a spectral shape that is the result of passing white noise through a first-order filter. With these assumptions the basic form of the model is defined by three power spectral densities that relate normalized gust intensities to spatial frequency. In terms of a temporal frequency, as

[8] J. K. Zbrozek, "Atmospheric Gusts—Present State of the Art and Further Research," *J. Roy. Aeron. Soc.*, 69, No. 649, 27–45 (Jan. 1965); Grant B. Skelton, "Investigation of the Effects of Gusts on V/STOL Craft in Transition and Hover," AFFDL-TR-68-85, Wright-Patterson Air Force Base, Ohio, Oct. 1968.

[9] C. R. Chalk, T. P. Neal, T. M. Harris, F. E. Pritchard, and R. J. Woodcock, "Background Information and User Guide for MIL-F-8785B(ASG), Military Specification—Flying Qualities of Piloted Airplanes," AFFDL-TR-69-72, Wright-Patterson Air Force Base, Ohio, Aug. 1969.

seen by an aircraft flying through the gust field, these spectra are given by

$$\Phi_{u_g}(\omega) = \sigma_u^2 \frac{2L_u}{\pi U_0} \frac{1}{[1 + (L_u\omega/U_0)^2]} \tag{11-36}$$

$$\Phi_{v_g}(\omega) = \sigma_v^2 \frac{L_v}{\pi U_0} \frac{1 + 3(L_v\omega/U_0)^2}{[1 + (L_v\omega/U_0)^2]^2} \tag{11-37}$$

$$\Phi_{w_g}(\omega) = \sigma_w^2 \frac{L_w}{\pi U_0} \frac{1 + 3(L_w\omega/U_0)^2}{[1 + (L_w\omega/U_0)^2]^2} \tag{11-38}$$

where $L_u$, $L_v$, $L_w$ is the scale length (ft); $U_0$ the aircraft's mean speed with respect to the air mass (ft/sec); $\omega$ the frequency (rad/sec); and $\sigma$ the standard deviation of gust velocity (the subscript indicates which orthogonal component).

Additional spectra are derivable from these basic spectra. One is the rolling gust spectrum which is derived from the apparent gradient of the $w_g$ gust across the wing span ($b$ ft):

$$\Phi_{p_g}(\omega) = \frac{\sigma_w^2}{4U_0L_w} \frac{\sqrt{\pi L_w/b}}{[1 + (20b\omega/\pi U_0)^2]} \tag{11-39}$$

Similar pitching and yawing gust spectra may also be derived in an analogous way, but the influences of pitching and yawing gusts on automatically controlled aircraft are very small and often completely neglected in design studies.

The procedure used to set the scale lengths for the spectra is based on extensive data fitting and the adjustment of the scales to make all three, i.e., $L_u$, $L_v$, and $L_w$, equal at an altitude of 1750 ft. The resulting scale lengths for clear air turbulence for the spectral forms, given above, below 1750 ft are

$$L_w = h \tag{11-40}$$

$$L_u = L_v = 145(h)^{1/3} \tag{11-41}$$

where $h$ and the $L$'s have dimensions of feet. The variation of $L_u$ and $L_v$ at low altitudes according to the $\frac{1}{3}$ power of altitude above ground level is a mechanism that forces the scales of the two horizontal gust components to be larger than the vertical scale. Although these formulas produce correct trends, there are little data available that can actually be used to substantiate the $h^{1/3}$ relationship.

The gust model that we have just discussed represents stationary, random gusts as stationary processes. However, the gusts actually encountered by a descending airplane are nonstationary because of the altitude dependence of the gust characteristics. Because the break points in the analytic expressions for the gust power spectra are nonlinear functions of altitude, the spectral characteristics of the gusts encountered by a descending airplane are actually nonlinear functions of time. In spite of these complicating factors, however, reasonably accurate performance calculations can still be made using Eqs. 11-32 to 11-35. This is justified as follows. Implicit in the calculation of variance of the output response of a linear system to a stationary random input is the assumption that the system has been operating for a sufficiently long (theoretically infinite) time so that all the transients have died out. For practical purposes, however, all that is really required is that steady-state conditions have been reached, i.e., that the system has been operating on the stationary input long enough so that the output has become approximately stationary. The time required for this to be the case depends, of course, on the system dynamics. For our airplane plus controller example, this time is relatively short—of the order of 5–10 sec (based on the settling time for step inputs).

As an analogous example, Fig. 11-17 shows a plot of the standard deviation in height error, $\sigma_h$, vs. time for an F4D-1 tracking a visual glide slope beam during a constant speed approach to an aircraft carrier.[10] To produce this plot, random beam motion was initiated at $t = 0$ sec. The

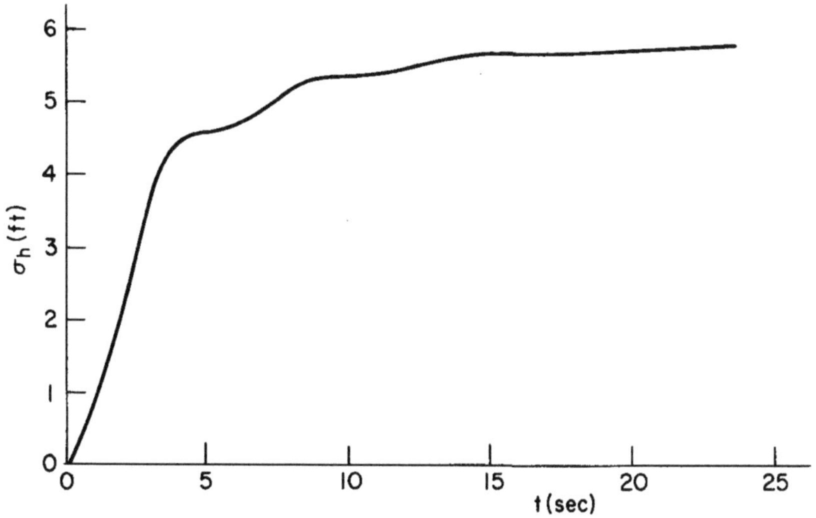


Fig. 11-17. Aircraft altitude excursions resulting from pilot's tracking the beam.

[10] T. S. Durand, and G. L. Teper, "An Analysis of Terminal Flight Path Control in Carrier Landing," Systems Technology, Hawthorne, Calif., Tech. Rept. 137-1(R), Aug. 1964.

plot shows that 80 percent of the steady-state airplane $\sigma_h$ is reached in 5 sec, and 90 percent in $7\frac{1}{2}$ sec.

Thus, although the mathematics requires a stationary input to be applied to a system for an infinite time, only the most recent 10 sec (or so) of input has any appreciable effect on the airplane's current condition. This observation has profound consequences, the most important of which is that the complete nonstationary gust input to the descending airplane is not required in order to estimate the statistics of the dispersion of the aircraft in height with respect to the glide path beam. Instead, only the gust characteristics corresponding to the last 5–10 sec preceding an altitude of interest need be considered. Because, typically, the airplane sink rate is about 10 ft/sec, 5–10 sec corresponds only to about a 50–100 ft altitude change. For such a small altitude change the altitude-dependent gust parameters cannot change very much during the time interval of interest. Therefore, the required short-time gust model is essentially a stationary input that is closely approximated by the gust model components evaluated at a particular altitude of interest.

If the altitude of interest were taken to be a decision height of, for example, 100 ft, then:

$$L_w = 100\ (\text{ft}) \tag{11-42}$$

$$L_u = L_v = 145(100)^{1/3} = 673.0\ (\text{ft}) \tag{11-43}$$

The breakpoints in the $w_g$ spectral density function, based on an aircraft speed of $U_0 = 228$ ft/sec, are

$$\text{poles at } \frac{U_0}{L_w} = \frac{228}{100} = 2.28\ \text{rad/sec} \tag{11-44}$$

$$\text{zeros at } \frac{U_0}{\sqrt{3}\,L_w} = \frac{228}{1.73 \times 100} = 1.32\ \text{rad/sec} \tag{11-45}$$

Similarly, the breakpoint for the $u_g$ spectral density function is at

$$\frac{U_0}{L_u} = \frac{228}{673} = 0.34\ \text{rad/sec} \tag{11-46}$$

Another aspect of the gust model is that the gust intensities along the various axes are related by the expressions:

$$\frac{\sigma^2_{u_g}}{L_u} = \frac{\sigma^2_{v_g}}{L_v} = \frac{\sigma^2_{w_g}}{L_w} \tag{11-47}$$

Thus the ratio of $\sigma_{u_g}$ to $\sigma_{w_g}$ will be

$$\frac{\sigma_{u_g}}{\sigma_{w_g}} = \sqrt{\frac{L_u}{L_w}} = \sqrt{\frac{673}{100}} = 2.59 \qquad (11\text{-}48)$$

The gust model also prescribes the probability of occurrence of gusts with a root mean square intensity, $\sigma_{w_g}$, at various altitudes, as is represented by $P(\sigma_{w_g})$, defined as the exceedance probability:

$$P(\sigma_{w_g}) = P_1\hat{P}(\sigma_{w_g}) \qquad (11\text{-}49)$$

where $P_1$ is the probability of occurrence of clear air turbulence; and $\hat{P}(\sigma_{w_g})$ is the probability of equaling or exceeding a given magnitude of $\sigma_{w_g}$, once clear air turbulence is encountered.

The probability of occurrence of clear air turbulence, $P_1$, at various altitudes is defined by the curve shown in Fig. 11-18.[11] It has been derived from a variety of measurements. According to Fig. 11-18, $P_1 = 0.8$ at 100 ft. The probability $\hat{P}(\sigma_{w_g})$ of equaling or exceeding a given $\sigma_{w_g}$ is depicted by the curve in Fig. 11-19.[12]

With these data the spectral characteristics of the gusts needed to enter the integrals of Eqs. 11-32 to 11-35 are defined.

Although, in principle, the root mean square values of the response variables of interest can be determined by using higher degree entries from a table of integrals similar to the one of Table 10-6, in practice we rely on a digital computer. The results of such computations for unit root-mean-square intensity gust inputs for the present example are given by

$$\sigma_{d_1} \equiv \frac{\sigma_d]_{w_g}}{\sigma_{w_g}} = 0.366 \qquad (11\text{-}50)$$

$$\sigma_{d_2} \equiv \frac{\sigma_d]_{u_g}}{\sigma_{u_g}} = 0.441 \qquad (11\text{-}51)$$

$$\sigma_{u_1} \equiv \frac{\sigma_u]_{w_g}}{\sigma_{w_g}} = 0.357 \qquad (11\text{-}52)$$

$$\sigma_{u_2} \equiv \frac{\sigma_u]_{u_g}}{\sigma_{u_g}} = 0.304 \qquad (11\text{-}53)$$

The total root mean square values of $d$ and $u$ are then found from,

$$\sigma_d = \sqrt{\sigma_{d_1}^2\sigma_{w_g}^2 + \sigma_{d_2}^2\sigma_{u_g}^2} \qquad (11\text{-}54)$$

[11] Chalk, Neal, Harris, Pritchard, and Woodcock, "Background Information and User Guide for MIL-F-8785B(ASG), Military Specification—Flying Qualities of Piloted Airplanes," AFFDL-TR-69-72, Wright-Patterson Air Force Base, Ohio, Aug. 1969.

[12] *Ibid.*

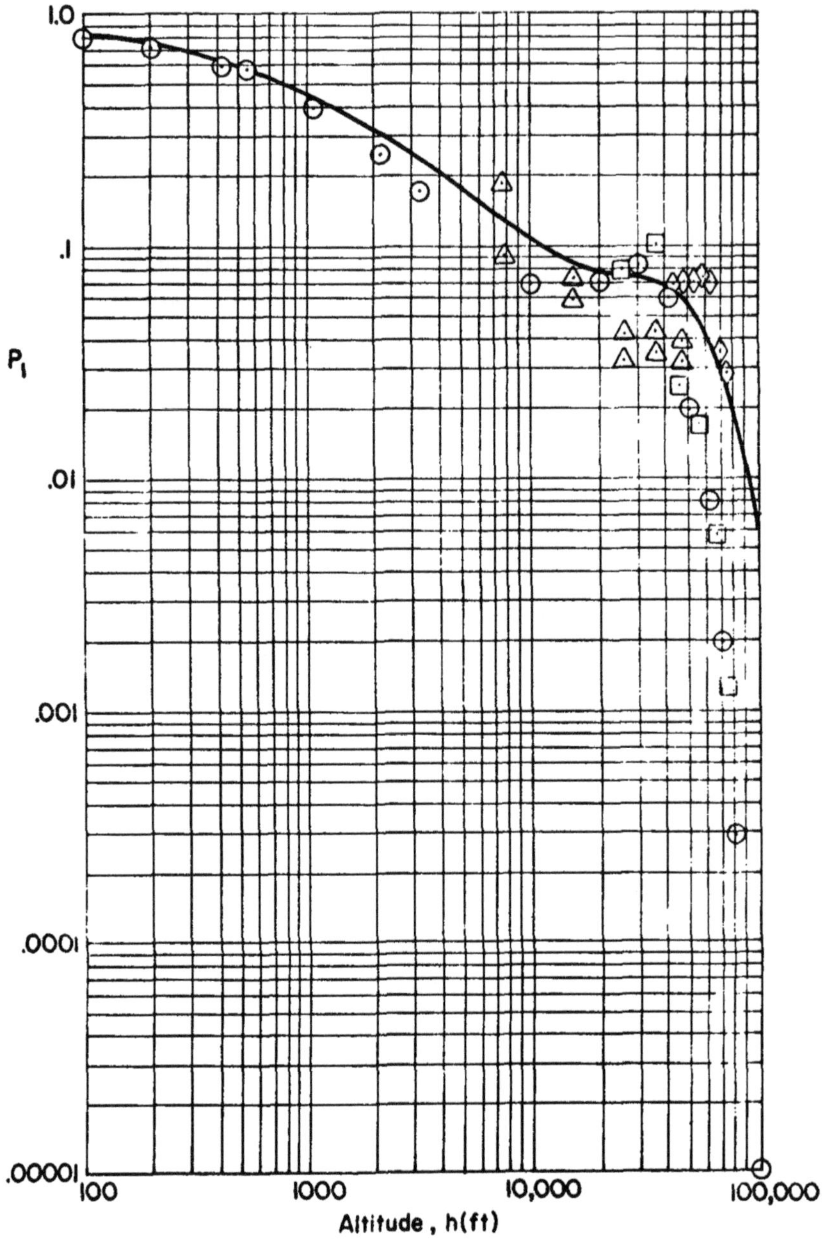


Fig. 11-18. Probability $P_1$ of encountering turbulence.

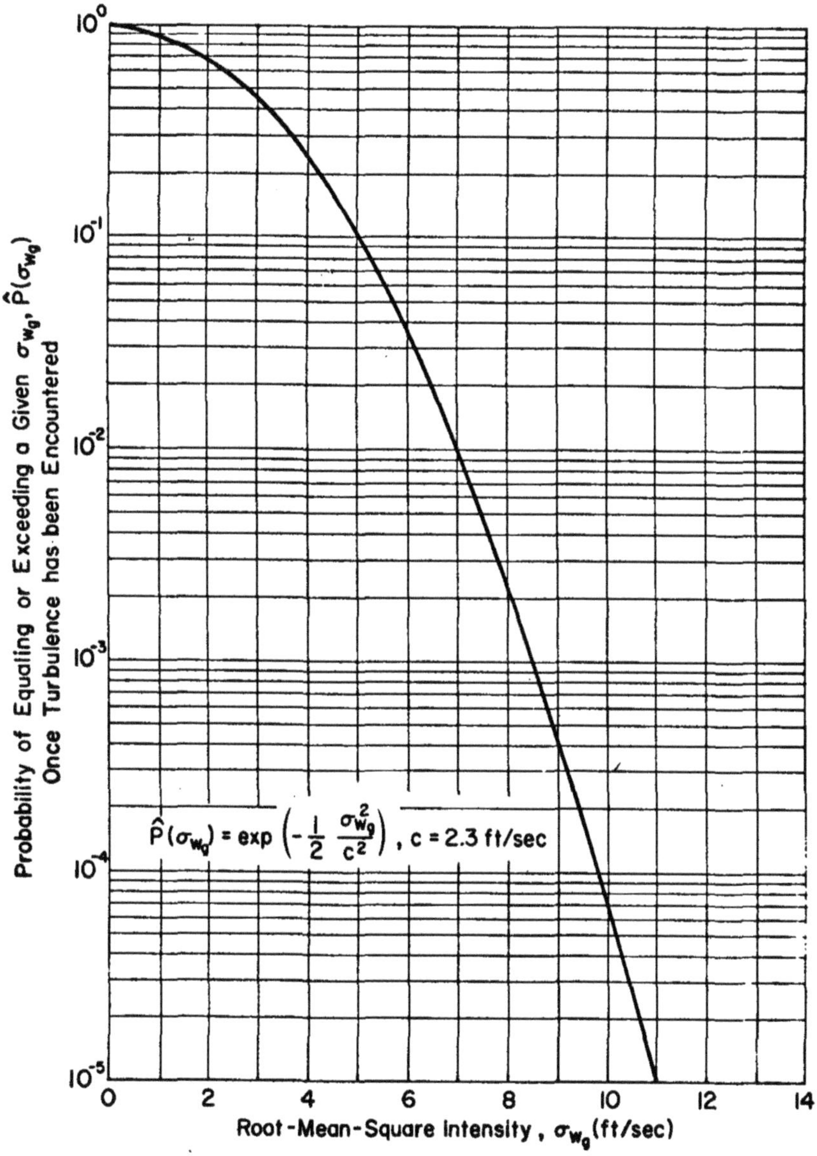


Fig. 11-19. Exceedance probability for $\sigma_{w_g}$.

and

$$\sigma_u = \sqrt{\sigma_{u_1}^2 \sigma_{w_g}^2 + \sigma_{u_2}^2 \sigma_{u_g}^2} \tag{11-55}$$

Equations 11-54 and 11-55, however, can be simplified by making use of the relation between $\sigma_{w_g}$ and $\sigma_{u_g}$ given for an altitude of 100 ft, that is:

$$\sigma_{u_g} = 2.59\sigma_{w_g} \tag{11-56}$$

Then, using this relationship and the values for unit root mean square inputs from Eqs. 11-50 to 11-53, the final results become

$$\sigma_d(\text{ft}) = 1.20\sigma_{w_g} \tag{11-57}$$

$$\sigma_u(\text{ft/sec}) = 0.865\sigma_{w_g} \tag{11-58}$$

These formulas indicate satisfactory regulation against disturbances by the multiloop feedback control system. Only small errors in deviation from the glide path are produced by the gust inputs, and it is not surprising that the fluctuating component of the aircraft's speed should be relatively large in the presence of turbulence and in the absence of a closed-loop speed control. This latter eventuality has, however, been acceptable in actual operations. In fact the low approach system analyzed here is very much superior in every respect to the minimum requirements recommended by RTCA SC-79,[13] even when its performance is evaluated in heavy turbulence.

This concludes the discussion of the longitudinal approach control system. We now turn to consideration of a different type of multiloop problem, one in which there is more than one control.

## 11-4. Lateral-Directional Multiloop Control System

The full power of the method of multiloop analysis that was presented in Section 3-5 has not yet been brought into play. This is because we have not had occasion to consider a feedback control system with more than one control point. Almost invariably, however, automatic control of the aircraft's lateral-directional motions involves the use of feedbacks to both the rudder and the ailerons. Thus the analysis of even the simplest lateral control examples requires the exercise of a method that is capable of dealing with this complication.

A simple example of a multipoint controller that is encountered in automatic flight control is a bank angle control system. Consideration of such a system will serve for the further exposition of the method of analysis

[13] "Standard Performance Criteria for Autopilot/Coupler Equipment," Paper 31-63/DO-118, Radio Technical Commission for Aeronautics, Washington, D.C., Mar. 14, 1963.

and of the means for satisfying requirements for completely automatic control of the aircraft's motions.

The primary functions of a bank angle control system are to turn the aircraft by introducing bank angle commands and to maintain the bank angle at the commanded level. An associated function may be to improve turn coordination. Indirect functions may include regulation of the yawing velocity and service as the inner loop of more elaborate heading and path control systems.

In order to accomplish bank angle control, one control loop clearly requires bank angle command inputs and bank angle feedback to generate an error that, after appropriate equalization, can be used to actuate the ailerons. Thus the simplest possible controller equation will be

$$\delta_a = G_\varphi(\varphi_c - \varphi) \tag{11-59}$$

On a few aircraft this type of control can be successfully applied by itself. More often, however, the dutch roll damping and, sometimes, frequency characteristics require modification by feedbacks to the rudder. Several possibilities for accomplishing these modifications have been discussed in connection with the material on essential feedbacks in Section 11-1. Presuming that only the dutch roll damping requires augmentation, we shall here use washed-out yawing velocity feedbacks to the rudder. The rudder controller equation will then be

$$\delta_r = -G_r r \tag{11-60}$$

A block diagram for the simplest system then appears as in Fig. 11-20. Using the rules of Section 3-5 the closed-loop system equations can be derived directly:

$$\varphi = \frac{G_\varphi(N^\varphi_{\delta_a} + G_r N^{r\ \varphi}_{\delta_r\delta_a})\varphi_c + (N^\varphi_{\beta_g} + G_r N^{r\ \varphi}_{\delta_r\beta_g})\beta_g + (N^\varphi_{p_g} + G_r N^{r\ \varphi}_{\delta_r p_g})p_g}{\Delta_{\text{sys}}} \tag{11-61}$$

$$\begin{aligned}\beta = \frac{1}{\Delta_{\text{sys}}} [G_\varphi(N^\beta_{\delta_a} &+ G_r N^{r\ \beta}_{\delta_r\delta_a})\varphi_c \\ &+ (N^\beta_{\beta_g} + G_r N^{r\ \beta}_{\delta_r\beta_g} + G_\varphi N^{\varphi\ \beta}_{\delta_a\beta_g} + G_r G_\varphi N^{r\ \varphi\ \beta}_{\delta_r\delta_a\beta_g})\beta_g \\ &+ (N^\beta_{p_g} + G_r N^{r\ \beta}_{\delta_r p_g} + G_\varphi N^{\varphi\ \beta}_{\delta_a p_g} + G_r G_\varphi N^{r\ \varphi\ \beta}_{\delta_r\delta_a p_g})p_g]\end{aligned} \tag{11-62}$$

$$r = \frac{G_\varphi N^r_{\delta_a}\varphi_c + (N^r_{\beta_g} + G_\varphi N^{\varphi\ r}_{\delta_a\beta_g}) + (N^r_{p_g} + G_\varphi N^{\varphi\ r}_{\delta_a p_g})p_g}{\Delta_{\text{sys}}} \tag{11-63}$$

where

$$\Delta_{\text{sys}} = \Delta_{\text{lat}} + G_r N^r_{\delta_r} + G_\varphi N^\varphi_{\delta_a} + G_r G_\varphi N^{r\ \varphi}_{\delta_r\delta_a} \tag{11-64}$$

There are five steps involved in discovering the closed-loop characteristics of this system.

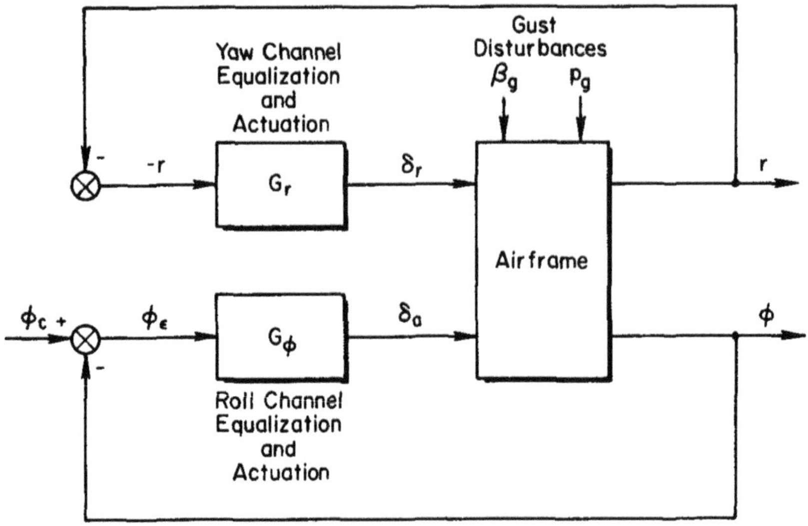


Fig. 11-20. Bank angle control system $(\varphi_c - \varphi) \rightarrow \delta_a$, $r \rightarrow \delta_r$.

1. The control channels are divided into two categories, inner and outer loops, reflecting the closure sequence to be used in the analysis.
2. The inner loop is closed with tentatively selected equalization and gains and the closed inner loop roots are found. These roots become the vehicle's *poles* for the outer loop closure. If the yaw rate channel is the inner loop, this calculation results in the zeros of $\Delta' = \Delta + G_r N^r_{\delta_r}$; whereas, if the roll channel is selected for the first closure, $\Delta'$ will be $\Delta + G_\varphi N^\varphi_{\delta_a}$.
3. By using the same gain and equalization selected above, the coupling loop is closed. The closed-loop roots resulting from this closure become the vehicle *zeros* for the outer loop closure. If the yaw rate channel is the inner loop, then for the bank angle outer loop these zeros are those of $N^\varphi_{\delta_a} + G_r N^{r\ \varphi}_{\delta_r \delta_a}$.
4. The outer loop is closed in a conventional single loop manner around the modified outer loop vehicle transfer function determined by Steps 2 and 3.
5. Possible repetitions of Steps 2–4 may be required with different equalizations and gains if the result of Step 4 should be unsatisfactory.

Fundamentally, the five *analysis* steps summarized above represent one cut in a cut and try *synthesis* procedure. In such procedures a trial system is analyzed to determine its dynamic characteristics, which are then compared with dynamic performance objectives. Deficiencies revealed by the comparison are then, hopefully, eliminated or reduced by modifications that result in a new system for trial. This system is analyzed and assessed in its turn, and the iterations continue until the trial system characteristics are consonant with the performance objectives.

The number of iterations that may be required in order to determine the characteristics of the ultimate system depends on the designer's intuition, on his ability to transfer insights into analytical procedures, and on his capacity to draw new knowledge and understanding from the results of previous analyses. An analysis procedure cannot be a substitute for background and experience, but the procedure can be matched to a presumed background so as to achieve a balance. The background presumed as a match for the multiloop analysis method described here is an intimate knowledge of single-loop servo analysis techniques (Chapter 3), a detailed understanding of vehicle dynamics (Chapters 4–6), and a thorough appreciation of the changes in effective-vehicle dynamics caused by idealized (i.e., no sensor or servo lag) single-loop controllers (Chapters 7 and 8). Based on this background, the multiloop analysis procedure has been organized so as to promote clarity and insight at each step in the analysis. In fact, if full use of the background can be made, the procedure may form the basis for almost direct synthesis.

*Loop-Closure Sequence and General Closure Criteria*

The selection of a sequence for the loop closures is extremely important in eliminating or reducing iterations. From a purely analytical point of view the closure sequence is immaterial. But in *synthesis* the closure sequence can be very important. For example, in practical forms, certain loops are necessarily closed before others for which they provide parallel equalization; or some loop-closure sequences result in extreme variation in the controller adjustments as a function of the iteration number, whereas for other sequences there is practically no change from one iteration to the next, and so forth. For these reasons there are at least some preferred loop-closure sequences and sometimes even a uniquely desirable one. For the present example there are, of course, only two possibilities, but if only a heading to aileron feedback were added, the possible loop-closure sequences jump to six, and far more complex arrangements are easily possible.

Desirable loop-closure sequences, however, are not always easy to determine. Fortunately, for aircraft control systems a set of factors can be enumerated to provide the insight needed to construct a closure sequence which, in the practical sense of minimizing inner-outer loop interactions, will also minimize iterations in synthesis. To a certain extent, the enumeration of these factors is facilitated not only by experience with the mathematical analysis of aircraft flight control systems but it also reflects desirable procedures for "tuning" an automatic flight control system in simulation or in flight test. In those cases, the choice of a sequence in which adjustments are made to the physical system may have an important influence on the number of trials and the amount of time that may be required in order to produce satisfactory results.

The factors that enter into a choice of which loops should be closed first are summarized in general terms in Table 11-11.

Table 11-12 is intended to illustrate how consideration of the same

Table 11-11. Factors involved in selecting a unique closure sequence.

| Factor | Remarks |
|---|---|
| Command loop | For a given system analysis the command loop is ordinarily made the last, or outer, loop. |
| Relative bandwidth | The bandwidth of a given loop closure is measured roughly by the crossover frequency $\omega_c$. (If more than one crossover frequency exists, the largest is taken as $\omega_c$.) The general sequence of loop closures in a multiloop system should then be in order of decreasing $\omega_c$, e.g., $\omega_{c_{\text{inner loop}}} > \omega_{c_{\text{outer loop}}}$. |
| Closure characteristics of a loop as a single-feedback system | Those loops that would require extensive equalization as individual single loops will ordinarily be made outer loops, so that inner loops can relieve the equalization requirements. |
| Subsidiary feedbacks | Feedbacks intended to provide equalization for subsequent loops, or to suppress subsidiary degrees of freedom that have undesirable effects on subsequent loops, are invariably inner loops. |
| Interdependent loops | Some loops contain common elements in their open-loop transfer functions that make independent closures impossible or undesirable. These loops are closed either simultaneously or in close sequence. |
| Multimode character | Flight control systems are ordinarily a composite of several systems, one for each mode of operation. A common result is a variety of command loops in which outer loops in one system (corresponding to one mode of operation) may be inner loops of another. To minimize gain and/or equalization changes with mode of operation, it is desirable to have the closures for a given loop the same, or closely connected, for each of the several "systems" (operational modes). |
| Equalization economy | Practical considerations (e.g., equipment complexity, maximum signal/minimum increment of control or dynamic range, minimization of internal disturbance effects, etc.) ordinarily restrict equalization of signals obtained from any one sensor to no more than approximate integration, proportional, and rate. When more equalization is required, a subsidiary loop is ordinarily needed. |

Table 11-12. Factors involved in bank attitude system loop closure sequence.

| Factor | Remarks | Consequences |
|---|---|---|
| Command loop | Turn commands can theoretically be introduced via $\varphi_c$ or $r_c$, since $r = (g/U_0) \sin \varphi$ in steady-state coordinated maneuvers. If $\varphi_c$ is used, a steady-state yaw rate will exist in steady-state turns; therefore either an $r_c$ command or $\varphi$ crossfeed must be introduced in the $r$ channel or $G_r(0)$ should be very small. | Either $\varphi \to \delta_a$ or $r \to \delta_r$ can be outer loop. Washed-out $r \to \delta_r$ or a separate $r_c \to \delta_r$ is required if $\varphi_c$ is used. |
| Relative bandwidth | $r \to \delta_r$ is ordinarily used to increase the Dutch roll damping. $\varphi_c \to \delta_a$ must provide attitude regulation against external disturbances and command turn capability. Resulting bandwidths, $\omega_{c_{r(\text{loop})}}$ and $\omega_{c_{\varphi(\text{loop})}}$, are ordinarily about the same. | Either $\varphi \to \delta_a$ or $r \to \delta_r$ can be outer loop. |
| Closure characteristics as single loops and equalization economy | $r \to \delta_r$ loop when $\zeta_r$, $\omega_r$, and $1/T_r$ are positive is readily closed with little or no equalization; $\varphi \to \delta_a$ can be difficult to close with stable results because of $\omega_\varphi$, $\omega_d$ effects. | $r \to \delta_r$ inner loop; $\varphi \to \delta_a$ outer loop. |
| Subsidiary feedbacks | $r \to \delta_r$ suppresses $\omega_\varphi$, $\omega_d$ effects, thereby relieving equalization in the $\varphi$ loop and/or supplementing the action of $\varphi \to \delta_a$ as an attitude controller. | $r \to \delta_r$ inner loop; $\varphi \to \delta_a$ outer loop. |
| Multimode character | A yawing velocity loop, without $\varphi_c \to \delta_a$, is a very common control mode to augment Dutch roll damping. On high performance manned aircraft it is almost invariably required to operate separately, as well as in conjunction with other loops. | Inner loop should be $r \to \delta_r$. |

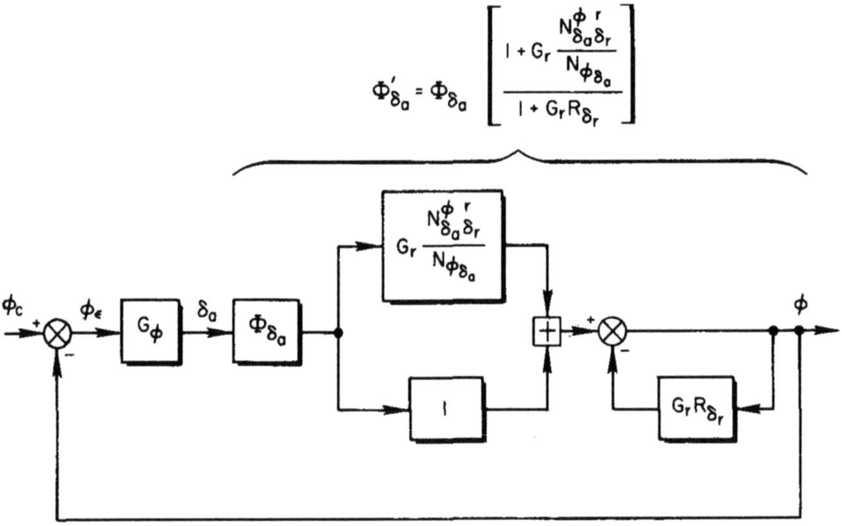


Fig. 11-21. Equivalent block diagram for bank angle control system representing the loop closure sequence $r \rightarrow \delta_r$, $(\varphi_c - \varphi) \rightarrow \delta_a$.

factors may be specifically taken into account in connection with the example of bank angle control so as to indicate which of the two loops should be the "inner" one and which the "outer" one. Based on the considerations summarized in Table 11-12, the yaw rate loop is made the inner loop and the bank angle loop becomes the outer loop. The appropriate equivalent system block diagram that also graphically represents the closure sequence is displayed as Fig. 11-21. This figure is a specific example of the type of block diagram shown in Fig. 3-40.

With a closure sequence determined, the next problem is to set up the closure criteria for the two loops. Only factors that represent engineering judgment can be delineated as generalities. They include the following considerations:

1. Inner loops intended as parallel equalization for a subsequent loop closure must be closed in such a way as to provide this function.
2. An inner loop for one system function may be an outer loop in another mode of operation. Its stability and response should be compatible with the desired performance in both operational modes, ideally without changes in controller settings.
3. The sensitivity factors of the various loops and their influence on the outer loops are strongly affected by the gains and equalization

that are selected. Ideally, these should be such that the effects of inner loop parameter variations on the outer loops are as small as possible.

Such considerations, taken in a context with overall system performance specifications, ordinarily serve as adequate guidelines for the selection of a specific gain and equalization for each loop in the sequence.

*Dynamic Characteristics of the Controlled Element*

The airframe characteristics that are to be used in this bank angle control system example are very similar to the ones for the straight wing aircraft of Section 6-4. If, however, the characteristics of that aircraft were to be used as they stand, it would be possible to achieve good bank angle control using a bank angle loop alone. This fact can be appreciated by examining Fig. 6-23, which shows that the dutch roll singularities are very nearly canceled by the roll numerator complex pair, so that an unequalized closure is seen to be possible. In order to provide a more interesting example of aircraft dynamics (i.e., with unfavorable $\omega_\varphi/\omega_d$ transfer function numerator, and therefore requiring a rudder loop to change the effective $\varphi/\delta_a$ transfer function numerator, as well as to provide dutch roll damping), the value of the effective aileron yawing moment derivative, $N'_{\delta_a}$, is changed from 0.395 to 5.11. With only this change, the aircraft transfer function quantities needed for our example, are

$$\Delta_{\text{lat}} = (s - 0.00135)(s + 1.78)[s^2 + 2(0.0243)(1.88)s + (1.88)^2] \quad (11\text{-}65)$$

$$N^r_{\delta_r} = -1.3617(s + 1.78)[s^2 + 2(0.00596)(0.292) + (0.292)^2] \quad (11\text{-}66)$$

$$N^\varphi_{\delta_a} = 27.276[s^2 + 2(0.0496)(2.06)s + (2.06)^2] \quad (11\text{-}67)$$

$$N^{\varphi\ r}_{\delta_a\delta_r} = -40.08(s + 0.0495) \quad (11\text{-}68)$$

Notice that the characteristic function and $r/\delta_r$ numerator are unchanged by the alteration in $N'_{\delta_a}$. They are the same as in the airplane of Section 6-4.

With the modification, $\omega_\varphi/\omega_d$ is now about 1.1 and $1/T_R\omega_d$ is approximately 0.95. Thus the single-loop roll control situation with a pure gain roll control system will be similar to the one depicted as Case 4A of Fig. 8-4, i.e., the closed-loop dutch roll mode will be *unstable* for otherwise desirable values of roll-loop gain.

The response of the aircraft alone to a 0.02 rad, 1 sec aileron pulse is shown in Fig. 11-22. This figure should be compared to Fig. 6-3(b) for the aircraft with the nominal $N'_{\delta_a}$ characteristics. There are several interesting differences in these figures, the most dramatic of them the sign reversal in the sideslip, $\beta$, and side acceleration, $a_{y_{c.g.}}$, oscillations. This stems from changes in the modal response ratios for the dutch roll mode because

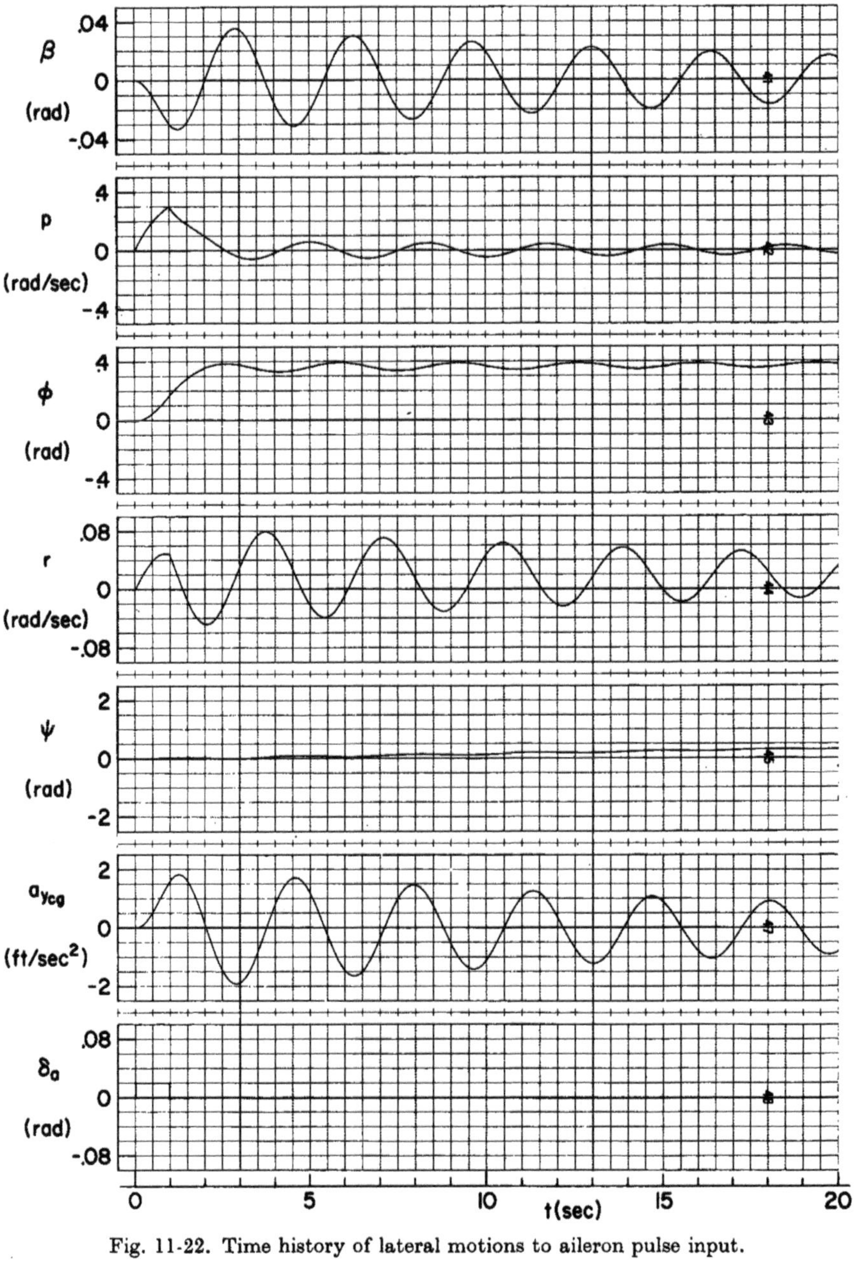


Fig. 11-22. Time history of lateral motions to aileron pulse input.

of the enlarged $N'_{\delta_a}$. In more physical terms the reversal can be explained as follows. Because the "favorable" aileron-induced yaw[14] is now so much larger, the modified airframe has an almost immediate yawing velocity, $r$, in Fig. 11-22, whereas the initial $r$ trace in Fig. 6-3(b) is essentially zero. Both aircraft have almost identical initial roll responses, as would be expected since they have the same aileron rolling moment characteristics. The sideslip rate is

$$\dot{\beta} = Y_v\beta + \frac{g}{U_0}\varphi - r \tag{11-69}$$

The bank angle contribution to $\dot{\beta}$ is positive for both aircraft in this transient. On the other hand, while the initial short time, yawing velocity is essentially nil for the unmodified airplane, it is sufficiently large for the altered aircraft to more than offset the $\varphi$ contribution to the sideslip rate. This accounts for the sign reversal in the $\beta$ and $a_{y_{c.g.}}$ oscillations when the two aircraft are compared. This difference in phasing of the oscillations is, of course, the transient response tip-off which indicates that the aircraft of Fig. 6-3(b) can easily be controlled by a single-loop roll controller whereas the airplane whose transient response is shown in Fig. 11-22 would be unstable with the same controller. An understanding of the difference between these two cases with the same characteristic function underscores the importance of studying the response to specific control and disturbance inputs. Much more needs to be known than the stability, as revealed by the characteristic function.

The actuators for both aileron and rudder in connection with the present example are assumed to be second-order systems with a 20 rad/sec undamped natural frequency and a damping ratio of 0.7. The yaw rate and roll sensor dynamics are assumed to be ideal pure gains in the frequency range of interest.

*Yaw Rate Loop Closure as a Yaw Damper*

The inner loop closure of washed-out yawing velocity to rudder has a system (or mode) identity of its own without the roll closure. This is as a yaw damper whose purpose is to damp the dutch roll without, at the same time, introducing uncoordinating yawing moments. Ideally, aileron-alone turns with this loop closed should exhibit a well-damped dutch roll and very little transient or steady-state sideslip. This may be accomplished by the pure lead-lag equalization of the yawing velocity feedback signal. As noted in Section 8-4, the inverse time constant of the washout, $1/T_{wo}$, should be made large in order to minimize the opposition to turning and, hence, to minimize the uncoordination. At the same time, the

[14] Aileron yawing moments are termed "favorable" if the initial yawing velocity motion is in the same direction as the intended roll.

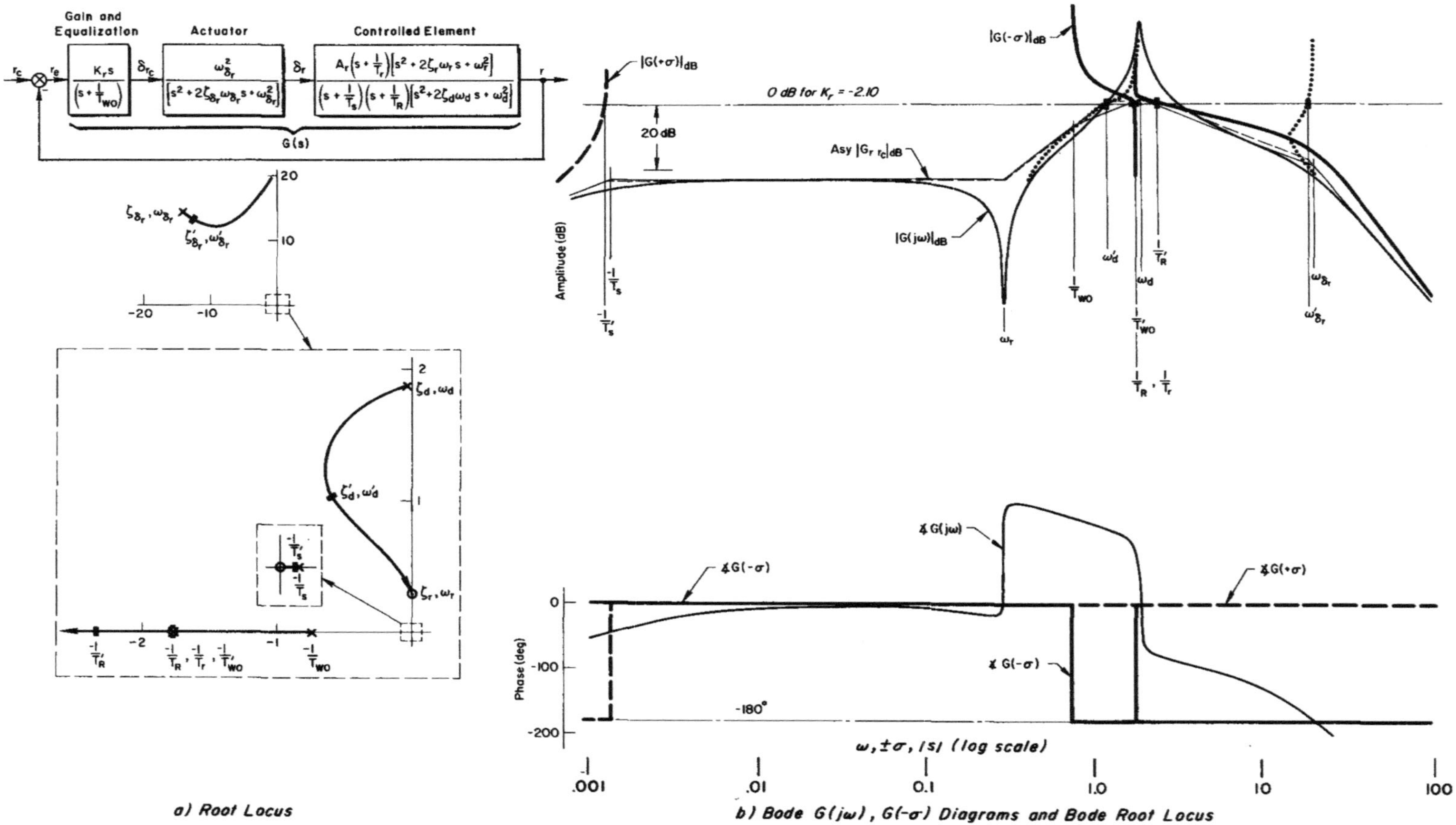


Fig. 11-23. System survey of washed-out yaw rate ($r \rightarrow \delta_r$) control system.

maximum obtainable closed-loop dutch roll damping is achieved when $1/T_{wo}$ is very small. The latter fact is best explained in terms of sensitivity. As shown in Fig. 8-11 the sensitivity vector, $S_1/T_{wo}$, for $1/T_{wo}$ variations is nearly parallel to the real axis. Thus the choice of a value for the inverse washout time constant is the subject of a compromise. This is the most fundamental consideration in connection with the yaw rate loop alone.

As criteria for the inner-loop closure, we shall assume a desired dutch roll damping ratio of approximately 0.5 and a closed-loop $r/r_c$ gain at low to mid-frequencies of less than $-20$ dB (the d-c gain will be zero because of the free $s$ in the washout numerator). These tentative criteria reveal only very slight opposition to turning, even in the transient.

A washout time constant, $T_{wo} = 1/0.75$ sec, is compatible with these desires, as is demonstrated by the system survey of Fig. 11-23. There it will be noted that for a gain of $K_r = -2.10$ ($K_r/T_{wo} = 1.57$ rad per rad/sec) the closed-loop dutch roll damping ratio is a bit greater than 0.5, while the mid-frequency amplitude ratio is down about $-23$ dB from the 0 dB line. With these same closure characteristics, the closed-loop lag introduced by the washout, $1/T'_{wo}$, forms a very close dipole with the numerator singularity at $-1/T_r$, the roll subsidence is slightly increased to $1/T'_R$, and the spiral mode time constant is slightly decreased (made more negative). With the selected gain $K_r = -2.10$, there is a phase margin of approximately 90 degrees. This is very large, more than sufficient to make up for higher frequency lags neglected in the analysis, as well as to provide for aircraft parameter uncertainties and controller tolerances.

In addition to these favorable system characteristics, the response of the aircraft plus yaw damper system to an aileron input is generally excellent, as is indicated in Fig. 11-24. The dutch roll is here seen to be well damped; the bank angle response has negligible overshoot; the steady-state sideslip is zero; and the transient sideslip is small and is balanced between positive and negative values.

As an aside, we should note that the transient sideslip might be further reduced by using relatively simple combinations of feedbacks and crossfeeds. The major sources of the sideslip are the aileron induced yawing acceleration, $N'_{\delta_a}\delta_a$, and the yawing accelerations due to rate of roll, $N'_p p$, and to yawing velocity, which derives from the aircraft alone plus the yaw damper. In principle, there would be no transient sideslip if the effective $\beta/\delta_a$ numerator were made zero. This result might be accomplished using aileron to rudder crossfeeds and roll rate feedback, either separately or in appropriate combinations.

If an aileron to rudder crossfeed with the transfer function $G_{\delta_a}^{\delta_r}$ were used, e.g., the effective $\beta/\delta_a$ numerator would then be

$$N_{\delta_a}^{\beta}]_{\mathrm{eff}} = N_{\delta_a}^{\beta} + G_r N_{\delta_a \delta_r}^{\beta\, r} + G_{\delta_a}^{\delta_r} N_{\delta_r}^{\beta} \qquad (11\text{-}70)$$

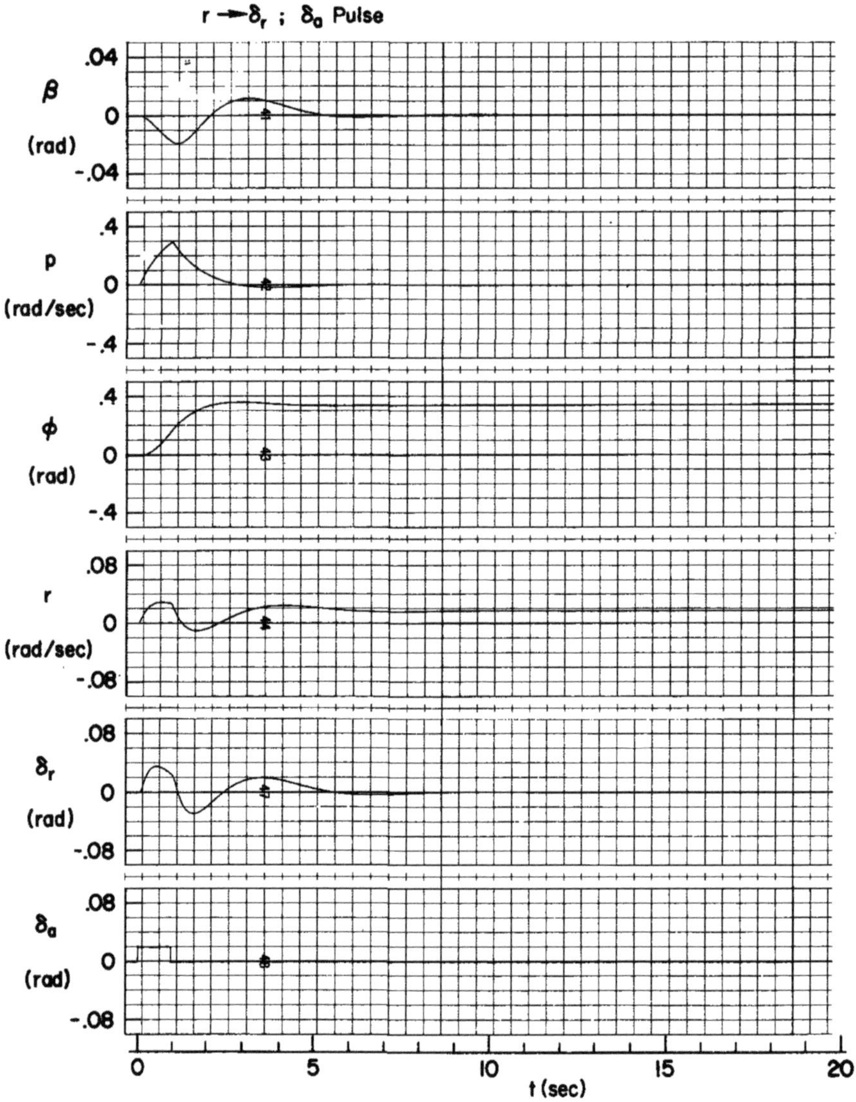


Fig. 11-24. Time history of aileron-induced lateral motions with washed-out yaw rate control system.

For this to be zero over all frequencies would require the aileron to rudder crossfeed transfer function to be

$$G_{\delta_a}^{\delta_r} = -\frac{N_{\delta_a}^{\beta} + G_r N_{\delta_a \delta_r}^{\beta\ r}}{N_{\delta_r}^{\beta}} \tag{11-71}$$

This expression can be very complex. The numerator and denominator are of the same degree. (Even if the servo dynamics were neglected in the present example, $G_{\delta_a}^{\delta_r}$, would be a ratio of fourth-order polynomials.) A simple approximation, however, using a lag-lead filter will often be effective.[15]

The yaw damper, analyzed here, is also an effective regulator system for the suppression of the effects of gust disturbances. This is shown graphically by the time histories for a step sideslip given in Fig. 11-25. Notice that the aircraft takes up a new sideslip (really a new inertial side velocity, since $\beta = v/U_0$) with very little oscillation; and that it quickly arrives at a new steady-state condition in which both bank angle, $\varphi$, and yawing velocity, $r$, are zero.

*Yaw Rate Loop Closure as Equalization for Roll Loop*

The effects of the washed-out yaw rate inner loop on the effective aircraft poles has already been described completely. Because of the coupling numerator, $N_{\delta_r \delta_a}^{r\ \varphi}$, however, the effective numerator of the $\varphi/\delta_a$ transfer function will be changed as well.

With a view toward equalization of the roll loop, the numerator $N_{\delta_a}^{\varphi'}$ of the $(\varphi/\delta_a)'$ transfer function (with the yaw rate loop closed), should exhibit complex singularities with large negative real (well-damped) parts. This is because the location of the $\omega_\varphi'$ zeros will determine the limiting parameters of the closed outer-loop dutch roll mode. Furthermore, the $\omega_\varphi'$, $\omega_d$ quadratic dipole should, ideally, have only a small effect on the closed-loop responses. Such a requirement implies that these transfer function quadratic factors should be very similar, so that they essentially cancel. Since, any effects of the washout mode are not desired in the roll response, the factor corresponding to that mode should also be associated as one member of a nearly canceling dipole pair by the time the roll loop is closed. Most basic is the desire that the inner loop be mechanized in such a way as to impose a minimum of additional equalization requirements on the outer roll loop.

The system survey for the coupling loop closure is shown in Fig. 11-26. It is the result of this step in the analysis which determines whether or

[15] "Automatic Flight Control Systems for Piloted Aircraft," BuAer Rept. AE-61-4, Vol. 6, Apr. 1956; R. L. Stapleford, D. T. McRuer, L. G. Hofmann, and G. L. Teper, "A Practical Optimization Design Procedure for Stability Augmentation Systems, AFFDL-TR-70-11, Wright-Patterson Air Force Base, Ohio, Oct. 1970.

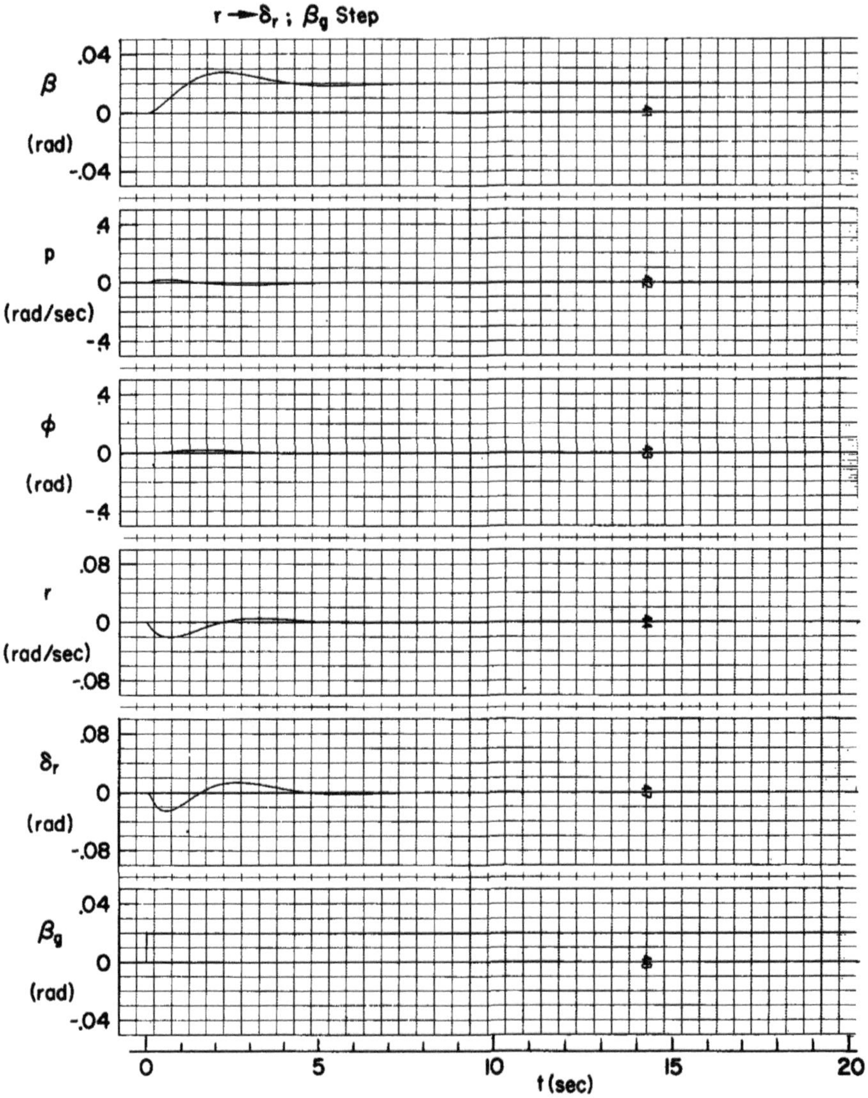


Fig. 11-25. Time history of sideslip gust-induced lateral motions with washed-out yaw rate control system.

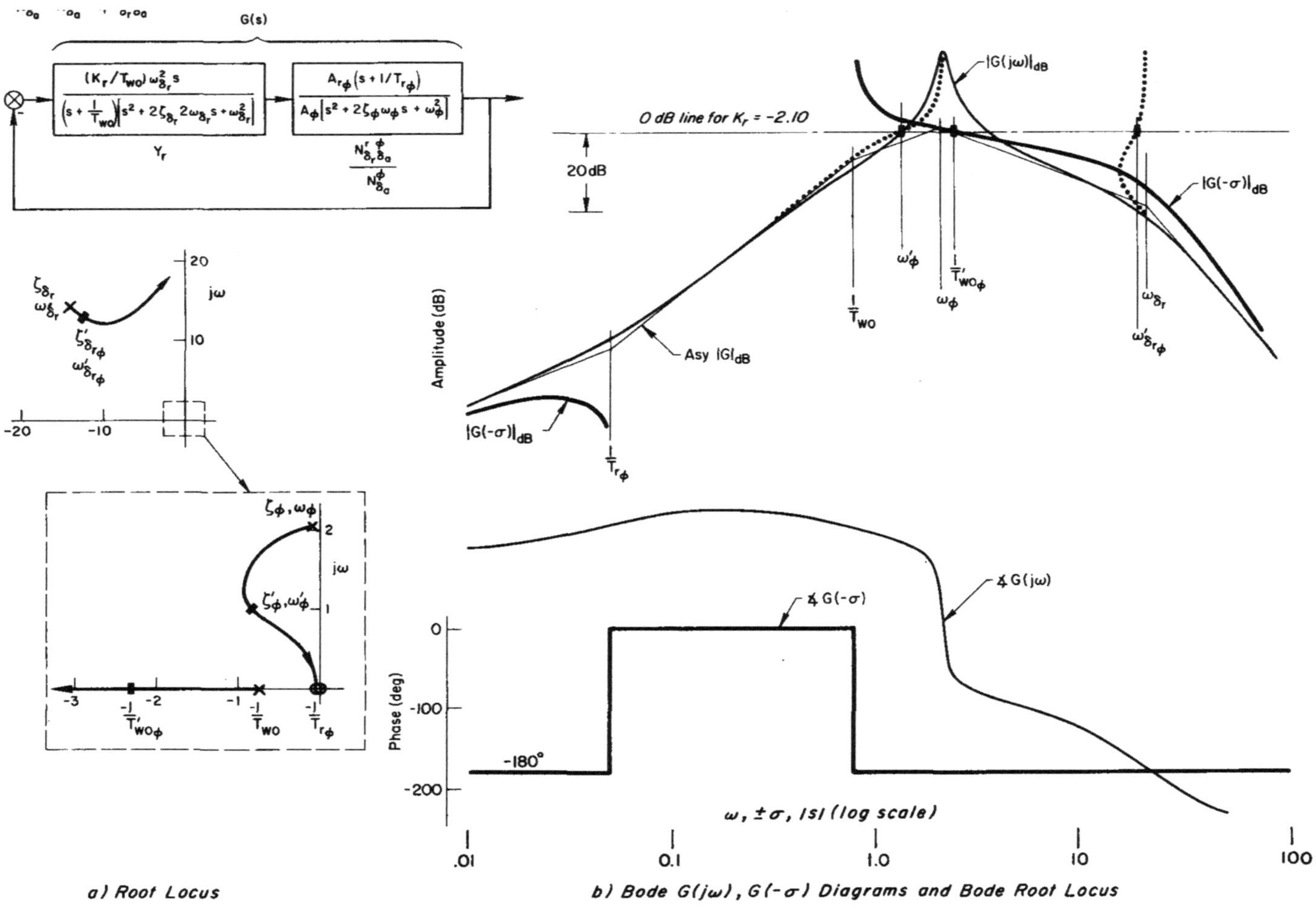


Fig. 11-26. System survey of effects of washed-out yaw rate ($r \rightarrow \delta_r$) feedback on bank angle to aileron numerator ($N^{\varphi}_{\delta_a}$).

not the yaw rate loop closure provides satisfactory equalization for the roll loop. The adjustable parameters, i.e., $K_r$ and $T_{wo}$, however, have already been determined in connection with the yaw rate loop closure so that this coupling numerator closure is, as we now present it, simply a consequence of the previously selected values. Actually, the coupling numerator considerations should play a role in selecting the appropriate value of $K_r$ and $T_{wo}$. If the values tentatively determined for the inner loop closure are not also satisfactory for the numerator closure, they may have to be changed.

Figure 11-26 shows that the roll numerator complex singularities move along a locus very similar to the one for the dutch roll singularities in Fig. 11-23. In fact, $\omega'_\varphi$ is almost identical to $\omega'_d$, although $\zeta'_\varphi$ is somewhat greater than $\zeta'_d$. These similar migrations reflect the inherent, almost identical high frequency characteristics of the respective open loops noted near the end of Section 3-5 (e.g., Eq. 3-77). Similarly, the singularity in the coupling numerator corresponding to the mode introduced by the washout moves at low gains in Fig. 11-26 in the same way as the washout singularity in Fig. 11-23. At higher gains it moves in the same way as the $1/T'_R$ singularity in Fig. 11-23. The net result is that the washout factor in the denominator of the effective $\varphi/\delta_a$ transfer function is very nearly canceled by an equivalent numerator factor. To give these remarks numerical precision, the effective aircraft transfer function, with the yaw rate loop closed, is given by

$$\left.\frac{\varphi}{\delta_a}\right]' = \frac{N^{\varphi'}_{\delta_a}}{\Delta'}$$

$$= \frac{27.276\overbrace{(s+2.313)}^{1/T'_{wo}}\overbrace{[s^2+2(0.635)(1.298)s+(1.298)^2]}^{\zeta'_\varphi,\,\omega'_\varphi}}{\underbrace{(s-0.001267)}_{1/T'_s}\underbrace{(s+1.764)}_{1/T'_R}\underbrace{(s+2.373)}_{1/T'_{wo}}\underbrace{[s^2+2(0.604)(1.203)s+(1.203)^2]}_{\zeta'_d,\,\omega'_d}}$$

$$\times \frac{\overbrace{[s^2+2(0.6915)(18.07)s+(18.07)^2]}^{\zeta'_{\delta_{r\varphi}},\,\omega'_{\delta_{r\varphi}}}}{\underbrace{[s^2+2(0.6928)(18.24)s+(18.24)^2]}_{\zeta'_{\delta_r},\,\omega'_{\delta r}}} \qquad (11\text{-}72)$$

Note that the servo actuator characteristics also appear in both the denominator and the coupling numerator closures as the $\zeta'_{\delta_r}$, $\omega'_{\delta_r}$, and $\zeta'_{\delta_{r\varphi}}$, $\omega'_{\delta_{r\varphi}}$ quadratics factors, respectively. This introduces still another dipole pair similar to the washout pair.

From this we see that the yaw rate loop closure provides satisfactory equalization for the bank angle closure. All of the features desired for the yaw rate loop closure have now been accomplished, so that we can proceed to the bank angle outer loop.

Before we continue, however, this is a good point at which to remark that, in practice, one would have just begun at this stage. Practical yaw dampers must operate effectively over a wide range of flight conditions, so that many more cases should be considered. From such a set of flight conditions, appropriate gain and/or washout time constant adjustments with flight condition are determined. These adjustments may be made either on a programmed basis or using self-adaptive devices. Furthermore, the yaw damper design is seldom as straightforward as is depicted here, in that the aircraft characteristics are ordinarily far less tractable. Some of the phenomena that may make the task more difficult have been described in Section 8-4.

*Closure of the Bank Angle Loop*

To a first approximation, the effective transfer function for $\varphi/\delta_a$ that is created when the rudder loop is closed is the transfer function of a simple second-order system. That is, when the washout dipole and the $\omega'_\varphi$, $\omega'_d$ and the $\omega'_{\delta_{r_\varphi}}$, $\omega'_{\delta_r}$ quadratic factors are assumed to cancel, and when the distinction between the spiral factor and a singularity at the origin is neglected,

$$\left.\frac{\varphi}{\delta_a}\right]' = \frac{N^{\varphi'}_{\delta_a}}{\Delta'} \doteq \frac{27.276}{s\underbrace{(s + 1.764)}_{1/T'_R}} \qquad (11\text{-}73)$$

The bank angle loop could then readily be closed with a pure gain controller. This is, indeed, one final conclusion of our multiple-loop example problem. However, $1/T'_R$ and the aircraft gain will both vary with flight condition. Consequently, the closed-loop system characteristics will either vary with flight condition or gain compensation will be needed in the roll loop. To reduce such variations and gain compensation, we shall consider a simple alternative system for which the dynamic performance is less configuration-sensitive. This system will make use of lead equalization in the feedback and a very high gain closure of the bank angle loop. The closed loop $\varphi/\varphi_c$ transfer function will then invariably approximate the inverse of the feedback equalization. This equalization can be held constant over a wide range of flight conditions and, as long as the total loop gain remains relatively large, the $\varphi/\varphi_c$ dominant characteristics will remain substantially the same.

Mechanization of this system can be accomplished using either passive equalization or an additional sensor. For the sake of concreteness, we shall assume that the additional sensor, a roll rate gyro, is the means

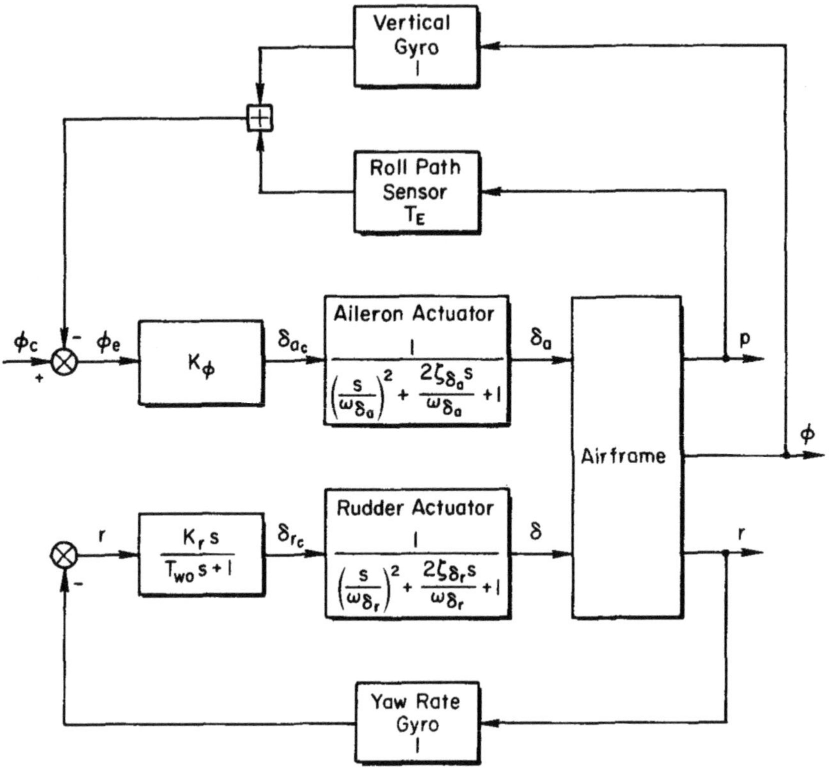


Fig. 11-27. Complete bank angle control system.

selected. Then the complete bank angle control system will be as illustrated in the block diagram of Fig. 11-27. Notice that the roll rate gyro sensor dynamics are presumed to be evident only at such high frequencies that their effect may be neglected at the major frequencies of interest. Because our attention is confined to straight and level horizontal flight, the rolling velocity is just the derivative of the bank angle; therefore, the net roll feedback of rolling velocity and bank angle corresponds to the transformed expression $(T_E s + 1)\varphi$.

A system survey of the outer loop for the bank angle control system is displayed in Fig. 11-28. For the purpose of constructing this diagram, the roll rate gain, $T_E$, is taken to be 1.25. This gain introduces a lead at 0.8 rad/sec in series with the $(\varphi/\delta_a)'$ characteristics of Eq. 11-72. Also introduced in series is the quadratic lag because of the aileron servo. The several dipole effects are readily apparent in the Bode root locus. The

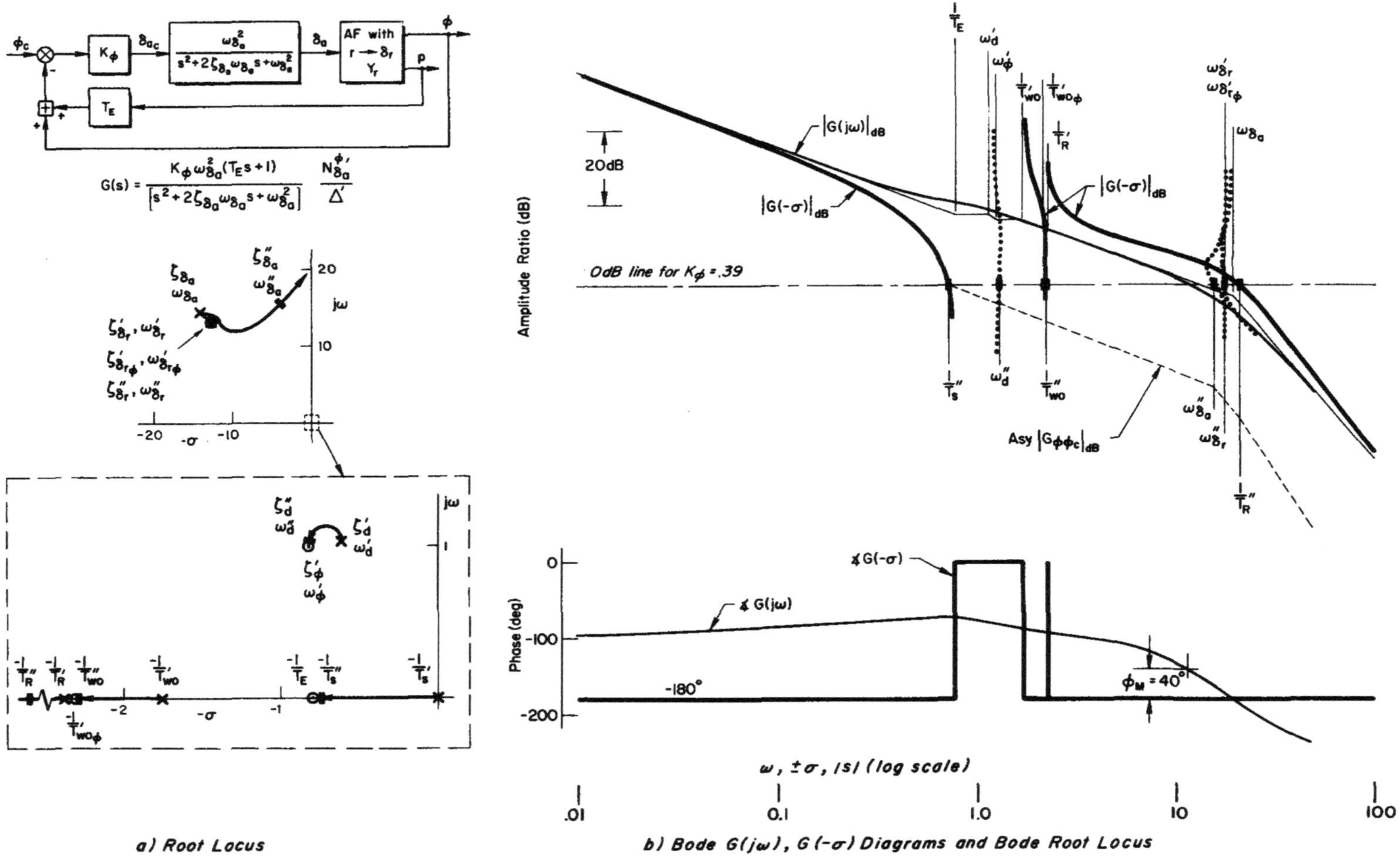


Fig. 11-28. System survey of bank angle command ($\varphi \rightarrow \delta_a$) control system with washed-out yaw rate inner loop.

fundamental character of the Bode plot for the outer loop, however, is its very close approximation to a pure integrator characteristic over an extraordinarily wide frequency range. As a consequence, the loop can be closed with no stability problem with a very wide range of gains. In fact, the stability limit occurs at a crossover frequency somewhat larger than 20 rad/sec.

In order to keep the closed-loop characteristic dominant mode, a first-order system with time constant close to $T_E$, the open loop amplitude ratio at the lead breakpoint, $1/T_E$, should be fairly large. For the nominal 0 dB line shown with a gain of $K_\varphi = 0.39$ degree (aileron)/degree (bank), the amplitude ratio at $\omega = 1/T_E$ is about +22 dB. The phase margin is 40 degrees, which is about as low as should be permitted. Notice that a factor of 2 gain reduction will only reduce the closed-loop dominant mode inverse time constant $1/T_s''$ from about 0.74 rad/sec to 0.68 rad/sec (8 percent decrease), while increasing the phase margin to 70 degrees causes a 75 percent increase. Consequently, there is a good deal of latitude in selecting the final value of gain for this closure without seriously degrading the closed-loop characteristics in response to command inputs. This flexibility is a very important feature, for, in a practical controller, many other flight conditions require investigation before the final gains are set. The ideal situation with respect to this loop would be to have either a fixed gain or, at most, a gain that is only discretely varied at one or two points in the flight envelope.

For the nominal gain ($K_\varphi = 0.39$) the closed-loop system roots are indicated by the factored characteristic function as follows:

$$\Delta'' = \underbrace{(s+0.73771)}_{1/T_s''}\underbrace{(s+2.308)}_{1/T_{wo}''}\underbrace{(s+21.185)}_{1/T_R''}\underbrace{[s^2+2(0.6253)(1.3236)s+(1.3236)^2]}_{\zeta_d'',\omega_d''}$$
$$\times \underbrace{[s^2+2(0.2418)(16.14)s+(16.14)^2]}_{\zeta_{\delta_a}'',\omega_{\delta_a}''}\underbrace{[s^2+2(0.6911)(18.036)s+(18.036)^2]}_{\zeta_{\delta_r}'',\omega_{\delta_r}''} \qquad (11\text{-}74)$$

All but four of these poles are paired with similar zeros for the bank angle command case. This is graphically demonstrated by the asymptotic plot of the closed-loop function $|G_{\varphi_c}^{\varphi}|_{dB}$ shown as dashed lines in Fig. 11-27. Only the pole at $1/T_s''$ (approximately $1/T_E$), $1/T_R''$, and the $\zeta_{\delta_a}''$, $\omega_{\delta_a}''$ servo characteristics are important in determining the closed-loop characteristics. Because the other three singularities occur at such high frequencies, the $1/T_s''$ inverse time constant is really the only one of significant interest in the basic rigid body, closed-loop system response to a bank angle command, $\varphi_c$.

The transient, closed-loop, system response characteristics to commands and disturbances are shown in Fig. 11-29. As already noted, for a step bank angle command [Fig. 11-28(a)] the bank angle response is very smooth and is closely similar to the response of a first-order system with a time constant of 1.25 sec. The transient sideslip in achieving the steady-state bank angle is relatively small, initially favorable (i.e., tending to help the aircraft bank) and then slightly adverse. This same characteristic is exhibited by the yawing velocity time history that shows an initial high positive rate of yaw which reduces to zero, then goes slightly negative, and finally builds up to the positive yaw rate corresponding to the steady-state bank angle. The dutch roll mode is very well damped and essentially is not present in either the bank angle or rolling velocity traces. The higher frequency oscillation seen in rolling velocity and aileron reflects the closed-loop servo dynamics.

The ability of the system to regulate against gust disturbances is demonstrated in Figs. 11-29(b) and 11-29(c) which depict responses to rolling and side gusts, respectively. For both of these inputs the bank angle and rolling velocity are hardly affected, remaining unperturbed. This is certainly to be expected because of the enormous outer loop bandwidth of about 12 rad/sec attained with the nominal gain used here for illustration. In fact, the bandwidth is so large that the aileron response is a close replica of the rolling gust disturbance and thereby nullifies the gust's effect almost directly. The side gust disturbance of Fig. 11-29(c) primarily affects the yawing responses. In fact, the responses shown here are very similar to the ones shown in Fig. 11-25, where only the yaw rate control system was engaged. The primary reason for this is the essential decoupling between roll and yaw at dutch roll frequencies and the relatively small value of effective dihedral, $L'_\beta$, associated with this straight-wing aircraft. On configurations with more wing sweep, the effective dihedral could perhaps be orders of magnitude larger. The system responses shown here are those of a very docile, almost ideal, airframe control system combination.

In order to conclude our discussion of this multiloop analysis example, it is worth remarking that the system sensitivity to either inner- or outer loop parameter variations will, in general, be very small for all of those characteristics that result in the nearly canceling dipole pairs. The key characteristics left over are the inner- and outer loop gains and, of course, the outer loop equalization time constant $T_E$. An equalization change would have its most direct effect on the time constant of the bank angle response to a command, as has already been described. The gains have their major influence on the time scale rather than upon response shape, at least for nominal variations about their selected values. In the inner loop, for example, it will be recalled (see Fig. 11-23) that the dominant

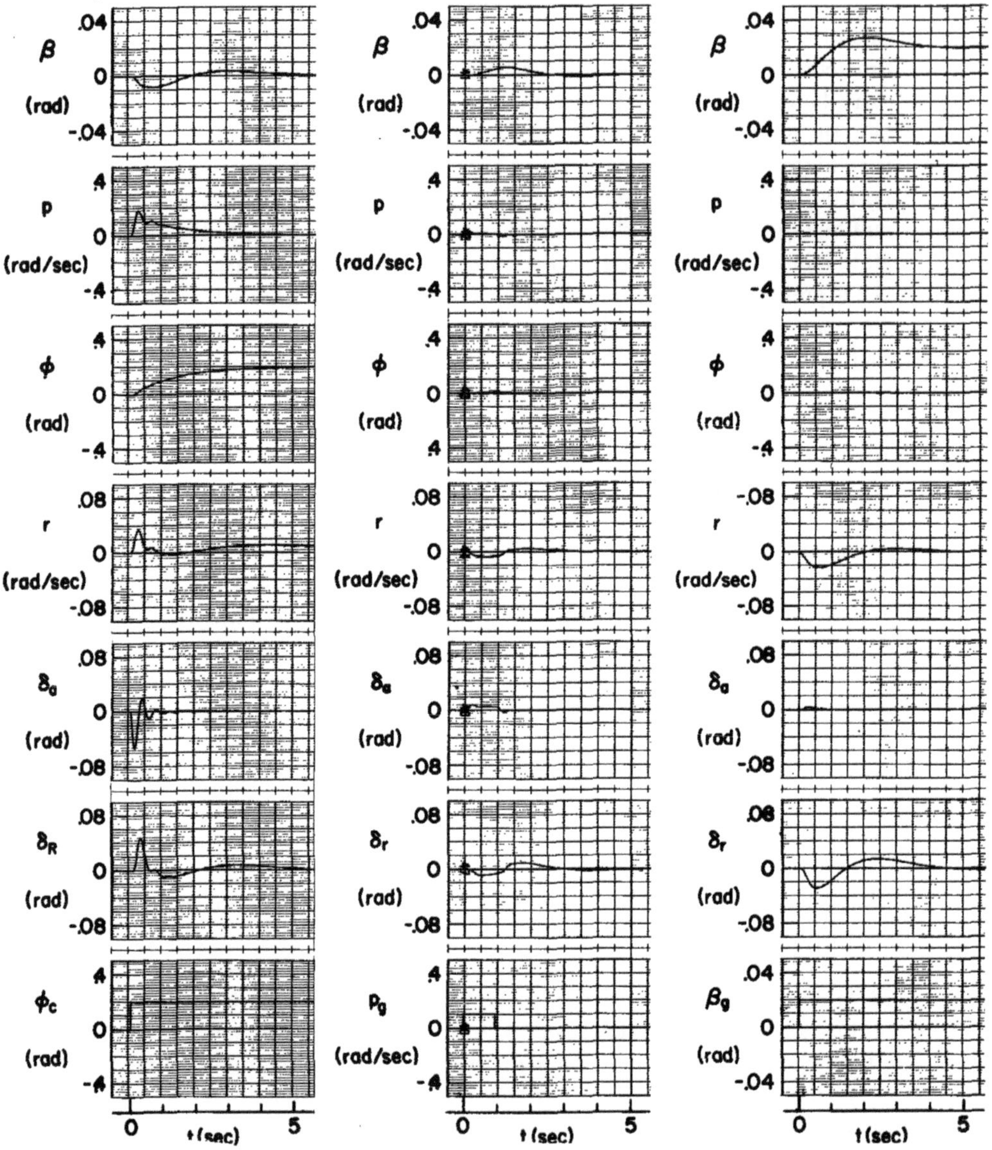


(a) Step bank angle command

(b) Rolling gust disturbances

(c) Side gust disturbance

Fig. 11-29. Bank angle control system response.

dutch roll mode damping ratio was not far from constant for a fairly wide range of gains about the nominal, whereas the undamped natural frequency and, hence, the response time scales, varied more directly with gain. For the outer loop, variations in the displacement gain, $K_\varphi$, will have relatively little effect on the $\varphi$ response to $\varphi_c$ as long as the amplitude ratio at the $1/T_E$ breakpoint is large. There is, however, a direct connection between system bandwidth and $K_\varphi$. In fact, this relationship is almost as pure as that for the simple $\omega_c e^{-\tau s}/s$ crossover model described in Section 9-5. It may even be a fitting end to this discussion of quite a complex system to point out that the outer loop characteristics of the bank angle control system described by three control loops, acting on two points of control application, and having a transfer function defined by ninth-order differential equation, is nevertheless approximated very well by the extraordinarily simple crossover characteristic, $\omega_c e^{-\tau s}/s$!

## 11-5. Conclusion

The last two sections have presented typical examples of the analysis of multiloop flight control systems. These examples have been representative of the type of problems that comprise possibilities for study in connection with the automatic control of aircraft. But they were treated analytically, of course, within the restrictions that have been imposed here. In design practice much more complicated analyses may be required, both in their degree of complexity and in scope.

It would now be conventional, for instance, to consider a system for control of the longitudinal motions in landing approach with feedbacks to three control points: the elevator, the direct lift flaps, and the throttle, instead of the elevator alone, as in our example. Future aircraft configurations may even add to this number of controls.

Greater complexities are inherent in phenomena whose description we have been compelled to insist is beyond the scope of this work. Such phenomena include:

Large angle maneuvering
Airframe structural flexibility effects
Sensor dynamics
Aircraft and controller nonlinearities

Of these items, all except the last are subjects for the supplemental bibliography that follows the Appendices. As we have already had occasion to point out, all of the effects of these complex phenomena are most usually studied in connection with detailed analyses and simulations of a candidate feedback control system after it has been selected as a tentative winner in its competition with alternatives. It has been primarily to the task of preliminary analyses of promising candidate systems to enter the competition that the discussion presented here has been addressed. This task, quite evidently, is one that must be accomplished with skill and expedition, since no level of subsequent effort can have the effect of rectifying mistaken decisions concerning the bases of the system design. The understanding of automatic flight control problems and solutions conveyed by the results that have been presented in this and in previous chapters, however, should enable the reader to circumvent or overcome many practical difficulties in satisfying the requirements for the feedback control of aircraft.

*APPENDICES*
*BIBLIOGRAPHY*
*INDEX*

APPENDIX A

# *STABILITY DERIVATIVES AND TRANSFER FUNCTION FACTORS FOR REPRESENTATIVE AIRCRAFT*

To apply the methods and techniques described in the text to the design of automatic flight control systems for real aircraft, it is necessary to have numerical data on the stability derivatives. In Chapters 5 and 6, for illustrative purposes, such data have been presented for a specific, conventional straight-wing airplane and for a tilt-duct VTOL aircraft in hovering flight. Later, in Chapters 7 and 8, a wide variety of generic analyses of closed-loop dynamics were introduced to show how the open-loop dynamics of the vehicle might influence the choice of which loops should be closed or the compensation that might be required. Although the characteristics illustrated there were not explicitly associated with particular aircraft, the reader can be assured that they were suggested by real features of the performance of actual vehicles.

Some readers of this volume may be working on projects in which stability and control aerodynamicists have made the necessary measurements or estimates, or they may themselves be stability and control aerodynamicists. If not, however, students of the subject will need a source of data in order to exercise the skills that they have acquired and to deepen their understanding. It is the purpose of this appendix to provide one such source.

Here the reader will find, conveniently arranged, stability derivatives and transfer function factors for nine aircraft. The notation employed has been defined in Chapters 4, 5, and 6. The choice of the aircraft has been somewhat arbitrary; it has been dictated primarily by the ready availability to the authors of unrestricted data. We have attempted, however, to provide a representative selection of both historical and modern aircraft of a very wide range of configurations. The reader will quickly note that in some cases the data are extensive, covering a large number of flight conditions; other cases are more briefly discussed.

The data presented in this appendix have been collected from very diverse sources over a long period of time. In a few cases the original source of the data is now unclear. In many cases the data have been altered because of internal inconsistencies or physical improbabilities revealed by the attempt to use them. For these reasons we wish to make it clear that the data are only *nominally* representative of the several aircraft configurations. In particular, the manufacturers of the aircraft cannot be held accountable for this information, nor would they be bound to concur in any conclusions with respect to their aircraft that might be derived from its use.

The data presented here may be made to serve at least two useful purposes: analysis and simulation. The stability derivative data are basic. They may be used to set up an analog or digital computer simulation of the linearized equations, or they may be used to compute the gust input numerators or some coupling numerators that are not tabulated here. On the other hand, most readers will find it convenient for many purposes to start with the tabulations of transfer function factors. These permit the methods and techniques, described in the text, to be applied relatively easily.

The aircraft represented are the following:

| | |
|---|---|
| Conventional jet-propelled: | |
| Straight wing | F-89 |
| Tailless delta | F-106B |
| Delta | A4D (A-4) |
| Transports: | |
| Propeller | C-47 (DC-3, R4D) |
| Jet | DC-8 |
| VTOL | |
| Tilt wing | XC-142 |
| Tilt duct | VZ-4 |
| Single-rotor helicopter | H-19 |
| Vintage biplane: | |
| Bristol fighter | F.2B |

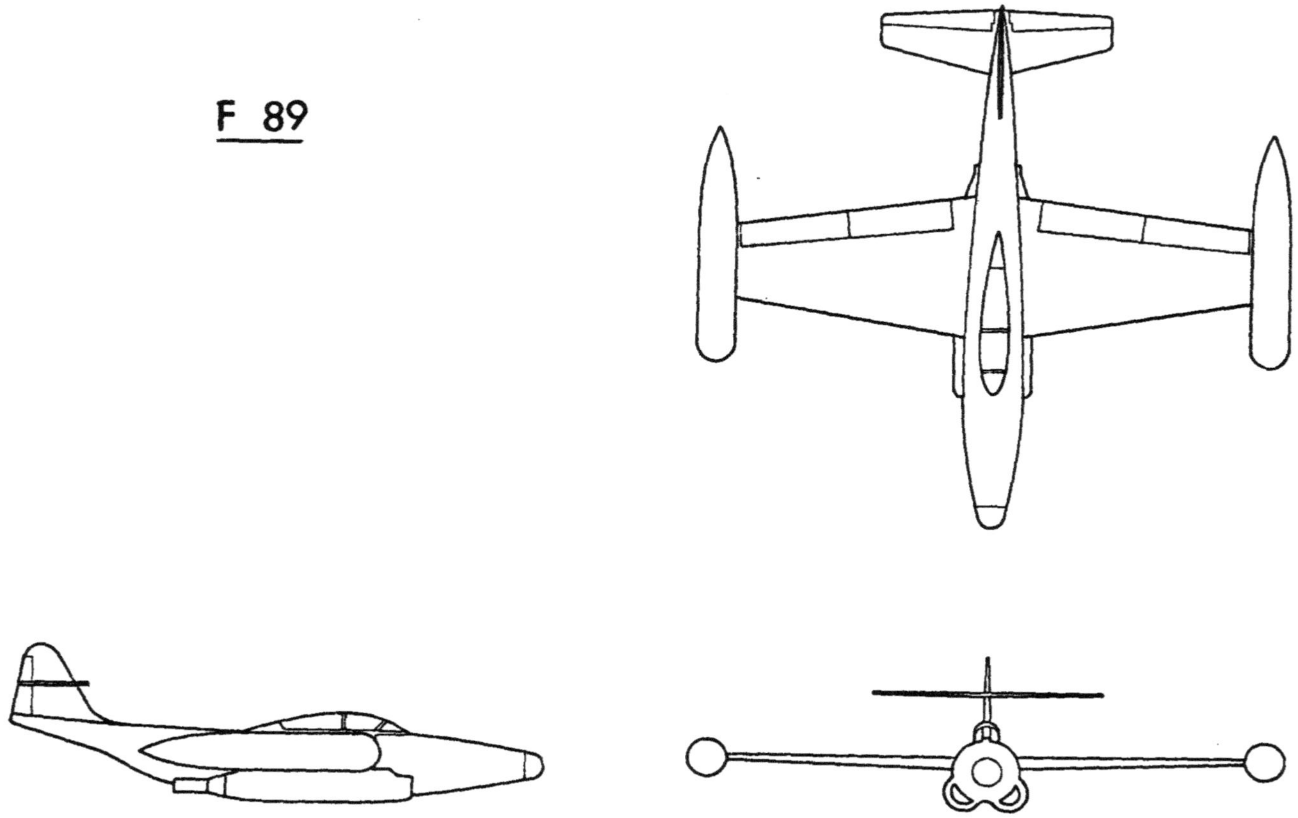


Fig. A-1. The F-89.

Table A-1. The F-89 (Fig. A-1)

| A. Geometrical and inertial parameters[a] | Flt. cond. 8901 | B. Longitudinal dimensional derivatives[a] | Flt. cond. 8901 | C. Lateral dimensional derivatives[a] | Flt. cond. 8901 |
|---|---|---|---|---|---|
| $h$ (ft) | 20,000 | $h$ (ft) | 20,000 | $h$ (ft) | 20,000 |
| $M$ (—) | 0.638 | $M$ (—) | 0.638 | $M$ (—) | 0.638 |
| $a$ (ft/sec) | 1,037 | $T_u$ (1/sec)[b] | — | $Y_v$ (1/sec) | −0.0829 |
| $\rho$ (slugs/ft$^3$) | 0.00127 | $X_u$ (1/sec) | −0.0097 | $Y_\beta$ [(ft/sec$^2$)/rad] | −54.7 |
| $V_{T_0}$ (ft/sec) | 660 | $X_w$ (1/sec) | 0.0016 | $Y_{\delta_a}$ [(ft/sec$^2$)/rad] | 0 |
| $\bar{q} = \rho V^2/2$ (lb/ft$^2$) | 276 | $X_{\delta_e}$ [(ft/sec$^2$)/rad] | 0 | $Y^*_{\delta_a}$ [(1/sec)/rad] | 0 |
| $W$ (lb) | 30,500 | $Z_u$ (1/sec) | −0.0955 | $Y_{\delta_r}$ [(ft/sec$^2$)/rad] | 7.65 |
| $m$ (slugs) | 947 | $Z_{\dot{w}}$ (—) | — | $Y^*_{\delta_r}$ [(1/sec)/rad] | 0.0116 |
| | | $Z_w$ (1/sec) | −1.43 | $L_\beta$ (1/sec$^2$) | −4.77 |
| | | $Z_{\delta_e}$ [(ft/sec$^2$)/rad] | −69.8 | $L_p$ (1/sec) | −1.70 |
| | | $M_u$ (1/sec-ft) | 0 | $L_r$ (1/sec) | 0.178 |
| | | $M_{\dot{w}}$ (1/ft) | −0.0013 | $L_{\delta_a}$ (1/sec$^2$) | 27.3 |
| | | $M_w$ (1/sec-ft) | −0.0235 | $L_{\delta_r}$ (1/sec$^2$) | 0.666 |
| | | $M_q$ (1/sec) | −1.92 | $L'_\beta$ (1/sec$^2$) | −4.55 |
| | | $M_{\delta_e}$ (1/sec$^2$) | −26.1 | $L'_p$ (1/sec) | −1.70 |
| | | | | $L'_r$ (1/sec) | 0.172 |
| | | | | $L'_{\delta_a}$ (1/sec$^2$) | 27.3 |
| | | | | $L'_{\delta_r}$ (1/sec$^2$) | 0.576 |
| | | | | $N_\beta$ (1/sec$^2$) | 3.55 |
| | | | | $N_p$ (1/sec) | −0.0025 |
| | | | | $N_r$ (1/sec) | −0.0957 |
| | | | | $N_{\delta_a}$ (1/sec$^2$) | −0.615 |
| | | | | $N_{\delta_r}$ (1/sec$^2$) | −1.38 |
| | | | | $N'_\beta$ (1/sec$^2$) | 3.38 |
| | | | | $N'_p$ (1/sec) | −0.0654 |
| | | | | $N'_r$ (1/sec) | −0.0893 |
| | | | | $N'_{\delta_a}$ (1/sec$^2$) | 0.395 |
| | | | | $N'_{\delta_r}$ (1/sec$^2$) | −1.36 |

## D.

| Elevator longitudinal transfer function factors | | Flt. cond. 8901 |
|---|---|---|
| Mach No., $M$ (—) | | 0.638 |
| Altitude, $h$ (ft) | | 20,000 |
| Weight, $W$ (lb) | | 30,500 |
| $\Delta_{long}$ | $\zeta_{sp}$ | 0.493 |
| | $\omega_{sp}$ | 4.27 |
| | $\zeta_p$ | 0.0714 |
| | $\omega_p$ | 0.0630 |
| $N^{\theta}_{\delta_e}$ | $A_\theta$ | −26.1 |
| | $1/T_{\theta_1}$ | 0.0098 |
| | $1/T_{\theta_2}$ | 1.372 |
| | d-c gain | −4.85 |
| $N^{u}_{\delta_e}$ | $A_u$ | −0.112 |
| | $1/T_{u_1}$ | 1.44 |
| | $1/T_{u_2}$ | −7251.4 |
| | d-c gain | 15920 |

## E.

| Aileron lateral transfer function factors | | Flt. cond. 8901 |
|---|---|---|
| Mach No., $M$ (—) | | 0.638 |
| Altitude, $h$ (ft) | | 20,000 |
| Weight, $W$ (lb) | | 30,500 |
| $\Delta_{lat}$ | $1/T_s$ | −0.00135 |
| | $1/T_R$ | 1.78 |
| | $\zeta_d$ | 0.0247 |
| | $\omega_d$ | 1.88 |
| $N^{p}_{\delta_r}$ | $A_p$ | 27.3 |
| | $1/T_{p_1}$ | 0 |
| | $\zeta_p$ | 0.047 |
| | $\omega_p$ | 1.86 |
| | d-c gain | 0 |
| $N^{\varphi}_{\delta_a}$ | $A_\varphi$ | 27.3 |
| | $\zeta_\varphi$ | 0.047 |
| | $\omega_\varphi$ | 1.86 |
| | d-c gain | −11075 |

## F.

| Rudder lateral transfer function factors | | Flt. cond. 8901 |
|---|---|---|
| Mach No., $M$ (—) | | 0.638 |
| Altitude, $h$ (ft) | | 20,000 |
| Weight, $W$ (lb) | | 30,500 |
| $\Delta_{lat}$ | $1/T_s$ | −0.00135 |
| | $1/T_R$ | 1.78 |
| | $\zeta_d$ | 0.0247 |
| | $\omega_d$ | 1.88 |
| $N^{p}_{\delta_r}$ | $A_p$ | 0.576 |
| | $1/T_{p_1}$ | 0 |
| | $1/T_{p_2}$ | 2.56 |
| | $1/T_{p_3}$ | −2.89 |
| | d-c gain | 0 |
| $N^{\varphi}_{\delta_r}$ | $A_\varphi$ | 0.576 |
| | $1/T_{\varphi_1}$ | 2.56 |
| | $1/T_{\varphi_2}$ | −2.89 |
| | d-c gain | 500 |

## G.

| Lateral coupling numerators | | Flt. cond. 8901 |
|---|---|---|
| Mach No., $M$ (—) | | 0.638 |
| Altitude, $h$ (ft) | | 20,000 |
| Weight, $W$ (lb) | | 30,500 |
| $N^{p\,\beta}_{\delta_a\delta_r}$ | $A_{p\beta}$ | 0.316 |
| | $1/T_{p\beta_1}$ | 0 |
| | $1/T_{p\beta_2}$ | 118 |
| $N^{p\,\beta}_{\delta_a\delta_r}$ | $A_{r\beta}$ | 0.00458 |
| | $\zeta_{r\beta}$ | −0.0706 |
| | $\omega_{r\beta}$ | 19.9 |
| $p/s_r$ $N_{\delta_a\delta_r} = N^{\varphi r}_{\delta_a\delta_r}$ | $A_{\varphi_r}$ | −37.4 |
| | $1/T_{\varphi_r}$ | 0.0537 |
| $N^{\beta\,a'_y}_{\delta_a\delta_r}$ | $A_{\beta_{a'_y}}$ | 3.02 |
| $l_x = 0$ | $1/T_{\beta_{a'_{y_1}}}$ | 0.114 |
| $l_z = 0$ | $1/T_{\beta_{a'_{y_2}}}$ | 2.70 |
| c.g. | $1/T_{\beta_{a'_{y_3}}}$ | — |

Table A-1 (Continued)

D.

| Elevator longitudinal transfer function factors | | Flt. cond. 8901 |
|---|---|---|
| Mach No., $M$ (—) | | 0.638 |
| Altitude, $h$ (ft) | | 20,000 |
| Weight, $W$ (lb) | | 30,500 |
| | $A_w$ | −69.8 |
| | $1/T_{w_1}$ | 248.5 |
| $N^w_{\delta_e}$ | $\zeta_w$ | 0.0713 |
| | $\omega_w$ | 0.068 |
| | d-c gain | −1110 |
| | $A_{\dot{h}}$ | 69.8 |
| | $1/T_{\dot{h}_1}$ | 0.00640 |
| $N^{\dot{h}}_{\delta_e}$ | $1/T_{\dot{h}_2}$ | 19.8 |
| | $1/T_{\dot{h}_3}$ | −17.03 |
| | d-c gain | −2080 |
| | $A_{a_z}$ | −69.8 |
| $N^{a_z}_{\delta_e}$ | $1/T_{a_{z_1}}$ | 0 |
| | $1/T_{a_{z_2}}$ | 0.0064 |
| | $1/T_{a_{z_3}}$ | 19.8 |
| c.g. | $1/T_{a_{z_4}}$ | −17.03 |
| | d-c gain | 0 |

E.

| Aileron lateral transfer function factors | | Flt. cond. 8901 |
|---|---|---|
| Mach No., $M$ (—) | | 0.638 |
| Altitude, $h$ (ft) | | 20,000 |
| Weight, $W$ (lb) | | 30,500 |
| | $A_r$ | 0.395 |
| | $1/T_{r_1}$ | 1.65 |
| $N^r_{\delta_a}$ | $\zeta_r$ | −0.828 |
| | $\omega_r$ | 2.65 |
| | d-c gain | −539 |
| | $A_\beta$ | −0.395 |
| | $1/T_{\beta_1}$ | 0.0496 |
| $N^\beta_{\delta_a}$ | $1/T_{\beta_2}$ | −6.23 |
| | $1/T_{\beta_3}$ | — |
| | d-c gain | −14.4 |
| | $A_{a_y}$ | 21.6 |
| $N^{a_y}_{\delta_a}$ | $1/T_{a_{y_1}}$ | 0.0496 |
| | $1/T_{a_{y_2}}$ | −6.23 |
| | $1/T_{a_{y_3}}$ | — |
| c.g. | $1/T_{a_{y_4}}$ | — |
| | d-c gain | 788 |

F.

| Rudder lateral transfer function factors | | Flt. cond. 8901 |
|---|---|---|
| Mach No., $M$ (—) | | 0.638 |
| Altitude, $h$ (ft) | | 20,000 |
| Weight, $W$ (lb) | | 30,500 |
| | $A_r$ | −1.36 |
| | $1/T_{r_1}$ | 1.78 |
| $N^r_{\delta_r}$ | $\zeta_r$ | 0.00596 |
| | $\omega_r$ | 0.292 |
| | d-c gain | 24.3 |
| | $A_\beta$ | 0.0116 |
| | $1/T_{\beta_1}$ | −0.00373 |
| $N^\beta_{\delta_r}$ | $1/T_{\beta_2}$ | 1.75 |
| | $1/T_{\beta_3}$ | 118 |
| | d-c gain | 1.05 |
| | $A_{a_y}$ | 7.65 |
| $N^{a_y}_{\delta_r}$ | $1/T_{a_{y_1}}$ | −0.00511 |
| | $1/T_{a_{y_2}}$ | 1.70 |
| | $1/T_{a_{y_3}}$ | 2.57 |
| c.g. | $1/T_{a_{y_4}}$ | −2 47 |
| | d-c gain | −49.5 |

G.

| Lateral coupling numerators | | Flt. cond. 8901 |
|---|---|---|
| Mach No., $M$ (—) | | 0.638 |
| Altitude, $h$ (ft) | | 20,000 |
| Weight, $W$ (lb) | | 30,500 |
| $N^{r\,a'_y}_{\delta_a\delta_r}$ | $A_{ra'_y}$ | 3.02 |
| $l_x = 0$ | $1/T_{ra'_{y_1}}$ | −4.09 |
| $l_z = 0$ | $\zeta_{ra'_y}$ | 0.279 |
| c.g. | $\omega_{ra'_y}$ | 2.29 |
| $N^{p\,a'_y}_{\delta_a\delta_r}$ | $A_{pa'_y}$ | 209 |
| $l_x = 0$ | $1/T_{pa'_{y_1}}$ | 0 |
| $l_z = 0$ | $1/T_{pa'_{y_2}}$ | 2.57 |
| c.g. | $1/T_{pa'_{y_3}}$ | −2.47 |
| $N^{\varphi\,\beta}_{\delta_a\delta_r}$ | $A_{\varphi\beta}$ | 0.316 |
| | $1/T_{\varphi\beta}$ | 118 |
| $N^{\varphi\,a_y}_{\delta_a\delta_r}$ | $A_{\varphi a_y}$ | 209 |
| | $1/T_{\varphi a_{y_1}}$ | 2.57 |
| c.g. | $1/T_{\varphi a_{y_2}}$ | −2.47 |

[a] The data in A, B, and C are for body-fixed stability axes.

[b] $T_u = \dfrac{1}{a_m}\dfrac{\partial T}{\partial M}$.

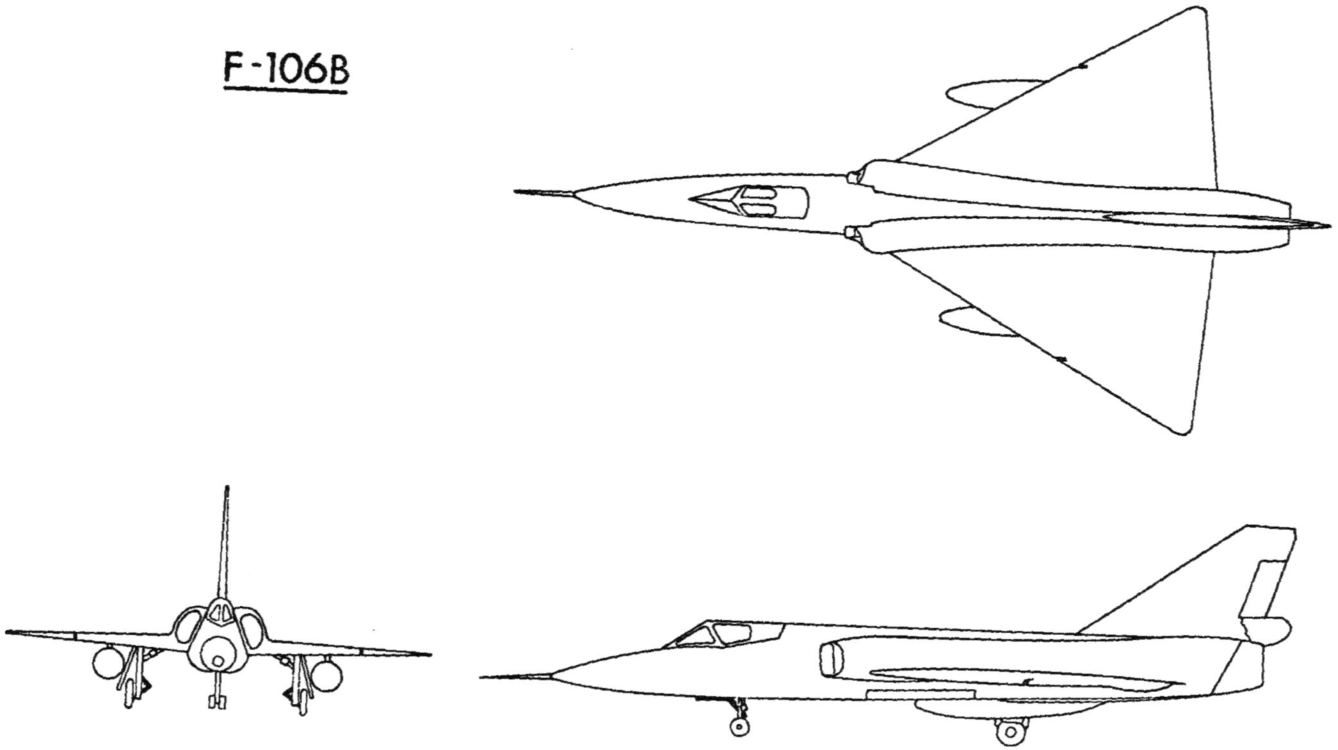


Fig. A-2. The F-106B.

Table A-2. The F-106B (Fig. A-2)
A. Geometrical parameters

| | Flight condition | | | | | | | | | | | |
|---|---|---|---|---|---|---|---|---|---|---|---|---|
| | 1 | 2 | 3 | 4 | 5 | 6 | 7 | 8 | 9 | 10 | 11 | 12 |
| $h$ (ft) | 20,000 | 20,000 | 20,000 | S.L. | S.L. | 20,000 | S.L. | 20,000 | 40,000 | 20,000 | 40,000 | 40,000 |
| $M$ (—) | 0.755 | 0.755 | 0.755 | 0.2 | 0.4 | 0.4 | 0.9 | 0.9 | 0.9 | 1.4 | 1.4 | 2.0 |
| $a$ (ft/sec) | 1,037 | 1,037 | 1,037 | 1,116 | 1,116 | 1,037 | 1,116 | 1,037 | 968 | 1,037 | 968 | 968 |
| $\rho$ (slugs/ft$^3$) | 0.001267 | 0.001267 | 0.001267 | 0.002377 | 0.002377 | 0.001267 | 0.002377 | 0.001267 | 0.000587 | 0.001267 | 0.000587 | 0.000587 |
| $V_{T_0}$ (ft/sec) | 785 | 785 | 785 | 223.2 | 446.4 | 414 | 1004.4 | 933 | 871 | 1,450 | 1,355 | 1,936 |
| $\bar{q} = \rho V^2/2$ (lb/ft$^2$) | 392 | 392 | 392 | 59.3 | 237.2 | 108.6 | 1,199 | 551 | 223 | 1,332 | 549 | 1,100 |
| $W$ (lb) | 35,000 | 30,000 | 28,000 | 25,500 | 29,776 | 29,776 | 29,776 | 29,776 | 29,776 | 29,776 | 29,776 | 29,776 |
| Mass (slugs) | 1,090 | 931 | 870 | 791.9 | 924.7 | 924.7 | 924.7 | 924.7 | 924.7 | 924.7 | 924.7 | 924.7 |
| $I_x$ (slug-ft$^2$) | 25,490 | 18,744 | 15,809 | 15,800 | 18,634 | 18,634 | 18,634 | 18,634 | 18,634 | 18,634 | 18,634 | 18,634 |
| $I_y$ (slug-ft$^2$) | 195,156 | 185,300 | 177,645 | 160,783 | 177,858 | 177,858 | 177,858 | 177,858 | 177,858 | 177,858 | 177,858 | 177,858 |
| $I_z$ (slug-ft$^2$) | 215,262 | 198,707 | 187,115 | 170,301 | 191,236 | 191,236 | 191,236 | 191,236 | 191,236 | 191,236 | 191,236 | 191,236 |
| $I_{xz}$ (slug-ft$^2$) | 4947.1 | 5310.9 | 6015.4 | 5,727 | 5,539 | 5,539 | 5,539 | 5,539 | 5,539 | 5,539 | 5,539 | 5,539 |
| $x_{c.g.}/\bar{c}$ | 0.29 | 0.305 | 0.26 | 0.305 | 0.305 | 0.305 | 0.305 | 0.305 | 0.305 | 0.305 | 0.305 | 0.305 |
| $\alpha_{trim}$ (degrees) | 4.42 | 4.04 | 3.88 | 18.0 | 4.9 | 11.0 | 2.0 | 2.7 | 5.4 | 1.2 | 2.70 | 1.2 |
| $C_{L_\alpha}$ (1/rad) | 2.7 | 2.7 | 2.7 | 2.5 | 2.54 | 2.55 | 2.86 | 2.94 | 3.00 | 2.46 | 2.58 | 1.84 |
| $\theta_0$ (degrees) | 4.42 | 4.04 | 3.88 | 18.0 | 4.9 | 11.0 | 2.0 | 2.7 | 5.4 | 1.2 | 2.7 | 1.2 |
| $U_0$ (ft/sec) | 784 | 784 | 784 | 212 | 445 | 406 | 1,004 | 933 | 868 | 1,450 | 1,355 | 1,936 |
| $W_0$ (ft/sec) | 60.6 | 55.4 | 53.1 | 73 | 40 | 80 | 36 | 44 | 82 | 30 | 64 | 40 |

### B. Lateral dimensional derivatives[b]

| | Flight condition | | | | | | | | | | | |
|---|---|---|---|---|---|---|---|---|---|---|---|---|
| | 1 | 2 | 3 | 4 | 5 | 6 | 7 | 8 | 9 | 10 | 11 | 12 |
| $h$ | 20,000 | 20,000 | 20,000 | S.L. | S.L. | 20,000 | S.L. | 20,000 | 40,000 | 20,000 | 40,000 | 40,000 |
| $M$ | 0.755 | 0.755 | 0.755 | 0.2 | 0.4 | 0.4 | 0.9 | 0.9 | 0.9 | 1.4 | 1.4 | 2.0 |
| $Y_v$ | −0.207 | −0.239 | −0.259 | −0.126 | −0.237 | −0.109 | −0.561 | −0.277 | −0.112 | −0.423 | −0.182 | −0.217 |
| $Y_\beta$ | −163 | −188 | −204 | −29.6 | −106 | −46.1 | −564 | −259 | −98.0 | −615 | −247 | −420 |
| $Y^*_{\delta_a}$ | 0.0799 | 0.0926 | 0.100 | 0.0492 | 0.0865 | 0.0443 | 0.175 | 0.108 | 0.0523 | 0.0470 | 0.0287 | 0.0128 |
| $Y^*_{\delta_a}$ $(\delta_r = -1.5\delta_a)$ | 0.0279 | 0.0323 | 0.0351 | 0.00720 | 0.0208 | 0.0105 | 0.0747 | 0.0494 | 0.0245 | 0.0118 | 0.0153 | −0.00128 |
| $Y^*_{\delta_r}$ | 0.0347 | 0.0402 | 0.0435 | 0.0280 | 0.0438 | 0.0225 | 0.0669 | 0.0392 | 0.0185 | 0.0235 | 0.00898 | 0.00940 |
| $Y^*_{\delta_r}$ $(\delta_a = 0.17\delta_r)$ | 0.0483 | 0.0559 | 0.0606 | 0.0364 | 0.0585 | 0.0301 | 0.0967 | 0.0575 | 0.0274 | 0.0315 | 0.0139 | 0.0116 |
| $L'_\beta$ | −6.61 | −8.78 | −10.1 | −20.0 | −22.3 | −19.2 | −51.2 | −27.6 | −18.9 | −116 | −55.1 | −60.5 |
| $L'_p$ | −1.69 | −2.30 | −2.74 | −1.22 | −2.35 | −1.08 | −5.14 | −1.89 | −1.23 | −4.25 | −2.05 | −2.69 |
| $L'_r$ | 1.22 | 1.64 | 1.91 | 3.51 | 2.86 | 2.12 | 4.56 | 2.59 | 1.60 | 2.63 | 1.36 | 2.65 |
| $L'_{\delta_r}$ | 7.06 | 9.51 | 11.1 | 2.08 | 6.17 | 2.97 | 19.5 | 11.1 | 5.07 | 7.31 | 4.18 | 5.23 |
| $L'_{\delta_r}$ $(\delta_a = 0.17\delta_r)$ | −0.540 | −0.882 | −1.30 | −0.175 | −0.475 | −0.353 | 1.61 | −0.975 | −0.817 | 1.11 | −0.271 | 0.784 |
| $L'_{\delta_a}$ | −44.7 | −61.1 | −73.0 | −13.3 | −39.1 | −19.5 | −105 | −71.2 | −34.7 | −36.4 | −26.2 | −26.1 |
| $L_{\delta_a}$ $(\delta_r = -1.5\delta_a)$ | −55.3 | −75.4 | −89.7 | −16.4 | −48.3 | −23.9 | −135 | −87.9 | −42.1 | −47.4 | −32.5 | −34.0 |
| $N'_\beta$ | 5.07 | 5.42 | 5.68 | −0.192 | 2.17 | 0.506 | 16.0 | 7.50 | 2.79 | 18.9 | 7.78 | 11.1 |
| $N'_p$ | −0.0307 | −0.0527 | −0.0787 | −0.0351 | −0.0582 | −0.0261 | −0.135 | −0.0442 | −0.0301 | −0.113 | −0.0512 | −0.0684 |
| $N'_r$ | −0.472 | −0.498 | −0.513 | −0.199 | −0.472 | −0.218 | −1.27 | −0.627 | −0.263 | −0.823 | −0.364 | −0.376 |
| $N'_{\delta_r}$ | −2.55 | −2.68 | −2.75 | −0.505 | −1.63 | −0.792 | −6.16 | −3.28 | −1.44 | −2.90 | −1.45 | −1.91 |
| $N'_{\delta_r}$ $(\delta_a = 0.17\delta_r)$ | −3.41 | −3.70 | −3.95 | −0.696 | −2.23 | −1.08 | −8.77 | 4.87 | −2.09 | −4.18 | −2.23 | −2.53 |
| $N'_{\delta_a}$ | −5.09 | −6.03 | −7.01 | −1.12 | −3.51 | −1.69 | −15.4 | −9.34 | −3.85 | −7.54 | −4.59 | −3.64 |
| $N'_{\delta_a}$ $(\delta_r = -1.5\delta_a)$ | −1.28 | −2.02 | −2.89 | −0.362 | −1.07 | −0.500 | −6.13 | −4.42 | −1.69 | −3.19 | −2.41 | −0.778 |

Table A-2 (Continued)
C. Aileron lateral transfer function factors

| | | Flight condition | | | | | | | | | | | |
|---|---|---|---|---|---|---|---|---|---|---|---|---|---|
| | | 1 | 2 | 3 | 4 | 5 | 6 | 7 | 8 | 9 | 10 | 11 | 12 |
| Mach No., $M$ | | 0.755 | 0.755 | 0.755 | 0.2 | 0.4 | 0.4 | 0.9 | 0.9 | 0.9 | 1.4 | 1.4 | 2.0 |
| Altitude, $h$ | | 20,000 | 20,000 | 20,000 | S.L. | S.L. | 20,000 | S.L. | 20,000 | 40,000 | 20,000 | 40,000 | 40,000 |
| c.g. | | 29 | 30.5 | 26 | 30.5 | 30.5 | 30.5 | 30.5 | 30.5 | 30.5 | 30.5 | 30.5 | 30.5 |
| Weight | | 35,000 | 30,000 | 28,000 | 25,500 | 29,776 | 29,776 | 29,776 | 29,776 | 29,776 | 29,776 | 29,776 | 29,776 |
| Trim $\alpha$, $\alpha_0$, degrees | | 4.42 | 4.04 | 3.88 | 18.0 | 4.9 | 11.0 | 2.0 | 2.7 | 5.4 | 1.2 | 2.7 | 1.2 |
| $\Delta_{lat}$ | $1/T_s$ | −0.0170 | −0.0166 | −0.0164 | 0.169 | 0.032 | 0.080 | −0.004 | −0.006 | 0.001 | 0.010 | 0.010 | −0.003 |
| | $1/T_R$ | 1.60 | 2.19 | 2.62 | 0.592 | 2.09 | 0.678 | 5.03 | 1.84 | 1.05 | 4.39 | 1.97 | 2.76 |
| | $\omega_D$ | 2.37 | 2.47 | 2.54 | 2.42 | 2.01 | 2.00 | 4.34 | 3.01 | 2.12 | 4.73 | 3.22 | 3.57 |
| | $\zeta_D$ | 0.164 | 0.175 | 0.178 | 0.162 | 0.233 | 0.162 | 0.224 | 0.159 | 0.129 | 0.116 | 0.095 | 0.074 |
| $p/\delta_a$ | d-c gain | −5.35 | −5.00 | −4.86 | 1.43 | 3.71 | 2.82 | −7.60 | −13.2 | 117 | 0.754 | 2.50 | −1.45 |
| | $A_{p_a}$ | −44.7 | −61.1 | −73.0 | −13.3 | −39.1 | −19.5 | −105 | −71.2 | −34.7 | −36.4 | −26.2 | −26.1 |
| | $1/T_{p_a}$ | −0.00310 | −0.00282 | −0.00270 | −0.041 | −0.006 | −0.014 | −0.001 | −0.002 | −0.003 | −0.0005 | −0.001 | −0.0003 |
| | $\omega_{p_a}$ | 2.44 | 2.53 | 2.61 | 1.24 | 2.08 | 1.48 | 4.96 | 3.37 | 2.22 | 6.59 | 4.19 | 4.43 |
| | $\zeta_{p_a}$ | 0.171 | 0.181 | 0.186 | 0.298 | 0.245 | 0.192 | 0.260 | 0.191 | 0.132 | 0.147 | 0.101 | 0.112 |
| $\varphi/\delta_a$ | d-c gain | 1,737 | 1,782 | 1,806 | −37.3 | −633 | −202 | 7,120 | 8,320 | −34,300 | −1,650 | −2,250 | 4,200 |
| | $A_{\varphi_a}$ | −45.1 | −61.5 | −73.5 | −13.6 | −39.4 | −19.8 | −106 | −71.7 | −35.0 | −36.6 | −26.4 | −26.2 |
| | $\omega_{\varphi_a}$ | 2.43 | 2.53 | 2.61 | 1.27 | 2.08 | 1.49 | 4.95 | 3.37 | 2.22 | 6.58 | 4.18 | 4.43 |
| | $\zeta_{\varphi_a}$ | 0.172 | 0.182 | 0.188 | 0.277 | 0.245 | 0.188 | 0.262 | 0.191 | 0.132 | 0.148 | 0.102 | 0.112 |

| | | | | | | | | | | | | | |
|---|---|---|---|---|---|---|---|---|---|---|---|---|---|
| $r/\delta_a$ | d-c gain | 69.2 | 70.8 | 71.7 | −4.40 | −43.3 | −14.5 | 218 | 280 | −1,240 | −36.0 | −53.0 | 69.2 |
| | $A_{r_a}$ | −5.09 | −6.03 | −7.01 | −1.12 | −3.51 | −1.69 | −15.4 | −9.34 | −3.85 | −7.54 | −4.59 | −3.64 |
| | $1/T_{r_a}$ | 0.591 | 0.672 | 0.716 | 0.430 | 0.949 | 0.406 | 2.64 | 0.837 | 0.411 | 2.68 | 0.569 | 1.26 |
| | $\omega_{r_a}$ | 1.88 | 1.97 | 1.99 | 2.32 | 1.87 | 2.15 | 1.40 | 1.87 | 1.98 | 1.31 | 2.04 | 1.36 |
| | $\zeta_{r_a}$ | 0.254 | 0.318 | 0.348 | 0.110 | 0.250 | 0.109 | 0.700 | 0.242 | 0.157 | 0.506 | 0.324 | 0.410 |
| $\beta/\delta_a$ | d-c gain | 7.41 | 7.57 | 7.65 | −1.52 | −7.70 | −2.77 | 18.1 | 24.5 | −114 | −1.16 | −1.87 | 2.67 |
| | $A_{\beta_a}$ | 0.0799 | 0.0926 | 0.100 | 0.049 | 0.086 | 0.044 | 0.175 | 0.108 | 0.052 | 0.047 | 0.029 | 0.013 |
| | $1/T_{\beta_{a_1}}$ | −0.295 | −0.281 | −0.271 | −61.7 | −1.57 | −45.4 | −0.116 | −0.231 | −0.445 | −0.042 | −0.006 | −0.045 |
| | $1/T_{\beta_{a_2}}$ $(\omega_{\beta_a})$ | 2.35 | 3.56 | 3.83 | (0.543) | (3.91) | (0.548) | 4.78 | 1.70 | 2.26 | 3.88 | 2.05 | 2.35 |
| | $1/T_{\beta_{a_3}}$ $(\zeta_{\beta_a})$ | 20.6 | 18.0 | 20.3 | (0.898) | (0.580) | (0.759) | 67.9 | 56.3 | 10.8 | 145 | 117 | 243 |
| $a_y/\delta_a$ (c.g.) | d-c gain | −1,143 | −1,354 | −1,484 | 56.6 | 856 | 147 | −10,000 | −6,250 | 11,200 | 783 | 502 | −1,090 |
| | $A_{a_a}$ | 62.9 | 72.9 | 78.9 | 11.6 | 38.8 | 18.7 | 176 | 101 | 45.8 | 68.3 | 39.0 | 24.8 |
| | $1/T_{a_{a_1}}$ $(\omega_{a_a})_1$ | (1.79) | (2.19) | (2.40) | (3.68) | (1.59) | (2.97) | −0.243 | −0.729 | (1.64) | −0.079 | −0.115 | −0.061 |
| | $1/T_{a_{a_2}}$ $(\zeta_{a_a})_1$ | (0.792) | (0.819) | (0.881) | (0.073) | (0.828) | (0.0837) | 6.07 | 1.52 | (0.184) | 3.38 | 2.19 | 2.17 |
| | $1/T_{a_{a_3}}$ $(\omega_{a_a})_2$ | (0.935) | (0.925) | (0.950) | (0.461) | (1.54) | (0.440) | −3.47 | −1.63 | (0.677) | −5.62 | −3.07 | −5.93 |
| | $1/T_{a_{a_4}}$ $(\zeta_{a_a})_2$ | (−0.358) | (−0.427) | (−0.511) | (0.959) | (0.0622) | (0.910) | 4.05 | 3.35 | (0.657) | 7.39 | 3.41 | 6.88 |
| $a_y/\delta_a$ (cockpit) | d-c gain | −1,143 | −1,354 | −1,484 | 56.6 | 856 | 147 | −10,000 | −6,250 | 11,200 | 783 | 502 | −1,090 |
| | $A_{a_a}$ | −176 | −237 | −288 | −52.5 | −154 | −76.1 | −446 | −301 | −138 | −186 | −129 | −126 |
| | $1/T_{a_{a_1}}$ | −0.324 | −0.336 | −0.337 | −1.12 | −0.589 | −0.719 | −0.197 | −0.275 | −0.291 | −0.071 | −0.083 | −0.055 |
| | $1/T_{a_{a_2}}$ | 0.481 | 0.551 | 0.584 | 0.334 | 0.574 | 0.289 | 1.51 | 0.567 | 0.277 | 1.28 | 0.536 | 0.859 |
| | $\omega_{a_{a_2}}$ | 2.53 | 2.61 | 2.69 | 1.30 | 2.11 | 1.42 | 5.25 | 3.60 | 2.25 | 6.66 | 4.22 | 4.75 |
| | $\zeta_{a_{a_2}}$ | 0.110 | 0.109 | 0.108 | 0.541 | 0.219 | 0.326 | 0.093 | 0.115 | 0.114 | 0.078 | 0.067 | 0.050 |

Table A-2 (Continued)
D. Rudder lateral transfer function factors

| | | Flight condition | | | | | | | | | | | |
|---|---|---|---|---|---|---|---|---|---|---|---|---|---|
| | | 1 | 2 | 3 | 4 | 5 | 6 | 7 | 8 | 9 | 10 | 11 | 12 |
| Mach No., $M$ | | 0.755 | 0.755 | 0.755 | 0.2 | 0.4 | 0.4 | 0.9 | 0.9 | 0.9 | 1.4 | 1.4 | 2.0 |
| Altitude, $h$ | | 20,000 | 20,000 | 20,000 | S.L. | S.L. | 20,000 | S.L. | 20,000 | 20,000 | 20,000 | 40,000 | 40,000 |
| c.g. | | 29 | 30.5 | 26 | 30.5 | 30.5 | 30.5 | 30.5 | 30.5 | 30.5 | 30.5 | 30.5 | 30.5 |
| Weight | | 35,000 | 30,000 | 20,000 | 25,500 | 29,776 | 29,776 | 29,776 | 29,776 | 29,776 | 29,776 | 29,776 | 29,776 |
| $q$, lb/ft² | | 392 | 392 | 392 | 59 | 237 | 109 | 1,199 | 551 | 223 | 1,332 | 549 | 1,100 |
| $\Delta_{lat}$ | $1/T_s$ | −0.0170 | −0.0166 | −0.0164 | 0.169 | 0.032 | 0.080 | −0.004 | −0.006 | 0.001 | 0.010 | 0.010 | −0.003 |
| | $1/T_R$ | 1.60 | 2.19 | 2.62 | 0.592 | 2.09 | 0.678 | 5.03 | 1.84 | 1.05 | 4.39 | 1.97 | 2.67 |
| | $\omega_D$ | 2.37 | 2.47 | 2.54 | 2.42 | 2.01 | 2.00 | 4.34 | 3.01 | 2.12 | 4.73 | 3.22 | 3.57 |
| | $\zeta_D$ | 0.164 | 0.175 | 0.178 | 0.162 | 0.233 | 0.162 | 0.224 | 0.159 | 0.129 | 0.116 | 0.095 | 0.074 |
| $p/\delta_r$ | d-c gain | 0.390 | 0.365 | 0.355 | 0.757 | 0.522 | 0.915 | −78.1 | −0.115 | 9.04 | 0.095 | 0.260 | −0.163 |
| | $A_{p_r}$ | 7.06 | 9.51 | 11.1 | 2.08 | 6.17 | 2.97 | 19.5 | 11.1 | 5.07 | 7.31 | 4.18 | 5.23 |
| | $1/T_{p_{r_1}}$ | −0.00314 | −0.00286 | −0.00275 | −0.043 | −0.006 | −0.015 | −0.351 | −0.782 | −0.003 | −0.0005 | −0.001 | −0.0003 |
| | $1/T_{p_{r_2}}$ $(\omega_{p_r})$ | (1.64) | (1.72) | (1.79) | 1.88 | 1.85 | 1.97 | −0.001 | −0.002 | 1.54 | 5.13 | 3.35 | 3.10 |
| | $1/T_{p_{r_3}}$ $(\zeta_{p_r})$ | (0.0633) | (0.0701) | (0.0736) | −2.64 | −2.05 | −2.34 | 0.565 | 0.829 | −1.69 | −5.29 | −3.40 | −3.58 |
| $\varphi/\delta_r$ | d-c gain | −125 | −128 | −129 | −19.3 | −87.8 | −65.0 | 11.9 | 74.9 | −2,650 | −207 | −234 | 474 |
| | $A_{\varphi_r}$ | 6.86 | 9.32 | 10.9 | 1.92 | 6.03 | 2.81 | 19.3 | 11.0 | 4.94 | 7.25 | 4.11 | 5.19 |
| | $1/T_{\varphi_{r_1}}$ $(\omega_{\varphi_r})$ | (1.67) | (1.74) | (1.81) | −2.98 | −2.12 | −2.49 | −0.406 | −0.808 | −1.74 | −5.34 | −3.45 | −3.61 |
| | $1/T_{\varphi_{r_2}}$ $(\zeta_{\varphi_r})$ | (0.0468) | (0.0545) | (0.0582) | 1.99 | 1.85 | 2.02 | 0.555 | 0.823 | 1.55 | 5.13 | 3.37 | 3.10 |

| | | | | | | | | | | | | | |
|---|---|---|---|---|---|---|---|---|---|---|---|---|---|
| $r/\delta_r$ | d-c gain | −5.04 | −5.16 | −5.23 | −2.33 | −6.09 | −4.70 | 0.224 | 2.44 | −95.7 | −4.54 | −5.51 | 7.80 |
| | $A_{r_r}$ | −2.55 | −2.68 | −2.75 | −0.505 | −1.63 | −0.792 | −6.16 | −3.28 | −1.44 | −2.90 | −1.45 | −1.91 |
| | $1/T_{r_{r_1}}$ | −0.436 | −0.431 | −0.427 | 0.442 | 2.18 | 0.430 | 5.57 | 2.00 | 0.735 | 4.28 | 1.43 | 2.71 |
| | $\omega_{r_r}$ $(1/T_{r_{r_2}})$ | (0.349) | (0.365) | (0.375) | 2.48 | 0.680 | 1.74 | (0.383) | 0.190 | 0.674 | 0.592 | 0.736 | 0.430 |
| | $\zeta_{r_r}$ $(1/T_{r_{r_3}})$ | (2.00) | (2.71) | (3.28) | 0.214 | 0.415 | 0.243 | (0.0062) | 0.582 | 0.502 | 0.435 | 0.610 | 0.384 |
| $\beta/\delta_r$ | d-c gain | 0.0351 | 0.0233 | 0.0172 | −0.351 | −0.560 | −0.416 | 0.402 | 0.640 | −8.39 | −0.044 | −0.070 | 0.436 |
| | $A_{\beta_r}$ | 0.0347 | 0.0402 | 0.0435 | 0.028 | 0.0438 | 0.023 | 0.067 | 0.039 | 0.019 | 0.024 | 0.009 | 0.009 |
| | $1/T_{\beta_{r_1}}$ | −0.00111 | −0.000727 | −0.000530 | −0.259 | −0.033 | −0.086 | −0.004 | −0.009 | −0.021 | −0.003 | −0.004 | −0.010 |
| | $1/T_{\beta_{r_2}}$ | 1.56 | 2.11 | 2.53 | 0.690 | 2.07 | 0.780 | 5.08 | 1.85 | 1.07 | 4.36 | 1.98 | 2.74 |
| | $1/T_{\beta_{r_3}}$ | 89.5 | 83.8 | 81.1 | 41.3 | 50.4 | 60.0 | 104 | 97.8 | 103 | 130 | 184 | 215 |
| $a_y/\delta_r$ (c.g.) | d-c gain | 21.6 | 27.2 | 30.7 | 17.0 | 79.1 | 28.7 | −159 | −129 | 839 | 60.9 | 29.4 | −165 |
| | $A_{a_r}$ | 27.3 | 31.6 | 34.2 | 6.57 | 19.6 | 9.52 | 67.2 | 36.6 | 16.2 | 34.1 | 12.2 | 18.2 |
| | $1/T_{a_{r_1}}$ $(\omega_{a_r})_1$ | 0.00614 | 0.00656 | 0.00679 | (1.02) | −0.067 | −0.333 | −0.004 | −0.010 | −0.034 | −0.012 | −0.011 | −0.012 |
| | $1/T_{a_{r_2}}$ $(\zeta_{a_r})_1$ | 1.53 | 2.04 | 2.44 | (0.986) | 2.05 | 0.904 | 6.89 | 1.85 | 1.08 | 4.29 | 1.99 | 2.74 |
| | $1/T_{a_{r_3}}$ $(\omega_{a_r})_2$ | −3.30 | −3.41 | −3.49 | (1.21) | −2.42 | −1.16 | −5.65 | −3.92 | −2.44 | −5.37 | −4.59 | −5.65 |
| | $1/T_{a_{r_3}}$ $(\zeta_{a_r})_2$ | 3.92 | 4.16 | 4.29 | −(0.243) | 3.25 | 1.88 | 5.17 | 4.59 | 2.88 | 6.16 | 5.02 | 5.99 |
| $a_y/\delta_r$ (cockpit) | d-c gain | 21.6 | 27.2 | 30.7 | 17.0 | 79.1 | 28.7 | −159 | −129 | 839 | 60.9 | 29.4 | −165 |
| | $A_{a_r}$ | 6.40 | 16.6 | 23.3 | 4.71 | 11.8 | 5.60 | 24.9 | 16.5 | 8.03 | 7.89 | 0.784 | 2.32 |
| | $1/T_{a_{r_1}}$ | 0.0646 | 0.00670 | 0.00693 | −0.300 | −0.061 | −0.135 | −0.004 | −0.010 | −0.031 | −0.012 | −0.010 | −0.012 |
| | $1/T_{a_{r_2}}$ | −8.81 | −6.05 | −5.56 | 0.402 | −6.36 | 0.428 | −16.1 | 2.69 | 0.844 | 3.35 | 1.33 | 2.50 |
| | $(\omega_{a_{r_1}})$ $1/T_{a_{r_3}}$ | (3.05) | (2.99) | (3.08) | −5.38 | (2.17) | −5.47 | (5.81) | −8.13 | −5.89 | −21.5 | −51.0 | −34.4 |
| | $(\zeta_{a_{r_1}})$ $1/T_{a_{r_4}}$ | (0.890) | (0.730) | (0.663) | 3.26 | (0.949) | 3.55 | (0.703) | 3.39 | 3.35 | 8.69 | 10.9 | 8.54 |

[a] The data in A are for body-fixed centerline axes. $S = 695$; $b = 38.13$; $c = 23.755$; $x_{a_{lat}} = 17.5$; $z_{a_{lat}} = -3.35$.

[b] The data in B are for body-fixed centerline axes. Static aeroelastic corrections are included. $\delta_a$ and $\delta_r$ derivatives include the effects of crossfeed shown.

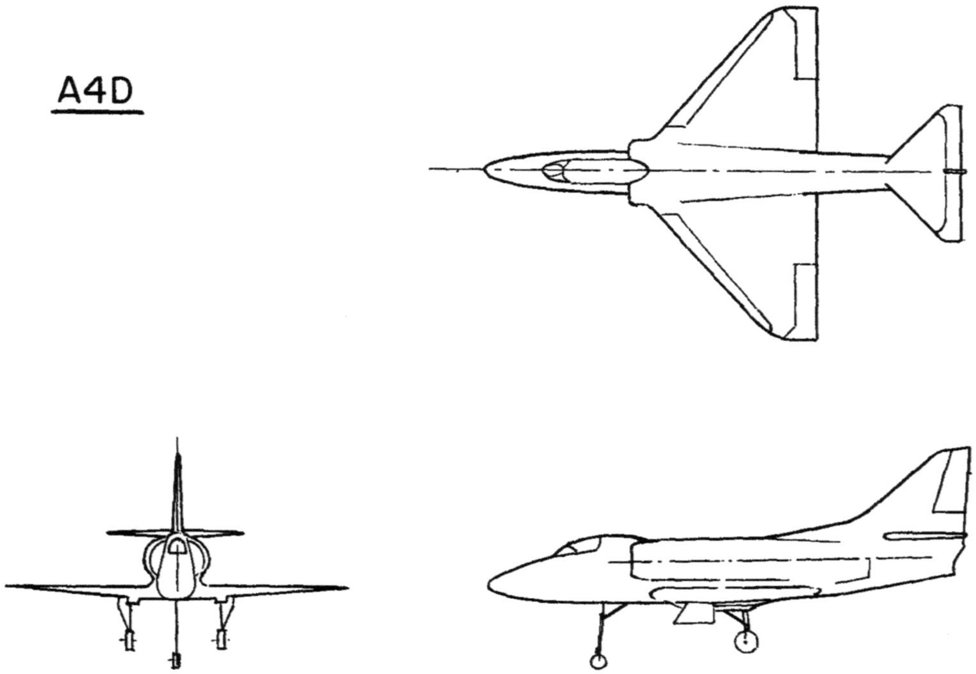


Fig. A-3. The A4D (A-4).

Table A-3. The A4D (Fig. A-3)

A. Geometrical and inertial parameters[a]

| | Flight condition | | | | | | | |
|---|---|---|---|---|---|---|---|---|
| | 1 | 2 | 3 | 4 | 5 | 6 | 7 | 8 |
| $h$ (ft) | 0 | 0 | 0 | 15,000 | 15,000 | 15,000 | 35,000 | 35,000 |
| $M$ (—) | 0.2 | 0.4 | 0.85 | 0.4 | 0.6 | 0.9 | 0.5 | 0.7 |
| $a$ (ft/sec) | 1116 | 1116 | 1116 | 1057 | 1057 | 1057 | 973 | 973 |
| $\rho$ (slugs/ft$^3$) | 0.00238 | 0.00238 | 0.00238 | 0.001496 | 0.001496 | 0.001496 | 0.000738 | 0.000738 |
| $V_{T_0}$ (ft/sec) | 223 | 447 | 950 | 423 | 634 | 952 | 487 | 681 |
| $\bar{q} = \rho V^2/2$ (lb/ft$^2$) | 59.2 | 237 | 945 | 134 | 301 | 678 | 87 | 171 |
| $W$ (lb) | 22,058 | 17,578 | 17,578 | 17,578 | 17,578 | 17,578 | 17,578 | 17,578 |
| $m$ (slugs) | 658 | 546 | 546 | 546 | 546 | 546 | 546 | 546 |
| $I_x$ (slug-ft$^2$) | 16,450 | 8,010 | 8,070 | 8,190 | 8,000 | 8,050 | 8,590 | 8,030 |
| $I_y$ (slug-ft$^2$) | 29,300 | 25,900 | 25,900 | 25,900 | 25,900 | 25,900 | 25,900 | 25,900 |
| $I_z$ (slug-ft$^2$) | 35,220 | 29,300 | 29,230 | 29,140 | 29,300 | 29,260 | 28,730 | 29,250 |
| $I_{xz}$ (slug-ft$^2$) | −5,850 | −446 | 1,159 | −1,994 | 37 | 1,040 | −3,460 | −891 |
| $x_{c.g.}/\bar{c}$ | 0.25 | 0.25 | 0.25 | 0.25 | 0.25 | 0.25 | 0.25 | 0.25 |
| $\alpha_{trim}$ (degrees) | 19.5 | 4.7 | 0.4 | 8.9 | 3.4 | 0.7 | 13.0 | 5.9 |
| $\gamma_0$ (degrees) | 0 | 0 | 0 | 0 | 0 | 0 | 0 | 0 |

Table A-3 (Continued)

B. Longitudinal dimensional derivatives[b]

| | Flight condition | | | | | | | |
|---|---|---|---|---|---|---|---|---|
| | 1 | 2 | 3 | 4 | 5 | 6 | 7 | 8 |
| $h$ (ft) | 0 | 0 | 0 | 15,000 | 15,000 | 15,000 | 35,000 | 35,000 |
| $M$ (—) | 0.2 | 0.4 | 0.85 | 0.4 | 0.6 | 0.9 | 0.5 | 0.7 |
| $T_u$ (1/sec) | −0.001347 | −0.000297 | 0 | −0.000052 | 0.000225 | 0.00303 | 0.000489 | 0.000907 |
| $X_{u_{aero}}$ (1/sec) | −0.0813 | −0.01558 | −0.0308 | −0.01476 | −0.01288 | −0.0635 | −0.01736 | −0.01612 |
| $X_u$ (1/sec) | −0.0826 | −0.01588 | −0.0308 | −0.01482 | −0.01266 | −0.0605 | −0.01687 | −0.01521 |
| $X_w$ (1/sec) | −0.0312 | −0.00379 | −0.0368 | −0.0371 | −0.00588 | −0.034 | −0.0338 | −0.0245 |
| $X_{\delta_e}$ [(ft/sec²)/rad] | — | — | — | — | — | — | — | — |
| $X_{\delta_T}$ [(ft/sec²)/rad] | 0.00432 | 0.00582 | 0.00769 | 0.00374 | 0.00414 | 0.00707 | 0.00201 | 0.00238 |
| $Z_{u_{aero}}$ (1/sec) | −0.26 | −0.1442 | −0.1134 | −0.1518 | −0.1012 | −0.1346 | −0.1291 | −0.1013 |
| $Z_u$ (1/sec) | −0.26 | −0.1442 | −0.1134 | −0.1518 | −0.1012 | −0.1346 | −0.1291 | −0.1013 |
| $Z_{\dot{w}}$ (—) | −0.001681 | −0.0022 | −0.00382 | −0.001385 | −0.001616 | −0.00265 | −0.000797 | −0.000902 |
| $Z_w$ (1/sec) | −0.307 | −0.873 | −2.23 | −0.52 | −0.818 | −1.475 | −0.211 | −0.454 |
| $Z_{\delta_e}$ [(ft/sec²)/rad] | −7.07 | −40.71 | −188.0 | −22.93 | −56.92 | −101.3 | −15.81 | −33.3 |
| $Z_{\delta_T}$ [(ft/sec²)/rad] | 0 | 0 | 0 | 0 | 0 | 0 | 0 | 0 |
| $M_{u_{aero}}$ (1/sec-ft) | −0.00285 | −0.000433 | 0.001108 | −0.000467 | −0.000407 | −0.01131 | −0.000549 | 0.000513 |
| $M_u$ (1/sec-ft) | −0.0029 | −0.000441 | 0.001108 | −0.000468 | −0.0004 | −0.01122 | −0.000533 | 0.000542 |
| $M_{\dot{w}}$ (1/ft) | −0.000646 | −0.000672 | −0.001308 | −0.000482 | −0.000556 | −0.000902 | −0.000253 | −0.000303 |
| $M_w$ (1/sec-ft) | −0.0102 | −0.01972 | −0.0502 | −0.01338 | −0.02 | −0.0378 | −0.00759 | −0.0109 |
| $M_q$ (1/sec) | −0.48 | −0.988 | −2.93 | −0.67 | −1.07 | −1.934 | −0.389 | −0.592 |
| $M_{\delta_e}$ (1/sec²) | −2.21 | −11.53 | −64 | −7.4 | −19.42 | −33.8 | −5.26 | −11.33 |
| $M_{\delta_T}$ (1/sec²) | 0.000152 | 0.0001617 | 0.000243 | 0.0001181 | 0.0001309 | 0.000224 | 0.0000637 | 0.0000753 |

C. Elevator longitudinal transfer function factors

| | | Flight condition | | | | | | | |
|---|---|---|---|---|---|---|---|---|---|
| | | 1 | 2 | 3 | 4 | 5 | 6 | 7 | 8 |
| Mach No., $M$ (—) | | 0.2 | 0.4 | 0.85 | 0.4 | 0.6 | 0.9 | 0.5 | 0.7 |
| Altitude, $h$ (ft) | | 0 | 0 | 0 | 15,000 | 15,000 | 15,000 | 35,000 | 35,000 |
| c.g. (% $\bar{c}$) | | 0.25 | 0.25 | 0.25 | 0.25 | 0.25 | 0.25 | 0.25 | 0.25 |
| Weight, $W$ (lb) | | 22,058 | 17,578 | 17,578 | 17,578 | 17,578 | 17,578 | 17,578 | 17,578 |
| $\Delta_{\text{long}}$ | $\zeta_{sp}$ $(1/T_{sp_1})$ | 0.317 | 0.348 | 0.436 | 0.286 | 0.304 | (−0.0783) | 0.1883 | 0.225 |
| | $\omega_{sp}$ $(1/T_{sp_2})$ | 1.558 | 3.11 | 7.35 | 2.45 | 3.69 | (0.121) | 1.941 | 2.77 |
| | $\zeta_p$ $(1/T_{p_1})$ | 0.0871 | 0.0706 | 0.226 | 0.0439 | 0.0867 | 0.344 | 0.0488 | 0.1177 |
| | $\omega_p$ $(1/T_{p_2})$ | 0.152 | 0.0905 | 0.0696 | 0.098 | 0.0635 | 6.23 | 0.0861 | 0.0752 |
| $N^{\theta}_{\delta_e}$ | $A_\theta$ | −2.21 | −11.53 | −64 | −7.4 | −19.42 | −33.8 | −5.26 | −11.33 |
| | $1/T_{\theta_1}$ | 0.0482 | 0.01519 | 0.0287 | 0.00308 | 0.01187 | 0.0578 | −0.00544 | 0.00909 |
| | $1/T_{\theta_2}$ | 0.309 | 0.804 | 2.08 | 0.49 | 0.76 | 1.365 | 0.21 | 0.428 |
| | d-c gain | −0.585 | −1.779 | −14.51 | −0.194 | −3.19 | 7.23 | 0.215 | −1.014 |
| $N^{w}_{\delta_e}$ | $A_w$ | −7.07 | −40.7 | −188 | −22.9 | −56.9 | −101.3 | −15.81 | −33.3 |
| | $1/T_{w_1}$ | 70.2 | 127.6 | 326 | 137.1 | 218 | 319 | 162.2 | 232 |
| | $\zeta_w$ | 0.217 | 0.0782 | 0.245 | 0.069 | 0.0888 | 0.519 | 0.0917 | 0.1091 |
| | $\omega_w$ | 0.1894 | 0.101 | 0.0626 | 0.1067 | 0.0711 | 0.0583 | 0.0918 | 0.0696 |
| | d-c gain | −317 | −669 | −915 | −622 | −1140 | 298 | −772 | −863 |
| $N^{\dot{h}}_{\delta_e}$ | $A_{\dot{h}}$ | 7.07 | 40.7 | 188 | 22.9 | 56.9 | 101.3 | 15.81 | 33.3 |
| | $1/T_{\dot{h}_1}$ | −4.04 | −9.58 | −24.5 | −7.72 | −12.29 | −19.84 | −5.29 | −9.59 |
| | $1/T_{\dot{h}_2}$ | −0.0777 | 0.00242 | 0.0268 | −0.0208 | 0.00519 | 0.0554 | −0.0511 | −0.0023 |
| | $1/T_{\dot{h}_3}$ | 4.71 | 10.61 | 27.4 | 8.44 | 13.37 | 21.8 | 5.74 | 10.2 |
| | d-c gain | 10.46 | 10 | −3380 | 31.1 | −48.5 | −2427 | 24.5 | 7.5 |
| $N^{a_z}_{\delta_e}$ | $A_{a_z}$ | 12.81 | 63 | 388 | 43.6 | 117.9 | 203 | 31.5 | 68.7 |
| | $1/T_{a_{z_1}}$ | 0 | 0 | 0 | 0 | 0 | 0 | 0 | 0 |
| | $1/T_{a_{z_2}}$ | −0.0772 | 0.00242 | 0.0268 | −0.0208 | 0.00519 | 0.0554 | −0.051 | −0.0023 |
| | $\zeta_{a_z}$ | 0.0471 | 0.0422 | 0.0461 | 0.0346 | 0.0345 | 0.0365 | 0.0202 | 0.026 |
| c.g. | $\omega_{a_z}$ | 3.25 | 8.1 | 18.05 | 5.86 | 8.91 | 14.69 | 3.9 | 6.88 |
| | d-c gain | 0 | 0 | 0 | 0 | 0 | 0 | 0 | 0 |

Table A-3 (Continued)

D. Lateral dimensional derivatives[b]

| | Flight condition | | | | | | | |
|---|---|---|---|---|---|---|---|---|
| | 1 | 2 | 3 | 4 | 5 | 6 | 7 | 8 |
| $h$ (ft) | 0 | 0 | 0 | 15,000 | 15,000 | 15,000 | 35,000 | 35,000 |
| $M$ (—) | 0.2 | 0.4 | 0.85 | 0.4 | 0.6 | 0.9 | 0.5 | 0.7 |
| $Y_v$ (1/sec) | −0.1026 | −0.243 | −0.575 | −0.1476 | −0.228 | −0.363 | −0.0864 | −0.1221 |
| $Y_\beta$ (ft/sec²)/rad | −22.9 | −108.6 | −547 | −62.4 | −144.8 | −345 | −42 | −83.2 |
| $Y_{\delta_a}$ (ft/sec²)/rad | −0.606 | −2.49 | −7.41 | −0.892 | −2.37 | −5 | −0.416 | −0.979 |
| $Y^*_{\delta_a}$ (1/sec)/rad | −0.00272 | −0.00557 | −0.0078 | −0.00211 | −0.00373 | −0.00525 | −0.000855 | −0.001437 |
| $Y_{\delta_r}$ (ft/sec²)/rad | 3.70 | 15.83 | 85.3 | 8.92 | 25.1 | 52.3 | 7.28 | 14.11 |
| $Y^*_{\delta_r}$ (1/sec)/rad | 0.0166 | 0.0354 | 0.0898 | 0.0211 | 0.0395 | 0.0549 | 0.01497 | 0.0207 |
| $L_\beta$ (1/sec²) | −3.21 | −26.5 | −127.3 | −14.02 | −35 | −87.2 | −6.4 | −21.35 |
| $L_p$ (1/sec) | −0.412 | −1.695 | −3.84 | −0.987 | −1.516 | −2.5 | −0.534 | −0.816 |
| $L_r$ (1/sec) | 0.0317 | 0.913 | 2.13 | 0.608 | 0.875 | 1.391 | 0.288 | 0.523 |
| $L_{\delta_a}$ (1/sec²) | 1.875 | 16.53 | 63.7 | 8.76 | 21.3 | 36.7 | 5.09 | 11.74 |
| $L_{\delta_r}$ (1/sec²) | 0.1284 | 6.99 | 40.9 | 2.8 | 9.96 | 24.1 | 1.164 | 4.73 |
| $L'_\beta$ (1/sec²) | −4.47 | −27.3 | −117.6 | −16.29 | −34.9 | −81.6 | −9.03 | −22.7 |
| $L'_p$ (1/sec) | −0.396 | −1.699 | −3.83 | −1.004 | −1.516 | −2.49 | −0.562 | −0.819 |
| $L'_r$ (1/sec) | 0.1455 | 0.948 | 1.93 | 0.717 | 0.872 | 1.273 | 0.404 | 0.56 |
| $L'_{\delta_a}$ (1/sec²) | 2.0 | 16.53 | 64.4 | 8.95 | 21.3 | 37.1 | 5.4 | 11.77 |
| $L'_{\delta_r}$ (1/sec²) | 0.617 | 7.36 | 37 | 3.75 | 9.92 | 21.9 | 2.25 | 5.27 |
| $N_\beta$ (1/sec²) | 2.8 | 14.49 | 72.7 | 8.21 | 18.73 | 46.4 | 5.44 | 11.3 |
| $N_p$ (1/sec) | −0.111 | 0.0392 | 0.1975 | 0 | 0.0398 | 0.1244 | 0 | 0.00845 |
| $N_r$ (1/sec) | −0.296 | −0.624 | −1.481 | −0.4 | −0.565 | −0.957 | −0.24 | −0.321 |
| $N_{\delta_a}$ (1/sec²) | −0.024 | 0.319 | 2.49 | −0.1641 | 0.478 | 1.49 | −0.1087 | 0.0837 |
| $N_{\delta_r}$ (1/sec²) | −1.272 | −6.43 | −28.6 | −3.64 | −8.3 | −17.38 | −2.44 | −4.69 |
| $N'_\beta$ (1/sec²) | 3.54 | 14.9 | 68 | 9.32 | 18.69 | 43.4 | 6.52 | 11.99 |
| $N'_p$ (1/sec) | −0.0452 | 0.0651 | 0.0456 | 0.0687 | 0.0379 | 0.0359 | 0.0676 | 0.0334 |
| $N'_r$ (1/sec) | −0.32 | −0.638 | −1.404 | −0.449 | −0.564 | −0.911 | −0.288 | −0.338 |
| $N'_{\delta_a}$ (1/sec²) | −0.357 | 0.0671 | 5.05 | −0.777 | 0.504 | 2.81 | −0.759 | −0.275 |
| $N'_{\delta_r}$ (1/sec²) | −1.374 | −6.54 | −27.1 | −3.9 | −8.29 | −16.6 | −2.71 | −4.85 |

E. Aileron lateral transfer function factors

| | | Flight condition | | | | | | | |
|---|---|---|---|---|---|---|---|---|---|
| | | 1 | 2 | 3 | 4 | 5 | 6 | 7 | 8 |
| Mach No., $M$ (—) | | 0.2 | 0.4 | 0.85 | 0.4 | 0.6 | 0.9 | 0.5 | 0.7 |
| Altitude, $h$ (ft) | | 0 | 0 | 0 | 15,000 | 15,000 | 15,000 | 35,000 | 35,000 |
| c.g. ($\%\ \bar{c}$) | | 0.25 | 0.25 | 0.25 | 0.25 | 0.25 | 0.25 | 0.25 | 0.25 |
| Weight, $W$ (lb) | | 22,058 | 17,578 | 17,578 | 17,578 | 17,578 | 17,578 | 17,578 | 17,578 |
| $\alpha_{trim}$ (degrees) | | 19.5 | 4.7 | 0.4 | 8.9 | 3.4 | 0.7 | 13.0 | 5.9 |
| $\Delta_{lat}$ | $1/T_s$ | 0.0655 | 0.00931 | 0.00438 | 0.00509 | 0.00595 | 0.00595 | −0.000577 | 0.00441 |
| | $1/T_R$ | 0.563 | 1.707 | 3.81 | 1.016 | 1.535 | 2.48 | 0.562 | 0.842 |
| | $\zeta_d$ | 0.0502 | 0.1115 | 0.1203 | 0.0949 | 0.0885 | 0.0964 | 0.0734 | 0.0625 |
| | $\omega_d$ | 1.894 | 3.87 | 8.29 | 3.06 | 4.34 | 6.61 | 2.55 | 3.47 |
| $N^{p}_{\delta_a}$ | $A_p$ | 2.0 | 16.53 | 64.4 | 8.95 | 21.3 | 37.1 | 5.4 | 11.77 |
| | $1/T_{p_1}$ | 0 | 0 | 0 | 0 | 0 | 0 | 0 | 0 |
| | $\zeta_p$ | 0.1209 | 0.1148 | 0.1214 | 0.0954 | 0.0924 | 0.0977 | 0.0695 | 0.0663 |
| | $\omega_p$ | 1.665 | 3.89 | 8.84 | 2.82 | 4.43 | 7.07 | 2.3 | 3.39 |
| | d-c gain | 0 | 0 | 0 | 0 | 0 | 0 | 0 | 0 |
| $N^{r}_{\delta_a}$ | $A_r$ | −0.357 | 0.0671 | 5.05 | −0.777 | 0.504 | 2.81 | −0.759 | −0.275 |
| | $1/T_{r_1}$ | −1.074 | 17.47 | 5.57 | −1.778 | 4.86 | 4.26 | −1.294 | −3.01 |
| | $\zeta_r$ | 0.644 | −0.0945 | −0.1411 | 0.548 | −0.283 | −0.223 | 0.531 | 0.464 |
| | $\omega_r$ | 1.439 | 3.90 | 2.45 | 1.975 | 2.93 | 2.28 | 1.382 | 2.78 |
| | d-c gain | 5.99 | 75 | 146.9 | 111.6 | 122.6 | 96.4 | −888 | 143 |
| $N^{\beta}_{\delta_a}$ | $A_\beta$ | −0.00272 | −0.00557 | −0.0078 | −0.00211 | −0.00373 | −0.00525 | −0.000855 | −0.001437 |
| | $1/T_{\beta_1}$ | 0.1874 | −2.8 | −0.1621 | −368 | −0.727 | −0.1674 | −887 | −191.6 |
| | $1/T_{\beta_2}$ ($\zeta_\beta$) | 1.266 | 3.61 | 4.14 | (0.935) | 1.708 | 2.69 | (0.835) | (0.871) |
| | $1/T_{\beta_3}$ ($\omega_\beta$) | −132 | 13.59 | 649 | (0.583) | 136.4 | 535 | (0.33) | (0.811) |
| | d-c gain | | | | | | | | |
| $N^{a_y}_{\delta_a}$ | $A_{a_y}$ | −0.606 | −2.488 | −7.406 | −0.89 | −2.365 | −5.0 | −0.416 | −0.98 |
| | $1/T_{a_{y_1}}$ ($\zeta_{a_{y_1}}$) | 0.177 | (−0.034) | −0.196 | 0.512 | 1.872 | −0.214 | (0.0163) | (−0.0056) |
| | $1/T_{a_{y_2}}$ ($\omega_{a_{y_1}}$) | 1.07 | (3.21) | 4.227 | 0.57 | 4.191 | 2.747 | (9.112) | (5.94) |
| | $1/T_{a_{y_3}}$ ($\zeta_{a_{y_3}}$) | (−0.064) | (0.706) | −16.82 | (0.023) | (−0.896) | −11.81 | (0.871) | (0.926) |
| c.g. | $1/T_{a_{y_4}}$ ($\omega_{a_{y_3}}$) | (4.198) | (1.81) | 18.02 | (7.96) | (2.223) | 12.68 | (0.317) | (0.661) |

Table A-3 (Continued)

F. Rudder lateral transfer function factors

| | | Flight condition | | | | | | | |
|---|---|---|---|---|---|---|---|---|---|
| | | 1 | 2 | 3 | 4 | 5 | 6 | 7 | 8 |
| Mach No., $M$ (—) | | 0.2 | 0.4 | 0.85 | 0.4 | 0.6 | 0.9 | 0.5 | 0.7 |
| Altitude, $h$ (ft) | | 0 | 0 | 0 | 15,000 | 15,000 | 15,000 | 35,000 | 35,000 |
| c.g. (% $\bar{c}$) | | 0.25 | 0.25 | 0.25 | 0.25 | 0.25 | 0.25 | 0.25 | 0.25 |
| Weight, $W$ (lb) | | 22,058 | 17,578 | 17,578 | 17,578 | 17,578 | 17,578 | 17,578 | 17,578 |
| $\alpha_{\text{trim}}$ (degrees) | | 19.5 | 4.7 | 0.4 | 8.9 | 3.4 | 0.7 | 13.0 | 5.9 |
| $\Delta_{\text{lat}}$ | $1/T_s$ | 0.0655 | 0.00931 | 0.00438 | 0.00509 | 0.00595 | 0.00595 | −0.000577 | 0.00441 |
| | $1/T_R$ | 0.563 | 1.707 | 3.81 | 1.016 | 1.535 | 2.48 | 0.562 | 0.842 |
| | $\zeta_d$ | 0.0502 | 0.1115 | 0.1203 | 0.0949 | 0.0885 | 0.0964 | 0.0734 | 0.0625 |
| | $\omega_d$ | 1.894 | 3.87 | 8.29 | 3.06 | 4.34 | 6.61 | 2.55 | 3.47 |
| $N^{p}_{\delta_r}$ | $A_p$ | 0.617 | 7.36 | 37 | 3.75 | 9.92 | 21.9 | 2.25 | 5.27 |
| | $1/T_{p_1}$ | 0 | 0 | 0 | 0 | 0 | 0 | 0 | 0 |
| | $1/T_{p_2}$ | −2.55 | −3.12 | −4.15 | −2.89 | −3.28 | −4.24 | −2.17 | −3.06 |
| | $1/T_{p_3}$ | 2.53 | 3.03 | 4.43 | 2.65 | 3.21 | 4.35 | 1.999 | 2.91 |
| | d-c gain | 0 | 0 | 0 | 0 | 0 | 0 | | 0 |
| $N^{r}_{\delta_r}$ | $A_r$ | −1.374 | −6.54 | −27.1 | −3.9 | −8.29 | −16.6 | −2.71 | −4.85 |
| | $1/T_{r_1}$ | 0.931 | 1.864 | 3.83 | 1.295 | 1.707 | 2.58 | 0.852 | 1.15 |
| | $\zeta_r$ | −0.34 | −0.0592 | 0.307 | −0.198 | −0.0796 | 0.0754 | −0.28 | −0.235 |
| | $\omega_r$ | 0.668 | 0.639 | 0.469 | 0.656 | 0.610 | 0.564 | 0.528 | 0.63 |
| | d-c gain | −4.32 | −20.9 | −19.97 | −45 | −30.7 | −21.1 | 305 | −49.6 |
| $N^{\beta}_{\delta_r}$ | $A_\beta$ | 0.0166 | 0.0354 | 0.0898 | 0.0211 | 0.0395 | 0.0549 | 0.01497 | 0.0207 |
| | $1/T_{\beta_1}$ | −0.000527 | −0.00964 | −0.0001478 | −0.021 | −0.00649 | −0.000944 | −0.01869 | −0.01081 |
| | $1/T_{\beta_2}$ | 0.482 | 1.716 | 3.81 | 1.032 | 1.538 | 2.49 | 0.579 | 0.845 |
| | $1/T_{\beta_3}$ | 83.0 | 185.4 | 304 | 185.4 | 210 | 303 | 181.2 | 234 |
| $N^{a_y}_{\delta_r}$ | $A_{a_y}$ | 3.70 | 15.83 | 85.3 | 8.91 | 25.09 | 52.26 | 7.28 | 14.11 |
| | $1/T_{a_{y1}}$ | −0.063 | −0.019 | −0.0031 | −0.034 | −0.0145 | −0.0055 | −0.031 | −0.0218 |
| | $1/T_{a_{y2}}$ | 0.434 | 1.72 | 3.82 | 1.041 | 1.54 | 2.49 | 0.591 | 0.848 |
| | $1/T_{a_{y3}}$ | −2.054 | −5.17 | −9.61 | −4.025 | −5.13 | −7.69 | −2.88 | −3.911 |
| c.g. | $1/T_{a_{y4}}$ | 2.40 | 5.80 | 11.03 | 4.471 | 5.68 | 8.61 | 3.17 | 4.243 |

[a] The data in A are for body-fixed stability axes. $S = 260$ ft$^2$; $b = 27.5$ ft; $c = 10.8$ ft.

[b] The data in B and D are for body-fixed stability axes.

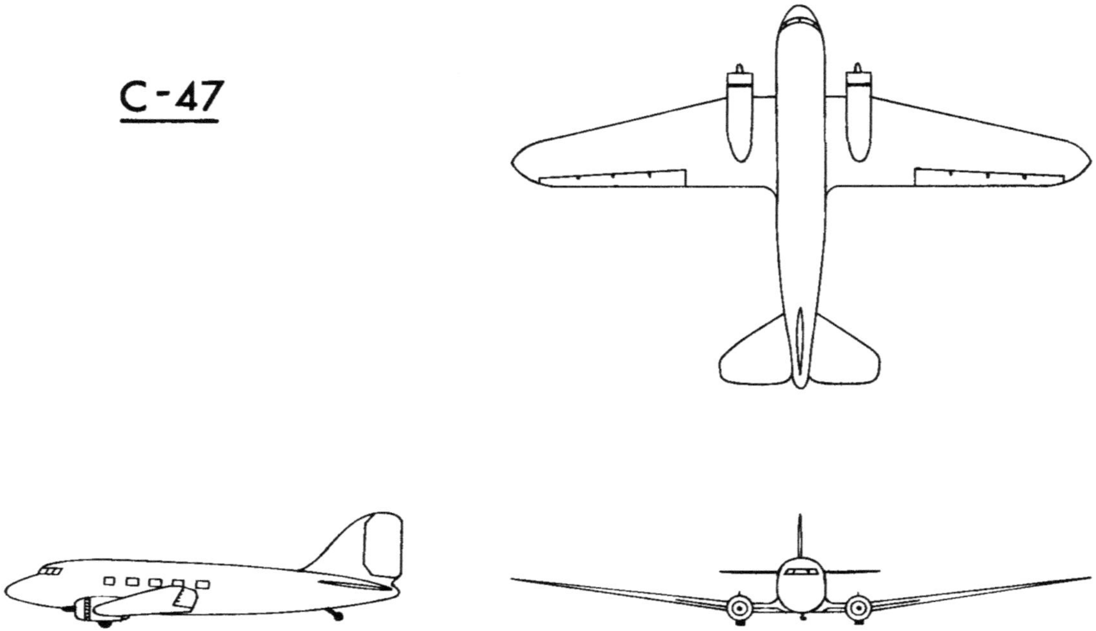


Fig A-4. The C-47 (DC-3, R4D).

Table A-4. The C-47 (Fig. A-4)

A.

| Geometrical and inertial parameters[a] | Flt. cond. 4701 (approach) |
|---|---|
| $h$ (ft) | 1,000 |
| $M$ (—) | 0.122 |
| $a$ (ft/sec) | 1113 |
| $\rho$ (slugs/ft$^3$) | 0.002309 |
| $V_{T_0}$ (ft/sec) | 136 |
| $\bar{q} = \rho V^2/2$ (lb/ft$^2$) | 21.3 |
| $W$ (lb) | 23,000 |
| $m$ (slugs) | 714 |
| $I_x$ (slug-ft$^2$) | 61,887 |
| $I_y$ (slug-ft$^2$) | 89,500 |
| $I_z$ (slug-ft$^2$) | 139,200 |
| $I_{xz}$ (slug-ft$^2$) | −4689 |

B.

| Longitudinal dimensional derivatives[a] | Flt. cond. 4701 (approach) |
|---|---|
| $h$ (ft) | 1,000 |
| $M$ (—) | 0.122 |
| $T_u$ (1/sec) | — |
| $X_{u_{aero}}$ (1/sec) | −0.0647 |
| $X_u$ (1/sec) | −0.0647 |
| $X_w$ (1/sec) | 0.115 |
| $X_{\delta_e}$ [(ft/sec$^2$)/rad] | — |
| $Z_{u_{aero}}$ (1/sec) | −0.476 |
| $Z_u$ (1/sec) | −0.476 |
| $Z_{\dot{w}}$ (—) | — |
| $Z_w$ (1/sec) | −1.235 |
| $Z_{\delta_e}$ [(ft/sec$^2$)/rad] | −0.712 |
| $M_{u_{aero}}$ (1/sec-ft) | — |
| $M_u$ (1/sec-ft) | — |
| $M_{\dot{w}}$ (1/ft) | −0.00845 |
| $M_w$ (1/sec-ft) | −0.011 |
| $M_q$ (1/sec) | −2.20 |
| $M_{\delta_e}$ (1/sec$^2$) | −4.56 |
| $M_\alpha$ (1/sec$^2$) | −1.496 |
| $M_{\dot{\alpha}}$ (1/sec) | −1.1492 |

C.

| Lateral dimensional derivatives[a] | Flt. cond. 4701 (approach) |
|---|---|
| $h$ (ft) | 1,000 |
| $M$ (—) | 0.122 |
| $Y_v$ (1/sec) | −0.0593 |
| $Y_\beta$ [(ft/sec$^2$)/rad] | −8.0648 |
| $Y_{\delta_a}$ [(ft/sec$^2$)/rad] | — |
| $Y^*_{\delta_a}$ [(1/sec)/rad] | — |
| $Y_{\delta_r}$ [(ft/sec$^2$)/rad] | 5.488 |
| $Y^*_{\delta_r}$ [(1/sec)/rad] | 0.0405 |
| $L_\beta$ (1/sec$^2$) | −2.05 |
| $L_p$ (1/sec) | −6.65 |
| $L_r$ (1/sec) | 2.61 |
| $L_{\delta_a}$ (1/sec$^2$) | 4.06 |
| $L_{\delta_r}$ (1/sec$^2$) | — |
| $L'_\beta$ (1/sec$^2$) | −2.12 |
| $L'_p$ (1/sec) | −6.63 |
| $L'_r$ (1/sec) | 2.65 |
| $L'_{\delta_a}$ (1/sec$^2$) | 4.06 |
| $L'_{\delta_r}$ (1/sec$^2$) | 0.073 |
| $N_\beta$ (1/sec$^2$) | 0.911 |
| $N_p$ (1/sec) | −0.423 |
| $N_r$ (1/sec) | −0.458 |
| $N_{\delta_a}$ (1/sec$^2$) | 0.184 |
| $N_{\delta_r}$ (1/sec$^2$) | −0.956 |
| $N'_\beta$ [1/sec$^2$) | 0.983 |
| $N'_p$ (1/sec) | −0.200 |
| $N'_r$ (1/sec) | −0.547 |
| $N'_{\delta_a}$ (1/sec$^2$) | 0.047 |
| $N'_{\delta_r}$ (1/sec$^2$) | −0.958 |

D.

| Elevator longitudinal transfer function factors | | Flt. cond. 4701 (approach) |
|---|---|---|
| Mach No., $M$ (—) | | 0.122 |
| Altitude, $h$ (ft) | | 1000 |
| $\Delta_{long}$ | $1/T_{sp_1}$ | 1.25 |
| | $1/T_{sp_2}$ | 3.31 |
| | $\zeta_p$ | 0.201 |
| | $\omega_p$ | 0.201 |
| $N^{\theta}_{\delta_e}$ | $A_\theta$ | −4.55 |
| | $1/T_{\theta_1}$ | 0.114 |
| | $1/T_{\theta_2}$ | 1.19 |
| | d-c gain | −3.64 |
| $N^{u}_{\delta_e}$ | $A_u$ | −0.082 |
| | $1/T_{u_1}$ | 2.40 |
| | $1/T_{u_2}$ | −920 |
| | d-c gain | 1074 |

E.

| Aileron lateral transfer function factors | | Flt. cond. 4701 (approach) |
|---|---|---|
| Mach No., $M$ (—) | | 0.122 |
| Altitude, $h$ (ft) | | 1000 |
| $\Delta_{lat}$ | $1/T_s$ | −0.043 |
| | $1/T_R$ | 6.57 |
| | $\zeta_d$ | 0.323 |
| | $\omega_d$ | 1.10 |
| $N^{p}_{\delta_a}$ | $A_p$ | 4.06 |
| | $1/T_{p_1}$ | 0 |
| | $\zeta_p$ | 0.312 |
| | $\omega_p$ | 1.02 |
| | d-c gain | 0 |
| $N^{\varphi}_{\delta_a}$ | $A_\varphi$ | 4.06 |
| | $\zeta_\varphi$ | 0.312 |
| | $\omega_\varphi$ | 1.02 |
| | d-c gain | −12.4 |

F.

| Rudder lateral transfer function factors | | Flt. cond. 4701 (approach) |
|---|---|---|
| Mach No., $M$ (—) | | 0.122 |
| Altitude, $h$ (ft) | | 1000 |
| $\Delta_{lat}$ | $1/T_s$ | −0.043 |
| | $1/T_R$ | 6.57 |
| | $\zeta_d$ | 0.323 |
| | $\omega_d$ | 1.10 |
| $N^{p}_{\delta_r}$ | $A_p$ | 0.0726 |
| | $1/T_{p_1}$ | 0 |
| | $1/T_{p_2}$ | 0.779 |
| | $1/T_{p_3}$ | −36.3 |
| | d-c gain | 0 |
| $N^{\varphi}_{\delta_r}$ | $A_\varphi$ | 0.0726 |
| | $1/T_{\varphi_1}$ | 0.779 |
| | $1/T_{\varphi_2}$ | −36.3 |
| | d-c gain | 6.02 |

G.

| Lateral coupling numerators | | Flt. cond. 4701 (approach) |
|---|---|---|
| Mach No., $M$ (—) | | 0.122 |
| Altitude, $h$ (ft) | | 1000 |
| $N^{p\ \beta}_{\delta_a \delta_r}$ | $A_{p\beta}$ | 0.164 |
| | $1/T_{p\beta_1}$ | 0 |
| | $1/T_{p\beta_2}$ | 24.4 |
| $N^{r\ \beta}_{\delta_a \delta_r}$ | $A_{r\beta}$ | 0.00191 |
| | $\zeta_{r\beta}$ | −0.238 |
| | $\omega_{r\beta}$ | 22.0 |
| $N^{p/s\ r}_{\delta_a\ \delta_r} = N^{\varphi\ r}_{\delta_a \delta_r}$ | $A_{\varphi_r}$ | −3.89 |
| | $1/T_{\varphi_r}$ | 0.0169 |
| $N^{\beta\ a_y}_{\delta_a \delta_r}$ | $A_{\beta_{a_y}}$ | −0.260 |
| | $1/T_{\beta_{a_{y_1}}}$ | 0.377 |
| | $1/T_{\beta_{a_{y_2}}}$ | −31.1 |
| c.g. | $1/T_{\beta_{a_{y_3}}}$ | — |

Table A-4 (continued)

D. Elevator longitudinal transfer function factors

| | | Flt. cond. 4701 (approach) |
|---|---|---|
| Mach No., $M$ (—) | | 0.122 |
| Altitude, $h$ (ft) | | 1000 |
| $N^{w}_{\delta_e}$ | $A_w$ | −0.712 |
| | $1/T_{w_1}$ | 873 |
| | $\zeta_w$ | 0.096 |
| | $\omega_w$ | 0.335 |
| | d-c gain | −415 |
| $N^{\dot{h}}_{\delta_e}$ | $A_{\dot{h}}$ | 0.712 |
| | $1/T_{\dot{h}_1}$ | 0.018 |
| | $1/T_{\dot{h}_2}$ | 34.5 |
| | $1/T_{\dot{h}_3}$ | −31.1 |
| | d-c gain | −80.3 |
| $N^{a_z}_{\delta_e}$ c.g. | $A_{a_z}$ | −0.712 |
| | $1/T_{a_{z_1}}$ | 0 |
| | $1/T_{a_{z_2}}$ | 0.0177 |
| | $1/T_{a_{z_3}}$ | 34.5 |
| | $1/T_{a_{z_4}}$ | −31.1 |
| | d-c gain | 0 |

E. Aileron lateral transfer function factors

| | | Flt. cond. 4701 (approach) |
|---|---|---|
| Mach No., M (—) | | 0.122 |
| Altitude, $h$ (ft) | | 1000 |
| $N^{r}_{\delta_a}$ | $A_r$ | 0.0474 |
| | $1/T_{r_1}$ | 1.35 |
| | $1/T_{r_2}$ | −1.48 |
| | $1/T_{r_3}$ | −10.3 |
| | d-c gain | −2.83 |
| $N^{\beta}_{\delta_a}$ | $A_\beta$ | −0.0474 |
| | $1/T_{\beta_1}$ | 0.377 |
| | $1/T_{\beta_2}$ | −31.1 |
| | $1/T_{\beta_3}$ | — |
| | d-c gain | −1.63 |
| $N^{a_y}_{\delta_a}$ c.g. | $A_{a_y}$ | 0.382 |
| | $1/T_{a_{y1}}$ | 0.377 |
| | $1/T_{a_{y2}}$ | −31.1 |
| | $1/T_{a_{y3}}$ | — |
| | $1/T_{a_{y4}}$ | — |
| | d-c gain | 13.2 |

F. Rudder lateral transfer function factors

| | | Flt. cond. 4701 (approach) |
|---|---|---|
| Mach No., $M$ (—) | | 0.122 |
| Altitude, $h$ (ft) | | 1000 |
| $N^{r}_{\delta_r}$ | $A_r$ | −0.958 |
| | $1/T_{r_1}$ | 6.66 |
| | $\zeta_r$ | 0.00811 |
| | $\omega_r$ | 0.270 |
| | d-c gain | 1.36 |
| $N^{\beta}_{\delta_r}$ | $A_\beta$ | 0.0404 |
| | $1/T_{\beta_1}$ | 6.83 |
| | $1/T_{\beta_2}$ | 24.2 |
| | $1/T_{\beta_3}$ | −0.0888 |
| | d-c gain | 1.73 |
| $N^{a_y}_{\delta_r}$ c.g. | $A_{a_y}$ | 5.49 |
| | $1/T_{a_{y1}}$ | 1.00 |
| | $1/T_{a_{y2}}$ | 6.57 |
| | $\zeta_{a_y}$ | −0.684 |
| | $\omega_{a_y}$ | 0.283 |
| | d-c gain | −8.50 |

G. Lateral coupling numerators

| | | Flt. cond. 4701 (approach) |
|---|---|---|
| Mach No., $M$ (—) | | 0.122 |
| Altitude, $h$ (ft) | | 1000 |
| $N^{r\ a_y}_{\delta_a \delta_r}$ c.g. | $A_{ra_y}$ | 0.260 |
| | $1/T_{ra_{y_1}}$ | −10.5 |
| | $\zeta_{ra_y}$ | 0.0418 |
| | $\omega_{ra_y}$ | 0.880 |
| $N^{p\ a_y}_{\delta_a \delta_r}$ c.g. | $A_{pa_y}$ | 22.3 |
| | $1/T_{pa_{y_1}}$ | 0 |
| | $1/T_{pa_{y_2}}$ | −0.408 |
| | $1/T_{pa_{y_3}}$ | 0.986 |
| $N^{\varphi\ \beta}_{\delta_a \delta_r}$ | $A_{\varphi\beta}$ | 0.164 |
| | $1/T_{\varphi\beta}$ | 24.4 |
| $N^{\varphi\ a_y}_{\delta_a \delta_r}$ c.g. | $A_{\varphi a_y}$ | 23.3 |
| | $1/T_{\varphi a_{y_1}}$ | −0.408 |
| | $1/T_{\varphi a_{y_2}}$ | 0.986 |

[a] The data in A, B, and C are for body-fixed stability axes. $S = 987\ ft^2$; $b = 95$ ft; $c = 11.5$ ft; $\gamma_0 = 0$ degrees.

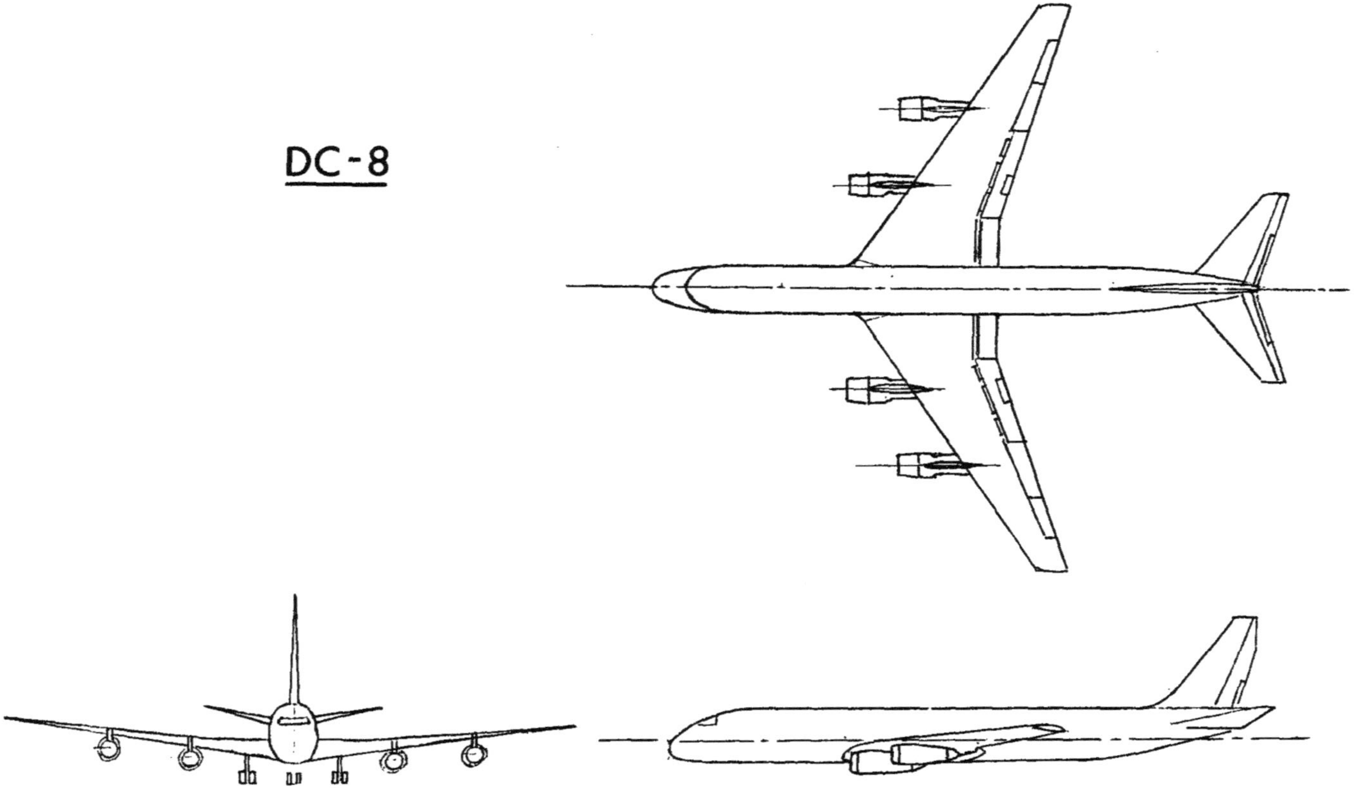


Fig. A-5. The DC-8.

**Table A-5. The DC-8 (Fig. A-5)**
**A. Geometrical and inertial parameters[a]**

| | Flight condition | | | |
|---|---|---|---|---|
| | 8001 (approach) | 8002 (holding) | 803 (cruise) | 8004 $V_{NE}$ |
| $h$ (ft) | 0 | 15,000 | 33,000 | 33,000 |
| $M$ (−) | 0.219 | 0.443 | 0.84 | 0.88 |
| $a$ (ft/sec) | 1,117 | 1,058 | 982 | 982 |
| $\rho$ (slugs/ft$^3$) | 0.002378 | 0.001496 | 0.000795 | 0.000795 |
| $V_{T_0}$ (ft/sec) | 243.5 | 468.2 | 824.2 | 863.46 |
| $\bar{q} = \rho V^2/2$ (lb/ft$^2$) | 71.02 | 163.97 | 270.0 | 296.36 |
| $W$ (lb) | 190,000 | 190,000 | 230,000 | 230,000 |
| $m$ (slugs) | 5,900 | 5,900 | 7,143 | 7,143 |
| $I_x$ (slug-ft$^2$) | 3,090,000 | 3,110,000 | 3,770,000 | 3,770,000 |
| $I_y$ (slug-ft$^2$) | 2,940,000 | 2,940,000 | 3,560,000 | 3,560,000 |
| $I_z$ (slug-ft$^2$) | 5,580,000 | 5,880,000 | 7,130,000 | 7,130,000 |
| $I_{xz}$ (slug-ft$^2$) | 28,000 | −64,500 | 45,000 | 53,700 |
| $x_{c.g.}/\bar{c}$ | 0.15 | 0.15 | 0.15 | 0.15 |
| $\theta_0$ (degrees) | 0 | 0 | 0 | 0 |
| $U_0$ (ft/sec) | 243.5 | 468.2 | 824.2 | 863.46 |
| $W_0$ (ft/sec) | 0 | 0 | 0 | 0 |
| $\delta_F$ (degrees) | 35 | 0 | 0 | 0 |

**B. Longitudinal dimensional derivatives[b]**

| | Flight condition | | | |
|---|---|---|---|---|
| | 8001 | 8002 | 8003 | 8004 |
| $h$ (ft) | 0 | 15,000 | 33,000 | 33,000 |
| $M$ (−) | 0.219 | 0.443 | 0.84 | 0.88 |
| $T_u$ (1/sec) | −0.000595 | −0.0000846 | 0.000599 | 0.000733 |
| $X_{u_{aero}}$ (1/sec) | −0.02851 | −0.00707 | −0.0145 | −0.0471 |
| $X_u$ (1/sec) | −0.0291 | −0.00714 | −0.014 | −0.0463 |
| $X_w$ (1/sec) | 0.0629 | 0.0321 | 0.0043 | −0.0259 |
| $X_{\delta_e}$ [(ft/sec$^2$)/rad] | 0 | 0 | 0 | 0 |
| $Z_{u_{aero}}$ (1/sec) | −0.2506 | −0.1329 | −0.0735 | 0.0622 |
| $Z_u$ (1/sec) | −0.2506 | −0.1329 | −0.0735 | 0.0622 |
| $Z_{\dot{w}}$ (−) | 0 | 0 | 0 | 0 |
| $Z_w$ (1/sec) | −0.6277 | −0.756 | −0.806 | −0.865 |
| $Z_{\delta_e}$ [(ft/sec$^2$)/rad] | −10.19 | −23.7 | −34.6 | −38.6 |
| $M_{u_{aero}}$ (1/sec-ft) | −0.0000077 | −0.000063 | −0.000786 | −0.00254 |
| $M_u$ (1/sec-ft) | −0.0000077 | −0.000063 | −0.000786 | −0.00254 |
| $M_{\dot{w}}$ (1/ft) | −0.001068 | −0.00072 | −0.00051 | −0.00052 |
| $M_w$ (1/sec-ft) | −0.0087 | −0.0107 | −0.0111 | −0.0139 |
| $M_q$ (1/sec) | −0.7924 | −0.991 | −0.924 | −1.008 |
| $M_{\delta_e}$ (1/sec$^2$) | −1.35 | −3.24 | −4.59 | −5.12 |

Table A-5 (Continued)
C. Elevator longitudinal transfer function factors

| | | Flight condition | | | |
|---|---|---|---|---|---|
| | | 8001 | 8002 | 8003 | 8004 |
| Mach No., $M$ (−) | | 0.219 | 0.443 | 0.84 | 0.88 |
| Altitude, $h$ (ft) | | 0 | 15,000 | 33,000 | 33,000 |
| c.g. (% $\bar{c}$) | | 15 | 15 | 15 | 15 |
| Weight, $W$ (lb) | | 190,000 | 190,000 | 230,00 | 230,000 |
| $\Delta_{\text{long}}$ | $\zeta_{sp}$ | 0.522 | 0.434 | 0.342 | 0.325 |
| | $\omega_{sp}$ | 1.619 | 2.40 | 3.15 | 3.59 |
| | $\zeta_p$ $(1/T_{p_1})$ | 0.0606 | 0.0310 | 0.241 | (−0.0708) |
| | $\omega_p$ $(1/T_{p_2})$ | 0.1635 | 0.0877 | 0.0243 | (0.108) |
| $N_{\delta_e}^{\theta}$ | $A_\theta$ | −1.338 | −3.22 | −4.57 | −5.1 |
| | $1/T_{\theta_1}$ | 0.0605 | 0.01354 | 0.01436 | 0.0493 |
| | $1/T_{\theta_2}$ | 0.535 | 0.675 | 0.725 | 0.76 |
| | d-c gain | −0.618 | −0.666 | −8.14 | 1.939 |
| $N_{\delta_e}^{u}$ | $A_u$ | −0.641 | −0.761 | −0.1489 | 1.00 |
| | $1/T_{u_1}$ | 1.08 | 1.279 | 0.816 | 0.449 |
| | $1/T_{u_2}$ | −35.3 | −72.7 | —879 | 279 |
| | d-c gain | 348 | 1598 | 18,257 | −1272 |
| $N_{\delta_e}^{w}$ | $A_w$ | −10.19 | −23.7 | −34.6 | −38.6 |
| | $1/T_{w_1}$ | 33.0 | 65.0 | 110.2 | −0.0364 |
| | $\zeta_w$ $(1/T_{w_2})$ | 0.0781 | 0.037 | 0.1362 | (0.0827) |
| | $\omega_w$ $(1/T_{w_3})$ | 0.1798 | 0.0947 | 0.0511 | (115.5) |
| | d-c gain | −155.3 | −312 | −1706 | −136 |
| $N_{\delta_e}^{\dot{h}}$ | $A_{\dot{h}}$ | 10.19 | 23.7 | 34.6 | 38.6 |
| | $1/T_{\dot{h}_1}$ | −3.75 | −5.95 | −8.24 | −8.63 |
| | $1/T_{\dot{h}_2}$ | −0.00182 | −0.000026 | 0.0107 | 0.0531 |
| | $1/T_{\dot{h}_3}$ | 4.83 | 7.29 | 9.59 | 100.9 |
| | d-c gain | 4.79 | 0.614 | −5000 | 1811 |
| $N_{\delta_e}^{a_z}$ c.g. | $A_{a_z}$ | −10.19 | −23.7 | −34.6 | −38.6 |
| | $1/T_{a_{z_1}}$ | 0 | 0 | 0 | 0 |
| | $1/T_{a_{z_2}}$ | −3.75 | −5.95 | −8.24 | −8.63 |
| | $1/T_{a_{z_3}}$ | −0.00182 | −0.000026 | 0.0107 | 0.0531 |
| | $1/T_{a_{z_4}}$ | 4.83 | 7.29 | 9.59 | 100.9 |
| | d-c gain | 0 | 0 | 0 | 0 |

Table A-5 (Continued)
D. Lateral dimensional derivatives[b]

| | Flight condition | | | |
|---|---|---|---|---|
| | 8001 | 8002 | 8003 | 8004 |
| $h$ (ft) | 0 | 15,000 | 33,000 | 33,000 |
| $M$ (—) | 0.219 | 0.443 | 0.84 | 0.88 |
| $Y_v$ (1/sec) | −0.1113 | −0.1008 | −0.0868 | −0.0931 |
| $Y_\beta$ [(ft/sec²)/rad] | −27.1 | −47.2 | −71.5 | −80.4 |
| $Y_{\delta_a}$ [(ft/sec²)/rad] | 0 | 0 | 0 | 0 |
| $Y_{\delta_a}^*$ [(1/sec)/rad] | 0 | 0 | 0 | 0 |
| $Y_{\delta_r}$ [(ft/sec²)/rad] | 5.79 | 13.48 | 18.33 | 20.12 |
| $Y_{\delta_r}^*$ [(1/sec)/rad] | 0.0238 | 0.0288 | 0.0222 | 0.0233 |
| $L_\beta$ (1/sec²) | −1.335 | −2.68 | −4.43 | −5.05 |
| $L_p$ (1/sec) | −0.95 | −1.233 | −1.18 | −1.289 |
| $L_r$ (1/sec) | 0.612 | 0.391 | 0.336 | 0.35 |
| $L_{\delta_a}$ (1/sec²) | −0.726 | −1.62 | −2.11 | −2.3 |
| $L_{\delta_r}$ (1/sec²) | −0.1848 | 0.374 | 0.559 | 0.63 |
| $L'_\beta$ (1/sec²) | −1.328 | −2.71 | −4.41 | −5.02 |
| $L'_p$ (1/sec) | −0.951 | −1.232 | −1.181 | −1.29 |
| $L'_r$ (1/sec) | 0.609 | 0.397 | 0.334 | 0.346 |
| $L'_{\delta_a}$ (1/sec²) | −0.726 | −1.62 | −2.11 | −2.3 |
| $L'_{\delta_r}$ (1/sec²) | 0.1813 | 0.392 | 0.549 | 0.612 |
| $N_\beta$ (1/sec²) | 0.763 | 1.271 | 2.17 | 2.47 |
| $N_p$ (1/sec) | −0.1192 | −0.048 | −0.01294 | −0.00744 |
| $N_r$ (1/sec) | −0.268 | −0.252 | −0.23 | −0.252 |
| $N_{\delta_a}$ (1/sec²) | −0.0496 | −0.0365 | −0.0519 | −0.0615 |
| $N_{\delta_r}$ (1/sec²) | −0.39 | −0.86 | −1.168 | −1.282 |
| $N'_\beta$ (1/sec²) | 0.757 | 1.301 | 2.14 | 2.43 |
| $N'_p$ (1/sec) | −0.124 | −0.0346 | −0.0204 | −0.01715 |
| $N'_r$ (1/sec) | −0.265 | −0.257 | −0.228 | −0.25 |
| $N'_{\delta_a}$ (1/sec²) | −0.0532 | −0.01875 | −0.0652 | −0.0788 |
| $N'_{\delta_r}$ (1/sec²) | −0.389 | −0.864 | −0.01164 | −1.277 |

Table A-5 (Continued)
E. Aileron lateral transfer function factors

| | | Flight condition | | | |
|---|---|---|---|---|---|
| | | 8001 | 8002 | 8003 | 8004 |
| Mach No., $M$ (–) | | 0.219 | 0.443 | 0.84 | 0.88 |
| Altitude, $h$ (ft) | | 0 | 15,000 | 33,000 | 33,000 |
| c.g. (% $\bar{c}$) | | 15 | 15 | 15 | 15 |
| Weight, $W$ (lb) | | 190,000 | 190,000 | 230,000 | 230,000 |
| $\Delta_{lat}$ | $1/T_s$ | −0.013 | 0.00649 | 0.00404 | 0.00447 |
| | $1/T_R$ | 1.121 | 1.329 | 1.254 | 1.356 |
| | $\zeta_d$ | 0.1096 | 0.1061 | 0.0793 | 0.0855 |
| | $\omega_d$ | 0.996 | 1.197 | 1.495 | 1.589 |
| $N^{p}_{\delta_a}$ | $A_p$ | −0.726 | −1.62 | −2.11 | −2.30 |
| | $1/T_{p_1}$ | 0 | 0 | 0 | 0 |
| | $\zeta_p$ | 0.223 | 0.1554 | 0.1072 | 0.1094 |
| | $\omega_p$ | 0.943 | 1.166 | 1.515 | 1.620 |
| | d-c gain | 0 | 0 | 0 | 0 |
| $N^{\varphi}_{\delta_a}$ | $A_\varphi$ | −0.726 | −1.62 | −2.11 | −2.30 |
| | $\zeta_\varphi$ | 0.223 | 0.1554 | 0.1072 | 0.1094 |
| | $\omega_\varphi$ | 0.943 | 1.166 | 1.515 | 1.620 |
| | d-c gain | 44.5 | −177.9 | −428 | −395 |
| $N^{r}_{\delta_a}$ | $A_r$ | −0.0532 | −0.01875 | −0.0652 | −0.0788 |
| | $1/T_{r_1}$ | 0.998 | 1.589 | 1.644 | 1.757 |
| | $\zeta_r$ | −0.656 | −0.727 | −0.392 | −0.345 |
| | $\omega_r$ | 1.242 | 2.23 | 1.323 | 1.269 |
| | d-c gain | 5.66 | −12.0 | −16.57 | −14.59 |
| $N^{\beta}_{\delta_a}$ | $A_\beta$ | 0.0532 | 0.01875 | 0.0652 | −0.0788 |
| | $1/T_{\beta_1}$ | −2.75 | −7.9 | −1.036 | −0.704 |
| | $1/T_{\beta_2}$ | 0.203 | 0.197 | 0.291 | 0.404 |
| | $1/T_{\beta_3}$ | — | — | — | — |
| | d-c gain | 2.05 | −2.35 | −1.733 | −1.467 |
| $N^{a_y}_{\delta_a}$ c.g. | $A_{a_y}$ | $-1.19 \cdot 10^{-7}$ | −0.885 | $-9.54 \cdot 10^{-7}$ | −6.33 |
| | $1/T_{a_{y_1}}$ | −2.75 | −7.9 | −1.036 | −0.704 |
| | $1/T_{a_{y_2}}$ | 0.203 | 0.197 | 0.291 | 0.404 |
| | $1/T_{a_{y_3}}$ | — | — | — | — |
| | $1/T_{a_{y_4}}$ | $1.21 \cdot 10^{7}$ | — | $4.89 \cdot 10^{6}$ | — |
| | d-c gain | −55.6 | 111.1 | 123.9 | 117.9 |

Table A-5 (Continued)
F. Rudder lateral transfer function factors

| | | Flight condition | | | |
|---|---|---|---|---|---|
| | | 8001 | 8002 | 8003 | 8004 |
| Mach No., $M$ (—) | | 0.219 | 0.443 | 0.84 | 0.88 |
| Altitude, $h$ (ft) | | 0 | 15,000 | 33,000 | 33,000 |
| c.g. (% $\bar{c}$) | | 15 | 15 | 15 | 15 |
| Weight, $W$ (lb) | | 190,000 | 190,000 | 230,000 | 230,000 |
| $\Delta_{lat}$ | $1/T_s$ | −0.013 | 0.00649 | 0.00404 | 0.00447 |
| | $1/T_R$ | 1.121 | 1.329 | 1.254 | 1.356 |
| | $\zeta_d$ | 0.1096 | 0.1061 | 0.0793 | 0.0855 |
| | $\omega_d$ | 0.996 | 1.197 | 1.495 | 1.589 |
| $N^p_{\delta_r}$ | $A_p$ | 0.1813 | 0.392 | 0.545 | 0.612 |
| | $1/T_{p_1}$ | 0 | 0 | 0 | 0 |
| | $1/T_{p_2}$ | 1.028 | 1.85 | 2.43 | 2.57 |
| | $1/T_{p_3}$ | −2.13 | −2.56 | −3.01 | −3.150 |
| | d-c gain | 0 | 0 | 0 | 0 |
| $N^{\varphi}_{\delta_r}$ | $A_{\varphi}$ | 0.1813 | 0.392 | 0.545 | 0.612 |
| | $1/T_{\varphi_1}$ | 1.028 | 1.85 | 2.43 | 2.57 |
| | $1/T_{\varphi_2}$ | −2.13 | −2.56 | −3.01 | −3.15 |
| | d-c gain | 27.5 | −150.5 | −353 | −324 |
| $N^r_{\delta_r}$ | $A_r$ | −0.389 | −0.864 | −1.165 | −1.277 |
| | $1/T_{r_1}$ | 1.124 | 1.335 | 1.276 | 1.377 |
| | $\zeta_r$ | −0.0743 | −0.0451 | −0.0619 | −0.0475 |
| | $\omega_r$ | 0.339 | 0.330 | 0.323 | 0.323 |
| | d-c gain | 3.46 | −10.19 | −13.68 | −12.00 |
| $N^{\beta}_{\delta_r}$ | $A_{\beta}$ | 0.0238 | 0.0288 | 0.0222 | 0.0233 |
| | $1/T_{\beta_1}$ | −0.0559 | −0.01475 | −0.00726 | −0.00637 |
| | $1/T_{\beta_2}$ | 1.141 | 1.297 | 1.217 | 1.323 |
| | $1/T_{\beta_3}$ | 16.47 | 30.2 | 52.6 | 55.0 |
| | d-c gain | 1.725 | −1.346 | −0.912 | −0.707 |
| $N^{a_y}_{\delta_r}$ | $A_{a_y}$ | 5.79 | 13.48 | 18.33 | 20.1 |
| | $1/T_{a_{y_1}}$ | −0.819 | −0.0347 | −0.01883 | −0.01746 |
| | $1/T_{a_{y_2}}$ | −0.1077 | 1.535 | 1.122 | 1.231 |
| | $1/T_{a_{y_3}}$ $(\zeta_{a_y})$ | (0.994) | −1.157 | −1.418 | −1.494 |
| c.g. | $1/T_{a_{y_4}}$ $(\omega_{a_y})$ | (1.078) | 1.147 | 1.723 | 1.819 |
| | d-c gain | −41.0 | 77.0 | 83.5 | 76.9 |

**Table A-5 (Continued)**
**G. Lateral coupling numerators**

| | | Flight condition | | | |
|---|---|---|---|---|---|
| | | 8001 | 8002 | 8003 | 8004 |
| Mach No., $M$ (−) | | 0.219 | 0.443 | 0.84 | 0.88 |
| Altitude, $h$ (ft) | | 0 | 15,000 | 33,000 | 33,000 |
| | $A_{p_\beta}$ | −0.01727 | −0.0466 | −0.0470 | −0.0536 |
| $N^{p\ \beta}_{\delta_a \delta_r}$ | $1/T_{p\beta_1}$ | 0 | 0 | 0 | 0 |
| | $1/T_{p\beta_2}$ | 17.21 | 30.5 | 53.4 | 56.0 |
| | $A_{r\beta}$ | −0.00127 | −0.00054 | −0.00145 | −0.001837 |
| $N^{r\ \beta}_{\delta_a \delta_r}$ | $\zeta_{r_\beta}$ | −0.0672 | −0.0656 | 0.0317 | 0.0507 |
| | $\omega_{r_\beta}$ | 5.52 | 13.39 | 8.2 | 7.79 |
| $N^{p/s\ r}_{\delta_a\ \delta_r}$ | $A_{\varphi_r}$ | 0.292 | 1.408 | 2.49 | 2.99 |
| $N^{\varphi\ r}_{\delta_a \delta_r}$ | $1/T_{\varphi_r}$ | 0.061 | 0.0567 | 0.0439 | 0.0464 |
| | $A_{\beta_{a_y}}$ | 0.308 | 0.253 | 1.195 | 1.586 |
| $N^{\beta\ a_y}_{\delta_a \delta_r}$ | $1/T_{\beta_{a_y 1}}$ | −2.75 | −7.90 | −1.036 | −0.704 |
| c.g. | $1/T_{\beta_{a_y 2}}$ | 0.203 | 0.1967 | 0.291 | 0.404 |
| | $1/T_{\beta_{a_y 3}}$ | — | — | — | — |
| | $A_{r_{a_y}}$ | −0.308 | −0.253 | −1.195 | −1.586 |
| $N^{r\ a_y}_{\delta_a \delta_r}$ | $1/T_{r_{a_{y_1}}}$ | −1.532 | −2.94 | −1.281 | −1.192 |
| c.g. | $\zeta_{r_{a_y}}$ | 0.359 | 0.317 | 0.593 | 0.645 |
| | $\omega_{r_{a_y}}$ | 1.1 | 1.86 | 1.519 | 1.536 |
| | $A_{p_{a_y}}$ | −4.21 | −21.8 | −38.7 | −46.3 |
| $N^{p\ a_y}_{\delta_a \delta_r}$ | $1/T_{p_{a_{y_1}}}$ | 0 | 0 | 0 | 0 |
| c.g. | $1/T_{p_{a_{y_2}}}$ | −0.871 | −1.184 | −1.413 | −1.482 |
| | $1/T_{p_{a_{y_3}}}$ | 1.18 | 1.445 | 1.651 | 1.744 |
| $N^{\varphi\ \beta}_{\delta_a \delta_r}$ | $A_{\varphi_\beta}$ | −0.01727 | −0.0466 | −0.0470 | −0.0536 |
| | $1/T_{\varphi_\beta}$ | 17.21 | 30.5 | 53.4 | 56.0 |
| | $A_{\varphi_{a_y}}$ | −4.21 | −21.8 | −38.7 | −46.3 |
| $N^{\varphi\ a_y}_{\delta_a \delta_r}$ | $1/T_{\varphi_{a_{y_1}}}$ | −0.871 | −1.184 | −1.413 | −1.482 |
| c.g. | $1/T_{\varphi_{a_{y_2}}}$ | 1.18 | 1.445 | 1.651 | 1.744 |

[a] The data in $A$ are for body-fixed stability axes. $S = 2600$ ft$^2$; $b = 142.3$; $c = 23$ ft; $\gamma_0 = 0$ degrees.
[b] The data in $B$ and $D$ are for body-fixed stability axes.

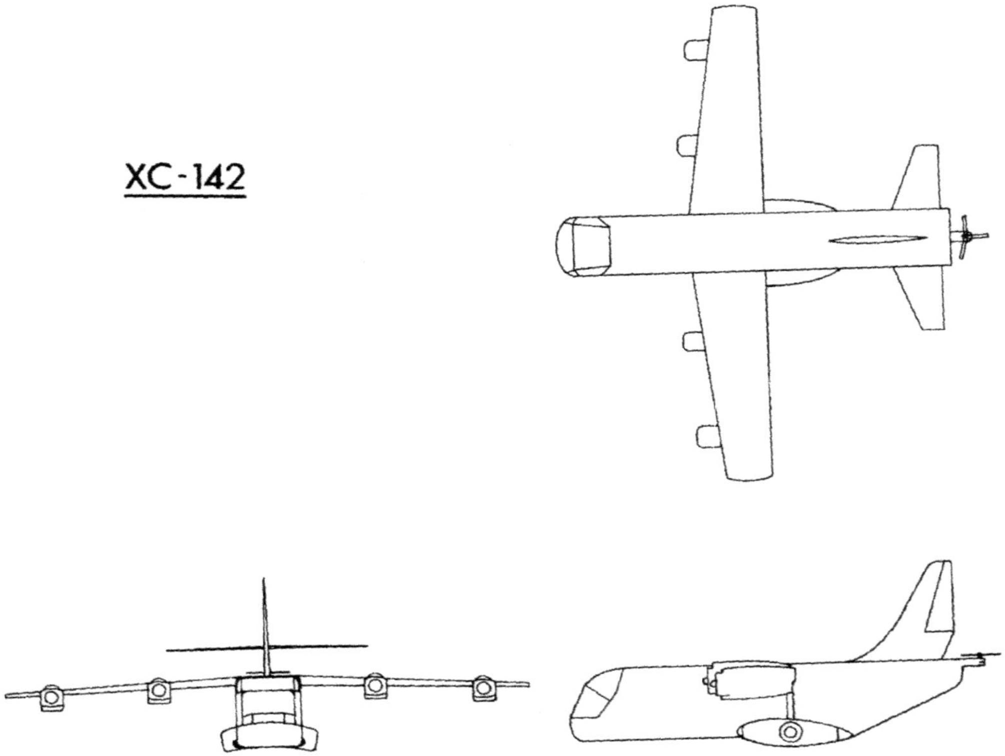


Fig. A-6. The XC-142.

Table A-6. The XC-142 (Fig. A-6)
A. Geometrical and inertial parameters[a]

| | Flight condition | | |
|---|---|---|---|
| | 1420 Hover | 1421 60 KTS | 1422 120 KTS |
| $h$ (ft) | 0 | 0 | 0 |
| $M$ (—) | 0 | 0.0906 | 0.1812 |
| $a$ (ft/sec) | 1,117 | 1,117 | 1,117 |
| $\rho$ (slugs/ft$^3$) | 0.002378 | 0.002378 | 0.002378 |
| $V_{T_0}$ (ft/sec) | 1.0 | 101.28 | 202.56 |
| $\bar{q} = \rho V^2/2$ (lb/ft$^2$) | 0 | 12.2 | 48.8 |
| $W$ (lb) | 37,474 | 37,474 | 37,474 |
| $m$ (slugs) | 1,163.8 | 1,163.8 | 1,163.8 |
| $I_x$ (slug-ft$^2$) | 173,000 | 173,000 | 173,000 |
| $I_y$ (slug-ft$^2$) | 122,000 | 122,000 | 122,000 |
| $I_z$ (slug-ft$^2$) | 267,000 | 267,000 | 267,000 |
| $I_{xz}$ (slug-ft$^2$) | 7,000 | 7,000 | 7,000 |
| $x_{c.g.}/\bar{c}$ | 0.20 | 0.20 | 0.20 |
| $i_w$ (degrees) | 90 | 14.5 | 1.25 |
| $\theta_0$ (degrees) | 0 | 0 | 0 |
| $U_0$ (ft/sec) | 1.0 | 101.28 | 202.56 |
| $W_0$ (ft/sec) | 0 | 0 | 0 |

B. Longitudinal dimensional derivatives[b]

| | Flight condition | | |
|---|---|---|---|
| | 1420 Hover | 1421 60 KTS | 1422 120 KTS |
| $h$ (ft) | 0 | 0 | 0 |
| $M$ (—) | 0 | 0.0906 | 0.1812 |
| $X_u$ (1/sec) | −0.21 | −0.196 | −0.22 |
| $X_w$ (1/sec) | 0 | 0.035 | 0.060 |
| $X_{\delta_e}$ [(ft/sec$^2$)/in.][c] | 0 | 0.124 | 0.120 |
| $X_{\delta_T}$ [(ft/sec$^2$)/rad][d] | 0 | 73 | 130.0 |
| $Z_u$ (1/sec) | 0 | −0.278 | −0.15 |
| $Z_{\dot{w}}$ (—) | 0 | 0 | 0 |
| $Z_w$ (1/sec) | −0.065 | −0.592 | −0.85 |
| $Z_{\delta_e}$ [(ft/sec$^2$)/in.][c] | 2.58 | 3.12 | 4.58 |
| $Z_{\delta_T}$ [(ft/sec$^2$)/rad][d] | −119.0 | −130 | −97 |
| $M_u$ (1/sec-ft) | 0.0073 | 0.0045 | 0.01 |
| $M_{\dot{w}}$ (1/ft) | −0.00127 | −0.00127 | −0.00127 |
| $M_w$ (1/sec-ft) | 0.0003 | −0.0002 | −0.0095 |
| $M_q$ (1/sec) | −0.085 | −0.486 | −0.89 |
| $M_{\delta_e}$ [(1/sec$^2$)/in.][c] | 0.765 | 0.87 | 1.195 |
| $M_{\delta_T}$ (1/sec$^2$-rad)[d] | 0.26 | −3.71 | −5.08 |

Table A-6 (Continued)
C. Elevator longitudinal transfer function factors[c]

| | | Flight condition | | |
|---|---|---|---|---|
| | | 1420 Hover | 1421 60 KTS | 1422 120 KTS |
| Mach No., $M$ (−) | | 0 | 0.0906 | 0.1812 |
| Altitude, $h$ (ft) | | 0 | 0 | 0 |
| c.g. (% $\bar{c}$) | | 20 | 20 | 20 |
| Weight, $W$ (lb) | | 37,474 | 37,474 | 37,474 |
| $\Delta_{\text{long}}$ | $\zeta_{sp}$ $(1/T_{sp_1})$ | −0.373 | −0.109 | 0.624 |
| | $\omega_{sp}$ $(1/T_{sp_2})$ | 0.570 | 0.411 | 1.60 |
| | $\zeta_p$ $(1/T_{p_1})$ | (0.0650) | (0.552) | 0.306 |
| | $\omega_p$ $(1/T_{p_2})$ | (0.722) | (0.940) | 0.352 |
| $N^{\theta}_{\delta_e}$ | $A_\theta$ | 0.762 | 0.866 | 1.19 |
| | $1/T_{\theta_1}$ | 0.0663 | 0.225 | 0.241 |
| | $1/T_{\theta_2}$ | 0.210 | 0.566 | 0.798 |
| | d-c gain | 0.694 | 1.26 | 0.715 |
| $N^{u}_{\delta_e}$ | $A_u$ | −24.5 | 0.124 | 0.120 |
| | $1/T_{u_1}$ | 0.0663 | 0.667 | 1.32 |
| | $1/T_{u_2}$ | — | −13.5 | −12.6 |
| | $1/T_{u_3}$ | — | 14.9 | 15.6 |
| | d-c gain | −106 | −189 | −98.0 |
| $N^{w}_{\delta_e}$ | $A_w$ | 2.58 | 3.12 | 4.58 |
| | $1/T_{w_1}$ | 0.834 | 28.7 | 53.7 |
| | $\zeta_w$ | −0.228 | 0.319 | 0.642 |
| | $\omega_w$ | 0.531 | 0.303 | 0.172 |
| | d-c gain | 39.7 | 94.1 | 22.7 |
| $N^{\dot{h}}_{\delta_e}$ | $A_{\dot{h}}$ | −2.58 | −3.12 | −4.58 |
| | $1/T_{\dot{h}_1}$ | 0.731 | 0.0562 | 0.198 |
| | $1/T_{\dot{h}_2}$ $(\zeta_{\dot{h}})$ | (−0.386) | −3.72 | −6.00 |
| | $1/T_{\dot{h}_3}$ $(\omega_{\dot{h}})$ | (0.562) | 4.47 | 7.16 |
| | d-c gain | −39.0 | −33.3 | 122 |
| $N^{a_z}_{\delta_e}$ $x_a = 18$ ft Pilot | $A_{a_z}$ | −11.1 | −12.5 | −16.8 |
| | $1/T_{a_{z1}}$ | 0 | 0 | 0 |
| | $1/T_{a_{z2}}$ | −0.293 | 0.0547 | 0.198 |
| | $\zeta_{a_z}$ | 0.660 | 0.178 | 0.110 |
| | $\omega_{a_z}$ | 0.428 | 2.07 | 3.42 |
| | d-c gain | 0 | 0 | 0 |

Table A-6 (Continued)
D. Throttle longitudinal transfer function factors[a]

| | | Flight condition | | |
|---|---|---|---|---|
| | | 1420 Hover | 1421 60 KTS | 1422 120 KTS |
| Mach No., $M$ (–) | | 0 | 0.0906 | 0.1812 |
| Altitude, $h$ (ft) | | 0 | 0 | 0 |
| c.g. (% $\bar{c}$) | | 20 | 20 | 20 |
| Weight, $W$ (lb) | | 37,474 | 37,474 | 37,474 |
| $\Delta_{long}$ | $\zeta_{sp}$ | −0.373 | −0.109 | 0.624 |
| | $\omega_{sp}$ | 0.570 | 0.411 | 1.60 |
| | $\zeta_p$ $(1/T_{p_1})$ | (0.0650) | (0.552) | 0.306 |
| | $\omega_p$ $(1/T_{p_2})$ | (0.722) | (0.940) | 0.352 |
| $N^{\theta}_{\delta_T}$ | $A_\theta$ | 0.411 | −3.54 | −4.96 |
| | $1/T_{\theta_1}$ | −0.0457 | 0.1409 | −0.1173 |
| | $1/T_{\theta_2}$ | 0.210 | 0.567 | 0.755 |
| | d-c gain | −0.258 | −3.24 | 1.374 |
| $N^{u}_{\delta_T}$ | $A_u$ | −13.24 | 73.0 | 130 |
| | $1/T_{u_1}$ | −0.0457 | 0.708 | 0.289 |
| | $\zeta_u$ | — | 0.1876 | 0.487 |
| | $\omega_u$ | — | 1.163 | 1.707 |
| | d-c gain | 39.62 | 798 | 342 |
| $N^{w}_{\delta_T}$ | $A_w$ | −119 | −130 | −97 |
| | $1/T_{w_1}$ | 0.721 | 3.63 | 11.92 |
| | $\zeta_w$ | −0.375 | 0.1541 | −0.0104 |
| | $\omega_w$ | 0.571 | 0.332 | 0.220 |
| | d-c gain | −1,831 | −594 | −175 |
| $N^{\dot{h}}_{\delta_T}$ | $A_{\dot{h}}$ | 119 | 130 | 97 |
| | $1/T_{\dot{h}_1}$ | 0.722 | −0.725 | −1.739 |
| | $1/T_{\dot{h}_2}$ $(\zeta_{\dot{h}})$ | (−0.373) | −0.1355 | −0.242 |
| | $1/T_{\dot{h}_3}$ $(\omega_{\dot{h}})$ | (0.570) | 1.827 | 3.55 |
| | d-c gain | 1,831 | 267 | 453 |
| $N^{a_z}_{\delta_T}$ $x_a = 18$ ft Pilot | $A_{a_z}$ | −126.4 | −66.2 | −7.78 |
| | $1/T_{a_{z1}}$ | 0 | 0 | 0 |
| | $1/T_{a_{z2}}$ | 0.707 | −1.133 | −4.60 |
| | $1/T_{a_{z3}}$ $(\zeta_{a_z})$ | (−0.375) | −0.1258 | −0.237 |
| | $1/T_{a_{z4}}$ $(\omega_{a_z})$ | (0.559) | 2.47 | 17.07 |
| | d-c gain | 0 | 0 | 0 |

Table A-6 (Continued)
E. Lateral dimensional derivatives[b]

| | Flight condition | | |
|---|---|---|---|
| | 1420 Hover | 1421 60 KTS | 1422 120 KTS |
| $h$ (ft) | 0 | 0 | 0 |
| $M$ (—) | 0 | 0.0906 | 0.1812 |
| $Y_v$ (1/sec) | −0.015 | −0.0945 | −0.175 |
| $Y_\beta$ [(ft/sec²)/rad] | −0.015 | −9.58 | −35.5 |
| $Y_{\delta_a}$ [(ft/sec²)/in.][e] | — | — | — |
| $Y_{\delta_a}^*$ [(1/sec)/in.][e] | — | — | — |
| $Y_{\delta_r}$ [(ft/sec²)/in.][f] | 0 | 0.248 | 0.94 |
| $Y_\delta^*$ [(1/sec)/in.][f] | 0 | 0.00245 | 0.00464 |
| $L_\beta$ (1/sec²) | −0.0006 | −0.724 | −1.93 |
| $L_p$ (1/sec) | −0.235 | −0.533 | −0.85 |
| $L_r$ (1/sec) | −0.025 | 0.395 | 0.582 |
| $L_{\delta_a}$ [(1/sec²)/in.][e] | −0.285 | −0.1663 | −0.192 |
| $L_{\delta_r}$ [(1/sec²)/in.][f] | 0.0706 | −0.081 | 0.0966 |
| $L'_\beta$ (1/sec²) | −0.000616 | −0.715 | −1.91 |
| $L'_p$ (1/sec) | −0.235 | −0.539 | −0.855 |
| $L'_r$ (1/sec) | −0.0335 | 0.382 | 0.559 |
| $L'_{\delta_a}$ [(1/sec²)/in.][e] | −0.285 | −0.167 | −0.193 |
| $L'_{\delta_r}$ [(1/sec²)/in][f] | 0.0622 | −0.0871 | 0.0913 |
| $N_\beta$ (1/sec²) | −0.00037 | 0.237 | 0.630 |
| $N_p$ (1/sec) | 0 | −0.123 | −0.094 |
| $N_r$ (1/sec) | −0.21 | −0.342 | −0.58 |
| $N_{\delta_a}$ [(1/sec²)/in.][e] | 0 | −0.0085 | −0.0215 |
| $N_{\delta_r}$ [(1/sec²)/in.][f] | −0.21 | −0.148 | −0.134 |
| $N'_\beta$ (1/sec²) | −0.000386 | 0.218 | 0.580 |
| $N'_p$ (1/sec) | −0.00617 | −0.137 | −0.116 |
| $N'_r$ (1/sec) | −0.211 | −0.332 | −0.565 |
| $N'_{\delta_a}$ [(1/sec²)/in.][e] | −0.00748 | −0.0129 | −0.0266 |
| $N'_{\delta_r}$ [(1/sec²)/in.][f] | −0.208 | −0.150 | −0.132 |

Table A-6 (Continued)
F. Aileron lateral transfer function factors

| | | Flight condition | | |
|---|---|---|---|---|
| | | 1420<br>Hover | 1421<br>60 KTS | 1422<br>120 KTS |
| Mach No., $M$ (−) | | 0 | 0.0906 | 0.1812 |
| Altitude, $h$ (ft) | | 0 | 0 | 0 |
| c.g. (% $\bar{c}$) | | 20 | 20 | 20 |
| Weight, $W$ (lb) | | 37,474 | 37,474 | 37,474 |
| $\Delta_{lat}$ | $1/T_s$ | 0.1911 | 0.1185 | 0.1236 |
| | $1/T_R$ | 0.389 | 0.886 | 1.161 |
| | $\zeta_d$ | −0.265 | −0.0292 | 0.1699 |
| | $\omega_d$ | 0.225 | 0.683 | 0.914 |
| $N^{p}_{\delta_a}$ | $A_p$ | −0.285 | −0.1668 | −0.1927 |
| | $1/T_{p_1}$ | 0 | 0 | 0 |
| | $\zeta_p$ $(1/T_{p_2})$ | (0.0131) | 0.411 | 0.418 |
| | $\omega_p$ $(1/T_{p_3})$ | (0.212) | 0.555 | 0.977 |
| | d-c gain | 0 | 0 | 0 |
| $N^{\varphi}_{\delta_a}$ | $A_\varphi$ | −0.285 | −0.1668 | −0.1927 |
| | $\zeta_\varphi$ $(1/T_{\varphi_1})$ | (0.0131) | 0.411 | 0.418 |
| | $\omega_\varphi$ $(1/T_{\varphi_2})$ | (0.212) | 0.555 | 0.977 |
| | d-c gain | −0.211 | −1.047 | −1.536 |
| $N^{r}_{\delta_a}$ | $A_r$ | −0.00748 | −0.0129 | −0.0266 |
| | $1/T_{r_1}$ | −0.764 | 0.793 | 1.056 |
| | $\zeta_r$ | 0.505 | −0.813 | −0.454 |
| | $\omega_r$ | 0.771 | 1.19 | 0.960 |
| | d-c gain | 0.903 | −0.296 | −0.215 |
| $N^{\beta}_{\delta_a}$ | $A_\beta$ | 0.00748 | 0.0129 | 0.0266 |
| | $1/T_{\beta_1}$ | −1,228 | −5.62 | −1.605 |
| | $1/T_{\beta_2}$ | 0.210 | 0.265 | 0.462 |
| | $1/T_{\beta_3}$ | — | — | — |
| | d-c gain | −513 | −0.391 | −0.1643 |
| $N^{a_y}_{\delta_a}$ | $A_{a_y}$ | 0.1507 | −0.065 | −0.285 |
| $x_a = 18$ ft | $1/T_{a_{y_1}}$ | 0.1449 | −5.36 | −1.152 |
| $z_a = 1.0$ ft | $1/T_{a_{y_2}}$ | 1.207 | 0.389 | 0.603 |
| Pilot | $\zeta_{a_y}$ | −0.448 | −0.1224 | 0.0817 |
| | $\omega_{a_y}$ | 1.048 | 1.165 | 1.876 |
| | d-c gain | 7.69 | 3.74 | 5.82 |

Table A-6 (Continued)
G. Rudder lateral transfer function factors

| | | 1420<br>Hover | 1421<br>60 KTS | 1422<br>120 KTS |
|---|---|---|---|---|
| Mach No., $M$ (—) | | 0 | 0.0906 | 0.1812 |
| Altitude, $h$ (ft) | | 0 | 0 | 0 |
| c.g. (% $\bar{c}$) | | 20 | 20 | 20 |
| Weight, $W$ (lb) | | 37,474 | 37,474 | 37,474 |
| $\Delta_{lat}$ | $1/T_s$ | 0.1911 | 0.1185 | 0.1236 |
| | $1/T_R$ | 0.389 | 0.886 | 1.161 |
| | $\zeta_d$ | −0.265 | −0.0292 | 0.1699 |
| | $\omega_d$ | 0.225 | 0.683 | 0.914 |
| $N^v_{\delta_r}$ | $A_v$ | 0.0622 | −0.087 | 0.0913 |
| | $1/T_{v_1}$ | 0 | 0 | 0 |
| | $\zeta_v$ $(1/T_{v_2})$ | (0.00725) | 0.444 | (−1.586) |
| | $\omega_v$ $(1/T_{v_3})$ | (0.331) | 1.245 | (1.422) |
| | d-c gain | 0 | 0 | 0 |
| $N^\varphi_{\delta_r}$ | $A_\varphi$ | 0.0622 | −0.0871 | 0.0913 |
| | $\zeta_\varphi$ $(1/T_{\varphi_1})$ | (0.00725) | 0.444 | (−1.586) |
| | $\omega_\varphi$ $(1/T_{\varphi_2})$ | (0.331) | 1.245 | (1.422) |
| | d-c gain | 0.0396 | −2.76 | −1.717 |
| $N^r_{\delta_r}$ | $A_r$ | −0.208 | −0.1504 | −0.1318 |
| | $1/T_{r_1}$ | 0.394 | 0.863 | 1.15 |
| | $\zeta_r$ | −0.291 | −0.281 | −0.0663 |
| | $\omega_r$ | 0.244 | 0.557 | 0.456 |
| | d-c gain | −1.303 | −0.821 | −0.263 |
| $N^\beta_{\delta_r}$ | $A_\beta$ | 0.208 | 0.00245 | 0.00464 |
| | $1/T_{\beta_1}$ | 0.326 | −0.308 | −0.0244 |
| | $1/T_{\beta_2}$ | 9.52 | 0.587 | 1.076 |
| | $1/T_{\beta_3}$ | — | 62.0 | 28.8 |
| | d-c gain | 171.9 | −0.560 | −0.0293 |
| $N^{a_y}_{\delta_r}$ | $A_{a_y}$ | −3.81 | −2.37 | −1.524 |
| | $1/T_{a_{y_1}}$ | 0.0796 | −0.204 | −0.0506 |
| $x_a = 18$ ft | $1/T_{a_{y_2}}$ | 0.399 | 0.765 | 1.110 |
| $z_a = 1.0$ ft | $\zeta_{a_y}$ | −0.397 | −0.0377 | −0.0746 |
| Pilot | $\omega_{a_y}$ | 0.283 | 0.861 | 1.664 |
| | d-c gain | −2.58 | 5.61 | 1.980 |

**Table A-6 (Continued)**
**H. Lateral coupling numerators**

| | | Flight condition | | |
|---|---|---|---|---|
| | | 1420 Hover | 1421 60 KTS | 1422 120 KTS |
| Mach No., $M$ (−) | | 0 | 0.0906 | 0.1812 |
| Altitude, $h$ (ft) | | 0 | 0 | 0 |
| | $A_{p\beta}$ | −0.0599 | −0.00041 | −0.000894 |
| $N^{p\ \beta}_{\delta_a\delta_r}$ | $1/T_{p_{\beta_1}}$ | 0 | 0 | 0 |
| | $1/T_{p_{\beta_2}}$ | — | 59.0 | 31.8 |
| | $A_{r_\beta}$ | −1.929 | −0.000032 | −0.000123 |
| $N^{r\ \beta}_{\delta_a\delta_r}$ | $\zeta_{r\beta}$ | — | −0.0398 | 0.000833 |
| | $\omega_{r_\beta}$ | — | 15.55 | 5.99 |
| $N^{p/s\ r}_{\delta_a\ \delta_r}$ | $A_{\varphi_r}$ | 0.0599 | 0.0240 | 0.0278 |
| $= N^{\varphi\ r}_{\delta_a\delta_r}$ | $1/T_{\varphi_r}$ | 0.015 | 0.0898 | 0.1479 |
| $N^{\beta\ a_y}_{\delta_a\ \delta_r}$ | $A_{r_{a_y}}$ | −0.0599 | 0.000159 | 0.00132 |
| $x_a = 18$ ft | $1/T_{r_{a_{y_1}}}$ | 0 | −5.76 | −0.37 |
| $z_a = 1.0$ ft | $1/T_{\beta_{a_{y_2}}}$ $(\zeta_{\beta_{a_y}})$ | −580 | −0.0399 | (−0.1797) |
| Pilot | $1/T_{\beta_{a_{y_3}}}$ $(\omega_{\beta_{a_y}})$ | — | −130.3 | (6.14) |
| $N^{r\ a_y}_{\delta_a\delta_r}$ | $A_{r_{a_y}}$ | 0.0599 | 0.0208 | 0.00286 |
| $x_a = 18$ ft | $1/T_{r_{a_{y_1}}}$ | 0.790 | 1.599 | 4.10 |
| $z_a = 1.0$ ft | $\zeta_{r_{a_y}}$ | −0.495 | −0.452 | −0.409 |
| Pilot | $\omega_{r_{a_y}}$ | 0.782 | 1.445 | 3.36 |
| $N^{p\ a_y}_{\delta_a\ \delta_r}$ | $A_{p_{a_y}}$ | 1.078 | 0.390 | 0.320 |
| $x_a = 18$ ft | $1/T_{p_{a_{y_1}}}$ | 0 | 0 | 0 |
| $z_a = 1.0$ ft | $\zeta_{p_{a_y}}$ | 0.26 | 0.0408 | −0.041 |
| Pilot | $\omega_{p_{a_y}}$ | 0.0289 | 0.748 | 1.615 |
| $N^{\varphi\ \beta}_{\delta_a\delta_r}$ | $A_{\varphi_\beta}$ | −0.0599 | −0.000408 | −0.000894 |
| | $1/T_{\varphi_\beta}$ | — | 59.0 | 31.8 |
| $N^{\varphi\ a_y}_{\delta_a\ \delta_r}$ | $A_{\varphi_{a_y}}$ | 1.078 | 0.390 | 0.32 |
| $x_a = 18$ ft | $\zeta_{\varphi_{a_y}}$ | 0.26 | 0.0408 | −0.041 |
| $z_a = 1.0$ ft | $\omega_{\varphi_{a_y}}$ | 0.0289 | 0.748 | 1.615 |
| Pilot | | | | |

[a] The data in $A$ are for body-fixed centerline axes. $S = 534$ ft$^2$; $b = 67.5$ ft; $c = 8.07$ ft; $\gamma_0 = 0$ degrees.

[b] The data in $B$ and $E$ are for body-fixed centerline axes; thrust corrections are included.

[c] $\delta_e$ = inches of scissors (horizontal tail and tail prop contributions included).

[d] $\delta_T$ = radians of main prop blade angle (static governor effects included).

[e] $\delta_a$ = inches of lateral stick (includes aileron and differential main prop blade angle); positive $\delta_a$ gives negative $\dot{p}$.

[f] $\delta_r$ = inches of pedal (includes rudder, aileron, and differential main prop blade angle); positive $\delta_r$ gives negative $\dot{r}$.

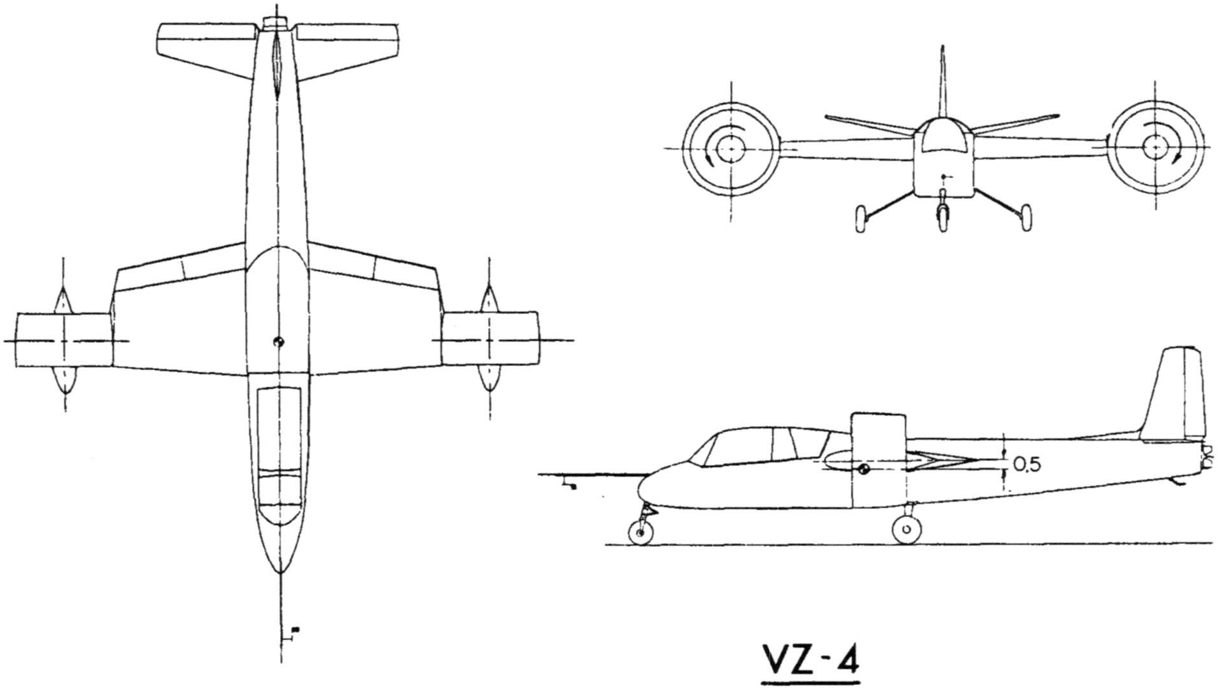


Fig. A-7. The VZ-4.

Table A-7. The VZ-4 (Fig. A-7)
A. Longitudinal derivatives

| $U_0$, ft/sec | 0 | 58.8 | 73.0 | 126.6 |
|---|---|---|---|---|
| $X_u$ | −0.137 | −0.130 | −0.140 | −0.210 |
| $X_w$ | 0 | −0.084 | 0.120 | 0.015 |
| $X_q$ | 0 | 0 | 0 | 0 |
| $X_{\delta_T}$ | 0 | 0.342 | 0.442 | 0.914 |
| $X_{\delta_e}$ | 0 | 0 | 0 | 0 |
| | | | | |
| $Z_u$ | 0 | −0.248 | −0.285 | −0.345 |
| $Z_w$ | −0.137 | −0.526 | −0.39 | −0.718 |
| $Z_q$ | 0 | 0 | 0 | 0 |
| $Z_{\delta_T}$ | 1.0 | −0.940 | −0.906 | −0.406 |
| $Z_{\delta_e}$ | 1.08 | 1.00[a] | 1.00[a] | 1.00[a] |
| | | | | |
| $M_u$ | 0.0136 | 0.0128 | 0.01205 | 0.0107 |
| $M_w$ | 0 | −0.032 | −0.046 | −0.082 |
| $M_{\dot{w}}$ | 0 | 0 | 0 | 0 |
| $M_q$ | −0.0452 | −0.858 | −1.065 | −1.839 |
| $M_{\delta_T}$ | 0 | 0 | 0 | 0 |
| $M_{\delta_e}$ | 1.0[a] | 0.775 | 0.775 | 0.775 |
| | | | | |
| $W$, lb | 3100 | 3100 | 3100 | 3100 |
| $I_y$, slug-ft$^2$ | 1790 | 1790 | 1790 | 1790 |

Table A-7 (Continued)

B. Longitudinal exact and approximate factors[b]

| | | Speed, $U_0$ (ft/sec) | | | |
|---|---|---|---|---|---|
| | | 0 | 58.8 | 73.0 | 126.6 |
| Denominator | $\omega_{sp}$ | $1/T_{sp_1} = 0.137$ | 1.40 (1.49) | 1.90 (1.94) | 3.404 (3.24) |
| | $\zeta_{sp}$ | $1/T_{sp_2} = 0.824$ | 0.459 (0.464) | 0.375 (0.374) | 0.374 (0.406) |
| | $\omega_p$ | 0.731 | 0.492 (0.457) | 0.399 (0.395) | 0.316 (0.315) |
| | $\zeta_p$ | −0.439 | 0.234 (0.378) | 0.216 (0.275) | 0.346 (0.375) |
| Stick | $1/T_{\theta_1}$ | 0.137 | 0.0757 | $\omega = 0.287$ | 0.224 |
| | $1/T_{\theta_2}$ | 0.137 | 0.539 | $\zeta = 0.820$ | 0.598 |
| | $1/T_{u_1}$ | 0.137 (0.137) | 343.1 (325.0) | −150.8 (−151.5) | −1564.0 (−1570.0) |
| | $1/T_{u_2}$ | — | 0.420 (0.441) | 0.456 (0.454) | 0.6510 (0.650) |
| | $1/T_w$ | $= 1/T_{sp_2}$ | 46.4 (45.8) | 57.6 (56.5) | 100.0 (97.5) |
| | $\omega_w$ | $= \omega_p$ | 0.377 (0.360) | 0.361 (0.370) | 0.299 (0.300) |
| | $\zeta_w$ | $= \zeta_p$ | 0.168 (0.181) | 0.191 (0.189) | 0.349 (0.350) |
| Throttle | $1/T_{h_1}$ | | −4.095 (−4.71) | −3.66 (−4.32) | −6.81 (−7.71) |
| | $1/T_{h_2}$ | [at $U_0 = 0$, $w = -\dot{h}$] | −0.218 (−0.215) | −0.1548 (−0.156) | 0.0700 (0.060) |
| | $1/T_{h_3}$ | | 5.302 (4.71) | 5.017 (4.32) | 8.79 (7.71) |
| | $1/T_w$ | | 1.36 (1.35) | 1.58 (1.62) | 3.24 (3.37) |
| | $\omega_w$ | [at $U_0 = 0$, $w = -\dot{h}$] | 0.550 (0.549) | 0.495 (0.505) | 0.326 (0.319) |
| | $\zeta_w$ | | −0.258 (−0.239) | −0.241 (−0.276) | −0.634 (−0.648) |
| | $1/T_{\theta_1}$ | — | 0.288 (0.267) | 0.264 (0.291) | 0.0924 (0.930) |
| | $1/T_{\theta_2}$ | — | | | |
| | $1/T_u$ | — | −0.688 (−0.712) | −0.645 (−0.878) | −0.0982 (−0.0914) |
| | $\omega_u$ | — | 2.029 (1.98) | 2.17 (1.84) | 3.46 (3.60) |
| | $\zeta_u$ | — | 0.5676 (0.594) | 0.427 (0.567) | 0.383 (0.360) |
| | $1/T_h$ | $= 1/T_{sp_2}$ | 0.580 | 0.425 | 1.25 |
| | $\omega_h$ | $= \omega_p$ | 1.335 | 1.81 | 3.20 |
| | $\zeta_h$ | $= \zeta_p$ | 0.104 | 0.254 | 0.247 |

C. Lateral derivatives[c,d]

| $U_0$, ft/sec | 1.0 | 58.8 | 73.0 | 126.6 |
|---|---|---|---|---|
| $Y_v$ | −0.14 | −0.2895 | −0.2945 | −0.333 |
| $Y_p$ | 0 | 0 | 0 | 0 |
| $Y_r$ | 0 | 0 | 0 | 0 |
| $Y_{\delta_a}$ | 0 | −24.9 | −26.8 | −30.04 |
| $Y_{\delta_r}$ | 1.017 | 1.85 | 2.29 | 5.31 |
| $L_v$ | −0.0122 (−0.0123) | −0.0224 (−0.0236) | −0.0216 (−0.0241) | −0.0136 (−0.0158) |
| $L_p$ | −0.271 (−0.273) | −0.455 (−0.467) | −0.497 (−0.508) | −0.67 (−0.677) |
| $L_r$ | 0 (0.0825) | 1.75 (1.848) | 0.911 (1.01) | 0.659 (0.807) |
| $L_{\delta_a}$ | 0.69 (0.696) | 0.5013 (0.5055) | 0.5208 (0.525) | 0.972 (0.979) |
| $L_{\delta_r}$ | −0.185 (−0.119) | −0.141 (−0.0442) | −0.13 (−0.0102) | −0.15 (0.126) |
| $N_v$ | 0 (0.000885) | 0.0081 (0.0098) | 0.01 (0.0121) | 0.0174 (0.0184) |
| $N_p$ | 0 (0.0197) | 0.0605 (0.0940) | 0.0535 (0.0900) | 0.0109 (0.0596) |
| $N_r$ | −0.656 (−0.662) | −0.655 (−0.788) | −0.723 (−0.796) | −1.13 (−1.19) |
| $N_{\delta_a}$ | 0 (−0.0500) | 0.003 (−0.0333) | 0.0041 (−0.0336) | 0.0133 (−0.0571) |
| $N_{\delta_r}$ | −0.539 (−0.531) | −0.78 (−0.777) | −0.962 (−0.961) | −2.204 (−2.213) |
| $I_{xz}/I_x$ | −0.1246 | −0.1246 | −0.1246 | −0.1246 |
| $I_{xz}/I_z$ | −0.07188 | −0.07188 | −0.07188 | −0.07188 |

Table A-7 (Continued)

D. Lateral exact and approximate factors[b,c]

| | | Speed, $U_0$ (ft/sec) | | | |
|---|---|---|---|---|---|
| | | 1.0 | 58.8 | 73.0 | 126.0 |
| Denominator | $\omega_d$ | 0.669 (0.704) | 0.866 (0.758) | 0.914 (0.92) | 1.59 (1.54) |
| | $\zeta_d$ | −0.347 (−0.355) | 0.1655 (0.035) | 0.205 (0.21) | 0.421 (0.436) |
| | $1/T_B$ | 0.888 (0.914) | 1.242 (1.477) | 0.957 (1.023) | 0.796 (0.804) |
| | $1/T_s$ | 0.653 (0.662) | 0.0159 (0.0141) | 0.265 (0.190) | 0.0620 (0.0583) |
| Ailerons/Differential Thrust | $1/T_{\varphi_1}$ | 0.14 (0.14) | 0.407 (0.464) | $\omega_\varphi = 1.13$ ($1/T_{\varphi_1} = 1.269$) | $\omega_\varphi = 1.65$ (1.70) |
| | $1/T_{\varphi_2}$ | 0.6564 (0.661) | 1.712 (1.764) | $\zeta_\varphi = 0.968$ ($1/T_{\varphi_1} = 1.161$) | $\zeta_\varphi = 0.595$ (0.595) |
| | $1/T_{r_1}$ | 0 (0) | −0.646 (−0.671) | −0.606 (−0.614) | −0.7188 (−0.70) |
| | $1/T_{r_2}$ | 0.140 (0.14) | 0.982 (0.979) | 1.02 (0.886) | 1.51 (1.42) |
| | $1/T_{r_3}$ | 0 (0) | 6.329 (7.32) | 8.24 (9.25) | 9.01 (9.85) |
| | $1/T_{v_1}$ | 0.657 | −0.611 | −0.577 | −0.754 |
| | $1/T_{v_2}$ | 447.0 | 0.595 | 0.846 | 1.35 |
| | $1/T_{v_3}$ | | 1.192 | 0.944 | 0.5911 |
| Rudder tail jet | $1/T_{\varphi_1}$ | 0.267 (0.368)[e] | 1.03 (1.054) | 1.90 (1.98) | 2.39 (2.71) |
| | $1/T_{\varphi_2}$ | 1.008 (0.775)[e] | 33.5 (33.0) | 99.5 (94.0) | −15.73 (−14.2) |
| | $1/T_{r_1}$ | 0.886 (0.910) | 1.17 (1.185) | 1.188 (1.191) | 1.15 (1.187) |
| | $\omega_r$ | 0.674 (0.699) | 0.802 (0.804) | 0.793 (0.792) | 0.642 (0.661) |
| | $\zeta_r$ | −0.348 (−0.352) | −0.274 (−0.303) | −0.259 (−0.252) | −0.147 (−0.1322) |
| | $1/T_{v_1}$ | −1.75 | −0.793 (0.79) | −0.452 (−0.447) | −0.205 (−0.289) |
| | $1/T_{v_2}$ | 0.994 | 1.27 (1.25) | 0.968 (0.954) | 0.896 (−0.979) |
| | $1/T_{v_3}$ | 2.22 | 25.47 (25.96) | 31.43 (31.9) | 53.93 (53.97) |

[a] Normalized. (Note $M_{\delta_e}/Z_{\delta_e}$ changes with forward speed due to shift from jet to tail control. Values quoted are approximate.)

[b] ( ) denotes approximate factors.

[c] Primed derivatives shown in parentheses.

[d] ($W = 3100$ lb; $I_x = 1990$ slug-ft$^2$; $I_z = 3450$ slug-ft$^2$; $\alpha_W = 12$ degrees)

[e] Condition of validity not satisfied.

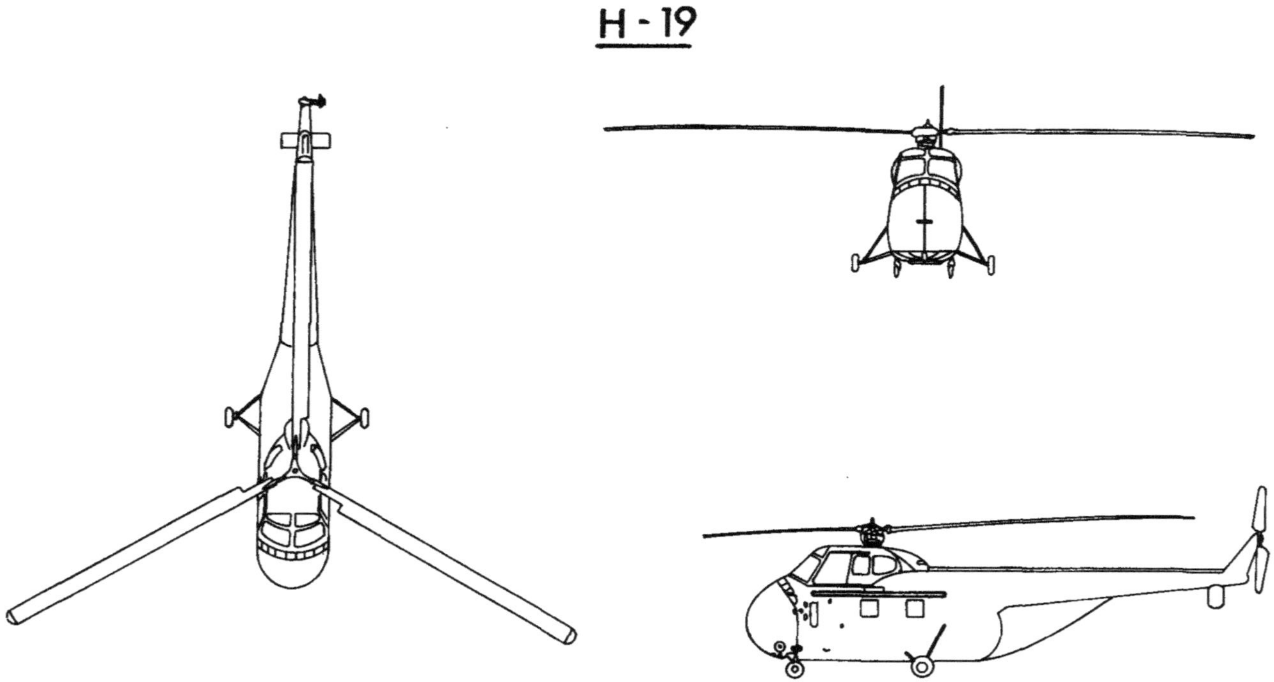


Fig. A-8. The H-19.

Table A-8. The H-19 (Fig. A-8)
A. Geometrical and inertial parameters[a]

| | Flight condition | | | |
|---|---|---|---|---|
| | 1901 | 1902 | 1903 | 1904 |
| $h$ (ft) | 0 | 0 | 0 | 0 |
| $M$ (—) | 0 | 0.0452 | 0.0643 | 0.104 |
| $a$ (ft/sec) | 1117 | 1117 | 1117 | 1117 |
| $\rho$ (slugs/ft$^3$) | 0.002378 | 0.002378 | 0.002378 | 0.002378 |
| $V_{T_0}$ (ft/sec) | 1 | 50.4 | 71.8 | 116.4 |
| $\bar{q} = \rho V^2/2$ (lb/ft$^2$) | 0.00119 | 3.02 | 6.12 | 16.1 |
| $W$ (lb) | 6400 | 7000 | 7000 | 7000 |
| $m$ (slugs) | 198 | 217 | 217 | 217 |
| $I_x$ (slug-ft$^2$) | 2118 | 1755 | 1755 | 1755 |
| $I_y$ (slug-ft$^2$) | 9640 | 9430 | 9430 | 9430 |
| $I_z$ (slug-ft$^2$) | 7840 | 7840 | 7840 | 7840 |
| $I_{xz}$ (slug-ft$^2$) | 0 | 0 | 0 | 0 |
| $\gamma_0$ (degrees) | 0 | 0 | 0 | 0 |

B. Longitudinal dimensional derivatives[a]

| | Flight condition | | | |
|---|---|---|---|---|
| | 1901 | 1902 | 1903 | 1904 |
| $h$ (ft) | 0 | 0 | 0 | 0 |
| $V_{T_0}$ (ft/sec) | 0 | 50.4 | 71.8 | 116.4 |
| $X_u$ (1/sec) | −0.0284 | −0.0394 | −0.0438 | −0.0525 |
| $X_w$ (1/sec) | 0 | −0.00151 | 0.00513 | 0.0207 |
| $X_{\delta_e}$ [(ft/sec$^2$)/rad] | 32.2 | 33.0 | 32.0 | 30.6 |
| $X_{\delta_T}$ [(ft/sec$^2$)/ ] | 0 | −4.02 | −3.04 | −0.88 |
| $Z_u$ (1/sec) | 0 | −0.141 | −0.0638 | 0.0151 |
| $Z_{\dot{w}}$ (—) | — | — | — | — |
| $Z_w$ (1/sec) | −0.69 | −0.79 | −0.80 | −0.81 |
| $Z_{\delta_e}$ [(ft/sec$^2$)/rad] | 0 | 39.4 | 56.6 | 92.0 |
| $Z_{\delta_T}$ [(ft/sec$^2$)/ ] | −358 | −284 | −291 | −304 |
| $M_u$ (1/sec-ft) | 0.00609 | 0.00805 | 0.00654 | 0.00612 |
| $M_{\dot{w}}$ (1/ft) | 0 | 0 | 0 | 0 |
| $M_w$ (1/sec-ft) | 0 | −0.0011 | −0.00171 | −0.00231 |
| $M_q$ (1/sec) | −0.610 | −0.944 | −0.984 | −1.004 |
| $M_{\delta_e}$ (1/sec$^2$) | −6.65 | −7.47 | −7.35 | −7.10 |
| $M_{\delta_T}$ (1/sec$^2$) | 0 | 1.26 | 0.902 | 0.425 |

Table A-8 (Continued)
C. Elevator longitudinal transfer function factors

| | | Flight condition | | | |
|---|---|---|---|---|---|
| | | 1901 | 1902 | 1903 | 1904 |
| Altitude, $h$ (ft) | | 0 | 0 | 0 | 0 |
| $V_{T_0}$ (ft/sec) | | 0 | 50.4 | 71.8 | 116.4 |
| Weight, $W$ (lb) | | 6400 | 7000 | 7000 | 7000 |
| $\Delta_{\text{long}}$ | $\zeta_{sp}$ $(1/T_{sp_1})$ | (0.69) | 0.995 | 0.996 | 0.908 |
| | $\omega_{sp}$ $(1/T_{sp_2})$ | (0.874) | 0.967 | 0.992 | 1.05 |
| | $\zeta_p$ | −0.250 | −0.159 | −0.106 | −0.0433 |
| | $\omega_p$ | 0.473 | 0.473 | 0.418 | 0.380 |
| $N_{\delta_e}^{\theta}$ | $A_\theta$ | −6.65 | −7.47 | −7.35 | −7.10 |
| | $1/T_{\theta_1}$ | −0.00109 | 0.00305 | 0.0153 | 0.0249 |
| | $1/T_{\theta_2}$ | 0.69 | 0.797 | 0.813 | 0.843 |
| $N_{\delta_e}^{u}$ | $A_u$ | 32.2 | 33.0 | 32.0 | 30.6 |
| | $1/T_{u_1}$ | 0.69 | 0.788 | 0.809 | 0.874 |
| | $\zeta_u$ | 0.118 | 0.174 | 0.180 | 0.187 |
| | $\omega_u$ | 2.58 | 2.71 | 2.74 | 2.68 |
| $N_{\delta_e}^{w}$ | $A_w$ | | 39.4 | 56.6 | 92.0 |
| | $1/T_{w_1}$ | | −8.71 | −8.35 | −7.94 |
| | $\zeta_w$ $(1/T_{w_2})$ | | 0.0393 | 0.0922 | (0.181) |
| | $\omega_w$ $(1/T_{w_3})$ | | 0.263 | 0.0955 | (−0.163) |
| $N_{\delta_e}^{\dot{h}}$ | $A_{\dot{h}}$ | | −39.4 | −56.6 | −92.0 |
| | $1/T_{\dot{h}_1}$ | | −0.0761 | 0.00522 | 0.0560 |
| | $\zeta_{\dot{h}}$ | | 0.171 | 0.179 | 0.183 |
| | $\omega_{\dot{h}}$ | | 2.76 | 2.75 | 2.75 |
| $N_{\delta_e}^{a_z}$ c.g. | $A_{a_z}$ | | 39.4 | 56.6 | 92.0 |
| | $1/T_{a_{z1}}$ | | 0 | 0 | 0 |
| | $1/T_{a_{z2}}$ | | −0.0761 | 0.00522 | 0.0560 |
| | $\zeta_{a_z}$ | | 0.171 | 0.179 | 0.183 |
| | $\omega_{a_z}$ | | 2.76 | 2.75 | 2.75 |

Table A-8 (Continued)
D. Throttle longitudinal transfer function factors

| | | Flight condition | | | |
|---|---|---|---|---|---|
| | | 1901 | 1902 | 1903 | 1904 |
| Altitude, $h$ (ft) | | 0 | 0 | 0 | 0 |
| $V_{T_0}$ (ft/sec) | | 0 | 50.4 | 71.8 | 116.4 |
| Weight, $W$ (lb) | | 6400 | 7000 | 7000 | 7000 |
| $\Delta_{long}$ | $\zeta_{sp}$ $(1/T_{sp_1})$ | (0.69) | 0.995 | 0.996 | 0.908 |
| | $\omega_{sp}$ $(1/T_{sp_2})$ | (0.874) | 0.967 | 0.992 | 1.05 |
| | $\zeta_p$ | −0.250 | −0.159 | −0.106 | −0.0433 |
| | $\omega_p$ | 0.473 | 0.473 | 0.418 | 0.380 |
| $N^{\theta}_{\delta_T}$ | $A_\theta$ | | 1.26 | 0.902 | 0.425 |
| | $1/T_{\theta_1}$ | | 0.0220 | 0.0227 | 0.0113 |
| | $1/T_{\theta_2}$ | | 1.03 | 1.35 | 2.49 |
| $N^{u}_{\delta_T}$ | $A_u$ | | −4.02 | −3.04 | −0.88 |
| | $1/T_{u_1}$ | | 1.03 | 1.35 | 6.35 |
| | $\zeta_u$ | | 0.094 | 0.149 | 0.533 |
| | $\omega_u$ | | 3.19 | 3.09 | 2.45 |
| $N^{w}_{\delta_T}$ | $A_w$ | | −284 | −291 | −304 |
| | $1/T_{w_1}$ | | 0.976 | 0.979 | 1.03 |
| | $\zeta_w$ | | −0.22 | −0.191 | −0.160 |
| | $\omega_w$ | | 0.495 | 0.456 | 0.437 |
| $N^{\dot{h}}_{\delta_T}$ | $A_{\dot{h}}$ | | 284 | 291 | 304 |
| | $1/T_{\dot{h}_1}$ | | 0.966 | 0.905 | 0.805 |
| | $\zeta_{\dot{h}}$ | | 0.503 | 0.483 | 0.501 |
| | $\omega_{\dot{}}$ | | 0.0152 | 0.127 | 0.251 |
| $N^{a_z}_{\delta_T}$ c.g. | $A_{a_z}$ | | −284 | −291 | −304 |
| | $1/T_{a_{z_1}}$ | | 0 | 0 | 0 |
| | $1/T_{a_{z_2}}$ | | 0.966 | 0.905 | 0.805 |
| | $\zeta_{a_z}$ | | 0.503 | 0.483 | 0.501 |
| | $\omega_{a_z}$ | | 0.0152 | 0.127 | 0.251 |

Table A-8 (Continued)
E. Lateral dimensional derivatives[a]

| | Flight condition | | | |
|---|---|---|---|---|
| | 1901 | 1902 | 1903 | 1904 |
| Altitude, $h$ (ft) | 0 | 0 | 0 | 0 |
| $V_{T_0}$ (ft/sec) | 0 | 50.4 | 71.8 | 116.4 |
| Weight, $W$ (lb) | 6400 | 7000 | 7000 | 7000 |
| $Y_v$ (1/sec) | −0.0731 | −0.096 | −0.1045 | −0.122 |
| $Y_\beta$ [(ft/sec²)/rad] | −0.0731 | −4.85 | −7.51 | −14.2 |
| $Y_{\delta_a}$ [(ft/sec²)/rad] | 32.2 | 21.3 | 24.8 | 27.9 |
| $Y_{\delta}^*$ [(1/sec)/rad] | 32.2 | 0.423 | 0.346 | 0.239 |
| $Y_{\delta_r}$ [(ft/sec²)/rad] | 17.3 | 15.8 | 15.8 | 16.7 |
| $Y_{\delta_r}^*$ [(1/sec)/rad] | 17.3 | 0.314 | 0.220 | 0.143 |
| $L'_\beta$ (1/sec²) | −0.052 | −3.65 | −5.19 | −8.43 |
| $L'_p$ (1/sec) | −3.18 | −4.97 | −5.01 | −4.81 |
| $L'_r$ (1/sec) | 0.804 | 1.00 | 1.00 | 1.00 |
| $L'_{\delta_a}$ (1/sec²) | 29.2 | 31.7 | 34.7 | 36.8 |
| $L'_{\delta_r}$ (1/sec²) | 9.78 | 12.3 | 12.3 | 13.0 |
| $N'_\beta$ (1/sec²) | 0.0352 | 1.71 | 2.42 | 3.79 |
| $N'_p$ (1/sec) | 0.22 | 0.338 | 0.276 | 0.201 |
| $N'_r$ (1/sec) | −1.1 | −1.1 | −1.1 | −1.1 |
| $N'_{\delta_a}$ (1/sec²) | 0 | 0 | 0 | 0 |
| $N'_{\delta_r}$ (1/sec²) | −13.53 | −13.80 | −13.9 | −14.4 |

Table A-8 (Continued)
F. Aileron lateral transfer function factors

| | | Flight condition | | | |
|---|---|---|---|---|---|
| | | 1901 | 1902 | 1903 | 1904 |
| Altitude, $h$ (ft) | | 0 | 0 | 0 | 0 |
| $V_{T_0}$ (ft/sec) | | 0 | 50.4 | 71.8 | 116.4 |
| Weight, $W$ (lb) | | 6400 | 7000 | 7000 | 7000 |
| $\Delta_{lat}$ | $1/T_s$ | 0.732 | 0.162 | 0.116 | 0.0821 |
| | $1/T_R$ | 3.43 | 5.09 | 5.10 | 4.87 |
| | $\zeta_d$ | 0.155 | 0.344 | 0.317 | 0.276 |
| | $\omega_d$ | 0.609 | 1.33 | 1.58 | 1.96 |
| $N^p_{\delta_a}$ | $A_p$ | 29.2 | 31.7 | 34.7 | 36.8 |
| | $1/T_{p_1}$ | 0 | 0 | 0 | 0 |
| | $\zeta_p$ $(1/T_{p_2})$ | (0.081) | 0.429 | 0.364 | 0.296 |
| | $\omega_p$ $(1/T_{p_3})$ | (1.04) | 1.33 | 1.58 | 1.97 |
| $N^\varphi_{\delta_a}$ | $A_\varphi$ | 29.2 | 31.7 | 34.7 | 36.8 |
| | $\zeta_\varphi$ $(1/T_{\varphi_1})$ | (0.081) | 0.429 | 0.364 | 0.296 |
| | $\omega_\varphi$ $(1/T_{\varphi_2})$ | (1.04) | 1.33 | 1.58 | 1.97 |
| $N^r_{\delta_a}$ | $A_r$ | 7.55 | 11.4 | 10.4 | 16.5 |
| | $1/T_{r_1}$ | — | — | — | — |
| | $\zeta_r$ | 0.117 | 0.103 | 0.119 | 0.136 |
| | $\omega_r$ | 2.09 | 1.74 | 1.90 | 2.15 |
| $N^\beta_{\delta_a}$ | $A_\beta$ | 32.2 | 0.423 | 0.346 | 0.239 |
| | $1/T_{\beta_1}$ | 1.11 | 2.85 | 3.65 | 4.50 |
| | $1/T_{\beta_2}$ $(\zeta_\beta)$ | 0.295 | 0.375 | 0.334 | 0.218 |
| | $1/T_{\beta_3}$ $(\omega_\beta)$ | 5.37 | 4.30 | 3.68 | 3.22 |

Table A-8 (Continued)
G. Rudder lateral transfer function factors

| | | Flight condition | | | |
|---|---|---|---|---|---|
| | | 1901 | 1902 | 1903 | 1904 |
| Altitude, $h$ (ft) | | 0 | 0 | 0 | 0 |
| $V_{T_0}$ (ft/sec) | | 0 | 50.4 | 71.8 | 116.4 |
| Weight, $W$ (lb) | | 6400 | 7000 | 7000 | 7000 |
| $\Delta_{lat}$ | $1/T_s$ | 0.732 | 0.162 | 0.116 | 0.0821 |
| | $1/T_R$ | 3.43 | 5.09 | 5.10 | 4.87 |
| | $\zeta_d$ | 0.155 | 0.344 | 0.317 | 0.276 |
| | $\omega_d$ | 0.609 | 1.33 | 1.58 | 1.96 |
| $N^{p}_{\delta_r}$ | $A_p$ | 9.78 | 12.3 | 12.5 | 13.0 |
| | $1/T_{p_1}$ | 0 | 0 | 0 | 0 |
| | $1/T_{p_2}$ | 0.313 | (1.58) | (1.85) | (2.39) |
| | $1/T_{p_3}$ | −0.283 | (−1.55) | (−1.85) | (−2.37) |
| $N^{\varphi}_{\delta_r}$ | $A_\varphi$ | 9.78 | 12.3 | 12.5 | 13.0 |
| | $1/T_{\varphi_1}$ | 0.313 | 1.58 | 1.85 | 2.39 |
| | $1/T_{\varphi_2}$ | −0.283 | −1.55 | −1.85 | −2.37 |
| $N^{r}_{\delta_r}$ | $A_r$ | −13.5 | −13.8 | −13.9 | −14.4 |
| | $1/T_{r_1}$ | 3.11 | 4.73 | 4.82 | 4.69 |
| | $\zeta_r$ | −0.0561 | −0.000756 | −0.00938 | −0.0198 |
| | $\omega_r$ | 0.524 | 0.534 | 0.530 | 0.547 |
| $N^{\beta}_{\delta_r}$ | $A_\beta$ | 17.3 | 0.314 | 0.220 | 0.143 |
| | $1/T_{\beta_1}$ | −0.00872 | −0.00281 | −0.000922 | −0.000396 |
| | $1/T_{\beta_2}$ ($\zeta_\beta$) | (0.518) | 5.26 | 5.17 | 4.88 |
| | $1/T_{\beta_3}$ ($\omega_\beta$) | (4.89) | 44.8 | 64.1 | 101.4 |

[a] The data in A, B, and E are for body-fixed stability axes.

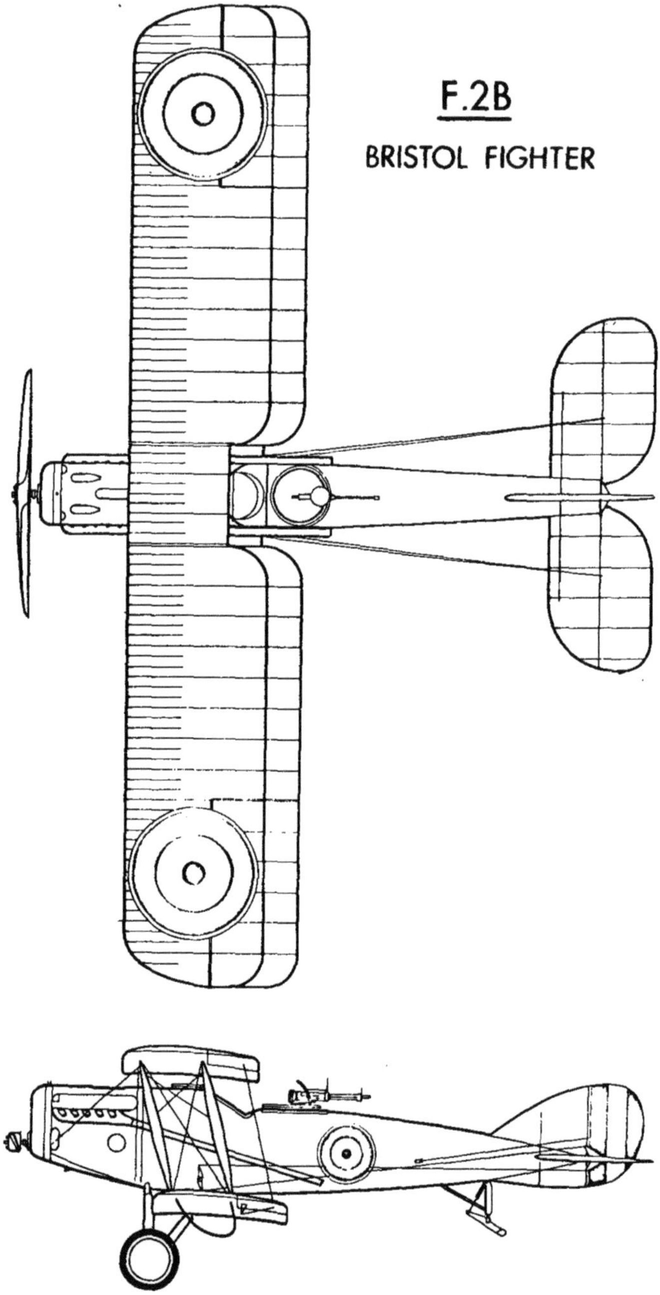


Fig. A-9. The F.2B, the Bristol Fighter.

Table A-9. The F.2B, the Bristol Fighter (Fig. A-9)

A. Geometrical and inertial parameters for the Bristol Fighter[a]

| | Flight condition 0001 | Flight condition 0002 |
|---|---|---|
| $h$ (ft) | 6000 | 6000 |
| $M$ (—) | 0.126 | 0.0895 |
| $a$ (ft/sec) | 1094 | 1094 |
| $\rho$ (slugs/ft$^3$) | 0.0020 | 0.0020 |
| $V_{T_0}$ (ft/sec) | 138 | 98 |
| $\bar{q} = \rho V^2/2$ (lb/ft$^2$) | 19.05 | 9.6 |
| $W$ (lb) | 3090 | 3090 |
| $m$ (slugs) | 96 | 96 |
| $I_x$ (slug-ft$^2$) | 2600 | 2600 |
| $I_y$ (slug-ft$^2$) | 1925 | 1925 |
| $I_z$ (slug-ft$^2$) | 4100 | 4100 |
| $I_{xz}$ (slug-ft$^2$) | 0 | 0 |
| $x_{c.g.}/\bar{c}$ | 0.33 | 0.33 |
| $\alpha_{trim}$ (degrees) | 3 | 10 |
| $\theta_0$ (degrees) | −7 | +1.5 |
| $U_0$ (ft/sec) | 138 | 96.5 |
| $W_0$ (ft/sec) | 7.24 | 17 |
| $\gamma_0$ (degrees) | −10 | −8.5 |

B. Longitudinal dimensional derivatives for the Bristol Fighter[b]

| | Flight condition 0001 | Flight condition 0002 |
|---|---|---|
| $h$ (ft) | 6000 | 6000 |
| $M$ (—) | 0.126 | 0.0895 |
| $X_u$ (1/sec) | −0.0471 | −0.032 |
| $X_w$ (1/sec) | 0.1765 | 0.248 |
| $X_{\delta_e}$ [(ft/sec$^2$)/rad] | 0 | 0 |
| $Z_u$ (1/sec) | −0.353 | −0.400 |
| $Z_{\dot{w}}$ (—) | 0 | 0 |
| $Z_w$ (1/sec) | −2.06 | −1.40 |
| $Z_{\delta_e}$ [(ft/sec$^2$)/rad] | 0 | 0 |
| $M_u$ (1/sec-ft) | +0.000684 | +0.00186 |
| $M_{\dot{w}}$ (1/ft) | −0.00768 | −0.00588 |
| $M_w$ (1/sec-ft) | −0.016 | −0.0108 |
| $M_q$ (1/sec) | −2.24 | −1.28 |
| $M_{\delta_e}$ (1/sec$^2$) | −12 | −5.8 |
| $M_\alpha$ (1/sec$^2$) | −2.21 | −1.06 |
| $M_{\dot{\alpha}}$ (1/sec) | −1.06 | −0.576 |

C. Lateral dimensional derivatives for the Bristol Fighter[c]

| | Flight condition 0001 | Flight condition 0002 |
|---|---|---|
| $h$ (ft) | 6000 | 6000 |
| $M$ (—) | 0.126 | 0.0895 |
| $Y_v$ (1/sec) | −0.176 | −0.12 |
| $Y_\beta$ [(ft/sec$^2$)/rad] | −24.3 | −11.8 |
| $Y_{\delta_a}$ [(ft/sec$^2$)/rad] | 0 | 0 |
| $Y^*_{\delta_a}$ [(1/sec)/rad] | 0 | 0 |
| $Y_{\delta_r}$ [(ft/sec$^2$)/rad] | 0 | 0 |
| $Y^*_{\delta_r}$ [(1/sec)/rad] | 0 | 0 |
| $L'_\beta$ (1/sec$^2$) | −4.50 | −4.11 |
| $L'_p$ (1/sec) | −7.17 | −4.88 |
| $L'_r$ (1/sec) | 2.06 | 2.36 |
| $L'_{\delta_a}$ (1/sec$^2$) | 26.1 | 12.7 |
| $L'_{\delta_r}$ (1/sec$^2$) | 0 | 0 |
| $N'_\beta$ (1/sec$^2$) | 1.55 | 0 |
| $N'_p$ (1/sec) | −0.436 | −0.52 |
| $N'_r$ (1/sec) | −0.341 | −0.152 |
| $N'_{\delta_a}$ (1/sec$^2$) | −1.66 | −0.805 |
| $N'_{\delta_r}$ (1/sec$^2$) | 1.30 | 0.63 |

**Table A-9 (Continued)**
**D. Elevator longitudinal transfer function factors**

| | | Flight condition | |
|---|---|---|---|
| | | 0001 | 0002 |
| Mach No., $M$ (—) | | 0.126 | 0.0895 |
| Altitude, $h$ (ft) | | 6000 | 6000 |
| c.g. (% $\bar{c}$) | | 33 | 33 |
| Weight, $W$ (lb) | | 3090 | 3090 |
| $\Delta_{\text{long}}$ | $\zeta_{sp}$ $(1/T_{sp_1})$ | (2.08) | 0.956 |
| | $\omega_{sp}$ $(1/T_{sp_2})$ | (3.25) | 1.69 |
| | $\zeta_p$ | 0.197 | 0.103 |
| | $\omega_p$ | 0.183 | 0.279 |
| $N^{\theta}_{\delta_e}$ | $A_{\theta}$ | −12.0 | −5.8 |
| | $1/T_{\theta_1}$ | 0.0785 | 0.109 |
| | $1/T_{\theta_2}$ | 2.03 | 1.32 |
| | d-c gain | −8.38 | −3.75 |
| $N^{u}_{\delta_e}$ | $A_u$ | 86.9 | 98.6 |
| | $\zeta_u$ | 0.518 | 0.571 |
| | $\omega_u$ | 3.00 | 1.63 |
| | d-c gain | 3428 | 1178 |
| $N^{w}_{\delta_e}$ | $A_w$ | −1656 | −568 |
| | $1/T_{w_1}$ | — | — |
| | $\zeta_w$ | 0.163 | 0.128 |
| | $\omega_w$ | 0.288 | 0.362 |
| | d-c gain | −603 | −334 |

Table A-9 (Continued)
E. Aileron lateral transfer function factors

| | | Flight condition | |
|---|---|---|---|
| | | 0001 | 0002 |
| Mach No., $M$ (—) | | 0.126 | 0.0895 |
| Altitude, $h$ (ft) | | 6000 | 6000 |
| c.g. (% $\bar{c}$) | | 33 | 33 |
| Weight, $W$ (lb) | | 3090 | 3090 |
| $\alpha_{trim}$ (degrees) | | 3 | 10 |
| $\Delta_{lat}$ | $1/T_s$ | −0.000836 | 0.0508 |
| | $1/T_R$ | 7.07 | 4.62 |
| | $\zeta_d$ | 0.215 | 0.273 |
| | $\omega_d$ | 1.44 | 0.886 |
| $N^p_{\delta_a}$ | $A_p$ | 26.1 | 12.7 |
| | $1/T_{p_1}$ | 0.0278 | −0.00865 |
| | $\zeta_p$ | 0.158 | 0.575 |
| | $\omega_p$ | 1.14 | −0.444 |
| | d-c gain | −76.4 | 0.152 |
| $N^\varphi_{\delta_a}$ | $A_\varphi$ | 26.3 | 12.7 |
| | $\zeta_\varphi$ $(1/T_{\varphi_1})$ | 0.216 | (0.563) |
| | $\omega_\varphi$ $(1/T_{\varphi_2})$ | 1.14 | (−0.463) |
| | d-c gain | −2788 | −18 |
| $N^r_{\delta_a}$ | $A_r$ | −1.66 | −0.805 |
| | $1/T_{r_1}$ | 0.637 | 13.0 |
| | $\zeta_r$ $(1/T_{r_2})$ | (14.1) | 0.261 |
| | $\omega_r$ $(1/T_{r_3})$ | (−0.512) | 0.319 |
| | d-c gain | −622 | −5.82 |
| $N^\beta_{\delta_a}$ | $A_\beta$ | 3.03 | 2.96 |
| | $1/T_{\beta_1}$ | 0.0656 | −0.00547 |
| | $1/T_{\beta_2}$ | 9.73 | 4.89 |
| | $1/T_{\beta_3}$ | — | — |
| | d-c gain | −157 | −0.431 |

Table A-9 (Continued)
F. Rudder lateral transfer function factors

| | | Flight condition | |
|---|---|---|---|
| | | 0001 | 0002 |
| Mach No., $M$ (—) | | 0.126 | 0.0895 |
| Altitude, $h$ (ft) | | 6000 | 6000 |
| c.g. (% $\bar{c}$) | | 33 | 33 |
| Weight, $W$ (lb) | | 3090 | 3090 |
| $\alpha_{trim}$ (degrees) | | 3 | 10 |
| $\Delta_{lat}$ | $1/T_s$ | −0.000836 | 0.0508 |
| | $1/T_R$ | 7.07 | 4.62 |
| | $\zeta_d$ | 0.215 | 0.273 |
| | $\omega_d$ | 1.44 | 0.886 |
| $N^{p}_{\delta_r}$ | $A_p$ | 2.67 | 1.49 |
| | $1/T_{p_1}$ | 0.0266 | −0.00800 |
| | $1/T_{p_2}$ | 2.33 | 1.84 |
| | $1/T_{p_3}$ | — | — |
| | d-c gain | −13.5 | −0.119 |
| $N^{\varphi}_{\delta_r}$ | $A_\varphi$ | −0.160 | 0.0165 |
| | $1/T_{\varphi_1}$ | 3.05 | 1.79 |
| | $1/T_{\varphi_2}$ | −12.5 | 93.3 |
| | d-c gain | −495 | 14.9 |
| $N^{r}_{\delta_r}$ | $A_r$ | 1.30 | 0.630 |
| | $1/T_{r_1}$ | 7.16 | 4.79 |
| | $\zeta_r$ | 0.248 | 0.200 |
| | $\omega_r$ | 0.381 | 0.527 |
| | d-c gain | −110 | 4.55 |
| $N^{\beta}_{\delta_r}$ | $A_\beta$ | −1.30 | −0.621 |
| | $1/T_{\beta_1}$ | −0.0383 | −0.176 |
| | $1/T_{\beta_2}$ | 7.13 | 4.64 |
| | $1/T_{\beta_3}$ | — | — |
| | d-c gain | −28.9 | 2.76 |

Table A-9 (Continued)
G. Lateral coupling numerators

| | | Flight condition | |
|---|---|---|---|
| | | 0001 | 0002 |
| Mach No., $M$ (—) | | 0.126 | 0.0895 |
| Altitude, $h$ (ft) | | 6000 | 6000 |
| $N^{p\ \beta}_{\delta_a \delta_r}$ | $A_{p\beta}$ | −33.9 | −7.88 |
| | $1/T_{p\beta_1}$ | 0.0284 | −0.00860 |
| | $1/T_{p\beta_2}$ | — | — |
| $N^{r\ \beta}_{\delta_a \delta_r}$ | $A_{r\beta}$ | −1.78 | −1.37 |
| | $\zeta_{r\beta}$ | 4.41 | 1.89 |
| | $\omega_{r\beta}$ | — | — |
| $N^{p/sr}_{\delta_a\ \delta_r}$ | $A_{\varphi_r}$ | 33.9 | 8.00 |
| $= N^{\varphi\ r}_{\delta_a \delta_r}$ | $1/T_{\varphi_r}$ | 0.176 | 0.120 |
| $N^{\varphi\ \beta}_{\delta_a \delta_r}$ | $A_{\varphi_\beta}$ | −33.7 | −7.92 |
| | $1/T_{\varphi_\beta}$ | — | — |

[a] The data in A are for body-fixed chord axes; engine off. $S = 405$ ft$^2$; $b =$ 40 ft; $c = 5.17$ ft.
[b] The data in B are for body-fixed chord axes; engine off.
[c] The data in C are for body-fixed chord axes.

# APPENDIX B

# *ELEMENTS OF PROBABILITY*

Because inputs to which automatic control systems for aircraft will be subjected are often either very complicated or are not known in detail, it may be necessary to consider performance of the system in response to commands and disturbances that are considered to be *random* variables. Other applications of the theory of random variables to the synthesis of automatic flight control systems have to do with the assessment of design qualities, in particular with the estimation of likelihood of failure of the system or its components.

The study of random variables, quantities that may take any values of a specified *set* with a specified probability, comprises the science of statistics. Although it is not feasible here to give an account of more than a minute part of the subject, we can aspire to expose just those elements of the underlying theory that our examples in the main text require.[1]

Probability as a concept is treated first. This is immediately followed by a discussion of how certain probabilities may be computed if other probabilities are known, as in the addition and multiplication laws of probability. Random variables are then defined, and consideration of the probability that a random variable will take certain values introduces the first probability distribution and density functions. Next, the most important averages or "statistics" of random variables are defined. These include the mean, mean square, and the characteristic function. Finally, the binomial, Poisson, uniform, exponential, and Gaussian distributions are described, derived, and put to use in elementary illustrative examples.

Although there is some inherent difficulty in defining probability in a completely satisfactory manner, for nearly every engineering use *empirical probability* is a suitable interpretation of the concept.

Imagine or make a large number of experiments that may materialize, under what are taken to be identical conditions, in two or more outcomes

[1] Useful introductory references include: A. C. Aitken, *Statistical Mathematics*, Oliver and Boyd, Edinburgh and London, 8th edn., 1957; M. J. Moroney, *Facts from Figures*, Penguin Books, 3rd edn., 1956; D. A. S. Frazer, *Statistics, An Introduction*, Wiley, New York, 1958; G. J. Hahn, "Probability and Statistics," Chapter 6 in H. Chestnut, *Systems Engineering Tools*, Wiley, New York, 1965; H. H. Goode, R. E. Machol, *Systems Engineering*, McGraw-Hill, New York, 1957; W. R. Bennett, "Methods of Solving Noise Problems," *Proceedings IRE*, 44, 609–638 (1956).

or *events*. (An event is, generally, something concerning which it can be determined whether or not it occurred.) The set of all possible events comprises the sample space, $\mathcal{S}$, and, corresponding to each *basic* possible outcome of the experiment we associate a sample point, $\zeta_i$, in this space. An event may then correspond to a single sample point or to a *set* of such sample points. For example, the outcome "three dots showing" of the roll of a single die is an event and a sample point. On the other hand, the event "three or fewer dots showing" comprises three sample points, since it corresponds to three basic possible outcomes of the experiment.

Now the results of certain experiments, repeated a large number of times, may show, or at least be believed to show, *statistical regularity*. If the outcome $A$ occurs $n_A$ times in $N$ trials and the ratio $\Pr(A) = n_A/N$ appears to approach a limit as the number of trials, $N$, becomes very large, $\Pr(A)$ may be taken to be the *probability* of the event $A$.[2] (The simplest example is the flipping of a coin. We may take it that in each trial there are only two possible outcomes: heads or tails. The result of a large number of trials will be, for instance, that the number of heads divided by the number of trials is very close to $\frac{1}{2}$.) It follows directly from the definition of probability that the probability has two properties, namely:

$$\begin{aligned} \Pr(A) &\geqq 0 \\ \Pr(\mathcal{S}) &= 1 \end{aligned} \tag{B-1}$$

Consider further, and more generally, that there may be four identifiable outcomes of an experiment each one of which is a distinct possibility:

the event, $A$ only, occurs $n_A$ times in $N$ trials
the event, $B$ only, occurs $n_B$ times in $N$ trials
the events, $A$ and $B$, occur together $n_{AB}$ times in $N$ trials
the event, neither $A$ nor $B$, occurs $n_{\overline{A+B}}$ times in $N$ trials

and that this takes care of all the possibilities. Then, for a large number of trials $N$:

$$\text{the probability of the event } A = \Pr(A) \doteq \frac{n_A + n_{AB}}{N}$$

$$\text{the probability of the event } B = \Pr(B) \doteq \frac{n_B + n_{AB}}{N}$$

$$\text{the probability of the event } A \text{ and } B = \Pr(AB) \doteq \frac{n_{AB}}{N}$$

[2] The difficulty with empirical probability as a concept resides in the postulated limit. It is impossible to show satisfactorily that the limit should exist and be unique; also, the limiting process does not behave in the same way as the familiar mathematical limiting process for a sequence.

the probability of the event $A$ or $B = \Pr(A + B) \doteq \dfrac{n_A + n_B + n_{AB}}{N}$

the probability of the event neither $A$ nor $B$

$$= \Pr(\overline{A + B}) \doteq \frac{n_{\overline{A+B}}}{N}$$

$$= 1 - \Pr(A + B)$$

$$\doteq 1 - \frac{n_A + n_B + n_{AB}}{N}$$

the probability of the event $A$ given that event $B$ has occurred

$$= \Pr(A/B) \doteq \frac{n_{AB}}{n_B + n_{AB}}$$

Total probability is the probability of any one of several mutually exclusive outcomes, say, $A$, $B$, or $C$, where each of the mutually exclusive events is a set of sample points or elements. The set of all possible outcomes (space) can be represented as the area inside a circle in *Euler* or *Venn diagram.* Then the fractions of the area marked $A$, $B$, and $C$ represent fractions of the total number of trials that resulted in the event $A$, $B$, or $C$. See Fig. B-1(a). On the other hand, if the outcomes are not mutually exclusive, areas on the Euler diagram may still represent the fractions of the area of the circle (sets of sample points) defined by Eq. B-2.

Reasoning from either the algebraic expressions (Eq. B-2) or the geometrical representation (Fig. B-1), we find that

$$\Pr(A + B) = \Pr(A) + \Pr(B) - \Pr(AB) \tag{B-3}$$

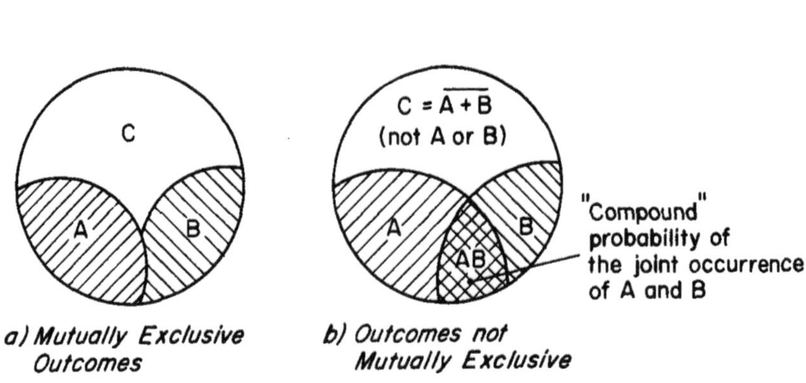


Fig. B-1. Euler diagrams.

This is called the addition law of probabilities. If the event $A$ and the event $B$ are *mutually exclusive*, the compound or joint probability $\Pr(AB) = 0$, and the addition law is simply:

$$\Pr(A + B) = \Pr(A) + \Pr(B) \tag{B-4}$$

In the case in which the event $A$ and the event $B$ may occur together, the conditional probability, $\Pr(A/B)$, can be expressed in terms of the probability of the event $B$ and the compound or joint probability, $\Pr(AB)$:

$$\Pr(A/B)\Pr(B) = \frac{n_{AB}}{n_B + n_{AB}} \frac{n_B + n_{AB}}{N} = \Pr(AB) \tag{B-5}$$

or

$$\Pr(A/B) = \frac{\Pr(AB)}{\Pr(B)}$$

It is not necessary for the events to be simultaneous in order to apply this definition of conditional probability. When the events $A$ and $B$ are *independent*,

$$\Pr(A/B) \equiv \Pr(A)$$

and

$$\Pr(AB) = \Pr(A/B)\Pr(B) = \Pr(A)\Pr(B) \tag{B-6}$$

This is called the multiplication law.

The addition law and the multiplication law are easily generalized to cases of more than two independent events. Thus for three independent events, $A$, $B$, and $C$,

$$\begin{aligned} \Pr(A + B + C) &= \Pr(A) + \Pr(B) + \Pr(C) - \Pr(AB) - \Pr(AC) \\ &\quad - \Pr(BC) + \Pr(ABC) \\ &= 1 - \Pr(\bar{A})\Pr(\bar{B})\Pr(\bar{C}) \qquad \text{(B-7)} \\ &= 1 - [1 - \Pr(A)][1 - \Pr(B)][1 - \Pr(C)] \end{aligned}$$

$$\Pr(ABC) = \Pr(A)\Pr(B)\Pr(C) \tag{B-8}$$

and so forth. Equation B-6 or B-8 is the mathematical statement of the definition of independence, i.e., $A$ and $B$ are independent when Eq. B-6 applies, or $A$, $B$, and $C$ are independent when Eq. B-8 applies.

An example of events that are not mutually exclusive but which are independent might be the drawing of an ace or a spade (on a single draw) from a standard deck of 52 playing cards.

$$\text{Pr (spade)} = \frac{13}{52} = \frac{1}{4}$$

$$\text{Pr (ace)} = \frac{4}{52} = \frac{1}{13}$$

The probability of drawing the card that is both a spade and an ace is

$$\text{Pr (spade ace)} = \text{Pr (spade) Pr (ace)} = \frac{1}{4} \times \frac{1}{13} = \frac{1}{52}$$

and this illustrates the multiplication law for independent events. Finally the probability of a spade *or* an ace is:

$$\begin{aligned} \text{Pr (spade + ace)} &= \text{Pr (spade)} + \text{Pr (ace)} - \text{Pr (spade ace)} \\ &= \frac{13}{52} + \frac{4}{52} - \frac{1}{52} \\ &= \frac{16}{52} = \frac{4}{13} \end{aligned}$$

This illustrates the operation of the addition law.

Consider next an example in which the events are neither mutually exclusive nor independent. Two coins are tossed simultaneously. There are four possible basic outcomes (sample points). They are head-head, head-tail, tail-head, and tail-tail. We ask first what is the probability that at least one head occurs. By counting the outcomes in which at least one head occurs:

$$\text{Pr (head)} = \frac{3}{4} = \text{Pr (tail)}$$

Also, the joint probability of a head and a tail is

$$\text{Pr (head tail)} = \frac{1}{2}$$

and the conditional probability of a head, given that a tail has occurred, is

$$\text{Pr (head/tail)} = \frac{2}{3}$$

From the addition law, the probability of a head or a tail is

$$\Pr(\text{head} + \text{tail}) = \frac{3}{4} + \frac{3}{4} - \frac{1}{2} = 1$$

and the relationship between the conditional and joint probabilities is illustrated by the calculation:

$$\Pr(\text{head/tail}) \Pr(\text{tail}) = \Pr(\text{head tail})$$

$$\frac{2}{3} \times \frac{3}{4} = \frac{1}{2}$$

Note that here $\Pr(\text{head tail}) \neq \Pr(\text{head}) \Pr(\text{tail})$, i.e., the events, "at least one head" and "at least one tail," are *not independent* in simultaneous tosses of two coins.

## Random Variables

It is often more interesting to consider some function of the outcomes of an experiment, rather than the outcomes themselves. If, then, we assign to every sample point (element), $\zeta_i$, of the sample space (set), $\mathcal{S}$, a number $X(\zeta)$ according to some rule, the function $X(\zeta)$, defined on the sample space, is a random variable,[3] provided that $X(\zeta) \leq x$ is an event and that $X(\zeta) = \pm\infty$ are events but that the probability of their occurrence is zero. Note that the outcome itself falls under the definition of a random variable, since the functional dependence may allowably comprise the identity function. Thus outcomes or functions of outcomes are random variables.

## Distribution and Density Functions

Now define, for any number $-\infty < x < \infty$, the probability that $X(\zeta) \leq x$ as the real-valued first probability *distribution function* of the random variable $X(\zeta)$:

$$\Pr(X(\zeta) \leq x) = P_1(x) \tag{B-9}$$

The first probability distribution function has the properties:

$$\begin{aligned} P_1(-\infty) &= 0 \\ P_1(\infty) &= 1 \\ P_1(x_1) &\leq P_1(x_2);\ x_1 < x_2 \end{aligned} \tag{B-10}$$

[3] The use of the words "random variable" for a *function* is rooted in precedent.

If $n(x)$ is the number of times, in an experiment repeated $N$ times, in which $X(\zeta) \leqq x$, then for a very large number of repetitions:

$$P_1(x) \doteq \frac{n(x)}{N} \tag{B-11}$$

The derivative of the first probability distribution function is called the *first probability density function*, $p_1(x)$.

$$\frac{dP(x)}{dx} = p_1(x) \tag{B-12}$$

Then the probability element $p_1(x)\,dx = \Pr\,[x < X(\zeta) < x + dx]$. As often happens, the distribution function may have step-form discontinuities, and there are delta functions in its derivative. For the case of so-called discrete random variables, the density functions actually consist of a sum of weighted delta functions:

$$p_1(x) = \sum_i p_i\,\delta(x - x_i) \tag{B-13}$$

The probability that a function $X(\zeta)$ lies between two values, say $x_1$ and $x_2$, is obtained by integrating the density function between those limits.

$$\Pr\,[x_1 < X(\zeta) < x_2] = \int_{x_1}^{x_2} p_1(x)\,dx \tag{B-14}$$

and clearly, from the definition of a random variable,

$$\Pr\,[-\infty < X(\zeta) < +\infty] = \int_{-\infty}^{\infty} p_1(x)\,dx = 1 \tag{B-15}$$

The names distribution function and density function come from the analogy to a *unit* mass distributed along the line that is the $x$ axis. The *mass element* in a small distance, $dx$, is $(dM/dx)\,dx$, where $dM/dx$ is the mass per unit length, or density. Then the total mass to the left of a given point $x_1$ is obtained by integrating the density function $dM/dx$, or

$$M(x) = \int_{-\infty}^{x} \left(\frac{dM}{dx}\right) dx \tag{B-16}$$

Now $M(-\infty) = 0$, while $M(\infty) = 1$, as in the probability distribution function.

**Expectation, Mean, and Variance**

In the theory and application of probability, the average, mean, or mathematical expectation of a random variable has a surpassing importance. As every schoolboy knows, the way in which to find the average value of a random variable, $X_i$, which takes a number of discrete values, is to multiply each value by the number of times, $n_i$, it occurs, form the sum, and divide by the total number of observations, $N$:

$$\text{average of } X = E[X] = \frac{X_1 n_1 + X_2 n_2 + \cdots X_n n_n}{N} \tag{B-17}$$

According to our interpretation of probability, however, a fraction such as $n_1/N$ is simply the probability, $p_1$, of the event $x_1$, so that Eq. B-17 could be rewritten:

$$E[X] = X_1 p_1 + X_2 p_2 + \cdots + X_n p_n = \sum_{i=1}^{n} X_i p_i \tag{B-18}$$

In the case of a continuous random variable, the probability that the variable lies in the narrow range between $x$ and $x + dx$ is the probability element $p_1(x)\,dx$. Now, considering the integral as the limit of a sum,

$$E[X] = \int_{-\infty}^{\infty} x p_1(x)\,dx \tag{B-19}$$

When the expected value of some function of $x$, say $g(x)$, is wanted:

$$E[g(X)] = \int_{-\infty}^{\infty} g(x) p_1(x)\,dx \tag{B-20}$$

Those averages in which $g(X) = X^n$ are of particular importance. They are called the *moments* of the distribution, and have the form:

$$m_n = E[X^n] = \int_{-\infty}^{\infty} x^n p_1(x)\,dx \tag{B-21}$$

From the defining properties of an allowable distribution function, the zero order moment, $m_0$, always equals 1. The first moment, $m_1$, is the arithmetic mean; the second moment, $m_2$, is the mean square; and so forth.

In terms of physical analogies, the mean is simply the centroid of the density function, while the mean square value is the radius of gyration of the density function. These two moments are illustrated for a rectangular density function in Fig. B-2.

If the mean value is subtracted from all values of $X$, the moments derived from the results become *central moments*. Thus,

$$\mu_n = E[(X - m_1)^n] = \int_{-\infty}^{\infty} [x - m_1]^n p_1(x)\,dx \tag{B-22}$$

The first central moment, $\mu_1$, is zero. The second central moment is

$$\begin{aligned} \mu_2 &= \int_{-\infty}^{\infty} [x - m_1]^2 p_1(x)\,dx \\ &= m_2 - [m_1]^2 \\ &= E[X^2] - \{E[X]\}^2 = \sigma^2 \end{aligned} \tag{B-23}$$

$\mu_2 = \sigma^2$ is the *variance*. The square root of the variance is the *standard*

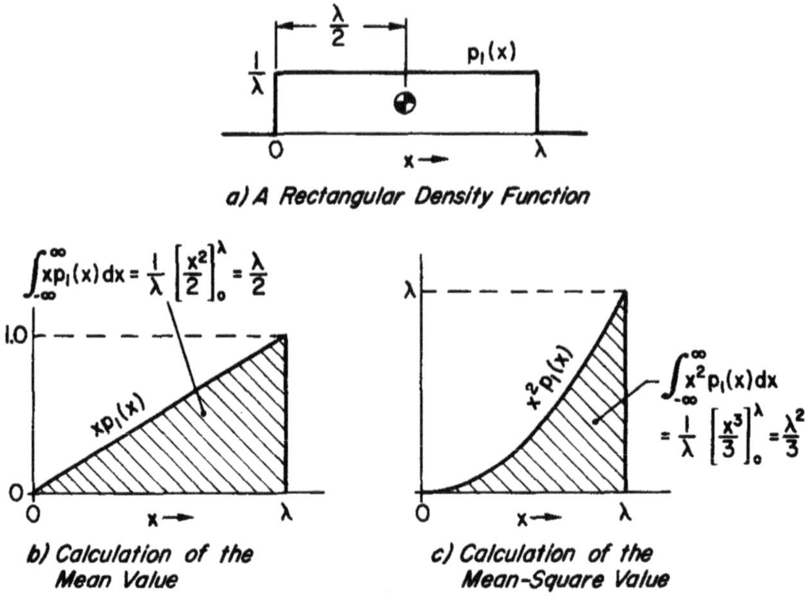


Fig. B-2. Mean and mean-square values for the rectangular density function.

*deviation* of the distribution for the ensemble $X_1 \ldots, X_n, \ldots$. When the mean is zero, the variance and the mean square are identical.

The mean and the variance (or standard deviation) are often the most significant quantities that may characterize a distribution.

**Characteristic Function**

The characteristic function is the expected value of the complex conjugate Fourier transform of a random variable $X$.

$$\varphi(\omega) = E[e^{+j\omega X}] = \int_{-\infty}^{\infty} e^{+j\omega x} p_1(x)\,dx \tag{B-24}$$

By expanding the exponential function in a series,

$$\begin{aligned}\varphi(\omega) &= \int_{-\infty}^{\infty} p_1(x)\left[1 + j\omega x + \frac{(j\omega x)^2}{2!} + \cdots + \frac{(j\omega x)^n}{n!} + \cdots\right] dx \\ &= \int_{-\infty}^{\infty} p_1(x)\,dx + j\omega\int_{-\infty}^{\infty} xp_1(x)\,dx + \frac{(j\omega)^2}{2!}\int_{-\infty}^{\infty} x^2p_1(x)\,dx \\ &\quad + \frac{(j\omega)^3}{3!}\int_{-\infty}^{\infty} x^3p_1(x)\,dx + \frac{(j\omega)^4}{4!}\int_{-\infty}^{\infty} x^4p_1(x)\,dx + \cdots\end{aligned} \tag{B-25}$$

The integrals of this last expression may be recognized as the central moments of the distribution defined by the density function $p_1(x)$. The characteristic function, therefore, contains information reflecting the character of *all* the central moments of the distribution. In fact, if the expression on the right-hand side of Eq. B-25 is compared to the Maclaurin series expansion for the characteristic function,

$$\varphi(\omega) = \varphi(0) + \varphi'(0) + \varphi''(0)\frac{\omega^2}{2!} + \cdots + \varphi^{(n)}(0)\frac{\omega^n}{n!} + \cdots \qquad \text{(B-26)}$$

it may be seen that the central moments of the distribution can be computed from derivatives of the characteristic function evaluated at $\omega = 0$.

$$E(X^n) = \int_{-\infty}^{\infty} x^n p_1(x)\,dx = (j)^{-n} \frac{d^n}{d\omega^n}\varphi(\omega)\Big]_{\omega=0} \qquad \text{(B-27)}$$

Another use of the characteristic function is in the computation of the first probability density function of the sum of two random variables. The characteristic function of the sum of two independent random variables is the product of the characteristic functions of the individual variables.

$$\varphi(\omega) = \varphi_1(\omega)\varphi_2(\omega) \qquad \text{(B-28)}$$

Then the first probability density function for the sum is the inverse complex conjugate Fourier transform:

$$\begin{aligned} p_1(x_1 + x_2) &= \mathcal{F}^{-1}[\varphi(-\omega)] = \mathcal{F}^{-1}[\varphi_1(-\omega)\varphi_2(-\omega)] \\ &= \frac{1}{2\pi}\int_{-\infty}^{\infty} e^{-j\omega x}\varphi(\omega)\,d\omega \end{aligned} \qquad \text{(B-29)}$$

This inverse transform may be written as a convolution integral

$$p_1(x_1 + x_2) = \int_{-\infty}^{\infty} p_{1_1}(t_1)p_{1_2}(x - t_1)\,dt_1 \qquad \text{(B-30)}$$

**Examples of Distribution Functions**

While the conditions on the first probability distribution function are so broad as to admit of a very large number of interesting possibilities, the number of different distribution functions often encountered in engineering design work is quite small. It includes the binomial (or Bernoulli), Poisson, uniform, exponential, and Gaussian (or normal) distributions. The first two of these are discrete distributions in which the

**Table B-1. Properties of some probability distribution functions**

| NAME | DOMAIN | PROBABILITY DENSITY FUNCTION |
|---|---|---|
| Binomial or Bernoulli | $x = s$<br>$s = 0,1,2,\ldots n$ | $p_1(x) = \sum_{s=0}^{n} \binom{n}{s} p^s q^{n-s}\,\delta(x-s)$<br>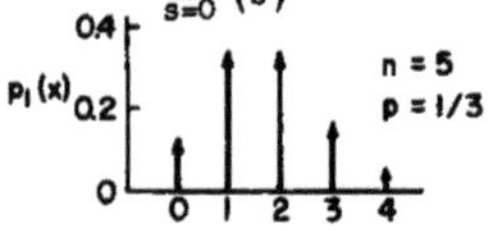 |
| Poisson | $x = s$<br>$s = 0,1,2,\ldots n$ | $p_1(x) = e^{-m} \sum_{s=0}^{\infty} \frac{m^s}{s!}\,\delta(x-s)$<br>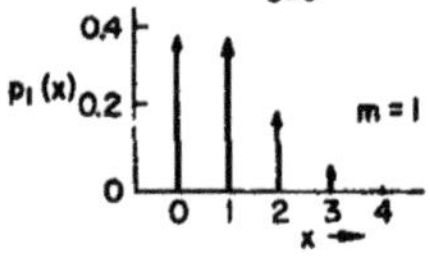 |
| Uniform | $m - \frac{\lambda}{2} \leq x \leq m + \frac{\lambda}{2}$ | $p_1(x) = 1/\lambda$<br>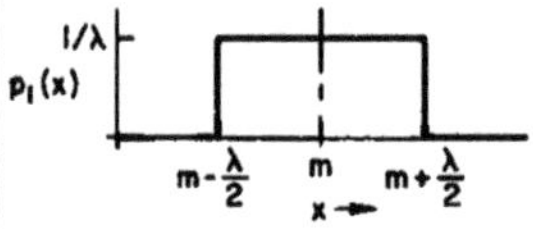 |
| Exponential | $0 \leq t \leq \infty$ | $p_1(x) = ce^{-ct}$<br>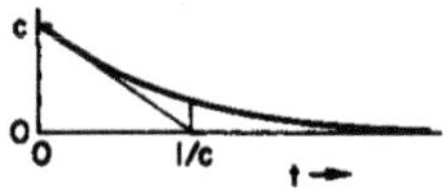 |
| Gaussian or Normal | $-\infty < x < \infty$ | $p_1(x) = \frac{1}{\sqrt{2\pi}\,\sigma} \exp - \left[ \frac{(x - m_1)^2}{2\sigma^2} \right]$<br>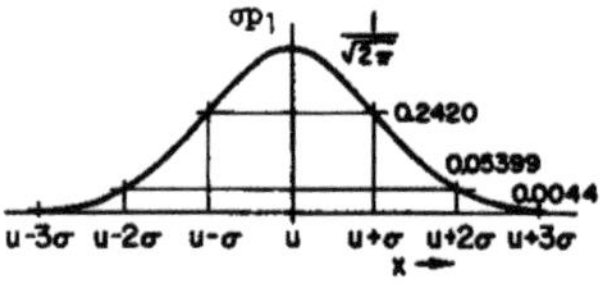 |

| PROBABILITY DISTRIBUTION FUNCTION | RESTRICTIONS ON PARAMETERS | MEAN | VARIANCE | CHARACTERISTIC FUNCTION |
|---|---|---|---|---|
| $\int_{-\infty}^{\infty} p_1(x)dx$ <br> 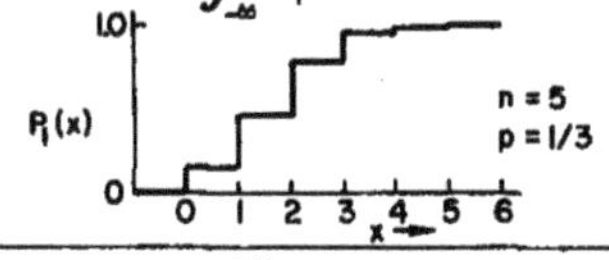  | $0 < p < 1$ <br> $q = 1 - p$ | $np$ | $npq$ | $\left(q + pe^{j\omega}\right)^n$ |
| $\int_{-\infty}^{\infty} p_1(x)dx$ <br> 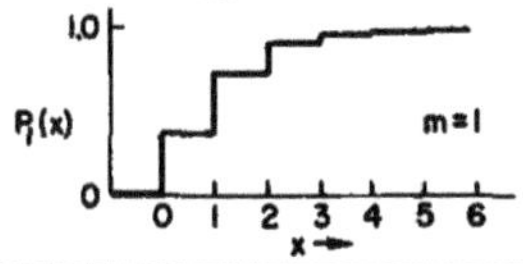  | $0 < m < \infty$ | $m$ | $m$ | $e^{m\left(e^{j\omega} - 1\right)}$ |
| $\int_{-\infty}^{\infty} p_1(x)dx = \begin{cases} 0 & ; x < m - \frac{\lambda}{2} \\ \frac{x - m + \lambda/2}{\lambda} & ; m - \frac{\lambda}{2} \leq x \leq m + \frac{\lambda}{2} \\ 1 & ; x > m + \frac{\lambda}{2} \end{cases}$ <br> 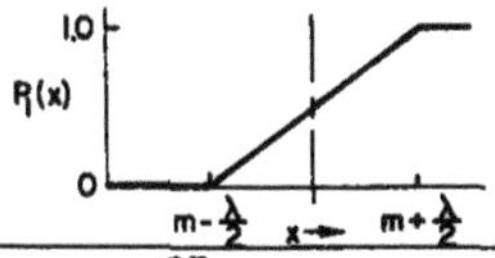  | $-\infty < m < \infty$ <br> $0 < \lambda < \infty$ | $m$ | $\frac{\lambda^2}{12}$ | $\frac{2}{\lambda t} \sin\left(\frac{\lambda t}{2}\right) e^{j\omega m}$ |
| $P_1(x) = \int_{-\infty}^{\infty} p_1(x)dx = 1 - e^{-ct}$ <br> 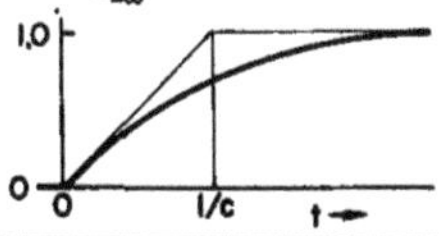  | $0 < \frac{1}{c} < \infty$ | $\frac{1}{c}$ | $\frac{1}{c^2}$ | $\frac{1}{\left(1 - j\frac{\omega}{c}\right)}$ |
| $P_1(x) = \frac{1}{2} + \text{erf}\left(\frac{x - m_1}{\sigma}\right)$ <br> 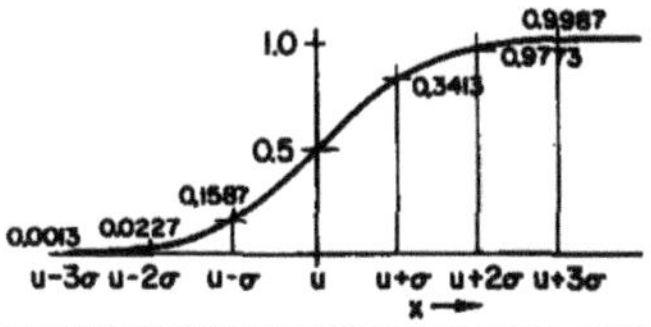  | $-\infty < m_1 < \infty$ <br> $0 < \sigma < \infty$ | $m_1$ | $\sigma^2$ | $\exp\left(jm_1\omega - \frac{\sigma^2\omega^2}{2}\right)$ |

density function is a sum of delta functions. The next three are continuous distributions. Some of the characteristics of these one-dimensional distributions are summarized in Table B-1. In order to illustrate the types of random variables to which each of these distributions is appropriate, we shall discuss each of them in turn.

*The Binomial Distribution*

The binomial distribution was discovered by Jacob Bernoulli and was published posthumously in 1713. It was the first probability distribution to be discovered. It is useful for the description of the number of times a given event, say $A$, will occur in a given number of trials, say $n$, when the probability of success on each trial is the same and the trials are independent of one another. It may, for example, be used to find the number of ground to air missiles required, on the average, to score at least one hit on an airplane, or, alternatively, it may be used to describe the probability that an item of equipment will satisfactorily complete a mission. It is of surpassing importance in the theory of sampling inspection.

Consider a random experiment with two possible outcomes, $A$ (success) and $\bar{A}$ (failure). Define $\Pr(A) = p$ and $\Pr(\bar{A}) = q = 1 - p$. Then consider the experiment repeated 1, 2, . . . $n$ times with the outcome on each trial independent of previous results.

On the first trial the result may be a failure or a success, and their respective probabilities are $q$ and $p$. After the second trial, we may have observed two failures (probability $= q^2$, from the multiplication law), a failure and a success (probability $= qp$), a success and a failure (probability $= pq$), or two successes (probability $= p^2$). Similarly, we could count the outcomes after three trials and then describe the results after the third trial. The results might be arranged as shown in Table B-2. If we consider the probabilities of these occurrences and now disregard the precise sequence in which successes and failures might occur, we can also arrange the probabilities in a table. Once the pattern is clear (Pascal's triangle), the table can be extended very easily (Table B-3). Of course, terms on any given line represent the terms in the expansion of the binomial $(q + p)^n$. (Hence the name of the distribution.)

More generally, the number of equally likely sequences of $s$ successes and $n - s$ failures in $n$ trials is

$$\binom{n}{s} = n!/[s!\,(n - s)!],$$

and the probability that the number $x = s$ (a specified number of successes):

$$\Pr(x = s) = \binom{n}{s} q^{n-s} p^s \qquad \text{(B-31)}$$

**Table B-2. Sequences of failures and successes**

| | |
|---|---|
| 1st trial | $\bar{A}$ or $A$ |
| 2nd trial | $\bar{A}\bar{A}$ or $\bar{A}A$ or $A\bar{A}$ or $AA$ |
| 3rd trial | $\bar{A}\bar{A}\bar{A}$ or $\bar{A}\bar{A}A$ or $\bar{A}A\bar{A}$ or $A\bar{A}\bar{A}$ or $\bar{A}AA$ or $A\bar{A}A$ or $AA\bar{A}$ or $AAA$ |

(This formula is only valid for integer values, $s = 0, 1, 2, \ldots n$.) Then a partial table of the distribution of the probabilities for any choice of a number of successes, $s$, and number of trials, $n$, might appear as follows:

| Number of successes, $s$ | 0 | 1 | 2 | 3 | $n$ |
|---|---|---|---|---|---|
| $\Pr(x = s)$ | $q^n$ | $nq^{n-1}p$ | $\binom{n}{2}q^{n-2}p^2$ | $\binom{n}{3}q^{n-3}p^3$ | $p^n$ |

This is just one row in the triangle table. Since $(q + p)^n = 1$, the sum of all the probabilities is one, as it should be.

The density function of the binomial distribution may be written:

$$p_1(x) = \sum_{s=0}^{n} \binom{n}{s} p^s q^{n-s}\, \delta(x - s) \qquad \text{(B-32)}$$

There is a binomial distribution for each value of $n$ and $p$. These are known as the parameters of the distribution. Extensive tables of the distribution have been calculated.[4] This, however, is very tedious when the number of trials, $n$, is large. For many practical purposes, the binomial distribution can be successfully approximated. When the probability of either success or failure is small, the binomial distribution can be approximated by the Poisson distribution (see below). On the other hand, if $p \doteq q$, the continuous Gaussian distribution may be used as an approximation to the discrete binomial distribution by fitting the mean and the variance, provided only that the number of trials, $n$, is reasonably large.

**Table B-3. Pascal's triangle**

| | | | | | | | | | |
|---|---|---|---|---|---|---|---|---|---|
| 0 trials | | | | | 1 | | | | |
| 1 trial | | | | $q$ | | $p$ | | | |
| 2 trials | | | $q^2$ | | $2qp$ | | $p^2$ | | |
| 3 trials | | $q^3$ | | $3q^2p$ | | $3qp^2$ | | $p^3$ | |
| 4 trials | $q^4$ | | $4q^3p$ | | $6q^2p^2$ | | $4qp^3$ | | $p^4$ |

[4] "Tables of the Binomial Probability Distribution," National Bureau of Standards, Applied Mathematics Series 6, U.S. Government Printing Office, Washington, D.C., 1950.

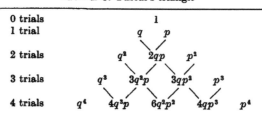

As an example of the application of the binomial distribution to problems in aircraft subsystem reliability, consider the question of the relative reliability of single- and twin-engined airplanes. Suppose the probability that an engine will complete a four-hour mission is $p = 0.9900$, and further that engine failures are independent events even in multi-engined airplanes. Then, from Eq. B-16, for a single-engined airplane during a single mission (trial):

$$\text{Pr (success)} = \binom{n}{s} p^s\, q^{n-s} = \binom{1}{1} p^1\, q^0$$
$$= 0.9900$$

Similarly, for a twin-engined airplane that requires both engines to complete its flight (two successes in two trials):

$$\text{Pr (success)} = \binom{n}{s} p^s\, q^{n-s} = \binom{2}{2} p^2 q^0$$
$$= 0.9801$$

On the other hand, if a twin-engined airplane can complete its flight on only one engine (at least one success in two trials), using the addition law for mutually exclusive events:

$$\begin{aligned}\text{Pr (success)} &= \text{Pr (one success)} + \text{Pr (two successes)}\\ &= \binom{2}{1} p^1 q^{2-1} + \binom{2}{2} p^2 q^0\\ &= 2pq + p^2 = 1 - q^2 = 1 - (1 - p)^2\\ &= 1 - 0.0001 = 0.9999\end{aligned}$$

This shows the powerful advantage of *redundancy*.

*The Poisson Distribution*

For the purpose of estimating the number, $s$, of independent events, $A$, that will occur when the average rate per unit time, length, area, or volume at which they occur is known, we use the Poisson distribution, named after Simeon Denis Poisson, the French mathematician. The events are isolated events in a continuum. Thus, e.g., the Poisson distribution may characterize the number of cars arriving at an intersection in 1 min, the number of bacteria on a slide, or the number of flaws in a casting.

If $\beta$ is the "density" of the event, $A$, (the average rate per unit time, for example), then the probability that the event occurs in the interval between $T$ and $T + \Delta T$ is proportional to $\Delta T$, and the constant of proportionality is $\beta$. Pr [event occurs between $T$ and $T + \Delta T$] $= \beta\, \Delta T$.

Now, from the multiplication law, Eq. B-6,

Pr [no event between 0 and $T + \Delta T$] $= P_0(T + \Delta T)$
$=$ Pr [no event between 0 and $T$] Pr [no event between $T$ and $T + \Delta T$]
or

$$P_0(T + \Delta T) = P_0(T)\,[1 - \beta\,\Delta T] \tag{B-33}$$

On rearranging,

$$\frac{P_0(T + \Delta T) - P_0(T)}{\Delta T} = -\beta P_0(T) \tag{B-34}$$

in the limit, or, as $\Delta T \rightarrow 0$:

$$\frac{dP_0(T)}{dT} + \beta P_0(T) = 0 \tag{B-35}$$

The initial condition is $P_0(0) = 1.0$ (the probability that no event occurs in zero time) and the solution is

$$P_0(T) = e^{-\beta T} \tag{B-36}$$

Similarly Pr [1 event between 0 and $T + \Delta T$] $= P_1(T + \Delta T)$. This probability is the sum of the probability that there is one event in the interval, 0 to $T$, and none in $\Delta T$, together with the probability that there is no event in the interval, 0 to $T$, and that there is one in $\Delta T$.

$$P_1(T + \Delta T) = P_1(T)[1 - \beta\,\Delta T] + P_0(T)\beta\,\Delta T \tag{B-37}$$

$$\frac{P_1(T + \Delta T) - P_1(T)}{\Delta T} = -P_1(T)\beta + P_0(T)\beta \tag{B-38}$$

Dividing by $\beta$, letting $\Delta T \rightarrow 0$, rearranging and substituting for $P_0$, we obtain

$$\frac{1}{\beta}\frac{dP_1(T)}{dT} + P_1(T) = e^{-\beta T} \tag{B-39}$$

which has the solution,

$$P_1(T) = \beta T e^{-\beta T} \tag{B-40}$$

By successively considering the probability that 2, 3, . . . , $n$ events will occur in the interval $T + \Delta T$, we should find that

$$P_2(T) = \frac{(\beta T)^2}{2}\,e^{-\beta T}, \qquad P_3(T) = \frac{(\beta T)^3}{6}\,e^{-\beta T}, \qquad P_n(T) = \frac{(\beta T)^n}{n!}\,e^{-\beta T}$$

These may be recognized as successive terms in the expansion

$$e^{-m}e^{m} = e^{0} = 1 = e^{-m}\left[1 + \frac{m}{1!} + \frac{m^2}{2!} + \cdots \frac{m^n}{n!}\right] \tag{B-41}$$

where $m = \beta T$. Notice that the sum of all the probabilities is one. For the Poisson distribution then,

$$\Pr(x = s) = \frac{e^{-m} m^s}{s!} \tag{B-42}$$

and the density function is given by the expression:

$$p_1(x) = e^{-m} \sum_{s=0}^{\infty} \frac{m^s}{s!}\, \delta(x - s) \tag{B-43}$$

Again $s$ takes only integer values. The term $m$ is the "frequency parameter" of the Poisson distribution which represents the number of events in a typical *finite* interval, i.e., $m = \beta T$, e.g., where $\beta$ is the "density" of the events and $T$ is a convenient interval, say 1 sec or 1 year.

An example of the application of the Poisson distribution is the prediction of the number of cavalrymen likely to be killed by horse kicks in the course of a year. The assumption that each such event is independent and occurs randomly in time, but with a constant overall frequency, suggests that the Poisson distribution may be appropriate. Data collected by von Bortkiewicz[5] for 10 Army Corps for 20 years (200 readings) shows a total of 122 deaths from this cause. There are, therefore, 122/200 = 0.61 deaths per corps in a year. This is the "frequency parameter," $m$, of the distribution. Considering the probability of $s = 0, 1, 2, 3$, and 4 deaths occurring in any one corps in any one year we can construct a table of probabilities of the event, $\Pr(s = x) = e^{-0.61}(0.61)^s/s!$

| Number of deaths in one year per corps, $s$ | 0 | 1 | 2 | 3 | 4 |
|---|---|---|---|---|---|
| Probability, $\Pr(x = s)$ | 0.543 | 0.331 | 0.101 | 0.021 | 0.003 |
| Frequency expected in 200 readings | 109 | 66.3 | 20.2 | 4.1 | 0.6 |
| Actual number of corps-years the number $s$ deaths occurred | 109 | 65 | 22 | 3 | 1 |

It may be observed that the Poisson distribution fits the actual data and might be used for extrapolation.

We have had occasion to mention that the calculation of the binomial distribution is tedious when the number of trials, $n$, becomes large. In

[5] L. von Bortkiewicz, *Das Gesetz der kleinen Zahlen*, B. G. Teubner, Leipzig, 1898.

some cases, the Poisson distribution is a suitable approximation. Recall that the binomial distribution is characterized by the expression

$$\Pr(x = s) = \binom{n}{s} p^s q^{n-s} = \frac{n!}{s!\,(n-s)!} p^s (1-p)^{n-s} \qquad \text{(B-44)}$$

In the Poisson distribution the parameter, $m$, is the average number of occurrences (successes) in a typical finite interval of interest. Take the interval to extend over the number of independent trials in the case of the binomial distribution and therefore let $m = np$ the average number of successes in $n$ trials. Substituting $p = m/n$ and rewriting Eq. B-39, we find that

$$\begin{aligned}
\Pr(x = s) &= \binom{n}{s}\left(\frac{m}{n}\right)^s\left(1 - \frac{m}{n}\right)^{n-s} \\
&= \frac{n!}{s!\,(n-s)!}\frac{m^s}{n^s}\left(1 - \frac{m}{n}\right)^{n-s} \\
&= \underbrace{\frac{n(n-1)\cdots(n-s+1)}{n\quad n\quad \cdots\quad n}}_{\text{there are } s \text{ of these factors}}\frac{(n-s)!}{(n-s)!}\frac{m^s}{s!}\left(1 - \frac{m}{n}\right)^{n-s} \\
&= 1\left(1 - \frac{1}{n}\right)\cdots\left(1 - \frac{s-1}{n}\right)\frac{m^s}{s!}\left(1 - \frac{m}{n}\right)^{n-s} \qquad \text{(B-45)}
\end{aligned}$$

If we let $n$ become very large while $s$ remains small, so that $n - s \doteq n$, and further recognize that $\lim_{n\to\infty} [1 - (m/n)]^n = e^{-m}$:

$$\lim_{n\to\infty} 1\left(1 - \frac{1}{n}\right)\cdots\left(1 - \frac{s-1}{1}\right)\frac{m^s}{s!}\left(1 - \frac{m}{n}\right)^{n-s} = (1)\frac{m^s}{s!}e^{-m} \qquad \text{(B-46)}$$

or finally

$$\Pr(x = s) = \binom{n}{s}\left(\frac{m}{n}\right)^s\left(1 - \frac{m}{n}\right)^{n-s} \doteq \frac{m^s}{s!}e^{-m} \qquad \text{(B-47)}$$

This last expression, of course, is the one that defines the Poisson distribution. The approximation is useful when $n$ is as small as 20, provided that $m = np \leqq 5$. For larger values of $m = np$ the approximation will only be useful for still larger values of $n$.

Consider now some continuous one-dimensional distributions.

*The Uniform Distribution*

A uniform distribution is appropriate to the description of a continuum of events that are equally likely in a given interval, say from $m - (\lambda/2)$ to $m + (\lambda/2)$. By "equally likely" we mean that the probability element

$p_1(x)\,dx = \Pr[x_1 < X < x_1 + dx_1]$ is a constant. The first probability density function is therefore a constant over the interval, and zero elsewhere. Since we have to arrange that

$$\Pr[-\infty < X < \infty] = \int_{-\infty}^{\infty} p_1(x)\,dx = 1 \tag{B-48}$$

$p_1(x) = 1/\lambda$. Then the distribution function $P_1(x)$, is a cutoff ramp function rising from 0 to one while $[m - (\lambda/2)] < x < [m + (\lambda/2)]$.

An example of the application of the uniform distribution would be to the calculation of the probability, disregarding the actually discrete nature of the problem, that a wheel of fortune would stop on any particular number.

### *The Exponential Distribution*

Another important continuous distribution is the exponential distribution. It is often used in estimating the reliability of systems or components that are subject to a constant hazard.

Consider a component that has survived to the time, $t$. Letting $F$ symbolize the event failure and $S$ the event survival we find that the conditional probability of failure in a time, $\Delta t$, given that the component has survived up to time, $t$, from Eq. B-5, is

$$\Pr(F/S) = \frac{\Pr(FS)}{\Pr(S)} \tag{B-49}$$

The conditional probability of failure is proportional to the hazard, $c$, and the interval, $\Delta t$.

$$\Pr(F/S) = c(\Delta t) \tag{B-50}$$

Suppose that the cumulative probability distribution for failure as a function of time were known. Call it $P_1(t)$. Then the probability of the component's surviving to the time, $t$, is

$$\Pr(S) = \Pr(\bar{F}) = [1 - P_1(t)] \tag{B-51}$$

and the derivative $dP_1(t)/dt = p_1(t)$ times $\Delta t$ is the probability element which defines the probability that the component will fail in the time, $\Delta t$. This is the same thing as surviving up to time, $t$, and then failing.

$$\Pr(FS) = \frac{dP_1}{dt}(\Delta t) \tag{B-52}$$

Substituting in Eq. B-35, we get

$$c(\Delta t) = \frac{(dP_1/dt)(\Delta t)}{[1 - P_1(t)]} \tag{B-53}$$

or

$$\frac{1}{c}\frac{dP_1}{dt} + P_1(t) = 1 = u(t); \qquad t \geqq 0$$

We may now solve this equation for the unknown distribution function

$$P_1(t) = [1 - e^{-ct}]; \qquad t \geqq 0 \tag{B-54}$$

and the corresponding density function is

$$p_1(t) = \frac{dP_1}{dt} = ce^{-ct}; \qquad t \geq 0 \tag{B-55}$$

The exponential functions that appear here give the distribution its name.

In applications to reliability calculations, the inverse of the parameter, $c$, which has the dimensions of time, often goes by the name "mean time before failure" (MTBF). This is its physical significance in such applications. Note from Table B-1 that the mean of the exponential distribution is $1/c$.

More generally, the exponential distribution governs the *time* between occurrences of *independent* random events that occur at a constant average rate. In this respect the exponential distribution bears a close relationship to the Poisson distribution. Recall from Eq. B-31, used in deriving the Poisson distribution, that for independent events occurring randomly in time at a constant average rate:

$$\Pr\,[\text{no event between 0 and } t] = P_0(t) = e^{-\beta t} \tag{B-56}$$

Similarly,

$$\Pr\,[\text{no event between } t \text{ and } t + \Delta t] = e^{-\beta \Delta t} \tag{B-57}$$

and therefore,

$$\Pr\,[\text{at least one event between } t \text{ and } t + \Delta t] = 1 - e^{-\beta \Delta t} \tag{B-58}$$

Since, for a very small $\Delta t$ (approaching $dt$),

$$e^{-\beta \Delta t} \doteq 1 - \beta\,\Delta t$$

therefore,

$$1 - e^{-\beta dt} = \beta\, dt \tag{B-59}$$

and this is the probability of *exactly* one event in the interval between $t$ and $t + dt$. Finally, the product of the probabilities of no events in the interval between 0 and $t$, and exactly one event in the interval between $t$ and $t + dt$, yields the probability element $(e^{-\beta t})\beta\, dt$, which, integrated from 0 to $T$, shows the probability of exactly one event's occurring more than $T$ seconds after its predecessor and less than $T + dt$ seconds afterward.

$$\text{Pr [one even between } T \text{ and } T + dt] = \int_0^T \beta e^{-\beta t}\, dt$$
$$= 1 - e^{-\beta T} \qquad \text{(B-60)}$$

Thus the intervals, $T_i$, between the events in a Poisson process are seen to be governed by the exponential distribution. The mean time between the events $T_{av} = 1/\beta$, the inverse of the "density" of the events.

To return briefly to the question of engine reliability used to illustrate the binomial distribution, one might have asked: "How was it determined that the probability of the engine's surviving a four-hour flight, $p = 0.9900$?" Under conditions approximating a constant hazard, it means that, on the average, there are 400 engine operating hours (during short missions) between random failures. Alternatively, the mean time between failures $1/c = 400$ hr. The cumulative probability of failure during the mission is governed by the exponential distribution. From Eq. B-37 and B-40 the probability that the engine will survive (not fail) up to a time, $t$ = mission time = 4 hr, is

$$\Pr(S) = \Pr(\bar{F}) = 1 - [1 - e^{-ct}] = e^{-ct}$$
$$\doteq 1 - ct = 1 - \tfrac{4}{400} = 0.9900 \qquad \text{(B-61)}$$

*The Gaussian Distribution*

By far the most important probability distribution in science and engineering is the normal or so-called Gaussian distribution. (It was actually discovered by Abraham De Morive as an approximation to the binomial distribution and was published by him in 1733, 60 years before Carl Friedrich Gauss used it in his astronomical calculations.) It can be derived very simply.[6]

The reader may recall that following some formulations of statistical mechanics, in the theory of communication, the entropy or information, $H$, of a discrete set of probabilities, $p_1, p_2 \cdots p_n$ is defined in such a way that

$$H = -\sum_{i=1}^{n} p_i \ln p_i \qquad \text{(B-62)}$$

Furthermore, the entropy is a measure of *disorder* or information (choice). For a continuous (single-dimensional) probability distribution, the analogous expression is

$$H = -\int_{-\infty}^{\infty} p_1(x) \ln p_1(x)\, dx \qquad \text{(B-63)}$$

[6] C. E. Shannon and W. Weaver, *The Mathematical Theory of Communication,* University of Illinois Press, Urbana, Ill., 1949, pp. 54–56.

We may inquire as to the first probability density function that makes the entropy (disorder, randomness) a maximum, subject to the condition that the standard deviation of the variable $x$ be a constant, $\sigma$. The square of the standard deviation or "variance" has been defined in terms of the density function in Eq. B-23, repeated here,

$$\sigma^2 = \int_{-\infty}^{\infty} x^2 p_1(x)\, dx \tag{B-64}$$

and from Eq. B-15,

$$\int_{-\infty}^{\infty} p_1(x)\, dx = 1 \tag{B-65}$$

Finding the maximum value of the integral function $H$, subject to the constraints (Eqs. B-48 and B-49) is a standard problem in the calculus of variations.[7] It is done by adjoining the constraints (each multiplied by a Lagrange multiplier) to the payoff function and maximizing the sum. The problem then becomes to find

$$\max_{p(x)} \int_{-\infty}^{\infty} [-p(x) \ln p(x) + \lambda x^2 p(x) + \nu p(x)]\, dx \tag{B-66}$$

A necessary condition for the maximum is that the Euler equation be satisfied. The Euler equation is: $\partial F/\partial y - (d/dx)(\partial F/\partial y') = 0$ where $F(x, y, y')$ is the integrand. In this case, where there are no derivatives of $p(x)$ in the integrand, the Euler equation is simply:

$$\frac{\partial}{\partial[p(x)]} [-p(x) \ln p(x) + \lambda x^2 p(x) + \nu p(x)] = 0 \tag{B-67}$$

$$-1 - \ln p(x) + \lambda x^2 + \nu = 0$$

or

$$p(x) = e^{\nu - 1} e^{\lambda x^2}$$

The *constants* $\nu$ and $\lambda$ are then chosen so as to satisfy the equations of constraint. An appropriate choice yields

$$p_1(x) = \frac{1}{\sqrt{2\pi}\,\sigma} e^{-(x^2/2\sigma^2)} \tag{B-68}$$

which is the first probability density function for the Gaussian distribution with a zero mean value. Its integral is the first distribution function

$$P_1(x) = \tfrac{1}{2} + \operatorname{erf}\left(\frac{x}{\sigma}\right) \tag{B-69}$$

where

$$\operatorname{erf}\left(\frac{x}{\sigma}\right) = \frac{2}{\sqrt{\pi}} \int_0^{x/\sigma} e^{-u^2}\, du \tag{B-70}$$

[7] L. E. Elsgolc, *Calculus of Variations*, Addison-Wesley, Reading, Mass., 1962.

is the so-called error function.[8] More generally, the Gaussian distribution is defined by

$$p_1(x) = \frac{1}{\sqrt{2\pi}\,\sigma} \exp - \left[\frac{(x - m_1)^2}{2\sigma^2}\right] \tag{B-71}$$

$$P_1(x) = \tfrac{1}{2} + \operatorname{erf}\left(\frac{x - m_1}{\sigma}\right) \tag{B-72}$$

which allows for the case in which the mean value of $x$ is not necessarily zero.

We have seen that, in a sense, the Gaussian distribution is the one associated with the maximum degree of randomness. Otherwise, subject to fairly general conditions, it can be shown that the first probability density function, $p_1(x)$, of the sum of a large number, $N$, of independent random variables tends to the probability density function of the normal (Gaussian) distribution. For this reason, if we are observing a random variable whose fluctuations are due to a large number of independent causes, and this is indeed often the case, the Gaussian distribution is likely to be the appropriate one characterizing the variable of interest.

For essentially the same reason, the Gaussian distribution is also an approximation to the discrete binomial and Poisson distributions if the mean $np$ or $m$ is large.

To fit a Gaussian distribution to a discrete one derived from plotting numbers of cases in given categories, so as to keep the total probability equal to one, the ordinates of the normal probability density curve are multiplied by the total number of observations, $N$.

$$y = \frac{N}{\sqrt{2\pi}\,s} e^{-(x-\bar{x}/2s)^2} \tag{B-73}$$

where the standard deviation is

$$s = \left(\frac{\sum_i x_i^2}{N} - \bar{x}^2\right)^{1/2} \doteq \sigma \tag{B-74}$$

and where $\bar{x}$ is the mean of all the observations, $\bar{x} \doteq m_1$. In the cases of the binomial and Poisson distributions, the standard deviation, of course, is the square root of the known variance (see Table B-1).

Figure B-3 shows several normal density functions with different means and standard deviations. Since the area under each curve must be one, they are higher at the mean value if the standard deviation is small. The points of inflection on each curve are located one standard deviation from the mean.

[8] I. N. Sneddon, *Special Functions of Mathematical Physics and Chemistry*, Oliver and Boyd, Edinburgh and London, 1961, p. 13, for example.

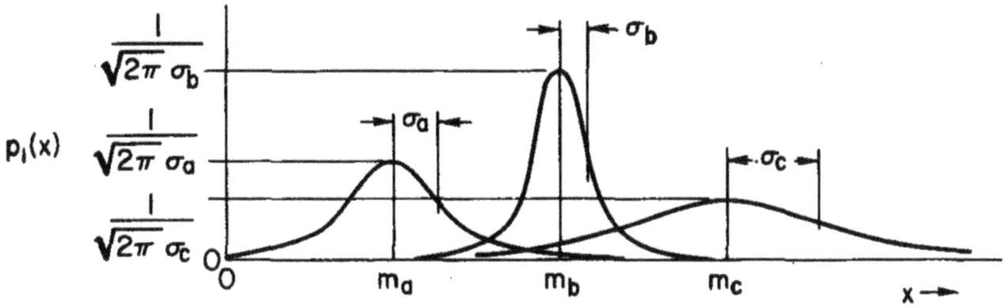


Fig. B-3. Normal density functions.

In order to make numerical calculations, the normal density function is usually cast into the *standard form* in which the mean is zero and the standard deviation is unity. Deviations from the mean are then measured in units of standard deviations, $t = (x - m_1)/\sigma =$ [standard deviations]. The standard form is

$$z(t) = \frac{1}{\sqrt{2\pi}} e^{-t^2/2}$$

This function is found tabulated in a number of places.[9]

One very useful configuration of tabulated values is in terms of

$$\Pr[(x - m_1) > t] = 1 - \int_{-\infty}^{t} z(t)\, dt$$

which represents the probability of a random variable, known to be governed by the normal distribution, exceeding a given positive value, greater than the mean, by an amount expressed as a number of standard deviations, $t$. This probability is the hatched area under the curve in Fig. B-4. A very short table of such probabilities is presented in Table B-4.

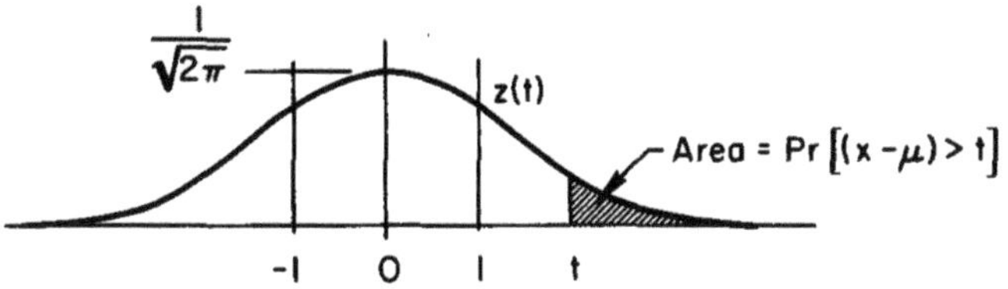


Fig. B-4. The standard form of the normal probability density function and $\Pr[(x - m_1) > t]$.

[9] See, e.g., A. Hald, *Statistical Tables and Formulas*, Wiley, New York, 1952, p. 33, where it is designated $\varphi(u)$; or M. M. Abramowitz and I. Stegun, eds., *Handbook of Mathematical Functions*, National Bureau of Standards, Applied Mathematics Series 55, U.S. Government Printing Office, Washington, D.C., 1964, pp. 966–972, where it is designated $Z(x)$.

Table B-4. Probability of a variable exceeding a value $t$ standard deviations greater than the mean

| $t$ standard deviations | $\Pr[(x - m_1) > t]$ | $t$ standard deviations | $\Pr[(x - m_1) > t]$ |
|---|---|---|---|
| 0.00 | 0.50000 | 2.00 | 0.02275 |
| 0.50 | 0.30854 | 3.00 | 0.00135 |
| 1.00 | 0.15866 | 4.00 | 0.00003 |
| 1.50 | 0.06680 | 5.00 | $3 \times 10^{-7}$ |

Since the Gaussian curve is symmetrical, Table B-4 may also be used to calculate the probability that a variable will be less than the mean by an amount that exceeds $t$ standard deviations; or, indeed, if the probability of a value either greater *or* less than the mean by the specified number of standard deviations is wanted, it is *twice* the values in the table.

$$\Pr[x \gtrless m_1 \pm \sigma] \doteq \frac{1}{3}, \quad \Pr[x \gtrless m_1 \pm 2\sigma] \doteq \frac{1}{22}, \quad \Pr[x \gtrless m_1 \pm 3\sigma] \doteq \frac{1}{370}$$

As a simple example of the applications of the normal distribution, consider that in typical aircraft carrier landing operations the mean sink rate of the airplanes at the instant of contact with the deck is $\bar{V}_I = 14.0$ ft/sec, that the standard deviation in the sink rate is 2.33 ft/sec, and that the sink rate is normally distributed about the mean value. If a sink rate of 21.0 ft/sec corresponds to the ultimate strength of the landing gear, in what fraction of all landings will the aircraft be damaged by a hard landing?

$$\Pr[V_I > 21] = \Pr[V_I > 14 + 3(2.33)]$$

From Table B-4, this is the probability that the sink rate will be more than the sum of the mean value and *three* standard deviations, i.e., $\Pr[V_I > 21] = 0.00135$, or the sink rate will be excessive once in 740 landings.

These few facts on the Gaussian and other distributions allow us to make a large number of interesting calculations concerning probabilities, but the material presented in this appendix is only introductory and is not a substitute for a more rigorous study of the subject.

# *BIBLIOGRAPHY*

## I. Additional Aspects of Aircraft Dynamics

### *Steady rolling flight, Steady turning flight, Internal angular momenta effects, Structural flexibility*


Abzug, M. J. "Effects of Certain Steady Motions on Small-Disturbance Airplane Dynamics," *J. Aeron. Sci.*, 21, No. 11, 749–752, 762 (Nov. 1954).

Blakelock, J. H. "Inertial Cross-Coupling," Chapter 5, and "Structural Flexibility," Chapter 8, in *Automatic Control of Aircraft and Missiles*, Wiley, New York, 1965.

Etkin, B., *Dynamics of Flight*, Wiley, New York, 1959, pp. 304–307.

Kolk, W. R. "Roll Coupling," Chapter 7, and "Aero-Structure-Control Interactions," Chapter 11, in *Modern Flight Dynamics*, Prentice-Hall, Englewood Cliffs, N.J., 1961.

McRuer, D. T., "A Feedback-Theory Analysis of Airframe Cross-Coupling Dynamics," *J. Aerosp. Sci.*, 29, No. 5, 525–533 (May 1962).

Paiewonsky, B. H., "The Effects of Engine Angular Momentum on an Airplane's Longitudinal and Lateral-Directional Dynamic Stability," WADC TR 56-225, June 1956.

Pass, H. R., Pearce, B. F., Siskind, R. K., and Wolkovitch, J., "Topics on Flexible Airplane Dynamics," ASD-TDR-63-334, Parts I through IV, July 1963.

Pearce, B. F., Johnson, W. A., and Siskind, R. K., "Analytical Study of Approximate Longitudinal Transfer Functions for a Flexible Airframe," ASD-TDR-62-279, Apr. 1962.

Phillips, W. H., "Effect of Steady Rolling on Longitudinal and Directional Stability," NACA TN-1627, 1948.

Pinsker, W. J. G., "A Preliminary Note on the Effects of Inertia Cross-Coupling on Aircraft Response in Rolling Maneuvers," RAE Tech. Note Aero 2419, 1955.

Pinsker, W. J. G., "Charts of Peak Amplitudes in Incidence and Sideslip in Rolling Maneuvers Due to Inertia Cross Coupling," RAE Tech. Note 2604, Apr. 1958.

Pinsker, W. J. G., "Critical Flight Conditions and Loads Resulting from Inertia Cross-Coupling and Aerodynamic Stability Deficiencies." RAE Tech. Note Aero 2502, Mar. 1957; also AGARD Rept. 107.

Rhoads, D. W. and Schuler, J. M., "A Theoretical and Experimental Study of Airplane Dynamics in Large-Disturbance Maneuvers," *J. Aeron. Sci.*, 24, No. 7, 507–526, 532 (July 1957).

Schuler, J. M., "Analytical Study of Airplane Dynamics and Tail Loads in Rolling Pull-Out Maneuvers," WADC TR 56-403, Sept. 1956.

Schuler, J. M., "Flight Evaluation of an Automatic Control System for Stabilizing the Large Uncontrolled Motions of Airplanes in Stalled Flight," Cornell Aeronaut. Lab. Rept. TB-1132-F-2, Oct. 1959.

Seamans, R. C., Jr., *et al.*, "Recent Developments in Aircraft Control," *J. Aeron. Sci.*, 22, No. 3, 145–164 (Mar. 1955).

Stapleford, R. L., Hofmann, L. G., Best, J. J., Wezeman, C. D., and Johnson, W. A., "Transfer Function Approximations for Large Highly Coupled Elastic Boosters with Fuel Slosh," NASA CR-464, Apr. 1966.

Thomas, H. H. B. and Price, P., "A Contribution to the Theory of Aircraft Response in Rolling Manoeuvres Including Inertia Cross-Coupling Effects, RAE Rept. Aero 2634, Apr. 1960; also ARC R & M 3349.

Waymeyer, W. K. and Sporing, R. W., "An Industry Survey on Aeroelastic Control System Instabilities in Aerospace Vehicles," IAS Paper 62-47, Jan. 1962.

Welch, J. D. and Wilson, R. E., "Cross-Coupling Dynamics and the Problems of Automatic Control in Rapid Rolls," *J. Aeron. Sci.*, 24, No. 10, 741–754 (Oct. 1957).

## II. Components

### *Gyros and accelerometers*

Ahrendt, W. R. and Savant, C. J., *Servomechanism Practice*, McGraw-Hill, New York, 1960.

Ausman, J. S., "Theory of Inertial Sensing Devices," in G. R. Pitman, Jr., ed., *Inertial Guidance*, Wiley, New York, 1962, pp. 72–91.

*Automatic Flight Control Systems for Piloted Aircraft*, Northrop Aircraft, BuAer Rept. AE-61-4VI, Bureau of Aeronautics, Navy Dept., Apr. 1956, pp. 2-93 thru 2-113 and Appendix, "Equations of the Gyroscope."

Becker, L., "Gyro Pickoff Indications at Arbitrary Plane Attitudes," *J. Aeron. Sci.*, 18, No. 11, 718–724 (Nov. 1951).

Broxmeyer, C., *Inertial Navigation Systems*, McGraw-Hill, New York, 1964.

Collette, J. G. R., "Analysis and Design of Space Vehicle Flight Control Systems, Vol. XI—Component Dynamics," NASA CR-830, July 1967.

Hopkin, H. R. and Dunn, R. W., "Theory and Development of Automatic Pilots, 1937–1947, RAE Rept. IAP 1459, Aug. 1947.

Lichtenstein, B., *Gyros, Platforms, Accelerometers*, 5th ed., Kearfott Div., General Precision, Little Falls, N.J., June 1962.

MacDonald, W. R., "Acceleration Transducers of the Force Balance Type," in *Proc. International Flight Test Instrumentation Symposium* (1960), Pergamon, New York, 1961, pp. 15–23.

Richardson, K. I. T., *The Gyroscope Applied*, Hutchinson and Company, London, 1954.

Savant, C. J., Jr., Howard, R. C., Solloway, C. B., and Savant, C. A., *Principles of Inertial Navigation*, McGraw-Hill, New York, 1961.

Savet, P. H., ed., *Gyroscopes: Theory and Design with Application to Instrumentation, Guidance and Control*, McGraw-Hill, New York, 1961.

Scarborough, J. B., *The Gyroscope, Theory and Application*, Interscience, New York, 1958.

Slater, J. M., *Inertial Guidance Sensors*, Reinhold, New York, 1964.

### *Air data sensors*

Chaffois, J., *Pitot-Static Probes for Subsonic and Supersonic Aircraft* ["Sondes Anémo-Barométriques pour Avions Subsoniques et Supersoniques," *Techniques et Sciences Aéronautiques et Spatiales* (1962) 1, pp. 7–17], RAE Library Trans. 1032, Nov. 1963. Translated by L. H. Townend, edited by A. A. Woodfield.

Coulthard, W. H., *Aircraft Instrument Design*, Pitman, New York, 1952, Chapter 1.

Emerson, F. M., Gardner, F. H., Gruenwald, G. D., Olshausen, R., and Sloma, L. V., "Study of Systems for True Angle of Attack Measurement," WADC TR 54-267, May 1955.

Goodman, D. G., Salter, C., and Warsap, J. H., "A New Design of Pitot-Static Tube with a Discussion of Pitot-Static Tubes and Their Calibration Factors, National Physical Laboratory Aero Rept. 1013, May 7, 1962.

Gracey, W., "Recent Developments in Pressure Altimetry," *J. Aircraft*, 2, No. 3, 161–165 (May–June 1965).

Hilliard, E. E., "Supersonic Wind Tunnel Tests of Several Pitot-Static Probes," AEDC TR 65-192, Aug. 1965.

Mallard, S. R., "Calibration Tests of a Litton Conical Air Data Probe at Mach Numbers of 2 to 8," AEDC TDR 62-186, Oct. 1962.

Muller, J. F., "Systematic Determination of Simplified Gain Scheduling Programs, *J. Aircraft*, 4, No. 6, 529–534 (Nov.–Dec. 1967).

Smetana, F. O., "Design and Tests of Aerodynamic Static Pressure Compensators for Four Service Aircraft," WADC TR 59-383, May 1959.

Teigen, M. J., "Dimensional Variables Affecting Calibration Characteristics of Null-Seeking Differential-Pressure and Vane Angle of Attack Sensors," WADC TR 59-388, June 1959.

### *Actuators*

Ahrendt, W. H. and Savant, C. J. Jr., *Servomechanism Practice*, 2nd edn., McGraw-Hill, New York, 1960.

Anderson, B. W., *The Analysis and Design of Pneumatic Systems*, Wiley, New York, 1967.

"Basic Research and Development in Fluid Power Control for the United States Air Force," ASD-TDR-62-3, Jan. 1962.

"Bibliography on Servomechanisms, Bulletin of Automatic and Manual Control Abstracts," Ministry of Aviation TIL/BIB/52/37, Apr. 1964.

Blackburn, J. F., Reethof, G., and Shearer, J. L., *Fluid Power Control*, Technology Press of Massachusetts Institute of Technology jointly with Wiley, New York, 1960.

Cataldo, R. S., *Analysis of Electrohydraulic Valves and Systems*, ISA Preprint ISA-11, presented at the Joint Automatic Control Conference, Massachusetts Institute of Technology Sept. 7–9, 1960.

Click, G. E., Henry, A. P., and McRuer, D. T., "Design Study of Multi-Function Hydraulic Actuating Devices," WADC TR 56-418, Aug. 1956.

Collette, J. G. R., "Analysis and Design of Space Vehicle Flight Control Systems, Vol. XI—Component Dynamics," NASA CR-830, July 1967.

"Electric Input Hydraulic Servovalve," SAE ARP 490, July 15, 1958.

Gibson, J. E. and Tuteur, F. B., *Control System Components*, McGraw-Hill, New York, 1958.

Graham, D., "Magnetic Clutches Add Muscle to Electronic Servos." *Space/Aeronaut.* 31, No.4, 138-141 (Apr. 1959).

Kegel, A. G. and Axelby, G. S., "Actuators," in E. M. Grabbe, S. Ramo, and D. E. Wooldridge, eds., *Handbook of Automation Computation and Control*, Vol. 3, *Systems and Components*, Wiley, New York, 1961.

May, K. D., "Advanced Valve Technology," NASA SP-5019, Feb. 1965.

Nightingale, J. M., "Hydraulic Servo-Valve Design," *Machine Design*, 27, No. 1, 191 (1955).

Proceedings of the First Piloted Aircraft Powered Surface Control System Symposium, Bureau of Aeronautics, Oct. 1949.

Report of the Second Piloted Aircraft Flight Control Symposium, Bureau of Aeronautics Rept. AE-61-5, June 1952.

Shearer, J. L., "Dynamic Characteristics of Valve Controlled Hydraulic Servomotors," *Trans. ASME*, 76, No. 8, 895–903 (Aug. 1954).

Shearer, J. L., "Study of Pneumatic Processes," *Trans. ASME*, 78, No. 2, 239–249 (1956).

Smyth, R. K., "Operation of Autopilot Servo Actuators," *Applied Hydraulics and Pneumatics*, 12, No. 3, 110–113 (Mar. 1959).

"The Hydraulic System," Northrop Aircraft, Bureau of Aeronautics Rept. AE-61-4 IV, Mar. 1953.

Truxal, J. G., ed., *Control Engineer's Handbook*, McGraw-Hill, New York, 1958, Sections 12, 14, 15, and 16.

Walters, R., *Hydraulic and Electro-Hydraulic Servo Systems*, Chemical Rubber Company Press, Cleveland, Ohio, 1967.

Willitt, A. A. J., "Developments in Aircraft Hydraulics Systems," *Aeroplane and Astronautics*, 100, No. 2574, 174-176 (Feb. 17, 1961).

## III. Descriptions of Complete Flight Control Systems and Related Documents

Abrams, C. R., "Final Report on the General Electric Self-Adaptive Flight Control System," NADC-ED-6455, June 1964.

Andeen, R. E., "Self-Adaptive Autopilots," *Space/Aeronaut.*, 43, No. 4, 46–52 (Apr. 1965).

Arnzen, H. E., "Flight Controls: A Look into the Future," *Space/Aeronaut.*, 33, No. 2, 46–49 (Feb. 1960).

"Automatic Flight Control Systems for Piloted Aircraft," Northrop Aircraft, Bureau of Aeronautics Rept. AE-61-4VI, Apr. 1956.

"Automatic Pilot Installations," SAE ARP 419, Nov. 15, 1957.

"Automatic Pilots," SAE AS 402A, Issued Aug. 1, 1947, revised Feb. 1, 1959.

Bodner, V. A. and Kozlov, M. S., "Stabilization of Flying Craft and Autopilots," (Stabilizatsiya Letatel'nykh Apparatov I Autopiloty, Gosudarstvennoye Nauchno-Teknicheskoye Izdatel'stvo, Oborongiz, Moskva, 1961), Foreign Technology Div. Transl. FTD-TT-62-811, July 31, 1962.

Boskovich, B. and Kaufmann, R. E., "Evolution of the Honeywell First-Generation Adaptive Autopilot and Its Applications to F-94, F-101, X-15, and X-20 Vehicles," *J. Aircraft*, 3, No. 4, 296–304 (July-Aug. 1966).

Buxton, E. R., "Prediction Servo Techniques of Smooth Automatic Flare Control," *Space/Aeronaut.*, 30, No. 5, 32, 33, 35, 36, 37 (Nov. 1958).

Cannon, C. H., "Military and Civil All Weather Landing Systems for C-141," *J. Aircraft*, 3, No. 6, 529–534 (Nov.–Dec. 1966).

Coulthard, W. H., *Aircraft Instrument Design*, Pitman, New York, 1952, Chapter 18.

"Current Autopilot Block Diagrams," prepared for SAE A-18 Committee, Oct. 1959,

Cushman, R. H., "Vanguard Control Demonstrates Minimum Hardware Approach. Parts I and II," *Automatic Control*, 25–33 (June 1958); 16–20 (July 1958).

Dawson, J. W., Harris, L. P., and Swean, E. A., "Dynamic Response of Two Aircraft-Autopilot Systems to Horizontal Turn Commands," Massachusetts Institute of Technology, Dynamic Analysis and Control Lab. Rept. 94, Jan. 31, 1955.

Ehlers, H. L., "Helicopter Automatic Flight Control Systems," North American Rockwell Corp., Autonetics Div., Oct. 1967.

Ehlers, H. L., "Technical Considerations in the Design of Gust Alleviation Control Systems," North American Aviation, Autonetics Div., Rept. X7-933/301, Apr. 17, 1967.

Eldridge, W., "An Investigation of Ways of Improving the Coordinated Turn Maneuver of Jet Transport Aircraft," Boeing Airplane Co. Document No. D6-5998, Dec. 9, 1960.

Graham, D. and Lathrop, R. C., "Automatic Feedback Control and All-Weather Flying," *Aeronautical Eng. Rev.*, 14, No. 10, 70–85 (Oct. 1955).

Holahan, J., "Electronic Pilot Automates Fighter in Aerial Attack, Ground-Controlled Intercept, All-Weather Flight, Low Approach," *Aviation Age*, 28, No. 6, 102–108 (Dec.1957).

Jex, H. R., Ashkenas, I. L., and Peters, R. A., "An Application of Airframe-Controller Optimization Techniques to a Supersonic Missile," *Proc. IAS Symposium on Vehicle Systems Optimization*, Garden City, N.Y. Nov. 28–29, 1961, pp. 115–126.

Klass, P. J., "Sperry Innovates Design in Autopilot for Jet Transport," *Aviation Week*, 65, No. 22, 76–86 (Nov. 26, 1956).

Kramer, K. C., "A-7A AFCS: A Flight-Proved High-Gain System," *J. Aircraft*, 3, No. 5, 454–461 (Sept.–Oct. 1966).
McDonald, E. H. and Farris, J. A., "The X-20 Flight Control System Development," Systems Engineering Group SEG-TDR-64-8, June 1964.
McKinley J. L. and Bent, R. D., "Electronic Navigation Equipment," Chapter 20; "Autopilot for Light Aircraft," Chapter 21; and "Autopilot for Jet Airliners," Chapter 22; in *Electricity and Electronics for Aerospace Vehicles*, McGraw-Hill, New York, 1961.
Mallery, C. G. and Neebe, F. C., "Flight Test of General Electric Self-Adaptive Control," *J. Aircraft*, 3, No. 5, 449–453 (Sept.–Oct. 1966).
Monroe, W. R., "Improving the Dynamic Response of Airplanes by Means of Electric Equipment," *AIEE Trans.*, 72, Part. 2, 441–447 (Jan. 1953).
Mueller, L. J., "Problems Unique to VTOL Automatic Flight Control," *J. Aircraft*, 2, No. 5, 357–360 (Sept.–Oct. 1965).
"P-3 Automatic Pilot—Attitude Control for Jet Aircraft," *Digest* (a publication of the U.S. Navy), Mar. 1951, pp. 9–17.
Richardson, K. I. T., *The Gyroscope Applied*, Philosophical Library, New York, 1954, Section 3, Chapter 5 and Section 4, Chapter 2.
Seacord, C. L., "Flight Control for Manned Spacecraft," *Space/Aeronaut.*, 40, No. 6, 72–80 (Nov. 1963).
Stambler, J., "Boeing 707 Flight Control System Tailored for Safety," *Aviation Age*, 27, No. 5, 34–41 (May 1957).
Williams, C. A., *Aircraft Instrument Control Systems*, Odhams Press, New York, 1963.

# INDEX

www.ingramcontent.com/pod-product-compliance
Ingram Content Group UK Ltd.
Pitfield, Milton Keynes, MK11 3LW, UK
UKHW022109140225
455125UK00007B/160

9 780691 600383